International Conference

RADAR-82

18—20 October 1982

Organised by the
Electronics Division of the Institution of Electrical Engineers

in association with the
Institute of Electrical and Electronics Engineers
(Aerospace and Electronic Systems Society)

with the support of the
Convention of National Societies of
Electrical Engineers of Western Europe (EUREL)
Institute of Mathematics and its Applications
Institution of Electronic and Radio Engineers
Royal Aeronautical Society
Royal Institute of Navigation

Venue
The Royal Borough of Kensington and Chelsea Town Hall,
Hornton Street, London W8, UK

Organising Committee

K F Slater (*Chairman*)
P K Blair
Dr J Clarke
Professor J R Forrest
K L Fuller
R N Lord*
Dr K Milne OBE
M F Radford
M H A Smith
F K Spokes
C M Stewart
P D L Williams

Overseas Corresponding Members

Colonel E Arkoumaneas, Greece
E P W Attema, Netherlands
Professor M H Carpentier, France
Dr A L Gilardini, Italy
Professor P Gudmandsen, Denmark
R T Hill, USA**
Professor G F Lind, Sweden
Dr T Lund, Norway
Professor S Okamura, Japan
Dr W D Wirth, Federal Republic of Germany

**representing the Institute of Electrical and Electronics Engineers (Aerospace and Electronic Systems Society)
*representing the Institution of Electronic and Radio Engineers

The cover design is a reproduction of a radar display of the Spithead Review of the Fleet on the occasion of Her Majesty The Queen's Silver Jubilee in 1977.

The radar display is reproduced by permission of the Ministry of Defence (PE) at the Admiralty Surface Weapons Establishment, Portsdown. Crown copyright reserved.

Published by the Institution of Electrical Engineers, London and New York, ISBN 0 85296268 1 ISSN 0537-9989.

Produced by MULTIPLEX techniques ltd, Orpington, Kent.

Contents

The Institution of Electrical Engineers is not, as a body, responsible for the opinions expressed by individual authors or speakers.

Page No

Radar Systems I

1 'Adaptive radar in remote sensing using space, frequency and polarization processing'
Dr D T Gjessing and J Hjelmstad
Royal Norwegian Council for Scientific and Industrial Research, Norway

7 'Intrapulse polarization agile radar'
Dr M N Cohen and E S Sjoberg
Georgia Institute of Technology, USA

12 'On search strategies of phased array radars'
W Fleskes
Forschungsinstitut für Funk und Mathematik, FGAN, Federal Republic of Germany

15 'Design and performance considerations in modern phased array radar'
E R Billam
Admiralty Surface Weapons Establishment, UK

20 'A new generation airborne synthetic aperture radar (SAR) system'
Dr J R Bennett and R A Deane
MacDonald, Dettwiler and Associates Ltd, Canada

Radar Systems II

24 'Results from a new dual band, dual purpose radar for sea surface and aircraft search'
P D L Williams
Racal-Decca Ltd, UK

30 'The Dolphin naval surveillance radar'
Dr J Blogh
Plessey Radar Ltd, UK

36 'AN/APS-134(V) maritime surveillance radar'
J M Smith
Texas Instruments Inc, USA

509 'Modular survivable radar for battlefield surveillance applications'
Dr E L Hofmeister, W E Szczepanski, R F Oot, Dr D C Dalpe and Dr M E Davis
General Electric Co, USA

41 'A British AEW radar system'
Dr J Clarke
Royal Signals and Radar Establishment, UK
Dr J King
Marconi Avionics Ltd, UK

Sequential Detection and MTI

46 'Sacrifices in radar clutter suppression due to compromises in implementation of digital doppler filters'
J W Taylor Jr
Westinghouse Electric Corporation, USA

51 'A comparison between noncoherent and coherent MTIs'
Dr F F Kretschmer Jr, Mrs F C Lin and B L Lewis
Naval Research Laboratory, USA

56 'Study of weather clutter rejection with moving target detection (MTD) processor'
Bao Zheng, Peng Xueyu and Zhang Shouhong
Northwest Telecommunication Engineering Institute, People's Republic of China

61 'Reliable single scan target acquisition using multiple correlated observations'
Dr R A Dana
Mission Research Corporation, USA
D Moraitis
Hughes Aircraft Co, USA

66 'A simplified sequential detection scheme'
M C Jackson
Marconi Research Centre, UK

Contents

Page No **Adaptive Processing Techniques**

71 'False alarm control in automated radar surveillance systems'
Dr W G Bath, Ms L A Biddison, S F Haase and E C Wetzlar
Johns Hopkins University, USA

76 'Performance evaluation of some adaptive polarization techniques'
Professor D Giuli and Dr M Gherardelli
University of Florence, Italy
Professor E Dalle Mese
University of Pisa, Italy

82 'Impact of extremely high speed logic technology on radar performance'
Dr E K Reedy and R B Efurd
Georgia Institute of Technology, USA
M N Yoder
Office of Naval Research, US Navy, USA

87 'Superresolution using an active antenna array'
U Nickel
Forschungsinstitut für Funk und Mathematik, FGAN, Federal Republic of Germany

92 'A fast beamforming algorithm for large arrays'
Dr E K L Hung and Dr R M Turner
Department of Communications, Canada

HF/VHF Radar

97 'HF sky-wave backscatter radar for over-the-horizon detection'
G R Nelson and Dr G H Millman
General Electric Co, USA

101 'HF ground-wave radar for sea-state and swell measurement; theoretical studies, experiments and proposals'
Professor E D R Shearman, Dr L R Wyatt, G D Burrows, M D Moorhead and D J Bagwell
University of Birmingham, UK
Dr W A Sandham
Formerly with University of Birmingham, now with British National Oil Corporation, UK

107 'Experimental studies of the performance of an MF/HF ground-wave radar on a coastal site of irregular contour'
Professor E D R Shearman, Dr D C Cooper, Dr K Kumar, D J Bagwell and M D Moorhead
University of Birmingham, UK

110 'Propagation effects on a VHF radar'
F Christophe and Dr P Golé
Office National d'Etudes et de Recherches Aérospatiales, France

Radar Systems III

115 'A barrier radar concept'
J Marshall, C Ball and I Weissman
Riverside Research Institute, USA

120 'Search and target acquisition radar for short range air defence systems. A new threat environment—a new solution'
Dr J O Winnberg
Telefonaktiebolaget L M Ericsson, Sweden

*125** 'CW multi-tone radar ranging using DFT techniques'
L C Bomar, Dr W J Steinway, S A Faulkner and Ms L L Harkness
Georgia Institute of Technology, USA

130 'A fixed-beam multilateration radar system for weapon impact scoring'
Dr S Gaskell
RCA Corporation, USA
M Finch
Ministry of Defence, UK

134 'Instrumentation and analysis of airborne pulse-doppler radar trials'
Dr J Clarke, E B Cowley and I W Scroop
Royal Signals and Radar Establishment, UK
Dr K Clifton and Dr J King
Marconi Avionics Ltd, UK

* *The text of this contribution was not available at the time of publication.*

Contents

Page No **Coherent Radar Processing**

138 'Optimum pulse doppler search radar processing and practical approximations'
V G Hansen
Raytheon Co, USA

144 'Resolution of ambiguous radar measurements using a floating bin correlator'
E R Addison and Dr E L Frost
Westinghouse Electric Corporation, USA

149 'Optimising the integration aperture for a high PRF CW surveillance radar'
R A Hall
Marconi Avionics Ltd, UK

154 'Performance comparison of MTI and coherent doppler processors'
Dr D C Schleher
Eaton Corporation, USA

159 'A spatially-variant autofocus technique for synthetic-aperture radar'
Dr M R Vant
Department of Communications, Canada

Multi-Site Radar Operation

164 'Problems of data processing in multiradar and multisensor defense systems'
Dr H Ebert
AEG-Telefunken, Federal Republic of Germany

169 'Association of multisite radar data in the presence of large navigation and sensor alignment errors'
Dr W G Bath
Johns Hopkins University, USA

174 'Active array receiver studies for bistatic/multistatic radar'
Dr J G Schoenenberger
Formerly with University College London, now with Racal-Decca Ltd, UK
Professor J R Forrest
University College London, UK
Dr C Pell
Royal Signals and Radar Establishment, UK

179 'Coherent multi-static radar: stochastic signal theory and performance evaluation'
Dr A Wernersson
National Defence Research Institute, Sweden

183 'Multistatic tracking and comparison with netted monostatic systems'
Dr A Farina
Selenia SpA, Italy

Radar Clutter: Sea

188 'Bistatic sea clutter return near grazing incidence'
Dr G W Ewell and S P Zehner
Georgia Institute of Technology, USA

193 'Sea clutter statistics'
Dr J Maaløe
Technical University of Denmark, Denmark

198 'Amplitude and temporal statistics of sea spike clutter'
I D Olin
Naval Research Laboratory, USA

203 'A radar sea clutter model and its application to performance assessment'
K D Ward
Royal Signals and Radar Establishment, UK

Contents

Page No **Air Traffic Control I: Secondary Radar**

208 'Monopulse secondary surveillance radar—principles and performance of a new generation SSR system'
M C Stevens
Cossor Electronics Ltd, UK

215 'Decoding-degarbling in monopulse secondary surveillance radar'
Professor G Marchetti and Professor L Verrazzani
University of Pisa, Italy

220 'Evaluation of angular discrimination of monopulse SSR replies in garble condition'
Dr G Benelli and Dr M Fossi
University of Florence, Italy
Dr S Chirici
Whitehead Motofides, Italy

225 'Integral SSR antenna having independently optimized sum and difference beams'
P T Muto
Electronic Navigation Research Institute, Ministry of Transport, Japan
T Izutani, S Itoh, H Yokoyama and H Takano
Nippon Electric Co, Japan

230 'Secondary radar performance prediction'
B E Willis
Ministry of Defence, UK
B Pugh and S Strong
British Aerospace PLC, UK

Simulation and Data Processing

235 'Generic tracking radar simulator'
Dr W K McRitchie, P I Pulsifer and G A Wardle
Defence Research Establishment Ottawa, Canada

240 'Simulation of radar returns from land using a digital technique'
Dr J R Morgan, P E Sherlock and D J Hill
Ferranti Computer Systems Ltd, UK

245 'Radar electromagnetic environment simulation'
J F Michaels
Republic Electronics Inc, USA

250 'An equipment for simulating airborne radar video'
T Snowball, T R Berry and A M Pardoe
Royal Signals and Radar Establishment, UK

254 'The automatic track while scan system used within the searchwater airborne maritime surveillance radar'
M Symons
Thorn EMI Electronics Ltd, UK

259 'Automatic integration of data from dissimilar sensors'
W I Citrin, R W Proue and J W Thomas
Johns Hopkins University, USA

Target Recognition

262 'Recognition of targets by radar'
Dr N F Ezquerra and Ms L L Harkness
Georgia Institute of Technology, USA

266 'Digital signal processing of scattering data from nonlinear targets'
J Y Hong and Professor E J Powers
University of Texas at Austin, USA

271 'Radar spectroscopy'
Dr P J Moser
Naval Research Laboratory, USA
Professor H Uberall
Catholic University of America, USA

274 'Classification of ships using an incoherent marine radar'
Dr J Maaløe
Technical University of Denmark, Denmark

Contents

Page No **Low Probability of Intercept Radar and Passive Operation**

278 'The impact of waveform bandwidth upon tactical radar design'
C H Gager
Mitre Corporation, USA

283 'Radar assisted passive DF tracking'
Dr R S Farrow
Admiralty Surface Weapons Establishment, UK

288 'A filtering technique of passive radar in hyperbolic coordinate system'
Hungcun Chang, Zhuoying Wang and Yiyen Feng
Tsinghua University, People's Republic of China

291 'Deghosting in an automatic triangulation system'
Dr G van Keuk
Forschungsinstitut für Funk und Mathematik, FGAN, Federal Republic of Germany

Air Traffic Control II

296 'The multiradar tracking in the ATC system of the Rome FIR'
Dr G Barale, Dr G Fraschetti and Dr S Pardini
Selenia SpA, Italy

301 'Methods for radar data extraction and filtering in a fully automatic ATC radar station'
Dr E Giaccari
Selenia SpA, Italy

306 'Presentation and processing of radar video map information'
Dr R J G Edwards
Ministry of Transport, New Zealand

311 'ASMI-18X an airport surface surveillance radar'
J D Holcroft and S J Martin
Racal-MESL Radar Ltd, UK

316 'Detection of hazardous meteorological and clear-air phenomena with an air traffic control radar'
D L Offi, W Lewis and T Lee
Federal Aviation Administration Technical Center, USA

Signal Processing

321 'Matched filtering using surface-acoustic-wave convolvers'
Dr D P Morgan, D R Selviah, D H Warne and Dr J J Purcell
Plessey Research (Caswell) Ltd, UK

326 'A fiber optic pulse compression device for high resolution radars'
Dr E O Rausch, R B Efurd and M A Corbin
Georgia Institute of Technology, USA

331 'New polyphase pulse compression waveforms and implementation techniques'
B L Lewis and Dr F F Kretschmer Jr
Naval Research Laboratory, USA

336 'A digital high-speed correlator for incoherent-scatter radar experiments'
Dr H J Alker
Electronics Research Laboratory, University of Trondheim, Norway

341 'A retrospective detection algorithm for extraction of weak targets in clutter and interference environments'
R J Prengaman, R E Thurber and Dr W G Bath
Johns Hopkins University, USA

346 'A Kalman approach to improve angular resolution in search radars'
Professor E Dalle Mese and Dr G De Fina
University of Pisa, Italy
Dr V Sacco
Segnalamento Marittimo ed Aereo SpA, Italy

Contents

Page No	Antennas: Low Sidelobe Antennas
351	'Reduced cost low sidelobe reflector antenna systems' Dr N Williams and Dr D J Browning *ERA Technology Ltd, UK* P Varnish *Admiralty Surface Weapons Establishment, UK*
355	'On the performance degradation of a low sidelobe phased array due to correlated and uncorrelated errors' Dr J K Hsiao *Naval Research Laboratory, USA*
360	'Minimisation of sidelobes from a planar array of uniform elements' G J Halford and W J McCullagh *Admiralty Surface Weapons Establishment, UK*
365	'The in-situ calibration of a reciprocal space-fed phased array antenna' Dr E K L Hung, N R Fines and Dr R M Turner *Department of Communications, Canada*
	Radar Returns from Weather and Land
370	'Multiple-parameter-radar techniques and applications for precipitation measurements: a review' S M Cherry, M P M Hall and J W F Goddard *Rutherford Appleton Laboratory, UK*
375	'The Federal Aviation Administration weather radar research and development program' D E Johnson *Federal Aviation Administration, USA*
380	'Land clutter study: low grazing angles (backscattering)' Dr J W Henn and D H Pictor *British Aerospace PLC, UK* Dr A Webb *Royal Signals and Radar Establishment, UK*
385	'Millimeter wave land clutter model' N C Currie and S P Zehner *Georgia Institute of Technology, USA*
	Antennas: Beam Forming and Steering
390	'Beamforming for a multi-beam radar' J M Chambers *Plessey Radar Ltd, Plessey Co Ltd, UK* R Passmore and J Ladbrooke *Plessey Electronic Systems Research Ltd, Plessey Co Ltd, UK*
394	'An X-band microstrip phased-array antenna with electronic polarization control' Dr C H Hamilton *AEG-Telefunken, Federal Republic of Germany*
399	'Results from an experimental receiving array antenna with digital beamforming' Dr U Petri *AEG-Telefunken, Federal Republic of Germany*
403	'Beam forming with phased array antennas' W Sander *Forschungsinstitut für Funk und Mathematik, FGAN, Federal Republic of Germany*
408	'Optical fibre networks for signal distribution and control in phased array radars' Professor J R Forrest, F P Richards and Dr A A Salles *University College London, UK* P Varnish *Admiralty Surface Weapons Establishment, UK*

Contents

Page No **Novel Applications of Radar**

413 'Portable FMCW radar for locating buried pipes'
Dr A D Olver, Dr L G Cuthbert, Dr M Nicolaides and Dr A G Carr
Queen Mary College, University of London, UK

419 'A novel method of suppressing clutter in very short range radars'
Dr A Al-Attar, D J Daniels and H F Scott
British Gas, Engineering Research Station, UK

424 'Cable radar for intruder detection'
A C C Wong and P K Blair
Standard Telecommunication Laboratories Ltd, UK

429 'Coupling mechanism for guided radar'
Dr P W Chen
ESL Inc, USA
Dr G O Young
TRW Inc, USA
Dr R K Harman
Formerly with Computing Devices Co, now with Senstar Corporation, Canada

Radar Tracking Systems

434 'A new family of Selenia tracking radars; system solutions and experimental results'
Dr T Bucciarelli, Dr U Carletti and Dr M D'Avanzo
Selenia SpA, Italy
Professor G Picardi
Selenia SpA and also University of Rome, Italy

439 'An X-band array signal processing radar for tracking targets at low elevation angles'
A Pearson, Dr P Barton and W D Waddoup
Standard Telecommunication Laboratories Ltd, UK
R J Sherwell
Admiralty Surface Weapons Establishment, UK

444 'Tracking radar electronic counter-countermeasures against inverse gain jammers'
S L Johnston
International Radar Directory, USA

Poster Displays — Radar Systems

448 'A new broadband array processor'
K M Ahmed and Dr R J Evans
University of Newcastle, Australia

453 'Estimation of ship's manoeuvres with a navigation radar'
Professor G F Lind
Lund Institute of Technology, University of Lund, Sweden

458 'Measuring target position with a phased-array radar system'
G A van der Spek
Physics Laboratory TNO, Netherlands

464 'Automatic detectors for frequency-agile radar'
Dr G V Trunk and P K Hughes II
Naval Research Laboratory, USA

469 'A novel 35 GHz 3-D radar for flight assistance'
G M Ritter
Siemens AG, Federal Republic of Germany

Contents

Page No **Poster Displays — Signal Processing**

473 'Suboptimum clutter suppression for airborne phased array radars'
Dr R Klemm
Forschungsinstitut für Funk und Mathematik, FGAN, Federal Republic of Germany

477 'Ambiguity functions of complementary series'
Squadron Leader J A Cloke
Royal Air Force, Ministry of Defence, UK

482 'MTI-filtering for multiple time around clutter suppression in coherent on receive radars'
S Carlsson
Royal Institute of Technology, Sweden

486 'The Gram-Schmidt sidelobe canceller'
Dr T Bucciarelli, Dr M Esposito, Dr A Farina and Dr G Losquadro
Selenia SpA, Italy

491 'Ground clutter suppression using a coherent clutter map'
Dr J S Bird
Department of Communications, Canada

496 'An experimental adaptive radar MTI filter'
Y H Gong
Chengdu Radio Engineering Institute, People's Republic of China
J E Cooling
University of Technology, Loughborough, UK

501 'The use of a multi-level quantiser in plot extraction'
P N G Knowles
Plessey Electronic Systems Research Ltd, UK

505 'Improved coherent-on-receive radar processing with dynamic transversal filters'
R L Trapp
Johns Hopkins University, USA

List of Authors

Page No

Addison, E R 144
Ahmed, K M 448
Al-Attar, A 419
Alker, H J 336

Bagwell, D J 101, 107
Ball, C 115
Bao Zheng 56
Barale, G 296
Barton, P 439
Bath, W G 71, 169, 341
Benelli, G 220
Bennett, J R 20
Berry, T R 250
Biddison, L A 71
Billam, E R 15
Bird, J S 491
Blair, P K 424
Blogh, J 30
Browning, D J 351
Bucciarelli, T 434, 486
Burrows, G D 101

Carletti, U 434
Carlsson, S 482
Carr, A G 413
Chambers, J M 390
Chen, P W 429
Cherry, S M 370
Chirici, S 220
Christophe, F 110
Citrin, W I 259
Clarke, J 41, 134
Clifton, K 134
Cloke, J A 477
Cohen, M N 7
Cooling, J E 496
Cooper, D C 107
Corbin, M A 326
Cowley, E B 134
Currie, N C 385
Cuthbert, L G 413

Dalle Mese, E 76, 346
Dalpe, D C 509
Dana, R A 61
Daniels, D J 419
D'Avanzo, M 434
Davis, M E 509
Deane, R A 20
De Fina, G 346

Ebert, H 164
Edwards, R J G 306
Efurd, R B 82, 326
Esposito, M 486
Evans, R J 448
Ewell, G W 188
Ezquerra, N F 262

Farina, A 183, 486
Farrow, R S 283
Finch, M 130
Fines, N R 365

Page No

Fleskes, W 12
Forrest, J R 174, 408
Fossi, M 220
Fraschetti, G 296
Frost, E L 144

Gager, C H 278
Gaskell, S 130
Gherardelli, M 76
Giaccari, E 301
Giuli, D 76
Gjessing, D T 1
Goddard, J W F 370
Golé, P 110
Gong, Y H 496

Haase, S F 71
Halford, G J 360
Hall, M P M 370
Hall, R A 149
Hamilton, C H 394
Hansen, V G 138
Harkness, L L 262
Harman, R K 429
Henn, J W 380
Hill, D J 240
Hjelmstad, J 1
Hofmeister, E L 509
Holcroft, J D 311
Hong, J Y 266
Hsiao, J K 355
Hughes II, P K 464
Hung, E K L 92, 365
Hungcun Chang 288

Itoh, S 225
Izutani, T 225

Jackson, M C 66
Johnson, D E 375
Johnston, S L 444

King, J 41, 134
Klemm, R 473
Knowles, P N G 501
Kretschmer Jr, F F 51, 331
Kumar, K 107

Ladbrooke, J 390
Lee, T 316
Lewis, B L 51, 331
Lewis, W 316
Lin, F C 51
Lind, G F 453
Losquadro, G 486

Maaløe, J 193, 274
Marchetti, G 215
Marshall, J 115
Martin, S J 311
McCullagh, W J 360
McRitchie, W K 235
Michaels, J F 245
Millman, G H 97
Moorhead, M D 101, 107
Moraitis, D 61
Morgan, D P 321

Page No

Morgan, J R 240
Moser, P J 271
Muto, P T 225

Nelson, G R 97
Nickel, U 87
Nicolaides, M 413

Offi, D L 316
Olin, I D 198
Olver, A D 413
Oot, R F 509

Pardini, S 296
Pardoe, A M 250
Passmore, R 390
Pearson, A 439
Pell, C 174
Peng Xueyu 56
Petri, U 399
Picardi, G 434
Pictor, D H 380
Powers, E J 266
Prengaman, R J 341
Proue, R W 259
Pugh, B 230
Pulsifer, P I 235
Purcell, J J 321

Rausch, E O 326
Reedy, E K 82
Richards, F P 408
Ritter, G M 469

Sacco, V 346
Salles, A A 408
Sander, W 403
Sandham, W A 101
Schleher, D C 154
Schoenenberger, J G 174
Scott, H F 419
Scroop, I W 134
Selviah, D R 321
Shearman, E D R 101, 107
Sherlock, P E 240
Sherwell, R J 439
Sjoberg, E S 7
Smith, J M 36
Snowball, T 250
Stevens, M C 208
Strong, S 230
Symons, M 254
Szczepanski, W E 509

Takano, H 225
Taylor Jr, J W 46
Thomas, J W 259
Thurber, R E 341
Trapp, R L 505
Trunk, G V 464
Turner, R M 92, 365

Uberall, H 271

van der Spek, G A 458
van Keuk, G 291

List of Authors

Page No

Vant, M R 159
Varnish, P 351, 408
Verrazzani, L 215

Waddoup, W D 439
Ward, K D 203
Wardle, G A 235
Warne, D H 321
Webb, A 380

Weissman, I 115
Wernersson, A 179
Wetzlar, E C 71
Williams, N 351
Williams, P D L 24
Willis, B E 230
Winnberg, J O 120
Wong, A C C 424
Wyatt, L R 101

Yiyen Feng 288
Yoder, M N 82
Yokoyama, H 225
Young, G O 429

Zehner, S P 188, 385
Zhang Shouhong 56
Zhuoying Wang 288

ADAPTIVE RADAR IN REMOTE SENSING USING SPACE, FREQUENCY AND POLARIZATION PROCESSING

Dag T. Gjessing and Jens Hjelmstad

Royal Norwegian Council for Scientific and Industrial Research, Environmental Surveillance Technology Programme

1 INTRODUCTION

The radar scientist is facing a challenging and inspiring future, one of matching new technological achievements to important applications in the field of remote detection and identification.

This contribution is dedicated to the following concept: Most of the existing detection/identification systems do not make optimum use of all the a priori information on the object of interest that one generally is in possession of. Knowing something about the geometrical shape of the object on which our attention is focused (distribution of scattering centers and motion pattern), an optimum illumination and detection system which adapts itself to this target against a terrestrial background through an adverse propagation medium can be designed.

We are thus faced with the consideration of three filter functions: the transmission medium between the observation platform and the target, the terrestrial background against which the target is viewed, and the target itself. The more detailed information we require about the particular target per unit time, the more widebanded must our radar illuminator be.

In this brief contribution we shall concentrate on the multifrequency radar system. As we shall see, this system lends itself directly to simple computer control in a manner which is very familiar to the computer scientist.

Having structured the illumination in the time domain for optimum coupling to the target, it remains to shape the phasefront in space so as to obtain maximum coupling to the particular reflecting structure of interest by making use of a matrix antenna (two-dimensional broadside array).

Finally, we can manipulate the polarization properties of our transmit/receive system so as to investigate the polarization characteristics (the symmetry properties) of the target.

In this brief presentation we shall lean heavily on earlier contributions from the authors' laboratory (1,2,3,4,5), highlight these and describe a multi-frequency polarimetric radar system which presently is being developed by the authors. We shall present results from simple mathematical models and offer some preliminary experimental verifications.

2 BASIC PHYSICAL CONCEPTS: FORMULATION OF THE PROBLEM

As introduced above, we shall be considering four signature domains:

a) By measuring the correlation properties $R(\Delta F)$ in the frequency domain of the waves scattered back from the illuminated area (target against background) we obtain information about the longitudinal distribution of the scatterers. Specifically it can be shown (1,2,3,4,5) that if we describe the distribution in range (longitudinally) of the scatterers by the delay function $f(z)$ which has dimension field-strength and is the square root of the scattering cross-section $\sigma(z)$, then the correlation function in the frequency domain $R(\Delta F)$ is the Fourier transform of $f(z)$. A measure of $R(\Delta F)$ is obtained, as we know, by multiplying the scattered field-strength of frequency F by the complex conjugate of the field-strength at frequency $F+\Delta F$.

Thus we have

$$E(F)\ E^*(F+\Delta F) \sim R(\Delta F) \sim FT\{R(\Delta z)\} \qquad (2.1)$$

Note that if the target is illuminated with two frequencies spaced ΔF apart, then irregularities in the target with scale size $\Delta z = c/(2\Delta F)$ contribute to the scattered field.

b) By measuring the spatial correlation properties of the field scattered back from the target in a plane normal to the direction of propagation, i.e. transversely, we obtain information about the transverse distribution of the scatterers. If then the x and y directions are orthogonal to the direction of propagation z (direction from radar to target), then we measure the field-strength at the points x and $x+\Delta x$ in exactly the same way as above where we were dealing with different frequencies.

It can here readily be shown (page 18 of reference 2) that

$$E(x)\ E^*(x+\Delta x) \sim R(\Delta x) \sim FT\{\sigma(x/R)\} \qquad (2.2)$$

where $\sigma(x)$ is the transverse distribution of the scatterers over the scattering body in the x direction and R is the distance to the target. This, of course, is the same as saying that the spatial auto-correlation of the transverse field-strength is the Fourier transform of the angular power spectrum of the scattered wave (angle of arrival spectrum).

c) By measuring the temporal distribution of the scattered field (the power spectrum) information about the motion pattern of the target is obtained through the well-known Doppler relationship

$$f = \frac{1}{2\pi}\ \vec{K}\cdot\vec{V} \qquad (2.3)$$

where f is the Doppler frequency, $\vec{K}$ is the vector difference $\vec{k}_i-\vec{k}_s$ between the wavenumber $\vec{k}_i$ of the illuminating (incident) wave and $\vec{k}_s$ the wavenumber of the scattered wave. $\vec{V}$ is the velocity of the scattering element.

Thus if we are dealing with a target (such as the sea surface) composed of many scattering centers or facets which have different velocity, we obtain information about the velocity distribution of scale size Δz by illuminating the target with two frequencies with frequency difference $\Delta F = c/(2\Delta z)$ and by measuring the temporal variation (power spectrum) of the quantity

$$W(\omega) \sim V(F,t)\ V^*(F+\Delta F,t) \qquad (2.4)$$

Note that the velocity of the scattering element of scale size Δz is obtained from equation (2.3) by noting that it is the wavenumber $\Delta K = 2\pi/\Delta z$ of the difference frequency ΔF that enters into the Doppler equation in this case.

Hence the Doppler frequency is given by

$$f = \frac{2\Delta F}{c} \cos \phi$$

For details the reader is referred to ref (5), page 15.

d) By measuring the distribution of the scattering centers (the $\sigma(z)$, $\sigma(x)$ and $\sigma(y)$ functions) for each element of the scattering matrix, we obtain information about the symmetry characteristics of the target.

Figure 2.1 finally sums up the basic concepts of the general adaptive (inverse scattering) radar system, whereas figure 2.2 shows the radar signature (correlation function in the frequency domain) of some idealized scattering objects. Note that the spatial correlation function is obtained by changing the abscissa of figure 2.2 from ΔF to $R\lambda/\Delta x$.

Before we proceed to give experimental verifications of the simple mathematical models based on first principle physics, let us consider the basics of polarimetry in relation to a multi-frequency radar system. Single frequency pulsed polarimetric radars have recently received considerable attention (6-9); in this very space limited presentation, we shall confine ourselves to introducing the basic concepts involved in an experimental investigation which is in progress at the authors' laboratory aiming at increasing the target identification potential of our multi-frequency radar system. Note that the present radar (see figure 3.1) makes use of six correlated computer controlled frequency synthesizers which will give 15 different frequency spacings (can couple to 15 different target scales). The radar has two transmitters and two receivers with polarization control so as to enable us to determine the frequency covariance function

$$R(\Delta F) \sim V(F)\ V^*(F+\Delta F)$$

simultaneously for 15 frequency separations and for the three polarization combinations (horizontal/horizontal (S_{11}), horizontal/vertical (S_{12})and vertical/vertical (S_{22}). The radar can also operate with adaptive polarization basis, that is, the radar optimizes its detection and identification capability by transmitting the optimal elliptic polarization.

After having completed the preliminary tests with this system, the program will be expanded to include six spaced receivers so as to allow us to determine the spatial correlation properties of scattered field with 15 spatial separations.

In order to introduce this concept, figure 2.3 is presented. Here we have "modelled" a target in the form of an airplane making use of seven isotropic and polarization invariant scattering centers. Note that the scattering matrix elements merely are represented by their moduli. The actual system, however, measures their relative phase thus allowing detailed analysis of the target's symmetry properties within each time/space resolution in an arbitrary polarization basis.

Figure 2.3 refers to the longitudinal case where the target is viewed head on with 15 coherent frequency spacings. The idealized delay function is shown and the corresponding frequency covariance function $R(\Delta F)$.

In exactly the same way, the spatial correlation of the scattered field can be calculated, giving a $R\lambda/\Delta x$ abscissa rather than a ΔF abscissa.

Having introduced the multi-domain adaptive radar designed primarily for environmental surveillance applications, we shall present a somewhat more general target configuration and give the $R(\Delta F)$ signatures for the three pertinent polarization configurations.

Figure 2.4 shows the simplified air plane target viewed head on with our 15 frequency spacings and three polarization combinations. Note that as in figure 2.3 the target has seven scattering centers. However, as a means of illustrating the polarization issue, we have selected four different classes of scattering in figure 2.4 with the following notations:

-- scatterers for horizontal/horizontal S_{11}
¦ scatterers for vertical/vertical S_{22}
| odd-bounce scatterers
< even-bounce scatterers

Note that to the extent that the target has polarization sensitive scattering centers, we can draw conclusions regarding detailed aircraft dimensions, not merely overall length as in the case of the simplest scheme illustrated in figure 2.2.

3 A BRIEF DESCRIPTION OF THE RADAR SYSTEM

The adaptive polarimetric multi-frequency radar system shown in figure 3.1 is made from a conventional 960 channel microwave communications link. The system structure is, as indicated in the figure, very simple. The six frequencies in the 50 - 90 MHz band delivered from a common crystal oscillator are up-converted to the 6 GHz band giving a total output power of 200 mW (33 mW per frequency) to the transmitting antenna. Two different antennas were used for the various applications: a horn antenna with an aperture of 30 x 30 cm and a paraboloid with diameter 1.20 m.

The receiving antenna was positioned adjacent to the transmitter (110 dB isolation). Transmitting and receiving antennas are identical. The backscattered signal is down-converted to the IF band (50 - 90 MHz) by means of a 6 GHz source. This source is x-tal-controlled and common to both the up- and down convertors. Upon amplification and filtering, the six separate VHF receiver frequencies are mixed with the corresponding six frequencies from the transmitting synthesizers. The resulting 6 voltages giving 15 different frequency pairs are then multiplied and the products (the 15 covariances V(F,t) V*(F+ΔF,t) resulting) are subjected to 15 sets of Fast Fourier Transform filters, thus producing 15 power spectra (Doppler spectra).

The essence of these processes is illustrated in figure 4.1.

4 EXPERIMENTAL VERIFICATIONS

The multi-frequency radar system was used for three different investigations:

- Measurement of directional ocean wave spectra (wave intensity, velocity and direction)
- Investigation of ship signatures against a sea background
- Classification of air targets (F-16 aircraft)

A brief highlighting of the results of these investigations will now be given.

4.1 Directional ocean wave spectra

Illuminating the sea surface from a cliff 50 m above the sea, the ocean wave spectra (wave intensity and wave velocity) was determined for various azimuth directions and for 15 different ocean wavelengths in the interval from some 5 m (couples to ΔF = 27 MHz) to 150 m (corresponding to ΔF = 1 MHz). For each frequency separation ΔF the power spectrumn of the frequency covariance function V(F,t) V*(F+ΔF,t) was computed.

Examples of such power spectra (Doppler spectra) are shown in the upper part of figure 4.1. In the lower part of the figure two curves are plotted: The curve marked with crosses gives the wave intensity (wave height) spectrum for ocean wavelengths ranging from 18 to 150 m. The curve marked with points gives the Doppler shift as a function of ocean wavelength obtained from the power spectra shown in the upper part of the figure. Note that the theoretical Doppler shift (phase velocity of gravity waves given by:

$$V = \sqrt{\frac{gL}{2\pi}}$$ where L is the ocean wavelength

(ref 4)) is also shown. The systematic shift of the experimental points towards higher velocity is probably due to tidal currents on which the wave motion is superimposed. We obtain one such set of spectra for each azimuth direction of the antenna system.

4.2 Radar signature of a ship against a sea clutter background

Let us focus our attention on ship targets obtained experimentally. Figure 4.2 shows the time-record of the frequency covariance function V(F,t) V* (F+ΔF,t). To illustrate the essential features, three values of ΔF (out of the total ensemble of 15 frequencies) have been selected. We see that when the radar beam illuminates the sea surface only, the time-record shows an irregular structure. When the ship enters the radar beam, a Doppler signature which is proportional to the frequency separation ΔF is clearly shown.

Before we change the subject from sea surface targets to aircraft, let us present another two sets of experimental results on ship signatures against a sea clutter background. With reference to figure 4.2 above, figure 4.3 shows the Doppler shift for the various frequency separations caused by the ship and by the sea surface, respectively. Note that the ship gives a linear (non-dispersive) relationship between frequency separation and Doppler shift whereas the sea surface gives results which are in reasonably good agreement with the theoretical relationship for deep water gravity waves.

Finally, figure 4.4 shows the normalized correlation as a function of frequency separation for a particular cargoship and also for the sea surface background. Note that the sea surface signature is expanded by a factor 10^3 relative to the signature of the ship. Note also that there is a remarkably good agreement between experimental results and the theoretical ones if the assumption is made that the scatterers are distributed evenly over the scattering body 37.5 m long.

4.3 Radar signature of a rigid airplane

Finally, we shall present a very brief highlightening of the radar signature investigations which were carried out for various types of air targets. In this brief presentation we shall focus our attention on a particular rigid aircraft, namely the F-16 fighter airplane. Tests were also performed on airplanes of comparable size, but different structure. These produced drastically different signatures both as regards the Doppler spectrum (flutter and vibrations superimposed on a translatory motion) and as regards the frequency covariance function. Figure 4.5 shows the signature for the F-16 airplane. Note that by-and-large the agreement with theory when the assumption is made that the scatterers are distributed in a Gaussian manner is reasonably good. There is clear evidence, however, of a small number of scattering centers which dominate over the Gaussian distribution. For details the reader is referred to ref (3).

5 ACKNOWLEDGEMENT

The authors would like to acknowledge the very valuable services rendered by Dr Dagfin Brodtkorb of A/S Informasjonskontroll who laboriously and conscientiously developed the computer software and played a very active role in the planning of the data analysis.

The multi-frequency radar is based on a 960 channel radio relay link generously provided by Mr Standahl of A/S Elektrisk Bureau, Division NERA.

Chief Scientist Karl Holberg of the Norwegian Defence Research Establishment, the previous affiliation of the authors, has been a continuous source of stimulation and encouragement, and we are also indebted to Mr Per Ersdal of the Norwegian Defence Communication Administration who provided antennas and test facilities.

REFERENCES

1. Gjessing, D.T., 1978, "Remote surveillance by electromagnetic waves for air-water-land", Ann Arbor Publishers Inc., Ann Arbor, USA.

2. Gjessing, D.T., 1981, "Adaptive radar in remote sensing", Ann Arbor Publishers Inc., Ann Arbor, USA.

3. Gjessing, D.T., Hjelmstad, J., Lund, T., 1982, "A multifrequency adaptive radar for detection and identification of objects. Results on preliminary experiments on aircraft against a sea clutter background", IEEE Trans., AP-30, 3, 351-365.

4. Gjessing, D.T., Hjelmstad, J., Lund, T., "Directional ocean spectra as observed with a Multi-Frequency CW Doppler Radar System", Submitted for publication.

5. Gjessing, D.T., 1981, "Adaptive techniques for radar detection and identification of objects in an ocean environment", IEEE Journal of Ocean Engineering, OE-6.1, 5-17.

6. Boerner, W.-M., Chan, C.-Y., Mastoris,P., El-Arini, M.B., 1981, "Polarization dependence in electromagnetic inverse problems", IEEE Transactions on Antennas and Prop., AP-29.

7. Huynen, J.R., 1970, "Phenomenological theory of radar targets", PhD Dissertation, Drukherij Bronder-Offset, N.V. Rotterdam.

8. Poelmann, A.J., 1979, "Reconsideration of the target detection criterion based on adaptive antenna polarization", Tijdschrift van het Nederlands Electronica en Radiogenootschap, 44.

9. Deschamps, G.A., 1951, "Geometrical representation of the polarization of a plane electromagnetic wave", Proc. IRE, 39.

10. Bass, F.B., Fuks, I.M., 1979, "Wave scattering from statistically rough surfaces, Pergamon, New York.

11. Plant, W.J., 1977, "Studies of backscattered sea return with a CW dual-frequency X-band radar", IEEE Trans. Antennas and Propagation, AP-25, 1, 28-36.

12. Weissmann, D.E., Johnson, J.W., 1977, "Dual frequency correlation radar measurements of the height statistics of ocean waves", IEEE Journal of Oceanic Eng. OE-77, 74-83.

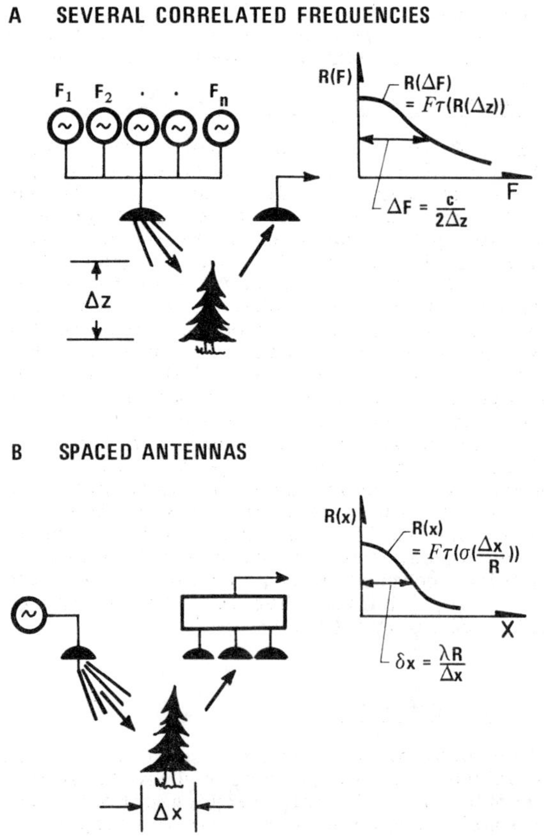

Figure 2.1 The basic concepts of the general adaptive (inverse scattering) radar system

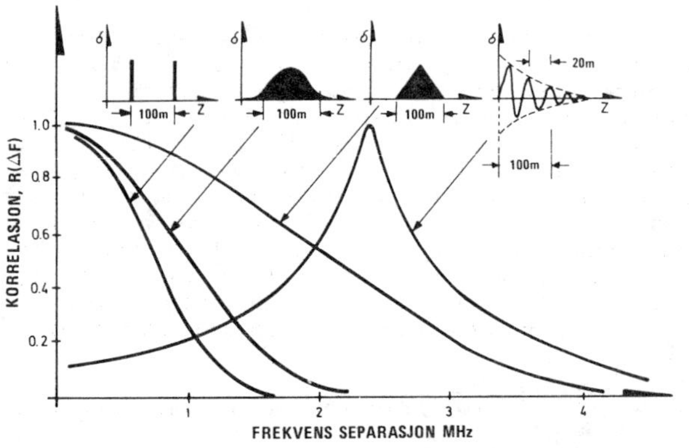

Figure 2.2 The radar signature (correlation function) of some simple scattering objects

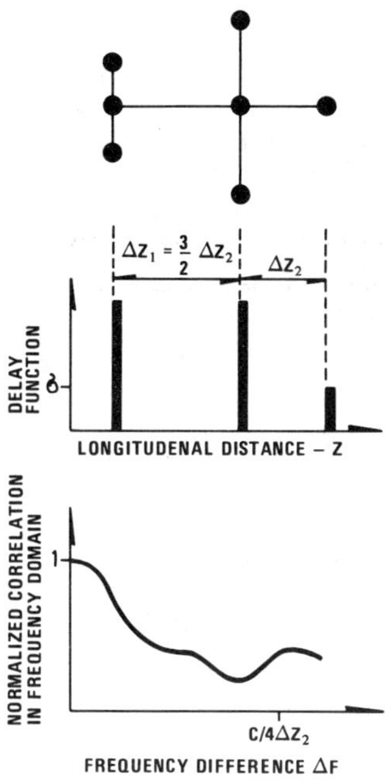

Figure 2.3 An idealized target consisting of isotropic polarization invariant scattering senters and the corresponding multi-frequency radar signature; the correlation function R(ΔF) is measured

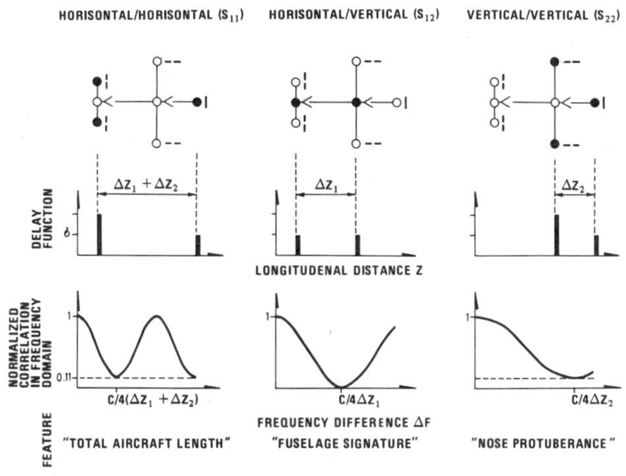

Figure 2.4 Measuring the frequency covariance function R(ΔF) for the three appropriate polarization combinations, information can in principle be obtained about the detailed longitudinal target dimensions.

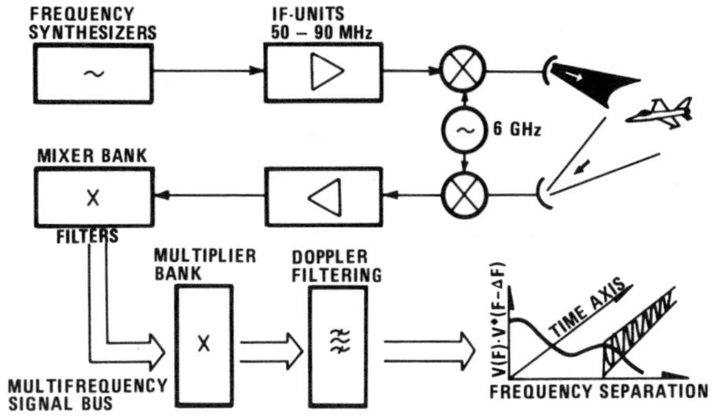

Figure 3.1 The multi-frequency radar system

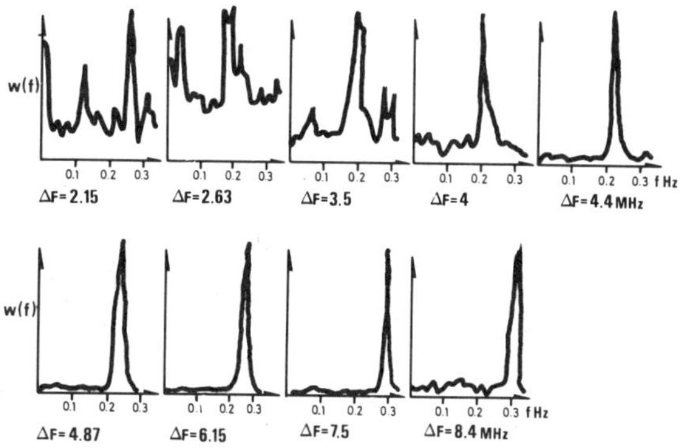

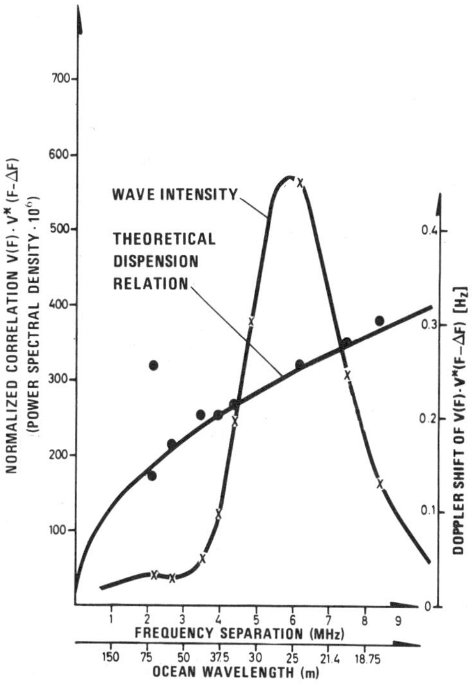

Figure 4.1 Wave intensity and wave velocity plotted to the basis of ocean wavelength (frequency separation). Note that the upper set of curves gives the Doppler spectra for various ocean wavelengths (frequency separation ΔF). (For details see ref (4))

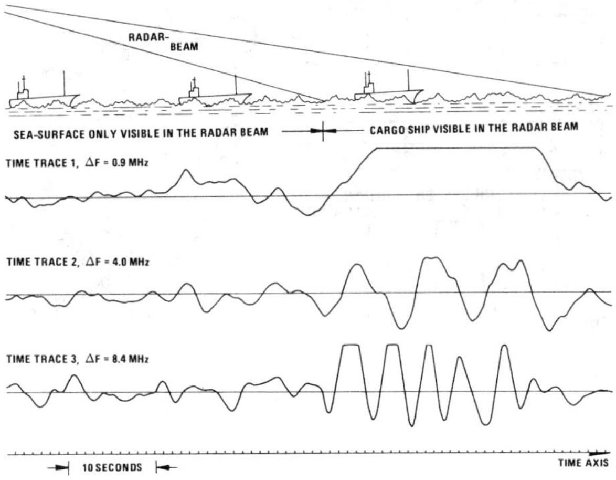

Figure 4.2 Time-record of the covariance function in the frequency domain v(F,t) V (F+ΔF,t) for three different values of the frequency separation ΔF

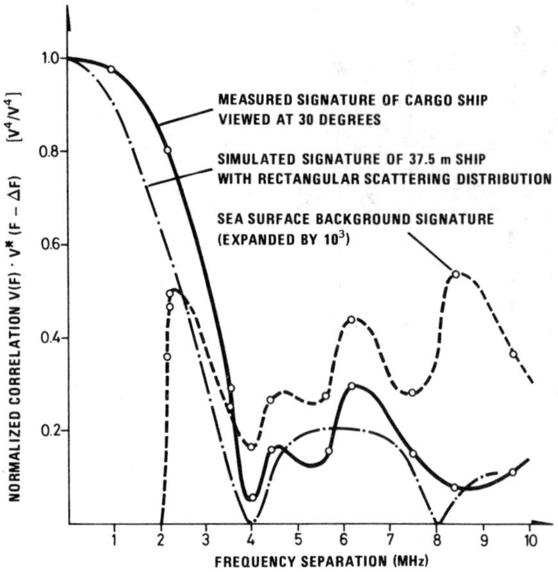

Figure 4.4 Normalized correlation in the frequency domain plotted as a function of frequency separation for a 37.5 m ship and for the sea surface background. Note the good agreement between measured and simulated signatures.

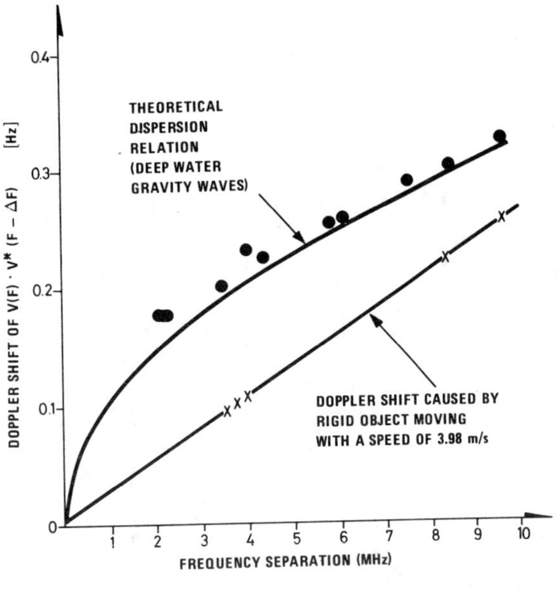

Figure 4.3 Doppler shift as a function of frequency separation for a rigid target and sea clutter, respectively. Note that deep sea gravity waves are dispersive: There is a square root relationship between phase velocity and wavelength. A rigid body moving at constant velocity gives a Doppler shift that is proportional to the frequency separation and the velocity of the object.

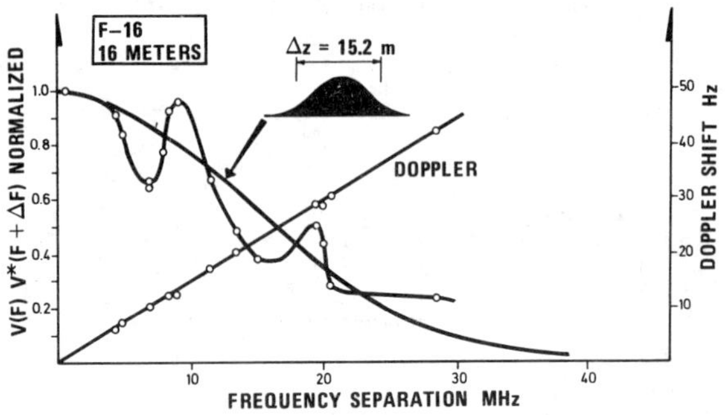

Figure 4.5 Longitudinal distribution of scattering centers for F-16 aircraft. There are a few dominant scattering centers and, apparently, many smaller ones that are distributed along the min axis of the aircraft (for details see ref (4))

INTRAPULSE POLARIZATION AGILE RADAR

M.N. Cohen and E.S. Sjoberg

Georgia Institute of Technology, USA

INTRODUCTION

The Intrapulse Polarization Agile Radar (IPAR) system developed by the Georgia Institute of Technology Engineering Experiment Station in conjunction with the United States Naval Sea Systems Command employs a new and unique form of pulse compression. As opposed to encoding the RF carrier with either phase or frequency modulation as has been done classically, IPAR utilizes polarization modulation to imprint the pulse compression coding on the carrier. As a consequence of this novel coding method, IPAR exhibits many unique features such as the ability to use a variety of RF waveforms, including single frequency, frequency agililty or wideband noise, the potential to exploit high range resolution polarization characteristics of targets and clutter, and insensitivity to Doppler shift.

The IPAR system has been implemented in an X-band radar designed and constructed at Georgia Tech. Traveling wave tubes amplify the independent horizontal and vertical polarization components of the RF energy which is then transmitted via a dual-mode feed as right or left circular polarization. The heart of the IPAR system is a high speed digital processor implemented in TTL and ECL technology. Binary codes up to 32 bits in length with bit rates up to 100 MHz may be generated and processed in real time.

BACKGROUND

Various signal processing techniques have proven so useful that they are currently considered indispensable in radar system design. Of these techniques, resolution improving processing--such as synthetic aperture imaging and pulse compression--have proven to be among the richest in terms of breadth and depth of application. The primary applications of pulse compression include: (1) S/N ratio enhancements due to the compression gain, (2) clutter reduction due to the resulting decrease in range cell size, (3) lowered probability of intercept (LPI) as well as other favorable electronic counter-countermeasure (ECCM) characteristics, again due to the compression gain, (4) imaging capability to provide detailed mapping as well as target characterization for identification purposes, and (5) waveform flexibility to implement a system capable of a multitude of functions (such as detect, track, identify).

Historically, pulse compression was first achieved utilizing linear frequency modulated (LFM) waveforms. Recently, as a result of the ascendancy of digital technology, biphase and polyphase coding of the carrier RF as well as discretely stepped (inter- and intra-pulse) frequency modulation have come to the fore.

The nucleus of the IPAR concept is that pulse compression can be achieved utilizing coding of the polarization of the transmitted electromagnetic wave as opposed to the classical methods of coding the phase or frequency of the carrier. The resulting waveforms have potential for power and resolution enhancement similar to that of the conventional compression techniques, but the IPAR waveforms exhibit certain strengths and sensitivities which are significantly different from those of the classicial techniques, especially when reflected by complex scatterers.

Polarization

Radar signals may be polarized in the sense that the E-field modulation can be constrained to a particular plane in 3-dimensional space or the plane of modulation can be varied in a particular manner. For example, two orthogonal linear polarizations are vertically (V) polarized and horizontally (H) polarized electromagnetic waves, where the plane of polarization is defined as that perpendicular to which no E-field modulation can be detected. Electromagnetic waves with elliptical polarizations are generated by transmitting sinusoidal H and V polarized waves simultaneously which differ in amplitude and/or phase. If the amplitudes of the H and V sinusoids are equal and are out of phase by 90°, then the polarization is said to be circular. If, in addition, the H component lags the V component by 90°, the resultant wave's plane of polarization varies according to the right hand rule in the direction of propagation, and the wave is said to be right circularly (RC) polarized. When the H component leads the V component, the wave is left circularly (LC) polarized. RC and LC are orthogonal polarizations.

The circular polarization of a reflected wave is affected by the nature of the reflecting surface. In particular, the circular polarization of a wave reflected from an odd-bounce scatterer is opposite in sense from the circular polarization of the incident wave (e.g., RC incident yields LC reflected and vice versa). As can be deduced from the odd-bounce case, even-bounce reflectors leave circular polarizations unchanged (e.g., RC incident results in RC reflected). The phenomenon underlying these effects is that on a single bounce the relative phase between the H and V components of the wave remains unchanged (since both are flipped by 180° in phase) while the direction of the propagation is reversed. The net result is a change in handedness of the electromagnetic wave, just as one would observe a change in handedness of the rotation of a transparent clock when it is viewed from front and back.

The phenomena just described have been exploited to achieve rain and snow clutter reduction.[1] The most straightforward application has been to utilize a circularly polarized antenna, say RC, to detect aircraft in rain. Since raindrops are essentially spherical, they act as odd bounce reflectors and thus return LC polarized waves. An aircraft, on the other hand, is a complex scatterer that will return some energy in each of the circular polarizations. A right circularly polarized antenna will reject the predominately LC polarized rain return and accept only the RC polarized rain returns. The percentage of even-bounce reflectance that the aircraft exhibits thus determines the achievable target to clutter enhancement.

Pulse Compression on Polarization Modulation

The IPAR system is capable of switching between right and left circular polarization on an intrapulse basis at rates of up to 100 MHz. Utilizing separate H and V ports on receive, IPAR downconverts the two channels separately, and accurately measures the phase between them. Let ϕ_{HV} represent the phase difference between the H and V channels (that is, $\phi_{HV} = (\phi_H - \phi_V)$), and recall that a signal is RC polarized if $\phi_{HV} = +90^{\circ}$ or LC polarized if $\phi_{HV} = -90^{\circ}$. For convenience, let $S\phi = \sin\phi$. Then $S\phi_{HV} = 1$ for RC signals and $S\phi_{HV} = -1$ for LC signals, thus giving a natural correspondence between binary codes and intrapulse polarization modulation. IPAR utilizes this correspondence to construct well-behaved polarization codes from the class of well-known, well-behaved binary codes.

Figure 1(a) depicts a natural choice for an IPAR transmit waveform. According to the correspondence described above, the chosen code represents a 13-bit Barker code, and it makes possible a pulse compression ratio of 13 to 1. Figures 1(b)-(e) represent four of the possible reflected waveforms. Figure 1(b) represents the echo from a flat plate, or any other idealized odd-bounce scatter. As explained above, the polarization of the echo in this case is always opposite in sense from the polarization of the transmitted wave. Figure 1(c) represents the return from an idealized even-bounce scatterer; thus the returned polarizations are the same as those transmitted. Figures 1(d) and 1(e) represent likely returns from complex scatterers where each was generated assuming a simple probabilistic model. In 1(d), it was assumed that the echo source was predominantly odd-bounce so that 75 percent of the time the polarization sense was changed, 12.5 percent of the time the polarization remained the same, and 12.5 percent of the time the sense was sufficiently mixed so as not to exceed a threshold either way. In 1(e), it was assumed that the scatterer had an equal mix of odd-bounce and even-bounce scatterers so that neither type predominated. Thus, the model assumed 33.3 percent polarization change, 33.3 percent retention of polarization sense, and 33.3 percent lack of sufficient bias to exceed a threshold either way (denoted by "0").

When receiving any of these four "possible" returns, the IPAR processor computes $S\phi_{HV}$ on a subpulse basis and passes the results through a matched filter (i.e., correlator). The compressed returns are depicted in Figure 2. Figure 2(a) represents the transmit waveform's autocorrelation, which is precisely the output of the processor in response to an even-bounce return (Figure 1(c)). Note that the resolution is $\tau/2$ rather than $T/2$ (a 13 to 1 improvement), and that the peak signal voltage is T/τ or 13 times the nominal uncompressed level. Figure 2(b) represents the output in response to an odd-bounce scatterer. Its power characteristics are precisely the same as those of 2(a), although there is a change of sign in voltage. As the reflecting scatterer gets more complex, the degree to which the transmitted code is retained in the reflected waveform decreases. Figure 2(c) still shows a recognizable peak in response to the "predominantly odd-bounce" reflection, but Figure 2(d) depicts the noise-like return expected from a scatterer that shows no predominance.

As with any scheme of compressing on frequency or phase modulation, there are particular strengths and weaknesses inherent in the specific compression process used. For the case of compressing on polarization modulation as in the IPAR system, the following observations are pertinent:

(1) Since only <u>relative</u> phase is important for compression, IPAR is totally insensitive to Doppler shift and can be implemented without a stable LO to provide coherency (i.e., IPAR can be implemented in a noncoherent system).

(2) Since man-made targets tend to be composed of fewer simple scatterers than most clutter, IPAR should, theoretically, provide improved stationary target detection capabilities (analogous to the MTI utility of binary phase codes).

(3) The added information contained in the polarization matrix of man-made targets may allow IPAR to provide enhanced target identification capabilities.

(4) Effective compression on polarization of complex extended targets may require very fine range and cross-range resolutions to isolate individual scatterers.

(5) ECCM - New equipment may be needed to counter polarization coded radar. The full ECCM impact of this technique has not been evaluated.

As noted above, IPAR's pulse compression coding is contained in the relative phase between the H and V polarized components of the echo pulse, and thus compression is totally insensitive to Doppler shift. That is, IPAR represents the first all-velocity compression scheme to be invented. LFM exhibits a range-Doppler ambiguity that makes it difficult to interpret the compressed signal returns from high velocity targets. Binary phase coded signals exhibit a loss in processing gain at any appreciable Doppler shifts which makes that compression coding technique difficult to use with moving targets. IPAR represents a truly all-velocity compression scheme that will faithfully perform detection, location, and track independent of target velocity and without the need to employ banks of Doppler filters.

Since absolute carrier phase is of no importance in an IPAR system, various RF carriers can be used. In particular, the RF may just as well be wideband modulated noise or a frequency hopped carrier instead of a single narrow band frequency. These possibilities hold a great deal of promise for LPI operation, ECCM, and the selective decorrelation of clutter. While the sensitivity of polarization coding to target composition necessitates detailed analyses of the waveform for various scenarios, this sensitivity also suggests many potentially useful characteristics. For example, if the subpulse length can be reduced to enable range resolution on the order of one foot or so, it seems likely that many man-made objects would appear to be made up of individual, simple scatterers--in which case, compression would be almost totally effective. On the other hand, many types of clutter such as leaves, windblown rain and snow, and field backscatter would still appear to be made up of many independent scatterers--thus giving more noise-like returns. This difference in composition should allow substantial target to clutter enhancement.

Similarly, in addition to the usual gains one achieves against multipath by employing pulse compression (by "gating out" double bounce returns on a range resolution basis), an IPAR type process would take advantage of the depolarizing effect of the second bounce to further reduce multipath interference, especially when the "ground" is the surface of a somewhat turbulent sea.

<u>IPAR IMPEMENTATION</u>

IPAR was designed and built as a demonstration and data collection system. The goals were first to establish that the technique would work and second to collect data which could be used to quantify its strengths and weaknesses relative to classical radar techniques.

Some of the design goals were to:

(1) Achieve a 100 MHz real-time correlation and compression rate;

(2) Correlate and compress using noncoherent techniques and a wide bandwidth noise RF carrier;

(3) Permit growth to computer control and data analysis;

(4) Incorporate the ability to range gate and replay correlation data for quick on-site observation and analysis.

Figure 3 shows a block diagram of the resulting system. Functionally, the system may be viewed as two subsystems: the RF subsystem and the IPAR digital processor subsystem. The basic system operation is described in the following paragraphs.

The IPAR system includes a two-channel master oscillator power amplifier (MOPA) radar transmitter. The two channels are connected to the vertical and horizontal inputs of a dual-polarized antenna feed. Whenever a 90° phase relationship exists between the two channels, the resultant transmission is circularly polarized. By switching the phase relationship to a negative 90°, the opposite sense circular polarization is transmitted. By switching in accordance with an advantageous coding sequence, a pulse is transmitted which is coded right and left circular.

The phase shift for one channel as referenced to the other is achieved at a low power level due to the speed limitations of high powered phase shifters. For this reason, the IPAR system was built with two high power amplifier channels. The phase shift of one channel is accomplished at low power using a mixer, which, for the devices used, could conceivably modulate the polarization at rates in excess of 2 GHz. The mixer (i.e., phase shifter or phase modulator) has the characteristic that a high level signal (i.e., +1) at the IF port causes zero phase shift from the input to output; whereas a low level signal (i.e., -1) causes a phase shift of 180°. With a permanent 90° phase shift added to the 0°, 180° phase shift combinations, the result is -90°, 90°. Hence, the required phase relationships are generated for the right and left circular polarization-coded transmissions, which may be at a high data rate utilzing a non-coherent carrier, even noise.

As in transmission, two receive channels are employed. The signals in the two receive channels are translated down in frequency, retaining the relative phase between the two signals. Finally, the two signals are applied to a phase detector. The polarity and amplitude of the resultant bi-polar video signal will be similar to the coding impressed on the transmission (and stored in the correlator) provided that the return is of high fidelity and not noise-like. The high speed correlator samples the returning signal at a maximum rate of 100 MHz. Future VHSIC and Gallium Arsenide advances could increase this rate considerably. Once the video signal has been quantized and input to the correlator, the remainder of the operation proceeds like conventional biphase modulated pulse compression techniques.

The prototype IPAR digital processor was designed to provide a flexible, complete and self-contained radar signal correlation processor. It may be operated in a stand-alone mode, under front panel control, or connected to and controlled by a minicomputer or microcomputer. The processor is capable of correlating received signals in real time with the stored reference code at selected rates up to 100 MHz. The operator has freedom to independently determine the transmitted code, subpulse width, PRF, and range gate setting to achieve his objectives.

Figure 4 presents a simplified block diagram of the digital processor and its interfaces to the external world. The slower TTL processor (6 MHz) may be considered the host or control portion of the unit which interfaces with the front panel and the control computer (when provided), and determines the operational mode of the processor. The high speed ECL processor interfaces with the radar generating the transmit and coding signals, receives and quantizes the video from the radar, performs correlation processing in real-time and stores data from the range gate position. A list of basic capabilities is presented below.

IPAR Digital Processor Capabilities

Subpulse Length:	10 - 160 ns (100 - 6.25 MHz)
PRF:	500 - 8000 Hz plus manual
Range Gate:	0 - 15 km, 0.75 m Maximum accuracy
Codes:	Any binary code up to 32 bits long
Correlation Processor:	True correlation for 32 bit code length Pseudo correlation for 1 - 31 Bit Code Lengths
Digital Threshold:	Adjustable
Self Test/ Calibration:	Built-in

Field Test Results

Local field measurements have been conducted at Georgia Tech to verify system operation and theoretical predictions. Some of the conclusions from those qualitative tests are listed below:

(1) Odd bounce (trihedral) targets produced a negative correlation.

(2) Even bounce (dihedral) targets produced a positive correlation.

(3) Tree targets produced correlations which change from positive to negative in an unpredictable manner. The rate of change and duration of positive and negative correlations appeared to be on the order of the time required for clutter to decorrelate using standard fixed frequency radars.

(4) When using the swept frequency source, odd and even bounce targets produced results similar to those observed during fixed frequency operation.

(5) When using the noise source, odd and even bounce targets produced results similar to those observed using fixed frequency.

(6) When using either the swept frequency or noise source, tree clutter did not appear to correlate to a significant level and changed from positive to negative values very rapidly. No actual measurements of the decorrelation time of the tree clutter were made.

(7) The IPAR correlation function showed that the system did compress target signals to a single bit width.

(8) Multiple targets in different resolution cells, but separated by less than the total pulse length, correlated separately and to a level inversely proportional to their separation distance. The dominant (largest) target correlated to the full 32 bit level while smaller targets correlated to a level determined by their separation from the largest scatterer. This result is as expected and is due to the one bit plus sign quantization used in the IPAR system.

In addition, preliminary experiments involving multipath reduction, Doppler-invariant pulse compression and ECCM properties have been performed. Qualitative results of these tests indicate that the IPAR technique behaves according to theoretical predictions.

FUTURE DIRECTIONS

Plans are being developed to perform an extensive data collection and analysis program in mid-1982 which should quantify the performance the IPAR technique. In addition, an advanced IPAR is being designed which includes several significant advances over the current IPAR:

(1) Embedded microprocessor controller;
(2) Six-bit A/D conversion of the phase video at 100 MHz;
(3) 64-bit code length and true correlation for all codes;
(4) Signal integration capability;
(5) Pseudo real-time processing of 256 range bins.

The embedded microprocessor controller will allow pulse-to-pulse code agility as well as provide a signal integration capability and an automatic digital tape recording facility. By employing six-bit A/D converters, small scatterers separated by more than one subpulse length from a larger scatterer should be resolved from the larger reflector, which will enable true target polarization profiling. The code length will be expanded to 64 bits, and a true correlation will be formed for all possible codes including those with embedded zeros. The extended code length and correlation capability is obtained at the expense of real-time operation. Instead, 256 range cells will be sampled and processed during each pulse repetition interval.

The IPAR system has evolved from a concept to a demonstrated working radar system. When the full potential of the IPAR concept has been investigated, it may prove to be a significant advance in radar technology.

REFERENCES

1. M. Skolnik, Introduction to Radar Systems, McGraw Hill, Inc., New York, New York, 1980.
2. B. C. Appling, E. S. Sjoberg, E. E. Martin, Intrapulse Polarization Agile Radar, Final Technical Report, Project A-2343, contract number SCEEE-NAVSEA/79-2, Georgia Institute of Technology, July 1981.
3. M. N. Cohen, E. E. Martin, E. S. Sjoberg, Intrapulse Polarization Agile Radar Development Program, Final Technical Report, Project A-2343. Contract SCEEE/81-2, Georgia Institute of Technology, December, 1981.

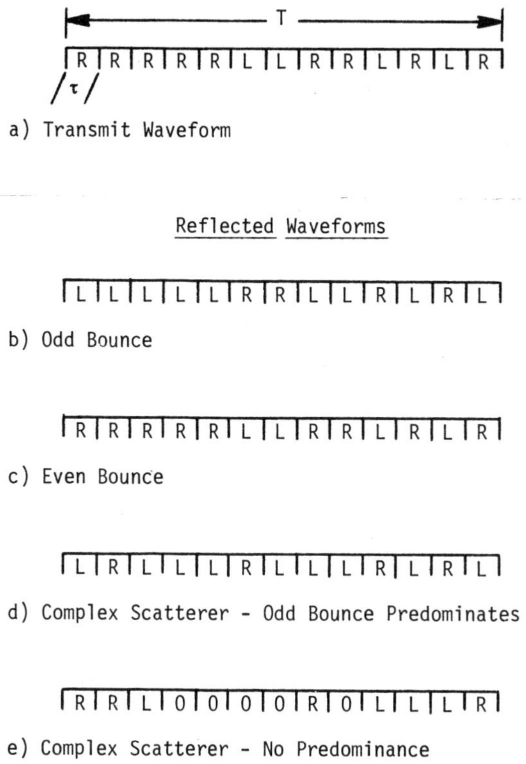

Figure 1. Simulated IPAR Transmit and Receive Waveforms.

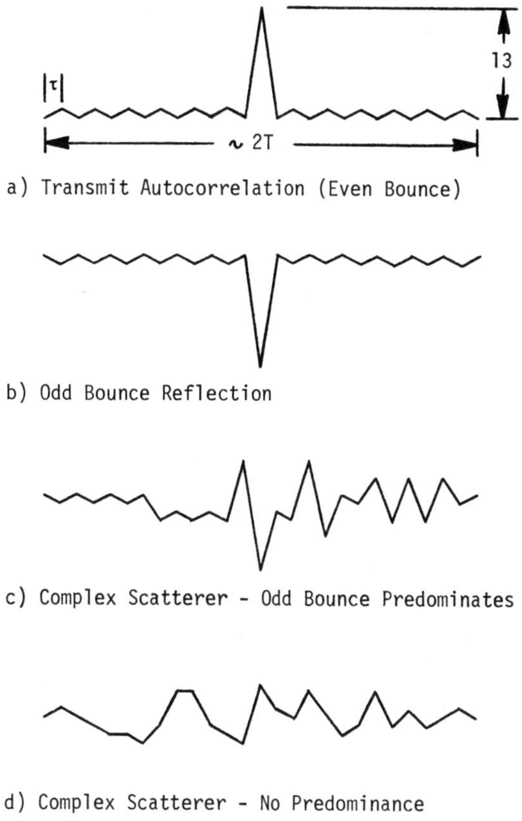

Figure 2. Compressed Output

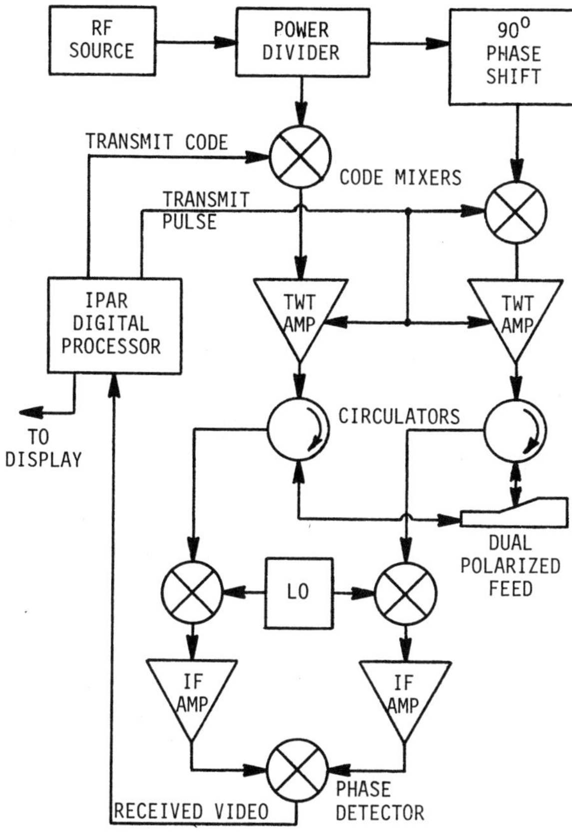

Figure 3. IPAR Block Diagram.

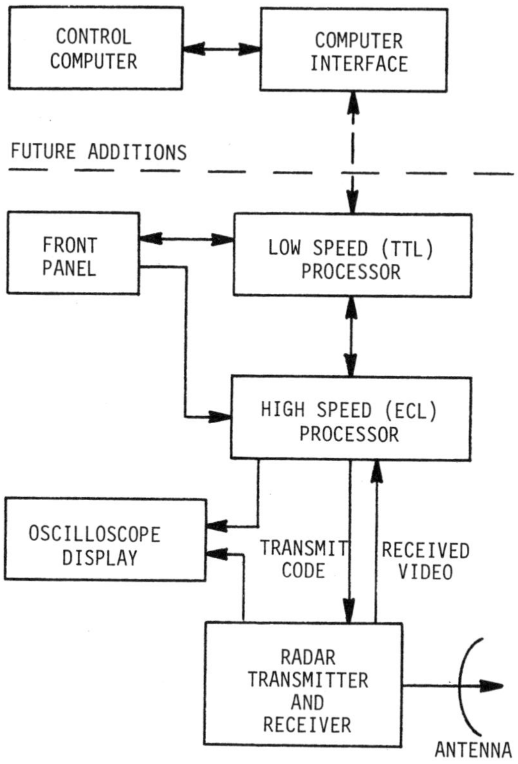

Figure 4. IPAR Digital Processor Block Diagram.

ON SEARCH STRATEGIES OF PHASED ARRAY RADARS

W. Fleskes

FGAN - Forschungsinstitut für Funk und Mathematik, Federal Republic of Germany

INTRODUCTION

The beam agility of a phased array radar system allows that the system can be seen as two independent devices, one for searching the surveillance area and the other for automatic tracking of the objects found (Skolnik (1)). This is in contrast to systems with a rotating antenna where searching and tracking activities use the same fan beam or pencil beam if elevation scanning is also done. As a consequence the track-while-scan mode of operation has to be used, which demands for a radar plot at each revolution during the whole time the target is in the surveillance area.

If the tracking activity can be decoupled from the searching, the latter has to provide only the first radar plot of a new target. An object once detected is then handed over to the tracking mode where by special means like adaptive scan-intervals, narrower correlation gates, other detection and position finding methods etc. the track can be conducted very precisely (Fleskes, v. Keuk (2)). Furthermore tracking is possible at longer ranges where automatic tracking by the track while scan mode breaks down due to missing plots and large update times. With computer controlled tracking low single pulse detection probability can be compensated by short scan intervals.

OPTIMISATION OF SEARCH EFFORT ALLOCATION

If the probability that a target appears at a specific azimuth-elevation direction is not the same for all directions then it seems to be quite naturally to spend more effort for those directions where this probability is higher. The question how much more in relation to the probability can be answered by the theory of optimal search, which provides for instance methods to maximise the probability of detection P_e under the constraint of a limited search effort N, which in this case means the total amount of radiated energy and which should be measured in units of illumination density or number of radar pulses. P_e is the probability for at least one detection, denoted sometimes cumulative detection probability in contrast to the single pulse detection probability in the following termed P_d.

Stationary Targets

Suppose there are k distinct direction-cells, each having a probability p_i for the target to be in cell number i and the search effort for that cell should be z_i, with i = 1, ... k and

$$\sum z_i = N$$

Furthermore there must be a detection function b(z) which relates a search effort z to the probability of detection. From Stone (3) this can be expressed by

$$b(z) = 1 - e^{-az}$$

where a is a constant to adjust the units and to describe the effectiveness of the sensor. E.g.

$$a = -\ln(1 - P_d) \tag{1}$$

if P_d is detection probability for a single pulse or one unit of effort. With this we have

$$P_e = \sum p_i\,(1 - e^{-az_i}), \quad z_i \geq 0 \tag{2}$$

Let the p_i be arranged in decreasing order and $k' \leq k$, then (2) is maximised with respect to the z_i by

$$\begin{aligned} z_i &= N/k' + (\ln(p_i) - 1/k' \sum_j \ln(p_j))/a, & i \leq k' \\ z_i &= 0 & i > k' \end{aligned} \tag{3}$$

The last equation indicates, that it may happen that no effort has to be placed in some cells where the p_i are too low, so that the bounded effort should be concentrated on these cells where the searching is more promising. The first equation describes a uniform distribution which is modified inversely proportional to the sensor effectiveness. The modification for a cell i is done according to the difference between $\ln(p_i)$ and their average.

So if the search frame time for the whole surveillance area is N seconds, then using z_i seconds search time in cell i will maximise the detection probability for stationary targets. That means the fraction

$$f_i = z_i/N, \quad \sum f_i = 1, \tag{4}$$

of the search frame time is spended in cell number i. These fractions could be used as a probability distribution from which samples can be drawn for random searches, but a more detailed analysis for random searches is given in the next section.

To optimise the time until first detection of a target by N discrete radar pulses one has to find an optimal sequence in the cell numbers i, i = 1,...k, which may be difficult and tedious due to the enormous number of possibilities. This problem does not arise for random searches.

Random Search

Phased array radars are not forced to use a fixed regular search pattern but can realize any pattern and at last random ones, where each radar pulse is drawn as a sample from prescribed distribution function. These irregular patterns may be useful for specific operational reasons. Rewriting (2) with (1) one obtains for the detection probability

$$P_e = \sum p_i(1 - (1 - P_d)^{z_i}),$$

where in this case the z_i are random variables. Therefore it only makes sense to maximise the expectation of P_e. This can be written as

$$E\left[P_e\right] = \sum_{z_i}\sum_i p_i(1-(1-P_d)^{z_i})\; p(z_1,\ldots z_k)$$

where $p(z_1, \ldots z_k)$ is the multinomial density depending on parameters $f_1, \ldots f_k$. The summation over z_i for i = 1, ...k under the condition

$\sum z_i = N$ yields

$$E\left[P_e\right] = \sum p_i(1 - (1 - P_d f_i)^N) \tag{5}$$

This is maximised with respect to the f_i by

$$f_i^* = (1 - p_i^{-1/(N-1)}(k'-P_d)/\sum_j p_j^{-1/(N-1)})/P_d \quad (6)$$

where $k' \leq k$ is again the member of cells with a probability p_i high enough to be searched. Using the optimal parameters f_i^* one obtains (assuming $k' = k$)

$$E\left[P_e^*\right] = 1 - (k-P_d)^N / (\sum p_i^{-1/(N-1)})^{N-1} \quad (7)$$

which can be compared with a uniform distributed random search, for which with $f_i = 1/k$ and from (5)

$$E\left[P_e\right] = 1 - (k-P_d)^N / k^N$$

The latter can be shown to be always less or equal than (7) and indicates that using a uniform random search, the expectation of P_e does not depend on the target distribution. Therefore the uniform random search serves in the following for comparision. For reasonable values of N and P_d there is only little difference between (4) and (6).

SIMULATION METHOD FOR MOVING TARGETS

To optimize search effort allocation one needs the distribution of the targets to be detected. This distribution depends not only on the scenario but also on the search effort allocation itself, because for a phased array system we have to deal only with targets not yet detected. Fig. 1 shows a typical surveillance area of a phased array antenna. In the following only one plane with fixed azimuth will be considered and it is assumed that the targets enter the surveillance area at maximum range with a uniform distribution for altitude and velocity and move radially without further accelerations towards the radar center. During the time of this movement a number of radar pulses can be radiated and used to detect the target. The distribution in the elevation angle ϑ of the undetected targets can be observed and this situation is depicted in Fig. 2.

Cosecant-Square Distribution

The distribution of all targets, detected and undetected, integrated over the time from entering to leaving the surveillance area is with R_m, h_m maximum-range and -altitude and

$$\vartheta_0 = \text{arctg}\,(h_m/R_m)$$

$$p(\vartheta) = R_m/(2\,h_m \cos^2(\vartheta)) \qquad 0 \leq \vartheta \leq \vartheta_0$$

$$p(\vartheta) = h_m/(2\,R_m \sin^2(\vartheta)) \qquad \vartheta_0 < \vartheta \leq \pi/2$$

The latter equation is the well known Cosecant-Square distribution which is derived here by describing the scenario statistically. These density functions can be used for a first optimisation step by calculating appropriate values p_i and using (6), but this involves all targets and not the undetected ones only.

Simulation of Radially Moving Targets

For simulation of the scenario using a random search one has to determine currently the next time the considered azimuth plane gets a next radar pulse, whose elevation value is then drawn as a sample from the search distribution for the considered azimuth plane. This can be done by a random generator producing exponentially distributed numbers until the target is detected, and after detection a next target is simulated. By observing the elevation positions of the undetected targets we can obtain their distribution and generate a more effective distribution for the search directions. Since the undetected targets are influenced by the search pattern itself the simulation has to be repeated several times until the observed distribution is stable, which can be confirmed by an identity-test for two distribution functions.

CONCLUSIONS

Fig. 3 shows the distance the targets travel in the surveillance area before detection for different single pulse detection probabilities P_d, which are assumed to be constant with range. As far as the average search frame is concerned we have assumed 50 different azimuth planes and a dwell-time of 10 ms per pencil-beam position. The beam-width (gaussian shaped) was assumed to be 2°. Consequently 50 x 10 ms will be the mean time difference between two consecutive search activities in the considered azimuth plane.

In comparision to an uniform random search (curve A) the optimised search effort allocation (curve B) exhibits earlier detection which can be seen as an extension of the average operational range of the radar system. To achieve a similar extension by using a more powerful system with higher energy radiation would be costly due to the radar equation.

REFERENCES

1. Skolnik, M.I., 1970, "Radar Handbook", McGraw-Hill Book Company, London, U.K.

2. Fleskes, W., van Keuk, G., 1980, "Adaptive Control and Tracking with the ELRA Phased Array Radar Experimental System", IEEE 1980 International Radar Conference Proceedings, Washington, U.S.

3. Stone, L.D., 1975, "Theory of Optimal Search", Academic Press, New York, U.S.

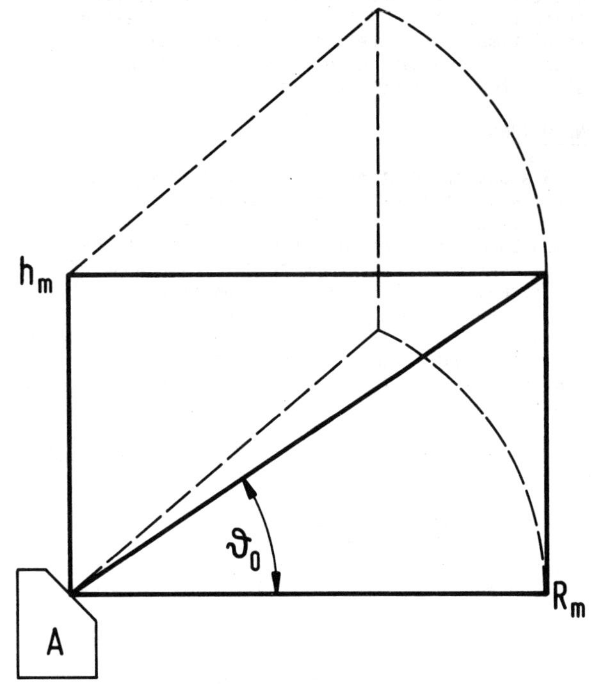

Fig. 1 Surveillance area of a phased array antenna A

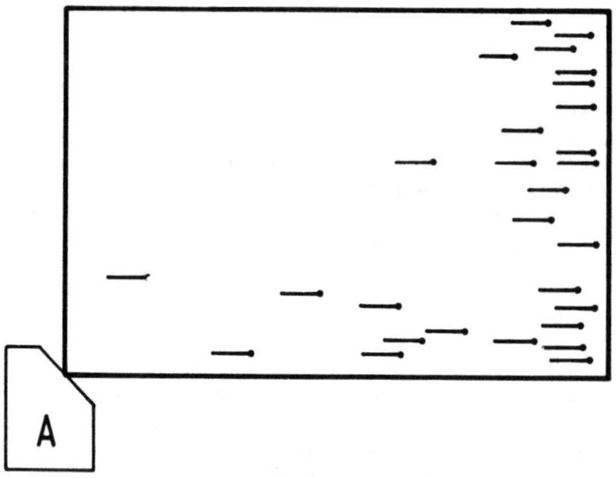

Fig. 2 Radially approching targets

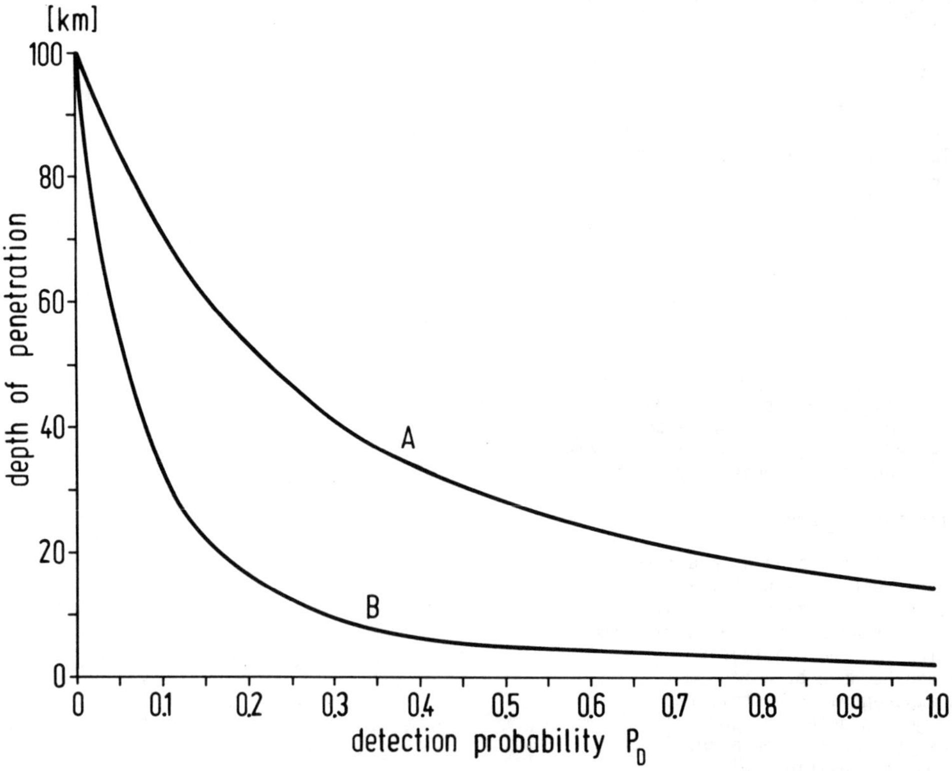

Fig. 3 Average Penetration depth for targets entering at a maximum range of 100 km with velocity of 300 m/s.

DESIGN AND PERFORMANCE CONSIDERATIONS IN MODERN PHASED ARRAY RADAR

E R Billam

Admiralty Surface Weapons Establishment, UK

INTRODUCTION

Phased array radars, with a fully steerable pencil beam, have been in existence since the 1950's and their virtues are generally appreciated. Nevertheless the impact of this type of radar on the air defence and air traffic control field has been slight. A reason for this is that conventional fixed rotation rate radars have been both good enough and cheaper. A conventional radar in this context is simply one which does not produce a fully agile pencil beam. However, the increasingly stringent performance requirements for radars has led to increasing sophistication and cost, even in nominally conventional designs. At the same time advances in computers, digital signal processing and solid state microwave components now offer the possibility of phased array radar designs which will not only allow a much greater realisation of the potential of this type of radar than has hitherto been the case but which could be competitive in cost to conventional radars.

Although the impact of modern technology can be felt in all aspects of phased array radar design, it is particularly significant that recent advances in monolithic microwave integrated circuits now make it possible to think in terms of an active solid state array in which a transmitter and receiver could be associated with each radiating element. This is very appropriate to solid state implementation on transmit since solid state devices are inherently low power devices. It permits high duty factor waveforms and represents a more effective use of transmitter power in that the loss from high power r.f. phase shifters and power dividing networks is avoided. An active array also allows the possibility of antenna array signal processing for both jammer and multipath suppression while programmable waveform generation and signal processing techniques allow the adaptive control of transmitted and compressed pulse length, pulse repetition interval and dwell time.

Given the parameter variability and adaptivity which are now in principle possible, questions arise as to the choice of parameters for optimum energy management and the sensitivity of performance to variations from the optimum values. This paper addresses some of these questions. The answers are illustrated by their application to the design of a long range air defence radar.

FRAME TIME OPTIMISATION

Fully agile electronic scanning allows step-scan surveillance in which the beam is directed in a given sequence to pre-determined beam directions. These can be represented as beam positions in u-v co-ordinates (sine θ space) as shown in Fig.1. The frame time, T_f, is in principle the time taken to look in all beam positions. In practice, the surveillance may be carried out in a far from simple manner, for example there may be a number of concurrent range-dependent surveillance modes, and so it is more appropriate to regard the frame time as the interval between successive looks in a given beam position (for a given surveillance mode), ie the surveillance data rate.

Cumulative Probability of Detection

A particularly significant consequence of a fully agile beam is that a single detection in the surveillance mode can lead to track initiation. On the occurrence of a detection the relevant beam position can be revisited until either a track is confirmed or a false alarm is declared. Thus the cumulative probability of detection is effectively the cumulative probability of track formation. The significance of this is illustrated in Fig.2, in which are plotted cumulative probability of detection curves for several values of ΔR, the radial distance moved between successive looks ($\Delta R = v.T_f$). A 'look' corresponds to a number of pulses sent in a particular direction, the returns from which are integrated. In Fig.2, the probability of detection curve is also plotted. This curve is characterised by R_o, the range at which the probability of detection (per look) has a value of 50%. The parameter $\Delta R/R_o$ is of particular significance as the curve of cumulative probability of detection scales linearly with R_o for a given value of $\Delta R/R_o$.

We define the surveillance range R_c to be the range at which the cumulative probability of detection has a value of 90%. R_c can also be thought of as a function of the number of samples, or looks, per unit distance and the value of (R_c/R_o) can be thought of as a 'sampling gain'. This gain is equal to the equivalent increase in mean power which would be required to give the same increase in R_c (which is proportional to R_o) as that caused by a reduction in ΔR, for a given dwell time. Curves of sampling gain are plotted against $\Delta R/R_c$ in Fig.3 for a range of values of N, the number of pulses per dwell, and for a Swerling 1 fading model. Note that ΔR is now normalised by R_c.

Surveillance Frame Time

First, we consider the case of surveillance only, ie no tracking load. The radar mean power is distributed in some manner between the given beam positions. We may write for the ith beam position

$$P(i) = \frac{PT_d(i)}{T_f}$$

where P is the mean radar power and P(i) and $T_d(i)$ are the ith beam position mean power and dwell time respectively. We thus have for a given beam position a number of combinations of dwell time and frame time varying from short dwell time and short frame time (small R_o, high sampling gain) to long dwell time and long frame time (large R_o, low sampling gain) all of which correspond to the same mean power. The question now arises as to the possibility of optimising the frame time, ie of choosing a value of frame time which will give the highest value of R_c for a given target closing radial velocity. Light is thrown on this question by Fig.4 in which, for a Swerling 1 fading model and for incoherent integration, R_c is plotted as a function of frame time for closing velocities of 125, 250, 500 and 1000 m/s and for the other parameter values indicated. The variation of R_c with T_f is shown as a dashed line. It can be seen from Fig.4 that the variation of frame time produces broad maxima in R_c for each value of velocity. At the maxima the rate of increase of integration gain with frame time is equal to the rate of decrease of sampling gain.

The question of frame time optimisation for surveillance can be restated in perhaps operationally more relevant terms. Instead of keeping the mean power constant and computing the effect of frame time variation on R_c, we can specify R_c and ask how the system sensitivity required to achieve this varies with frame time. The question can be expressed in even more general terms by substituting $\Delta R/R_c$ for frame time.

For a given target, the reduction in sampling gain caused by an increase in frame time is accompanied by an increase in the integration gain. The difference between them (in dB) represents the change in system sensitivity required to maintain R_c. This has been computed for Swerling 1 and 2 fading models and a plot of the required system sensitivity vs R/R_c is given in Fig.5. Because the ordinate is the required system sensitivity (for which we could substitute with less generality the mean power), then the minimum of the curve represents the optimum value since the less system sensitivity (or mean power) required the better. This type of plot was first presented by Whitfield (1).

We see from Fig.5 that the optimum values of $\Delta R/R_c$ are 4% for the Swerling 1 case (incoherent integration) and 20% for both the Swerling 2 case (incoherent integration) and for coherent integration (Swerling 1). However these values are associated with quite broad minima in the required power.

We may observe from Fig.5 that at the optimum values of $\Delta R/R_c$ the required mean power is lower than that for the Swerling 1 case by 0.7 dB for incoherent integration. The Swerling 2 case corresponds to the use of frequency agility on the assumption that this produces pulse-to-pulse decorrelation of fading.

The Effect of Tracking Load

The time spent tracking reduces the surveillance performance by virtue of either an increase in the frame time or a reduction in dwell time. Clearly, the tracking load is highly dependent on the scenario; nevertheless it is possible to make some general observations.

In the first place, it can be appreciated that the appropriate strategy for minimising the effect of tracking load is to attempt to maintain the frame time at its optimum value by reducing the dwell time, if sufficient pulses per dwell are available. The effect of this can be seen in Fig.6 in which R_c is plotted as function of tracking load, defined as the proportion of time devoted to tracking. Note that a tracking load of 50% corresponds to a reduction in R_c of only 10%, which would require a compensating increase in mean power of only 1.8 dB to restore R_c to its value for no tracking load. Such a tracking load is very high and implies that as much power is put into the relatively few beam positions containing targets as in all other beam positions for surveillance.

The tracking load is characterised by a fairly high probability of detection per look with a small number of range cells to be processed on each track update. The requirement to maintain a high probability of detection for targets at all ranges means in effect that the dwell time in the tracking mode is proportional to R^4. This in turn means that the dwell time requirements for targets at long range dominate the tracking load, much more so than variations in radar cross section or update rate. The converse of this is that the exclusion of some long-range targets by, say, a small reduction in instrumented range, can dramatically reduce the tracking load. For example, if we assume targets to be uniformly distributed by area then it can be shown that a reduction in instrumented range of 10% would cause a reduction in tracking load to a little more than half its initial value (assuming all targets have the same size and update rate).

CHOICE OF FREQUENCY BAND

Let us consider an active phased array radar which has a transmitter and a receiver at each element in the array. It would be reasonable to assume in the following that the cost of the radar will be dominated by the cost of the transmit-receive (T/R) modules and that the number of elements, N_e, will be fixed by the number of T/R modules which can be afforded. The following expressions for detection performance have been derived (P_e is element power and λ is wavelength).

Surveillance

(a) In the Clear

$$R^4 \propto P_e N_e^2 \lambda^2$$

This is the usual power-aperture product relationship. For fixed N_e, R^4 is now proportional to λ^2.

(b) Jamming

$$R^4 \propto P_e N_e^2$$

Here for fixed N_e, R^4 is independent of λ. Error sidelobes have been assumed ($\sim 1/N_e$).

(c) Rain Clutter

$$R^2 \propto N_e \lambda^4$$

$$R^2 \propto N_e \lambda^8 \qquad \text{(3-pulse MTI)}$$

Tracking

(a) In the Clear

$$R^4 \propto P_e N_e^4 \lambda^2$$

For tracking the performance depends on both the signal-to-noise ratio and the beamwidth. Note that for fixed N_e, R^4 is proportional to λ^2, ie tracking performance improves with wavelength. This is contrary to conventional wisdom.

(b) Jamming

$$R^4 \propto P_e N_e^4$$

(c) Rain Clutter

$$R^2 \propto N_e^2 \lambda^4$$

$$R^2 \propto N_e^2 \lambda^8$$

An example of the significance of these expressions is given in Fig.7 in which for a number of frequency bands the relative system sensitivity (R^4) in the clear is plotted against array diameter for a number of elements assumed to be fixed. This shows the effect on range performance of constraints on array size. For instance, the reduction in array diameter at L-band from about 9 m to 7.5 m reduces the tracking performance to that of a 4 m S-band array.

WAVEFORMS

A consequence of the separation of the surveillance and tracking functions in phased array radar is that specialised waveforms for the two functions can be employed. Since only a rough estimate of range is required in surveillance (or even none at all) a fairly long pulse length can in principle be used. In practice, as with conventional radar, the length of the compressed pulse is influenced by the need to minimise clutter returns. In this connection the absence of scanning modulation should enhance MTI/Doppler processing against ground clutter and consequently should, as far as ground clutter is concerned, allow a somewhat longer pulse length.

The proportion of the scanned volume in which clutter is likely to be experienced is, for at least medium and long range applications, quite small. This can be observed in Fig.8 which shows the computer variation of clutter power with range for a moderate rainfall. It can be seen that, as a consequence of the use of a pencil beam, the clutter falls off at relatively short range for the second and higher beam-position rows. Where surveillance regions are effectively clutter-free the use of long compressed pulses, possibly of the order of several micro-seconds, is desirable on several grounds:

(a) It is appropriate to the use of fairly long dwells (many pulses per dwell), which in turn allows flexibility in energy management.

(b) The associated reduction in the number of range cells could lead to an improvement in sequential detection performance.

(c) The signal processing load is reduced, particularly for the application of some computation-intensive adaptive antenna array techniques.

In the tracking mode only a small number of range cells in the vicinity of the expected position of the target need be employed. Even for high bandwidth waveforms the average processing load could be acceptable. Since range ambiguity is not of primary concern, high prf waveforms are appropriate.

It is desirable for tracking to maintain an adequately high signal-to-noise ratio (and consequently probability of detection) for parameter estimation. The adaptive control of tracking dwell time, based on the estimation of signal-to-noise ratio, is envisaged. This amounts to matching the energy per dwell to the target range and radar cross section. A lower probability of false alarm (Pfa) than is customary can be tolerated in both the tracking and surveillance modes.

A LONG RANGE AIR DEFENCE RADAR

The multiple target tracking capability of a multifunction phased array radar for application to weapon systems is one which cannot be matched by conventional radars. An intriguing question, however, is the extent to which a phased array radar employing modern technology could be a cost effective competitor to conventional radars for applications in which high -accuracy multiple target tracking is not required, for example in long range air defence. We now treat this case as an illustration of some of the concepts and results presented in this paper.

The basic requirement for the radar is taken to be a 90% probability of track formation of 300 km against a target with an RCS of 1 m^2 and a velocity of 500 m/s, with surveillance cover to a height of 30 km and an elevation of 30°. Applying the expressions given earlier, the desirability of modest array dimensions indicates the choice of S-band. For R_c = 300 km and v = 500 m/s the optimum frame time for a Swerling 1 target is 24 s. At the optimum value of $\Delta R/R_c$ of 0.04, R_c/R_o = 1.06. Thus the required R_o (average) is equal to 300/1.06, ie 283 km (R_o on peak of beam = 330 km).

The low-elevation instrumented range, R_s, is taken to be 400 km (a R_s/R_o value of 1.4, close to the optimum value), and a transmitted pulse length of 1 ms is chosen, giving a duty factor of 27.5%. The use of such a high duty factor waveform requires the use of one or more short-range surveillance modes to cover the dead zones in range (150 km for a 1 ms pulse length).

We choose a circular array diameter of 4 m, which corresponds to about 5000 elements at 3 GHz. An assumption of uniform illumination on transmit and a -45 dB Taylor distribution on receive give an equivalent 3 dB beamwidth of 1.67°. The transmit and receive gains are 41 dB and 38.6 dB respectively.

We are now in a position to draw what amounts to a vertical coverage diagram for our radar and this is shown in Fig.9. Each beam shown corresponds to a beam-position row as shown in Fig.1. Three surveillance modes are illustrated. The medium and short range modes have maximum instrumented ranges of 150 and 60 km, and transmitted pulse lengths of 380 μs and 150 μs respectively. The instrumented ranges for the higher elevation beams are matched to the required height cover as shown.

The number of pulses per beam which are shown are those which will produce the required R_c (P_d for the shorter range modes) and which will give the required optimum frame time of 24 s. The frame times of the individual modes are 20.8, 2.9 and 0.5 seconds respectively. We can conclude that the effect of the medium and short range modes on long range performance is fairly marginal. Thus the use of such multiple surveillance modes ('range zoning') is an effective way of exploiting the high duty factor capability of solid state power sources.

It should not be necessary to use MTI in the long-range surveillance mode. In the presence of rain clutter in the lowest beam-position row it would be preferable to extend the medium-range mode as appropriate (see Fig.8).

The long-range mode in Fig.9 corresponds to a single pulse R_o (average) of 178 km. Given a receiver noise figure of 4.5 dB, losses of 9 dB (excluding propagation loss) and a Pfa of 10^{-4}, this can be obtained with an element peak power of only 0.5 W, which is readily achievable in GaAs monolithic implementation. It would correspond to a peak power of 2.5 kW and a mean power of 713 Watts (overall duty factor of 28.5%). A power of 0.62 W would allow the same surveillance performance with a tracking load of 30%. Note that sequential detection has not been assumed, for which an effective gain in power of 4 dB is claimed for a Pfa of 10^{-4} (2).

At this point it is worth noting that for a conventional radar with a maximum instrumented range and an R_o both of 450 km and an antenna rotation period of 12 s, the 90% cumulative probability of detection ranges against our target are 170 km and 106 km for track initiation criteria of 2 out of 4 and 3 out of 4 respectively (Swerling 1, Pfa = 10^{-6}).

The corresponding Swerling 2 case has also been computed. Here the optimum value of $\Delta R/R_c$ dictates a frame time of 120 s, 200-300 pulses per beam in the lower beam positions and corresponding dwell times of about 1 s! However, the range gate straddling loss which becomes significant when the radial distance moved by the target during the dwell time becomes comparable to the range cell size is taken into account, the optimum value of $\Delta R/R_c$ will be somewhat lower than the value indicated so far.

We have assumed up to now a fixed-face phased array radar of which 4 would be needed to give 360° azimuth cover. For a long-range application, however, a single rotating phased array is likely to be a more cost-effective solution. A power increase of somewhat less than 6 dB would preserve the surveillance performance (the scanning loss would be less), and although the tracking performance would be degraded it should still be quite adequate. Note that the surveillance data rate, which could be optimised, would not be the same as the rotation period of the antenna.

REFERENCES

1. Whitfield G.R., 1966, 'Optimisation of Surveillance Radar', Proc. IEE, 1277-1280.

2. Wirth W.D., 1981, 'Signal Processing for Target Detection in Experimental Phased Array Radar ELRA', Proc. IEE, 120 Pt F, 311-316.

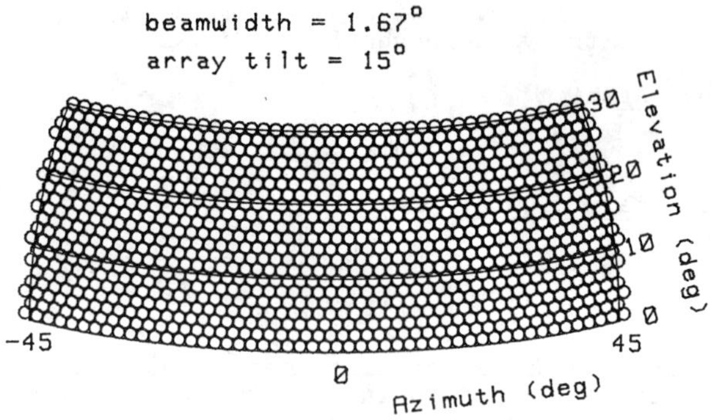

Fig.1 Surveillance beam positions for ± 45° azimuth and 0-30° elevation cover, plotted in u-v co-ordinates (sine θ space).

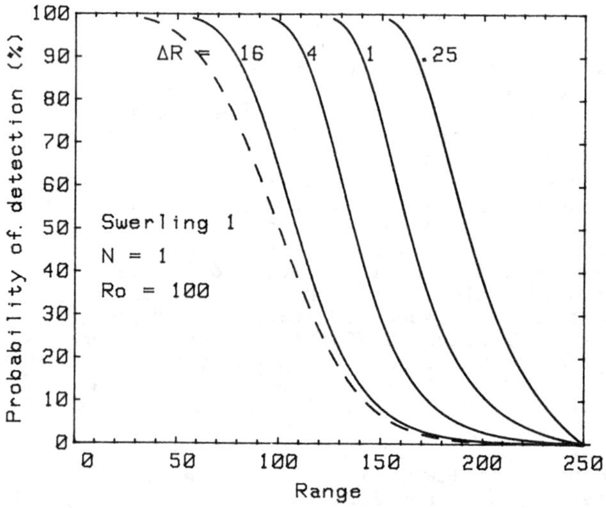

Fig.2 Probability of detection and cumulative probability of detection curves for values of $\Delta R/R_o$ of 0.25, 1, 4, and 16 per cent. Pfa = 10^{-4}

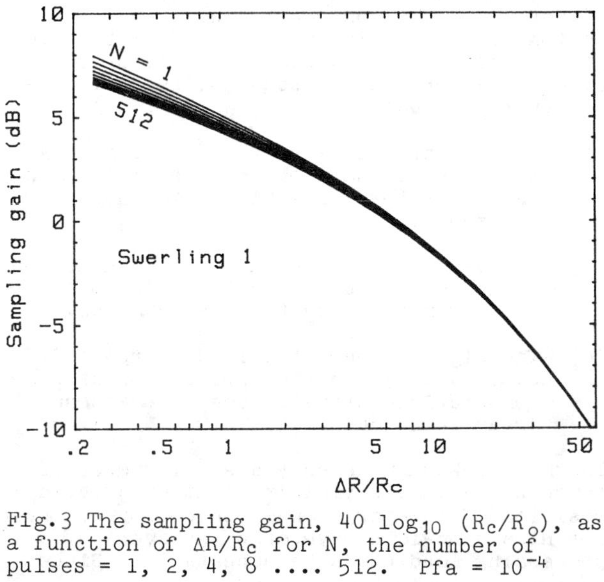

Fig.3 The sampling gain, 40 $\log_{10}$ (R_c/R_o), as a function of $\Delta R/R_c$ for N, the number of pulses = 1, 2, 4, 8 .... 512. Pfa = 10^{-4}

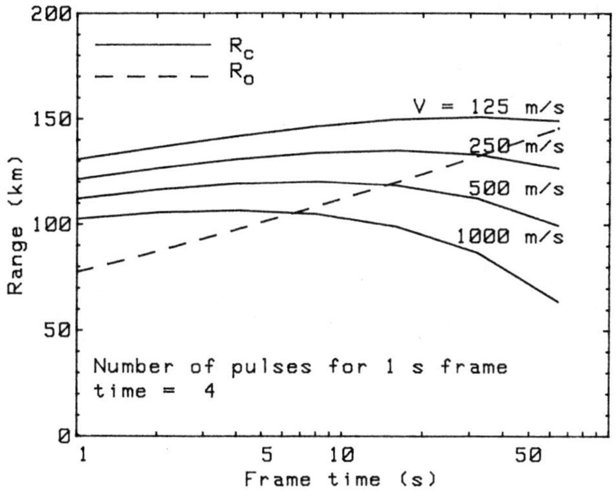

Fig.4 Variation of R_c and R_o with frame time for radial velocities of 125, 250, 500 and 1000 m/sec. Pfa = 10^{-4}

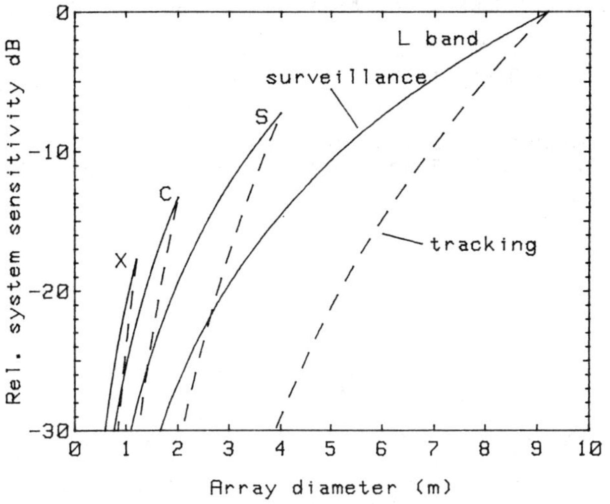

Fig.7 Surveillance and tracking performance ($40 \log_{10} R$, where R is the range for specified performance relative to that for 9 m L-band array).

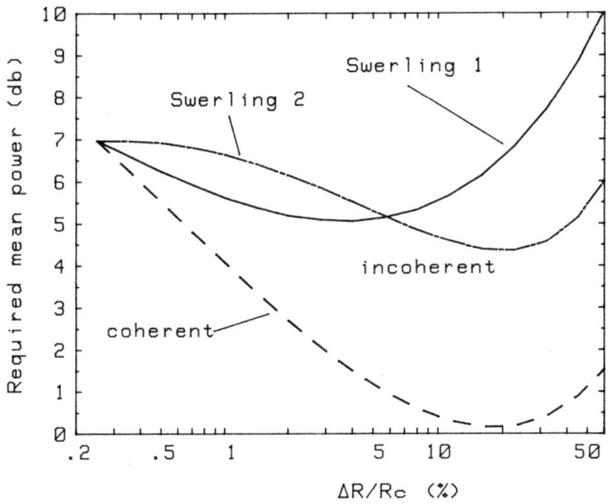

Fig.5 The relative system sensitivity (mean power) required to achieve a given R_c as a function of $\Delta R/R_c$. Pfa = 10^{-4}

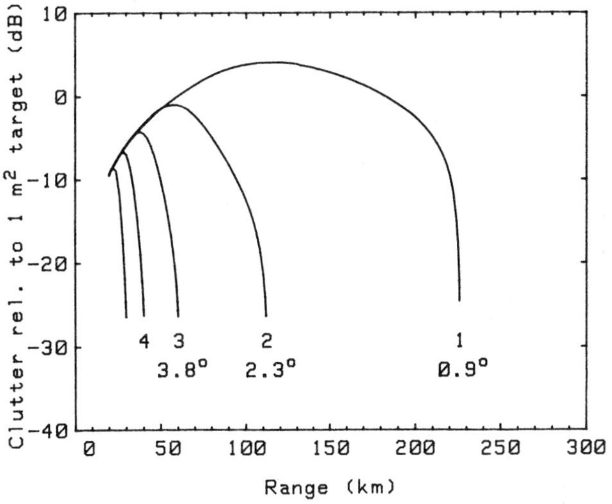

Fig.8 Rain clutter vs range for beam-position rows 1 to 5. Beam axis elevation angles are given. Rainfall rate 4 mm/h, 0-3000 m. Pulse width 1 μs. Beamwidth 1.67°.

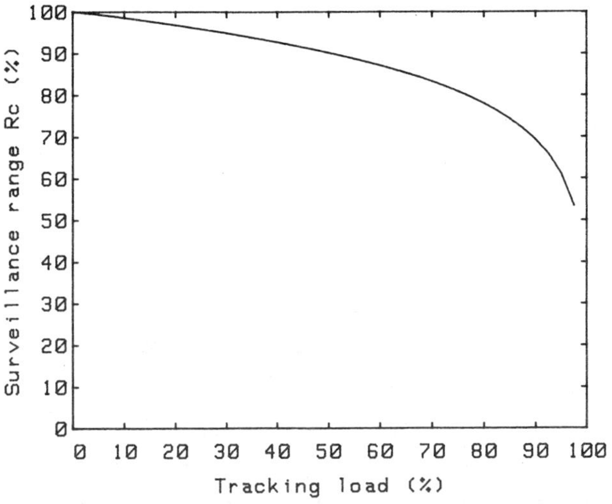

Fig.6 Surveillance range, R_c, versus tracking load.

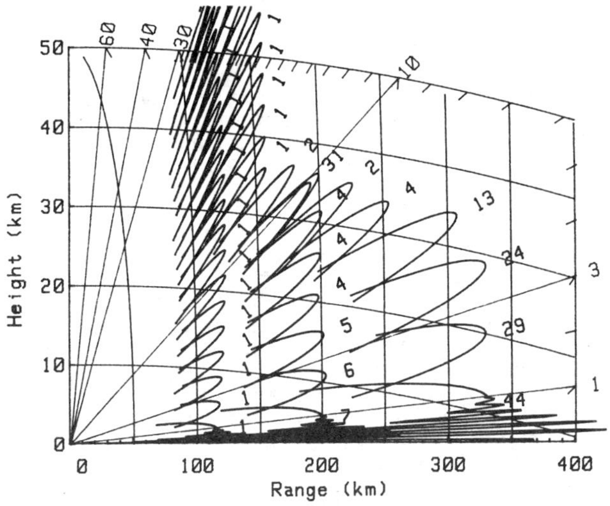

Fig.9 'Vertical coverage diagram' for long range air defence radar. Contours at 50% P_d level for beam position rows.

A NEW GENERATION AIRBORNE SYNTHETIC APERTURE RADAR (SAR) SYSTEM

J.R. Bennett and R.A. Deane

MacDonald, Dettwiler and Associates Ltd.
3751 Shell Road, Richmond, B.C., Canada, V6X 2Z9

1. INTRODUCTION

With the launching of the spaceborne Seasat-A, L-band SAR in 1978 and the subsequent clarity of the digitally processed Seasat imagery, interest in SAR as a remote sensing instrument for resource related applications has considerably intensified. The unique properties of SAR, including day or night all-weather operation, controlled, coherent illumination source and large dynamic range combine to make SAR sensors highly competitive and cost effective in several applications. Chief among these are:

1. Ice classification and migration surveillance in support of resource extraction and transportation in Arctic regions.
2. Mineral resource exploration.
3. Land use mapping.
4. General cartography in rain forest/jungle regions.
5. Maritime surveillance.

Perhaps the most immediate and pressing requirement for airborne SAR systems exists in the Canadian north where offshore oil and gas exploration and drilling operations are constantly threatened by the annual fluctuations of the ice pack and the seasonal variations in ice flow thickness, location and movement. Experiments with Seasat as well as airborne SAR and SLAR have clearly demonstrated the utility of radar sensors in ice type detection and classification, especially in winter conditions where darkness and cloud cover normally prevail and the fine structure of the ice pack is further obscured to visible sensors by a layer of snow.

In response to the requirement for radar support of arctic energy extraction, the Canadian Government Department of Energy, Mines and Resources, acting through its remote sensing division, the Canada Centre for Remote Sensing, is funding the research and development of an advanced, high-performance, airborne synthetic aperture radar system, optimized for civilian reconnaissance and resource related remote sensing applications. Contracts have been awarded to MacDonald, Dettwiler and Associates Ltd., together with their subcontractor in the microwave and radio frequency aspects, Canadian Astronautics Ltd., and work is well underway towards the delivery of the first prototype SAR in early 1984.

The full prototype SAR will be a two-frequency transmit, four channel receive system with two polarizations received for each transmitted frequency. The first delivered system will have a single C-band transmitter and two receive channels with provision for the later addition of an X-band transmitter and two X-band receive channels.

The main features of the radar are:

- dual mode, resolution/swath trade-off capability,
- on-board real-time processing and display,
- wide bandwidth motion compensation for operation on small (12,000 lb.) turbo prop aircraft,
- centralized operator console interaction, testing and control,
- raw SAR data recording for off-line, precision digital processing on the ground in addition to the airborne hard-copy imagery.

Future options include:

- on-board processed data recording and playback,
- realtime processed data downlink to surface terminals,
- surface terminals with track and display capability for ship-borne and land-based installations.

2. PERFORMANCE OF THE RADAR

The major parameters of the C-band radar are shown in Table 2-1:

TABLE 2-1 RADAR PARAMETERS

Radar frequency	5.3 GHz
Peak transmit power	50 KW
Average transmit power	>100 W
Antenna beamwidths	
Azimuth	3°
Elevation	25°
Target reflectivity:	
at 25 km range	-40 dB
at 70 km range	-29 dB

The above average transmitted power is at the maximum transmitted pulse repetition rate, which is 325 pps for a turbo prop aircraft, or 650 pps for a jet aircraft. The radar has been designed to operate in a variety of aircraft, with particular emphasis on small, fuel-efficient turbo prop systems. The exact performance depends on the particular aircraft and selected antenna, but is typically that shown in Table 2-2 for a small aircraft.

TABLE 2-2 AIRCRAFT DATA

Aircraft altitude	sea level to 11000 m (36000 ft)
Ground speeds (m/s)	
turbo prop -	75 - 160
jet -	150 - 320
Radar weight	~ 500 kg
Average wind speed	0 to 50 m/s
Superimposed gust speeds	0 to 10 m/s

A unique feature of this radar is the motion compensation subsystem. The radar covers very wide swath widths at moderately high resolution and necessitates far more accurate motion compensation that has been previously used. The full swath covers 4096 range cells and each of these range cells is individually motion compensated in the SAR digital processor to cope with the high wind speeds and turbulences common in the arctic.

A second unique feature of the radar is that real-time multilook processing is used to reduce image speckle. In this respect, civilian requirements tend to be somewhat different from military applications. In most military SARs, the main objective is maximum detectability of small point targets hidden in a background of nuisance clutter, whereas in many civilian applications, such as ice field mapping, it is this very terrain clutter which is of most interest. The military systems often select single look imagery at the expense of image speckle, whereas the civilian radar seeks to maintain an even grey scale for similar types of ice all over the imagery to aid identification of the type of ice (first year, multiyear, blocks, ridges, flows and similar). Although it would be technically feasible to adapt the processor to one look imagery, the present design is optimized for multilook processing with particular emphasis on adaptable range dependent gains, such as sensitivity time control (STC), to obtain equal image intensity for similar terrains at near and far ranges. In-flight displays assist the radar operator to select the correct STC law using data from measurements on actual incoming echoes.

Table 2-3 shows some of the imagery specification options which are available in real-time on a small turbo prop aircraft. This imagery not only checks the correct operation of the system but also considerably assists navigation in mapped areas of the world. The imagery is of sufficient quality to hand directly to an end user after a flight, and is fully annotated (lat, long, distance, gain setting, etc.).

TABLE 2-3 AIRBORNE IMAGERY OPTIONS

Nominal operational altitude	10000 m
High resolution mode	
- range plane resolution	5 m
- azimuth resolution	6 m
- ground swath (km)	10 to 29
Low resolution mode	
- range plane resolution	18 m
- azimuth resolution	9 m
- ground swath (km)	10 to 75
Ground range/slant range	conversion available while airborne
Number of looks summed	7
Hard copy imagery	Full or ¼ swath

3. CONCEPTUAL BLOCK DIAGRAM

The overall conceptual block diagram of a single channel of the prototype radar is shown in Figure 1. In the selected aircraft, the antenna is a dual polarized linear waveguide array mounted on a dual axis servo controlled platform and suspended in a radome under the fuselage. The servo controls stabilize the antenna in the yaw and roll planes, controlled by the angular outputs from a laser gyro inertial reference unit (IRU).

A synthetic array radar measures the range, amplitude and phase of radar echoes and requires a stable phase reference (10 MHz STALO in Figure 1). This STALO frequency is multiplied up to the intermediate frequency (IF) of 300 MHz where range pulse encoding is superimposed. A non-linear chirp waveform was selected for this application because surface acoustic chirp/dechirp devices of adequate dynamic range, stability and low side lobe performance are readily available. Alternative SAW pairs are used to obtain the two different range resolutions rather than filtering high resolution data to avoid loss of signal-to-noise ratio in the low resolution mode.

The transmitter's high power amplifier is a TWT with its own stabilized power supply. The pulse repetition frequency is controlled from the motion compensation computer which aims to maintain an equal sample spacing on the ground of successive transmitted pulses at all operational speeds of the aircraft.

The receiver is based on a low noise GAS FET amplifier and uses the SAW dechirp unit to range compress the received echoes. Its output provides baseband in-phase and quadrature (I,Q) complex signals. The receiver not only includes the STC functions mentioned above, but also has optional automatic gain control (AGC) of relatively long time constant to assist the maintenance of overall average grey scale levels when major terrain reflectivity changes occur, such as at land/sea boundaries. The analogue I,Q receiver outputs are digitized and stored until the appropriate motion compensation data has been computed. If the radar is fitted to a jet aircraft, a presummer filter is required to reduce data rates to practical levels for in-flight processing and recording. The presummer is omitted for a turbo prop aircraft.

The motion compensation computer interfaces to the inertial reference unit and converts body axis motions into earth axis motions in order to compute ground speed and drift velocities. The laser gyro platform contains three axis accelerometers operating in the body axis of the aircraft and are ideal for determining short term motions of the antenna relative to the ground. By integrating these accelerations and evaluating the changes of range of the antenna phase centre from the ground points, the motion compensation phase unit individually corrects the motion induced Doppler on each of the 4096 range cells for every received range line. This necessitates the use of high speed dedicated pipeline digital phase shifters (complex multipliers), and the control of the start range for the digitization window. For side gusts of 10 m/s, a synthetic aperture time of 6 seconds and range gates separated by 4 m, a target would move through 15 range gates without such range gate control applied at the A/D sampler.

The motion compensated data is compressed in azimuth to obtain the specified azimuth resolutions, using time domain matched filtering of the azimuth phase histories. The input data is first filtered into seven separate sub-apertures of the beam, as in a beam sharpening radar, and each of these sub-apertures are independently compressed and detected with their own matched filter. The detected look data is then recombined in the look summation module and formatted for display or for onward transmission to the ground via a high speed communications downlink.

The image recorder is a hard-copy display using dry silver paper to show 2000 pixels across track. The operator can actively control the level of saturation in the A/D converter, the grey level settings for this display and the range dependent gain controls such as the STC function. He can reconfigure the radar in a few seconds to change the resolution/swath width combination or the direction of the antenna pointing (port or starboard).

In some installations, it is required to record radar data immediately after motion compensation for later processing on the ground. Such processing is to different resolutions and numbers of looks from that of the airborne processor. The data rate is high, necessitating a high density digital recorder (HDDT). The formatter for this unit merges annotation, gain settings and motion information with the data to assist later ground processing. MDA's Generalized SAR software processor is used for the ground processing.

Finally, mention might be made of two important control loops within the radar processor for some installations. An auto clutter lock tracker measures the antenna illumination deviation from zero doppler and adjusts the antenna yaw angle to restore pointing orthogonal to the time track. An autofocus tracker, which provides a tenfold improvement in the measurement of along track velocity, is also incorporated. In military systems, a Doppler radar often provides this needed improvement, but a SAR also contains the necessary information and can be substituted for the Doppler system. Expected accuracies are 0.15% ground speed or better.

4. CONCLUSION

The radar system described above contains many features which have not so far been available in civilian SARs. Many are based on our experience both in the ground processing of Seasat data and in the last two years of flights of a real-time airborne processor. The features of this new radar are aimed at providing a much improved interface to the operator, the automation of many features, the provision of relatively wide swaths at high resolution and the full motion compensation required to cater to the high wind speeds and turbulence conditions found over the Bering and Beaufort sea areas of the arctic.

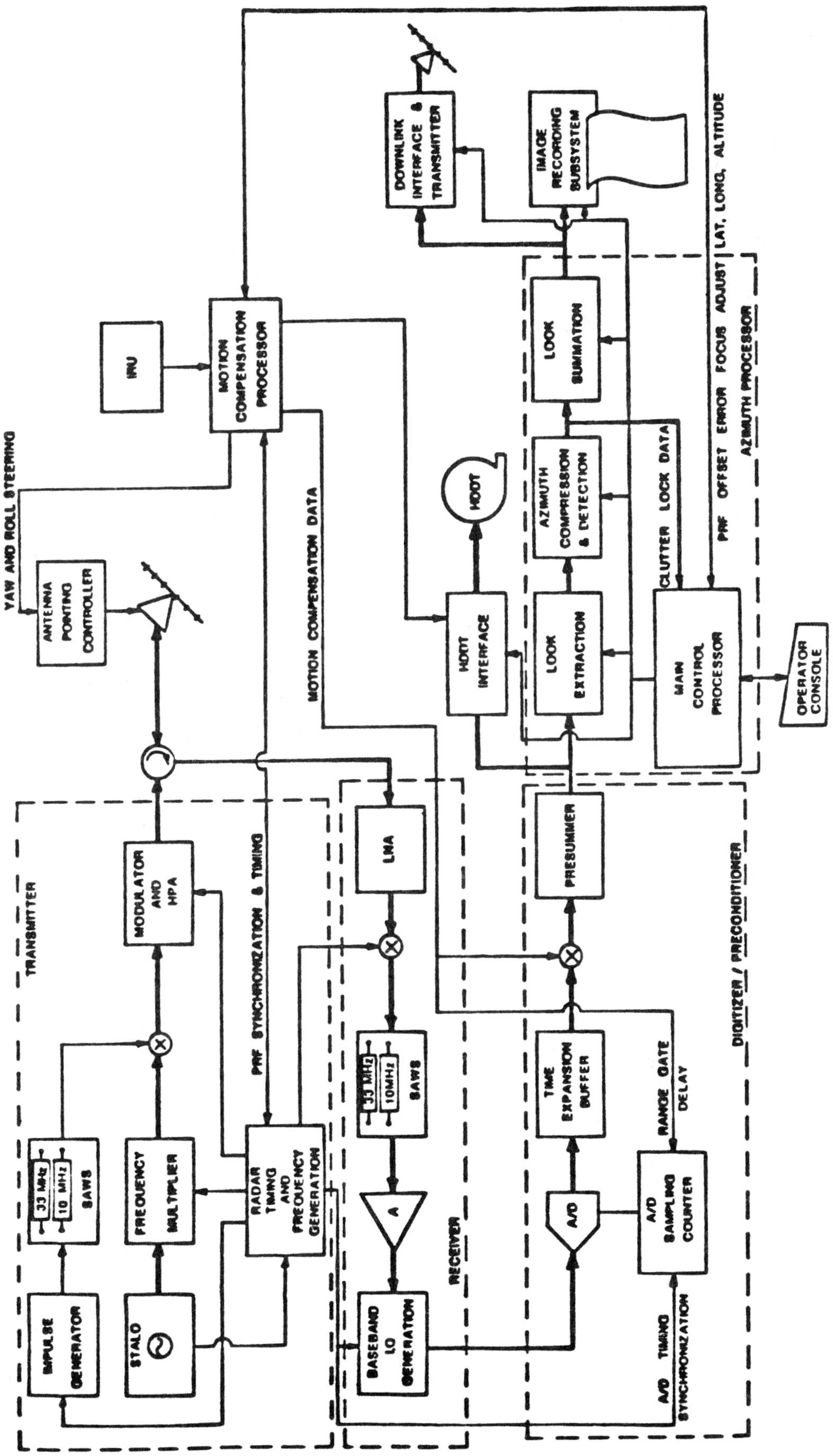

FIGURE 1 Block Diagram of the MDA SAR System

RESULTS FROM A NEW DUAL BAND, DUAL PURPOSE RADAR FOR SEA SURFACE AND AIRCRAFT SEARCH

P D L Williams

Racal-Decca Limited, UK

1. INTRODUCTION

The task of carrying out efficient surveillance of local air space and sea surface has traditionally been carried out by several radars aboard most naval ships. In today's stringent economic climate such multiplicity of radar equipment may not always be fully justified and on smaller vessels consideration of top weight and space at the mast head must also be borne in mind.

On balance the tasks of navigation, anti collision and target search for surface targets is best carried out at a wavelength of 3.2cm whilst search for aircraft, and other airborne targets out to the same range but in the larger volume up to 10,000ft, or more, is best carried out at a longer wavelength with a larger aperture aerial. The fundamental form of radar equation is given as (1) below.

$$R = \left[\frac{P\ A\ ts\ \sigma}{S/N)\ Am\ (\mathrm{Sin}\ Em - \mathrm{Sin}\ Eo)} \times \frac{1}{4\pi\ kT\ NFL}\right]^{4} \quad \text{--1}$$

This form is discussed by Barton (1) in Berkowitz and is attributed to an earlier worker in Sperry Gyroscope. The equation states that for a given target of RCS, σ to be detected in a given search time ts, in a given volume of sky set by the aerial elevation limits Em and Eo and azimuth sector Am; all that matters is transmitter mean power, aerial aperture radar losses L and receiver noise factor, subject of course to the usual considerations of probability of detection versus false alarm rate as a function of detector threshold. It is interesting to note that neither spatial resolution nor wavelength appear in this form of radar equation and it is only when target detection in a hostile environment is considered that these factors become important.

The choice of carrier frequency varies considerably for surface and airborne targets in various environments when comparing radar equipment, either with fixed aperture aerials or those of equal beamwidths and a range of calculations have been given previously by Williams (2).

For a wide variety of reasons, (3) frequency agility and diversity is also an attractive option for this class of equipment and this paper describes results obtained from a dual beam dual frequency system which could be considered as the ultimate in thinned spread spectrum!

2. EQUIPMENT

A prototype of the equipment being considered has been previously described in detail by the author (4) and was based on two separate inexpensive marine radars which had previously been used in early dual frequency radar trials (5). Fig. 1 shows the twin aerials on the turning gear and the general parameters of this present equipment are given in Table 1.

The lower beam aerial with its 0.8 degree azimuth beam width is primarily for use in producing a high resolution display to enable the navigational aspects of a civil marine radar to be fulfilled ref. (6). The upper beam being used on a lower frequency but needing a larger aperture aerial produces

Fig. 1. TWIN AERIALS AND TURNING GEAR OF THE DUAL BAND RADAR

good radar performance in the much greater volume of air space and Fig. 2, illustrates the vertical coverage of forcast performance against a $10m^2$ target with the parameters of both sets given in Tables 1 and 2.

Parameter	Carrier Wavelength 3.2cm	10cm
Transmitter Peak Power	75kw	30kw
(	0.05μs	0.05μs
Pulse length choice (	0.25μs	0.25μs
(	1 μs	1 μs
(	3300Hz	3300Hz
Corresponding PRF (	1650Hz	1650Hz
(	825Hz	825Hz
Receiver		
Type, superhet with preamplifier and logarithmic l.F. amp		
Noise Factor (overall)	6dB	3.5dB
Bandwidth	5 or 15MHz	5 or 15MHz
Aerials		
Horizontal Physical aperture	2.8m	3.75m
Vertical Physical aperture	0.29m	0.5m
Polarisation	H	H
GAIN	33dB	27dB
Horizontal Beam Width	0.85°	2°
Vertical Beam width	15°	32°
Estimated losses	6dB	6dB

TABLE 1. Essential Parameters of dual Band Radar

The performance forcast is for the equipment operating on long pulse under which conditions a free space range estimate is given on Table 2.

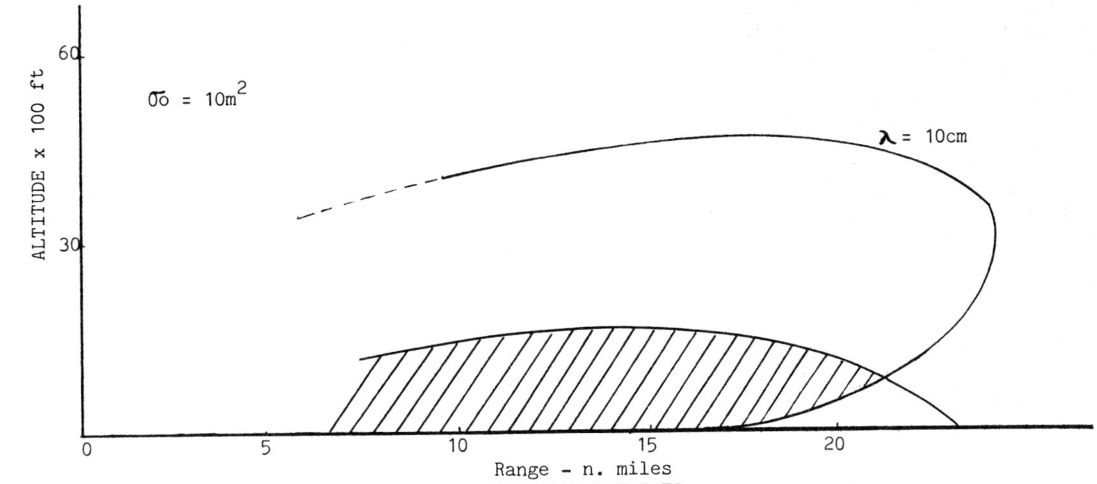

FIG. 2. INDIVIDUAL COVERAGE DIAGRAM FOR RADAR ELEMENTS

	3.2cm	10cm
Single channel detection probability	50%	50%
False alarm probability	10^{-6}	10^{-6}
Hits per beam width	6	14
Required S/N	7dB	4dB
Free space range on $1m^2$	13.5 n.m.	13.1 n.m.
Free space range on $10m^2$	23.3 n.m.	24 n.m.

TABLE 2. DETAILS OF RANGE FORCAST

The area of cover common to both radar elements where the full diversity gain operates is shown hatched on Fig.2.

3. ENVIRONMENTAL DIAGRAMS

The background against which most targets have to be detected is usually dominated by sea and precipitation clutter compared to system thermal noise in a benign military environment Figs. 3 and 4 are an indication of such background for the two wavelengths being considered with the following notes:- precipitation back scatter co-efficients are taken from Nathanson (7) with the simplistic model that rain fills the vertical beam in both equipments, in practice the high beam rain back scatter will be much less due to the low probability of rain extending vertically beyond 10,000'. The sea clutter background is dominated by the aerial height and spatial resolution of the two radars concerned. It will be seen that the S Band radar enjoys a lower clutter background than the 3.2cm set but suffers a shorter transition range against small low lying surface targets. For these diagrams an aerial height of 20 metres above sea level has been chosen with a sea state of 3.

4. PERFORMANCE AGAINST SURFACE TARGETS

In general, for modest scanner heights of the order of 10-20 metres the radar horizon and transition range limits long range performance on all but the smallest targets which become lost in receiver noise. However, the 3.2cm element of the radar being described has a performance over 40dB more than that required to meet the marine radar specification (6).

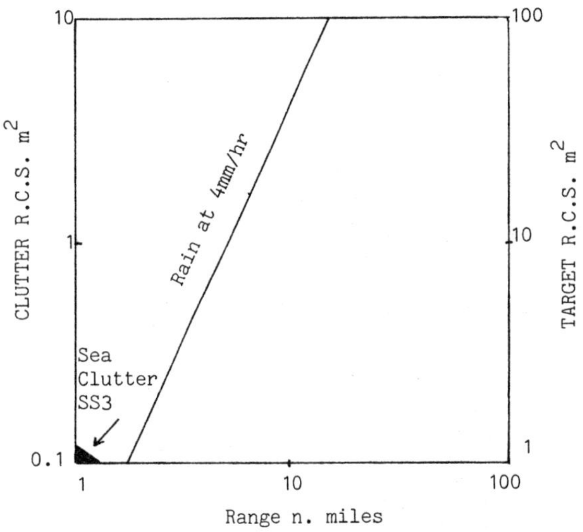

FIG. 3 ENVIRONMENTAL DIAGRAM FOR 10cm ELEMENT ON 1μS PULSE WITH A 10dB TARGET TO CLUTTER MARGIN

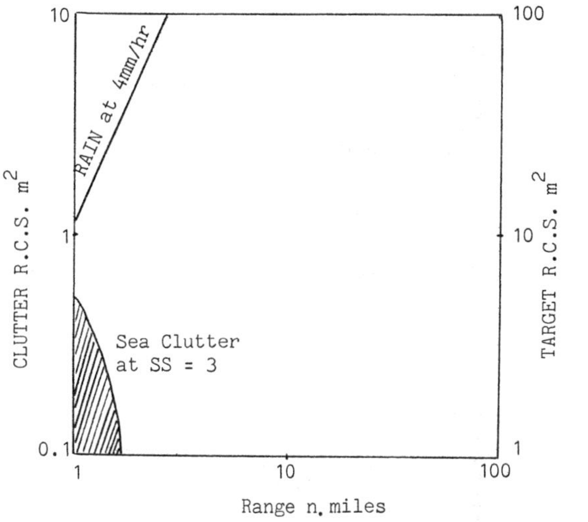

FIG. 4 ENVIRONMENTAL DIAGRAM FOR 3.2cm ELEMENT ON 1μS PULSE WITH A 10dB TARGET TO CLUTTER MARGIN

The diversity aspects of the two equipments used to detect and track surface targets have previously been studied in detail and are reported in (4) and (5), as an illustration Figs. 5 to 8 show the position, absolute and relative performance of a ship travelling a distance of 10 miles with the aerials so positioned above sea level, as to put the transition range for both equipments beyond 10 miles. It will be seen that the differential performance fluctuates even for data derived from smoothing the 50% probability paint over ten aerial scans to produce each data point.

5. BALLOON TRIALS

Initial flight trials were carried out with a tethered balloon covered with a metallic reflective coating and from the scanning radar an average over 8 runs of the actual pulses per beam width received by each channel in a beam dwell time are shown in Fig. 9 when the balloon was so positioned as to present an elevation of 3°.

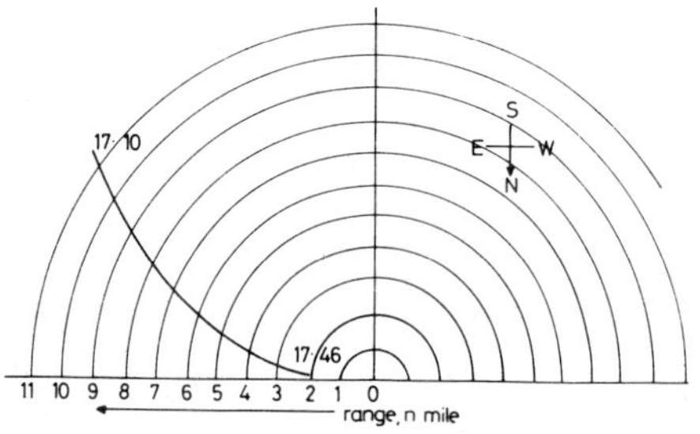

Fig. 5. PLOT OF SHIP'S TRACK FOR 36 MIN.

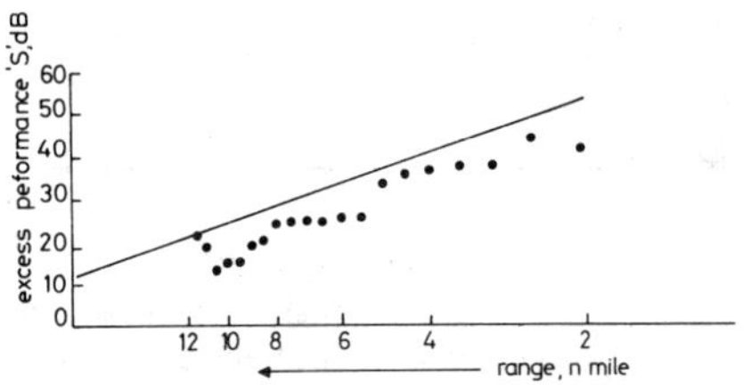

Fig. 6. S-BAND PERFORMANCE AT P_d = 50%

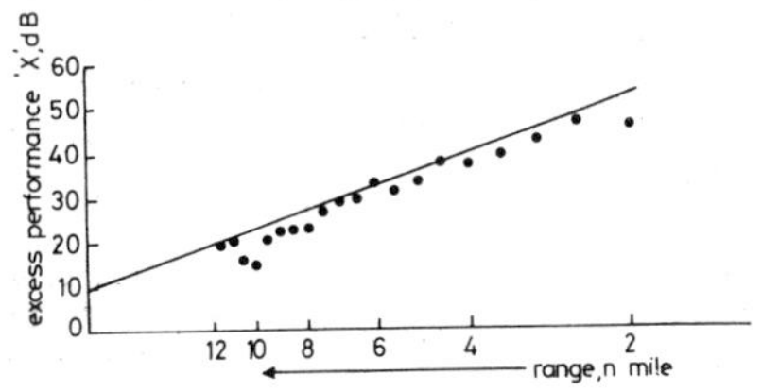

Fig. 7. X-BAND PERFORMANCE AT P_d = 50%

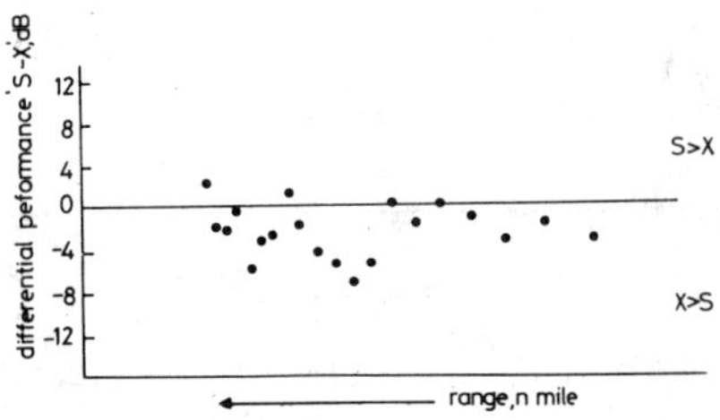

Fig. 8. S-X

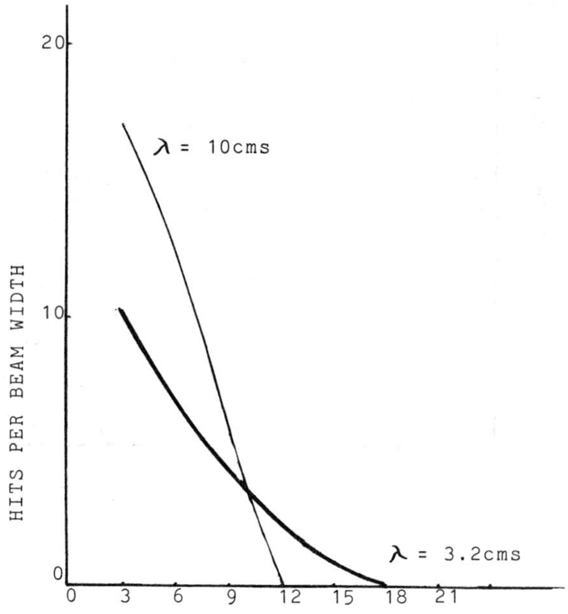

FIG. 9. AVERAGE HITS PER BEAM WIDTH ON TETHERED BALLOON IN 6 QUANTISED AMPLITUDE LEVELS

Fig. 9. Average hits per beam width on each carrier for balloon trial at range of half a mile and elevation of 3°.

Results indicated the lower hits per beam width expected from the 3.2cm radar but amplitude levels in excess of the 10cm carrier in the region of common coverage shown in Fig. 2. Theoretical results were not matched by the balloon trials so that the performance of the present equipment has been improved by 6 to 8dB over that described in Ref. 4.

6. AIRCRAFT TRIALS

The first series aircraft trials were carried out with a twin engine turbo prop aircraft, the British Aerospace - Jetstream. Horizontal level flights were carried out both in bound and out bound and these are shown for the individual radar sub systems in Figs. 10 and 11. A further set of flight trials were carried out down to as near sea level as the pilot felt safe viz: 50'. The recorded coverage from these trials is shown in Figs. 12 and 13. The next series of trials were carried out with a medium size helicopter and the coverage to 10,000' is shown against the predicted coverage for a medium 10 square metre target in Fig. 14.

Classification of RCS of aircraft.

During work up trials at Biggin Airfield a number of radar observations were made on such aircraft as were using the airport or in the controlled airways over flying the airport. The best detection range for all the 36 aircraft were noted and details of actual planes gained from either observations as they took off or landed or by use of the aircraft radio with help from air traffic controllers when possible. The three groups were represented by the following examples in Table 3.

Small Aircraft	Medium Aircraft	Large Aircraft
Cesna 150	Cesna 310	737
Cesna 172	Aztec	747
Cheetah	Jetstream	Tristar
Banduranti	Citation	
Piper Navajho		

TABLE 3. Examples of aircraft in three Classes.

The ranges of the three classes were examined and grouped separately from the above table and the final analysis gave a remarkably good fit between the aircraft groups and corresponding detection ranges:- as shown in Table 4.

No. of Samples in Batch	Best range n. miles	Standard Deviation of range	Class
6	9.08	1.74	Small
15	12.73	2.88	Medium
15	29.6	7.53	Large

TABLE 4. Mean of best detection range of individual aircraft by class

Having regard to the fact that these chance aircraft were flying at altitudes ranging from a few thousand feet for local traffic from the airport to those at much higher altitudes over flying Biggin Hill and of course presented themselves at various aspects to the radar, the correlation of detection range and group encouraged the median value RCS of each group to be calculated from the group range means. The results are given in Table 5, and were later compared to estimates made by Skolnik (8) in his review of aircraft RCS values, albeit with many qualifying remarks.

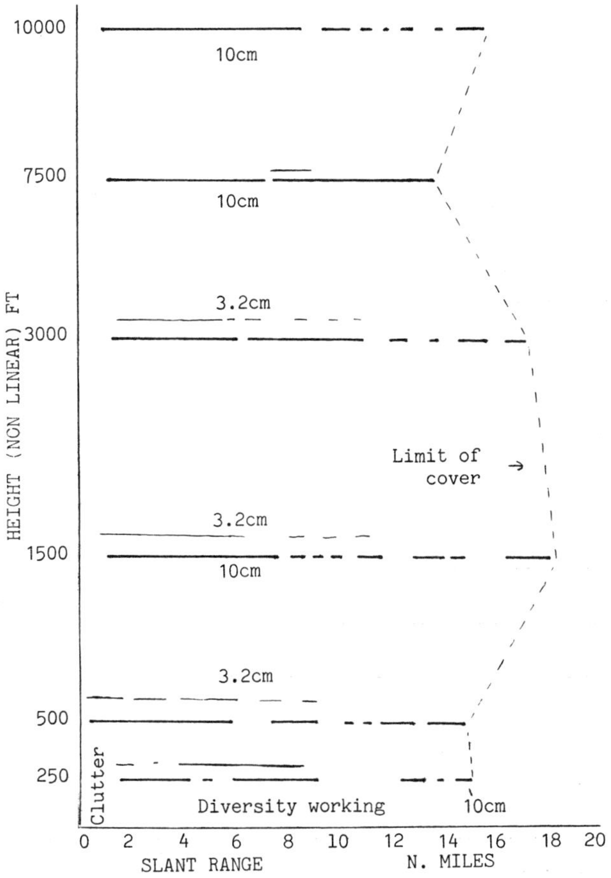

FIG. 10. TRIALS RESULTS ON JETSTREAM AIRCRAFT APPROACHING "HEAD ON"

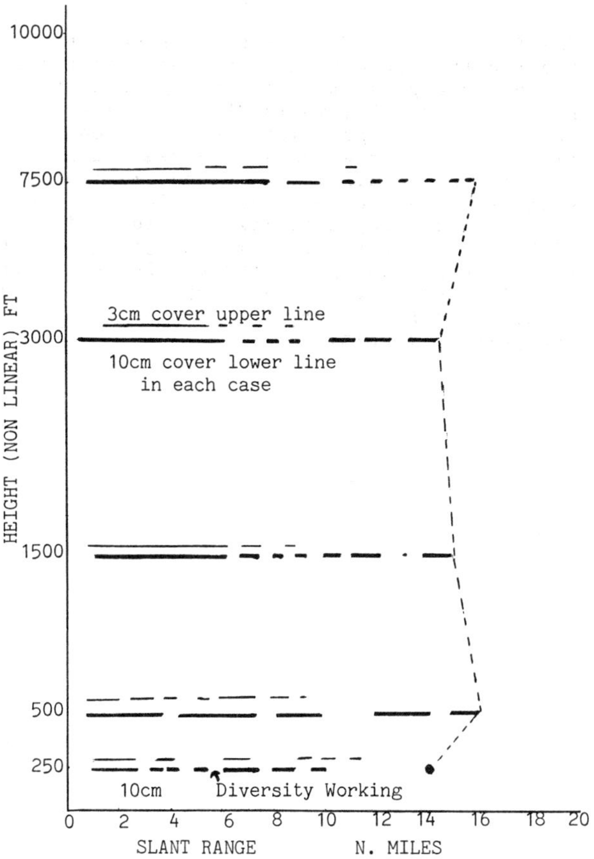

FIG.11 TRIALS RESULTS ON JETSTREAM AIRCRAFT RECEDING - TAIL ASPECT

Clearly, the means of each group could be shifted by a factor of 2 or 3 each way dependent on the make up of each sample population, nevertheless, the results are presented as all coming from one equipment over a limited period so that variables were not as wide as a large review of many separate trials over widely varying conditions.

Class	Calculated R.C.S. - m^2	Skolnik's estimate
Small	0.2	1
Medium	0.8	2
Large	21	40
Largest	200	100

TABLE 5. Calculated R.C.S. of the mean value of each aircraft group.

7. SUMMARY AND CONCLUSION

The radar described has evolved from two modest surveillance equipments and has fulfilled the aim of providing a high resolution surface mapping radar for shipboard navigation, whilst the air cover has been achieved in accordance with the theoretical forcasts to the accuracy of R.C.S. known values. The advantage of dual band diversity has been clearly seen for a wide range of shipping targets and, to a limited extent, for aircraft particularly well inside the area of common cover but from the limited trials so far, diversity at the cover boundaries has to date not been any more successful than predicted, in that it does not raise detection from 0% to 50% but does raise it from 50% to 100% over a few aerial scans.

The flexibility of the system is such that if desired the 3.2cm element may be operated on a 50 nano-sec pulse to yield a high quality high resolution display,whilst the 10cm element operating on a 1μs pulse provides good solid air cover thus meeting a very real dual operational requirement from one mast head equipment. Suitable video combining circuits have been developed to mix the video after CFAR processing to yield maximum target sensitivity in sea and preciptation clutter in order to drive both operational displays and Automatic Radar Plotting Aid or ARPA anti collision (or engagement) displays.

REFERENCES

1. Barton,D.K. The Radar Equation. Chapter 2 of Modern Radar, ed. R.S. Berkowitz, John Wiley 1965.

2. Williams, P.D.L. Limitations of Radar Techniques for the Detection of Small Surface Targets in Clutter. The Radio and Electronic Engineer, Vol. 45, No. 8, 379-389, August, 1975.

3. Barton, D.K. ed. Frequency, Agility and Diversity, Radars Vol.6. 1977.

4. Williams, P.D.L. Results from an Experimental Dual Band Search Radar. The Radar and Electronic Engineer, Vol. 51, No.11/12, Nov/Dec 1981.

5. Williams, P.D.L. Medium Term Fading of Radar Targets at Sea with Special Reference to Operation at 3cm and 10cm. IEE Proc. Col. 127, PT,F, No. 3. June 1980.

6. Croney, J. Civil Marine Radar, Chapter 31 of Radar Handbook ed. M.I. Skolnik. McGraw Hill 1970.

7. Nathanson, F.E. Radar Design Principles McGraw-Hill, New York, 1969.

8. Skolnik, M.I. Introduction to Radar Systems. Second Edition, McGraw Hill 1980.

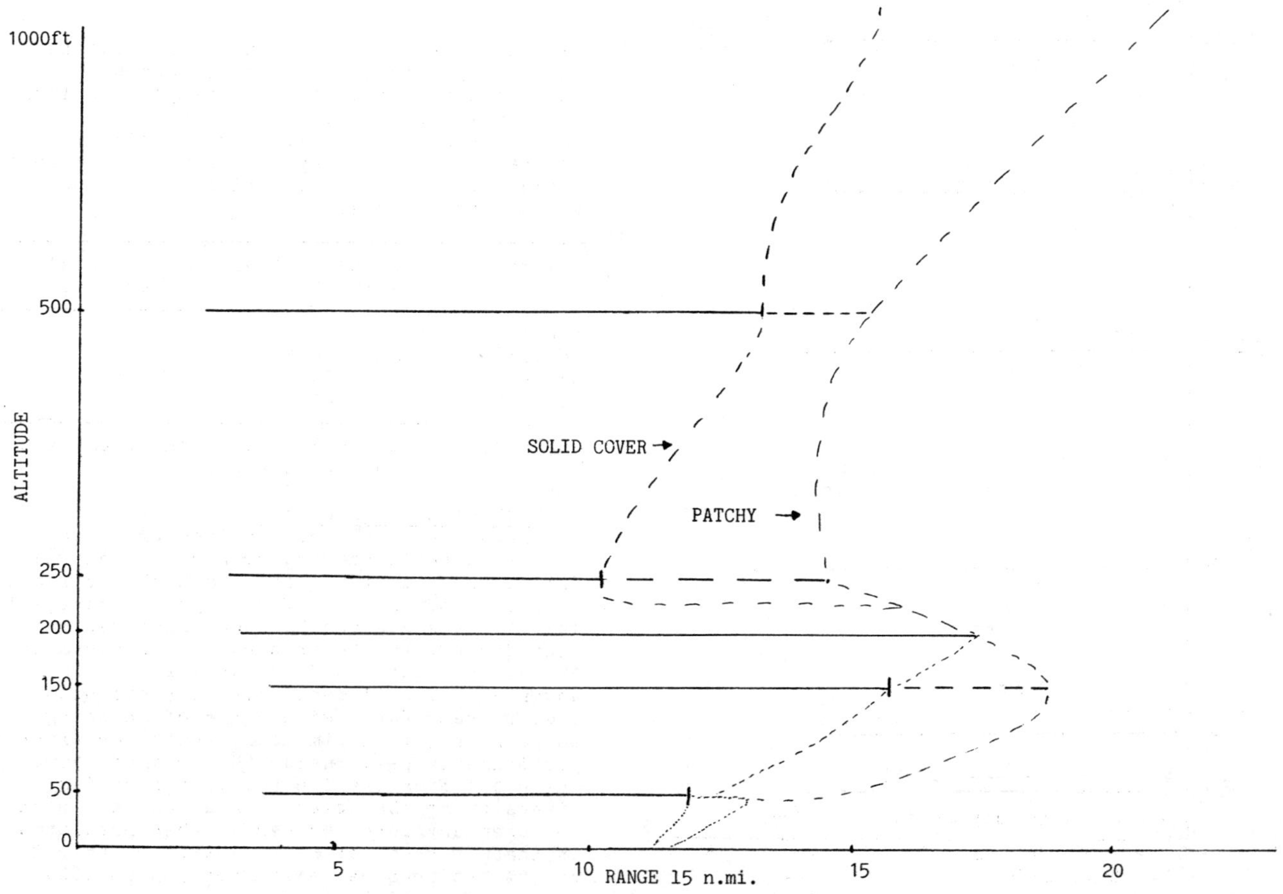

FIG. 12 LOW ALTITUDE TRIALS

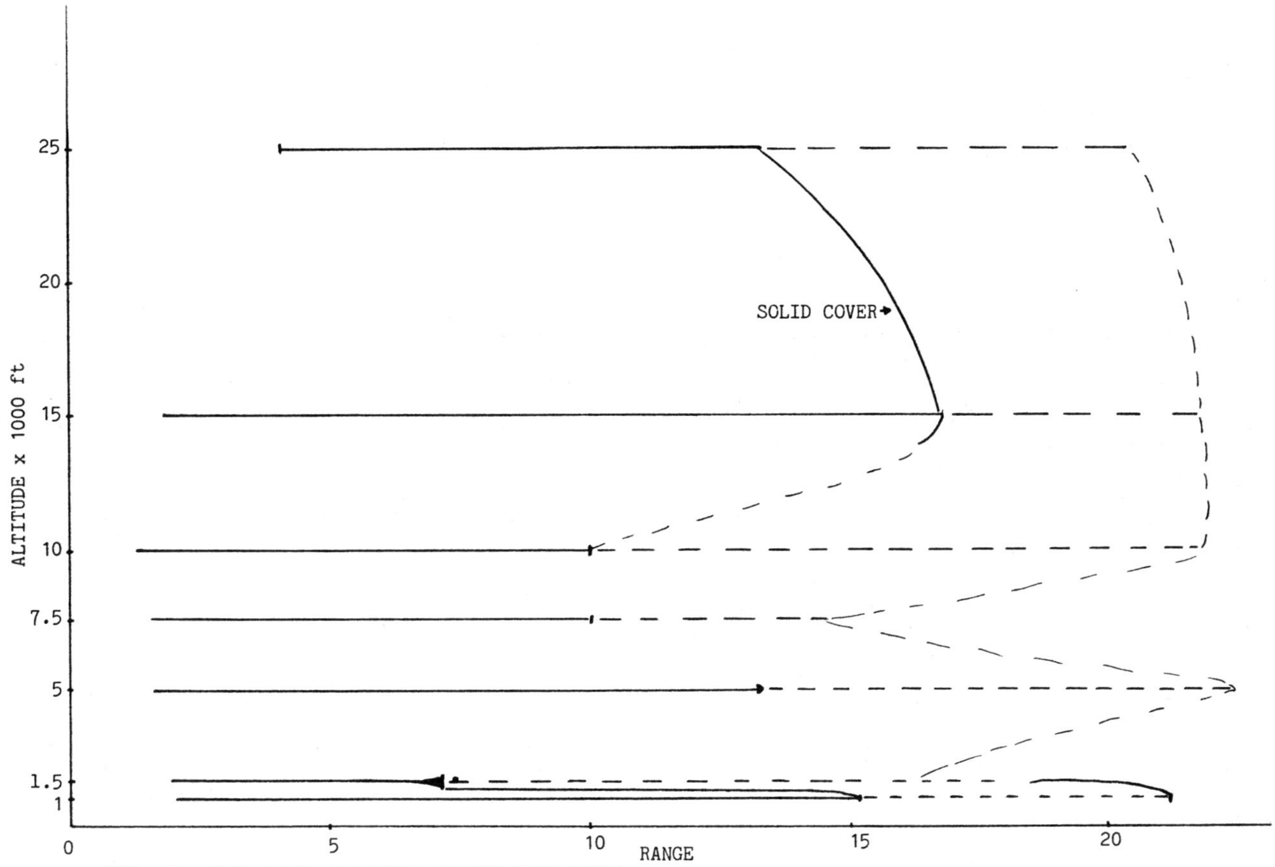

FIG. 13 CUMULATIVE JETSTREAM RESULT HIGH LEVEL

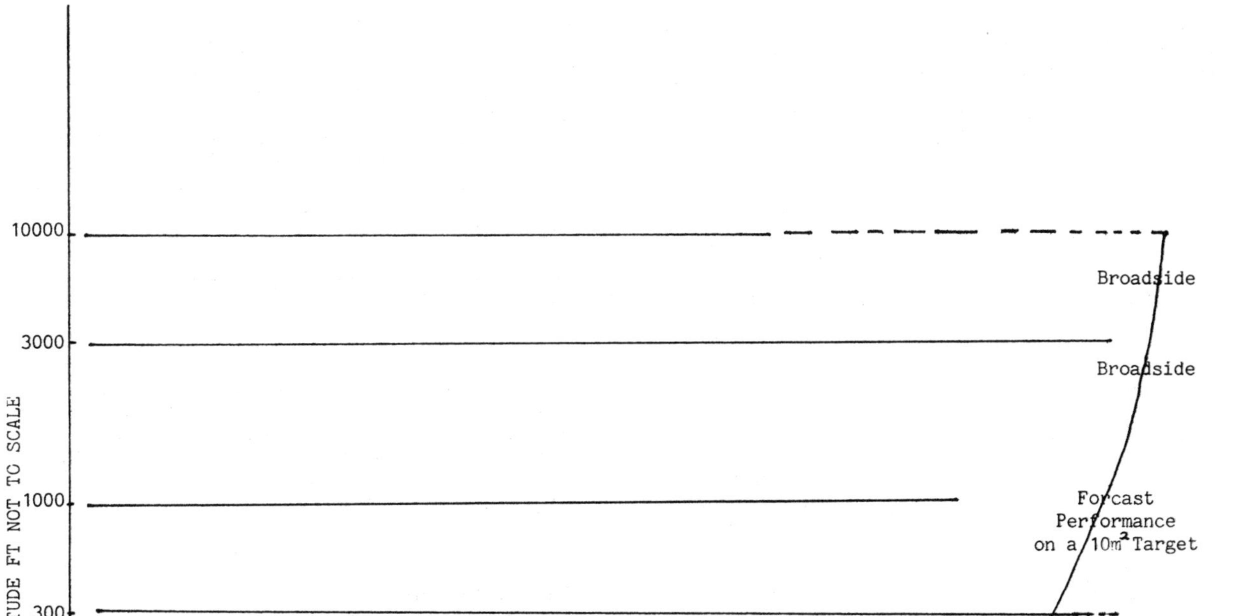

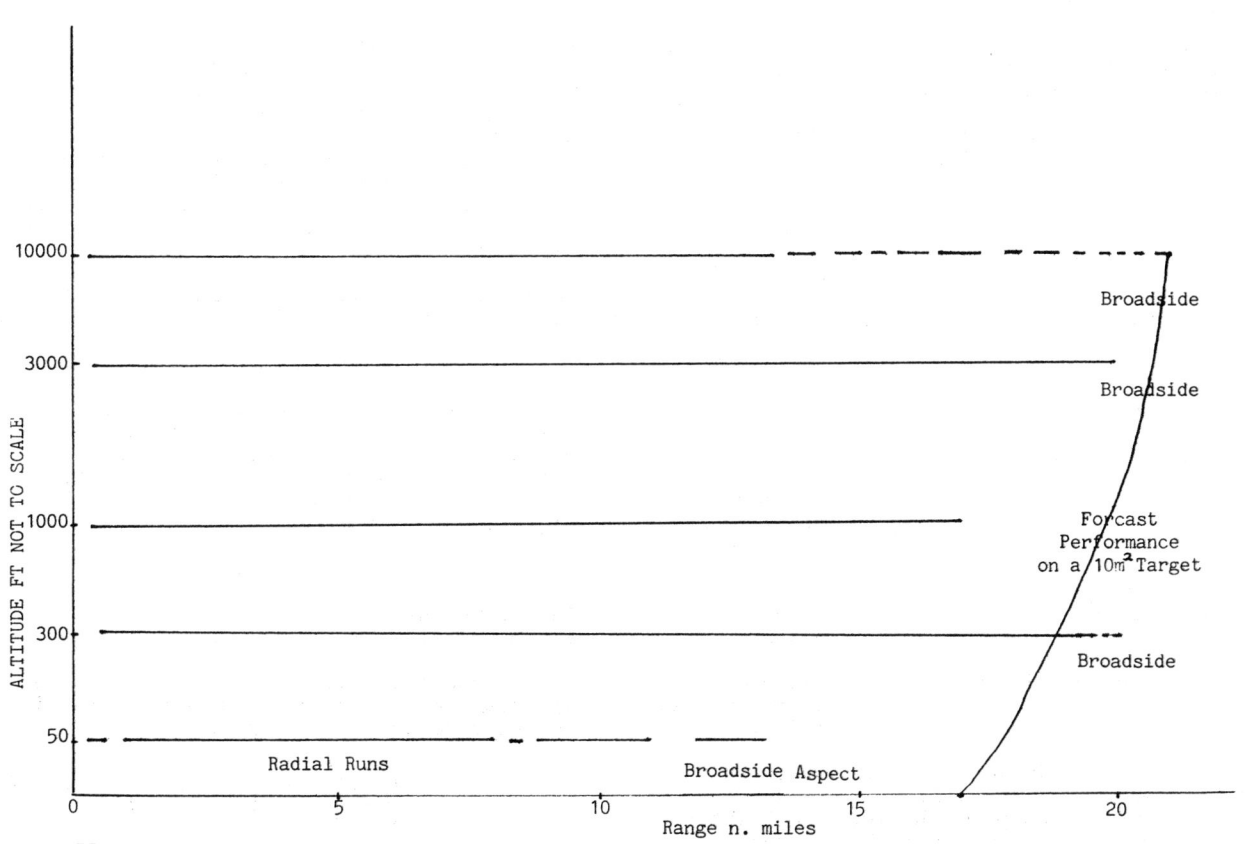

FIG. 14. MEASURED PERFORMANCE ON A MEDIUM BATTLEFIELD HELICOPTER

THE DOLPHIN NAVAL SURVEILLANCE RADAR

J. Blogh

Plessey Radar Limited, UK

INTRODUCTION

The development of anti-ship missiles with a capability of being launched from aircraft, surface ships or submarines constitutes a significant threat to warships at sea. These missiles have the ability to approach their targets from sea-skimming height to an approach angle in excess of 70°, with velocities ranging from sub-sonic to Mach 2+ and with a capability of performing terminal high 'g' manoeuvres. Missile systems exist which may defeat this threat at ranges in excess of a few kilometres, but there is a requirement for a complementary gun based defence system, suitable for installation on small vessels of around 200 tons and upwards, which can destroy the anti-ship missile at close range as a last ditch defence.

The Dolphin radar has been developed specifically to perform the target indication function for a close in weapon system (c.i.w.s.) and to perform general surveillance on the smaller classes of ship. It is designed to provide high accuracy target range and bearing information at a high up-date rate to the weapon system to enable the tracking radar to acquire its target in a minimum time.

RADAR REQUIREMENTS

The definition of the integrated weapon system resulted in the following target indication/surveillance radar requirement.

Detection Range, (a)	$\sigma = 0.1m^2$ Elevation coverage	8km 0° to 70°
(b)	$\sigma = 2m^2$ Elevation coverage	30km 0° to 10°
Radar Coverage		See Figure 1
Data Rate		1Hz
False Alarm Rate		10^{-6}
Azimuth Accuracy		0.5° r.m.s. (Missile targets) 0.3° r.m.s. (Surface targets)
Range Accuracy		50 metres r.m.s.
Masthead Weight (including i.f.f. antenna)		approximately 500kg.

The defined requirement must be satisfied in the presence of sea and rain clutter, chaff and both noise and pulse jamming. A light-weight compact mechanical design is required to enable the system to be installed in 200 ton craft.

DESIGN PHILOSOPHY

Certain key elements must be considered in the design of a radar system for a specific application. These include radar frequency, antenna parameters, transmitted waveform and modulation, clutter suppression techniques and immunity to electronic countermeasures. The selection of radar parameters generally involves compromises in an attempt to obtain the optimum performance against the spectrum of operational environments.

Radar Frequency

Whilst radar frequency band affects detection in clear conditions, antenna beamwidths, clutter rejection capability, and performance in an e.c.m. environment, it is the latter three aspects which for Dolphin dictated the frequency selection. Considering the azimuth accuracy requirement, basic radar angular accuracies for a scanning beam radar of one tenth of a beamwidth are achievable. To obtain the overall azimuth accuracy specified, not more than 0.15° can be allocated to the radar measurement directly resulting in a maximum azimuth beamwidth of 1.5°. To achieve this at 'S' Band requires an antenna aperture of 4.5 metres which is inconsistent with the desired masthead weight. The choice of frequency therefore lies between C Band (5.5GHz) and X Band (9.3GHz). The maximum width of antenna, constrained by masthead weight, is considered to be 2.5 metres consistent with a 1.5° beamwidth at C Band. At X Band an antenna azimuth aperture of 2.5 metres must also be employed to enable the required detection range to be achieved without an unacceptable increase in transmitter power.

Rain echo is a function of radar frequency according to the relationship: (Katzin (1)).

$$\sigma_{rain} \propto f^4 \times \theta az \times \theta el \quad \ldots\ldots (1)$$

where f is radar frequency, θaz is the azimuth beamwidth and θel is the elevation angle over which rain is illuminated. For a fixed antenna azimuth aperture and a constant value of θel defined either by the coverage requirement or the rain ceiling we get:

$$\sigma_{rain} \propto f^3 \quad \ldots\ldots (2)$$

Therefore the rain echo at X Band is approximately 7dB greater than at C Band.

The standard deviation σ_c of clutter power spectra due to internal clutter fluctuations and antenna scanning is given by:

$$\sigma_c^2 = \left(\frac{2\sigma_v}{\lambda}\right)^2 + \left(\frac{0.265\ fr}{n}\right)^2 \quad \ldots\ldots (3)$$

where σ_v is the standard deviation of the clutter spectrum
λ is the radar wavelength
f_r is the radar p.r.f.
n is the number of hits/beamwidth.

For a fixed antenna aperture n is proportional to λ and hence

$$\sigma_c \propto 1/\lambda \qquad \ldots\ldots (4)$$

For an m.t.i. system employing a double delay canceller the improvement factor I is inversely proportional to the fourth power of σ_c. Hence I is 9dB higher at C Band than X Band and taking account of the reduced clutter return at C Band, the clutter residue after m.t.i. processing is approximately 16dB lower than at X Band. This assumes that equipment instabilities do not limit the m.t.i. performance.

The relationship between sea clutter echo and radar frequency is not well defined and the assumption is made that they are independent. For a fixed antenna azimuth aperture therefore the echo at X Band is 2.3dB lower than at C Band due to the smaller area of clutter illuminated. As in the case of rain clutter the m.t.i. improvement factor is 9dB higher at C Band giving an overall gain of 6.7dB when compared to X Band. Furthermore, the sea clutter transition range is proportional to frequency as illustrated in Figure 2, (Skolnik (2)) resulting in a much reduced sea clutter echo beyond the C Band transition range.

Figure 3 illustrates the detection performance of the Dolphin radar which operates in C Band with a 3 pulse m.t.i. canceller against a $0.1m^2$ target. Using similar radar parameters at X Band the detection range fails to meet the requirement. Table 1 makes the comparison in the presence of 4mm/hour rain with the severe effect on detection of the target at X Band. The use of circular polarisation may be considered to combat rain clutter but the benefits are unquantifiable as the missile target may suffer a loss of echoing area approaching that of the rain clutter.

TABLE 1 - Low target detection range in rain for similar C and X band system.

	Detection range km				
Target ht. m.	2	5	10	20	50
C logarithmic video	-	-	-	-	-
C m.t.i. processing	5.0	7.6	10.3	14.3	12.5
X logarithmic video	-	-	-	-	-
X m.t.i. processing	4.0	2.6	4.5	1.8	1.8

Radar frequency influences radar performance in jamming as follows:-

(a) For a given antenna azimuth aperture the lower the radar frequency the wider the main antenna lobe. Jammers with a capability for effective main lobe jamming only will therefore be effective over a wider azimuth angle at the lower radar frequency and will produce more effective screening in a stand-off jammer role.

(b) The ability to determine the direction of the jammer by Jamming Strobe Extraction techniques will be enhanced at the higher frequencies.

(c) Whilst greater jammer power can be generated at lower frequencies limitations on jamming antenna dimensions will normally reduce the effective radiated power at these frequencies.

(d) Radar instantaneous bandwidth is independent of operating band. Operational agile bandwidths of radars increase with an increase of transmission frequency and thus a jammer must spread its power over a wider bandwidth at the higher frequency. This results in less jamming power entering the radar receiver with an accompanying reduction in jammer effectiveness.

In optimising the radar performance for an environment of rain and sea clutter, and active and passive jamming within the constraints of antenna system weight, C Band was selected for the Dolphin application.

Elevation Coverage

Simultaneous roles of target indication up to 70° combined with general surveillance for detection of air and surface targets are enhanced by the use of a dual beam antenna system, a single 70° fan beam not having the necessary gain at low elevation angles. A typical coverage diagram, Figure 3, illustrates the dual beam advantage. Further benefits of this approach include the absence of clutter in the high beam and the possibility of jamming entering only one of the two beams, particularly if each beam operates on a different frequency. Dual beams in 2-D radars can be fed concurrently or alternatively. In the former case the transmitter power in each pulse is divided between the two beams. Dual receivers are required to avoid a collapsing loss and to retain the data as to which beam detected a target, useful information to minimise target acquisition by a tracker. The Dolphin radar transmits groups of pulses in each beam alternately, with all target information being received in one prior to the other transmitting. A single receiver chain is multiplexed between the two beams whilst allowing target beam identification to be retained. This method allows a very high group pulse repetition frequency to be employed in the high beam compatible with its short range requirement and yielding an improved m.t.i. performance.

Transmitter Selection

The need for clutter rejection and high performance in an e.c.m. environment dictates the use of a wideband coherent transmitter. Pulse to pulse phase stability of 0.6° is required to obtain the necessary clutter cancellation. Within pulse phase ripple of 3° ensures that pulse compression time sidelobes of less than -35dB with respect to the peak of the compressed pulse are achieved. To obtain these stabilities a travelling wave tube transmitter is employed with e.c.c.m. advantages. Firstly instantaneous changes of transmitter frequency over a wide bandwidth is possible. Secondly immediate switch off of r.f. transmissions can be achieved for sector blanking or if required to defeat anti-radiation missiles. Thirdly the peak power of such tubes is an order of magnitude lower than magnetron transmitters reducing their probability of detection by e.s.m. equipment. The use of a grided t.w.t. allows the alteration of transmitted pulse lengths by changing the grid and r.f. drive waveforms. Long transmitter pulses inherent in t.w.t. operation require to be compressed on reception to produce operationally acceptable range resolution and accuracy and to minimise the range extent of clutter illumination. The benefits of the short compressed pulse are obtained by utilising high processing speeds in the doppler processing and target detection circuitry. Transmission of long pulses extends the minimum range of the radar. The interspersion of short uncoded pulses into the transmission pattern will reduce the radar minimum range to an acceptable level with a small percentage of the transmitter power being diverted for this purpose.

Doppler Processing

Extraction of moving targets from relatively low velocity clutter or chaff involves the application of either m.t.i. in which a filter is used to reject the interference, or pulse doppler in which a filter band is generated and the outputs are accepted of the filters in which wanted targets are expected. Pulse doppler processing is generally employed where high values of sub-clutter visibility are required. An examination of the operational environment often shows that acceptable performance can be achieved with a less complex m.t.i. processor typically a three pulse canceller. In the choice of doppler processor it must be considered that for e.c.c.m. reasons ideally pulse-to-pulse frequency agility should be employed. Pulse doppler systems require that a relatively long burst of constant frequency pulses are transmitted, typically eight or sixteen pulses whereas a three pulse canceller requires three pulses at constant frequency. As an example using the Dolphin radar parameters a stand-off jammer at a range in excess of 110km, equivalent to three pulse repetition intervals, would not be able to jam the radar with repeater jamming when the Dolphin radar is operating in the "burst agile" mode. This produces a change of frequency after each three pulse burst.

Doppler processing must remove the effects of platform motion. Use of m.t.i. cancellers require that the position of the filter notch must adapt to the environment to perform effectively.

EQUIPMENT DESCRIPTION

The Dolphin radar is illustrated in Figure 4, with the corresponding block diagram in Figure 5. It consists of a dual beam primary radar antenna system with i.f.f. antenna attached all mounted on a two axis stabilised turning gear. The antenna system is fed from the transmitter cabinet containing the t.w.t. power amplifier and duplexer. The r.f. drive waveform is generated in the Signal Processing Rack which houses the r.f. and i.f. processing circuits and the digital circuits for m.t.i. and c.f.a.r. processing. After integration signals are fed to the Distribution and Antenna Control Rack for plot extraction prior to handing over the target data to the weapon system. This rack also houses the servo processors and power amplifiers for control of the stable platform.

Antenna System

The linear array high beam antenna consists of a row of 52 dipoles housed in a corner reflector, the vertical aperture of which produces the required elevation beamwidth. The dipoles are fed from an air spaced strip-line corporate distribution network. The electrical path length from the input of the network to each radiating element is constant thereby producing an equiphase front across the radiating aperture over the frequency agile bandwidth. The network generates a Taylor amplitude distribution across the array designed for 29dB first sidelobes. The main beam antenna utilises an identical feed system although the vertical aperture of the corner reflector is altered to illuminate a single curvature reflector. This reflector is shaped to produce a $cosec^2$ beam in elevation with a vertical beamwidth of 14°. Both antennae normally produce horizontal polarisation with circular polarisation being switchable on the main beam. This is produced by means of a Lerner polariser which rotates over the main beam feed aperture. The arrays are fed by means of a passive duplexer which filters the transmission energy to the appropriate beam as a function of radar frequency. Attached to the high beam is an i.f.f. antenna consisting of two rows of ten radiating elements fed from a printed distribution network. The i.f.f. beam has an electronic squint of 8° lagging the primary radar. Targets may therefore be interrogated after detection by the primary radar whilst ensuring that the full i.f.f. beamwidth is utilised.

The antenna system is mounted on the stabilised turning gear which rotates the antenna system at 60 r.p.m. The platform operates over $\pm 10^{\circ}$ in pitch and $\pm 25^{\circ}$ in roll with a minimum period of 5 seconds. The stabilisation accuracy of the platform is 29'r.m.s. under these conditions. The weight of the complete above decks equipment is 520kg.

Stabiliser Servo System

The stabiliser servo system utilises ships attitude information from the ships master vertical reference system to control the two axis mount in roll and pitch maintaining the turning axis of the antenna vertical. Comparision of the vertical reference data with the stabiliser angles produces an error signal which after filtering is fed to switching mode solid state amplifiers. These have sufficient power handling capability to drive the platform servo motors. The servo motors are permanent magnet d.c. torque motors each fitted with a tacho generator and electro-mechanical brake.

Transmitter

The two stage amplifier in the transmitter uses a coupled cavity high gain travelling wave tube as the power stage. The driver stage is a solid stage amplifier. The t.w.t. is grid modulated giving flexibility of pulse widths and p.r.f.'s, and operates in the depressed collector mode to achieve 25% efficiency. The high stability required for m.t.i. is achieved by using a method of quantum charging the e.h.t. storage capacitors.

The quantum of energy is generated by a 14KHz inverter. As the requirements of the power supply change ... p.r.f. ... the ... quanta ... is ... voltage ... Additional ... thode to ... iser. ... al from ... thode ... is used ... tor ... essing ... on a ... fast ... ieved by ... feeding ... trans- ... e of

... r is ... net. A ... eralised ... and ... gers ... and the ... from the ... primary ... er to ... er.

... 70MHz ... hin the ... uration ... h imparts ... wave- ... by ... on, centred at 70MHz, is selected by amplitude gating. The coded pulse is mixed with derivatives of the Frequency Agile Synthesiser to produce the transmission waveform.

The frequency synthesiser provides a specific frequency from the channels available over the agility band. To minimise the frequency switching time, the unit uses direct synthesis by selective mixing of the derivatives of four low noise crystal controlled oscillators. Change of frequency can be accomplished within 5μs and the spectral noise of the source is less than -77dB with respect to the carrier, in a 1Hz bandwidth, 1KHz from the carrier.

Receiver channel. The received signal is converted down to a final intermediate frequency of 70MHz by mixing with the local oscillators used in the transmission chain, in order to maintain the correct phase reference. The signals are subject to limiting prior to compression in a weighted s.a.w. compressor. A main lobe pulse width of 125ns and time sidelobes of less than -36dB with respect to the main lobe are achieved.

The compressed i.f. signal is converted to video signals in quadrature phase detectors by reference to a 70MHz oscillator.

Variable phase shifter. Operation of the m.t.i. fast loop, to control the m.t.i. notch position is achieved by altering the phase of the 70MHz reference signal. This is controlled by the introduction of delay into the signal path. Each bit of a parallel seven bit word representing the observed phase shift, controls an i.f. switch which introduces a delay into the circuit. The delays are binomially weighted with the least significant bit equating to 2.8°.

Processing of Digital Signals

The processing of the received signals provides a digital message comprising range and bearing of each target, for subsequent processing by the threat assessment function.

Cancellation of returns from clutter by accurate positioning of an m.t.i. rejection notch at the mean clutter velocity is maintained by adaptive feedback loops.

The m.t.i. processor consists of quadrature dual loop cancellers including two adaptive control loops.

A slow closed loop controls the frequency of the phase reference oscillator, using an offset error derived from the clutter background and averaged over several pulse intervals. This removes the doppler offset from the clutter surrounding the radar.

A high speed closed loop system operating on the digitally controlled phase shifter following the reference oscillator, uses an offset which is updated at range bin rate to compensate for clutter velocities which vary as a function of range.

The doppler offset is estimated by a phase error algorithm which compares the signals at the input and output of the first delays of the m.t.i. canceller.

The control signal to the fast loop is obtained by integrating the error signal at the output of the phase error algorithm in a 16 bit accumulator. The phase of the 70MHz reference signal is shifted by the estimated error signal in the digital phase shifter at a 10 megabit rate. The control signal of the slow loop is obtained by sampling the above error and averaging over 8 pulse intervals.

Optimum performance is obtained at close range when the clutter rejection notch is controlled by the slow loop. In operation of the fast loop, a compromise is made between improvement factor and the loop response time to a block of clutter at a new velocity. In the interest of response speed, a small degradation of improvement factor is accepted.

A parallel channel which does not undergo doppler processing provides visibility of targets with low radial velocities.

Both channels are subject to constant false alarm rate circuitry and correlation across the antenna beam dwell, but utilisation of the non m.t.i. processed channel is restricted by control from an adapting clutter map.

Extraction of range and bearing in a plot message format, from the real time radar returns is undertaken in non-saturable hardware, which maintains the identity of the plots i.e., antenna beam and processing channel.

CONCLUSIONS

The Dolphin radar has been designed specifically to meet the target indication requirement of a shipborne close in weapon system. Radar parameters have been selected to enable this function to be achieved in the presence of severe clutter and e.c.m. Evaluation equipments have been manufactured for integration into the weapon system and for subsequent trials, with an equipment having successfully undergone full environmental tests to naval specifications.

Deployment of the Dolphin radar with its associated weapon system should greatly enhance the survivability of naval craft subject to attack from anti-ship missiles.

ACKNOWLEDGEMENTS

The author wishes to thank Mr. G. Bardanzellu, Marketing Director of Contraves AG. and Mr. J. Hakes, Managing Director of Plessey Radar Limited, for permission to publish this paper. The Dolphin radar was developed against a specification from Contraves Italiana Spa. for the Seaguard close in weapon system.

REFERENCES

1. Katzin, M., January 1957, "On the mechanisms of radar sea clutter", Proceedings of the IRE.

2. Skolnik, M., "Radar Handbook", P24-21, McGraw Hill Book Company.

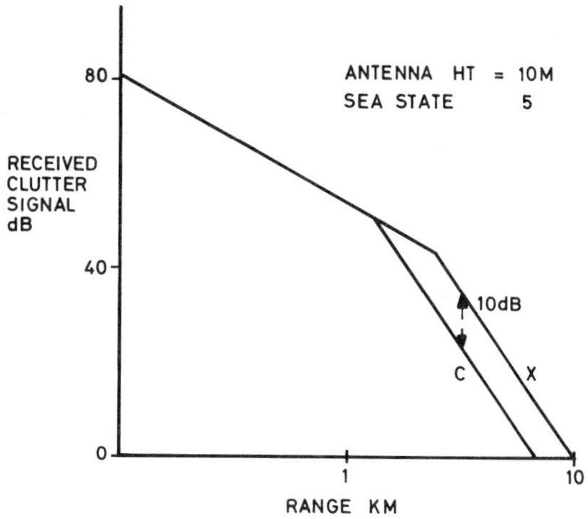

Figure 2 Sea clutter signal versus range

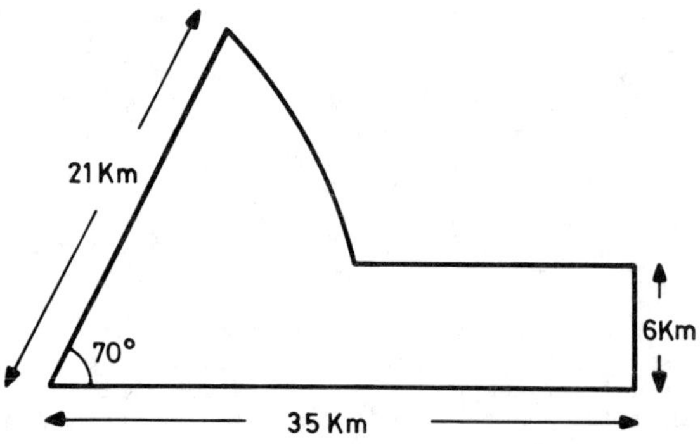

Figure 1 Instrumented range requirement

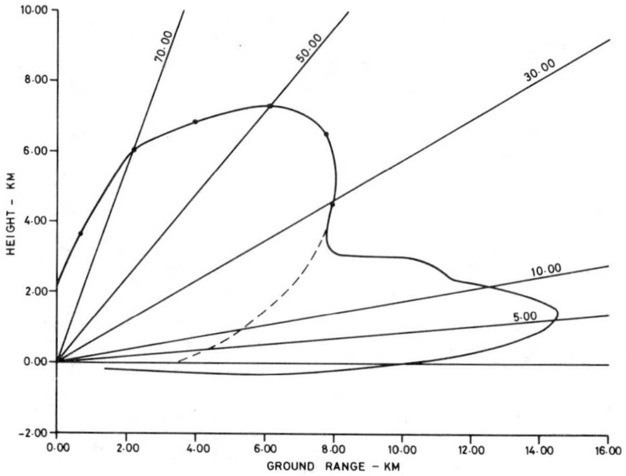

Figure 3 Dolphin coverage diagram $0.1m^2$ target

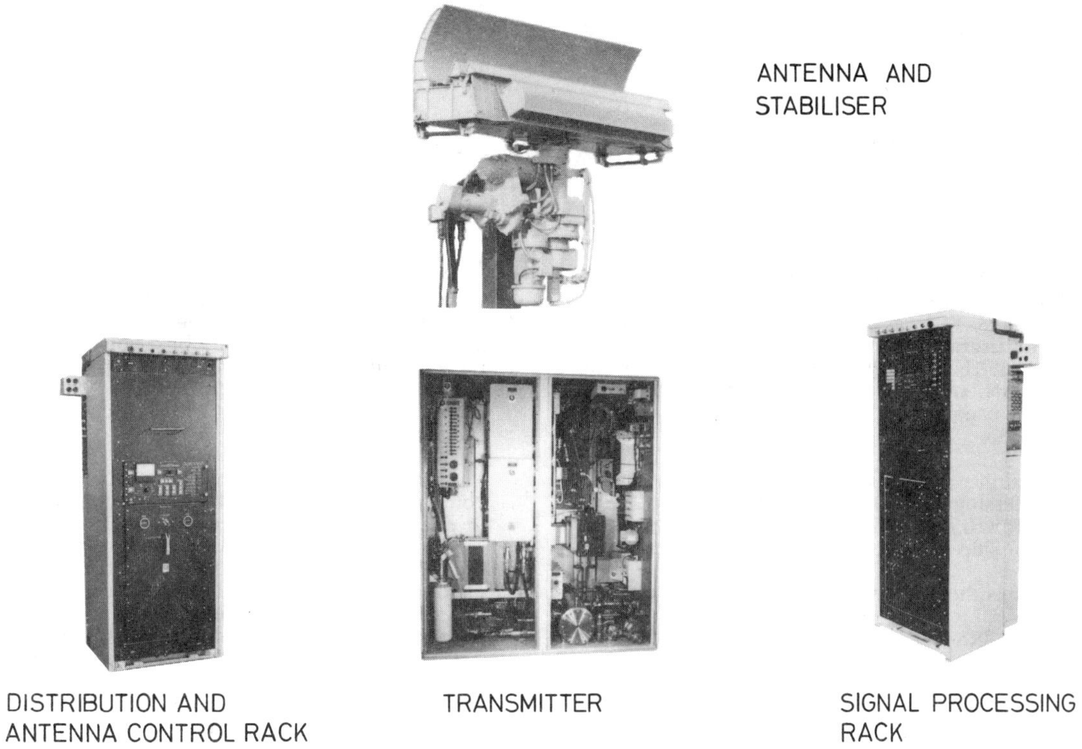

Figure 4 Dolphin radar hardware

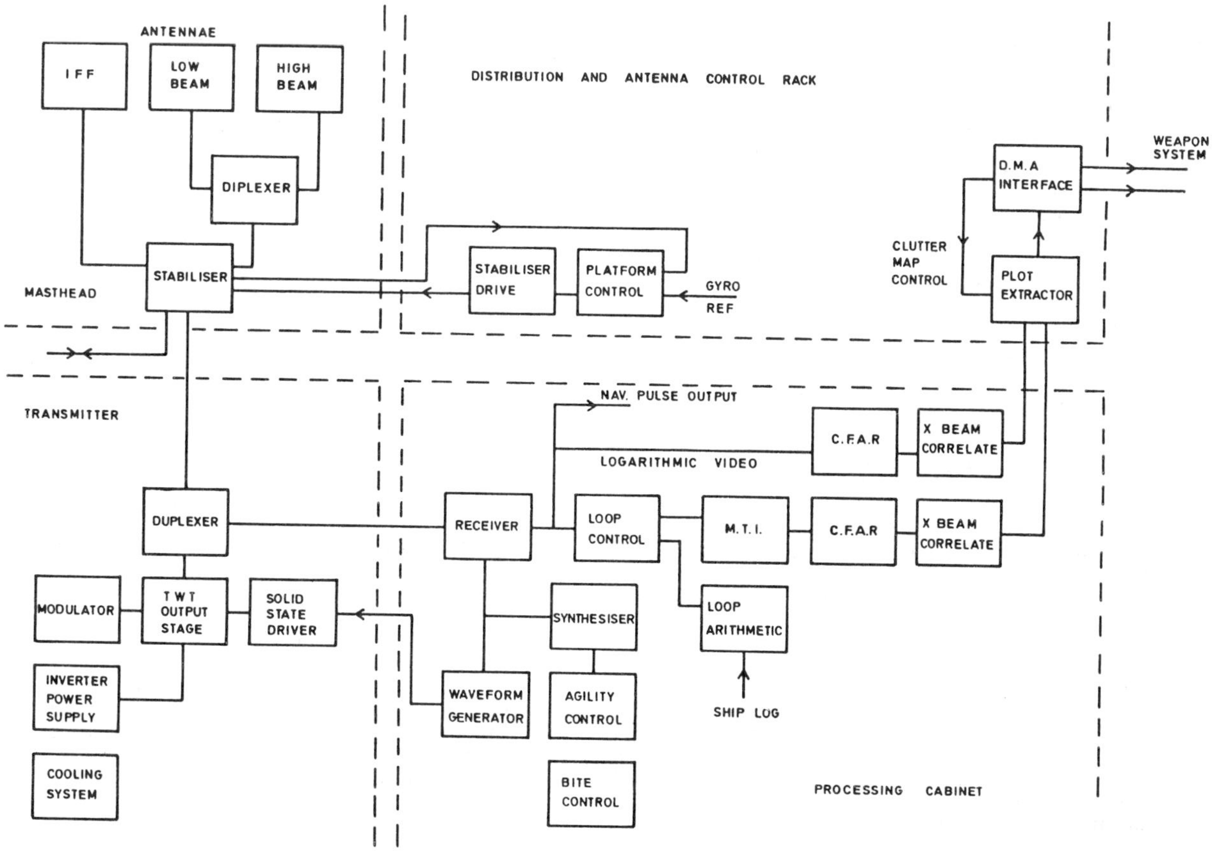

Figure 5 Dolphin radar block diagram

AN/APS-134(V) MARITIME SURVEILLANCE RADAR

James M. Smith

Texas Instruments Incorporated, USA

INTRODUCTION

The AN/APS-134(V) Radar System is the international successor to the U.S. Navy's AN/APS-116 Periscope Detecting Radar. The AN/APS-116 was originally designed to provide the S-3A Viking aircraft with periscope detecting capability in high sea states. The AN/APS-134 incorporates all of the features of the AN/APS-116 while improving performance and adding new capabilities, including a unique maritime surveillance mode.

Improved operability and reliability are achieved by replacing analog processing and scan conversion with an all digital design. A growth mode that produces two-dimensional radar images for ship classification is in development.

Figure 1 identifies the units comprising the radar system.

DESIGN CONSIDERATIONS

The inherent problem of the maritime surveillance radar is detecting small targets in the sea-clutter environment. The combination of extremely high resolution coupled with scan-to-scan (S/S) processing offers the only proven means of overcoming this limitation. The AN/APS-134 incorporates both features to provide optimum antisubmarine and surface detection capability. The transmitted waveform is frequency modulated ("chirped") in a linear manner over a 500-MHz bandwidth. This permits the high average power necessary for long-range periscope detection in the noise environment. The received signal is pulse-compressed, using state-of-the-art surface acoustic wave devices, from a 0.5-microsecond (μs) to a 2.5-nanosecond (ns) pulse. This effectively reduced the sea surface area illuminated by the radar beam. Since the surface sea clutter return is a function of this area, the clutter is reduced by the pulse compression ratio (23 dB). Average clutter return, however, is only part of the problem. Clutter spikes of 3 or more seconds compete with target returns to confuse the radar operator. To combat the spiky nature of the sea clutter return and further reduce the average value, fast scan processing is employed to time-decorrelate the clutter. The radar returns are processed on an S/S basis over a period of several seconds. The digital processing circuitry integrates the correlated target signals while rejecting the uncorrelated clutter.

Examples of the foregoing discussion are shown in Figure 2, top row. These are actual photographs taken during shore testing of the radar at an altitude of 1,200 feet. Figure 2 at top left shows the uncompressed 0.5-μs pulse. The sea state is 3-plus and the clutter extends to beyond 20 nmi. In Figure 2, top center, the clutter spikes mask the target returns. Figure 2, top right, shows the result of S/S processing. The top return is from an attack periscope. The lower target is the combination search periscope and ESM mast of a second submarine. The bottom returns are from near-range land (shore) mass.

SYSTEM DESCRIPTION

The design, at the system level, ensures installation flexibility. This is accomplished by providing both on- and off-line control and display options. The basic radar set, comprising modified AN/APS-116 units [detailed description by Smith and Logan (1)], remains the same for any configuration. The signal data converter (SDC) is the interface between the control commands and the radar set, and the input/output channels are tailored for the particular requirement.

In the on-line version, the SDC supplies composite radar video with 512 active lines for use by external multipurpose displays. Radar control is effected through a MIL-STD-1553B serial data bus. For off-line installations, the SDC interfaces with the IP-1385/APS-134(V) Radar Control/Display (RC/D) to provide a complete radar system. The operation of the self-contained system is described as follows.

The SDC, with the RC/D, provides processed radar video. It controls radar operation by performing aircraft and radar interface functions. Hence, the SDC is divided into two functional sections (plus a power supply). These are the digital scan converter (DSC) and the radar set control (RSC) sections. The DSC section accepts radar video and processes it by pulse-to-pulse (P/P) and S/S integration in the fast and intermediate scan speed modes. The processed data is stored in memory locations (write addresses) selected by the address generator. Read address from the generator recalls data from the memory for display. The data is converted to an analog signal and combined with synchronization pulses to provide a composite video signal for transmission to the display. The RSC section contains aircraft and radar interfaces and performs processing and timing functions. The processing and timing function provides the synchronization and processing needed to operate the radar and DSC. The timing portion provides clocks and triggers to the DSC for synchronization with radar timing. The processor stores and distributes the radar and DSC commands and performs arithmetic operations on the data from the aircraft systems, display control panel, and the antenna azimuth encoder. The results are used by the DSC.

OPERATIONAL CHARACTERISTICS

Three operational modes are provided, and they are discussed in the following paragraphs.

Mode I is optimized for the low-altitude detection of periscope-size targets of limited exposure time. The compressed pulse equates to a range resolution of less than 1.5 feet, which closely matches the physical dimension of periscope and snorkel classes of targets. The rapid scan antenna rotates at 150 rpm to provide the necessary S/S processing gain in a short time. Maximum range in this mode is 32 nmi.

Mode II is the navigation and weather avoidance mode and is designed for maximum sensitivity against land and other distributed targets. The techniques associated with the fast-scan modes emphasize point target detection while rejecting distributed or area targets (i.e., sea clutter). Therefore, this mode employs a slow antenna rotational rate and an uncompressed waveform. Maximum range is 150 nmi.

Mode III provides high altitude maritime surveillance to ranges of 150 nmi. Again, the use of pulse compression and fast-scan processing allows detection of patrol/fishing boat-size targets and larger under all sea conditions. These targets are continuously exposed, and longer integration times (i.e., S/S processing intervals) can be employed. Thus, the antenna rotational rate can be slowed to 40 rpm to minimize processing losses.

Table 1 is a listing of the system parameters of the AN/APS-134.

TABLE 1. AN/APS-134(V) system parameters

ANTENNA	
Radar	
Gain	35 dB (nominal)
Azimuth beamwidth	2.4 degrees
Elevation beamwidth	4.0 degrees
Sidelobes	Down at least 20 dB
Polarization	Vertical
IFF (integral)	Standard IFF interrogator sets monopulse compatible
Scan speed	
Mode I	150 rpm
Mode II	6 rpm
Mode III	40 rpm
TRANSMITTER	
Peak power	500 kW
Average power	500 W
Frequency	
Mode I, III (linear FM sweep)	9.5 to 10.0 GHz
Mode II (random frequency agility)	9.6 to 9.9 GHz
PRF (four-pulse stagger)	
Mode I	2,000 pps
Mode II, III	500 pps
Pulsewidth	0.5 μs
RECEIVER	
Noise figure (system)	4.5 dB
Intermediate frequency	
Mode I, III	1300 MHz
Mode II (dual conversion)	1300 MHz and 100 MHz
IF bandwidth	
Mode I, III	500 MHz
Mode II	2.4 MHz
Pulse compression gain (Mode I, III)	23.0 dB
Compressed pulsewidth (Mode I, III)	2.5 ns
AGC	Three loops, CFAR

SIGNAL PROCESSING

Considerable analysis aided choices of modes/parameters and postdetection processing. Basically, this centered on the use of either P/P or S/S integration, or a combination of both. In the slow-scan antenna mode, P/P integration results in reduced signal-to-noise (S/N) ratio requirements at the input to the detector for a given detection probability and false alarm probability. Pulse-to-pulse integration does not improve performance in the sea clutter case because clutter returns are highly correlated for short time intervals (e.g., milliseconds) between pulses. Scan-to-scan integration over longer time periods (e.g., seconds) can improve both S/N and signal-to-clutter (S/C) ratio. The processing gain or degradation in S/N or S/C depends primarily on the time interval and number of samples for integration.

Figure 3(A) shows a generic block diagram of the signal processing chain although not all elements of this chain are used in each mode of operation.

The digitizer converts the analog video to digital data either by two-level digitization (thresholding) or multiple-level digitization (quantization). Range stretching is a method of reducing the number of range resolution elements to an acceptable value for subsequent processing and display. The stretching consists of taking the peak value in a number of successive range cells as representative of that group, which is then treated as one stretched range cell. The total range stretch is determined by the range resolution inherent in the radar signal structure and the fact that the digital scan converter can accommodate only 512 stretched range cells. Figure 3(A) indicates that the total range stretch can be apportioned in the processing chain to optimize tradeoffs between memory requirements and/or processing rate and performance.

An additional signal processing feature of the AN/APS-134 is multilevel processing, which allows preservation of a background clutter map. This enhances the ability of the radar operator to discriminate between debris (or other nonsignificant returns) and small targets of interest. The clutter (or ocean) map is generated by allowing the large-area returns to integrate within the digital scan converter memory to a preset amplitude (Level 80), which represents approximately one-third of the display dynamic range. Point targets are allowed to integrate to the full memory depth (Level 256) and thus be displayed at the maximum intensity. The presentation is an underlay representing the ocean surface with target returns appearing as intensified spots. Both surface and subsurface currents can be distinguished along with predominant wave structures. The relationship of the point targets to the clutter map provides the basis for operator discrimination. An additional benefit of the clutter-mapping technique is the ease of detecting calm or smooth areas within the ocean structure. These are readily apparent and can be investigated for possible contamination from oil spills or leakage.

Figure 2, lower row, highlights the signal processing features of the radar. These photographs were all taken during shore tests at an altitude of 1,300 feet. Ocean coverage is approximately 170 to 290 degrees. Range scale is 32 nmi and sea state is between 2 and 3.

Figure 2 at lower left represents the slow scan navigation mode. Sea clutter is visible over the entire 32-nmi range. The larger targets are fishing boats.

Figure 2, lower center, shows fast scan with S/S processing using threshold detection. The clutter returns are completely decorrelated and the targets are presented against a "black" background. Although useful for the unattentive operator, it is difficult to distinguish between small targets of interest and debris.

Figure 2, lower right, shows fast scan with S/S and multilevel (shades-of-gray) processing. The clutter map is clearly visible, yet the smallest of targets is integrated to a higher level. Note specifically the navigation buoy at 260 degrees and 13-nmi range. This target is barely visible in the threshold-detected picture and undistinguishable from clutter in the first photograph. Although the reproduction process degrades the quality of these pictures, it is still possible to see the ocean surface structure in the clutter map. Note the current ridges in the shallow areas near land. Real-time observation of these conditions over several hours showed remarkable stability in the clutter patterns.

PERFORMANCE

Considerable analysis has been done in calculating and predicting detection ranges. Optimization routines with the basic AN/APS-116 parameters were used to develop new digital processing algorithms and antenna rotation rates. The system loss budget contains all installation losses including radome. Final results of the analysis are shown in Table 2. This performance data has been confirmed by company field testing and customer flight qualification and evaluation testing.

TABLE 2. AN/APS-134(V) detection performance

Target Type	Radar Cross Section (m^2)	Detection Range (nmi)	Sea State	Altitude (feet)
Periscope	1	22	3	2,500
Snorkel	10	38	4	8,000
Patrol/fishing boat	100	65	5	20,000
Small transport or destroyer	1,000	105	5	20,000
Cruiser	3,300	140	5	20,000
Aircraft carrier or weather	>5,000	>150	5	25,000

Probability of detection (P_d) = 0.5

Probability of false alarm (P_{FA}) = 10^{-6}

RADAR IMAGING GROWTH

The AN/APS-134 is compatible with the inverse synthetic aperture radar (ISAR) imaging technique under development by Texas Instruments for the U.S. Navy. This processing generates an image of a recognizable nature on surface ships for use in long-range ship classification.

Previous attempts to image ships using conventional synthetic aperture techniques were largely unsuccessful, primarily owing to ship motion, which defocused the image. The U.S. Naval Research Laboratory developed the ISAR concept, which actually relies on ship motion.

The AN/APS-134 pulse compression chain and its attendant coherent operation provide the two key parameters of image generation.

The orthogonal axes used for the display of ISAR images represent directions parallel to the line of sight (range) and perpendicular to the line of sight (cross range). High resolution in the range dimension is obtained from the pulse compression. Radial velocity, and its effect on the returned signal phase, is used for resolution in the cross-range direction. Ship motion is the sole source of the radial velocities.

ISAR images are essentially maps indicating the location and reflectivity of the component scattering elements comprising a target. The positions of these elements are displayed in Cartesian coordinates proportional to range and cross range. Range locations are determined by the round-trip time of the signals. Cross-range measurement is dependent on target motion, with the radial velocity of the target being proportional to its cross-range location. Amplitude of the returned signals is plotted as intensity and indicates the radar cross section (reflectivity) at that location.

Generating the cross-range location requires selecting a reference point on the target and measuring target rotational motion relative to that point. A fundamental prerequisite for imaging is that any relative motion between the radar and the reference point must be compensated so that only the effects of target rotation remain. This motion compensation is achieved through highly precise range tracking and doppler tracking circuit functions to stabilize the reference point in both dimensions.

The type of image generated is determined by the nature of the ship motion and the location of the radar with respect to that motion. Figure 4 depicts yaw motion with the radar as shown. In this case, a plan view is produced. Figure 5 represents pitch motion and generates the profile view. In the true environment, ship motion is a combination of yaw, pitch, and roll (also producing a profile view). These combined motions generate images that are isometric in appearance, which enhances the ability of the interpreter to recognize and classify the ship.

The ISAR image is basically a range-doppler plot generated from a time aperture of coherent radar video. Figure 3(B) is a block diagram illustrating the three basic functions that comprise ISAR range-doppler processing: data presum, doppler processing, and display buffering.

The presum provides coherent integration and data reduction, providing the bandwidth of the target data is much less than the radar pulse repetition frequency (PRF). The presum, therefore, reduces the effective PRF and performs integration that otherwise would be done by the doppler processor. This effective PRF reduction is important because it allows additional time to generate the required frequency data.

Spectral analysis is performed in the doppler processor. For each sample range, the doppler processor converts time domain data cells to the frequency domain. This function can be performed by various techniques, the most common being the fast Fourier transform (FFT). The FFT is an efficient algorithm when the required image update rate is on the order of the time aperture. As the image update rate increases, the FFT becomes less attractive. Each time the image is updated, a complete time aperture of data must be processed. This means that the same data is processed over and over. An FFT system that would provide an output for every time input would require $(N/2)\log_2 N$ complex multiplications to produce each updated range cell output, where N is the number of samples comprising an aperture. This means that each range cell would require $N^2 \log_2 \sqrt{N}$ complex multiplications per time aperture.

A unique doppler processor has been developed by Texas Instruments that produces an updated output for each time input and requires only N complex multiplications per range cell update. Each range cell, therefore, requires N^2 complex multiplications per time aperture. This technique produces a *real-time continuous presentation* of range-doppler imagery and is called a continuous Fourier transform (CFT).

The processor transforms 256 points for 256 range cells at a rate of 256 operations per second. The doppler filter resolution is equal to the reciprocal of the effective aperture time τ_A. This implies that PRF and number of pulses presummed can be used to control the doppler filter resolution while keeping the number of filters constant. Thus, the complexity of the doppler processor and display circuitry is minimized. Providing the ISAR operator with control of τ_A allows doppler resolution and physical size of the image in the doppler dimension to be adjusted to compensate for ship motion at the time of observation.

The display buffer stores image data and converts it to a form compatible with a multipurpose display system.

CONCLUSION

The AN/APS-134 radar system is an evolutionary improvement of the AN/APS-116 system. It greatly enhances the periscope detection mission and provides the high-altitude capability required of modern maritime surveillance aircraft. An optional multiple-target tracking feature allows for long-term maintenance of the surface plot. Physical characteristics are detailed in Table 3.

TABLE 3. AN/APS-134(V) physical characteristics

Unit	Weight (pounds)	Dimensions (H × W × D inches)	Power Dissipation (watts)
Transmitter	174	13 × 22 × 21	3,495
Synchronizer-exciter	36	10 × 10 × 18	155
Receiver-pulse compressor	48	11 × 13 × 16	149
Power supply	44	9 × 12 × 19	365
Antenna	62	36 × 41 × 27	221
Signal data converter	50	8 × 15 × 20	310
Waveguide pressurizer	15	13 × 12 × 5	
Total	429		4,695
Optional			
Radar control/display	98	16 × 23 × 23	300
Total	527		4,995

When available, addition of ISAR ship imaging will allow real-time target classification to take place at standoff ranges that are beyond the llthality of current and projected surface-to-air weapons.

Current users of the AN/APS-134 include the Federal Republic of Germany in the BR-1150 "Atlantic" update, the Royal New Zealand Air Force P-3B "Orion" update (in production), and the United States Navy for specialized applications. A new wideband, low-profile, linear array antenna, suitable for helicopter bottom-fuselage mounting, is planned. This will allow installation of the radar in the larger antisubmarine helicopters.

REFERENCES

1. Smith, J.M., and Logan, R.H., 1980, *IEEE Trans. AES-16*, No. 1.

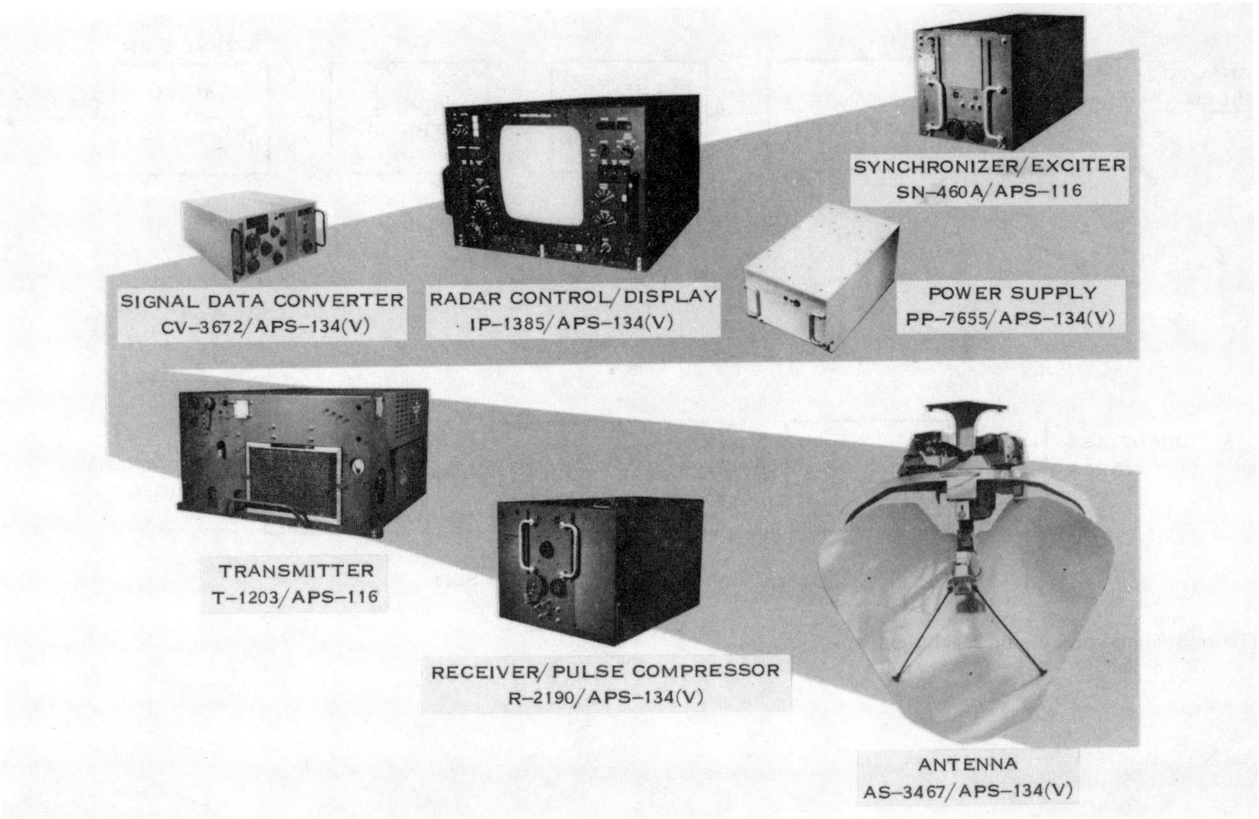

Figure 1. AN/APS-134(V) Maritime surveillance radar

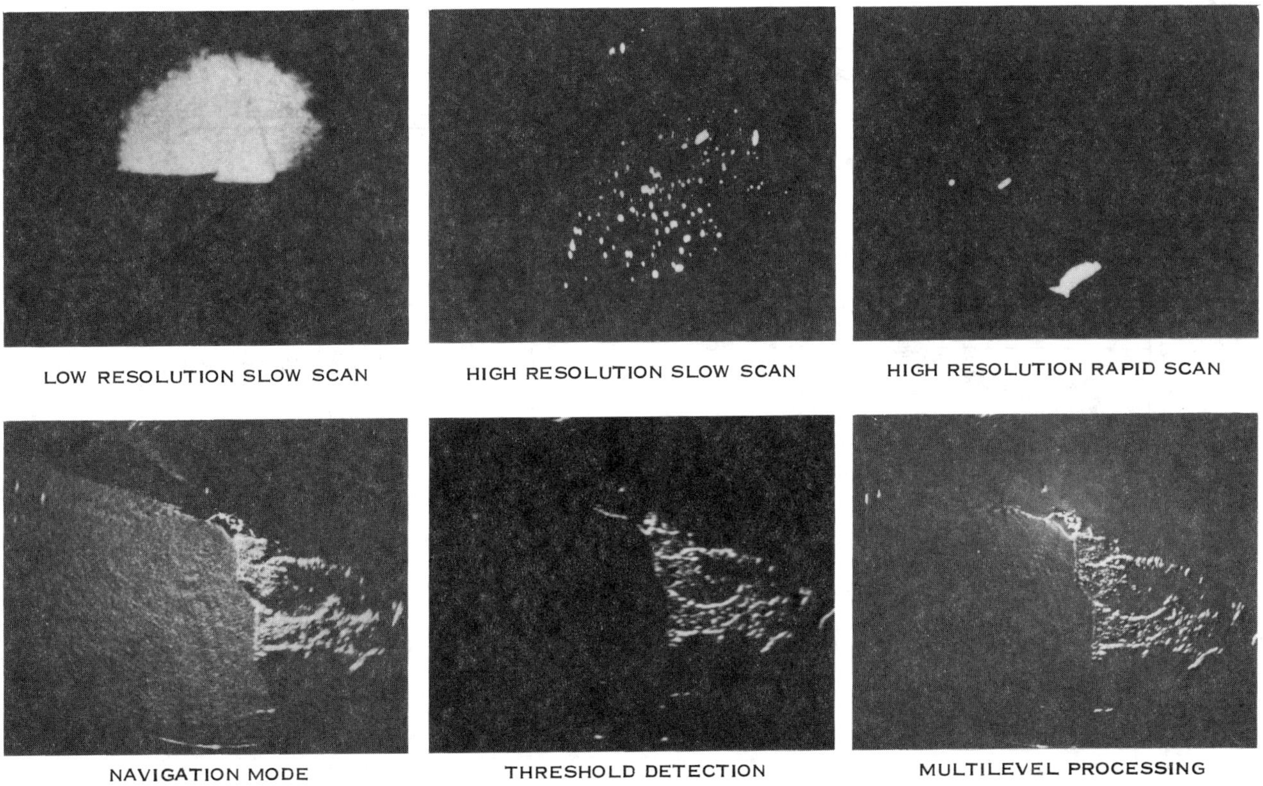

Figure 2. Examples of signal processing features

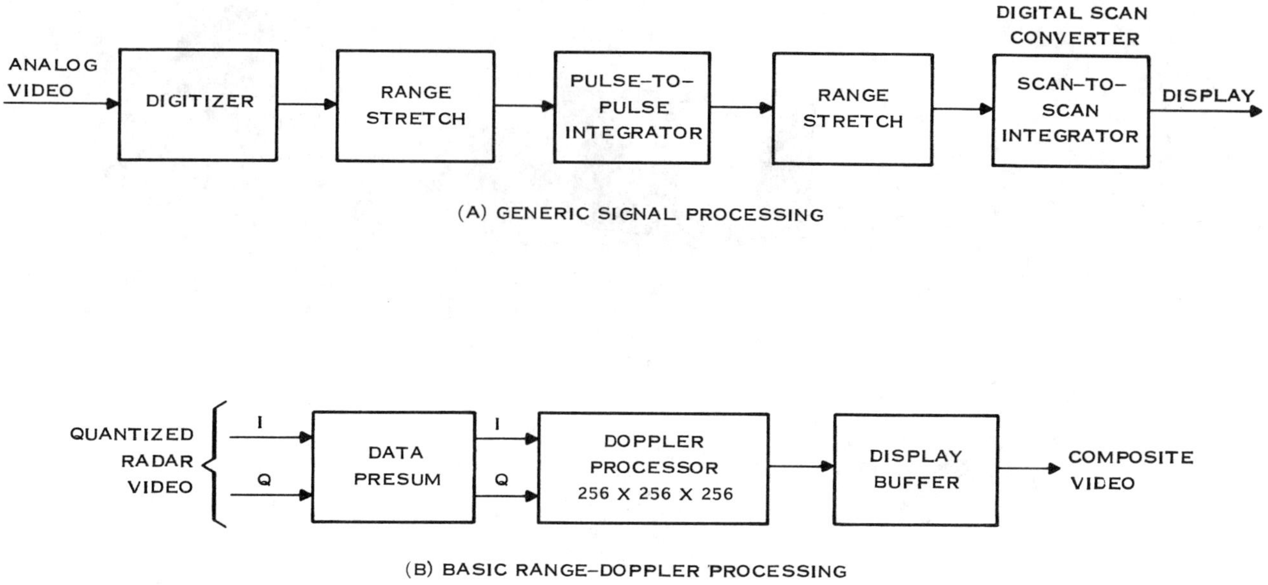

Figure 3. Signal processing block diagrams

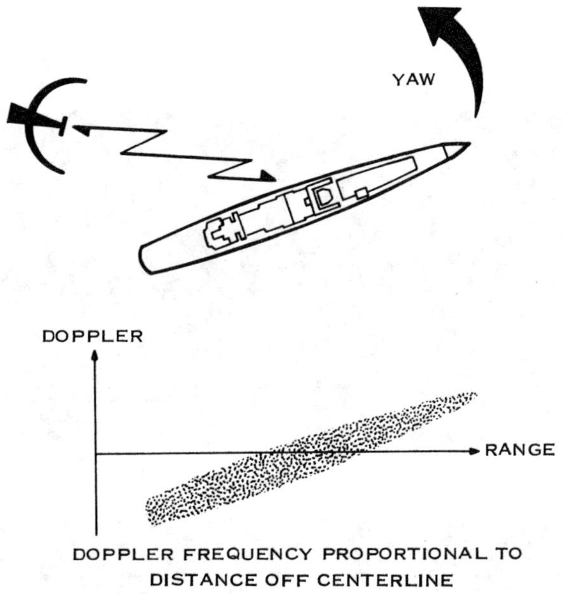

Figure 4. Yaw motion imaging

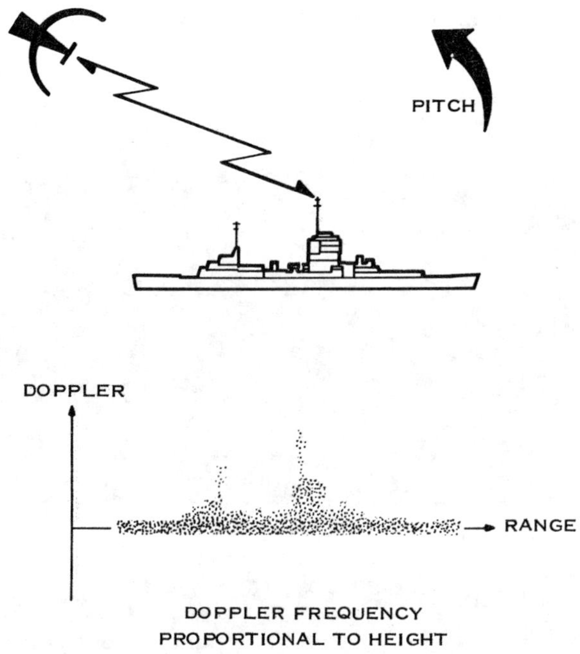

Figure 5. Pitch motion imaging

A BRITISH AEW RADAR SYSTEM

J. Clarke[1], J. King[2]

1. Royal Signals and Radar Establishment, UK; 2. Marconi Avionics, UK

INTRODUCTION

The British requirement for a new Airborne Early Warning (AEW) aircraft will soon be fulfilled with the introduction into service of the NIMROD AEW MK 3 shown in Fig. 1. The AEW Mission System Avionics (MSA) will provide airborne surveillance, detection, tracking and recognition of airborne and maritime targets and will link into the NATO Air Defence Environment, particularly the UK segment. The prime sensor of this system is the surveillance radar which provides long range detection for both low and high flying aircraft and maritime targets, and tracking of all types especially those manoeuvring and crossing.

The British AEW concept at Marconi Avionics has been backed by UK Ministry of Defence funding since the mid-1960's and, following evaluation of alternatives, the decision was made to further develop the AEW Nimrod System currently in production. Whilst final flight proving trials of the total MSA remain to be completed, there is an extensive background of subsystem trials which began in 1977. The performance of the radar subsystem was proven and the data handling system (including Mission System Software) has been established. Performance of all the individual aspects of NIMROD AEW have been proven in 4 development aircraft, and current flight trials of the complete System are underway in 2 further aircraft.

DESIGN RATIONALE

The radar subsystem design adopted is coherent pulse compression radar (Fig 2) operating in two modes; medium prf pulse-Doppler optimised for the detection of airborne targets and a low prf pulse mode to detect slow moving/stationary maritime targets.

The pulse-Doppler mode provides aircraft detection in a high clutter environment down to low radial velocities which allows continuous automatic tracking of manoeuvring and crossing targets. The prf has been chosen to allow the detection of crossing targets, provide a large detection region free of clutter and to ease range ambiguity resolution in a dense target environment. Target velocity is extracted to provide initialisation information for the tracking algorithm within the Data Handling System, and to allow discretes and overland vehicles to be blanked from the tracking system.

Adoption of this medium prf approach has led to the transmitter-receiver spectral purity (coherence) requirement being mid-way between the relative ease of low prf and the difficulty of high prf techniques. The signal processing system utilises modern digital circuitry and microprocessors to ensure a low false alarm rate and to provide the Data Handling System a single report per target from each scan; this contains range, azimuth, range-rate and height together with associated IFF/SSR responses.

Range and Doppler ambiguities are resolved by correlating the returns from bursts of different prfs over the target dwell time. Both range and velocity correlators are designed to provide true range and Doppler from only two burst returns to minimise the ambiguous blind speeds and ranges within the radar surveillance volume. Height finding is achieved by amplitude comparison of correlated returns, using dual receiver channels operating on two vertically stacked beams. Pulse compression techniques have been adopted to increase the effective mean power of the radar whilst employing a relatively low peak power. In this instance a non-linear chirp utilising surface acoustic wave (SAW) techniques has been implemented to provide pulse compression with low sidelobes for all possible Doppler shifts of interest.

TRANSMISSION SYSTEM

Coherent pulses, with non-linear FM chirp are generated at the receiver IF by impulsing a SAW expander in phase synchronism with the IF coherent reference (COHO). These pulses are then upconverted to the transmission frequency using the local oscillator signal generated in the microwave frequency generator (MFG) according to the transmission channel selected.

The rf pulses are amplified to a high power level within the transmitter initially by a travelling wave tube amplifier (TWTA) driver which feeds a high power travelling wave tube stage (TWTS) for amplification to the final level. The output is fed to the high power fore/aft switch for transmission to either of the aerial systems.

ANTENNA SYSTEM

A full 360° azimuthal coverage is obtained without interference from the airframe by mounting the antennas on the extreme front and rear of the fuselage, each giving a 180° azimuth cover.

The antennas shown in Fig 3, are dual frequency, (Primary and Secondary Radar) wide bandwidth, twist reflecting Cassegrain structures chosen for compactness of minimum swept volume, low inertia, and low sidelobes. Polarisation twisting has been employed to eliminate the subreflector aperture blockage. Two feeds are utilised to provide twin primary beams in elevation for height finding. As Fig 4 shows, each antenna is supported on a twin gimbal scanner structure which is electrically driven. The scanner control system, with inputs from the aircraft Navigation Sub-system, provides a ground stabilised uniform azimuth scan rate at constant aerial boresight elevation, with respect to the horizon. The aerial boresight is pitch, roll and yaw compensated for normal patrol manoeuvre and turbulence.

Separate fore/aft microwave transmit/ receivers are mounted local to the scanner system. Each of these units accepts the high power pulse from the transmitter and splits the power into the two elevation beams. For each beam, the output is fed through a duplexer and azimuth and elevation rotating joints to the antenna feed for transmission. In the reception path two microwave receivers are used at each aerial to provide low noise amplification (utilising parametric amplifiers and FETs) of the upper and lower beam duplexer outputs. Image suppression mixers are used to mix the signal down to a fixed IF using the local oscillator signal generated in the microwave frequency generator. The fore/aft receiver outputs are fed back to the centre of the aircraft to the main Receivers.

RECEPTION SYSTEM

The microwave receiver outputs are fed to the central IF Receivers which amplify, pulse compress and mix down to base band the received signals from each beam. A SAW pulse compression line (forming a conjugate match pair with the expander used to generate the transmission pulse) is employed for pulse compression. The use of non-linear FM allows a heavy weighting for range-sidelobe reduction without suffering any system losses. An output from the IF coherent reference (COHO) is used to mix the compressed pulses using inphase and quadrature components such that second IF rejection takes place. In the pulse-Doppler mode the COHO reference signal is offset in frequency by a prediction of the peak clutter doppler return to ensure that the clutter spectrum is centred at zero frequency in the baseband inphase and quadrature receiver outputs.

Upper and lower beam inphase and quadrature outputs are quantised at intervals corresponding to the compressed pulse width by four high dynamic range digitisers. These provide a range gated sample complete with phase information suitable for digital analysis and further processing.

DOPPLER SIGNAL PROCESSING

The signal processing operates entirely with digital information subjecting the signal to the following sequential processes using hard-wired (but reconfigurable) digital processing and microprocessor techniques:

a) Cancellation of the zero Doppler signal component to remove ground clutter and discrete returns.

b) Fourier transformation from time to frequency samples utilising a hard-wired Fast Fourier Transform incorporating Taylor Weighting to contain spectral broadening. The remaining clutter and discrete returns are thus concentrated into well defined range-Doppler regions.

c) Adaptive thresholding of signal returns in each velocity bin to ensure target detection above surrounding signals with a low false alarm rate.

d) Peak sensing of detections in both range and Doppler to provide one detection per target per burst.

e) Range and Velocity correlation over returns from several bursts to resolve target range and velocity.

f) Beam sharpening to establish the peak azimuth return during the aerial dwell.

g) Comparison of upper and lower beam amplitudes for target height evaluation.

h) Association of primary and secondary radar returns.

i) Conversion of target reports into a geographical frame of reference and in addition to perform the various plot filtering and control functions available to the operator.

The digital circuits used standard TTL and CMOS devices. The boards have been laid out by computer-aided-design to reduce costs, ease documentation and to suit from the outset Automatic Test Equipment. A typical rack containing several Line Replaceable Units (LRUs) is shown in Fig 5 and the special test connector may be seen at the front of each unit whereas the normal connectors are at the rear. The LRUs are mounted horizontally in mating trays.

The output of the radar signal processing consists of a single report per target; reports are ordered in ground stabilised azimuth and consist of range, azimuth, range-rate/radial velocity and height, together with associated IFF/SSR responses.

PULSE MODE (LOW PRF)

To enhance the maritime capability for the detection of slow moving or stationary ships whilst flying over sea, a supplementary pulse mode has been incorporated in the AEW Nimrod system. This mode utilises large sections of the pulse-Doppler hardware, configured to optimise the detection performance in clutter, by using a fairly simple low prf pulse technique. Pulse to pulse frequency agility provides clutter decorrelation when integrating the returns over the dwell, using a two-pole feed-back technique. Adapative thresholding circuits provide target detection over clutter with low false alarm rates which are processed and reported to the Data Handling System, as in the main mode.

DATA HANDLING AND DISPLAY

To minimise the risks of dependence on a single mainframe computer, the design of the data processing has placed emphasis on distributing the processing using micro-processors. The central computer (a GEC 4080M with Coral 66 real time software) has the task of automatic track initiation and track maintenance, though manual areas may be established by the operator. Separate procedures for manoeuvring and straight-line tracks are utilised for increased accuracy.

The display console, Fig 6, provides each operator with the entire data base available, from which he can select to display plots/ tracks of interest on his 12" diameter synthetic tactical screen; symbology has chosen to ease operator interpretation in a dense target situation. Full data on tracks of interest is available on the co-located tabular TV display.

BUILT IN TEST

An important feature of any equipment of this complexity is automatic built-in test. Each unit has high-integrity sensors and comparators to immediately detect and identify malfunction. A hierarchical reporting scheme is employed to relay BIT information to the data processor and thence to the operators. Hence by being alerted to a degraded unit, the operator may reconfigure the equipment or adjust the flight profile to ameliorate its effects.

CLOSE

The AEW Nimrod currently in production is a powerful surveillance system. With the implementation of medium prf pulse-Doppler and modern digital techniques in the air surveillance role, it will detect high and low altitude aircraft independent of the clutter environment.

Careful consideration during the design and development of the radar has been given to ensuring maximum detection performance against manoeuvring and crossing targets and to the minimising of blind speeds and ranges. Performance trials using a prototype radar sub-system mounted in a converted Ministry of Defence(PE) Comet aircraft has proved the design concept for operation both over sea and over land.

ACKNOWLEDGEMENTS

The enormous development programme summarised here has, of course, been the work of a large team of engineers. The authors therefore merely represent their colleagues at Marconi Avionics and the Ministry of Defence. Acknowledgement is also given to the co-prime contractor British Aerospace and the numerous sub-contractors who have supported this programme with such dedication.

Figure 1 Nimrod AEW Mk. 3

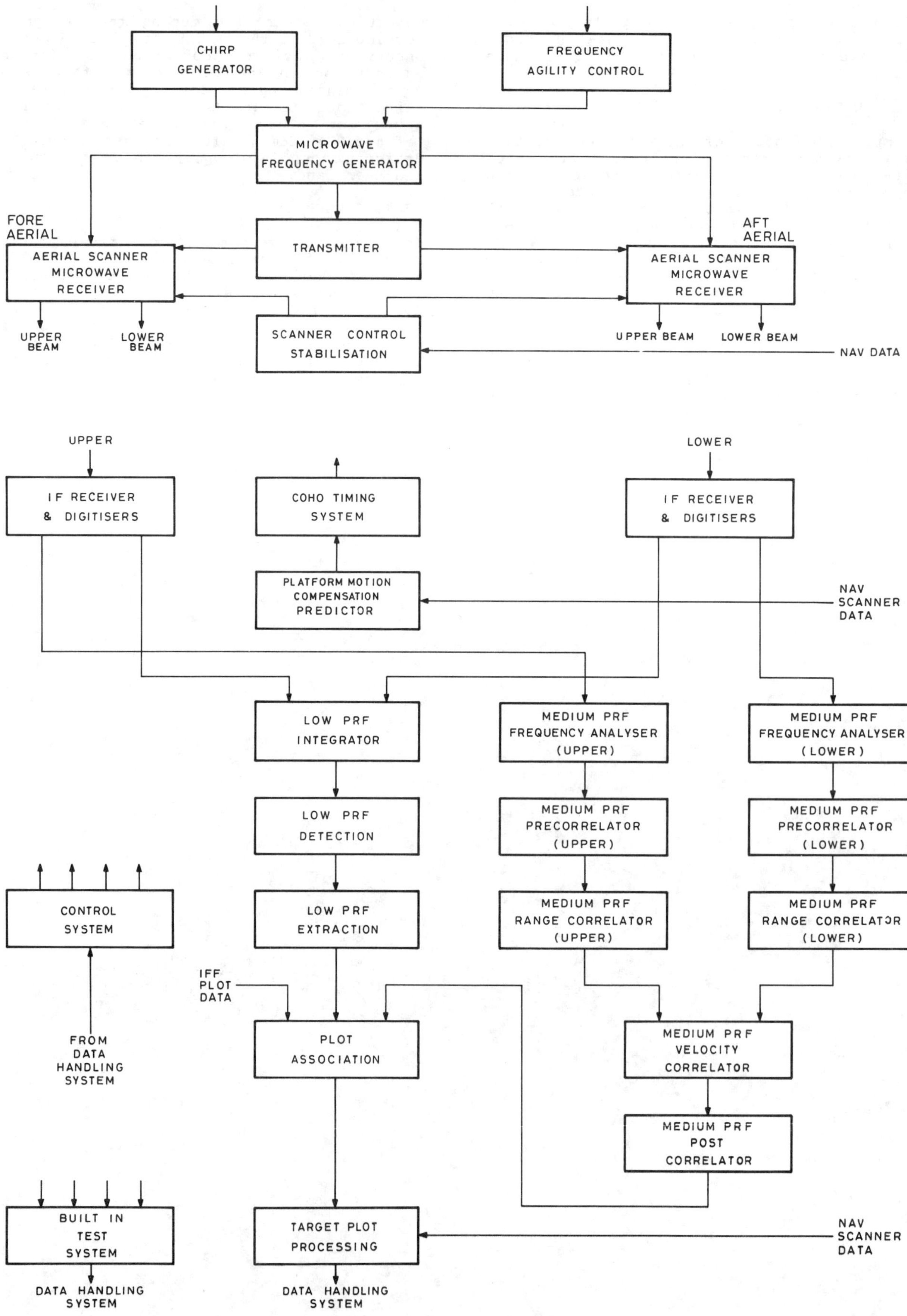

Fig.2 Radar Subsystem Block Diagram

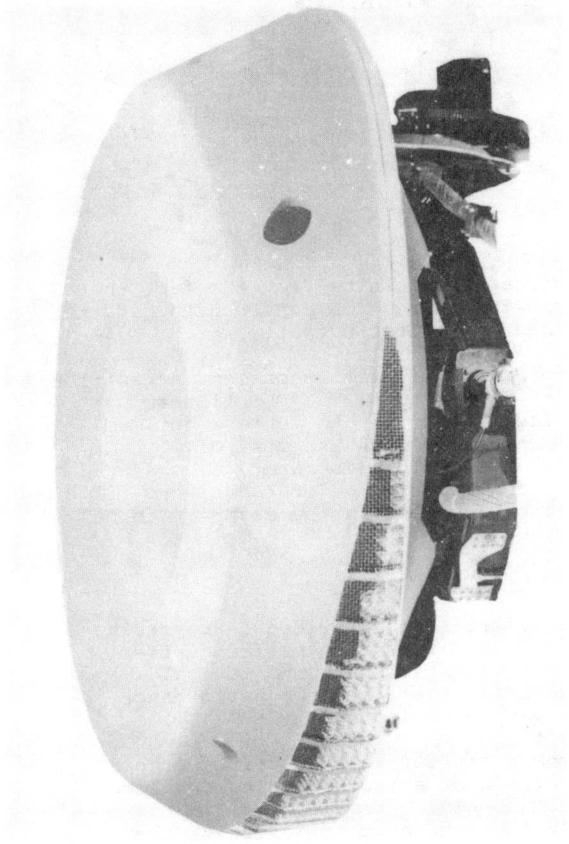

Figure 3 Cassegrain antenna

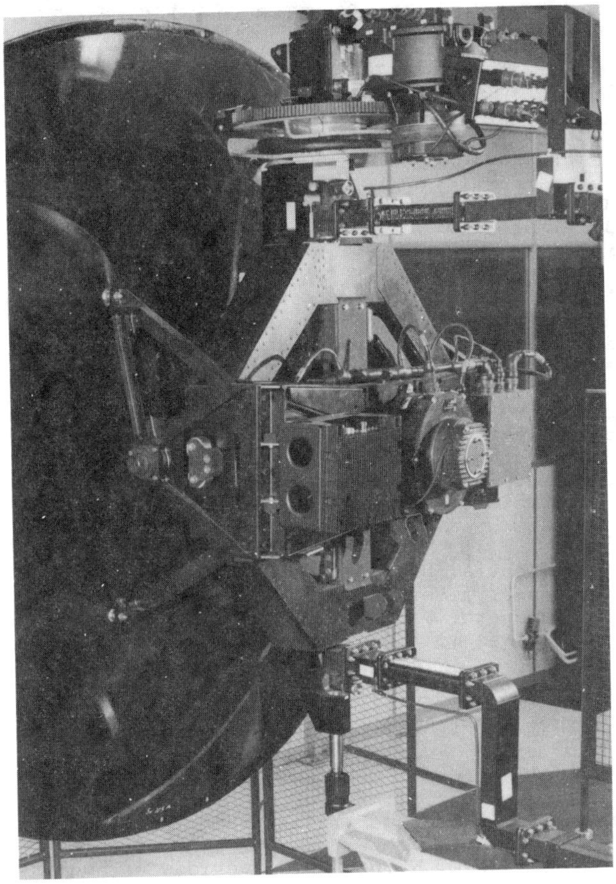

Figure 4 Light weight scanner

Figure 5 Digital LRU rack

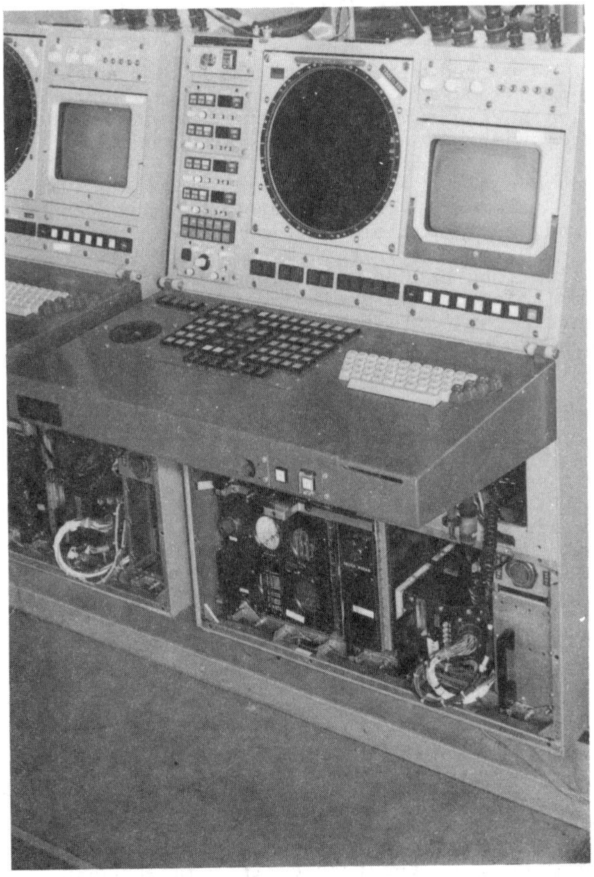

Figure 6 Console

SACRIFICES IN RADAR CLUTTER SUPPRESSION DUE TO COMPROMISES IN IMPLEMENTATION OF DIGITAL DOPPLER FILTERS

J. W. Taylor, Jr.

Westinghouse Electric Corporation
Defense & Electronic Systems Center
Box 1897
Baltimore, Maryland 21203 USA

Digital doppler filtering is employed in most modern radars to discriminate between desired echoes from aircraft and undesired clutter echoes from terrain, sea, rain, and chaff. The most widely-utilized filters are non-recursive or finite impulse response (FIR) types; they recover quickly from clutter transients created by change in radar frequency or beam position, and are less vulnerable to pulse interference from other radars or jammers.

The potential performance of such filters is often sacrificed in order to reduce cost of the hardware. The impact of three common compromises will be discussed.

Fundamental Concepts

Digital FIR filtering in radar involves:

1. Collection of echo samples in digital form at a multiplicity of precisely spaced times following each radar transmission.

2. Storage of N interpulse periods of such data.

3. Weighted summation of the N echo vectors from a given time (range) location to form a filtered output.

4. Repeated weighted summation of the same N vectors to form M different filtered outputs, each having a frequency response which depends upon the chosen weighting coefficients.

In general, the only limitation on the number of different filter outputs (M) which can be produced from the N data samples is the speed of the computer, which must make MN multiplications and accumulations. Complete freedom to choose weighting coefficients and interpulse periods permits each FIR filter to achieve optimum response characteristics for the available number of data samples (N)

Fast Fourier Transform (FFT) is a computational algorithm reported by Cooley and Tukey (1)(2) in 1965 which requires less computational effort, but it imposes undesirable constraints on filter response. Specifically, if the number of data samples (N) is a power of two, and the time interval between samples is constant, about $(N/2) \log_2 N$ complex multiplications and $N \log_2 N$ complex additions are required to produce N filter outputs. When N is a large number, the FFT provides substantial cost advantage over the more general FIR procedure, which requires nearly N^2 multiplications and N^2 additions.

The FFT is a clever computational technique of sequentially combining progressively larger weighted sums of data samples so as to produce N outputs of filtered data. In other words, the weighting coefficients are not optimum, because they cannot be defined independently for each filter; the reduction in number of multiplications results from the fact that a single multiplication can furnish data to more than one filter output.

In summary, FFT filters represent a restricted class of FIR filters. The FFT filters sacrifice the ability to achieve optimum performance to reduce the cost of implementation. This is rarely a good trade of cost and performance of radar echo filters, as will be apparent in the examples to be presented.

Performance Objectives

Doppler filters are provided in radars to suppress two different types of clutter interference: stationary and moving. Consequently, there are two performance objectives which may conflict.

Stationary clutter is produced by echoes from terrain or structures. Its doppler spectrum is highly predictable, in many cases being generated primarily by the motion of a mechanically-scanned antenna. For ground-based radars, the spectrum is symmetrical about zero doppler. The examples to be discussed are implemented on ground based radars with mechanically rotating antennas.

One objective is to provide maximum detectability of aircraft over ground clutter. An optimum group of filters can be defined mathematically, if one assumes that the amplitude of the clutter interference is known and that the aircraft doppler is randomly distributed across the doppler band. With a finite number of data samples (N), there is a limit as to how closely one can approach the optimum filter response; the more pulses, the better the approximation. These suboptimal filters are often called "optimum filters employing N pulses."

In reality, ground clutter amplitude varies widely from one spatial location to another, so it is difficult to justify the "optimum" label to a filter optimized for a specific clutter-to-noise ratio. One design approach is to optimize for a level of clutter which generates clutter residue equal to noise at the filter output, due to limited dynamic range and instabilities in the radar. However, this results in different optimum filters for different radars, even if their antennas' scanning produce identical clutter spectra.

The difficulty of defining an optimum filter should not obscure the fact that the filter response at low doppler frequencies dictates the ability of the radar to suppress heavy ground clutter interference. A common measure of this capability is the attenuation of clutter relative to noise, which may either include or exclude radar instability. This is a useful measure for comparing the ability of different filters or different radars to suppress ground clutter.

Moving clutter created by rain or chaff is generally several orders of magnitude less intense than ground clutter, so less suppression is required. However, this suppression is required over a wide band of doppler frequencies, because the clutter is borne by the wind; the mean doppler of this interference and the spread due to wind shear are variables, as well as RMS echo intensity. The ideal filter must adapt its doppler response to the variable rain or chaff spectrum. The objective is to provide maximum detectability of aircraft in rain or chaff clutter.

One approach to creating an adaptive filter is to generate a multiplicity of filters with overlapping rain rejection bands, each with its own detection process to control false alarm rate (CFAR). Those filters which do not reject the rain are desensitived by their CFAR processes. The resulting composite sensitivity to aircraft, as a function of its doppler, varies with the interference conditions.

The ideal filters for such a configuration are ones having rain clutter rejection bands as wide and as deep as possible. The width determines the degree of overlap; at least one filter must be free from rain interference, regardless of its mean doppler, but it is desirable to have several filters with full sensitivity in order to improve the probability of detecting aircraft.

The depth of the rain rejection band, represented by the highest doppler sidelobe of an individual filter, dictates the rainfall rate which the filter can tolerate under the most unfavorable wind speed conditions. With a finite number of data samples, increasing the depth of the rain rejection band sacrifices its width. The usual compromise is to design the depth to cope with a specified rainfall rate; and the maximum width is achieved when all doppler sidelobes have equal amplitude.

Whatever one's criteria for judging a filter to be "optimum", the degree to which one can approach this ideal is dictated by compromises. The examples which follow illustrate how these compromises affect the filter characteristics, causing them to degrade from the ideal performance objectives.

Multiple Filter Bank Examples

One of the most well-known applications of these concepts to ground-based radars is the Moving Target Detector (MTD), built by the Lincoln Laboratory of the Massachusetts Institute of Technology for use on Airport Surveillance Radars (ASR). Two generations of MTD processors have been built: the first employing FFT algorithm; and the second, the general FIR.

The examples were selected because published information provided adequate detail of the implementation compromises, and because MTD has been well-publicized in the radar literature.

The first generation MTD employed a cascade combination of 3-pulse canceller and an 8-pulse FFT, requiring a burst of ten pulses with fixed interpulse periods to create a single output from each filter. The composite doppler sidelobes were lowered to some degree by adding to each FFT output one-quarter of the output of the neighboring filters. Figure 1, illustrates the capability of this FFT process in suppressing both ground and rain clutter.

Muehe (3) shows the performance of these 10 pulse filters in ground clutter (40 dB above noise) to be essentially equivalent to that achievable with 8 pulse FIR filters. Requiring 25% more data and time to achieve equivalent results has several negative impacts: wasted power, increased memory cost, increased vulnerability to pulsed interference from other radars, and decreased ability to fill in blind speeds by repetition of the pulse bursts with different interpulse periods during the time that the beam dwells on a target.

In figure 1 and subsequent figures, the ordinate represents gain of the filter at any doppler frequency relative to gain to noise. The coherent gain provided an optimum speed target is not a significant benefit of doppler filtering; post-detection integration of N pulses would produce essentially equivalent detectability, averaged over all dopplers. The most significant characteristic of any filter is its ability to suppress clutter relative to noise (0dB in subsequent figures), which is termed MTI Improvement Factor when a single doppler filter is employed.

Although suppression of ground clutter is dependent only upon the filter characteristics in the vicinity of zero doppler (the shape of the ground clutter rejection notch), suppression of rain is dependent on the doppler sidelobes of the filter across a wider band of rain dopplers. The rain rejection band of the filter of Figure 1a is quite shallow; a narrow spectrum of rain echoes, moving at the speed corresponding to the worst doppler sidelobe, would be attenuated only 5 dB relative to noise. Cascading a three-pulse canceller with the FFT filters improved the suppression of ground clutter but degraded the suppression of rain clutter, because the canceller has maximum response at half the pulse repetition frequency.

The second generation MTD was developed to achieve much better performance at comparable cost by employing 8-pulse FIR filters rather than 10-pulse FFT filters. Figures 2a and 3a show the lower doppler sidelobes achieved in these filters. Note that, in contrast to Figure 1, the choice of independent sets of weights for these two filters allows their zeroes to be located closer to optimum for each individual filter. The fact that the doppler sidelobes are unequal means that the rejection notch at the level of the worst doppler sidelobe is not as wide as possible; but substanital improvement in ability to suppress rain or chaff was achieved by the FIR filters.

Second generation MTD implemented the filters which attenuate ground clutter as a cascade combination of 2-pulse canceller and 7-pulse FIR filters. This reduces the number of multiplications required to form the filtered echoes and ensures that the output is not influenced by bias in the input data.

Second generation MTD restricted their weighting coefficients to 5-bit integers, tabulated in Table 1. The filters implemented by both Lincoln Laboratory and Westinghouse retain all the bits of the weighted echoes in the digital filter output, because both organizations have encountered performance degradation if the bits of lesser significance in the products are eliminated by truncation or roundoff. Consequently, the number of bits of weights affects the number of bits required at the output of the multiplier and in the accumulator.

These compromises in the hardware implementation of MTD II restricts the freedom of the filter designer to place zero responses at the most desirable doppler frequencies. This impairs his ability to achieve doppler sidelobes of equal amplitude across the rain rejection band and to maximize the suppression of ground clutter by shaping the bottom of the rejection notch.

A new Westinghouse radar provides an 8-pulse FIR filter (Figure 4a) with peak response of virtually the same doppler frequency/ PRF as Figure 2a and 3a (27 vs. 25 and 29% respectively). The weighting coefficients are implemented with more bits than Lincolns Lab's design (12 vs. 5), allowing the creation of both a deeper rain rejection band having doppler sidelobes of nearly equal amplitude and a ground clutter rejection notch of more optimum shape.

The ability of these filters to suppress rain clutter under unfavorable wind conditions is indicated by their doppler sidelobes relative to noise : 32 dB in the Westinghouse filter and 15 and 26 dB in the two MTD filters. The Westinghouse filter (Figure 4a) provides the same width of rain rejection band at the -26dB level as MTD filter 2 (Figure 3a), but the 12 bit weights provide 6 dB lower doppler sidelobes.

The ability of these filters to suppress ground clutter is illustrated by the spectra superimposed on Figures 2b, 3b and 4b, which provide expanded views of the low doppler region. The clutter input spectrum, illustrated by dashed lines, is generated by a Gaussian beam scanning at a rate which creates 31 hits per beamwidth. The clutter residue spectra, illustrated by dot-dash lines, is the sum (in dB) of the input clutter at any doppler frequency and the filter response at that frequency. The ratio of the total power in the input spectrum to that in the output spectrum is computed and printed at the top of each Figure; this is the attenuation of ground clutter power relative to noise, which each filter could provide if employed in a perfectly stable radar having a signal processor dynamic range greater than these numbers.

Comparison of Figures 2b & 4b shows that the 12 bit weighting coefficients provide a better shape of ground clutter rejection notch, producing a more symmetrical clutter residue spectrum of lower amplitude. The degree of symmetry of the residue spectrum is a good indicator of the degree to which the attenuation of clutter has approached the theoretical limit imposed by other parameter choices. Proper location of two zeros within the clutter spectral region allows the spectrum to be broken into three lobes, and the least clutter residue power results when the three lobes have nearly equal amplitude. When the constraint is imposed that one zero be at zero doppler, to totally eliminate any effect of A/D bias, the best clutter attenuation is achieved when the clutter residue spectrum is symmetrical, having two lobes of equal amplitude as shown in Figure 4b. The 5 bit weights used in the MTD II filters do not provide ability to locate the zeroes precisely where desired, and the resulting unsymmetrical clutter residue spectra, shown in Figures 2b and 3b, contain more power than necessary.

The second-generation MTD represents a recognition within Lincoln Laboratory of the unnecessary degradation of performance caused by compromises in their first-generation implementation. Similarly, the Westinghouse implementation which has been described represents a recognition of the remaining deficiencies of the second-generation MTD which can be avoided by use of currently-available digital multipliers. The examples represent three evolutionary stages, not three alternative implementations at a common time.

Finally, it must be recognized that achievement of low doppler sidelobes in practical hardware is dependent upon the degree to which the response of the hardware conforms to the theoretical. The precisely predictable response of digital components must be accompanied by appropriate accuracy of the analog components which form the in-phase (I) and quadrature (Q) digital echo data. For example, lower doppler sidelobes demand more accurate gain and phase balance between I&Q data. Figure 5 illustrates the effect of a 4% gain unbalance on the filter response, and a phase error of 4% of a radian would create similar effect. These tolerances, representative of the best available hardware, are acceptable for the doppler sidelobes of the first and second generation MTD filters, but they are incompatable with the much lower sidelobes of the Westinghouse filters.

Methods of automatically compensating for gain and phase unbalance, which change slowly with time and temperature, are available (5) but are rarely implemented. Inclusion of these automatic compensation devices is a necessary additional cost to achieve superior filters, better able to attenuate wind-borne rain and chaff interference.

Single Filter (MTI) Examples

Moving target indicator (MTI) is a single doppler filter having a symmetrical response about zero doppler. Symmetry prevents imbalance in I and Q gain from having any effect on velocity response, permitting MTI's to be implemented with only one component of the echo vector to reduce cost. The penalty is an increase in the echo strength required to achieve a desired detection probability, in either clutter or noise environments.

Digital MTI's are often implemented with 3-bit binomial weights (1, -3, 3, -1 for a four pulse MTI) which are far from optimum, particularly when pulse-to pulse variation in interpulse period is used to eliminate blind speeds. Optimum weights (4) can increase MTI Improvement Factor by 3-8 dB, reducing the aircraft echo required to achieve a desired detection probability in heavy clutter.

Figure 6 illustrates the benefits of the optimum 4-pulse MTI implemented in the latest versions of a Westinghouse radar, compared to the I-only binomially-weighted 4-pulse MTI of the prior model. Sensitivity is improved 4 dB in weak clutter and up to 9 dB in heavy clutter, even though both MTI's have the same dynamic range.

Summary

There can be vast differences in the capability of different doppler filters to detect moving targets in clutter, even if the filters process the same number of echoes with the same number of bits of dynamic range. Common compromises to reduce hardware cost which cause substantial degradation of doppler filter performance:

- o The FFT algorithm.
- o A small number of bits of weighting coefficients.
- o Omission of automatic compensation for gain and phase imbalance in I&Q data, or omission of I or Q entirely in MTI.

As the cost of digital hardware continues to drop, the cost penalty of better filters becomes less significant than their performance benefits.

REFERENCES

1. Cooley, J. and Tukey, J., 1965, "Algorithm for the Machine Calculations of Complex Fourier Series", Math of Comput., 19, 297-301.

2. Cochran, W., Cooley, J. ..., 1967, "What is the Fast Fourier Transform", IEEE Trans., AU-15, 45-55.

3. Muehe, C., 12 Apr 75, "Digital Signal Processor for Air Traffic Control Radars", MIT Lincoln Laboratory Industrial Liaison Report.

4. Taylor, J. and Brunins, G., 1975, "Long-Range Surveillance Radars for Automatic Control Systems", The Record of the IEEE 1975 International Radar Conference, 319.

5. Churchill, F., Ogar, G. and Thompson, B., 1981, "The Correction of I and Q Errors in a Coherent Processor", IEEE Trans., AES-17(I), 131-136.

TABLE 1 - Weight coefficients

Echo Number	MTD II* Filter 1	MTD II* Filter 2	Westinghouse Filter 1
1	- 3 + j3	+ 2 + j3	186 + j0
2	+ 4 + j7	+ 8 - j6	- 84 - j541
3	+11 - j5	-12 - j13	-984 + j282
4	- 2 - j15	-15 + j15	521 + j1270
5	-12 - j1	+13 + j12	1207 - j652
6	- 2 - j8	+ 6 + j8	-583 - j842
7	+ 4 + j2	- 3 - j2	-407 + j365
8	*	*	144 + j118

* MTD II weights are applied to the 7 outputs of a two-pulse canceller.

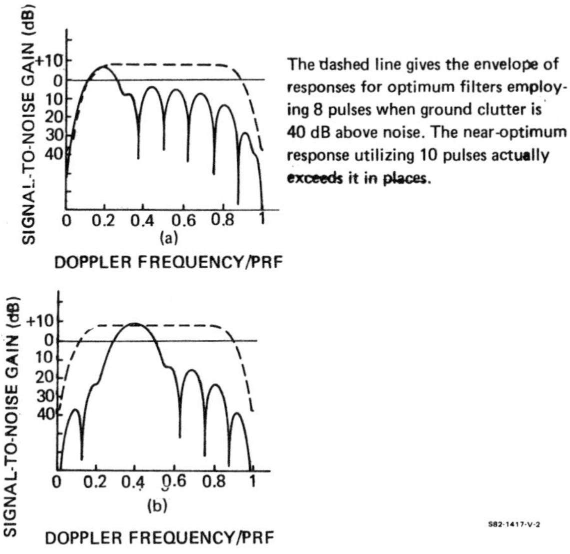

Figure 1. First Generation MTD Filters Using FFT Algorithm

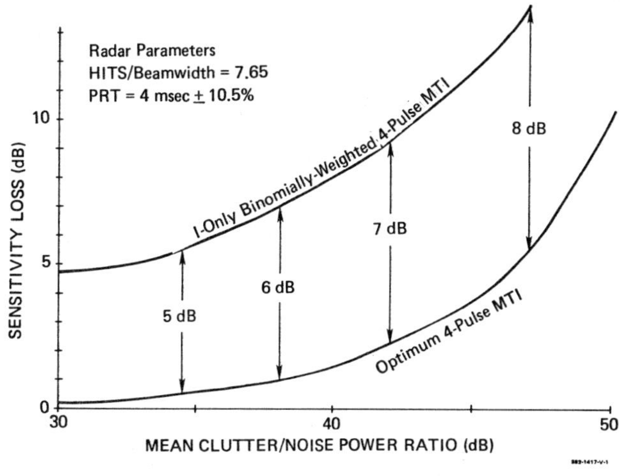

Figure 6. Optimum MTI Detectability Of Aircraft in Distributed Clutter

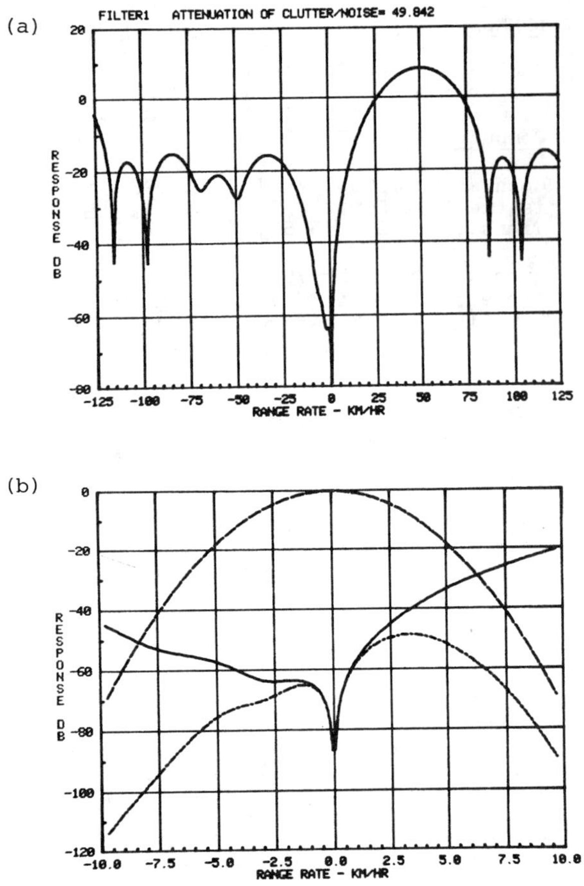

Figure 2. MTD II 8-Pulse FIR Filter 1 Using 5 Bit Weights

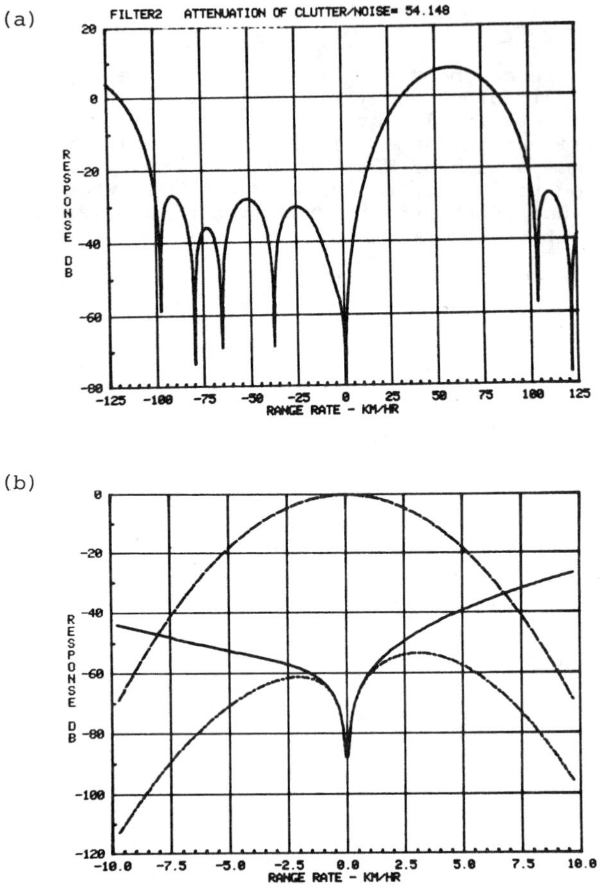

Figure 3. MTD II 8-Pulse FIR Filter 2 Using 5 Bit Weights

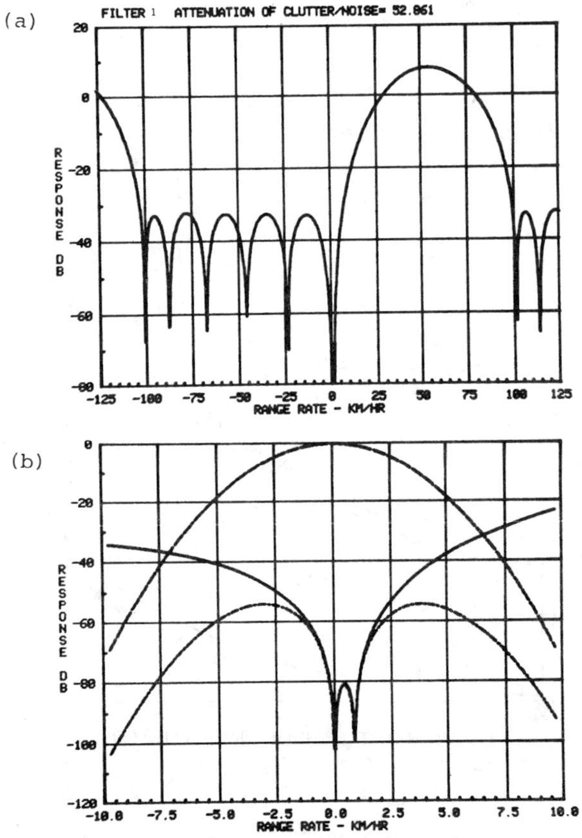

Figure 4. Westinghouse 8-Pulse FIR Filter 1 Using 12 Bit Weights

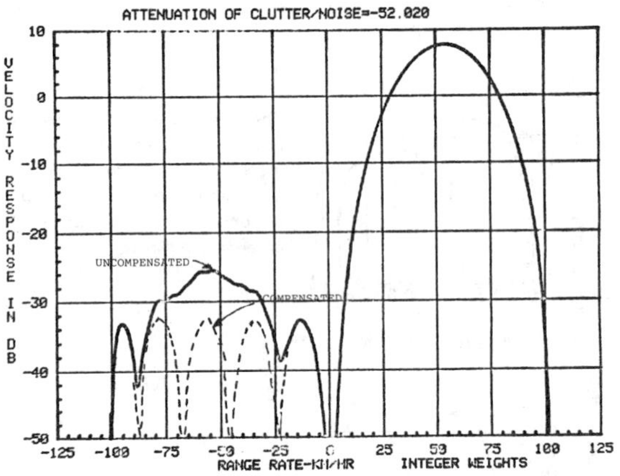

Figure 5. Effect of 4% Unbalance in I/Q Gain

A COMPARISON BETWEEN NONCOHERENT AND COHERENT MTI'S

Frank F. Kretschmer, Jr., Feng-ling C. Lin and Bernard L. Lewis

Naval Research Laboratory, USA

INTRODUCTION

Noncoherent MTI differs from coherent MTI in that the MTI processing is performed at video after an envelope or square law detector (Skolnik (1), Nathanson (2), Steinberg (3), Schleher (4), and Emerson (5)). In the coherent MTI, processing is done at i.f. or more commonly in the complex video inphase I and quadrature Q channels.

The noncoherent MTI is similar in many ways to the coherent MTI but differs in some important respects. With the noncoherent MTI, if the pulse-to-pulse changes in the amplitude of the clutter return are small the clutter will be heavily attenuated by the MTI filter following the detector. Since the phase of the clutter is not used, the noncoherent MTI is capable of centering the cancellation notch on clutter having a non-zero average radial velocity. In the presence of clutter, a target having a different radial velocity than the clutter will cause pulse-to-pulse amplitude variations and the target plus clutter signal will not be as heavily attenuated by the MTI filter. Since the target signal is clutter referenced it will be canceled in the absence of clutter unless the target scintillates.

An examination of the existing literature indicated that little analytical work has been performed on the noncoherent MTI, particularly based on the envelope detector which is of interest in this paper. Prior work (3,4) considered the envelope detector to behave approximately as a square law detector and based the results on arguments relating to the spectral characteristics of the clutter signal after a square law detector.

In this paper, a statistical approach is taken, in both the analysis and the Monte Carlo simulations, to evaluate the noncoherent envelope detector MTI. Clutter attenuation and improvement factors based on generalized definitions are computed and compared to the coherent and the noncoherent square -law detector MTI's.

COMPARISON OF COHERENT AND NONCOHERENT MTI

A. Coherent MTI

First we consider the clutter attenuation for a 2-pulse coherent MTI. The successive clutter returns are designated by the complex video vectors consisting of the I and Q values, or equivalently the amplitude and phase which are computed from $(I^2 + Q^2)^{1/2}$ and $\tan^{-1}(Q/I)$ respectively.

For a two pulse canceler the complex residue signal is

$$R = C_1 - C_2 \qquad (1)$$

where C_1 and C_2 are successive complex clutter return signals from a given range cell separated in time by the interpulse period T.

The average residue power C_o is taken to be

$$C_o = \overline{|R|^2} = \overline{|C_1 - C_2|^2}$$

$$= \overline{|C_1|^2} + \overline{|C_2|^2} - 2\mathrm{Re}\,\overline{C_1C_2^*}. \qquad (2)$$

The cross correlation term Re $\overline{C_1C_2^*}$ may be determined from the I and Q components as

$$R_e\,\overline{C_1C_2^*} = \overline{C_{1I}C_{2I}} + \overline{C_{1Q}C_{2Q}} = 2\rho\sigma^2 \qquad (3)$$

where it is assumed that the I and Q components of each clutter return have the same variance σ^2 and each have the same correlation coefficient ρ. Substituting (3) in (2) results in

$$C_o = \overline{|R|^2} = 4\sigma^2 - 4\rho\sigma^2 = 4\sigma^2(1-\rho). \qquad (4)$$

Letting the input clutter power be

$$C_{in} = \overline{|C_1|^2} = \overline{|C_2|^2} = 2\sigma^2, \qquad (5)$$

the clutter attenuation or cancellation ratio is

$$CR = C_{in}/C_o = 1/[2(1-\rho)] \qquad (6)$$

This result may be generalized in terms of matrix notation as follows. Let W denote the column matrix of weights applied to the successive clutter samples C_i and let M_c denote the covariance matrix of the clutter which is given by

$$M_c = \overline{CC^t} \qquad (7)$$

where C^t denotes the complex conjugate of the transposed matrix C. Letting T denote the transpose operation, the residue power is

$$C_o = \overline{(W^TC)(W^TC)^t} = W^T\overline{C\,C^t}W^* = W^TM_cW^* \qquad (8)$$

The clutter cancellation ratio may be expressed in general form as

$$CR = 2\sigma^2/(W^TM_cW^*). \qquad (9)$$

The MTI improvement factor I is determined by the ratio of the output target-to-clutter ratio divided by the input target-to-clutter ratio where the input target is averaged over all velocities if no a priori knowledge is available. The average target response, or target enhancement factor, is equivalent to normalizing the filter response to white noise so that the output noise power is equal to the input noise power. It follows that

$$I = \overline{(T_o/C_o)/(T_{in}/C_{in})} = \overline{(T_o/T_{in})}/(C_{in}/C_o)$$

$$= (W^TW^*)\cdot(CR) = 2\sigma^2(W^TW^*)/(W^TM_cW^*) \qquad (10)$$

where we have made use of (9) and the fact that the white noise response of the filter is given by W^TW^*.

B. Noncoherent MTI

For the 2-pulse noncoherent MTI, the output

clutter residue is

$$R = |C_1| - |C_2| \tag{11}$$

and the output clutter power is

$$C_o = \overline{R^2} = \overline{(|C_1| - |C_2|)^2}$$

$$= \overline{|C_1|^2} + \overline{|C_2|^2} - 2\overline{|C_1||C_2|} \ . \tag{12}$$

From Middleton (6) and Lawson and Uhlenbeck (7) the cross correlation term is given by

$$\overline{|C_1||C_2|} = 2\sigma^2(\pi/4)\,{}_2F_1\,(-1/2,-1/2,1,\ \rho^2) \tag{13}$$

where ${}_2F_1$ (·) above is the Gaussian hypergeometric function given in (6) by

$${}_2F_1\ (\cdot)$$

$$=1 + \ (1/4)\rho^2 \ +(1/64)\rho^4 \ +(1/256)\rho^6 \ +\ldots \ . \tag{14}$$

The above stated references use an identity given by

$${}_2F_1\ (\cdot) = (4/\pi)E(\rho) - (2/\pi)(1-\rho^2)\ K(\rho) \tag{15}$$

where K and E are complete elliptic integrals of the first and second kinds respectively. These functions are tabulated (Abramowitz and Stegun (8)) and are also available as a subroutine on the Naval Research Laboratory (NRL) Advanced Scientific Computer (ASC) digital computer.

From (12), (13) and (15), the resultant output clutter may be written as

$$C_o = 4\sigma^2\ [1 - F(\rho)] \tag{16}$$

where,

$$F(\rho)=(\pi/4)\ {}_2F_1(\cdot)=E(\rho)-[(1-\rho^2)/2]K(\rho). \tag{17}$$

Comparison of (16) with the expression for the coherent MTI in (4) shows the expressions are the same if one interchanges ρ and F(ρ). This result generalizes for higher order cancelers so that for the noncoherent canceler the output clutter may be expressed in a similar way to (8) as

$$C_o = W^T M_N W^* \tag{18}$$

where M_N is the covariance matrix of the noncoherent MTI which is the same as M_c in (8) except that ρ is replaced by F(ρ). In either case, it is noted that ρ is a function of time separation which is equal to the Fourier transform of the clutter power spectral density and is dependent on the interpulse intervals.

From the previous results, the cancellation ratio for the noncoherent MTI is given by

$$CR = 2\sigma^2/(W^T M_N W^*)\ . \tag{19}$$

A discussion of the improvement factor for the noncoherent MTI is deferred to section E.

C. Comparison of Cancellation Ratios

At this juncture, we compare the cancellation ratios of the coherent and noncoherent MTI for a clutter spectrum which is assumed to be Gaussian and given by

$$G(f) = 1/(2\pi\sigma_s^2)^{1/2}\ \exp(f^2/2\sigma_s^2) \tag{20}$$

where σ_s denotes the standard deviation of the clutter spectrum. The correlation function ρ(τ) is the Fourier transform of G(f) and is given by

$$\rho(\tau) = \exp[-2\pi^2(\sigma_s\tau)^2] \tag{21}$$

where τ is the time delay variable.

Making use of these expressions, and eqs. (8) and (19), the cancellation ratio was computed for the coherent and noncoherent MTI and the results are shown in Fig. 1 for various order cancelers where the standard binomial weighting was used. In Fig. 1 the abscissa denotes the spectral width σ_s which is normalized to the pulse repetition frequency (prf). Fig. 1(a) shows the comparison for the 2 and 3 pulse cancelers, Fig. 1(b) shows the comparison for the 4 and 5 pulse cancelers.

For the 2-pulse canceler, it is shown that the noncoherent MTI is approximately 3 dB better than the coherent MTI in contrast to prior beliefs. For higher order cancelers the noncoherent clutter attenuation is seen in Fig. 1 to be generally worse than the coherent MTI except for the 3-pulse canceler whose curves cross over for σ_s/prf approximately equal to 0.07.

The clutter attenuation of the coherent and noncoherent MTI's were also simulated on a digital computer using Monte Carlo techniques, which are described later, and excellent agreement was obtained.

D. Spectral Spreading Evaluation

Next, we examine the relationship between the coherent and noncoherent cancelers in terms of a spectral spreading factor. That is, for a given σ_s/prf in Figs. 1(a) and 1(b) and the associated noncoherent clutter attenuation, we find the multiplicative factor for σ_s/prf which results in the corresponding coherent canceler having the same cancellation ratio as the noncoherent canceler.

This is plotted in Fig. 2 where it is seen that there is no simple relationship which may be stated in regard to the spectral spreading factor.

the spectrum corresponding to F(ρ) consists of a summation of Gaussian terms having unequal variances and that the resultant spectrum is non-Gaussian, in contrast to the output spectrum of the square law detector. Hence, the spectral spreading factor is an equivalence only in terms of the resultant value of the cancellation ratio.

E. Improvement Factor Comparison

Because of the nonlinearity of the envelope detector in a noncoherent MTI system it is not appropriate to separately determine the doppler-averaged target response and the clutter attenuation and combine these factors as was discussed for the coherent MTI whose improvement factor was given by (10). To circumvent this problem we define the output target-to-clutter power ratio as

$$T_o/C_o = [(T+C)_o - C_o]/C_o\ . \tag{22}$$

The improvement factor using the generalized definition becomes

$$I = \overline{[(T+C)_o - C_o]/T_{in}} \cdot (C_{in}/C_o) \tag{23}$$

where averaging is over all target velocities. Due to the nonlinearity of the noncoherent MTI, the improvement factor will in general be dependent on the input target-to-clutter ratio.

Therefore, to determine I, computer simulations were performed using Monte Carlo techniques. For the different order cancelers successive clutter samples, corresponding to the returns from a given range cell on successive sweeps, were generated on the computer. The correlation between successive sweeps was specified and the MTI residue was computed for each trial consisting of N sweeps for an N-pulse MTI. On each trial the first return was taken to have a Rayleigh distributed amplitude and a uniformly distributed phase, and successive returns were correlated with the first return.

The MTI was then simulated by applying the binomial weighting to the correlated complex samples, or to the amplitude of the samples, for the coherent and nonocherent cases respectively. For different values of σ_s/prf, 3000 independent trials were run and the output residue powers were averaged. The cancellation ratios were computed for the two cases, where the input clutter power for each was taken as the average of the input ($I^2 + Q^2$) value which is $2\sigma^2$. The output clutter power for the coherent case was taken as the average residue, again computed from ($I^2 + Q^2$), averaged over the 3000 independent trials. The noncoherent residue power was computed by averaging the square of the residue. Thus, both systems are computed with common input and output terminals.

The target signal was added to the clutter signal with a random initial phase angle and a specified phase shift corresponding to the target's doppler. The average target output power as determined from the generalized definition (22) was determined for 10 different target velocities uniformly spaced across the prf interval. Excellent agreement was obtained with the known improvement factor for the coherent MTI case. The results for the coherent and noncoherent MTI are plotted in Fig. 3 for an input clutter-to-target ratio of 20 dB. The results were very nearly the same for all clutter-to-target ratios above 10 dB. It is seen that the improvement factors for the 2-pulse canceler are nearly the same while the noncoherent MTI improvement factor degrades relative to the coherent MTI as the number of pulses increases. A comparison with the improvement factor computed from the relation in reference (3), for a square law device, is shown in Fig. 4 where it is seen that the improvement factor for the square law detector is equal to or greater than for the envelope detector. It was found that the noncoherent MTI target enhancement factor is approximately 3 dB less than the coherent MTI for a clutter-to-target ratio of 20 dB.

SUMMARY AND CONCLUSIONS

A comparison between coherent and noncoherent MTI using an envelope detector has been made based on statistical analysis and on computer simulations using Monte Carlo techniques. This approach differs from prior investigations which base the results on the envelope detector behaving approximately as a square law detector from which it is argued that the standard deviation of the power spectral density of the clutter is increased by $\sqrt{2}$ due to the self-convolution of the clutter spectra.

The results of the analysis in this paper indicate that in general, the envelope detector cannot be regarded as a square law detector which increases the standard deviation of the clutter spectrum by $\sqrt{2}$. The equivalent spectral spread for the envelope detector case was found to differ from $\sqrt{2}$ depending on the the number of pulses used in the MTI as well as on the correlation of the returned clutter signals. It was noted that the spectrum after envelope detection is not Gaussian since the correlation function consists of higher order terms than ρ^2 which cannot be ignored. Thus, the equivalent spectral spreading of the input Gaussian clutter spectrum prior to envelope detection is an equivalency only in terms of the value of the cancellation ratio. It was also found that the clutter attenuation for the 2-pulse noncoherent canceler using an envelope detector is 3 dB better than the coherent canceler in contrast to some prior conceptions.

In terms of improvement factors, it was found that the two noncoherent MTI's and the coherent MTI are nearly the same for the 2-pulse canceler. As the number of pulses increases, the improvement factors become more unequal. The coherent MTI has the largest improvement factor, followed in order by the square-law and envelope detector MTI's.

REFERENCES

1. Skolnik, M., 1962, "Introduction to Radar Systems", McGraw-Hill, Inc., New York.

2. Nathanson, F., 1969, "Radar Design Principles", McGraw-Hill, Inc., New York.

3. Steinberg, B., 1965, (Chapters 1,2, Part VI,) "Modern Radar Analysis, Evaluation and System Design", (edited by R. Berkowitz), John Wiley and Sons, New York.

4. Schleher, D., (ed), 1978, "MTI Radar", Artech House.

5. Emerson, R., "Some Pulsed Doppler MTI and AMTI Techniques", (This paper appears in reference 4).

6. Middleton, D., 1960, "Introduction to Statistical Communication Theory", McGraw-Hill, New York.

7. Lawson, J., and Uhlenbeck, G., (eds.), 1950, "Threshold Signals", MIT Radiation Laboratory Series, McGraw-Hill, New York.

8. Abramowitz, M., and Stegun, I., (eds.), 1964, "Handbook of Mathematical Functions", National Bureau of Standards, Washington, D. C.

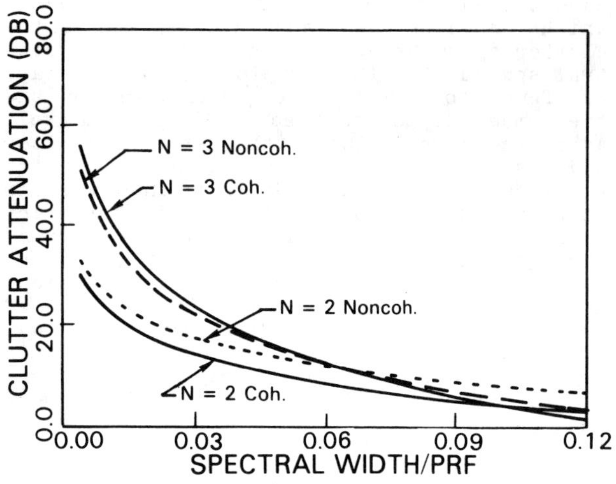

FIGURE 1(a) Clutter attenuation of coherent and noncoherent (envelope detector) MTI's, (2 and 3 pulse canceler)

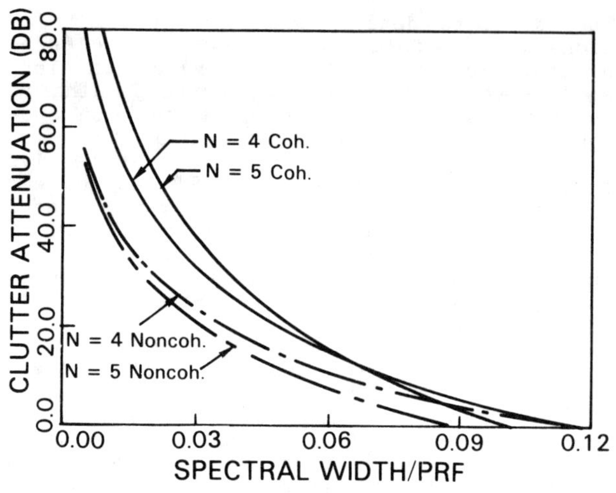

FIGURE 1(b) Clutter attenuation of coherent and noncoherent (envelope detector) MTI's, (4 and 5 pulse canceler)

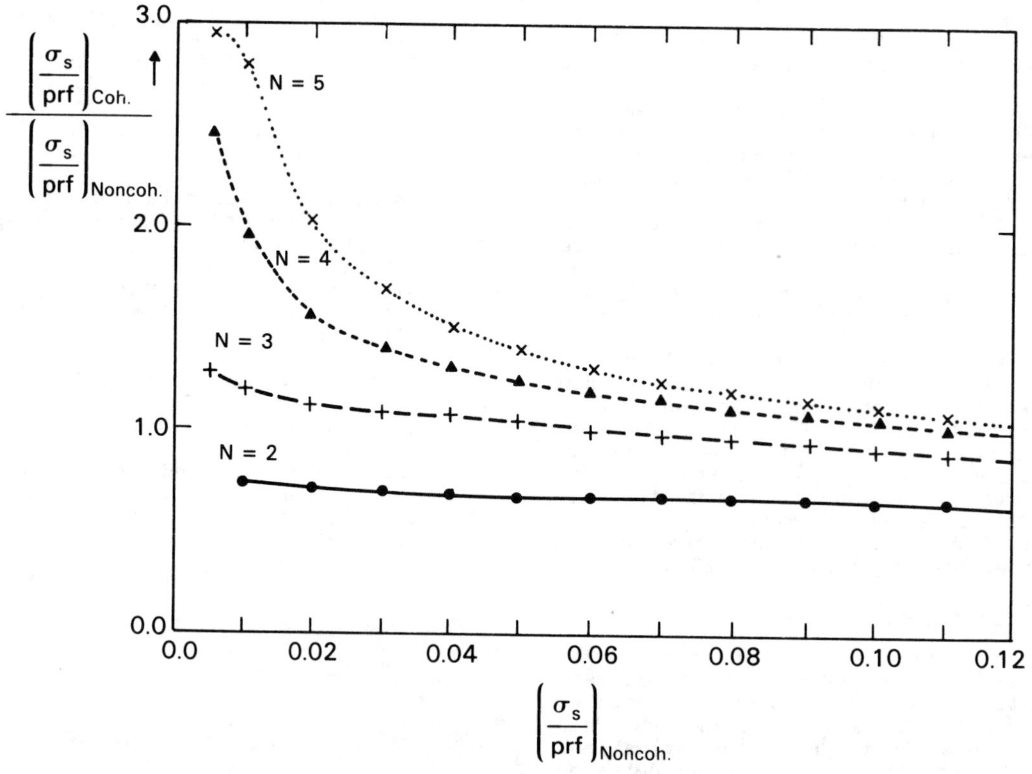

FIGURE 2 Spectral spreading factor for noncoherent MTI using an envelope detector

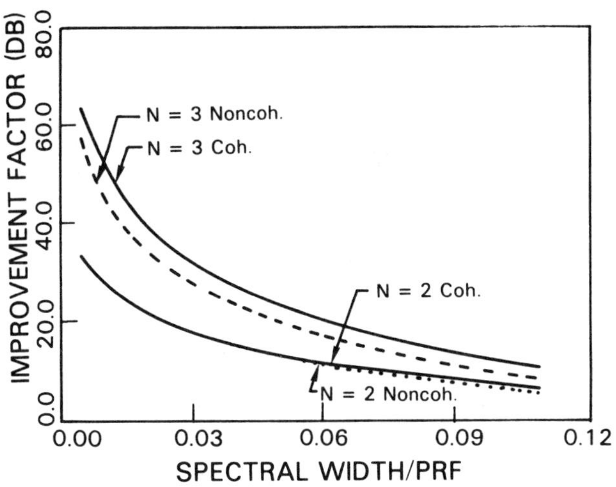

FIGURE 3(a) Improvement factors of coherent and noncoherent (envelope detector) MTI's, (2 and 3 pulse canceler)

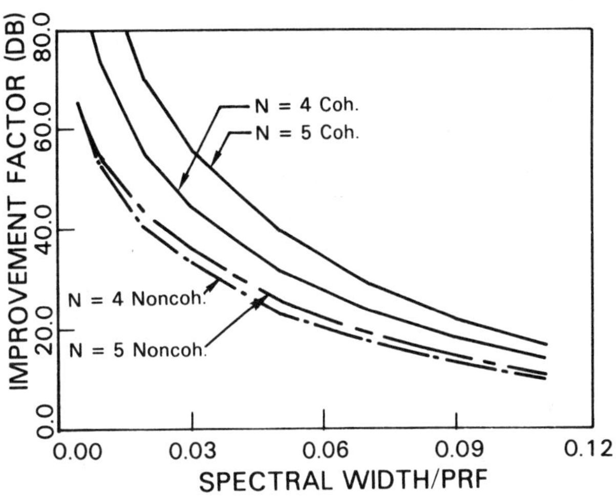

FIGURE 3(b) Improvement factors of coherent and noncoherent (envelope detector) MTI's, (4 and 5 pulse canceler)

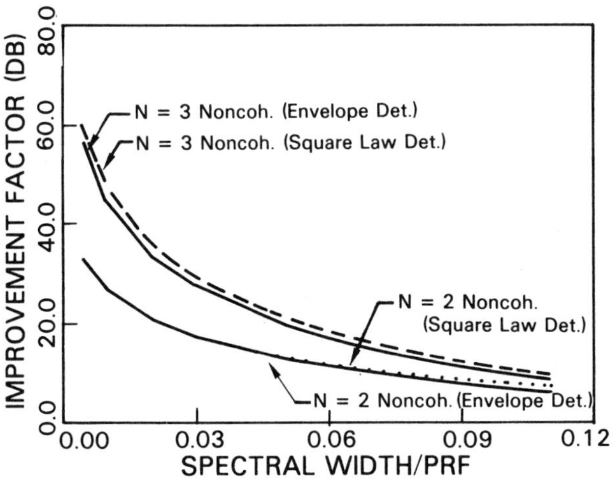

FIGURE 4(a) Improvement factors for noncoherent MTI's using an envelope and a square law detector, (2 and 3 pulse canceler)

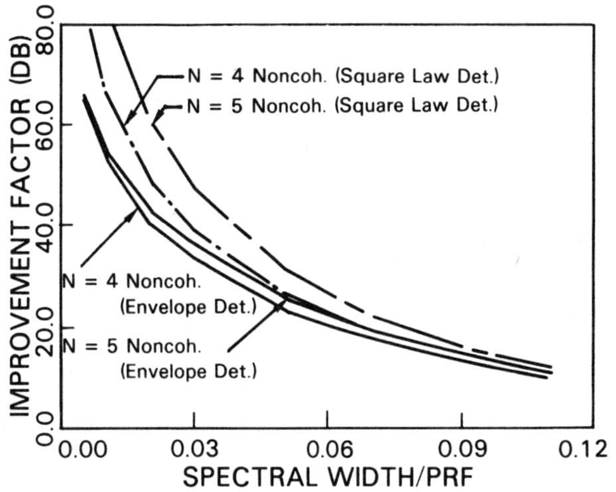

FIGURE 4(b) Improvement factors for noncoherent MTI's using an envelope and a square law detector, (4 and 5 pulse canceler)

STUDY OF WEATHER CLUTTER REJECTION WITH MOVING TARGET DETECTION (MTD) PROCESSOR

Bao Zheng, Peng Xueyu, Zhang Shouhong

Northwest Telecommunication Engineering Institute, Xi'an China

INTRODUCTION

The MTD, consisting mainly of a bank of narrow-band filters and individual CFAR circuits, has been widely studied in recent years. Usually, it works together with a clutter map to detect the tangential aircraft and to control the output threshold of each narrow-band filter. This greatly improves the ability to discriminate the target from ground clutter.

Using narrow-band filters and a CFAR circuits is very efficient in rejecting weather clutter and chaff jamming too. In this paper analyses is given from the view point of self-adaptivity to explain the function of these circuits. Filter bank has multiple output. Usually, for simpler MTD equipment, speed measure is not neccessary, so the multiple outputs are often combined into a single by using a special interface. The interface can be of various kinds. Mainly, they are classified into two:"greater-selecting" circuit (i.e. analog-OR) and analog-AND.

THE RESPONSE CHARACTERISTIC UNDER THE INFLUENCE OF WEATHER CLUTTER

Weather clutter is generally a kind of extended clutter with Rayleigh distribution. The CFAR circuit used in this case is often a cell averaging CFAR by which the clutter can be normalized. Thus it is equivalent to dividing the gain of each channel by the standard deviation of its clutter (noise) output. For ease of explanation, we can exchange the position between the dividing operation and detection. Thus the block diagram of the overall system is shown in Fig 1. Obviously, the exchange above has no effect upon the overall system performance. The only assumption is that the clutter of each channel has been normalized before detection.

Under the influence of colored noise (clutter), because each channel has its own normalization, the gains of different channels will change differently. In the channel with stronger clutter, the gain will decrease noticeably. This is equivalent to making the colored clutter "white" in frequency domain by means of open-loop adaptivity. As for target echo, because of its narrow time-width, it has no great influence upon gain control. The two kinds of filter characteristic groups referring to these before and after gain control are shown in Fig 2. It shows that the overall system may be constructed adaptively into "notch" filter to the clutter.

The signal is usually mono-frequency. So one of the channels can be adjusted to match or nearly match it. Then the output of this channel has higher signal-clutter (noise) ratio. However, target speed can't be predicted, we can't get output from a fixed channel. To keep this match, it would be better to combine every channel after detection with an interface. In desiging the interface the interference between channels should be reduced as much as possible. The details will be discussed later.

Fig 2 (b) shows an example of a characteristic group after gain control. When using an interface to combine all channels, the combined characteristic is based on and predominantly determined by it. In order to get a satisfactory response characteristic, obviously, the number of filters should not be too few, the band width of each filter should match the spectrum width of the clutter, and its side-lobe level should be low enough.If clutter is colored, the gain-fall only happens in one or two filters, the pass-band of which is coincident with the clutter spectrum. The other filters are not much effected by clutter, so that the overall characteristic has wider band and higher gain. It would no doubt contribute to increasing average signal gain.

EQUIVALENT AVERAGE IMPROVEMENT FACTOR

The improvement factor is one of the most important figure-of-merit of MTI filter. The system shown in Fig 1 is defferent from the ordinary MTI filter. It involves nonlinear processing,so that the improvement of S/C ratio should be redefined and reexplained. In addition, clutter and noise should not be considered separately.

Suppose the input clutter-noise power density spectrum is $C(f)+N_0$. If the Doppler shift is given, it is easy to calculate the improvement of each channel. Taking the k th channel as an example, according to the usual definition, the improvement factor before detection is

$$I_K=\left\{\frac{|H_K(f)|^2}{\int_0^{F_r}[C(f')+N_0]\,|H_K(f')|^2df'}\right\}\left\{\int_0^{F_r}[C(f')+N_0]df'\right\} \quad \ldots\ldots \quad (1)$$

where the first factor on the right hand side is signal power gain, the second is clutter-noise power attenuation. If the amplitude of the input signal is A, then, before detection, the signal-to-clutter (noise) power ratio is

$$\left(\frac{S}{C+N}\right)_K=\frac{A^2|H_K(f)|^2}{2\int_0^{F_r}[C(f')+N_0]|H_K(f')|^2\,df'} \quad \ldots\ldots \quad (2)$$

let J=C+N —— clutter-noise power before

detection.

If the target speed is unknown, the improvement factor makes sense only when taking its average over range of the target speeds. Since the output finally comes from the interface, it is not significant if the average is made only in one channel. Once signal frequency drifts off the pass-band, its output will fall down, but in another channel, the output will go up. Therefore, the behaviour of the interface must be taken into account in making the average.

When combining multiple channels, no matter what kind of interface is used, the correlation among all channels has great influence upon the combined output. The correlation coefficient between two arbitrary channels (number k and i) is determined by their weight coefficient $\mathbf{W}_K$, $\mathbf{W}_i$ and the covariance matrix of clutter (noise), $\mathbf{R}$ i.e.

$$\rho_{Ki} = \frac{\mathbf{W}_K^T \mathbf{R}\, \mathbf{W}_i^*}{(\mathbf{W}_K^T \mathbf{R}\, \mathbf{W}_K^* \cdot \mathbf{W}_i^T \mathbf{R}\, \mathbf{W}_i^*)^{1/2}} \quad \cdots\cdots \quad (3)$$

Calculation shows that, as long as the characteristic of the filter has very low side-lobe level, especially in such case that C/N is not too large, the correlation between non-adjacent channels is negligible. The correlation between adjacent channels is also small. So, we can treat each channel as nearly statistically independent. This will make calculations much easier. Before detection, there are 2M degrees of freedom including the real and imaginary part, and every clutter-noise variance is 1.

Based on the assumptions above, the calculation becomes very easy if square-law detection and analog-AND interface are used. In this case, interface output has a χ^2-distribution. Its probability density function is

$$P(V) = \frac{1}{2}\left(\frac{V}{\lambda}\right)^{\frac{M-1}{2}} \exp\left(-\frac{\lambda+V}{2}\right) I_{M-1}(\sqrt{\lambda V}), \quad V>0 \qquad (4)$$

here

$$\lambda = A^2 \,|H_{el}(f)|^2$$
$$= A^2 \sum_{K=0}^{M-1} \frac{|H_K(f)|^2}{\int_0^{F_r} [C(f')+N_0]\, |H_K(f')|^2\, df'} \quad \cdots \quad (5)$$

On taking an average over signal speed, λ should replaced by $A\, \overline{|H_{el}(f)|^2}$.

As mentioned above, under this special condition, the distribution of interface output is related to $\overline{|H_{el}(f)|^2}$. It is reasonable to use it instead of the first factor in equation(1) in order to find the single-channel average improvement factor. However, here the function of the interface has been takem into account, so it is called "modified single-channel average improvement factor", $\overline{I}_{M1}$, i.e.

$$\overline{I}_{M1} = \overline{|H_{el}(f)|^2} \int_0^{F_r} [C(f')+N_0]\, df'$$

$$= \overline{\left\{ \sum_{K=0}^{M-1} \frac{|H_K(f)|^2}{\int_0^{F_r} [C(f')+N_0]\, |H_K(f')|^2\, df'} \right.}$$

$$\times \int_0^{F_r} [C(f')+N_0]\, df' \quad \cdots \quad (6)$$

Having got the modified single-channel average improvement factor, we should further find S/J ratio loss caused by the interface, called "interface loss" for short. Because non-linear processing is involved, interface loss is related to the detecting condition. This means that interface loss should be calculated with a given false alarm probability P_{fa} and probability of detection P_d. If the input clutter has a Gaussian distribution, no matter whether there is a signal or not, the distribution of the single-channel detecting output can be easily found. For example, in equation (4) when M=1, the result is the distribution of output with squar-law detection. From this distribution and given P_{fa}, P_d, we can find the S/J ratio which is required to meet the detecting condition.

In a similar way, from the distribution given by equation (4), and with the same P_{fa} and P_d, we can find the $(S/J)_M$ ratio which is required by the output of an interface with M channels. Compared with the single-channel case, it is increased by the interface loss L.

Subtracting L from $\overline{I}_{M1}$, which is given by equation (6), gives I, the equivalent improvement factor of the overall system.

Another kind of interface——the "greater-selecting" circuit——will now be discussed.

If only clutter exists, and the correlation among all channels are neglected, then after linear detecting, the probability density of the greater-selecting output is

$$P(V) = M(1-e^{-V^2/2})^{M-1}\, V e^{-V^2/2}, \quad V>0 \quad \cdots\cdots \quad (7)$$

To calculate false alarm probability, we are mainly concerned with the "tail" of density function. In this case, equation (7) can be rewritten as

$$P(V) \approx M V e^{-V^2/2} \qquad (V \gg 1) \qquad \cdots\cdots \quad (8)$$

It shows that the probability density of the "tail" is about M-times as much as it is in the single channel.

When signal exists, under usual detection conditions, for example, when $P_{fa} = 10^{-6}$, $P_d = 90\%$, it is known that the detectable target has to have a great enough S/J ratio. If the signal frequency is just located at the center of one of the channels, the detection output in this channel will be much greater than those in other channels (fluctuation should be taken into account). Its distribution has an approximate Gaussian shape. Its mean is approximately equal to the signal amplitude before detection, and the variance is 1. For these points, the average of the "greater-selecting" output is

$$A|H_{e2}(f)| = \mathop{Max}_{K=0,1,\cdots M-1}\left[\frac{A|H_K(f)|}{\left\{\int_0^{F_r}[C(f')+N_o]|H_K(f')|^2 df\right\}^{1/2}}\right] \qquad \cdots\cdots \quad (9)$$

$$f=f_0,\ f_1,\ \cdots,\ f_{M-1}$$

For the frequencies which are just at the intersection of two characteristics in Fig 2(b), since two channels have the same mean and variance, the "greater-selecting" output would be mainly determined by them. If the variance is 1, the correlation coefficient between two channels is ρ , it can be shown that the mean of "greater-selection" output will increase by $\sqrt{(1-\rho)/\pi}$. The probability density curve no longer has a Gaussian shape, its peak drifts toward the bigger amplitude side, this is advantageous to detecte the target. If the calculation is still based on the Gaussian distribution, the equivalent mean will be a bit greater. As for the frequencies which are neither channel centers nor intersection of two characteristics, the result would lie between the above two. As mentioned previously, it is difficult to find the distribution of "greater-selecting" output exactly. In fact, under the condition of ordinary operation, using the M values of the center of every channel as standards, and making the Lagrange polynomial interpolation, we can find a mean value curve more satisfactorily and express it as $A|H'_{e2}(f)|$.

Taking a speed-average the greater-selecting output has a Gaussian distribution with mean $A|H'_{e2}(f)|$ and variance 1, So, using $|H'_{e2}(f)|$ to replace the first factor (signal gain factor) in equation (1), we can find the modified single-channel average improvement factor $\bar{I}_M$. Based on the two distributions and given P_{fa} and P_d, the interface Loss L can be calculated. Taking M=8, $P_{fa} = 10^{-6}$, P =0.9, we can find the equivalent improvement factor I as shown in table 1 refer to two different interfaces.

In the previous calculations, we neglected the correlation between different channels. For the analog-AND interface, this leads to optimistic results; but for the greater-selecting interface,a little bit pessimistic.

In addition, further explanations about MTD operation under white noise background should be given. Take greater-selecting interface as an example. If signal frequency is at the center of one of the channels,then the ideal coherent integration gain of this channel is 10logM. The weighting used to reduce side-lobe will cause some mismatch loss, and also get a multiple channel signal gain. Besides, using multiple channel interface to give output will cause interface loss. It should be noticed that when ordinary amplitude detection is used, since white noise is statistically independent of time, we can take non-coherent integration over M pulses to increase S/N ratio. The integration gain obtained by this way is equal to the coherent integration gain minus the detection loss. Table 2 shows the improvement factor comparision of two situations V.S. different values of M, when P_{fa} =10 , P_d=0.9.

Obviously, if the value of M is not big enough,The effect of coherent integration is not noticeable. This means that under non-coherent intergration condition, the detection loss is not noticeable. MTD in this case is not helpful.

From the discussion above, we can also see that to calculate the improvement factor of MTD exactly is quite difficult, but to estimate its approximate value is easy. Again taking the greater-selecting output as an example, the interface loss has been discussed in detail before. It is related to the detecting condition. As for the single channel modified improvement factor, we can divide it into several parts. In ideal situation, for a given target doppler, which is lies far off the clutter spectrum, the maximum value of improvement factor is obtained,and is equal to C/N(clutter-to-noise ratio), multiplied by coherent processing gain, and divided by mismatch loss, which is similar to the white noise background situation. Then, according to the control state of the filter characteristic group [see Fig 2(b)], the gain fall caused by speed-averaging can be estimated. Considering all the factors above, we can roughly estimate the equivalent improvement factor of the system.

CONCLUSION

Using a filter bank and a normalized circuit we can get equivalent adaptive effects. This works well in rejecting weather clutter. To construct this kind of detector, the following points should be noticed. Firstly, it has to have perfect equivalent frequency response, i.e. wide pass band and high gain. Thus, the number of filters should not be too few, its width should roughly match the spectrum width, and there should be certain amount of overlaping between adjacent characteristics. Besides each filter has to have a lower side-lobe level. At the same time we should be careful not to accept a large mismatch loss under white noise background. To get narrow-band FIR filter, the order of filter must high enough. With the lower order (2-3 order), and using clutter map control, we can get higher rejection performance on ground clutter. As to weather clutter,which has an undetermined Doppler shift, it doesn't work well. The reason is that under certain circumstances, there is a wide notch on its equivalent frequency characteristic, and the gain in the pass-band is low.

As far as the type of interface is concerned, we hope, of caurse, that the interface will give certain amount of gain. But the more important is it to avoid the fluctuation increment caused by adding multiple channel clutter to the signal. So the greater-selecting interface is a little bit better than the analog-AND.

To reduce the interface loss, we should use fewer channels, however, this is in conflict with the requirement of equivalent frequency characteristic. The latter should have the first priority. But the overlaping between adjacent characteristics should be small in order to reduce the number of channels. For some shortrange search radar, the frequency is below S band, but its PRF is high. Comparatively, the Doppler shift of weather clutter is low, and, the spectrum of weather clutter is rather narrow, so relatively it is always located at the lower frequency end. In this case, combining those filters, that clutter can not reach, into a wide-band high pass filter, we can get a good performnace with a simple equipment.

REFERENCE

1. W.H. Drury, Improved MTI Radar Signal Processor, AD-A010478.
2. T. Irabu, et al., IEEE 1980 International Radar Conference. PP. 311-316.
3. R.E. Ziemer, et al., IEEE Trans. Vol. AES-16, May, 1980.

TABLE 1 - Comparison of equivalent improvement factor I with different interface type. Clutter relative frequency shift is 1/8, power spectrum of clutter has Gaussian shape, its relative standerd deviation is 60/1024

Interface Type	Weight Type	Clutter-to-Noise Ratio (C/N)	
		30dB	20dB
Greater-Selecting	$W(n)=0.5-0.5\cos(2\pi n/M)$	31dB	21.5dB
	$W(n)=1-0.5\cos(2\pi n/M)$	22.7dB	19.5dB
χ^2-sum.	$W(n)=0.5-0.5\cos(2\pi n/M)$	30.9dB	22.9dB
	$W(n)=1-0.5\cos(2\pi n/M)$	21.8dB	20.2dB

TABLE 2 - Improvement factor in different operating system under white noise background

Type	M=8	M=16	M=32
MTD (Hanning Weight)	6.8dB	9.7dB	12.5dB
Normal detection (non-coherent integration)	7.2dB	9.6dB	11.7dB

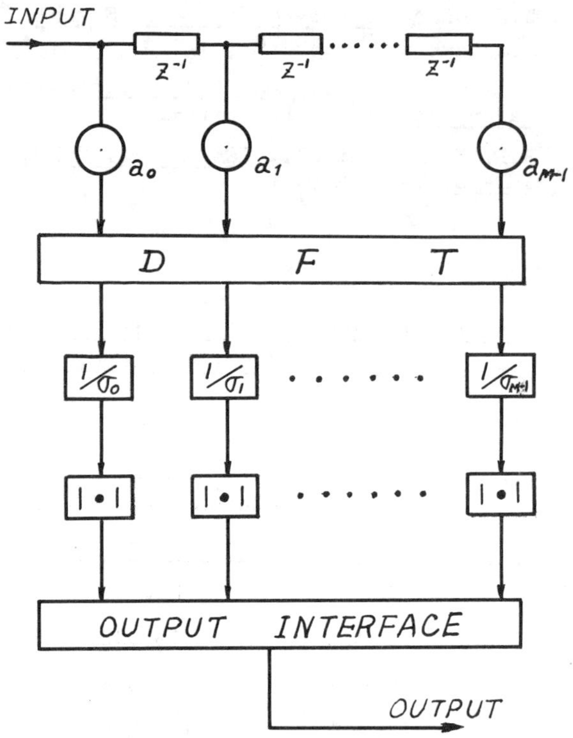

Figure 1 The block diagram of the overall system.

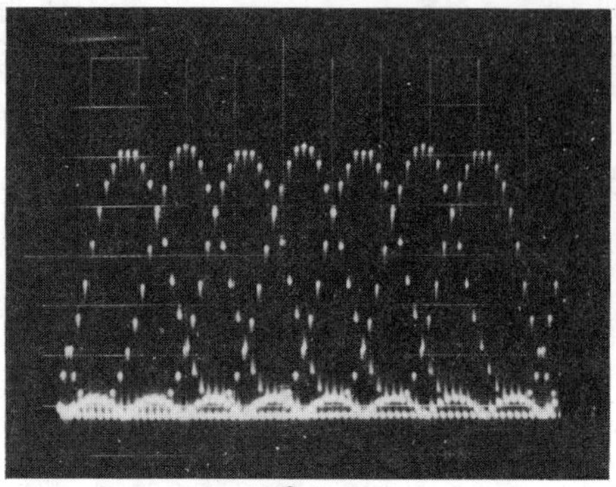

(a)

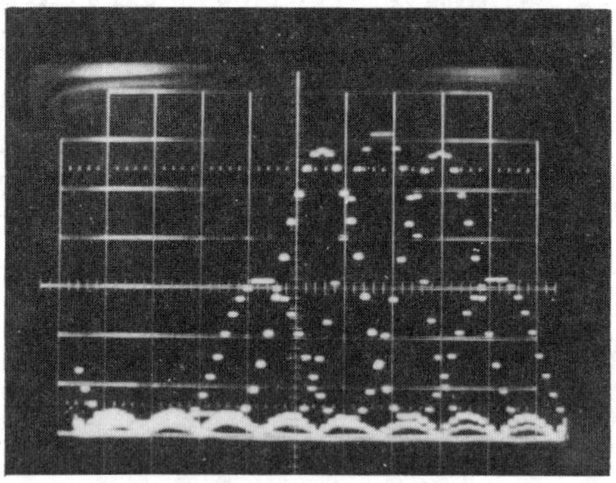

(b)

Figure 2 Filter characteristic groups.
(a) before control.
(b) after control.

RELIABLE SINGLE SCAN TARGET ACQUISITION USING MULTIPLE CORRELATED OBSERVATIONS

R. A. Dana

Mission Research Corporation
Santa Barbara, California, USA

D. Moraitis

Hughes Aircraft Company
Fullerton, California, USA

INTRODUCTION

The state of current radar technology allows new and innovative use of radar resources. A desirable characteristic of an air defense radar is quick, reliable acquisition of new targets. The conventional medium-to-long range track-while-scan radar is characterized by antenna rotation speeds of 4-6 RPM and pulse repetition frequencies (PRFs) which are unambiguous to the maximum range of interest. Such a radar, when automated, typically requires detections on m-out-of-n opportunities to acquire new targets, where each opportunity occurs on a different scan. For the Swerling I target model, the cross section decorrelates scan-to-scan resulting in missed detections. This lower probability of detection, as shown in Lindeberg, Margulies, and Smyton (1), results in acquisition times that could be on the order of minutes from the time the first detection is obtained. By utilizing the flexibility offered by current radar technology, including adaptive, reactive computer control of waveforms, energy level, and electronic beam pointing, new targets can be reliably acquired in a matter of milliseconds from the time of the initial search detection. The technique requires that each new search detection be followed by additional observations (opportunities) on the same scan. By exploiting the correlated nature of the target cross section within a scan, each new search detection can be resolved in favor of a true air target or be rejected as a false alarm. The additional observations, for the case of ambiguous range radar designs, characteristic of maneuver-following high data-rate radars, can also be used to resolve range and/or Doppler ambiguities.

An analytic expression for the probability of acquiring a Swerling I target on three observations during a scan is derived and presented in an easy to evaluate form that includes the effects of cross section decorrelation and variable energy and detection thresholds on successive observations. Acquisition performance results are presented which show the effects of target cross section decorrelation during the scan and increased energy or reduced detection thresholds on the second and third observations.

ACQUISITION ALGORITHM

The target acquisition algorithm is characterized by sequential reactive beam scheduling and processing of multiple observations in a single scan. A single observation (dwell) is made at each individually scheduled beam position. Each new search detection is followed by a "verify" and "track" dwell on the same scan. The search waveforms are characterized by non-monopulse, range ambiguous waveforms with multiple PRFs and transmission frequencies to accommodate system requirements on visibility and data rate. The verify beam, electrically steered to the position of the search detection, is used to reduce false alarms, obtain an initial monopulse angular measurement and resolve range and/or Doppler ambiguities. It has been shown by Dana and Moraitis (2) that in order to maximize the probability of acquisition, the transmission frequency of the verify beam should be the same as that of the search beam on which the initial detection occurred. The monopulse information obtained on the verify beam is used to point the beam for the (same scan) track beam and thereby reduce the beam profile loss and the off-axis monopulse error of the second angular measurement. The data extracted from the track dwell is used to check and refine the target's unambiguous range and range rate calculated from the data of the first two dwells. Again, the transmission frequency is held constant. Upon detection on all three dwells, a target track is established with reliable range, range rate, azimuth, and elevation information for updating on the subsequent scan. If the target is not detected on the verify and track dwells, it is rejected as a false alarm. It will be shown that the time between observations, determined by the time it takes to process the target report data and schedule beams, should be less than the decorrelation time of the target cross section in order to maximize the acquisition probability.

PROBABILITY OF n DETECTIONS ON n DWELLS DURING A SCAN

The probability of detecting a target n times on n dwells during one scan when the target has a radar cross section that is correlated within a scan but is independent from scan-to-scan is given by

$$P_n = \int_0^\infty da_1 \cdots \int_0^\infty da_n \, f(a_1, \cdots, a_n) \, Q(a_1,t_1) \, Q(\sqrt{\Delta_2}\, a_2,t_2) \cdots Q(\sqrt{\Delta_n}\, a_n,t_n) \quad (1)$$

where $f(a_1, \cdots, a_n)$ is the joint probability density function (pdf) of the n target amplitudes normalized to the r.m.s. noise voltage, t_i is the detection threshold used on the ith dwell normalized to the r.m.s. noise voltage, and $Q(\cdot,\cdot)$ is Marcum's Q function. Implicit in Equation (1) is the assumption that the target cross section remains constant during a dwell and that decorrelation occurs only during the inter-dwell period. The factors Δ_i $(i = 2, \cdots, n)$, defined as the ratio of the mean target power of the ith dwell to the mean target power of the first dwell are included to allow the use of variable waveform parameters such as the number of pulses coherently integrated per dwell or the pulsewidth on subsequent dwells after the initial detection. In addition, the target information gained from the first i-1 dwells can be used to develop range, Doppler, and/or angular gates on the ith dwell and thereby allow the use of lower detection thresholds on the ith and subsequent dwells. The target information can also be used to reduce processing losses (e.g. beam profile loss) on the ith dwell which can be accounted for in the value of Δ_i.

Constant Cross Section During a Scan

Equation (1) cannot, in general, be reduced to a closed form expression for arbitrary first and second order statistics of the target cross section. However, if the cross section remains constant during all n dwells of a scan, then the joint pdf of the n target amplitudes reduces to

$$f(a_1, \cdots, a_n) = f(a_1)\, \delta(a_2-a_1) \cdots \delta(a_n-a_1) \qquad (2)$$

where $\delta(\cdot)$ is the Dirac delta function. The probability P_n then collapses to

$$P_n = \int_0^\infty da_1\, f(a_1)\, Q(a_1,t_1)\, Q(\sqrt{\Delta_2}a_1,t_2) \cdots Q(\sqrt{\Delta_n}a_1,t_n) \qquad (3)$$

This form of P_n with a single integral is easily evaluated using numerical quadrature techniques for arbitrary n and $f(a_1)$.

A Swerling I (SWI) target has a cross section which is constant during a scan and exponentially distributed from scan-to-scan. The received voltage from the target then has an amplitude a which is Rayleigh distributed with the pdf

$$f(a) = (a/\langle S\rangle)\exp(-a^2/2\langle S\rangle) \qquad (4)$$

where

$$\langle S\rangle = \frac{1}{2}\int_0^\infty a^2 f(a)\, da \qquad (5)$$

is the mean signal-to-noise ratio (SNR). Subsequently, a target whose amplitude obeys Equation (4) will be referred to as a Rayleigh target. A closed form expression for P_2 and a SWI target is given in reference (2).

Independent Cross Section From Dwell-to-Dwell

The probability of n detections on n dwells is also easy to evaluate if the target cross section is uncorrelated from dwell-to-dwell. This situation will occur if different transmission frequencies of sufficient separation (relative to target length) are utilized on each dwell. The joint pdf of the n target amplitudes then becomes

$$f(a_1, \cdots, a_n) = f(a_1)\, f(a_2) \cdots f(a_n) \qquad (6)$$

when the target amplitude is statistically stationary. The probability P_n in this case reduces to

$$P_n = P_1(\langle S\rangle,t_1)\, P_1(\Delta_2 \langle S\rangle, t_2) \cdots P_1(\Delta_n \langle S\rangle, t_n) \qquad (7)$$

where $P_1(\langle S\rangle,t)$ is the single dwell detection probability of a target with mean signal-to-noise ratio $\langle S\rangle$ using a detection threshold of t.

A Swerling II (SWII) target has a Rayleigh distributed amplitude and a cross section that varies independently from dwell-to-dwell. For a Rayleigh target,

$$P_1(\langle S\rangle,t) = P_{FA}^{\;1/(1+\langle S\rangle)} \qquad (8)$$

where the detection threshold is related to the probability of false alarm P_{FA} as

$$t = \sqrt{-2\ln(P_{FA})} \qquad (9)$$

Rayleigh Target with Cross Section Decorrelation

Intermediate between a SWI and a SWII target is a target that has a Rayleigh amplitude distribution and a cross section that decorrelates somewhat between dwells. The derivation of P_3 for this target is summarized here with the probability P_2 being a limit of the resulting expression. The calculation of P_3 requires the joint pdf $f(a_1,a_2,a_3)$ of the target amplitude at the times of the three dwells. This is obtained by considering first the joint pdf of the in-phase (I) component x and the quadrature-phase (Q) component y of the received voltages from the target. It is easy to show, for an amplitude that has a Rayleigh distribution and a phase ϕ that is uniformly distributed on the interval $(0,2\pi]$, that the I and Q components

$$x = a\cos\phi$$
$$\text{and} \quad y = a\sin\phi \qquad (10)$$

are independent and normally distributed with variance $\langle S\rangle$ and zero means. It is then assumed that at three different times the I and Q components of the received voltage are jointly normal. Thus for the I components,

$$f(\vec{x}) = \frac{\exp\left[-\frac{1}{2}\vec{x}\,\mu^{-1}\vec{x}^T\right]}{\sqrt{(2\pi)^3|\mu|}} \qquad (11)$$

where $\vec{x} = (x_1,x_2,x_3)$ and where the components of the symmetric covariance matrix are

$$\mu_{i,j} = E\left[x_i\, x_j\right] = \langle S\rangle\, \rho_{i,j} \qquad (12)$$

An expression identical to Equation (11) holds for the joint pdf of the Q components. The additional assumptions that $\vec{x}$ and $\vec{y}$ are Markov processes and that the dwells are equally spaced in time will be made for mathematical convenience. As a consequence, the components of the covariance matrix become

$$\mu_{1,1} = \mu_{2,2} = \mu_{3,3} = \langle S\rangle$$
$$\mu_{1,2} = \mu_{2,3} = \langle S\rangle\rho \qquad (13)$$
$$\mu_{1,3} = (\langle S\rangle\rho)^2$$

Upon expansion of Equation (11), the joint pdf of $\vec{x}$ becomes

$$f(x_1,x_2,x_3) = \frac{\exp\left\{-\dfrac{x_1^2 + (1+\rho^2)\,x_2^2 + x_3^2 - 2\rho\,(x_1x_2+x_2x_3)}{2\langle S\rangle(1-\rho^2)}\right\}}{\left[2\pi\langle S\rangle\right]^{3/2}(1-\rho^2)} \qquad (14)$$

with an identical expression holding for $f(y_1,y_2,y_3)$. Noting that the I and Q components are independent, it is then an easy matter to make the change of variables

$$\begin{aligned} x_i &= a_i \cos\phi_i \\ y_i &= a_i \sin\phi_i \end{aligned} \qquad (i = 1,\ 2,\ 3) \qquad (15)$$

and integrate out the phase dependence to give the joint pdf of the target amplitude at three different times as

$$f(a_1,a_2,a_3) = \frac{a_1\,a_2\,a_3\exp\left\{-\dfrac{a_1^2 + (1+\rho^2)\,a_2^2 + a_3^2}{2\,\langle S\rangle\,(1-\rho^2)}\right\}}{\langle S\rangle^3\,(1-\rho^2)^2}$$

$$\cdot\, I_o\left[\frac{\rho a_1a_2}{\langle S\rangle(1-\rho^2)}\right] I_o\left[\frac{\rho a_2a_3}{\langle S\rangle(1-\rho^2)}\right] \qquad (16)$$

where $I_o(\cdot)$ is the modified Bessel function. The probability of detecting a Rayleigh target on three dwells is now calculated by substituting Equation (16) into Equation (1). Two of the remaining three integrals (those over a_1 and a_3) are given in closed form by Nuttall (2, integral 45). The resulting expression for P_3 is

$$P_3 = \langle S\rangle^{-1}\int_0^\infty da_2\,a_2\exp\left[-a_2^2/2\langle S\rangle\right]$$

$$\cdot\, Q\left[\rho a_2/\sqrt{m+1},\ t_1/\sqrt{m+1}\right] Q(\sqrt{\Delta_2}\,a_2, t_2)$$

$$\cdot\, Q\left[\rho\sqrt{\Delta_3}\,a_2/\sqrt{\Delta_3\,m+1},\ t_3/\sqrt{\Delta_3\,m+1}\right] \qquad (17)$$

where

$$m = \langle S\rangle\ (1-\rho^2) \qquad (18)$$

The integral over a_2 cannot, to the authors knowledge, be obtained in closed form. It is however, easy to evaluate numerically. The probability of detecting a Rayleigh target on two dwells can be obtained from P_3 by noting that $Q(a,0)$ is unity and hence that

$$P_2 = P_3\ (t_3 = 0) \qquad (19)$$

The resulting integral is given in closed form by Nuttall (2, integral 59).

PROBABILITY OF ACQUISITION

The probability of acquisition (P_{ACQ}) for a Rayleigh target and for an acquisition algorithm that requires detections on an initial search (S) dwell followed by detections on a verify (V) dwell and a track (T) dwell during the same scan is given by Equation (17). Three issues on the acquisition performance of this algorithm that are addressed by this analysis are: 1) the effects of lowered detection thresholds on the V and T dwells; 2) the effects of increased energy on the V and T dwells; and 3) the effects of target cross section decorrelation between the dwells.

Effects of Lowered Thresholds

It is desireable to maintain a small probability of false alarm on the initial search dwell in order to minimize the number of false alarms that are verified. However, because the probability of false acquisition is the product $P_{FA,S}\,P_{FA,V}\,P_{FA,T}$, and because the initial search detection allows the use of range gates and perhaps Doppler gates on the V and T dwells, the P_{FA} can be increased on the latter two dwells to improve acquisition performance. The probability of acquiring a SWI target is shown in Figure 1 versus the mean SNR of the search dwell, SNR_S. The mean SNR of each of the dwells is equal and $P_{FA,S}$ is set at 10^{-6}. The P_{FA} of the V and T dwells is set at 10^{-6}, 10^{-4}, or 10^{-2} giving an overall probability of false acquisition of 10^{-18}, 10^{-14}, or 10^{-10} respectively. The asterisks in the figure give the probability of detection for the search dwell alone which defines the maximum (optimal) acquisition performance (i.e. the probability of acquisition is limited by the probability of obtaining the initial search detection). It can be seen from the figure that when the P_{FA} on the V and T dwells is set at 10^{-2} that the acquisition performance is very near this limit. Note that for sufficiently high SNR on the search beam, lowering of the thresholds does not significantly improve the acquisition probability.

Effects of Increased SNR

Another method that can be used to improve the acquisition performance of three dwell algorithm is to increase the energy on target on the V or T dwells. This can be accomplished, for example, by increasing the pulsewidth or by increasing the number of pulses per dwell that are transmitted and coherently integrated. The effects of increased energy are shown in Figure 2 where P_{ACQ} is plotted for a SWI target. The P_{FA} of each dwell is set at 10^{-6} and the SNR of the V or T dwells is allowed to vary relative to the initial search dwell. Near optimum acquisition performance, again shown by the asterisks, is achieved when the energy of the V and T dwells is 3 dB higher than the energy of the search dwell. Note again that for high SNR on the search beam, increasing the energy on V and T dwells does not significantly improve the acquisition performance.

Effects of Cross Section Decorrelation

Finally, the effect of inter-dwell cross section decorrelation on the acquisition performance of a Rayleigh target is shown in Figure 3. For this case, the three dwells have equal energy and thresholds. The probability of acquisition is shown for the correlation coefficient ρ, defined in Equation (12) and (13), ranging from 0 (SWII target) to 1 (SWI target). To achieve a P_{ACQ} value of 0.7, the SWII target requires 3.5 dB more SNR on each dwell than does the SWI target. These results show that the acquisition performance of a Rayleigh target approaches that of a SWII target for correlation coefficients which are less than 0.6. Even a small amount of cross section decorrelation between dwells significantly degrades acquisition performance. For example when ρ equals 0.9, the Rayleigh target requires 2 dB more SNR on each dwell than a SWI target requires to achieve a P_{ACQ} of 0.7.

CONCLUSIONS

The flexibility offered by today's modern radar allows reliable target acquisition on a single scan. Each new search detection can be readily resolved in favor of a new target acquisition or be rejected as a false alarm by the use of sequential, multiple observations. Near optimum acquisition performance can be achieved by keeping the time between observations less than the decorrelation time of the target cross section and increasing the energy and/or reducing the target detection thresholds on the second and third observations. Effects of partial decorrelation can be offset by energy and/or threshold adjustment on the second and third observations. For the case in which the amplitude of the search detection is high, substantial cross section decorrelation can be tolerated without impacting acquisition performance.

ACKNOWLEDGEMENT

This work was initiated while R.A. Dana was with the Hughes Aircraft Company, Fullerton, California, USA.

REFERENCES

1. G.E. Lindeberg, A.S. Margulies, and P.A. Smyton, "Performance of Automatic Track Initiation Logic in Specific Target Environments", Proc. AGARD, 252, 13.

2. Dana, R.A. and Moraitis D., 1981, "Probability of Detecting a Swerling I Target on Two Correlated Observations", IEEE Trans., AES-17, 727-730.

3. Nuttall, A.H., 1972, "Some Integrals Involving the Q-Function", Technical Report NUSC 4297, Naval Underwater Systems Center (AD743066).

4. Papoulis, A., 1965, Probability, Random Variables, and Stochastic Processes, McGraw-Hill Book Company, New York.

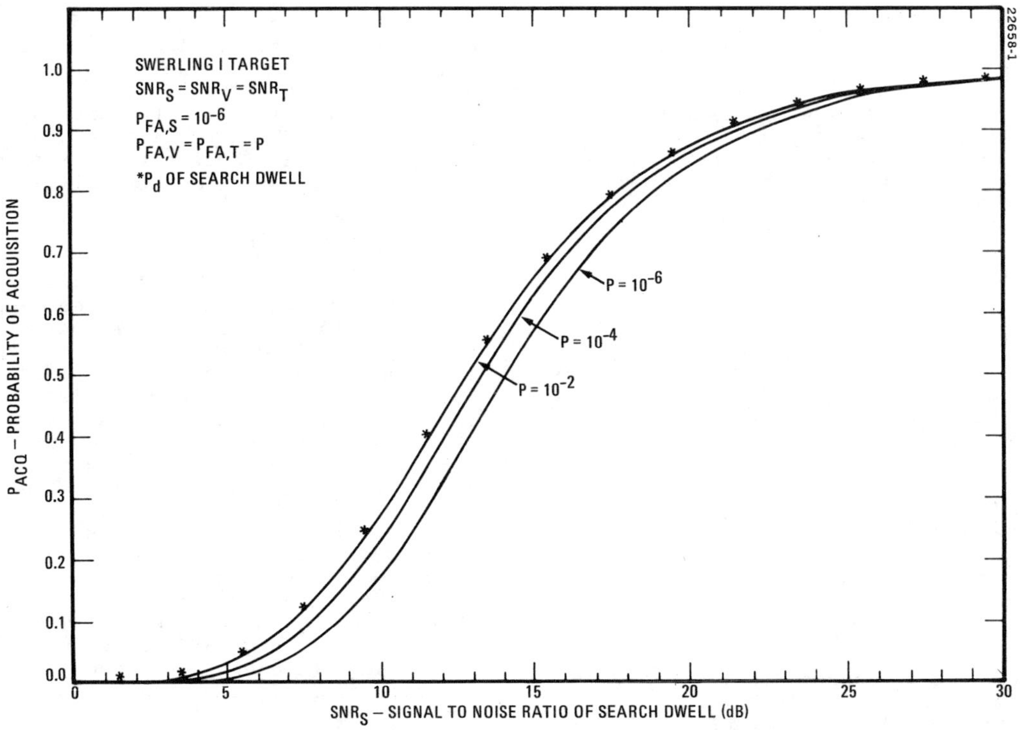

Figure 1. The Effect on Acquisition Performance of Lowered Detection Thresholds on the Verify and Track Dwells

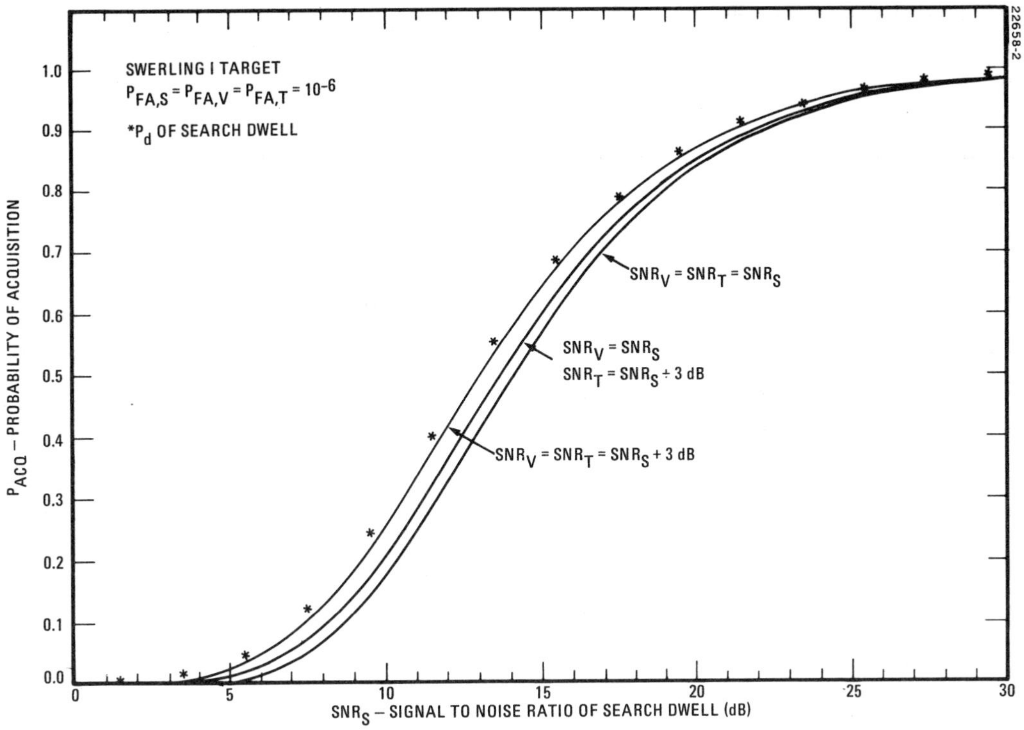

Figure 2. The Effect on Acquisition Performance of Increased Energy on the Verify and Track Dwells

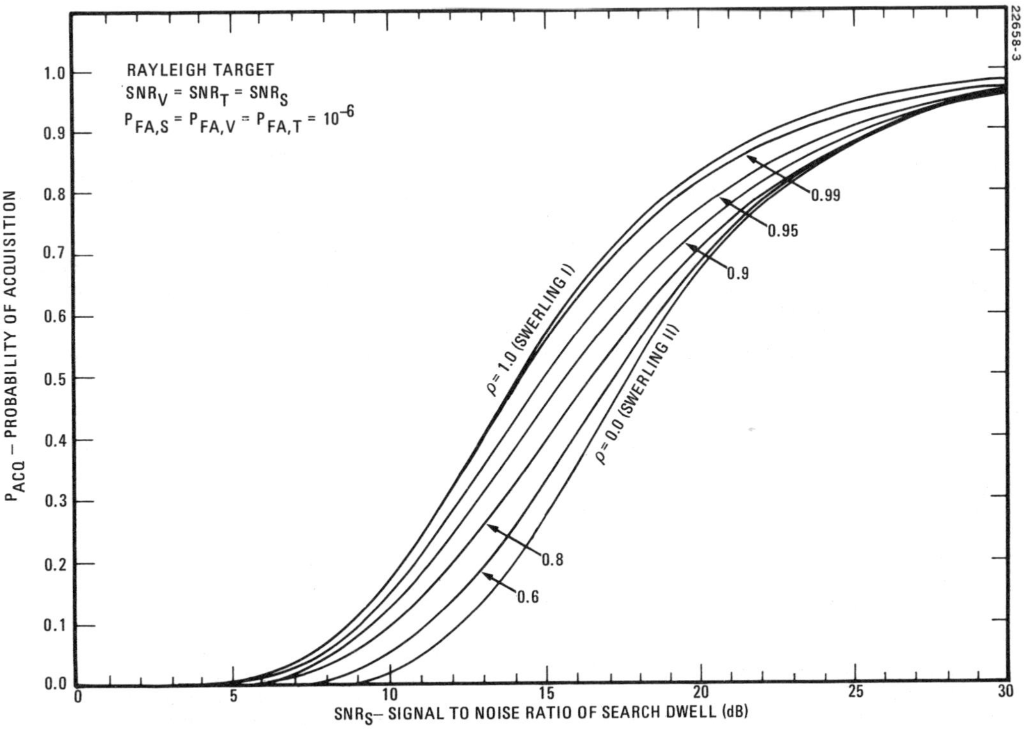

Figure 3. The Effect of Inter-Dwell Cross Section Decorrelation on the Probability of Acquisition

A SIMPLIFIED SEQUENTIAL DETECTION SCHEME

M.C. Jackson

Marconi Research Centre

Introduction

With phased-arrays, radars which adapt their scan pattern to circumstances can be built, and sequential detection, S D, which involves varying the dwell time, becomes an obvious possibility.

With a fixed dwell-time a constant number of returns in each range cell are simply added together, and the total is then compared with a threshold set far enough above noise to avoid false alarms. The sum builds up erratically over the whole of the dwell, as shown in Fig. 1 for a very large number of pulses and for two sizes of target or no target. The trend is obvious relatively early and in some cases the final result could be anticipated with little risk of error by having a pair of thresholds like those shown. The saving depends on how clear-cut the evidence is, and is small in borderline cases. The key idea in SD is the use of a pair of thresholds, which depend on the number of the pulse, to make a decision as soon as the evidence justifies it.

There is a statistical theory of such processes due to A. Wald, who coined the name, Ref. 1. Wald's thresholds differ markedly from those shown as they are roughly parallel with a dead-band between, and indefinitely long runs are possible because they never meet. This classical theory is based on distinguishing between just two possibilities, such as no target or some 'design' target, and is optimised (in the sense of giving the quickest decision on average) only for them. Moreover for radar the best procedure depends on the target fade type (e.g. Swerling model) as well as on its size.

For radar the need is for quick decisions when there is no target, giving economical 'search', combined with good sensitivity to all types of target. Some compromise design must be chosen and in conventional radars, where an 'ideal' receiver involves similar considerations, the solution is usually to ignore such subtleties and use a simple integrator, accepting whatever performance is thus obtained. Integration can be coherent, using spectral analysis, or incoherent after envelope detection, either linear or square-law.

A simple example of Wald's theory

The classical SD theory places the thresholds where the ratio of the 'likelihoods' of the set of observations under the two hypotheses, i.e. the relative probabilities of getting the actual data with the design target or with no target, has values depending on how many errors are acceptable, i.e. on the specified detection and false alarm probabilities, Pd and Pfa.

When the observations are independent the overall likelihoods are obtained by multiplication but it is more convenient to work with log-likelihoods which just add. In radar this means summing some measure of signal strength and comparing with thresholds which rise with the pulse number, as described above.

Only an idealised radar case, a known-phase sinewave in narrow-band Rayleigh noise, has been fully worked out. Although of no practical importance, it is instructive.

The ideal procedure is to produce a video signal, DC in Gaussian noise, by mixing with an in-phase local oscillator, and then to add sample values directly. In this case the signals are bipolar.

By a suitable choice of units and regarding the pulse number as a continuous variable the design target's size can be eliminated and a dimensionless diagram can be produced on which the only variations are the vertical positions of the thresholds which depend on Pd and Pfa (Fig. 2). In effect both coordinates represent energy, in terms of mean noise.

The following points should be noted.

1) The trend-lines showing the average behaviour of the no-target and design target sums are straight lines leaving the origin with slopes of 0 and 1 respectively. In practice the sums stray from these lines in a random-walk fashion, depending on the noise on each occasion.

2) The thresholds are parallel lines with slope ½. In principle a test could go on indefinitely and although the chance of very long runs becomes vanishingly small another stopping rule is needed to eliminate unacceptable cases.

3) The dismiss threshold, B, intercepts the y axis at a slightly negative value which depends only weakly on Pfa (via 1-Pfa) so that Pfa has virtually no effect on the pulses used when there is no target.

4) The alarm threshold, A, intercepts at a much larger positive value which depends mainly on Pfa, and consequently the cost of making a detection depends on Pfa in the usual way. This alarm line need not go through the threshold point of the equivalent standard radar because they depend on the specification in different ways.

   The big gain in SD comes from demanding small Pfa, e.g. 10^{-6}, whilst accepting much larger risk of missed detections.

Di Franco and Rubin, Ref. 2, analyse the performance of this system, so it can be compared with an ordinary radar of the same specification.

Fig. 3 shows how the transmission requirements of a search radar, in terms of the energy which would be received from the design target, depend on the specified Pfa and Pd. Only the standard radar is affected by Pfa, so the saving depends on this and is typically 10dB. Pd has a minor effect but very high values become costly.

Fig. 4 compares the detection characteristic, Pd:SNR, with that for a standard processor of the same specification. The curves cross at the design point, 80% in this example, but the slope is much less for the sequential system because the fluctuations of run length accentuate the statistical nature of detection.

In Fig. 5 the effect of the presence of a target on the average transmission requirement, i.e. the number of pulses, is shown. There is a peak for targets a few dBs below the design strength, where the demand is roughly the same as for a standard radar (the exact relationship depends on the design Pd and Pfa). For stronger or weaker targets the demand is less because the thresholds in Fig. 2 are crossed earlier. Unfortunately targets 10dB below design have a marked effect even though they are not detected, and good MTI is therefore needed.

Practical aspects

In practice the echo phase is not known beforehand and at some stage the signal amplitude must be found by envelope detection. Usually Doppler shift makes even the frequency uncertain and the best procedure is to use coherent integration, i.e. spectral analysis, before rectification, but this is tricky when the number of pulses is unknown. Incoherent integration after rectification is more practical but then the best detector law depends on the target fade model.

This is also the case for incoherent integration of a fixed number of pulses but as the exact detector law makes only a few tenths of a dB difference, most systems are nominally linear. Linear detection is also the most natural for digital systems and is compatible with MTI and sidelobe cancellers so it was chosen for this study.

In principle thresholds based on likelihood ratios can be found using the Rice distribution but this is difficult and a more empirical approach was used. With incoherent integration there is no universal diagram like Fig. 2 but something like Fig. 6 is always obtained. The main changes are that all signals are positive as even noise alone has a DC component, and that the threshold lines are curved. The dismiss threshold rises from the pulse axis some way from the origin and at least that many pulses are always used.

Beside being difficult to calculate, these thresholds are awkward to implement and the performance seems impossible to predict, and so a more empirical approach was used. Other reasons for this are the following limitations of the Wald theory when applied to radar.

1) The theory treats the pulse number as a continuous variable so that the process can stop exactly on the threshold. With incoherent integration only a modest number of pulses should be used to avoid integration loss so the thresholds will be overshot to some extent, i.e. their effective positions will be slightly further apart.

2) The theory applies only to a single range cell. With multiple cells the process cannot stop until the last of them has been cleared, which is delayed by statistical straggling.

3) The cost of detecting a design target is of less importance than obtaining good sensitivity to a range of targets.

Accordingly, for simplicity, a pair of straight lines which intersected after a chosen number of pulses were used, Fig. 7, and the performance was studied by simulation. By moving the thresholds of the mean number of pulses with no target, the sensitivity, and Pfa can be adjusted empirically.

As the ideal alarm threshold is nearly straight and only the top end of it is important, the linear approximation is reasonable. The line can be estimated by drawing the trend line for some nominal target and taking a line of intermediate slope intersecting it a few, say 5, noise standard deviations above the no-target line. The converging dismiss threshold is a convenient way of cutting-off excessively long runs, and it should also reduce the straggling with multiple cells.

With the threshold thus linked together Pfa will have some effect on the no-target demand, so it was set using tests of 10^{-7} trials with noise only. Pfa = 10^{-6} was adopted for the remaining work, although in the event changes by orders of magnitude had only small effects.

Statistics of the run lengths obtained during this phase allow the straggle effect to be assessed. Fig.8 is an example in which at least 20 pulses were used about once in a thousand trials, whereas the mean was barely 2. Thus 1000 cells would incur a loss of ~10dB, which about equals the gain offered by SD.

Fig. 9 shows a typical detection characteristic. The sequential system used 1.5 pulses when there was no target, and a theoretical curve for a standard radar was using 12 pulses is shown for comparison. The resemblance to Fig. 4 is striking although the system details are quite different.

Fig. 10 shows how the average number of pulses used depends on the target strength. This shows a maximum and again agrees with the ideal model results in Fig. 5.

Evidently the precise threshold arrangements are not crucial, at least for a single range cell, as a gain of ~10dB is obtained with no bother. However, multicell simulations show that the straggle effect eliminates the SD gain at about 1000 cells. The cure for this is to desensitise the system at short range so that the holdups only occur in a few outer cells, as demonstrated by Wirth, Ref. 3. This can be implemented quite simply in the proposed system.

If the sensitivity is varied the set of thresholds obtained form a neat pattern, Fig. 11. The alarm thresholds form a curved envelope and in each case the important part is near the point of tangency so it is possible to use a common set of values which depend only on the pulse number and are not adjusted with range. In contrast the dismiss thresholds form a fan, with the cut-off pulse number and the sensitivity depending on the slope of the line used. If the origin is moved to the point of this fan the sensitivity can be increased, in real time, by multiplying the threshold by a constant slightly less than unity for each new cell. In dBs:range the sensitivity law is rather peculiar, but this does not matter.

Simulating this system is rather lengthy and only one case, with the dismiss intercept starting at -1 noise unit, was fully investigated. This was chosen to use a modest number of pulses to reduce integration loss whilst keeping small the overhead of any pulses used to initialise an MTI canceller. The range dependence used reduced the sensitivity by ~4½dB between maximum and 70% range, which is slower than $1/R^4$. The maximum number of pulses allowed at extreme range was varied from 5 to 500 to cover systems of different sensitivity.

Fig. 12 summarises the results obtained by plotting the mean no-target pulse demand against the single-pulse SNR for Pd=½ at extreme range. The corresponding curve for ordinary incoherent integration is also shown.

Evidently the 10dB gain which is possible with 1 range cell is reduced to 1dB or less at 1000 cells. However when range dependence was used ~5dB were recovered, i.e. the gain was ~6dB which would be well worth while. This was for a first trial system which had not been optimised. A few exploratory trials showed that the gain obtained is proportional to the amount of taper used, but it must be weighed against the loss of short range sensitivity.

Conclusions

Sequential detection is a signal processing technique which could be useful in phased array search radars which allow adaptive beam scheduling.

The salient characteristics are:

1) For a single range cell typical energy savings are ∿10dB for Pfa = 10^{-6}. The actual energy demand is virtually independent of Pfa so very low values can be used without penalty.

2) This gain is obtained only when there is no target so the technique would not be used in tracking. Even undetectably small targets call for more energy so good MTI is needed.

3) With multiple range cells the performance falls because of statistical straggling. By desensitising at short range (range dependence) the effective number of cells is reduced and useful performance is obtained even with 1000 of them.

A practical scheme using incoherent integration and allowing MTI and range dependence has been devised, and shows promise on simulation.

References

1. Wald, A. Sequential Analysis John Wiley & Sons, 1947.

2. Di Franco and Rubin, Radar Detection Chap. 19 Prentice-Hall, 1968.

3. Wirth, W.D. Fast and efficient target search with phased array radars Proc. IEEE International Radar Conf. 1975 p. 198.

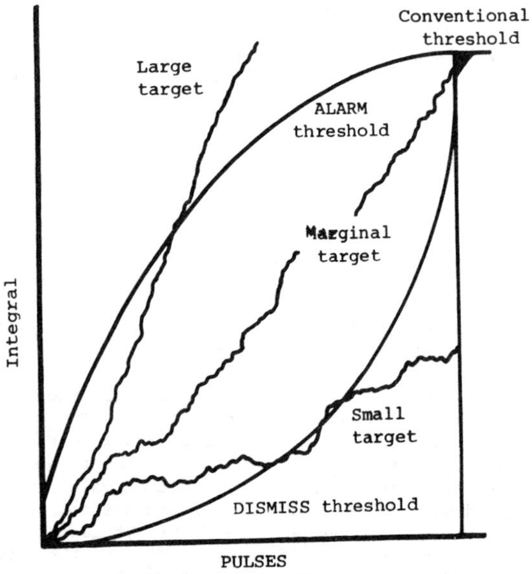

Fig. 1 The idea behind sequential, detection

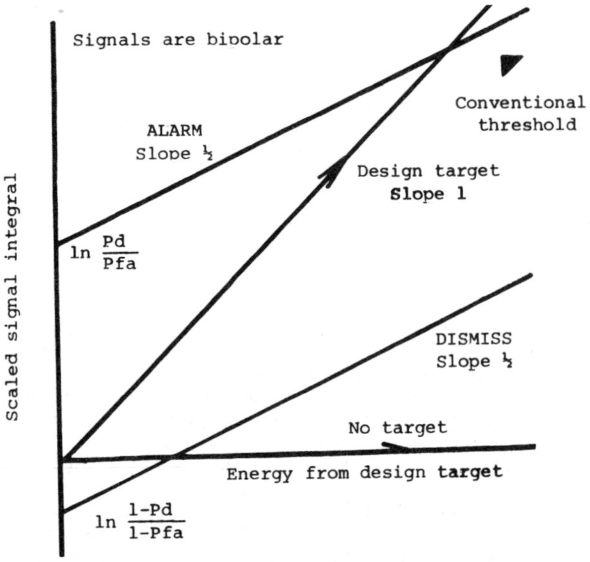

Fig. 2 Thresholds when target phase known

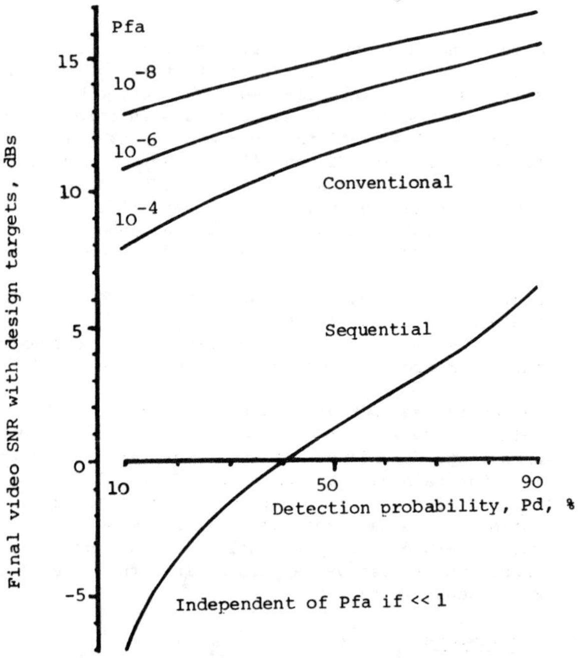

Fig. 3 Energy requirements when phase known

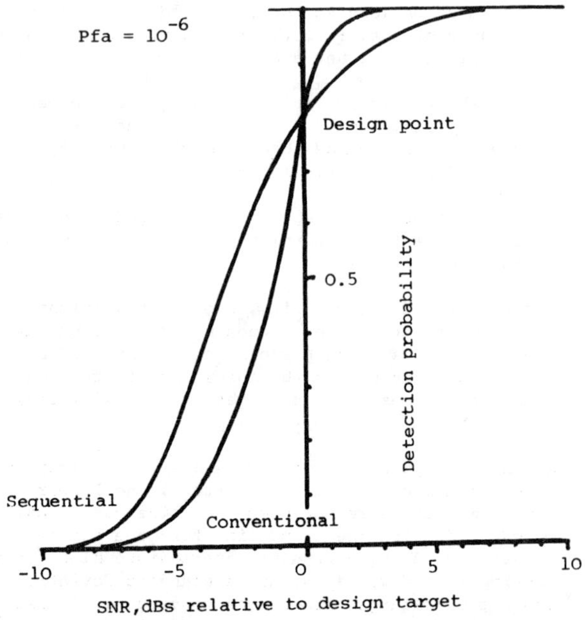

Fig. 4. Detection characteristics

"THIS WORK HAS BEEN CARRIED OUT WITH THE SUPPORT OF PROCUREMENT EXECUTIVE, MINISTRY OF DEFENCE."

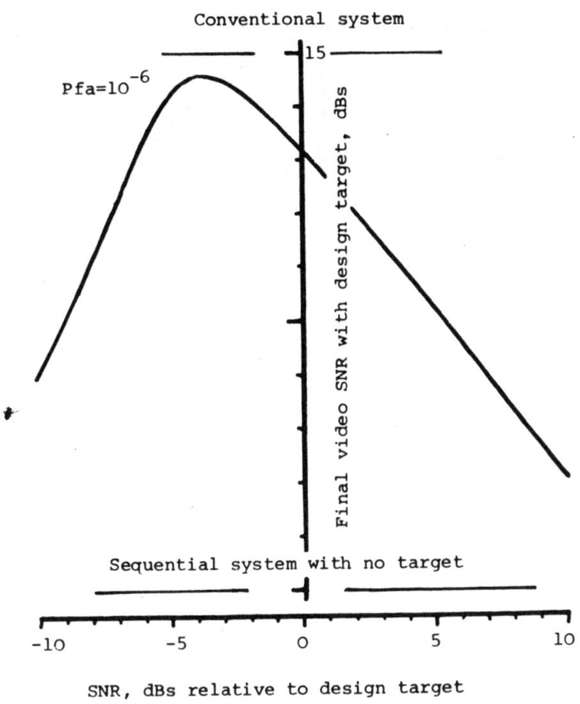

Fig. 5. Energy demand when target present

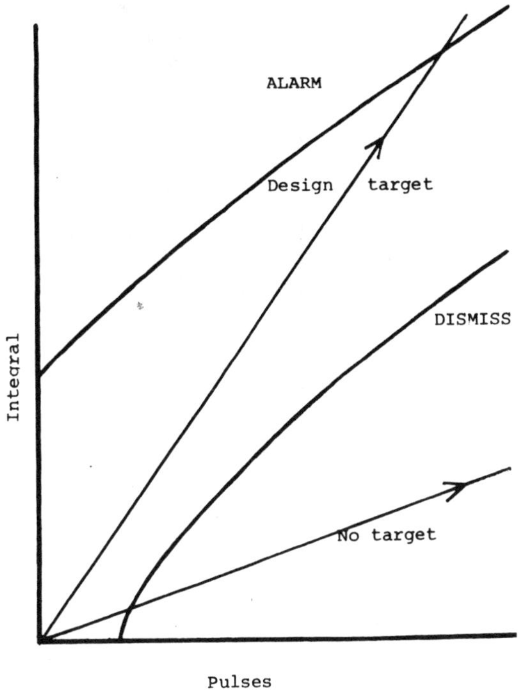

Fig. 6. Thresholds for incoherent integration

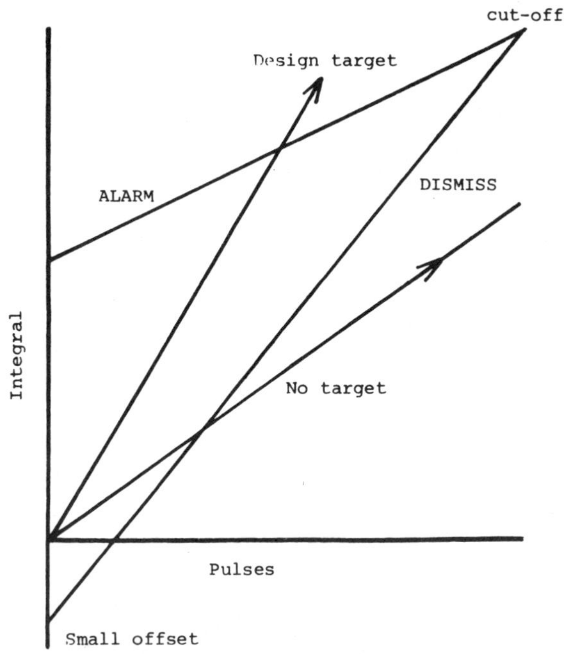

Fig. 7. Simple thresholds adopted

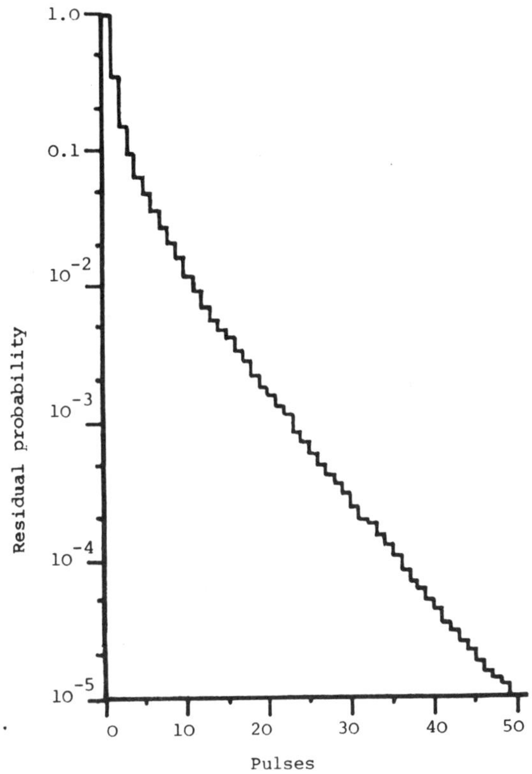

Fig. 8. Probability of long runs

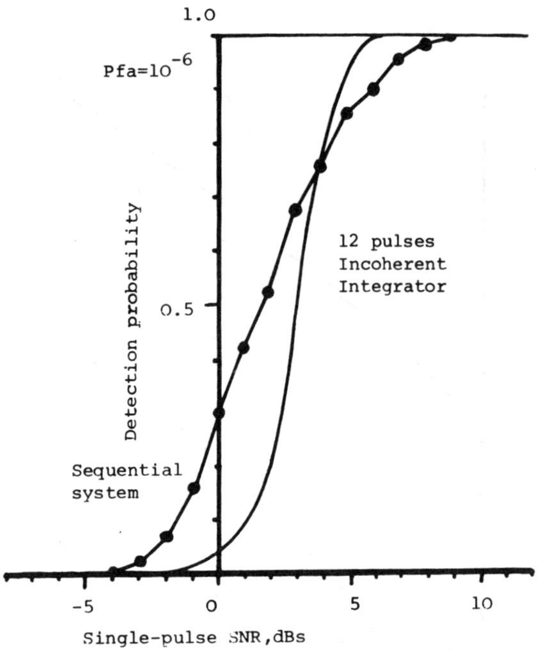

Fig. 9. Detection characteristics

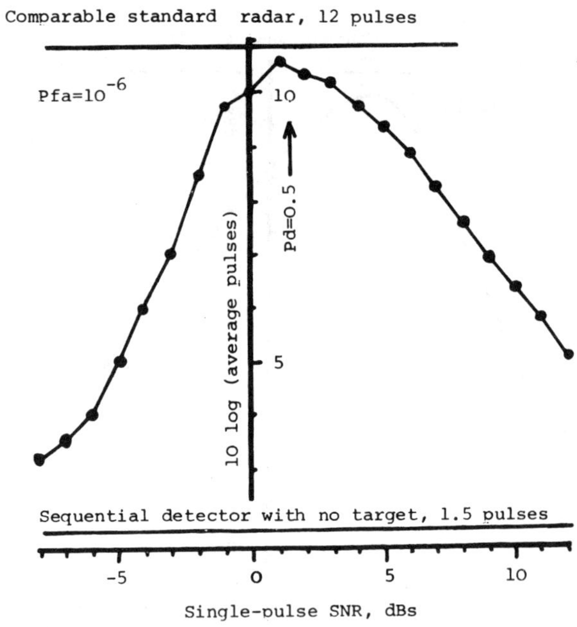

Fig. 10. Pulses used when target present

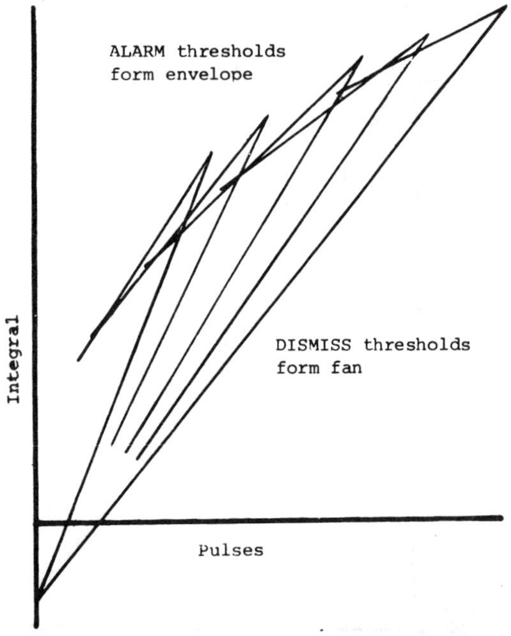

Fig. 11. Thresholds for range dependence

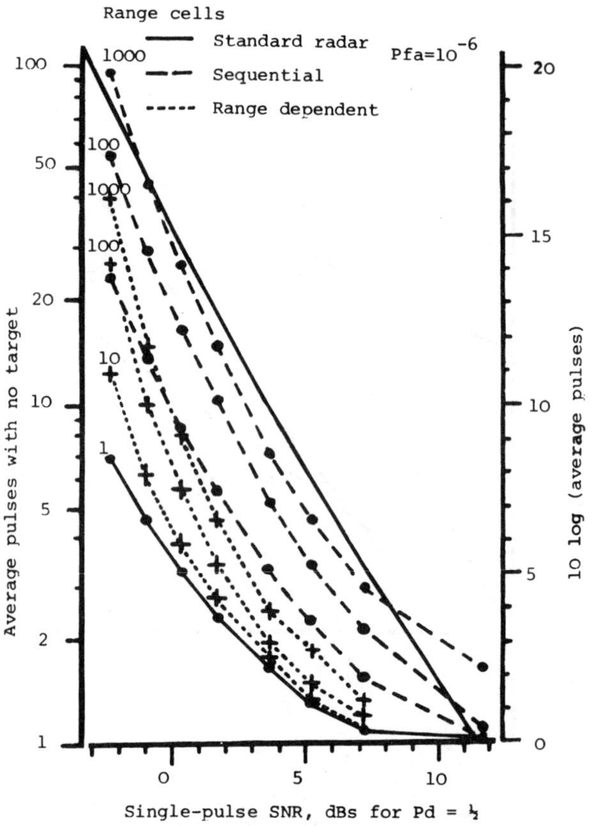

Fig. 12. Performance found by simulation

FALSE ALARM CONTROL IN AUTOMATED RADAR SURVEILLANCE SYSTEMS

W. G. Bath, L. A. Biddison, S. F. Haase, E. C. Wetzlar

The Johns Hopkins University, Applied Physics Laboratory, Laurel, Maryland

INTRODUCTION

Early radar systems relied upon a human operator to make all detection and tracking decisions, sometimes aided by track symbology and computer rate computations. However, these manual surveillance systems are increasingly inadequate because the operator is limited in tracking capacity, reaction time and accuracy.

Fully automatic detection and tracking systems have recently been developed to avoid the limitations of manual detection and tracking. These systems take radar IF or video signals as input, automatically produce a display of track symbology for the decision maker and produce computer files of accurate track data. The human operator does not have to detect and track each target, but instead, acts as a monitor of the automatic process adjusting its parameters to optimize performance.

Although automation produces tremendous improvements in tracking capacity, accuracy, and reaction time, the greatest challenge in radar automation is to control false alarms due to clutter or jamming in order to produce a reliable air surveillance picture with a low number of false tracks. This is a crucial design requirement in any automated radar system because excessive numbers of false tracks cause time and resources to be wasted trying to identify and react to nonexistent targets and eventually destroy the user's confidence in the system.

The sources of potential false alarms are well known. Clutter reflections from land, weather and the sea can be expected, as well as radio frequency interference (RFI) from friendly sources. In military systems signals from unfriendly sources will also be likely. While an automated radar system obviously cannot overcome basic theoretical radar waveform and power limitations in clutter and jamming, it must prevent excessive false alarms in these environments while simultaneously maintaining enough sensitivity to detect a target when it is theoretically detectable.

Ideally one would build radars with sufficient range and doppler resolution to make processed clutter returns significantly smaller than target returns. In practice, this is seldom, if ever, possible and when confronted with very small targets, even radar systems currently considered high resolution will have to deal with clutter detections. Simple techniques such as suppressing clutter (and attenuating targets) with STC are becoming increasingly dangerous.

Figure 1 illustrates an automated detection and tracking (ADT) false alarm management strategy. Hardware adaptive video processors employing high dynamic range digital constant false alarm rate (CFAR) processors reduce the false alarm rate to a level where computer processing in possible. Software false alarm processing is based upon quantitative measures of activity developed by the computer for each local region and qualitative descriptions of the environment (e.g., sea, land, etc.) provided by an operator. These activity and environment signals are combined to provide three levels of false alarm control:

1. Closed-loop contact acceptance screening which utilizes activity/environment descriptors and contact statistics to eliminate contacts which are unlikely to be related to real targets.
2. Fixed clutter point removal through bifurcated air/clutter tracking.
3. A track formation and update process which tailors track formation and promotion decisions to the local environment, local activity levels, and individual contact statistics.

The following sections describe these hardware and software techniques giving measured examples of system performance.

HARDWARE FALSE ALARM CONTROL

As shown in Figure 1, a digital signal processor - termed a radar video converter (RVC) is installed in the radar, or in the case of an integrated system, in each component radar. Each RVC samples and A/D converts the radar video, then using adaptive decision processes, produces a contact (set of digital words describing target location and video statistics) for each target detection. Although RVC design is unique for each individual radar, the generic approach is to use a high dynamic range (60-80 dB) digital processor which adapts its detection decisions to the local signal background and to control signals from tracking computers identifying gross clutter regions and known target locations.

Figure 2 illustrates a simplified, generic RVC. Both noncoherent logarithmic and coherent MTI (when available) videos are processed in parallel with final selection or combination of the videos' detections made under computer control. Predetection azimuth log-batch integration is employed in the 2D RVC's. The batch integration of log video is a robust process which makes the final detection decision insensitive to aberrations (such as RFI) which occur on only a few dwells. The constant false alarm rate (CFAR) technique, which uses a log-mean-level estimator to establish a threshold, is effective for Rayleigh noise, jamming, and clutter. It is also useful against non-Rayleigh clutter through the addition of a slow feedback loop which monitors the number of threshold crossings and adjusts the threshold offset in spikey or azimuth-correlated clutter regions. Additional threshold controls are received from the computers in known clutter regions or in the vicinity of known targets. The centroiding process makes the final hardware detection decision using a M out of N binary integrator. For a 3D radar this is usually a 2 out of 2 requirement to eliminate asynchronous RFI while for a 2D radar, it is usually a 3 out of 5 batch requirement which allows detection even if 1 or 2 batches correspond to prfs near blind speeds. All contiguous detections are then combined to form a contact with a position estimated by a center-of-mass or a leading edge technique. The relevant contact statistics (detection amplitude, local noise level, general noise level and detection angular extent) are computed for each contact and forwarded to the tracking computers for use in false alarm management algorithms.

SOFTWARE FALSE ALARM CONTROL

Real-Time Characterization of Environment

To optimize performance in an ADT system, it is necessary to provide more sensitivity control than that provided by the radar adaptive thresholding technique alone. In a real operating environment, there are regions of nonhomogeneous and correlated clutter which should be localized so that special techniques can be used to maintain good performance in the cluttered regions without sacrificing performance in adjacent uncluttered regions.

The first step toward providing this capability is to allow the operator to enclose the region of interference with a simple geometric shape such as a ring around the ship or a range-bearing extent window. The region can then be labeled with a choice of several environment designations, such as land, sea, rain, chaff, or RFI. so that special processing can be used within those areas of increased radar activity. The environment description affects not only radar video selection and RVC sensitivity, but also other processes such as contact acceptance screening and track promotion.

Because the operator provides only a gross, qualitative description of the environment, automatic quantitative environment characterization and sensitivity control is required as well. Activity is automatically measured and sensitivity is automatically controlled in discrete regions of radar coverage known as monitor sectors. These sectors are fixed in space and are chosen as small as possible within the requirements for statistical significance. For each monitor sector two measures of activity are maintained for each radar: an Input Monitor Level (IML) and a Track Monitor Level (TML). As the radar environment changes, the IML and TML gradually control the system sensitivity to avoid false tracks and processing overloads.

The IML system is a slow responding feedback system which monitors the average number of new saved contacts (potential new tracks) and controls that value by modulating the contact acceptance screening. The average number of new saved contacts is a measurable quantity which is relatively unaffected by the number of real targets or fixed clutter points in the sector. Thus, sensitivity is not lost in heavy air traffic regions or in regions with easily recognized clutter.

The IML is adjusted by comparing the average number of new saved contacts to one of a catalog of threshold values to determine whether the IML should be incremented in response to an activity increase. If not incremented, then an estimate is made of the average number of new saved contacts at the next lower value of the IML by counting the number of additional contacts that would be accepted if the IML were decremented. If this estimate is less than the threshold for the next lower value of the IML, then the IML is decremented. This double threshold process prevents the IML from "dithering" back and forth about its desired value and perturbing the tracking functions.

Although the IML system and the clutter tracking capability do much to control the false track level in the system, there is still a need for an even faster responding monitor system for either mildly correlated activity or for a sudden increase in the contact activity, such as when a deceptive jammer is turned on. This need is fulfilled by the TML system, which controls system sensitivity by affecting such mechanisms as tentative track entry, track promotion, the number of scans a saved contact is retained, whether to allow ungated contacts to update tracks, whether to allow severe target maneuvers, etc.

The TML is determined as the greatest of three separate monitor levels: one based upon the average number of new Saved Contacts, one based upon the average number of new Tentative Tracks to detect correlated activity, and one based upon a sudden increase in the number of ungated contacts entering the system. In contrast to the IML, the TML can make sudden changes to any level and is intended to reduce system sensitivity until the IML has had a chance to respond.

An important feature of the monitor level system is the automatic response to a sudden increase in contact activity, such as when a radar is first turned on (Figure 3). The TML immediately jumps to its maximum value due to a sudden increase in ungated contact activity and automatic track formation is momentarily disabled. The IML gradually responds until the Saved Contact activity is reduced to the proper level. The TML then drops and track formation is allowed, with perhaps some reduced sensitivity consistent with the Saved Contact and new Tentative Track activity present in the system.

Contact Acceptance Screening

In order to regulate the number of false reports entering the system, contact acceptance screening is employed prior to the track processing stages. Contact acceptance screening consists of sequences of tests based upon the sensor measured statistics associated with each contact, to determine whether a contact should be allowed into the system. The test sequence for a particular contact may include up to five types of tests: a signal-to-noise test, a local noise-to-system noise test, a maximum amplitude test, a minimum amplitude test and a number-of-elscans or bearing extent test.

The applicability and order of the tests are determined by the surrounding operator-defined environment. The severity of the test is dynamically adjusted in response to the local detection activity (reflected in the IML). The track gates sent to the radars also have an impact upon the contact acceptance screening. These are small sectors automatically defined about known track locations to avoid rejecting contacts corresponding to existing tracks. If a contact falls within a track gate, then with high probability it represents a return from a legitimate target. Such gated contacts are accepted even if they fail contact acceptance tests, but are tagged as candidates for updating only firm tracks.

Performance examples of the contact acceptance screening are provided in Figures 4 and 5, assuming that track-gated returns describe legitimate targets and ungated contacts represent clutter or interference.

The first example consists of heavy land clutter whose returns are so heavy that the contact history plot describes the area topographically. The empirical contact parameter probability density functions (pdf's) are obtained for the detections within a low elevation land environment sector. By observing the differences between the pdf's of the ungated contacts and those of the track-gated contacts (Figure 4), it was determined that the local noise-to-system noise ratio (NL/N) test was the most effective discriminant for heavy land clutter. A contact in a land environment is rejected if NL/N exceeds a threshold set by environment and IML.

In a more homogeneous environment, such as rain or chaff, a signal-to-noise test is also useful. The signal-to-noise test eliminates contacts whose peak amplitude to local noise ratio is less than a threshold controlled by environment and IML. Figure 5 illustrates the effectiveness of using the S/N test to distinguish rain clutter from real target signals.

In general, analysis of empirical contact statistics such as those in Figures 4 and 5 has shown that there is a great deal to be gained by contact acceptance tests based on matching measured contact statistics to the contact statistics characteristic of a given environment and to the contact statistics characteristic of real targets. The closed-loop IML system provides a rational way of regulating this contact acceptance process.

Clutter Point Tracking

Contacts which pass acceptance screening are candidates to associate with existing tracks, to associate with clutter points or to form new tracks or clutter points. The system tracks clutter points by simultaneously establishing both a tentative air track and a companion tentative clutter point. The progress of both the air track and the associated clutter point are monitored and prior to the air track becoming firm, a decision is made whether the data is clutter or a real target, and the companion track is eliminated. The decision is based on whether the air track (with its high bandwidth filter) or the clutter point (with its limited velocity, low bandwidth filter) has more accurately followed the target. The approach has significant advantages over clutter "mapping" techniques which can have long transients in responding to ship motion or changing clutter conditions.

Bayesian Track Promotion

A track begins as a saved contact, which is a contact that remains in the system after all tracks and clutter points have been updated. If a contact associates with the saved contact during the next few scans of the radar, then both a tentative track and a clutter track are formed. Tentative tracks which continue to receive updates are promoted first to Assumed, and then to Firm track status if the companion clutter track is dropped. Tracks can be dropped at any step of the promotion process if they cease to receive updates or if they are declared to be clutter points.

An attempt is made to minimize the possibility of false updates by using as small an association window as possible around the track prediction. Large deviation associations are handled by establishing an internal provisional track in addition to the original track, (i.e., bifurcation). During the next few scans, the update sequences of the original and provisional tracks are used to decide which is "real" (i.e., whether the high deviation association is a legitimate high-g maneuver or simply a false alarm).

The final line of defense against false tracks is track promotion. The system uses a Bayesian method by which promotion is retarded for tracks which are potentially false, while quick reaction is guaranteed for targets in clear or light clutter environments. This is accomplished by assigning a likelihood L_i to each contact based upon 4 factors:

1. The target strength as compared to the local background.
2. The relative strength between the local background and the system background (measured at maximum range).
3. The TML for the monitor sector containing the target,
4. The accuracy of the track prediction (statistical distance).

The use of the first three factors is tailored to the environment definition provided by the operator so that track promotion is accelerated for contact statistics unlikely to occur naturally in a given environment, and is accelerated in environments which the TML indicates are being well controlled.

Using Bayes rule, the sequence of likelihoods for a particular track can then be used to estimate the probability that a track is real:

$$P\ (\text{track real} \mid \text{measurements}) = \frac{LP_0}{LP_0+1-P_0}$$

where $L = \prod_{i=1}^{n} L_i$ = the likelihood product since track initiation

P_0 = the prior probability that the track is real (assumed 1/128)

Track promotion is based upon the probability being greater than a threshold, or equivalently upon the likelihood product being greater than a threshold. When no contact occurs to associate with a particular track, a miss is recorded, which is assigned a likelihood less than unity and causes the likelihood product to decrease. Tracks are dropped whenever the likelihood product is less than a threshold. Fewer misses are required to drop a track in the earlier stages of track promotion corresponding to smaller values of the likelihood product.

For a single radar, the track promotion process can be represented by a Markov chain in which the states represent the value of the log likelihood sum (Reference 1). Assuming that false tracks are updated predominantly by one radar, analysis of the Markov chain provides a strategy for controlling the IML. Since the interest is on false tracks, the probability associated with each update is that of at least one false contact appearing in the track association window. The total number of chains is equal to the maximum possible number of tracks, which is essentially the number of radar resolution cells in the monitor sector. The assumption is made that the number of false tracks present is small, so the chains can be assumed to not interact. Solving for the steady state probabilities, it is possible to show that

$$F \approx a\, s^4$$

where F = average number of false firm tracks

s = average number of saved contacts

a = a scale factor which depends upon the maximum possible number of tracks

This relationship has been used to set up control thresholds for changing the IML based upon the average number of saved contacts being formed in order to obtain preset false track goals.

FALSE ALARM CONTROL EXAMPLE

An automatic detection and tracking system has been built and the false alarm control is quantitatively described by an information flow diagram such as Figure 6. In an average radar scan time (6 seconds) about 6 million A/D samples are taken in the A/D converters. 392 hardware detections are produced from these samples including a few real target detections and many more false alarms from clutter and RFI. Most of these contacts can be safely eliminated by contact acceptance screening and clutter point tracking leaving 80 possible legitimate detections. The track formation, update and promotion processes identify only about 20 residual contacts corresponding to real targets currently in track and about 6 new tentative tracks. One out of every 15 tentative tracks eventually becomes firm and is made available to the command, control and weapons systems. Thus only about one out of every 1000 contacts received from the hardware actually corresponds to a new target detection (i.e., eventually leads to a firm track). The huge volume of information which must be discarded to produce a reliable track picture is a graphic illustration of the necessity for sophisticated false alarm control in automated radar surveillance systems.

The average number of false Firm tracks was measured for operation under land, sea, rain, and RFI clutter environments. The false track rates were found to be low and easily managed by the operator.

SUMMARY AND CONCLUSIONS

Radar automation demands careful attention to false alarm control since a high false track rate can make an automatic system unusable. Reliable false alarm control requires automation of the "mental signal processing" a human would perform if he had the opportunity to examine the radar video in detail and at his leisure. This translates into a requirement for high dynamic range digital processors; software which automatically adapts its contact acceptance, track formation, track update and track promotion decisions to the changing radar environment; and, considerable exchange of environment, activity and tracking data between the hardware and software.

REFERENCES

1. Casner, P. G. and Prengaman, R. J., 1977, "Integration and Automation of Multiple Colocated Radars," IEEE International Radar Conference, London, 145.

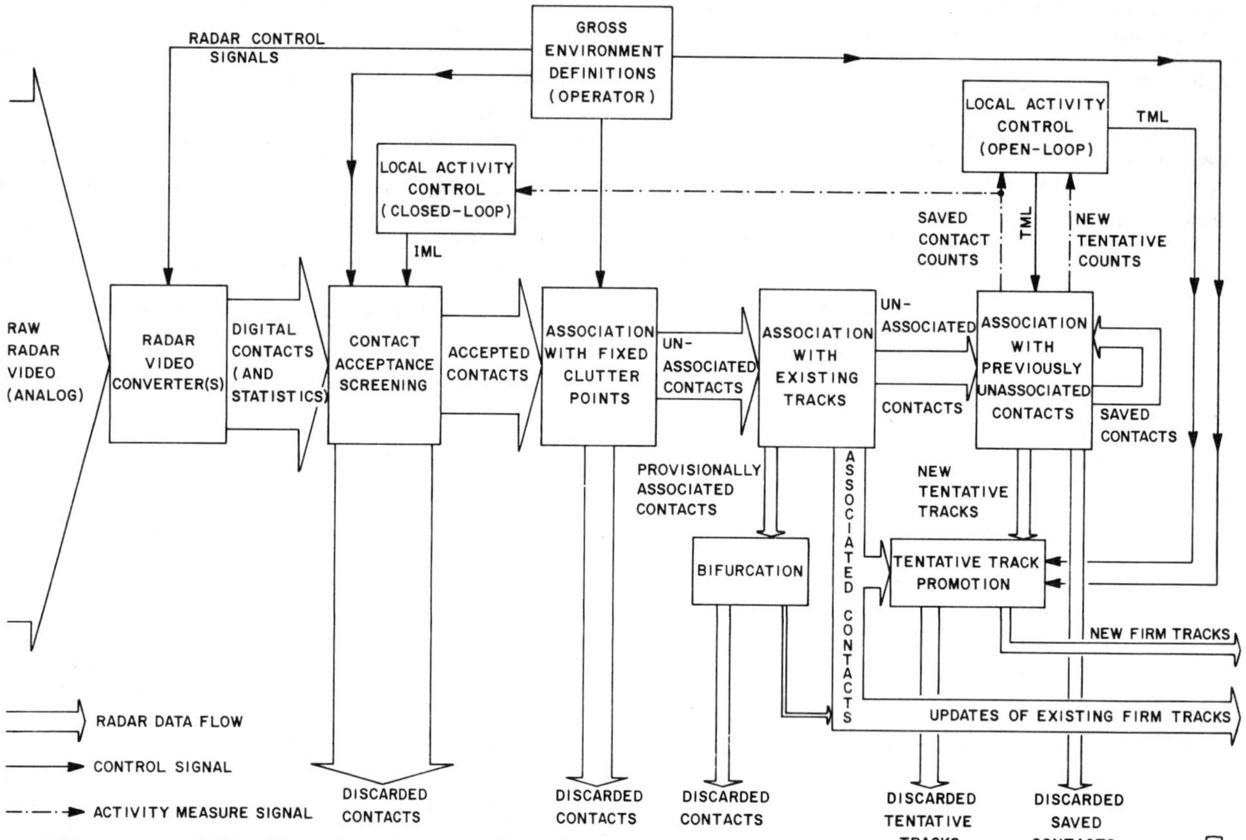

Figure 1. Management of False Alarms in an Automated Radar System.

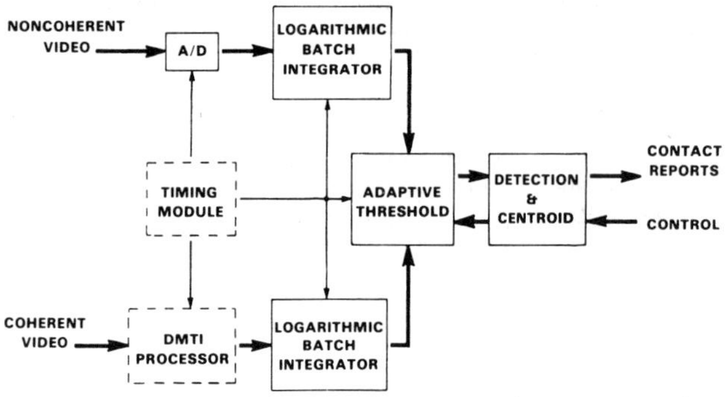

Figure 2. Simplified Diagram of a Typical Radar Video Converter.

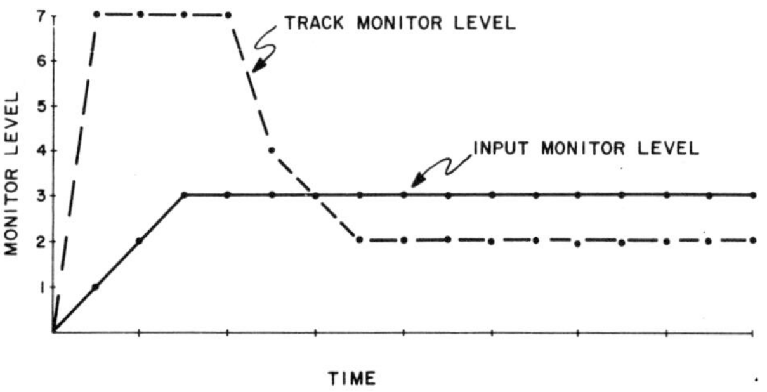

Figure 3. Response of the System Monitor Levels to a Sudden Increase in Radar Detection Activity.

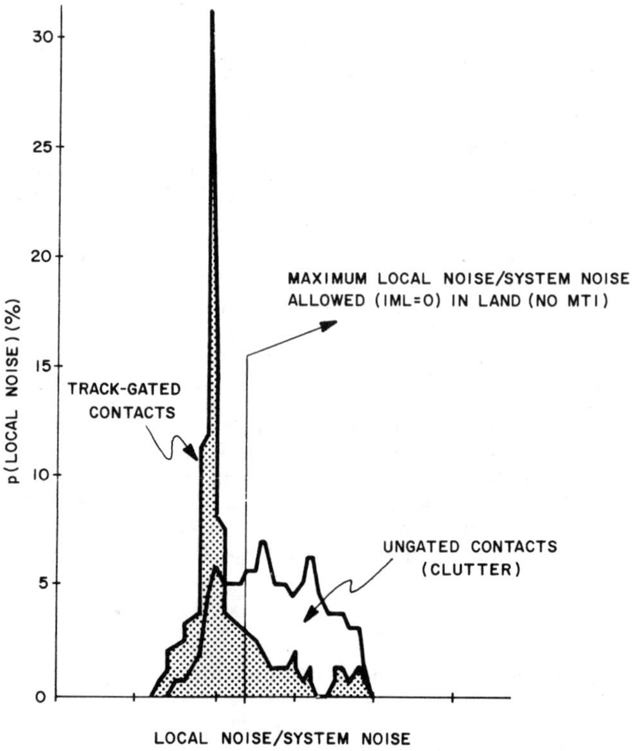

Figure 4. Effectiveness of Local Noise/System Noise Ratio Test in Land Clutter.

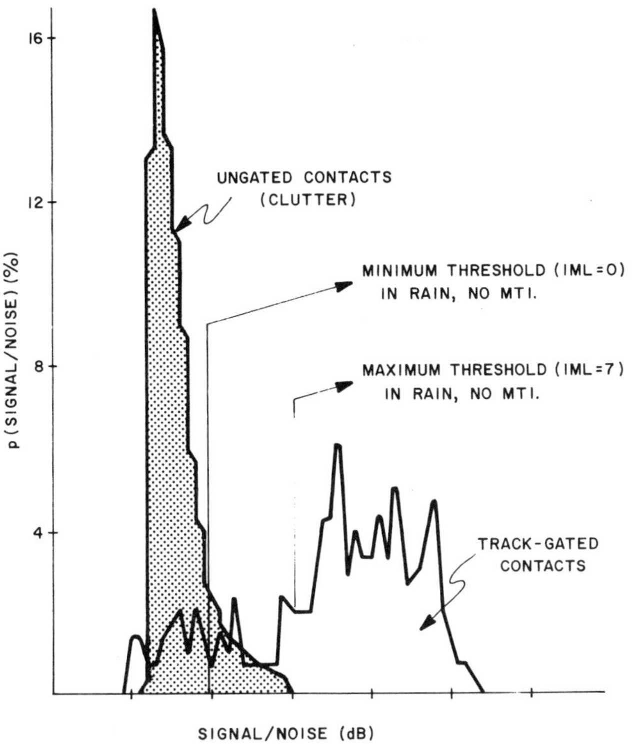

Figure 5. Effectiveness of Signal/Noise Ratio Test in Rain Clutter.

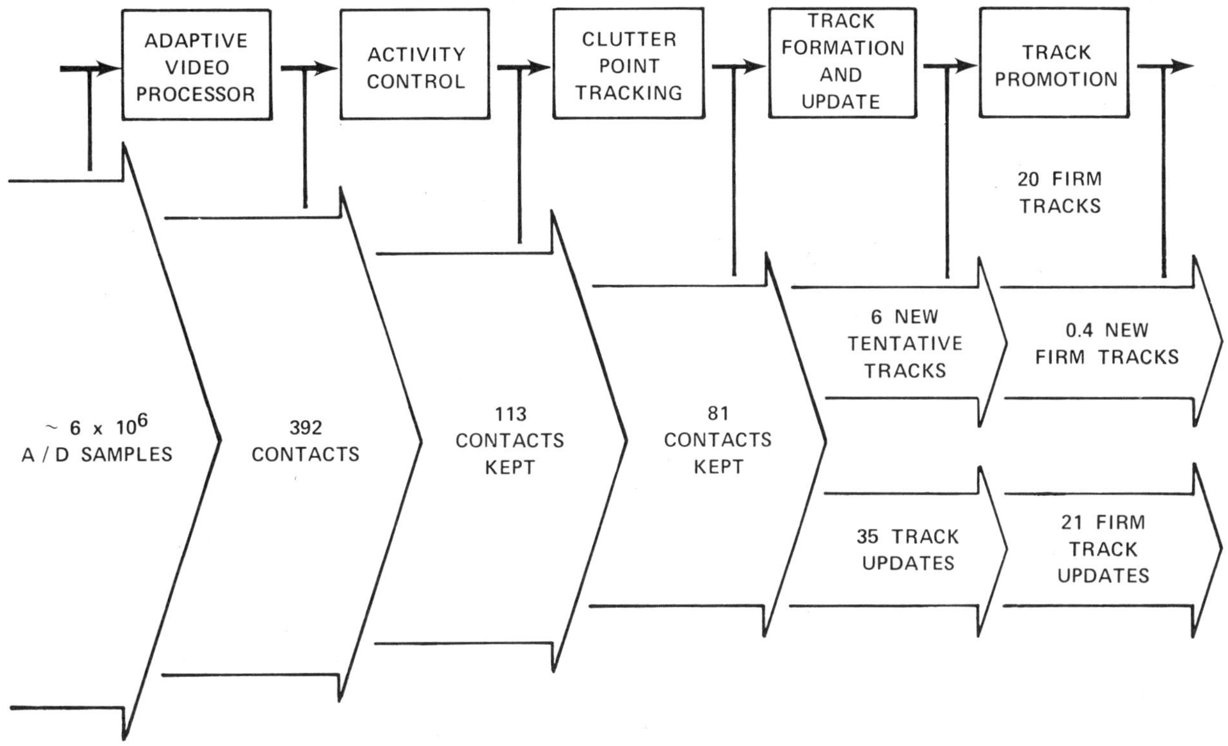

Figure 6. Example of Information Flow in a Severe Clutter/RFI Environment.

PERFORMANCE EVALUATION OF SOME ADAPTIVE POLARIZATION TECHNIQUES

D.Giuli°, M.Gherardelli°, E.Dalle Mese°°

°University of Florence, Italy - °°University of Pisa, Italy

INTRODUCTION

Owing to the necessity of improving radar performance, many studies have recently been devoted to optimally exploiting the polarization state of target echoes (1-16). Polarization, added to time, frequency and bearing can be considered a further, not sufficiently explored, dimension of the radar signal space. It is indeed envisaged, Huynen (1), Poelman (2), McCormick (3), that radar polarization techniques can be successful in:

i) improving the ratio of the target signal power to the unwanted signal power;

ii) discrimination, classification and identification of targets;

iii) decorrelating radar echoes through polarization agility;

iv) synthesis of polarization coded waveforms for target resolution;

v) reducing multipath effects;

vi) radar analysis of precipitation microstructure.

Polarization techniques always increase the complexity of radar systems, in particular if polarization has to be controlled also at transmit. Therefore, the actual improvement of radar performance, which can be gained through these techniques, has to be carefully evaluated.

This paper is concerned with evaluating radar performance in case i), i.e. when antenna polarization is controlled, so as to improve the ratio of the target signal power to the unwanted signal power. Three different cases of unwanted signals are considered: background noise, rain clutter and intentional interference (jamming). In the first case the improvement in detecning targets in the presence of background white Gaussian noise is evaluated when adapting transmit and/or receipt polarization to the target polarization properties. For this purpose a suitable analytical target model is devised which accounts for the dependence of echo strength and fluctuation from polarization. Explicit formulae of the target detection probability are thus derived as a function of target parameters. With reference to some particular target parameters some numerical results are shown which evidence the improvement obtainable when controlling antenna polarization at transmit and/or receipt.

In the second case, which refers to target in the presence of rain clutter, improvements in signal-to-clutter ratio are evaluated when using an adaptive polarization canceller (APC) proposed by Nathanson (4). The performance of the APC is evaluated through simulations based on suitable models for both the rain clutter signals and the receiver operation.

In the third case, which refers to target in the presence of jamming, the improvements in the signal-to-interference ratio are evaluated when using a symmetric adaptive polarization canceller (SAPC) proposed by D.Giuli (5,6). The performance of SAPC is evaluated, through numerical simulations, for single or multiple jamming sources generating polarized uncorrelated white-noise signals.

This work has been supported by National Council of Research, under Contract n. 81.00202, and by Selenia Company, Roma, Italy.

POLARIZATION CONTROL TO IMPROVE DETECTION OF TARGETS EMBEDDED IN BACKGROUND NOISE

Low-reflectivity targets embedded in background noise require the use of higher sensitivity radars. In principle, such a radar improvement can be achieved by controlling antenna polarization and adapting it to the polarization state of the target. Given a target, optimum antenna polarizations at transmit and/or receipt should be selected in order to maximize the echo strength or target detection probability. Since polarization properties of a given target under radar coverage are generally unknown "a priori", the capabilities of polarization control techniques cannot be fully exploited, even if good results can be expected through suboptimal techniques, such as the "virtual adaption" polarization technique (Poelman (7)).

However, to evaluate maximum detection gain, which can ideally be achieved by controlling antenna polarization, we analyze target detection by supposing that target polarization properties are known. For this analysis we resort to a target model based on a general theorem on target decomposition (see reference (1)). According to this theorem any target scattering matrix T(t) can be decomposed as:

$$T(t)=a(t)T_o+T_N(t) \quad \ldots\ldots\ldots\ldots \quad (1)$$

where $a(t)T_o$ is the scattering matrix of the "effective target" and $T_N(t)$ is the scattering matrix of the "noise target". These two matrices are independent and $T_N(t)$ is given by:

$$T_N(t)=b_1(t)T_{N1}+b_2(t)T_{N2} \quad \ldots\ldots\ldots\ldots \quad (2)$$

where

$$T_{N1}=\begin{bmatrix}1 & 0\\ 0 & -1\end{bmatrix} \; ; \quad T_{N2}=\begin{bmatrix}0 & 1\\ 1 & 0\end{bmatrix} \quad \ldots\ldots \quad (3)$$

Based on eq. (1) we develope a target model for pulse-to-pulse representation of the signal echo. Therefore we consider a discrete-time representation of T(t), with $t=iT_s$ (integer i) and T_s equal to the pulse repetition period. This model specifies the characteristics of the processes $a(t)$, $b_1(t)$ and $b_2(t)$ when referring to the same target and at different pulses during antenna dwell-time. The developed model is also simple enough to allow for analytical evaluations, while being physically sound and general enough to cover the most significant actual cases. Here we summarize our assumptions ($t=iT_s$):

- All the involved random sequences are supposed to be

locally stationary, Gaussian, zero mean, and with uncorrelated real and imaginary parts.
- The sequence a(t) is decomposed as

$$a(t)=A+a_o(t) \qquad (4)$$

where A is a Gaussian random variable, with zero mean and variance σ_A^2, constant during the antenna dwell-time, while $a_o(t)$ is a white sequence independent of A.
- The two sequences $b_1(t)$ and $b_2(t)$ are white sequences.
- The variances of the sequences a(t) and $a_o(t)$ are, respectively, σ_a^2 and $\sigma_{a_o}^2$; while the two sequences $b_1(t)$ and $b_2(t)$ have equal variance σ_b^2.

A detailed discussion of the physical meaning of the above hypotheses is contained in Dalle Mese and Giuli (8). Notice that our assumptions lead to a target model which is given by the sum of two independent targets according to the following relation:

$$T(t)=AT_o+T_F(t) \qquad (5)$$

with

$$T_F(t)=a_o(t)T_o+T_N(t) \qquad (6)$$

The first and the second term in (5) correspond, respectively, to a Swerling I model (scan-to-scan fluctuating target) and a Swerling II model (pulse-to-pulse fluctuating target). The received signal can thus be written as

$$s(t)=s_o+s_F(t)+n(t) \qquad (7)$$

where s_o and $s_F(t)$ are the two target signal components, while n(t) is the background noise supposed white, Gaussian, with variance σ_n^2 and independent of antenna polarizations. The target signal components are given by:

$$s_o=A(T_o\underline{f})\cdot\underline{g} \qquad (8)$$

$$s_F=a_o(t)(T_o\underline{f})\cdot\underline{g}+(T_N(t)\underline{f})\cdot\underline{g} \qquad (9)$$

where $\underline{f}$ and $\underline{g}$ represent the antenna polarization vectors at transmit and receipt respectively.

Eight parameters generally specify the above target model: five parameters characterize T_o and the three variances σ_A^2, $\sigma_{a_o}^2$ and σ_b^2 specify the statistical nature of targets. In order to reduce the number of independent parameters, in numerical computations we used a simplified version of our model by taking

$$T_o=\begin{bmatrix} \frac{1+\lambda}{2}+\frac{1-\lambda}{2}\cos2\psi & \frac{1-\lambda}{2}\sin2\psi \\ \frac{1-\lambda}{2}\sin2\psi & \frac{1+\lambda}{2}-\frac{1-\lambda}{2}\cos2\psi \end{bmatrix} \qquad (10)$$

The scattering matrix T_o given in (10) is obtained from the scattering matrix T'_o,

$$T'_o=\begin{bmatrix} 1 & 0 \\ 0 & \lambda \end{bmatrix} \qquad (11)$$

with a rotation of the angle ψ around the antenna-target axis. Without any loss we assume $0\leq\lambda\leq1$. The new model is specified by the same three statistical parameters and by the two parameters λ and ψ. It is also more convenient to refer to the following parameters instead of σ_A^2, $\sigma_{a_o}^2$, σ_b^2:

- polarization degree: $\gamma_p=(\sigma_A^2+\sigma_{a_o}^2)/(\sigma_A^2+\sigma_{a_o}^2+\sigma_b^2)$ (12)
- fluctuation degree: $\gamma_f=(\sigma_{a_o}^2+\sigma_b^2)/(\sigma_A^2+\sigma_{a_o}^2+\sigma_b^2)$ (13)
- polarized fluctuation degree : $\gamma_{pf}=\sigma_{a_o}^2/(\sigma_{a_o}^2+\sigma_A^2)$ (14)

It is our opinion that the simplified model describes a major class of targets. We note also that the simplified model embodies, as a particular case, the target model proposed by Poelman (9).

The proposed target models allow for computation of target detection probability, P_D once the target parameters and antenna polarizations are given. When using the incoherent optimum receiver reported in Fig. 1, explicit formulae of P_D are indeed found (8,10). It is also possible to determine the couple of transmission and reception antenna polarizations which maximize the average echo power; similarly the optimum antenna polarization in reception can be determined once the polarization in transmission is fixed (reference (8)).

In Fig. 2 and 3 some curves are reported which have been obtained with the simplified target model. They report the gain of transmission power, G, as a function of the angle, Φ, of the antenna polarizations supposed to be linear. The gain G is evaluated with respect to the case of horizontal linear polarizations, used both in transmission and reception, at the parity of P_D, P_{fa} (False alarm probability) and N (number of integrated pulses). Fig. 2 refers to the case of changing polarizations both in transmission and reception while keeping them equal to each other. Fig. 3 compares a curve obtained as in Fig. 2 (PR=PT) with that obtained by only changing polarization at receipt while keeping the horizontal polarization in transmission. Both figures refer to an asymmetric target (γ=0.2), with orientation angle ψ=60°, partially fluctuating (γ_s=0.3 in Fig. 2 and γ_s=0.5 in Fig. 3) and with no polarized fluctuation (γ_{sp}=0). The fixed detection parameters are $P_{fa}=10^{-6}$ and N=10. From Fig. 2 it can be inferred that the detection gain obtainable by controlling antenna polarization is higher for low-probability target detection; while Fig. 3 shows that a much higher detection gain can be achieved by controlling antenna polarization both in transmission and in reception, rather than only in reception.

ADAPTIVE POLARIZATION RECEPTION TO IMPROVE DETECTION OF TARGETS EMBEDDED IN RAIN CLUTTER

Raindrop backscattering often reduces to the maximum extent the capability of the radar for a correct detection of true target. In this situation optimum choice of antenna polarization is one possible method for rain clutter rejection. To this end circular polarizations are generally used, even if some factors can limit the obtainable rain clutter cancellation, such as (see Nathanson (4) and Peebles (11)):
- non-spherical shape of larger raindrops;
- different attenuation of horizontally and vertically polarized waves propagating through rain;
- not uniform space distribution of rain;
- ground reflections with different strength for horizontal and vertically polarized components;
- ellipticity of antenna polarization.

Owing to the above limiting factors a slightly elliptical polarization, with ellipticity degree varying according to the range of the range cell, is the ideal choice for rain clutter rejection, once a fixed circular polarization is used in transmission. Based on this principle, Nathanson (4) proposed the use of an adaptive polarization canceller (APC). The analytical model of the APC is reported in Fig. 4. Its operation is based on using a dual-polarization receivers. The two reception signals, $S_1(t)$ and $S_2(t)$, are received through two circularly polarized channels with "same" and "opposite" rotation sense of polarization with respect to that in transmission. The APC performs a coherent linear combination (at intermediate or video frequency) of the signals $S_1(t)$ and $S_2(t)$ according to eq. (15):

$$S(t)=S_1(t)+W(t)S_2(t) \quad . \; . \; . \; . \; . \; . \; . \; . \; . \qquad (15)$$

The closed-loop circuit of the APC operates so as to force to zero the component of S(t) correlated with $S_2(t)$. The operation is performed along each sweep thus affecting all the range cells therein contained. The clutter cancellation ratio is ideally given by

$$C_r = \frac{E\left\{|S_1^2(t)|\right\}}{E\left\{|S^2(t)|\right\}} = \frac{1}{1-\mu^2} \quad . \; . \; . \; . \; . \; . \; . \; . \; . \qquad (16)$$

where the correlation coefficient μ is defined as:

$$\mu = \frac{|E\{S_1(t)S_2(t)^*\}|}{\left[E\left\{|S_1^2(t)|\right\}\cdot E\left\{|S_2^2(t)|\right\}\right]^{\frac{1}{2}}} \quad . \; . \; . \; . \; . \qquad (17)$$

Actually, it turns out that the value of C_r is lower than that given by (16), owing to the following impairments in the APC circuit operation:

- signals decorrelation due to asymmetry of the reception channel;
- non stationarity of the input signals;
- stability constraints;
- limited accurateness of the closed-loop estimate of the signal correlation coefficient.

In order to evaluate performance of the APC when using it for rain clutter cancellation, as well as to optimize the APC structure, a numerical program has been developed to simulate the APC operation. This program is based on suitable models of both the APC circuit and the rain clutter signals (16). In particular, rain clutter dual-polarization signals are simulated by resorting to both analytical models and results of experiments published in the literature (17-21). As far as e.m. scattering by rain is concerned this model accounts, in the frequency range 2-10 GHz, for:

- rainfall rate;
- mean canting angle of raindrops;
- degree of common alignement of raindrop axis;
- raindrop scattering coefficients;
- size distribution of raindrops.

In simulating the rain clutter signals the target decomposition in eq. (1) is also adopted to account for their partial polarization.

Simulation results confirm in many cases the effectiveness of the APC for rain clutter cancellation. As an example, in Fig. 5 the average signal power is reported as a function of range at both the two inputs and the output of the APC. The curves in Fig. 5 refer to the case of uniform spatial distribution of rain and rainfall rate of 2.5 mm/hr, with the low-pass filter in the loop implemented by an integrator with memory given by 9 consecutive range cells (range cell dimension =75 m). The curves are obtained by averaging 100 independent runs for 50 consecutive range cells starting at 20 Km. The target is located at the 30-th range cell and it is assumed as completely fluctuating from one run to another. As it can be inferred from Fig. 5 an additional average cancellation of 9 dB is achieved through the APC, to be added to 24 dB of average cancellation obtained by using circular polarization, which is enough to evidence the presence of target.

ADAPTIVE POLARIZATION RECEPTION TO IMPROVE DETECTION OF TARGETS EMBEDDED IN INTENTIONAL INTERFERENCE

To enhance targets embedded in intentional interference, techniques based on controlling antenna polarization can be used. If antenna polarization is fixed at transmit, an adaptive choice of polarization at receipt can be effective only if the following conditions are met: a) the intentional interference has a high degree of polarization; b) polarization of intentional interference differs enough from target echo polarization; c) fast polarization adaptation can be achieved to counteract signal non stationarity. These conditions are generally met if i) the intentional interference is a background disturbance generated by a single jamming source; ii) the adaptive polarization receiver, depicted in Fig. 6 and proposed in (5,6),is adopted. Its operation is based on using a symmetric adaptive polarization canceller (SAPC) formed by two APC's which still perform cancellation according to the scheme in Fig. 4, for each of the two orthogonally polarized reception channels. Then, the cancelled output with the weakest background disturbance is selected. The device for channel selection performs logarithmic amplification and low-pass filtering of the two cancelled outputs, before their comparison by means of a comparator with hysteresis. Too frequent channel changes not connected with non-stationary behaviour of jamming source are thus avoided.

Since the SAPC ideally achieves the same cancellation ratio on both the orthogonally polarized channels, the channel selection device allows for maximum rejection of interference. Using circular polarization at transmit, together with the horizontal and vertical linear polarization at the two reception channels, it should allow for bounded reduction of target echo power, at least when disturbance signal powers differ enough from one reception channel to another (5). In the case of multiple independent jamming sources the proposed receiver performs worse, since condition a) is no longer necessarily met. However, a good cancellation ratio can still be achieved if the polarization of one source is slightly different from the others as well as if one source is stronger than the others.

The proposed adaptive polarization receiver is based on a concept analogous to that of the conventional single side-lobe canceller (SSLC). However, the former presents the following advantages with respect to the latter: a) it is capable of jamming cancellation within the antenna mainlobe; b) it is capable of high side-lobe cancellation also for jamming bearing far away from the antenna boresight (in this situation the SSLC performance can be degraded by a signal decorrelation caused by the displacement of the antenna centers of phase); c) mainlobe channel symmetry allows for additional jamming reduction through fast adaptive switching of the orthogonally polarized channels.

The performance of the SAPC has been evaluated for

mainlobe interference through numerical simulations, and assuming that each source independently generates a white, Gaussian, zero-mean signal. As an example, in Fig. 7, the behaviour of the SAPC is reported in the case of a single jamming source employing a linear polarization. It is assumed that the polarization angle keeps a constant value (α=15°) up to the 40-th range cells; then it shows an abrupt change (α=75°) and keeps constant again. The target has been located at the 60-th cell embedded with jamming (jamming-to-signal ratio= +4 dB). Fig. 7 compares the two input signals of the SAPC with the cancelled output after channel selection, and shows the good performance of the SAPC which yields a cancellation ratio of 25 dB.

CONCLUSIONS

The results obtained in evaluating radar performance when using some polarization techniques have been discussed. It has been evidenced that the target embedded in background noise can be enhanced mainly by controlling antenna polarization both in transmission and in reception and in particular under conditions of low target detection probability. Techniques based on an adaptive polarization receiver are less complex while appearing effective in reducing rain clutter or intentional interference. To become confident with the results obtained through analysis and simulations, it would be advisable to achieve some experimental validation. This is even more necessary due to the scarce availability of short-term dual-polarization radar measurements. To this purpose a radar measurement set has been designed to collect measurement data from dual-polarization radars (Biffi Gentili et al. (22)). These measurements are first planned for validation of the APC performance in the presence of rain clutter. Successive analogous measurements will be executed for validation of the SAPC performance when used for jamming cancellation.

REFERENCES

1. Huynen, J.R., 1978, "Phenomenological Theory of Radar Targets", in "Electromagnetic Scattering", (P. I.E. Uslenghi Editor), Academic Press, New York.

2. Poelman, A.J., 1980, "A Study of Controllable Polarization Applied to Radar", Military Microwave Conference, London, England.

3. McCormick, G.C. and Hendry, A., 1975, Radio Science, 10, 421-434.

4. Nathanson, F.E., 1975, "Adaptive Circular Polarization", IEEE Int.Radar Conference, Washington, USA.

5. Giuli, D., 1980, "On Using the Polarization Diversity in Radar Systems" (in Italian), Selenia Company, Rome, Italy, Report No. RT-80064/In.

6. Giuli, D., 1981, "Adaptive Polarization for the Cancellation of Intentional Interference in a Radar System", Selenia Company, Rome, Italy, Patent (Pending), No. 48438/A81.

7. Poelman, A.J., 1980, "A Study of Controllable Polarization Applied to Radar", Military Microwave Conference, London, England.

8. Dalle Mese, E. and Giuli, D., 1981, "Effects of Antenna Polarization Control on Detection of Targets in Clear" (in Italian), Selenia Company, Rome, Italy, Report No. RT 81127/In.

9. Poelman, A.J., 1976, "Reconsideration of the Target Detection Criterion Based on Adaptive Antenna Polarizations", AGARD Conference Proc. "New Devices, Techniques, and Systems in Radar", The Hague, Holland.

10. Dalle Mese, E., and Giuli, D., 1982, "Detection Probability of a Partially Fluctuating Target", (submitted for publication).

11. Peebles, P.Z., 1975, "Radar Rain Clutter Cancellation Bounds Using Circular Polarization", IEEE Int.Radar Conference, Washington, USA.

12. Poelman, A.J., 1976, IEEE Trans., AES-12, 774-785.

13. Poelman, A.J., 1975, IEEE Trans., AES-11, 660-663.

14. Ioannidis, G.A. and Hammers, D.E., 1979, IEEE Trans., AP-27, 357-363.

15. McCormick, G.C. and Hendry, A., 1975, Radio Science, 10, 421-434.

16. Gherardelli, M. and Giuli, D., 1981, "Numerical Simulation of an Adaptive Polarization Receiver", (in Italian), CNR Special Project, FUB ATC Report, No. III/A/3.3.

17. Marshall, J.S. and Palmer, W.M., 1948, J.Meteorology, 5, 165.

18. Oguchi, T., 1977, Radio Science, 12, 41-51.

19. Fang, D.J. and Lee, F.J., 1978, Comsat Technical Rev., 8.

20. McCormick, G.C. and Hendry, A., 1976, Radio Science, 11, 731-740.

21. Hendry, A. and McCormick, G.C., 1974, Journal Res. Atmos., 189-200.

22. Biffi Gentili, G., Fossi, M., Gherardelli, M., Giuli, D. and Giaccari, E., 1982, "A System for Dual-Polarization Radar Measurement and Analysis of Precipitation Echoes", Conference Proceedings "Multiple-Parameter-Radar Measurements of Precipitation", Bournemouth, England.

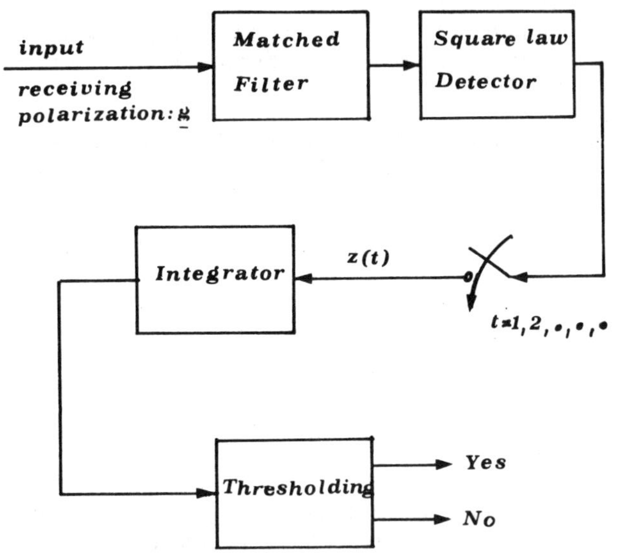

Figure 1 Incoherent detection scheme with single controllable polarization at receipt

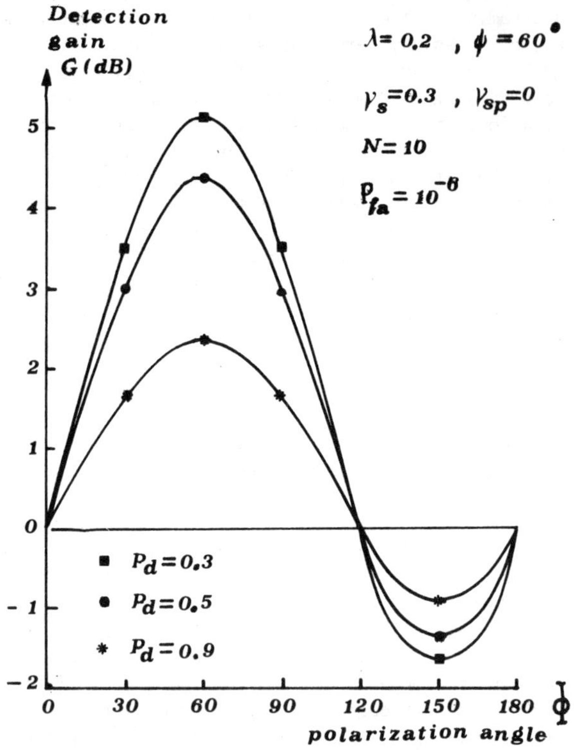

Figure 2 Detection gain as a function of linear polarization angle for a fluctuating target obtained by controlling the polarization at both receipt and transmit (PR=PT)

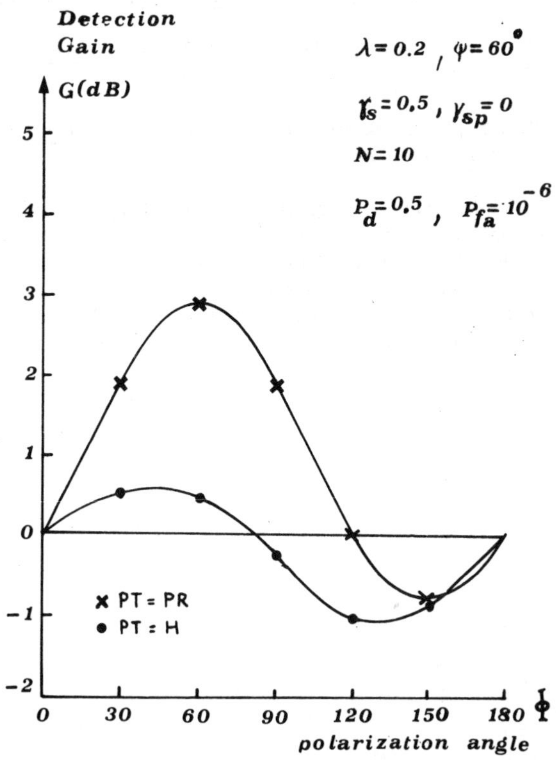

Figure 3 Detection gain as a function of linear polarization angle for a fluctuating target obtained bycontrolling the polarization only at receipt (PT=H) or at both receipt and transmit (PR=PT)

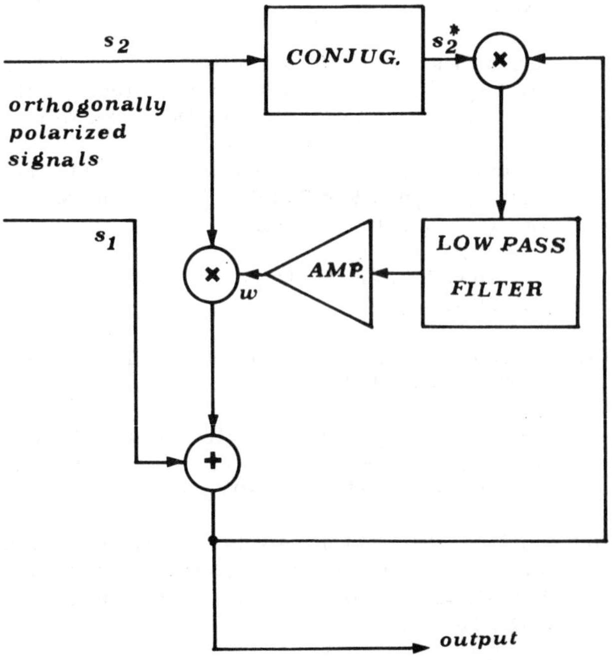

Figure 4 Analytical model of the adaptive polarization canceller

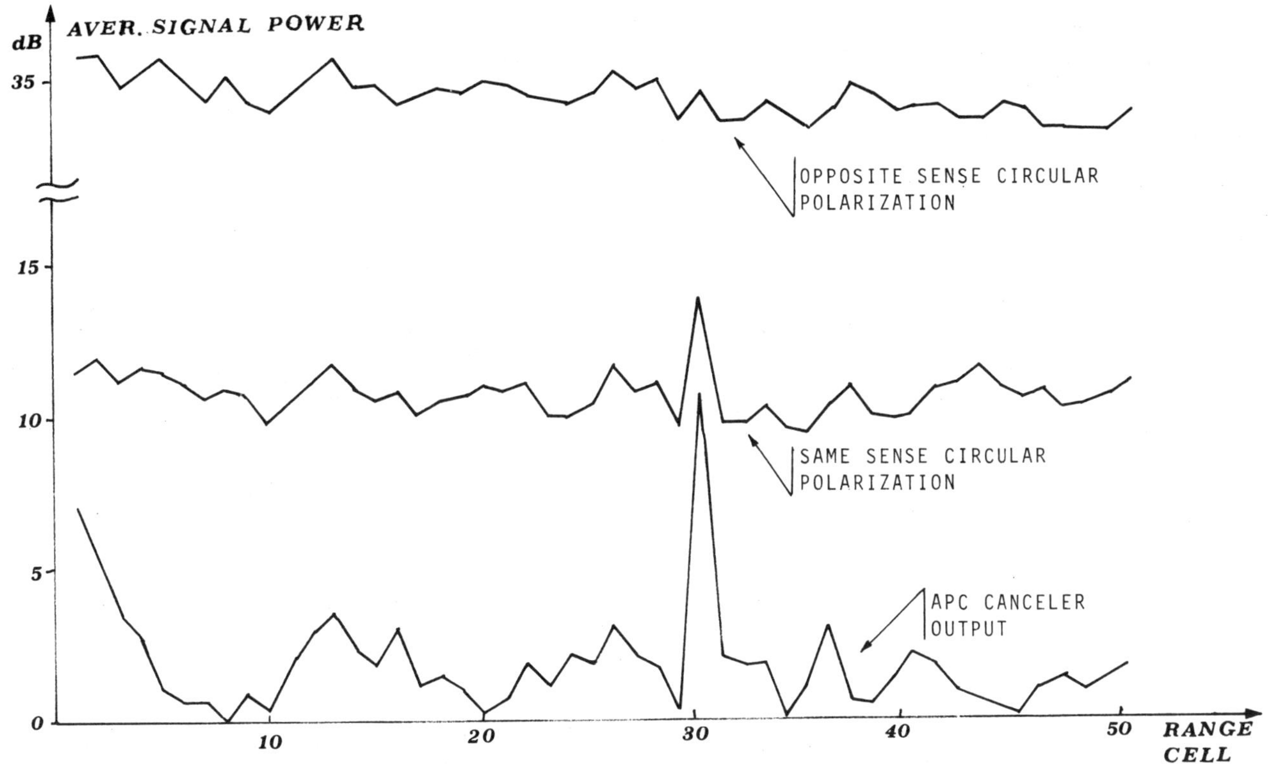

Figure 5 Average signal powers of the APC inputs and output as a function of range (light-rain clutter)

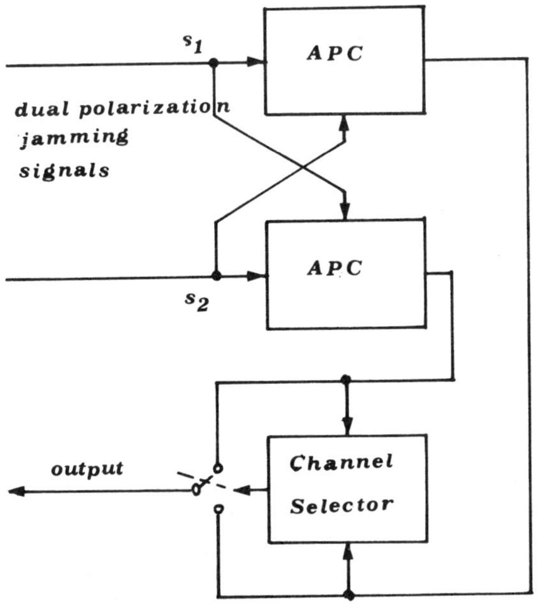

Figure 6 Scheme of the symmetric adaptive polarization cancellers

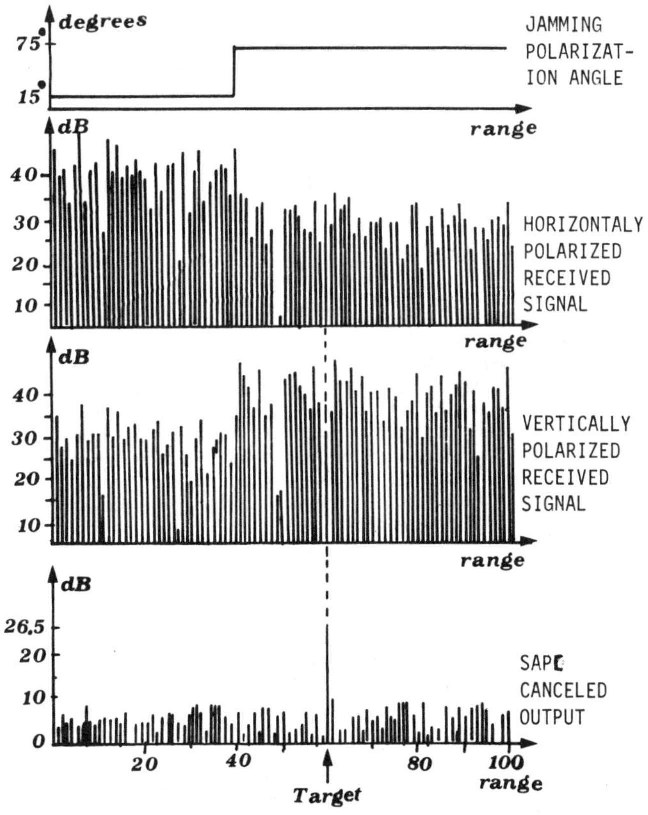

Figure 7 Performance of SAPC in the presence of jamming with abrupt change of linear polarization angle

IMPACT OF EXTREMELY HIGH SPEED LOGIC TECHNOLOGY ON RADAR PERFORMANCE

E. K. Reedy and R. B. Efurd

Georgia Institute of Technology, USA

M. N. Yoder

U.S. Navy, Office of Naval Research, USA

INTRODUCTION

Application of high speed and high density digital electronics to radar system designs has received much attention from researchers throughout the evolution of digital electronics. The chief limitation of the digital approach to radar system design has been and remains the extreme difference between the throughput rates of the digital devices and the required throughput rate for broadband, multiple-range-gated radar signals. Previous examples of digitial processors for radar signals offer less than optimal performance both in their ability to digitize the input data or adequately process the input data that could be digitized. Although this fundamental problem remains, the current rapid change in the price and performance of digitial electronics will substantially alter the boundary between practical and impracticial designs.

From June to December 1981, the Radar and Instrumentation Laboratory of the Georgia Institute of Technology Engineering Experiment Station, working under Office of Naval Research contract number N00014-81-K-0429, performed a research program to investigate radar system performance improvements based on exploiting very large scale integrated circuit and the emerging very high speed integrated circuit (VLSIC/VHSIC) technologies.

The study objectives were to identify functional areas of radar design (old and new) that merit consideration for adaptation of extremely high speed digital technology, synthesize system concepts that exploit this technology to provide improve performance, analyze expected performance limitations of these concepts, rank the system concepts according to potential performance improvement and operational need, and develop an experimental plan for verifying the feasibility of selected concepts.

RESEARCH BACKGROUND

The rapid advances in electronic technology have significantly improved weapon system performance in friendly environments, but the proliferation of such systems and the increasing emphasis on electronic warfare may well render such systems useless on any realistic electronic battlefield. The adversary may exploit emissions from radar and communication systems to determine the composition, location, and intent of friendly forces; then employ electronic countermeasures to degrade radar performance and disrupt communications to either cover his retreat or increase the effectiveness of his attack. The probability of war on an electronic battlefield implies a need for improvements in radar system performance accompanied by a reduction of radar system vulnerability to enemy interception and jamming.

Concepts for simultaneously improving performance and reducing the vulnerability of radar systems have been formulated and a few such systems have been developed. These concepts are generally based on the use of advanced antenna technology and/or wide bandwidth systems that use waveform modulation techniques to achieve pulse compression or low probability of intercept (LPI) for increased performance and reduced vulnerability. The very high speed signal processing technology necessary to realize the full potential of these concepts on a large scale is still being developed. Emerging large scale and high speed integrated circuit technologies offer the potential for improving the performance of existing radar systems and provide the stimulus for development of new radar system concepts. The U.S. Department of Defense (DOD) initiated a program in 1979/1980 to produce advanced digital technology and hardware components tailored for the rigorous demands of the military environment. The acronym VHSIC (Very High Speed Integrated Circuits was applied to this program. The research effort reviewed in this paper was not directly related to the VHSIC program, but the impetus and motivation for this effort was derived from the VHSIC and other related technology programs.

Office of Naval Research personnel recognized the need to assess the impact of high speed and large capacity digital circuits on Navy effectiveness. As a subset of this effectiveness assessment, the Navy contracted with the Engineering Experiment Station of the Georgia Institute of Technology to investigate the effect of high speed and high capacity digital circuits on the design of radars and other high technology sensors.

The Navy's effectiveness is a function of a complex mix of system performance, availability, tactics, strategy, adversary capability, initial cost, total cost, etc. the exact effect of radar design on this integrated effectiveness is impossible to assess. In this circumstance, an inexact design strategy becomes part of the experience and bias of radar designers, and individual radar designs are judged against this experience and bias. Unfortunately, the radar design is typically limited by the designer's background in a benign laboratory or manufacturing environment. The resulting design does not normally adequately include the harshness of the real environment, the level of training of the operators/repairers, or the Navy management's perceptions of the ingredients that lead to an effective system.

Besides these general problems, the electronic industry readily generates its own specialized problems for military systems. The commercial life of electronic parts is historically less than ten years, but the life of military systems can be thirty years or more. In some cases, the design period for a complex weapon system can exceed the commercial life of an electronic part which (worst case) leads to a design for which there is no parts supply. The commercial electronics industry is driven by intense competition in both price and performance. In the near future the rate of innovation is not likely to slow down.

If the above outline is taken as a reasonable statement of some of the major elements of the radar designer's problems, then what does the designer community do to cope with these problems? More specifically, how does the designer community cope with the forced evolution in digital circuits sponsored by the services under the title VHSIC? VHSIC represents a step function in capability but the designer community's general response will exhibit time lags as each element of community becomes aware of the technology. Moreover, the designer community will be biased by training and experience to use the VHSIC capability to increase the technical capabilities of radar with the implicit belief that this will necessarily enhance the effectiveness of the Navy weapon systems in which these radars reside.

The objective of the effort reported in this paper was to investigate innovative uses of extremely high speed integrated circuits (primarily digital) in radar. The objective was unrestricted in the innovation sense while tightly coupled to radar design. This effort was directed not at the creation of new circuits, devices or technologies, but rather to the use of high speed digital electronics to improve the form, fit, and function of radars. Because form and fit are specifics of particular radars and their transport, function was the dominant theme of this study. Function generally contains elements of technical performance, convenience, cost, etc. This research concentrated on the search for those areas where high speed/density circuits make or could make significant changes in any of these subcategories of radar function.

TECHNOLOGY SURVEY

The probable technologies for high speed electronics for the 1980s and 1990s are: (1) Silicon, (2) Gallium Arsenide, and (3) Josephson Junctions. The silicon technology is of immediate interest to DOD, and the VHSIC program bears witness to this interest. The Gallium Arsenide (GaAs) technology is of interest because it has a four or five to one speed advantage over silicon and GaAs can be used as a substrate for building both electronic and optical devices. The GaAs speed advantage leads to Microwave Integrated Circuit (MIC) devices while its joint electronic/optics capability leads to integrated electro-optics devices. The Josephson Junction (JJ) devices exhibit gate speed on the order of 10 picoseconds, but they require cooling to a temperature within a few degrees of absolute zero.

Based on these well known facts, the probable evolution of high speed electronics is:

(1) For the period of 1981 to 1990, silicon IC will be the overwhelmingly dominant technology for all low speed (i.e., less than 1 GHz) integrated circuits.

(2) In the 1981 to 1990 time period, GaAs will capture the low power devices in the middle microwave frequencies (i.e., 1 to 15 GHz). GaAs will receive a large boost because of its joint use in MICs and electro-optics, particularly for communications purposes.

(3) In the 1990s, JJs will make their initial market penetration. This penetration will probably be in the super computer arena and then only for fixed location computers.

The speed-density goals in the VHSIC program are well publicized and are: 5×10^{11} Gate-Hertz per square centimeter of silicon for VHSIC Phase 1 and 1×10^{13} Gate-Hertz per square centimeter of silicon for VHSIC Phase 2. The gate-hertz per square centimeter measure of performance used by the VHSIC program is sometimes called the functional throughput rate (FTR). VHSIC success will have two prime effects: (1) many functions or function extensions which were prohibited by speed will be available, and (2) many applications which were prohibited by volume or power will be practical. In addition, there may be a decrease in the dollar-to-function ratio which will increase the use of this technologogy in applications which are not primarily driven by technical requirements.

In the commercial market devices that represent the state-of-the-art in high speed digital electronics are the digital correlator (100 megabits/second) and the analog-to-digital converter (8 bits at 75 MHz) manufactured by TRW, Inc. These devices enable the collection and processing of digital data from a limited number of range gates at the rates corresponding to video or IF frequencies of most of deployed radars. The signals from those radars with small range resolution cells (i.e., broad bandwidth waveforms) could not be processed by these devices.

The limits on both the data taking circuits and the data processing circuits have been or will be improved; but, for some combination of peak data rate or total data to be processed per unit time either the single device speed or the general processor data throughput rate will become the limiting factor. By the mid to late 1980's time frame, it should be reasonable to consider radar designs with peak data rates per range resolution cell in the interval of 0.1 to 1.0 gigahertz for a number of range cells between 10 and 1000.

SPECIFIC APPLICATIONS, DIGITAL COHERENT-ON-RECEIVE

Potential applications of high-speed/high-density electronics explored were: (1) adapative correlation processing (digital coherent-on-receive and matched filter performance with a non-stabilized transmitter), (2) intrapulse polarization agile radar (IPAR), (3) multistatic systems, and (4) small, lightweight remotely piloted vehicle (RPV) sensors. Only the digital coherent on receive (DCOR) technique is described in this paper.

A coherent radar requires that the phase of the radar carrier frequency be "known" for each transmitted pulse and used during listening time for signal processing. Knowledge of the carrier phase can be obtained by: (1) maintaining control of the transmitted phase relationship between the RF carrier and a reference RF signal, or (2) measuring the phase relationships between an uncontrolled RF carrier and a reference RF signal. The first approach is called coherent, and there is a fixed relationship between the carrier phase at time t_o and any later time $t_o + t$. The second approach is called coherent-on-receive, and there is a measured relationship between the phase of the transmitted RF carrier and a reference RF signal, as well as the received RF carrier and the same reference RF signal.

Due to the lack of phase stability from one pulse interval to the next, a fully coherent radar is difficult if not impossible to develop with magnetron or similar power oscillator tubes. Otherwise, power oscillator tubes are desirable for transmitters because: (1) they cost less than power amplifier tubes for a given power output, (2) they generally have longer field life, and (3) they require simpler and more reliable modulator systems. It is, therefore, very desirable to achieve the advantages of a coherent radar with a noncoherent power oscillator tube (magnetron). Incorporation of coherent-on-receive capability in existing operational non-coherent systems allows improved performance, i.e., detection of various targets, without the need to replace the expensive transmitter section.

CLASSICAL COHERENT-ON-RECEIVE

Figure 1 is the block diagram of the classical implementation of a coherent-on-receive system. For the cases in which an existing non-coherent system is modified for coherence, the transmitter, duplexer, antenna, and receiver mixer can usually remain intact. These are usually costly items and are often the most expensive parts of a system.

A stable local oscillator (STALO) is added to the system since the typical local oscillator (LO) is usually not designed for coherent operation. During a transmit pulse, a sample of the transmitted signal is mixed with a sample of the STALO to provide an IF signal which is in phase with the transmitted signal on that particular pulse. A coherent oscillator (COHO) is turned on at the same time the IF reference sample is injected into it. This causes the COHO to begin oscillating at the IF frequency, in phase with the reference. The COHO is allowed to oscillate at the IF frequency for most of the interpulse period to provide a reference for the received signals. The COHO is turned off just prior to the next transmit pulse so that it can begin again at a new phase that is determined by the injected signal.

The received signal is first mixed with the STALO. The resulting IF signal is mixed with both in phase (I) and quadrature (Q) components of the COHO signal to provide I and Q representations of the received signal for processing.

Figure 2 shows the block diagram of a typical coherent double delay line canceller for a moving target indication (MTI) system. A double delay line canceller could be employed to incorporate an MTI capability into a coherent-on-receive modification of a power oscillator radar.

One serious limitation in MTI performance associated with the classicial approach described here is related to the fact that the COHO usually drifts in phase (frequency) during the interpulse period. If the drift were consistent from one pulse to the next, there would be no degrading effect in performance; however, some pulse-to-pulse variation is unavoidable. It is estimated that approximately a 20 dB limit on improvement factor would result at a range of 30 km; performance would degrade even further at longer ranges. MTI performance is a function of range.

PHASE ERROR MEMORY COHERENT-ON-RECEIVE COHO CORRECTION

An improvement over the injection COHO approach involves the use of a stable CW COHO and a system which measures the phase between the CW COHO and the sampled IF and makes the appropriate correction. Figure 3 is the block diagram of such a system. The two sample-and-hold (S/H) and analog-to-digital (A/D) converter units sample the phase (I and Q) between the COHO and the transmit reference. The phase logic determines the unambiguous phase difference and provides a digital number to the digital-to-analog (D/A) converter to indicate the phase correction to be made to the COHO signal to place it in phase with the transmitted sample. This phase correction is by means of an analog phase shifter in the COHO line. With this system, the problems associated with the drifting injection COHO are avoided. The limiting item is the stability of the D/A converter and analog phase shifter.

PHASE ERROR MEMORY COHERENT-ON-RECEIVE VIDEO CORRECTION

A further improvement in the phase-error-memory approach can be made with a system such as that shown in Figure 4. In this system, the phase between the COHO and the transmit reference is measured in the same way as with the previous system. The phase correction is performed digitally on a cell-by-cell basis in the digital phase corrector unit. This eliminates the instability of the analog phase shifter. In addition to a phase correction, this system allows correction of pulse-to-pulse amplitude scintillation, which is one of the limiting factors in a non-coherent sytem.

The digital phase corrector must convert the received I and Q signals from rectangular to polar coordinates, apply the appropriate phase and amplitude correction, and convert the new (corrected) vector to rectangular coordinates. Figure 5(a) depicts the vector operations involved. The mathematical operations are given by:

$$I' = A' \cos(\phi + \Delta\phi)$$

$$Q' = A' \sin(\phi + \Delta\phi)$$

where

$A = (I^2 + Q^2)^{1/2}$ = original amplitude

$A' = A + \Delta A$ = corrected amplitude

$\phi = \tan^{-1} Q/I$ = original phase

$\Delta\phi$ = phase correction.

Figure 5(b) is the block diagram of the digital phase corrector. The correction must be applied to the received video on a range cell by range cell basis. For a 0.5 microsecond magnetron pulse width, a range cell is approximately 250 ns in length. The circuitry which provides the speeds required for this process consists of random access memory and adders. The trigonometric functions, squaring, square roots, and multiplications would be implemented using "look-up tables" in a random access memory (RAM). Since the RAM would lose its information any time the power is turned off, the RAM would be loaded from programmable read-only-memory (PROM) when system power is turned on.

PROCESSING AT IF

A concept in which the analog IF signals are sampled and converted to digital form for processing may soon be practical. Figure 6 shows the block diagram of such a concept. The IF of the many radars is centered at 30 MHz and A/D conversion rates of 75 MHz are available, allowing for sampling at a rate that is at least consistent with the Nyquist sampling criteria.

The task of the phase processors is to compute the difference in phase between the two sampled signals supplied as inputs. At least two techniques are available for determining this phase. One operates in the time domain and the other operates in the spectral domain. The time domain process involves determining a "best fit" expression associated with each of the sampled IF waveforms and performing a cross-correlation to determine the amount of phase shift required to achieve a correlation peak. The spectral domain process involves performing a complex discrete Fourier transform (DFT) and analyzing the results to determine the relative phase. If the response resides in only one output bin, then the signals are in phase. If the response is shared between the two bins, then the relative phase is related to the ratio of signal levels in the two bins.

Once the phase and amplitude errors have been determined, the received signal can be processed in a way similar to that of the video correction system. For each range cell, the phase and amplitude of the signal are corrected (digitally) before the signal can be applied to the signal processor (MTI filter). The correction is applied to either the digitized receiver IF signal or to the digitized COHO signal, depending on which of the two techniques would provide the best performance.

COMPARATIVE PERFORMANCE TRADEOFF

Table I lists the major parameters that limit MTI improvement factor for coherent-on-receive systems such as those described. A summary of the performance limits would indicate that an improvement factor limit of approximately 20 dB would be achieved at 30 km range for the classical implementation of a system. The primary contributors to this limit are the COHO drift and the sea clutter spectrum. An improvement factor of 30 to 35 dB should be achievable using the digital phase error memory approach; the actual performance depends specifically on the radar's magnetron characteristics and the sea clutter spectrum.

System detection performance was estimated by assuming the following sea clutter and radar system characteristics and the sea clutter spectrum:

Average reflectivity (σ_o)	= -30 dB
RMS internal clutter motion	= 1 m/s
Antenna beamwidth (azimuth)	= 1.5°
Pulse length	= 0.5 s
Subclutter visibility	= 30 dB (phase error correction)
	= 20 dB (injection COHO)
Carrier frequency	= 5.825 x 10^9 Hz
Pulse repetition frequency	= 650 Hz

For the injection COHO system, a 1 m^2 moving target could be detected at about 5 km. With the digital phase error correction system, the same target could be detected at 50 km.

SUMMARY

This paper is a condensed review of a radar system design concept explored in the full technical report[1]. This full report explores several design concepts (i.e., radar systems, remotely piloted vehicles, multistatic sensors, etc.) All explorations addressed the impact of high-speed and high-density digital electronics on the design concepts/strategies which should be employed by the designer. A central finding was that precision and accuracy requirements currently imposed on the RF, IF, and video circuits of radar systems could be partially transferred to the digital circuits and that these transfers are expected to produce both technical performance and cost advantages. The DCOR explicitly discussed in this paper is an example of the process of transferring the precision and accuracy requirements to the digital circuits and the technical performance advantages realized by this transference. The price and performance of near-future digital electronics will make digital video and digital IF techniques the techniques of choice.

TABLE I
COHERENT-ON-RECEIVE
MTI IMPROVEMENT
FACTOR LIMITS

QUANITY	AMOUNT	LIMITS OF I FACTOR
Freq. Var.	10 KHz	42 dB
Amp. Var.	1%	40 dB
Phase Noise	1%	40 dB
Amp. Quant.	8 bits	46.9 dB
Sea Clutter Spectrum	1.0 m/s	20 dB
Range Dep. Phase Var.	10 Km	40 dB
Phase Quant.	9.25 m rad	40.7 dB

REFERENCES

1. Reedy, E. K., Efurd, R. B., Scheer, J. A., Sjoberg, E. S., Cohen, M. N., and Hayes, R. D., 1982, "Impact of Extremely High Speed Logic Technology on Radar System Performance," Final Technical Report on Contract No. N00014-81-K-0429, Georgia Institute of Technology, Engineering Experiment Station.

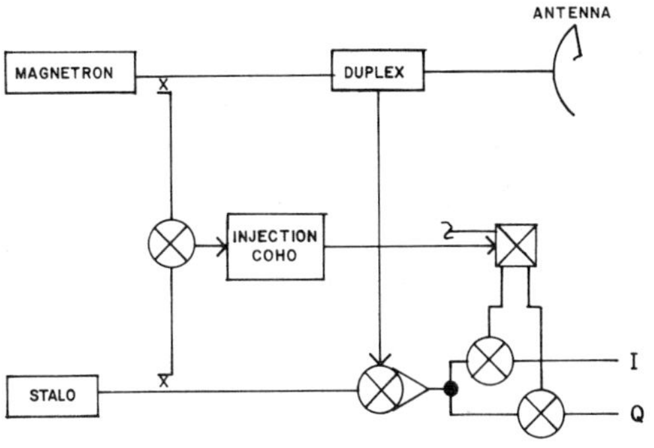

FIGURE 1. CLASSICAL COHERENT ON RECEIVE

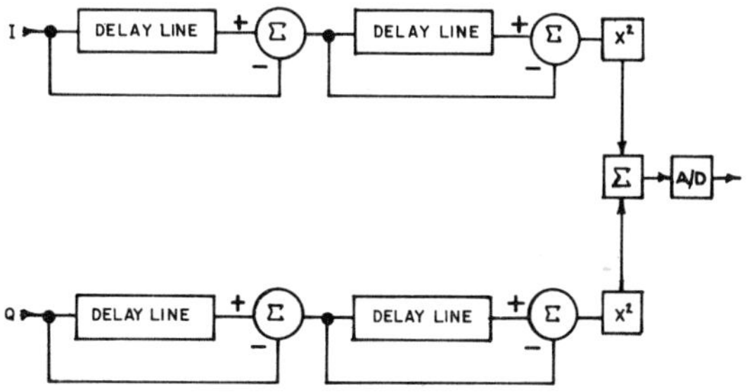

FIGURE 2. DOUBLE DELAY VECTOR CANCELLER

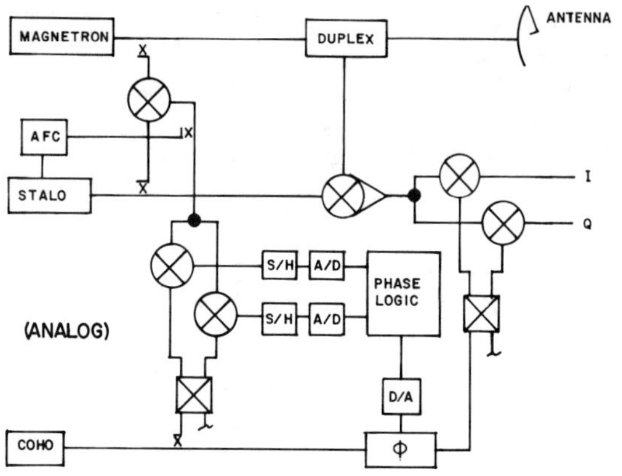

FIGURE 3. PHASE ERROR MEMORY COHERENT-ON-RECEIVE

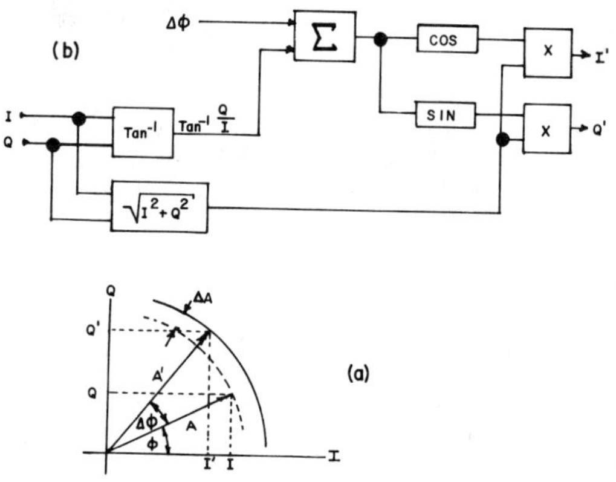

FIGURE 5. DIGITAL PHASE CORRECTION

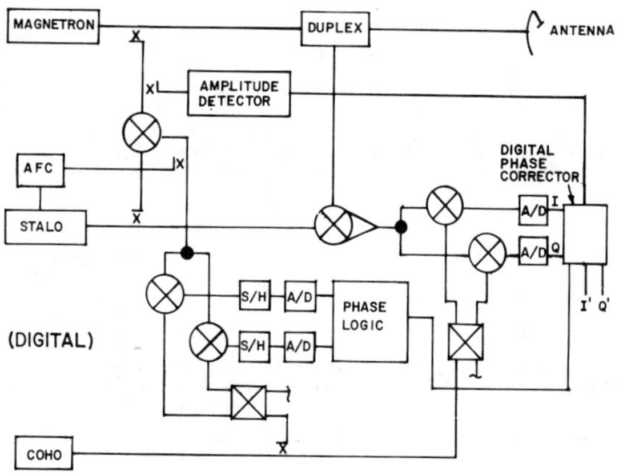

FIGURE 4. PHASE ERROR MEMORY COHERENT-ON-RECEIVE

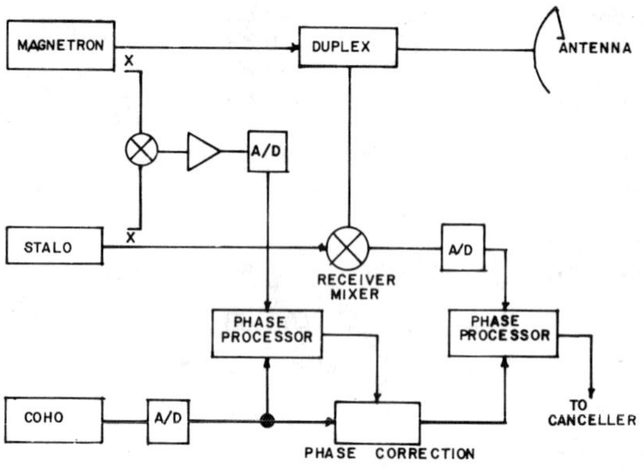

FIGURE 6. DIGITAL COHERENT-ON-RECEIVE (IF)

SUPERRESOLUTION USING AN ACTIVE ANTENNA ARRAY

U. Nickel

Forschungsinstitut für Funk und Mathematik (FFM) of FGAN, F.R. Germany

INTRODUCTION

The limited resolution of radar antennas is the reason for several errors. Besides tracking errors for formations of targets or jammers, we have multipath and glint errors, which are caused by insufficient resolution. An active antenna array offers the potential for angular superresolution if the sequence of spatial samples of the received waves is available. From spectral analysis and other fields superresolution methods like the maximum-entropy method, Capon's maximum-likelihood method, or spectra generated by eigenvectors are well known. These methods can also be applied for angular resolution with some modifications. Essentially these methods generate a peaky estimate of the angular spectrum. The resolution task then is still left to the user who has to interpret the spectrum. A decision rule for the detection and resolution of peaks with these superresolution methods is to date not known. In addition these methods are only applicable for stationary stochastic signals. The important case of superposed pure sinusoids is not in this class. For automatic signal processing fixed algorithms with well defined properties are necessary. Such algorithms can be found by parameterising the signal with the desired parameters and by fitting this signal model directly to the measured data. The more we specialise the signal model, the better the achievable resolution will be. But the resolution may perhaps be much worse, if the data do not belong to the chosen model. This dilemma is common to all superresolution methods. The signal model should just comprise the essential features to give a powerful superresolution method. For radar with narrow-band receivers a point-target model seems appropriate. This model leads to angular spectral line fitting to the data.

RESOLUTION AS A DECISION PROBLEM

Model fitting is a decision problem. The formulation of radar resolution as a decision problem has been attacked by several authors e.g. Root (1), Ksienski and McGhee (2), Birgenheier (3), and others. The output of an antenna array is a time sequence of vectors. It is essential to formulate the decision problem for a given set not only of spatial but also temporal samples. The narrow-band point target signal model is still open for several fluctuations of the signal amplitude and phase. No assumptions should be made on these fluctuations because they may differ from one application to another. Therefore we consider the amplitudes and phases as a deterministic, but unknown sequence. Assuming measured data of the form signal plus pure receiver noise, we can formulate hypotheses (i.e. families of distributions of the data) and we have to decide from which class the data were taken from.

Multi-hypothesis test. It can be shown that a solution of this multi-hypothesis problem can be given by a sequence of likelihood-ratio tests of the following form: We test the hypothesis "the number of targets is $\leq$ M" against the alternative "the number of targets is $>$ M" starting with M=0 and increasing the number of targets M. Once we have accepted the hypothesis "the number of targets is $\leq$ M", we stop testing. The decision for the target number is M. In this way the multi-hypothesis test is sequential with respect to the number M. The overall procedure is shown in figure 1. A precise formulation of the hypotheses, the test problem, and the solution can be given and will be published elsewhere. The likelihood ratio test has the advantage of giving a fixed asymptotic error level, in contrast to the general Bayes approach. For further data processing, like tracking algorithms, this is important. If the test procedure is terminated at some stage M, before the decision has been made, one has at least the information that the number of targets is greater than M. This information may sometimes be useful. To compute the likelihood ratio, at each stage M a maximum likelihood parameter estimation has to be carried out, which essentially estimates the directions of the targets for the assumed number M. There are several applications where a test procedure is not needed, e.g.for multipath error reduction a two target model may be sufficient. The procedure is then simpler.

ESTIMATION OF THE DIRECTION

Signal model. Suppose we have given M point sources in the farfield of the antenna. For an array with elements at the positions x_i, y_i, i=1,...N, the complex sample ouput at the i-th element then can be written

$$z_i = \sum_{k=1}^{M} \beta_k e^{j\phi_k} e^{-j2\pi/\lambda(x_i u_k + y_i v_k)} + n_i$$

where u_k, v_k denote the direction cosines for azimut and elevation. λ is the wavelength and n the receiver noise sample ($j^2=-1$). In vector notation we can write this equation as

$\underline{z} = \underline{A}\,\underline{b} + \underline{n}$, where $b_k = \beta_k e^{j\phi_k}$, and the

transmission matrix $\underline{A}$ has the elements

$$e^{-j2\pi/\lambda(x_i u_k + y_i v_k)} \qquad (i=1,..N,\ k=1,..M)$$

(vectors and matrices are underlined).
Signal models other than $\underline{A}\underline{b}$ are also possible, the matrix $\underline{A}$ can have another form.

Maximum likelihood estimation

For white gaussian uncorrelated receiver noise maximum likelihood estimation leads to the minimisation of the mean squared error between the measured data and the signal model. For K data vectors $\underline{z}$ we have to minimise

$$\sum_{k=1}^{K} \| \underline{z}_k - \underline{A}\,\underline{b} \|^2 \qquad (1)$$

($\|..\|$ denotes the square norm for complex vectors.) The minimisation with respect to $\underline{b}$ has to be done for each $\underline{z}_k$, because the amplitudes and phases are assumed to be an unknown, deterministic sequence. This is a linear least squares problem and the solution can be written down at once: $\hat{\underline{b}}_k = (\underline{A}^* \underline{A})^{-1} \underline{A}^* \underline{z}_k$. ($..^*$ denotes complex-conjugate transpose). We are thus left with the minimisation of the function

$$Q(\underline{u},\underline{v}) = \sum_{k=1}^{K} \| \underline{z}_k - \underline{A} (\underline{A}^* \underline{A})^{-1} \underline{A}^* \underline{z}_k \|^2 \quad (2)$$

Equivalently we could maximise

$$S(\underline{u},\underline{v}) = \sum_{k=1}^{K} \underline{z}_k^* \underline{A} (\underline{A}^* \underline{A})^{-1} \underline{A}^* \underline{z}_k \quad (3)$$

Function (3) shows that this estimation is a generalisation of conventional beamforming which is the case M=1. We have to maximise the square of a vector of M simultaneous decoupled sum-beams. For the derivation of these functions we have assumed omnidirectional patterns of the antenna elements. If the single elements have a directional, but equal pattern, we would have to replace the matrix $\underline{A}$ by a matrix $\underline{A}\,\underline{C}$, with $\underline{C}$ a diagonal (MxM)-matrix, depending on the directions. The form of the functions (3), (4) shows that these functions are invariant under such left-side transformations $\underline{C}$ of the matrix $\underline{A}$ (as long as $\underline{C}$ is regular). Equal antenna element patterns therefore always lead to the same minimisation problem. As a consequence mutual coupling effects do not influence the estimation procedure, if the coupling effects are the same for all elements (e.g. for large arrays on a regular grid). One single snapshot $\underline{z}$ is in principle sufficient for this kind of resolution, if the signal-to-noise ratio is high enough. No assumptions have been made on the $\underline{b}_k$, they may be deterministic or stochastic. Direction finding by maximising (minimising) these functions may be considered as the optimum procedure, but it is very time consuming and for most radar applications of less value. The main problem of this kind of resolution is the maximisation (minimisation) of the function (2) or (3).

Suboptimum estimation

From equation(2),(3) we can derive a simple, suboptimum estimation algorithm. We can minimise Q for only one observation (K=1) with a gradient algorithm, but we use a new observation $\underline{z}$ for each iteration step. Thus the iteration proceeds in the direction of steepest descent, but the underlying function is time varying. This leads to a stochastic approximation algorithm

$$\underline{w}_{k+1} = \underline{w}_k - a_k\, \underline{G}(\underline{z}_k, \underline{w}_k) \; ; \; k=1,2,3... \quad (4)$$

where $\underline{w}_k = (\underline{u}_k, \underline{v}_k)$ and

$\underline{G}(\underline{w}_k, \underline{z}_k) = \text{grad}\, \underline{Q}\, (\underline{w}_k, \underline{z}_k)$

and a_k is a sequence of real numbers with

$\Sigma\, a_k = \infty$, $\Sigma\, a_k^2 < \infty$ to achieve

convergence with probability 1. A sequence $a_k = \text{const}/k$ has the desired properties. The double null tracker of White (4) is a special case of this algorithm. Conditions for convergence with probability 1 can be found in the literature. The convergence point is that set of directions $\tilde{\underline{w}}$ with $E\{\underline{G}(\tilde{\underline{w}},\underline{z})\} = \underline{0}$. If $\underline{G} = \text{grad}\, \underline{Q}$, convergence properties can therefore be studied by dicussing the function E{Q} for various fluctuations of the signal amplitude and phase. To ensure convergence, the iteration has to be bounded to a convergence region, which can be chosen to be the disk of the antenna beamwidth. This region is found by scanning the conventional sum-beam (the stage M=1 in the procedure of fig.1). The iteration (4) has the advantage that only one observation vector $\underline{z}$ has to be stored (in contrast to equ.(2) or (3)). In effect the gradient depends only on M simultaneous sum- and difference-beams in the directions $\underline{w}$. Figure 2 shows the computations for the gradient $\underline{G}$ for the case M=2 and a linear antenna. Without beamforming (which should be done by special analog or digital hardware) 8 complex multiplications and 4 additions are necessary for one iteration step for this example. For M=1 the algorithm simply tries to null the difference beam.

Experimental results

Computer Simulations. For the simulations shown here we used an antenna consisting of 192 elements and a diameter of 37λ . The location of the elements with a slight parabolic density tapering is shown in figure 3. Figure 4 shows the estimation of the azimut direction cosines u with a version of the stochastic approximation algorithm for two given targets. The elevation estimation looks quite similar. The 2 targets are located at $\tilde{\underline{u}}$ = (-0.35 BW/2, 0.35 BW/2), $\tilde{\underline{v}}$ = (-0.35 BW/2, 0.35 BW/2).The starting point of the iteration is always $\underline{u}$ =(0,0), $\underline{v}$ = (-0.9 BW/2, 0.9 BW/2). Figure 4(a) shows the estimation for uncorrelated targets with Rayleigh-fluctuating amplitudes, figure 4(b) with fixed amplitudes and a constant phase difference of 0 degree. The antenna output signal-to-noise ratio is 19.8dB. Large numbers of element outputs are often pre-processed by forming subgroups. We may consider these subgroups as new antenna elements with certain antenna patterns. The estimation procedure can then be applied for this element configuration. If the subgroups are chosen to be all equal, these new elements have equal patterns. The estimation therefore remains the same as mentioned above. We have only to use the center of gravity of the subgroups as element positions. Unequal subgroups affect the estimation. The transmission matrix $\underline{A}$ then needs a correction. Figure 4(c) shows the estimation using only subgroup outputs without any correction. The 192-element antenna was divided into 24 subgroups, each consisting of 8 elements as indicated in figure 3. This result shows that empirically the estimation by stochastic approximation is robust against small errors of the transmission matrix $\underline{A}$.

Measured 2 target signals. Further tests were made with an experimental setup. To measure 2-target configurations, a 6-element antenna at S-band with 2λ diameter was taken. 5 elements were located at the corners of a pentagon and one element at the center. The 12 element outputs (I and Q channels) were converted analog to digital with 8 bits and then processed by a desk computer. Targets were simulated by two transmit elements of 1m separation. These were located in front of the antenna at a distance of ca. 4m. Figure 5 shows the configuration. The 2 targets were simulated by 2 doppler-shifted pure sinusoids. Figure 6 shows the estimation. The above part shows the conventional azimut sum-beam patterns if only one of the 2 transmit elements is active (gross lines) and 3 patterns by superposition of the two sources for some random phase differences. Below is shown the azimut and elevation estimation of the directions by stochastic approximation with 30 iterations. The circle indicates the size of the 3 dB beamwidth. The estimated directions differ from the maxima of the single-target sum-beam patterns by 0.02 of the 3dB-beamwidth. The signal-to-noise ratio in this case was very high. The main errors were due to misadjustment, channel quantisation, and multipath effects.

TEST FOR THE NUMBER OF TARGETS

With the directions estimated by the sub-optimum stochastic approximation, we are no longer able to perform a likelihood-ratio test, which would give a fixed asymptotic error level. Nevertheless we can find a test statistic which uses the suboptimum estimation of the directions, for testing the hypotheses under consideration H : " $M \leq \hat{M}$ " against K :"$M > \hat{M}$". This statistic is the value of (2) $Q(\hat{\underline{u}}, \hat{\underline{v}})$, where $(\hat{\underline{u}}, \hat{\underline{v}})$ are found by stochastic approximation. In this case K can be small, e.g. K=2,3,4. $Q(\hat{\underline{u}}, \hat{\underline{v}})$ measures the residual energy after signal extraction.

Test at approximate error level. If the estimation is sufficiently accurate, we can approximate the distribution of Q by a χ^2 - distribution with 2K(N-M) degrees of freedom, because then Q is only a sum of squares of 2K(N-M) noise samples. This is because

$$(\underline{I} - \underline{A}(\underline{A}^* \underline{A})^{-1} \underline{A}^*)\, \underline{s} = \underline{0} \qquad (\underline{s} = \underline{A}\underline{b})$$

We are thus able to construct a sequence of tests at an approximate level of probability of error of the first kind for each stage $\hat{M}$ (i.e. $P\{Q > \eta_{\hat{M};\alpha}\} \leq \alpha$ if H_M is given). The sequential form of the multi-hypothesis test of fig.1 then yields that for the overall type-1-error probability for a given target number M, we have also $P\{\hat{M} > M\} \leq \alpha$, because if H_M is valid, we have :

$$P\{\hat{M} > M\} = P\{Q > \eta_{1;\alpha}\}P\{Q > \eta_{2;\alpha}\}..P\{Q > \eta_{M;\alpha}\} \leq \alpha . \qquad (5)$$

We have thus constructed a multi-hypothesis test which has approximately a given level of error $P\{\hat{M} > M\}$. For target resolution the computation of the set of thresholds $\eta_{M;\alpha}$ is essential, because these thresholds cannot be adjusted experimentally when the radar is in operation , as is often done for single target detection. Multiple target situations are in general too rare. In the case of Rayleigh-fluctuating targets with uniformly distributed phase differences we can even compute an approximation of the probability of detecting the correct number, because under H_M, $P\{Q > \eta_{L;\alpha}\}$ for L<M can be computed. If we set $P\{Q < \eta_{M;\alpha}\} = 1 - \alpha$ in equation (5), we get a lower bound of this overall probability of detection. Computer simulations showed that for 2 targets the computed probability of detection gave a reasonable good approximation for the observed probability of detection . For targets with fixed amplitudes the simulations gave a higher probability of detection, so that the computed probability in this case may also be considered as a lower bound. Figure 7 shows this computed probability of detection as a function of the output signal-to noise ratio (SNR) for the 192-element antenna and 2 given targets. The different curves are for different target separations, varying from 1.2 to 0.4 degrees in steps of 0.2 degrees. The 3dB beamwidth of the 192-element antenna is approximately 1.6 degrees. Averaging of the residual error was done with only 2 samples (the function Q of equ. (2) taken with K=2). The level α is 0.04 and 0.08. One can see that in this case a SNR of 20 dB gives a sufficiently high probability of detection. By the $1/R^4$-law we can reach such values for the SNR at 0.4 times the distance with a SNR of 3dB. Further averaging increases the probability of detection.

CONCLUSIONS

We have shown that using an active antenna array a resolution enhancement over the conventional 3-dB beamwidth by signal processing is in principle possible under the assumption of a point target model. The resolution procedure can be formulated as a sequence of direction estimations and hypothesis tests. The estimation of the directions by the stochastic approximation is a rather simple procedure and compatible with the common array signal processing, because only independent steerable sum- and difference-beams are needed. The procedure is flexible for modifications. A multihypothesis test of sequential form (with respect to M) for the number of targets can be constructed, which has an approximate level of error for overestimating the number of targets. Computations and simulations showed that to resolve 2 targets separated by 0.5 beamwidth a reasonable signal-to-noise ratio of ca. 20dB is sufficient.

REFERENCES

1. Root, W.L., 1962, IRE Trans. Military Electr.,April,197-204.

2. Ksienski, A.A. and McGhee, R.B., 1968, IEEE Trans. AES-4,No.3,443-455.

3. Birgenheier, R.A.,1972, Ph.D. dissertation University of California, Los Angeles.

4. White, W.D., 1974,IEEE Trans. AES-10,No.6, 835-852.

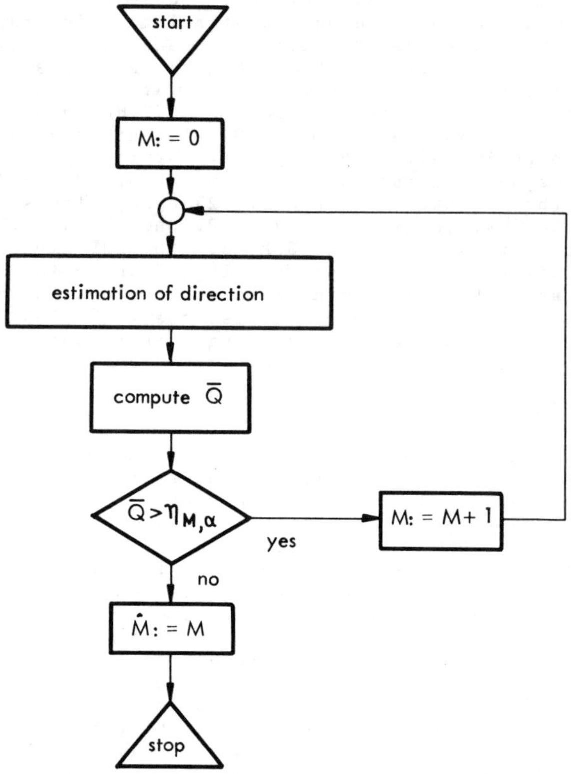

Figure 1. Resolution Algorithm

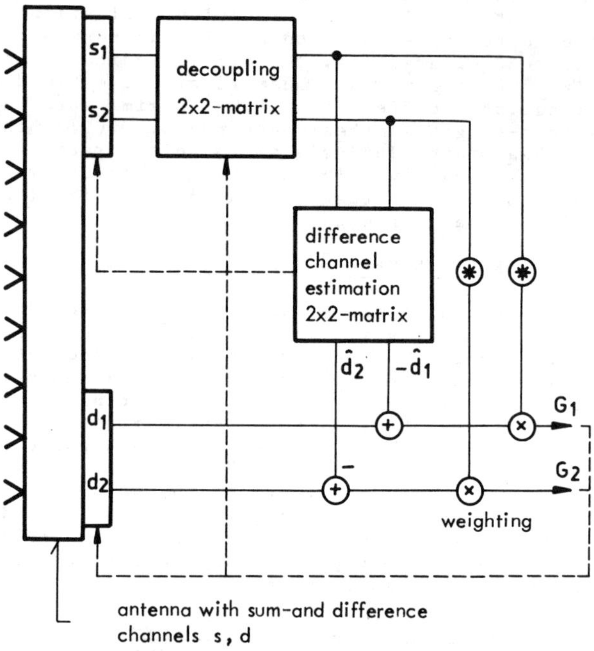

Figure 2. Computation of the Gradient

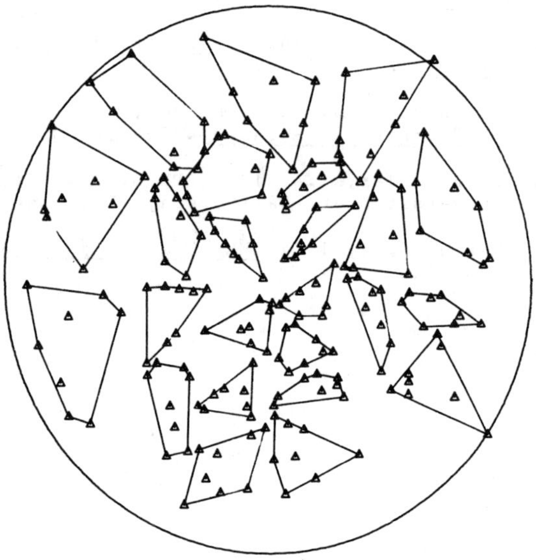

Diameter 32 λ

192 antenna elements arranged in 24 subgroups

Figure 3. Antenna Array

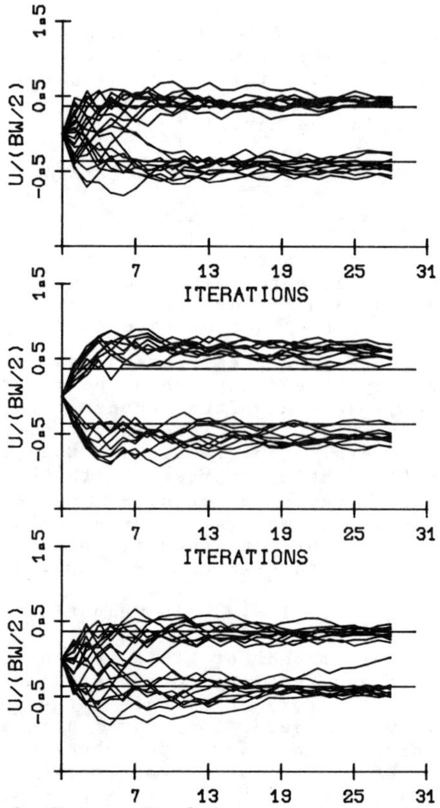

Figure 4. Computer Simulations

Figure 5. Arrangement of Measurement Equipment

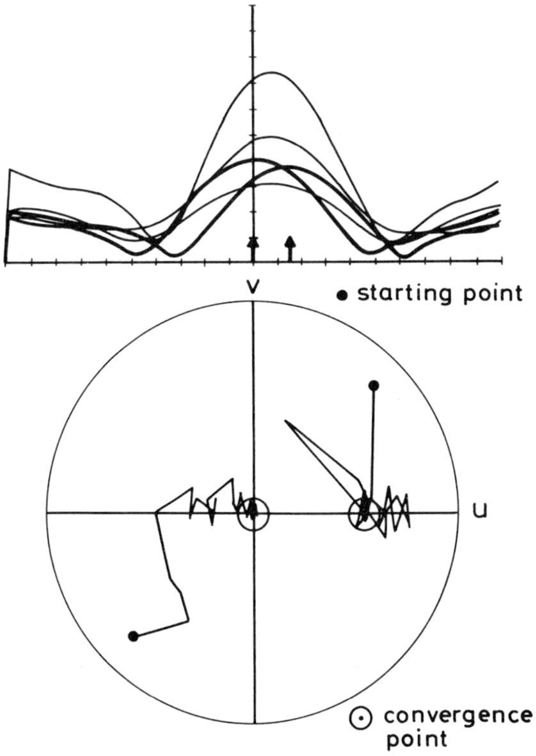

Figure 6. Estimation with measured Data

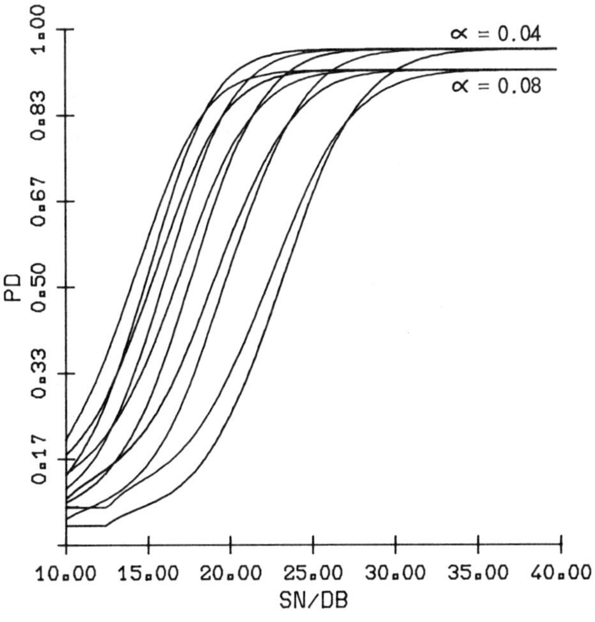

2 targets given at 0.5 BW separation

Figure 7. Computed Probability of Detection for 2 Targets

A FAST BEAMFORMING ALGORITHM FOR LARGE ARRAYS

Eric K.L. Hung and Ross M. Turner

Department of Communications, Canada

I. INTRODUCTION

Presented in this paper is a fast beamforming algorithm for array radars in which the number of array elements K is very large compared with the number of jammers L the radar is designed to suppress. This algorithm has several useful properties. First, it is very simple in design and programming. Second, it requires only about L samples of the noise environment to construct the weighting coefficients. Third, computation requirements are rather low. Given M samples of the noise environment, it requires approximately $(4M^2+6M+2)K$ real add and $(4M^2+8M+4)K$ real multiply operations. Most of these operations are in the form of vector operations and can be carried out efficiently with vector array processors. Fourth, it is effective in the suppression of many types of jammers. Fifth, it can be used in other applications such as the location of a target inside the main beam and the artificial placement of nulls in the radar beam pattern.

There are other techniques which modify the phases and amplitudes of array element output signals to produce a radar beam with jammer suppression properties. They include the coherent sidelobe cancellation (CSLC) technique of Howells and Applebaum (1,2), the least mean squares (LMS) technique of Widrow (3-5), and the sample matrix inversion (SMI) technique of Brennan and Reed (6). The CSLC and the LMS techniques converge too slowly in the case of unfavorable distributions in jammer powers and directions. The SMI technique in (6) has better performance in these jamming environments. Here, the noise covariance matrix must be non-singular to be invertable. The number of multiply and add operations in the construction of adapted weight vector is also proportional to K^3.

There are modified SMI techniques which construct adapted weight vectors in the case of singular noise covariance matrices. One of these techniques, described by Hudson in (7), adds a small positive quantity to each diagonal element of the matrix before inverting it. Another, also described by Hudson, uses Bühring's orthogonal projection method (8, 9). Accelerated covergence SMI techniques are available to calculate adapted weight vectors by iteration. Examples are Bühring's projection method (8, 9) and the recursive matrix inversion methods described by Hudson (7). The number of arithmetic operations in these techniques is proportional to MK^2. Significant reductions in the total number of arithmetic operations are realized whenever $M \ll K$.

The beamforming algorithm in this paper is a new technique for sampled data which yield a singular noise covariance matrix. The number of arithmetic operations is proportional to M^2K and is significantly smaller than those in the other techniques in (7) to (9) whenever $M \ll K$. Most of the arithmetic operations are in the form of vector operations. Additional reduction in computation time is realized if vector array processors are used.

II. ALGORITHM

A. Radar Receiver Output Power

The output signal of a K-element array antenna is represented here as a K-dimensional complex column vector $\underline{r} = (r_1, r_2, \ldots, r_k, \ldots, r_K)^T$, where r_k is the output of the k-th array element. The radar receiver output power is constructed as

$$P = \left|\underline{w}^T \underline{r}\right|^2 \qquad (1)$$

where $\underline{w} = (w_1, w_2, \ldots, w_K)^T$ is the weight vector for the antenna output signal and

$$\underline{w}^T \underline{r} = \sum_{k=1}^{K} w_k r_k \qquad (2)$$

The value of P given by (1) is proportional to $|\underline{w}|^2$. For the sake of convenience in the comparison of radar output powers measured with different weight vectors, the following condition is imposed,

$$|\underline{w}|^2 = 1 \qquad (3)$$

B. Construction of Adapted Weight Vector

The adapted weight vector $\underline{w}_A$ is constructed with a set of M noise vectors denoted by $\{\underline{u}_1, \underline{u}_2, \ldots, \underline{u}_M\}$. Initially an orthonormal basis $\{\underline{v}_1, \underline{v}_2, \ldots, \underline{v}_{M'}\}$, $M' \leq M$, is generated using the Gram-Schmidt orthogonalization method. The component of $\underline{w}_Q$ orthogonal to $\{\underline{v}_1, \underline{v}_2, \ldots, \underline{v}_{M'}\}$ is then calculated, renormalized to unit length, and identified as $\underline{w}_A$. There are twelve steps in the procedure. In Step 4, vector $\underline{u}'_m$ is the component of $\underline{u}_m$ orthogonal to all previously calculated $\underline{v}$-vectors. In Step 7, $\underline{u}'_m$ is normalized to unit length and identified as a new $\underline{v}$-vector if $|\underline{u}'_m|^2 > \Delta$, where Δ is a preassigned threshold.

Step 1 Set $M' = 0$

Step 2 Set $m=1$ and $\underline{u}'_1 = \underline{u}_1$.

Step 3 Proceed directly to Step 5.

Step 4 Calculate $\underline{u}'_m$ as

$$\underline{u}'_m = \underline{u}_m - \sum_{m'=1}^{M'} (\underline{v}_{m'}{}^{\dagger}\underline{u}_m)\underline{v}_{m'} \qquad (4)$$

Step 5 Calculate $|\underline{u}'_m|^2$

Step 6 If $|\underline{u}'_m|^2 \leq \Delta$, proceed directly to Step 8.

Step 7 Replace M' by $M' + 1$, set $m' = M'$ and calculate $\underline{v}_{m'}$ as

$$\underline{v}_{m'} = \underline{u}'_m / |\underline{u}'_m| \qquad (5)$$

Step 8 If $m=M$, proceed directly to Step 10.

Step 9 Replace m by m+1.

Step 10 Return to Step 4.

Step 11 Calculate $\underline{w}^{\circ}_Q$ as

$$\underline{w}^{\circ}_Q = \underline{w}_Q - \sum_{m=1}^{M'} (\underline{v}_m{}^T \underline{w}_Q)\underline{v}^*_m \qquad (6)$$

Step 12 Calculate $\underline{w}_A$ as

$$\underline{w}_A = \underline{w}_Q^\circ / |\underline{w}_Q^\circ| \qquad (7)$$

C. Computation Count

A rough estimate on the number of arithmetic operations in each step of the algorithm procedure is given below:

Step 1 Zero
Step 2 Zero
Step 3 Zero
Step 4 4(M-1)MK additions + 4(M-1)MK multiplications
Step 5 2MK additions + 2MK multiplications
Step 6 Zero
Step 7 2M'K multiplications + M' divisions + M' square root operations.
Step 8 Zero
Step 9 Essentially zero
Step 10 Zero
Step 11 8M'K additions + 8M'K multiplications
Step 12 2K additions + 4K multiplications + one division + one square root operation.

The sum total in the most unfavourable case of M'=M is $(4M^2+6M+2)K$ real additions + $(4M^2+8M+4)K$ real multiplications + (M+1) square root operations + (M+1) divisions. Given M, the number of add and multiple operations are proportional to K only.

III. COMPUTER SIMULATIONS

Two simulation experiments have been carried out to study the performance of this beamforming algorithm in the following jamming environments:

1. 3 narrow-band noise jammers
2. 1 broad-band noise jammer with 2% spread in frequency

Although this algorithm is written for large arrays, a choice K=10 was made. This value of K was large enough to demonstrate the properties of this algorithm, and yet small enough for visual inspection of changes in the array beam pattern.

A. Equations

The array chosen for study was a linear array as shown in Fig. 1. There were ten isotropic elements (i.e., K-10) spaced at a distance of half a wavelength (d=0.5λ). The quiescent beam pointing direction was arbitrarily chosen to be the broadside direction (θ_Q=0). The quiescent weight vector $\underline{w}_Q = (w_{Q1}, w_{Q2}, \ldots, w_{QK})^T$ was calculated as

$$w_{Qk} = \frac{1}{\sqrt{K}}, \quad k=1,2,\ldots,K \qquad (8)$$

Each vector in $\{\underline{u}_1, \underline{u}_2, \ldots, \underline{u}_M\}$ was calculated as a sum of a jammer vector and a white noise vector,

$$\underline{u}_m = \underline{j}_m + \underline{n}_m . \quad m=1,2,\ldots,M \qquad (9)$$

The jammer vector was calculated as

$$\underline{j}_m = \alpha_{1m}\underline{a}_1 + \alpha_{2m}\underline{a}_2 + \ldots + \alpha_{Lm}\underline{a}_L \qquad (10)$$

where

$$\alpha_{\ell m} = \sqrt{p_\ell}\, e^{j\phi_{\ell m}} \qquad (11)$$

and

$$\underline{a}_\ell = \left\{ e^{j\pi f_\ell \sin\theta_\ell}, e^{j\pi f_\ell 2\sin\theta_\ell}, \ldots, e^{j\pi f_\ell K\sin\theta_\ell} \right\}^T \qquad (12)$$

$$\ell = 1,2,3,\ldots,L.$$

Here, L was the number of jammers, $(p_1, p_2, \ldots p_L)$ were the jammer powers at the array elements, $(\phi_{1m}, \phi_{2m}, \ldots, \phi_{Lm})$ were random phases extracted from a population uniformly distributed in the range $(-\pi,\pi]$, $(f_1, f_2, \ldots, f_L)$ were jammer frequencies relative to the radar design frequency and were defined as $f_\ell = 2d/\lambda_\ell$, and $(\theta_1, \theta_2, \ldots, \theta_L)$ were the jammer directions off broadside. The components of the white noise vector $\underline{n}_m = (n_{m1}, n_{m2}, \ldots, n_{mK})$ were generated as complex Gaussian variates with the following properties:

$$E\left\{Re(n_{mk})\right\} = E\left\{Im(n_{mk})\right\} = 0 , \qquad (13)$$

and

$$var\left\{Re(n_{mk})\right\} = var\left\{Im(n_{mk})\right\} = 0.5\,\sigma^2 , \qquad (14)$$

where $Re(n_{mk})$ and $Im(n_{mk})$ were the real and imaginary components of n_{mk}, respectively, and σ^2 was the white noise power at the array elements.

Array gain in direction θ at frequency f was defined as

$$G(\theta,f) = \left|\underline{w}^T\underline{s}\right|^2 \qquad (15)$$

with $\underline{w}=\underline{w}_Q$ before adaptation and $\underline{w}=\underline{w}_A$ after adaptation. The components of signal vector $\underline{s} = (s_1, s_2, \ldots, s_K)^T$ were calculated as

$$s_k = e^{j\pi fk\sin\theta} , \quad k=1,2,\ldots,k , \qquad (16)$$

where f was the relative frequency at which the gain was calculated.

B. Simulation 1

This simulation studied the manner in which three narrow-band noise jammers with unequal strengths were suppressed as the number of noise vectors was increased. The parameters for the number of array elements, quiescent beam pointing direction, and frequency in the calculation of array gain were K=10, θ_Q=0°, and f=1.0. The jammer parameters were L=3, (p_1,p_2,p_3) = (1.0,0.1,0.01), (f_1,f_2,f_3) = (1,1,1), and $(\theta_1,\theta_2,\theta_3)$ = (-17,30,65) degrees. White noise power was $\sigma^2=10^{-4}$. The threshold in Step 6 of procedure was

$$\Delta = 2K\sigma^2 \qquad (17)$$

The results of this simulation are presented in Table I and Figs. 2 to 5. There was no improvement in the suppression of output jammer power when the number of noise vectors was increased from M=3 to M=4. In this study $|\underline{u}_4'|^2 < \Delta$ in Step 5 of the algorithm procedure. Consequently M'=3 for both M=3 and M=4.

Column 1 of Table I lists the number of noise vectors used in beamforming. For the sake of convenience in presentation, results obtained with the quiescent weight vector are identified with M=0. Column 2 of the table lists the array gain in the quiescent beam pointing direction. This gain decreased from $G(\theta_Q,f)$ = 10.00 dB to 9.66 dB as the number of noise vectors increase to M=3. The decrease in the value of $G(\theta_Q,f)$ was small in this simulation, because the jammers were purposely located outside the main lobe of the quiescent beam pattern. Columns 3 to 6 of the table list input jammer parameters. In column 6, p_ℓ/σ^2 was the ratio of jammer to white noise power at the array elements. This ratio was 40, 30, and 20 dBs for the jammers at -17°, 30°, and 65°, respectively. Column 7 lists the array gain in the jammer direction. In the quiescent beam pattern, the gains were $G(\theta_1,f_1) = -3.00$, $G(\theta_2,f_2) = -6.99$, and $G(\theta_3,f_3) = -9.95$ dBs for the jammers at -17, 30, and 65 degrees, respectively. After adaptation with one noise vector, $G(\theta_1,f_1)$ was reduced by 9.80 dB to -12.80 dB. At the same time, $G(\theta_2,f_2)$ and $G(\theta_3,f_3)$ were increased slightly. With two noise vectors, $G(\theta_1,f_1)$ and $G(\theta_2,f_2)$ were reduced to -29.50 and -29.72 dBs, corresponding to reductions of 26.50 and 22.73 dBs, respectively. Gain $G(\theta_3,f_3)$ was slightly above its value calculated with M=0. When the number of noise vectors was equal to the number of jammers, the array gains were $G(\theta_1,f_1) = -38.17$, $G(\theta_2,f_2) = -44.49$, and $G(\theta_3,f_3) = -23.84$ dBs. Column 8 lists the ratio of receiver output jammer to white noise power. Originally, they were 37.00, 23.01, and 10.05 dBs for the jammers at -17°, 30°, and 65°, respectively. With one noise vector, the strongest jammer seen by the array at the quiescent state was given the most suppression. With two noise vectors, the two strongest jammers were given the most suppression. After adaptation with three noise vectors, the power due to the strongest jammer was reduced to 1.83 dB above the white noise background, and the output powers due to the weaker jammers were submerged below this background.

In the array beam patterns in Figs. 2 to 5, the quiescent beam pointing direction is marked with an arrow above the array gain at θ_Q. The jammer positions are marked with arrows below the x-axis. The figures show that there was broadening of the main lobe as M increased. Suppression of the jammers were accomplished by reducing the peak values and the widths of the jammer-present sidelobes.

C. Simulation 2

The suppression on a broad-band noise jammer with 2% spread in frequency was studied in this simulation. The band jammer was positioned at -17° and was constructed as a sum of 21 equal strength narrow-band noise jammers with frequencies equally spaced in the range 99% to 101% of the radar frequency. The simulation parameters were K=10, $\theta_Q=0°$, L=21, M=0 to 5, $\{p_\ell=1;\ \ell=1,2,\ldots,L\}$, $\{f_\ell=0.989+0.001\ell;\ \ell=1,2,\ldots,L\}$, $\{\theta_\ell=-17°;\ \ell=1,2,\ldots,L\}$, $\sigma^2=10^{-4}$, and $\Delta=2K\sigma^2$.

The results on radar receiver output power relative to the white noise background are summarized in Table II. The results obtained with M=4 and M=5 were identical and are placed in the same column. In this table, $G(\theta_\ell,f_\ell)p_\ell$ was the output power due to the ℓ-th component of the broad-band jammer, and $G(\theta_\ell,f_\ell)\ p_\ell/\sigma^2$ was the output power relative to the white noise background. The summation over all the values of ℓ was the total output power due to the band jammer. This power was 50.2 dB above the white noise background in the case M=0. With M=4, it was reduced by 49.7 dB to 0.5 dB. The array gain pattern calculated with f=1 and M=2 and 4 are presented in Fig. 6. In the case of M=4, there was a significant overall increase in the sidelobe levels in regions far away from the jammer direction. The overall increase was significant because a small value of K was used in this simulation.

IV. ACKNOWLEDGEMENT

This work was supported by the Canadian Department of National Defence under Research and Development Branch Project 33C69.

V. REFERENCES

1. P.W. Howells, 1965, "Intermediate frequency sidelobe canceller", U.S. Patent 3 202 990.

2. S.P. Applebaum, 1976, "Adaptive arrays", IEEE. Trans. Antennas and Propagation, Vol. AP-24, pp. 585-598.

3. B. Widrow, 1971, "Adaptive filters I: Fundamentals", Stanford University, Electronic Labs., Syst. Theory Lab., Centre for Syst. Res., Rept. SU-SEL-66-126, Tech. Rep. 6764-6.

4. B. Widrow, 1971, "Adaptive filters", in Aspects of Network and System Theory, R.E. Kalman and N. DeClaris, Eds., New York: Holt, Rinehart, and Winston, Ch. 5.

5. B. Widrow, P.E. Mantey, L.J. Griffiths, and B.B. Goode, 1967, "Adaptive antenna systems", Proc. IEEE, Vol. 55, pp. 2143-2159.

6. I.S. Reed, J.D. Mallett, and L.E. Brennan, 1974, "Rapid convergence rate in adaptive arrays", IEEE Trans. Aerosp. Electron, Syst., Vol. AES-10, pp. 853-863.

7. J.E. Hudson, 1981, "Adaptive Array Principles, IEE publication, New York: Peter Peregrinus, Ch. 5.

8. W. Bühring, 1976, "Adaptive orthogonal projection for rapid converging interference suppression", Electronics Letters, Vol. 14, pp. 515-516.

9. W. Bühring, 1978, "Adaptive antenna with rapid convergence", IEE Conference on Antennas and Propagation, No. 169, pp. 51-54.

TABLE I

M	$G(\theta_Q,f)$ (dB)	ℓ	f_ℓ	θ_ℓ (degrees)	p_ℓ/σ^2 (dB)	$G(\theta_\ell,f_\ell)$ (dB)	$G(\theta_\ell,f_\ell)p_\ell/\sigma^2$ (dB)
0	10.00	1	1	-17	40	- 3.00	37.00
		2	1	30	30	- 6.99	23.01
		3	1	65	20	- 9.95	10.05
1	9.88	1	1	-17	40	-12.80	27.20
		2	1	30	30	- 4.58	25.42
		3	1	65	20	- 8.56	11.44
2	9.70	1	1	-17	40	-29.50	10.50
		2	1	30	30	-29.72	0.28
		3	1	65	20	- 9.20	10.20
3.4	9.66	1	1	-17	40	-38.17	1.83
		2	2	30	30	-44.49	<0
		3	1	65	20	-23.84	<0

Table I Simulation results of the suppression of three narrow-band noise jammers. Results obtained with the quiescent weight vector are identified with M=0. There was no difference between the results obtained with M=3 and M=4.

TABLE II

ℓ	f_ℓ	$G(\theta_\ell,f_\ell)p_\ell/\sigma^2$				
		M=0	1	2	3	4,5
1	0.990	5044	0.3526	0.7096	0.9140	0.0620
2	0.991	5042	0.3280	0.6265	0.8290	0.0611
3	0.992	5040	0.3031	0.5477	0.7463	0.0601
4	0.993	5038	0.2781	0.4733	0.6664	0.0589
5	0.994	5035	0.2533	0.4036	0.5895	0.0576
6	0.995	5033	0.2286	0.3387	0.5159	0.0561
7	0.996	5030	0.2044	0.2790	0.4461	0.0547
8	0.997	5027	0.1806	0.2244	0.3801	0.0532
9	0.998	5024	0.1577	0.1754	0.3184	0.0518
10	0.999	5020	0.1355	0.1321	0.2613	0.0505
11	1.000	5016	0.1144	0.0946	0.2090	0.0493
12	1.001	5013	0.0945	0.0633	0.1620	0.0483
13	1.002	5008	0.0760	0.0383	0.1204	0.0477
14	1.003	5004	0.0590	0.0198	0.0848	0.0473
15	1.004	5000	0.0439	0.0082	0.0553	0.0474
16	1.005	4995	0.0306	0.0036	0.0324	0.0480
17	1.006	4990	0.0196	0.0063	0.0164	0.0492
18	1.007	4985	0.0109	0.0166	0.0077	0.0510
19	1.008	4980	0.0047	0.0346	0.0065	0.0536
20	1.009	4974	0.0013	0.0606	0.0132	0.0569
21	1.010	4969	0.0010	0.0949	0.0283	0.0612
$\sum_{\ell=1}^{21} G(\theta_\ell,f_\ell)p_\ell/\sigma^2$		105267 (50.2 dB)	2.878 (4.6 dB)	4.351 (6.4 dB)	6.403 8.1 dB)	1.126 (0.5 dB)

Table II Simulation 2 results of the suppression of a broad-band jammer constructed as a sum of 21 equal strength narrow-band noise jammers with frequencies equally spaced in the range 99% to 101% of the radar frequency. The radar receiver output powers $\{G(\theta_\ell,f_\ell)p_\ell;\ \ell=1,2,\ldots,21\}$ were calculated with $p_\ell=1$ and $\theta_\ell=-17°$ for all values of ℓ. There was no difference between the results obtained with M=4 and M=5.

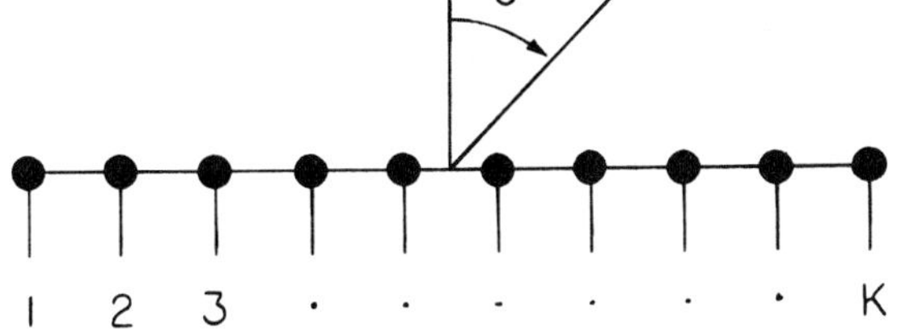

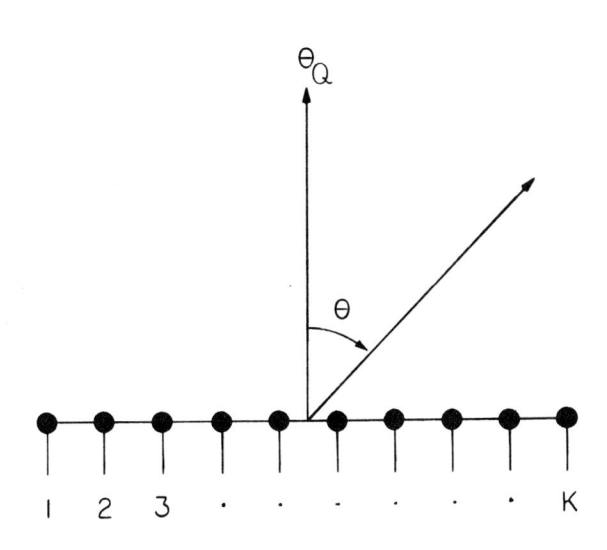

Figure 1 Array configuration in computer simulations. The broadside direction is identified as the zero degree direction. Angle θ is measured clockwise from the broadside direction.

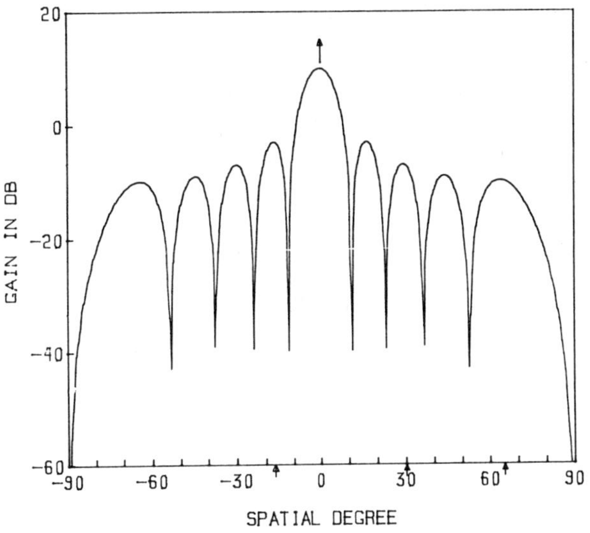

Figure 2 The quiescent beam pattern in Simulation 1. The beam pointing direction is marked with an arrow above the array gain at $\theta_Q=0°$. Jammer positions are marked with arrows below the x-axis.

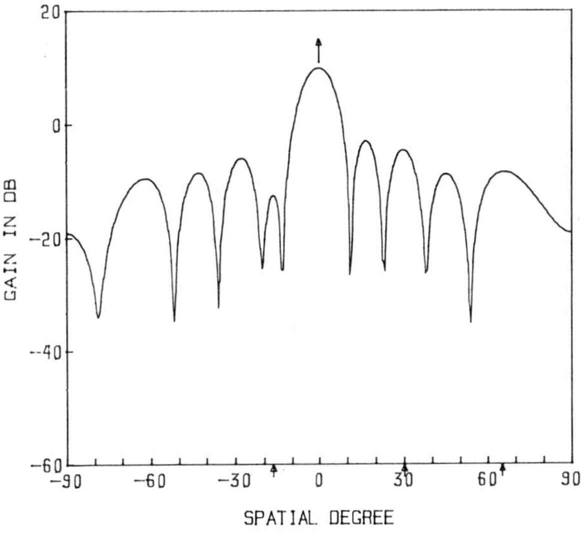

Figure 3 Simulation 1 array beam pattern calculated with one noise vector, M=1. Jammer powers in directions -17°, 30°, and 65° are 40, 30 and 20 dBs, respectively, above the white noise background.

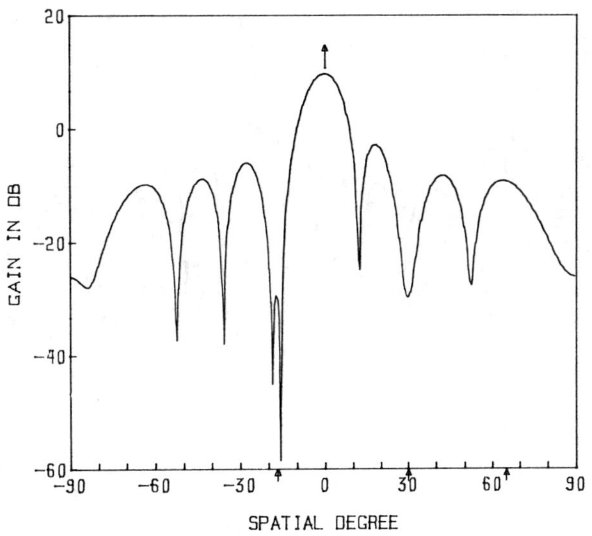

Figure 4 Simulation 1 array beam pattern calculated with two noise vectors, M=2.

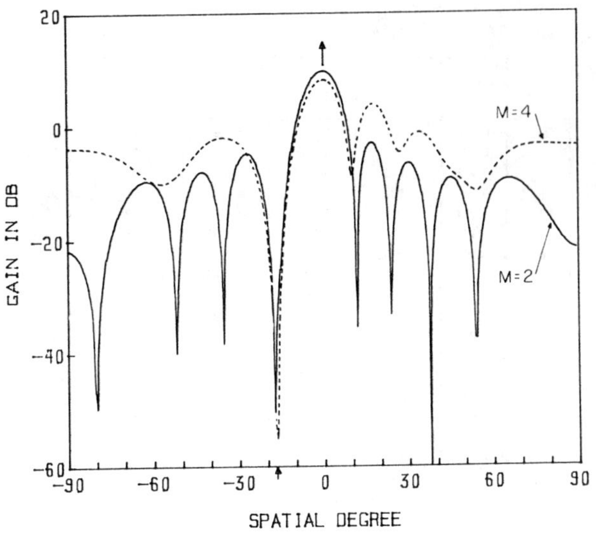

Figure 6 Simulation 2 array beam patterns calculated with M=2 and M=4. The band jammer is located at -17°.

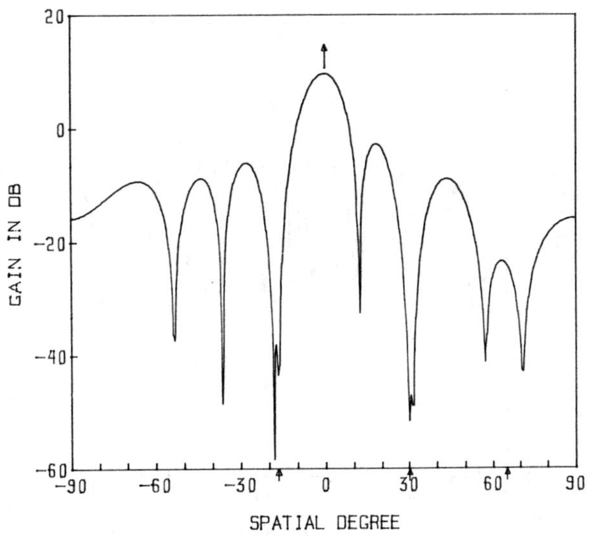

Figure 5 Simulation 1 array beam pattern calculated with three noise vectors, M=3.

HF SKY-WAVE BACKSCATTER RADAR FOR OVER-THE-HORIZON DETECTION

Gary R. Nelson and George H. Millman

General Electric Company, Syracuse, New York 13221, USA

ABSTRACT

The environmental factors that impact the design and performance of over-the-horizon backscatter (OTH-B) radars are reviewed in this paper along with the quantitative characteristics at high frequency (HF) of targets, external noise, and deleterious propagation effects. The performance as limited by these environmental factors is also examined. Key design considerations in the synthesis of cost effective systems are reviewed.

INTRODUCTION

HF, over-the-horizon backscatter (OTH-B) radars which take advantage of ionospheric refraction to achieve target detection at ranges up to 10 times that of conventional microwave, line-of-sight radars, are addressed in this paper. A comprehensive understanding of the highly dynamic characteristics of the ionospheric environment is essential to the effective design and application of OTH-B radars. While this type of radar is not new, only recently have the state of knowledge and technology matured to the point that such radars are operationally practical for a variety of applications. The key factors that affect performance and design are considered.

SYSTEM CONSIDERATIONS

OTH-B radar performance, e.g., detection and accuracy, is dependent on the signal-to-noise ratio (SNR) given by

$$\mathrm{SNR} = \frac{\overline{P}_t G_t G_r t \lambda^2 \sigma}{(4\pi)^3 R^4 K T_o N L_s L_p} \qquad (1)$$

where $\overline{P}_t$ is the average transmitted power, G_t and G_r are the gains of the transmitting and receiving antennas relative to an isotropic antenna, t is the coherent integration time, λ is the transmission wavelength, σ is the target cross section, R is the range to the target, K is Boltzmann's constant, T_o is the ambient temperature, N is the external noise level relative to KT_o, L_s is the system loss, and L_p is the two-way loss due to the ionospheric propagation.

External noise can originate from various sources. Atmospheric radio noise is predominant except at the high end of the HF band where galactic noise prevails. Man-made noise can be important near urban and/or industrial area. Figure 1 depicts the composite external noise levels for a typical mid-latitude location. The most significant feature is the increasing noise level with decreasing frequency.

Various types of targets, such as ships, aircraft and missiles, can be detected by OTH-B radars. Since the transmission wavelengths fall between 60 m (5 MHz) and 10 m (30 MHz), target scattering takes place in the resonance region. This results in relatively large cross sections; e.g., on the order of 10^4 -10^5 m^2 for ships and 10^2 -10^3 m^2 for aircraft. Backscatter from the ground and/or sea, although many orders-of-magnitude larger than that from ships and aircraft, can be resolved from the target reflections by virtue of its relatively low Doppler frequency shift and spread (typically less than 1-2 Hz). Reflections from field-aligned E-layer auroral ionization and F-layer irregularities could result in strong clutter for OTH-B radars sited at higher latitudes. Ionospheric clutter-to-noise ratios (CNRs) as high as 70 dB can exist (Millman, 1975) (1). This type of clutter cannot be readily resolved in Doppler since the combined Doppler shift and spread of ionospheric clutter could be on the order of ±40 Hz at 10 MHz (Millman, 1975) (1).

Meteor trail reflections give rise to noise and false targets. It is estimated that, in the 5-30 MHz frequency range, approximately 10^4 -10^5 sporadic, underdense meteor trails (line densities < 10^{14} electrons/m) per hour, could be present in the radar beam. Because of the extremely high rate, their reflections appear as additional background noise in the radar's 100-2800 km range window. Underdense "meteor noise" could at times be as high as 10-20 dB above the atmospheric/galactic noise level. Specular reflection from overdense meteor trails (line densities > 10^{14} electrons/m) can result in cross sections up to 10^6 -$10^7 m^2$ with time durations of 1-35 s and Doppler shifts on the order of 3-18 Hz. The rate for overdense meteors is relatively low - on the order of 200 meteors/h at night (Millman, 1978) (2).

Repetitively solving Equation (1) for the expected range of sunspot numbers over a full 11-year cycle, all seasons and all times days, and statistically presenting the results as a function of range is a useful way to determine the required detection energy ($\overline{P}_t G_t G_r t$) for a typical OTH-B radar. Figure 2 is the result of such an exercise for a mid-latitude sited radar where SNR ≥ 10 dB has been taken as the criterion of success. The results indicate that detection energies between 110 and 120 dBj are required for good availability. Meteor and ionospheric clutter effects were not included in the analyses. It is estimated that both effects the availability percentages by less than 10%.

Another measure of performance is the cumulative probability of detection. Figure 3 depicts this parameter over a 180° surveillance sector for a large jet aircraft size target and for a low sunspot number (SSN) of 20. For a high SSN of 120, the 0.9 cumulative probability of detection contour is extended out to a ground range of approximately 3000 km.

The range and cross range resolution is primarily limited by available bandwidth and antenna size considerations. Typical values range from 8-30 km and 1°-3° respectively. Doppler resolution is a function of the coherent integration time and ionospheric temporal stability. Typical values range between 2 and 0.1 Hz with the operational value selected based on mission considerations.

Localization measurements are errored by propagation anomalies such as ionization gradients (tilts) and uncertainty in ionospheric reflection height. A tilt which can impose an error both in angular position and ground range is present during sunrise/sunset transitions, with traveling ionosphere disturbances, and in the mid-latitude trough (Muldrew, 1965) (3). For an ionospheric tilt of 5° and ionospheric reflection height of 350 km, the maximum lateral deviation can be on the order of 63 km. In the radial direction, the ground range error can be on the order of 2.5 and 0.5 km at ranges of 750 and 3700 km, respectively (Millman, 1978) (4).

An ionospheric reflection height estimation error of 15 km results in a range error of about 28 and 9 km at a ground distance of 750 and 3700 km, respectively (Millman, 1978) (4).

DESIGN CONSIDERATIONS

A convenient starting point for OTH-B radar architecture considerations is Equation (1). The variables can be grouped into two categories; those set by nature (λ, σ, R, K, T_o, N, L_p) and those under the designer's control ($\overline{P}_t$, G_t, G_r, t, L_s).

The values of the former variables are fixed by the environment. The designer's task is to select a cost-effective combination of the latter which provides the desired level of performance in light of the characteristics of those set by nature. As previously indicated, the required detection energy, $\bar{P}_t G_t G_r t$, is on the order of 110-120 dBj. The question now is how best to select values for the constituent parameters.

One of the first decisions that must be made is the selection of the system configuration; i.e., monostatic or bistatic. A monostatic system implies that the transmit and receive functions share a common antenna, thus a single site is required. However, a bistatic system utilizes separate transmit and receive antennas resulting in two sites. In spite of the increased site demands, bistatic configurations are often preferred. By separating the transmit and receive sites by 50 km or more, 100% duty factor, continuous wave (CW) waveforms can be used. (Separations as close as 10-20 km may be possible if adaptive spatial nulling is used at the receive site to further attenuate the transmitted ground wave.) The high average powers required can be generated more cost effectively with CW waveforms than with low-duty factor (<10%) pulse waveforms. In addition, CW waveforms are significantly more compatible with the other HF band users than high peak power, low-duty factor pulse waveforms. This results from the fact that radar performance is a function of the average transmitted power while radio frequency interference (RFI) effects are primarily a function of peak power.

Bistatic configurations, and also monostatic systems with separate but colocated antennas, can take advantage of the fact that external noise at HF (See Figure 1) is much larger than that generated internal to the system. This allows the use of inefficient receive antenna elements which can be small relative to the operating wavelength. By proper design, the combination of element efficiency, transmission line losses and receiver noise figure results in a system noise figure which is still well below that presented by the external environment. Thus, large numbers of simple receive elements can be line arrayed to provide the desired azimuthal resolution.

Figure 4 depicts the required array length as a function of azimuthal beamwidth, and frequency. It is noted that, for operation at 5 MHz, the array must be approximately 1830-m long to form a 2° beam. Assuming a 30-MHz upper operating frequency, about 300 elements would be required to prevent grating lobes when scanning. This type of an array can be implemented relatively inexpensively using short monopoles (6 m), dipoles, or small loops arrayed in front of a backscreen or in combinations to produce a unidirectional pattern.

Radiation considerations lead to quite different conclusions for the transmit antenna. Element efficiency is important since it is desirable to radiate as much of the HF power generated by the transmitters as is economically practical. Efficient arrays require electrically long elements, i.e., $\lambda/2$ dipoles at 5 MHz are 30.5-m long. Furthermore, in order to operate over wide bandwidths and at high powers leads to large, mechanically complex elements. Fortunately, it is possible to restrict the overall length of the transmit array(s) to minimize the number of these costly elements by utilizing a short transmit array. This array is made up of relatively few elements to form a broad transmit beam filled with narrow receive beams formed by a long receive array of many simple elements.

Bistatic OTH-B radars employ long receive arrays with directive gains from 25 dBi to 35 dBi to achieve the required azimuthal resolution, and short transmit arrays from 0.25 to 0.1 as long as the receive array, with gains from 15 dBi to 25 dBi. Several independent transmit arrays, each operating over a different frequency subband, may be required to cover efficiently the entire operating band. Since it is desirable to keep the azimuthal resolution essentially constant with frequency, variable aperture weighting is necessary on both transmit and receive.

Another key antenna consideration is elevation angle coverage. The tradeoff is between antennas with shaped, broad elevation patterns which illuminate the required propagation angles (3°- 35°) to cover the desired ranges and propagation modes of interest and antennas with narrow beams which can be elevation steered to maximize performance for the prevailing ionospheric conditions. Narrow elevation beams enhance performance by virtue of their increased vertical aperture (gain). More important, however is that the narrow elevation steered beam could be very useful in selecting the desired propagation mode while rejecting undesired multipath modes and/or elevation angle regimes where ionospheric or ground clutter originates.

Figure 5 is a plot of the difference in the elevation angles between the principal propagation modes as a function of radar range. Assuming operation beyond 1850 km, discrimination between elevation angles down to a few degrees is required. Figure 6 depicts the required physical antenna height as a function of the elevation discrimination angle. It is evident that the required height ranges between 61 m (30 MHz) and 366 m (5 MHz) for a discrimination angle of 5°. Thus, high structures are necessary to provide mode discrimination over the entire frequency band.

The cost of a large HF antenna is proportional to its length and the square of its height. Since antenna gain is proportional to effective area, it is obvious from cost considerations that it is preferable to increase aperture by increasing its length rather than height. Thus, the only consideration which might dictate narrow, elevation steered beams is mode/clutter discrimination. There is insufficient experimental data to date to verify if the resulting performance enhancement is worth the additional cost to provide elevation steering.

The next variable under design control is the average transmit power. Depending on the application, average powers ranging from 200 to 2000 kW are required. Typically, elemental transmitters are employed to feed the transmit antenna array so that generation of the required high average power is distributed among several amplifiers, i.e., 4 to 20. The elemental transmitters must operate over the required several octave frequency band. This can be achieved with distributed or band tuned amplifiers employing high-power vacuum tubes. Band tuned amplifiers are generally preferred because of their higher efficiency. The state-of-the-art in high-power transistors currently precludes the use of solid-state devices in the final stage of such HF transmitters for OTH-B radar application.

The system loss, L_s, includes all losses internal to the radar such as transmission line loss, scan loss, beam shape factor loss, waveform mismatch loss and Doppler filter straddling loss. It is, for the most part, not cost effective to attempt to reduce it to less than about 10 dB.

The effective coherent integration time, t, is generally determined by the Doppler or velocity resolution necessary to discriminate the weak desired target signals from the much stronger ground/sea clutter. Natural ground, i.e., trees, sea movements, and ionospheric temporal instabilities limit the clutter spectrum to a maximum Doppler consistent with motion up to a few tens of km/h. Aircraft have Dopplers consistent with radial velocities from essentially zero up to 2800 km/h. Effective coherent integration times ranging from 1 to 10 s, depending on the instantaneous operating frequency and target radial velocity, are required in order to resolve the target from the much stronger clutter. For slow-moving ship targets, effective coherent integration times up to 60 s may be necessary at times.

With regard to ground/sea clutter, the subclutter visibility, SCV (dB), required for an OTH-B radar is derived from the relationship

$$SCV = SNR_m + NCR + FM + CTR \qquad (2)$$

where SNR_m is the maximum desired signal-to-noise ratio, NCR is the noise-to-clutter ratio residue margin for less than 1-dB SNR degradation, FM is the fade margin and CTR is the clutter-to-target ratio which is defined by σ_o + CCS - σ_t where σ_o is the clutter backscatter coefficient, CCS is the clutter cell size and σ_t is the target cross section. For a typical set of parameters, i.e., SNR_m = 20 dB, NCR = 6 dB, FM = 12 dB, σ_o = -29 dBsm, CCS = 94 dB, and σ_t = 23 dBsm, SCV is on the order of 80 dB.

In addition to coherent integration, other necessary signal processing functions are pulse compression, Doppler processing (coherent integration), clutter blanking, background normalization, peak detection, parameter estimation and data compression. These functions must be accomplished over a wide range of operating parameters (such as waveform bandwidths, waveform repetition frequencies and coherent integration times) with a dynamic range in excess of 100 dB to achieve the required SCV. This can be achieved with networks of high-speed, special-purpose digital processors under the dynamic control of a host computer. To provide the high dynamic range, 32-bit floating point operation is generally utilized through the coherent functions up to envelop detection.

After signal processing, the candidate "hits" are passed to a data processor which initiates tracks, maintains tracks, assesses the propagation evnvironment, registers tracks from radar to geographic coordinates, formats information for display, supports operator interactions with data bases being automatically processed, assesses the significance of the radar derived information and communicates the results to users. The data processing functions can be accomplished in one or more general-purpose computers with a total capacity of several million instructions per second. For the operator display functions, computer driven, digital displays with high resolution and interactive graphics are required.

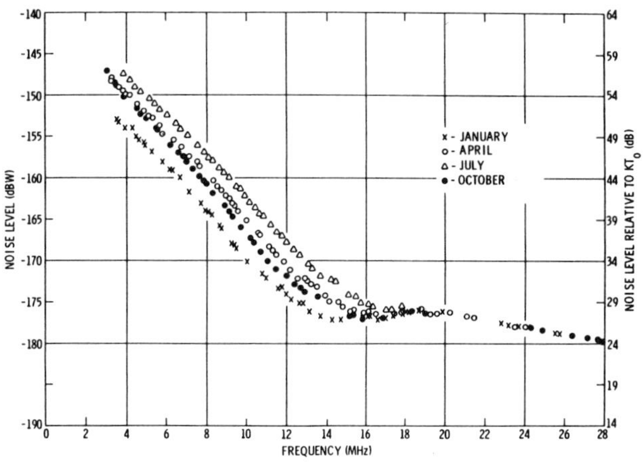

Figure 1 Environmental noise as a function of frequency

CONCLUSIONS

As a result of extended range detection possible via ionospheric refraction, an OTH-B radar can provide surveillance of an area as large as 7 x 10^6 km^2. It can provide all altitude detection of targets from the earth's surface up to ionospheric heights. Principal applications of interest are those requiring surveillance of areas which are not cost-effectively accessible by other radar sensors, e.g. remote ocean areas and politically unavailable land regions. Propagation effects limit the availability, resolution and accuracy of OTH-B radars; their performance is less than that achievable by microwave radars. Nevertheless, their ability to "see" into the line-of-sight shadow zone at ranges up to 10 times that of microwave radars makes OTH-B radars attractive candidates for many applications.

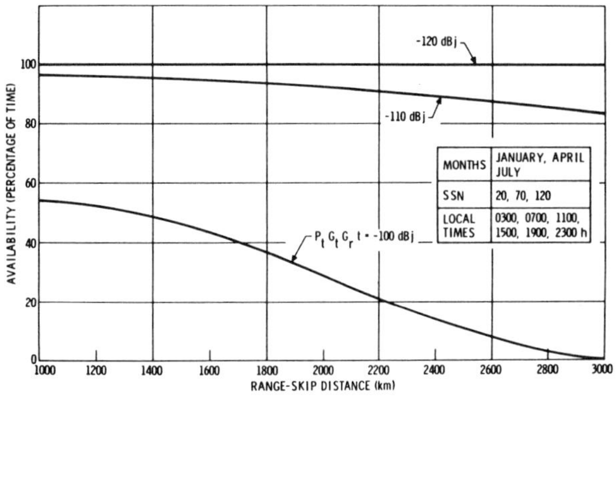

Figure 2 OTH radar availability due to propagation effects

REFERENCES

1. Millman, G.H., "An Evaluation of HF Ionospheric Backscatter Echoes", General Electric Technical Information Series Report No. R75EMH19, November 1975.

2. Millman, G.H., "HF Scatter from Overdense Meteor Trails", Aspects of Electromagnetic Wave Scattering in Radio Communications, AGARD Conference Proceedings No. 244, pp. (6-1)-(6-14), September 1978b.

3. Muldrew, D.B., "F-Layer Ionization Troughs Deduced from Alouette Data", Journal of Geophysical Research, Vol. 70, pp. 2635-2650, June 1, 1965.

4. Millman, G.H., "Ionospheric Propagation Effects on HF Backscatter Radar Measurements", Proceedings of the Symposium of the Effect of the Ionosphere on Space and Terrestrial Systems, Naval Research Laboratory, pp. 211-218, 1978a.

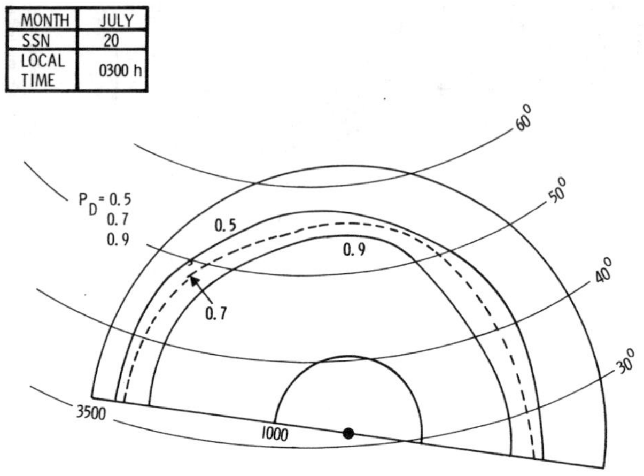

Figure 3 Cumulative probability of detection

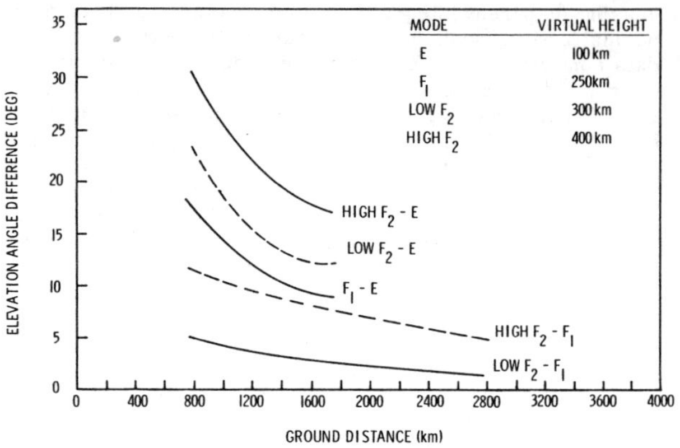

Figure 5 Elevation angle differences between various one-hop propagation modes

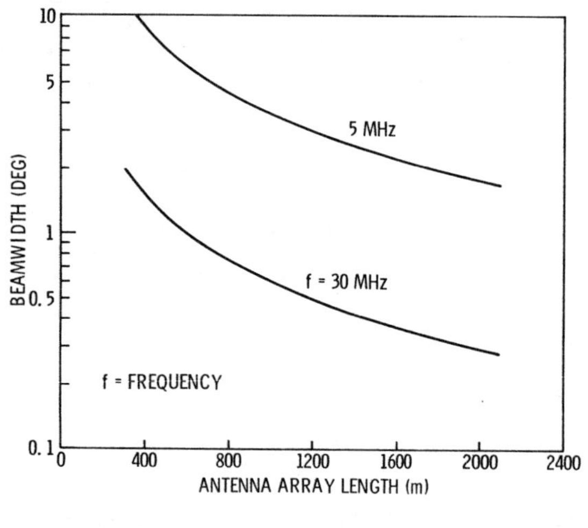

Figure 4 Antenna aperture characteristics (30-dB Taylor Weight)

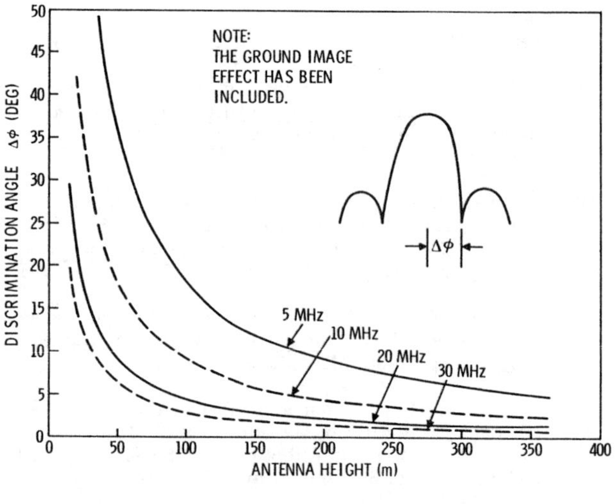

Figure 6 Elevation angle discrimination as a function of antenna height and frequency

HF GROUND-WAVE RADAR FOR SEA-STATE AND SWELL MEASUREMENT; THEORETICAL STUDIES, EXPERIMENTS AND PROPOSALS

E.D.R. Shearman* L.R. Wyatt* G.D. Burrows* M.D. Moorhead* D.J. Bagwell* and W.A. Sandham+

* Department of Electronic & Electrical Engineering, University of Birmingham

+ Formerly at the above, now at British National Oil Corporation, Glasgow

INTRODUCTION

A capability for sensing remotely from a terrestrial site the characteristics of sea-waves, surface-currents and surface-winds over an area would clearly be a very useful enhancement of the point-sampling techniques using buoys and ships which are in use at present. Microwave radar, operated from a coastal site or platform for such purposes, has a range limited to 10 km or so by line-of-sight considerations. In contrast, ground-wave radar, that is radar operating in the MF or HF band using vertical polarization, gives a capability for sensing sea-state up to ranges of the order of 200 km. In addition, these dekametric radio waves scatter resonantly from sea-waves of dekametric wavelength, which are those of most interest to man because they carry the major part of the sea-wave energy and, therefore, influence and damage man-made structures.

The basic first-order mechanism of scattering of dekametric waves by the sea, is the Bragg resonant effect originally identified by D.D. Crombie [1]. In this process, the outward propagating radio waves interact selectively with those radially-approaching and receding sea-wave components whose wavelength is one half of the radio wavelength. (Sea-wave components of different lengths or different directions scatter the radio waves in directions other than back towards the radar). Because the sea-waves are moving, the radar echoes from, respectively, the approaching and receding sea-wave components, display positive and negative Doppler shifts.

The work of Long and Trizna[2] and of Maresca and Barnum[3] has shown that with a knowledge of the directional properties of sea-waves launched by the wind, one may deduce from the ratio of the amplitude of the approach and return radar echoes, the direction of the sea-surface wind.

Also, from these first-order radar returns, as has been shown by Crombie[4] and by Barrick and Evans[5], one may deduce the radial component of surface-current. This arises because if the water body as a whole moves away from the radar, the Doppler shift of the approaching waves and of the receding waves will both be offset by an amount corresponding to the velocity of the water on which the waves move. Using two radars, looking in different directions at a particular sea area, it is possible from these offsets to map the vector current flow.

These two techniques, for wind-measurement and surface-current measurement respectively, have been well-substantiated and their capability for sensing surface-wind associated with weather systems and estuarial current flow, has been validated.

A more challenging task, but one of very great interest to offshore-technologists, is the measurement of sea-state. The parameters sought here are the mean wave-height and two spectral descriptors, the non-directional wave-height spectrum and the directional wave-height spectrum. With these parameters, the response of ships and oil-rigs to the complex pattern of sea-waves, may be computed statistically.

Hasselmann[6] and Barrick[7] laid the theoretical foundation from which the second-order components in the Doppler radar echo spectrum may be used to determine the wanted wave-height parameters. In this paper, we review the form that a radar should take to measure such parameters most effectively. In particular we compare pulse and FM radar technique for this application and report some initial results using FM technique and their use to determine wave-properties. The radar antenna directivity considerations which are needed in the design of a radar for sea-state sensing are also relevant in determining the accuracy of measurement of surface-wind and surface-current by the first-order scattering processes discussed above.

The physical processes of scattering and the background of oceanographic description of the sea-surface, will be presented only in outline here, as they have been discussed in more detail in a recent survey-paper[8]. This survey-paper also discusses the associated sky-wave radar technique, by which coverage of the sea-surface in a zone from 1,000 to 3,000 km from the radar-site, is achieved by the use of ionospherically-reflected waves.

THE SEA-SURFACE AS A RADAR-SCATTERER

The most common descriptor of the sea-surface used by oceanographers is the non-directional wave-height spectrum, either in terms of wave frequency or wavelength, the two being simply related through the dispersion relationship between wave-velocity and wavelength. Figure 1 illustrates a typical wave-spectrum, where the high frequency continuum is due to wind-waves locally generated, and the discrete component at lower frequency is due to swell propagating in from a distant storm area. A fuller description of the sea-surface can be given by a two-dimensional representation, in which the energy incident in each azimuthal direction is plotted for each frequency component. This is the directional wave-height spectrum.

Strong back-scattered echoes are received from the sea-surface by the first-order Bragg resonant diffraction process already mentioned involving radially-travelling sea-wave components, whose wavelength is one half a radio wavelength (see Figure 2a). Weaker second and higher order scattered echoes are received (A) by Bragg resonant diffraction from harmonics of radially-travelling non-sinusoidal sea-waves (Figure 2b), (B) by Bragg resonant diffraction from radially-travelling sea-wave components arising from non-linear interactions between co-directional (Figure 2c) or crossing sea-waves (Figure 2d) and (C) by successive electromagnetic scattering from two sea-wave components travelling orthogonally (Figure 2e). Each of these backscattered echoes is Doppler-shifted, so that a continuous spectrum of echo-components is received, coded in Doppler frequency in a manner related to the velocities of the sea-waves responsible.

Figure 3 shows an example of the resulting form of Doppler echo spectrum from a 7.5 km x 10 km resolution cell on the sea-surface at a range of 40 km obtained with the Birmingham University radar at Angle, Pembrokeshire. The various spectral features are indicated in the Figure and related to the processes A, B and C referred to above.

RADAR CONSIDERATIONS

In HF ground-wave radar the range achieved for sea-state sensing depends on the ground-wave attenuation in addition to normal spatial spreading. At 2 MHz the one-way attenuation over sea to 200 km range is about 4 dB. At 30 MHz, however, the one-way attenuation has risen to 53 dB at 200 km and to 17 dB at 50 km.

A second factor which governs achievable range is the noise level, which at these frequencies is controlled by external natural and man-made noise, the former being subject to diurnal and seasonal variations. Thus at 2 MHz the summer night-time level in a short unipole is governed by atmospheric noise and is 60 dB greater than kT_oB while in the day-time throughout the, atmospheric noise is attenuated in the D layer of the ionosphere and the received noise is governed by man-made noise at a level of 45 dB greater than kT_oB. At 30 MHz the noise level would be expected to be governed by galactic noise and to be in the neighbourhood of 18 dB greater than kT_oB throughout the year. In addition, the congestion of the HF band also implies that interference from other transmissions sharing the channel allocation is likely, geographical separation not being sufficient to guarantee immunity.

It is against this background that the requirements of a sea-state radar must be judged and implementation chosen. We are concerned with maximising the signal/noise ratio of received sea-echo and displaying its Doppler spectrum in the manner shown in Fig. 3. To obtain data of this quality, a Doppler resolution of the order of 2.5-5 mHz is needed, corresponding to coherent integration of the radar echoes over 200-400 seconds. In addition, incoherent averaging of the power spectra obtained from such individual coherent integration periods, or 'dwells', is required to reduce the sampling error involved in a single spectral estimate of a Gaussian process such as sea echo at HF. Typically 13 such spectra are averaged in the present work, overlapping of these dwells being permissible, since the weighting utilized reduces the effect of the correlation of overlapping data.

Barrick(7) has shown that for first-order Bragg scattering, the amplitude of the scattered echo from the sea may be characterised by a back-scattering cross-section per unit area of sea, σ_0, which is related to the magnitude of the Bragg-matched component of the sea-wave spectrum.

Barrick has evaluated σ_0 for a non-directional sea-wave spectrum of the Phillips equilibrium form, which represents the spectral distribution of wave height variance for a 'saturated' wind-sea. This is a sea for which the wind has been blowing for long enough and over a sufficient distance to build up a stationary spectral distribution, with a low-frequency cut off of $f = g/2\pi u$, where u is the wind-speed, and with a spectral distribution,

$$s(f) = \beta(2\pi)^{-4} g^2 f^{-5} \qquad f > g/2\pi u$$

$$= 0 \qquad f < g/2\pi u.$$

With these assumptions the observed value of received echo power from a patch of sea, ds, is shown to be given, for a monostatic radar, by

$$dP_R = \frac{P_T G_T G_R \lambda^2}{(4\pi)^3 R^4} \cdot F^4 \cdot \sigma_0 \cdot dS \quad \ldots \ldots \ldots \quad (1)$$

with $\sigma_0 = 0.02$. Here P_T is the transmitted power, λ the radio wavelength, R the range, F the Norton ground-wave amplitude attenuation factor relative to free-space propagation and G_T and G_R the transmitting and receiving antenna gains. With Barrick's definition of σ_0, G_T and G_R must be defined as 'equivalent free-space gains', 6 dB less than the measured value over a reflecting ground).

To focus attenuation on the capabilities and limitations of HF radar, we consider a particular experimental system being developed, having a broad-beam logarithmic periodic transmitting antenna of gain 10 dB (G_T = 4 dB by the above definition) and a family of 7° beams for reception, of gain 18 dB (G_R = 12 dB). The mean transmitting power is taken as 250 W and the values of F are obtained by inspection of the recently computed revisions of the CCIR field-strength curves[8]. For a monostatic radar, ds becomes $R.\Delta R.\Delta\theta$, where $\Delta\theta$ is the narrower of the transmit and receive beams, 7° in this case, and ΔR is the range resolution. The range-dependent terms thus yield $dP_R \propto F^4 R^{-3}$.

Fig. 4 illustrates the capability of this radar at the three frequencies for which the receiving beams have the above parameters. The arrows along the abscissa indicate the ranges at which the sea-echo to noise ratio falls to 50 dB, the value considered desirable for Doppler spectra of the quality of Fig. 3, suitable for inversion to yield wave spectra. The computed values are for a single 200 s dwell; some improvement in range would be expected for incoherent averaging of a number of spectra.

The effect of antenna beamwidth on radar observations is demonstrated by Fig. 5, (after Sandham[9]). This concerns the use of the observed ratio between approach and recede Bragg spectral line amplitude to deduce wind direction. The curve for beamwidth 10° shows the relationship between Bragg line ratio and wind direction assuming that the spreading function of the wind waves about the mean wind direction is given by the Munk relationship,

$$G(\theta) = \epsilon + (1-\epsilon)\cos^4(\theta/2),$$

when $\epsilon = 0.004$. The remaining curves show the effect of broadening the beam so that an average of the ratio over the beam is performed. We deduce that little error is introduced by the use of a 20° beamwidth.

RADAR TECHNIQUE

The various HF radar techniques which can be used to extract desired oceanographic and meteorological data have been discussed elsewhere[10]. For current measurement it has been demonstrated[5] that the use of omnidirectional antennas and direction-finding technique is practicable. Doppler resolution is used to separate the first-order Bragg echoes returned by patches having different radial velocities before their direction is identified. (The technique is related to that described for ship-tracking in a companion paper[11]). The use of omnidirectional antennas makes smaller the gains G_T and G_R in equation 1, reducing the range, while the need to use the highest HF frequencies to maximise the Doppler shift also reduces the achievable range to some 50 km.

Beam-forming systems are required when wave measurements are to be made, as the shape of the second-order Doppler continuum must be resolved and this would be irretrievably smeared by superposition of the continua from other directions, possibly current-shifted. A beam-forming system has, therefore, been adopted for our experiment. It seems probable that operation at more than one radio frequency is desirable to determine wave spectra, and accordingly, the system under development has a frequency range 6-30 MHz with 7 beams covering a 90° sector.

Previously-described systems have employed pulse technique and, therefore, to achieve the mean power requirement already discussed, the short pulses required for high range resolution and the limited pulse repetition frequency set by range ambiguity considerations would dictate a high pulse power.

In the present experiments, frequency-modulated technique is used, using the principles described elsewhere[12]. To permit transmission and reception from a common site, the saw-tooth frequency-modulated carrier is interrupted with a pseudo-random binary sequence, this enabling echo reception to be achieved in the carrier-off intervals. (The system is termed 'FMICW', frequency-modulated interrupted continuous wave). With this system the duty ratio can approach 50% (compared with a typical 1% for a pulse system), permitting a modest peak-power requirement, more practical for a transportable antenna system. An added advantage of the FMICW scheme is that high range resolution can be obtained by increasing the deviation only. The high data rates associated with sampling of short pulses do not occur because the sweep repetition rate can be kept low, being governed only by the Nyquist rate for Doppler shifts of a few tenths of a Herz. Other advantages result from the mismatch of the radiated waveform to CW interference[12].

With an FMICW system, care is needed with the dynamic range of receiver, which must handle echoes from all ranges simultaneously, and with the chopping waveforms used for transmission and reception. That these problems have been solved may be seen from the quality of the data being obtained with the prototype system.

EXPERIMENTAL RESULTS

The prototype FMICW radar discussed in the previous section was evaluated at 9 MHz and 12 MHz in a preliminary experiment at a field site at Angle, Pembroke[11]. The transmitted signal was a (chopped) sawtooth FM carrier of deviation 20 kHz. The transmitting antenna used was a $\lambda/4$ resonant monopole. The received sawtooth FM waveform was 'de-ramped' using a sawtooth FM carrier as a first local oscillator. The resulting beat-frequency echo waveform was translated to audio frequency (0-150 Hz), monitored on a spectrum analyzer and also recorded on analogue tape. The echo spectrum was in the form of lines spaced by 2.5 Hz (the sweep repetition frequency) the amplitude of successive lines being proportional to the sea echo from successive range resolution cells. With higher spectral resolution each 'line' showed the double structure and second-order continuum of sea-echo, Fig. 3 being an example.

Data of this type obtained at 9 MHz and 12 MHz was analyzed by the model-fitting technique of Wyatt and compared with available in situ measurements and wave predictions. The technique can be used to analyse the region in the spectrum between the first-order line and the $2^{\frac{1}{2}}$ singularity. This region corresponds to interactions between one wave considerably longer than the first-order wave and one wave of the same order. In the model-fitting method, short wind-wave and long-wave directional wave-height spectra are assumed and the corresponding radar Doppler spectrum is computed. The long-wave parameters are then varied in an orderly fashion until the difference between computed and observed radar spectrum in this region is minimized. The process can sometimes be accelerated since swell components isolated in frequency from the rest of the spectrum may be identified rapidly in frequency and direction by a linearized inversion technique due to Lipa and Barrick[13].

Fig. 6a shows the resulting radar-derived non-directional long-wave spectrum. Unfortunately, the nearest wave-rider buoy yielding non-directional wave spectra was at the Isles of Scilly some 170 km from the radar resolution cell. Fig. 6b shows for comparison the wave-rider buoy spectrum at this location. The dotted lines in each Figure show predictions (from a wave-model with meteorological inputs) at the two locations, from which it is seen that the spectral shape was expected to be similar but with a slightly higher sea-state at the Isles of Scilly. Although there is clearly no substitute for local measurements as a comparison standard, the evidence shows that the radar-derived spectrum differed from the wave-model prediction in similar respects to the buoy-measurement from its corresponding prediction.

An additional piece of evidence is that a light-vessel instrumented for wave-measurement, located only 45 km from the radar point, measured a significant wave-height (obtained by a weighted average over the spectrum as a whole) of 2.4 metres compared with 2.47 metres from the wave-model and 2.7 metres from the radar result. The measurements made at a radio frequency of 9.25 MHz gave a value for the significant wave-height of 2.4 metres.

For a preliminary feasibility study of wave spectral-measurement, these results are considered to be very promising. A second experiment using a wave-directional buoy at the location of the radar resolution cell has recently been carried out and an extended programme of comparisons is now being planned.

PROPOSALS FOR FUTURE WORK

The installation at Angle, Pembroke, already discussed, is to be used in an extended study of sea-state measurement over the next eighteen months. The radar system is a prototype FMICW equipment and will be used in conjunction with a logarithmic periodic transmit antenna and 15-element receiving broadside arrays, each yielding 7 beams. An on-line computer incorporating an array processor will shortly be installed. This is planned to yield Doppler spectra on-line and to provide maps of radar-deduced wind-directions and currents. Digitised tapes will also be produced, which will be analysed off-line (at least initially) to deduce sea-state and spectra by the model-fitting technique already described and other methods under study.

With a view to applications studies at other locations, a more advanced FMICW radar using a commercial receiver with built-in single-loop digital sweep-frequency synthesizer is being developed. Back-synthesis technique is used to generate the transmitted carrier. This radar will be used with a transportable antenna system, either as a stand-alone system, or as a second radar viewing the same area as that surveyed by the Angle radar so as to obtain vector data.

ACKNOWLEDGEMENTS

The authors acknowledge the valuable assistance of Mr. B. Fursman and T. Facer at the site at Angle, and that of Mr. D. Eccles and his colleagues at the SERC Rutherford Appleton Laboratory, who are developing the on-line computer installation and software for the FMICW radar.

REFERENCES

1. CROMBIE, D.D., 1955, Nature, 175, p.682.

2. LONG, A.E., and TRIZNA, D.B., 1973, IEEE Trans. on Antennas and Propagation, AP-21, 5, p.680.

3. MARESCA, J.W., and BARNUM, J.R., 1977, IEEE Trans. on Antennas and Propagation, AP-25, 1, p.132.

4. CROMBIE, D.D., 1972, "Resonant backscatter from the sea and its application to physical oceanography", IEEE Conf. Publn. 72 CHO 660-1 OCC, p.173.

5. BARRICK, D.E., EVANS, M.W., and WEBER, B.L., 1977, Science, 198, p.138.

6. HASSELMANN, K., 1971, Nature, Phys. Sci., 229, p. 16.

7. BARRICK, D.E., 1972, "Remote sensing of the troposphere", V.E. Derr (Ed.), (US Govt. printing Office, Washington, D.C.).

8. ROTHERAM, S., 1981, IEE Proc., 128, Pt.F, 5, p.285.

9. SANDHAM, W.A., 1980, "Remote sensing of the ocean surface using MF/HF radar", Ph.D. Thesis, University of Birmingham.

10. SHEARMAN, E.D.R., 1981, "Area surveillance of sea-state, surface current and surface wind by dekametric radar technique", IERE Conf. Proc., 51, p.101.

11. SHEARMAN, E.D.R., COOPER, D.C., KUMAR, K., BAGWELL, D.J., and MOORHEAD, M.D., 1982, "Experimental studies of the performance of an MF/HF ground-wave radar on a coastal site of irregular contour", this Conference Publication.

12. SHEARMAN, E.D.R., and UNSAL, R.R., 1980, "Compatibility of high frequency radar remote sensing with communications", IEE Conf. Publn., 188, p.103.

13. LIPA, B., BARRICK, D., 1980, RADIO SCIENCE, 15, p.843.

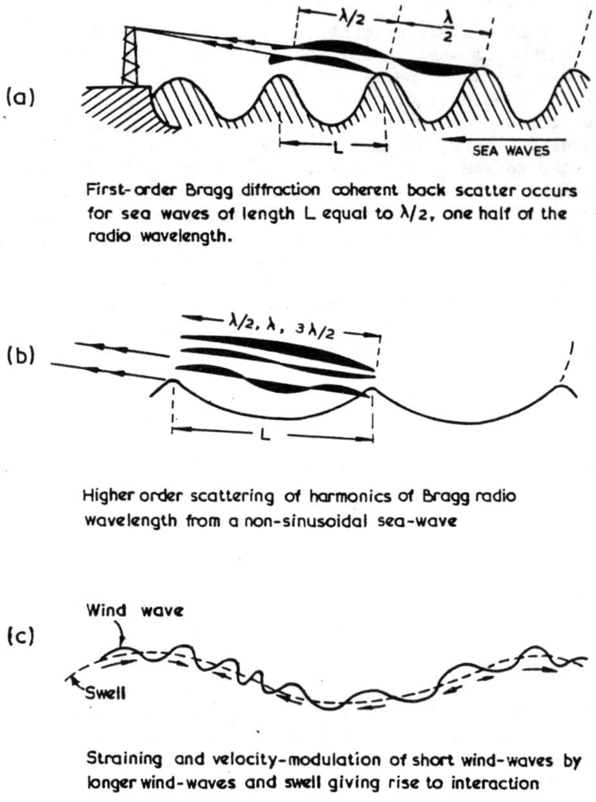

(a)
λ/2
λ/2
L
SEA WAVES
First-order Bragg diffraction coherent back scatter occurs for sea waves of length L equal to λ/2, one half of the radio wavelength.
(b)
λ/2, λ, 3λ/2
L
Higher order scattering of harmonics of Bragg radio wavelength from a non-sinusoidal sea-wave
(c)
Wind wave
Swell
Straining and velocity-modulation of short wind-waves by longer wind-waves and swell giving rise to interaction

FIG. 2 (a-c).

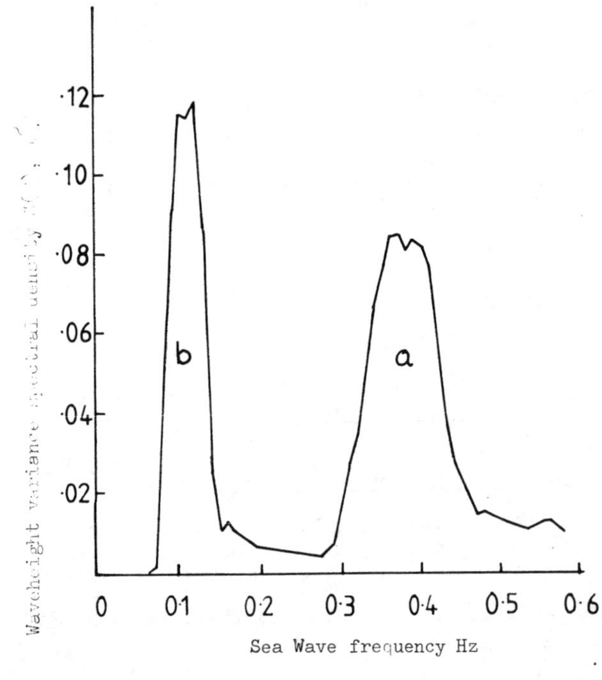

·12
·10
·08
·06
·04
·02
b
a
0
0·1
0·2
0·3
0·4
0·5
0·6
Sea Wave frequency Hz

FIG.1 Typical waveheight spectrum measured with buoys during the JONSWAP experiment, 1969, showing (a) wind-wave and (b) swell contributions. [Wave energy per Hz per m² of ocean = ρgS(f)]

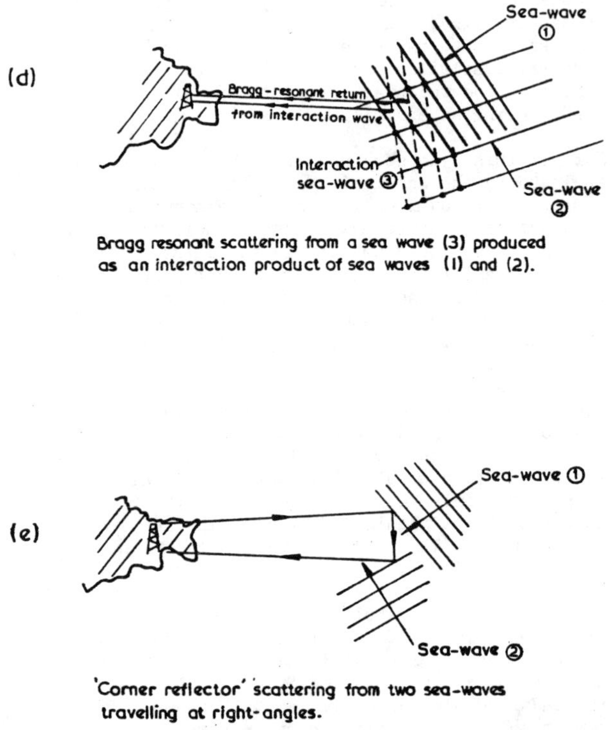

(d)
Sea-wave ①
Bragg-resonant return from interaction wave
Interaction sea-wave ③
Sea-wave ②
Bragg resonant scattering from a sea wave (3) produced as an interaction product of sea waves (1) and (2).
(e)
Sea-wave ①
Sea-wave ②
'Corner reflector' scattering from two sea-waves travelling at right-angles.

FIG. 2 (d-e)

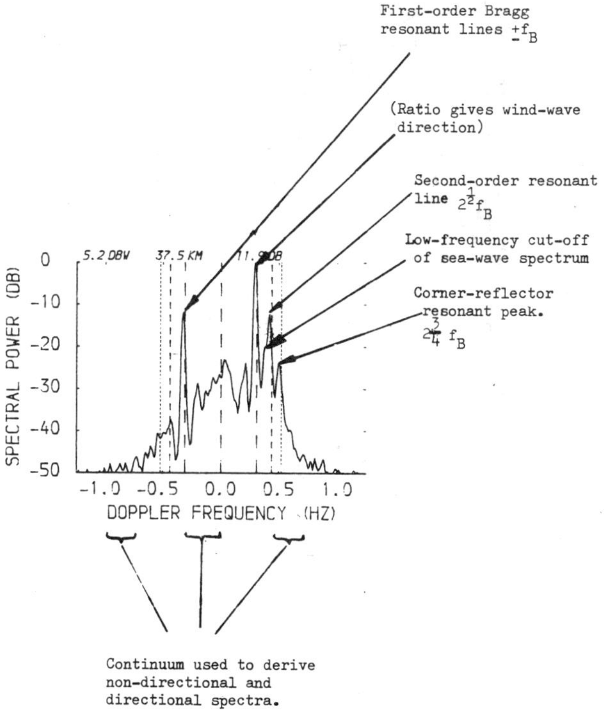

FIG. 3 Doppler sea-echo spectrum for range 37.5 km recorded at 9 MHz with the FMICW radar at Angle on 25.2.82. Dynamic range > 50 dB. First-order and second order features indicated.

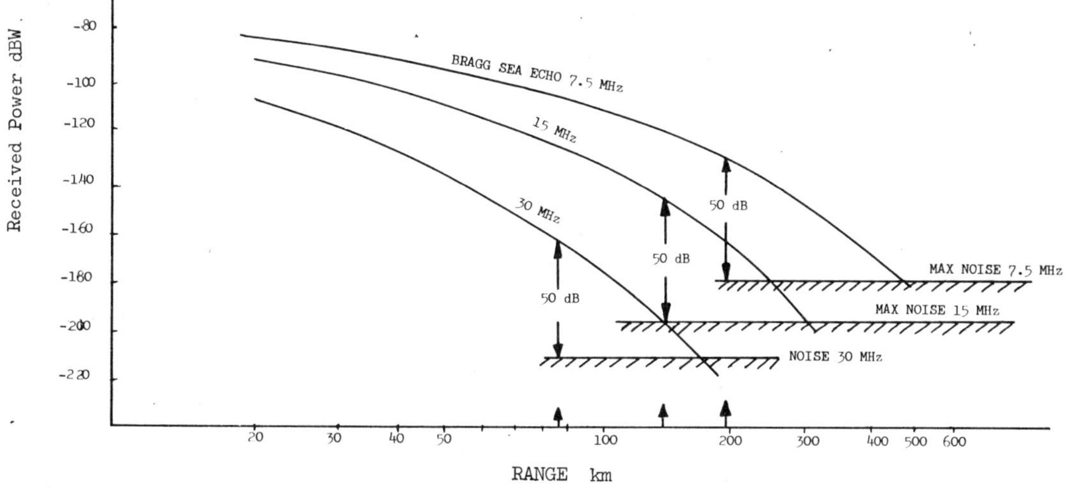

FIG.4 Bragg sea-echo and noise power vs range for three operating frequencies. Parameters of experimental system.

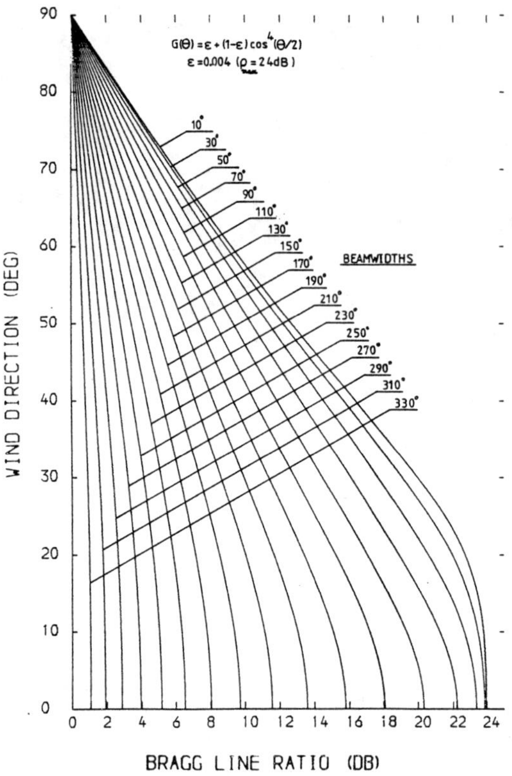

FIG. 5 Finite antenna beamwidth effect on Bragg line ratio for spread parameter S = 4.

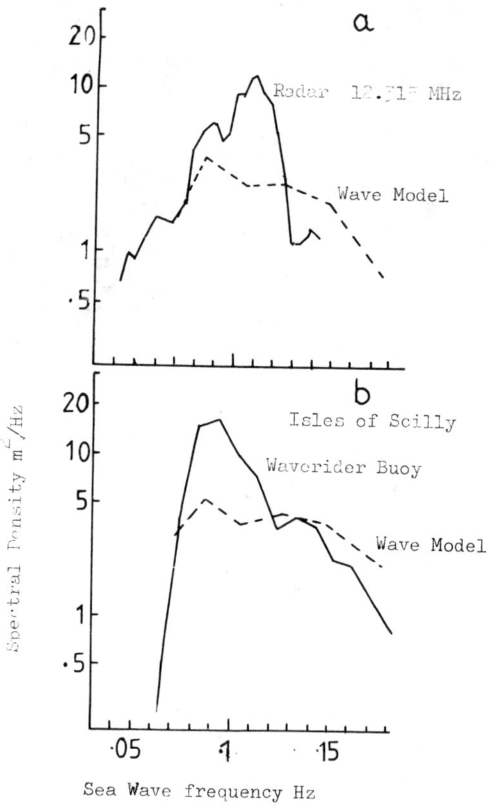

FIG. 6 Sea wave spectra measured by (a) the radar and (b) a waverider buoy at the Isles of Scilly, compared with local wave model spectra.

EXPERIMENTAL STUDIES OF THE PERFORMANCE OF AN MF/HF GROUND-WAVE RADAR ON A COASTAL SITE OF IRREGULAR CONTOUR

E.D.R. Shearman, D.C. Cooper, K. Kumar, D.J. Bagwell and M.D. Moorhead

Department of Electronic and Electrical Engineering, University of Birmingham

INTRODUCTION

An MF/HF ground-wave radar launching vertically polarized waves from a coastal site has been used for a number of research investigations in remote sensing of sea-state and ship-tracking (1)(2). Early investigations were conducted using omnidirectional transmitting and receiving antennas and in order to provide directional information from the received signals a wideband antenna array has recently been installed. The elements of the array are 2-30 MHz active loops and in one configuration eight elements are used to form four beams at ±14° and ±38° from the array normal. The beams are formed by networks of cables with weighting designed to provide -30 dB sidelobe levels. The geographical location of the radar and the beam positions are shown in Figure 1.

For accurate directional measurements the beam patterns should be known precisely and the irregular coastal contours and unknown ground parameters at the site are likely to have an important effect on these patterns. In this paper we present antenna patterns experimentally measured (using an early unweighted beam-former) and show that simple methods for estimating the site and propagation losses only give an approximate explanation for the observed perturbations in the patterns.

The results of experiments on the tracking of a ship using a pulse-Doppler radar operating at 1.95 MHz are also presented. A feature of MF/HF radar is the form of the sea-clutter spectrum, this being dominated by two Bragg-resonant spectral lines with much reduced energy at other frequencies. This makes it possible to detect and track ships whose speeds are such that the resulting Doppler frequencies differ from those of the Bragg lines (2).

Thus the processing of the received signals in two adjacent receive antenna beams involves range gating and spectral analysis in each channel followed by target detection and finally direction indication using the amplitude of the target in each channel and the known beam shapes.

RADIATION PATTERN MEASUREMENTS AND SIMPLE ESTIMATION OF SITE AND PROPAGATION LOSSES

Antenna pattern measurements were made using a transmitter carried by a ship navigating a circular track, centred on the radar and of radius 46.3 km. The ship's position was reported at intervals of 15 mins and position of the ship at intermediate times were interpolated from the reported data.

The directional patterns for each of the loop elements was recorded and some variation between the results for the various elements was noted. However, the average behaviour of the elements patterns did not correspond to the radiation pattern measurements given by the manufacturers for this type of loop and hence a simple approach to the estimation of the propagation losses for the ship-to-radar path was adopted to determine whether the discrepancy would be expected.

The radiation pattern of four unweighted beams was also measured in addition to the directional pattern of the individual loops. In this section we report the measured patterns and their agreement with the computed results. As shown in Fig. 1, the beam centres of three beams to the east of the St. Ann's Head are in the clear and are expected to be unaffected by the coastal site irregularities other than the local ground variation. The extent to which the ground in front of the loops affects the radiation pattern will be discussed later in the section.

The radiation pattern, for beams slewed away from the array normal for uniformly excited unweighted elements, is computed using standard equations and a simple program. A very good agreement between the measured and computed patterns existed for the above mentioned three beams and their first side lobes. However, the fourth beam which passes over a series of headlands, not surprisingly, showed a very poor agreement. Two sets of beam patterns are shown in Fig. 2 and 3, showing good and poor agreement respectively. We note that the disagreement shown in Fig. 3 could be due to the presence of irregular terrain and varying extent of ground in different directions. A terrain feature could explain the presence of a "dip" in the radiation pattern at about 60°, which corresponds to traversal of the maximum width of the peninsular in Fig. 1.

To estimate the propagation loss and ground attenuation, the length of land paths in different directions was determined using appropriate maps. For short distances in front of the loops an approximate expression (3) was used to calculate the attenuation coefficient. The average value of the ground constants for the area of interest were taken as 3×10^{-3} S/m and $\epsilon_r = 4$. The ground-wave propagation reported here takes place over a mixed path of ground and sea. In certain directions there are several discontinuities as a result of different ground constants. Millington's method (4)(5) is used to estimate the propagation losses. The curves of Rotherham (6) are used as the required source of data for all-ground and all-sea propagation respectively.

The theoretically-estimated and experimentally-measured radiation patterns for one of the loops is shown in Fig. 4. The general pattern of the two curves is as anticipated, showing a "dip" at about 60°; however, the perturbation on the measured pattern in the vicinity of the boresight can not be explained simply on the basis of ground losses. It is believed that this is caused by the irregular cliff contour and the presence of a bunker of significant size between the cliff and the loop. The "dip" in the pattern at 60° is due to the interposition of the full width of the St. Ann's Head peninsular seen on the map of Fig. 1, while the peak at 70° is attributable to the narrower isthmus joining the peninsular to the mainland.

The measured radiation pattern of all the eight loops clearly demonstrate the presence of dip and it is not surprising to find it in the beam pattern of Fig. 3. Thus it can be concluded that the shape of the radiation pattern is modified by variable attenuation of the radio waves due to different land paths over the distant peninsular in different directions, while the effect of computed attenuation over a small piece of land immediately in front of the loops is not very significant. To explain the perturbations due to the irregular cliff-line adjacent to the loops, a full three-dimensional modelling analysis is clearly required.

RESULTS OF THE TRACKING EXPERIMENT

To track a target in range and azimuth, four beams at $\pm 14^{\circ}$ and $\pm 38^{\circ}$ on either side of the boresight are used. The radiation patterns of the two central beams which are most extensively used for this experiment are shown in Fig. 5. The range of a target is determined by the time difference between the transmit and receive pulse. The accuracy of the range detection is a function of the resolution cell. In this section we will describe the method used for finding the direction of the target. In principle, the amplitude of the backscattered signal is recorded in two overlapping beams and used to estimate the direction of the incoming signal. The angular accuracy is best when the target appears almost in the middle of the two beams and preferably within their 3 dB points. The ratio of the signal levels in two beams will give the true direction of the incoming signal only when a single target is present and the background clutter or noise is negligible. The angular accuracy is a function of signal/clutter ratio and the measurement error falls as the ratio increases. Using Fig. 5(a) of Cooper and Longstaff [7] it can be shown that for (signal/clutter) > 10 dB the measurement error $\propto$ (signal/clutter)$^{-0.4}$. For constant echoing area of the target, (signal/clutter) $\propto R^3/R^4 \propto R^{-1}$ assuming clutter filling the beam. Thus it can be concluded that the measurement error $\propto$ (range)$^{0.4}$.

For comparison purposes the measured errors are normalised to 40 km assuming the (range)$^{0.4}$ relationship. As a result of the above process the mean and R.M.S. errors are 2.8° and 4.8° respectively. The R.M.S. error is consistent with the results given in reference 7 for a signal/clutter ratio of about 15 dB, this being typical of the experimentally observed signal/clutter ratio at the range of 40 km.

CONCLUSIONS

The directional characteristics of a broadside array of loop antennas on a coastal site have been studied experimentally and theoretically. The directional pattern was modified by an irregular coast-line and local terrain features and also by the interposition of a more distant headland in certain directions. The gross features of the modification of the directional pattern by the headland were satisfactorily explained by computation of the attenuation introduced by the mixed paths. The smaller perturbations introduced by the local features could not be so explained and clearly require a full three-dimensional model treatment. In spite of the irregular coast-line the main lobe and first side lobes for the complete array were in good agreement with free-space computed values down to the -20 dB level, except where propagation over the headland took place.

Tracking of ships using amplitude comparison between two overlapping antenna beams has been satisfactorily demonstrated. Increased tracking accuracy could have been obtained by using measured values for the beam directional patterns.

ACKNOWLEDGEMENT

The authors wish to thank Mr. B. Fursman and T. Facer at the experimental site for their assistance throughout.

REFERENCES

1. Shearman, E.D.R., 1980, The Radio and Electronic Engineer, 50, No.11/12, 611-623.

2. Sandham, W.A., Shearman, E.D.R., and Bagwell, D.J., 1978, "Remote sensing of sea-state and surface winds in the Irish and Celtic Seas using MF radar" IEE Conf. Publn. Antennas and Propagation, No.169, 1-5.

3. Jordan, E.C., and Balmain, K.G., 1968, "Electromagnetic Waves and Radiating Systems", Prentice Hall Inc.

4. Millington, G., 1949, IEE Proc., Part III, 96, 53-64.

5. Stokke, K.N., 1945, Telecommunication J., 42, 157-163.

6. Rotheram, S., 1981, IEE Proc., 128, 285-295.

7. Cooper, D.C., and Longstaff, I.D., 1972, IEE Proc., 119, 305-311.

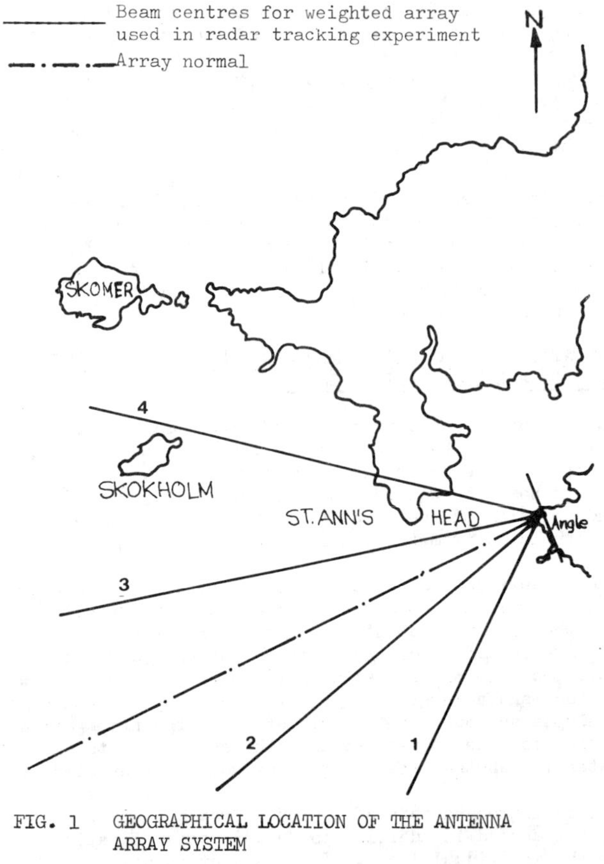

FIG. 1 GEOGRAPHICAL LOCATION OF THE ANTENNA ARRAY SYSTEM

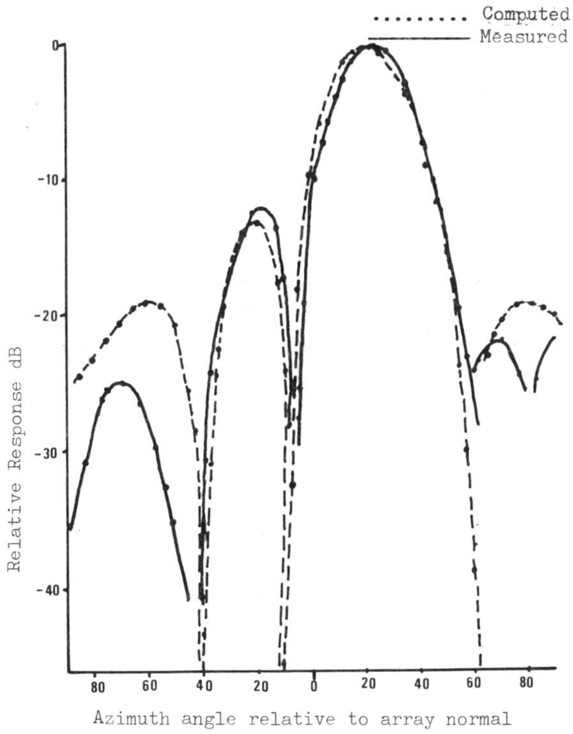

FIG. 2 DIRECTIONAL PATTERNS FOR BEAM 3 (Unweighted array)

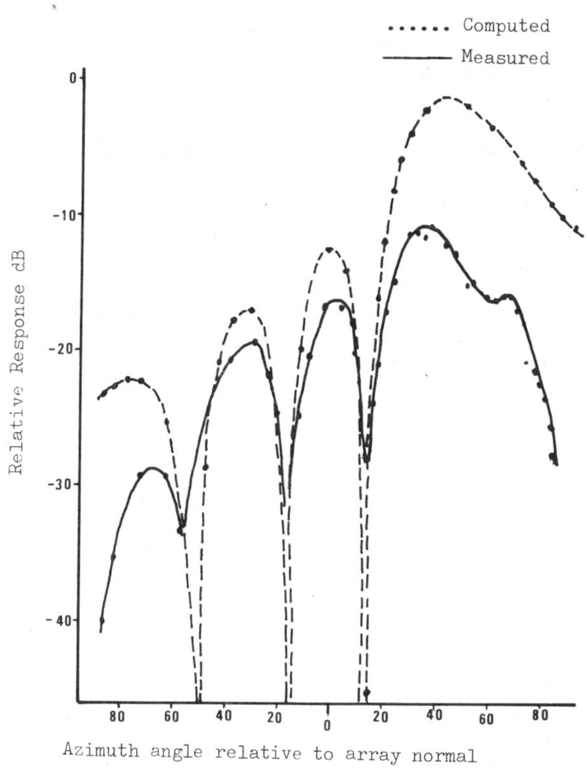

FIG. 3 DIRECTIONAL PATTERNS FOR BEAM 4 (Unweighted array)

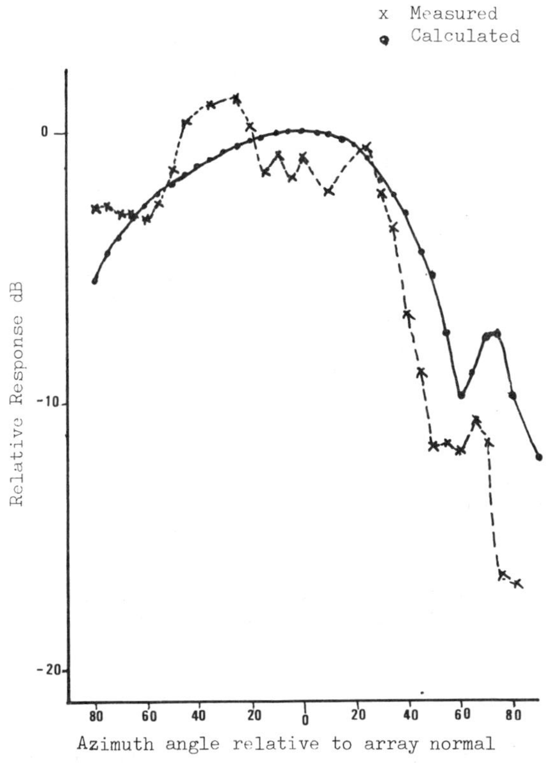

FIG. 4 RADIATION PATTERN OF A LOOP ELEMENT (Loop 3 from North end)

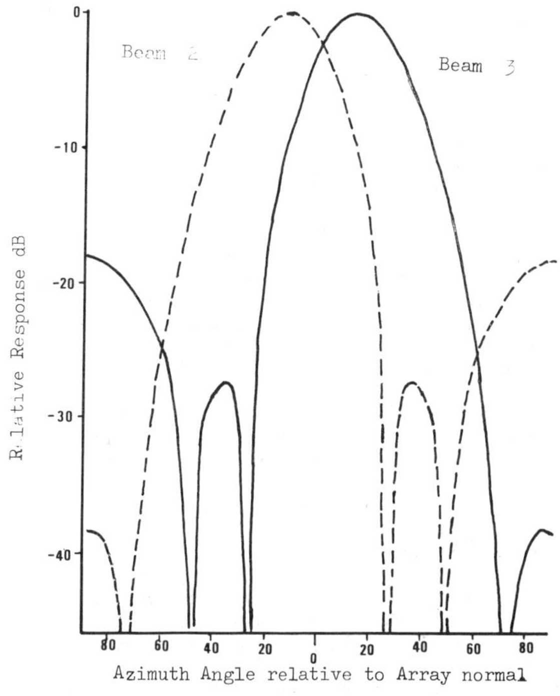

FIG. 5 RADIATION PATTERNS FOR BEAMS 2 AND 3 (weighted array) USED IN TRACKING EXPERIMENT

PROPAGATION EFFECTS ON A VHF RADAR

F Christophe and P Golé

Office National d'Etudes et de Recherches Aérospatiales (ONERA), France

INTRODUCTION

Since the first systems were built, the radar designers have required components operating at ever higher frequencies, in order to obtain better distance and angular resolutions.

Nevertheless, the VHF band offers various advantages for air surveillance, such as non ambiguous pulse-doppler capability, improved power balance in the presence of active or passive countermeasures, etc...

Therefore, ONERA has carried out the study of the VHF propagation effects, which may affect the performance of such systems.

INFLUENCE OF THE GROUND

Large Scale Effects

Range forecast . A qualitative illustration of the difference between conventional microwave and VHF systems as regards to the large scale ground effects is given by the curves in fig. 1 and 2 -from reference (1)- which show the electric field variation as a function of distance for constant transmitter and receiver heights over a smooth spherical homogeneous earth.

The well-known region of interference between the reflected and direct rays extends almost to the radio horizon in the case of microwave, with a sharp cut-off at this distance (fig. 1) whereas the field at VHF decreases in a smoother way from the last interference lobe, which is much closer to the transmitter, at a distance approximately proportional to the frequency (fig. 2).

The mathematical modelisation of the field may be made separately in three regions -interference, intermediary and diffraction- according to the important litterature which has been developed in the last four decades and is useful for predicting power budgets in various radio applications (see specially (1), (2)).

As a first step, we developed a computer program, based upon the residue series of reference (1) for evaluating the field attenuation as a function of soil parameters, radar and target heights, for distances close to the radio-horizon or greater ; at smaller ranges the interfering rays can be easily calculated, and both models have been connected in the intermediary region.

A fair agreement with the well-known CCIR curves (3) was obtained, and this program has been confronted to measured values. The experiment consisted in measuring the electric field received on-board an airplane flying above the ocean as a function of its height and distance to the transmitter. The airplane position was precisely known and fed as an input parameter to the computer program : the flight profile, mainly located in the intermediary and diffraction regions, is represented in fig. 3 where the data points correspond to 30s intervals. The field values measured during the flight are shown in fig. 4 (solid lines). These values are expressed as the attenuation relative to the free space field. The curve in dashed line shows the computed values. Though the above models do not account for atmospheric or surface inhomogeneities, it may be seen that a good agreement is obtained between theoretical and measured values, thus allowing to use this program in the range estimation of a given VHF radar located in a flat place. The particular case of reliefs surrounding the radar location will not be treated here - a good prevision of this effect can be calculated, as shown in (4)- and the lower diffraction losses behind obstacles at VHF than at microwaves will only be repeated.

Low elevation defocusing. For a radar system capable of measuring the elevation of the targets, some vertical extent of the antenna is required, which may be sufficient to separate the direct and reflected rays from a target in the interference region ; in the opposite case, the system has to deal with the well-known problem of target and image in the beamwidth, with its many proposed remedies.

At VHF, since the intermediate and diffraction regions may be within range of the radar, the effect on elevation measurement has to be evaluated there in a different way.

According to reference (1), the received field is proproportional to :

$$A = \sum_{s=0}^{\infty} F_s(D)\ f_s\ (h_1)\ f_s\ (h_2) \qquad (1)$$

which can be usually separated within the diffraction region into:

$$A = F\ (D)\ f(h_1)\ f(h_2) \qquad (2)$$

where h_1 and h_2 are the transmitter and receiver heights, F (D) is the ground wave attenuation at distance D, and f (h) is the height gain factor.

Figures 5 and 6 illustrate the amplitude and phase values of the height gain factor as a function of height for propagation over sea water and medium ground respectively, at 5 m wavelength.

As could be expected the amplitude increases with height, with the exception of a minimum occuring above seawater for a height of 10 m, revealing a preponderant surface wave at low altitude.

Calculation of the focal plane elevation response of the vertical extent antenna to a target in these regions is now feasible by Fourier transform of the complex gain factor, giving both apparent elevation of the target, and sidelobes level.

It has to be pointed out that, depending on the median height of the antenna, and on the parameters of the soil, different wavefront tilts can be obtained : for example an antenna at an altitude of 60 m above sea water should see an incident wavefront at elevation 0.4° from any target in the region allowing the approximation of equation (2).

The experimental confirmation of this effect, actually intended by ONERA, may be altered by a distorted refraction index profile at low altitudes, acting in a similar way.

Near Ground Effects

In the case of a large array, that we shall consider now for good angular resolution, each radiating element has poor directivity at VHF, hence illuminating the nearby ground and possibly causing multipath interference. Ground scattering when in the near field zone of the array creates uncorrelated amplitude and phase fluctuations at each element, thus principally rising the sidelobes.

Since theoretical estimation of this multipath effect does not seem practical for locations on rough soils, when the low elevation angle regularization is no longer possible at short distances, experiments were conducted in the following manner with emphasis on phase perturbations by scatterers around VHF ground antennas.

In a similar experiment as the one described before, a transmitting antenna was mounted aboard an aircraft with inertial trajectography capability, moving along various flight paths over the sea, and the signal was received some 100 km away on various antennas -vertical monopole on 5 to 10 m masts- on a vegetation-covered irregular hill on the shore. This geometry allows for separation between the previously mentioned large scale effect of the sea, and the scatterers -trees, rocks,...- near the antennas, when associated with constant altitude and distance flight paths.

Interferometric measurements were performed, the phase of the received signal at a master antenna being compared with the signal received at five other antennas -see fig. 7- with interferometric arms of 5 to 50 wavelengths at various orientations.

After compensation of the theoretical fringe movement deduced from the trajectography data of the aircraft, the remaining phase fluctuations are attributed to propagation effects ; an example is given fig. 8, corresponding to the previously shown flight path.

The decorrelation of these phase errors and the slowness of their variations for large angular motion of the transmitter, confirms the stated multipath interference near each antenna.

The R.M.S. phase fluctuations for the 5 interferometers during this flight was 20°, the instrumental error being negligible.

If an equal uncorrelated contribution at each antenna is assumed, its R.M.S. value is then 14°, which may characterize the random effect, at the experimental site, of the scatterers surrounding the chosen vertical monopole antennas ; no clear dependence of the height of the mast appears in this case between 5 to 10 m.

Related experiments -reported in (5)- allowing for phase perturbation measurements on the link between two such ground antennas on the same test site, yielded similar results.

Introducing phase errors of this order could be acceptable in a phased array, depending on the required sidelobes level and the number of radiating elements as can be found for example in (6). In the opposite case, the array should be located on a smoother ground.

FALSE-ALARM PRODUCING PHENOMENA

As is well-known, radars are affected by various false-alarm inducing phenomena according to the frequency. It is only intended here to show the relative importance of such phenomena and compare them with the values for conventional radars at microwave frequencies.

Resonant Sea Clutter

The HF band is now in current use for remote sea-state sensing, since a resonant Bragg diffraction occurs for radio wavelengths and incident angle tuned to the swell (7) ; in such cases, a Doppler shift of the echo is produced by swell propagation.

Similar effects, mainly in the lower part of the band, are expected at VHF.

Conventional "Angels"

Ducts. There are several laws relating the superrefractive layer-thickness to the maximum wavelength that can be propagated within the duct with very little attenuation. They show that, while a layer thickness of only several meters is sufficient to provide ducting at microwave frequencies, a thickness of several hundred meters is needed for the same results in the case of VHF which is very seldom the case (8), thus deeply decreasing the probability of occurence of such phenomenon, and of the associated blind elevation angles.

Tropospheric inhomogeneities. The relevant parameter to consider in this case is the radar cross-section per unit volume η . It is generally admitted (9) that it decreases as $\lambda^{-1/3}$. η should therefore be 2 to 10 times smaller at VHF than at microwaves. However, tropospheric turbulence nevertheless has a very low radar cross-section compared to practical applications.

Hydrometeors and insects. It is generally admitted that in this case, the radar cross-section decreases as λ^{-4}, which means that at VHF these conventional "angels" are not detected.

Ionospheric inhomogeneities. The radar cross-section of incoherent backscattering by ionospheric inhomogeneities is proportional to the number of electrons contained in the radar pulse-volume and depends strongly on the magnetic latitude. It also appears that it is negligible compared to echoes involved on practical applications and would not lead to false-alarm problems in medium latitudes (10).

Meteor Trails

Among all the phenomena investigated here, near specular reflections on meteor trails are those producing the higher radar cross-section in the VHF band, and thus echoes from these trails are a source of false alarms for such applications as surveillance radars.

These echoes have been studied since the first radars were built, but their resolution usually did not allow for direct observation of the details. While developing a VHF test array with improved 4-D resolution capability, whose description is out of the scope of this paper, ONERA has detected several times very strong echoes from overdense meteor trails. Fig. 9 shows a three-dimensional representation of a meteor trail echo which was detected at an elevation of about 35° and a distance of 150 km (parts A-C). Part D of this figure shows the time variation of the echo which lasted for about 3s, with strong periodic fluctuations showing that a beat occurs between two nearly equally reflecting parts of the trail. It was often suggested that such beating resulted when the trail is separated into two parts by a high altitude wind-shear. The relative drift velocity of the two trails was supposed to explain this doppler beat effect (11). Part A of fig. 9 where both parts of the trail have actually been angularly and Doppler separated, confirms such an explanation.

According to reference (12), the average meteor flux incident on the earth's atmosphere corresponding to line densities greater than 10^{14} electrons/meter is $1.6.10^{-12} m^{-2} sec.^{-1}$, which, for an omnidirectional array with a maximum range of several hundred kilometers may result in an average number of false-alarm producing echoes of the order of 20 meteors/min, depending on the shape of the Doppler filters, and of the antenna beams.
As a conclusion, overdense meteor trails may produce very strong echoes which may result in a high false alarm rate. However there is a need for more data concerning the actual number and strength of echoes detectable at VHF by a given system.

CONCLUSION

The study which has been carried out on the propagation effects upon a VHF ground radar, shows advantages and disadvantages when compared with conventional microwaves systems, none of which seems decisive.

REFERENCES

1. Bremmer, H., 1949, "Terrestrial Radio Waves", Elsevier Publ. Co.

2. Bullington, 1946 "Radio Propagation at frequencies above 30 Megacycles. National Electrics Conference, Chicago.

3. CCIR, 1959, "Atlas of Ground-Wave Propagation Curves for frequencies between 30 and 10000 MHz". UIT, Geneva.

4. Ott, R.H., 1979, Ground Wave Propagation over irregular homogeneous terrain, AGARD CP-269.

5. Christophe, F., 1981, Transport of a phase reference by VHF propagation above vegetation covered ground. 2nd ICAP, York, U.K..

6. Steinberg, B.D., 1976, Principles of Aperture-Array and System Design, Wiley.

7. Lipa, B., Barrick, D., 1979, Ocean Swell parameters from narrow beam HF radar sea echo. AGARD CP-263.

8. Hall, P.M., 1979, Effects of the troposphere on radio communications. P. Pengrimms Ltd, U.K.

9. James, P.K., 1980, A review of radar observations of the troposphere in clear air conditions. Radio Science, 15, 151-175.

10. Towle, D.M., 1980, VHF and UHF radar observation of equatorial F region ionospheric irregularities and background density. Radio Science 15, 87-94.

11. Greenhow, J.S, 1952, Characteristics of Radio Echoes from Meteor Trails. III The behaviour of the electron trails after formation Proceedings of the Physical Society, 65, 3B, 169-181.

12 Millman, G.H., 1977, H.F. Scatter from overdense meteor trails, AGARD CP-244.

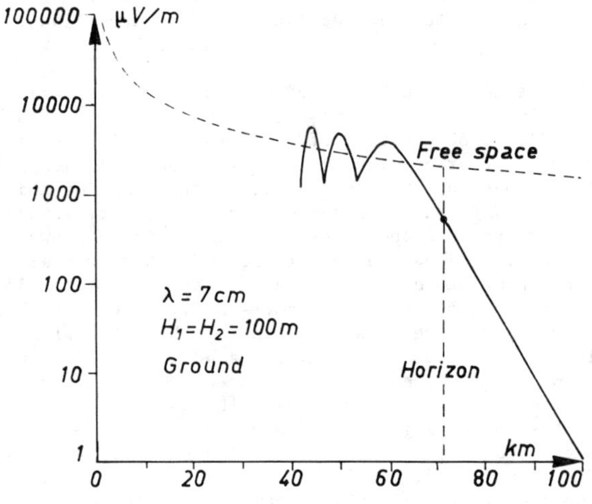

Fig. 1 : (From Bremmer [1])

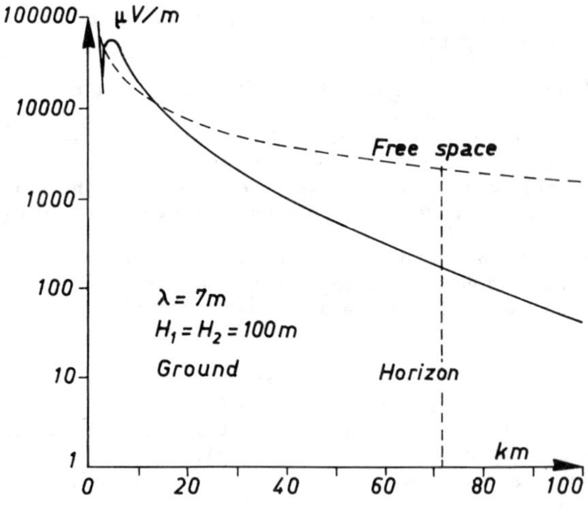

Fig. 2 : (From Bremmer [1])

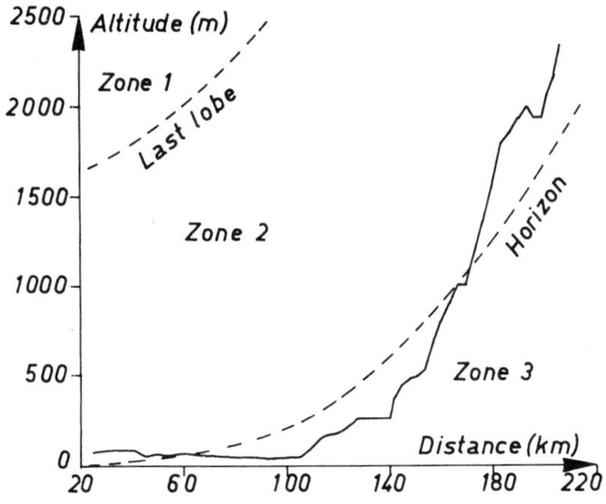

Fig. 3 : Flight profile

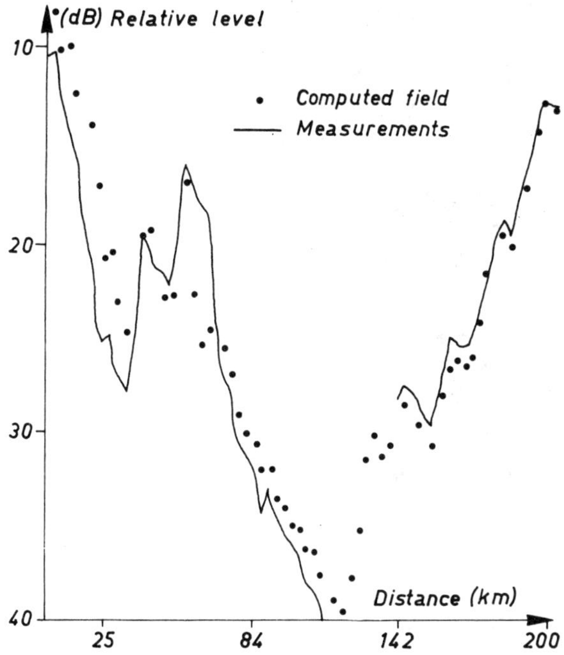

Fig. 4 : Attenuation relative to free space

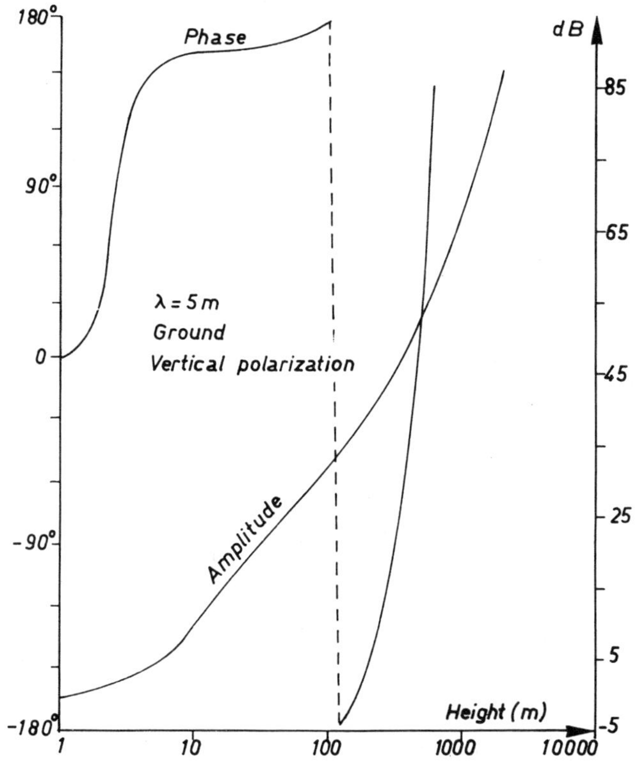

Fig. 5 : Complex height gain factor

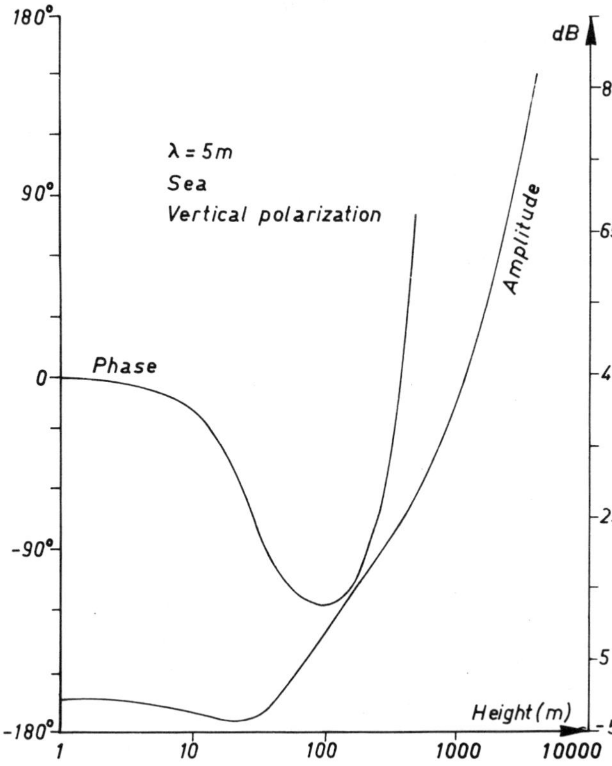

Fig. 6 : COmplex height gain factor

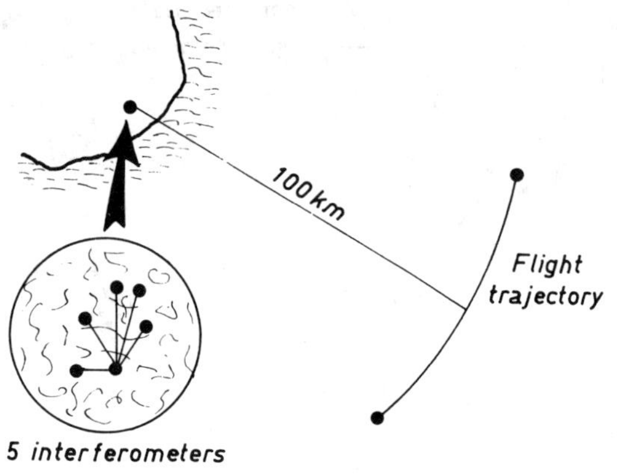

Fig. 7 : Interferometric measurements

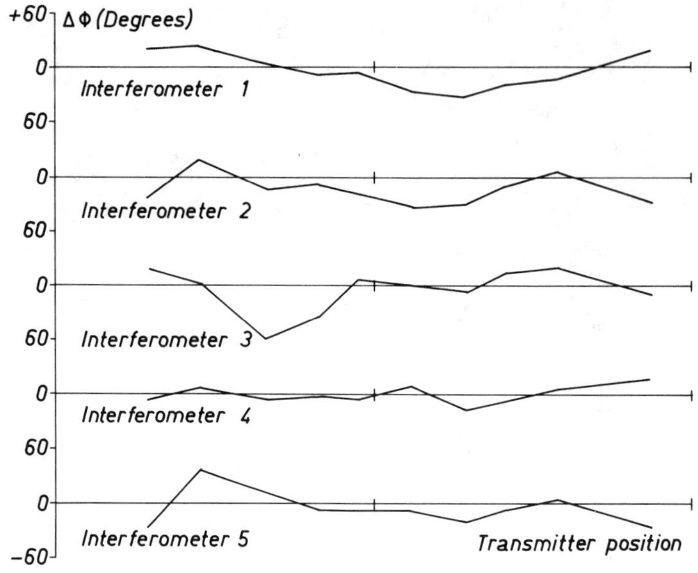

Fig. 8 : Phase fluctuations

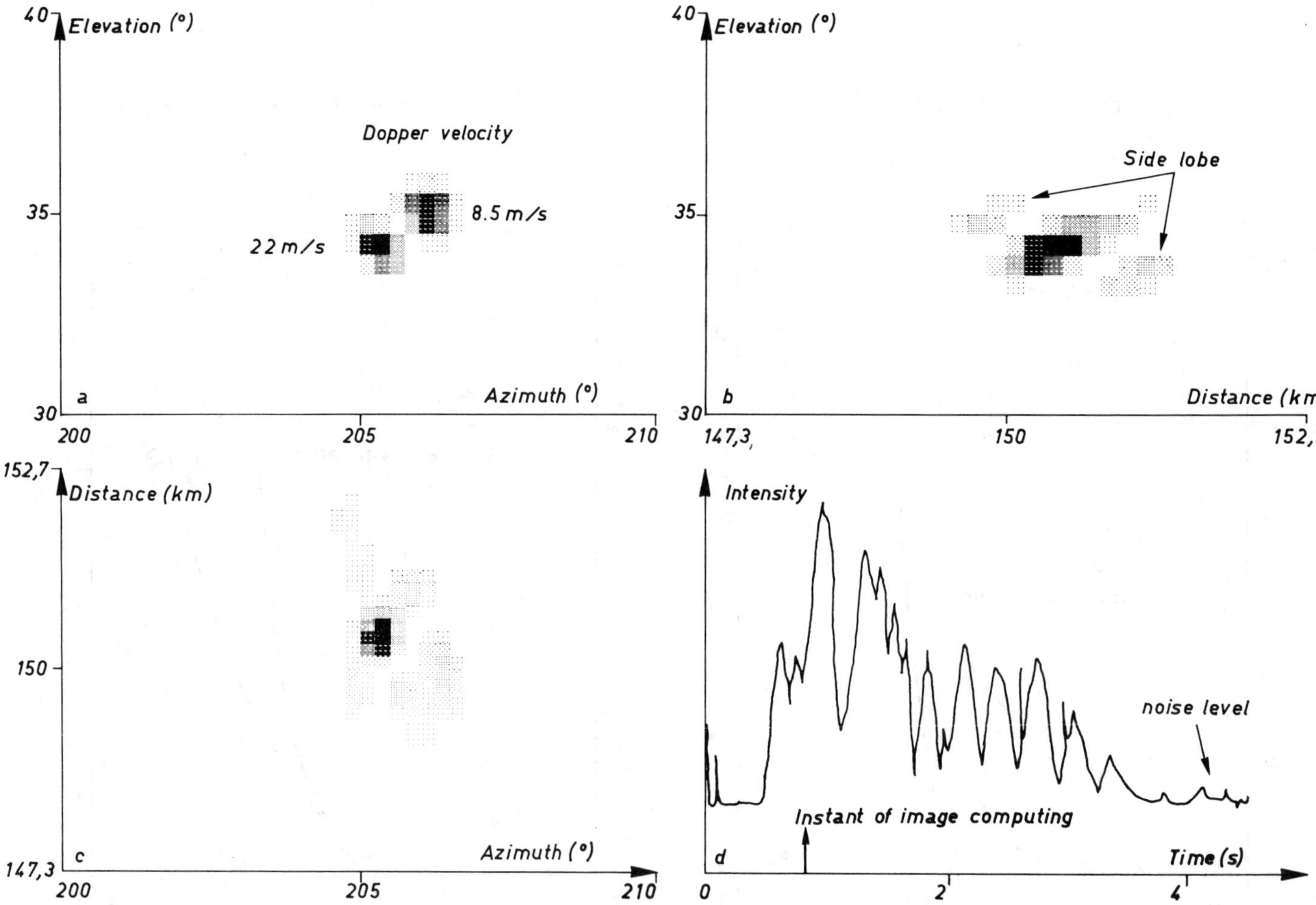

Fig. 9 : VHF detection of an overdense meteor trail

A BARRIER RADAR CONCEPT

J. Marshall, C. Ball, I. Weissman

Riverside Research Institute, New York, N.Y.

INTRODUCTION

There exist many situations, both military and civil, in which it is desired to monitor an extended perimeter to provide detection and warning of intrusions. Typically, such regions may be remotely located and have an extent of hundreds or even thousands of kilometers. Primary considerations in establishing such a warning system involve suitability for unmanned operation, high reliability, low power consumption, ease of equipment transportability and set up, and minimal cost. This paper describes one approach for meeting these requirements, namely, a low power, light-weight radar which can be quickly set up and operated on batteries for extended periods of time to detect airborne intruders. By keeping the equipment procurement and operating costs sufficiently low, it becomes practical to employ a multiplicity of such radars to provide an unbroken intrusion fence over the desired perimeter.

As shown in Figure 1, each radar establishes a single transmitted fan beam extending vertically from horizon to horizon. This beam is generated by a two-face array antenna built in an A-frame configuration, and is shaped, through phasing of the array elements, to concentrate the transmitter power in a manner consistent with the expected operating altitude ceiling of the targets of interest. The angular width of this beam in the dimension transverse to the fan is dependent on the radar transmission frequency and the antenna aperture dimension, but is typically wide enough so that a target at the maximum altitude or range will require tens of seconds to pass through the beam. Thus a large number of independent samples of radar data will be available, to provide many opportunities for target detection. Short range targets will of course pass through the beam more rapidly; however such targets are also more readily detected because of the R^{-4} characteristic of the received power variation.

On reception, the fan is divided into a number of narrower receive beams. The number of these is determined by the transmitter frequency and available aperture. In typical designs this number might lie within the range of 4 to 16 beams total. Formation of these beams results in the most efficient use of the aperture available, and also provides capability to establish the angular location of detected targets. Since these beams are non-scanning, there is no requirement for use of variable phase shifters. Instead, the outputs of the array elements are combined in a suitable matrix which provides both the phase shifts for beam formation and suitable amplitude tapering for sidelobe control. In typical cases it may be feasible to establish the phase relations required by use of appropriately-cut fixed lengths of coaxial cable.

In order to provide more precise localization of targets, monopulse receive beams may be implemented. When the radar is in its normal surveillance mode, only the monopulse sum beam is processed. A detection in this beam results in immediate processing of a subsequent set of radar transmissions to verify the detection. If this verification is positive, the monopulse off-boresight difference data is processed to localize the target in angle. By use of periodic looks as the target crosses the beam, an estimate of the target trajectory and speed may be formed. Figure 2 indicates the matrix connections used to form the monopulse beams.

SELECTION OF RADAR SYSTEMS PARAMETERS

A fundamental goal in the design is to keep the total power consumption as low as possible in order to allow operation from batteries (or, in some cases, solar cells) in remote areas. To this end, transmitter power is minimized by exploiting the short ranges associated with the vertical barrier. Also the use of coherent signal integration provides an efficient match to the transmitted waveforms. The integration is conveniently performed by a fast Fourier transform (FFT) processor, which also serves the function of enhancing the target detection in the ground clutter which may be encountered in the lower beams. The maximum time over which the returns may be coherently processed is limited by the processor losses resulting from the change in Doppler shift as the target moves across the beam. To prevent these losses from being excessive, the maximum allowable coherent processing time is of the order of $(\lambda R)^{1/2}/V$, where λ is the transmitted wavelength, R the target range, and V the component of target speed normal to the fan. Since this time is a function of range, a mixture of coherent and non-coherent integration is used for targets at shorter ranges, where the returns are stronger, and fully coherent processing is used only at the longer ranges. In a typical case a 256 point FFT might be used at the longer ranges, with 4 consecutive 64 point FFT's at shorter ranges.

The number of range cells which must be processed is determined by the maximum detection range required and the range resolution selected. Detection range will usually be limited by the horizon in the lower beams. While this range can be extended by placing the radars on towers, this procedure may not always be feasible in the projected applications. Favorable topographic features of course can and should be exploited in siting the radars. The range resolution is dictated in part by the requirement that a target having maximum range rate remain within the same range cell for the entire integration (coherent plus non-coherent) interval, and in part by the requirements of clutter reduction in order to minimize the region in which low altitude targets will be obscured by ground clutter. As shown in Figure 3, this region lies along or near the center-line of the beam in the case of targets penetrating normal to this line. In practical cases the beam will be sufficiently wide that the target will be detectable above the clutter for a considerable time duration after it enters the beam, and again before it leaves.

Figure 4 shows the average transmitter power requirement for each coherent transmission for detection of objects having radar cross sections representative of typical airborne targets. These curves pertain to an assumed range of 34 kilometers, and show the variation with both antenna size and transmitted wavelength. Operation in bands B through E (1.0 meter to 0.1 meter wavelength) has been investigated. While the higher frequencies require somewhat less power for a given antenna size, most practical applications considered have favored the longer wavelength (bands B or C) because of the smaller number of radiating elements required and the resulting saving in transceiver modules and processing load. Solid-state modules capable of delivering hundreds of watts of peak power are readily available in these bands, enabling use of solid-state circuitry throughout to maximize the reliability of the radar.

MINIMIZATION OF SYSTEM POWER

The reduction in peak transmitter power brought about by use of coherent processing in effect shifts much of the burden to the processor. In a typical application the processor must compute a 256 point FFT within an integration period of perhaps 0.1 second for each of several hundred range cells and for up to 16 simultaneous beams. The resulting processor throughput rate may be equivalent to between 10 and 100 million operations per second. Study of the processor requirements indicates that such throughput can be achieved by use of hard-wired pipeline processors with a separate processing module for each beam. However, the electrical power, as well as weight and cost, of the processor increase linearly with the number of beams which must be processed. A typical processor, having capability to service 16 beams, may require as much as 300 watts input power. This estimate is based on a hypothetical processor design which utilizes CMOS technology wherever possible. However, certain components, notably high capacity RAM's and high-speed multiplier-accumulators, are not at present commercially available in CMOS, and it is these components which largely determine the power consumption.

As is evident from the preceeding discussion, when many beams must be processed the processor power can become the dominant contributor to the overall system power drain. Reduction of this power to a level consistent with battery operations (with possible solar cell replenishment) is a major goal of the design. One effective means of accomplishing this result is through operation of the radar at a low duty. If the fan beam is made sufficiently wide in the transverse dimension, many hundreds of coherent processing intervals may be available during the transit time of a long-range target. This number is more than is actually required, and can be reduced substantially and still permit a high probability of detecting the target by operating the radar intermittently, typically by using only 5 to 10 percent of the available coherent transmission intervals. However, targets crossing the barrier at short ranges could pass through undetected at such a low transmission rate, and hence require additional consideration. Fortunately, much less processing is required for these targets, since extensive coherent integration is not needed for detection. The main consideration is the ability to detect them in the clutter background, and this ability can be provided by conventional MTI canceller techniques.

The signal processor which has evolved from consideration of all these factors is shown in Figure 5. FFT processing is performed in the range cells at the longer ranges (256 points) and at intermediate ranges (64 points, with 4 consecutive groups non-coherently combined). A three-pulse canceller is used at the short ranges. A special waveform, described subsequently, is used to allow these various processing modes to be intermixed. The outputs of both the FFT and MTI processors are sampled by a microprocessor controller, which perform thresholding and target detection functions, directs generation of verification looks and processing of monopulse data, and formulates target reports for verified targets. The microprocessor operates continually; however the power drain of this unit is quite modest.

Since the radar waveform is transmitted intermittently, at low duty, it is no longer necessary to provide a separate processor module for each beam. Instead, the data received in all beams on each transmission may be entered into buffer storage and processed sequentially by a single module. This procedure provides reduced cost and a marked improvement in the processor weight and power consumption. In a representative configuration the processor power drain can be reduced to about 15 watts, instead of the several hundred watts required with operation at 100 percent duty.

REPRESENTATIVE BARRIER RADAR DESIGN

Table I lists parameters of a barrier radar configured to detect an intruding aircraft with 10 m^2 RCS travelling at Mach 1 speed. The waveform designed for this particular application is shown in Figure 6. It comprises a pair of interleaved waveforms: a triplet of unchirped narrow pulses is transmitted at a sufficiently high data rate to insure detection of targets passing through the beam at ranges between 0.15 and 4.2 km. MTI processing is used with this waveform. A higher energy chirped waveform is used with a 10 MHz frequency offset to provide detection at longer ranges. This waveform is transmitted as a burst of 256 consecutive pulses for FFT processing with a waiting period between bursts. The duty is sufficient to provide from 3 to 24 independent looks at any target penetrating the barrier at ranges between 4.2 and 34 km. The range resolution is 450 meters for the chirped pulses and 150 meters for the unchirped triplets.

Table II provides estimates of the system weight and power drain. These figures apply to the radar equipment itself, and do not include communications and batteries or other prime power provisions. (The weight of the batteries will of course depend on the required period of operation without recharging.)

TABLE I

Representative barrier radar parameters

Transmitter Frequency	400 MHz
Peak Power	2 kW
Average Power (Long-Term)	3 W
Antenna Aperture	1 m^2 per face
Maximum Detection Range	35 km (10 m^2 RCS)
Range Resolution	450 m (med. and long range) 150 m (short range)
Waveforms	27 μs chirped pulse (10: 1 PC, 256 pulse burst) 1 μs unchirped pre-pulse triplet
Processing	256 point FFT (long range) 64 point FFT (med. range) 3 pulse MTI (short range)
System Noise Temperature	725° k
System Losses	10.5 dB
Total Power Consumption (exclusive of communication)	45 W

TABLE II

Power and weight estimates

	Power (W)	Weight (kg)
Antenna		
Radiators	—	25
Support Structure	—	90
Transceiver Modules	10	10
Exciter	5	10
Other Receiver Elements	10	15
A/D Converters	5	5
Signal Processor	15	15
Total	45	170

An overall block diagram of the radar is shown in Figure 7 and a brief description of its features is provided here. The antenna requirements are compatible with use of microstrip array technology. This type of construction results in a light-weight, inexpensive, readily manufactured antenna structure. Solid state transceiver modules, one module per receive beam, connect directly to the rear of the antenna structure. These modules contain the transmitter power transistors, the TR switch and isolation circuitry, and the receiver preamplifiers which are needed to achieve low noise performance and overcome losses in the beam forming matrix which is connected to the preamplifier outputs. This matrix provides the sum and two monopulse difference outputs for each of the beams. Additional amplification and down-conversion to i.f. is performed after the matrix, followed by compression of the chirped pulses using SAW delay lines, and coherent demodulation into in-phase (I) and quadrature (Q) components. The I and Q demodulator outputs are sampled once per range bin, A/D converted, and sent to the signal processor. An exciter unit provides r.f. drive signals at a level sufficient to drive the power amplifier stages, and also provides the local oscillator and I and Q demodulator reference signals. The generation of the transmitter waveforms is initiated by commands from the microprocessor controller in the signal processor unit.

OTHER CONSIDERATIONS

A communications link, operating at the radar site, is an obvious requirement for most of the contemplated barrier applications. While communications requirements have not been considered in detail in this paper, the use of on-site processing of the radar data to formulate target and system status reports implies that low data rate transmission (probably less than 10 bits per second) will be adequate. A 600 km HF data link operating at this information rate will probably consume not more than a few watts.

Other issues, such as site security and weather protection, must of course be addressed in the formulation of the system design. High reliability, a paramount consideration is enhanced by use of all solid-state components, fixed beam positions, and by the low-power operation contemplated. Modularity of construction allows rapid maintainence through replacement of faulty modules when failures do occur. It also permits growth to a higher power capability, should this be required for lower RCS targets. Provisions will be made for each radar to continually monitor its own operational status and transmit this information on the data link along with target reports.

The barrier radar, while at present only a concept which has not yet been submitted to thorough study or detailed design, nevertheless appears to be an attractive option for providing intrusion monitoring in a variety of applications.

ACKNOWLEDGEMENT

The concepts described here were developed under internal research funding provided by the Riverside Research Institute.

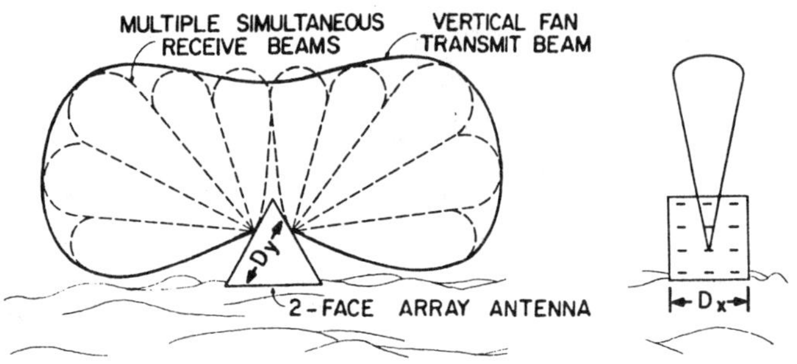

Figure 1 Barrier radar concept

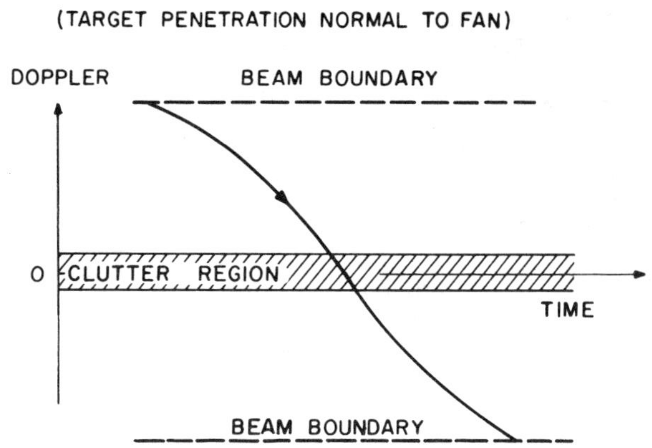

Figure 3 Target Doppler history

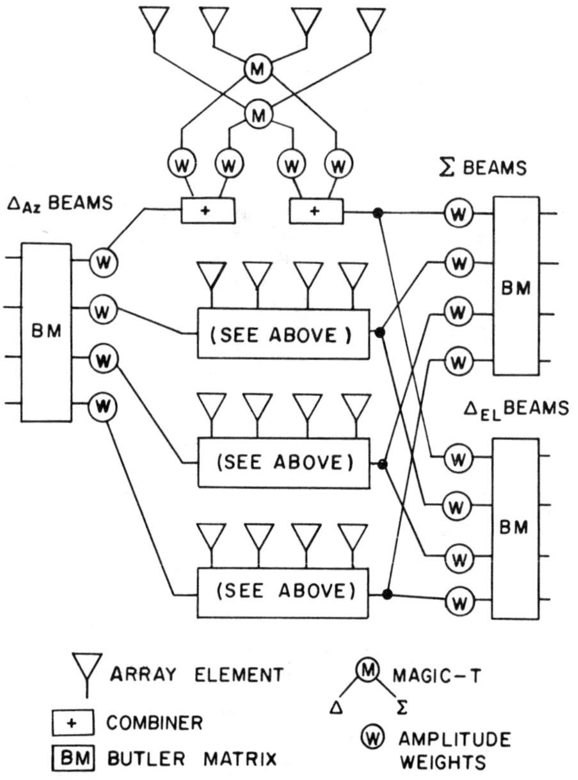

Figure 2 Beam forming matrix

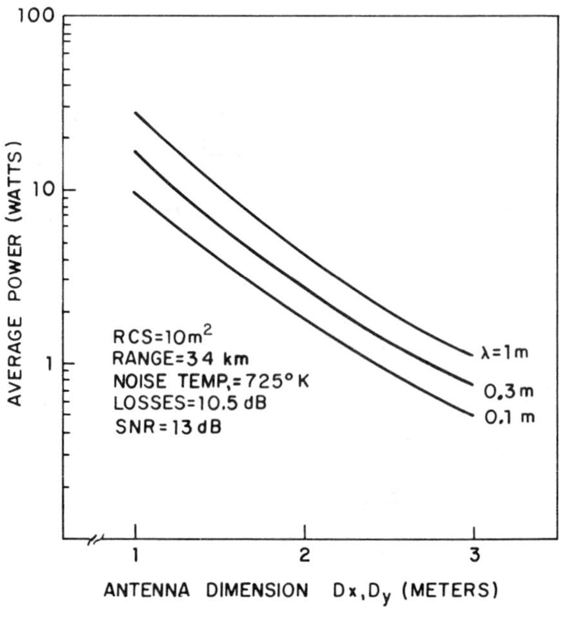

Figure 4 Transmitter power requirement (Average over a single coherent processing interval)

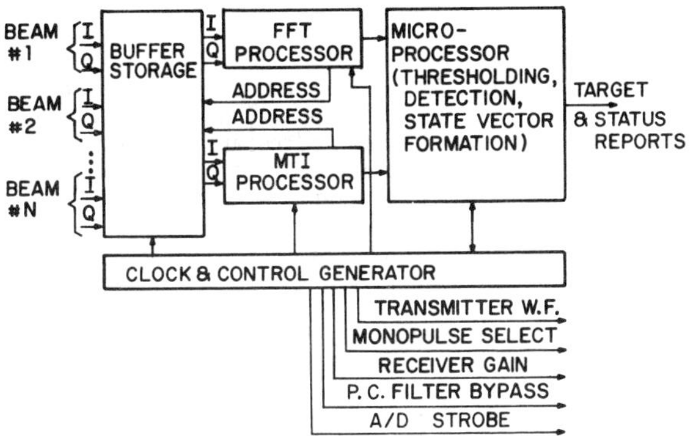

Figure 5 Signal processor

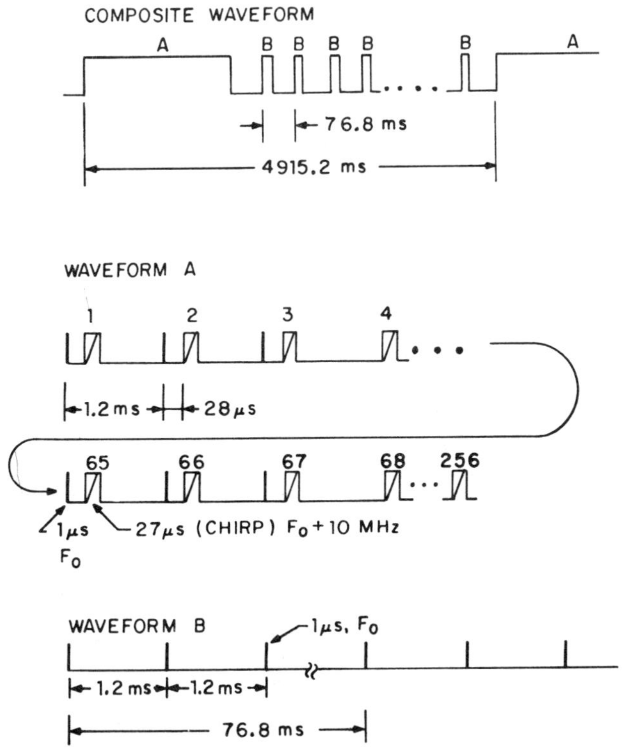

Figure 6 Transmitted waveform

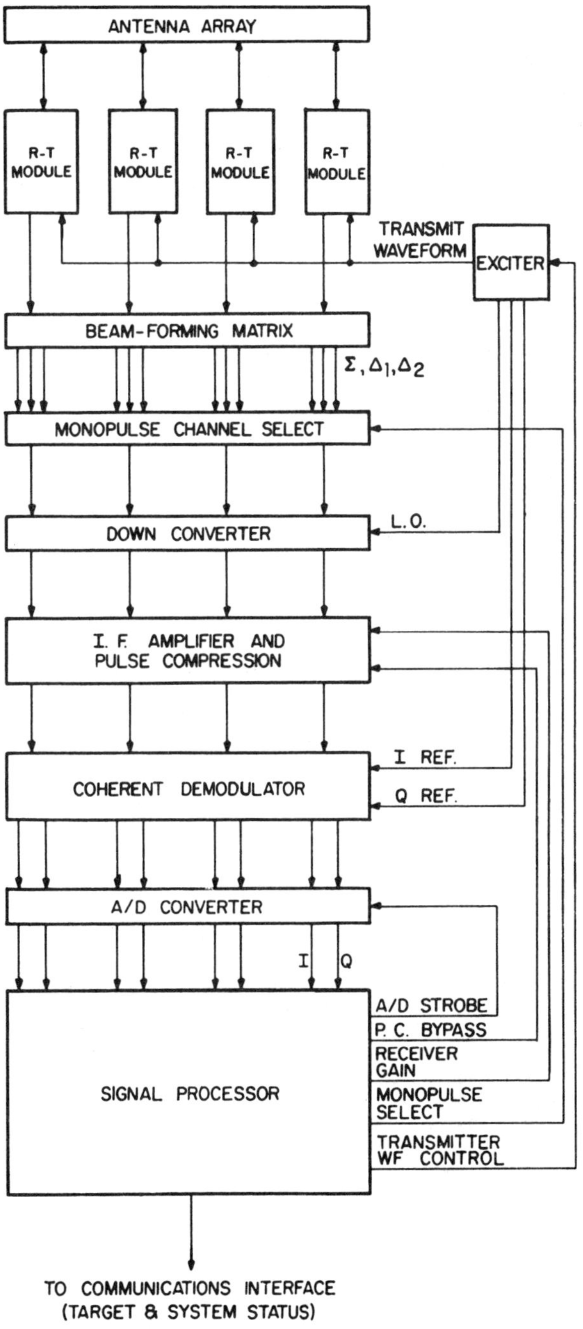

Figure 7 Barrier radar block diagram

SEARCH AND TARGET ACQUISITION RADAR FOR SHORT RANGE AIR DEFENCE SYSTEMS. A NEW THREAT ENVIRONMENT - A NEW SOLUTION

J.O. Winnberg

Telefonaktiebolaget L M Ericsson, Defence and Space Systems Division, Sweden

INTRODUCTION

The threat envionment of air defence systems is constantly changing. Developments in aircraft, weapons and ECM technology are used to form new tactics, aiming at exploiting weaknesses in the existing air defence. Such developments are made necessary by the high effectivness of present anti-aircraft missiles and gun systems.

Parallelling the development of the threat there is a development of air defence systems. Particularly the sensor and control functions have benefitted from the advances in solid state and digital technology.

In this paper we shall present a new radar sensor, designed to meet the demands of the future threat environment and to serve as search and target acquisition radar in short range air defence systems. The design makes use of several technology advances, such as planar array phase controlled antenna, solid state transmitters, frequency agility combined with doppler processing and a multi-mode, software controlled signal processing.

The paper is organized as follows. We begin by briefly considering the development of the threat environment. This is followed by an investigation of the consequential requirements on the air defence system. In the ensuing chapter we present the general system principles of the radar, followed by an account, in some detail, of the design. We end with a chapter on the data and performance, both as calculated and as verified with a prototype system.

DEVELOPMENT OF THE THREAT ENVIRONMENT

In this section we shall examine the threat environment, particularly as it affects radar system design.

A modern anti-aircraft defence contains, apart from the weapon itself, several functions. These are target detection, threat evaluation and weapon allocation, acquisition and, finally, fire control. If allowed to operate on its own terms, such a defence would be highly effective. This forces the development of tactics and technology to deny the air defence adequate target data and adequate time to react upon them.

Denial of target data is obtained in several ways:

a) low-level approach. Purposefully using the terrain for masking and using the advantage of radar clutter and multipath interference at very low level, the attacking aircraft can remain undetected until very close to its target area.

b) jamming. Using stand-off and escort jammers to reduce search radar range and self-screening jamming to eliminate range measurement, an advanced attacker can virtually blind radars that do not have proper jamming resistance.

c) anti-radiation weapons (ARW). Both anti-radiation missiles carried by the attacking aircraft and harassment drones launched by ground forces can force the shut-down of search radars.

To be effective, all these methods rely on electronic surveillance to reveal the position and electronic parameters of the air defence radars. Anti-radiation weapons in particular must be able to home in on the signal emitted by its target.

Reduction of the time available for the air defence to respond is obtained by:

a) fast, manoeuverable aircraft. At lowest level, the speed will be up to transsonic.

b) weapons allowing very short aiming and delivery times (Standoff weapons, however, we consider to be targets for the air defence themselves).

c) saturation. Many aircrafts in coordinated attack will limit the time available for the air defence to handle each target.

d) jamming. Various forms of deception jamming will create false targets, producing a saturation especially of the threat evaluation and weapon allocation function.

Apart from the developments concerning conventional attack aircraft, a new target category has entered the threat environment: the armed helicopter. When moving at full speed, this is essentially a very low level attack aircraft. Particularly anti-armour helicopters use also another tactic, hovering at prechoosen positions and concealed by terrain and vegetation. If a search radar is to detect and distinguish this target among ground clutter and moving vehicles, it must include the design features necessary for this purpose.

REQUIREMENTS ON AIR DEFENCE SEARCH RADARS

Two different kinds of search radar can be used in an air defence system: a central search radar providing target data and overall combat control for several firing units or a local search radar included in the firing unit itself. In each air defence application there is some optimum combination of these two types. The elaboration of this point is, however, beyond the scope of the present paper. We shall in the following section consider the requirements on the local search radar belonging to a short range firing unit. By short range we mean up to about 6 km intercept range.

The range performance of a local search radar must be adequate whereas a surplus range will be not only costly but in some sense degrading. This is because an unnecessarily high output power will increase weight, power consumption and cooling requirements (reduced mobility, bulkier equipment) as well as increasing the probability of intercept by ESM and warning receivers and anti-radiation weapon seekers. This fundamental fact being stated, we list other requirements as they follow from the threat environment of the previous chapter.

In order to provide target data, the radar must have

a) low level performance. Terrain masking is avoided mainly by choosing a proper site for the firing unit. The weapon itself, it must be remembered, must also have a free line of sight to the target. The main demand then is one on mobility such that the best sites can be reached.
The radar itself should as far as possible be immune to multipath interference and have a very high subclutter visibility.

b) anti-jamming performance. Since the jamming environment can be very dense, the jamming resistance should be by basic design rather than by operator skill. Also it must work together with clutter suppression.

c) anti-radiation weapons protection. By use of a low power density in the space, time and frequency domains and a confusing radiation scheme, the seeker of the weapon should be denied an adequate signal for lock-on and tracking.

In order to reduce the reaction time of the air defence system to its minimum, the radar must provide

a) high search rate.

b) accurate and comprehensive target data. The fast and correct threat evaluation requires both target position and height. Data must provide the accuracy needed to calculate, from succesive scans, speed, heading and diving angle. The time needed for the fire control system to acquire the target depends on data accuracy and on whether elevation angle is provided.

c) multiple target capability. To a high degree this is a matter of correct threat evaluation, such that the optimum engagement sequance can be choosen.

d) resistance to deception jamming. This is achieved by emitting signals with too low a power density to analyze and too complicated a pattern to imitate.

Finally, when protecting armoured units, the radar must be able to detect and distinguish hovering helicopters.

SYSTEM PRINCIPLES

In the previous chapter we analyzed the requirements on a short range search radar. We shall now describe the system principles we have choosen in order to meet these requirements.

The demands on mobility and on lowest possible power density speak for a low peak power. Adequate range performance can be obtained by high antenna gain and high duty factor. The high antenna gain is compatible with the intended search volume only if beam position can be shifted. This, together with the great advantage of target elevation data speaks strongly for a 3-D radar. Added advantages are then increased jamming resistance and greatly reduced ESM signature due to the narrow main beam.

For a lightweight radar with moderate range performance, we have choosen a planar array, phase controlled antenna where each antenna element is fed by a solid state transmitter-, receiver- and phase shifter module. The very high duty cycle of a solid state transmitter fits well with the requirement for low peak power. Adequate range resolution is obtained by pulse compression, yielding a wide pulse spectrum and complicated signal and thus improved deception jamming-, ESM- and ARW resistance.

The modules are fed, through a power dividing network, by a radio frequency generator (RFG) and solid state driver amplifier. The fast frequency shifting of the RFG is the single most important factor in giving high resistance to jamming ESM and ARW threats. The fast shifting is combined with full spectral purity performance, making possible the unhindered combination of clutter rejection and frequency agility.

The signal processor contains software controlled arithmetic units, allowing different modes of operation in different threat- and clutter environments. In particular a special mode for detection and characterization of hovering helicopters is included.

When mounted on a AA vehicle the radar can automatically compensate for platform velocity and, using gyro signals, stabilize target data and beam elevation electronically.

The radar scan pattern is controlled by a processor. Thus an adaption is possible for each given application.

DESIGN

In this chapter we describe the design of the prototype system.

The overall system block diagram is shown in fig 5. Physically, the prototype consists of two units, radar unit (fig 1 and 2) and signal processing unit. These units are indicated in the block diagram.

The radar unit consists of antenna, transmitter-receiver modules, power divider, receiver and RFG.

The RFG generates a number of preselected frequencies. Switching time is short enough not to reduce significantly radar duty cycle even when switching from pulse to pulse.

The receiver is a heterodyne with high IF (above 1 GHz). It contains a low noise RF amplifier as well as quadrature video channels.

The antenna is of planar array slotted waveguide type. Each of the sixteen waveguide elements are fed by its own module.

A module, shown in fig 3 and as block diagram in fig 4, contains a three-bit reciprocal diode phase shifter, a transmitter amplifier chain and a low noise receiver amplifier. All amplifiers use GaAs FET's.

The signal processing starts, after the A/D-converters, with a video conditioning and velocity compensation unit. This is needed both for helicopter detection and since the prototype is intended for use on a moving platform.

It further consists of software controlled arithmetic units that perform the MTI processing and video correlation. Different modes can be used for different clutter situations. A special mode is used for detection and analysis of hovering helicopters.

The signal processing unit also contains a processor for target data extraction, tracking and threat evaluation. The target is tracked in three coordinates by a Kalman filter. The filter estimates target position and velocity, giving high precision acquisition data in all coordinates and at any time.

The scanning patterns of the prototype system is shown in fig 6. The search volume is covered in two antenna revolutions. In the first one only the lowest lobe position is used. Either the helicopter detection mode or one of the two wide rejection band modes can be used. During the second revolution the rest of the search volume up to 30° elevation is scanned with five fixed lobe positions. The helicopter detection mode is of course not used at these elevations.

DATA AND PERFORMANCE

The technical data of the prototype are given in table 1. Future production versions will differ mainly in a higher output power.

The range performance is shown in fig 6 and other performance figures are given in table 2. These values, which are based on calculations and extensive simulations, agree well with the values obtained from the testing of the prototype system.

TABLE 1 - Technical data.

Frequency	X-band
System bandwidth	> 5%
Peak power	30 W
PRF	10 kHz, staggered
Pulse length, compressed	1 μs
Gain	33 dB
Beam width, azimuth	2.4^{o}
elevation	5°
Max beam deflection	$\pm 30^{o}$
Antenna revolution rate	40 rpm
MTI Improvement Factor	45 dB

TABLE 2 - Performance.

The following target data are assumed:

Target cross section	3 m^2 (fixed-wing)
	0.3 m^2 (hovering helicopter)
Probability of detection	50% single scan
False alarm rate	1 per 10 s
Range: Lobe 1	12 km (3 m^2)
	7 km (0.3 m^2)
Lobe 2-4	9.5 km
Lobe 5-6	8.5 km

Acquisition data accuracy (2σ-value)

Range	50 m
Azimuth	0.5^{o}
Elevation	1°

SUMMARY

We began by analyzing the demands on a short range search and target acquisition radar in the future threat environment. We found the following important requirements

* Mobility
* Clutter rejection
* Low-level capability
* Jamming resistance
* ESM protection
* ARW protection
* Threat evaluation: elevation, velocity
* Acquisition data: real-time, elevation
* Helicopter detection

We have described a radar system designed for these demands:

Mobility

The system is a low power, solid state, X-band radar consisting of two main units with a total weight around 125 kg. It can be integrated in any anti-aircraft vehicle, with compensation for vehicle motion, roll and pitch. It can be mounted in a separate small radar vehicle or, as a freestanding unit, be transported by a small vehicle or helicopter.
In the latter case, the units are lightweight enough to be handled and mounted manually.

Clutter Rejection

The radar is fully coherent with high subclutter visibility and several modes for different clutter environments.

Low Level Capability

The choice of X-band itself minimizes the problem of multipath interference. Furthermore, the narrow beam hits the ground only in its lowest position. Frequency agility eliminates blind elevations above the first maximum.

Jamming Resistance

Frequency agility (the same frequency is never repeated more than five pulses) combined with MTI, pulse compression, narrow main lobe, low sidelobes and high receiver IF together make the radar almost immune to jamming.

ESM and ARW Protection

A very low power density through low peak power, coded pulse, low sidelobes and frequency agility makes it very difficult to detect and track the radar.

Threat Evaluation

Targets are tracked in three dimensions making complete evaluation possible including elevation and diving angle.

Acquisition Data

A Kalman tracking filter gives accurate real-time data in all three dimensions and in velocity.

Helicopter Detection

The radar has a special facility to detect and distinguish hovering helicopters.

A prototype system has been built and is by now thoroughly tested. It's performance is in good agreement with the one expected from theory and computer simulations.

Figure 1 Radar Unit. Front view

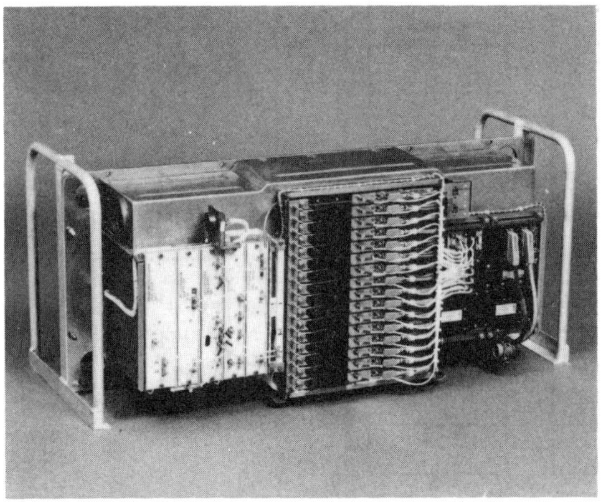

Figure 2 Radar Unit. Back view

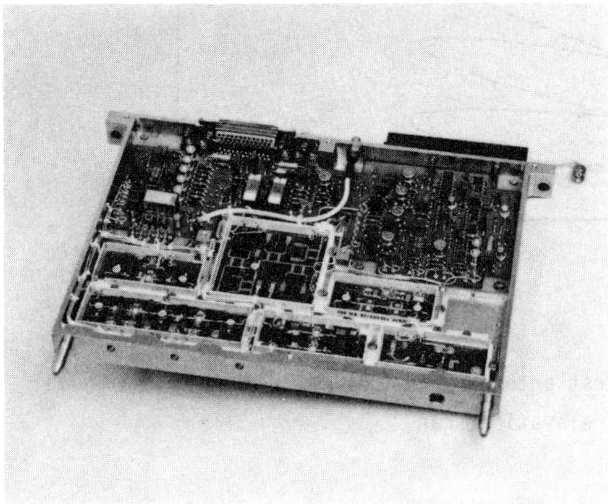

Figure 3 Transmitter-, receiver and phase shifter module

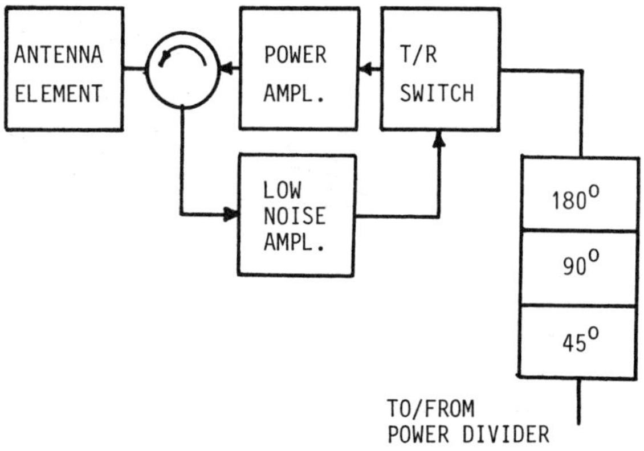

Figure 4 TR module block diagram

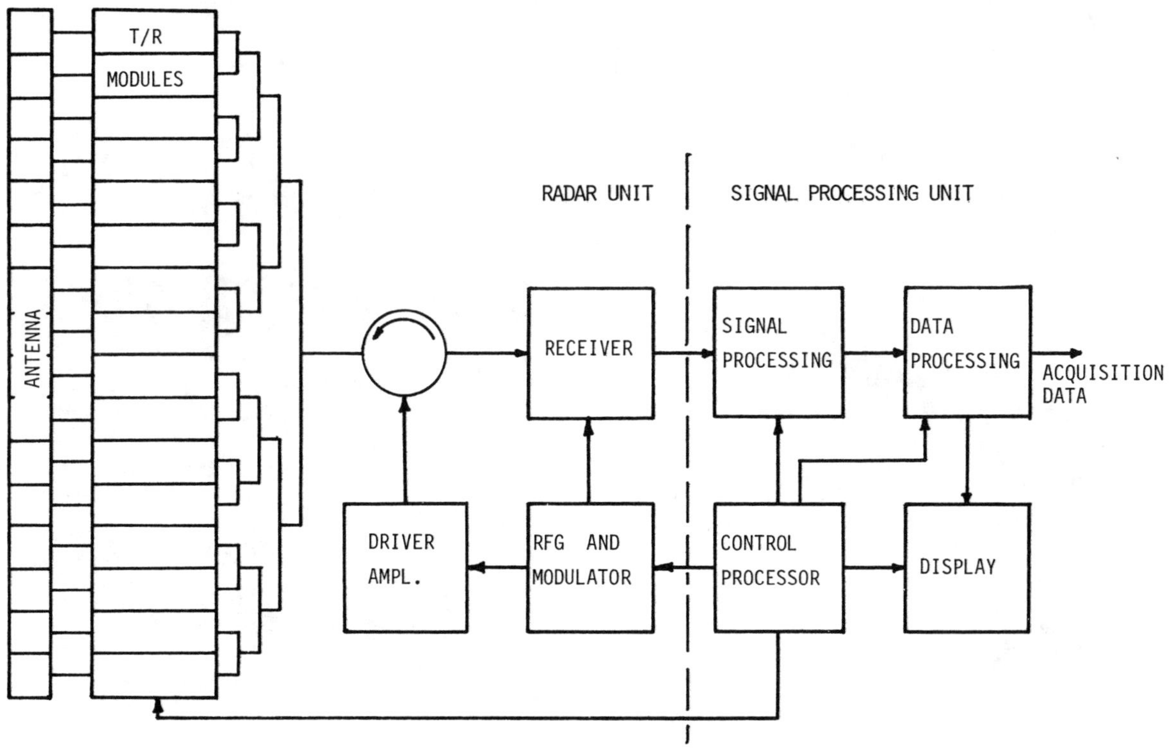

Figure 5 System block diagram

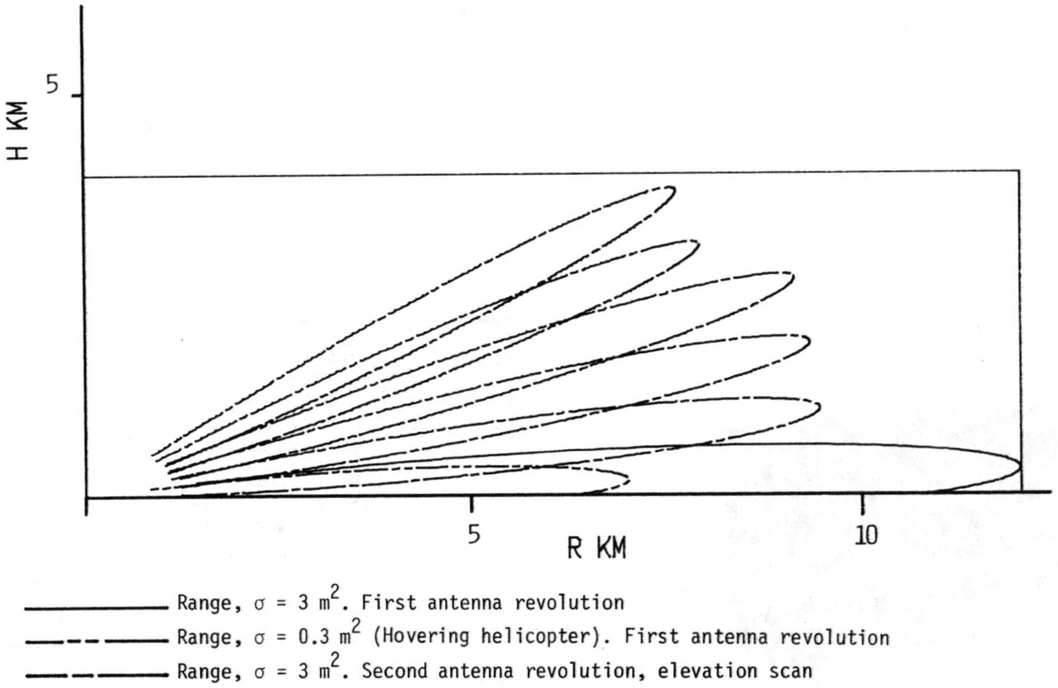

Figure 6 Search pattern and coverage diagram

A FIXED-BEAM MULTILATERATION RADAR SYSTEM FOR WEAPON IMPACT SCORING

S. Gaskell

RCA Corporation, Government Systems Division, US

M. Finch

Ministry of Defence, UK

INTRODUCTION

The impact scoring of practice bombs is necessary for the evaluation of air-crew proficiency. It is usually achieved by observing with telescopes or televisions the dust plume or water splash resulting from impact. Apart from the obvious drawback of relying on highly variable secondary phenomena, which may be masked by the designated target, the scoring can be accomplished only in conditions that afford good visibility. This paper describes a radar scoring system that determines the impact point by direct measurement and extrapolation of the bomb trajectory; by using radar, scoring can be achieved in any operational weather conditions, day or night. The radar requirements are within current state of the art and most of the required hardware is proven, off-the-shelf design.

The United Kingdom scoring ranges have widely varying topography with land and/or sea targets. For both radar and visual scoring, the most difficult observation problem occurs with sea ranges having targets dispersed over a wide area; the sea range considered in this paper uses three tethered barges as bombing targets and requires bomb impact detection out to 10 km. The operating conditions pertinent to the system design can be summarized as follows

a. Aircraft Attack Profile

Speed	400 to 550 knots
Level	45 m to 300 m altitude (high drag bomb)
	150 m to 600 m altitude (low drag bomb)
Dive	5° to 30° dive angle (low drag bomb)
Toss	Release at 6 km from target (low drag bomb)

b. Ordnance (practice bombs)

3 kg high drag bomb (minimum drag coefficient = 1.66)
13 kg low drag bomb (mininum drag coefficient = 0.112)

c. Target Characteristics (tethered barges)

Offshore distance 6.0 km (maximum)
3 km radius scoring circle (maximum)
Radar reflectors on mast

d. Environment

Sea state	$\geq$ 5 (5% probability)
	$\geq$ 3 (50% probability)
Rain rate	$\leq$ 4mm/hr
Tidal variations	6 m
Birds	
No electronic warfare	

e. Communication (ground-to-air)

Score is relayed orally to pilot $\leq$ 30 s after impact

Bomb radar cross section as a function of the aspect angle was required for the radar design. Extensive measurements were made on both bombs for all aspect angles and fin orientation, and this data was used in the computer simulation for evaluation of system performance.

The measure of performance is the circular error probable (CEP) for the distribution of impact prediction errors. For a probability density function p(x,y) of impact prediction error, the CEP is defined as the radius of the circular region S such that,

$$\iint_S p(x,y)\, dx\, dy = 0.5$$

The design goal for the system was a CEP of 3 m.

SYSTEM CONCEPT

Figure 1 illustrates the bomb scoring system. Two independent 2D monopulse radars (range and elevation angle) are used, each having a different transmit frequency in the same band. Each radar has a fixed, linear-array, narrow-band antenna radiating a horizontal fan beam; the common azimuth coverage, between the 3 dB points of the two beams, covers the designated target area. The beams are at the same height (ground level to beam center). Each radar obtains a sequence of range and angle measurements on the bomb as it falls through the beam, and takes range measurements on the target.

The digitized data from the two radars is transmitted via modems and cables to the central computer in the range control building. The bomb data from each radar is smoothed in a mid-point, least-squares polynomial filter, where the mid-point is the estimated time at beam center (zero elevation angle). The estimated ranges, the range rates, and the altitude rate are combined with the known beam height and radar locations to give an estimate of the state vector (position and velocity) of the bomb at beam center. The impact point relative to the target is obtained by trajectory extrapolation. The result is shown on an alphanumeric display and relayed orally to the pilot.

SYSTEM ANALYSIS

The principal tool in these studies was a detailed computer simulation of the system from which the CEP values were determined by Monte-Carlo techniques. Analytical and computer simulation parametric studies provided essential insights for the system design, specifically for the radar beam characteristics and signal processing and also for the sensitivity of impact prediction errors to measurement errors.

Computer Simulation

From input data on target location, aircraft attack profile, and bomb type, the aircraft flight path and bomb trajectory are computed. The bomb trajectory computation uses input data on the bomb drag characteristics.

The antenna tower locations, beam height, and azimuth and elevation beamwidths are input data. The beam and antenna characteristics, including sidelobes, are calculated from the input sidelobe level for a Taylor sum weighting and a Bayliss difference (monopulse) weighting in elevation.

When the bomb is between the 3 dB points of each sum beam, there is a sequence of computations at the radar data rate. (The radar data rate equals the pulse repetition interval multiplied by the number of pulses integrated.)

The sequence is as follows:

a. True range and angle

b. Aspect angle and table look-up of radar cross section

c. Signal level, including lobing

d. Clutter cross section (sea, rain) and resultant interference signal

e. Signal-to-interference (S/I) ratio and comparison with input threshold

f. Multipath elevation angle error

g. Target position and aircraft position, velocity, and signal level

h. Measured range and angles obtained as algebraic sum of true value, input residual random and bias errors, multipath error, S/I related random errors. The S/I related errors are randomly selected from a Gaussian population whose standard deviation is determined by the S/I ratio.

The measured data is input to the smoothing algorithms and the state vector at beam center is estimated. The trajectory prediction algorithm determines the estimated impact point and the impact prediction error. The CEP is calculated by repetition of the last part of the program and by random selection of Gaussian distributed errors from which the probability distribution of impact prediction errors, p(x, y), is determined.

Beam Characteristics

The azimuth beamwidth is determined by the area of the target complex. The height of the beam and the elevation beamwidth are determined primarily by the need to avoid receiver saturation from aircraft returns and to minimize elevation multipath error. The radar cross section of the aircraft is 30-40 dB above the bomb. For the minimum aircraft altitude (45 m) at a maximum range of 10 km, optimum results were obtained with a beamwidth of 3 mrad and a beam center height of 20 m. This narrow beamwidth minimizes the clutter cross section and lobing, but puts an upper limit on the bomb vertical velocity, depending on the interval between measurements and the number of data points required for effective smoothing.

Signal Processing

For the large azimuth beamwidth (~1 rad) required to cover the target area, the resulting signal-to-clutter ratio for the bomb is about -30 dB for rain clutter or sea clutter. Since the power spectral densities of the rain and sea clutter will, in general, have different mean and standard deviation values, and because the bomb must also be discriminated from the aircraft and birds, the optimum signal processing would be a bank of doppler filters that are matched to the clutter environment. A near-optimum solution is provided by a digital MTI (moving target indicator) filter followed by FFT (fast fourier transform) signal processing, as shown by McAuley(1). The FFT provides pulse integration gain against wideband (noise) interference.

When more than one object is in the beam at the same time, a "ghosting" problem may arise. Data is available before each mission from which the bomb trajectory can be predicted, permitting the correct association of range and angle measurements for each object in the beam.

Sensitivity Factors

The impact prediction errors are determined by the errors in the state-vector estimate at beam center propagated along the extrapolated trajectory. For given measurement errors, the variance of the state-vector errors is inversely related to the number of data points. For this system, the ratio is about 10:1 between the high drag bomb (~1 s in beam) to the low drag bomb (~0.1 s in beam). The sensitivity factor relates the errors in the state vector estimate at beam center to the impact prediction errors. In general, they are unity for position errors and equal to t_f, the trajectory extrapolation time, for velocity errors. The exception is the height and height rate errors (noting that errors in the estimated time at beam center are equivalent to height error). The sensitivity factors in this case are $\cot\gamma$ for height and $t_f \cot\gamma$ for height rate, where γ is the bomb trajectory angle at impact. For level delivery at low altitude and maximum speed, $\cot\gamma \sim 5$.

Two important conclusions for the system design, confirmed by the computer simulation, are:

a. The most significant errors are in the determination of beam height, elevation angle, estimated time at beam center, and estimated vertical velocity at beam center.

b. Impact prediction errors, and therefore CEP values, are a maximum for minimum values of trajectory impact angle; the performance of the system can be bounded by considering these delivery conditions.

RADAR DESIGN

Figure 2 is a simplified block diagram of the radar, and Table 1 gives the pertinent characteristics. The rationale behind the selection of the radar parameters and the hardware implementation are the topics of this section of the paper.

TABLE 1 — Radar characteristics

Transmitter	
Frequency	9.1 GHz/9.3 GHz (fixed)
Peak power	250 kW
Duty cycle	0.001
PRF	5000 sec^{-1}
Pulsewidth	0.2 μs
Antenna	
Type	Resonant-slotted linear array
Length	13 m (nominal)
Gain	32 dB
Polarization	Horizontal
Sum Beamwidth (half-power)	3 mrad (elevation); 1220 mrad (azimuth)
Sidelobe level (elevation)	-35 dB
Boresight null depth	-35 dB
Receiver	
Type	2-channel monopulse, pulse-to-pulse coherent
Noise figure	4 dB
Gain (manual gain control)	77 dB
IF bandwidth	6 MHz
IF frequency	60 MHz
Data Processing	
Analog-to-digital conversion	10 MHz, 8 bits (I and Q)
Clutter rejection	3-pulse MTI with manual clutter lock 64-point FFT

Antenna Characteristics

The antenna is a resonant-spaced, slotted waveguide array that consists of 16 uniformly illuminated elements and 8 sum/difference beamformers between symmetric elements. Taylor weighting is used for the sum pattern and Bayliss weighting for the difference pattern to give -35 dB sidelobes. The selected X-band frequencies give an array length of about 13 m for an elevation beamwidth of 3 mrad; the next higher frequency (16 GHz) would reduce the array length and reduce the Fresnel region, but would result in severe signal attenuation in rain. For this array, the range R of the unfocused Fresnel region for an antenna length D ($R < 2D^2/\lambda$) is $\leq$11 km. Focusing is used to reduce this value to $\leq$3 km, which establishes the lower limit of the radar range. The 16-element configuration gives grating sidelobes at acceptable angular displacements.

Horizontal polarization was selected for the relative ease of manufacture of the antenna. A horizontally polarized array has straight symmetric slots in the broad face of the waveguide, giving easier control of the factors affecting sidelobe level compared with a vertically polarized array which has tilted slots on the narrow face of the waveguide. Tests on elements of the proposed array have confirmed the predicted characteristics. The reflection coefficients of sea water for horizontal and vertical polarization differ by less than 0.5 dB for the <0.4° grazing angles obtained with this system configuration.

Transmitter

A 250 kW X-band (magnetron) transmitter of modern design and proven reliability was selected. A PRF of 5000 and 64-pulse integration gives an adequate number of returns for the low drag bomb released at high altitude (minimum time in beam). The corresponding narrow pulse width of 0.2 μs minimizes clutter cross-section, but is wide enough, in conjunction with the wide azimuth beamwidth, to avoid the problem of "spiky" (non-Rayleigh) sea clutter returns as discussed by Nathanson(2).

The unambiguous velocity interval is 81 m/s, and foldover is obtained for aircraft and low drag bombs. For the specified operating conditions, this foldover did not invalidate doppler discrimination in that the aircraft, bomb and clutter appeared in separate doppler (FFT) filters.

Receiver/Signal Processing

The receiver is a solid-state, two-channel monopulse superheterodyne type, and is pulse-to-pulse coherent with a stable local oscillator (STALO) and a phase-locked coherent oscillator (COHO). The IF (intermediate frequency) signal in each receiver channel is coherently detected, and the in-phase (I) and quadrature-phase (Q) are each sampled at a 10 MHz rate. Four hundred video samples are obtained over a 6 km range gate corresponding to the diameter of the safety circle around the target. Since the read-time (40 μs) is one-fifth of the pulse repetition interval (PRI), the data is buffered to enable it to be processed sequentially at a single three-pulse digital MTI and a 64-point FFT rather than four MTIs and FFTs.

The receiver and digital components are linear over the maximum expected clutter-to-noise ratio of 40 dB in order to prevent spectral spreading. A remotely controlled digital phase shifter is used in conjunction with a clutter meter to displace the peak of the clutter spectra to the MTI notch. The FFT is a 64-point pipeline processor with filter weighting for -40 dB sidelobes. The output magnitude, $(I^2 + Q^2)^{1/2}$, is obtained for each range cell at each doppler (400 x 64) for each channel. The process time of 12.5 ms gives the data rate for the system. The two receiver channel outputs are combined to give the sum and difference signals and then the sum signal is input to a threshold detector.

Figure 3 shows the most severe clutter environment, i.e., 4 mm/hr rain and sea-state 5 in comparison with the lowest average bomb cross-section of -20 dBsm. The mean ($\overline{v}$) and the standard deviation (σ_v) of the clutter spectra are given. For these conditions, the rain clutter return would be displaced to the MTI notch, and the theoretical improvement factor is 31.4 dB. In practical radars, the effective improvement factor (I_e) is calculated from the theoretical improvement factor and from the cancellation ratio (CR) resulting from radar instability. Schleher(3) has published data indicating CR values of 30-35 dB for magnetron transmitters having STALO and COHO. A CR of 38 dB has been measured with this MTI and transmitter; for this analysis, a CR of 35 dB is assumed.

The doppler diference between the rain clutter, sea clutter, and bomb return signals will place them in separate FFT filters. Leakage into adjacent filters is suppressed by the -40 dB sidelobes. The decorrelated signals of receiver thermal noise and rain clutter are reduced by 16.6 dB in the FFT.

The total interference cross section is shown in Figure 3, and the resultant signal-to-interference ratio is the difference between the bomb cross section and the interference cross section. A signal-to-interference detection threshold of 14 dB is required for a 99% probability of detection, based on a two-second false alarm time (twice the maximum time of a bomb in the beam) and a 40 μs range gate. The number of independent noise samples over the false alarm time is 3.125×10^4, giving a false alarm probability of 1.6×10^5. From the computer simulation, the aspect angle changes for the bomb in the beam have a correlation time of the order of one second, so that a steady target may be assumed. The threshold is remotely controlled from monitored alarm counts.

Special Features

The antenna support tower is a 27-m trussed tower designed for an operating wind speed of 30 knots at 10°C and a survival wind speed of 120 knots. The tower and foundation tilt at operating wind speed is calculated to be not greater than 0.05 mrad.

The antenna waveguide and beamformer are bolted into a fabricated aluminum H-structure to give stiffness for handling and structural loading. The antenna is pivoted at the top of the tower and fastened at the bottom with adjustable jack screws. Alignment of the radiating surface with the vertical plane will be < 0.1 mrad, and the beam displacement errors due to environmental changes in the antenna and beamformer will be < 0.05 mrad.

The measurement errors due to deflection of the fan beam by atmospheric refraction must be corrected in the central computer. It is well known that in coastal regions the refractivity lapse rate dN/dZ (where N = refractivity and Z = altitude) may be much greater than for the standard model atmosphere; evaporation ducts may develop. The measured lapse rate can be used for accurate refraction correction as shown by Ghiloni(4). It will be determined from measurements obtained by three microwave refractometers mounted at the top, center, and base of the antenna support towers. The data will be relayed directly to the central computer.

SYSTEM PERFORMANCE

Results of extensive analyses made with computer simulation have shown that an impact prediction CEP of ≤3m can be met with the high-drag bomb for most operational conditions. A CEP range of 1-4m was obtained for the most adverse conditions, namely sea-state 1 (specular multipath), 4 mm/hr rain and an aircraft speed of 550 knots. The maximum CEP was obtained for the lowest release altitude of 45m. For the low drag bomb and the above conditions of sea-state, rain and aircraft speed the CEP is 3-9m for release altitudes of 150-600m and 30° dives. These higher CEP values for the low drag bomb are primarily due to reduced time in the beam. Lower CEP values could be achieved by providing separate antennas with wider elevation beamwidths.

DEVELOPMENT AND APPLICATION

The system that has been described in this paper was developed for a specific application. The fundamental concept of determining impact (or launch) point by passage of the projectile through a fixed beam is applicable to land or sea ranges for bombs, shells, and rockets. Such configurations have been developed for the land and land-sea ranges of the United States.

REFERENCES

1. McAuley, R.J., 1972, "A Theory of Optimal MTI Digital Signal Processing - Supplement 1," Technical Note 1972-14, Lincoln Laboratory, Massachusetts Institute of Technology, USA.

2. Nathanson, F. E., 1979, "Radar Design Principles," McGraw-Hill Book Company, New York, USA.

3. Schleher, D. C., 1978, "MTI Radar," Artech House, Boston, USA.

4. Ghiloni, J.C., 1973, "Millstone Hill Radar Propagation Study - Instrumentation," Technical Report 507, DDC AD-775140/7 Lincoln Laboratory, Massachusetts Institute of Technology, USA.

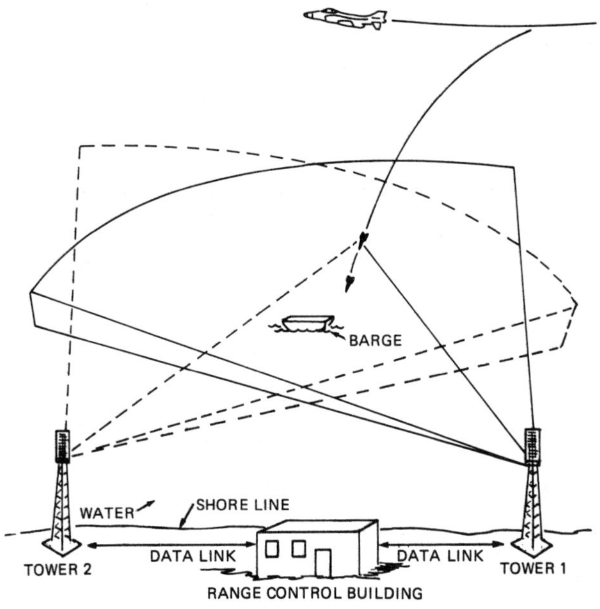

Figure 1 Radar bomb scoring system

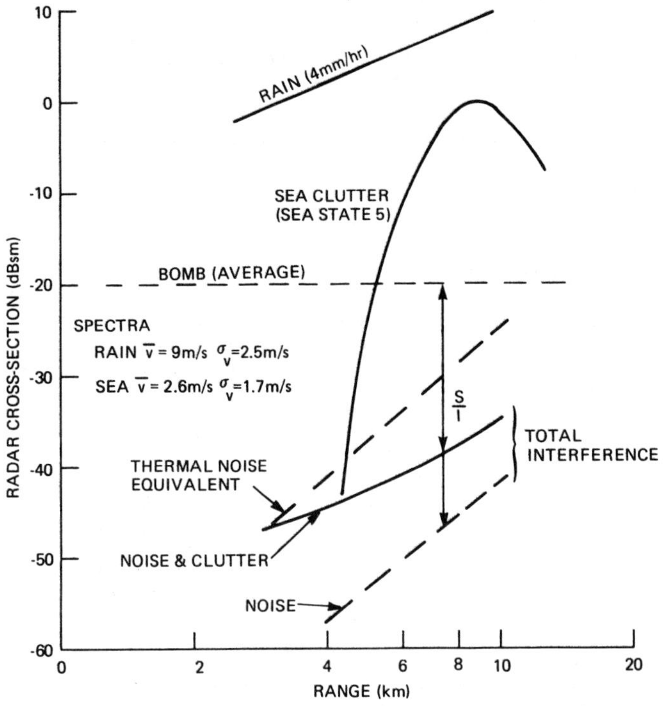

Figure 3 Radar environment and signal processing output

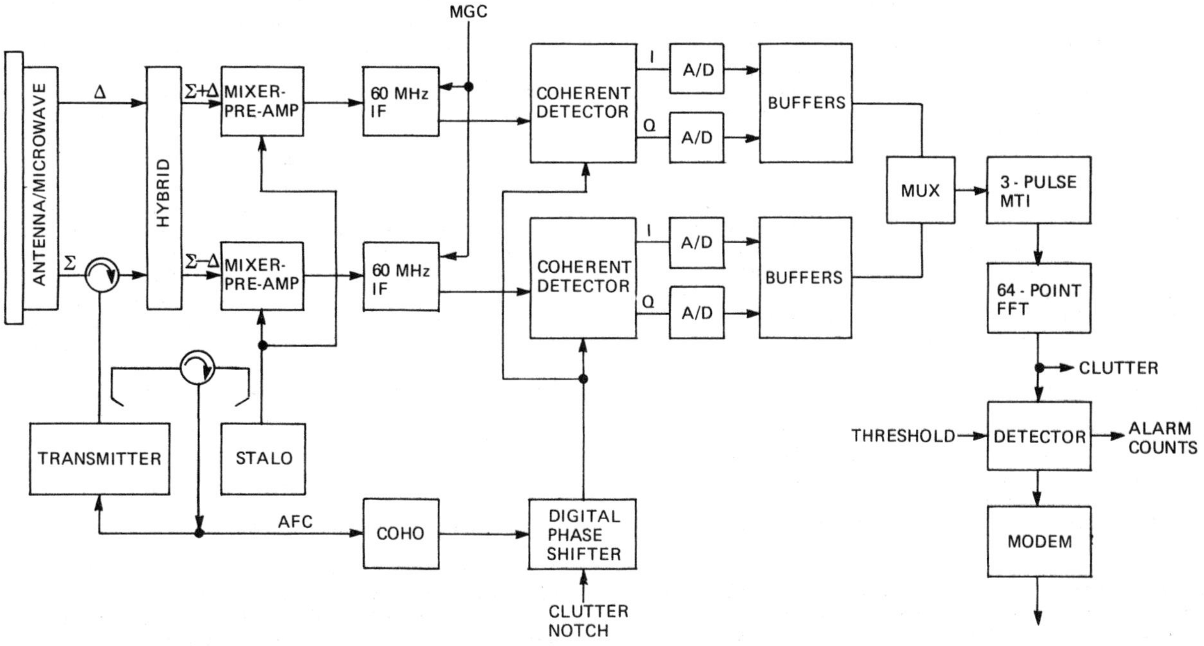

Figure 2 Radar block diagram

INSTRUMENTATION AND ANALYSIS OF AIRBORNE PULSE-DOPPLER RADAR TRIALS

J Clarke[1], K Clifton[2], E B Cowley[1], J King[2], I W Scroop[1]

1. Royal Signals and Radar Establishment, UK; 2. Marconi Avionics, UK

INTRODUCTION

The development of a pulse-doppler radar generally involves an advance in the state of the art of either the antenna, transmitter, waveform, signal processing, or of several of these aspects. Often, therefore, design assumptions have to be made based upon extrapolations of previous experience or upon an interpretation of published work. In such programmes, it is necessary to instrument an early prototype radar and by way of data analysis, to confirm the validity of the assumptions that were made and to characterise fully the performance of the radar. This paper will present a practical account of a modern radar instrumentation package and describe associated analysis tools that have been prepared and proven.

The class of airborne pulse-doppler radar of interest here is a scanning surveillance coherent radar which uses a fairly high duty ratio transmitter. These radars are often, but not necessarily, ambiguous in range and undertake a spectral analysis of each range cell. The dwell time on target will contain one or more periods of coherent signal processing.

Modern pulse-doppler radars comprise, in essence, a high gain low-sidelobe antenna, RF and IF amplifiers, an Analogue-Digital converter and a Spectrum Analyser followed by various plot extraction circuits. Since it is the plot extraction processing which is of greatest interest, particularly false alarm rate control, the data recording must be undertaken at various stages in the signal processing chain. The spectral analysis is generally performed by a Fast Fourier Transform routine (FFT) which provides the amplitude of each Range and Velocity resolution cell. The system described in this paper records data at the output of the FFT at full radar resolution and also at the output of the plot extraction circuits. To do this it needs to record binary data at a rate of at least 40 M bits per second.

INSTRUMENTATION RECORDER

Two instrumentation packages have been designed and manufactured. The first utilises two SE Labs. TDR 10 recorders and has the facility to switch between the recorders without any data loss, thereby allowing trials of any length to be recorded in entirety. The second package utilises a single Honeywell 101 recorder with a co-axial tape spool system which together with its interface circuits provides an easily transportable recording system in a self-contained rack. Both recorders use Phase Code Modulation (PCM) for recording digital data and interface circuits compress the data into well defined blocks to permit identification, synchronisation and deskewing. The twin deck recorder is shown in Fig 1.

The two instrumentation packages are very similar in specification. They both record the data onto a 2800 metre 28 track High Density Digital Recorder (HDDR) magnetic tape at 3 m/sec which allows 15 minutes of trials data to be available for subsequent analysis from each tape. Each tape track has a maximum data rate in excess of 2 M bits/sec giving a maximum total data rate of about 60 M bits/sec. In current applications the amplitude of each Range-Velocity (R-V) cell is recorded as a 12 bit binary number, with 2 cells recorded simultaneously onto 24 tracks of the tape at 1.6 MHz word rate. The remaining tape tracks are used for recording data parity, scanner and navigation data, target plot parameters (from the output of the plot extractor circuits) and system time reference.

TAPE COPYING AND TRANSCRIPTION

Since airborne trials are very expensive, the HDDR trials tapes are copied at full density and with unchanged format before use in order to minimise the risk of damage to master tapes. This copying is performed at half the original recording speed (ie 1.5 m/sec) on a dedicated tape copying station utilising 2 laboratory tape transports. The lower recording speed improves the quality of the recording in preparation for use in the next process. After copying the master tapes are archived.

The (copy) tapes are then transcribed, via a specially configured PDP-15 computer onto 7 track computer compatible tape. The data is demultiplexed by a special hardware interface into a form acceptable to this 18 bit computer, which then stores selected data on a 7 track Ampex TMA tape recorder. The 7 track tapes have a recording density of about 30 k bits/metre/track and are 800 metres in length; this compares with about 500 k bits/metre/track and 2800 metres for the HDDR tapes. Thus only 3 secs of full resolution data can be transferred to each computer tape, although provision exists for transfer of contiguous data to successive computer tapes. It is necessary therefore to provide a facility to select data, in time and azimuth sector, for detailed analysis at this processing stage. Another constraint on the amount of data that can be transcribed is that, due to the limited speed of the PDP-15 and the 7 track tape recorders, the transcription process can only be performed at a HDDR trials tape replay speed of 0.07 m/sec which means that an entire HDDR tape would take over 12 hours to transcribe.

Alternatively, specialised hardware is available to reprocess the recorded spectral data through adaptive thresholding, range and velocity correlators, and plot extractors into an identical format to the recorded plot data. This enables different hardware, firmware and software algorithms together with signal processing techniques to be developed, optimised and assessed.

The specialised hardware has 3 basic displays:

1) Amplitude (A) vs Range (R)

2) Amplitude vs Velocity (V)

3) PPI type Range/Azimuth (RA)

In addition, it is of interest to know that such displays have also been used on board trials aircraft as well as an R-V display which can be sliced at a selectable amplitude level. These four airborne displays have proved to be invaluable for taking measurements and photographic records during flight to aid subsequent analysis.

Plot data which has been either recorded or reprocessed from spectral data is reformatted, condensed and then sent down a standard link to a main frame computer to be recorded onto magnetic disc or tape. This process effectively reduces the data by a ratio of 4000:1. The link has been adapted to handle peak data rates up to 250 Kbaud; this type of transcription is therefore achieved in real time (3 m/sec).

The relationship of these tape copying and transcription facilities to the analysis procedures described below is shown in Figure 2.

SPECTRAL DATA ANALYSIS

Suitable analysis software has been written in CORAL 66, comprising over 100,000 program statements and designed to allow characterisation of the data in many different ways. The software has been structured in a modular format to provide high flexibility of use and to minimise the difficulties of adding extra facilities which are required as the analysis proceeds. All the analysis options can be performed for any selected azimuthal sector, for any number of antenna scans and for any selected section of range-velocity space.

An important group of software packages perform the analysis of distributed clutter. The analysis options provided include the ability to:

- compute the mean or median clutter level for each velocity resolution cell
- plot histograms of signal amplitude
- fit a number of specific analytic probability density functions to the data
- construct range-velocity (R-V) contour maps of mean returned power from distributed and discrete clutter

The R-V contour map is of particular interest since it can be used to give either a general overall view of the radar performance (eg clutter levels and their R-V spatial distributions) or to examine at full radar resolution the R-V appearance of specific areas of clutter or of discrete echoes. An example of the output obtained is shown in Fig 3; the high clutter level from velocities corresponding to the main beam are apparent as well as the high clutter "line" at all velocities over a narrow range band. (This "altitude line" arises from clutter illuminated at normal incidence beneath the radar.) The clutter statistical modelling package currently allows Rayleigh, Log-Normal, Weibull and Log-Weibull distributions to be fitted to the radar data. However, the package is written in a format that enables other distributions to be added to the alternatives with minimal effort. Examples are shown in Fig 4 which examine the fit of models to some measured data.

Other software packages provide the ability to simulate and evaluate a variety of plot extractors. This involves the facility to model various detection schemes employing fixed and adaptive threshold levels, and to examine peak sensing methods that isolate targets from associated skirt responses in all three axes of range, velocity and azimuth. Finally, with radars that are ambiguous in range and/or velocity, ambiguity resolution algorithms may be tested. At all stages in the modelling, comprehensive listing and graphical output provisions permit efficient monitoring of the processes which are aimed at obtaining high probability for target plots coincidently with low probability for false plots. Schemes which appear promising in this software simulation may be built in hardware for further evaluation and optimisation on the "plot extractor specialised hardware" indicated in Fig 2.

PLOT DATA ANALYSIS

With the plot data held on disc or tape at the main analysis computer, the analyst can select boundaries on parameters such as azimuth, range, velocity and scan number.

All the data for a given timeslice is superimposed and displayed in aircraft or ground referenced format. Fig 5 shows an example of a tape sample of 300 scans in a 70° sector. At this stage the analyst can estimate the false alarm rate (very low in Fig 5), identify targets of interest and look for any notable effects of clutter breakthrough. The data is then reduced by a series of extraction algorithms to focus on a particular target area, a reduction of the order 400:1. These extracted parameters are then checked to ensure positive identification of co-operative targets. The detection probability based on blip scan ratio is then calculated (averaged over a selected number of scans) and plotted against a chosen parameter. Fig 6 indicates the kind of display obtained for this analysis option; the 'best fit' smooth curve added by the computer to this reduced data may be seen. In a similar fashion, any of the detailed characteristics of plots from a specific target, such as height report or range-rate report, may be examined throughout the selected period; reduced data may be plotted automatically or listed on a line-printer.

All recorded radar and platform parameters may be similarly examined for accuracy and stability throughout the trial, since aircraft navigation sensors and radar scanner angles are also generally considered important.

This analysis system was set up to allow basic analysis of data from a trials tape to be assessed within hours of the trials aircraft landing and the results immediately fed back into sortie planning. This maximises the efficiency and effectiveness of a flight trials programme.

CLOSE

Sophisticated recording facilities and automated analysis tools of the nature described here are now considered essential for the efficient conduct of airborne radar flight trials. This is particularly pertinent to pulse-Doppler radars which are ambiguous in range. Study of both the clutter characteristics and plot details allows confirmation of the performance of key components (eg antenna), determination of complete system behaviour, prediction of performance in non-standard conditions, and optimisation of the radar configuration.

ACKNOWLEDGEMENTS

We are grateful for the support and contributions of our colleagues, associated contractors and suppliers.

Fig 1 Recorder station in aircraft

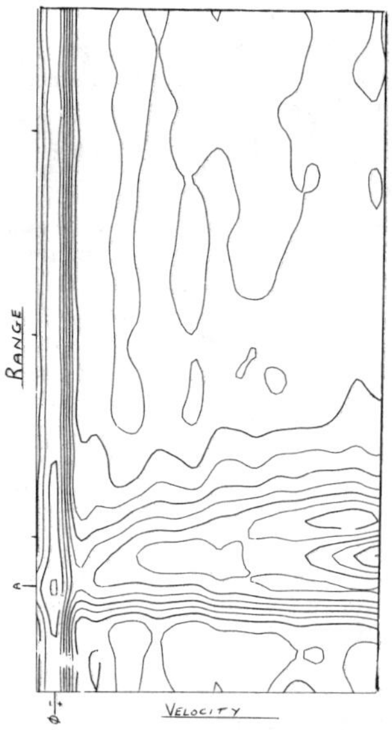

Fig 3 Clutter contours in range – velocity

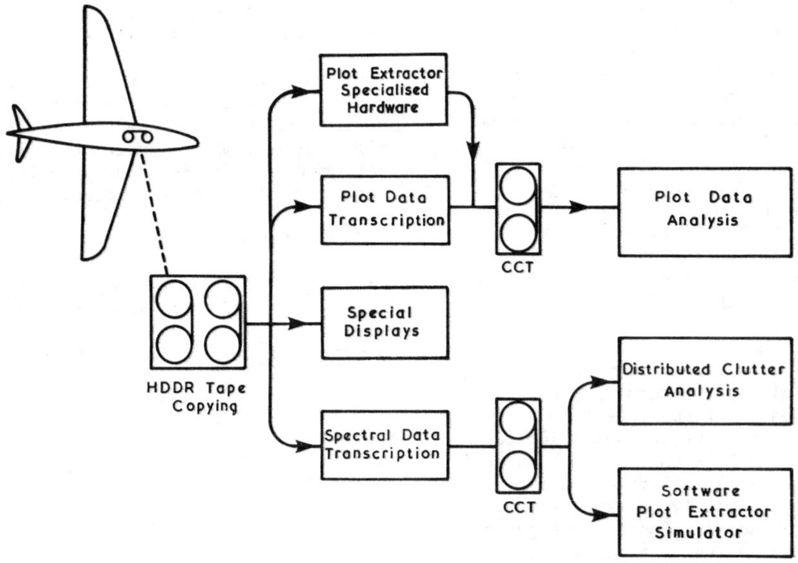

Fig 2 Data processing flow diagram

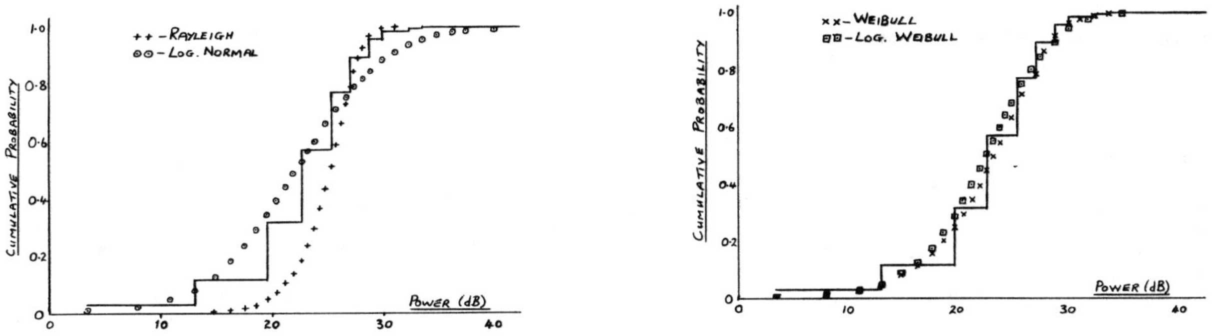

Fig 4 Clutter statistical modelling

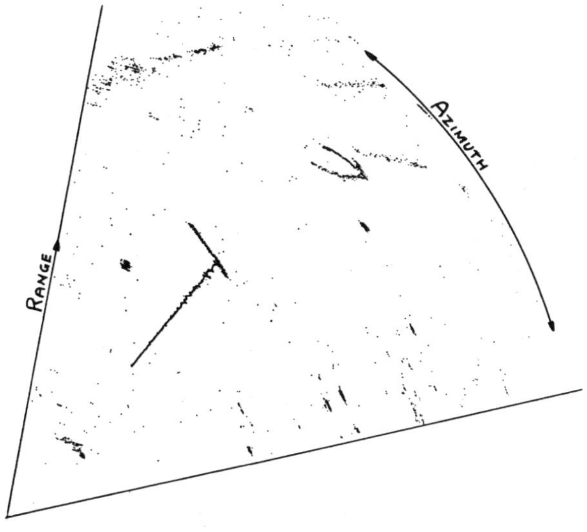

Fig 5 Superimposed plots (ground referenced)

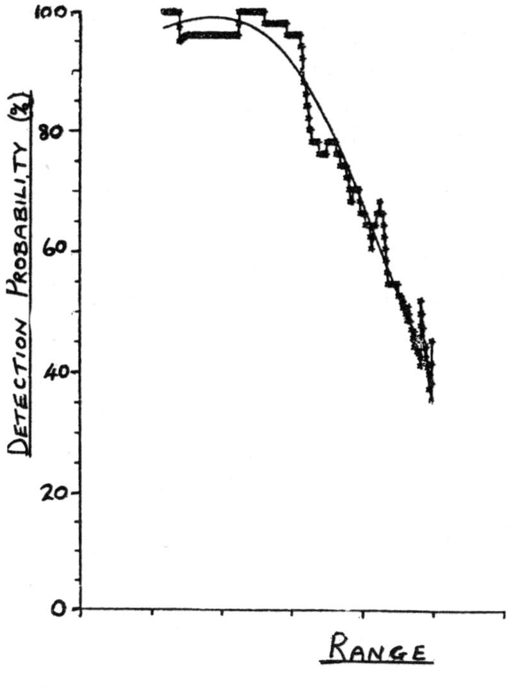

Fig 6 Observed probability of detection

OPTIMUM PULSE DOPPLER SEARCH RADAR PROCESSING AND PRACTICAL APPROXIMATIONS

V. Gregers Hansen

Raytheon Company, Wayland, Mass. 01778, USA

1. INTRODUCTION

When a target return with a known Doppler shift is received in a background of clutter and noise, which are characterized as Gaussian random processes with known spectrum, the classical detection theory [1] can be used to derive the characteristics of the optimum coherent radar processor. The optimum processor maximizes the signal-to-interference ratio (SIR) at its output and represents the generalized matched filter.

For an unknown or random target Doppler shift the theory no longer produces results which are practically useful (except for the rather trivial 2-pulse case [2]) and the maximum likelihood ratio strategy must be used to guide the design effort. It has been shown that the maximum likelihood strategy provides results which are very close to the optimum average likelihood ratio approach [3]. The maximum likelihood receiver can be implemented as a Doppler filter bank, where each filter is separately optimized against the specified clutter model at the particular Doppler frequency.

In practice the optimization of the individual filters is made difficult because the processor must be designed to operate against a range of clutter scenarios with varying power level, mean Doppler shift, spectral spreading and possibly spectral shape. Thus unless adaptive techniques are to be considered, the individual filters must be designed as a compromize between several conflicting requirement usually based on a somewhat empirical procedure. Such design procedures have been described for several recent applications [4,5].

For a given practical filter implementation the maximum likelihood theory can be used as a reference to determine how well this particular design meets the performance requirements for all variations in the clutter model. Such a comparison will indicate the extent to which an adaptive approach might be used to improve the processor performance.

Theoretical results for the performance of the maximum likelihood processor are usually presented as the signal-to-interference improvement, I_{SIR}, as a function of target Doppler. Interference here refers to the sum of clutter and thermal noise. For the actual processor the I_{SIR} is calculated for each of the Doppler filters as a function of target Doppler. The closeness of these results gives a direct indication of how well the practical filters are matched to the particular clutter model considered.

When performance against a wide range of clutter models must be considered, the comparison of many such curves can become difficult to interpret. A simplified measure of performance can be obtained from such results by calculating the average value of I_{SIR} across the unambiguous Doppler interval. This average SIR improvement will be denoted $\overline{I_{SIR}}$. For the practical Doppler filter bank the average is calculated by using the maximum value of I_{SIR} at each target Doppler frequency.

This type of averaging is quite similar to that which is routinely employed to obtain the standard MTI improvement factor [6]. Indeed, if a single filter is used by the coherent processor it is found that $\overline{I_{SIR}}$ is equal to the improvement factor.

The use of the average I_{SIR} as a performance measure is primarily valid for low PRF radars where multiple PRFs are used during the time on target to eliminate blind speeds. For high PRF pulse Doppler radars the actual response curves must be considered to assess performance versus target Doppler.

In Section 2 a set of curves are derived which show the optimum value of $\overline{I_{SIR}}$ as a function of the spectral width of a Gaussian clutter spectrum and for different values of the number of pulses N in a coherent processing interval (CPI).

As N becomes large the maximum likelihood detector can be implemented as the cascade of a whitening filter and a coherent integrator. Then the optimum $\overline{I_{SIR}}$ simply becomes the product of the MTI improvement factor of the whitening filter (which is independent of target Doppler) and the coherent processing gain of a coherent integrator. The results of such an analysis which provides an upper bound on $\overline{I_{SIR}}$, is presented in Section 3.

Finally in Section 4 the performance of some practical coherent processors using Chebyshev filter bank designs, is determined and comparisons are made with the corresponding optimum results for the maximum likelihood processor.

2. OPTIMUM PULSE DOPPLER RADAR PROCESSING

A block diagram showing a single Doppler filter of a pulse Doppler processor, as considered in this paper, is shown in Figure 1. The signal, noise and clutter inputs to the Doppler processor are defined at the output of the IF filter. This filter would usually be implemented as an approximation to a single pulse (white noise) matched filter. For a CPI consisting of N pulses the signal return is represented as a complex vector:

$$\underset{\sim}{s}(f_d)^T = \sqrt{P_{Si}} \cdot e^{j\phi}\{ e^{j\theta}, e^{j2\theta},\dots,e^{jN\theta}\} \quad (1)$$

where P_{Si} is the per pulse input signal power, ϕ is a random signal phase, and $\theta = 2\pi f_d T$ is the pulse-to-pulse phase shift due to the target Doppler shift f_d in conjunction with the interpulse period T. The accompanying thermal noise is:

$$\underset{\sim}{n}^T = \{n_1, n_2, \ldots, n_N\} \quad (2)$$

where by definition $E\{|\underset{\sim}{n}_i|^2\} = P_{Ni}$ represents the thermal noise power level. Finally the clutter return is:

$$\underset{\sim}{c}^T = \{c_1, c_2, \ldots, c_N\} \quad (3)$$

where $E\{|c_i|^2\} = P_{Ci}$ is the clutter power. The total input to the Doppler filter is:

$$\underset{\sim}{x} = \underset{\sim}{s}\,(f_d) + \underset{\sim}{c} + \underset{\sim}{n} \quad (4)$$

and the per pulse power is :

$$P_X = E\{|\underset{\sim}{x}_i|^2\} = P_S + P_C + P_N \quad (5)$$

The input signal-to-interference ratio (SIR) is:

$$(SIR)_{IN} = \frac{P_S}{P_C + P_N} \quad (6)$$

The Doppler filter has the complex weights

$$\underset{\sim}{w}^T = \{w_1, w_2, \ldots, w_N\} \quad (7)$$

and the filter output therefore is

$$\underset{\sim}{y} = \underset{\sim}{x}^T \underset{\sim}{w}^* \quad (8)$$

The corresponding output power is

$$P_Y = E\{|\underset{\sim}{y}|^2\} = E\{(\underset{\sim}{x}^T\underset{\sim}{w}^*)^{T*}(\underset{\sim}{x}^T\underset{\sim}{w}^*)\}$$
$$= \underset{\sim}{w}^T\, E\{\underset{\sim}{x}^*\underset{\sim}{x}^T\}\, \underset{\sim}{w}^* \quad (9)$$

The expected value in (9) represents the covariance matrix of the input which can be written

$$\underset{\sim}{M_X} = P_{Si}\underset{\sim}{M_S}\,(f_d) + P_{Ci}\underset{\sim}{M_C} + P_{Ni}\underset{\sim}{M_N} \quad (10)$$

assuming statistical independence of signal, clutter and thermal noise. By definition

$$\underset{\sim}{M_S}(f_d) = \frac{1}{P_S}\{\underset{\sim}{s}\;(f_d)^*\underset{\sim}{s}\;(f_d)^T \quad (11)$$

The i-jth element of $\underset{\sim}{M_S}(f_d)$ is $e^{j\theta(i-j)}$.

$$\underset{\sim}{M_N} = \frac{1}{P_N}\, E\{\underset{\sim}{n}^*\underset{\sim}{n}^T\} = \underset{\sim}{I} \text{ (identity matrix)} \quad (12)$$

$$\underset{\sim}{M_C} = \frac{1}{P_C}\, E\{\underset{\sim}{c}^*\underset{\sim}{c}^T\} \quad (13)$$

The ij-th element of M_c can be determined from the correlation function $\rho_C(\tau)$ of the clutter returns using the argument (i-j)T.

The noise gain of the Doppler filter is

$$G_N = \frac{P_{No}}{P_{Ni}} = \frac{\underset{\sim}{w}^T P_{Ni}\, I\underset{\sim}{w}^*}{P_{Ni}} = \underset{\sim}{w}^T\underset{\sim}{w}^* = \sum_{i=1}^{N} |\underset{\sim}{w}_i|^2 \quad (14)$$

The output SIR ratio is

$$(SIR)_{OUT} = \frac{P_{So}}{P_{Co} + P_{No}} = \frac{P_{Si}\underset{\sim}{w}^T\underset{\sim}{M_S}(f_d)\underset{\sim}{w}^*}{P_C\underset{\sim}{w}^T\underset{\sim}{M_C}\underset{\sim}{w}^* + P_{Ni}G_N} \quad (15)$$

and the SIR improvement therefore is

$$I_{SIR}(f_d) = \frac{(SIR)_{OUT}}{(SIR)_{IN}}$$

$$= \frac{\underset{\sim}{w}^T M_S(f_d)\underset{\sim}{w}^*}{\underset{\sim}{w}^T\underset{\sim}{w}^*} \cdot \frac{\underset{\sim}{w}^T\underset{\sim}{w}^*(P_{Ci} + P_{Ni})}{\underset{\sim}{w}^T(P_{Ci}\underset{\sim}{M_C} + P_{Ni}\underset{\sim}{I})\underset{\sim}{w}^*} \quad (16)$$

Corehent Processing Gain (G_C) — Normalized Clutter Cancellation (C)

This expression has been arranged to identify I_{SIR} as the product of the coherent processing (or integration) gain G_C (against thermal noise) and the normalized clutter cancellation C. The latter is equivalent to the definition of the MTI improvement factor.

The maximum SIR improvement is achieved if the weight vector is determined as as follows [7] :

$$\underset{\sim}{w}_{OPT}(f_d) = (P_C\underset{\sim}{M_C} + P_N\underset{\sim}{I})^{-1}\,\underset{\sim}{s}\;(f_d)^* \quad (17)$$

The average SIR improvement for the optimum processor is then determined by:

$$\overline{I_{SIR}} = \frac{1}{PRF}\int_0^{PRF} I_{SIR}(f_d)\, df_d \quad (18)$$

where $\overline{I_{SIR}}\,(f_d)$ is given by (16) and the filter weights are a function of f_d as shown in (17). The value of $\overline{I_{SIR}}$ was determined by a numerical integration assuming a Gaussian clutter Doppler spectrum of the form

$$S_C(f_d) = k\;.\;e^{-\frac{f_d^2}{2\sigma_f^2}} \quad (19)$$

where σ_f is the standard deviation of the width of the Doppler spectral lines. It was further assumed that $P_{Ni} << P_{Ci}$ so that $\overline{I_{SIR}}$ is not limited by thermal noise.

The results obtained for several values of N are shown in Fig.2. A universal abscissa axis is obtained by defining the relative clutter spectral spread as $\sigma_f T$ where T is the PRF interpulse period. These results are used in Section 4 for comparison with the performance of practical implementations of Doppler filter banks.

3. UPPER BOUND FOR LARGE N

When the number of pulses in a CPI is large the optimum Doppler processor can be configured as the combination of a whitening filter and a coherent integrator. The whitening filter has the transfer function

$$H_W\;(f_d) = \sqrt{\frac{1}{U_C\;(f_d)}} \quad (20)$$

where $U_C\;(f_d)$ is the clutter spectrum at the filter input. Across one unambiguous PRF interval we can write

$$H_W(f_d) = \sqrt{\frac{1}{rep_{1/T}\, S_C(f_d)}} \qquad (21)$$

where $S_C(f_d)$ is the spectrum of the Doppler spreading of the clutter returns. The function $rep_{1/T}$ is defined as [8]:

$$rep_{1/T}\, S_C(f_d) = \sum_{j=-\infty}^{\infty} S_C(f_d + j/T) \qquad (22)$$

The average normalized clutter cancellation due to the whitening filter is

$$C_{AV} = \frac{P_{Ci}}{P_{Co}} \cdot \frac{P_{No}}{P_{Ni}} \qquad (23)$$

We find

$$P_{Ci} = \int_0^{PRF} rep_{1/T}\, S_C(f)\, df = \int_{-\infty}^{\infty} S_C(f)\, df \qquad (24)$$

$$P_{Co} = \int_0^{PRF} rep_{1/T}\, S_C(f) \frac{1}{rep_{1/T} S_C(f)}\, df = PRF \qquad (25)$$

$$P_{Ni} = 1 \text{ (by definition)} \qquad (26)$$

$$P_{No} = \int_0^{PRF} \frac{T}{rep_{1/T} S_C(f)}\, df \qquad (27)$$

Thus

$$C_{AV} = T^2 \int_{-\infty}^{\infty} S_C(f)\, df \int_0^{PRF} \frac{1}{rep_{1/T} S_C(f)}\, df \qquad (28)$$

The coherent integrator which follows the whitening filter provides a coherent gain which is

$$G_C = 10 \log_{10} N \qquad (29)$$

at every signal Doppler frequency, and the average value of I_{SIR} is therefore

$$\overline{I_{SIR}} = C_{AV} \cdot G_C \qquad (30)$$

Assuming

$$S_C(f) = \frac{1}{\sqrt{2\pi}\sigma_f} \exp\left(\frac{-f^2}{2\sigma_f^2}\right) \qquad (31)$$

so that for $P_{Ci} = 1$ we have

$$C_{AV} = T^2 \int_0^{PRF} \frac{1}{rep_{1/T}\, S_C(f)}\, df \qquad (32)$$

This expression was evaluated numerically for several values of $\sigma_f T$ and the results are given in Table 1.

Table 1. Upper Bound on Average Clutter Cancellation

$\sigma_f T$	.07	.08	.1	.12	.14	.2
C_{AV} (dB)	85.2	61.0	33.5	19.4	11.6	2.8

For large N we therefore find

$$(\overline{I_{SIR}})_{OPT} = C_{AV}\text{ (dB)} + 10 \log_{10} N \qquad (33)$$

In Fig. 3 this asymptotic result is shown together with the values calculated in Section 2, using N as the independent variable. A good agreement between the exact calculation and the asymptotic result can be seen.

4. PRACTICAL PULSE DOPPLER PROCESSORS USING CHEBYSHEV WEIGHTED FILTERS.

Chebyshev filter designs are attractive for pulse Doppler processors due to their uniform Doppler sidelobes and ease of design. The weights of a Chebyshev filter for arbitrary N and sidelobe level, SLL, can be determined as described in [9]. In this Section Chebyshev designs for N = 6, 9, and 16 are considered, and a Doppler sidelobe level of 68 dB is specified. These filter designs will provide a maximum clutter suppression of approximately 60 dB.

Figure 4 shows the response of the corresponding Doppler filter bank for N=9. A total of 9 parallel filters are used to cover the unambiguous Doppler frequency interval using uniform spacing. The uniform spacing of the filters and a total number of filters equal to the number of pulses in the CPI are characteristic of DFT (or FFT) implementations of pulse Doppler processors. When general FIR (finite impulse response) filters are implemented as shown in Fig. 1, no such restrictions apply and an arbitrary number and spacing of the filters may be chosen. As discussed below, the use of a larger number of filters may be desirable to ensure a more uniform performance against clutter with an unknown Doppler shift.

The response curves shown in Fig. 4 are normalized to the noise gain of the filters (eq. (14)) and therefore represent the coherent filter gain of the individual filters as given by the first factor of eq. (16). By determining the normalized clutter cancellation for each filter as given by the second factor of eq. (16) the SIR improvement versus Doppler frequency can be calculated for any specified clutter model. Examples of the results of such calculations are shown in Figures 5 and 6 for the Chebyshev filter bank using N = 9 and assuming a Gaussian clutter spectrum with zero average Doppler shift and relative spectral spreads of $\sigma_f T = .05$ and .1, respectively. Only the maximum envelope for the nine filters are shown and a clutter-to-noise ratio of 100 dB was assumed in these and the following calculations to prevent thermal noise from limiting the performance in these graphs. The average I_{SIR} is obtained by averaging this response curve (in power) across the Doppler interval. The corresponding values of $\overline{I_{SIR}}$ are shown as broken lines in the two figures.

These calculations were repeated for a number of discrete values of $\sigma_f T$ and the resulting curves of $\overline{I_{SIR}}$ are shown in Figures 7 through 9 for each of the three Chebyshev filter bank designs.

Also shown in broken line in each of these graphs are the optimum results as derived in Section 2. The average $\overline{I_{SIR}}$ obtained with a given Doppler filter bank depends to some extent on the mean clutter Doppler shift. The best performance is obtained when the clutter spectrum is centered on the rejection region of one of the Doppler filters. Due to the finite number of filters in the Doppler filter bank it is possible for the clutter spectrum to be up to 1/2 of the filter-to-filter Doppler separation away from these optimum locations. Curves of $\overline{I_{SIR}}$ are therefore shown in Figures 7 through 9 for each of these two (best and worst) locations of the clutter spectrum. As the number of pulses per CPI (and therefore also the number of Doppler filters) increase from 6 to 16, the variation in the performance is seen to become negligible. This effect can be explained by the change in the ratio between the Doppler rejection region and the Doppler separation between the individual filters as listed in Table 2.

Table 2. Ratio of Doppler Rejection Region to Filter Separation

	N = 6	9	16
Doppler Rejection Region	.22	.414	.65
Filter Separation	.167	.111	.0625
Ratio	1.3	3.7	10.4

This effect is further illustrated by Figure 10 which shows $\overline{I_{SIR}}$ as a function of the mean Doppler shift of the clutter returns for N = 6 and $\sigma_f T = .05$. The maximum and minimum value are of course equal to those which can be read from Figure 7 for $\sigma_f T = .05$. A significant variation of performance is noted against clutter with an unknown Doppler shift. To reduce this variation to the order of 5 dB, the number of Doppler filters would have to be increased to approximately 12 parallel filters.

Except for this effect of the clutter Doppler shift the Chebyshev filter designs are seen to provide a performance which is within a few dB of the optimum value.

5. CONCLUSIONS

The performance criterion of average signal-to-interference ratio improvement ($\overline{I_{SIR}}$) has been used in this paper to compare optimum likelihood ratio pulse Doppler processors with practical approximations. The practical implementations were based on digital Chebyshev filter designs which were found to provide a performance which in many cases is within a few dB of the optimum. However, whenever the number of pulses in a coherent processing interval is small and clutter with an unknown Doppler shift must be suppressed, significant devictions from the optimum occurs for certain mean Doppler shifts of the clutter. This performance degradation can be avoided either by implementing a larger number of digital Doppler filters or by increasing the number of pulses used in each coherent processing interval.

REFERENCES

1. C.W. Helstrom, Statistical Theory of Signal Detection, Oxford: Pergamon Press, 1968.

2. L.A. Wainstein and V.D. Zubakov, Extraction of Signals from Noise, New York: Dover, 1970.

3. L.E. Brennan, I.S. Reed and W. Sollfrey, "A comparison of average-likelihood and maximum likelihood ratio tests for detecting radar targets of unknown Doppler frequency", IEEE Tr. Inf. Th., Vol. IT-14, pp. 104-109, January 1968,

4. C.E. Muehe et. al., "New techniques applied to air-traffic control radars", Proc. IEEE, Vol. 62, pp. 716-723, June 1974.

5. V.G. Hansen, "Clutter suppression in search radar", 1977 IEEE Conf. on Decision and Control, New Orleans, LA, 7-8 Dec. 1977.

6. W.W. Shrader, "MTI radar", Ch. 17 in Radar Handbook, M. Skolnik, Ed., McGraw-Hill, 1970.

7. L.J. Spafford, "Optimum radar signal processing in clutter", IEEE Trans. Information Theory, Vol. IT-14, pp. 734-743, Sept. 1968.

8. P.M. Woodward, Probability and Information Theory, With Application to Radar. London: Pergamon, 1953.

9. H.R. Ward, "Properties of Dolph-Chebyshev weighting functions". IEEE Tr. Aerospace and Electronic Systems, Vol. AES-9, pp. 785-786, Sept. 1973.

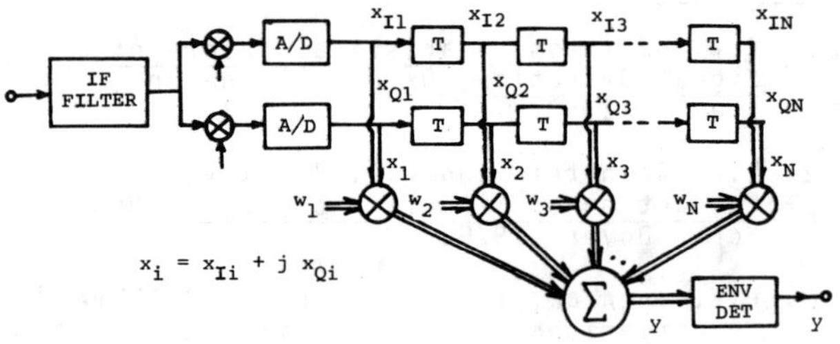

Figure 1 - Implementation of Each Digital Filter in a Doppler Filter Bank.

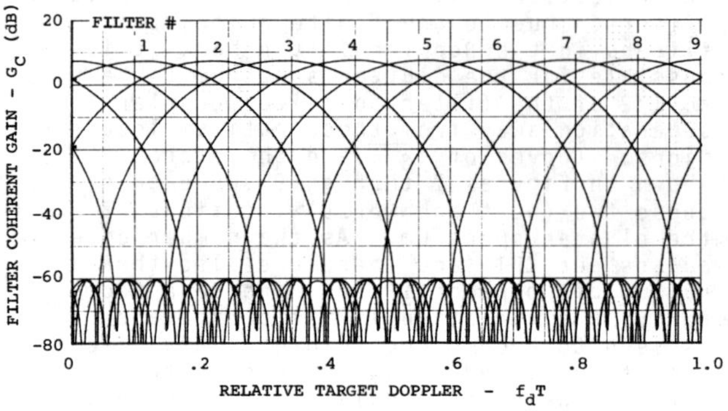

Figure 4 - Chebyshev Doppler Filter Bank with 68 dB Sidelobes for N = 9 Pulses.

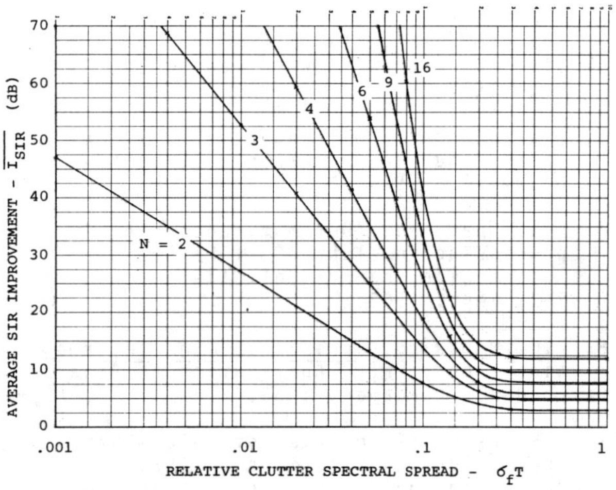

Figure 2 - Optimum Average SIR Improvement for Gaussian Clutter Spectrum.

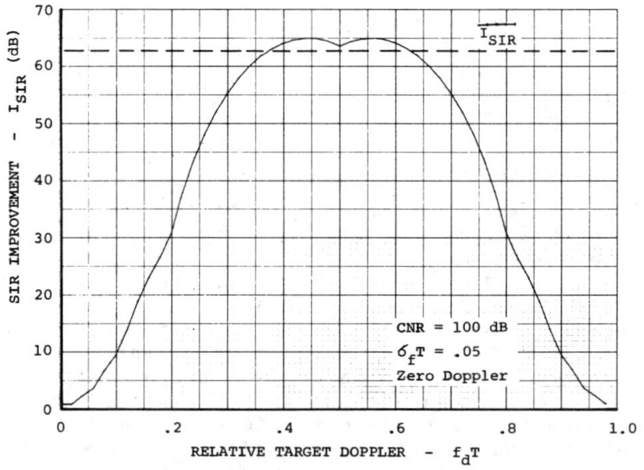

Figure 5 - Curve of SIR Improvement for Zero-Doppler Gaussian Clutter Spectrum With σ_fT = .05 and N = 9 Pulses.

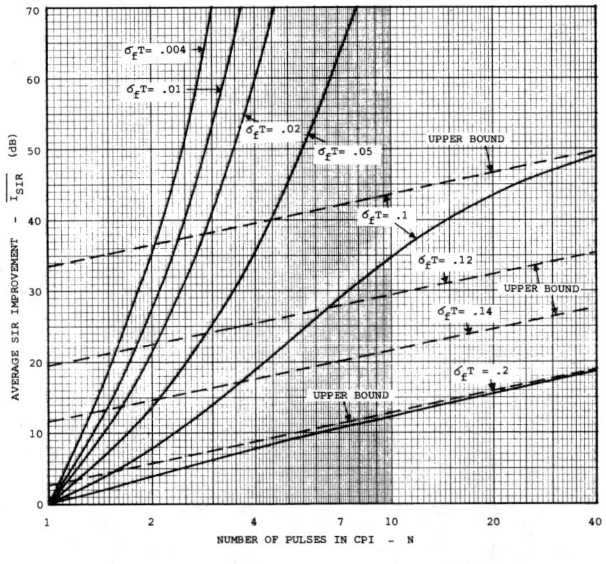

Figure 3 - Optimum Average SIR Improvement Compared With Upper Bound.

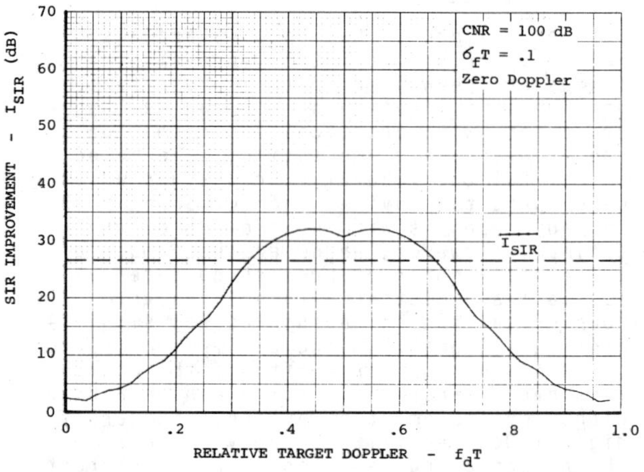

Figure 6 - Curve of SIR Improvement for Zero-Doppler Gaussian Clutter Spectrum With σ_fT = .1 and N = 9 Pulses.

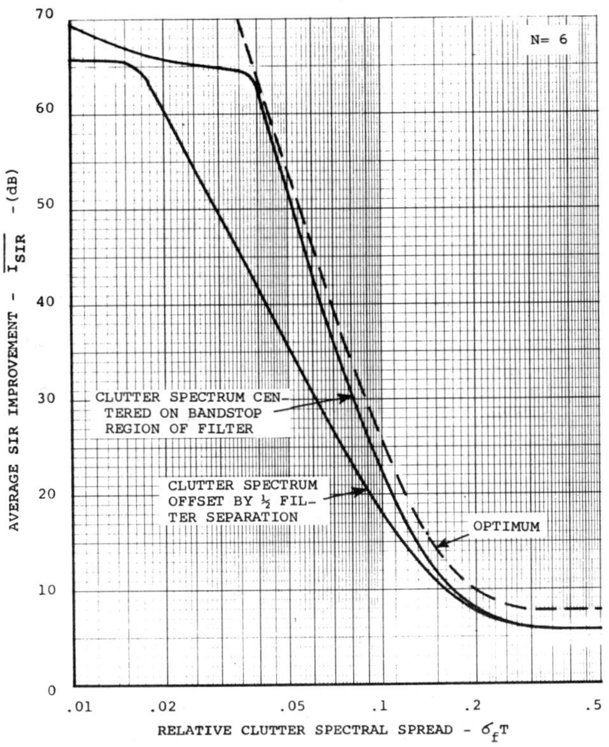

Figure 7 - Average SIR Improvement Curves for 68 dB Chebyshev Filter Bank and N = 6 Pulses.

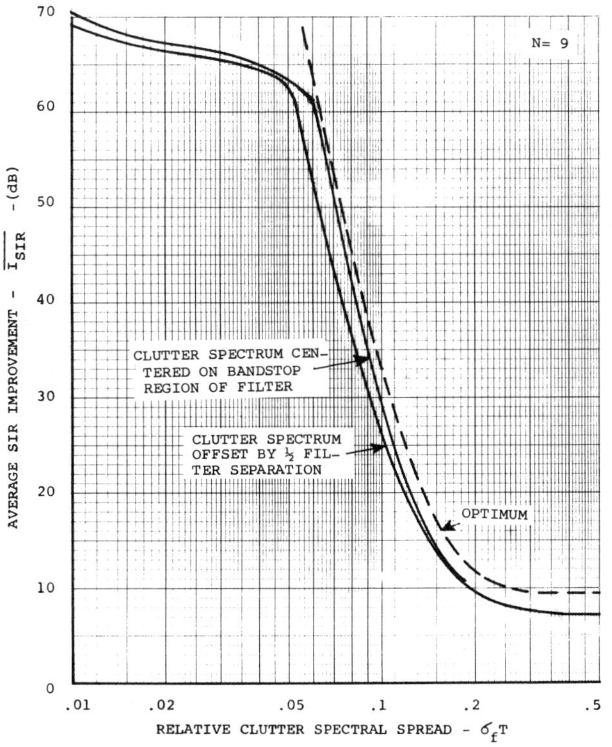

Figure 8 - Average SIR Improvement Curves for 68 dB Chebyshev Filter Bank and N = 9 Pulses.

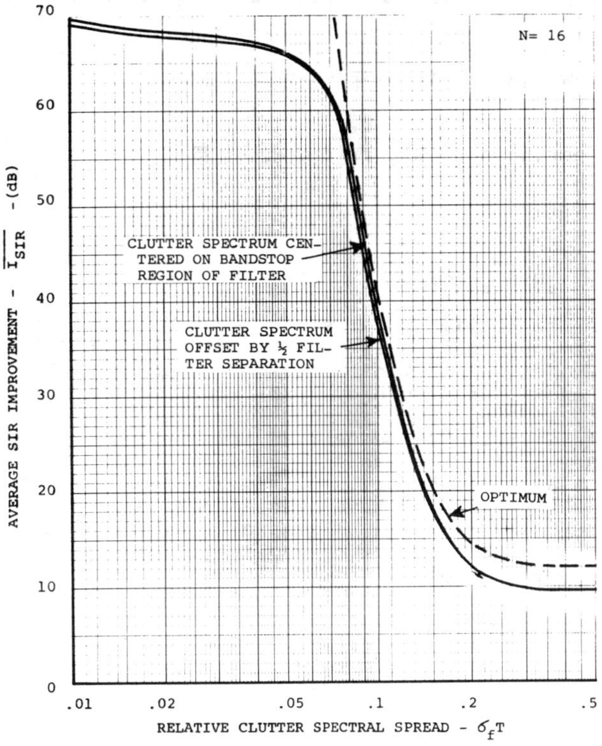

Figure 9 - Average SIR Improvement Curves for 68 dB Chebyshev Filter Bank and N = 16 Pulses.

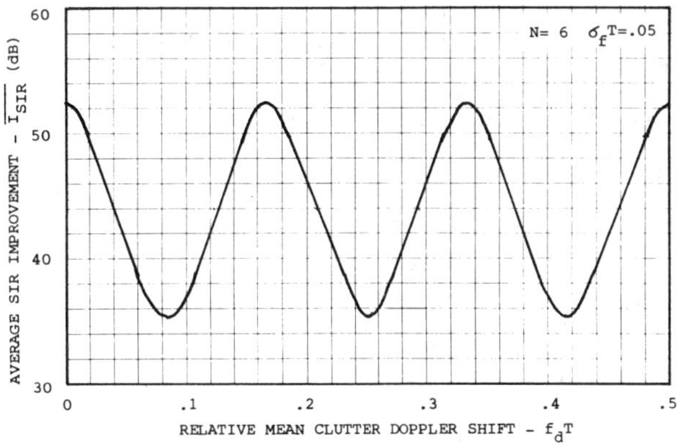

Figure 10- Average SIR Improvement for N = 6 as a Function of Mean Doppler Shift of Clutter.

RESOLUTION OF AMBIGUOUS RADAR MEASUREMENTS USING A FLOATING BIN CORRELATOR

Edwin R. Addison, Edward L. Frost

Westinghouse Electric Corp., USA

1. INTRODUCTION

Pulse doppler radar systems may be characterized by their ability to provide target range and/or velocity estimates. Whether the pulse repetition frequency (PRF) characterization is by low, medium or high PRF, one or both of the estimates must be derived from ambiguous data. Resolution of such ambiguous or aliased data is accomplished by implementation of multiple PRF systems and have been addressed by a number of authors (1-7). The aliasing problem is common to a number of applications other than radar (8-10).

While the advent of digital signal processing provides the means of resolving the ambiguities, the development of suitable algorithms must ensue. The Chinese Remainder Theorem (11), can be employed to yield unambiguous measurements by comparing outputs allocated to fixed integer numbered bins using integer arithmetic to modulo to the correct bin number. Targets straddling two or more bins or the assignment of an incorrect bin number will generally yield incorrect parameter values.

This paper sets forth an ambiguity resolution technique using multiple PRF data and utilizes a sliding floating point window or "floating bin" to correlate ambiguous centroided doppler measurements. This technique has the advantage that false targets are much less prevalent than in classical techniques. In addition, the same technique may be employed to resolve ambiguous range wherein centroided range measurements are moduloed with the pulse repetition interval (PRI) associated with each PRF.

Section 2. describes the algorithm. Section 3. derives performance and design equations for the algorithm. Sections 4. and 5. give simulation results for typical doppler and range correlators, respectively.

2. THE ALGORITHM

The reporting of target range and velocity parameters requires more than a simple parameter measurement when the parameter value is beyond the basic measurement interval. This situation occurs in pulse doppler radar systems when the expected target range or doppler exceeds the pulse repetition interval (PRI) or the pulse repetition frequency (PRF) respectively. The classical treatment of this problem is to resolve the ambiguous measurements using multiple PRF's which are related by a known factor and then applying the Chinese remainder theorem to the several measurements. This technique works well when the parameter measurement is in discrete increments (or bins) such that the number of bins in each of the PRFs in an integer and the set of integers have a basis which is relatively prime so that the expected parameter region can be resolved by the desired number of PRFs. Unfortunately, the relatively prime basis for the number of bins is difficult to meet in more than a one parameter domain as is the selection of a single bin for the parameter measurement when the bins are straddled. Parameter estimation techniques can be used to quantize the domain into bins and the ambiguity resolved. In practical situations, however; the parameter measurement is subject to error, thus the correct bin is not always selected and the resolved parameter value is incorrect.

These problems can be alleviated by a very similar technique which accounts for the measurement error. The ambiguous parameter measurements are moduloed toward larger values until their differences lie within some value, Ω, of each other. The most recent measurement is centered in Ω and the ambiguity resolved when a modulo of the remaining measurements falls within the width Ω of a modulo of the most recent measurement. Ω is then a "floating bin". Letting $(\alpha_i;\ i = 1, n)$ be the ambiguous parameter values for each of the PRFs $(F_i,\ i = 1, n)$, it is desired that the following conditions hold:

$$\Omega > |\alpha_1 + \delta\alpha_1 + K_1F_1) - (\alpha_i + \delta\alpha_i + K_iF_i)|;\quad i = 2, n \quad \text{.........................(2-1)}$$

where $\delta\alpha_i$ are the measurement error of the α_i respectively. But

$$\alpha_i = \alpha - K_iF_i\ ;\ i = 1, n \quad \text{...........(2-2)}$$

where α is the true target parameter value. Then

$$\Omega > |\delta\alpha_1 - \delta\alpha_i|\ ;\ i = 2, n \quad \text{..........(2-3)}$$

The measurement errors are independent random variable having zero mean and variance σ_α^2. Thus Ω is also a random variable having zero mean and variance $2\sigma_\alpha^2$. As a result, the selection of an optimum binwidth is solely dependent upon achieving a desired probability of correct ambiguity resolution.

3. DESIGN PERFORMANCE EQUATIONS

This section takes a more general approach to the correlation algorithm discussed in section 2. to take into account both range and doppler correlation as well as any other signal processing application requiring the resolution of ambiguous measurements. Equations for the probability of correlation and false correlation are developed.

3.1 Problem Restatement

Consider a set of positive real numbers $(F_1, F_2, \ldots F_n)$ which we shall call bases and a parameter α. Consider also a set of real numbers $(\alpha_1, \alpha_2, \ldots \alpha_n)$ each given by:

$$\alpha_i = \mathrm{mod}\ (\alpha, F_i) \quad \text{(3-1)}$$

We shall call α_i the ambiguous value of the parameter α with respect to the base F_i. Suppose an estimator is available to compute estimates $\hat{\alpha}_i$ of α_i. Let the probability density of the errors be $p_i\ (x)$. We desire to obtain an estimate $\hat{\alpha}$ of the parameter α.

The problem stated above is mathematically equivalent to doppler resolution in a pulse doppler radar system if (F_i; i = 1, n) is the PRF set, and α is the true target doppler. The set ($\hat{\alpha}_i$; i = 1, n) is the set of ambiguous measurements of the true doppler α. The problem is also mathematically equivalent to range resolution in a multiple range gate system. In this case, (F_i; i = 1, n) is the set of PRIs corresponding to the PRFs, and ($\hat{\alpha}_i$; i = 1, n) is the set of range measurements. The parameter α is then the true target range.

For the purpose of this problem we make the following assumptions:

A. $\alpha_\ell < \alpha < \alpha_u$ where α_ℓ and α_u are the lower and upper limits

$\alpha_\ell < 0$

$\alpha_u > \max F_i$; i = 1, n

B. $E\ (X_i) = 0$

$E\ (X_i^2) = \sigma_i^2$

$E\ (X_i\ X_j) = 0;\ i \neq j$

where X_i is a random variable having probability density $p_i(x)$.

C. $p_i(x)$ is symmetric about 0.

3.2 The Floating Bin Correlator

We may attempt to find a value for $\hat{\alpha}$ by choosing a pair of bases F_i and F_j, successively adding (or subtracting) F_i to $\hat{\alpha}_i$ and F_j to $\hat{\alpha}_j$, and comparing the intermediate results against each other with respect to a "floating binwidth", W_{ij}. This may be expressed by:

$$|(\hat{\alpha}_i + K_1 f_i) - (\hat{\alpha}_j + K_2 f_j)| < W_{ij} \quad \text{(3-2)}$$

The quantities K_1 and K_2 are integer values (modulo number) defined over the intervals $I_1 = (\frac{\alpha_\ell}{F_i} - 1, \frac{\alpha_u}{F_i} + 1)$ and $I_2 = (\frac{\alpha_\ell}{F_j} - 1, \frac{\alpha_u}{F_j} + 1)$. Any algorithm that uses one or more pair of bases from (F_1, F_2 . . . F_n) to find a value for $\hat{\alpha}$ in this manner shall be called a "floating bin correlator".

3.3 Probability of a Correlation

Consider using F_i and F_j and any floating bin correlator to find a value for $\hat{\alpha}$. It is of interest to know the probability of a correlation at the true value α. Consider the random variable Y_{ij} given by:

$$Y_{ij} = X_i - X_j \quad \text{(3-3)}$$

From assumptions, one can write the density function $P_{ij}(y)$ of Y_{ij} as

$$P_{ij}(y) = \int_{-\infty}^{\infty} P_i(y-x)\ P_j\ (x)\ dx \quad \text{(3-4)}$$

If $P_i\ (x)$ and $P_j(x)$ are normal densities, then

$$P_{ij}(y) = \frac{1}{\sqrt{2\pi\ (\sigma_i^2 + \sigma_j^2)}}\ \exp\ -\left[\frac{y^2}{2\ (\sigma_i^2 + \sigma_j^2)}\right] \quad \text{(3-5)}$$

The probability of a correlation at α; P_{ij} is then

$$P_{ij} = 2 \int_0^{W_{ij}} P_{ij}(y)\ dy \quad \text{(3-6)}$$

If $P_i(x)$ and $P_j(x)$ are normal densities as in equation (3-5), then equation (3-6) becomes

$$P_{ij} = 2\ \mathrm{erf} \left[\frac{W_{ij}}{\sigma_i^2 + \sigma_j^2}\right] \quad \text{(3-7)}$$

If a floating bin correlator is being designed for a certain minimum probability of correlation at α, P_α, then the minimum required binwidth, W_{min}, is given by

$$W_{min} = F_{ij}^{-1}\ (P_\alpha) \quad \text{(3-8)}$$

where $F_{ij}\ (W) = 2\int_0^W P_{ij}\ (y)\ dy$

Consider a floating bin correlator where the designer requires that m pair of the bases be used and that they will correlate to α. If P_c is the total probability of correlation at α and the binwidths were selected by equation (3-8), then

$$P_c > P_\alpha^m \quad \text{(3-9)}$$

If m of n pair are required to correlate, then

$$P_c > \sum_{k=0}^{n-m} \binom{n}{k}\ P_\alpha^{n-k}\ (1-P_\alpha)^k \quad \text{(3-10)}$$

Equations (3-9) and (3-10) are useful for designing a floating bin correlator to achieve a desired probability of correlation.

3.4 Probability of False Correlations and Selection of F

In order to minimize the possibility of false correlations, it is necessary to select the elements of (F_i; i = 1, n) such that they are relatively prime over the region of interest (α_ℓ, α_u). It would be desired that these elements satisfy

$$|K_1F_i - K_2F_j| > \varepsilon \; ; \; i,j < n, \; i \neq j \; \ldots (3\text{-}11)$$

where K_1 and K_2 belong to the intervals I_1 and I_2 as defined in equation (3-2). The value of ε may be selected by the designer based on the following:

Suppose that somewhere during the correlation process, the true location of α in an overlapping modulo of F_i and F_j differ by an amount Δ. Then, the probability of a false correlation, P_{fc}, is given by

$$P_{fc} = \int_{-W_{ij}}^{W_{ij}} P_{ij} \; (y + \Delta) \; dy \; \ldots\ldots\ldots\ldots (3\text{-}12)$$

If $P_{ij}(x)$ is a normal distribution, (3-12) can be rewritten as

$$P_{fc} = \text{erf}\left[\frac{\Delta - W_{ij}}{\sqrt{\sigma_i^2 + \sigma_j^2}}\right] + \text{erf}\left[\frac{\Delta + W_{ij}}{\sqrt{\sigma_i^2 + \sigma_j^2}}\right]$$

$$\equiv G_{ij} \; (\Delta) \; \ldots\ldots\ldots\ldots\ldots\ldots\ldots\ldots\ldots (3\text{-}13)$$

From a design viewpoint, P_{fc} is often given, and the minimum value of Δ must be found. This may be written

$$\varepsilon = \Delta_{min} = G_{ij}^{-1} \; (P_{fc}) \; \ldots\ldots\ldots\ldots\ldots\ldots (3\text{-}14)$$

The total probability of false correlation for a pair is the sum over all such modulos. However, in a practical sense, the smallest Δ term will dominate and equation (3-13) is approximately valid. If a floating bin correlator is used with m pair of correlations, then the total probability of false correlation, P_{FC}, is given by

$$P_{FC} < \left[\frac{W^*}{|\alpha_u - \alpha_\ell|}\right]^{m-1} \cdot P_{fc}^m \; \ldots\ldots\ldots (3\text{-}15)$$

where W* is the window for correlation across pairs. If m > 1, P_{FC} is very small since all m pair would have to correlate at the same false value.

Equations (3-15) may be solved to obtain P_{fc} from a specified P_{FC}. Then equations (3-14) and (3-11) may be employed to select an acceptable set (F_i; i = 1, n). This is a trial and error process. Choosing "relatively prime" sets works well as previously discussed.

3.5 Probability of Cross Correlations

Consider a situation where K other targets are present in addition to the parameter α. This introduces the possibility of cross correlations when using a floating bin correlator. On a given correlation using bases F_i and F_j, the expected number of cross correlations, E_{xc}, is given by

$$E_{xc} = \begin{pmatrix}\text{Target exposure} \\ \text{per } F_i \text{ modulo}\end{pmatrix} \times \begin{pmatrix}\text{Fraction of } F_i \\ \text{that overlaps } F_j\end{pmatrix} \times \text{(no. of modulos)}$$

$$= K \; \frac{2\,W_{ij}}{F_i} \cdot \frac{F_i}{F_j} \cdot \frac{|\alpha_u - \alpha_\ell|}{F_i}$$

$$= \frac{2\,W_{ij}\,K|\alpha_u - \alpha_\ell|}{F_i\,F_j} \; \ldots\ldots\ldots\ldots (3\text{-}16)$$

If a floating bin correlator with m pair of correlations is used, the probability of cross correlation, P_{xc}, is approximately limited by

$$P_{xc} < \left(\frac{2\,W^* E_{xc}}{|\alpha_u - \alpha_\ell|}\right)^m \; \ldots\ldots\ldots\ldots\ldots (3\text{-}17)$$

4. DOPPLER CORRELATION EXAMPLE

A representative floating bin doppler correlator was designed and simulated for an airborne MPRF radar application. This section describes the design and performance.

The velocity limits used in the design of the doppler correlator are V^+_{max} = 3500 fps and V^-_{max} = -3500 fps. A minimum of 3 target detections are required for a target report and at least two correlations are needed in order to prevent false correlations at incorrect doppler frequencies.

The doppler correlator was designed for a 99% probability of a true doppler correlation during a target dwell for a target at 10 dB S/N. There are at least 2 detections on which a correlation may be done. Since each detection requires 2 correlations, this translates to a required probability of 99% for each individual correlation. Assuming a normal distribution, a binwidth of W = 2.75σ, where σ is the rms error of the centroiding algorithm at S/N = 10 dB is sufficient. For a MPRF radar using a 64 point, 60 dB weighted FFT, the optimum binwidth is 400 Hz.

On a given look of processing, doppler is correlated using the filter with the largest target response that has been detected in each range gate. The correlator is designed to handle a maximum of 32 targets and cross correlations simultaneously.

The doppler correlator will eliminate cross correlations as follows. If only 3 looks detect and doppler does not correlate, the target is reported regardless.

The algorithm is implemented as follows. Two pair of looks are correlated for each target. The detection in the current look is correlated with the look containing the most recent detection of the same target. A second correlation is done between the current look and the second most recent detection. A comparison table is set up by successively adding (and subtracting) the PRF to the centroided frequency for the 2 looks being correlated. The table is set up to cover the velocity range of ±3500 fps. A comparison is done among the words in the table. The smallest difference is less than the binwidth specified then there is no correlated doppler. The MBC position must be added to this doppler and the correlated doppler from both pair must agree.

A detailed simulation was used to evaluate the performance of the doppler correlator. The simulation evaluated targets that range from -3500 fps to +3500 fps uniformly in doppler. A full dwell of processing was simulated for each target. Statistics on the probability of correlation were obtained and tabulated. Various S/N ratios were looked at. The simulation was used to obtain the binwidth.

Table 4-1 shows the simulation results for 10 dB and 15 dB targets. The result is the number of valid doppler reports per dwell. It represents the probability of a valid doppler report at least once within a target dwell.

The performance of the doppler correlator as a false target eliminator is also of interest. It was shown by calculation that 99.997% of all false targets will not correlate in doppler. Thus, almost all 3-of-n cross correlations are eliminated. A cross correlation that is 4-of-n will be reported without doppler, but chances are it will not repeat itself on future scans.

Another performance degrader is the presence of an interfering signal. Targets are never lost due to the presence of an interfering signal, however, several other effects may occur: 1) the interfering signal may prevent a doppler correlation, 2) doppler may correlate to the interfering signal frequency or 3) two doppler reports may be reported within a dwell, both the target and the interfering signal frequency. Table 4-2 shows the simulated performance with an interfering signal based on the assumption that the interfering signals are always smaller than the target. If the interfering signals are larger than the target, then the doppler will correlate to the signal frequency.

TABLE 4-1 - Doppler correlation performance

S/N = 10 dB Binwidth = 400 Hz

No. of Correct Pair Correlations	=	44425	96.0%
No. of Incorrect Pair Correlations	=	241	.5%
No. of Missed Pair Correlations	=	1178	2.6%
No. of Multiple Pair Correlations	=	1431	3.1%
No. of Correct Valid Doppler	=	21608	94.3%
No. of Incorrect Valid Doppler	=	0	.0%
No. of Invalid Doppler	=	1314	5.7%
No. of Valid Doppler/Dwell	=	6814	99.5%

S/N = 15 dB Binwidth = 400 Hz

TABLE 4-1 - Doppler correlation performance continued

No. of Incorrect Pair Correlations	=	20	.0%
No. of Missed Pair Correlations	=	165	.4%
No. of Multiple Pair Correlations	=	775	1.7%
No. of Correct Valid Doppler	=	22739	99.2%
No. of Incorrect Valid Doppler	=	0	.0%
No. of Invalid Doppler	=	183	.8%
No. of Valid Doppler/Dwell	=	6842	99.9%

TABLE 4-2 - Effects of an interfering signal

	10 dB Target	15 dB Target
Probability of No Doppler	2.0%	0.4%
Probability of Reporting Signal Doppler	0.1%	0.0%
Probability of Reporting Both Doppler	0.3%	1.0%

5. RANGE CORRELATION EXAMPLE

A floating bin range correlator was simulated as a second example. A medium PRF radar was assummed with the average PRF containing 64 range cells and an upper limit of the unambiguous range to be resolved of 1024 range cells. Several PRFs were used with the requirement that centroided target detections in at least 3 of the looks (PRFs) line up within the binwidth for correlation.

The range centroids are an estimate of the ambiguous position of a target in range. The targets amplitude in neighboring range gates (RG) is used to estimate the position in fractions of a RG. The fraction is added to the RG number with the detection to give the targets range centroid. The rms error for the minimum detectable target was such that a floating binwidth of 0.6 range cells was appropriate. The simulation results are discussed below:

The range correlator is implemented as follows. Correlation begins when the new set of ambiguous centroids have been determined. Any range cell may correlate if there exists a range centroid in that cell for the current look and there are (k-1) or more range centroids in prior looks with the same range centroid as the current plus or minus half the binwidth.

In other words the binwidth acts as a window which is placed around each of the range centroids of the current look. The window is then moved to the same range cell number for each of the past looks and if there is another centroid within the window then there is a correlation between the current look and this look. The window may cross range cell boundaries and be comparing range centroids in adjacent range cells, since a target detection may slip a range gate for one of the looks.

Each range cell of the current look is correlated to the corresponding range cells of the previous looks. To correlate a range cell of the current look with one of the older looks, the ambiguous centroids from: the range cell to be correlated, the previous range cell, and the next range cell of the older look are compared. The current look's centroid is used as a reference, it is subtracted from each of the older PRF's centroids and if the absolute value of any of the differences is less than the binwidth/2 then that look correlates with the current look, the correlation result is then set to one; zero otherwise. All of the older look's correlation results are added to give a total number of correlations, if this total is equal to or greater than k-1 then a target is reported for this range cell.

Results were obtained from a Monte Carlo simulation of the binwidth correlator and a discrete correlator. The percentage of false correlations verses the number of targets in a dwell is shown. Eight mutually prime PRFs were used and the k criterion for a correlation was 3 out of 8 for range cells between 0 and 511 and 4 out of 8 for range cells between 512 and 1024. The number of targets in a dwell (8-looks) was specified then the target ranges were chosen at random. A different Gaussian random variable was added to each of the ranges and the ambiguous value was calculated for each look. Seven percent of the PRFs were blanked to represent mainbeam clutter blanking. Each ambiguous range was then unfolded out to 1024 and placed into a correlator memory. All range cells passing the target criterion (k or more correlations) were checked to the original target list. If the range cell's centroid was equal to any true target range plus or minus half a binwidth then a true correlation was declared. If the centroid was not equal to any target plus or minus half the binwidth then a false correlation was made. A count was kept of all of the false correlations. A percentage of the number of false correlations to total number of targets in the dwell is shown in Figure 5-1. The simulation of the discrete case was similar except that the target range was rounded up to a whole number and no gaussian noise was added in.

It can clearly be seen that the binwidth range correlator has a much higher resistance to false correlations as the number of targets in a dwell increases.

6. CONCLUSION

Resolution of ambiguous range and doppler measurements in pulse doppler radar systems using a floating bin correlator has been described and demonstrated by simulation. Results show that this method is better than conventional approaches in that the number of false targets produced is significantly lower while simultaneously providing a high probability of correlation. In addition, this correlation may be achieved in real time.

REFERENCES

1. F. J. Taylor, "Adaptive Pulse Doppler Ambiguity Resolution," IEEE Trans. on Aerospace and Electronic Systems, AES-12, No. 2, pp. 98-103 (1976).

2. F. J. Taylor, "Pulse Doppler Ambiguity Resolution," IEEE Trans. on Aerospace and Electronic Systems, AES-8, No. 1, pp. 591-595 (1972).

3. J. M. Colin, "RACINE: An Unambiguous Pulse Doppler Radar Process," IEEE 1975 International Radar Conference, pp. 69-72 (1975).

4. T. Hair and M. G. Cross, "Some Advances in Pulse Doppler Processing," IEEE 1975 International Radar Conference, pp. 63-68 (1975).

5. R. L. Mitchell, "Resolution in Doppler and Acceleration with Coherent Pulse Trains," IEEE Trans. on Aerospace and Electronic Systems, AES-7, No. 4, pp. 630-636 (1971).

6. R. L. Mitchell, Radar Signal Simulation, Artech House, Inc., Dedham, Mass. (1976).

7. B. H. Cantrell, "A Short-Pulse Area MTI," NRL Report 8162, AO-47419, available from National Technical Information Service, Springfield, Va. (1977).

8. L. J. Pinson, "A Quantative Measure of the Effects of Aliasing on Raster Scanned Imagery," IEEE Trans. on Systems, Man, and Cybernetics, SMC-8, No. 10, pp. 774-778 (1978).

9. R. LeGault, "The Aliasing Problems in Two-dimensional Samples Imagery," Perception of Displayed Information, edited by L. M. Bikerman, Plenum Press, Inc., New York, N.Y., chapter 7 (1973).

10. E. Masry, "Alias-Free Sampling: An Alternative Conceptualization and Its Application," IEEE Trans. on Information Theory, IT-24, No. 3, pp. 317-324 (1978).

11. D. H. Mooney and W. A. Skillman, "Pulse Doppler Radar," in Radar Handbook, edited by M. Skolnik, McGraw-Hill Book Co., New York, N.Y., pp. 19-16 (1970).

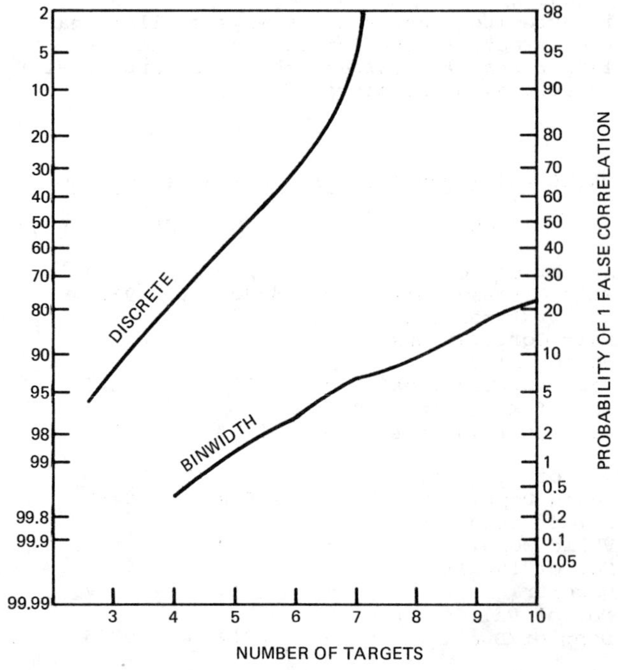

Figure 5-1. Performance of Floating Bin Range Correlator

OPTIMISING THE INTEGRATION APERTURE FOR A HIGH PRF CW SURVEILLANCE RADAR

R.A. Hall

Marconi Avionics Limited, U.K.

INTRODUCTION

Radar system performance is often modelled upon the principle of performing an analysis (1) of a signal in thermal noise, then, by use of the range equations, reduce the signal-to-noise ratio by amounts each assigned to a radar parameter (e.g. waveguide loss etc). These losses may be divided into two groups, firstly those which are constant and secondly those that vary as a function of time, i.e. for a scanning radar, beamshape. These time dependent effects modulate returned target signals and are responsible for defining the maximum window over which a scanning radar can usefully dwell on a target.

Consider now a typical application of a narrow beam radar in an airborne surveillance role. To keep the antenna within reasonable physical limits the transmitted carrier frequency will need to be x-band or higher. Doppler information will be extracted using a DFT technique, so, the PRF will then need to be at least the maximum doppler shift expected.

Using this DFT technique two windows are formed. The window, or aperture, over which the frequency analysis is made and the overall window, as just introduced, limited by signal modulations. Due to available signal processing modules however, the coherent window over which the DFT is performed is a fixed quantity. Therefore the overall window width can be fully utilized either by summing, in series, a number of signal doppler returns after the DFT or by the parallel use of DFT's, to increase the coherent window size.

Integration is thus being performed over the returned windows non-coherently for the series case and coherently when in parallel.

It has been shown by Marcum (1), and others, that it is the antenna gain pattern half-power points which indicate the maximum usable window for integration.

This paper will consider integration of cw returns in the doppler frequency domain and include directly in the analysis effects of time dependent losses which include:

1) Swerling I fading model
2) Hanning amplitude weighting
3) Eclipsing in the time domain from the high PRF
4) Beamshape as modelled from a typical experimental pattern.

and will, by quantifying detection performance against mean signal-to-noise ratio, show the optimum window (for a particular set of parameters) over which to integrate. For general usage, the mathematics have been presented enabling performance evaluation under different conditions. It is shown also that the beamshape and Hanning weighting function, which effectively modulate the integration, reduce to a single scaling factor which modifies the S/N directly. This factor, henceforth called the beam-factor, greatly simplifies the inclusion of these modulations within performance calculations.

CONCEPT

In a point surveillance role, signals are exchanged via a single scanning beam. Aperture time available is, therefore, limited to that defined by the beamwidth and scan rate. Within this aperture the target return is modulated by the scanning of the beam, any weighting function used during the sampling period, and eclipsing in the time domain due to the high PRF.

System noise bandwidth of a CW radar is usually defined by the coherent integration or aperture time. So, increasing the overall aperture will improve the probability of detecting a signal in noise.

However, amplitude modulations due to scanning will mean energy, both incident from the target and transmitted by the radar, is severly reduced as targets are scanned across. Thus an optimal aperture width will therefore exist that maximises the effective energy transfer between the scanning radar and the target.

Transmitted Modulation I.C.W.

In the high PRF CW system considered, the transmission of signals is continuously interupted, enabling the same aerial to be used both to transmit and receive. In this paper a single range gate is assumed, although multiple gates can be used. Duty ratio is thus 1:1. As a direct consequence of this duty ratio, unambiguous range is small, and considerable eclipsing of the returned signal occurs. However, with this unambiguous range being small (e.g. <1 n. mile), compared to usual detection ranges of many nautical miles, it can be inferred that each scan will provide a return uncorrelated with respect to the eclipsing. This property is used later to determine the average loss under eclipsing over many looks.

Weighting Function

Due to the discontinuous nature of the aperture period, e.g. a rectangular function in time, each doppler filter in the frequency domain is a sinc function.

High sidelobes (typically -13 dB), severely reduce resolution in the contiguous series of the (DFT) doppler bank, so in the system described in this paper, Hanning weighting of the sample aperture is utilised to reduce the first sidelobes by a further 18 dB.

Note: Hanning is equivalent to a $\cos^2$ amplitude modulation over the returned sample window.

Beamshape

The beamshape can be considered as another weighting function, although the weighting is now on the integrator. It is shown in Appendix 2 that the weighting of the integrator, for a constant amplitude signal, can be reduced to a single factor, ρ_B, which modifies the S/N ratio directly. This is a very convenient result as it simplifies the effects of beamshape upon the performance algorithm. The application of Hanning is incorporated simply as changes in value of this beam factor.

COHERENT INTEGRATION

Under coherent integration one variate of signal plus noise is considered.

The integration or aperture time is increased comensurate with constraints of beamshape.

From reference 2, probability of detecting a Swerling type target using a square law envelope detector is given by:

$$Pd(s) = \int_{qn}^{\infty} (1+s)^{-1} \exp(-x/1+s)\, dx \qquad (1)$$

with $$PFA = \int_{qn}^{\infty} \exp(-x)\, dx \qquad (2)$$

Having obtained the probability distribution of signal plus noise, then to find the total probability of detection of a target, this must be averaged over the probability distribution of receiving that signal

$$PD(s') = \int Pd(s')\, P_E(s')\, ds' \qquad (3)$$

with $P_E(s')$ the PDF of the eclipsing function in Appendix 1A.

Note: s' is in the decibels

Noise bandwidth is inversely proportional to the sample window or aperture time. So this performance, as measured in S/N ratio, is increased in proportion; the probability of detection PDF being unchanged. This direct proportionality is the rationale behind usage of the term "coherent integration".

The radar system has two main parameters which perturb the return signal as a function of time during its operation. They are firstly beamshape and secondly Hanning weighting.

It can be shown that optimum performance obtainable occurs when the beamshape and weighting function are matched. This is unrealistic for practical reasons, although results indicate optimum performance is obtained under conditions of nearest match.

NON-COHERENT INTEGRATION

The basic sample window is now significantly smaller than the beamwidth. After allowance for any doppler shift, integration is applied upon the series of returned signal plus noise samples. In other words, after the FFT the filters are summed prior to the detection threshold being applied. Phase is not preserved between returns hence integration is non-coherent.

Under complete correlation between return amplitudes, the probability of detecting a signal using a square law detector-integrator was derived by Marcum (1). In addition the effects of Swerling fading and eclipsing, assumed decorrelated on a scan-by scan basis, are included by the method of Appendix 1.

$$PD(s') = \int_{0}^{\infty} Pd(s')\, F(s')\, ds' \qquad (4)$$

with F(s') = combined PDF of fading and eclipsing (see Appendix 1)

and Pd(s') the PDF of integrated returns.

SYSTEM IMPLEMENTATION

For surveillance, a contiguous bank of filters over the complete doppler band is required. Here, this is accomplished by an DFT technique. A continuous band of 1024 points over the doppler band is provided by the sampling rate supplied by PRF of the signal.

Required doppler band	= 100kHz
Sampling rate - (PRF)	= 200kHz
Aperture time for 1024 points	= 10.24ms
With bandwidth	= 97.6Hz

For longer transforms, more points are taken within the doppler band, thereby reducing the noise bandwidth in proportion. However, 10.24ms is the standard aperture time in the proposed system and therefore all integrations will be a multiple of this. A Block schematic illustrating ICW doppler radar processing is presented in Figure 1. The two possible integration methods being included.

Using the integration and perturbation algorthms outlined in the appendices, curves of probability of detection versus signal-to-noise ratio were computed.

The following are the set of parameters used:

Beamwidth(-3dB)	=	2.2^o
Scan rate	=	60° per sec
Aperture time	=	10.24ms
Aperture window	=	0.61^o
Overall false-alarm rate	=	1 per 100 secs
Tx-Rx duty ratio	=	1:1

Figures 2 and 3 indicate performance in S/N ratio, for a fixed Pd, under both integration modes. For illustration a complete range of integration aperture windows has been shown.

Conclusion

Examination of figure 2 shows, for the set of system parameters considered, optimum performance under non-coherent integration of doppler returns is acheived when 3 sample windows are integrated. A complete set of Pd curves show a slight change in integration gain with Pd, however it is not significant enough to change the optimum integration number. Coherent integration shows an interesting result in that Hanning sidelobe weighting allows a greater usable integration window than either the non-Hanned coherent case or under non-coherent integration (with or without Hanning). Figure 3 illustrates the integration window can be almost doubled.

Comparison of the two integration systems show a greater detection performance under the coherent mode. However, implementation results in significant differences in cost, complexity and physical size; the coherent case being greater in all three. This is basically because faster processing is required.

These are important points which need consideration for cost effectiveness under particular system utilisation.

Acknowledgements

The author wishes to thank:

1. Mr. C.R. Hardaker for his valuable assistance in producing computer solutions to the integrals shown.

2. Procurement Executive, Ministry of Defence for support in the carrying out of this work.

References

1. Marcum, J.I., 1960, IRE TRANS IT-6

2. Schwartz, M., 1956, IRE TRANS Vol IT-2

APPENDIX 1

Fading and eclipsing

If the probability of detection for a stationary signal in noise is known, and the signal allowed to vary as some probability distribution. Then the resultant probability of detection is given by

$$PD(s') = \int_{-\infty}^{\infty} Pd(s')\ P\ (s'-s_o')ds' \qquad (1.1)$$

i.e. statistical averaging

Now if two (or more) independent variates produce variations in the signal, their combined effects can be reduced to a single PDF by convolving

Under fading and eclipsing

$$P\ (s') = P_E(s')\ *\ P_F(s')$$
$$= \int_{-\infty}^{\infty} P_E(s\ ')\ P_F\ (s'-s\ ')\ ds\ ' \qquad (1.2)$$

1A Eclipsing PDF

With the one range gate and a 1:1 mark¦space ratio, the output signal amplitude varies as a function of range according to a sawtooth law.

Since the sawtooth period is small in comparison with absolute target range it can be assumed all positions on it are equally probable.

ie, $P_E\ (A) = K$ (constant) (1.3)

But $\int_0^1 P_E\ (A) = dA = 1$

e.g. $K = 1$

$$P_E\ (A) = 1 \qquad (1.4)$$

Therefore to express in dB, let 20 log A = s'

which gives $P_E(s') = \dfrac{A}{8.686}$ (1.5)

1B Fading

Variation in target echoing area is assumed Swerling (type I)

$$P_F\ (P)\ dp = \frac{1}{Po} \exp \left(\frac{-P}{Po}\right) dp \qquad (1.6)$$

Where P = return power

Let $P = A^2$, and normalising mean power to unity, gives:

$$P_F\ (A) = 2A \exp\ (-A^2) \qquad (1.7)$$

or in dB with $S = 20 \log_{10}A$

$$P_F\ (s') = \frac{A^2}{4.343} \exp\ (-A^2) \qquad (1.8)$$

APPENDIX 2

Using the method given by Marcum (1), the characteristic function of a single variate of signal plus noise is

$$C_x(\varepsilon,s) = (\varepsilon+1)^{-1} \exp(-s) \exp(s/\varepsilon+1) \quad (2.1)$$

Let the jth variate be attenuated by the factor B_j, thus

$$S_j = B_j S$$

Where S = S/N ratio before modulation

Hence the characteristic function for the sum Y, of N independently weighted signal plus noise samples, is:

$$C_Y(\varepsilon,s) = \prod_{j=1}^{N} (\varepsilon+1)^{-1} \exp(-SB_j) \exp(SB_{j/\varepsilon+1})$$

$$= (\varepsilon/1)^{-1} \exp(-NS/\rho_B) \exp(NS/\rho_B(\varepsilon+1))$$

$$\text{which gives } \rho_B = N \Big/ \sum_{j=1}^{N} B_j \quad (2.2)$$

This parameter, ρ_B, henceforward called the Beam Factor, is seen to be an extension of the collapsing ratio as defined by Marcum. Indeed, it reduces to Marcum ratio under similar conditions.

Let N samples consist of M noise and P equal signal plus noise samples

$$\rho_B = \frac{M + P}{(M.Bm + P.Bp)}$$

However $Bm \rightarrow 0$ and $Bp \rightarrow 1$

So, $\rho_B = \frac{M + P}{P} = \rho_c$ i.e. collapsing ratio

APPENDIX 3

Beamshape or the aerial gain function, is taken as measured on a experimental aerial system and was found to be, over the main-beam, essentially a Gaussian function to well below the -3dB points; then reducing at a progressively greater rate to the first null

$$G_1(\theta) = \exp -0.568(\theta)^2 \qquad 0<\theta<1.47 \quad (3.1)$$

$$G_2(\theta) = G_1(\theta) \cos^2 \frac{\pi}{2.5} (\theta-1.47) \qquad 1.47<\theta<2.5 \quad (3.2)$$

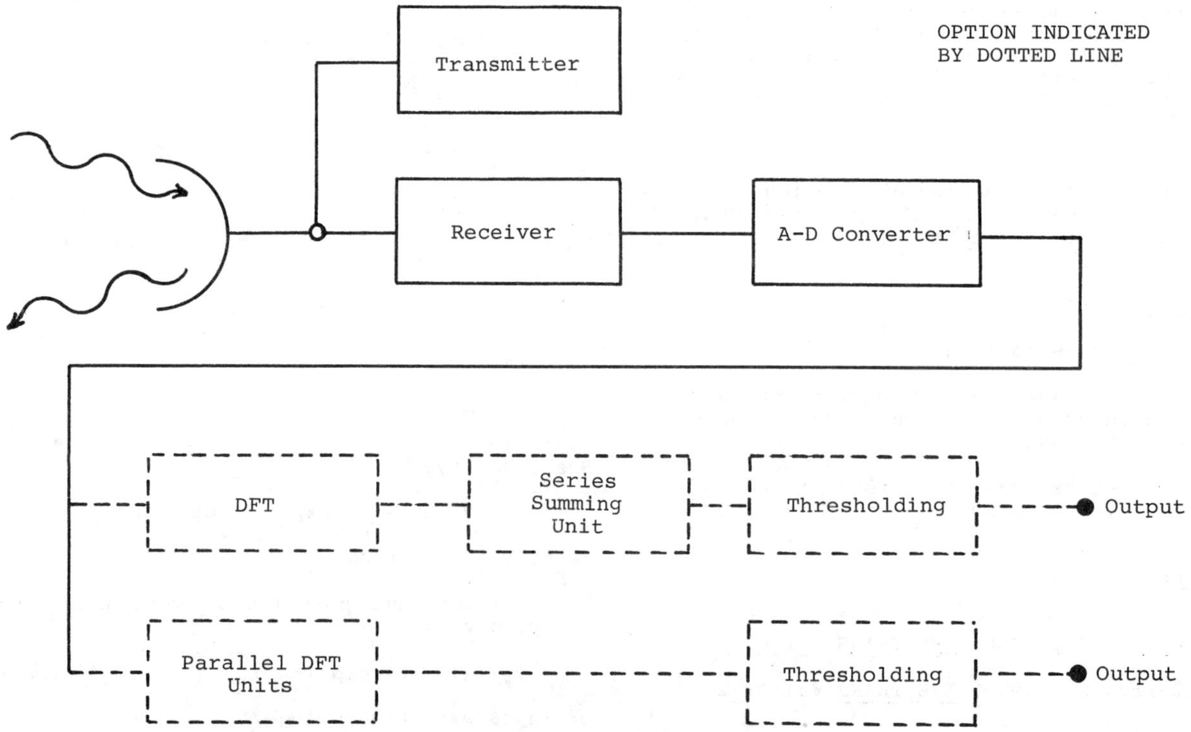

Figure 1 Surveillance Radar Simplified Block Schematic

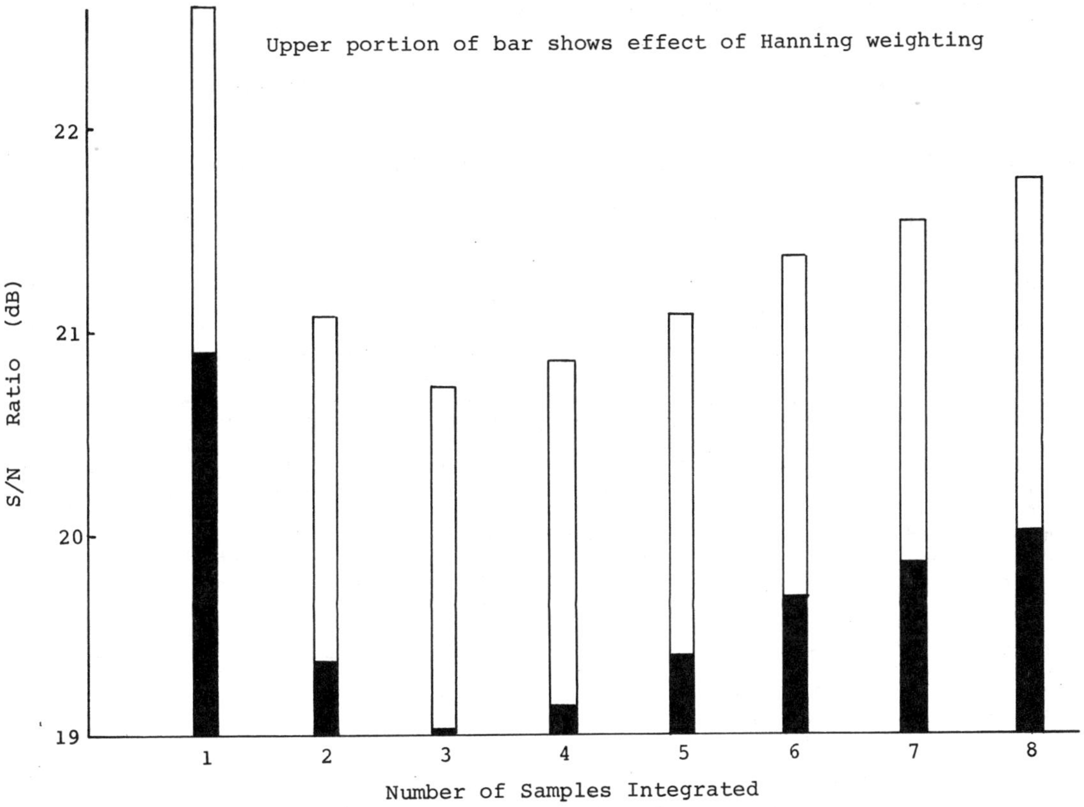

Figure 2 Mean S/N ratio for Pd of 0.5 under non-coherent integration

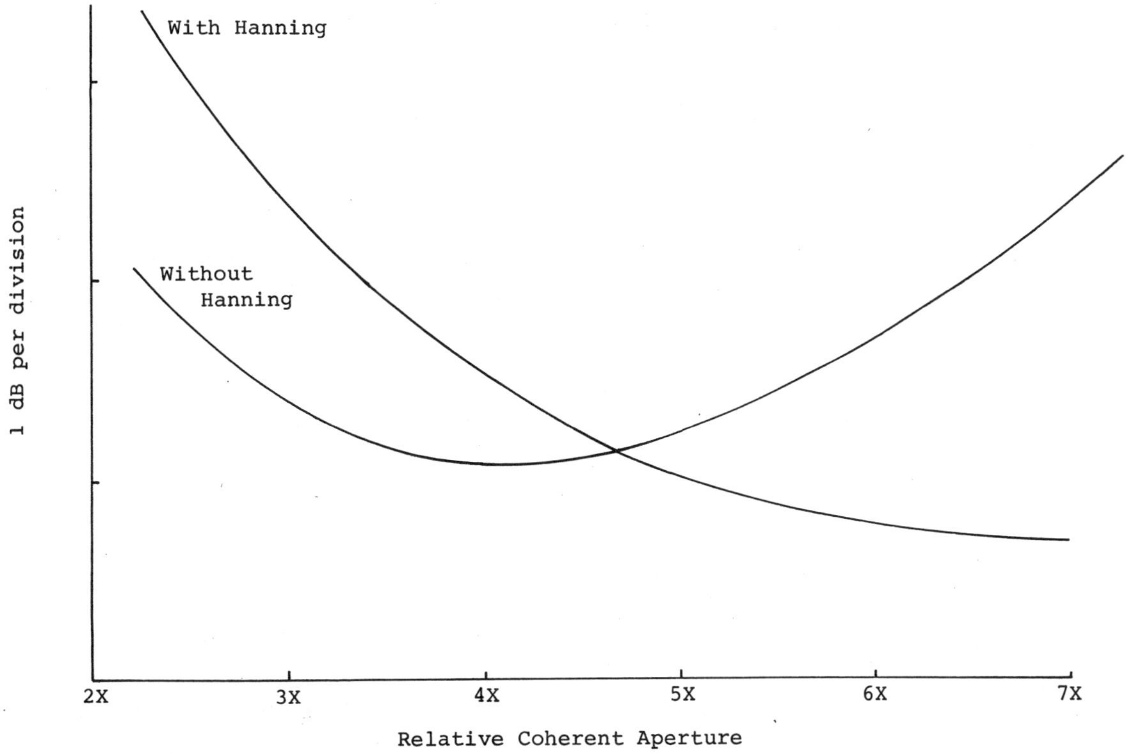

Figure 3 Relative S/N ratio for Pd of 0.5 under coherent integration

PERFORMANCE COMPARISON OF MTI AND COHERENT DOPPLER PROCESSORS

D. C. Schleher

Eaton Corporation, AIL Division, Deer Park, New York, 11729, USA

INTRODUCTION

MTI and coherent doppler processors are utilized in radar systems to detect targets imbedded in clutter. The relative detection performance of these processors is of interest to the radar designer. Several recent papers [1-6] have presented the performance of MTI's followed by both coherent and incoherent integrators in the presence of receiver noise. While this is of theoretical interest, there would be no purpose to use an MTI unless it was desired to extract targets from a clutter background. Hence this paper considers the problem of detecting targets in both a clutter and noise interfering background. Three basic doppler processing configurations are considered: the optimum transversal filter, an MTI followed by a weighted FFT coherent integrator, and an MTI followed by an incoherent integrator.

Performance comparisons between the various doppler processing configurations is provided for the same number of pulses in a radar dwell and various clutter-power to receiver-noise-power ratios. A nine-pulse optimum transversal filter and a two-pulse MTI followed by an eight-pulse coherent or incoherent integration process are considered. The numerical results presented in the paper can be readily extended to higher order MTI's or to a different number of processed pulses using the analytic formulations provided.

The fundamental problem considered in this paper is the processing of n-coherent received pulses which contain either background interference (radar clutter plus receiver noise) or a doppler shifted target in background interference. Using statistical detection theory [7] or maximizing the signal-to-interference power ratio [8] allows the complex weights for the optimum transversal filter to be found under the assumption that all signal and interference parameters are known. Since the target doppler shift is not known in most applications, a bank of n equally spaced contiguous transversal filters can be used to cover the doppler frequency band of interest. Each filter is then optimum at its center frequency but mismatched otherwise [9].

An almost optimum alternative doppler processor uses an MTI followed by a bank of coherent integrators [10]. This two-stage cascaded filter uses the MTI to enhance the signal-to-clutter ratio and the coherent integrator to primarily enhance the signal-to-noise ratio. Also, only one of the m-pulses used in the MTI processing is effective in building up the signal-to-noise ratio so that the performance of this processor can be expected to be inferior to that of a transversal filter [9].

A processor extensively used in practice is an MTI followed by an incoherent integrator. The detection loss with this type of processor can be quite severe. The MTI in addition to reducing the number of pulses available for integration also correlates the receiver noise, which when incoherently integrated, causes a detection loss that increases with the order of the MTI employed. This problem is further compounded by any residual clutter not eliminated by the MTI, which in the extreme, completely negates the effect of the integrator.

Performance Analysis of Doppler Processors

The improvement factor for the transversal filter depicted in Figure 1 is given by:

$$I = \frac{W^T M_s W^*}{W^T M_x W^*} \quad (1)$$

where M_s and M_x are the signal and interference covariance matrices and W* is the complex conjugate weight vector [11]. For Gaussian distributed interference the optimum weights, which are obtained by maximizing (1), are given by:

$$W_o = M_x^{-1} S^* \quad (2)$$

where M_x^{-1} is the inverse of the interference covariance matrix and S* is the complex conjugate of the doppler shifted signal vector. The improvement factor for the optimum weights can be found from:

$$I_o = \mathrm{Trace}\,[\,M_x^{-1} M_s\,] \quad (3)$$

The probability of detection is then equal to [12]

$$P_d = Q\,\{[2P_s/P_n\,(1+F/I_o)]^{1/2},\ [2\ln(1/P_{fa})]^{1/2}\} \quad (4)$$

where F is the clutter-to-noise power ratio, P_s and P_n are the signal and noise powers, P_{fa} is the probability of false alarm and Q (.,.) is Marcum's "Q" function.

The cascaded MTI and FFT coherent integrator is depicted in Figure 2. The improvement factor for this configuration can be found from:

$$I = \frac{|H(f)|^2 \int_0^{f_r} S_i(f)\, df}{\int_0^{f_r} |H(f)|^2 S_i(f)\, df} \quad (5)$$

where $S_i(f)$ is the normalized clutter-plus-noise power spectral density, f_r is the radar's PRF and:

$$H(f) = \sin^{m-1}(\pi x) \frac{\sin[\pi(x-k)]}{n \sin\left[\frac{\pi(x-k)}{n}\right]} \quad (6)$$

Shaping of the integrators response can be accomplished by appropriate postcombining of adjacent filter outputs [11] or by time weighting the integrators input signal. The probability of detection for the cascaded binomial MTI and FFT coherent integrator can be found by substituting equation (5) into equation (4).

The MTI followed by incoherent integration is depicted in Figure 3. The characteristic function of the integrator output can be obtained by decorrelating the MTI output using an orthogonal transformation and is given by [3,4]:

$$\phi(u) = \exp\left[\sum_{i=1}^{n} u\, g_k^{\,2}/(1-2u\lambda_k)\right] \prod_{k=1}^{n} 1/(1-2u\lambda_k) \quad (7)$$

where λ_k are the eigenvalues of the covariance matrix at the output of the MTI and g_k are the projections of the signal vector onto the kth eigenvector. An expression for the probability of false alarm can be developed from equation (7) with $g_k=0$ as:

$$P_{fa} = \sum_{k=1}^{n} \left\{ \prod_{\substack{\ell=1 \\ \ell \neq k}}^{n} \frac{\lambda_k}{\lambda_k - \lambda_\ell} \right\} \exp\left(-\frac{V_T}{2\lambda_k}\right) \quad (8)$$

from which the thresholds V_T can be determined. Equation (7) can be used to develop a cumulant generating function from which the cumulants can be determined as:

$$\chi_\ell = \sum_{k=1}^{n} (2\lambda_k)^\ell \, (\ell-1)! \, (\ell g_k^{\,2} + 1) \quad (9)$$

Equation (9) can be used in a Gram-Charlier series to determine the probability of detection. A three term series with Edgeworth grouping was found to provide good accuracy.

Doppler Processor Performance Comparison

The doppler processing configurations analyzed consist of: a nine-pulse optimum transversal filter, a single canceller MTI followed by an eight-pulse FFT coherent integrator, and a single canceller followed by a square law detector and an eight-pulse incoherent integrator. The above configurations all process nine-input pulses and numerical results in the form of detection curves are presented for $P_{fa} = 10^{-6}$ and Gaussian shaped clutter autocorrelation functions.

Detection curves for the transversal filters, coherent FFT integrators and incoherent integrators are generally evaluated at $f_d = f_r/4$. This doppler frequency provides a response which is equivalent to the average response of a single canceller MTI when all velocities are equally likely. In addition, several other detection curves are provided which allow detection losses to be estimated. A detection curve for Hamming weighting of the FFT coherent integrator allows the effect of integrator weighting to be determined. Detection curves for MTI-coherent integrator performance (with Hamming weighting and no weighting) at crossover between adjacent doppler filters allows doppler filter straddling losses to be estimated. Detection curves for eight-pulse incoherent integration allows the integration loss introduced by the MTI into the incoherent integration process to be determined. A detection curve for a double canceller MTI-coherent FFT integrator allows the relative detection improvement by a higher order MTI in a heavy clutter situation to be determined.

Figure 4 shows the detection performance of the various doppler processors for a narrow clutter spectrum ($\sigma_c T = .01$). The single canceller MTI improvement factor for this spectrum is 43 dB. The detection loss of the optimum transversal filter is of the order of 0.2 dB between no clutter (F = 0) and a clutter-to-noise power ratio of F = 20,000. The detection performance in heavy clutter (F = 20,000) of the MTI-coherent integrator at maximum response is essentially that of an eight-pulse coherent integrator indicating that the processor was effective in removing the clutter. The effect of the MTI and clutter on the incoherent integrator results in a loss of 1.1 dB with just noise alone and 4.3 dB when the clutter residual power equals the noise power.

Figure 5 depicts the detection performance of the various doppler processors for a medium width clutter spectrum ($\sigma_c T = .05$) which provides a single canceller MTI improvement factor of 29 dB. The detection loss of the optimum transversal filter is of the order of 0.4 dB between no clutter (F=0) and a clutter-to-noise power ratio of F=1000. The peak detection performance of the MTI-coherent integrator combination for F=1000 is within the order of 0.15 dB of an eight-pulse coherent integrator in noise. This indicates good performance in clutter. However, at the crossover between adjacent doppler filters the composite performance (considering detection probabilities in both filters) is down almost 2 dB from the peak. Using Hamming weighting of the coherent integrator response considerably smooths the overall doppler response but the overall performance for F=1000 is uniformly down by the order of 1.4 dB from that of an eight-pulse coherent integrator in noise. The detection loss of the MTI-incoherent integrator is again 1.1 dB for noise alone and 3.5 dB when residual clutter power is equal to noise power.

Figure 6 depicts the detection performance of the various doppler processors for a wide clutter spectrum ($\sigma_c T=.1$). The single canceller MTI improvement factor is 23 dB. Performance of the optimum transversal filter is again good over strong clutter (F=1000) with a reduction the order of 0.8 dB with respect to noise only performance (F=0). The performance of the single canceller-coherent integrator for F=1000 is poor; down the order of 2.7 dB with respect to an eight-pulse coherent integrator in noise. This indicates that a better MTI filter is required. Using a double canceller instead of a single canceller reduces the detection loss for F=1000 to within 0.5 dB of an eight-pulse coherent integrator in noise. Detection loss for the MTI-incoherent integrator range from 1.1 dB for noise alone to 7.5 dB for F=1000. This loss is plotted in Figure 7 as a function of residual clutter power (F/I). The curve indictes that residual clutter power must be made significantly less than one if good detection performance is to be realized using the MTI-incoherent integrator doppler processor.

The performance comparison for the cases considered can be summarized as:

- Good detection performance is provided by optimum complex weight transversal filters even in heavy clutter environments.
- The performance of cascaded MTI-coherent integrators is inherently somewhat less than that of transversal filters for an equal number of pulses processed.
- Cascaded MTI-coherent integrators have an average doppler filter straddling loss that is the order of 1 to 1.5 dB.
- Coherent integrator weighting is required to provide a uniform overall doppler processing response.
- Higher order MTI filters in combination with coherent integrators are required for wide clutter spectrums to provide good detection performance.
- MTI-incoherent integrator doppler processors are inherently lossy. For a single canceller MTI with eight-pulse incoherent integration losses range from 1.1 dB for noise alone to 3.5 to 4 dB when residual clutter equals noise, to over 7.5 dB when residual clutter is 5 times noise power.

1. Dillard G., "Signal-to-Noise Ratio Loss in an MTI Cascaded with Coherent Integration Filters," IEEE International Radar Conference, April 1975.

2. Trunk G., "MTI Noise Integration Loss," NRL Report No. 8132, July 1977.

3. Kanter I., "A Generalization of the Detection Theory of Swerling," IEEE EASCON Conference, September 1974.

4. Dillard G. and Richard J., "Performance of an MTI Followed by Incoherent Integration for Nonfluctuating Signals," IEEE International Radar Conference, April 1980.

5. Hall W. and Ward H., "Signal-to-Noise Loss in Moving Target Indicator," IEEE Proceedings, February 1968.

6. Weiss M. and Gertner I., "Loss in Single-Channel MTI with Post-Detection Integration." IEEE Trans. Vol AES-18, No. 2, March 1982.

7. Brennen L. and Reed I., "Optimum Processing of Unequally Spaced Pulse Trains for Clutter Rejection," IEEE Trans. Vol AES-4, No. 3, May 1968.

8. Brooks L. and Reed I., "Equivalence of the Likelihood Ratio Processor, the Maximum Signal-to-Noise Ratio Filter, and the Wiener Filter," IEEE Trans. Vol AES-8, No. 5, September 1972.

9. Andrews G., "Comparison of Radar Doppler Filtering Techniques," NRL Report 7811, October 1974.

10. McAulay, R., "A Theory for Optimal Moving Target Indicator," AD 751923, October 1974.

11. Schleher D. C., "MTI Radar," Artech House, 1978.

12. Schleher D. C., "MTI Detection Performance in Rayleigh and Log-Normal Clutter," IEEE International Radar Conference, April 1980.

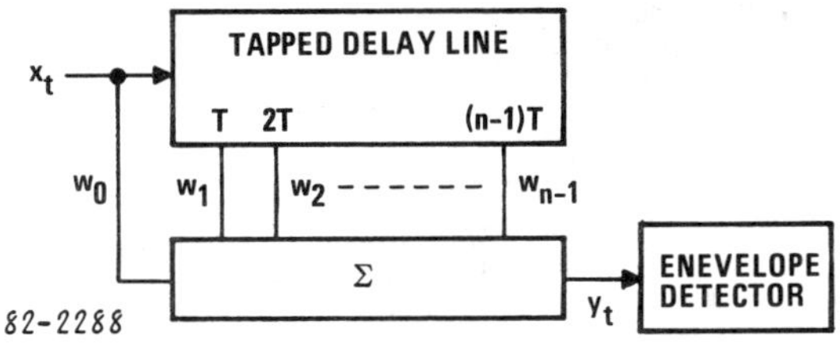

Figure 1 Optimum transversal filter processor

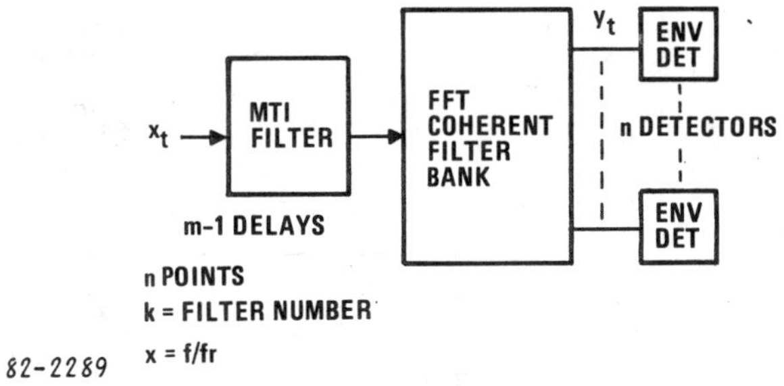

Figure 2 Cascaded MTI and coherent integrator

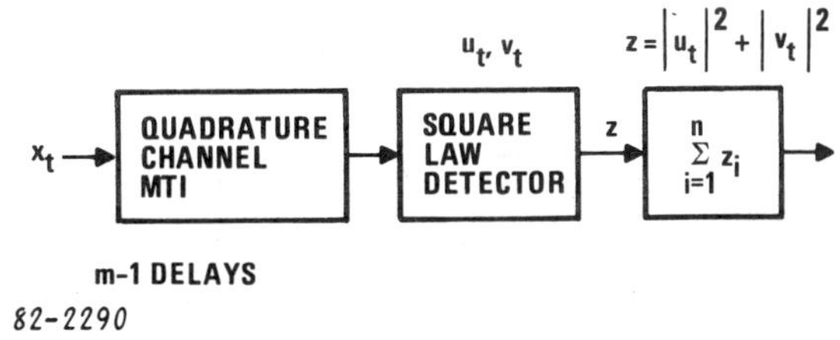

Figure 3 MTI and Incoherent Integrator

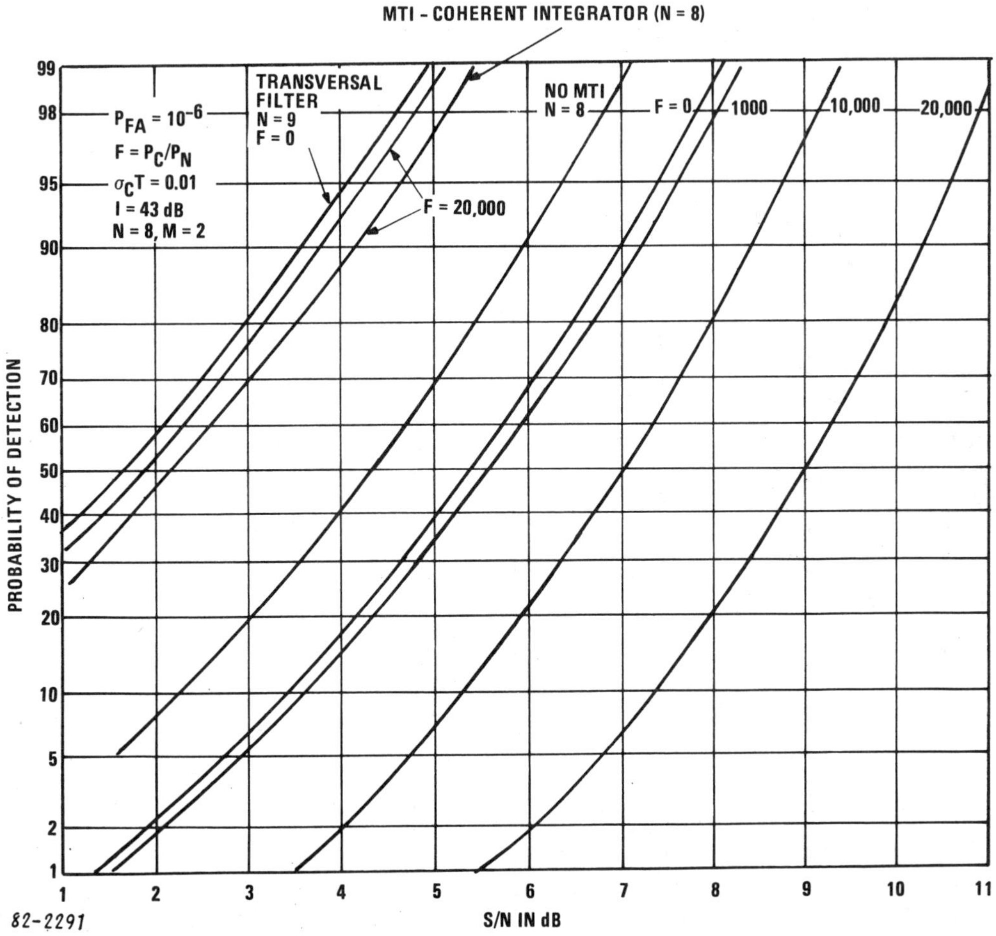

Figure 4 Probability of detection for single canceller quadrature MTI with video integration (N=8)

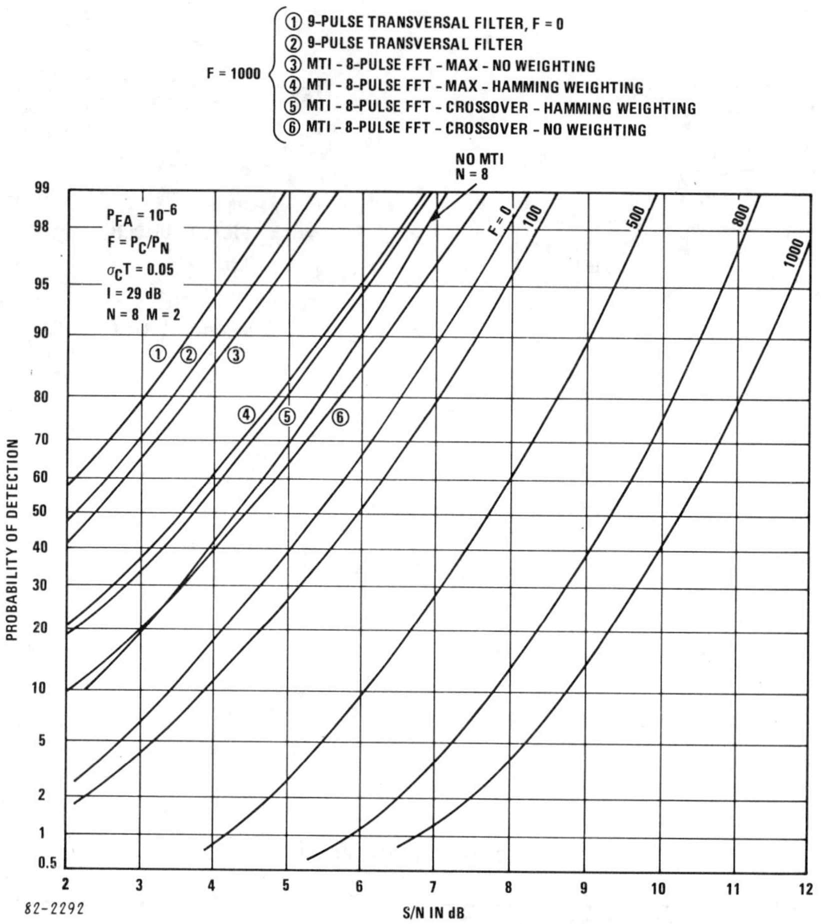

Figure 5 Probability of detection for single canceller quadrature MTI with video integration (N=8)

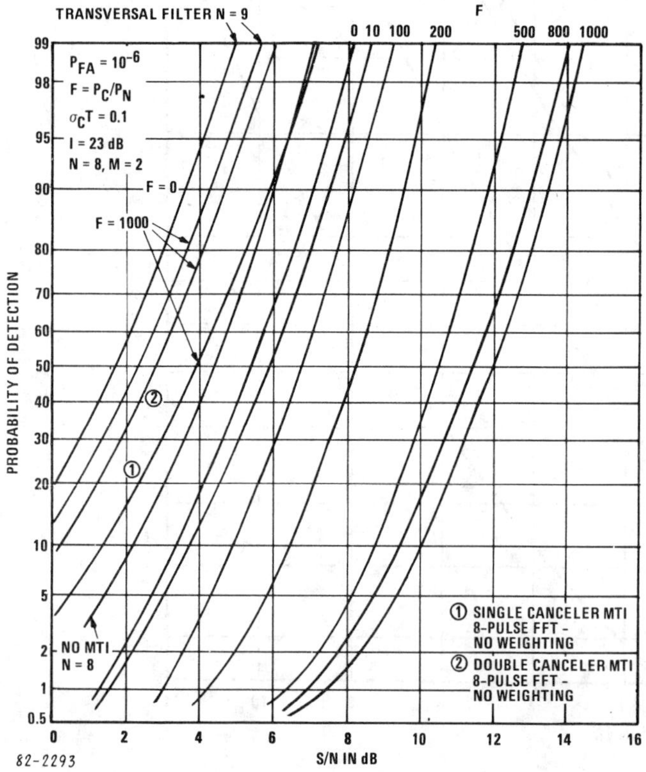

Figure 6 Probability of detection for single canceller quadrature MTI with video integration (N=8)

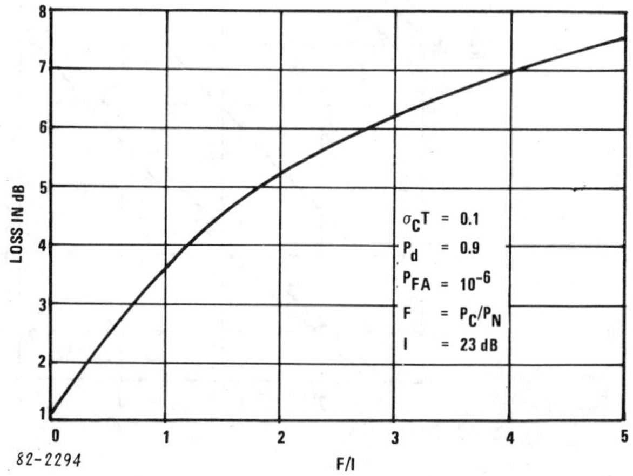

Figure 7 Video integration loss with single canceller MTI

A SPATIALLY-VARIANT AUTOFOCUS TECHNIQUE FOR SYNTHETIC-APERTURE RADAR

Malcolm R. Vant

Department of Communications, Canada

INTRODUCTION

The quality of synthetic-aperture radar (SAR) images, acquired from long range, depends critically on the ability of the motion compensation system to correct for deviation from a uniform vehicle velocity. Uncorrected errors in path and velocity cause the image to be smeared and defocussed.

This paper describes a technique for automatically removing cross-range phase errors, left by the motion compensation system, from nominally focussed SAR images that have been stored in complex number format. These phase errors, normally caused by medium to long term motion compensation system drift or target motion, are assumed to be quadratic in form and spatially-variant. It is assumed that any significant range delay errors have been removed previously by other methods.

To correct for the phase errors, the nominally focussed complex image is convolved with a small time-bandwidth product (TBWP) auxiliary focussing function. Since in general, the TBWP of this function is smaller than that of the matched filter for the original SAR signal, considerable computational savings can be realized over techniques that require the reprocessing of the SAR signal.

DERIVATION OF THE FOCUSSING FUNCTION

After range compression the SAR signal is assumed to be the summation of the reflections from many scatterers. The i'th scatterer gives rise to a signal of the form:

$$x_i(m) = A_i B(n-n_i)\,\mathrm{rect}[(m-m_i)/N_1]\,\exp[-j(4\pi/\lambda)a_2(m-N_1/2)^2\Delta x^2]\,, \qquad (1)$$

where A_i is a complex reflectivity constant for this scatterer; $B(n-n_i)$ describes the range compression impulse response, n and m are the range and azimuth sample numbers, m_i and n_i denote the scatterer's position in cross-range and range samples, respectively; N_1 is the number of pulses in the synthetic-aperture; Δx is the cross-range sample spacing; λ is the radar wavelength; and a_2 is the cross-range quadratic phase constant. This signal is convolved with the nominal impulse response of its matched filter, h_1, (or equivalent processing) to obtain the nominally focussed image y_i of x_i:

$$y_i = x_i * h_1\,. \qquad (2)$$

The term nominally focussed is used because an imperfect estimate, $\Delta\hat{x} \neq \Delta x$, of the sample spacing was used in the calculation of the impulse response h_1. Only when the true impulse response, based on the correct value of Δx, is used, can a truly focussed image be obtained.

The true impulse response can be shown to be a convolution of h_1 with an auxiliary function h_2, which is given by

$$h_2(m) = 1/(N_1\Delta\hat{x})[\lambda\Delta\tilde{x}^2/(8a_2|\Delta\hat{x}^2-\Delta\tilde{x}^2|)$$
$$\exp\{-j4\pi a_2 m^2\Delta\hat{x}^2\Delta\tilde{x}^2/[\lambda(\Delta\hat{x}^2-\Delta\tilde{x}^2)]\}$$
$$\{[C(f_+)-C(f_-)]+j[S(f_+)-S(f_-)]\},$$
$$m = -N_2/2 \ldots N_2/2-1\,, \qquad (3)$$

where, $$f_{+,-} = [\lambda(\Delta\hat{x}^2-\Delta\tilde{x}^2)/(a_2\Delta\tilde{x}^2\Delta\hat{x}^2)]^{\frac{1}{2}}$$
$$\{2a_2N_1\Delta\hat{x}^2/\lambda \pm 4ma_2\Delta\hat{x}^2\Delta\tilde{x}^2/[\lambda(\Delta\hat{x}^2-\Delta\tilde{x}^2)]\,, \qquad (4)$$

$\Delta\tilde{x}$ is our test value of Δx, $C(\cdot)$ and $S(\cdot)$ are the Fresnel Integrals, and N_2, is the length of $h_2 \cdot N_2$ must be sufficiently large that almost all the energy of h_2 is contained in the interval $|m| \leq N_2/2$. In order to completely specify h_2, the best estimate of $\Delta\tilde{x}$ must be found.

If it is assumed that y_i is corrupted with white noise, then the best value, in the maximum likelihood sense, of $\Delta\tilde{x}$, and hence of h_2, is found by convolving the nominally focussed image, y, with trial values of h_2, and evaluating from the resulting output a parameter that measures the goodness of the estimate.

In the implementation described here, a maximum likelihood ratio test, similar to that of Van Trees (1), and based on the following assumptions, was used:

1) the output from the h_1 and h_2 convolution filters is a Normal random variable with mean μ_1, and either variance $\sigma_0^2 > \sigma_1^2$ when a strong target is present, or variance σ_1^2 when one is not;

2) the strong target, when present, is surrounded by many weak ones;

3) x_i is a Normal random variable;

4) the noise added to y is white; and

5) h_1 and h_2 are linear systems.

The normalized sufficient statistic used for the test was the ratio of the standard deviation to the mean of the absolute values of the amplitudes of the data being analyzed, i.e., σ_r/μ_r, where $\mu_r \simeq \mu_1$. The threshold on σ_r/μ_r was set at unity, so that the best estimate of $\Delta\tilde{x}$ was accepted as valid if the maximum value of σ_r/μ_r, over all $\Delta\tilde{x}$'s tested, was greater than unity.

ALGORITHM

The algorithm used to implement the autofocussing procedure is as follows. A large block, e.g., (N_2+100)x100, of complex data is selected and transferred to fast memory. First, the focussing parameter, FP = σ_r/μ_r, is computed. Then, the auxiliary function $h_2(m)$ is calculated for the first value of $\Delta\tilde{x}$, and a copy of the input, nominally focussed image data, y, is convolved with h_2. Next, the FP value is computed for this data, and the process, starting with the computation of h_2, is repeated until a set of $\tilde{\Delta x}$'s, centred on $\hat{\Delta x}$ has been tested. Upon completion of the testing, the FP vs. Δx vector is scanned for the maximum value of FP, FP_{MAX}. Then, if $FP_{MAX}>1$, the corresponding value of Δx is accepted as the maximum likelihood estimate of Δx, Δx_{ML}, and the data are reprocessed with an h_2, which has been computed using $\tilde{\Delta x} = \Delta x_{ML}$. If $FP_{MAX} \leq 1$, the original input data are returned unaltered. In the case $FP_{MAX}>1$, the output data are also rescaled to ensure that the mean of the data are unaffected by the processing.

For spatially-variant autofocussing, the operations described above are done first, and the Δx_{ML} obtained is assigned the label Δx_{DEF}(Δx default). Then, the large block is segmented into a number of equal size areas, e.g. 100-10x10 sub-blocks, and the previously described processing is applied to each. However, the spatially-variant procedure differs in one respect: if for a sub-block $FP_{MAX} \leq 1$, the data is convolved with an h_2 based on the Δx_{DEF} value, and is not returned unchanged.

An algorithm incorporating a simple routine to automatically converge on the peak of the FP vs. $\tilde{\Delta x}$ function, instead of scanning every increment of Δx, was also tested. It was found that, except under the very special circumstances of small errors and high initial FP, the FP vs. $\tilde{\Delta x}$ functions exhibited too many local maxima to allow the peak finding algorithm to be used.

Other focussing algorithms, which depend on the correlation of images taken from different look angles ('map matching'), and require only one iteration to estimate Δx, have been informally suggested by Howard (2) and Bennett (3). However, a shortcoming of such algorithms is that they require a high edge-contrast-ratio (high FP) for good results, see Novak (4). This means that the scene to be focussed must be highly structured and have a high initial FP value, i.e. it must be close to being properly focussed in the first place. Whereas, the technique proposed here requires only that FP be high after focussing. See the RESULTS section for an example.

In situations where map-matching is applicable, it can be combined with our technique to improve processing efficiency, i.e. map-matching can be used to find the optimum Δx, and an h_2 function, as described here, can be used to fine focus the image.

COMPUTATIONAL EFFICIENCY

A comparison, based on a fast-convolution implementation, of an iterative solution using h_1 versus an iterative solution using the auxiliary function h_2, shows that for $N_2 \ll N_1$ many more real multiplies are required when h_1 is used. For an image size of 100x100, and typical values of $N_1/N_2=8$, the use of the auxiliary function is approximately 8 times more efficient than the straightforward convolution of equation (2).

The number of equivalent real multiplies required to compute each point of the impulse response, for each test value of Δx, is 144 for h_2, vs. 8 for h_1. But, because N_1 is typically much bigger than N_2, the total number of multiplies required to compute all of h_2 is only about twice the number required for h_1. Also, the effort required to compute either h_1 or h_2, for each test value of Δx, is typically small compared to that required for the fast convolution of either function with the data.

If the auxiliary function algorithm is implemented on hardware with a limited size memory, e.g. most minicomputer-array processor systems, considerable input/output time is saved over the approach of equation (2). Normally, the array of complex words required for focussing with the auxiliary function method will fit entirely on a 64K word array processor memory page, whereas the array required for focussing using the approach of equation (2) will not. Therefore, the latter technique requires the time consuming movement of the entire array into and out of memory for the calculation of each test value of Δx.

In general, the map-matching technique, where applicable, requires many more real multiplies than each loop of the method proposed here. However, it requires only one loop to obtain Δx_{ML}. For errors bounded by ± 250/TBW% in Δx, the iterative h_2 method is competitive in terms of computation load down to image sizes of 25x25.

RESULTS

The auxiliary function algorithm proposed here was used to obtain the results shown in Figures 1-10.

Figure 1 shows a SEASAT SAR image of uniform land clutter, with a cross-range resolution of 7m, a TBWP≃2000 and an error in Δx of 1%. The 100x100 area shown was segmented into 25-20x20 blocks and each was focussed independently. The 4 blocks, for which FP>1, returned Δx_{ML} estimates of between 4.110 and 4.117, compared to the correct value of 4.114. Therefore, the error was 0.1% versus a desired accuracy of 0.025%, i.e. 1/(2TBWP). This size of error is still tolerable in most cases, but if not, the FP threshold can be raised to decrease the variance in Δx_{ML}. Figure 2 shows the focussed image.

Figures 3 to 6 show focussing results obtained using data from an L-Band airborne SAR with a very poor motion compensation system. Figure 3 shows a 100x100 image of dock structures. It was produced using the supposedly correct $\hat{\Delta x}$. Note the 'spikey' appearance in the image. This 'spikeyness' usually indicates a spatially-variant, quadratic, or high order, phase error. Figure 4 shows the same image after it was focussed using one 100x100 block. The major peaks have improved, but there is still some spikeyness. Figure 5 is the same area focussed as 500, independent 20 (cross-range) x 1 (range) cell sized blocks. Note that a new strong peak has appeared near the previous one, and the image has a much smoother overall appearance. Figure 6 shows the sub-blocks for which FP>1. The spread of Δx_{ML} values obtained was 1.29 to 1.44, compared to the nominal value of 1.42. In comparison to the SEASAT scene, this spread of values does not represent the variance of the estimator, but instead is due primarily to real variations in the data.

Figures 7 and 8 are defocussed and focussed images of the same locale, acquired using the same L-Band airborne SAR. An error in Δx of 50% was deliberately introduced in the processing of the defocussed image in Figure 7. The TBWP was approximately 120. A value of N_2=121 was used in the formation of the focussed scene in Figure 8. This N_2 is too short for perfect correction. Note however, that despite the very low contrast in the initial data, the correct focus was obtained. It would seem, from the very low contrast in the defocussed image, that map-matching techniques would not succeed in focussing the data in Figure 7.

Figures 9 and 10 show typical FP vs $\Delta\tilde{x}$ plots for uniform land clutter and ship type targets, respectively. The FP vs $\Delta\tilde{x}$ function exhibits many local maxima in both plots. As mentioned previously, this type of behaviour means that a simple peak-finding algorithm cannot be used. It can also be seen from Figure 9, that for scenes with inherently low contrast, i.e. low FP_{MAX}, it is impossible to estimate Δx_{ML} reliably.

REFERENCES

1. Van Trees, 1968, Detection, Estimation, and Modulation Theory Part I, John Wiley, New York, U.S.A., 29.

2. Howard, J., 1981, "Design Principles for Radar Systems and Signal Processing", Course Notes, Mark Resources, Marina Del Ray, California, U.S.A.

3. Bennett, J., 1981, Personal communication.

4. Novak, L., 1978, IEEE Trans. Aerosp. Electron, Syst., 14, 641-648.

ACKNOWLEDGEMENTS

This work is supported by the Department of National Defence, Research and Development Branch. The author would like to acknowledge the contribution of Dr. D. Hughes of W.E. Thorp Associates who did all the programming for this project.

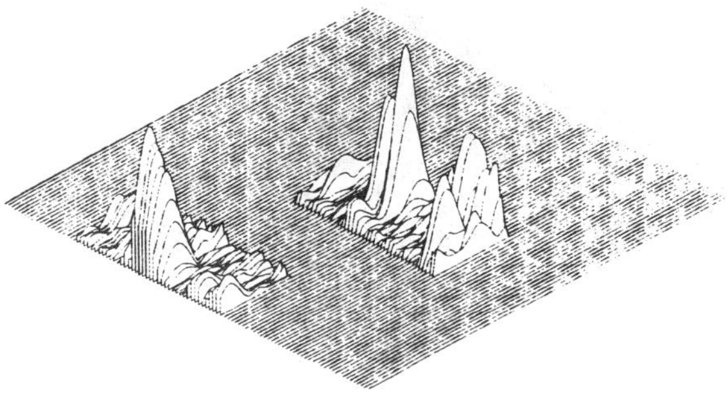

Figure 2 Segments of the SEASAT image that could be automatically focussed.

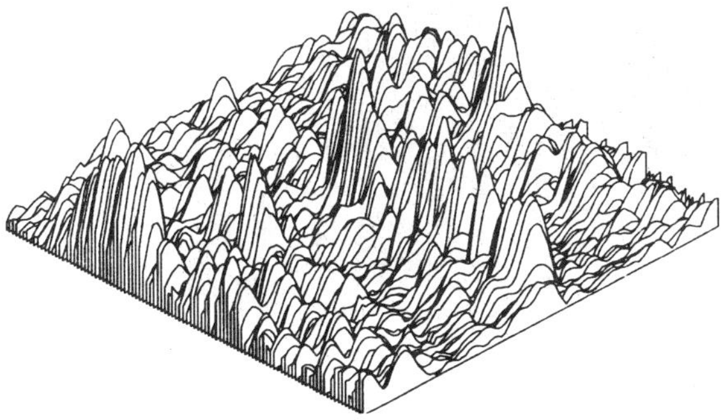

Figure 1 Defocussed SEASAT SAR image, processed with a velocity error of 1%.

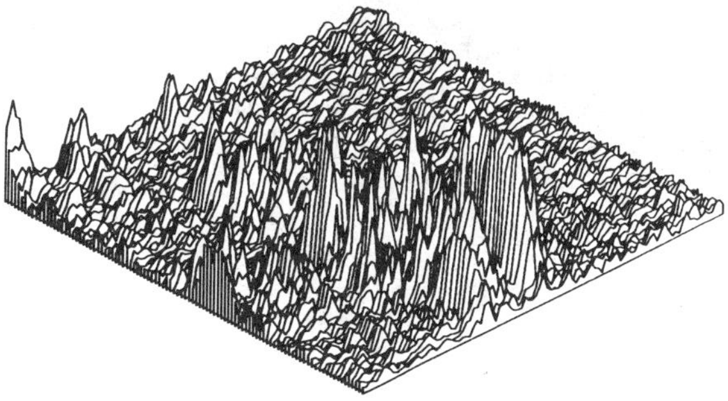

Figure 3 A poorly motion compensated L-band airborne SAR image of docks.

Figure 4 The previous image automatically focussed using 1 100x100 block.

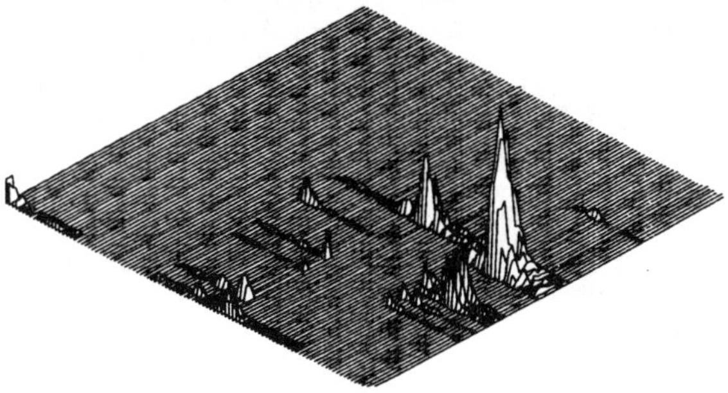

Figure 6 The blocks in Figure 5 for which the threshold was satisfied.

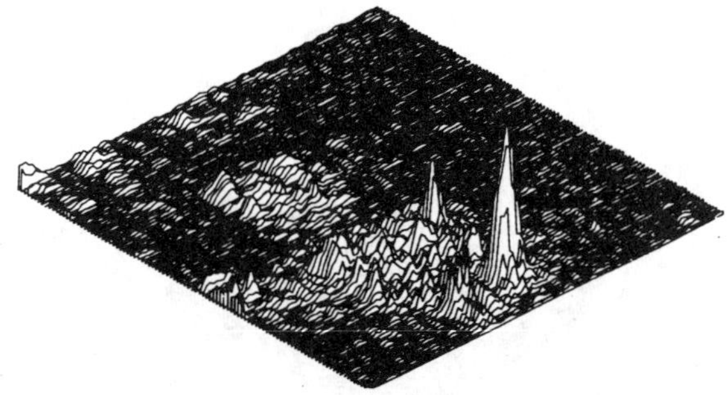

Figure 5 The previous image automatically focussed using 500-20x1 blocks.

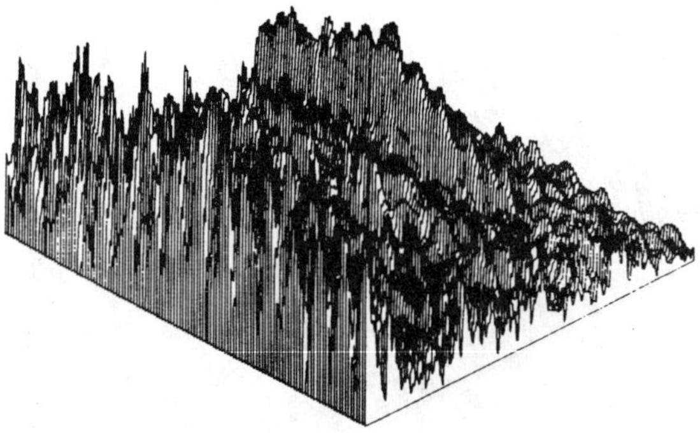

Figure 7 A defocussed L-band airborne SAR image, processed with a velocity error of 50%.

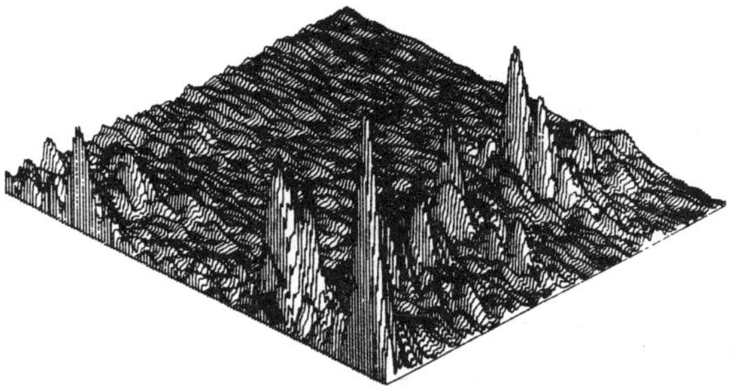

Figure 8 The image of Figure 7 after automatic focussing.

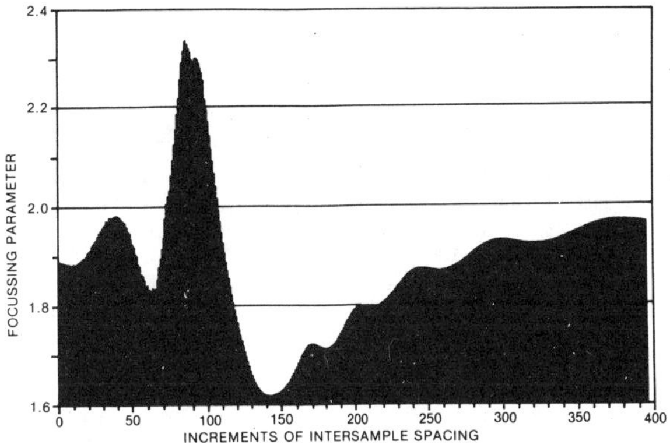

Figure 10 A plot of the focussing parameter vs. intersample spacing for a block with FP>1.

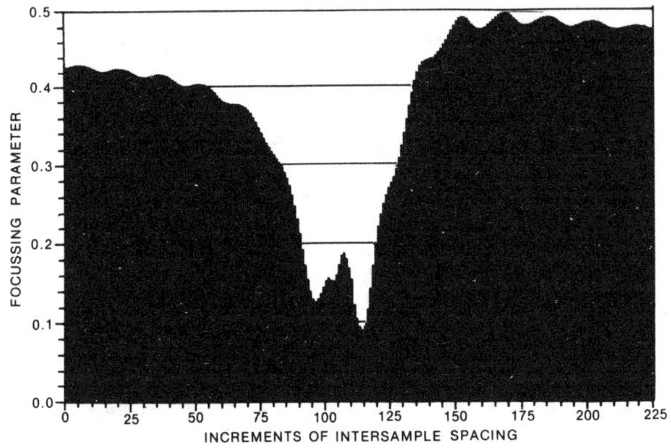

Figure 9 A plot of the focussing parameter vs. intersample spacing for a block with FP<1.

PROBLEMS OF DATA PROCESSING IN MULTIRADAR AND MULTISENSOR DEFENSE SYSTEMS

H. Ebert

AEG-Telefunken, Federal Republic of Germany

INTRODUCTION

The editorial of the February issue 1981 of the "Defense Electronics" Magazine is headed: "Will New Sensors and Computers Sound the Death Knell of Radar?". The background of this discussion is the well-known fact that in defense systems radars of various types cooperate with other sensors (e. g. infrared cameras, receivers for passive radio reconnaissance, laser systems) as well as with a large number of computers and computerized displays and finally with "effectors" (e.g. weapon systems). In this environment, radar still accomplishes the function of a primary sensor, however, other sensors contribute valuable additional information to support radar reconnaissance. In addition, the operational application of radar in military systems is considerably more reliable if, instead of individual radars, a larger number of radar facilities is used which are designed to fulfill special partial tasks and to mutually support each other within the overall system. Therefore, major defense systems always include a larger number of cooperatively used radars.

It does not need any further explanation that the interaction of many radars and sensors inperatively requires the aid of computers because of the abundance of information to be evaluated. It is, however, of great importance to know that combining many radars and sensors and connecting them with installations for information display such as computerized displays and finally coupling them with weapon systems of different types implies transition from the function of individual equipments to collective functioning in the form of a complex overall system. Within this system which, on the side of the sensor, is designed as a multiradar/multisensor configuration, computers ensure the tasks of signal processing, the organization of information display as a necessary prerequisite for taking command and control (C^2) and mission decisions by man and finally the communication between the different components. Such a system is therefore called a computerized command/control/communication system (C^3 system). The consideration of ADP problems in multiradar and multisensor systems thus leads directly to an analysis of the ADP funktions in typical computerized C^3 systems.

EXAMPLES OF COMPUTERIZED MULTISENSOR/ MULTIRADAR SYSTEMS

Computerized C^3/Weapon System for a War Ship

Systems of the basic configuration as described above are either already available for mission purposes or are being implemented in present development projects. A first example to be given here is a short description of an ADP supported C^3/ weapon system for a war ship which has been implemented in the FRG for the frigate F122. Fig. 1 shows a survey of functions:
Typical equipments of the sensor area are the long-range radar, the tracking radar, the sonar for under-water target detection and the electronical reconnaissance equipment. Typical equipments of the effector area are the gun, the missile system HARPOON for surface target engagement, the ASMD = Anti Ship Missile Defense system for engagement of air targets in the short-distance range, the missile system SEASPARROW for engagement of air targets and the ECM equipment. Typical equipments of the communication area are the message handling units and the Link 11 = communication means to other ships as well as to the accompanying helicopter. All the equipments mentioned above are connected to a central computer system enabling them to exchange data. Furthermore, multi-function displays are coupled with the central computer which permit analogue radar image display as well as a synthetical situation display. They serve for the execution of the typical operational functions on a war ship such as the support of the formation command and control, the surface weapon mission control,the under-water target engagement (mission control of helicopters), the air situation acquisition, the ECM equipment mission and the tracking. The equipments of the sensor, the effector and the data transmission area dispose of their own ADP subsystems to fulfill their respective mission tasks, that means their functioning is independent of the central computer. The central computer complex thus mainly supports the C^3 functions as well as the interaction of the individual systems.

Mobile Anti-Aircraft C^3 System for Land Forces

A typical example for a mobile multiradar C^3 system is the so-called Army Anti-aircraft Reconnaissance and Combat System HflaAFüSys which is at present under development in the FRG. Its task is combat command and control against aircraft flying in medium and low altitudes. This task is subdivided into the following subareas:
- early target recognition with the highest possible range up to ground proximity
- acquisition, composition and long-distance transmission of target information by many sensors
- instruction of weapon carriers
- air situation display and preparation of decision aids for the tactical command and control
- recognition, localization and defense against enemy ECM measures.

Fig. 2 gives a survey of the system structure: The reconnaissance is performed by the subsystems LÜR = air surveillance radar, TÜR = low-flying aircraft surveillance radar and LKWFlaAFü = anti-aircraft reconnaissance vehicle. The LÜR is a mobile 3D reconnaissance radar developed by AEG-TELEFUNKEN with pencil-beam antenna diagram the antenna being turned mechanically in the azimuth direction and electronically in the elevation direction. The TÜR is a mobile pulse doppler radar for the acquisition of low-flying aircraft developed by Siemens. C^3 functions are ensured by the LKWFlaFü vehicle (= combat centre of the anti-aircraft regimental commander) and the TPZFlaFü tank (= C^3 vehicle of the battery commander). Available weapon systems are the anti-aircraft gun tank GEPARD as well as the anti-aircraft rocket tank ROLAND.

COMPONENTS AND FUNCTIONS OF A COMPUTERIZED MULTIRADAR/MULTISENSOR C^3 SYSTEM

Basic Structure of a Computerized C^3 System

The fundamental components and functions whose interaction constitute a computerized multiradar/multisensor C^3 system may be deduced from the above mentioned examples:

COMPUTERIZED MULTIRADAR/MULTISENSOR C^3 SYSTEM

- Sensors/detectors/signal processing/ message acquisition
- Signal/message transmission
- Message processing, display, evaluation
- Decision finding, command + control
- Inition, control of effectors
- Data transmission from/to other C^3 centers

Basically, a computerized C^3 system may be considered as an integration of procedures and components of communication technique and informatics. Man is included in the system as decision carrier since he still assumes the tasks of decision finding and command and control.

Tasks and Functions of the ADP Components

General. The tasks and functions of the ADP components of a C^3 system are subject to the components and special functions existing within the system. The following ADP function areas may be deduced from the general basic structure such as described above:

- ADP functions of the sensor area
- ADP functions of the effector area
- ADP functions of the C^2 area including information display
- ADP functions of the communication/data transmission area

The ADP functions of the effector area are normally specifically tailored to the task and the functioning of the respective effector. Within this consideration which is primarily orientated on the processing of sensor data, the ADP functions of the effector area will not be dealt with in detail.

ADP Functions of the Sensor Area. In the sensor area the digital radar data processing normally constitutes the major part. It starts already in the radar installations itself, since in modern radars digital methods are almost exclusively used for doppler filtering. Modern methods of digital radar signal and data processing have been described in detail in other sections, so that a short glance on the main functions is sufficient. If radar equipments are to be connected with computers, two important intermediate functions must be fulfilled which constitute a necessary prerequisite for further processing of radar data in computers. These are 1) the automatic radar target detection and 2) the automatic radar target tracking. ADP methods and circuits are used for both functions. The methods of radar target detection are closely orientated on the technique of the radar such as 2D radar, 3D stacked beam radar, 3D pencil beam radar, 2D pulse doppler radar. Due to the high real-time requirements, digital radar extractors have so far exclusively been composed of hardwired digital circuits. Recently, specially developed microprogrammable processors are also being used for this propose. The automatic radar target tracking is the machine execution of the task to recognize target movements from the positions of radar targets detected during subsequent rotations of the radar antenna. The automatic target tracking is normally part of the tasks of a central computer. Within the tasks which are being handled there, it is however the one which is most orientated on the real time, i. e. it requires the shortest reaction times. Furthermore a distinction must be made between monoradar and multiradar target tracking.

The principles of radar data processing can be transferred to the processing of signals delivered by other sensors. This applies especially when these sensors are able to supply location data or even movement data of detected targets. The real-time processing of sensor data in computers always requires the functions of automatic target detection and target localization.

ADP Functions of the Data Transmission Area. The selection of technical procedures and components for the data transmission is decisively influenced by the fact whether it is a "local" data transmission or one in the distant range. This does not only apply for the physical transmission media (wire, broad-band cable, light line, radio), but also for all other problems related with data transmission, such as automatic call establishment and clearance, protection against transmission errors and message losses, function control of data transmission units etc. Due to the large number of communication partners in modern military Systems, point-to point connections between two partners are no more sufficient, real "multiuser" systems are necessary. This results in the complex problem of controlling the "access" to the communication media which is solved by the so - called "access protocols".

The technique of data transmission between dislocated computers has now gone through a long development phase where a number of special solutions has been created. In the meantime they have been supersided in the civil area to a large extent by standardized techniques and interface conventions. A distinction is now being made between the techniques of computer networks over geographically long distances ($>$ 1 km) and local computer networks (max. distance several 100 m). Connections of computers over long distances have so far been called "computer networks". Recently, "local computer networks" is the term used for locally connected computer systems.

Modern techniques of computer communication practically use the packet switching technique only. This means that a message transmitted by a network station is fragmented into "packets" of a specified length prior to being transmitted (fig. 3). After transmission of the individual packets the communication system "reassembles" the packets into the original message and transfers it to the addressee. Both the remote and the local computer communication are ensured on the basis of hierarchically levelled procedures called "protocols". The so called ISO architecture model recommended by CCITT for international use demonstrates one of the most modern and clear forms of hierarchical protocol categorization. Fig. 4 shows its basic structure: a distinction is made between protocols of the higher levels for communication of processes in separate computers and protocols of the "transport function", i. e. of the communication system: The transport function is subdivided into a total of 4 levels, to which the following functional areas are assigned:

Level 4 = end-to-end transport control i. e. interface between computer tasks wanting to communicate with each other through the computer network
Level 3 = packet level or logic channel level in which virtual calls can be established by means of the packet switching technique
Level 2 = link level protocols
Level 1 = physical lines, radio connections.
Together, levels 1 to 3 form the well-known interface standard X25.

The data communication through radio is mostly ensured according to the so-called "polling technique", that means a central master station inquires the partners assigned to it according to a specified scheme. In the civil area the problem of a decentral computer communication through radio has been solved for the first time with the establishment of the "ALOHA" computer network on the Hawai Islands where communication via cable connections could not be used. A so-called "broadcast" technique is used in this case, where a station wanting to transmit transfers its message directed to a specific addressee through a commonly used transmission channel. All other network stations receive the transmitted message, but only the addressee evaluates it and acknowledges correct receipt. Collisions caused by transmissions by several stations are registered by the respective transmitter by non-appearance of the acknowledgement signal, transmission is then repeated. Such a "pure ALOHA" technique, however, only achieves a max. exploitation of the message channel used of approx. 18 %. An improvement is achieved when the station wanting to transmit checks the radio channel for any possible engagement by another transmission prior to starting its own transmission. The access protocols working according to this principle are called "carrier sense multiple access" protocols, and permitt a channel exploitation of up to 85 %. In the meantime, CSMA protocols are not only used for computer communication through radio, but also in local computer networks. Here, all network stations are connected through a common broadband cable.

ADP Functions of the C^2 Area. In the ADP area of the C^2 functions the information supplied by the different sensor systems is combined and correlated as far as possible. In presently implemented systems, the accomodation of ADP functions of the C^2 area in a central computer prevails. The system of a war ship described earlier is a typical example. More recent plannings are based on the progress of the computer technique and tend towards decentralization of ADP functions of the C^2 area. The mobile anti-aircraft C^3 system described earlier may serve as an example. In general, the ADP functions of the C^2 area may be subdivided into the following tasks:

- Establishment of communication between subsystems, especially, between sensors and effectors
- Organization of information display on computerized displays
- Preparation of decision aids e. g. C^2 strategies for interceptors, recommendations for engagement measures
- Preparation of background data, e. g. performance data of enemy ships
- Execution of certain real-time functions e. g. mono-/multiradar target tracking, communication with other C^2 centers
- Preparation of programs for operational scenarios
- Organization of reconfiguration measures in case of failure of parts of the system.

REQUIREMENTS FOR THE ADP COMPOMENTS OF FUTURE MULTIRADAR/MULTISENSOR SYSTEMS

The focal point of future requirements may be summarized as follows:

1) Requirement for "decentralization" of the ADP functions, i. e. for their distribution over a large number of computers for the following reasons:
° increase of failure safety of the overall system in case of failure of system components
° decrease of the system vulnerability
° maintenance of the function of components in case of failure of important system functions, e. g. of the C^2 area
° possibility of direct communication between components, e. g. between sensors and effectors
2) Increase of the number of integratable subsystems especially of sensors:
In future C^3 systems it is necesarry to evaluate information delivered by a considerably heigher number of sensors and in addition by sensors working according to different physical principles.
3) Increase of reaction time of the system:
In the future, military systems will only have a chance to survive if they are able to rapidly recognize threats and to immediately react by countermeasures. A necessary prerequisite is the largely automatic interaction of sensors and effectors by means of computers.
4) Automatic interaction i. e. computerized interconnection of different C^3 system.

ADP CONCEPTS FOR C^3 SYSTEMS INCLUDING DEVELOPMENT TENDENCIES OF COMPUTER TECHNIQUE

Analysis of the Requirements

The analysis of the requirements for ADP components in future Multiradar/Multisensor

C^3 systems shows two focal points for the requirements which up to now are not or incompletely fulfilled:
1) Distribution of ADP functions upon several or even upon a large number of computers
2) Inclusion of a flexible and efficient system for computer communication.

Development Tendencies of Computer Technique

When we analyze the development tendencies of computer technique, we find that they are in absolute conformity with the abovementioned requirements in future C^3 systems. Due to the increased possibilities of semiconductor technology expressed by monolithic computer modules with at present several 100 000 transistor functions per ship, miniund microcomputer systems can now be constructed whose performance are comparable to that of former large-size computers. A direct consequence of this possibility is the tendency towards decentralization of ADP functions, since a high computer performance can easily be installed directly at places where it is required and can even be integrated into other systems, such as radar and communication installations. The same technological progress also permits the use of cheap mini/microcomputers for the establishment of efficient communication systems, i. e. the tendencies towards distributed machine intelligence and interconnection of distributed computers by efficient communication systems are closely linked together.

Discussion of Detail Problems

The design of the ADP components of future C^3 systems requires the discussion of a large number of detail problems which may be classified as follows:
- problems of interaction of several computers
- topology of computer systems
- methods of computer communication

Interaction of Several Computers. The tendency towards decentralization of the ADP functions as a consequence of the hardware progress has already been mentioned. Due to the present "hardware enthusiasm", optimistic circles are already talking of a forthcoming replacement of the present large-size computers by "multi-microcomputers". Fundamentally, such tendencies comply with the requirements of the defense area for an increase of the failure safety or of the maintenance of subsystem functions in case of failures due to external influences. However, even computer experts often forget that the distribution of ADP functions on a large number of computers results in complex problems especially in the fields of system and operating software some of which have not yet or only insufficiently been solved. These problems mainly concern the interaction of computers such as classified below. A strict distintion must be made between a network of loosely coupled computers and a multiprocessor computer. A network of loosely coupled computers consists of coupling of independent computers through a system of direct or indirect connections. Each computer disposes of its own operating systen. Only the exchange of messages, i. e. of data blocks, requests etc. is possible between the computers. A multiprocessor computer is a self-contained computer with several processors and disposes of a single integrated operating system (1). The so-called principles of "sharing" and "interaction" are entirely fulfilled. This means that 1) all processors have access to all memory units, I/O channels and peripheral units and 2) programs running on individual processors are capable of mutual support on the level of jobs, tasks, data sets and data elements.

As compared with a network of loosely coupled computers the multiprocessor computer constitutes a much more efficient computer concept. This applies more particularly to the important requirement in the defense area for an automatic system reconfiguration in case of partial failures of the ADP system. The problem of the multiprocessor computer is the availability of an efficient multiprocessor operating system which enables to make an effective use of the multiple hardware facilities. Statistics on available multiprocessor computers show that in spite of great efforts this problem has not yet been solved for universal computer application. For this reason, the use of real multiprocessor computer configurations in the defense area will only be possible on a long - term basis. The recommendable objectives for further development of ADP systems for the defense area are the transition from mainly centralized and hierarchically structured ADP system configurations to the configuration of loosely compled computer networks of hierarchically adequate computers.

Topology of the Computer Network. Fig. 5 gives a survey of typical topology forms of computer networks:
1) Global bus (fig. 5a). All computers are coupled through a common ("global") data bus. In case of high data traffic, the bus rapidly constitutes a bottle neck (2).
2) Loop (fig. 5b)
3) Complete interconnection (fig. 5c). Each computer is connected directly with the other computers. This form of connection is recommended for high data traffic between the individual computers, but is very sophisticated.- Besides these direct computer connetions there are the forms of indirect computer connections through intermediate computers which are typical for remote coupling of computers (star connection, regular and irregular network). The topology form influences the scope and the efficiency of a computer network very decisively. The decision for a certain topology form or for mixed forms can only be taken as a function of the application.

Computer Communication. The techniques of computer communication in the civil area as described before may be incorporated directly into the concepts for future military and civil C^3 systems (e. g. air traffic control). The application of computer network techniques presents itself for the interconnection of complete C^3 systems or for the establishment of C^3 systems from dislocated units. This implies the use of the principles of packet switching and of virtual call connections. For the connection of local computers in C^3 systems, the use of the communication techniques in local computer networks, i. e. the procedures of packet switching and virtual call connections together with broadcast protocols for access controll, promizes an essential advantage over presently used methodes of data exchange between local computers. Development tendencies show that in local computer networks, glass-fiber lines will mainly be used as physical transmission media in the future.

CONCLUSION

The preceding considerations have demonstrated that the use of a larger number of sensors and of sensors working along different physical principles in a C³ system and their cooperative connection with effectors and C³ elements implies the necessity for an extremely sophisticated ADP system. It is the cooperation of computers which permits operational cooperative use of a larger number of sensors. The efficiency of multiradar/multisensor C³ systems is essentially depending on the efficiency of the ADP subsystem. Progress in the field of ADP is decisive for the further development and improvement of military Multiradar/Multisensor C³ systems. (For further details concerning C³ Systems refer to (3)).

REFERENCES

1. Enslow, P., 1977, Computing Surveys, 9, 103-129.
2. Ebert, H., 1980, El. Rechenanlagen, 22, 134-142.
3. Hofmann, H., Huber, R., und Molzberger, P., 1982, "Führungs- und Informationssysteme", Oldenbourg, München/Wien, Germany/Austria.

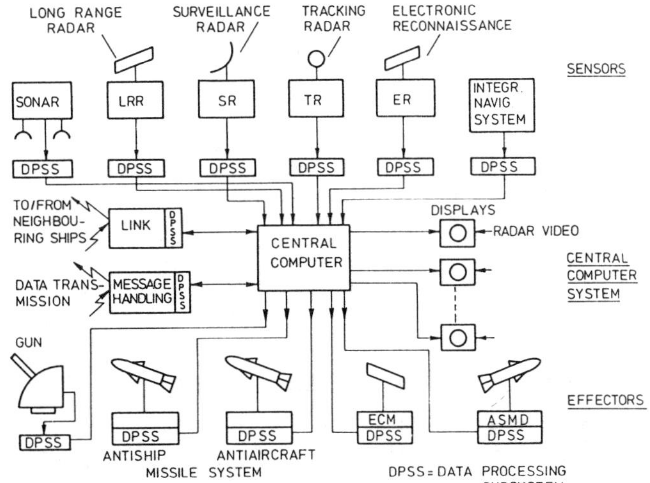

FIG. 1 COMPUTER AIDED GUIDANCE SYSTEM FOR WARSHIPS

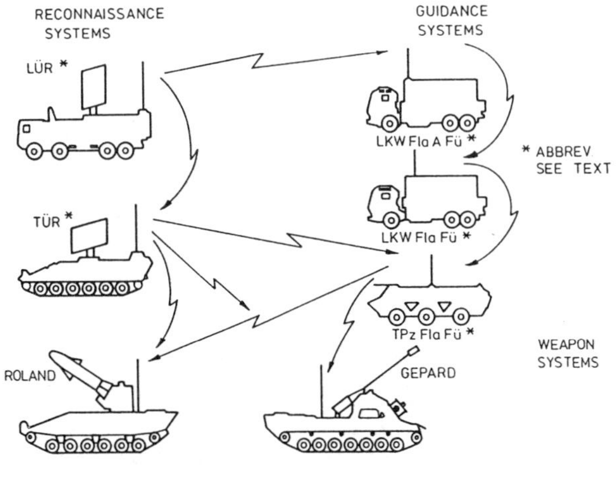

FIG. 2 STRUCTURE OF THE GERMAN "HFlaAFüSys" SYSTEM (MOBILE MULTIRADAR ANTIAIRCRAFT CCC SYSTEM)

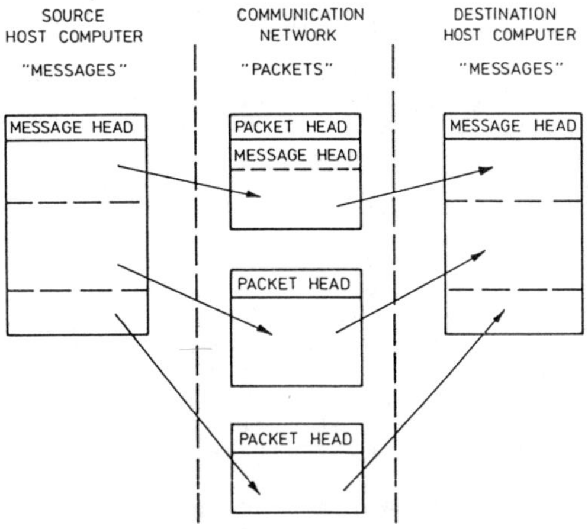

FIG. 3 FRAGMENTATION OF MESSAGES INTO PACKETS AND REASSEMBLING

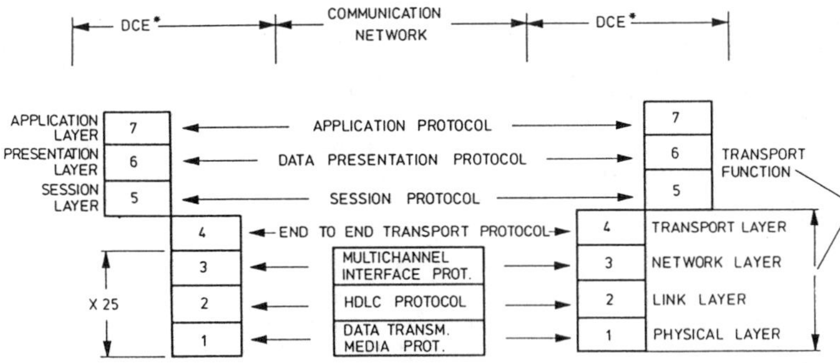

FIG. 4 HIERARCHICAL STRUCTURE OF THE ISO ARCHITECTURE MODEL

* DATA CIRCUIT TERMINATING EQU.

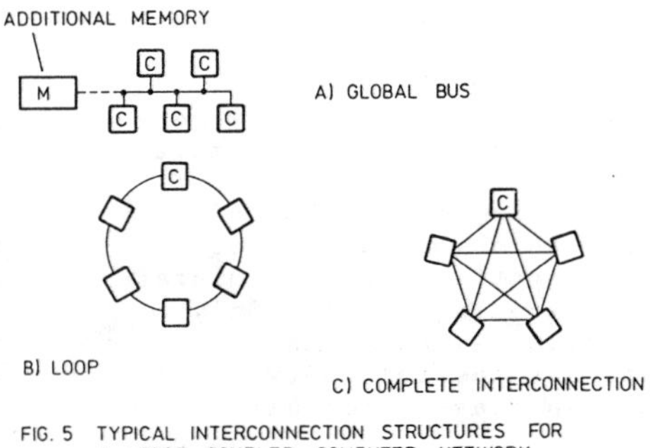

FIG. 5 TYPICAL INTERCONNECTION STRUCTURES FOR A LOOSE COUPLED COMPUTER NETWORK

ASSOCIATION OF MULTISITE RADAR DATA IN THE PRESENCE OF LARGE NAVIGATION AND SENSOR ALIGNMENT ERRORS

W. G. Bath

The Johns Hopkins University, Applied Physics Laboratory, Laurel, Maryland

ABSTRACT

This paper discusses a technique for associating radar track data from two or more moving platforms in the presence of very large navigation errors and sensor misalignments. In practice, these errors and misalignments can be so large as to render conventional association/correlation algorithms ineffective.

A novel approach to association is to view each platform's track file as a picture and apply two dimensional signal processing techniques to identify common features in the different pictures. This approach produces track association decisions which are unaffected by the size or direction of the navigation errors and by the size or sign of azimuth misalignments. This technique was tested using two 3D pencil beam surveillance radars separated by about 66 km. Navigation/alignment errors were artificially inserted into the track data bases and shown not to affect track association decisions, thus allowing an integrated surveillance picture to be produced even when navigation/alignment errors exceed typical target separations.

INTRODUCTION

There is increasing emphasis on "integrating" surveillance systems by combining measurements from two or more sensors to produce a single surveillance picture superior to the pictures produced by each component sensor individually. When the sensors are surveillance radars located at different sites or on different moving platforms, the resultant integrated surveillance picture will involve different radar-target geometries, different aspect angles of reflection and possibly different radar frequencies and waveforms as well. This diversity greatly reduces the likelihood that targets will go undetected due to natural clutter or man-made interference; and so improves the effectiveness of the radar surveillance system.

One of several issues in designing a distributed radar system (or in netting existing radars) is the data processing necessary to combine radar data from the different sites or platforms. Suppose each platform performs its own automatic detection and tracking, and tracks are exchanged via a communication link. The data processing compares tracks made locally with those received from other sites or platforms and decides which tracks correspond to the same target. This track association process eliminates duplicate tracks and allows an unambiguous data base to be formed.

Track association is complicated by site registration or platform navigation errors and by biases in the radar measurements (e.g., North misalignments). These errors and biases can prevent association of tracks which do correspond to the same target and cause misassociation of tracks corresponding to different targets. The problem is particularly severe when one or more radars are on moving platforms (e.g., aircraft or ships), when navigation systems are disrupted and when the radars have not been designed for bias-free measurements.

Previous studies of radar track association (e.g., [1-2]) have emphasized zero-mean Gaussian radar measurements errors as the limiting factor in association. In many practical situations, these classical measurement errors are dwarfed by registration/navigation errors and radar biases. This paper develops radar track association algorithms which can operate reliably even in the presence of large unknown registration/navigation errors and radar biases.

One aspect of multiple site radar integration which has received considerable attention is automatic registration (sometimes referred to as "gridlocking") in which registration/navigation errors and radar biases are estimated in real-time by comparing coordinates of associated tracks from different sites. A linearized or extended Kalman filter is typically used to estimate the biases [3-5]. In principle, automatic site registration eliminates the registration/navigation errors and the radar biases. However, automatic site registration algorithms presuppose the existence of correctly associated track pairs whose coordinates can be compared. Thus the problem of track association in the presence of large registration/navigation errors and radar biases must still be solved prior to automatic site registration.

SIMPLIFIED TRACK ASSOCIATION GEOMETRY

Figure 1 illustrates a simplified (flat-earth) two-dimensional geometry for the track association problem. Platform 1 is an angle Δ_1, out of North alignment. Platform 2 is an angle, Δ_2, out of North alignment. The two platforms are separated by a vector with length R_o and angle (relative to North) θ_o.

A target with coordinates (x,y), in a North aligned coordinate system located at Platform 1, will be seen by Platform 1 at coordinates:

$$\begin{bmatrix} x_1 \\ y_1 \end{bmatrix} = \begin{bmatrix} \cos\Delta_1 & \sin\Delta_1 \\ -\sin\Delta_1 & \cos\Delta_1 \end{bmatrix} \begin{bmatrix} x \\ y \end{bmatrix}$$

and will be seen by Platform 2 at coordinates

$$\begin{bmatrix} x_2 \\ y_2 \end{bmatrix} = \begin{bmatrix} \cos\Delta_2 & \sin\Delta_2 \\ -\sin\Delta_2 & \cos\Delta_2 \end{bmatrix} \left\{ \begin{bmatrix} x \\ y \end{bmatrix} - R_o \begin{bmatrix} \sin\theta_o \\ \cos\theta_o \end{bmatrix} \right\}$$

$$= \begin{bmatrix} \cos\Delta_2 & \sin\Delta_2 \\ -\sin\Delta_2 & \cos\Delta_2 \end{bmatrix} \begin{bmatrix} x \\ y \end{bmatrix} - R_o \begin{bmatrix} \sin(\Delta_2+\theta_o) \\ \cos(\Delta_2+\theta_o) \end{bmatrix}$$

The Platform 1 target coordinates may be expressed in terms of Platform 2 target coordinates as follows:

$$\begin{bmatrix} x_1 \\ y_1 \end{bmatrix} = \begin{bmatrix} \cos(\Delta_1-\Delta_2) & \sin(\Delta_1-\Delta_2) \\ -\sin(\Delta_1-\Delta_2) & \cos(\Delta_1-\Delta_2) \end{bmatrix} \begin{bmatrix} x_2 \\ y_2 \end{bmatrix} + R_o \begin{bmatrix} \sin(\Delta_1+\theta_o) \\ \cos(\Delta_1+\theta_o) \end{bmatrix}$$

This is the transformation one would use to transform remote data (Platform 2) into the local coordinate frame (Platform 1). Note that three biases must be known to effect the transformation:

$\Delta_1-\Delta_2$	:	difference in angular misalignments
R_o	:	magnitude of site separation
$\theta_o+\Delta_1$	:	angle of site separation in local (Platform 1) coordinates.

CONVENTIONAL CORRELATION ALGORITHMS

Most track correlation algorithms assume that navigation/registration errors and sensor misalignments are small. One then correlates tracks with similar positions and velocities using some measure of position/velocity "closeness" which is consistent with radar measurement accuracies, navigation or registration accuracies, site-target geometry, sensor alignment accuracy, target maneuver capability and site tracking filter accuracies. For the flat earth example described previously, one would typically base the correlation decision on five track parameters

x_i = x coordinate of track #i
y_i = y coordinate of track #i
a_i = altitude (above sea level) of track #i
s_i = speed of track #i
h_i = heading of track #i

Other parameters (such as altitude rate) and other coordinate systems (such as x rate and y rate instead of speed and heading) can be added or substituted as desired. The decision whether local track i and remote track j should be associated can then be based upon a statistical distance D_{ij} between the two tracks. D_{ij} is a quadratic form

$$D_{ij} = d_{ij}' P_{ij} d_{ij}$$

where d_{ij} is a 5x1 vector of transformed coordinate differences

$$d_{ij} = \begin{bmatrix} u_{ij} \\ v_{ij} \\ a_i - a_j \\ s_i - s_j \\ h_i - h_j - (\Delta_1 - \Delta_2) \end{bmatrix}$$

where

$$\begin{bmatrix} u_{ij} \\ v_{ij} \end{bmatrix} = \begin{bmatrix} x_i \\ y_i \end{bmatrix} - \begin{bmatrix} \cos(\Delta_1 - \Delta_2) & \sin(\Delta_1 - \Delta_2) \\ -\sin(\Delta_1 - \Delta_2) & \cos(\Delta_1 - \Delta_2) \end{bmatrix} \begin{bmatrix} x_j \\ y_j \end{bmatrix} + R_o \begin{bmatrix} \sin(\Delta_1 + \theta_o) \\ \cos(\Delta_1 + \theta_o) \end{bmatrix}$$

are the x and y track separations after coordinate transformation. The matrix P_{ij} is the [5x5] covariance matrix for the vector d_{ij}. This matrix can be computed from radar accuracies, individual site tracking filter variance reduction ratios, expected registration/ navigation and alignment accuracies, and the geometries involved [1,2,6]. Clearly, if two tracks have separations in each of the five coordinates which are comparable to the separations expected from radar measurements errors, then the statistical distance, D_{ij}, will be small. In particular, if tracks #i and #j do correspond to the same target and the tracking errors are Gaussian then D_{ij} has a chi-squared probability distribution with 5 degrees of freedom. Thus probabilities of correct and false association can be computed easily.

A further refinement of this technique is to pose the problem as a hypothesis test:

H_o: track #i and track #j correspond to two different targets.
H_1: track #i and track #j correspond to the same target.

or,

H_o: d_{ij} has density $f_o(d_{ij})$
H_1: d_{ij} has density $f_1(d_{ij})$

The most powerful test is then to threshold the likelihood ratio

$$L(d_{ij}) \triangleq \frac{f_1(d_{ij})}{f_o(d_{ij})}$$

Note that if the difference vector is assumed to be Gaussian when the two tracks do indeed correspond to a common target, then

$$f_1(d_{ij}) = \frac{\exp\left\{-\frac{1}{2} d_{ij}' P_{ij}^{-1} d_{ij}\right\}}{(2\pi)^{5/2} |P_{ij}|^{1/2}}$$

Thus if one assumes that the x,y, speed, heading and altitude differences are uniform across some possible ranges (perhaps specified by a coarse association process) then

$$lnL(d_{ij}) = -\frac{1}{2} d_{ij}' P_{ij}^{-1} d_{ij} + \text{constant terms}$$

and the most powerful test is just to threshold the statistical distance. If, on the other hand, there is more a prior information about the null density $f_o(d_{ij})$ (e.g., planes generally fly in formation or along established air corridors causing a natural "clumping" of speeds and headings even among different aircraft) then the more general likelihood ratio form allows one to account for this.

Note the implicit assumption that navigation /registration and radar misalignment errors are small compared to typical intertarget distances. If these errors are large, then P_{ij} will become large and the five dimensional ellipsoid

$$d_{ij}' P_{ij}^{-1} d_{ij} < T$$

will contain several aircraft pairs leading to false correlation decisions.

TRACK ASSOCIATION IN THE PRESENCE OF LARGE REGISTRATION/NAVIGATION AND ALIGNMENT ERRORS

The algorithm developed here will be described in heuristic terms. A rigorous derivation appears in [6].

When neither accurate navigation and sensor alignment nor accurate (steady-state) automatic registration data are available, the correlation decisions become much more difficult. Direct application of the approaches such as described above can result in numerous false correlations as the effective correlation windows are expanded to include the maximum possible navigation and alignment errors.

A practical approach to association with large navigation and alignment errors is to cross-correlate the two entire track files (local and remote) in the same manner the human eye does when studying pictures of local and remote tracking data. Consider two track bases with some fraction of the tracks held in common. By properly rotating and translating one track base, the two pictures can be made to coincide. This is most easily done by overlaying a transparency of one track base on top of the other, then sliding them around until they line up. While many different sliding operations may make one pair of tracks line up, if one sliding operation makes many more pairs line up than any other operation, those pairs are most likely to be the true associations.

The visual process of "lining up" two track bases can be automated using multidimensional cross-correlation functions. Consider first the case of an unknown translation in the horizontal plane (Figure 1). If one temporarily ignores velocity and altitude information and considers each track to be a dot (with finite size) in the horizontal plane, then the two dimensional cross-correlation function for the two track pictures is as shown in Figure 2. Whenever a given $(\Delta x, \Delta y)$ coordinate corresponds to the separation between some local track and some remote track, a spike occurs. The majority of spikes will not correspond to true pairings. However, the spikes which do correspond to the true pairings will all occur in the same place, producing a large recognizable spike. The pairings which contribute to this spike are selected as associations.

Now consider actual navigation and alignment errors (instead of a simple horizontal plane translation). The three largest errors will be assumed to occur in latitude, longitude and azimuth alignment. It can be shown that if a remote track base, misaligned in azimuth, is transformed into the local coordinate frame using incorrect latitude and longitude (but exact curved Earth equations), then this transformed track base and the local track base can be very nearly lined up by a simple horizontal plane translation and horizontal plane rotation. The cross-correlation approach is, then, directly applicable to actual navigation and alignment errors, provided it is extended to include a horizontal plane rotation as well as a translation. This is done as shown in Figure 3. A third dimension (azimuth rotation, $\Delta\theta$) is added. When the two-dimensional cross-correlation function is plotted for each $\Delta\theta$, the result is a three dimensional collection of spikes (shown as balls) with the spike corresponding to the real pairings dominating the surrounding spikes.

Velocity and altitude data can be used to make the real pairing spike more visible in the background of random association pairs. This is done by viewing the cross-correlation process as a composite hypothesis test with three unknown parameters (x translation Δx, y translation Δy, and azimuth rotation $\Delta\theta$). The approximate generalized likelihood ratio test is derived in [6] and shown to correspond to the cross-correlation process described above with the spikes weighted according to how well the two targets' velocities and altitudes agree. That is, the contribution of each possible local-remote track pair is weighted by the pair's velocity and altitude likelihood ratios. Since the real pairings will tend to have higher likelihood ratios than the false pairings, this will further enhance the composite spike corresponding to the collection of real pairings.

The association process, in the presence of large navigation errors and misalignments, can be summarized as follows:

1. Compute the three-dimensional cross-correlation array (Figure 3) using the local and remote track data bases.
2. Find the "center of mass" in the cross-correlation array (this will be referred to as the cluster point).
3. Select the track pairs at (and in the near vicinity of) the cluster point as the associated pairs.

This process is the mathematical version of what the human eye would do the "line-up" two track pictures.

A real time implementation of this process is possible by computing the cross-correlation array in an unconventional way. The key is that almost all the values in the array are zero. It is only necessary to compute those values which are nonzero. If one assumes a large upperbound on the navigation errors, then an entry into the cross-correlation array is made whenever a local and remote track are within this upper bound of each other. **But these situations are identified by any conventional correlation algorithm.** For example, a cartesian linked list structure provides an efficient way of identifying remote tracks. Whenever a remote track is close to a local track, the appropriate cells in the cross-correlation array are incremented by the weight determined by how well the tracks line up in speed, heading and altitude. After this has been done for all close pairs of tracks, the cross-correlation array contains the cross-correlation function.

After the cross-correlation array has been filled, the cluster point is computed. This centroiding operation is done only once per correlation cycle and so is not a significant contributor to computing time. The weighted average or "center of mass" centroiding techniques which has been applied to automatic target detection can be modified for this application. Basically, one selects the azimuth rotation which makes the picture the "sharpest" (i.e., concentrates as much of the cross-correlation function's mass as possible in one place). This defines an azimuth focal plane. The largest and second largest spikes in this plane are then center-of-mass centroided to obtain the x,y coordinates of the cluster point (Δx^*, Δy^*). The azimuth coordinate of the cluster point, $\Delta\theta^*$, is obtained by center-of-mass centroiding in the azimuth direction.

Since the vast majority of cross-correlation array cells are never addressed while filling up the array, and since the cluster point is found as the array is being filled **the only operation which requires addressing all elements in the cross-correlation array is zeroing out the array prior to starting the correlation process.**

EXPERIMENTAL EVALUATION OF TRACK ASSOCIATION ALGORITH USING SIMULTANEOUS RADAR DATA

The cross-correlation association algorithm described above has been evaluated using recorded simultaneous radar track data. Two S-band 3D pencil beam surveillance radars 66 km apart were used (Figure 4). The two radars were located near major airports and typically detected many targets of opportunity. Digital data collectors performed automatic plot extraction and recorded digitized plots on magnetic tape. These plot tapes were fed into two automatic detection and tracking (ADT) computer systems which assembled the plot into tracks. A description of the type of ADT systems used appears in [7]. The association algorithm, implemented on VAX-11/380 computer, took the two track files as input and produced track association decisions as output.

The experiment typically yielded 30-60 tracks per site depending on commercial air traffic. Of these 30-60 tracks, about 10-20 corresponded to common targets; the rest, due to radar blockage, geometry or horizon effects, were tracked only by a single radar.

A portion of the cross-correlation array computed for a single scan of radar track data from each site is shown in Figure 5. Three azimuth planes are shown corresponding to rotations of (-1), (-3/2) and (-2) times the radars azimuth beamwidth θ_a. The planes each consist of a matrix of cells with each cell 2 range resolution cells (width τ) on a side. The center ($\Delta\theta = -(3/2)\ \theta_a$) plane, is "in focus". The correlation array shows the registration errors ($\Delta x^* = -2\tau$, $\Delta y^* = +4\tau$) and groups together the track pairs to be associated. For comparison, Figure 6 shows an azimuth plane corresponding to an azimuth rotation ($\Delta\theta = +(1/2)\theta_a$) which is two beamwidths out of alignment. The picture is clearly out of focus.

It is interesting to note that even though the radar sites have been very accurately surveyed, it is necessary to perform both a rotation and translation to align the tracks. This is an illustration of the fact [3,5] that when an azimuth misalignment at local site is corrected by transforming data from a remote site, both a rotation and a translation of remote data are required.

As noted above, this cross-correlation process produces correlation decisions insensitive to large navigation/registration errors or sensor misalignments. If the navigation/registration errors are each exact multiples of the x and y cell size and if the azimuth misalignment is an exact multiple of the azimuth plane spacing, then one can easily prove mathematically that the correlation decisions are totally unaffected by these errors and misalignments. (The effect on Figure 5 would be simply to translate the correlation pattern in Δx, Δy, $\Delta\theta$ space unchanged.) Of more practical interest is the case where navigation errors are not exact multiples of cell spacing. This case is illustrated in Figure 7 in which artificial navigation errors of 9 range cells in x, 9 range cells in y, and 2.5 beamwidths in azimuth have been artificially inserted into one site's data. Note the fractional errors and biases have slightly "smeared" the pattern redistributing mass across cell boundaries. However the picture is still "in focus" and the computed parameters (Δx^*, Δy^*, $\Delta\theta^*$) very accurately remove the effects of the artificial errors as follows:

Effects of Artificially Induced Navigation Errors and Sensor Biases

	Error Applied	Change in Δx^*, Δy^* or $\Delta\theta^*$
x coordinate	9τ	8.8τ
y coordinate	9τ	8.9τ
θ rotation	$2.5\theta_a$	$2.7\theta_a$

Furthermore, no changes in the correlation decisions result from introducing the artificial errors.

Table I illustrates the accuracy (rms variations) of the parameters (Δx^*, Δy^*, $\Delta\theta^*$) over 250 separate association operations each spaced one antenna rotation period (8 seconds) apart. Table I indicates how accurately Δx^*, Δy^* and $\Delta\theta^*$ can be computed using only a **single look** at each of the two track files. Very stable association decisions were made throughout this 30 minute period.

One would certainly expect the cross-correlation process to work best when the fraction of tracks which are held in common by the two radars is high. However, the algorithm has proven effective even when only two or three targets are seen in common. Table II illustrates an example of this in which fractions of 12 common tracks are randomly discarded without impact on the remaining correlation decisions.

CONSIDERATIONS FOR A PRACTICAL REAL-TIME SYSTEM

A practical, real-time automatic association process based on this principle is currently being implemented in an AN/UYK-20 minicomputer. A cross-correlation array, 16 azimuth planes by 16 Δx cells by 16 Δy cells (4096 cells total) is used. The array is filled by a coarse association process. A running tally is kept of the most focused region in the array. The computed registration and alignment biases ($\Delta\theta^*$, Δx^*, Δy^*) are then used to transform one (e.g., the remote) track file. A conventional correlation algorithm (i.e., likelihood ratio test) is then applied to the two track files. The result is a set of correlation decisions corresponding directly to the cluster of spikes near the cluster point.

In order to accommodate the possibility of extremely large navigation/registration errors and sensor biases without using an exorbitant amount of computer memory, a two-step, iterated cross-correlation process is used in which the cell size and azimuth plane spacing are successively reduced. This enables one to make reliable correlation decisions even when navigation/registration errors are hundreds or thousands of range resolution cells. In particular, errors much larger than typical target spacings can be handled. Such errors would be catastrophic to a conventional correlation algorithm.

TABLE I Accuracy [1] of Parameters Δx^*, Δy^*, $\Delta\theta^*$ Computed by Cross-Correlation

Δx^*	0.4 range cell (τ)
Δy^*	1.0 range cell (τ)
$\Delta\theta^*$	0.24 beamwidth (θ_a)

[1]rms fluctuation over 250 single scan correlation attempts

TABLE II Change in Correlation Decisions and Parameters As Track Data Base Decreases

Modification	Δx^* Change	Δy^* Change	$\Delta\theta^*$ Change	Correlation Decision Changes
50% of correlated track pairs randomly eliminated	$.4\tau$	0	$-.1(\theta_a)$	None
75% of correlated track pairs randomly eliminated	$.4\tau$	0	$-.2(\theta_a)$	None

REFERENCES

[1] Kanyuck, A. and Singer, R., "Correlation of Multiple-Site Track Data", IEEE Trans., AES-6, March 1970, pp. 180-187.

[2] Stein, J. and Blackman, S., "Generalized Correlation of Multi-Target Track Data", IEEE Trans., AES 11, November 1975, pp. 1207-1217.

[3] Brickner, E., Castella, F., Cook, S., and Miller, J., "Multi-Sensor Tracking Study Final Report", Johns Hopkins University Applied Physics Laboratory Report to FAA, FP8-T-024, September 1975.

[4] Fischer, W., Muehe, C., and Cameron, A., "Registration Errors in a Netted Air Surveillance System", MIT Lincoln Labs Tehnical Note 1980-40, 2 September 1980.

[5] Miller, J., et. al., "Battle Group AAW Coordination Gridlock Analysis Report", Johns Hopkins University Applied Physics Laboratory, Fleet Systems Report FS-82-023-1,2, February 1982.

[6] Bath, W., "Automatic Correlation of Radar Track Data from Physically Separated Platforms in the Presence of Navigation Errors and Sensor Misalignments", Johns Hopkins University Applied Physics Laboratory, Fleet Systems Report FS-80-281, September 1980.

[7] Casner, P. and Prengaman, R., "Integration and Automation of Multiple Co-Located Radars", Proc. Radar - 77, London, October 1977, pp. 145-149.

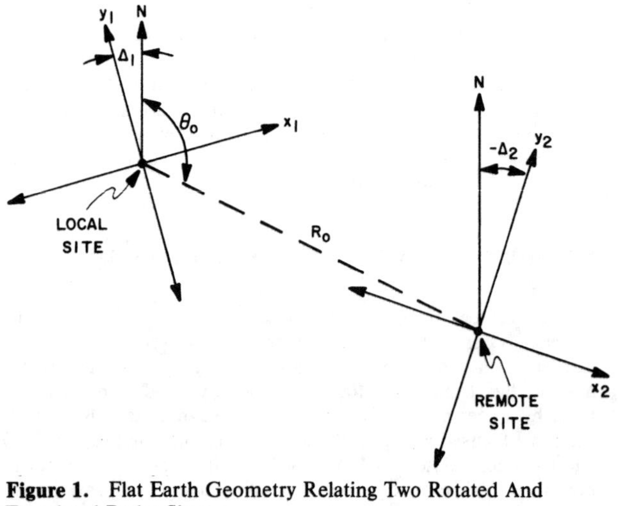

Figure 1. Flat Earth Geometry Relating Two Rotated And Translated Radar Sites

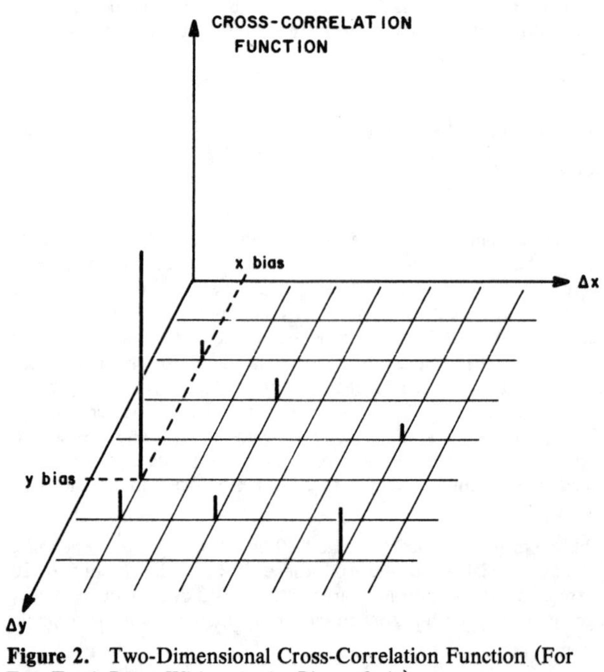

Figure 2. Two-Dimensional Cross-Correlation Function (For Two Track Bases With x And y Biases Only)

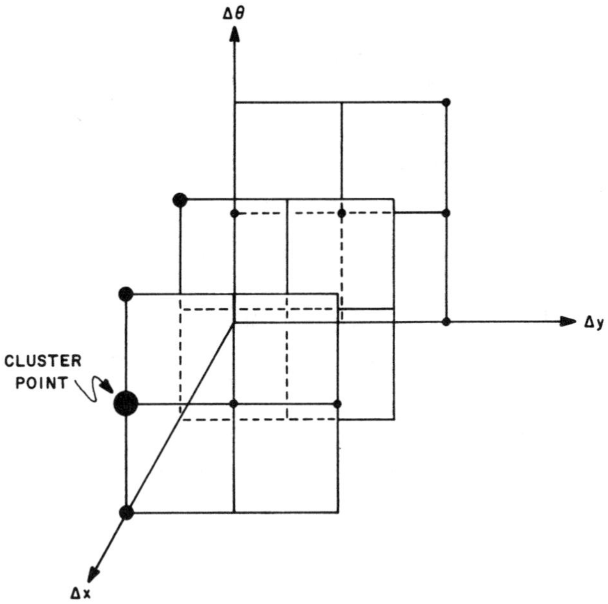

Figure 3. Three Dimensional Cross-Correlation Array (For Two Track Bases With Latitude And Longitude Errors Plus Azimuth Misalignment)

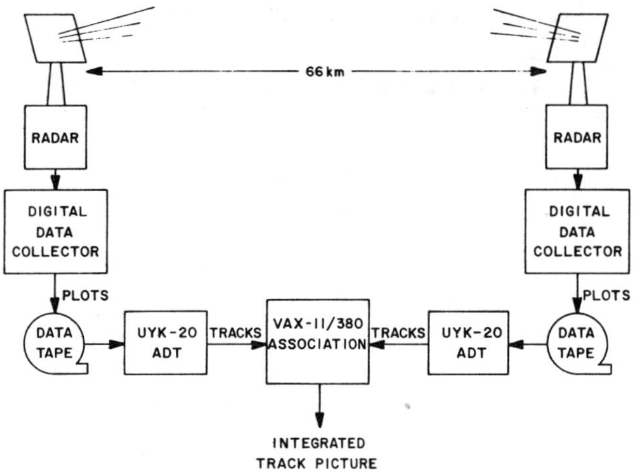

Figure 4. Experimental Setup For Testing Cross-Correlation Track Association Technique

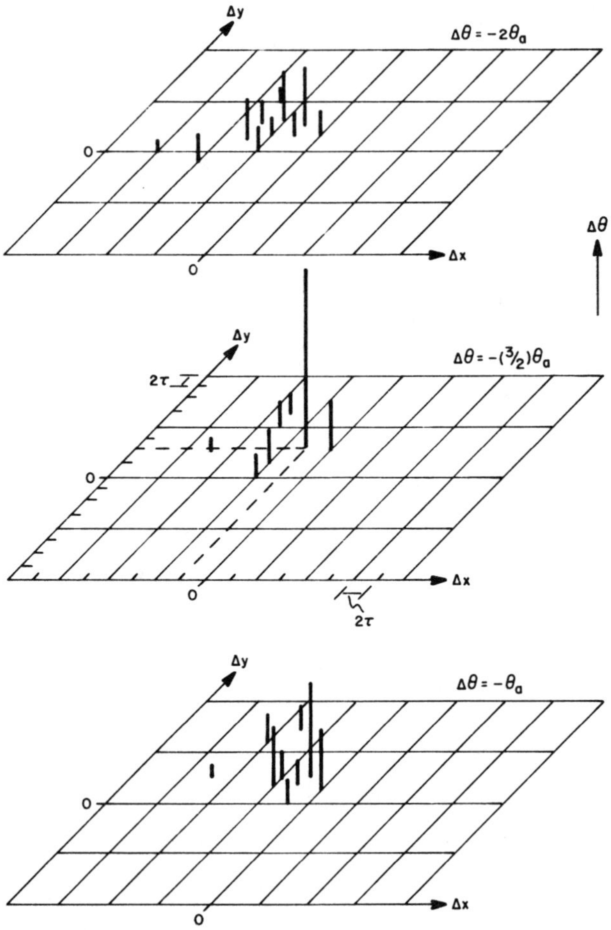

Figure 5. Small Portion of Three Dimensional Cross-Correlation Array. Computed from a Single Look at Two Track Files Formed from Radar Experiment Shown in Figure 4.

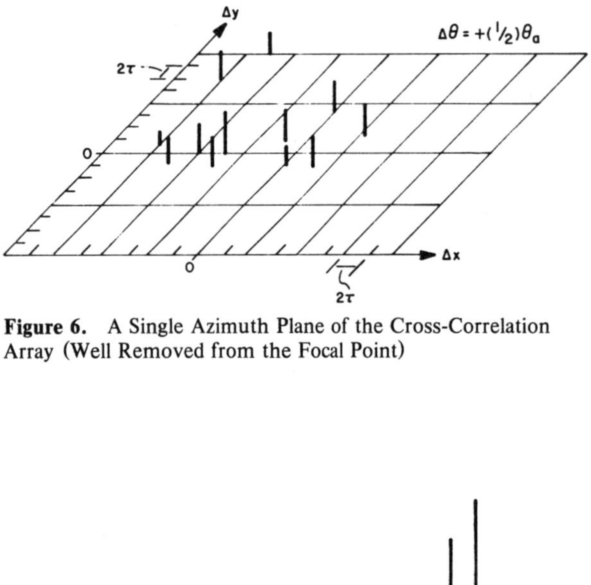

Figure 6. A Single Azimuth Plane of the Cross-Correlation Array (Well Removed from the Focal Point)

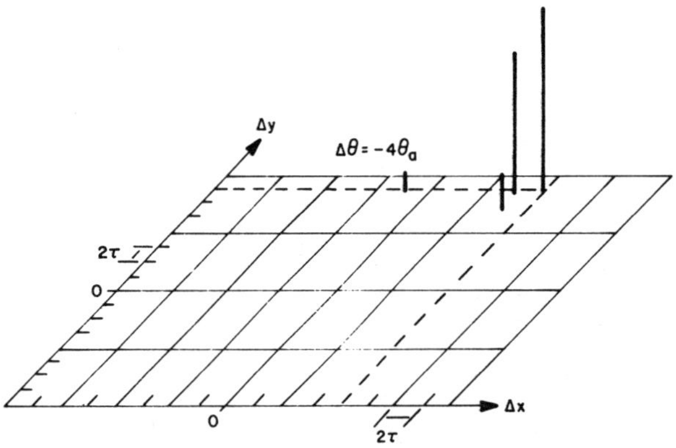

Figure 7. A Single Azimuth Plane of the Cross-Correlation Array After Insertion of Artificial Navigation Errors And Misalignments

ACTIVE ARRAY RECEIVER STUDIES FOR BISTATIC/MULTISTATIC RADAR

J.G. Schoenenberger[*+] and J.R. Forrest[*], C. Pell

[*]University College London, U.K., Royal Signals and Radar Establishment, U.K.
[*+]Now with Racal-Decca Ltd, U.K.,

INTRODUCTION

For several decades, surveillance performed by radar systems has been almost exclusively based upon the monostatic, or co-located transmitter and receiver, principle. This is true for civil and military applications in the airborne, shipborne or ground environment. However, over the last 5 to 8 years there has been a significant increase of interest in bistatic and multistatic systems. The bistatic radar principle is seen by many as offering important potential advantages over the monostatic counterpart, albeit sometimes with an attendant increase in complexity. Some of the more obvious advantages, such as a reduction in signal dynamic range and immunity to highly directional interference are well known, and have been summarised by Milne (1).

Applied research into bistatic radar has been actively pursued for several years by the Royal Signals and Radar Establishment and University College London. Some of the early experimental work conducted by RSRE with support from Plessey Electronic Systems Research has been reported by Pell et al (2) and concerned a ground-based experimental system operating over a baseline length of approximately 45km. Techniques for sensor synchronisation, communication and real-time bistatic data display and error correction were developed and assessed. Data were gathered and processed relating to a common aircraft target, simultaneously viewed both monostatically and bistatically.

The experimental bistatic radar systems research at UCL commenced with the development of a totally independent bistatic receiver utilising UHF transmissions from a civil air traffic control radar located some 25km away at London Heathrow airport. The work has been reported by Schoenenberger and Forrest (3), (4) and mainly concerned the techniques required for synchronising the receiver and providing accurate p.r.f. and transmitter pointing direction data. In order that the receiver remains totally independent, no link to the transmitter other than the usually transmitted radar signals was assumed. The p.r.f. and azimuth rotation signals were therefore synthesised locally at the receiver site. In addition, the well known bistatic display distortion was corrected in real-time by a fast microprocessor-based system, producing a transmitter-site origin PPI display. This system enabled an extremely versatile, totally independent, bistatic radar receiver to be evaluated, and some of the likely operational problem areas to be identified.

It is the aim of this paper to devote attention to beamforming, control and pattern optimisation for an active receive array, these techniques being regarded as vital to future systems.

BEAMFORMING FOR BISTATIC RECEIVE ARRAYS

In the original UCL system (4), the receiving antenna is a low gain, essentially omnidirectional in azimuth, element. Whilst this is convenient and achieves a satisfactory data rate and surveillance area when employed in conjunction with a narrow beamwidth, scanning illuminator antenna, only two spatial discriminants are accessible, viz. the range sum ellipse and illuminator azimuth. In addition to the limited detection performance usually resulting from this configuration, there is a high probability of clutter and spurious target detections resulting from illuminator sidelobe to receiver main beam coupling since only one-way sidelobe protection is afforded.

There is, therefore, considerable motivation in providing a third spatial discriminant-receiver azimuth. This can be achieved by providing angular selectivity at the receiver site with a narrow beam receiving antenna. Although Milne (1) has indicated several configurations of illuminator and receiver antennas, it is considered that in most applications, a multiple beam receiving antenna will be required, implemented as either a small number of electronically agile beams or a static set of contiguous beams covering the required surveillance sector. The non-uniform nature of the required receive beam steering makes a mechanically-steered antenna virtually impossible. Beamforming and control is a key area of research, particularly if an extension to angular selectivity in a further co-ordinate (e.g. elevation) is required, or when complex beam sets are required on reception for angle estimation refinement (e.g. amplitude interpolation or true monopulse).

Owing to the wide instantaneous azimuth coverage sector required, typically in excess of 90°, a reflector/multiple feed or a reflector/array feed hybrid were regarded as unlikely to be satisfactory. Array antennas are the only option since many techniques exist for the synthesis of multiple, controllable beams. Techniques available include:

i. Rotman Lenses possessing multiple input and output ports (5) perhaps employing microstrip or triplate structures (6).
ii. Analogue tapped delay line i.f. matrices.
iii. The 4-phase mixer or resistive i.f. matrix (7).
iv. The Blass/Maxson and Butler array matrices.
v. i.f. phase shifting by variable i.f. scanning.
vi. Digital, baseband beamforming.

Radford (8) has presented a most useful review of system applications of beam forming networks, including a consideration of multiple beam array antennas. Other relevant work concerning time-multiplexed, multi-beam synthesis is described by Fleskes et al (9).

The following aspects in regard to the different beamforming techniques were considered in the present work:

i. Theoretical and practical limits on achievable sidelobe levels, instantaneous signal bandwidth.
ii. Orthogonality constraints of formed beams, as defined by Allen (10).
iii. Ease of selection of arbitrary phase and amplitude weights or array element outputs for pattern optimisation and adaptivity.
iv. Ease of achieving the surveillance volume coverage.
v. Applicability to large one- and two-dimensional arrays.
vi. Implementational complexity, versatility and cost.

The synthesis of multiple static contiguous beams can be accomplished in one dimension (azimuth) by a fixed analogue beamforming matrix, such as the i.f. resistive matrix. However, whilst this technique is extremely efficient in terms of cost and complexity, it is inherently inflexible and pattern adaptivity or error compensation at an element level is not generally possible within the beamformer. Since adaptivity and, generally, the use of arbitrary element complex weights was required, the most attractive method available appeared to be digital beamforming at baseband. Although the technology required to implement a large scale system is not yet economically available, theoretical and experimental experience gained from a small scale exercise was considered vital. Generally, the number of real multiplications required per second in a digital beamformer are:

$$M = 4\,m\,n\,B\,\alpha$$

where m = number of elements or separate subarrays at beamformer input ports.
n = number of formed output beams
B = signal bandwidth (Hz)
α = factor required to meet the specified sampling criterion
(α = 2 for Nyquist sampling)

As an example, for m = 128 elements, n = 64 beams, B = 5MHz, and $\alpha = 2$, M becomes:

$$M \approx 10^{11}$$

This is a very large processing problem for todays technology, but may be tractable in the future.

There are, of course, structured algorithms such as the Fast Fourier Transform (FFT) that can be used to reduce the computational overhead, requiring $2N \log_2 N$ complex multiplications for the synthesis of N beams for an N element array, but the usage of arbitrary or rather custom complex weights for individual beams is precluded.

If the experimental system is limited, for convenience, to 8 elements generating 8 beams with a signal bandwidth of 300kHz and a sampling rate of 600kHz, M reduces to 38 million multiplications/second for the DFT which may be realised quite readily.

Theoretical Aspects of the Digital Beamforming Network

The formation of a single beam from an N-element array involves the following sequential operations:

1. Digitisation of the baseband signal in each element into in phase (I_i) and quadrature (Q_i) signals (i = 1 to N).

2. The I_i and Q_i signals are multiplied by complex weights, represented by I_{wi} and Q_{wi}. I_{wi} and Q_{wi} represent the required phase front and amplitude taper across the array. The resulting complex number for each element, $(I_i' + jQ_i')$ is obtained as:

$$I_i' = I_i\,I_{wi} - Q_i\,Q_{wi}$$

$$Q_i' = I_i\,Q_{wi} + Q_i\,I_{wi}$$

3. The $I_i' + jQ_i'$ signals from each element must be summed across the array to produce the sum signal $I_s + jQ_s$. I_s and Q_s are obtained as:

$$I_s = \sum_{i=1}^{N} I_i' \,, \quad Q_s = \sum_{i=1}^{N} Q_i'$$

4. In order to obtain a signal magnitude output from the array, $|I_s + jQ_s|$ must be determined. This involves squaring both I_s and Q_s, adding the squares, and taking the square root of the result to yield V_o the array voltage output:

$$V_o = (I_s^2 + Q_s^2)^{1/2}$$

Various errors within the beamforming network contribute to a degradation of overall performance:

1. Thermal noise
2. Quantisation errors in analog-to-digital conversion
3. Truncation errors
4. Errors in weight determination
5. Gain and Phase mismatches

Barton (11) has analysed such error sources in some detail. The errors may degrade performance either by limiting the linear dynamic range of the system or by providing extraneous sidelobes. For instance, the errors due to truncation and the effects of thermal and quantisation noise both limit the overall dynamic range of the beamforming network, the dynamic range due to thermal and quantisation noise being:

$$\frac{3N2^{2b_s}}{26}$$

For an N-element array with digitisation to b_s bits with the quantisation step equal to the r.m.s. noise voltage, this yields for an 8-element array (100 element array in parentheses) maximum dynamic ranges of:

b_s (bits)	maximum dynamic range (dB)
8	47.8 (58.8)
10	59.9 (70.8)
12	71.9 (82.9)

The errors due to weight definition (or quantisation), for quantisation of the weights to b_w bits yield rms sidelobe levels below the nose of the formed beam of:

$$\frac{3N}{2} \cdot 2^{2b_w}$$

for a uniformly illuminated array. The effect of these quantisation errors are illustrated for 8 and (100) element arrays:

b_w (bits)	r.m.s.sidelobe levels (dB)
6	46.9 (57.9)
8	59.0 (69.9)
10	71.0 (82.0)
12	83.0 (94.0)

Similarly, the r.m.s. sidelobe levels due to errors in gain and phase matching throughout the array may be expressed as:

$$\frac{4N}{(\delta^2 + k^2)}$$

where δ is the phase error in radians either between the in-phase and quadrature channels of any element or the element-to-element mismatch, and k is the ratio of gain error to mean channel gain. The effects of these errors may be illustrated by:

δ (degrees)	k (dB)	r.m.s.sidelobe levels (dB)
1	0.2	45.8 (56.7)
0.5	0.1	51.8 (62.8)
0.2	0.05	58.5 (69.5)
0.1	0.02	65.8 (76.8)

In order that a complete beamforming network should be implemented rapidly and inexpensively, it was decided to use an 8 element array and 8-bit analog-to-digital convertors. This implied a signal dynamic range of about 48dB at the beamformer output (and 42dB at the input, since 8 bit quantisation is really 7 bits plus sign).

Initial design performance aims were for -35dB sidelobes using a Taylor $\bar{n}$=5 amplitude taper in the complex weighting function, and 8-bit weight definition to keep the mean sidelobe level due to weight errors at under -59dB (i.e. under the thermal noise level). Initial aims for array element mismatch to be better than 1° in phase and 0.2dB in gain were also stipulated in order that extraneous sidelobes might be kept very low, at under -45dB r.m.s.

Design of the Array Receiver

The design of the receiver array circuitry is quite conventional (Fig. 1). Each of the eight elements comprises a four-element Yagi array aerial, r.f. bandpass filter, low noise (NF = 3dB) r.f. amplifiers, r.f. mixer, i.f. stages, quadrature phase detectors, sample/hold amplifiers and analog-to-digital convertors. The i.f. stages comprise a tuned amplifier for receiver bandwidth determination and a specially designed low phase-shift limiting stage. This limiting stage is required to reduce the overall signal dynamic range (66dB) to that used by the beamforming network (42dB) without the loss of phase information that often occurs with conventional limiters.

The phase detectors employed are bipolar transistor arrays used as two quadrant multipliers. These devices are used for their excellent linearity characteristics and ease of array calibration. The anti-aliasing filters ensure that any outputs from the quadrature phase detectors at or above the digital sampling frequency are attenuated to below the thermal noise level. The sample/hold amplifiers are used to enable inexpensive successive-approximation analog-to-digital convertors to be employed.

Some of the primary system parameters are:

r.f. frequency	600MHz
r.f. bandwidth (3dB)	20MHz
noise figure	< 5dB
i.f. frequency	70MHz
i.f. bandwidth (3dB)	250kHz
radar pulse length	4μs
sampling frequency (ADC)	400kHz

The worst errors are likely to occur in the i.f. limiters (when limiting) and the quadrature phase detectors. Results obtained to date indicate that the limiters introduce a differential phase shift of under 0.6° even under 30dB of limiting (which is more than required by the present system) and that the quadrature phase detectors maintain their phase and gain linearity to better than 0.1° and 0.1dB over the whole required dynamic range.

Long term drifts due to temperature and ageing have not yet been investigated, though these are not of great importance, since the complex weights used for beam control may also be used to correct for both gain and phase errors introduced both at r.f./i.f. and at baseband in the quadrature phase detectors (12).

Initial tests are therefore very promising and indicate that the desired gain and phase matching (0.2dB and 1° respectively) for the individual array element channels are likely to be achieved.

Design of Digital Beamformer

The processes to be performed to produce a single beam output have been described above in Section 2.1. For ease of sub-system design, the complete beamforming process may be pipelined. The element I_i and Q_i signals are sampled at regular intervals. At the start of each sample period, the data from each stage of the beamforming process is passed to the next stage. This incurs a real time (but constant) delay between the beamformer input and output signals. The process is illustrated in Fig. 2. The primary advantage of this type of system is that the amount of time available for performing each of the functions of the beamforming process may be quite long (up to the sampling period). If each process can be made considerably shorter than the sampling period, then a high sampling rate may be realised very simply by the provision of faster sample/hold and ADC circuits (if these are the limiting factors, as will often be the case).

The overall design of the beamformer is shown in the block diagram of Fig. 3. The complex multipliers each employ one LSI multiplier device, time-multiplexed to perform four multiplications, one addition and one subtraction in each sampling interval of 2.5μs. The complex weights are entered into the multiplier as required, the weighting functions for various beam positions and array amplitude tapers being stored in the "Weight Store". The stored beam position may be altered for every new sample (i.e. successive samples may be taken from totally differing beam positions if necessary). Access to the weight control lines is also made available to an external computer for adaptive weight control, if required.

The outputs of the eight complex multipliers are summed to produce the I_s and Q_s signals which are passed into the remaining sections of the beamformer as well as being made available to any external device. The square operation is performed by a single LSI multiplier device, the sum of I_s^2 and Q_s^2 is produced and the square root function is performed. The square root operation is, perhaps, the most difficult operation to perform accurately in 2.5μs. It is realised here with programmable read only memories (PROMs) as a look-up table and associated circuitry to decrease device count and complexity without compromising accuracy. The log function is produced in a similar fashion using PROMs, and the resulting digital linear and logarithmic signals are converted into analogue form for eventual display purposes, as well as being left in digital form for use by other signal processors.

No results on the performance of the digital beamformer are available, since it is currently under construction.

ASSESSMENT AND MEASUREMENT PROCEDURE PLANNED

Initial calibration and on-line error detection and correction may be performed most simply if a far field source at the radar frequency is used. Initial gain and phase mismatches should be adjustable to less than 0.1dB and 0.1° respectively.

The operation of the digital parts of the system may most simply be verified with simulated digital signals, and once correct operation has been determined, array radiation patterns may be measured using the far field calibration source, for various beam positions. All pertinent array parameters may be assessed using such a far field source enabling a prediction of the performance with the present transmitter to be made with confidence.

The beamformer output signals may be used with the existing bistatic display correction processor at UCL to provide a geometrically correct PPI display of the beamformer output (say over a $\pm 45^{\circ}$ sector) for all transmitter pointing angles. Alternatively, the digital output signals may be applied to an additional external processor such as a plot extractor.

CONCLUSIONS

This paper has attempted to summarise the current techniques research being conducted at UCL with RSRE support on digital beamforming and is considered to be an essential exercise to give the necessary theoretical and experimental understanding for any future large scale adaptive system. Its extension to a larger system is dependent upon the future availability and maturity of VLSI signal processing devices, but more efficient structures and algorithms are also likely to be needed.

At the time of submission of this paper, construction of the digital processor is nearing completion, the array antenna and analogue circuits have been constructed and detailed sub-system performance assessment is under-way. The full testing, optimisation and characterisation of the integrated system is expected to commence shortly.

ACKNOWLEDGEMENTS

The authors wish to extend their thanks to Professor K. Milne and Professor D.E.N. Davies for much practical advice, and also to Mr H.D. Griffiths and Mr A.D. Williams for their painstaking constructional work.

The authors have also appreciated the assistance of NATO (AGARD) in a collaborative research programme with the Hellenic Air Force Technology Research Centre (KETA), Athens.

REFERENCES

1. Milne, K., 1977, "Principles and Concepts of Multistatic Surveillance Radars", IEE Conference Publication No. 155, "Radar 77", October, 46-52.

2. Pell, C. et. al., 1981, "An Experimental Bistatic Radar Trials System", IEE Colloquium on Ground and Airborne Multistatic Radar, 4 December.

3. Schoenenberger, J.G., and Forrest, J.R., 1980, "Totally Independent Bistatic Radar Receiver with Real-time Microprocessor Scan Correction", IEEE International Radar Conference, "Radar 80", April, 380-386.

4. Schoenenberger, J.G., and Forrest, J.R., 1982, "Principles of Independent Receivers for use with Cooperative Radar Transmitters", The Radio and Electronic Engineer, 52, 2, February, 93-101.

5. Rotman, W. and Turner, R.F., 1963, "Wide Angle Microwave Lens for Line Source Applications", IEEE Trans-AP, AP-11.

6. Niazi, A.Y. et. al., 1980, "Microstrip and Triplate Rotman Lenses", Proc. 1980 Military Microwaves Conference, 3-12.

7. Hansen, R.C., 1966, "Microwave Scanning Antennas", III, Academic Press.

8. Radford, M.F., 1977, "System Applications of Beam Forming Networks", IEE Conference Publication No. 155, "Radar 77", October, 63-65.

9. Fleskes, W., and van Keuk, G., 1980, "Adaptive Control of Multiple Target Tracking with the ELRA Phased Array Radar", IEEE International Radar Conference, "Radar 80", April, 8-13.

10. Allen, J.L., 1961, "A Theoretical Limitation on the Formation of Lossless Multiple Beams in Linear Arrays", IRE Trans. AP, AP-9, July, 350-352.

11. Barton, P., 1980, "Digital Beam Forming for Radar", IEE Proc. 127, Part F, No.4, August, 266-277.

12. Churchill, F.E., Ogar, G.W., and Thompson, B.J., 1981, "The Correction of I and Q Errors in a Coherent Processor", IEEE Trans-AES, AES-17, No.1, January.

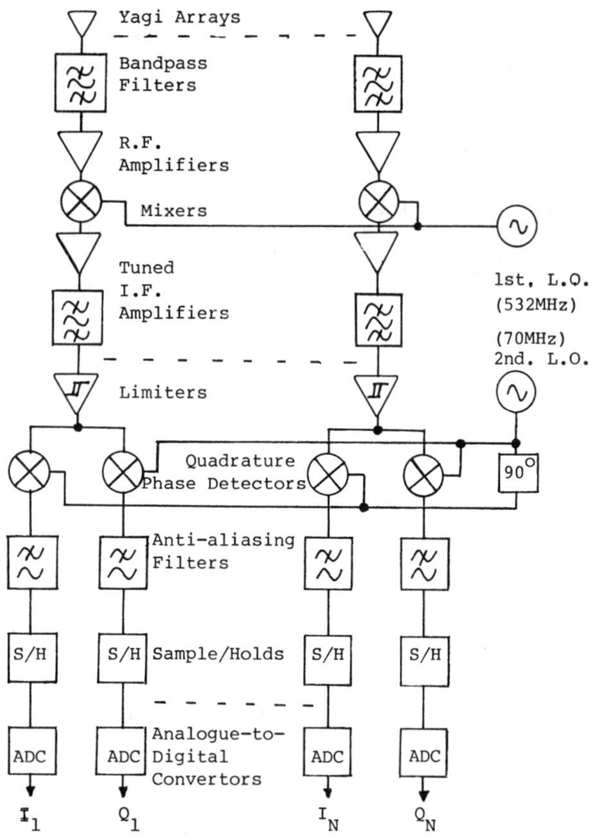

Figure 1. Block Diagram of Array Receiver.

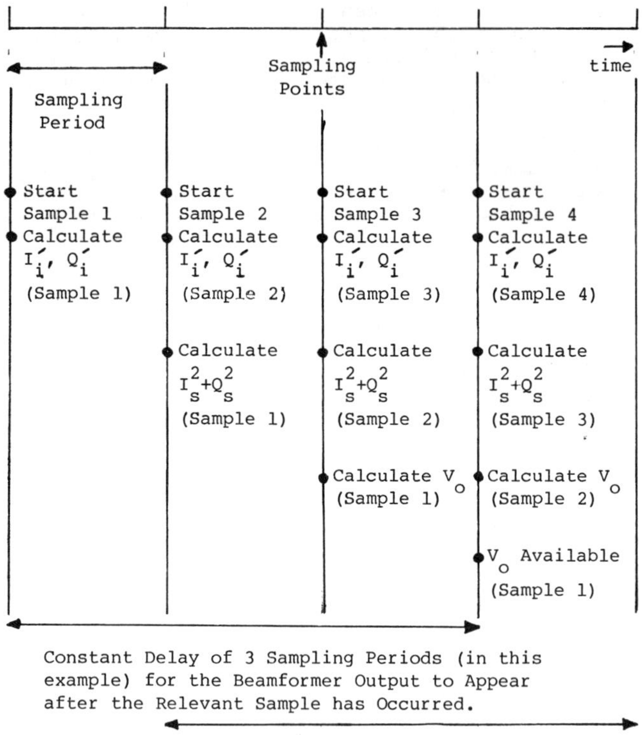

Figure 2. Pipelined Beamforming System (Example).

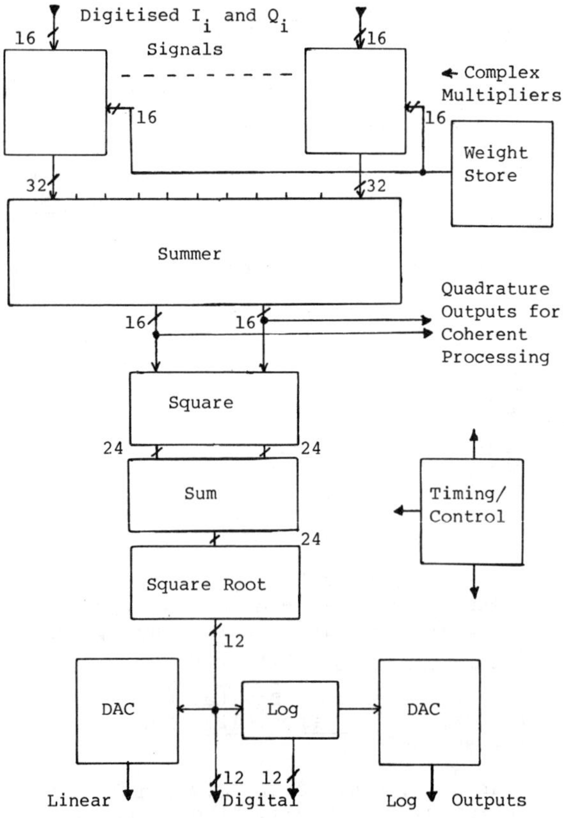

Figure 3. Block Diagram of Beamforming System.

Figure 4, Photograph of the Antenna Array.

COHERENT MULTI-STATIC RADAR: STOCHASTIC SIGNAL THEORY AND PERFORMANCE EVALUATION

Åke Wernersson

Nat. Defence Research Institute, 581 11 Linköping, Sweden

INTRODUCTION

The spatially distributed receivers of a multi-static radar system gives, compared with a monostatic or a bistatic radar, new possibilities of protection against jamming. The possibility of extracting information will also increase. Two main topics in this paper is signal modelling and the related influence of jamming. The paper is - for obvious reasons - nontechnical. Also, the presentation is semi-quantitative in the sense that only the order of magnitude has been studied.

The Geometrical Configuration

Figure 1 describes a quadrustatic configuration with three receivers and one transmitter. It is assumed, that the receivers have complete knowledge of the transmitted signal S_o (t). Hence, the time delay between the three receivers is always known with approximately the same accuracy as the resolution of the radar.

It should be observed that a jammer will introduce a different time delay than the scattered transmitted signal - compare Fig. 1. Together with the appropriate signal processing, this property can be used to obtain a "space-time" discrimination against repeated short pulse jamming. An essential improvement can be made if we have three instead of two receivers. In the former case some kind of "majority voting" can be included in the processing, say, by using the median of the three signals. One main topic in the paper is to study the possibilities of discrimination in the case of, say, pulse compression.

A SIMPLIFIED SIGNAL MODEL

Below, a simplified model is given for the transmitted signal, for the propagation, and for the scattering at the target. This model as well as the statistical dependence between the signals will be used later when we study jamming.

The Transmitted Radar Signal

Let S_o (t) be the video of the transmitted radar signal i.e. the modulating waveform. To obtain sufficient resolution, either a short pulse or some coded signal has to be used - compare pulse compression. For the sake of simplicity, regard S_o (t) as a discrete signal with M discretization points. S_o (t) can thus be considered as a vector of dimension M i.e.

$$S_o(t_1) = S_o = (s_1, \ldots, s_M)^* \tag{1a}$$

with all other components of S_o (t_1) identically equal to zero.

In the sequel it is assumed that S_o (t) is an approximately "Barker like" pseudorandom sequence with an amplitude normalization $|S_o(t)| \leq a$ for all t. More precisely, we assume that S_o (t) is "approximately orthogonal", AO, when time delayed. Using scalar products, the AO-property can be expressed as:

$$S_o(t)^* S_o(\tau) = \begin{cases} \leq M \cdot a^2 & \tau = t \\ c(\tau) \cdot a^2 & \tau \neq t \\ 0 & |\tau - t| > M \end{cases} \tag{1b}$$

where * stands for transposition and $c(\tau)$, at its most, has unit magnitude. With an obvious abuse of notations (and of the concept of an expected value) we thus assume that $E\, c(\tau) \approx 0$ and $\mathrm{Var}\, c(\tau) \leq 1$. The AO-property in Eq 1 is essential in Fig. 2 below. It should also be mentioned that the AO-property above is not identical with the properties postulated for PN sequences; compare chapter 7 in ref (1). One difference is that the autocorrelation function for PN sequences is defined for periodically continued sequences. The AO-property seems to be less restrictive than the PN-postulates.

Propagation and Scattering

Let $S_r(t)$ be the received videosignal at receiver r. To obtain a unique time index, the time reference is defined by the arrival of the transmitted wave form S_o (t) to the first receiver: $t=t_1$. Then, for two receivers, we have the simplified model

$$\begin{aligned} S_1(t) &= A_1\, S_o(t) + W_1(t) + B_1\, S_J(t) \\ S_2(t) &= A_2\, S_o(t-t_2) + W_2(t) + B_2\, S_J(t-t_J) \end{aligned} \tag{2}$$

where W is a white additive receiver noise and S_J is the signal from the jammer. A_r and B_r are used to model the signal amplitudes from the search bin and the jammer, respectively. Since all time scales are referenced with respect to receiver 1 the time delays t_2 and t_J appear clearly from Fig. 1.

Both receivers have the common propagation factor A_o from the transmitter to the target/ /search bin. One simple model for the A_r:s would thus be

$$A_r = A_o\, A_{or}\, e^{i\,\psi_r} \tag{3}$$

where A_{or} includes the scattering at the target and the propagation towards receiver r. A phase ψ has also been included in the model. The equation seems to be reasonable for one search bin, a fixed frequency, and for a signal S_o (t) that is shorter than the time constants of the fluctuations in A_o and A_{or}.

Statistical Dependence. In the simplified model in Eq 2 above the amplitude of the signals and the "degree of dependence" varies over very large ranges. To have knowledge of this dependence is important if we want to take full advantage of the new possibilities in designing the signal processing. In this paper we can at most discuss how the dependence affects the structural properties of the total system.

In conventional terminology, for a fixed search bin, A_1 and A_2 are proportional to the square root of the target cross section. It may be reasonable to assume that each of the A:s fluctuates in the same way as a monostatic cross section - for most bistatic angles geometrical optics describes the scattering. The Swerling models are certainly useful.

The next question is; how dependent are the fluctuations in A_1 and A_2? The answer to this question depends on the situation at hand. Consider first two receivers. Using Eq 3, A_1 and A_2 have a common factor A_o. If, in a specific situation, there is a propagation fading between the transmitter and the target there will be an obvious dependence between A_1 and A_2.

Another, trivial, type of dependence is the change of phase, ψ in Eq 3, due to Doppler shift. For a moving target, A_1 and A_2 does not cluster randomly. If several frequencies are used simultaneously we get a much more complicated structure of the dependence.

One portion of the received signal is from the jammer. What is the structure of the dependence? Any use of adaptive arrays methods , ref (2), assumes (silently) that there is a strong dependence between the signals when properly delayed. In fact most of the adaptive array theory assumes that every one of the waves arrive at all antenna elements with the same amplitude. In this case we does not have the same extreme case. In summary; the signals S_1 and S_2 should not be regarded as two independent samples of the same search bin. Most of the dependence is caused by the jammer.

A SIGNAL MODEL FOR THE JAMMER

It is reasonable to assume that a jammer retransmits a portion of the radar signal $S_o(t)$ as well as some true noise n_J. Hence, with weights g_1 and $1-g_1$ we have the preliminary model

$$S_J(t) = g_1 n_J(t) + (1-g_1) S_o(t-t'_J) \qquad (4a)$$

for the signal transmitted by the jammer. Several questions arises like, how does this signal S_J affect the multi-static radar system? Which are the degrees of freedom?

One Receiver - the Bistatic Radar

To describe the relations between the different signals it is convenient to use vector notations. The vector S_o in Eq 1a is the starting point in the illustration in Fig 2. From the AO-property in Eq 1b it follows that, say, the vector

$$S_o\ (t_1 + 1) = (S_2, \ldots, S_M, 0)^*$$

is AO to S_o. The same is true for all other time translations. Hence, compare Fig 2, the retransmitted portion of S_J is AO to S_o except for <u>one</u> time index were S_J is aligned with S_o.

Consider the set of ellipses with one focus at the transmitter and the other foci at the receiver. Elliptic coordinates are natural. If the jammer only retransmits the signal then the radar is disturbed in those search bins that are located on the ellipse that is passing through the locus of the jammer. Also, if the jammer retransmits , say, periodically delayed versions of the signal the radar is jammed on the search bins that are located on periodically distributed ellipses.

In the discussions above it was assumed that the jammer retransmits portions of the signal. Thus, it is natural to transmit a less harmful misleading signal S_M towards the the jammer. The generalization of Eq 4a is thus

$$S_J = g_1 n_J + (1-g_1) \left[g_2 S_o + (1-g_2) S_M \right] \qquad (4b)$$

where the weight g_2 is determined by the sideloobs of the transmitting antenna and the power in the transmission of S_M. Figure 2 suggests that the misleading signal should be such that S_M is <u>always</u> AO to S_o. Thus, when S_M dominates over S_o it is more efficient for the jammer to transmit noise n_J than to repeating the received signal.

All properties discussed above are essentially the same in the monostatic case as in the bistatic case. One difference is that long pulses or continuous transmission is always feasible in the bistatic case.

Several Spatially Distributed Receivers

The properties discussed above were limited to the time domain - the only degree of freedom there is with one receiver. Just two receivers, instead of one, gives new possibilities in discrimination against jamming. As a starting point we consider a "pulse compression radar with several parallel receivers".

Let $X(t)$ be the pulse compression signal vector defined as

$$\begin{aligned} X_1(t) &= S_o{}^* \ S_1(t) \\ X_2(t) &= S_o{}^* \ S_2(t + t_2) \end{aligned} \qquad (5)$$

where S_1 is according to Eq 2 for that particular model. The first equation is just conventional pulse compression but expressed as a vector product. The time translation in the second equation alignes the transmitted signal in the receivers. Equation 2 gives

$$\begin{aligned} X_1(t) &= A_1\, Ma^2 + S_o{}^* \ W_1 + B_1\, S_o{}^* \ S_J(t) \\ X_2(t) &= A_2\, Ma^2 + S_o{}^* \ W_2 + B_2\, S_o{}^* \ S_J(t+t_2-t_J) \end{aligned} \qquad (6)$$

The parameters of the noise are

$$E\ S_o{}^* \ W_r = 0 \ ; \ \mathrm{Var}\ S_o{}^* \ W_r = Ma^2 \sigma_W^2$$

where σ_W is equal to the standard deviation of each component in the observation noise $W(t)$. It can be shown that

$$E(S_o{}^* \ W(t) \cdot S_o{}^* \ W(t+T)) \approx 0; \ T \neq 0.$$

Hence, the moving average sequence $S_o{}^* \ W(t)$ has the AO-property with respect to t.

The structure of the compressed signal X is illustrated in Fig 3. The scattered portion and the jamming portion of the signal are given separately. The time axis is equivalent to the search bins along the transmitted pulse - compare Fig 1. In the figure it is assumed that only one large target (A_1, A_2) is present. The noise $S_o{}^* \ W_r$ is represented by the circles.

The illustrated jamming signal is assumed to contain both a retransmitted signal S_o and the noise n_J. The signal S_o gives two large spikes (B_1, B_2) that are separated by a time interval t_J. A repeating jammer will give an entire sequence of similar pairs of spikes. The noise n_J also appears at the two receivers with the same difference t_J in time. As a consequence, $X_1(t)$ and $X_2(t + t_2 - t_J)$ are strongly correlated. In the case of "quasi-stationary narrow band signals" there is a well developed linear theory and technique, ref (2,3), for resolving the incoming signals. These results are of greatest value but, as will be discussed later, the purpose of the multistatic receivers is a somewhat different problem viz to probe a set of search bins. It should also be remembered that Fig. 3 is an oversimplification in several cases.

Previously, the transmission of a misleading signal S_M towards the jammer was suggested. Combining Eq. 4b and Eq. 6 and using the covariances found above we can compare the effectiveness of different jamming signals. For a power limited jammer the effectiveness is proportional according to;

$S_M \sim 1, \quad n_J \sim \sqrt{M}$ while

$S_o \sim M$ for every M:th time index.

At least with more than one receiver the last type of jamming seems to be simple to reduce by signal processing - compare Fig. 3.

Several Different Signals

A natural design concept is to probe each search bin with several different types of signals. By using signals containing several frequencies some of the structure of the target can be revealed. Also, jamming of all the signals for a long time is more diffucult in a properly designed system. A preliminary signal model is discussed below and some "surprising" observations are made.

Consider Eq. 1a for K transmitted signals S_{ok} where the last subindex k is for each signal; $1 \leq k \leq K$. Hence, we have the signal matrix

$$\tilde{S}_o = [S_{o1}, \ldots, S_{oK}]$$

where each column represents one signal. In the same way as in Eq. 5 X_{rk} is introduced as the pulse compression signal matrix. Each of the signals k is delayed τ_k at transmission so as not to overlap in time. Expressed as a formal equation we have

$$X_{rk}(t) = \tilde{S}_o^* \, S_r(t + t_r + \tau_k)$$

where index r denotes the r:th receiver. We refrain from writing down the equations for the scattering matrix A_{rk} and the contribution from several jammers - they are obvious.

The AO-properties illustrated in Fig. 2 are those of a vector space. Hence, the AO-property is still meaningful if S_o is replaced by $\tilde{S}_o$. Some conditions are needed on the S_{ok}:s to make the AO-property hold. One unnecessary strong such condition is that the S_{ok}:s do not overlap in time and are transmitted in different frequency bands. Thus, at least in principle, it is possible to make a direct pulse compression on all signals simultaneously. Such an exaggerated pulse compression is perhaps optimal under certain Gaussian assumptions. For more realistic signals containing outliers and jamming more robust signal processing methods are necessary.

It is obvious that the properties illustrated in Fig. 3 are also valid if each of the X_r:s is replaced by a vector of dimension K. With several signals we have redundant observations at each receiver. This redundancy can be used for robustifying the signal processing. As a suggestion, Fig. 4 illustrates how the median for three parallel signals removes outliers in one signal without smoothing out true pulses. If more signals are available, it is very unlikely that a jammer will disturbe a majority of the signals in the same way and at the same time. For a reasonable number of signals the median can be computed as fast as the the mean. There is a large experience and a well established theory for the median, ref (4). We just mention that if the classical assumption of additive Gaussian noise is modified to an assumption of additive doubly exponential noise the median gives the optimal estimate.

SUMMARY

This paper aims at a signal theoretical understanding of the basic limitations for multistatic radar systems. Emphasis is on different space, time and signal configurations. Models and methods for performance analysis are equally important. The main findings and remarks are summarized by the figures and the following points:

1. A vector description of the time discretized signals gives a convenient formalism for also treating several dimensions.
2. Pulse compression is described as an approximate orthogonality, "AO", property of the sliding signal vector.
3. Using the AO-property misleading signals can be dessigned. If a misleading signal is transmitted towards the jammer, the jammer is more efficient transmitting noise than repeating the received signal.
4. The median is suggested as one robustifying step in this type of processing - Fig. 4.

All properties listed above are preliminary. Recent conference literature has been consulted but is not listed - a short list of references is given in the regular paper (5). Finally, it must be kept in mind that multi-static radar theory is an essentially unexplored field. Detailed modelling and non-linear signal processing seems to be much more important, when the degrees of freedom/dimensionality increases.

REFERENCES

1. Holmes, J.K., 1982, Coherent Spread Spectrum Systems, Wiley.
2. Monzingo, R. & Miller, T., 1980, Introduction to adaptive arrays, Wiley.
3. Gething, P.J.D., 1978, Radio direction-finding and the resolution of multicomponent wave-fields, P. Peregrinus.
4. Justusson, B., 1981, Median Filtering: Statistical Properties, Huang, T.S. (editor); Topics in Applied Physics, vol. 43, Springer.
5. Schoenenberger, J.G. & Forrest, J.R., 1982, Principles of independent receivers for use with co-operative radar transmitters, The Radio and Electronic Engineer, Vol. 52, No. 2, pp. 93-101.

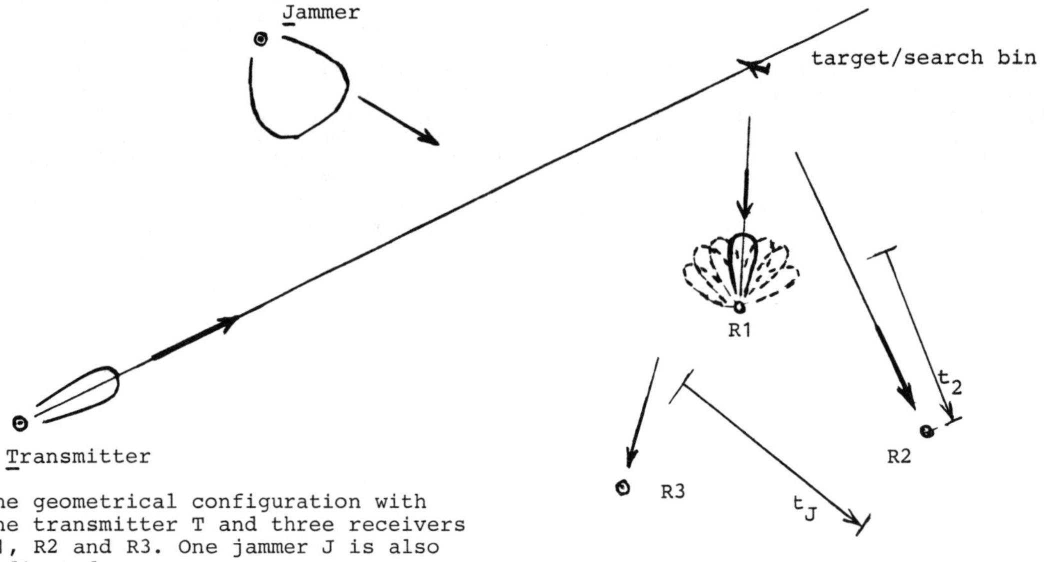

FIG.1 The geometrical configuration with one transmitter T and three receivers R1, R2 and R3. One jammer J is also indicated.

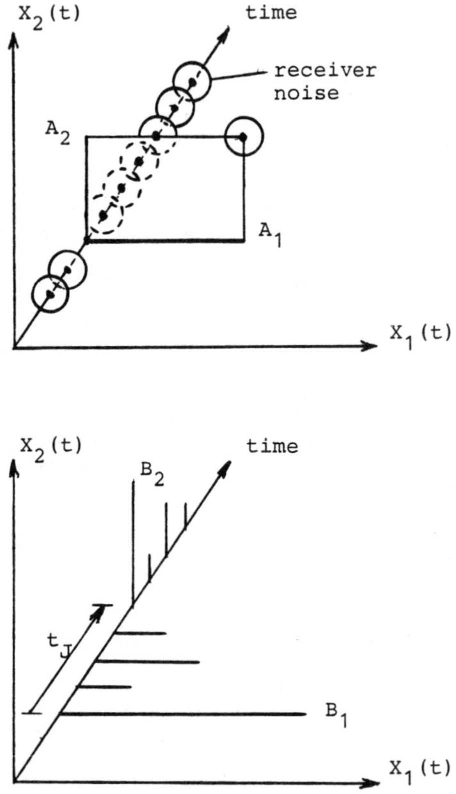

FIG.3 The two contributions to the pulse compression vector X. In the figure at the top one search bin contains a large target (A_1, A_2). The jamming is both a repeated signal of amplitude (B_1, B_2) and noise n_J. Note the time delay t_J between the two components.

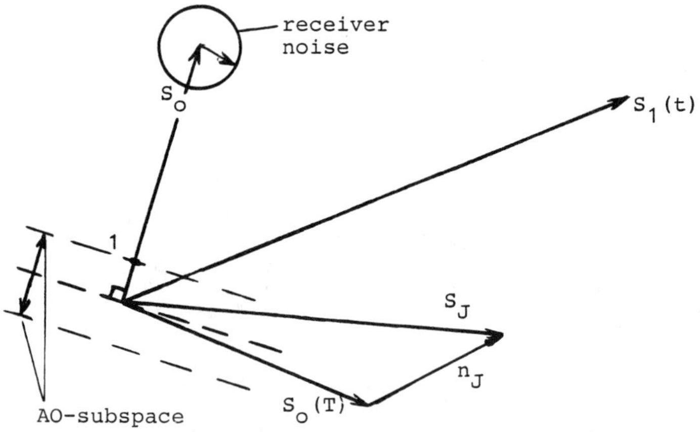

FIG.2 A vectorial illustration of the components S_o and S_J in the received signal $S_1(t)$. Due to the AO-property a time translated version of S_o is in the "AO-subspace". The signal S_J from a repeater jammer is also in this "AO-subspace" "most of the time".

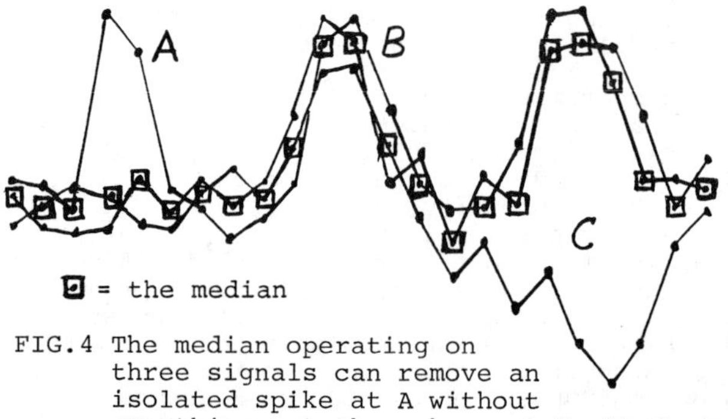

FIG.4 The median operating on three signals can remove an isolated spike at A without smoothing out the echoes at B. At C, two of the signals follows the echo while the third is disturbed.

MULTISTATIC TRACKING AND COMPARISON WITH NETTED MONOSTATIC SYSTEMS

A. Farina

Selenia S.p.A., I

INTRODUCTION

The main purpose of a tracking system is the estimation of target path parameters (position, velocity and, eventually, acceleration) and its evolution (track) in the near future from noise affected radar measurements (plots). The tracking system involves two basic functions: the plot-track association and the utilization of the associated plot for updating a recursive estimation of target parameters. The target tracking can be performed using plots coming from one or more radars having overlapping coverage (netted monostatic systems). Fuller details of this technique can be found in Farina and Pardini (1) and (2). Multistatic radar systems, having multiple transmitters, receivers and also monostatic sensors at different sites are similar in many respects to netted monostatic radar systems from the tracking problem point of view. Therefore, the usual tracking filters for netted monostatic systems can be extended to the multistatic tracking problem. Several advantages can be obtained over a single monostatic or bistatic system (e.g.: higher precision in the track estimation, reconfiguration capability in the case of failure of one or more sensors) at expense of a centralization of data processing which involves a spatial and time alignment of the sensors. Moreover, transmission lines for centralization of data and more processing resources than for monostatic case are also required.

In this paper, the multistatic tracking problem is considered and a general architecture for data processing is suggested together with a suitable selection of radar measurements, which increases the tracking filter performances. It is supposed in this study that the netted systems do not only overlap with a small intersection area but cover the same area totally. For reason of transparency the theory and application examples are outlined for a two-dimensional coverage area only, usually azimuthal, but corresponding results are also valid for three-dimensional considerations including an additional target direction measurement in elevation. Detailed calculations have been carried out for a bidimensional multistatic system with one transmitter and two receivers and a computer simulation has been performed in order to evaluate the tracking performance of this system. A comparison between multistatic tracking system and netted monostatic ones is also considered, the results are shown by means of constant accuracy contours of target position-finding.

ARCHITECTURE OF THE MULTISTATIC TRACKING SYSTEM

To determine the architecture of the tracking filter, mathematical model of target motion and equation of radar measurements should be established. The target path model is obtained with respect to a Cartesian co-ordinate reference system which is the only one that does not introduce ficticious accelerations; e.g. uniform straight line motion does not contain accelerations when projected on the rectangular axes x, y. The model of target path, as state equation, is derived when assuming that the target normally moves at constant velocity and turn or evasive manoeuvers may be considered as perturbation upon the straight lines. Therefore acceleration is a driving input for the state equation, which is usually linear. A simple way to model the unpredictable behaviour of acceleration is to consider a Gaussian process with zero mean, proper standard deviation and a correlation depending on the time duration of the manoeuver. Consider the state vector $\underline{s}^T_k$ (6 x 1) = $[x_k, \dot{x}_k, \ddot{x}_k, y_k, \dot{y}_k, \ddot{y}_k]$ where (x_k, y_k) are the target Cartesian co-ordinates sampled at time k. The state vector at k + 1 is given by the linear difference equation (2.2) of reference (2). The sampled radar measurements are polar in nature and can be the range ρ, the angle θ and the radial velocity $\dot{\rho}$ (if doppler filtering is available in the radar receiver) when monostatic radar is considered.

These measurements are affected by mutually independent additive white Gaussian noise n_ρ, n_θ, and $n_{\dot{\rho}}$ with zero mean and standard deviations σ_ρ, σ_θ and $\sigma_{\dot{\rho}}$, respectively. When multistatic sensor are emploied, if only one transmitter is considered, the following measurements can be available at the i-th radar receiver:

– the range sum : $\rho_i = \rho_T + \rho_{Ri}$

– the angle of the receiver beam: θ_i

– the radial velocity sum : $\dot{\rho}_i = \dot{\rho}_T + \dot{\rho}_{Ri}$

– the angle of the transmitted beam if it is of pencil beam type: θ_T

Figure 1 shows the set of measurements available in a multistatic system with a transmitter and two no-colocated receivers. Also the multistatic measurements are affected by additive noise having the same properties as that of the monostatic case.

In both monostatic and multistatic cases, the measurement equations can be set in the following vectorial form:

$$\underline{Z} = \underline{h}(\underline{s}) + \underline{N} \qquad (1)$$

where $\underline{z}^T$ (3 x 1) = $[\rho, \theta, \dot{\rho}]$ is the measurement vector while $\underline{h}$ (s) is a non-linear vector function of the state $\underline{s}_k$ and $\underline{N}^T$ (3 x 1) = $[n_\rho \; n_\theta \; n_{\dot{\rho}}]$ having zero mean and a diagonal variance matrix $\underline{R}$.

Stochastic filtering theory is a suitable mean to find the data processing structure. This data processing enables the best (in the least mean square sense) estimate $\hat{\underline{s}}_k$ of the target state by making use of the target model, of the measurement equation (1) and of the measurement set $\{\underline{z}_i; 0 \leq i \leq k\}$. Due to the non linearity of the measurement equation (1) with respect to the state $\underline{s}_k$, also the filtering algorithm is non linear and the optimal filter is hardly obtained.

An alternative approach refers to the block diagram of Fig. 2 which is mainly formed of a non linear zero memory filter performing a polar-to-Cartesian co-ordinate conversion of the measurements and a dynamic linear filter to reduce the measurement noise and to extrapolate the state estimation. As a matter of fact, from the polar measurements ρ, θ of the target position, the corresponding Cartesian coordinates x, y can be obtained. Moreover, if the measurements of two or more different radial velocity sums are available, it is possible to evaluate the corresponding rectangular velocity components $\dot{x}$, $\dot{y}$.

The noise sequences affecting the Cartesian measurements are still zero mean, but dependent on each other and having a new covariance matrix $\underline{P}$, different from $\underline{R}$, which depends on the position and velocity of target.

This measurement noise is then reduced using a linear Kalman filter. Let us come back to the Fig. 2 for a complete explanation of the data processing block diagram. Before the co-ordinate conversion, the measurements are aligned on a time scale to account for their time shift. This happens especially when radial velocities coming from monostatic sensors are used. A detailed explanation of the time alignment is given in (2). After the time alignment, the measurements, coming from the different sensors, are gathered in different sets from each of them it is possible to obtain, after the co-ordinate conversion, a complete measurement $\underline{Y}^T_j$ = $(x, y, \dot{x}, \dot{y})_j$ (j = 1, ... m).

As a matter of fact, the radar measurements coming from different sensors can be mixed in different ways. For sake of example, consider the bidimensional multistatic system shown in Fig. 1, where the following measurement are available:

– range-sums: $\rho_1 = \rho_T + \rho_{R1} \qquad \rho_2 = \rho_T + \rho_{R2}$

– angles: $\theta_T \quad \theta_1 \quad \theta_2$

– radial velocity sums: $\dot{\rho}_1 = \dot{\rho}_T + \dot{\rho}_{R1} \qquad \dot{\rho}_2 = \dot{\rho}_T + \dot{\rho}_{R2}$

they can be grouped in several ways to obtain the Cartesian measurements (x, y, $\dot{x}$, $\dot{y}$) of target after conversion. Examples of gathering of target position measurements to obtain (x, y) are: (ρ_1, θ_T);

$(\rho_1, \theta_1); (\rho_1, \rho_2$ and an angle); (θ_1, θ_2).

Among these it is possible to select some set of independent measurements such as (ρ_1, θ_1) and (ρ_2, θ_2) from which to obtain two sets of Cartesian co-ordinate $(x, y)_1$ and $(x, y)_2$ which will be compressed in a single set (x^*, y^*). The radial velocity measurement sums are, on the other hand, grouped in a single way in this example. Finally, the compression of data allows the combination of the measurements $(x, y, \dot{x}, \dot{y})_j$ (j = 1,2 .... m) into a single equivalent one $(x^*, y^*, \dot{x}^*, \dot{y}^*)$ which is processed through a linear Kalman filter. The data compression will be presented in the next section together with an off-line method to select data, thus reducing the amount of the on-line computation to perform co-ordinate transformation of several sets of measurements and their compression in a single datum.

DATA COMPRESSION AND MEASUREMENT SELECTION

To process the several data coming from the radar sensors two methods are available. The first has been described in the previous section and refers to the selection of independent sets of polar measurements which are transformed in Cartesian ones and after being compressed into a single equivalent measurement to be processed through a linear filter. Let us indicate with $\underline{Y}_j$ (j = 1,2 .... m) the Cartesian measurements to be compressed. They are affected by measurement noise having zero mean and covariance matrices $\underline{P}_j$ (j = 1,2 .... m). If "m" is two, the equation giving the compressed data $\underline{Y}^*$ is:

$$\underline{Y}^* = (\underline{P}_1^{-1} + \underline{P}_2^{-1})^{-1} \underline{P}_1^{-1} \underline{Y}_1 + (\underline{P}_1^{-1} + \underline{P}_2^{-1})^{-1} \underline{P}_2^{-1} \underline{Y}_2 \quad (2)$$

having the following covariance matrix:

$$\underline{P}^* = (\underline{P}_1^{-1} + \underline{P}_2^{-1})^{-1} \underline{P}_1^{-1} (\underline{P}_1^{-1} + \underline{P}_2^{-1})^{-1} + (\underline{P}_1^{-1} + \underline{P}_2^{-1})^{-1} \underline{P}_2^{-1} (\underline{P}_1^{-1} + \underline{P}_2^{-1})^{-1} \quad (3)$$

The equivalent measurement $\underline{Y}^*$, which is a linear combination of the known measurements $\underline{Y}_1$, and $\underline{Y}_2$, is obtained by means of the least-mean-square criterion. When "m" is greater than two, equation (2) is applied to $\underline{Y}^*$ and the third measurement $\underline{Y}_3$ thus obtaining a new compressed data, and so on.

To avoid this lot of on-line computation the following off-line data selection is suggested. Consider again the sets of Cartesian measurements $\underline{Y}_j$ (j = 1,2 .... m) and their covariance matrices $\underline{P}_j$. In each point of the controlled air-space it is possible to compare the variable $(\det \underline{P}_j)^{+1/2}$ (j = 1,2 .... m) and to select the measurement having the smallest value. The controlled airspace can therefore be divided into different contiguous regions in which different sets of measurements should be used. For example, consider again the bidimensional multistatic system shown in Fig. 1. Two sets of position measurements are compared (ρ_1, θ_T) and (θ_1, θ_2). From the first set, the following target Cartesian co-ordinates are obtained:

$$x = (a - 0.5\,\rho_1 \cos\theta_T) / (\frac{2a}{\rho_1} \cos\theta_T - 1)$$

$$y = (\frac{2a^2}{\rho_1} - 0.5\,\rho_1) \operatorname{sen}\theta_T / (\frac{2a}{\rho_1} \cos\theta_T - 1) \quad (4)$$

with exception of the set of the plane in which the two denominators are zero.

Let ρ_1 and θ_T be affected by mutually independent white Gaussian noises with zero means and variances σ_ρ^2 and σ_θ^2 respectively. If the noise is not too high, compared with the measurements, the noise sequences affecting x, y are still zero mean, white, but dependent on each other. The variances σ_x^2, σ_y^2 and covariance σ_{xy} can be evaluated by differentiating x and y of equations (4) with respect to ρ_1 and θ_T and then assuming that the following identitied hold:

$$\sigma_x^2 = E\{dx^2\},\ \sigma_y^2 = E\{dy^2\},\ \sigma_{xy} = E\{dx\,dy\} \quad (5)$$

Detailed evaluation of (5) are shown by Farina (3) p. 21. In a similar way, from the second set of measurements, the following target Cartesian co-ordinates are obtained:

$$x = (-a\,\mathrm{tg}\,\theta_1 + b\,\mathrm{tg}\,\theta_2 - c) / (\mathrm{tg}\,\theta_2 - \mathrm{tg}\,\theta_1)$$

$$y = [(b-a)\,\mathrm{tg}\,\theta_1\,\mathrm{tg}\,\theta_2 - c\,\mathrm{tg}\,\theta_1] / (\mathrm{tg}\,\theta_2 - \mathrm{tg}\,\theta_1) \quad (6)$$

and variances and covariance are evaluated.

The variable to be used to compare the two sets of measurements is the square root of the determinant of covariance matrix $\underline{P}$ of measurement errors:

$$\{\det \underline{P}\}^{1/2} = \{\sigma_x^2\,\sigma_y^2 - \sigma_{xy}^2\}^{1/2} \quad (7)$$

This variable is proportional to the area of an ellipse which represents a locus with measurement error having probability density of constant value. The smaller the area, the better is the accuracy of the measurements. See reference Farina and Hanle (4) to have further details on the meaning of this variable. The comparison is shown in Fig. 3 in terms of contours of constant value of the following ratio $\gamma = \{|\underline{P}(\rho_1, \theta_T)| / |\underline{P}(\theta_1, \theta_2)|\}$. It has been assumed that the three angular measurements θ_1, θ_2 and θ_T have the same standard deviation σ_θ = 0.003 rad while σ_ρ is equal to 100 m. In the figure, the dashed regions represent the areas in which the first set (ρ_1, θ_T) of measurement allows a better accuracy to be obtained with respect to the second set (θ_1, θ_2). Further curves of this type can be found in (3). The measurement selection can be made off-line and the corresponding regions can be stored in the tracking computer to select the sensors to be aimed at certain targets. In a more general case the previous data selection should take into account also the radial velocity measurements when they are available.

SIMULATED EXAMPLE OF MULTISTATIC TRACKING

Consider again the bidimensional multistatic system of Fig. 1 having one transmitter and two receivers. Tracking filters are developed for the following sets of data:

$$(1)\ \{\rho_1, \theta_T, \dot{\rho}_1, \dot{\rho}_2\} \quad (8)$$

$$(2)\ \{\theta_1, \theta_2, \dot{\rho}_1, \dot{\rho}_2\} \quad (9)$$

The target Cartesian co-ordinates x and y and the corresponding variances and covariances of measurement errors have been already obtained from the polar co-ordinates of position (ρ_1, θ_T) and (θ_1, θ_2). Similar computations can be carried out in order to obtain the Cartesian components $(\dot{x}, \dot{y})$ of target velocity from the polar measurements $(\dot{\rho}_1, \dot{\rho}_2)$.

The following equation holds:

$$\begin{bmatrix} (\cos\theta_T + \cos\theta_1) & (\operatorname{sen}\theta_T + \operatorname{sen}\theta_1) \\ (\cos\theta_T + \cos\theta_2) & (\operatorname{sen}\theta_T + \operatorname{sen}\theta_2) \end{bmatrix} \begin{bmatrix} \dot{x} \\ \dot{y} \end{bmatrix} = \begin{bmatrix} \dot{\rho}_1 \\ \dot{\rho}_2 \end{bmatrix} \quad (10)$$

If the coefficient matrix is not singular, the Cartesian co-ordinates $(\dot{x}, \dot{y})$ of target speed is easily obtained from the previous equation. The variances $\sigma_{\dot{x}}^2$, $\sigma_{\dot{y}}^2$ and covariance $\sigma_{\dot{x}\dot{y}}$ can be evaluated applying the same method described before. As a consequence, it is possible to obtain a measurement $\underline{Y}^T{}_k$ (4 x 1) = $[x_k\ \dot{x}_k\ y_k\ \dot{y}_k]$ linearly dependent on the target state $\underline{s}_k$:

$$\underline{Y}_k = \underline{H}\,\underline{s}_k + \underline{\xi}_k \quad (11)$$

$\underline{H}$ is a matrix (4,6) full of zeroes except the elements (1,1), (2,2), (3,4) and (4,5) which are equal to the unity. Finally, $\underline{\xi}_k$ is the measurement noise having zero mean and covariance matrix $\underline{P}_k$ expressed in terms of $\sigma_x^2, \sigma_y^2, \sigma_{xy}, \sigma_{x\dot{x}}$ and so on. Since the state equation and the measurement equation (11) are linearly dependent on the state $\underline{s}_k$, the Kalman filter can be used to obtain the estimate $\hat{\underline{s}}_k$.

A digital simulation has been performed in order to evaluate the performance of the two tracking filters which process the two sets of data (8) and (9) respectively. The results refer to two typical target paths shown in Fig. 3 and 4. The first is a straight line path with constant velocity, while the second path is formed of two straight lines separated by a centripetal acceleration portion. The path are sampled with a period Δ = 10 s. During the straight line portions of paths, the filtering errors are unbiased but with standard deviations depending on the measurement set processed and the target position. Fig. 5, which refers to the first path, shows the filtering error standard deviations of the target co-ordinate x. It is shown clearly, through the intersections of the two curves, the effects of different accuracy of the processed data. The simulation results relevant to the second path are shown in Figs. 6 and 7. The filter was run with and without radial velocity measurements ($\dot{\rho}_1$, $\dot{\rho}_2$). The filtering errors are biased during the accelerated portion of path, this bias can be reduced through the processing of the radial velocity measurements. Further simulation results are shown in reference (3).

COMPARISON BETWEEN MULTISTATIC AND NETTED MONOSTATIC SYSTEMS

This section affords briefly the problem of comparison between multistatic and netted monostatic systems from the tracking point of view. This comparison should be made in terms of tracking filters performance. But in this case it would be difficult to obtain general results, as a consequence the two netted systems will be compared in terms of data rates available to obtain new measurements and in terms of target position measurements accuracy. The data rate values establish the maximum bias of filtering errors for strongly manoeuvring targets. It can be noted that the non-synchronous sampling by netted monostatic system can lead to a reduced track quality, assuming the same mean value of data rate. On the other hand, the target position measurement accuracy determines the standard deviation of filtering errors during straight line portion of target path. This latter problem has been deeply analyzed in reference (4), therefore in this paper only some consideration will be included.

The comparison of the target position accuracy is performed in terms of Cartesian co-ordinates that are commonly used in the tracking sysyems. The comparison results are shown by means of constant accuracy contours. It is assumed that the monostatic sensor sites are coincident with the receiver sites of the corresponding multistatic systems.
The relative position of sites plays a key role in the comparison together with the accuracy of polar measurements which have been assumed equal for monostatic and bistatic sensors.
With reference to the usual multistatic system of Fig. 1, it is assumed that the following measurements of target position are available: 1st receiver (ρ_1 , θ_1); 2nd receiver (ρ_2 , θ_2). From these two couples of measurements it is possible to obtain the following two sets of Cartesian measurement of target position:

1st receiver: $\underline{Y}^T_{1,Mul} = (x, y)_{1,Mul}$

2nd receiver: $\underline{Y}^T_{2,Mul} = (x, y)_{2,Mul}$ (12)

The covariance matrices of these measurement are respectively $\underline{P}_{1,Mul}$ and $\underline{P}_{2,Mul}$. The measurements (12) can be compressed in an equivalent one $\underline{Y}^*_{N.Mul}$ having a covariance matrix $\underline{P}^*_{N.Mul}$ which are expressed through the equations (2) and (3). With reference to the same Fig. 1, the netted monostatic radar system consists of two sensors located at the receiver sites R1 and R2, they measure the following polar co-ordinates of target position: 1st monostatic ($\rho_{1,Mon} = 2\rho_{R1}$, θ_1); 2 nd monostatic ($\rho_{2,Mon} = 2\rho_{R2}$, θ_2). From these two couples of measuremens it is possible to obtain two sets of Cartesian measurement of target position:

1st monostatic: $\underline{Y}^T_{1,Mon} = (x, y)_{1,Mon}$

2nd monostatic: $\underline{Y}^T_{2,Mon} = (x, y)_{2,Mon}$ (13)

Each of these measurements $\underline{Y}_{1,Mon}$ and $\underline{Y}_{2,Mon}$ have the errors covariance matrices $\underline{P}_{1,Mon}$ and $\underline{P}_{2,Mon}$.
The two previous sets of Cartesian co-ordinate can be combined in an equivalent single measurement $\underline{Y}^*_{N.Mon}$ having covariance matrix $\underline{P}^*_{N.Mon}$ as done for the multistatic case.
The problem is now reduced to compare the square root of the determinants of the covariance matrices representing the error areas of the multistatic and netted monostatic systems, as done by means of the parameter γ .
It has been assumed that the standard deviations σ_ρ and σ_θ of monostatic and multistatic measurements are equal and independent of the target range. Constant accuracy contours of position finding accuracy are shown in Fig. 8 where it can be noted that the netted monostatic system has a slight advantage over the multistatic radar. However, in reference (4) it is shown that the assumption of range-dependent standard deviations σ_ρ and σ_θ allows to obtain more significant results.

CONCLUSIONS

The tracking function of the multistatic radar systems has been analyzed and a general data processing architecture has been suggested. Tracking performance have been obtained for a particular multistatic system. Multistatic radars compared to netted monostatic offer similar performance in terms of tracking accuracy.

REFERENCES

1 Farina, A., and Pardini, S., 1980, "Survey of radar data-processing techniques in air-traffic-control and surveillance systems".
Proc. IEE, 127, Pt. F, 190-204.

2 Farina, A., and Pardini, S., 1979, "Multiradar tracking system using radial velocity measurements".
Trans. IEEE on Aerospace and Electronic Systems, 15,555-563.

3 Farina, A., 1981, "Introduction to multistatic tracking function". Selenia Technical Report RT-81083 In.

4 Farina, A., and Hanle, E., 1982, "Comparison of target position accuracy between netted monostatic and bistatic radar systems".
Submitted for publication in Trans. IEEE on Aerospace and Electronic Systems.

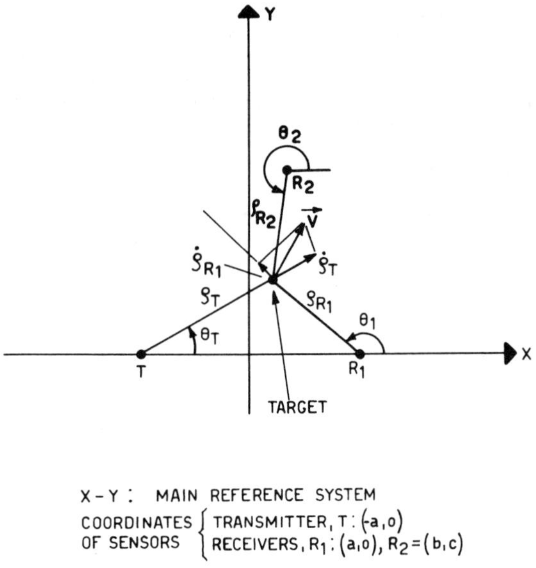

Fig. 1 — Multistatic radar measurements

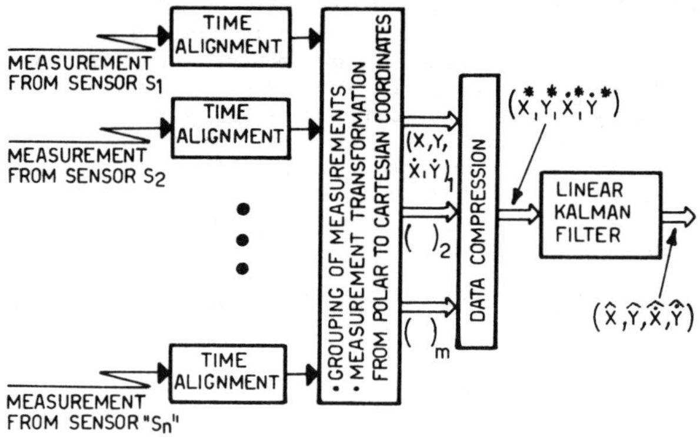

Fig. 2 — Multistatic tracking filter architecture

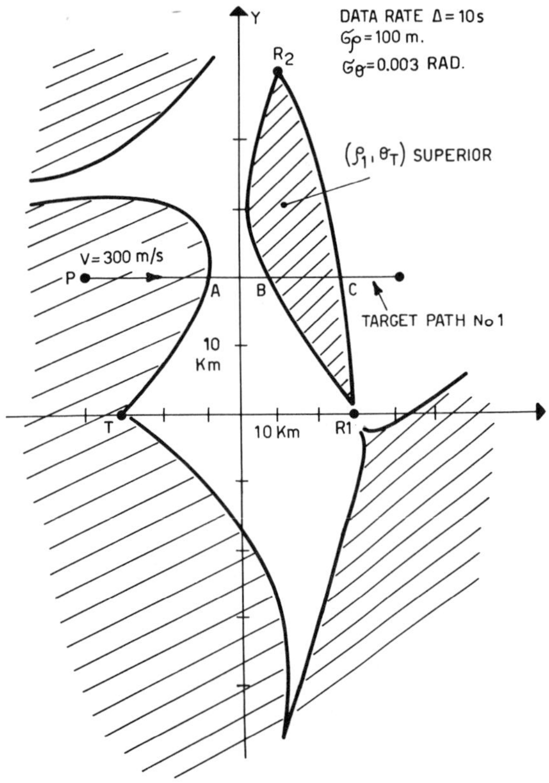

Fig. 3 — Accuracy of two sets (ρ_1, θ_T) and (θ_1, θ_2) of multistatic measurements

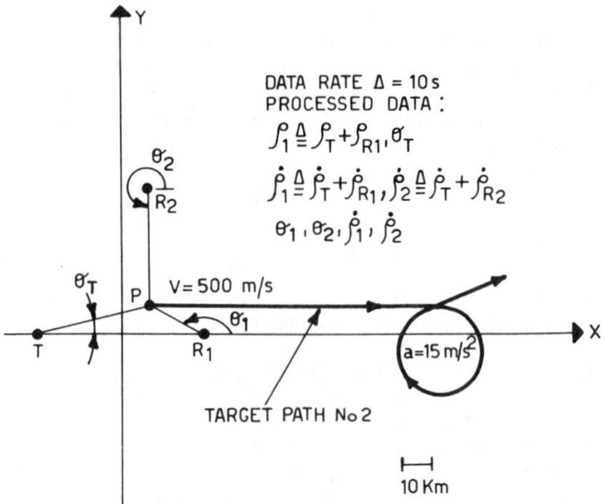

Fig. 4 — Simulated target path

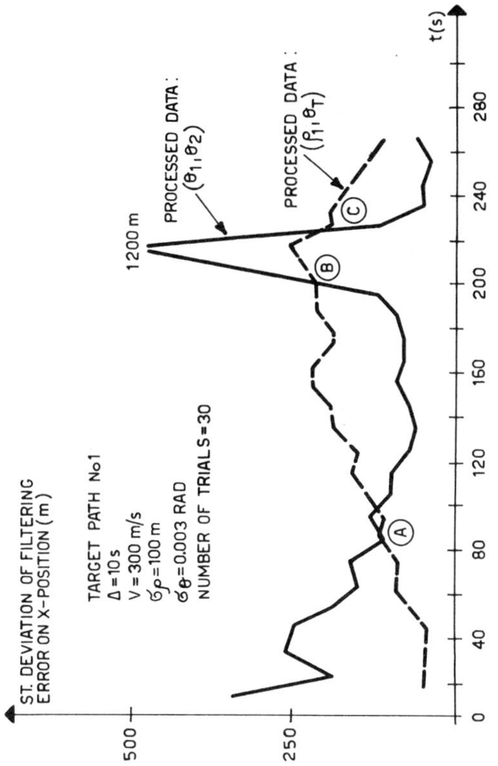

Fig. 5 — Standard deviation of filtering error on X-position (Path No. 1)

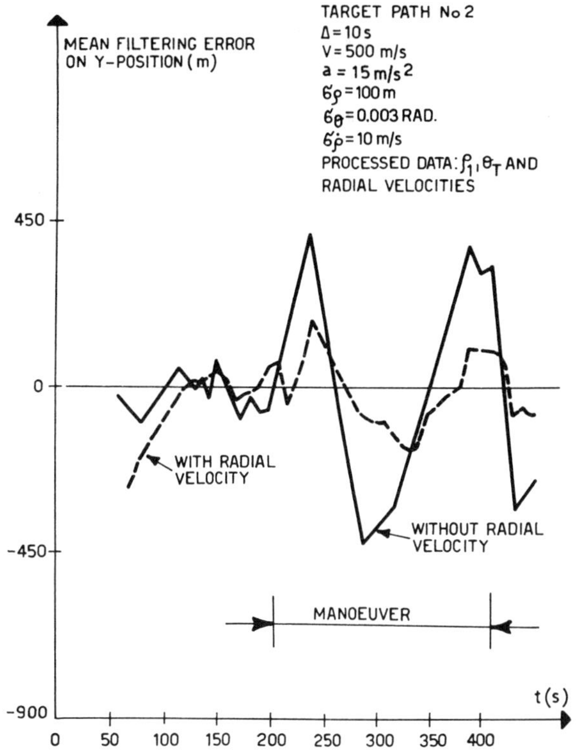

Fig. 6 — Mean filtering error on Y-position (Path No. 2)

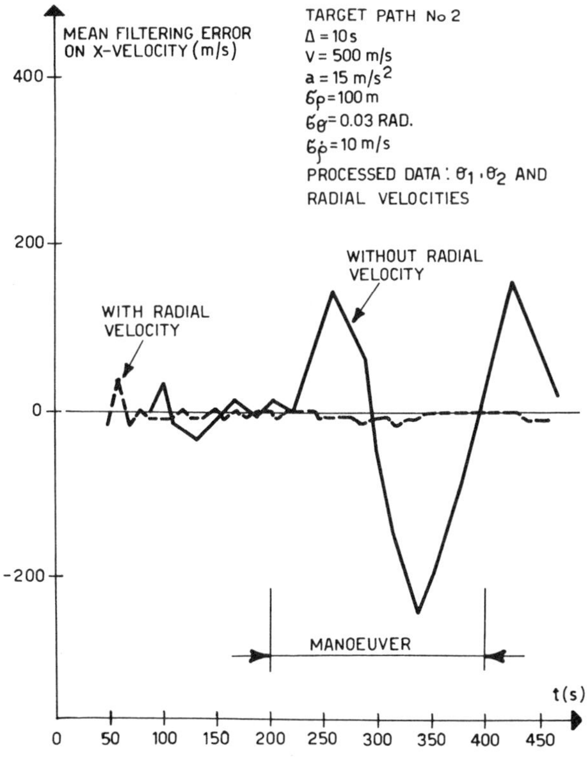

Fig. 7 — Mean filtering error on X-velocity (Path No. 2)

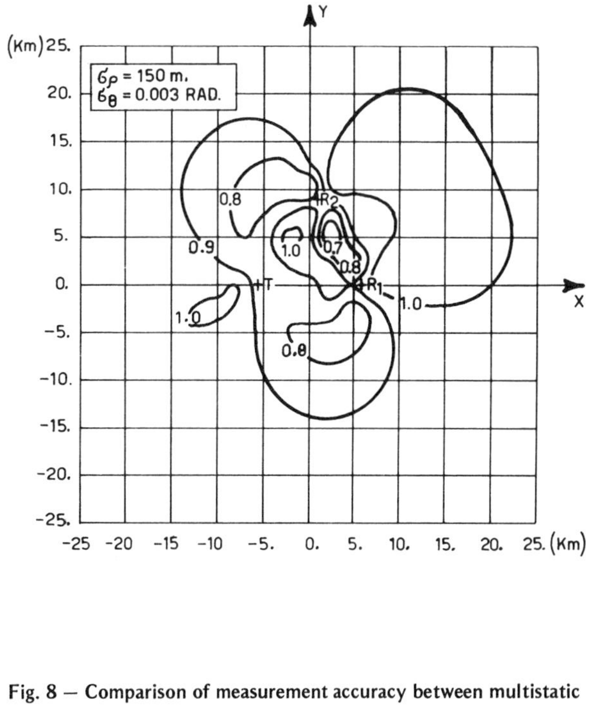

Fig. 8 — Comparison of measurement accuracy between multistatic and netted monostatic systems

BISTATIC SEA CLUTTER RETURN NEAR GRAZING INCIDENCE

George W. Ewell and Stephen P. Zehner

Engineering Experiment Station, Georgia Institute of Technology, USA

INTRODUCTION

The recent interest in bistatic radar, multistatic radar and electronic support measures (ESM) systems has led to a need for information concerning the bistatic reflectivity of clutter viewed by such systems. Unfortunately, little information is available describing land or sea clutter for a bistatic radar system [1], particularly when operating near grazing incidence. Hence, the development of realistic theoretical models describing bistatic reflectivity has been severely handicapped by the lack of experimental data. This paper presents a set of experimental data describing simultaneous bistatic radar cross section (BRCS) and monostatic radar cross section (MRCS) measurements of sea clutter taken with a moderately high resolution X-band radar system operating near grazing incidence for both horizontal and vertical polarizations over a range of bistatic angles from 23 to 85 degrees. These data are then used to examine the applicability of some possible bistatic reflectivity clutter models.

EQUIPMENT, CALIBRATION, AND EXPERIMENTAL PROCEDURES

The transmitter (and monostatic receiver) for this experiment was the X-band radar located at the Georgia Tech Field Site, Boca Raton, Florida. The bistatic receiver was located immediately north of the Boca Raton inlet, which is a distance of 3.5 km (1.9 nmi) south of the field site. Major system parameters are summarized in Table I, and a photograph of the receiving site is given as Figure 1. A sample of the transmitted pulse was radiated through a separate antenna toward the receiving site. This signal was received through a dedicated antenna (the small dish visible in Figure 1) and receiver, and used to establish a timing reference between the sites. Range-gated sample-and-hold circuits were used to sample both monostatic (at the transmitter site) and the bistatic (at the receiver site) logarithmic video return signals. The transmitting antenna was normally fixed at a constant bearing of 90 or 150 degrees, and the direction of the receiving antenna and range delay were varied to change the bistatic angle. Bistatic data were taken with the range gate sampling the peak of the signal formed by the intersection of the transmitting and receiving antenna beams, and monostatic data were acquired using a sampler positioned to sample the monostatic return corresponding to this intersection point. The resulting stretched signals were recorded on magnetic tape, along with voice annotation and time code, using instrumentation FM tape recorders.

Calibration of these recorded data was accomplished using several techniques. Known RF test signals were injected into the receiver front ends and varied in known increments over the dynamic ranges of the receivers while the resultant video signal was sampled and recorded on tape. This information was used to establish the transfer function of the receiver, data acquisition, and recording system at each site. Next, the received power from standard targets was used to establish levels corresponding to signals received from targets of known radar cross section. A rotating diplane (dihedral corner reflector) was used to calibrate both the monostatic and the bistatic systems, and a trihedral corner reflector was used as an additional aid in calibrating the monostatic system. The rotating diplane was oriented such that the seam was horizontal and the axis of rotation was vertical. This permitted observation of the specular cross section, which is approximately the monostatic cross section reduced by the square of the cosine of half

TABLE I - Characteristics of monostatic and bistatic systems

Parameter	Monostatic System	Bistatic Receiver
Frequency	9.38 GHz	9.38 GHz
Peak Power	250 kW	---
Pulse Width	0.2 µs	---
PRF	1000 pps	---
Antenna	Cut Paraboloid (1" Az x 3° El)	Paraboloid (1.5" x 1.5°)
Polarizations Used	HH or VV	HH or VV
Antenna Gain	38 dB	37 dB
IF Center Frequency	60 MHz	60 MHz
IF Bandwidth	5 MHz	20 MHz
IF Response	Logarithmic	Logarithmic
Dynamic Range	80 dB	75 dB
Antenna Height Above Mean Water Surface	22.9 m (75 ft)	8.2 m (27 ft)
Site Separation	3.5 km (1.9 nmi)	

the bistatic angle. The reduction from monostatic cross section was small, since the bistatic angle was kept small for the calibration data runs.

The next step in calibration involved using the radar equation to determine expected values of received power from these standard targets. A useful form of the bistatic radar equation is

$$P_r = \frac{P_t G_t G_r \sigma \lambda^2 F_t^2 F_r^2}{(4\pi)^3 R_1^2 R_2^2} \qquad (1)$$

where

- P_r = received power
- P_t = transmitted power
- G_t = transmitting antenna gain
- G_r = receiving antenna gain
- σ = target radar cross section
- λ = wavelength
- F_t = pattern propagation factor for transmitting antenna-to-target path
- F_r = pattern propagation factor for receiving antenna-to-target path
- R_1 = range from transmitter to target
- R_2 = range from target to receiver.

The propagation factors F are included to account for the effects of multipath returns. When the angle between the direct and reflected rays is small, the function F may be written as

$$F = |1 + \rho D \exp(-j\alpha)| \qquad (2)$$

where

- ρ = reflection coefficient of the ocean surface
- D = divergence factor to account for the curved reflecting earth
- α = phase differences between the direct and reflected waves at the target.

In the monostatic case $R_1 = R_2 = R$ and $F_t = F_r = F$, and Equation (1) reduces to which is the familiar monostatic radar equation:

$$P_r = \frac{P_t G_t G_r \sigma \lambda^2 F^4}{(4\pi)^3 R^4} \qquad (3)$$

Comparisons of measured and calculated received powers from the standard targets were used as measures of the quality of the radar cross section data. The expected value of received power was calculated using Equations (1) and (3), including the propagation factor given in Equation (2). Radar cross sections of the standard reflectors were calculated using measured physical dimensions; these radar cross sections were used with measured values of gain, transmitted power, and range to calculate expected values of received power.

The corner reflectors were elevated at a height which would place them near the peak of the interference lobe. Since accurate alignment of a corner reflector may be difficult from a small boat, values of received power equal to the largest value received (the value corresponding to the 0.99 point on the cumulative probability distribution) were associated with the received power from the reflectors. This procedure was justified due to the expectation that the +18 dBsm corner reflector would dominate the return from the small boat supporting the corner reflector (this was verified experimentally); then the maximum return would occur when the reflector was aligned with the receiver and when the reflector was also at the peak of the interference lobe.

Comparisons of the calculated and measured returns from numerous calibration runs conducted throughout the experiment showed excellent agreement between measured and calculated values; worst-case errors were less than ±2 dB, with ±1 dB differences being more typical. This excellent agreement between measured and calculated received powers from standard targets gives a high degree of confidence in data bracketed by such calibrations.

ANALYSIS OF DATA

Data describing monostatic and bistatic clutter returns were returned to Georgia Tech and used, along with calibration information, to generate distributions of received power at each of the receivers [2]. Since the multipath propagation factor is difficult to interpret in the case of distributed clutter returns, the free space radar equations

$$P_r = \frac{P_t G^2 \lambda^2 \sigma}{(4\pi)^3 R_1^4} \quad \text{(monostatic)} \qquad (4)$$

and

$$P_r = \frac{P_t G_t G_r \lambda^2 \sigma}{(4\pi)^3 R_1^2 R_2^2} \quad \text{(bistatic)} \qquad (5)$$

were used to convert received power to radar cross section.

Since illuminated cell sizes varied widely, the use of the normalized radar cross section per unit area, $\sigma°$, was selected as the measure of clutter reflectivity and was calculated by dividing the clutter cross section by the illuminated area. For the monostatic case, the calculation of effective illuminated area is rather straightforward and is given by

$$kR_1\left(\frac{c\tau}{2}\right)(BW) \qquad (6)$$

where

- k = a constant close to unity, dependent upon beamshape [3]
- BW = 3 dB antenna beamwidth, expressed in radians
- c = velocity of light
- τ = width of the transmitted pulse

for incidence angles near grazing.

Calculation of the effective illuminated area is more complex for the bistatic case, since the illuminated area is determined by the intersection of the two antenna beams and the portions of the constant delay ellipses illuminated by the transmitted pulse. Calculation of the illuminated area was accomplished using a digital computer, and the assumed geometry is shown in Figure 2. In Figure 2, the nominal intersection of the beams is indicated by the "aimpoint" and a clutter cell under consideration is indicated by the small square. The clutter return is obtained by integrating all of these incremental clutter areas which contribute to the return. It is assumed that the beamwidths of the antennas are BW_1 and BW_2. The delay at the bistatic receiver relative to the timing pulse received from the transmitter, δ, is

$$\delta = \frac{1}{c} (R_1 + R_2 - R_T) \tag{7}$$

and the coordinates of the aimpoint in rectangular coordinates are

$$(R_1 \cos\theta_1 + \frac{R_T}{2}, R_1 \sin\theta_1) \tag{8}$$

Then one may write

$$\theta_2 = \tan^{-1}(R_1 \sin\theta_1/R_1 \cos\theta_1 + R_T) \tag{9}$$

and

$$R_2 = (R_1^2 + R_T^2 + 2R_1R_T \cos\theta_1)^{1/2}. \tag{10}$$

If a rectangular pulse of width τ is transmitted, clutter elements will make a contribution if

$$c(\delta - \tau) < r_1 + r_2 - R_T < c\delta. \tag{11}$$

Next, by defining a factor Δ given by

$$\Delta = \begin{matrix} 1, \text{ if inequality (11) is satisfied} \\ 0, \text{ otherwise} \end{matrix}$$

then, from the previous discussion,

$$\text{area} = \sigma/\sigma°.$$

The area may be calculated by

$$\text{area} = R_1^2R_2^2 \int \Delta \frac{f_t(\alpha-\theta_1)}{r_1^2} \frac{f_r^2(\beta-\theta_2)}{r_2^2} dA \tag{12}$$

where

$$r_1 = ((x - R_T/2)^2 + y^2)^{1/2}$$

$$r_2 = ((x - R_T/2)^2 + y^2)^{1/2}$$

$$\alpha = \tan^{-1}(y/(x - R_T/2))$$

$$\beta = \tan^{-1}(y/(x + R_T/2))$$

and the f_t and f_r are the antenna pattern factors given by

$$f(\alpha) = (\sin(u))/u \tag{13}$$

where

$$u = 2.783\ \alpha/BW.$$

The integration was performed numerically starting at the aimpoint and increasing until either Δ went to zero or until u for both antennas was greater than π. The resulting values of illuminated area were used to convert values of bistatic cross section to bistatic radar cross section per unit area.

RESULTS

Data are presented as the ratio of median values, σ° monostatic/σ° bistatic, as a function of bistatic angle. Since the results are a function of look direction, environment, and polarization, results are separated by look direction and by day. Figures 3-5 show the ratio of median bistatic to monostatic σ° expressed in dB as a function of bistatic angle, indicating a rather abrupt decrease in BRCS from MRCS, which generally continues to decrease with increasing bistatic angle. The weather conditions during these measurements were typical of those normally encountered in southern Florida during the summer months; i.e., a fairly steady sea from the east or southeast with occasional local frontal activity and associated showers. Visual estimates of sea conditions were 1.2 - 1.8 m (4-6 ft) significant waveheight on August 3 and 4 and 0.9 m (3 ft) on August 5. The seas were somewhat disorganized on August 4, which may contribute to the relatively broad scatter of the data in Figure 5.

Amplitude distributions were for the most part nearly lognormal in shape. The lognormal standard deviation (LSD) was calculated as the difference of the cumulative distribution values in dB at the 50 and 84 percent points. Figure 6 is a scatter diagram of monostatic LSD versus bistatic LSD. There appears to be a tendency for the monostatic return to have a larger LSD than the bistatic, and for horizontal to have larger LSD than vertical polarization. Figure 6 is for a number of days and bistatic angles lumped together, since significant daily differences or dependence upon bistatic angle could not be identified.

BISTATIC CLUTTER MODELS

Perhaps the simplest bistatic clutter model might be one with limited quasi-specular monostatic scattering, with diffuse scattering largely independent of bistatic angle. The continued reduction in BRCS with increasing bistatic angle shown in Figures 3-5 does not support such a simplistic model.

Another relatively straightforward approach to modeling bistatic sea clutter might employ existing sea clutter models which are frequency-dependent [4,5] along with the monostatic-bistatic equivalence theorem [6]. Unfortunately, attempts to use such an approach (which must attempt to isolate multipath propagation factors and account for them separately) for bistatic angles of 30, 45, and 60 degrees predicted BRCS values substantially larger than observed during these experiments.

A third, and more complex approach is to assume that scattering arises from a number of locally correlated scattering areas (similar to the facets of some sea clutter theories [7]) and to allow these facets to have some distribution of slope in elevation and azimuth. Calculations were carried out for octagonal facets having normal distribution of slope about selectable median values with rms slope taken from Macdonald [8] and facet sizes from Schooley [9]. The result of one such set of calculations is shown in Figure 7 and closely approximates many features of the experimental measurements. By appropriate selection of facet size and slope distributions it would appear to be possible to closely reproduce experimental data of the type presented in

Figures 3-5, but confirmation of this modeling approach would require careful measurement of not only bistatic radar cross section, but also wave angular spectra and slope and scatterer size distributions as well. Such a set of measurements might yield useful insight into the mechanisms of low-angle monostatic sea clutter return in addition to providing an explanation for the observed nature of bistatic scattering from the sea surface.

REFERENCES

1. Pidegon, V.W., 1966, "Bistatic Cross Section of the Sea", IEEE Trans. Ant. and Prop., 3, pp. 405-408.
2. Ewell, G.W. and Zehner, S.P., 1980, "Bistatic Radar Cross Section of Ship Targets", IEEE Jrnl. Oceanic Eng., 4, pp. 211-215.
3. Barton, D.K., 1965, Radar System Analysis, Prentice-Hall, Inc., Englewood Cliffs, N.J., USA, pg. 96.
4. Ewell, G.W., Horst, M.M., and Tuley, M.T., 1979, "Predicting the Performance of Low-Angle Microwave Search Radars", Oceans 79 Conf. Rec., pp. 373-378.
5. Sittrop, H., 1977, "On the Sea Clutter Dependency on Windspeed", Radar '77, pp. 110-114.
6. Kell, R.E., 1965, "On the Derivation of Bistatic RCS from Monostatic Measurements", Proc. IEEE, 53, pp. 983-988.
7. Long, M.W., 1975, Radar Reflectivity of Land and Sea, Lexington Books, Lexington, Mass, USA, pp. 79-84.
8. Macdonald, F.C., 1956, "The Correlation of Radar Sea Clutter on Vertical and Horizontal Polarization with Wave Height and Slope", IRE Nat. Conv. Rec., pp. 29-32.
9. Schooley, A.H., 1962, "Upwind-Downwind Ratio of Radar Return Calculated from Facet Size Statistics of a Wind-Disturbed Water Surface," Proc. IRE, 50, pp. 456-461.

Figure 1 View of the bistatic receiving site located approximately 3.5 km south of the transmitting site. The large dish is the target receiving dish, while the smaller dish visible below was used to receive the synchronizing pulse.

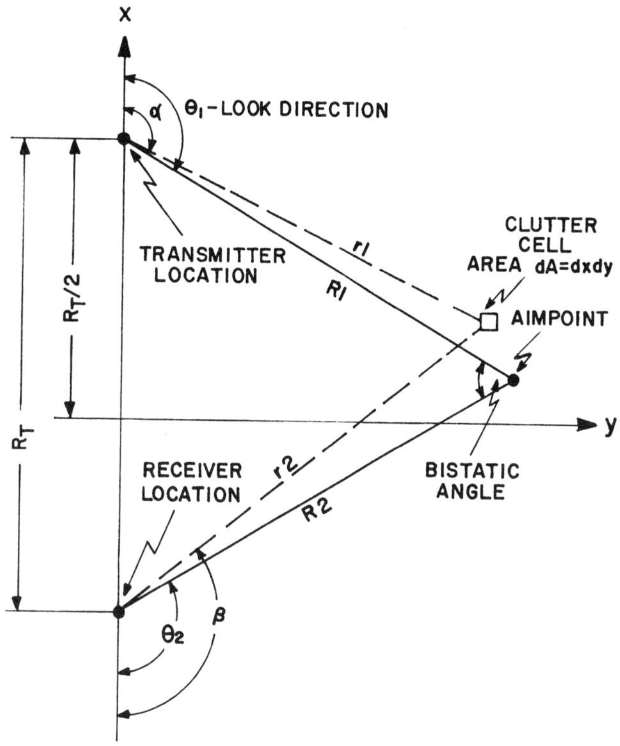

Figure 2 Geometry and nomenclature for the experiment and for calculation of bistatic illuminated area.

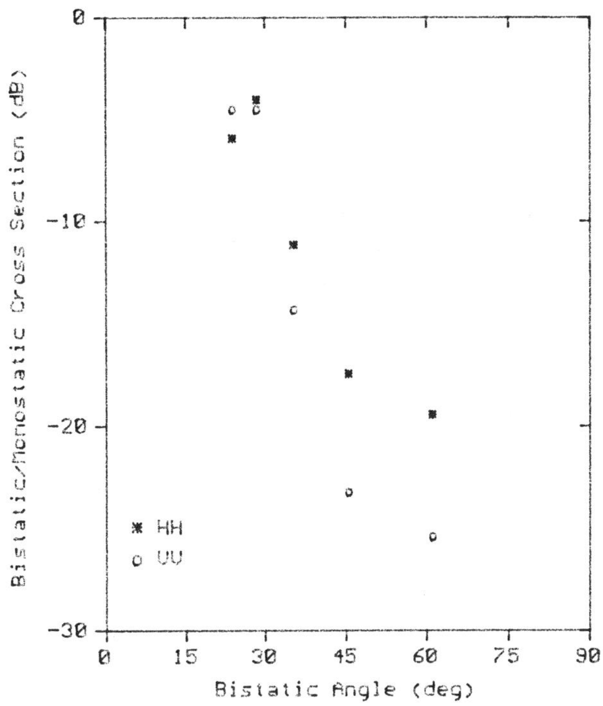

Figure 3 Ratio of bistatic to monostatic radar cross section per unit area as a function of bistatic angle for horizontal (HH) and vertical (VV) polarizations. Data of August 3 with transmitter look direction of 90 degrees.

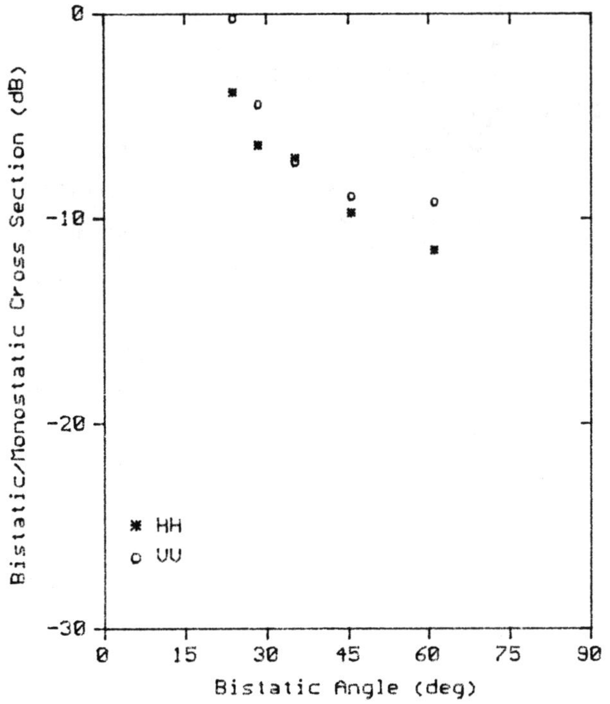

Figure 4 Ratio of bistatic to monostatic radar cross section per unit area as a function of bistatic angle for horizontal (HH) and vertical (VV) polarizations. Data of August 5 with transmitter look direction of 90 degrees.

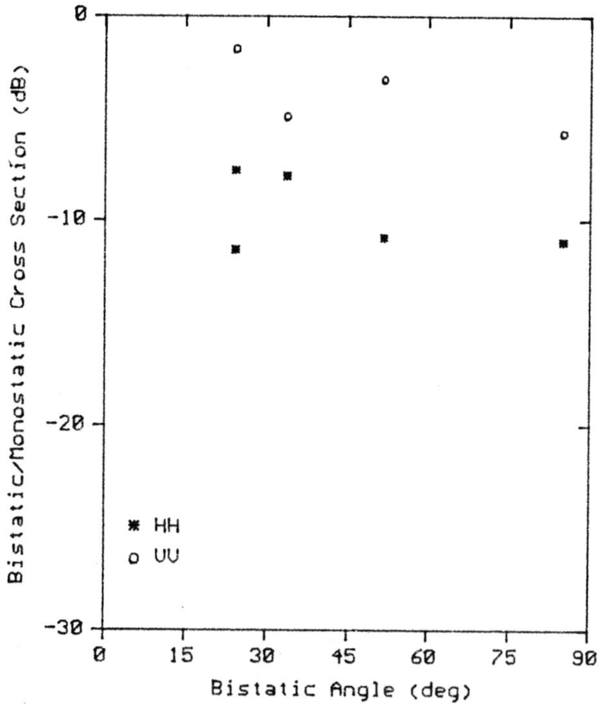

Figure 5 Ratio of bistatic to monostatic radar cross section per unit area as a function of bistatic angle for horizontal (HH) and vertical (VV) polarizations. Data of August 4 with transmitter look direction of 150 degrees.

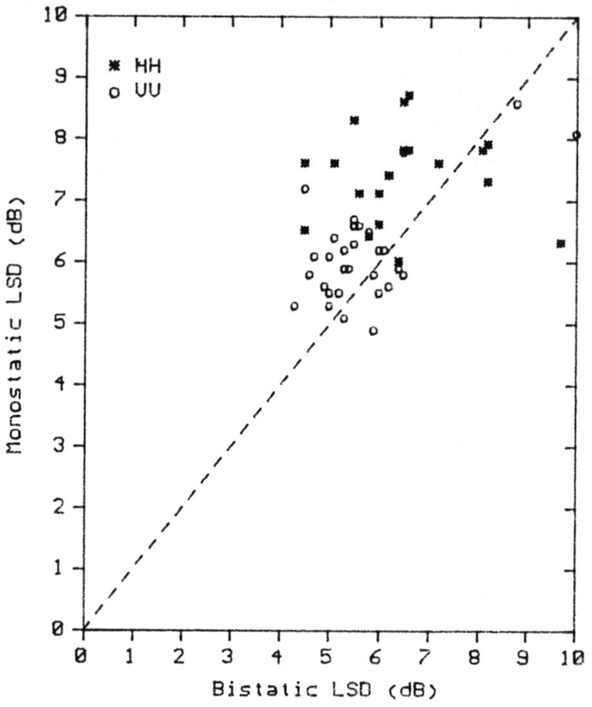

Figure 6 Scatter diagram for monostatic versus bistatic log normal standard deviations (LSD) in dB. Data for horizontal (HH) and vertical (VV) polarizations are included, and the dotted line is the loci of equal LSD.

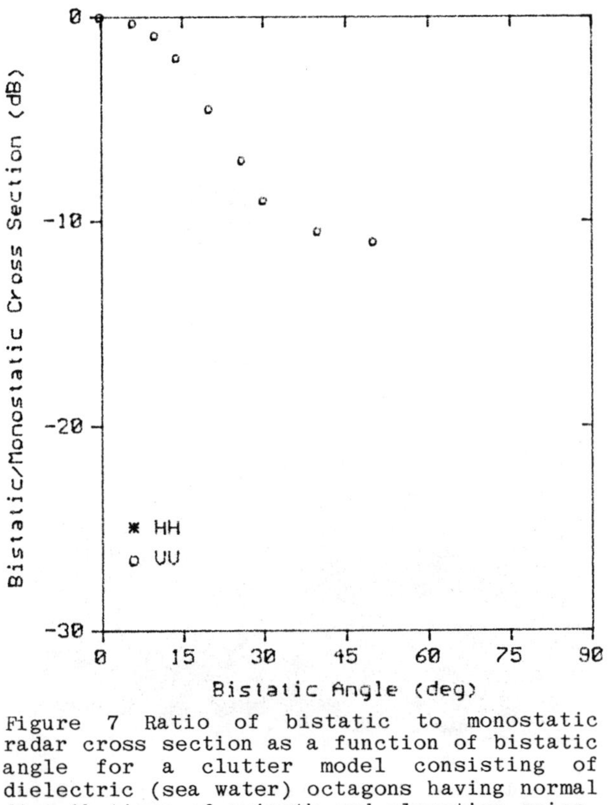

Figure 7 Ratio of bistatic to monostatic radar cross section as a function of bistatic angle for a clutter model consisting of dielectric (sea water) octagons having normal distributions of azimuth and elevation orientation. Parameter values were selected from the data of Macdonald [8] and Schooley [9] for conditions approximating those for the data set given in Figure 4.

SEA CLUTTER STATISTICS

J. Maaløe

Electromagnetics Institute, Technical University of Denmark, DK 2800 Lyngby, Denmark

INTRODUCTION

Radar signals backscattered from the sea surface are a limiting factor in any marine radar system as regards the detection of targets on or near the surface of the sea. With the aim of reducing this disturbing effect a number of analyses have been carried out on the statistical properties of sea clutter. The purpose of the analyses was

- o to calculate the thresholds exceeded by 50% and 5% of the sea clutter amplitudes
- o to compare the Log-normal and the Weibull distributions to that of the sea clutter amplitudes recorded
- o to study the sea clutter level dependence on range, pulse width and wind direction
- o to determine the spatial correlation of sea clutter amplitudes

Measurements

The measurements are carried out using a Racal-Decca X-band marine radar placed at a coastal site overlooking the English Channel off Dungeness, England [1]. The essential radar characteristics as regards the sea clutter analysis are listed in table 1.

Aerial:

- Height : 15 metres above mean sea level
- Polarization : Horizontal
- Azimuth resolution : 0.8^{o} at 3 dB
- Rotation speed : 20 rpm
- Gain : 32 dB
- Operating frequency: 9410 MHz

Transmitter:

- Type : Simple pulse
- PRF : 850 Hz, 1650 Hz, 3300 Hz *
- Pulse length : 1 µs, 0.25 µs, 0.05 µs

Receiver:

- Type : Superhet - log IF
- Bandwidth : 5 MHz or 18 MHz

The recordings were made during the month of October 1980. The weather was dry and the windspeed was about 11 knots corresponding to sea state 3 [2].

The radar signals are recorded at the output of the log-detector of the receiver. The recording system [3] used, stores the radar video signal together with the trigger, the antenna angle, and the heading marker as a composite video signal on a video recorder [IVC 601]. The bandwidth of the video recorder is 5 MHz, which is sufficient for recording of pulse widths larger than 200 nsec without deterioration. Later, the radar signals can be replayed either directly on a standard PPI or, after digitizing, into a computer for numerical analysis. The signals are digitized into 7 bits corresponding to a dynamical range of 42 dB. Hence, the amplitudes as presented in the next sections are divided into 128 arbitrary levels. The sampling rate in range, i.e. within a single A-scan, is 10 MHz, independent of the pulse width.

*The transmitter was externally triggered with a 1KHz PRF.

The sea clutter analysis is carried out for the ranges from 0.5 NM to 3.5 NM, using data from an up-wind as well as a cross-wind direction. The pulse widths used for the measurements are 250 nsec and 1 µsec respectively.

50% and 5% Amplitude Crossing Levels

The amplitude levels exceeded by 50% and 5% of the clutter samples are calculated for each of 10 antenna rotations over a sea surface sector corresponding to 225 metres in the range direction and 8.9 deg. in the azimuth direction. Further, the mean value and the standard deviation of these two crossing levels are calculated for the 10 antenna rotations. The calculations are repeated for each of the ranges selected (0.5 - 3.5 NM) and the results are shown in figures 1 to 4.

From the curves in figures 1 and 2, it appears that

- o the standard deviation of the 50% and 5% threshold levels are very small, hence the sea-clutter level may be estimated to a "reasonable" accuracy within 10 successive antenna rotations
- o the 50% and 5% threshold are larger for the up-wind direction than for the cross-wind direction
- o the tail of the distributions decreases with range

When using the 1 µs pulse width, figures 3 and 4, the 5% threshold appears to increase abruptly at a range of 1.5 NM. The phenomenon which has not yet been explained has also been observed in ref. [4]. Apart from this, the tendency is the same as that of figures 1 and 2 obtained for the 250 nsec pulse.

Log-normal and Weibull Distributions applied to the Sea Clutter

In order to study the sea clutter amplitude distributions, histograms representing the frequency of occurrence of the sea clutter amplitudes are calcluated for each of the ranges, the pulse widths and the wind directions under investigation. To each of the histograms a Log-normal or a Weibull distribution has been fitted [5]. The distributions are selected according to a distribution test, which defines the best fit as the one resulting in the least sum of the squared differences.

The density function corresponding to the Log-normal distribution is

$$f(x) = \frac{1}{x\cdot\beta\sqrt{2\pi}}\cdot\exp\left(-\frac{1}{2}\cdot\left(\frac{\ell nx-\alpha}{\beta}\right)^2\right) \quad , x \geqslant o$$

The Log-normal distribution is strongly skewed hence it is reasonable to expect it to represent the distribution of sea-clutter due to its infrequent occurrence of large amplitude values.

The density function of the Weibull distribution is

$$f(x) = \frac{C}{B}\left(\frac{x}{B}\right)^{C-1}\exp\left(-\left(\frac{x}{B}\right)^{C}\right) \quad , x \geqslant 0$$

where B is a scale parameter and C is a shape parameter. The skewness of the Weibull distribution is smaller than that of the Log-normal distribution hence to be expected as being a better choice for describing the sea-clutter in cases with more equally distributed amplitudes. As an example, figure 5 shows a histogram representing the frequency of occurrence of sea clutter amplitudes recorded at a range of 2 NM in the up-wind direction using a 250 nsec pulse. In addition, the Log-normal distribution determined as being the best fit to this set of measurements is shown for comparison.

The histograms calculated [5] indicate that the frequency of occurrence of large amplitudes decreases with increasimg range, i.e. the variance of the clutter amplitudes decreases with range. The tails of the histograms still remain rather wide out to a range of about 2 NM whereabout the sea clutter seems to disappear.

The distributions applied to the sea clutter amplitudes show that at close ranges, i.e. in a heavy clutter environment, the Weibull distribution appears as being the best fit. At ranges where the clutter dissolves into so-called clutter islands, the clutter amplitude tends to be better represented by a Log-normal distribution. At even larger ranges, where the clutter decreases and finally disappears, the amplitude distribution again approaches a Weibull distribution, due to the fact that this is the distribution of the receiver noise amplitude.

Sea Clutter Level as Function of Range, Pulse Width and Wind Direction

The mean values of the distributions fitted to the histograms are plotted in fig.6 as function of range for the two pulse widths and for the 2 azimuth directions. The curves show that

- the mean clutter level decreases with increasing range, until the receiver noise becomes the dominant source
- the clutter level appears to depend linearly on the logarithm of range
- the mean clutter level is greater in the up-wind direction than in the cross-wind direction
- the mean clutter level is greater for the long radar pulse (1 μs) than for the short radar pulse (250 ns)

The linear variation of the clutter level with the logarithm of range is to be expected from the radar equation and the fact that the data analysed are recorded at the output of a logarithmic amplifier. The analysis shows that the mean clutter level is greater in the up-wind direction than in the cross-wind direction. This is in accordance with the results obtained by G. Bishop [6]. The higher clutter level measured for the longer pulse is obviously due to the larger area of the resolution cell.

The standard deviation of the distributions does not seem to depend on the pulse width and the wind direction, but simply decreases with range.

Adaptive Video Detectors

The purpose of this section is to work out the relation between the threshold d of an adaptive video detector and the parameters of the Log-normal and the Weibull distributions, with the aim of maintaining a constant false alarm rate. Throughout the whole of this section, the false alarm probability P_{fa} is assumed to be 10^{-6}. This implies that noise will be mistaken for a signal on the average once out of a million when no target is present.

Firstly, the relation between the threshold and the parameters of the Log-normal distribution is derived.

At medium ranges where the distribution of sea-clutter amplitudes is best described by a Log-normal distribution, the distribution of the $y_j = \ell n\, x_j$ log-converted amplitudes are approximated by a Normal distribution with the mean and the variance

$$E(y) = \alpha$$

$$VAR(y) = \beta^2$$

The density function of y is

$$f(y) = \frac{1}{\beta \cdot \sqrt{2\pi}} \cdot \exp\left(-\left(\frac{y-\alpha}{\sqrt{2}\cdot\beta}\right)^2\right)$$

and the false alarm probability is

$$P_{fa} = \int_{d}^{\infty} \frac{1}{\beta \cdot \sqrt{2\pi}} \cdot \exp\left(-\left(\frac{y-\alpha}{\sqrt{2}\cdot\beta}\right)^2\right) dy$$

$$= \int_{\frac{d-\alpha}{\sqrt{2}\cdot\beta}}^{\infty} \frac{1}{\sqrt{2\pi}} \cdot \exp(-z^2)\, dz$$

$$= \frac{1}{2}\left(1 - \mathrm{erf}\left(\frac{d-\alpha}{\sqrt{2}\cdot\beta}\right)\right)$$

where erf(z) is the error function associated with the Normal distribution.

Tables of the error function do not cover values of the false alarm probability as low as 10^{-6}. However, for P_{fa} small, the error function can be represented by an asymptotic series, and for $P_{fa} \ll 1$ the first term is a satisfactory approximation [7].

$$P_{fa} = \frac{1}{2 \cdot \sqrt{\pi}} \cdot \frac{e^{-z^2}}{z} \quad , \quad z > 3$$

Using this equation it is found that

$$P_{fa} = 10^{-6} \Rightarrow z = 3.36 \Rightarrow d = \alpha + 4.75 \cdot \beta$$

Obviously, a different choice of P_{fa} will change d, i.e. reducing P_{fa} increases d, and vice versa.

Hence, the threshold level d can be made adaptive using the maximum likelihood estimators for α and β

$$\hat{\alpha} = \frac{1}{n} \sum_{i=1}^{n} \ell n\, x_i$$

$$\hat{\beta} = \frac{1}{n} \sum_{i=1}^{n} (\ell n\, x_i - \hat{\alpha})^2$$

i.e.

$$\hat{d} = \hat{\alpha} + 4.75 \cdot \hat{\beta}$$

Secondly, the relation between the threshold and the parameters of the Weibull distribution is derived.

At close range and at far range the clutter amplitudes obey a Weibull distribution. The false-alarm probability and the corresponding threshold level d, are simply related as

$$P_{fa} = \int_{d}^{\infty} \frac{C}{B} \left(\frac{x}{B}\right)^{C-1} \exp\left(-\left(\frac{x}{B}\right)\right) dx$$

$$= \exp\left(-\left(\frac{d}{B}\right)^C\right)$$

i.e.

$$d = B \cdot (-\ell n\, P_{fa})^{1/C}$$

or

$$d' = \ell n\, d = \ell n\, B + \frac{1}{C} \ell n(-\ell n\, P_{fa})$$

Using the estimators for B and C derived by Menon [8]

$$E(y) = \ln B - \frac{\gamma}{C} \quad , \quad \gamma = 0.57722$$

$$V(y) = \frac{\pi^2}{6C^2}$$

where $y = \ln x$ and γ is the Euler constant, it is found that

$$d' = \alpha + \frac{\sqrt{6}}{\pi} \cdot \beta (\ln(-\ln P_{fa}) + \gamma)$$

α and β are the first and second moments of the variable y. The values of the parameters α and β are estimated in a similar way as the parameters in the Log-normal distribution.

Hence, when Weibull distributed sea-clutter is present, the adaptive threshold is calculated using the equation

$$\hat{d} = \hat{\alpha} + \frac{\sqrt{6}}{\pi} \cdot \hat{\beta} (\ln(-\ln P_{fa}) + \gamma)$$

A false alarm probability of 10^{-6} results in the adaptive threshold

$$\hat{d} = \hat{\alpha} + 2.5 \cdot \hat{\beta}$$

This threshold dependence is similar to the one obtained for the sea-clutter obeying a Log-normal distribution, the only difference being the constant. From the equations expressing the adaptive threshold it appears that if sea-clutter amplitudes following a Log-normal distribution are erroneously assumed to obey a Weibull distribution, the false-alarm probability will increase.

Correlation Analysis of Sea Clutter Amplitudes

The purpose of the correlation analysis is to calculate the radial and the angular correlation of sea clutter amplitudes. The spatial correlations are determined out to a sample distance of 50 and are all estimated by [7]

1. calculating the power spectrum segments using the natural segmentation of the sea clutter data, e.g. A-scans
2. averaging the power spectrum estimates computed separately from the segments
3. calculating the inverse Fourier transform
4. smoothing the autocorrelation function, using a rectangular window with the length of 3 samples

The autocorrelation functions in the range direction are estimated on the basis of 3000 samples all obtained within a single antenna rotation. The data are collected from a sector consisting of 60 range samples and 50 A-scans, corresponding to 900 metres in the range direction and 6 degrees in the azimuth direction. Having selected the range interval size to 60 samples the correlation function is calculated for lags up to 50, for which lag the value of the correlation function is based on 500 samples.

The samples are obtained with a sampling rate of 10 MHz - independent of the pulse width - corresponding to a range resolution of 15 m. Hence, on the average the video signal is oversampled 2.5 times when using the short pulse and 10 times when using the long pulse.

The calculations are repeated for each of the 3 ranges selected: 0.5, 1 and 1.5 NM. An example of the autocorrelation functions [9] calculated in the range direction is plotted in figure 7. The results of the autocorrelation analysis carried out in the range direction may be summarized as follows:

- the decorrelation distance is approximately 3-5 samples for the 250 nsec pulse width and 10-20 samples for the 1 μsec pulse width, i.e. in both cases approximately equal to the range resolution, a result which agrees with the work carried out by Pidgeon [10]
- the decorrelation distance increases with range for both pulse widths
- the decorrelation distance is apparently independent of the wind direction
- when the autocorrelation functions are periodic, the period appears to be between 8 and 14 sample intervals.

The autocorrelation in the azimuth direction is estimated on the basis of 2100 samples, taken from three successive antenna rotations. During each antenna rotation, the data are collected from a sector consisting of 10 range samples and 70 A-scans, corresponding to 150 metres in range and 8.4 degrees in azimuth. The extent in azimuth is selected to be 70, hence the correlation at the maximum distance between samples - i.e. for a lag of 50 - can be estimated on basis of 600 samples.

The calculations are repeated for each of six ranges: 0.5,0.75,1,1.25,1.5 and 2 NM. An example of the autocorrelation functions [9] is plotted in figure 8. The results of the autocorrelation analysis carried out in the azimuth direction may be summarized as follows:

- the decorrelation distance varies from 35 A-scans (app. 6 beam widths) near the radar to 6 A-scans (app. 1 beam width) at larger distances
- the decorrelation distance is larger in the up-wind direction than in the cross-wind direction for both pulse widths
- the decorrelation distance is apparently independent of the pulse width selected
- when the autocorrelation function shows some periodicity, the period falls in the interval from 20 to 30 A-scans

From the radar characteristics (table 1) it appears that each resolution cell is hit by a pulse over six consecutive A-scans for each antenna rotation. Hence, the minimum decorrelation distance to be expected in the azimuth direction corresponds to six A-scans. The measurements prove that a video integration as well as a binary integration in azimuth over less than 6 A-scans will have no effect as regards the signal-to-clutter ratio. Further, the measurements indicate that to ensure a higher probability of detection of small targets close to radar, the detection has to be based on far more than ten azimuth integrations.

The decorrelation distance (No. of A-scans) in the azimuth direction decreases with range in a way which indicates that the physical decorrelation distance (in metres) is constant.

Conclusion

The radar return from a sea surface has been studied with the aim of investigating the statistical properties of sea clutter as a function of range, pulse length and wind direction. The measurements have been obtained with an X-band marine radar using two pulse lengths, 250 nsec and 1 μsec. The radar returns from both an up-wind direction and a cross-wind direction have been analysed.

The studies show that the sea clutter amplitudes as measured at the output of a log-detector do follow a Log-normal distribution or a Weibull distribution depending on the distance from the radar. It is further shown, that the parameters of the two distributions depend on the pulse length and the wind direction. The dependence on wind speed has not yet been analysed due to lack of data.

The correlation analysis indicates that the decorrelation distance in the range direction is between one and two resolution cells, increasing slightly with range. In the azimuth direction, the decorrelation distance varies from 35 A-scan (6 beam widths) close to the radar to 6 A-scan (1 beam width) far from the radar.

One information which is often missing is the transfer function of the radar receiver, which has a marked influence on the sea clutter amplitude distribution. Hence, the results presented in this paper only indicate the kind of stochastic behaviour of the sea clutter amplitude to be expected. The actual parameter values of the statistical distributions will have to be measured for each particular radar system under consideration.

Acknowledgement

The author wishes to express his appreciation to Racal-Decca Ltd. for making available their research station at Dungeness for the measurements.

References

[1] Maaløe, J, "Interim Report of a 2-month visit to U.K. to carry out radar sea clutter measurements," Electromagnetics Institute, Technical University of Denmark, November 1980, IR 220.

[2] Nathanson, F.E., "Radar design principles," McGraw-Hill Book Co., 1969.

[3] Maaløe, J., "Radar signal recording and analysing system," Electromagnetics Institute, Technical University of Denmark, October 1981, IR 253.

[4] Williams, P.D.L., "Results from an experimental dual-band search radar," The Radar and Electronic Engineer, Vol. 51, No. 11/12, pp 541-552, November/December 1981.

[5] Maaløe, J., "Sea-clutter amplitude statistics," Electromagnetics Institute, Technical University of Denmark, December 1981, R 249.

[6] Bishop, G., "Radar sea clutter," Royal Radar Establishment, Malvern, England.

[7] Schwartz, M and Shaw, L., "Signal processing," McGraw-Hill Book Co., 1975.

[8] Menon, M.V., ""Estimation of the shape and scale parameters of the Weibull distribution," Technometries, Vol. 5, No. 2, May 1963.

[9] Maaløe, J., "Correlation analysis of sea clutter amplitudes," Electromagnetics Institute, Technical University of Denmark, January 1982, R 250.

[10] Pidgeon, V.W., "Time, Frequency and Spatial Correlation of Radar Sea Return," Space Sys. Planetary Geol. Geophys., American Astronautical Society, May 1967.

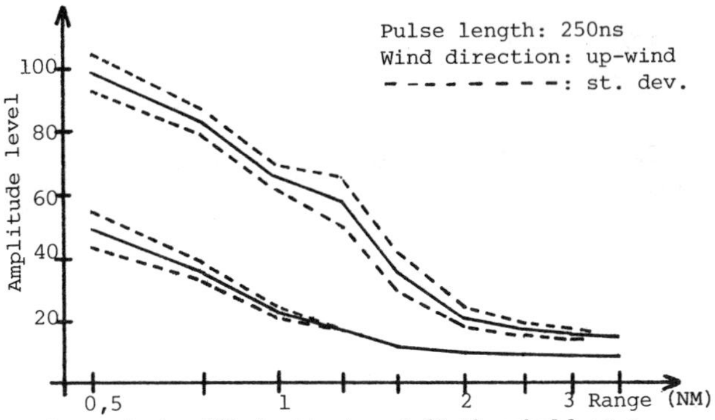

Figure 1 The 50% (median) and 5% threshold as a function of range.

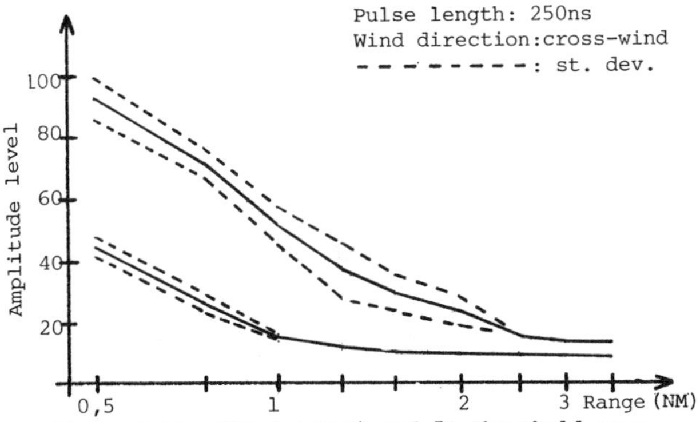

Figure 2 The 50% (median) and 5% threshold as a function of range.

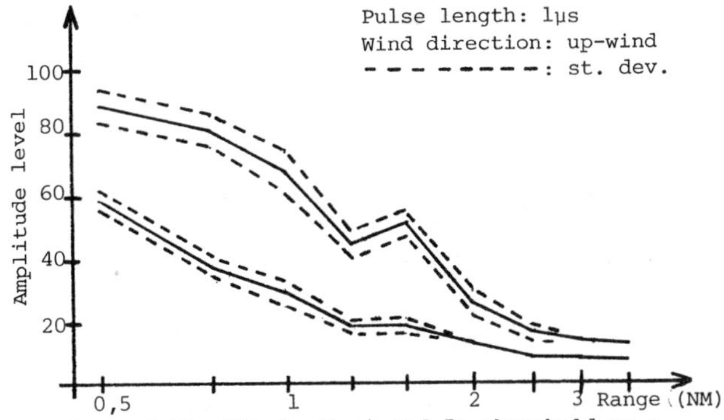

Figure 3 The 50% (median) and 5% threshold as a function of range.

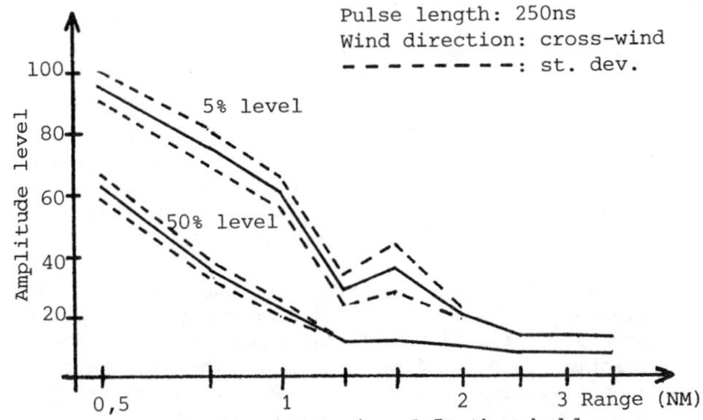

Figure 4 The 50% (median) and 5% threshold as a function of range.

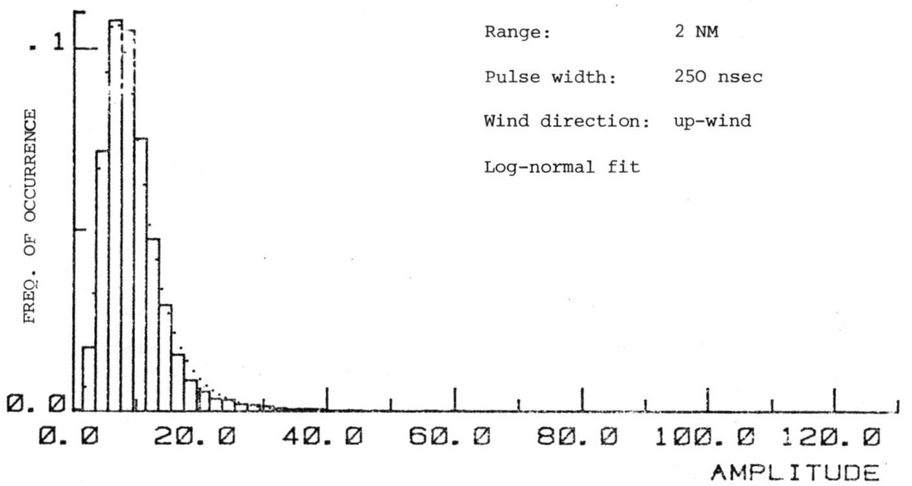

Figure 5 Histogram calculated from sea clutter amplitudes. The dotted curve is a Log-normal distribution applied to the histogram.

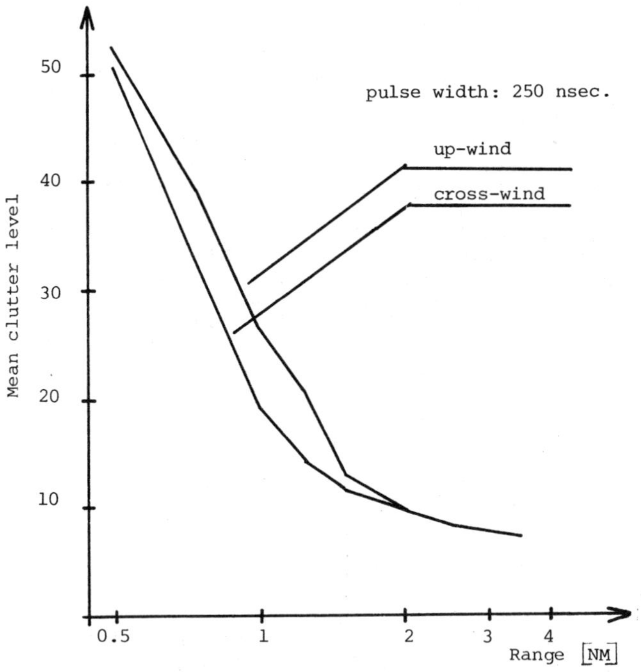

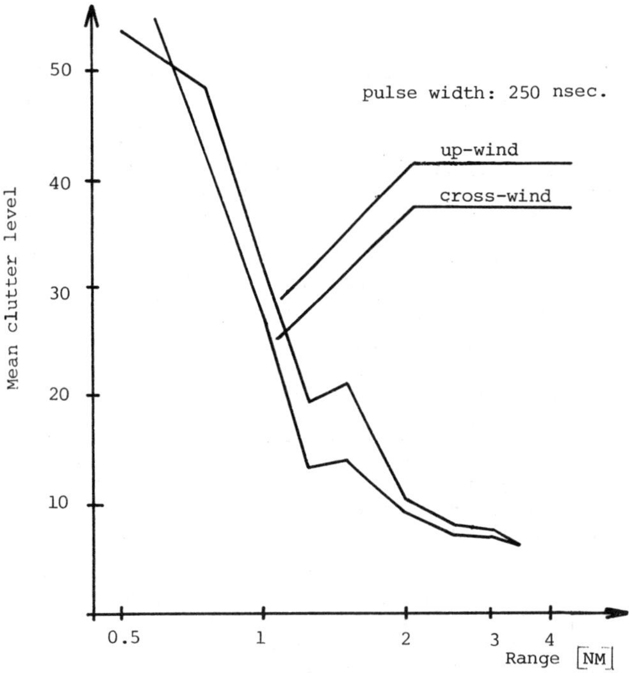

Figure 6 Mean sea clutter level as a function of range.

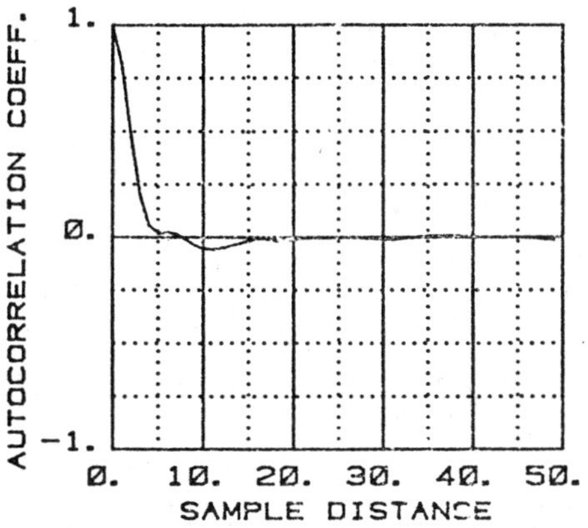

Figure 7 Correlation of sea clutter amplitudes in the range direction. The range is 1NM, the pulse width is 250 nsec, and the wind direction is up-wind.

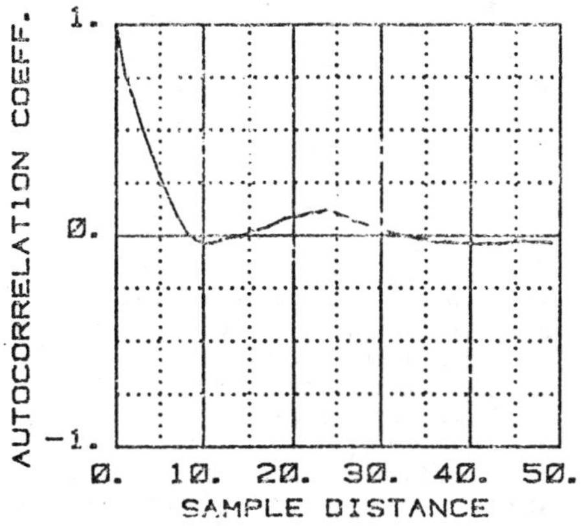

Figure 8 Correlation of sea clutter amplitudes in the azimuth direction. The range is 1 NM, the pulse width is 250 nsec, and the wind direction is cross-wind.

AMPLITUDE AND TEMPORAL STATISTICS OF SEA SPIKE CLUTTER

I. D. Olin

Radar Division, Naval Research Laboratory, Washington, D.C. 20375, USA

INTRODUCTION

Microwave radar backscatter from the sea surface is generally characterized by noise-like clutter signals which must usually be discriminated from the desired targets. Since the targets of interest are most often much smaller than the radar illuminated area, a convenient means of improving the target-to-clutter ratio for a pulsed radar is to decrease the illuminated area. This is more readily accomplished by a reduction in radar pulse width. However, as the range cell decreases below about 75 m, it is observed that the backscattered clutter peaks are approximately constant with respect to time. Only the period between the peaks increases with increased range resolution. Therefore, in order to determine a threshold for target detection, the statistics of such "spikey" clutter must be better understood. Thus, the amplitude statistics need to be recorded and useful distributions fitted. With high range resolution, however, the periods between the signal peaks can be substantial so that an analysis of the temporal statistics may also be useful in developing detection algorithms.

Sea clutter measurements were made by Hansen and Cavaleri (1) with a 3 cm radar using a pulse width of 40 ns and a pulse repetion frequency of 2000 Hz. The radar was installed on a platform site about 16 km offshore from Panama City, Florida at a water depth of about 30 m and at an antenna height of 15 m above mean sea level. These conditions approximated open sea conditions. A 1° pencil beamwidth was used and extensive measurements were made at 300 m range, corresponding to a grazing angle of 2.9^o. The equivalent illuminated patch was 31.6 m^2 (15dB>1m^2). Both horizontal and vertical polarizations were used and data were recorded under a variety of conditions.

DATA ACQUISITION

Radar calibration was accomplished by means of a corner reflector suspended so as to be range resolvable from any of the platform structure. Data were recorded on analog tape and later digitized for analysis. For this purpose sampling was at a rate of 250 Hz and a resolution of 8 bits. Figure 1 is typical of the data observed from a playback of the analog tape on a chart recorder and provides ample basis for characterizing the signals as "spikey." The marked difference between vertically and horizontally polarized backscatter is well known and has been noted by Long (2), and Lewis and Olin (3). Moreover, although SS-5 (Sea State 5) is depicted here, very similar envelopes, albeit with much reduced signal peaks also occurs in SS-2, or even in relatively calm water (3). In order to establish sufficient samples for analysis, data periods of at least 80s were used. Some were as long as 480s.

AMPLITUDE STATISTICS

The cumulative distributions of backscatter data from SS-2 and SS-5 taken upwind are depicted in figures 2 and 3 by the plotted data points (dots). The abscissa scale is in terms of the actual measured radar cross section (RCS). The normalized RCS can be obtained by subtracting 15dB from these values. The different character of the distributions for the two polarizations is clearly evident. Fitting a known distribution function to such data is important in extending the limited observations to a wide class of detection analyses. For this purpose a variety of functions have been advocated. (George (4), Trunk (5), Jakeman and Pusey (6) and Fay et al (7)). Universally, however, it is agreed that the data are non-Rayleigh. The Rayleigh form for radar cross section would also plot as a straight line on figures 2 and 3, but the slope would be much higher; i.e., the distribution "tails" would be shorter. The Weibull distribution, which has been suggested (7) also plots as a straight line, but accommodates a variety of slopes, including the special Rayleigh case. The density function for the Weibull, written in terms of a radar cross section parameter, R is:

$$p(R) = \frac{B}{A} R^{B-1} \exp\left[\frac{-R^B}{A}\right], \qquad (1)$$

from which the following probability distribution is determined:

$$P\left[R > R_t\right] = \exp\left[\frac{-R^B}{A}\right]. \qquad (2)$$

Often a slope parameter 1/B is defined, since it represents the slope of the straight line when plotted on Rayleigh statistical paper.

For B=1 this function reduces to the Rayleigh form (so called because the envelope detected signal voltages are Rayleigh distributed). Eqns. (1) and (2) also represent Weibull statistics for the signal voltages, p(v) and P[v], in which case B=2 defines Rayleigh statistics. As noted by Blake (8), this duality also occurs for the lognormal distribution.

The solid straight line for vertical polarization in figures 2 and 3 was fitted by eye and qualitatively estimated to fit the data points. Based on this line the parameters A, B of Eqn. (2) can be determined and are shown in Table 1. Also shown is the median radar cross section, R_m, derived from Eqn. (2).

TABLE 1 - Vertical Polarization Weibull Parameters

	A	B	R_m(dB>1m^2)
SS-2	0.065	0.622	-21.7
SS-5	0.228	0.495	-16.2

Clearly the horizontally polarized component cannot be fitted by any of the Weibull distributions. Another distribution suggested for high resolution clutter is the lognormal (4), defined by

$$P(R) = \frac{1}{\sqrt{2\pi}\,\rho R}\exp\left[-\frac{1}{2\rho^2}\left(\ln\frac{R}{R_m}\right)^2\right]. \qquad (3)$$

R_m is the median radar cross section and ρ is the standard deviation of ln R in natural units (if K equals ρ expressed in dB, then $\rho = 0.1K \ln 10$). The corresponding probability distribution is

$$P[R > R_t] = \frac{1}{2}\left[1 - \text{erf}\left(\frac{W_t - W_m}{\sqrt{2}\,K}\right)\right], \qquad (4)$$

where W_t and W_m are RCS expressed in dB>1m^2, corresponding to R_t and R_m.

A fit to the horizontally polarized data by the lognormal is indicated in figures 2 and 3 by the dashed line. The parameters of Eqn. (4) used in the plot are shown in Table 2.

TABLE 2 - Horizontal Polarization Lognormal Parameters

	W_m	K
SS-2	-29	9.0
SS-5	-25	8.5

Neither data point sets are fitted very well and this agrees with previous investigators (5) and (7) who noted the inadequacies of the lognormal distribution. It has been suggested (2) that the horizontally polarized component may have two independent sources of fluctuation. One possibility is to model the data using two independent Weibull distributions, the variates of which are simply summed. Thus given independent Weibull variates R_1, R_2 with respective distributions defined by (A_1, B_1) and (A_2, B_2) a combined variate $R=(R_1+R_2)$ is formed and the resulting distribution function plotted. Using a computer simulation with 10^4 samples from each distribution, fits to the horizontally polarized data were explored. Those shown by the solid curve in figures 2 and 3 represent the best fit qualitatively estimated. Parameters used for these are given in Table 3.

TABLE 3 - Horizontal Polarization "Double" Weibull Parameters

	A_1	B_1	R_{m1}(dB)	A_2	B_2	R_{m2}(dB)
SS-2	0.006	0.833	-29	0.163	0.172	-55
SS-5	0.034	0.625	-26	0.164	0.189	-50

While the fit to the experimental data is good, it doesn't necessarily confirm that the actual data are generated by two Weibull distributions. It is not surprising, for example, that with the four independent parameters available, a good fit is obtained.

TEMPORAL STATISTICS

There are considerable periods of low amplitude signals for both the horizontally and vertically polarized backscattered components, as indicated in figure 1. This is very different from lower resolution backscatter which tends to be more continuously noisy. To quantify the temporal characteristics of the spikey clutter the statistics of the inter-spike periods and the spike widths (in terms of time) were analyzed. To do this, however, first required formulating a suitable definition for a sea spike.

Sea-Spike Definition

Qualitatively, any signal rising above a background level can be termed a spike. Therefore, by establishing a threshold, periods above and below this threshold can be sorted and the desired statistics formulated. Figure 4 depicts a time sample of high resolution sea clutter backscatter on which an arbitrary amplitude threshold has been drawn. Sea spikes contain substantial high-frequency modulation components (3). As illustrated in the figure a sudden drop in signal level can be quickly followed by a rapid rise. This is distinct from the sustained low amplitude period indicated, so that some criterion is needed regarding the minimum period below threshold that must occur before declaring the end of the sea spike. In figure 4 this period is indicated as T_1. For those signals above threshold in the figure, the widths of the two spikes are indicated as W_1 and W_2. There will also be very short isolated periods in which a high amplitude spike will cross the threshold, as represented by T_2. These can be regarded as noise impulses and have been ignored in the analyses.

Temporal analyses of the sea spikes therefore involves the selection of three parameters: an amplitude threshold and periods T_1 and T_2. In conducting the analyses presented here $T_1 = 0.1$s and $T_2 = 0.01$s. 0.1s allows sufficient time for the high-frequency modulated components of a spike to be included before the end of the spike interval is declared. Thus the width of the spike on the left in figure 4 would be measured, but not recorded, until a below threshold crossing persisted for 0.1s. Similarly the spike width W_2 is defined.

Analyses Results

Table 4 lists the probability of the backscattered signal coming from a sea spike as a function of the threshold expressed in square meters of radar cross section. Sample lengths consisted of 300s of SS-2 data and 480s of SS-5 data. As expected the probability generally decreases with increasing threshold. An exception for vertical polarization in SS-5 is due to the splitting of a single spike into two widely separated spikes.

TABLE 4 - Probability of backscatter from a sea spike for horizontal and vertical polarization

	Sea State 2	Sea State 5
Threshold (m^2)	H/V pol.	H/V pol.
0.025	.07/.57	.36/.60
0.100	.03/.11	.15/.34
0.175	.01/.03	.11/.41
0.250	.01/.01	.08/.34

Table 5 lists average durations for spike width and inter-spike periods as a function of the same amplitude thresholds. Except for SS-5 using vertical polarization, little change in the average sea spike width occurred with changes in threshold. The average inter-spike period is a strong function of the threshold for both horizontal and vertical polarization in SS-2. It is a weaker function of threshold for horizontal polarization in SS-5 and changes very little for vertical polarization in SS-5.

In order to depict the distributions of the time intervals a threshold of 0.1 m^2 was selected. The results are shown in figures 5 and 6. In SS-2 the width of the sea spikes are mostly independent of polarization. However, the intervals between the spikes are longer for horizontal than for vertical polarization in both sea states, but especially in SS-2. It has been concluded in (1) and (3) that vertical polarization tended to profile the wave crests, whereas for horizontal polarization a particular wave crest alignment or range resolvable feature is necessary. Since measurements were made upwind, vertical polarization profiled all of the unshadowed wave crests. This is reflected in the much larger spike width in SS-5 when using vertical than horizontal polarization.

CONCLUSIONS

Amplitude and temporal statistics of sea backscatter using high resolution (40 ns pulse width) 3 cm radar from a deep water site at about 3° grazing incidence have been analyzed. Upwind amplitude data from the vertically polarized backscatter in both SS-2 and SS-5 are fitted by the Weibull distribution with slopes of 1.6 and 2.0 respectively, as compared with the Rayleigh distribution of RCS with a slope of 1.0. The RCS distribution of horizontally polarized backscatter is not well fitted by either the Weibull or the lognormal distribution. It appears to contain a break with different slopes for the low amplitudes than for the higher amplitude distribution tails. A distribution comprising two independent Weibull variates was qualitatively estimated to fit this data, although no conclusive evidence is offered that the Weibull is unique in this respect.

Temporal statistics of the spike width and inter-spike period have been analyzed. To accomplish this it was necessary to consider a definition for the presence of a sea-spike based on a specified threshold and which allowed for the oscillatory character usually observed in its fine structure. Results of this analysis indicate that except for a very low amplitude threshold, the probability of a backscattered signal originating from a sea spike in SS-5 for vertical polarization is 0.3-0.4, whereas for horizontal polarization it is about 0.1. In SS-2 for vertical polarization this probability is 0.01-0.1 and the corresponding probability for horizontal polarization is 0.01-0.03.

Given that a sea spike has occurred (based on a 0.1 m^2 threshold), the results indicate that in SS-5, using horizontal polarization, about 10% persist longer than 1s. Using vertical polarization this figure is approximately doubled. In SS-2, independent of polarization, only 10% of the spikes persist longer than about 1/2s.

The longest inter-spike periods occurred using horizontal polarization. In SS-2 occasional periods over 17s were observed. In SS-5 periods over 10s were observed. Using vertical polarization in both SS-2 and SS-5 only about 10% of the interspike periods persisted longer than 2½s.

ACKNOWLEDGMENTS

The author wishes to acknowledge the contributions of J.P. Hansen in acquiring the original data and P.A. Minthorn who furnished the machine compatible tapes used in the analyses presented here.

REFERENCES

1. Hansen, J.P. and Cavaleri, V.F., 1982, "High-Resolution Radar Sea Scatter Experimental Observations and Discriminants", NRL Report 8557.

2. Long, M.W., 1975, "Radar Reflectivity of Land and Sea", D.C. Heath and Company, Lexington, Mass.

3. Lewis, B.L. and Olin, I.D., 1980, "Experimental Study and Theoretical Model of High-Resolution Radar Backscatter from the Sea", Radio Science, 15, Number 4, 815-828.

4. George, S.F., 1968, "The Detection of Nonfluctuating Targets in Log-Normal Clutter", NRL Report 6796.

5. Trunk, G.V., 1976, "Non-Rayleigh Sea Clutter: Properties and Detection of Targets", NRL Report 7986

6. Jakeman, E., and Pusey, P.N., 1977, "Statistics of Non-Rayleigh Sea Echo", Radar-77, IEE Conference Publication Number 155, 105-109.

7. Fay, F.A., Clarke, J., and Peters, R.S., 1977, "Weibull Distribution Applied to Sea Clutter", Radar-77, IEE Conference Publication Number 155, 101-104.

8. Blake, L.V., 1980, "Radar Range-Performance Analysis", D.C. Heath and Company, Lexington, Mass..

TABLE 5 - Average duration in seconds of sea spike width and inter-spike period for horizontal and vertical polarization.

	Sea State 2		Sea State 5	
Threshold (m^2)	Width (H/V pol.)	Inter (H/V pol.)	Width (H/V pol.)	Inter (H/V pol.)
0.025	0.15/0.50	1.85/0.35	0.48/1.93	0.82/0.78
0.100	0.15/0.17	4.37/1.25	0.33/1.15	1.73/1.09
0.175	0.13/0.12	6.48/3.21	0.32/0.87	2.23/1.01
0.250	0.13/0.10	10.04/9.06	0.30/0.59	2.41/0.94

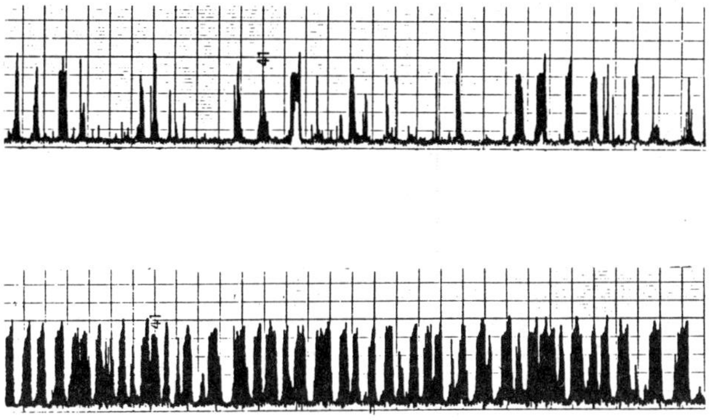

Sea State 5
abscissa: 10s per division
illuminated cell area: 31.6 m^2
pulse width: 40 ns

top trace: horizontal polarization
bottom trace: vertical polarization

Figure 1 High resolution sea clutter

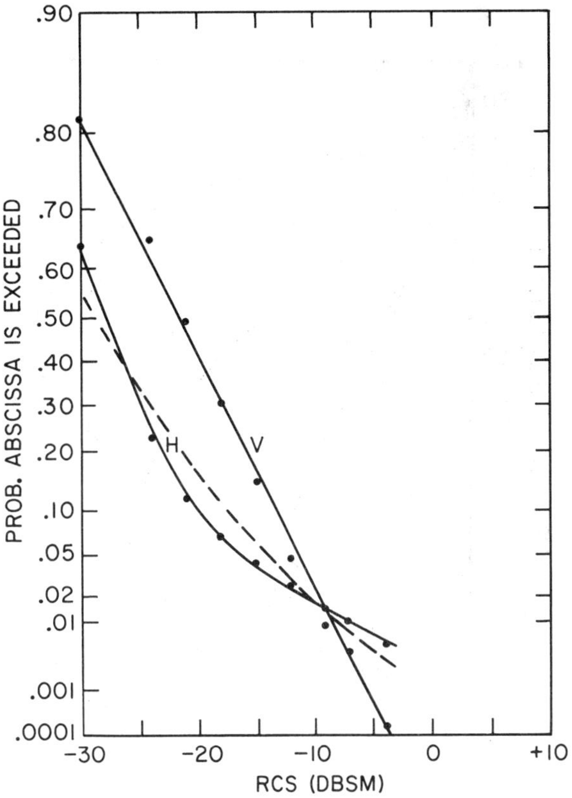

Figure 2 Sea state 2 amplitude distributions

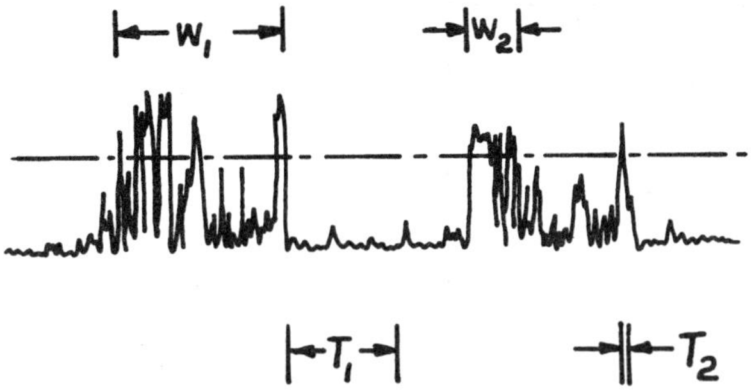

T_1 = minimum below threshold period
T_2 = minimum spike width period
W_1, W_2 = defined spike widths

Figure 4 Sea spike representation

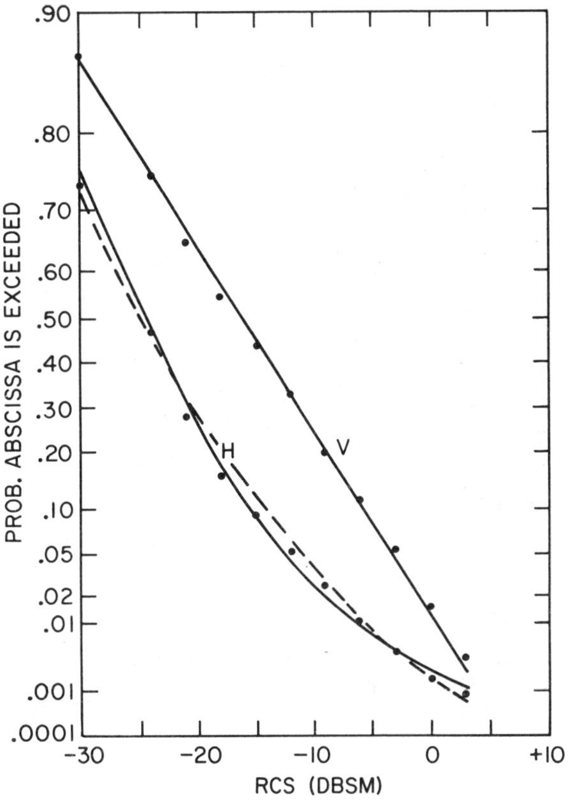

Figure 3 Sea state 5 amplitude distributions

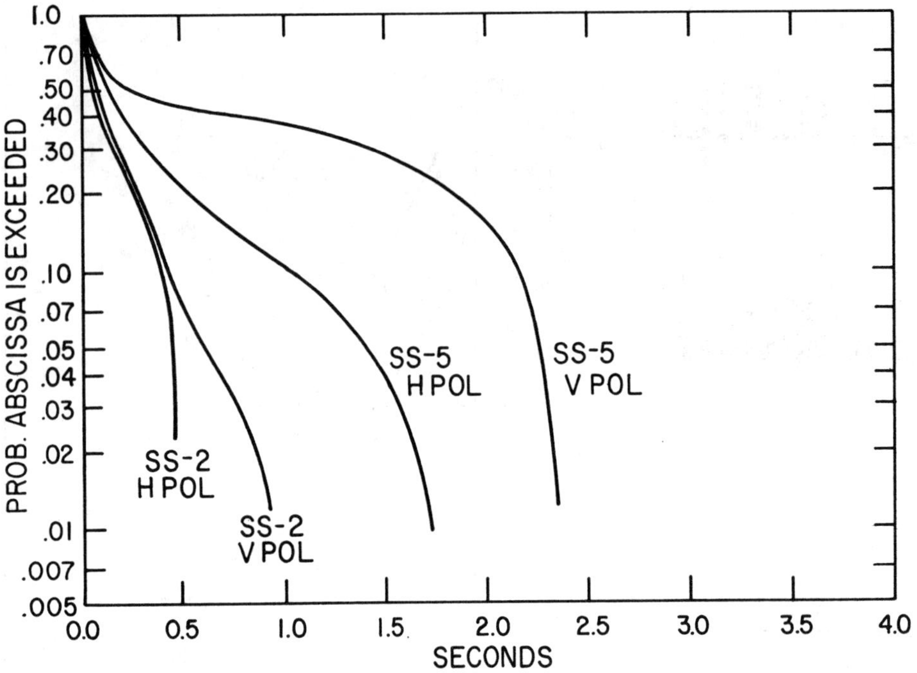

Figure 5 Sea spike width distributions

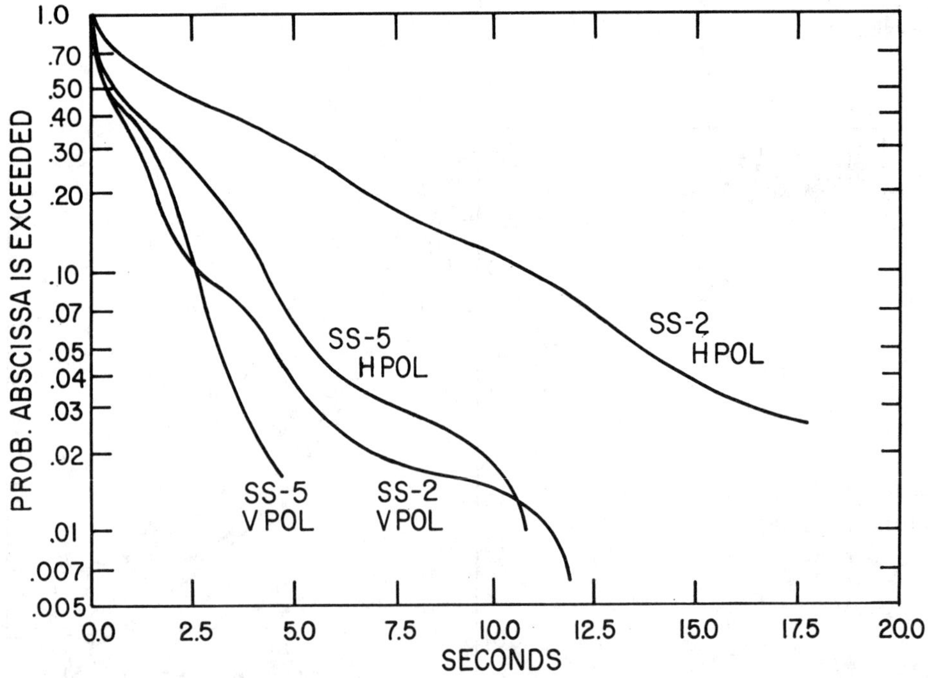

Figure 6 Inter-spike period distributions

A RADAR SEA CLUTTER MODEL AND ITS APPLICATION TO PERFORMANCE ASSESSMENT

K D Ward

Royal Signals and Radar Establishment, UK

1 INTRODUCTION

Radar returns from the sea surface, sea clutter, must be statistically modelled for radar performance to be predicted and for radar parameter and signal processing optimisations in a maritime environment. The modelling is usually split into two sections, mean power return per unit illuminated area (σ_0) and normalised statistical fluctuations. Measurements of σ_0 and empirical models for variation with environmental conditions and radar parameters are widely reported and in good general agreement (1-4). Fluctuation statistics have been extensively measured but without a unanimous description emerging. The subject of this paper is a model for the statistics of non-coherent sea clutter and its application to the assessment of radar performance.

For large illuminated patch sizes and high grazing angles ($> 10^o$) it is found that sea clutter obeys the central limit theorem and has Rayleigh distributed amplitude statistics. The range correlation is commensurate with the pulse length and, if the transmitter frequency is stepped by the pulse bandwidth each pulse (frequency diversity), the returns are independent from pulse to pulse. When the grazing angle is reduced and/or the radar resolution increased, this description no longer applies and the clutter is described as 'spiky'. The amplitude statistics have been modelled by Lognormal (5), Weibull (6) and K-distributions (7). The correlation properties have received less attention and no closed form expressions derived.

The deviations from noiselike statistics have a considerable impact on performance. Based on Rice's work (8) performance predictions in noise were calculated by Marcum and Swerling (9) and more recently extended into a reference book by Mayer and Meyer (10). A similar exercise is possible for spiky clutter if independence is assumed from pulse to pulse in the clutter echo. This has been performed for Lognormal (11, 12) and Weibull (13) distributions. Performance predictions including the effects of correlation have not been reported mainly due to the lack of a closed form description.

In this paper a model for sea clutter will be developed based upon airborne measurements. The model incorporates the frequency diversity, pulse to pulse correlation effects analytically and demonstrates the longer term and spatial correlation diagrammatically. The model is a compound distribution of two components each with differing correlation properties. The overall amplitude statistics are K-distributed and the ν parameter, which characterises the spikiness, has been measured against radar and environmental parameter variations. This leads, through scatter plots, to an empirical model for the ν value, limited at present to the radar pulse length used (30 nS). Section 2 describes the data collection and statistical model and section 3 develops the empirical model for ν.

The incorporation of the pulse to pulse correlation into the model allows performance predictions to be performed including these effects. Section 4 presents results derived from the model and shows that previous analysis based on independent samples gives optimistic predictions. Also, certain detection systems reported to have improvement over analogue integration in spiky clutter no longer show any advantage when the correlation effects are included. Section 5 contains a summary.

2 THE COMPOUND MODEL OF SEA CLUTTER

Sea clutter data have been recorded from an experimental, non-coherent, pulsed I-band radar capable of frequency diversity on a pulse-by-pulse basis. The pulse length is 30 nS, beamwidth 1.2^o and the transmitter polarization vertical or horizontal. The aircraft flew at various altitudes from 250' to 5000' and recorded data in a range profile form in a ground stabilised mode. Recordings were made at ranges from 22 miles down to 3 miles at 8 aspect angles, the first in the swell direction and thereafter at 45° intervals. Sea states ranging from 1 to 6 were encountered all in open sea conditions.

The envelope detected radar video was sampled at 17 nS intervals and 192 range samples from a ground stabilised range gate stored every echo. This form of recording has the advantage over a single range gate system in allowing spatial correlation to be investigated and also range time coupling due to sea wave motion.

The sea clutter data, on analysis, displayed two dominant components, with differing correlation times, contributing to the amplitude distribution. The fast varying component, which can be identified with the changing interference between scatterers, has a correlation time in the order of 10 ms and can be decorrelated by the use of frequency diversity. The slow varying component which can be associated with a bunching of scatterers, has a correlation time of the order of seconds and is unaffected by frequency diversity. The spatial correlation of the components is also different. The interference or speckle component has range correlation commensurate with the pulse length and is in all ways similar to noiselike clutter as its amplitude distribution confirms. The second component has considerable spatial correlation, depending upon aspect, which displays periodic effects and is coupled to the temporal correlation.

Plots of the correlated component demonstrate these effects. To extract the modulation, the clutter return is averaged over a time sufficient to give many independent samples to integrate out the speckle fluctuation, yet short enough for the 'bunching' to remain correlated. Figures 1 and 2 show range time plots of clutter where each trace is the integration of 200 ms of returns. The total plot is a time history of 10 seconds from a patch of sea over a range interval of 1/3 mile. Figure 1 is a typical V polarization recording at an upswell aspect in a medium to high sea state. The range time coupling and periodicity mentioned is evident. Figure 2 shows a H polarization recording in the same conditions and demonstrates the difference in long term and spatial correlation properties.

In general, plots of the type shown are a function of polarization, aspect angle, grazing angle and wave and swell condition of the sea. However the amplitude statistics all provide a good fit to one family of distribution. Also, because the correlation properties fall into two distinct components, the pulse to pulse correlation (and within beam correlation of a scanning radar) can be accurately described and performance predicted for fixed threshold systems where range correlation is not an active factor.

This develops as follows. Based on the two components described in the previous paragraphs, the overall amplitude sea clutter return (a) is represented as the product of two independent random variables.

$$a = xy \quad \ldots\ldots\ldots\ldots\ldots\ldots\ldots\ldots(1)$$

where x has a long correlation time and spatial and temporal structure and y is decorrelated by frequency diversity.

The results from the averaged clutter returns show that p(x), the probability density function of x, fits well to the chi distribution over a wide range of radar parameters and sea conditions.

$$p(x) = \frac{2x^{2\nu-1}}{\Gamma(\nu)} \exp -(x^2) \quad , \quad \nu > 0 \quad (2)$$

The value of ν depends on range, grazing angle, aspect angle, sea conditions and radar parameters. It has been found that ν generally falls within the range.

$$0 < \frac{1}{\nu} < 10 \quad \ldots\ldots\ldots\ldots\ldots\ldots\ldots(3)$$

The speckle component is well modelled by the Rayleigh distribution, which results in the overall amplitude, a, being distributed according to the K-distribution.

$$p(a) = \frac{4a^{\nu}}{\Gamma(\nu)} K_{\nu-1} \quad (2a) \quad \ldots\ldots\ldots(4)$$

where $K_\nu(z)$ is the modified Bessel function.

The K-distribution has been previously shown to be a good fit to the amplitude statistics of high resolution radar sea echo (7). Further work (14) develops a model based on a birth-death-immigration process leading to bunching of scatterers which is consistent with the present observations of clutter reported here. Of considerable importance to radar engineers is the ability to describe pulse to pulse correlation and its effect on performance. This is described in more detail in section 4.

3 AN EMPIRICAL MODEL FOR THE VARIATION OF ν PARAMETER

In order for a clutter model to be applied it is necessary to have a knowledge of how the various parameters in the model change with sea and radar parameters. As mentioned earlier, it is believed that σ_o, the mean echoing area, is understood. The next parameter to investigate is the shape or ν value. This provides information about the amplitude statistics and also some of the correlation properties.

The data base, used for the derivation of the empirical model, consists of about 300 recordings as described at the beginning of section 2. There is a predominance of V polarization recordings allowing a detailed investigation of this mode only. For H polarization the only real conclusion is that in general the ν value is lower than similar conditions with V polarization and the clutter therefore spikier and pulse to pulse correlation greater. Figure 3 shows a scatter plot of the ν parameter against grazing angle for the V polarization recordings. The delineated region shows the extent of H polarization recordings and the effect described.

Breaking down the Vertical recordings into groups defined by other parameters, reveals the trends shown diagrammatically in figure 4. The main conclusions are listed:-

1. All the results taken together show that a small grazing angle ϕ implies smaller ν.
2. There is no strong trend with sea state as seen in σ_o.
3. Aspect angle variation depends on swell and long wavelength sea wave content of sea spectrum. Considerable swell content implies aspect angle dependence as -
   (i) up and down swell, smaller ν
   (ii) across swell, larger ν
   (iii) between these directions, medium ν.
   Since a locally driven sea will have more long wavelength components for higher sea states, this aspect dependence often appears to relate to sea state.
4. At different ranges the across-range patch size is different. There is a strong trend for increased ν with increased patch size. (It must be noted that the along range resolution of all recordings is fixed at 30 nS.)

Measuring all the above trends leads to the following empirical relationship for V polarization sea clutter with a range resolution of 30 nS.

$$\log \nu = \frac{2}{3} \log \phi + \frac{5}{8} \log \ell - 1\left(\pm \frac{1}{3}\right) \quad ..(5)$$

where ϕ is the grazing angle in degrees
ℓ is the across-ranges resolution in metres
± 1/3 applies to the aspect angle dependence when it exists (see above (3)).

Comparing the empirical formula to the scatter of results gives an r.m.s. error of 0.24 in log ν.

4. PERFORMANCE PREDICTIONS USING THE COMPOUND DISTRIBUTION

The compound distribution of two components as described in section 2 can be written as

$$p(a) = \int_{\nu x} p_1(x)p_2(a|x)dx \quad \text{.........(6)}$$

where $p_1(x)$ is the pdf of the slow component (chi distribution)
$p_2(a|x)$ is the pdf of the amplitude, a, given a value of x (Rayleigh distribution of mean x).

For any fixed threshold system of processing where the integration is short compared to the correlation time of x, performance can be calculated assuming that x is constant and y = a/x is independent from pulse to pulse.

This leads to the expression for probability of false alarm, P_{FA}, for a linear analogue integrator,

$$P_{FA} = \int_{\nu x} p(x)P\left(\sum_i y_i > \frac{t}{x}\right)dx \quad \text{...(7)}$$

where t is the threshold, and
$P(\sum_i y_i > t/x)$ is the probability that the test statistic, Σy_i, is greater than t/x, assuming independent y_i.

A similar formula can be written for the probability of detection and also for other processing algorithms. The resulting performance will include pulse to pulse correlation effects. This method has been applied using numerical integrations and computations where a direct analysis is not possible. Results are presented here for a non-fluctuating target, single hit detection (figure 6) and analogue and binary integration of 10 returns (figure 7). The measure used for performance is the signal to r.m.s. clutter ratio required for 50% (or 90%) probability of detection at a given false alarm rate. This differs from signal to median clutter ratios quoted elsewhere (11, 12, 13) and is chosen since r.m.s. clutter can be directly related to σ_o, the usual clutter level measure. The difference between median and r.m.s. is not insignificant; for ν = 0.1, the r.m.s. to median ratio is 25 dB.

4.1 Single Hit Detection

As a preliminary to discussing the performance curves, it is instructive to consider the cumulative distribution of clutter. Figure 5 shows cumulative K and Weibull distributions plotted on Lognormal paper. Each K-distribution (ν = 0.1, 1, 10) is matched to a Weibull using the first two moments. The horizontal axis is threshold with respect to r.m.s. clutter to preserve the convention above. It can be seen that Weibull and K-distributions are very similar and both have negative second differentials on this plot. This implies less of a 'tail' to the distributions than Lognormal (a straight line on this paper). Perhaps more important is the relationship of the curves within a family. At high thresholds the probability of false alarm increases with the higher moments (as ν and α decrease). This is expected from the concept of 'spikiness'. However at lower thresholds the trend reverses (as it must, since the mean powers are the same). This has consequences for single hit and binary integration detection.

Figure 6 shows the single hit detection performance in K-distributed clutter. As expected, for low probability of false alarm (P_{FA}), the signal to clutter ratio (SCR) increases as the ν value decreases and the clutter becomes more spiky. For high P_{FA} (0.1), the SCR required decreases with ν for 90% Probability of Detection (PD) and wavers for 50% PD. This upset of the expected trend established for lower P_{FA} is due to the cumulative distribution effect explained in the above paragraph.

4.2 Pulse to Pulse Integration

Figure 7 compares performance using frequency diversity for three processing systems as follows:-

(i) Analogue integration - the radar video from a linear detector is summed over 10 echoes and applied to a threshold.

(ii) Binary integration - the radar video is applied to a threshold each echo and a decision is based on the number exceeding the threshold over the 10 echoes. In this case the criterion is that 6 or more must exceed the threshold.

(iii) Binary integration assuming independent clutter samples - the processing is the same as (ii) but a false assumption is made that independent samples of clutter are received each return. This is in contrast to (i) and (ii) where a compound distribution is assumed.

The results show that when the effects of correlation in clutter are taken into account, the performances of an analogue and binary system are very similar and both much degraded on the performance implied assuming independent samples. It should also be noted that a similar effect is observed for independent samples to that described under single hit detection at high P_{FA}.

Reference 11 shows that in Lognormal clutter that binary integration performance (median detection) is superior to analogue processing. The results presented suggest that this is a consequence of the assumption of independent samples and highlights the importance of correlation in performance predictions and therefore in clutter modelling.

5. SUMMARY

A compound form of the K-distribution has been used to describe the amplitude statistics and pulse-to-pulse correlation properties of high resolution, I-band, radar sea clutter. An experimental data base has been used to demonstrate qualitatively some of the spatial and larger term correlations and to derive an empirical expression for the variation of ν, the K-distribution shape parameter, with some radar and operating parameters. The sea clutter model was then applied to some fixed threshold processing predictions and results presented showing that binary integration gives very similar performance to analogue integration and that correlation has a very marked effect on performance and cannot afford to be ignored.

REFERENCES

1. Skolnik, M.I., "Introduction to Radar Systems", McGraw-Hill, 1980.

2. Nathanson, F., "Radar Design Principles", McGraw-Hill, 1969.

3. Guinard, N.W., and Daley, J.C., 1970, IEEE Proc. 58, 543-550.

4. Sittrop, H., 1977, Radar 77, 110-114.

5. Trunk, G.V., 1972, IEEE AES-8, 196-204.

6. Fay, F.A., Clarke, J., and Peters, R.S., 1977, Radar 77, 101-103.

7. Jakeman, E., and Pusey, P.N., 1976, IEEE AP-24, 806-814.

8. Rice, S.O., 1944, Bell. Syst. Tech. J. 26, 282, 1946, 24, 44.

9. Marcum, J.I., 1960, IRE Trans. IT-6.

10. Meyer, D.P., and Mayer, H.A., "Radar Target Detection", Academic Press, 1973.

11. Trunk, G.V., and George, S.F., 1970, IEEE AES-6, 620-628.

12. Schlerer, D.C., 1975, IEEE Int. Radar Conf., Washington DC, 262-267.

13. Schlerer, D.C., 1976, IEEE AES-12, no 6.

14. Jakeman, E., 1980, J. Phys. A. 13(2).

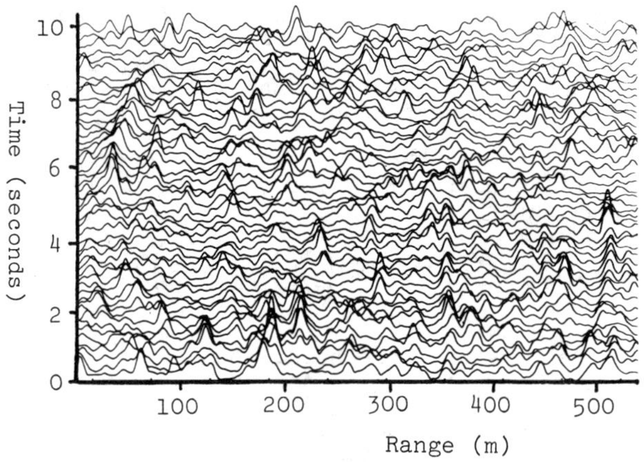

Figure 2 Range time plot of averaged sea clutter. (H pol, upswell aspect)

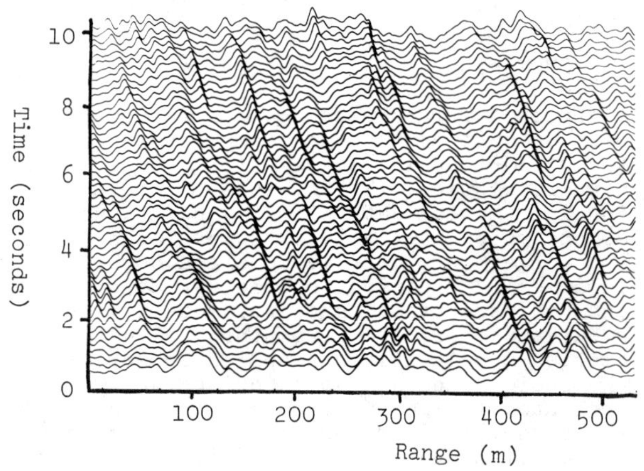

Figure 1 Range time plot of averaged sea clutter. (V pol, upswell aspect)

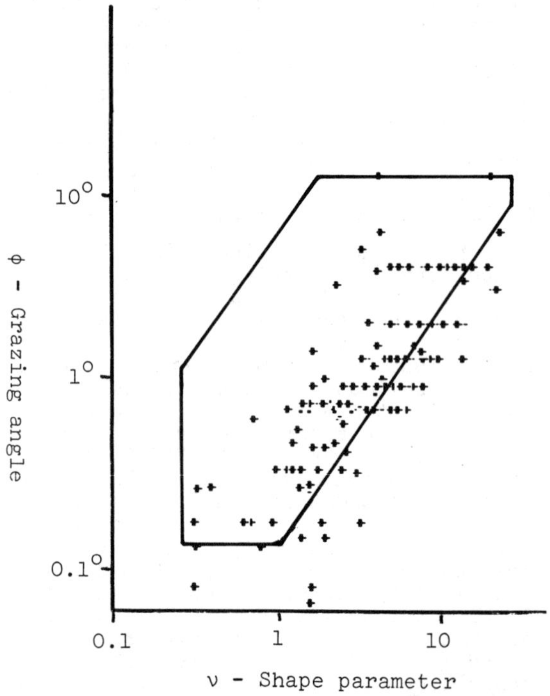

Figure 3 Scatter plot of ν vs grazing angle for V pol. (Spread of H pol is delineated region)

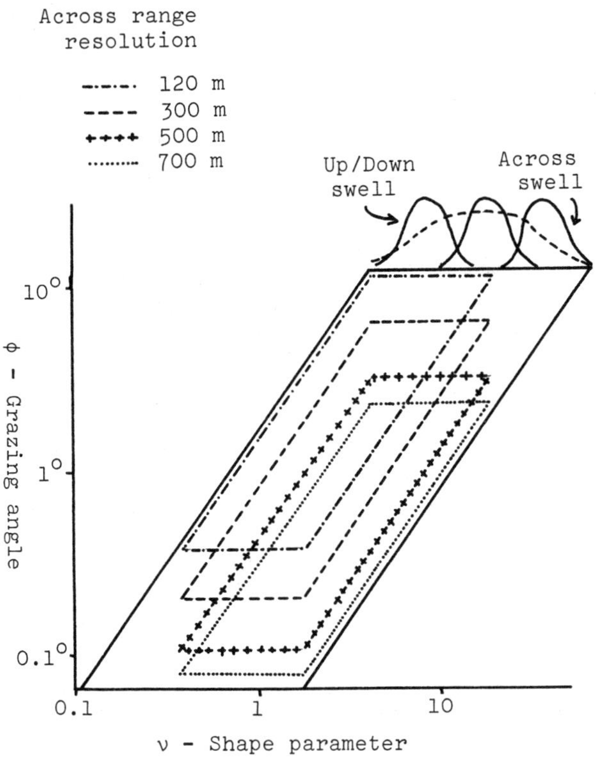

Figure 4 Schematic representation of scatter of ν for changing azimuth resolution and wave aspect.

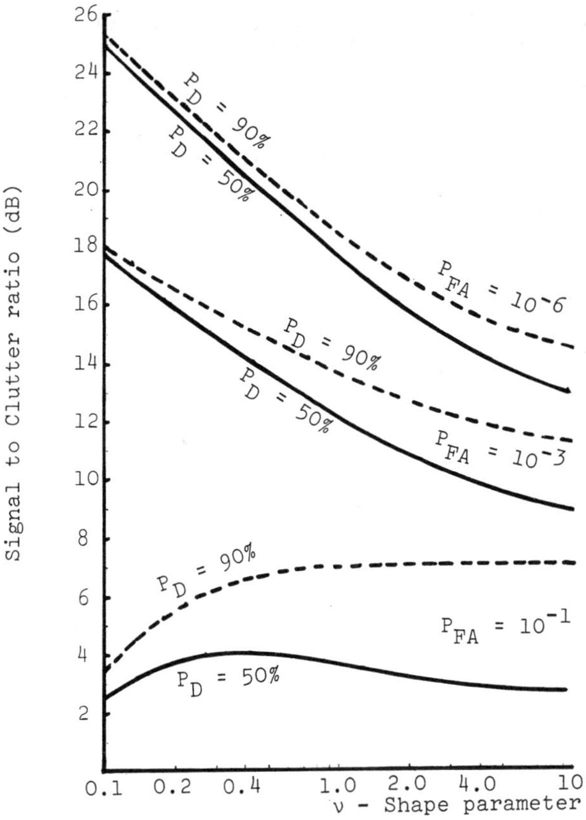

Figure 6 Single hit performance in K-distributed noise.

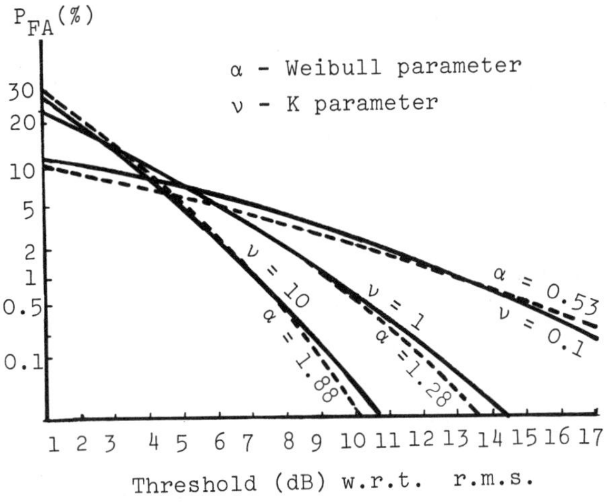

Figure 5 Cumulative Weibull and K-distributions.

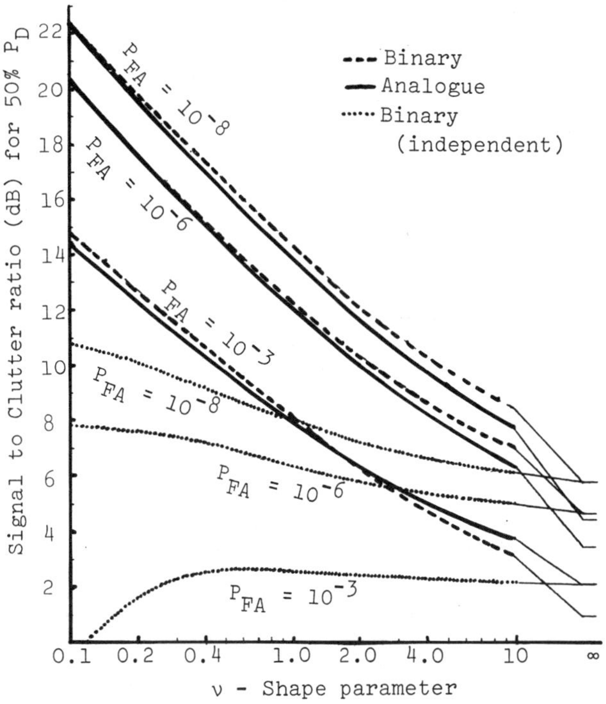

Figure 7 Performance for Binary and Analogue integration of 10 returns

MONOPULSE SECONDARY SURVEILLANCE RADAR
PRINCIPLES AND PERFORMANCE OF A NEW GENERATION SSR SYSTEM

M.C. Stevens

Cossor Electronics Ltd, Harlow, Essex, U.K.

INTRODUCTION

Secondary surveillance Radar (SSR) is becoming increasingly important as an A.T.C. tool and due to its many advantages over primary radar is becoming the principal sensor. This trend results in more stringent demands being placed on SSR performance and extends the requirement beyond that originally envisaged.

Although present day SSR works well in uncongested airspace free of interference conditions it can degrade significantly at times. Common problems are: (i) poor bearing measurement performance, so that aircraft have to be given wider separation; (ii) garble, which is a term used to describe a mutual interference condition which occurs when two aircraft are at a similar range so that their SSR replies interfere making the aircrafts' identities and flight levels difficult to determine; and (iii) false targets, which are caused by the presence of buildings and other reflecting objects and result in aircraft being detected at other than their true positions. SSR system problems are more fully described in reference 1.

The problems of poor bearing accuracy encouraged the consideration of the use of monopulse direction finding techniques, (reference 2). Not only does this approach prove effective but it also provides a source of additional signal quality information which enables other SSR system problems to be overcome. The use of monopulse data in SSR processing gives rise to what can be truely described as a quantum step in the improvement of SSR system performance.

SSR Signal Formats

A typical SSR ground station consists of a rotating antenna, a transmitter-receiver known as an interrogator and a signal processor called a plot extractor. The antenna has a narrow, high gain beam and as the antenna rotates pairs of pulses are transmitted at regular intervals. These pulses, called P1 and P3 see figure 1a, have spacings of either 8 µs or 21 µs corresponding to interrogation mode A and mode C. The two modes are, in effect, asking two questions of any aircraft receiving the pulses; mode A asks for a report of the aircraft's identity number and mode C for its flight level.

The pulses are detected by equipment, known as transponders carried on board aircraft. The transponders reply with coded messages with the format shown in figure 1b. The replies consists of two pulses F1 and F2, which are always present, and up to twelve intermediate pulses, which may or may not be present depending on the coded information being transmitted.

Monopulse Direction Finding

Figure 2 shows the three radiation patterns typical of a monopulse SSR antenna. The highly directional 'interrogate' or 'sum' pattern is the principal beam used on both the transmit and the receive paths. The difference pattern provides the monopulse function. The third pattern is used for side-lobe-suppression and is not discussed here.

The monopulse function can be best understood by considering the sum and the difference patterns of the antenna. Figure 3 shows these in more detail. The first requirement is to determine the bearing of aircraft to high accuracy and this is accomplished by measuring the ratio of the signal received in both the sum and difference channels. The ratio is converted to angle-units-from-beam-centre by means of a calibration table. Measurements can be made on both the left and right hand side of the beam centre and which applies is determined by the relative phase angle between the two signals. The bearing of the aircraft is found by adding (or subtracting) the angle within the beam to the direction of the antenna beam. Monopulse direction-of-arrival measurements are made on each individual pulse in a reply and these measurements can be averaged to increase accuracy.

The monopulse technique is useable over the full beamwidth over which replies are received. For a typical SSR antenna this beamwidth is about 4.5° wide which is about twice the 3 dB beamwidth.

Figure 4 shows an error scatter diagram plotted with the horizontal axis representing the position within the beam, and the vertical axis the error in the bearing measurement made on a reply received at that position. Each point on the diagram represents a separate reply. From the diagram one can judge the accuracy of the measurements at different positions as well as the repeatability of separate measurements made at the same position. Ideally all plots should fall on a single, straight horizontal line. Departure from a straight line is due to errors in the receivers used which distorted the ratio of the sum and difference channel signals. The scatter of points about the line is principally caused by the limited resolution in the analogue-to-digital convertors used to process the receiver signals For signals weaker than -70 dBm at the receiver mixers, the scatter increases further due to receiver noise influencing the received signal strength.

In contrast to monopulse SSR, a conventional, non-monopulse, SSR only operates on the sum channel signal and deduces aircraft bearing by calculating the direction of the centre of replies received as the antenna scans across the aircraft.

Reply Processing

The principal process in the decoding of SSR replies is the detection, by time discrimination of the pulses F1 and F2 and of the data pulses present. In previous SSR systems this was almost the only criteria applied and considerable problems arose when multiple, overlapping replies were present since replies could be missed or the data wrongly decoded. The problem is doubly significant since overlapping situations arise when aircraft are closely spaced and when good detection and decoding is most needed.

However, since in the monopulse SSR system the direction of arrival of each reply pulse is determined there is now an independent indication as to which reply an individual pulse belongs. Figure 5 shows the receiver waveforms for both sum and difference channels for an overlapping reply situation. It will be recalled that the ratio of sum to difference channel signal strengths is a measure of direction of arrival.

Solely from a timing point of view a reply decoder could find four replies in the reply signal. The next step in the monopulse decoder is to choose a reference pulse. Usually this is F1 but this choice will be inhibited if F1 could belong to more than one reply in which case F2 will be selected. If F2 also can belong to more than one reply then no valid reference is available and the reply cannot be decoded. This is more beneficial that it may appear at first consideration since this process eliminates the two "phantom" middle replies in figure 5 which only appeared because of the chance timing of pulses in the first and last replies.

Each pulse is now compared with the references. Since the first six pulses occur before the F1 pulse of the second reply, comparison is only performed with the first F1 (first reply reference). All pulses correlate with the reference, i.e. they have similar monopulse sum - difference ratios, and are declared as belonging to the first reply with high confidence - shown diagrammatically as solid vertical rectangles. The seventh pulse occupies the position of F1 of the last reply but could also be a data pulse belonging to the first reply, it cannot be used as the reference for the last reply so that the last F2 must be chosen for that purpose.

The seventh pulse is now compared with both references and correlation is found with the second reference so that a high confidence pulse is allocated to the last reply. A no-pulse (i.e. zero) is allocated to the first reply since it does not correlate with the first reference but does with the second reference. However caution is shown by labelling that decision with low confidence shown diagrammatically as a short, dotted rectangle.

The eighth pulse is treated similarly with a high confidence pulse being allocated to the first reply and a low confidence no-pulse allocated to the last reply.

In the above example the decoding process is easy to see but the process also works well in more complicated cases as shown in figure 6.

Target Reports

As the antenna rotates a number of replies are received from each aircraft. All replies from each aircraft are associated to form a target report. The criteria for association is,first, that each reply has the same range and bearing The second criteria is that there is no disagreement in the high confidence decisions between the replies. Where one reply has a low confidence decision and another a high confidence decision then the high confidence decision wins upon merging the replies.

It is normal to interrogate on two modes so as to obtain at least two mode A replies (identity data) and two mode C replies (flight level data). Since the different mode replies contain different data only replies on the same mode can be merged.

A complete target report is compiled for each aircraft containing data on range (deduced from the time of the reply following the interrogation), bearing, identity and flight level.

Tracking

Most replies received can be decoded with all data decisions achieving high confidence. After merging replies into target reports the number of low confidence decisions is further reduced. However, when two aircraft are close together it is probable that all replies from both aircraft will be garbling one another. This particular situation is one of those for which a system improvement is being demanded. The requirement is to resolve the low confidence data decisions. No further information is available from the current scan of the antenna but almost certainly the aircraft were seen on earlier rotations of the antenna and when they were were sufficiently separated as to not be in a mutual garble situation.

To make use of data obtained earlier each aircraft is tracked so that its new position can be predicted. By associating a target report with the track on that aircraft, previously obtained data can be used to confirm or correct the remaining low confidence decisions.

Fruit

Fruit (a somewhat tortuous acronym for False Replies Unsynchronised to Interrogator Transmissions) is the term used to refer to replies received from aircraft which have been triggered by other ground stations. The main problem posed by fruit replies is one of interference with the wanted synchronous replies from the aircraft.

Fruit can overlap wanted replies, to produce a form of garble and reduce decoding efficiency. In addition a fruit reply can become associated with a sequence of wanted replies to produce a code-swap error. This latter effect occurs in the following manner.

A fruit reply is received, say, following a mode A interrogation. On the next interrogation, which is on mode C a real reply is received which happens to be close enough to the apparent range and bearing of the earlier fruit reply for them to become associated and a target report compilation started. No comparison is made between the code data of the two replies since they were received on different interrogation modes. On the following interrogation, which is mode A again, a real mode A reply is received, but although it correlates in range and bearing with the earlier replies there is now code disagreement with the mode A reply data already obtained (from the fruit reply). The consequence is that the processor assumes that the latest reply is from a second aircraft and starts a second target report. The result is two target reports, one with a single, incorrect mode A reply and several mode C replies, and the other with several correct mode A replies and no mode C.

When target reports are associated with known aircraft tracks the above error becomes apparent and the correct target report produced.

Measured Performance

1. Bearing measurement. Sixteen aircraft tracks with ranges spreading from 20 miles to 208 miles have been analysed for track wander. The analysis used recorded reply data which allowed a conventional, non-monopulse, plot extractor to be simulated using the same data. Figure 7 gives a histogram of the standard deviations of the tracks for the two plot extractors. Figure 8 shows an example of one of the tracks.

2. Target detection. Table 1 gives measured results of aircraft detection probability for both the typical aircraft and the more stringent crossing aircraft case. The absolute values should be treated with caution since they include fades produced by the antenna system employed. Results are also influenced by the environment in which the equipment is operated which in this case is a typical location in England which suffers from relatively high traffic density and fruit rates. Table 1 also includes the detection probability achieved on a typical operational site using current, non-monopulse, equipment.

TABLE 1 - Aircraft detection probability

	Overall performance	Crossing track performance
Monopulse SSR	98.7%	98.0%
Current SSR	97.0%	90.0%

Figure 9 shows the detected positions of three aircraft with replies processed by conventional SSR and monopulse SSR techniques. It can be seen that where the tracks cross and reply garbling occurs the target detection efficiency of conventional SSR degrades, whilst the monopulse SSR maintains good performance.

3. Target data decoding. Table 2 gives the proportion of target reports containing complete, high confidence decoded data. The results are presented after reply correlation (target report declaration) as well as after further improvement realized by using track file information. The performance of mode C (flight level) is slightly less than that for mode A (aircraft identity) since allowance must be made for aircraft changing height which inhibits some low confidence situations being resolved.

TABLE 2 - Code performance

	After reply correlation	After tracking
Overall performance		
Mode A code complete	98.5%	99.7%
Mode C code complete	97.6%	98.7%
Crossing track performance		
Mode A code complete	95.3%	99.5%
Mode C code complete	92.0%	97.0%

ACKNOWLEDGEMENTS

The work described in this paper relates to development performed by Cossor Electronics Limited under contract to the United Kingdom Civil Aviation Authority and in collaboration with the Royal Signals and Radar Establishment. Special acknowledgement is made to Mr. J. Shaw of RSRE who processed some of the results and has made significant personal contributions to the programme. The development is based on original work performed by Lincoln Laboratory, U.S.A. and Cossor Electronics Ltd, with many enhancements.

REFERENCES

1. Stevens, M.C.,'Multipath and interference effects in secondary surveillance radar systems', Proc. IEE, 1981, Vol. 128 Pt.F, No 1, pp 1729-1735.

2. Stevens, M.C.,'Precision secondary radar', Proc. IEE, 1971, Vol. 118 No 12, pp 1729-1735.

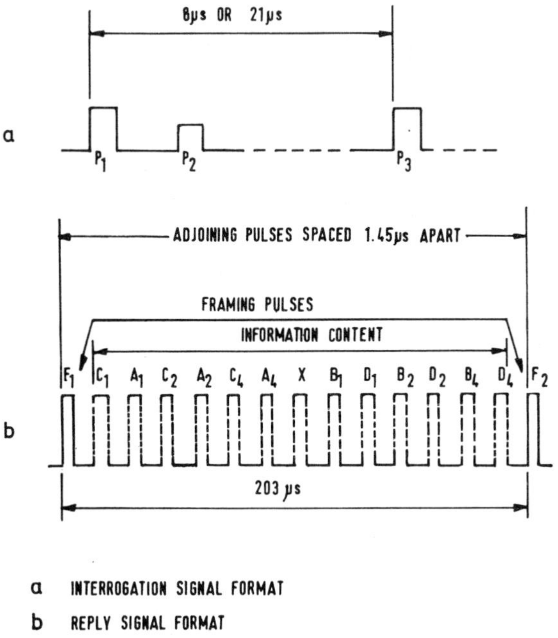

Figure 1 Interrogation and reply signal format

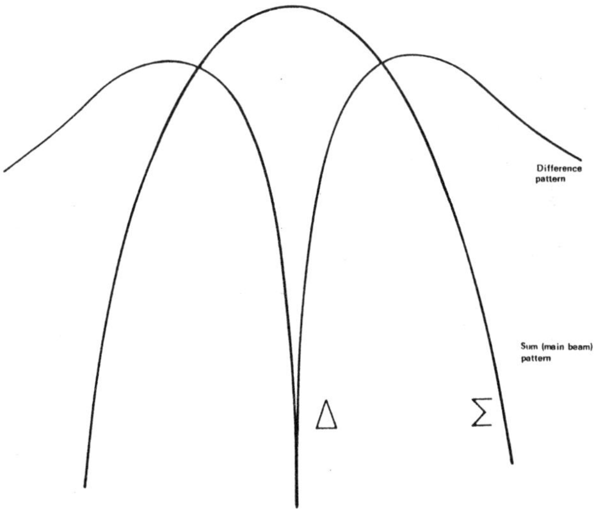

Figure 3 Main (sum) and difference antenna patterns

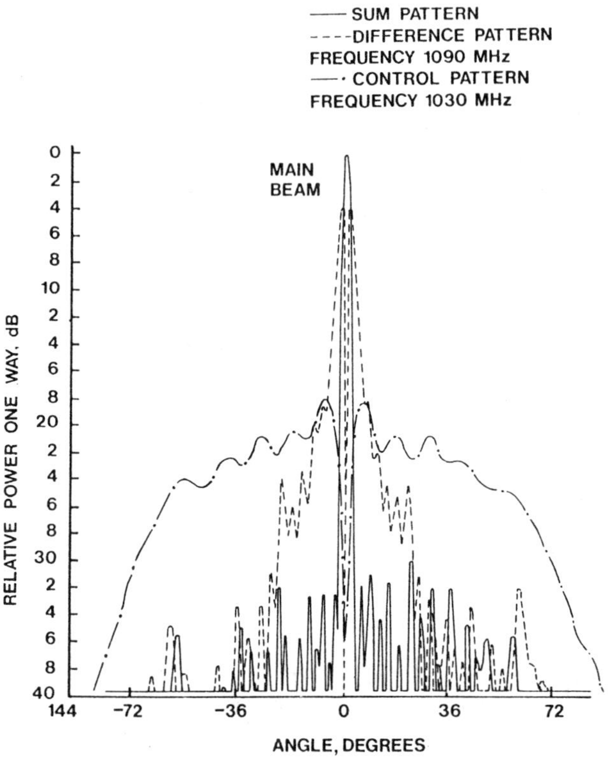

Figure 2 Three patterns of antenna

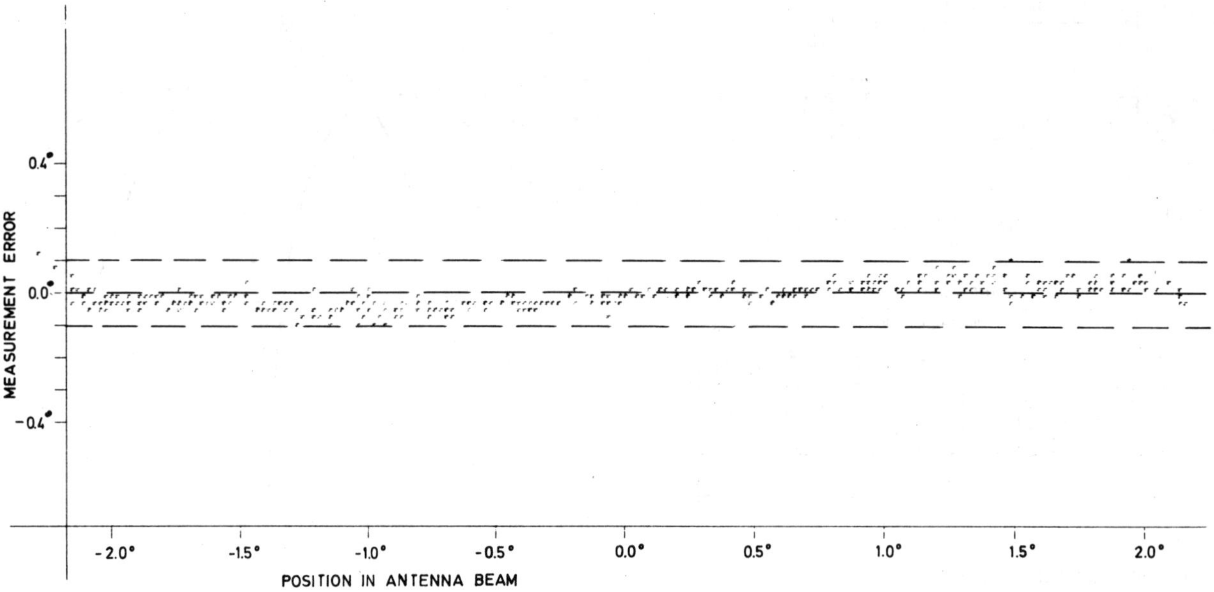

Figure 4 Monopulse Measurement errors across the antenna beam

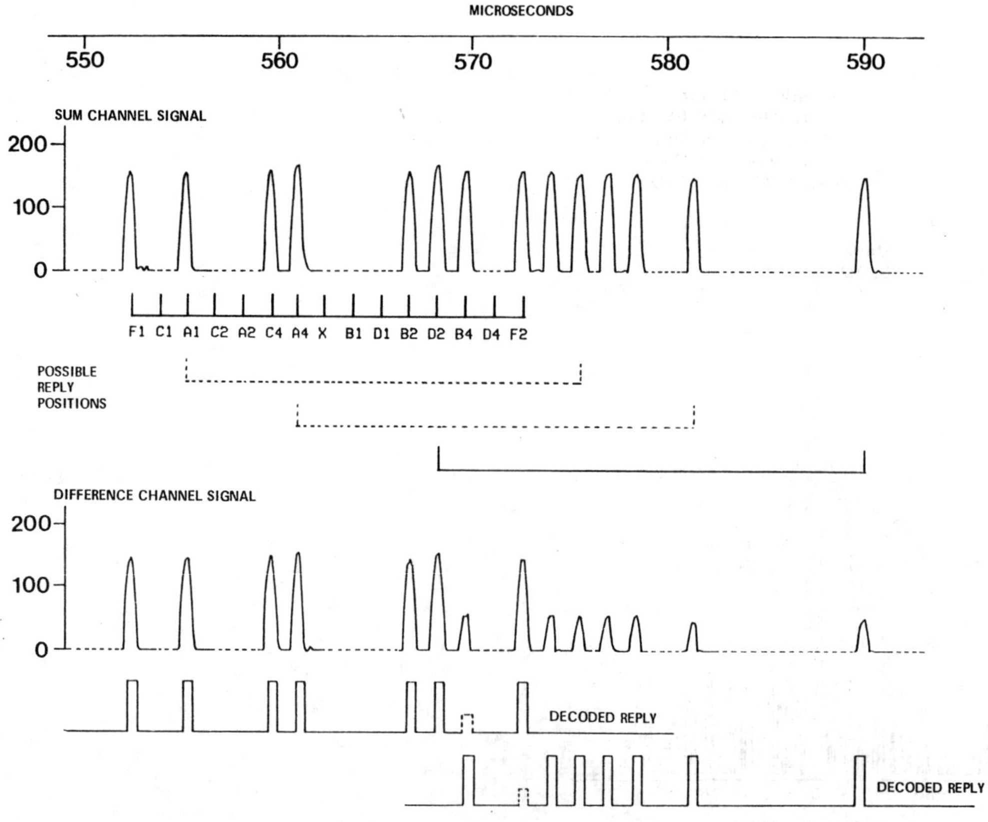

Figure 5 Sum and difference channel reply waveform

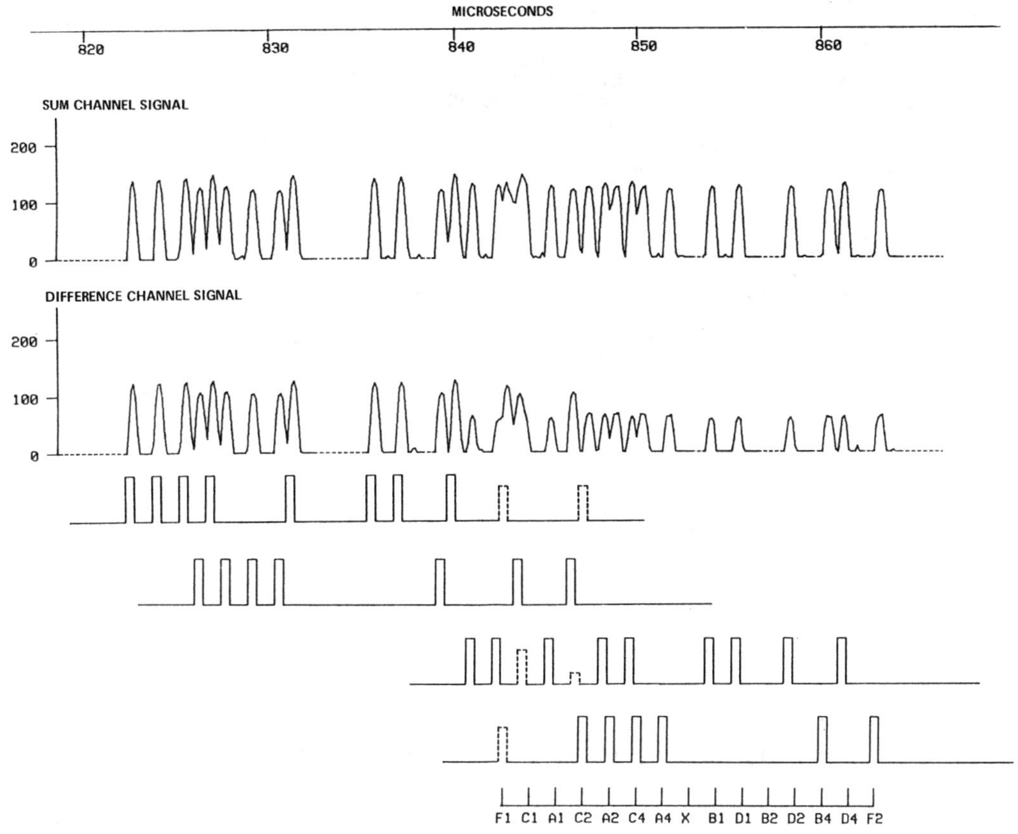

Figure 6 Sum and difference channel reply waveforms - complex example

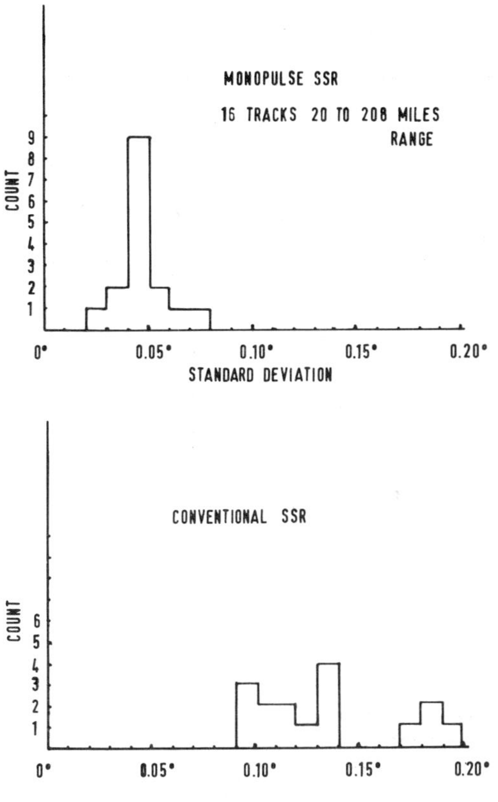

Figure 7 Histogram of bearing standard deviation

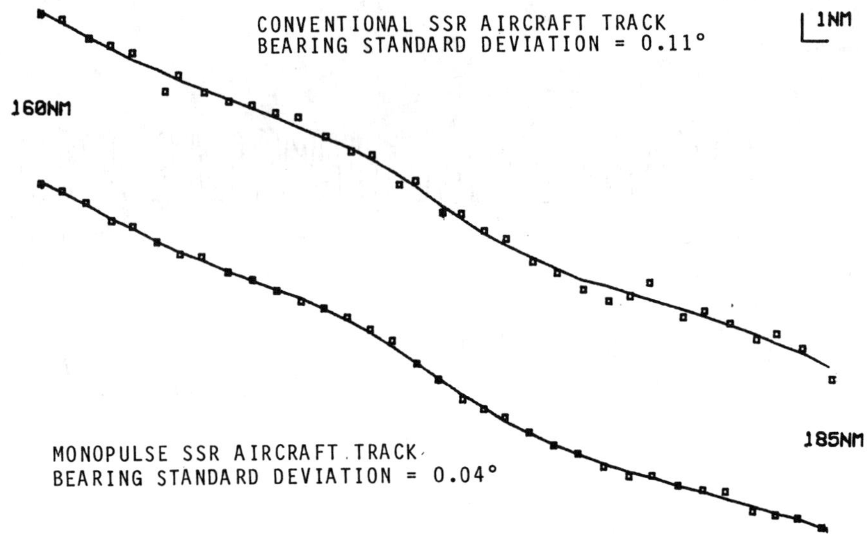

Figure 8 Same aircraft track as detected by conventional SSR and by monopulse SSR

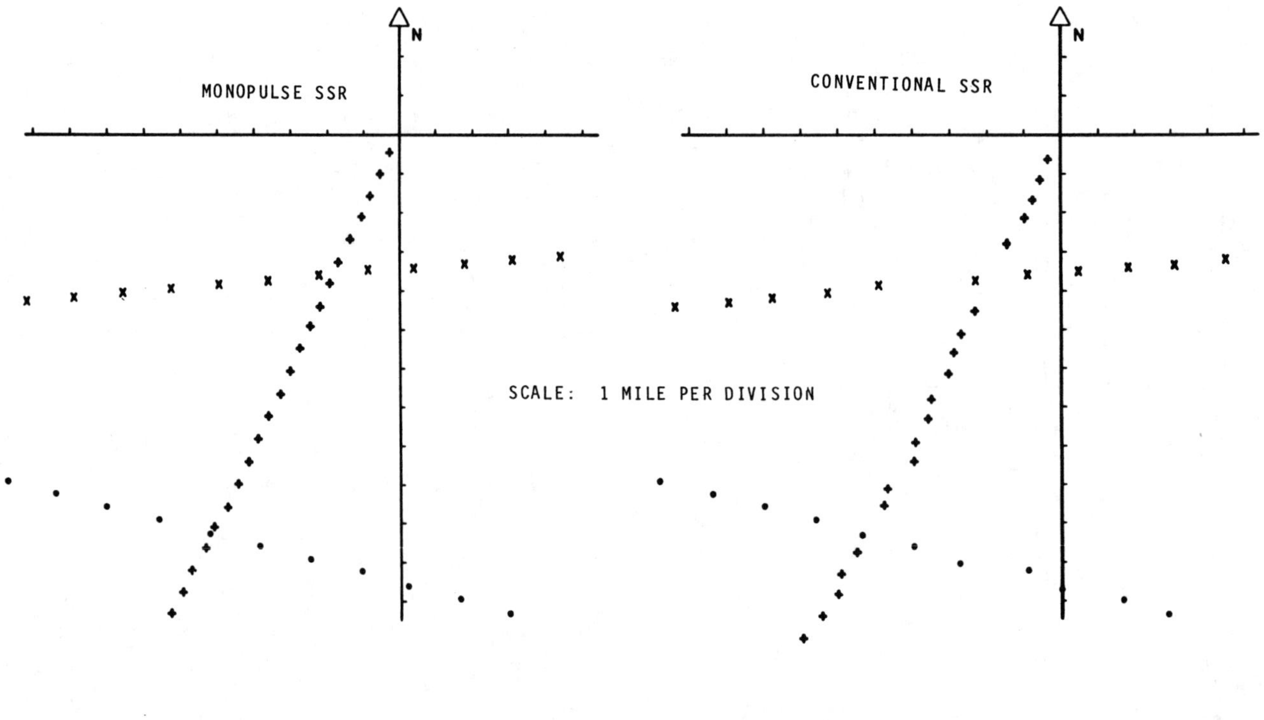

Figure 9 Crossing track performance for conventional SSR and monopulse SSR

DECODING-DEGARBLING IN MONOPULSE SECONDARY SURVEILLANCE RADAR

G.Marchetti and L.Verrazzani

Istituto di Elettronica e Telecomunicazioni, Pisa, Italy

INTRODUCTION

Secondary surveillance radar systems with monopulse target capability can perform their functions well even in synchronous code garbling conditions. For this purpose any reply detected is decoded by accompanying each information bit with a confidence bit (high or low). Then the surveillance processing is completed through a target report formation program and a report-to-track association routine.

Since in severe code garbling conditions just one wrong high confidence attribution is likely to cause a single false alarm report for each garbling aircraft, the target-to-track association algorithm may be seriously impaired. This drawback is overcome if two operating conditions are fulfilled:
- the probability of detecting code garbling is high and, whenever it is detected,
- the high confidence attribute is assigned to any code pulse only if its amplitude and the reference framing pulse amplitude overcome a prefixed confidence threshold level and the usual monopulse consistency tests are satisfied.

In this paper emphasis is placed on criteria for detecting code garbling and on a proposed confidence attribution strategy, which may be easily implemented.

GARBLING DETECTION

Secondary surveillance radar (SSR) replies can be successfully decoded,within any sweep time if each signalling pulse is detected and, simultaneously, its time-of-arrival is estimated. These combined operations will henceforth be denoted leading edge (LE) detection.

Whenever overlapping pulses are received, the true leading edge of the lagging pulse may be undetectable, and it must be deduced from the time-of-arrival of the trailing edge, since the signalling pulse duration is known.

To be more precise, the time-of-arrival of a pulse is the estimate of the time instant at the half amplitude point on the leading edge (or on the trailing edge, for a lagging overlapped pulse). This is the best choice because both the pulse duration and the reply coding frame are defined with reference to these time instants. Each LE, whether it is directly detected or not, is accompained by the corresponding monopulse estimate $\hat{\theta}$ and by the sidelobe flag (SL) which is set when a sidelobe pulse is received.

Replies are identified by searching the LE train for framing pulse pairs. Once a reply has been detected, each of the twelve code positions must be located to decide which of the code pulses exist and to detect potential garbling conditions. These occur whenever decoding windows of different replies overlap, and ambiguity arises whenever a LE is detected within the overlapped region. Ambiguity can be eliminated if there is a high probability of identifying garbling conditions. To meet this requirement two conditions are necessary:
i - each signalling pulse,connected with all replies to our interrogations,must be detected and its time-of-arrival must be correctly estimated,so that
ii - all framing pulse pairs and the corresponding decoding windows are recognized.

The first condition may be easily satisfied because, in the SSR system, the replies to our interrogations are received with a signal-to-noise ratio (SNR) which is no lower than approximately 15 dB (1). Thus, a minimum triggering level (MTL) can be chosen to ensure a LE detection probability that is practically equal to one, with a very low false-alarm rate. For example, MTL=12 dB is a fully suitable choice (2).

Condition (ii) can be fulfilled if we make due allowance for tolerance as to the shape and position of the reply pulses and for the jitter produced by noise in the pulse time-of-arrival estimation. A bracket declaration will be made, in the form of coincidence pulse, whenever two LE's are found to be spaced by $20.3 \pm \Delta F$ microseconds. Thus, to identify all F1, F2 pulse bracket pairs, ΔF must equal the sum of the maximum time tolerance as to framing pulse spacing (d=0.1 microseconds) and of the maximum error ε which, in the worst operating conditions, affects the pulse time-of-arrival estimate. ε depends on the working characteristics of the pulse detector-estimator, for a given SNR. Assuming that the detector- estimator used is the one analyzed in (2), ε cannot be larger than one half the rise or decay time of the signalling pulses. Since certain LE's may be deduced from the trailing edges, the maximum tolerated decay time equals 0.2 microseconds. Thus we obtain $\varepsilon \leq 0.105$ microseconds, if a second order intermediate frequency Butterworth filter is used with a 10 megahertz bandwidth. Consequently the coincidence window for identifying all pulse bracket pairs is $2\Delta F=0.41$ microseconds wide. The twelve decoding windows are located in the time slots, ΔC wide, spaced by $1.45\ \nu$ microseconds with respect to the LE of the reference framing pulse (ν is an integer ranging from 1 to 13, 7 excluded). Since the framing pulse F2 may be assumed as the reference, allowance must be made for a maximum pulse spacing tolerance of 0.15 microseconds so that $\Delta C=0.51$ microseconds must be used.

Let us consider two aircraft within the mainbeam of the directional ground antenna, that differ in slant range by a distance of $\delta<3.045$ kilometers so that their replies overlap. The probability P_G that a LE, pertaining to an information or bracket pulse of one reply, is detected within a decoding window of the other reply, is given by

$$P_G=[\sigma\sqrt{2}/(4d)]\sum_{k=1}^{4}(-1)^k\Big[\mathrm{erf}(x_k)+ +(1/\sqrt{\pi})\exp(-x_k^2)\Big] \quad , \quad \ldots\ldots\ldots\ldots\ldots\ldots \quad (1)$$

where:

$x_1=(\Delta C-\tau-d)/(\sigma\sqrt{2})$; $x_2=(\Delta C-\tau+d)/(\sigma\sqrt{2})$;

$x_3=(\Delta C+\tau-d)/(\sigma\sqrt{2})$; $x_4=(\Delta C+\tau+d)/(\sigma\sqrt{2})$;

$\sigma=(2\sigma_j^2+\sigma_1^2+\sigma_2^2)$, and:

σ_j=standard deviation of the transponder time jitter;

σ_i=standard deviation of noise induced error on time-of-arrival estimate of signalling pulses pertaining to one reply ($i=1,2$);

$\tau=2\delta/c$=time delay between the replies in ideal operative conditions, c being the velocity of light;

$\mathrm{erf}(x)$=Kramp-Laplace function.

In eq. (1) noise-induced error and transponder time jitter are assumed to be Gaussian independent processes with zero mean value.

P_G is a simmetrical function of τ with respect to $\tau=0$, and it is repeated at periods of 1.45 microseconds. Thus, examination of fig. 1 leads to the following remarks:

- the time jitter of transponders has unfavourable effects on degarbling, because P_G decreases at $\tau=0$ with increasing jitter, but P_G differs appreciably from zero within larger intervals of the relative time delay τ. Consequently, time jitter cause a synchronous garbling condition to last longer for a given flight configuration;
- on the basis of the delay between the framing pulses of the two overlapped replies it is best not to declare code garbling. On the contrary ambiguity must be detected on a pulse-by-pulse basis, in order to fully utilize the system's degarbling capabilities;
- whenever overlapped replies are detected, the reference bracket pulse of the lagging reply must be F2, to obtain reliable framing and monopulse references.

We have so determined the MTL, the coincidence window duration ΔF and the procedure for detecting ambiguity when garbling is produced by close targets. Code garbling may remain undetected whenever it is caused by a fruit reply. This occurs if garbling fruit has a SNR approaching the MTL, so that one or both its framing pulses may be missed and code pulses appear as isolated interferences. But the above described event is quite sporadic and the decoding techniques to be described below also give some protection from interferences of this kind.

THE DEGARBLING CAPABILITY OF THE REPLY DECODING SYSTEM

As previously stated, code extraction is initiated by the declaration of a bracket pair, which defines the possible information pulse locations (decoding windows). Since ambiguity is fairly common in garbling situations, the reply processor may not be able to decide whether or not a specific information pulse is present. Rather than force a possibly wrong decision to be made, the idea of confidence bits (high/low confidence) was developed at Lincoln Laboratory (3). Each information bit decision is accompained by a confidence bit. A pulse, whose LE falls within one of the decoding windows of a reply being decoded, is declared as:

- a high-confidence '1' (CONF=1, CODE=1), if it is received from the mainbeam (SL=0) and its monopulse sample correlates only with the reference monopulse sample $\hat{\theta}_R$ of that reply (MCT=1, CF=0);
- a low-confidence '1' (CONF=0, CODE=1), if SL=0, $\hat{\theta}$ correlates with $\hat{\theta}_R$ and also with the reference monopulse sample of another reply (MCT=1, CF=1);
- a low-confidence '0' (CONF=0, CODE=0), if SL=0 and $\hat{\theta}$ does not correlate with $\hat{\theta}_R$, or if the pulse is received from sidelobe (SL=1).

A high-confidence '0' (CONF=1, CODE=0) is declared whenever no LE falls within a decoding window. To be more precise, the monopulse correlation test is satisfactory if $|\hat{\theta}-\hat{\theta}_R|$ is lower than a prefixed value ξ, and it is always carried out to gain protection from isolated interfering pulses. The initial monopulse reference of a reply being decoded is the monopulse sample of the first framing pulse F1, except when that pulse is detected in a region that is potentially garbled by an earlier reply. As previously pointed out this procedure establishes good monopulse and framing time references only when two replies are garbling each other.

Note that the ambiguity is flagged by any low-confidence bit and it is detected only on the basis of monopulse correlation test.

A low confidence attribute must have the following meaning: the best guess is made concerning the information bit and its validation is deferred to further reply processing, i.e. the target report formation and the report-to-track association routines. The first routine attempts to combine all the

replies received from a transponder, within each beam dwell time, into a single target report. It operates on the basis of range, azimuth and code correlation tests. Code correlation is done by comparing only high-confidence information bits of a new reply with the high-confidence information bits of the code estimate in the target-report file. This code estimate is updated by adding the new high-confidence positions of the correlating reply. Consequently the probability P_W of a wrong high-confidence declaration assesses the effectiveness of the decoding-degarbling process. P_W is given by a quite complicated formula that requires numerical computation (1). The results show that the following basic limitations exist: if the monopulse correlation constant ξ equals a suitably small fraction of the mainbeam traverse aperture α_H, P_W may rise to unaccettable values ($P_W \simeq 0.25$) whenever:
- ambiguity arises from synchronous garbling replies radiated from airborne transponders at close bearings and
- one reply undergoes a deep fade due to an adverse flight attitude.

In this conditions information pulses of the weakest reply are too frequently assigned as high-confidence '1''s to the strongest reply. Decoding errors occur because the monopulse samples of the low amplitude pulses fail to correlate with the right reference monopulse sample and the ambiguity remains undetected. As a consequence, target reports may be formed with a false code estimate, which impair the surveillance process. The probability P_W is satisfactorily approximated by:

$$P_W \simeq 1-\mathrm{erf}(\xi/2\sigma_\theta) \text{ , ...} \qquad (2)$$

for any value of practical interest ($P_W<0.01$) for close aircraft. Here σ_θ is the standard deviation of the noise-induced error on $\hat{\theta}$ .Eq. (2) leads to the conclusion that 2ξ must be equal to a large fraction of the mainbeam traverse aperture α_H, in order to limit P_W below an acceptable value, in all system operating conditions. But, thus doing, the angular resolution capability of the monopulse receiver is not fully exploited at all. Indeed, the maximum value of σ_θ is $0.18\ \alpha_H$, for the minimum SNR of replies to our interrogations, and P_W is no higher than 0.01 if $2\xi=0.65$. This result confirm the relevant conclusion give above. To overcome the previously mentioned drawback we proposed (1) a mainbeam (SL=0) information pulse be assigned as a high-confidence '1' to a reply being decoded,whenever one of the following conditions is met:
- the pulse LE occurs at a time instant pertaining only to one of the decoding windows of that reply (W=0) and the monopulse correlation tests are successful (MCT=1; CF=0) or
- the pulse LE occurs in a time-slot common to another decoding window (W=1), the monopulse correlation tests are successful (MCT=1, CF=0) and both information pulse amplitude and reference pulse amplitude overcome a prefixed confidence threshold level.

This decoding-degarbling procedure entails minor modifications for the reply processor. Indeed, an amplitude comparator must be added to the pulse detector-estimator circuit and one extra bit (CTL) must be transmitted to the reply decoders which must exchange one more flag (W).

It should be noted that replies with a low SNR, when received in synchronoys garbling situations, will produce target reports with a low quality code estimate (many low confidence bits) because of the confidence threshold test. But these target reports do not impair the surveillance processing; indeed low-confidence attributes can be removed by means of the report-to-track association routine. Moreover, the confidence threshold test makes it possible to control the effects of deep fade in transponder radiation patterns.

We must point out that the proposed decoding-degarbling procedure performs as well as that of Lincoln Laboratory at ranges usually lower than 50 nautical miles. Indeed, in these operative conditions, there is sufficient fade allowance in both uplink and downlink power budgets to ensure that ambiguity is very likely to be detected and that, simultaneously, the confidence threshold test is satisfied.

Besides the confidence threshold level can be chosen in order to bring about a considerable reduction in the rate of target report formation with a false code estimate, at extended operative ranges. Thus, we can fully exploit the monopulse resolution capability of a monopulse receiver. This can be evaluated by means of the system angular resolution $\Delta\theta$ which is defined as the target azimuth separation necessary to obtain the following result: that the probability of correctly assigning high-confidence '1''s is equal to one half the probability that the same decision is made in the absence of garbling. Computations produces the following general result:

$$\Delta\theta=\xi .$$

Therefore ξ must be chosen to obtain the required azimuth resolution and,consequently, the confidence threshold is assessed in order to fix the specified upper bound to the probability P_W.

We are now in a position to state that in practice two replies, radiated from garbling aircraft differing in azimuth by 2ξ, produce the following results:
- each isolated information pulse is correctly decoded as a high-confidence '1';
- overlapped pulses, at worst, are decoded as low-confidence '1's for both garbling replies.

The modified decoding-degarbling strategy also works well for stochastic-response SSR systems (4) with target monopulse capability. In these systems code falsification due to code garbling prevents a correct target-report formation, even if replies are separated within any beam dwell time by means of the stochastic-response mechanism.

In conclusion, to obtain a good angular separability within an extended system operating range, the following design guidelines must be adopted:
- ξ is chosen to obtain the required azimuth separability,
- the design signal-to-noise ratio γ_D is obtained by solving eq. (2), assuming P_W to be equal to the prefixed upper bound, taking into account that σ_θ equals $\alpha_H/\sqrt{\gamma_D}$ (1); γ_D must be consistent with the uplink and downlink power budgets;
- the confidence threshold level is chosen to ensure a high probability (e.g. equal to 0.99) of its being overcome by a signalling pulse with the design SNR.

As an example, assuming $\Delta\theta$ is equal to 0.1 α_H and $P_W \leq 0.01$, we obtain $\gamma_D \simeq 26$ dB and a confidence threshold level of 25 dB (a pulse with a 25 dB SNR overcomes this level with a probability equal to 0.5). Four identical reply decoders (3),(5) are used to permit the handling of up to four overlapped replies. Each decoder receives the following data from a bus:
- LE=1 (0), pulse present (pulse absent);
- SL=1 (0), pulse is received from a sidelobe (from the mainbeam);
- CTL=1 (0), pulse is over the confidence threshold (is below that threshold);
- $\hat{\underline{\theta}}$, the monopulse sample for each pulse detected.

The reply decoders exchange two flags:
- W=1 (0), a decoding window is being explored (is not being explored);
- CF=1 (0), the monopulse correlation test is successful (fails or is not executed).

Each decoder is started (through a selector) by a bracket-pulse-pair coincidence pulse, and, when starting, it stores the CTL flag and the monopulse sample $\hat{\underline{\theta}}_R$ of the reference framing pulse. Afterwords, the decoder processes only the data occurring within each of the twelve decoding windows of the assigned reply. The operations performed to decide the confidence bit and the code bit are described in the flow chart of fig. 2. Operations start at the beginning of each decoding window and, in any case, they stop at the end of the window, when OW is reset to zero. Fig. 2 is self-explanatory in view of the previous descriptions and of the following specifications: the flag CTL° is the logical product of the stored reference CTL with the CTL of each incoming information pulse. Note that the CTL° flag is used to decide on a pulse-by-pulse basis if ambiguity occurs when a LE falls within overlapped decoding windows. Moreover, we must point out that operations stop before the end of a decoding window if a high-confidence '1' is decided on, but whenever a low-confidence decision is made, the remaining portion, if any (OW=1), of the decoding window is explored once in order to search for another LE. This repetition is suitable for fully exploiting the intrinsic degarbling capabilities of the monopulse receiver and of the pulse detector-estimator which is extensively described in (2).

CONCLUSIONS

The correct assignment of the high confidence bits to SSR decoded replies is of primary importance as far as the degarbling process is concerned. The decoding-degarbling strategy, based only on the monopulse correlation tests, works well within those operating ranges where there is a sufficient fade allowance in both uplink and downlink power budgets. But at extended ranges, decoding errors, i.e. uncorrect high-confidence bit assignment, may frequently occur, when garbling is present, because of adverse flight attitudes. This occurrence may impair the surveillance process since:
- if a track is wrongly initiated, several scans may be required to obtain aircraft identity and altitude;
- if a track has failed to correlate with a new target report for a specific number of consecutive scans, it is dropped and a false track may be initiated.

This drawback is overcome and the azimuth resolution capability, intrinsic in the monopulse receiver, is fully exploited if any information pulse is decoded as a high-confidence '1', whenever the usual monopulse consistency tests are satisfied, the ambiguity is detected on a pulse-by-pulse basis and, if it occurs, both the information pulse amplitude and the reference framing pulse amplitude overcome a suitable confidence-threshold level. Thus doing, the uncorrect attribution of high-confidence '1"s is limited by the monopulse correlation tests in the case of replies with SNR larger than the confidence level, and it is limited by thresholding when either or both garbling replies fade down.

The proposed decoding-degarbling strategy is suitable for extending the operative range of monopulse SSR systems, the stochastic-response system included, since it allows correct target report formation with full utilization of the angular separation capability of monopulse receiver.

REFERENCES

1. Marchetti, G., Picchi, G., and Verrazzani, L., 1982, "Reply Processing and Angular Resolution in Monopulse SSR Systems", ETS, Pisa, Italy.

2. Marchetti, G., Picchi, G., and Verrazzani, L., 1980, IEEE Trans. on AES, vol. AES-16, No. 3, 294-303.

3. Nelson, R.G., and Nuckols, J.H., 1977, "A Hardware Implementation of the ATCRBS Reply Processor Used in DABS", Project Report, FAA-RD-77-92.

4. Milosevic, Lj., 1976, L'onde électrique, Vol. 56, No. 12, 499-504.

5. Marchetti, G., Russo, F., and Verrazzani, L., 1979, "Architettura del Rivelatore-Elaboratore delle Singole Risposte per la Sezione SSR di un Sistems DABS", Congress Proceedings "Il Radar nel Controllo del Traffico Aereo", Roma, Italy.

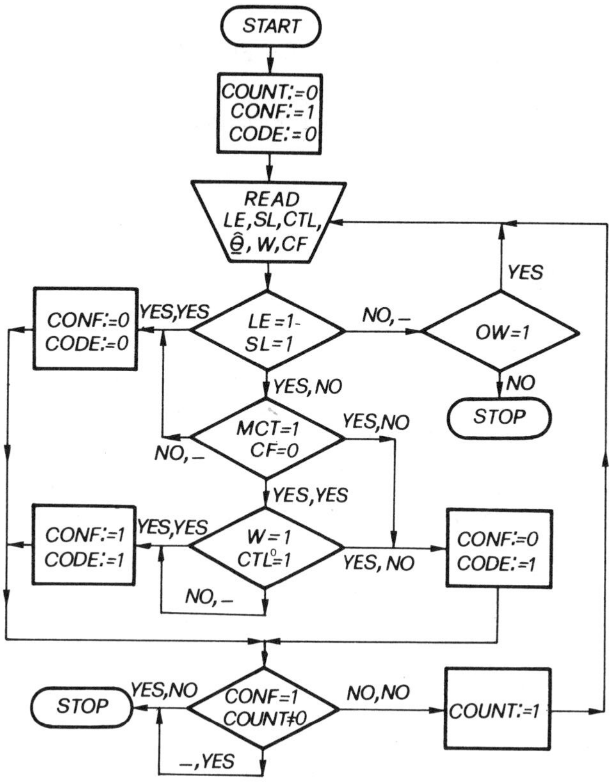

Figure 2 Flow chart of operations performed in each decoding window.

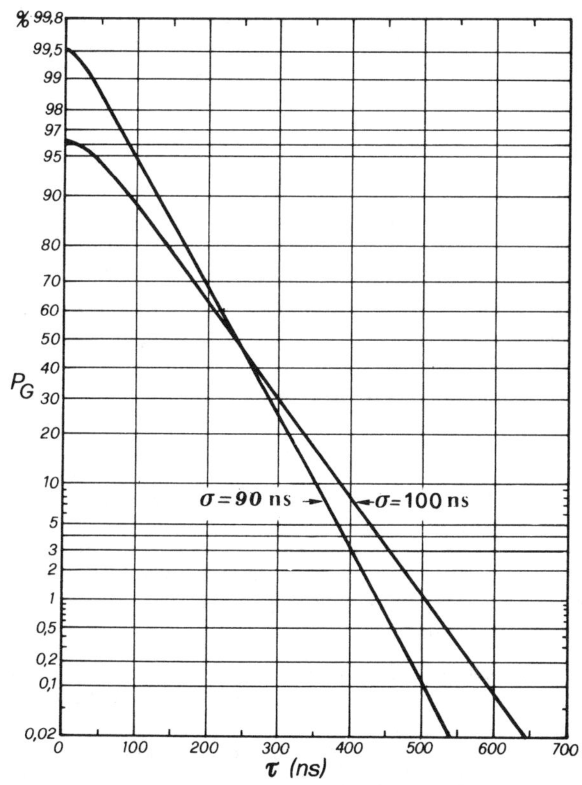

Figure 1 Garbling probability P_W vs. relative delay τ.

EVALUATION OF ANGULAR DISCRMINATION OF MONOPULSE SSR REPLIES IN GARBLE CONDITION

G. Benelli[o],M.Fossi[o],S.Chirici[oo]

[o]University of Florence, Italy. [oo]Whitehead Motofides,Leghorn,Italy.

INTRODUCTION

Modern systems for Air Traffic Control,such as DABS and ADSEL,utilize a monopulse sensor to obtain the azimuth estimate for aircraft. Even as regards the SSR functioning mode,monopulse sensors make possible some important improvements with respect to traditional systems,especially when there is inteference.In particular,the use of monopulse receivers allows a more reliable estimate of the angular positions of targets,at the same time permitting a reduction in the number of interrogations of the sensor during the dwell-time of the target.

A typical SSR monopulse receiver may be seen as consisting of three subsystems:the Reply-Time Processor (RTP),the Dwell-Time Processor (DTP),and the Scan Time Processor (STP),according to the three processing levels of the signal that those receivers implement.In particular,the RTP subsystem decodes the sequence of impulses received,providing a succession of reply reports in output,with the corresponding range,azimuth,and code estimates.The DTP analyzes the replies received during each dwell-time,for the purpose of associating the replies coming from one and the same aircraft,according to suitable criteria breifly mentioned here.The reply association algorithm is based on the consideration that owing to the short duration of dwell-time,the N consecutive replies from the same aircraft must have the same range,azimuth and code.A target is considered to be detected if at least L ($L \leq N$) replies correlate in range,azimuth and code.

The next subsystem,STP,analyzes the raw target reports received in consecutive scans and takes care of updating the tracks and eliminating false information.

Use of the monopulse receiver offers various advantages.As already mentioned,the accuracy of the azimuth estimate remains very high,even when there is interference;moreover,the availability of the azimuth estimate for each impulse,makes it possible to reduce the number of false alarms and false associations.

This paper describes the characteristics and analyzes the performance of an SSR monopulse receiver operating according to the DABS system protocols,limiting the research to the RTP and DTP sections and simulating conditions of operation corresponding to complete,permanent superimposition of the replies coming from two aircraft (garble).Thus it is supposed that the two aircraft have the same range,and the impulses of one reply are completely superimposed on the corresponding impulses of the other reply.The analysis is performed by suitable modeling : i)the hardware sections of the transmission and reception antennae and the monopulse receiver; ii)the software sections of RTP and DTP.Utilizing the devised models, a complete computer simulation of the system was performed.The probability of detecting and discrminating the two targets with correct codes is determined as a function of the relative power of the two signals,the azimuth separation and the signal-to-noise ratio.Some results are presented which show that the SSR monopulse section of DABS is capable of achieving satisfactory performance in critical synchronous garble situations.

DESCRIPTION OF THE SSR MONOPULSE RECEIVER

In this section the SSR monopulse receiversubsystems in relation to which the simulation was developed,are described.The block-diagram of this receiver is shown in fig.1.The receiver may be considered to be divided in three parts: i)Monopulse Receiver;ii)Reply-Time Processor;iii)Dwell-Time Processor (1).The analysis carried out is limited to a single antenna scan,and the process of target-track association is not taken into account.

As has already been stated,the analysis was conducted considering two aircraft under total garble conditions,that is ,assuming that the replies from the two aircraft are completely superimposed ,and on the other hand excluding all other possible causes of interference.

The receiver considered is of the monopulse amplitude-comparison type.As is well known, in such a system the information about the angular position of the target is obtained by a sensor consisting of two antennae with staggered beams,such that the amplitudes of the signals received depend on the angular position of the target with respect to the sensor (amplitude-sensitive antenna).In output from the antenna system,reference is made to the signals sum Σ and difference Δ , related to the replies from the two aircraft,to which Gaussian white noise is added.

We indicate with $G_{\Sigma}(\Theta)$ and $G_{\Delta}(\Theta)$ the tension gains of the sum and difference radiation pattern of the monopulse antenna for a certain off-boresight azimuth Θ;with A_1 and A_2,the amplitude of the impulses of the reply from the first and the second aircraft respectively . In addition, let $X_{n\Sigma}$ and $Y_{n\Sigma}$ be the noise components ,in phase and quadrature,of the sum channel; and $X_{n\Delta}$ and $Y_{n\Delta}$,those of the diffe rence channel.The complex envelopes of the sum Σ and difference Δ signals received may be written

$$\underline{\Sigma}=A_1G_\Sigma(\theta_1)+A_2G_\Sigma(\theta_2)\ \exp(j\varphi)+X_{n\Sigma}+jY_{n\Sigma}$$
$$\underline{\Delta}=A_1G_\Delta(\theta_1)+A_2G_\Delta(\theta_2)\ \exp(j\varphi)+X_{n\Delta}+jY_{n\Delta} \quad (1)$$

where θ_1 and θ_2 represent the off-boresight angles of the first and second aircfrat and φ the phase displacement of the signal from the first aircraft with respect to the signal from the second aircraft.This displacement is assumed to be an aleatory variable with uni form distribution between O and 2π or,alter natively,Gaussian distribution with standard deviation $\sigma \ll 2\pi$;these hypothesis express the two extreme conditions of low correlation or high correlation between the initial pha ses of the pairs of the replies from the two transponders of the aircraft.
The monopulse receiver processes the signals received by the two antennae,though two dis tinct channels,until they are brought to the amplitude levels required for the subsequent processing;this consists in determining,accor ding to a prefixed optimality criterfon,the optimum strategy for detecting the target and estimating its parameters,in particular its azimuth.

To this end, the sum signal received is sent into a threshold detector to detrmine whether in a certain position there is an impulse. An impulse is detected if

$$10\ \log_{10}|\underline{\Sigma}|^2/2\sigma^2 \geqslant S_1 \quad (2)$$

S_1 being the Threshold in dB and σ^2 the noi se variance.For each impulse detected,the re lated $\underline{\Sigma}$ and $\underline{\Delta}$ signals are sent to a monopul se processor which operates according to the criterion of the ADSEL receiver (2,3).This ty pe of receiver implements the hyphothesis that the phases of the carriers of the signals re ceived are unknowns and not necessarily the same in each channel.In output from the mono pulse processor the signals

$$S = \text{sign}\left\{\text{Re}\left[\underline{\Delta}/\underline{\Sigma}\right]\right\}$$
$$V = \ln\left\{\text{Mod}\ \underline{\Delta}/\underline{\Sigma}\right\} \quad (3)$$

are had .This parameters are used by the an gle estimator to estimate the angle θ_{es} from which comes the detected impulse.

We indicate with $E_T(x)$ the normalized antenna pattern,that is

$$E_T(\theta)=G_\Delta(\theta)/G_\Sigma(\theta) \quad (4)$$

The angle estimator utilizes the inverse fun ction $E_T^{-1}(x)$ to calculate the angle of origin of the impulse using the signal x=S V.

Finally ,every impulse detected is classified by main lobe or side lobe (RSLS function) ac cording to whether the following relationship is satisfied or not :

$$20\ \log_{10}\left\{|\underline{\Sigma}|/|\underline{\Delta}|\right\} > \text{RSLS} \quad (5)$$

where RSLS is the threshold value expressed in dB.

The next block (Reply Report Declaration) ta kes care of declaring the presence of a reply if it the following conditions related to the frame impulses F_1 and F_2 are verified :
-i) F_1 and F_2 have been detected;
-ii)F_1 and/or F_2 has been declared to origina te from the main lobe.
For each detected impulse falling between the two frame impulses,assignement to the reply under examination is attempted on the basis of the decoding algorithm illustrated in fig.2 (1).Note that a flag of high (H) or low (L) reliability is associated to each impulse pre sent in the reply (O or 1);this flag will la ter be utilized in the process of correlation of the individual replies in code.

The Dwell-Time section of the SSR receiver per forms the operation of associating the replies received into groups.A reply is associated to a group if it satisfies the azimuth and code correlation tests,since in our hyphothesis the range correlation test is assumed to be alwa ys satsfied.In the case where the reply does not correlate with any group,it starts a new group.Subsequent replies attempt correlation with all the existing groups in the order in which they are generated.The azimuth test is not done if the reply and/or the group with which it is attempting correlation does not have a reliable monopulse estimate.

The azimuth of the reply is estimated by means of the following expressions (1) :

$$\theta^j_{rp} = \theta^j_{bs} + \theta^j_{ob} \qquad j=1,2,\ldots,N \quad (6)$$

where j indicates the sweep to which the re ply belongs and N the total number of interro gations during the Dwell-Time of the target. The reference azimuth θ_{gr} for the group is set equal to that of the last reply in the group that is provided with a monopulse esti mate :

$$\theta^j_{gr} = \theta^{j-k}_{bs} + \theta^{j-k}_{ob} \quad (7)$$

where k-1 indicates the number of any last consecutive sweeps for which there is no mono pulse estimate in the group.

The azimuth test is successful if :

$$\left|\theta^j_{rp}-\theta^j_{gr}\right| = \left|-2(\theta_T/N)k+\theta^j_{ob}-\theta^{j-k}_{ob}\right| \leqslant \Delta\theta_{max} \quad (8)$$

where $\theta^j_{bs} - \theta^{j-k}_{bs} = -2(\theta_T/N)k$, havingassumed the direction of rotation of the sensor anten na to be counterclockwise,and indicating the angular dynamics of the monopulse estimator by $2\theta_T$.

Once azimuth correlation test has been done, the code correlation test is carried out.A re ply and a group satisfy the code correlation test if all the high-reliability bits turn out equal.This condition is set both for the mode A (identity) codes and for the mode C (altitude) codes.When a reply correlates with a group in azimuth and in code,both the azi muth and the code of the group are updated. The group's azimuth is initially set equal to that of the reply that starts it,and updating takes place according to the following procedures (1):

-i) if the group and /or the reply those not have a monopulse estimate ,we set

$$\theta_g^{n+1} = (\theta_g^n + \theta_{rp}^{n+1})/(n+1) \qquad (9)$$

where n indicates the number of replies in the group under examination before association of the last reply.

-ii) if the group only has replies whose monopulse estmate lies to the right of the direction of aim (boresight) of the antenna,and the monopulse azimuth estmate of the reply under consideration is also "right",we pose:

$$\theta_g^{n+1} = \theta_{rp}^{n+1} \qquad (10)$$

-iii) if the group only has replies from the right,and the reply in consideration comes from the left ,we pose :

$$\theta_g^{n+1} = (\theta_g^n + \theta_{rp}^{n+1})/2 \qquad (11)$$

In this last case,any further replies do not influence the final azimuth estimate of the group.

Once a reply has correlated with a group,the symbols of the code and the related reliability preceed to be updated according to table 1.

		New Reply			
		0H	0L	1L	1H
Group	0H	0H	0H	0L	1L
	0L	0H	0L	0L	1H
	1L	0H	0L	1L	1H
	1H	1L	1H	1H	1H

Table 1.Code Update Rules.

SIMULATION PARAMETERS AND RESULTS

To evaluate the performance of the SSR monopulse receiver described in the previous section, under the operating conditions specified in section 1,a computer simulation was performed of the entire system described.

Simulations were carried out for two different types of antenna.The first is a dummy antenna, whose sum pattern is a Gaussian curve and whose difference pattern is a straight line.The second (fig.3) is a real antenna of the type used for air traffic control.The results here presented refer to this real antenna.

In relation to the codes of the two aircraft, the results presented refer to average conditions characterized by codes having a number of impulses equivalent to 6 out of 12,of which half are positioned in the same locations for the two aircraft.

Below are described the principal parameters that take part in characterizing the simulation performed:

P_1 (P_2) power of the signal emitted by the transponder of the first (second) aircraft,estimated at the peak of the sum pattern of the monopulse antenna;

S_N power of the Gaussian white noise,estimated at the peak of the sum pattern o of the monopulse antenna;

θ angular separation between the two aircrafts;

θ_{rp} angular separation between two successive replies;

α beamwidth of the sum antenna (points at -3dB of the power gain);

RSLS declaration threshold of the side-lobe impulses;

S_1 azimuth correlation test window for decoding the replies;

$\Delta\theta_{max}$ azimuth correlation test window of the Dwell-Time Processor;

L minimum number of replies for the groups declared;

DP_1 declaration probability of the first aircraft with codes exact;

DP_2 declaration probability of the second aircraft with both codes exact;

DP_{12} declaration probability of the two aircraft with codes exact;

NH_1 (NH_2) average number of bits with high reliability in the codes of the first (second) aircraft;

NR_1 (NR_2) average number of replies in the groups which declare the first (second) aircraft;

NG average number of declared groups.

Some results obtained from the computer simulation of the described system are reported in the figs. 4-6.In fig.4 the probability DP_1 and DP_2 of declaring with exact codes the first and the second aircraft respectively as a function of the second aircraft power P_2 are reported.The power P_1 of the first aircraft is set equal to 15 dB and the angular separation θ is set equal to 4 degrees.The curves are drawn for two different values of the noise power S_N and the threshold RSLS.In fig.5 the probability DP_{12} of declaring in the same time both the aircfrat is shown,for the same parameter values of fig.4.Finally,in fig.6 the quantities NH_1,NR_1 and NG are reported.

From fig.4, it follows that the probability DP_1 of declaring the first aircraft is high if the signal power of the second aircraft is lower than 10 dB.Neverthless,such a probability is strongly dependent from the RSLS threshold and the signal-to-noise ratio,particularly when the signal powers of the two aircraft become comparable.

From fig.6 it is also clear that the average number NR_1 of the replies which form a declared group decreases until to minimum value equal to 2 ,which is the threshold of group declaration.

CONCLUSIONS

In this paper some results on the performance of a SSR monopulse receiver for Air Traffic

Control in a particulary hard garble condition obtained through a computer simulation,are reported.The results,showed in this paper and others,obtained by the computer simulation, permit to characterize the degarbilng capability of a SSR monopulse receiver and in particular to its subsystems.

This work has been supported from Selenia Company ,Italy.

REFERENCES

1. Gertz, J. L.,1977,"The ATCRBS Mode of DABS" MIT Lincoln Laboratory Report,ATC 65.
2. Stevensen, M. C.,1971,"Precision Secondary Radar", Proc. IEE , 118,pp.1729/1735.
3. Biffi Gentili, G., Chierchini, F.,and Giuli, D., 1978, "Studio sugli Effetti della Propagazione e della Interferenza sulla Stima Monopulse" ,National Council of Research Report II/A/4.8.

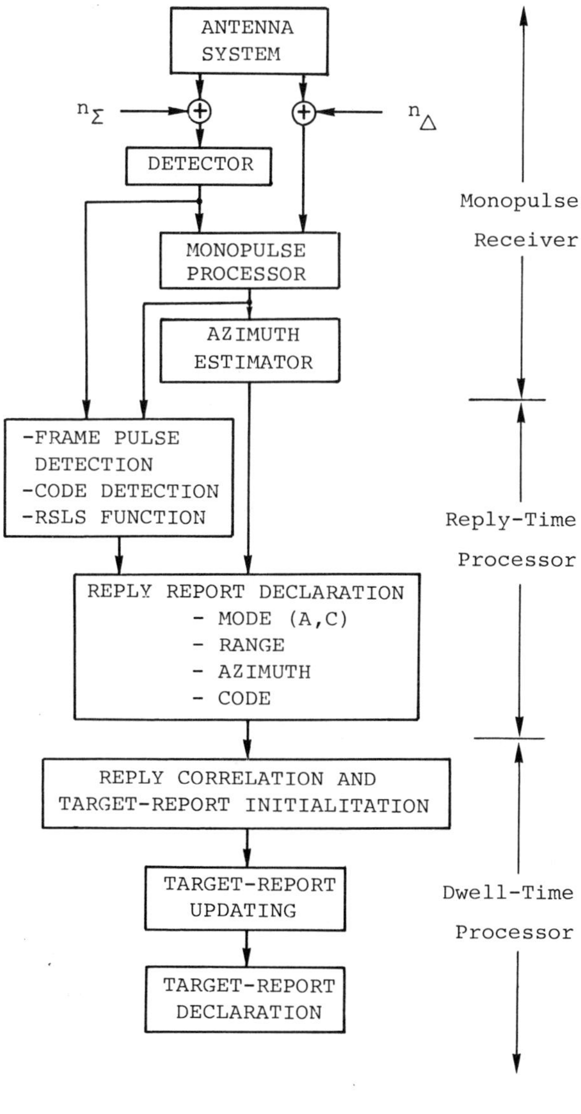

Figure 1.General block-diagram of the SSR monopulse receiver.

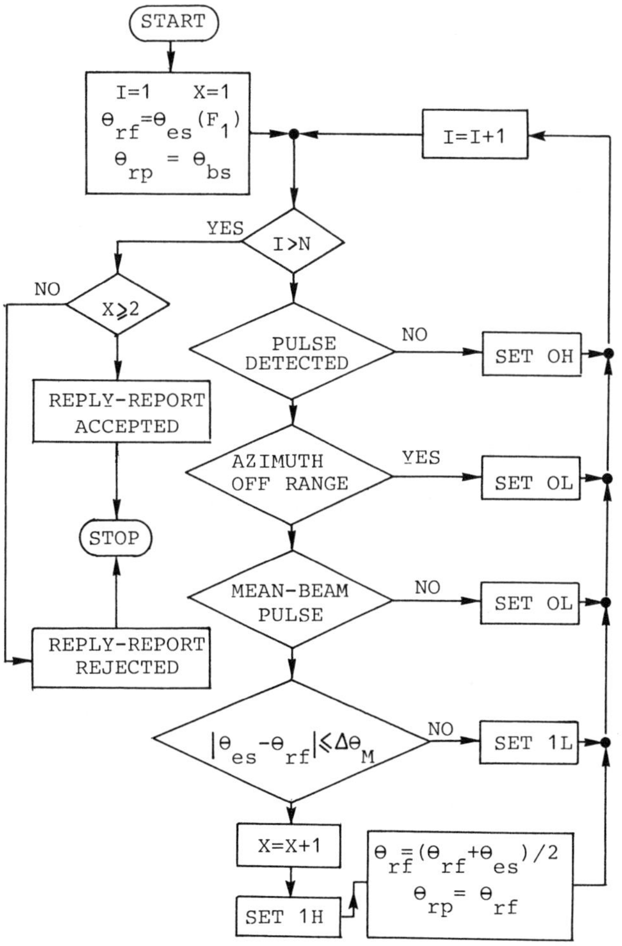

Figure 2.Reply decoding algorithm for the Reply Time Processor; I=position counter of the impulse in the current ply; W=counter of the high reliability impulses in the current reply.

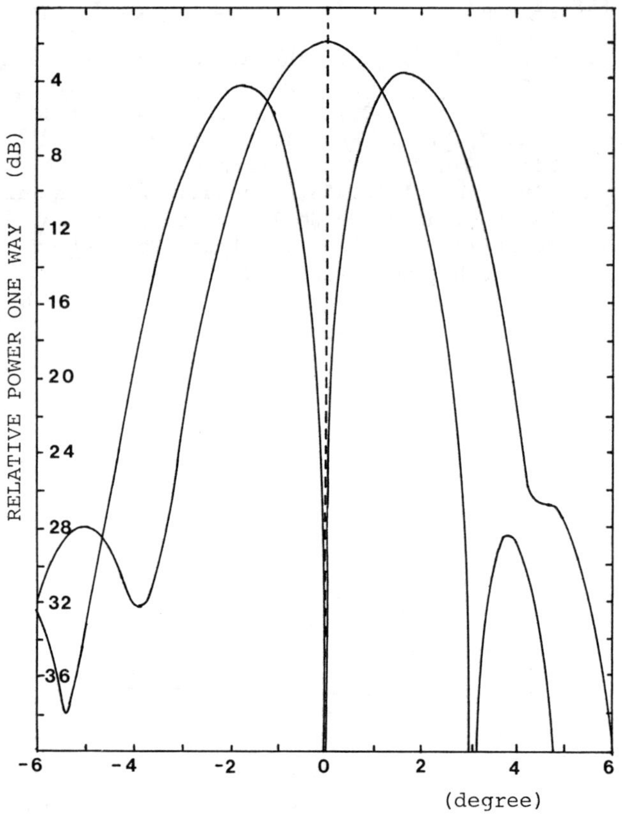

Figure 3. Sum and difference diagrams of the utilized antenna.

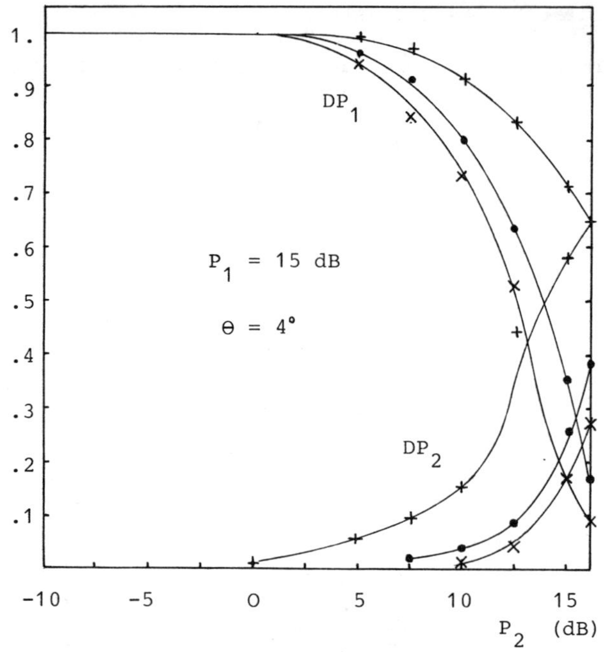

Figure 4. The declaration probability DP_1 and DP_2 of the two aircraft for the following parameter values :

++ S_N =-30 dB; RSLS=0 dB ; S_1=15dB
•• S_N =-30 dB; RSLS=3 dB ; S_1=15dB
XX S_N =-15 db; RSLS=3 dB ; S_1=15dB

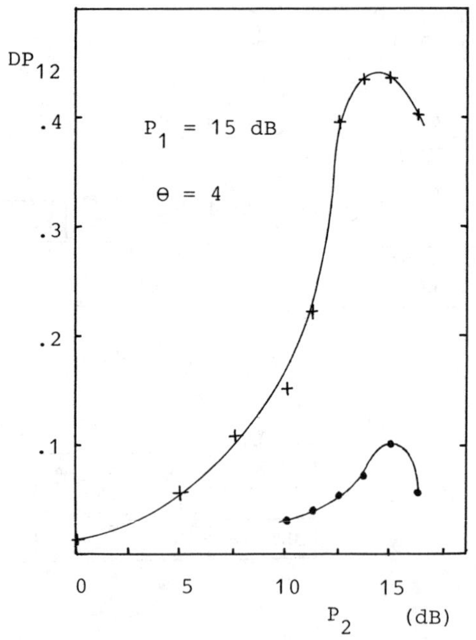

Figure 5. Declaration probability DP_{12} of the two aircraft for the following parameter values:

++ S_N=-30dB; RSLS=0dB; S_1=15dB
•• S_N=-30dB; RSLS=3dB; S_1=15dB

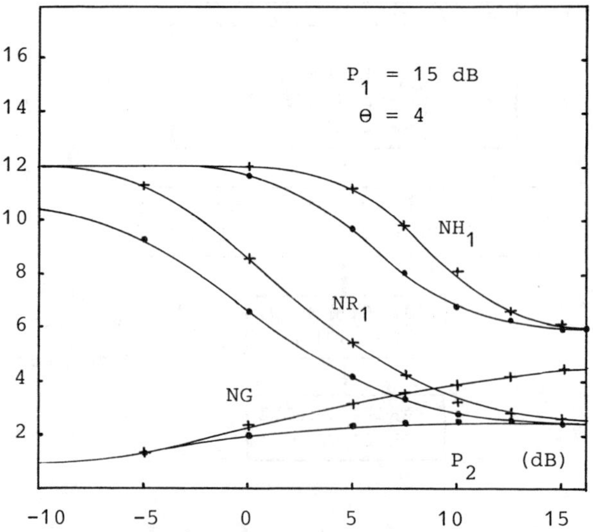

Figure 6. NH_1, NR_1 and NG for the following parameter values :

++ S_N=-30dB; RSLS=0dB; S_1=15dB
•• S_N=-30dB; RSLS=3dB; S_1=15dB

INTEGRAL SSR ANTENNA HAVING INDEPENDENTLY OPTIMIZED SUM AND DIFFERENCE BEAMS

P.T. Muto*, T. Izutani**, S. Itoh**, H. Yokoyama** and H. Takano**

* Electronic Navigation Research Institute, Ministry of Transport, Tokyo, Japan
** Nippon Electric Co., Radio Application Division, Tokyo, Japan

INTRODUCTION

Since Secondary Surveillance Radar (SSR) has recently played an important role for air traffic control, an improved SSR antenna is required which has low angle sharp cut-off characteristics in the elevation plane for considerable reduction of ground lobing and multipath effects, and has sum and difference beams for an accurate azimuth direction measurement using the monopulse technique.

An integral SSR antenna, which utilizes the reflector in common with the Air Route Surveillance Radar (ARSR) antenna, meets above mentioned requirements, having the advantages of low cost and applicability to existing radomes (1)(2). However, compared with a conventional hog-trough antenna, it has some design problems; restrictions for gain flatness in elevation and azimuth beamwidth.

First, this paper discusses fundamental requirements for the SSR antenna from the SSR system design point of view and sets the design goal for the desirable integral SSR antenna patterns, taking its practical application into consideration. Next, this paper shows the practical design of the SSR primary feed which is compatible with the dual beam ARSR feed. Reflector illuminations for sum and difference patterns in azimuth are independently optimized. In addition, measured secondary patterns of the integral SSR antenna whose reflector size is 14m wide by 8m high are presented.

SSR ANTENNA PATTERN CONSIDERATIONS

An integral SSR antenna which utilizes the reflector in common with the ARSR antenna has to realize the improved SSR pattern while retaining the ARSR antenna pattern. For this reason, some restrictions are imposed on the SSR antenna pattern formation.

Gain Variation in Elevation

The ideal elevation pattern of an SSR antenna required by an SSR system is a sector type (with a constant gain in elevation) pattern having sharp cut-off characteristics at low elevation angles. On the other hand, most of the existing ARSR antennas have a modified cosecant-squared pattern to boost the antenna gain at high elevation angles in order to compensate for the radar system gain reduced at short distances by the sensitivity time control (STC). As a result, the elevation pattern of an integral SSR antenna, as illustrated in Fig.1, has a shape similar to that of the ARSR antenna, and tends to have a reduced gain in the plateau at high elevation angles in relation to the beam nose gain.

The allowable limit to this gain reduction is determined by the condition which enables effective sidelobe suppression (SLS) and stable replies in the SSR system. The sidelobe gain of the interrogation antenna should be lower than the SLS antenna gain for a perfect suppression of the sidelobe of the interrogation antenna by the transponder SLS or the receiver SLS. At the same time, the interrogation antenna gain has to be higher than the SLS antenna gain by not less than 9 dB in order to assure correct replies of the transponder (3). Moreover, considering a margin of the order of 3 dB for the gain variation at high elevation angles by ground lobing, the difference in gain between the interrogation and the SLS antennas should be not less than 12 dB.

The beam nose gain of an integral SSR antenna estimated by pattern computation is approximately 27 dB. The SLS antenna which is used in combination with an integral SSR antenna is a vertical linear array antenna and is installed on top of the reflector. The vertical aperture size of the SLS antenna is approximately 2m to be housed in the existing radome. The estimated gain of this antenna is approximately 2 dB at high elevation angles. In view of these gain relationship of the antennas realizable, the allowable limit to the gain reduction at high elevation angles of the integral SSR antenna in relation to its beam nose gain to meet the minimum gain difference of 12 dB mentioned earlier is 13 dB as shown below.

A. Integral SSR antenna gain	:	27 dB
B. SLS antenna gain	:	2 dB
C. Gain difference minimum	:	12 dB
D. Allowable gain reduction	:	13 dB (A-B-C)

Effective Azimuth Beamwidth

The azimuth beamwidth of an integral SSR antenna to be installed in an ARSR antenna having an azimuth beamwidth of 1.2° would be approximately 1.8°, which is narrower than that of a conventional hog-trough antenna (2.4°). Moreover, the effective azimuth beamwidth of an integral SSR varies with the gain variation in elevation described earlier.

The effective azimuth beamwidth (blip width) of an integral SSR antenna and a conventional hog-trough antenna, estimated by using the existing SSR system parameters, as they vary with the radar range, are shown in Fig.2. The integral SSR antenna pattern characteristics used for this estimation are the following: beam nose gain : 27 dB, high elevation angle plateau level : -10 dB from the beam nose, azimuth beamwidth : 1.8°. In Fig.2, the target navigational condition is set at a constant altitude flight of 40 kft. As indicated in the figure, due to the system gain margin in the up-link and down-link, the blip width is determined by the up-link farther than 70 NM, and by the down-link within 70 NM. The blip width variation of an integral SSR antenna is roughly the same as that of a conventional hog-tough antenna farther than 70 NM, and the blip width of the former is reduced within 70 NM.

However, this reduction in the blip width is

preferable because it improves the azimuth accuracy and angular resolution at short distances. Therefore, the blip width variation of an integral SSR antenna is acceptable for the SSR system.

SSR PRIMARY FEED DESIGN

Azimuth Feed Optimization

Although the monopulse angular measurement technique is an effective means for improving the azimuth accuracy of the SSR system, careful design is required for obtaining good monopulse patterns because the reflector width 14m of an integral SSR antenna is basically determined by the required azimuth beamwidth 1.2° of the ARSR antenna.

To begin with, this reflector width is a little too large for the SSR sum pattern formation. In other words, if this reflector is used with ordinary tapered illumination of approximately -15 dB at the reflector edge, the SSR azimuth beamwidth would be expanded by the ratio of the frequencies used by the ARSR and the SSR to approximately 1.5°. In order to obtain the azimuth beamwidth of 1.8° acceptable for the SSR system mentioned earlier, deep tapered illumination to direct the first null direction of the primary pattern toward the reflector edge is required. This illumination distribution for the sum pattern can be realized by a feed composed of two to four elements.

With respect to the difference pattern, if one half of the above feed and the other half are excited in reverse phase to each other, the pattern obtained would have a small angle separation between the peaks of the difference pattern, large gain reduction due to spillover, and high sidelobes. This problem can be resolved by realizing an illumination suitable for the difference pattern independently from the sum pattern.

The azimuth feed used in this integral SSR antenna, as shown in Fig.3, is composed of three horizontal elements to optimize the illumination respectively for the sum and difference patterns. In this azimuth feed, all the three elements are excited for forming the sum pattern, while only the two elements on both sides are excited in reverse phase to each other for forming the difference pattern.

In this configuration, the illumination distribution for the difference pattern is optimized by adjusting the element spacing of the azimuth feed, while the illumination distribution for the sum pattern can be maintained as desired even if there is a little variation in the element spacing by varying the inter-element amplitude excitation. The monopulse angular measurement characteristics for three cases of the element spacing is obtained from the secondary pattern computation and shown in Fig.4. This secondary pattern computation is made by using the current distribution method where the integration is performed over the entire reflector area. As seen from Fig.4, the monopulse characteristics is optimum near the case B for the reflector under consideration. The calculated values of the sum and difference primary patterns in this optimum condition are shown in Fig.5. The illumination distribution obtained is a deep tapered distribution with a minimum of spillover for both the sum and the difference patterns, which realizes gain improvement and sidelobe reduction in the secondary patterns.

Elevation Beam Forming

The low angle sharp cut-off characteristics in elevation of the integral SSR antenna can be realized by utilizing the large vertical aperture (8m) of the ARSR reflector. The SSR primary feed is constructed by installing the above-mentioned 3-element azimuth feed immediately above and below the low beam horn of the dual beam ARSR. The SSR elevation pattern similar to that of the ARSR can be obtained by adjusting the amplitude ratio and the relative phase between the two azimuth feeds. Figure 6 shows the circuitry of the integral SSR feed.

EXPERIMENTAL RESULTS

The developed integral feed is shown in Fig.7. The SSR primary feed is composed of six cavity backed slot antennas; a pair of three azimuth elements above and below the low beam horn of dual beam ARSR feed. The dual beam ARSR antenna which includes this integral SSR feed is shown in Fig. 8.

The SSR elevation pattern measured at 1090MHz is shown in Fig.9. The measured power gain of this antenna is about 27dB. As expected, a good low angle sharp cut-off performance is obtained. The pattern slope at the nominal radar horizon is about 2dB/deg which demonstrates a significant improvement compared with that of a conventional hog-trough antenna; 0.05dB/deg. The plateau level in high elevation angle is 8 to 10dB down from its beam nose. This value satisfies the limitation of 13dB which is derived in the preceding section. The elevation service area of the SSR system which utilizes this antenna can be obtained up to about 45° which is the same as the existing system.

Figure 10 shows the azimuth patterns of the integral SSR antenna measured at 1090MHz. The measured azimuth beamwidth of the sum pattern is 1.9°. As a result of the independent optimization of the azimuth feed, desirable cross over level of sum and difference patterns are obtained due to the increase in gain of the difference pattern. The monopulse measurement angle expected from the measured pattern is about $\pm 1.5^{\circ}$ from the boresight. The effect of the independent optimization of the azimuth feed also appears in the low sidelobe characteristics of the difference pattern. Both sum and difference sidelobes are more than 24dB down from their peaks (3).

Measurements of the integral SSR antenna patterns at 1030MHz have also shown satisfactory characteristics.

The measured patterns of the dual beam ARSR antenna which incorporates the integral SSR feed are shown in Fig.11 and Fig.12. The patterns are measured at 1300MHz. The measured azimuth beamwidth is 1.2° for both the low beam and the high beam. It is observed that the integral feed does not cause any significant degradation of the ARSR patterns.

CONCLUSIONS

An integral SSR antenna for ARSR has been developed which has independently optimized sum and difference patterns in the azimuth plane.

Gain variation in elevation and associated

effective azimuth beamwidth variation that are characteristics of the integral SSR antenna are acceptable for the SSR system.

The SSR primary feed is composed of six array elements; a pair of three azimuth elements above and below the low beam horn of dual beam ARSR feed. Measured secondary patterns of the integral SSR antenna, the reflector size of which is 14m wide by 8m high, shows satisfactory characteristics;

- o excellent low angle sharp cut-off in elevation,
- o desirable plateau shaped pattern at high elevation angles,
- o desirable cross over level of the sum and difference patterns due to the difference beam gain improvement,
- o low sidelobes for both sum and difference azimuth patterns, lower than -24 dB.

Therefore, a monopulse direction finding system using this integral SSR antenna is promising.

ACKNOWLEDGMENT

The authers wish to express their gratitude to Mr. E. Yoshioka, Director, Evaluation Division and the members of the Air Traffic Control Laboratory of Electronic Navigation Research Institute for their helpful contributions and encouragements.

REFERENCES

1. Winter, C.F.,"Vertical Plane Pattern Calculations for Shaped Beam Reflectors with Multiple Feeds", IEEE Trans., AP-22, pp.495-497, May, 1974.

2. Chiba, T. and Suzuki, Y.,"Newly Developed Co-feed Type SSR Antenna," Conf. Proc., 11th European Microwave Conf., pp.551-555, 1981.

3. "International Standards and Recommended Practices, Aeronautical Telecommunication", Annex 10, ICAO, 1972.

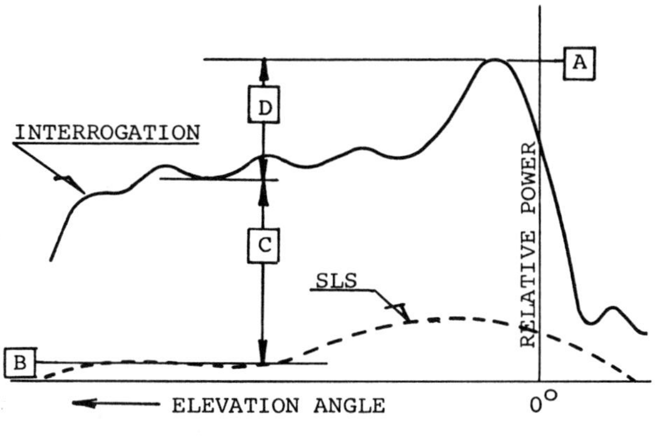

Fig.1 Required Integral SSR Pattern

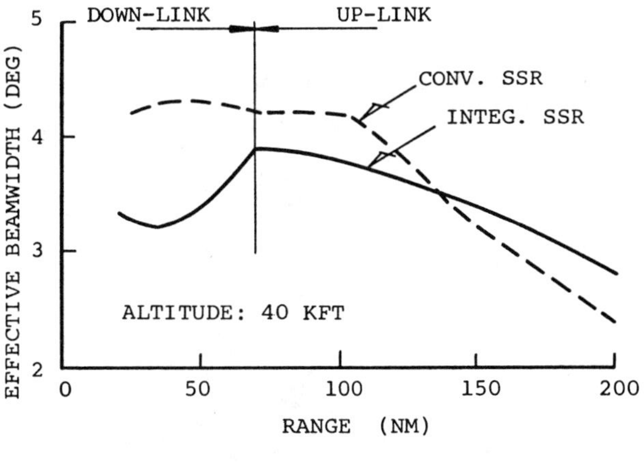

Fig.2 Comparison of Effective Beamwidth

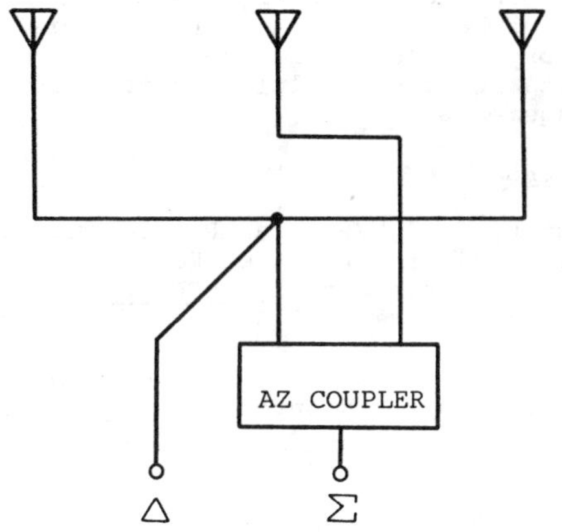

Fig.3 Azimuth Monopulse Feed Circuit

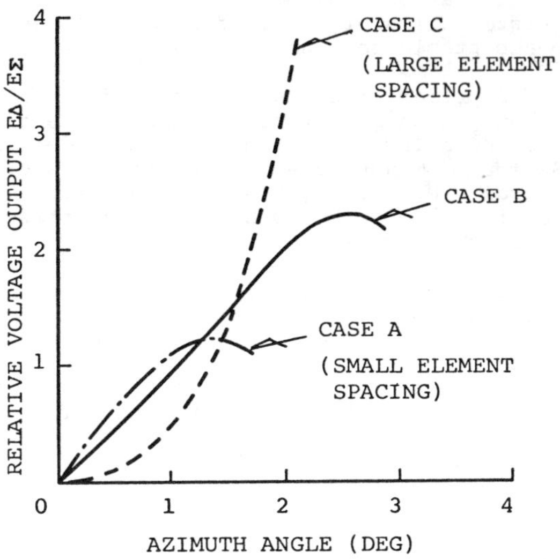

Fig.4 Monopulse Output Characteristics

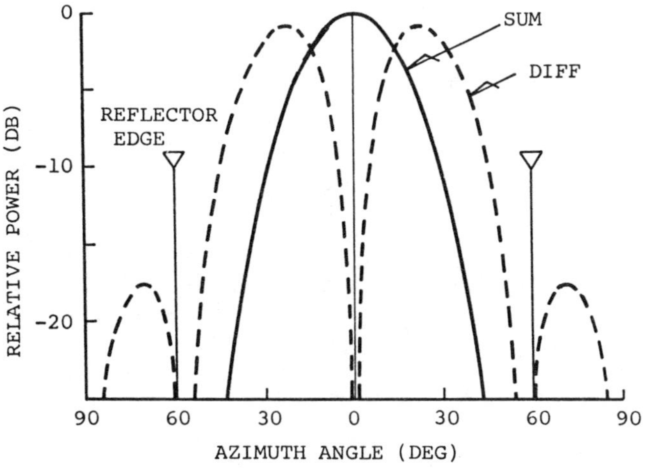

Fig.5 Optimized SSR Primary Patterns

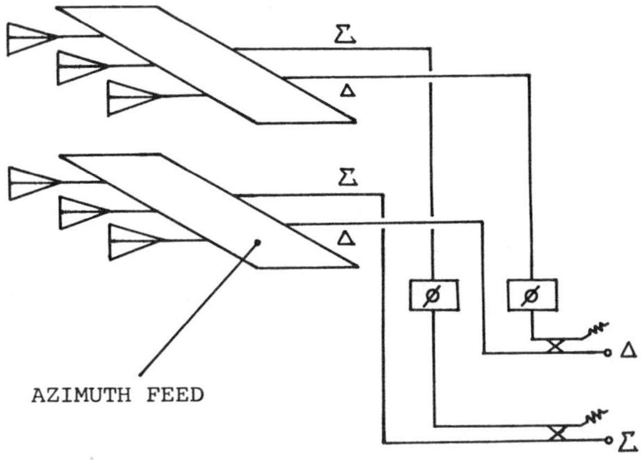

Fig.6 SSR Feed Circuit

Fig.7 Integral SSR Feed

Fig.8 Integral SSR/ARSR Antenna

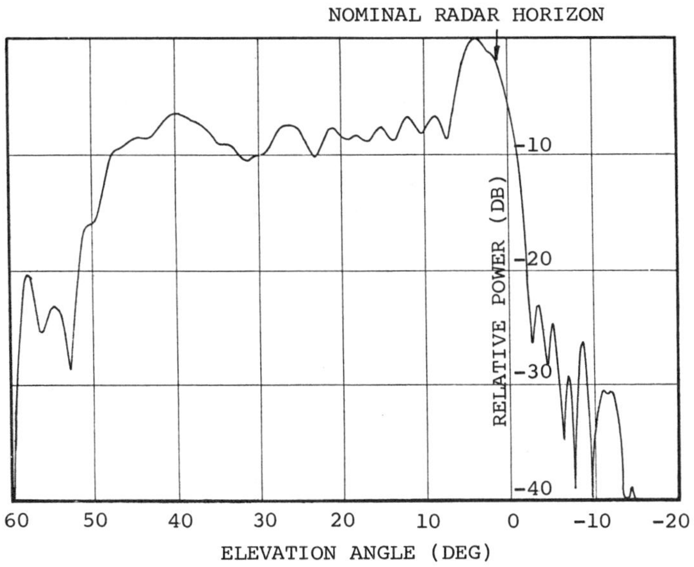

Fig.9 Measured SSR Elevation Pattern

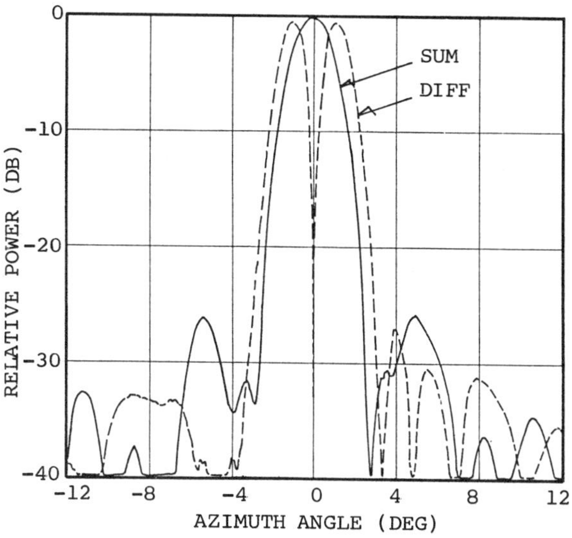

Fig.10 Measured SSR Azimuth Patterns

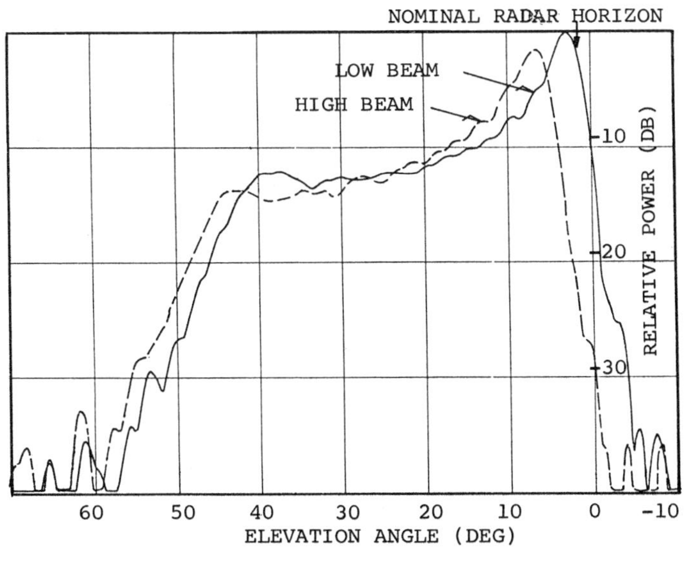

Fig.11 Measured ARSR Elevation Patterns

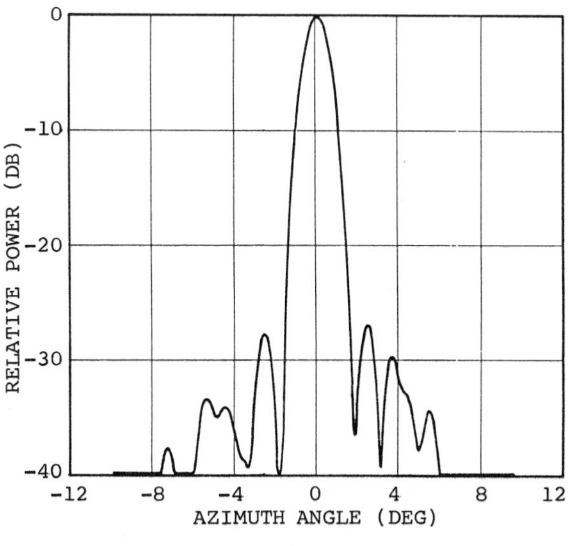

Fig.12 Measured ARSR Azimuth Pattern

SECONDARY RADAR PERFORMANCE PREDICTION

B.E. Willis*, B. Pugh** and S. Strong**

* Ministry of Defence, UK ** British Aerospace P.L.C., Dynamics Group, Bristol Division, UK

INTRODUCTION

Secondary radar is performing an increasingly important role in Air Traffic Control and identification. The system comprises an interrogator and antenna, normally associated with a ground primary radar, and a transponder and antenna(s) onboard the aircraft. Interrogations from the ground station, received by the aircraft, trigger reply signals which are received and processed by the interrogator. Aircraft identity or height information, coded on the reply, can then be presented on ATC displays. In wartime, the replies give the vital "friend or foe" decision for military aircraft.

For a system description see Honold (1). Accurate performance prediction for given combinations of aircraft and ground interrogators, "platform pairs", is important, to give confidence in reliable identification.

Performance is dependent not upon an aircraft echoing area, but upon the aircraft receiving sensitivity and EIRP; the positioning of the antenna(s) on the aircraft structure can markedly affect these terms.

Ground multipath can also impair performance.

The aims of the work reported were:

(a) To develop a calculation procedure which, when applied to a specified platform pair, will predict the extent of interoperability, by giving the secondary radar performance as a function of the aircraft variables (height and attitude). This will enable optimisation of interoperability.

(b) To validate the above procedure, by comparison with measured performance in flight trials.

(c) To optimise antenna positioning on airframes.

(d) To reduce the necessity for expensive flight proving trials.

The procedure has been named "MIISPEC" (merit of individual IFF system performance characteristics).

CAUSES OF PERFORMANCE DEGRADATION

Incorrect identification has various possible causes. Assuming that the airborne and ground equipment is fully serviceable and correctly set up, malfunctions may be due to:

Garbling

Garbling exists when two or more transponder replies are related in time such that the reply information combines to give an ambiguous interpretation of those replies. Since all aircraft within the interrogator antenna beam will reply to a given interrogation, in dense air traffic environments overlap of reply signals is probable. The signal processing logic is designed as far as practicable to recognise the wanted replies even if extraneous pulses are present; nevertheless malfunctions can occur.

Fruit

A transponder will reply to any valid interrogation which is received. Because an aircraft may be within range of many interrogators simultaneously, a given interrogator can receive many replies from a single aircraft which are triggered by other interrogators. This effect is called fruit (friendly replies unsynchronised in time). Fruit replies create confusion since the random arrival time produces an incorrect apparent range. Methods for discrimination between correct replies and fruit exist, based upon the variations in signal repetition frequencies between different interrogators.

Overinterrogation

Transponders can only generate a specified maximum number of reply pulses per second without overload. If the interrogation rate would cause this maximum to be exceeded, protection logic imposes a limit on the reply rate. Thus some of the interrogations do not generate replies. Of the rejected interrogations some are from the wanted interrogator, and therefore system operation is impaired.

Inadequate R.F. Link Margins

For correct operation, both the up-link and down-link must have adequate R.F. power margins, i.e. the received signal power must exceed the minimum permissible received value for the transponder or interrogator.

STUDY DEFINITION

A fully comprehensive performance assessment procedure would take account of all possible degradation causes. Garbling, fruit and overinterrogation can be important in some situations, but are dependent on the overall secondary radar environment in the area of interest. To obtain a comprehensive MIISPEC all environmental data would be modelled. The link margin is in contrast fundamental to correct operation even for a single platform pair. An unsatisfactory margin guarantees poor operation, even if other degradation causes are negligible.

This work therefore concentrates on the estimation of link margins and their effects on performance.

DEVELOPMENT OF METHOD

Basic Considerations

In principle, there is no difficulty in evaluating the link margin once the input terms in the link budget are known. In practice, however, there are considerable problems.

The signal margin is a function of many inputs, some of which are associated with the interrogator platform and others of which are associated with the transponder platform. It must be possible to obtain the margins for various interrogator-transponder combinations.

The outputs should be simple enough to be readily interpreted and used.

Many of the input values are difficult to quantify for specific cases.

The aircraft antenna radiation pattern is an important factor. Since angular regions with low gain cannot readily be eliminated, the orientation of the aircraft with respect to the propagation path must therefore be taken into account.

The input terms in the link budget are listed in Table 1. Then, all quantities in dB units, the link margin (M) is given by:

$$M = P_t - P_r + G_t + G_r - L_t - L_p - L_{po} - L_a - L_m - L_r \quad \ldots\ldots\ldots\ldots \quad (1)$$

This applies both for up-link and down-link, with the appropriate terms used.

Aircraft Antenna Gain (G_t or G_r)

Civil aircraft have one antenna, while military aircraft can have two antennas. Most aircraft with two antennas have nominal locations one on the top side and one on the underside of the body. The transponder is either continuously switched between the two antennas (one at a time) or operates in diversity, selecting the stronger received signal. System performance is thus determined by the higher gain value in any direction. Figure 1 shows a demonstration model of the radiation pattern given by two monopoles, one above and one below a flat 50λ diameter ground plane. This gives an approximate representation of an aircraft antenna pattern. Evidently the gain is markedly dependent on the orientations; there are deep minima vertically up and down, and near horizontal. Actual radiation patterns are affected by the aircraft configuration and are far more irregular.

Three methods of estimating aircraft antenna gain are possible.

Measurements. If of sufficient quality and comprehensiveness, measured patterns would be ideal, either from full-scale aircraft or reduced scale models. However, adequate information is rarely available.

Rigorous Mathematical Modelling. Such techniques as GTD (Geometrical Theory of Diffraction) can give very good predictions of the aircraft structure effects on the antenna radiation patterns. The problem is the high engineering and computing costs for comprehensive coverage.

Simplified Mathematical Modelling (SMM)

This technique has been developed in the Antenna Department of British Aerospace Dynamics Group, Bristol Division. It enables good estimates of the radiation patterns to be calculated with low cost. The perturbation caused by the aircraft structure are well predicted, as shown by comparisons with measured radiation patterns. It is intended to publish a description of this work in the near future.

Link budget evaluations to date have used antenna patterns calculated by SMM.

Ground Multipath

Two methods can be considered for evaluation.

Measurement. Ideally in-situ measurement should be used, but once again it is rare that such information will be available, particularly for transportable interrogators.

Calculation. Given the profile of the ground surface near the antenna and the ground electrical parameters (permittivity and conductivity), the multipath term can in principle be calculated. If the surface is assumed to be plane and smooth, this simplifies the calculation since only one reflected path is present.

Adequate ground multipath measurements have not been available for cases of interest, and the calculation method has been employed.

Polarisation Loss

This paper considers performance exclusive of polarisation loss.

CALCULATION TECHNIQUES

Programmes have been developed for evaluation of link budgets on a small desk-top computer. The terrain near the ground antenna is assumed to be a plane surface of specified slope and material permittivity. Antenna patterns are stored in numerical form.

The received signals via both the direct and reflected rays are calculated, and vectorially summed to give the total signal inclusive of multipath. The link margin is then obtained.

The aircraft attitude is specified by two angles, azimuth and tilt (Figure 2).

Azimuth is the angle in the horizontal plane between the aircraft forward longitudinal axis and the ground station direction.

Tilt is the angle between the yaw axis of the aircraft and a geographic perpendicular through the aircraft, measured in the vertical plane through the aircraft and the ground station. When the aircraft is nose-on or tail-on the tilt angle is purely a pitch angle. When the aircraft heading is 90 degrees with respect to the ground site, the tilt angle is purely an aircraft roll angle. For intermediate orientations the tilt angle is a composite of pitch and roll. Employing tilt angle simplifies the presentation of information, by removing the need for separate pitch and roll variables. For given aircraft attitude and other input values, link margin is calculated as a function of range at constant height.

MIISPEC PRESENTATION

Large amounts of performance data can be calculated, and a simplified presentation is needed to be of practical use. The number of variables displayed must be reduced to a minimum.

A given ground interrogation system will have maximum and minimum range limits, within which satisfactory operation is needed. The range values at which any failures occur are not essential as data. Thus GOOD performance is defined to be satisfactory at all ranges between the limits, BAD otherwise. This removes the range variable from display.

The numerical values of link margin are not important, since the essential question is: will the system work? A positive link margin is taken as GOOD, negative as BAD.

Aircraft height and attitude must be retained, since their effect on performance is useful information. The aim is therefore to present the combinations of aircraft height and attitude angles (azimuth and tilt, or equivalent) which give positive link margins. Several formats have been proposed.

(1) Contour Plot

Azimuth and tilt angle are presented as rectangular co-ordinates (Figure 3). For a given aircraft height, contours of zero link margin are drawn, which divide the azimuth-tilt angle plane into GOOD and BAD regions. The favourable aircraft attitudes are then readily visible. Several aircraft heights can be plotted on the same diagram (like contours on a map) to give a compact display. From Figure 3, at the given height the aircraft when broadside-on can tilt (i.e. roll) to about ±40 degrees with good performance. However, when tail-on the good region only covers tilt (i.e. pitch) angles from +10 to +70 degrees. This is for an upper antenna only; when the lower antenna coverage is included the bad regions are filled in.

(2) Compass Card

This gives a further simplification. A given span of tilt angles is taken, between limits which are not likely to be exceeded, and for each azimuth angle it is noted whether performance is good for all tilt angles within these limits. For a given height the data are thus reduced to good and bad azimuth sectors. These are displayed as white and black arcs respectively on a polar format (Figure 4).

(3) Roll - Azimuth Plot

The azimuth and tilt angle coordinate system is convenient for calculations and display with respect to a ground observer. The air-crew would normally use aircraft based co-ordinates (roll-azimuth-pitch). An alternative format developed gives GOOD-BAD contours in the roll-azimuth plane for various pitch angles. Polar coordinates are used to show more clearly the azimuth angle (Figure 5).

This format may be more convenient in some cases; the same information is contained as in the contour plots.

MEASUREMENT COMPARISONS

Calculated link margins for given platform pairs have been compared with the results of flight trials, in which the down-link received power was measured. Good agreement was obtained, generally within 2 dB. Figure 6 shows the comparison for a horizontal flight path towards the ground equipment.

Comparisons which the secondary radar system performance have also been made.

CONCLUSIONS

The method developed for secondary radar performance prediction (MIISPEC) has the following features:

(1) Inclusion of link budget terms specific to the aircraft and ground system types. These include the aircraft antenna patterns and ground multipath. The SMM method for antenna pattern prediction which is used has been shown to give satisfactory results.

(2) Calculation of link margins as functions of aircraft height and attitude.

(3) Presentation of results in convenient compact formats.

(4) Optimisation of platform pair performance is possible, based on the tilt angle concept and SMM.

(5) Good agreement with flight trial results.

(6) Low cost.

FURTHER WORK

At present, further studies are in progress on all other aspects of link margin which need to be taken into account.

REFERENCES

1. Honold, P., 1976, "Secondary Radar", Chapter 1., Heyden & Son Limited, London.

TABLE 1 - Link Budget Terms

		Up-Link		Down-Link	
1 Tx Output power	(P_t)	S		S	
2 Tx Feeder losses	(L_t)	S	Ground	S	Aircraft
3 Tx Antenna gain	(G_t)	S		C or S	
4 Path loss	(L_t)	C		C	
5 Polarisation loss	(L_{po})	A		A	
6 Atmospheric attenuation	(L_a)	N		N	
7 Ground multipath	(L_m)	C or S		C or S	
8 Rx Antenna gain	(G_r)	C or S		S	
9 Rx Feeder losses	(L_r)	S	Aircraft	S	Ground
10 Required Rx input power	(P_r)	S		S	

S - Values can be supplied from specs., measurements, etc.

A - Assumed zero

N - Values small, can be neglected

C - Can be calculated

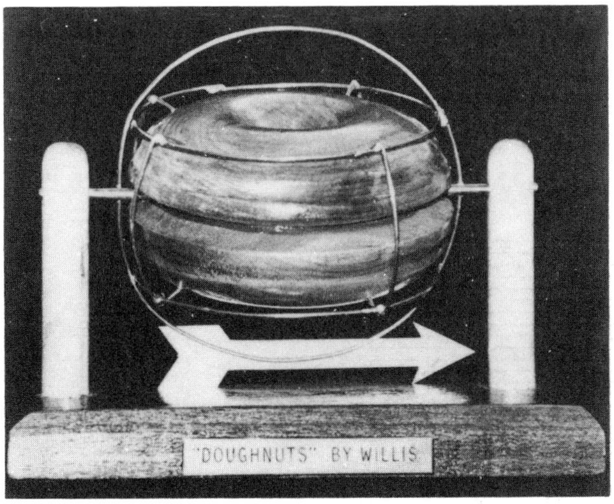

Figure 1 Antenna Pattern Model

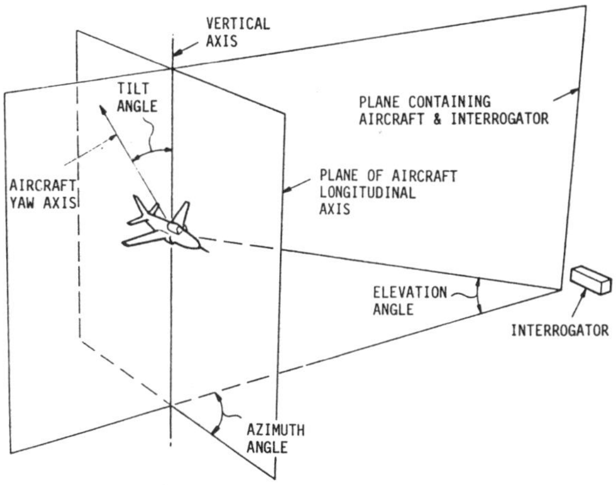

Figure 2 Definition of Angles

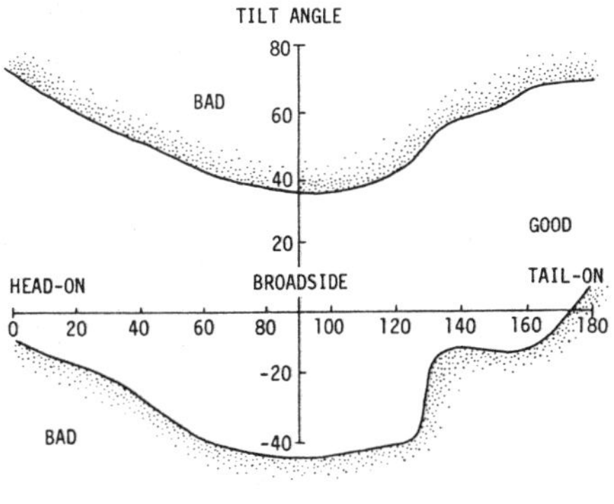

Figure 3 MIISPEC Contour Plot

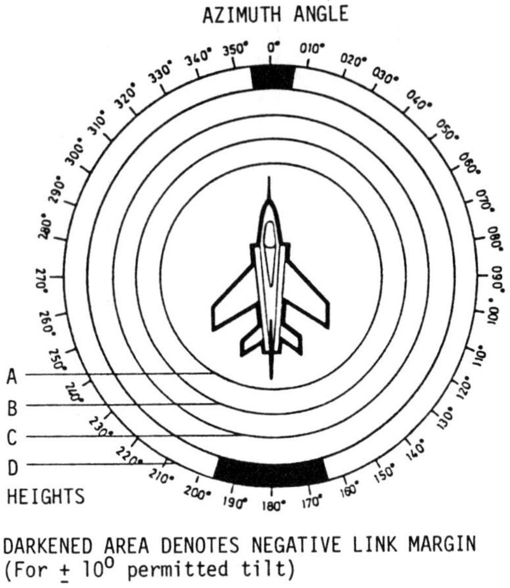

Figure 4 MIISPEC Compass Card Plot

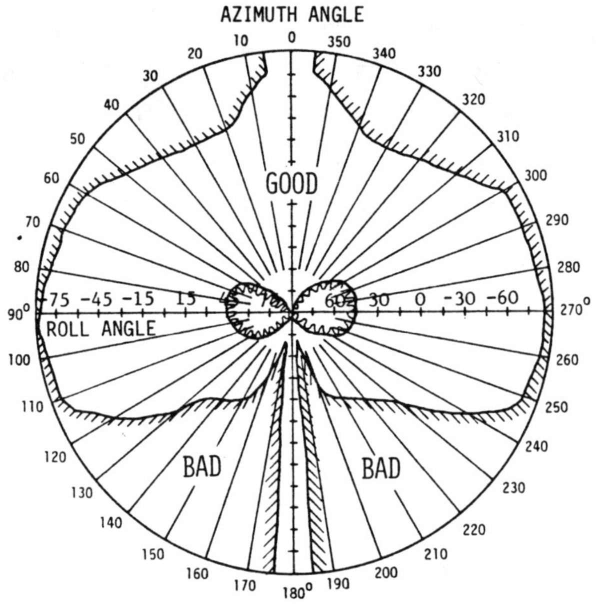

Figure 5 MIISPEC Roll-azimuth Plot

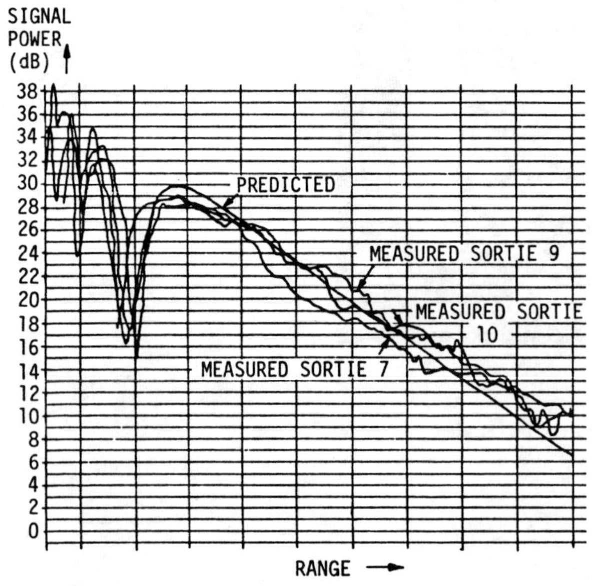

Figure 6 Measurement Comparison

GENERIC TRACKING RADAR SIMULATOR

W.K. McRitchie, P.I. Pulsifer and G.A. Wardle

Defence Research Establishment Ottawa, Canada

INTRODUCTION

This paper describes the development and operational capabilities of a generic tracking radar simulator (TRS) that is now in operation at Defence Research Establishment Ottawa. The TRS is part of the Radar/Countermeasures Simulation Facility, a block diagram of which is shown in Fig. 1. In this facility, target return pulses, jamming signals and clutter are generated under computer control and radiated through a two-dimensional array of dipole antennas into an anechoic chamber. Each element of the array is controlled by a PIN diode switch, which in turn is controlled by the ECM Simulator computer. A timing pulse from the radar indicates zero range so that it is not necessary for the radar to transmit an RF pulse. Delay circuitry is used to generate the low level target return pulses at the appropriate range.

The TRS receives the simulated target returns at the opposite end of the anechoic chamber and processes these signals to produce range tracking, antenna steering and display information. Since the radar operates autonomously to track the target(s), it responds to jamming techniques which could be in the form of noise, deception or active or passive decoys. In this way, the TRS is used to assess the effectiveness of countermeasures (CM) techniques against radar systems. It can also be used in the assessment of various radar design or CCM techniques in overcoming certain types of jamming. This is facilitated by the modular construction and the extensive use of digital circuitry in the signal processing subsystem of the TRS. Parallel processing with dedicated microcomputers permits relatively simple changes to the radar design through software control of many radar parameters.

DESCRIPTION OF THE TRS

Figure 2 is a simplified block diagram of the complete radar simulator, showing the key elements of the system. It is a hardware simulator designed primarily for laboratory use. Some parts of the signal processing subsystem are still under development and have not yet been incorporated (those units within the dashed box). Several types of radar can be simulated by the TRS including fire control radars for gun systems or air interceptors and active or semi-active missile seekers. These radars typically operate in the H/I/J frequency bands. The TRS employs a three channel monopulse antenna and positioner. The antenna consists of a four-horn feed and a parabolic reflector which is oblong, giving it an azimuth beamwidth of 3.5° and an elevation beamwidth of 6°.

Receiver

An expanded diagram of the receiver subsystem is shown in Fig. 3. To simplify the initial design, the three monopulse antenna output signals are multiplexed onto two down-conversion channels. At the inputs to the hybrid coupler, the RF signals are carefully phase matched so that the outputs are $\Sigma + \Delta$ and $\Sigma - \Delta$. Note that Δ alternates between the azimuth difference and elevation difference signals on successive pulses, and can be positive or negative depending on the target location: i.e., left, right, up or down with respect to the antenna boresight. From this point on, the relative phase angles of the two signals are unimportant, so that accurate phase matching through the mixer and IF stages is not required.

Both logarithmic and linear IF amplifier stages are used: log for angle tracking, linear for range tracking and display. The difference of the two log detector outputs yields $\log|\Sigma + \Delta| - \log|\Sigma - \Delta|$ which can be approximated by Δ/Σ provided $|\Delta| \ll \Sigma$. High speed sample and hold devices convert the detected pulses to d.c. error voltages which are used to drive the antenna when it is in track mode. Through proper timing of the sample commands, the sample and hold devices also provide de-interleaving of the azimuth and elevation difference signals. The sum of the two log detector outputs yields $\log|\Sigma^2 - \Delta^2|$ which is proportional to $\log(\Sigma)$ if $|\Delta| \ll \Sigma$. This output provides fast automatic gain control (AGC) for the linear IF amplifiers, the outputs of which are summed to give a detected pulse proportional to Σ. This is then used in the range tracking and display units.

The local oscillator and target return signals are derived from the same source, a direct synthesizer. With direct synthesis, frequency changes may be made in approximately 10 μs which is necessary for pulse to pulse frequency agility. At the same time it has the frequency stability that would be required for coherent processing techniques. Such devices are, however, inherently low frequency so it is necessary to upconvert the synthesizer output by muliplication and mixing stages to the desired frequency. Band pass filters are used at each stage to eliminate harmonics. The upconverted signal labelled "RF" in Fig. 3 is applied to the input of a PIN modulator which is triggered on at the appropriate time by the target generation unit.

Antenna Controller

The antenna control unit provides the interface between the radar receiver or hand control and the antenna servo motors. It is completely software programmable with its own dedicated microcomputer and uses non-linear control to vary the maximum antenna slew rate and acceleration.

There are three basic modes of operation of the antenna controller: automatic track, manual or hand control and search. In autotrack, the error voltages required to steer the antenna in the direction of the target are generated by the radar receiver. In manual mode, the joy-stick/hand control dictates the position of the antenna. The antenna controller compares the requested position with the actual position of the antenna to determine the appropriate antenna motion. In search mode, the controller defines the motion of the antenna such that it follows a pre-programmed search pattern, the parameters for which are loaded into the controller microprocessor from the host computer.

In conjunction with the new range tracking unit, the antenna controller will have the option of automatically acquiring and locking onto a target from its search mode. If prior tracking information is available, the controller can enter a reacquisition mode. This can be in the form of fine search about the last known target position, or a blind tracking mode in which the antenna is made to follow the predicted target position. Three types of search patterns have been programmed to date: unidirectional multi-level raster, bidirectional multi-level raster and spiral. Typical pattern variables that are programmed in at initial set-up are: search limits, antenna slew speed and acceleration, number of levels and direction of search for raster scan.

Digital Signal Processing

Prior to processing by the moving target indicator (MTI), constant false alarm rate (CFAR) and range tracking units, the video sum signal is digitized. The A/D converter samples at 10 MHz in search mode and 25 MHz in track mode. The sampling rate is controlled by the range tracker, which also synchronizes the processing of data in the MTI and CFAR units. With a minimum of two samples per target return pulse being required to replicate the input data in digital form, these sampling rates accommodate TRS pulse widths of 200 nsec in search and 100 nsec in track modes.

MTI. Noncoherent MTI processing is implemented digitally as a finite impulse response filter. Data from the A/D converter is read into high speed ECL RAM and combined in a multiplier/accumulator. The RAM is directed to store the incoming data for one to four PRI, thus allowing two, three, four or five pulse cancellation. Control of internal data paths is via a micro-sequencer and writeable control store. Communication with the TRS host computer is through a microprocessor controlled I/O board and serial interface. The operator can select the filter order and specify filter coefficients through the host computer terminal.

Range tracking. The digital range tracker presently in use affords manual target acquisition. The tracker input signal is formed by comparing the video sum signal with a fixed detection threshold level, which is manually adjustable. The comparator output is a TTL logic level signal. Using the TRS joystick a range gate can be positioned over the detected target. The range tracker can then track either the leading or the trailing target edge (operator selectable).

A fully automatic range tracker, now under development, will replace the existing unit as indicated in Fig. 2. Input data from the A/D converter can be MTI processed, CFAR processed, MTI and CFAR processed or processed for fixed threshold level crossings only. Four operating modes will be provided: search, acquisition, track, and reacquisition. In search mode the full range is divided into small segments or range bins. The host computer will set the size of each bin and the dwell time for each. Once a target has been detected in one of the bins, the acquisition mode initiates fine search to locate the target precisely within the bin. The search is either from near to far or from far to near as specified by the host computer or by a joystick control switch. When the target is located the track mode is automatically selected, and the leading edge, trailing edge or centre of the target return pulse is tracked. Reacquisition mode provides a partial search pattern to reacquire the target if lock is broken. Two search patterns are available in this mode: 1) the range gate searches far to near from the last known target position for a closing target or near to far for a receding one; and 2) the range gate will search alternately on one side of the last known target position and then on the other, each time moving one range gate width away.

CFAR. The CFAR processor, which is still under development, will provide an adaptive threshold level in the detection system of the new range tracker. To implement the adaptive threshold the video sum signal is sampled immediately before and after the current range bin. Up to 16 samples are taken on each side of the bin. Using a PROM lookup table these samples are converted to logarithmic form, facilitating geometric mean detection. The samples on each side of the bin are separately summed and the two results are either averaged or the greater of the two is selected, depending on the mode requested by the operator. This result is the detection threshold for the range bin. The "greatest of" technique lowers the increase in the false alarm rate at clutter edges. The number of samples taken and the technique requested will be communicated to the unit through the TRS host computer terminal.

Missile Guidance And Autopilot

For scenarios in which the tracking radar is a missile seeker, a hybrid computer simulation of the missile guidance and autopilot is used to calculate the missile aerodynamics. Three degrees of freedom of missile motion - translation in azimuth, elevation and range - are modelled by the analog computer program. Air friction, gravity and thrust factors are also incorporated. With the inclusion of missile guidance and autopilot functions in the simulator, missile/target engagements involving manoeuvering targets and countermeasures can be simulated in a closed-loop manner, with system effectiveness expressed by a missile miss distance.

PERFORMANCE AND CAPABILITIES

Table 1 summarizes the main features of the TRS and gives the parameter ranges or characteristic values for each item. Most of these features have been discussed earlier in this paper. Again, the wide range of options that are available in setting up the TRS is facilitated by the use of microprocessors in the antenna controller, the MTI unit, the CFAR unit and the range tracker. The wide range of PRF and pulse width options is made possible by the design of the central timing unit. This unit generates the zero range or synchronization pulse which is used by the target return pulse generator. The sync pulse is also used to initialize the sensitivity time control (STC) and to co-ordinate the PIN diode switch in the receiver front end, the frequency controller and the display. Thus, through independent control of the delay between sync pulses, it is possible to alter the PRF or to program various PRI stagger or jitter sequences.

CONCLUSIONS

The facility described herein makes use of advanced technology to simulate the generic characteristics of many different fire control and missile seeker type tracking radar systems. Through the use of dedicated microprocessor control, the TRS has a wide range of parameter values and signal processing options, such as software programmable angle and range search patterns, PRF parameters and multiple pulse cancellation MTI. In conjunction with the countermeasures simulation facility, the radar simulator is a valuable tool in assessing both countermeasures and counter-countermeasures effectiveness.

TABLE 1 - Principal features and characteristics of the tracking radar simulator.

Feature	Description/Parameter Range
Frequency Agility	pseudo random, pulse-to-pulse
Antenna Beamwidth	3.5 Az, 6° El
Angle Tracking	2 channel monopulse variant
Angle Search	uni- or bi-directional multi-level raster, spiral
Angular Accuracy	$\pm 0.3^{\circ}$ Az or El
STC	60 dB dynamic range
Lin/Log IF	lin: 50 dB AGC range; log: 80 dB dynamic range
PRF	0.2 to 6 KHz
PRI Jitter	pseudo random ± 5 or ± 50 μs ranges
PRI Stagger	4 programs of up to 16 PRI's each
PW	0.1 to 4 μs
Range Tracking	leading edge, trailing edge, centroid
Range Search	manual, programmable with auto acquisition
Detection Threshold	manually set, log CFAR
Range Gate Width	0.1 to 4.0 μs
Auto Acquis./Reacq.	coordinated range and angle search/lock-on
Range Tracking Rate	Mach 5 closing speed, 10G acceleration
Range Uncertainty	± 22.5 m
Digital MTI	2 to 5 pulse cancellation, up to 40 dB improvement factor
Display	A scope, B scope

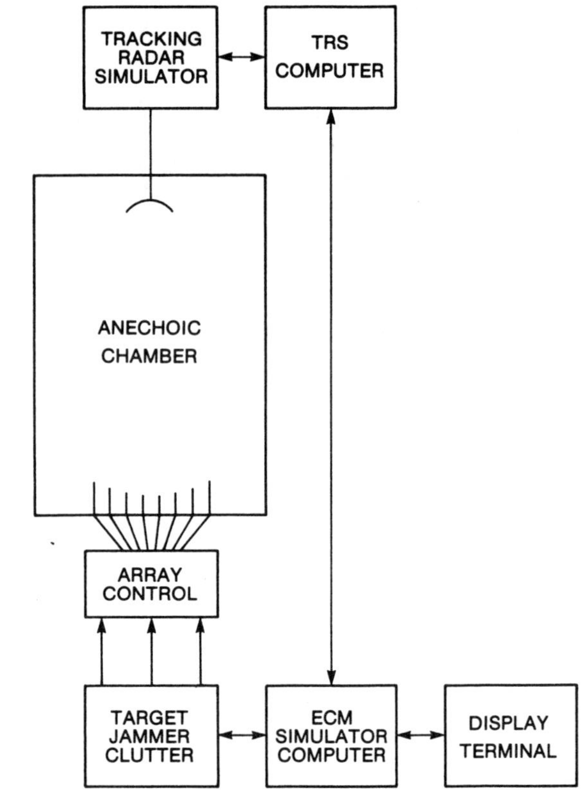

Figure 1 Radar/countermeasures simulation facility

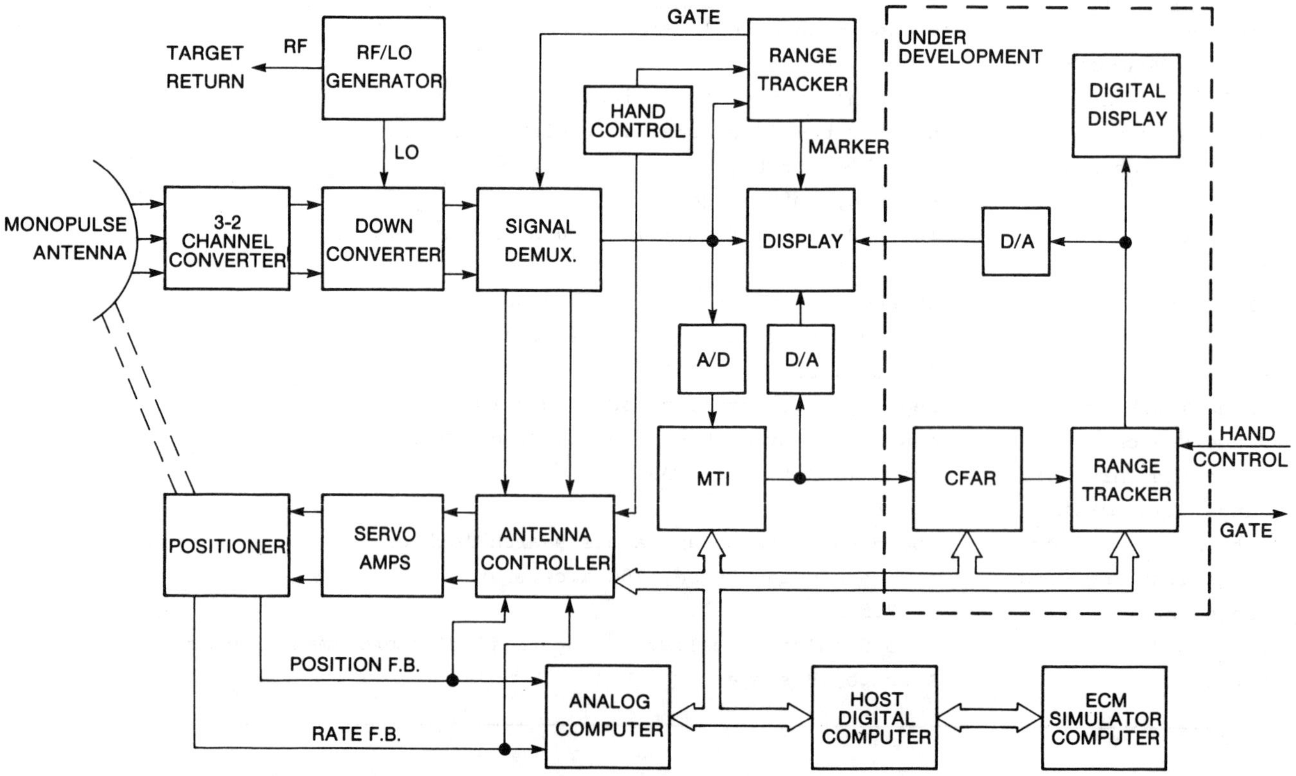

Figure 2 Tracking radar simulator overall system block diagram

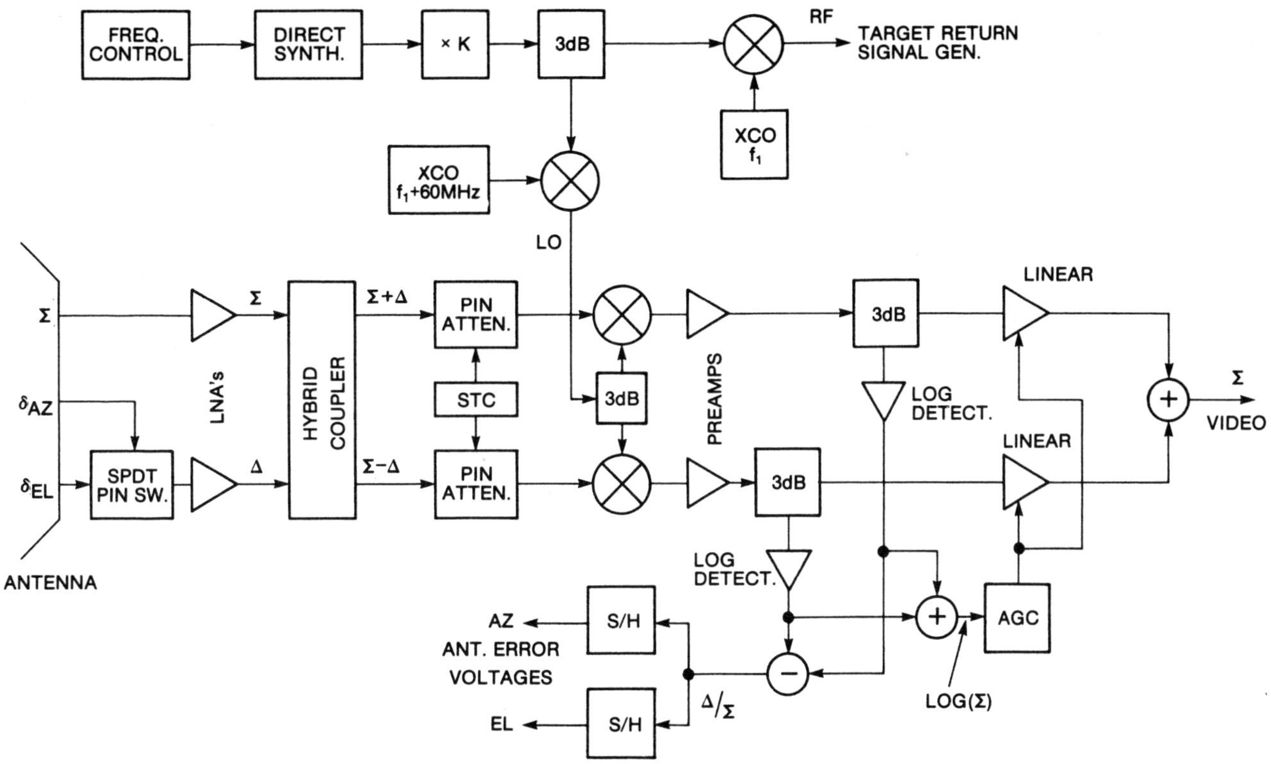

Figure 3 TRS receiver subsystem

SIMULATION OF RADAR RETURNS FROM LAND USING A DIGITAL TECHNIQUE

J.R. Morgan, P.E. Sherlock, D.J. Hill

Ferranti Computer Systems Ltd., UK

INTRODUCTION

Radar echo simulators have been used in training naval and air personnel for some time. These have used digital methods to display returns from targets, but when echoes from land masses have been required they have been produced using a flying spot scanner. The realism demanded of trainers has steadily increased and the radars to be simulated have also gained in complexity. Faced with the requirement to provide simulators for highly mobile airborne radars in which terrain has to be mapped in three dimensions so that targets can hide behind it, a solely digital method was developed. This land simulator package is linked to the target echo generator hardware and a host computer. It is also readily adaptable to simulate weather and chaff corridor effects.

REASONS FOR A SOLELY DIGITAL METHOD

The method often used to simulate terrain echoes has been the flying spot scanner, in which land is depicted on a transparent slide and scanned by a light beam. Light is transmitted through the slide where the radar pulse would be reflected by land, then detected by a light sensitive device which allows the echo to be shown at the appropriate point on the radar display.

This method has numerous deficiencies. The lack of three dimensional information makes the masking of targets by terrain difficult to simulate, as well as the associated shadowing of lower lying land by nearer ground. Representation of slant range for airborne radar systems is also a problem with flying spot scanners. The accuracy and resolution of the radar map are dependant on the slide and on the dynamics of the scanner. The latter tends to need frequent adjustment and the position of the terrain often drifts noticeably during a training exercise.

OUTLINE OF THE DIGITAL METHOD

The radar map is represented on an XY cartesian coordinate grid contained in three stores in the terrain simulator hardware. Two of these stores are 64K x 16 bits and the last one 64K x 3, into which the data are loaded at the beginning of an exercise. The area of the map can be varied, but is related to the resolution. The best resolution is 15 5/8 yards, corresponding to a map or playing area of 128 miles by 128 miles (1 mile = 2000 yards).

To hold each 15 5/8 yard square uniquely for a playing area of this size would require a large amount of store. To overcome this the map is modelled using general purpose 'building blocks' which can be used repeatedly in different parts of the playing area and hence reduce the amount of stored data. Two levels of building block are used as shown in Fig.1. The three stores are physically linked as shown in Fig.2. The first store holds the complete playing area and is addressed by the most significant eight X and Y coordinate bits, dividing it into 65,536 squares each with sides of 1/2 mile. These are termed "symbols". Ten of the data bits from this store form the most significant address of the second memory. This contains the large building blocks, of which there can be 1024 different types pointed to by the playing area store. The next three significant X and Y coordinate bits form the least significant address bits of this store, dividing each symbol into 64 squares, each with sides of 125 yards. These are the smaller building blocks termed "characters". Again, ten of the data bits constitute the most significant address of the third store card. Thus there are also 1024 possible characters. The three least significant X and Y coordinate bits form the remainder of the address. Each character is therefore composed of 64 squares with sides of 15 5/8 yards termed "pixels". The three bit data from this store allows eight different reflectivity levels for each small square. There are thus 1024 possible characters which can be combined in groups of 64 to produce 1024 unique symbols.

Height is also encoded into the map. It is divided into two parts. Base height, in steps of 1024 feet is stored in five of the remaining data bits of the playing area store. Finer height information is stored in the six other data bits of the symbol store, giving heights defined to character level.

Although there are only 1024 different symbols to fill the 65,536 locations, this is not such a great disadvantage as may appear at first sight. Naval training exercises obviously use maps with large areas of sea so any of the playing area squares consisting wholly of water will use the same symbol. Sea clutter is generated separately. Inland details are rarely visible from the sea. Many aircraft training exercises also contain areas of sea.

Sometimes a simulation can use an invented map. In this case it is relatively straightforward to generate the symbols and characters as the map can be adjusted to optimise their re-use. When an actual piece of land is to be represented then the problems are more involved. Usually, however, approximations can be made. Complete fidelity is unlikely to be necessary over the whole playing area so realism can be concentrated where it is needed, for example around harbours, estuaries and airfields. These areas tend to use a large number of symbols, but in contrast a large plain which did not need to be accurately represented could be simulated using very few symbols. Even hilly countryside uses only a handful of symbols when an approximate picture is sufficient. It is fairly easy to distribute the hills in a random manner. In practise it has been found that more symbols are required than characters, especially when the land is mountainous. The number of symbols can be easily increased to 2048 by adding on extra symbol memory, there being a spare data bit from the playing area store which can be used to switch between them. For certain types of scenario, in which the computer can exercise direct control over the immediately viewable playing area, it is possible to dynamically update the contents of the terrain memories transparent to the simulation. Consequently, the size and detail of the playing area need only be limited by the amount and speed of available backing store.

The playing area size and resolution are easily altered being set by a data word from the simulation computer. They can be set to multiplies of 128 miles and 15 5/8 yards respectively.

SIMULATOR OPERATION

The simulator operates in real time so that the stores are accessed as the radar pulse is 'transmitted'. The terrain part of the hardware has various links to the target echo circuits and to the simulation computer. Antenna movement is controlled by a microprocessor in the target echo hardware and by the computer. The latter informs the microprocessor of the type of scan selected which could be rotation through 360°, limited scan over a particular arc or lock and track modes, and the rate at which it is moving. The motion can be in elevation and azimuth. If necessary the dynamics of the antenna system can be simulated by an analogue circuit controlled by the microprocessor. The results of the antenna position calculations are required by various circuits in the simulation and as they become available, are stored in a separate RAM. Just before the next pulse transmission, this data with its destination address is sent out in a quick burst. The information specifically for the terrain simulation includes the sine and cosine of the bearing, the elevation, the cosine of the elevation and also the pulse length. For moving target indicating radars, a data word is sent which changes the intensity of the terrain reflectivity as a function of velocity of radar platform with respect to the terrain illuminated by the beam.

Another microprocessor controls the timing of the 'transmitted' pulses. It provides a wide range of radar type synchronisation signals and can be locked to external sources, for example the radar set in the simulator, or it can generate its own timing on being given the p.r.f. required. Two-way communication between the two microprocessors gives a flexible system which can be adapted to numerous situations.

Just before the beginning of a pulse transmission the computer informs the terrain simulator of the radar's position in the playing area and its height above sea level. The former data is loaded into the X and Y coordinate counters addressing the map stores, and the latter is used in the hardware downstream of the stores in calculations which determine if the terrain reflectivity accessed from the stores is actually visible in the radar beam. As the simulated radar pulse is transmitted the X and Y coordinate counters are incremented or decremented as appropriate, at rates which follow the path of the pulse as it is projected onto the ground. To achieve this, the basic clock frequency of 10.5MHz, whose period 95nS is the two way propagation time for a distance of 15 5/8 yards, is multiplied by the cosine of the antenna elevation or depression from its horizontal position to give the horizontally resolved rate at which the radar transmission crosses the ground. The X coordinate counter is clocked at a rate proportional to the product of this ground rate clock and the sine of the bearing, and the Y coordinate counter at a rate proportional to the ground rate clock and the cosine of the bearing. If a resolution lower than 15 5/8 yards is required then the signals clocking the counters are decreased in frequency in proportion to the change in resolution. For example with 62 1/2 yards the frequencies are divided by four.

The X and Y coordinate counters are not as straightforward as they appear at first sight, because the access times of the first two stores have been ignored in the discussions so far. Thus the X and Y address bits to the symbol memory must be delayed by the access time of the playing area memory, and the X and Y bits to the character memory by the sum of the access times of the playing area and symbol stores.

The cycle times of the playing area and symbol stores is 380nS, but if the radar pulse passes across a corner of a symbol or character square when the simulation is set to the highest resolution, two accesses could be required in less than this time. However, the shortest time between alternate accesses is 760nS so that on average the time can be met. The control circuitry associated with the coordinate counters evens out the closely spaced accesses by storing the results in an intermediate holding register ahead of time, and shifting them back to the correct separation. The access time of the character store is 95nS. Thus the reflectivity and height information is made available for all points over which the pulses passes.

The base height from the playing area store and the finer heights from the symbol memory are combined together, and along with the reflectivity of the pixel squares are passed to the next stages of processing. This is where the shadowing of lower lying terrain is simulated along with polar response and radar pulse length. The processing is divided into a chain of small sections each of which operate on the data within 95nS and pass the results onto the next section, effectively forming a long pipeline.

A target echo should be hidden by terrain whenever the elevation of the land from the antenna is greater than the target's, and the target is at a greater range. The elevation angle of each target echo as viewed from the antenna is stored in the target echo hardware, and the elevation angle of terrain is continually calculated as the radar pulse sweeps across the ground. Elevation and masking processes are performed as follows.

The ground rate generator used in the X and Y coordinate counters also increments a ground range counter, providing the horizontal range of each pixel. Since the terrain model is coded assuming a flat earth, a factor is subtracted from the height to correct for curvature of the earth, this being read from a PROM as a function of ground range. The difference between the corrected terrain height and the antenna height is determined, divided by the ground range, and the arctangent taken to yield terrain elevation. This is achieved by using logarithm PROMS, subtraction and then arc tangent-antilog PROMS.

The maximum elevation to date for the current transmission is stored in a register and each successively calculated elevation compared with it. If the new elevation is greater then the register is updated with the new value. If it is less, then that piece of terrain is in 'shadow' and is blanked out. The continually updated maximum elevation to date with range is linked across to the target echo hardware for comparison with the elevation of targets at that range.

The difference between the terrain elevation and the antenna elevation is now calculated and addresses a prom containing the vertical polar response. The data from this prom modifies the intensity of the pixel, such that only pixels actually illuminated by the beam contribute to the display.

The stream of reflectivity data up to this point represents the return from each portion of terrain for a time of 95ns. The pulse length is simulated next. Suppose the terrain echo from previous pixels is at a certain level. The reflectivity of the next pixel is added to this level and at the same time temporarily stored. After a time equal to the pulse length this data is read and subtracted from the combined level. Thus every 95ns a new return is being added to the total echo and an old return subtracted. The attenuation with slant range is simulated by incrementing a range counter as the pulse is travelling outwards and addressing a prom which attenuates the terrain echo by an appropriate amount. The simulation computer can alter the overall terrain level so that it can be set with respect to target echoes etc. One of the eight reflectivity levels is used to represent sea. When this code is detected sea clutter is selected instead of the land echo.

SEA CLUTTER

This effect is simulated by producing a series of random data bits which form arcs with programmable amounts of correlation from one radar pulse to the next, and controllable depths and number of arcs so that various sea states can be simulated. The arc intensities change in a random manner, but follow a Rayleigh curve, and the overall signal level can be altered. These parameters can be set differently for sectors of bearing so that the effect of wind direction can be taken into account. The sea clutter is attenuated with the third power of range.

The sea clutter or terrain echoes are combined with weather or chaff corridor returns if they are being simulated and sent across to the target echo hardware. There they are combined with the other radar simulation effects, converted to an analogue video signal and connected directly to a display or injected into the video processing stage of the radar set.

LIMITATIONS OF THE SIMULATOR

The simulator described is a cost effective way of producing terrain echoes, as returns from several million pixels are represented. However there are obviously some deficiencies. The size of the pixel squares on the radar screen at the shortest range scale determines one of the limitations. In converting the range and bearing of the radar pulse to cartesian coordinates an error is incurred which results in the pixels being displaced by up to plus or minus one pixel from their true positions. In addition they appear as curvilinear squares, and as the radar position alters they change position and shape. If the pixels appear large then this can be objectionable, but usually airborne radars have small screens and do not have very short range scales so there is no problem. The most appropriate resolution can obviously be chosen to minimise this effect commensurate with the playing area required. This could be made larger by using more store cards to hold the map.

The horizontal polar response is not simulated for the terrain returns, a thin pencil beam being assumed. To simulate it the reflectivity of adjacent terrain would be required which is not available using this technique. A rough approximation could be made by storing the outputs for pulses on various bearings and combining them in the correct proportions, but this could only easily be done with a 360° scan and with a slow moving radar vehicle. With narrow beam radars this is not a great loss.Polar response (with side lobes) is much more important with target echoes for realistic effects. The vertical polar response's main lobe is simulated, but not sidelobes. The position of a piece of land is only displayed in the correct position when it is in the centre of the beam, due to small angle approximations used in evaluating the in beam components. The error increases with increasing angle from the centre so sidelobes could produce undesirable results. Again this has been found to detract very little from the training value.

The strength of the echo should vary with the angle of incidence of the beam to the ground. Although this has not been simulated, it would be straightforward to add. The slope of the terrain is proportional to the arctangent of the difference in height between successive pixels. By adding this angle to the elevation angle which the terrain makes to the antenna the glancing angle of the pulse to the land is found. This can be used to modify the reflectivity accessed from the stores.

CLOUDS AND CHAFF CORRIDORS

The representation of clouds and chaff corridors poses similar problems to that of terrain and in the past have been simulated in the same way using flying spot scanners. The digital method described here can be adapted to simulate them. The same store arrangement is used, except that the height is split into two components, the height of the top and the height of the bottom of the cloud. They are thus represented by columns with pixel size cross-section. The reflectivity is constant in a column between the top and bottom heights. Each column has an attenuation factor so that targets behind clouds are attenuated depending upon cloud density and depth.

The cloud map is built up from building blocks and accessed in the same way as the terrain. The angles which both top and bottom heights make with respect to the antenna are calculated to determine what proportion of the cloud is within the vertical beamwidth of the aerial. The reflectivity of the pixels is modified to show the effects of pulse length and range attenuation, and then added into the terrain reflectivity.

The cloud attenuation values are accumulated as the pulse is transmitted and sent across to the target echo hardware, so as to attenuate all echoes in and behind the cloud.

During the course of an exercise the computer must periodically move the cloud about the playing area. It can also change its shape, reflectivity and attenuation. With chaff corridors it must lay the corridor as the aircraft flies along and alter its shape, reflectivity and attenuation quite rapidly. The computer therefore has a heavier workload for these effects than with terrain simulation.

SUMMARY

The simulator that has been developed overcomes many of the limitations and problems of flying spot scanners, yet is more cost effective than computer generated imagery techniques. By taking advantage of the fact that the terrain model can be constructed from re-useable building blocks the data can be stored in quite small semiconductor memories allowing real time processing to take place, so that airborne tracking radars can be simulated. The hardware is compact, consisting of ten double euro size printed circuit cards and requires very little external processing power.

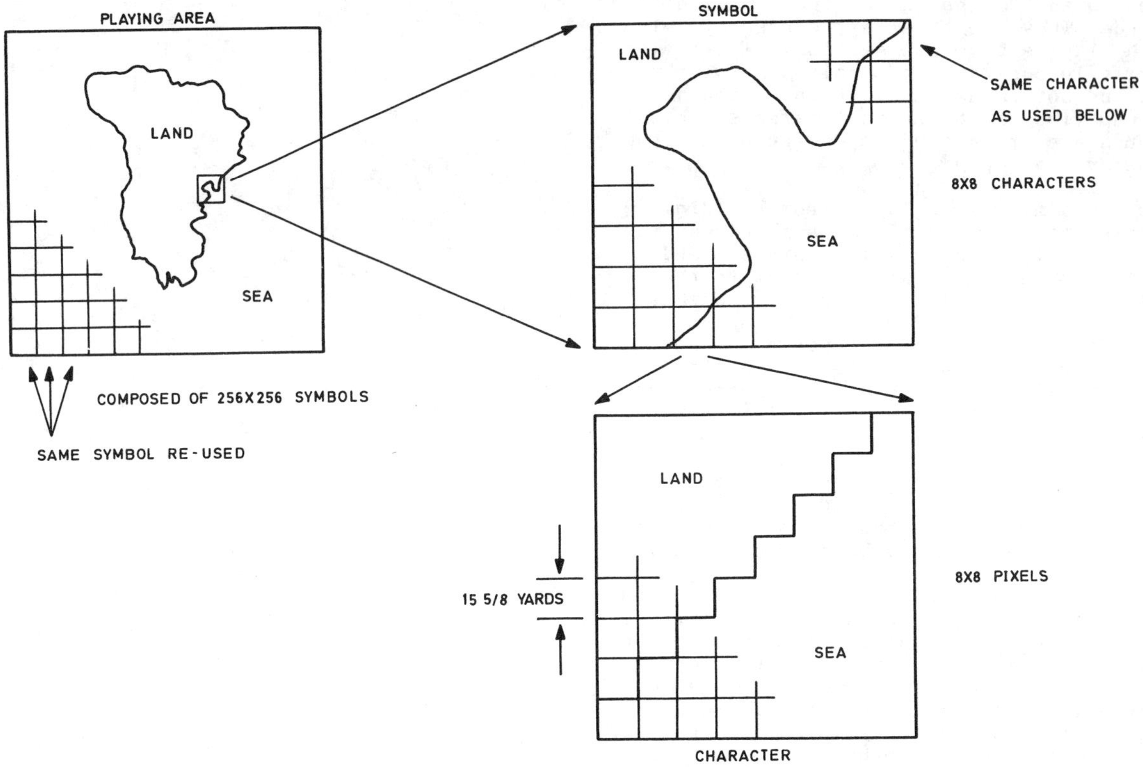

Figure 1 Terrain building block structure

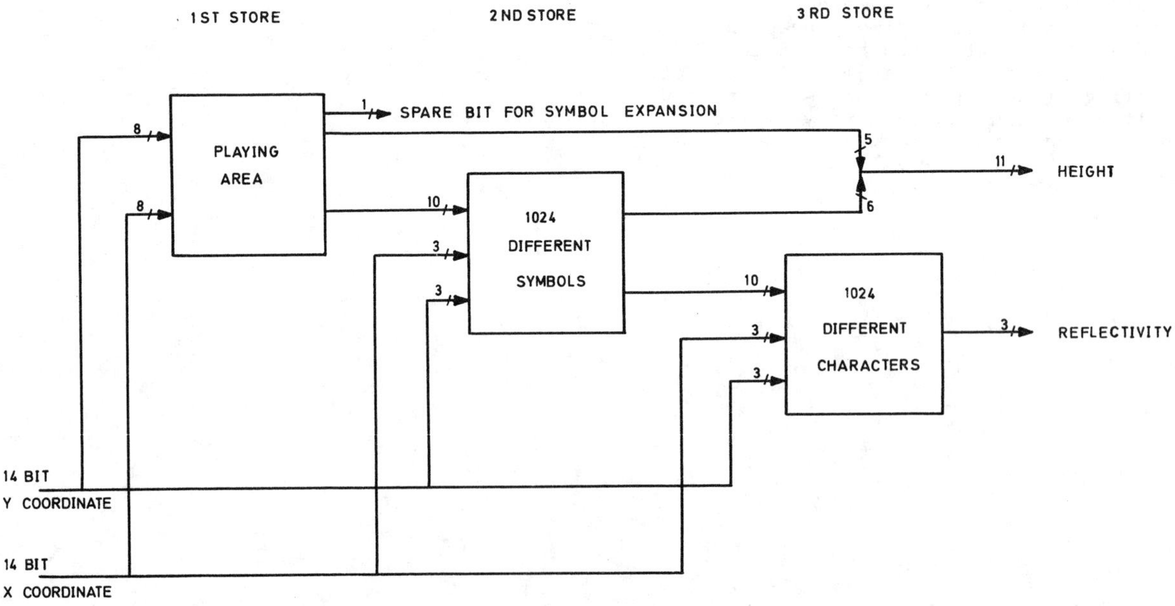

Figure 2 Physical configuration of the stores

RADAR ELECTROMAGNETIC ENVIRONMENT SIMULATION

John F. Michaels

Republic Electronics, Inc., USA

INTRODUCTION

Real world, real time testing and training with a radar system can be a formidable undertaking, to say the least. Consider the high cost of flights with chaff drops and jammers and the difficulty of obtaining repeatable data for system study and evaluation. Consider also the problem of effective training without having the ability to provide repeatable scenarios. Simulation provides a solution to these problems.

Today I'd like to discuss some of the factors which must be considered in order to provide a simulator for radar systems. These include: 1) The signals which represent multiple dynamic airborne targets with chaff, clutter and selected coherent and non-coherent ECM emanating from predesignated targets, 2) The models used for the generation of coherent synchronous signatures of aircraft/missile/ship targets, chaff, clutter, rain and jammers, including in each the effect of doppler, noise due to aircraft and antenna motion, the effect of time, frequency, spatial parameters and antenna pattern scan modulation. Descriptions of two simulators follow this discussion. The first simulator creates a real time coordinated and coherent electromagnetic environment at RF about a 3D "S" band radar, a 2D "L" band radar and their associated IFF equipment. The second simulator is capable of supplying the same environment while interfacing with many radars, and without restrictions on distance between radars and net control facilities.

A RADAR ELECTROMAGNETIC ENVIRONMENT MODEL

Radar Cross Section

There are two distinct types of targets - those that are short compared to the radar's pulse length and those that are long. A physically short signal effectively echoes a near exact replica of the transmitted signal. Long radar echoes include rain, clutter, and extended chaff drops. For long or extended signals, the radar return is no longer a simple replica of the transmitted signal, but a superposition of many returns from individual scattering centers. The RCS assigned to each type of target is based on measured mean values. This statistical mean is taken as a point of departure for all subsequent variation and is the operator entry into the system.

Scintillation Noise and Other Noise Components

The signal returned to the radar's antenna for even a physically very short target varies from reply to reply. The first important signal strength variation is the deviation from mean RCS caused by irregularities of the target itself. Minor changes in aspect angle can vary reply signal strength up to 20 dB, yet for all of its wild variations, the return signal strength is essentially free of step functions. For an aircraft in a "g" limited maneuver the return signal strength variations on a pulse to pulse basis (PRI of 2 milli-seconds) are minor, but in the 5 to 10 seconds scan-to-scan period, the return strength can vary to 40 dB correlation, pulse to pulse, and full decorrelation scan to scan, defined in literature as Swerling Case 1[1]. The probability-density function for the cross section σ is given by the exponential density function[1] (corresponding to a Rayleigh distribution of voltage).

$$P(\sigma)=\frac{1 \exp}{\sigma\,av.}\left(\frac{-0}{\sigma\,av.}\right)\ \sigma\geq 0$$

The second most important signal strength variation arises from multipath.[2] There are four possible multipaths considered on both transmit and receive. Path one, direct, is most significant because of least attenuation. Paths two and three, equal length, experience a single reflection. Path four, the longest, experiences two reflections and has a much lesser effect. This near equivalence of path one and the sum of two and three, has been amply demonstrated by various tracking test. In a multipath environment, two significant signals can be at any phase with equal probability, the addition can vary from the norm of unity (i.e., 1 volt) for the direct ray (path one) to a maximum of 2 volts and a minimum of 0 volts, a variation from 0 to +6 dB upward to infinity downward. The contribution from path four, however, prevents the zero (-infinity dB) and -10 dB is a realistic approximation.

In literature, Swerling I and II are often used as two distinct effects, some occurring under some circumstances, and the other under some conditions. Data recordings taken in the past have convinced us that both Swerling I and II are usually superposed. Special provisions were made to correlate replies from any one target, but decorrelated from another target within a radar beam width. Since correlated data variations are time dependent, the degree of change is based on elapsed time only and the radar PRI does not enter the generation of correlated noise.

Target Dynamics

In order to achieve the desired real time system coordination the target model must perform the following functions during each radar PRI: 1) Automatically acquire radar transmitted pulse characteristics, frequency, coding, and peak power level, 2) Determine the time delay for target range, 3) Determine target altitude in relation to antenna elevation angle, 4) Correlate target position with any interposed environmental model, 5) Calculate Doppler offset frequency as a function of slant range rate, 6) Output pulse frequency phase coherent with radar transmitter including any Doppler offset, 7) Calculate "target power" as a function of target RCS and radar transmitter peak power, 8) Determine antenna gain as a function of target position and antenna pointing angle, 9) Determine one-way (jammer) or two-way (Skin

Return) path loss as a function of slant range, 10) Determine atmospheric attenuation as a function of frequency and range if applicable, 11) Determine rain attenuation if applicable, 12) Output target pulse in real time synchronism at the proper power level and frequency.

Path Loss Factors

The radar range equation[3], calculates the power a radar receiver would receive. The simulation model accounts for variations in RCS, attenuation and radar system losses. Atmospheric losses as a function of frequency and range[4] (if applicable), and the attenuation of signal strength when in or behind rain[5] are also included.

Doppler

Two simulator calculations are required, the instantaneous slant range rate as a function of the present target velocity and heading and the Doppler offset frequency or phase, to output each moving target pulse phase coherent with the radar. In this system, phase shifters with digital networks are used and dedicated to each target in view. The phase is constant during each transmission period, so the spectral purity is not as good as using individual oscillators, but this approach handles a multiplicity of speeds and targets. Typically, a 4 bit phase shifter in an L band system yields a speed resolution of about ±10 knots for a simulated target speed from 0 to ±2,800 knots (±1/4% accuracy).

Extended Targets (Chaff, Rain, Sea Clutter)

Extended targets create an echo return that is a superposition of many returns from individual scattering centers. The range convolution is that of the radar pulse, a square wave, with that of the extended target shape (e.g., rain or chaff). Similarly, a convolution integral determines the combined effects of target and antenna pattern azimuthal variations.

For chaff, the variations are two fold. The vertical variations with time can best be described as a normal distribution with the upper 2-sigma anchored at the level of sowing and a mean fall rate dependent on the material employed. Typically the centroid fall rate is about 250 ft./minute[6]. Simultaneously the lateral spread, perpendicular to the direction of sow, increases at about the same rate.

Rain, although similar to chaff in certain aspects, has some significant differences. With its higher fall velocity, rain (a) reaches the ground in near zero time, and (b) has a uniform vertical profile from the ground up to the level of the forming cloud. Chaff has a uniformity along the line of sow but rain contains a near circular center with a density taper toward the edges not to zero but with a step function near its edge providing a definite profile.

Further, in the consideration of Doppler, the majority of the rain motion is vertical, and any lateral motion due to wind is uniform to all raindrops for the rain below the cloud, which is the majority of rain field in radar view, though not within the storm center of the raincloud. Rain within the center of the cloud, the region of turbulence, experiences large radial motion (up to 6m/sec versus 1m/sec for chaff[7]), but subtends only a small vertical angle, so it is neglected for search radar simulation.

For small targets, the multipath superimposition of two different signals was discussed. In the case of large extended targets there is a similar effect with many more reflectors delivering their one reply at slightly different times and phases. For small targets, the amplitude of the total simulated signal varies and a finer resolution is not necessary, since the receiver bandwidth is matched to the transmitted signal duration. For extended targets, however, the receiver can resolve such variations, although bandlimited to the signal bandwidth. Thus the random variations, (of +6 to -20 dB), as discussed previously under multipath, are again superimposed on the return signal. A noise pattern is generated with the data varied in a broadband manner with frequencies compatible with the receiver bandwidth. This high frequency noise (500 kHz for a 2 microsecond radar) is then superimposed in the amplitude patterns representing the convolutions of the target physical reflectivity profile and the scanning antenna pattern.

For simple pulse radars, the chaff/rain signal model is complete at this point. For pulse compression systems, scatterers at different ranges reply with different frequencies at the same time. For small targets a replication of the FM pulse suffices. For extended targets an additional high rate FM modulation is required to simulate the receipt of multiple frequencies while maintaining the pulse to pulse frequency reference. For simulation of extended targets, the modulating wave shape is modified to simulate the frequency centroid as a function of time during the return signal's duration (pulsewidth, plus target depth). The new centroid waveshape is further modulated with a signal, higher than the receiver bandwidth, whose peak to peak excursions correspond to the earliest and latest frequency in view at any given instant.

To generate a sea return signal, similar techniques are used. They differ from either the chaff or the rain (longitudinal or cylindrical symmetry) in their density patterns. While chaff or rain are isolated phenomena, sea clutter differs in that (a) it surrounds the radar (360°), and (b) it appears only at close-in ranges, less than 15NM from the ship's radar horizon for low antenna grazing angles.

Skolnik states [8] "It seems to have been well established that theories describing radar scattering from the ocean must take account of the small-wave structure (ripples, capillaries, facets) as well as the large-wave structure." Katzin's facet theory, supported by measurements taken by Schooley, appears to present the best model. He advanced the suggestion that, instead of droplets, the scattering elements are small patches, or facets, that overlie the main large-scale wave pattern.

The main reflecting patterns are known to be correlated to the wind (not wave) direction and appear as slowly moving reflecting strips perpendicular to the wind direction. Additionally, minor small targets appear to fill in the spaces between the main scatterers. Another wind-caused effect is that mean reflections are maximum into the wind, somewhat less alee, and a minimum cross wind, yielding an hourglass figure for the clutter simulation.

In forming a sea state pattern, the complexity is in

generating cross wind lines and minor scatterers as a function of wind direction, which correlate scan-to-scan but whose position drifts in minutes. Multiscatter effects are simulated in a manner similar to other extended targets.

ECM

To complete the electromagnetic environment model, the jammer platforms are replicated with independent ERP control and multiple operation modes. Among these are coherent range deceptions (including cover pulse) and inverse gain, noncoherent swept CW, spot and barrage noise, plus synchronous, and nonsynchronous pulses.

Scan Modulation and Antenna Gain

Actual antenna pattern data are used to amplitude modulate the simulated targets. The values of antenna gain relative to peak are stored in ROM with sufficient capacity dependent upon the desired resolution and system accuracy. For the usual target return loss the antenna gain is used both with the transmit link and with the receive link. For most jammer cases, it is only needed for the receive calculation.

A DELIVERED SIMULATOR (REES-201)

The REES-201, has the capability to provide 36 maneuverable targets, six chaff events and six jamming platforms in a simulated natural environment of background clutter and selected rain effects. The simulator interfaces with two radars plus the IFF system at the same time. Both radars operate independently of one another but see the same simulated environment. The basic concept for the REES-201 is indicated in Figure 1 and shows the simulator and its interconnection with the radar. The video display terminal and controller are common to the radar interfaces.

In order to achieve the degree of coherency necessary, certain information is required from the radar and is indicated in Figure 1. The STALO and COHO frequencies and their anticipated variations are necessary to recreate a return echo with the necessary degree of phase coherency. A radar trigger to synchronize the system, azimuth, elevation scan data and platform are also necessary.

Figure 2 shows the basic block diagram for the total REES-201 system. The video terminal is essentially the front panel of the radar simulator. The terminal is used to enter all data, control commands and scenario decisions. The following data is displayed on the video terminal: aircraft, jammer, chaff, rain, selected radar parameters, selected platform parameters, IFF, wind and clutter level.

Referring to Figure 2, static data (unvarying data) is entered and transferred to the 2D and 3D controllers. Any changes in static data entered by the operator will be transferred to the 2D and 3D controllers on entry. The dynamic data (varying parameters) is range ordered by the controllers as it is input initially and is kept in a range ordered file. The VDT screen format information and data is kept and updated in the central controller. The central controller recalculates the present and expected position (1 sec later) based on inputted heading and velocity of each object every second. The objects are then range ordered. While one CPU calculates this information, the other CPU transfers its previously calculated information to the 2D and 3D controllers. The 2D and 3D controllers store the received information in one of their two range ordered files. The 2D and 3D controllers take the computed present and expected position for each target each second and output a straight line approximation between these points with target data updated each 10 milliseconds to the radar. Analysis shows that the maximum position error is less than 70 feet, at worst case, for a Mach 3 target. Both processors receive appropriate radar antenna data. This information is used to determine those objects within a selected angle about the antenna boresight. Elevation angle gating is handled in hardware.

The control of the RF circuits is the responsibility of the digital interface circuitry under control of the microprocessors. The RF circuitry has been designed to correspond to the desired degree of simultaneity between radar and jammer returns. There are no conflicts between coherent target returns and sea clutter, chaff and rain with screening effects of the above considered as they effect target return amplitude.

In order to allow a minimum of two targets to occur simultaneously in time, two coherent RF channels are implemented in each radar interface. Figure 3 shows the signal processing that is required with one coherent target return, outputting simultaneous with either sea clutter, rain or chaff and one noncoherent jammer.

The jammer channel is separately VCO-derived with identical control elements as used in the coherent return channels. The combined outputs from each RF Processor, are inputted into the receiver system via a directional coupler to minimize any effects on received system sensitivity.

A RADAR NET ENVIRONMENT SIMULATOR

The previous section described a simulator capable of generating a real time coordinated and coherent environment about 2 independent primary and one secondary radar all located in the same vicinity. This section concerns a simulator that does all that has been previously described, and also generates this environment into many widely separated radars, all properly coordinated in real time. This has applicability for stressing a country's total engagement system and performing coordinated radar operator training involving the total net.

Figure 4 indicates the concept for the Radar Net Environment Simulator. Data is entered either manually or automatically via a magnetic tape unit and includes initial target longitude, latitude and altitude coordinates, velocity, heading, turn and altitude change rates, mean radar cross section, jammer and/or chaff data is applicable, weather, clutter and IFF data for friendlies. Individual site coverage parameters are resident in master controller memory. All data is updated each second and outputted to the appropriate radar site. Figure 5 is a block diagram of the master controller in synchronism with each site's central processor. Data is transmitted over full duplex dedicated communications lines to each radar site and target data is transformed to rho-theta and altitude coordinates appropriate to the site. At each site target data is updated each second identical to tasks performed at the master controller, and in synchronism with it. This data is further segmented appropriate to the particular radar or radars at the site and outputted each ten milliseconds in synchronism and phase coherency with each radar. Figure 6 depicts the receipt and processing of data for each radar. Because of the data processing techniques used there are no restrictions or distance between radars.

SUMMARY

By utilization of the techniques described, a total electromagnetic environment can be generated, manually or via magnetic tape control, phase coherent and real time coordinated, into a radar or radars that are part of a net in order to both stress the total radar/ data processing system and accomplish the desired level of radar operator readiness.

It should be noted also that the concept presented is not dependent on the type of radar, but rather three required data items: 1) Where the antenna is pointing 2) LO and COHO information 3) Synchronizing trigger. Given this information, a simulator can be developed for all types of radars.

REFERENCES

1. Skolnik, M.I., 1962, Intro. to Radar Syst., 51.
2. Nathanson, F.E., 1969, Rad. Des. Principles, 33.
3. Ibid., 41.
4. Ibid., 51.
5. Ibid., 195.
6. Van Brunt, L.B., 1978, Applied ECM, 385.
7. Nathanson, F.E., op. cit., 226.
8. Skolnik, M.I., 1970, Radar Handbook, 26-23.

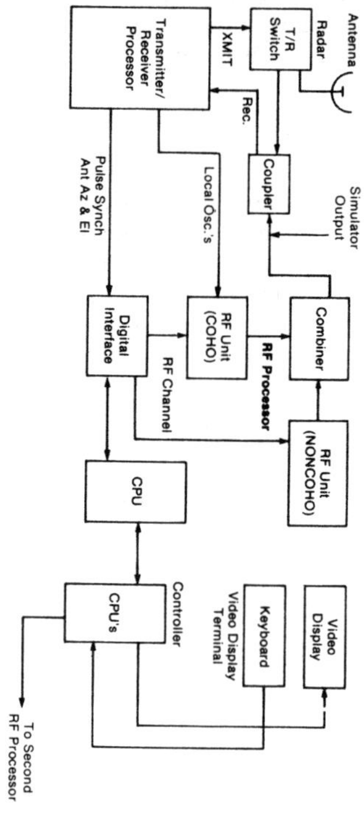

Figure 1: Radar Interconnection

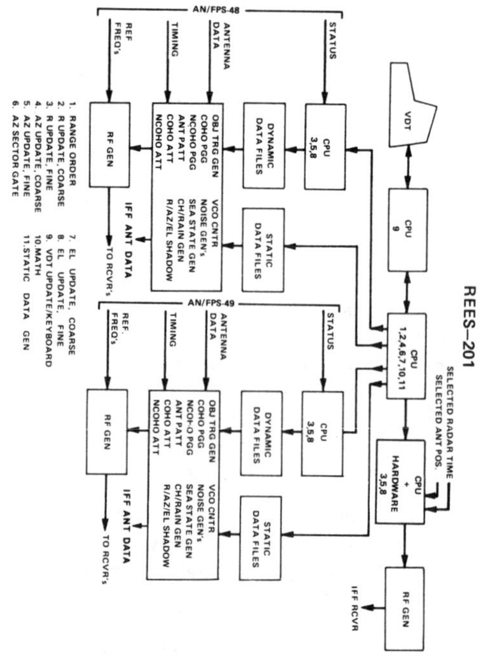

Figure 2: REES-201 System Block Diagram

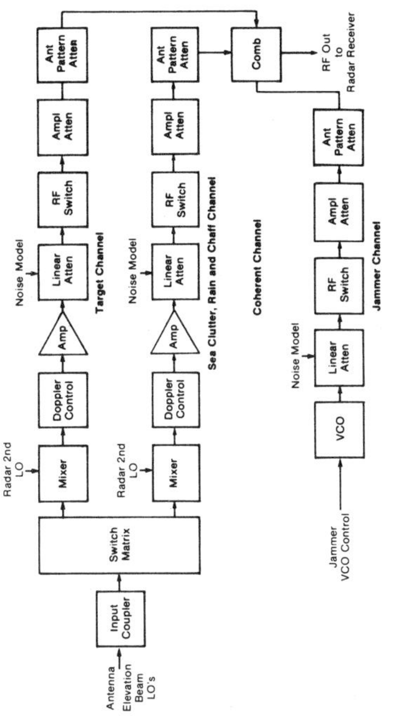

Figure 3: RF Block Diagram

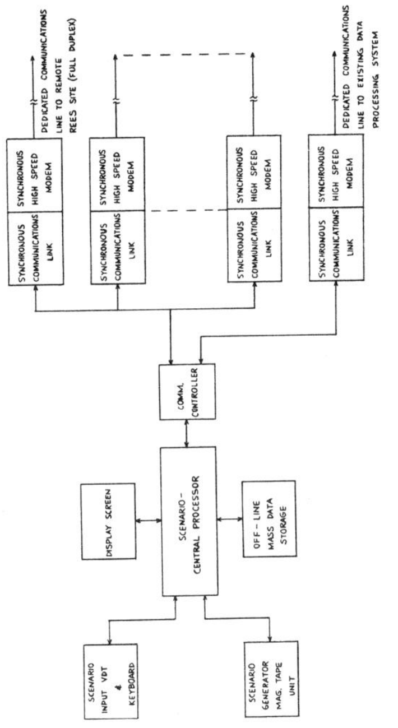

Figure 5: Control Center Master Controller/
Data Entry and Display

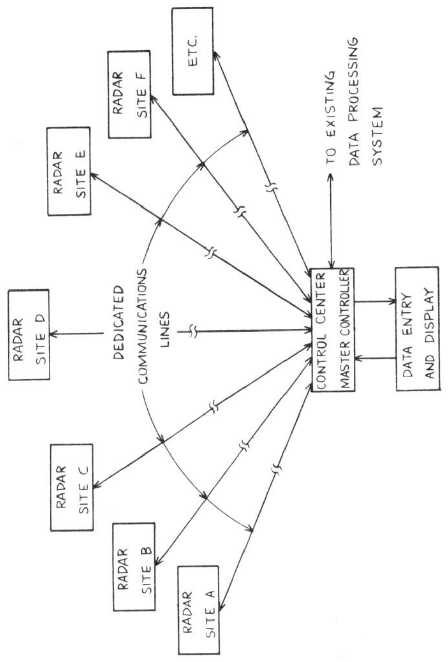

Figure 4: Radar Net Environment Simulator

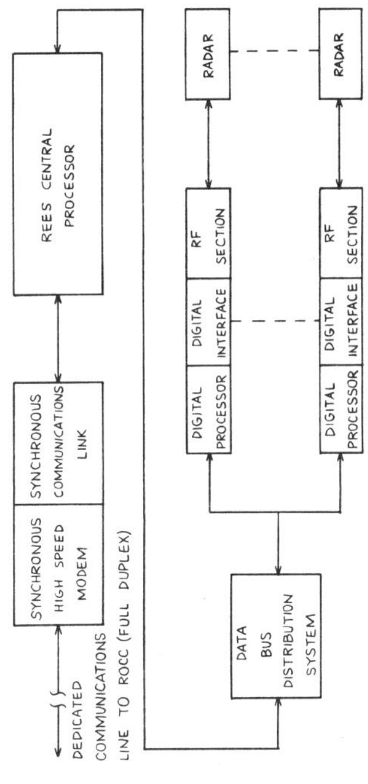

Figure 6: Radar Site RF Processor

AN EQUIPMENT FOR SIMULATING AIRBORNE RADAR VIDEO

T. Snowball, T. R. Berry, A. M. Pardoe

Royal Signals and Radar Establishment, St Andrews Road, Malvern, Worcs, UK

There has long been a market, albeit limited, for cheap flexible radar video simulators. With the emergence of digital scan converted airborne displays, this market is likely to increase significantly in the future. In the short term this is because the new display formats, made possible by scan conversion, require considerable laboratory investigation to determine their operational potential. In the longer term, the complexities introduced into the display system - scan to scan integration with moving origin, complex stabilisation etc will require a fairly sophisticated back-up for testing. Ideally, such a test facility should reproduce as near as possible, the radar characteristics which the scan converter is to accept.

For the reasons given above a versatile radar simulator, simple in concept and cheap to produce, has been developed and built at RSRE Malvern. The device is capable of generating video signals representative of a range of airborne, shipborne and ground based radars. It consists, basically, of a miniature commercial TV camera whose deflection coils are fed with ramps similar in principle to those used to generate conventional PPI displays, ie the scan pattern of the camera is polar (Rθ) in form rather than cartesian (XY) as in conventional TV.

The camera is used to 'look at' at 'suitable' film strip - where 'suitable' means some form of simulated radar picture. In this way, the video out of the camera can be made to resemble closely the video which would be produced by a radar scanning the area represented by the film. By moving the film past the camera, and by suitably controlling the various parameters which determine the deflection waveforms and the active area on the vidicon used, it is possible to simulate a wide range of corresponding radar and aircraft parameters. These include: aerial scan format (ie PPI or Sector scan); aerial rotation rate and direction; radar range and prf, aircraft heading and drift velocity.

The equipment also simulates the effect of radar shadowing due to relative heights and bearing between targets and aircraft.

INTRODUCTION

Considerable work has recently gone into the development of Radar Digital Scan Converters mainly in an attempt to overcome the problems of fading and flicker of conventional radar displays. Not surprisingly, the introduction of Digital Scan Converters with permanent frame stores and some form of digital processing has not only improved on conventional displays, but they have also made possible entirely new display formats. Very little work has been done so far to determine the potential of these new formats. Any laboratory experiments aimed at rectifying this neglect would require considerable quantities of representative radar video. This could be obtained by recording data in a series of flight trials. However this is a very expensive way to obtain data and is limited by the performance of the recording equipment, the available radars and the instrumentation fit of the trials aircraft. Such data, once obtained is inflexible, and any changes in format could well require a new flight.

In consequence we have developed a versatile but inexpensive simulator, the main advantages of which are that the radar video is easily adaptable and can represent a whole range of radar equipments and usage at very low cost.

The present system is based on the use of a cheap TV Vidicon camera, modified to scan in polar form, which looks at a moving strip film. This film can be made to be representative of the radar being simulated, but if it is to be used to show up defects in the scan converters, it is vital that any resulting imperfections on the display should be easily recognised. Consequently if we were to use truly representative radar pictures - which might consist of just ill defined white blobs on a black background - such imperfections could easily be overlooked. It has been found more suitable therefore to use a film strip made from a conventional ordnance survey map, modified to enhance those features on the map that give good radar returns - eg banks of rivers, coastlines, densely populated areas and power lines, etc.

SYSTEM

A simple block schematic of the simulator is shown in Fig 1. Digital electronics generates all the waveforms, with the final output as sawtooth waveforms to the TV camera deflection amplifiers.

Incremental range (δR) is generated in the Range Oscillator and forms the basic calibration unit for the deflection system of the vidicon and the film drive motor. Instantaneous range R is established by counting δR in the variable length Range Counter. Increasing the length of the Range Counter effectively increases the length of the radial scan in the camera, which is equivalent to increasing the range of the simulated radar video.

Incremental aerial rotation (δθ) is generated in the θ Oscillator and converted to instantaneous θ in the Aerial θ generator in a manner directly analogous to the Range Counter. The sense of rotation (clockwise or counter-clockwise) is achieved by making the Aerial θ generator count up or down. An initial value of fixed or variable origin may be fed into the Aerial θ generator to simulate either a fixed angular offset, or the effect of a changing pointing angle. The latter is particularly useful in simulating an airborne sector scanning radar whose aerial pointing angle is being slowly rotated relative to the aircraft axis. (Alternatively it can represent simply a change in aircraft track angle).

When simulating an all round scanning radar (PPI) the UP/DOWN count line to the Aerial θ generator, is held constant to give a continuous clockwise or anticlockwise rotation. When simulating a sector scan the UP/DOWN line is changed at the extremity of each scan. By adjusting the rotation rate under control of the θ generator and the duration of the count up and count down times angular coverage of the sweep may be determined.

The conversion from θ to sin θ and cos θ is performed in a read only memory (ROM) situated in the Aerial θ generator. Sin θ and cos θ are then multiplied by the incremental range (δR) to produce pulse trains whose frequencies represent the rate of increase of incremental X ($\delta X = \delta R \sin\theta$) and incremental Y ($\delta y = \delta R \cos\theta$). These pulse trains are counted in X and Y to produce instantaneous digital X and Y which in turn are applied to digital to analogue converters (DAC) to produce the X and Y sawtooth waveforms to the TV camera deflection amplifiers.

On the completion of a scan, determined by the Range Counter length, the Range osc is halted and waits for the next Tx pulse from the PRF generator.

In order to generate simulated Airborne Radar video it is essential to mechanise a realistic flight path by simulating both heading and drift motions. This could be achieved by providing the film with 2 axis motion. In practice, however, heading motion is many times greater than drift, and we take advantage of this to simplify the film drive so as to represent heading only; and simulate drift by electronic methods with the vidicon itself. The principle is illustrated in fig 2. The system is optically scaled so that the whole of the across-track dimension of the film falls on the vidicon target, but only a small part is actively scanned by the read-out beam at any time. Modifying the starting values of the X and Y counters effectively moves the scanned area across the vidicon target and hence simulates motion in the along and across track dimension of the film. The across track movement produces the required drift component, and the along track component acts as a fine adjustment on the mechanical film motions.

SHADOWING

Although specifically designed for generating video to help in developing radar displays, it is reasonable to ask if the simulator could be extended for other applications. One area of particular interest is for providing cheap simulators for training radar operators. The system so far described however does suffer from one major shortcoming which severely limits its use for training purposes. That is, the video output from the TV camera is independent of the simulated aircraft height and bearing relative to the target, since the camera is always looking at a fixed two dimensional film. In the real 3 dimensional world, because of radar shadowing, a radar produces a picture dependent upon aircraft height, aspect and flight path.

Consequently an additional feature has been incorporated which attempts to simulate shadowing. The technique used is based on the premise that good returns imply prominent objects, that is, we may relate terrain contours to signal amplitude. This is, at best, only a very gross approximation, nevertheless it gives us a system in which the effects of terrain are easy to mechanise and does meet the primary requirement that the picture will vary in a reproducable manner with aircraft height and bearing. The system is illustrated by fig 3 and fig 4. At each incremental range point the measuring signal is sampled and compared with the previous value, held in the previous result store. If the new value is equal to, or greater than, the previous one the target at that point is taken as not being in a shadow, and the output is made equal to the input. This situation is represented in fig 3 by R_O to R_N. If the current signal is found to be less than its predecessor however, as at $R_N + 1$, the target is taken as being in shadow, the output is clamped to zero, and the grazing angle of aircraft to the last used value derived from the expression

Grazing Angle =

$$\frac{\text{Aircraft Height - Signal Height}}{R}$$

(on the basis that Signal Height represents terrain). Using the stored value of this grazing angle the last used value of signal is progressively decremented at each range increment until it becomes less than the incoming value (represented by Rm), when the output is restored.

Presently the performance of the simulator is limited by the characteristics of the TV camera which is specifically designed for linear scanning at TV rates. The signal from the Vidicon is extracted by equalising the charge pattern on the target, by means of a scanning electron beam which in the case of TV, only scans each element once per frame, used, with a PPI scan however, the centre area of the circle is overscanned many times during each revolution - so that residual charge patterns can be established. Again, with the many variations of scanner rate, PRF and range that are possible it becomes difficult to maintain the light input/electron scan ratio within the automatic gain control range of a vidicon camera. So some modifications either to light input or other electrical parameters of the vidicon are necessary to cover a wider range of simulated radars.

Other more expensive scanning devices may be considered as a replacement for the vidicon such as flying spot scanners, but these are generally more bulky and complex.

Although the simulator so far built has been used exclusively to simulate airborne radar polar scanners, the principles involved could equally well be applied to represent sensors which have other scan patterns, IRLS, SLAR, SONAR for example.

The range of parameters covered with the present equipment are as follows:

PRF: 900 Hz to 27 KHz

RANGE: 25 μsec to 920 μs
(2 miles to 75 miles)

ROTATION: 1 rev/sec to 1 rev/24 sec

some of the above rates are together mutually exclusive, PRF and Range for example.

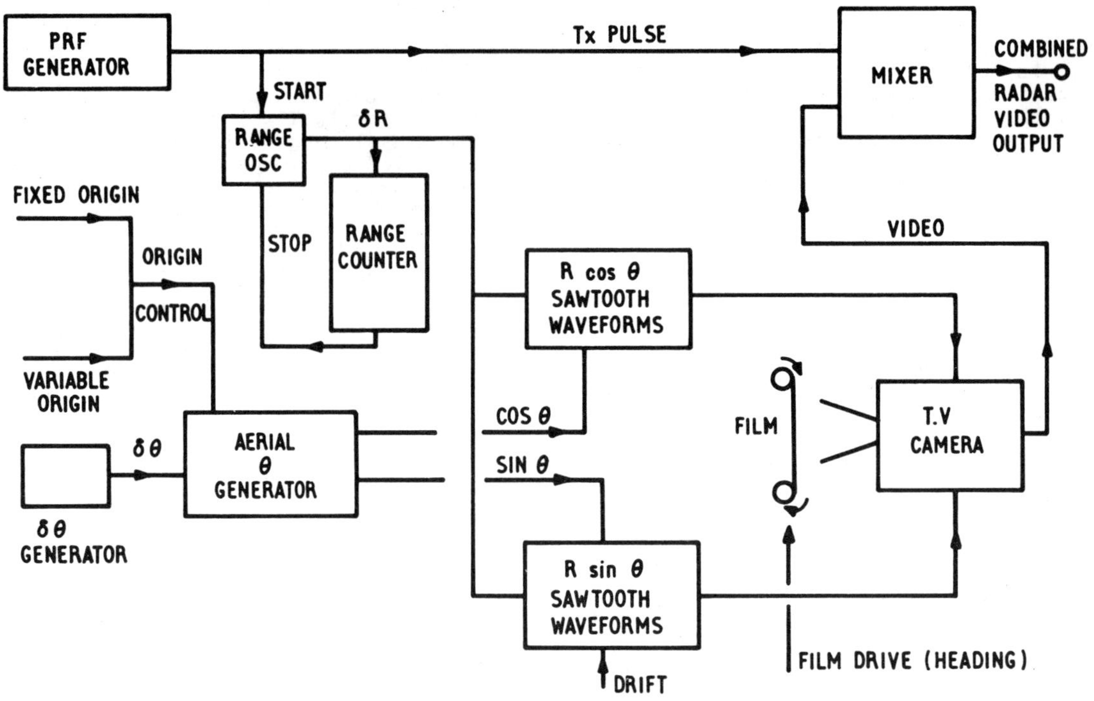

FIG.1 SIMPLIFIED SCHEMATIC OF RADAR VIDEO SIMULATOR

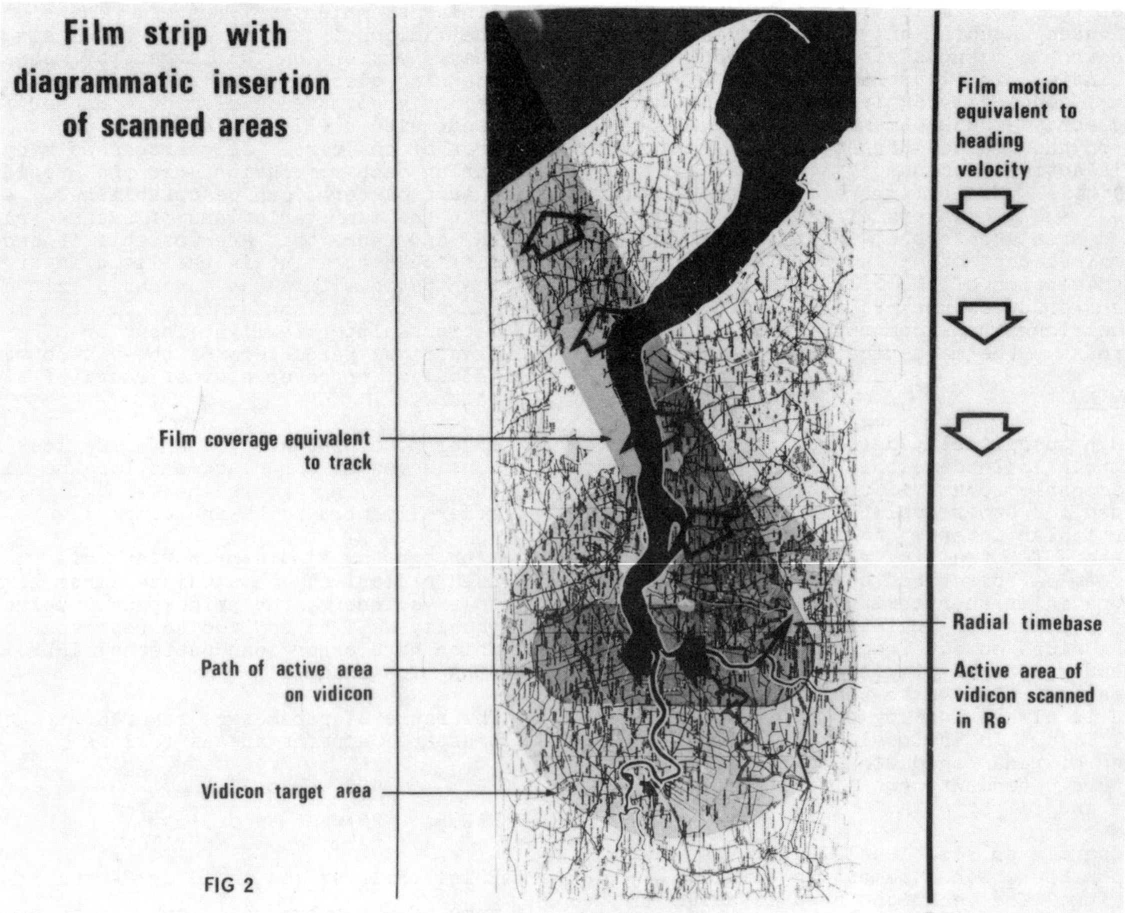

FIG 2

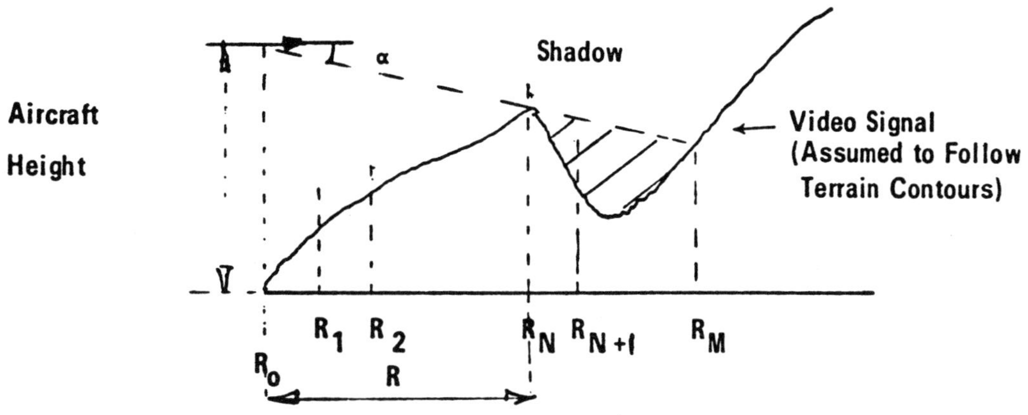

$$\text{SLOPE OF } \alpha = \frac{\text{HEIGHT} - \text{SIGNAL}}{R}$$

$R_o \rightarrow R_N$ New Signal $\geq$ Previous $\therefore$ Output Signal = New Signal

$R_N \rightarrow R_M$ New Signal $<$ Previous $\therefore$ In Shadow

i Output Clamped to Zero

ii Signal at R_N Decremented

At R_M New Signal $>$ Decremented $\therefore$ Restore Output

FIG 3 PRINCIPLE OF SHADOWING

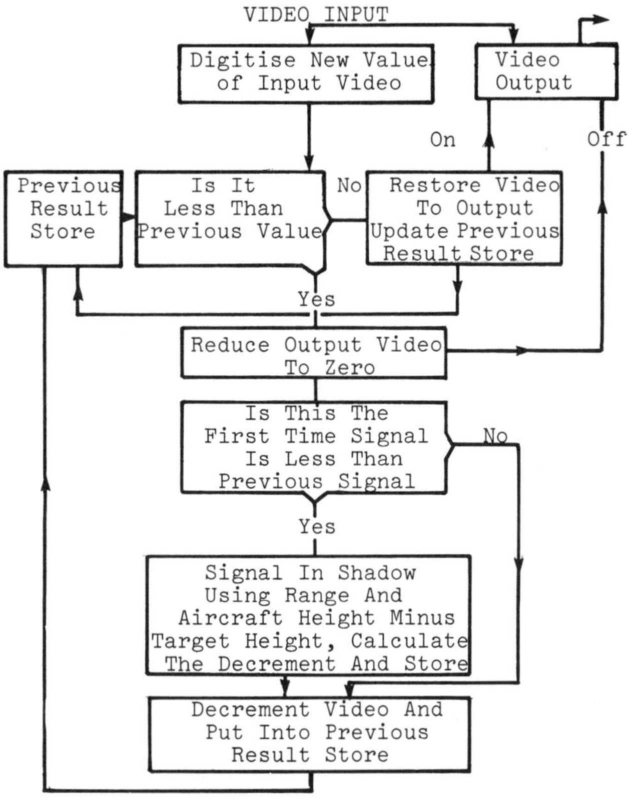

FIG 4 FLOW CHART

THE AUTOMATIC TRACK WHILE SCAN SYSTEM USED WITHIN THE SEARCHWATER AIRBORNE MARITIME SURVEILLANCE RADAR

M. Symons

THORN EMI Electronics Limited

INTRODUCTION

The Searchwater Radar System is an airborne surveillance radar used for detecting, tracking and classifying both small and large vessels on the surface of the sea.

The main facilities are the following.

The presentation of a ground stabilized North oriented Plan Position Indicator (PPI) display.

A choice of A scope or B scope high resolution displays.

Display of target file alphanumeric data and various symbolic markers.

A number of operator controls, including roll ball, buttons and switches.

An automatic track while scan facility for a multiplicity of targets, which will be the main subject of this paper.

THE BASIC BLOCKS OF THE SYSTEM

Before concentrating on the tracking facility, the basic blocks of the radar system as shown in figure 1 are briefly described here to serve as a background.

A pitch and roll stabilized scanning antenna is used to minimise the effects of the aircraft movement. In addition, the tilt of the antenna is controlled by the computer to optimize the radar performance in range allowing for aircraft height.

A special purpose electronic hardware block includes a signal processing system designed to enhance the detection of surface vessels in high sea states and at large ranges. The domains of space, frequency, time and amplitude are all exploited. The dimensions of the patch of sea illuminated at any instant are kept small in azimuth by employing a narrow aerial beamwidth and effectively small in range by processing a narrow pulse generated with pulse compression techniques. Frequency agility is employed and the total number of returns during the time taken for the aerial to scan over predetermined angular widths results in an improvement in discrimination of targets against clutter. A threshold immediately after the demodulation of the returned signal is continuously adjusted using a control signal derived from the local clutter level so that as nearly as possible a uniform constant low false alarm rate is presented to the operator. The combination of these and other techniques ensures that generally the display is fairly free from clutter and any new radar contact is readily apparent.

With use of both the computer and the hardware, the radar data is converted from its polar scan form to appear on a display in television raster format so that bright flicker free information is presented to the operator. Furthermore, the display presents the target returns plan corrected in range as a North oriented ground stabilized map. A higher resolution A scope or B scope display of operator selected localized areas can be presented to the operator without interruption of the main PPI display. The main display can also present target details from files selected from within the computer by the operator.

The radar controls are simple sequences of push button and roll ball operation and no measurements have to be taken directly from the screen.

The system also contains a real-time dedicated airborne digital computer which has a number of tasks. One of the tasks is to interpret the operator controlled keyboard inputs. A second is the correction for aircraft movement to provide the ground stabilized radar data for the display. A third is the display of the target file data. Another is the control of the high resolution A scope and B scope displays. There are various subsidiary requirements such as the antenna tilt control. However, the main task is that described in this paper, namely the continuous automatic track while scan for a multiplicity of targets. It was necessary that the tracking of vessels occurred with the aerial still scanning in order that the surveillance role of the radar remained uninterrupted. Furthermore, in order that the work load for the radar operator be kept to a minimum, it was also required that the tracking system be automatic. The software uses CORAL as its high level language and uses FIXPAC assembly code in areas where speed is essential.

The remaining block shown in figure 1 is a hardware interface unit for transferring aircraft navigational data such as aircraft position, speed and heading.

THE TRACKING PROBLEM

The concept of an automatic track while scan has been developed over a number of years and some ground based surveillance radar installations, in particular those for air traffic control, include its implementation. However, an air-to-ground tracking radar system has a number of significant differences which, in the past, prevented the development of an airborne automatic track while scan system. An airborne radar tracking low velocity targets (seaborne vessels) has a disadvantage not encountered by a zero velocity ground based radar tracking high velocity targets (aircraft). This disadvantage is that any navigational errors in the aircraft system, particularly in velocity, though small when compared with the motion of the aircraft, become effectively large when superimposed on the motion of the target. This does not impair the tracking itself but can affect the accuracy of the resulting speed estimate. Another hazard is that, while the sky presents a relatively clean background to a ground based radar, the sea presents an extremely poor background, especially under adverse meteorological conditions.

This latter problem of sea clutter was reduced by the use of a comparatively small moving two-dimensional window around the area of the moving target, together with a high resolution radar system and a hardware front end signal processor to minimise the amount of clutter entering the window. The window is moved under control of the computer as will be explained later. If more than one return appears in the window, the return nearest the centre of the window is taken to be the target plot.

Even after selecting this return nearest the centre, the estimation of the true target position still presents a problem for the tracker. The return differs from the true position due to a number of reasons. Firstly since the radar return signal has width, the target centre can vary from scan to scan because of positional quantization, noise supplementation, the radar return's fading characteristics and the target's change in relative aspect. Secondly there are the navigational errors in the aircraft system, as already mentioned. Thirdly in spite of a high degree of stabilization against the aircraft movement, including stabilization against pitch and roll, there are small residual errors affecting the scanner beam direction.

The method for solving this problem was to use the algorithm known as an $\alpha\beta$ tracker applying an adaptively weighted mix of past history of target returns together with the current plot return. This is a technique proposed over a number of years by various researchers including Sklansky (1), Marks (2), Benedict and Bordner (3)(4), and Quigley (5). Modern control theory shows that, for fitting a continuous track to a sequence of noisy plots, the Kalman filter is a better linear estimator (see, for example, Barham and Humphries (6)). The weaker but related $\alpha\beta$ tracker algorithm was adopted for economy in computer loading. It was possible to use this alternative since surface targets are relatively more sedate compared with highly manoeuvrable airborne targets.

THE TRACKER GATE

The moving window, referred to above, is a special piece of hardware known as a tracker gate, which is effectively moved by the computer controlling the timing for the hardware to accept the radar returns. The gate is used solely for tracking; the other radar facilities such as the surveillance PPI display do not use the gate. There is in fact more than one tracker gate but, by multiplexing the use of these hardware devices, a few gates are able to process a relatively large number of targets.

The gate consists of a two dimensional array of quantized cells (see figure 2) with a fixed number of cells in both range and azimuth. Each cell can take the value zero or one as a result of a threshold after the signal processor, where one represents the presence of a signal. A target will generally occupy more than one cell and the positions of the returns within the tracker gate are passed to the computer in the form of the start of target azimuth, the target width in azimuth cells, and the mid-range position. From this information, the computer deduces the azimuth and range displacements of the centre of a target from the centre of the tracker gate.

Although, under most conditions the tracker gate will contain a single target, in some situations there may be more than one target or false alarm in the gate. For the latter case, the computer scales the two dimensions of the return within the tracker gate to determine the distance of each return from the centre. The return nearest the centre is found using this scaled distance.

For this and other purposes, two running integrals, FR and FT, are produced over successive plots. FR and FT are measures of the jitter on the range and azimuth displacements in the tracker gate and are produced from a weighted sum of the differences between displacements for each current plot and the displacements from the previous plot. The values of FR and FT are used in turn to scale the dimensions of the instantaneous displacement of each return within the tracker gate. In subsequent tracking computations the actual displacement values are used without weighting.

Thus the distance Dn of a return n from the tracker gate centre is defined as

$$Dn = \left[\left(\frac{DRn}{FR}\right)^2 + \left(\frac{DTn}{FT}\right)^2\right]^{\frac{1}{2}}$$

where DRn and DTn are the displacements in range and azimuth respectively for return n. The return with the smallest value of Dn is taken to be the target nearest to the centre of the gate, and therefore the most probable plot.

Thus if returns vary enormously in relative azimuth compared with the range, then more significance will be given to the displacement from the gate centre in range.

THE $\alpha\beta$ TRACKER ALGORITHM

By mathematical analysis of the plot in the gate and of past plotted positions, a best fit piece-wise linear track is developed by giving a value for the speed and direction of the target. If a large number of past plotted positions appear to be on a relatively straight line, then a reasonably reliable velocity can be calculated. If, however, the target appears to be manoeuvring, then only the latest few of the past plotted positions can be used with consequent degradation of the estimates of speed and direction.

From the past trends of the target motion, the system predicts the most likely position F for the return of the next scan and places the tracker gate centred on position F accordingly. If however, the return appears at position P then there is a correction to the forecast position to produce an estimate E of the true position. The amount of correction is an adaptively weighted proportion of the error FP. The velocity vector is also corrected (with another proportion of the error FP) to form an updated estimate of the target speed and direction. This process is the basis of the $\alpha\beta$ tracking algorithm and the equations used are equivalent to

$$F(r+1) = E(r) + V(r).T$$
to forecast the position for the next scan.

$$E(r+1) = F(r+1) + KP.\left[P(r+1) - F(r+1)\right]$$
to estimate the true position.

$$V(r+1) = V(r) + KV.\left[P(r+1) - F(r+1)\right]/T$$
to estimate the velocity.

where
E(r) is the estimated position at scan r
F(r) is the forecast position at scan r
P(r) is the plot return position at scan r
V(r) is the estimated velocity at scan r
T is the time interval between the successive plot returns r and r+1
KP and KV are the weighting adjustment factors

There are in fact two sets of the above equations to adjust the estimated position and velocity components in an xy plan grid. Thus, there are parameters Ex, Ey, Vx and Vy in the equations used in the software. Consequently, the azimuth and range displacements in the tracker gate and the forecast position have to be translated from and to the radar tracker gate polar position and the xy grid.

The weighting factors KP and KV are adaptive and calculated from

$$KP = 2(2N-1)/N(N+1)$$

$$KV = 6/N(N+1)$$

where N is a quality index which increases in value by unity to reach an upper limit each time the forecast position is sufficiently accurate and otherwise decrements in value by unity to a lower limit.

It is the factors KP and KV which are the α and β values of the algorithm.

Thus the automatic track while scan sequence of processes is that as shown in figure 3. The first process is to forecast the position of the target at the next plot. This is performed in two stages. The first stage is to make a coarse prediction fairly soon after a plot using the current aircraft position and velocity information. Nearer the time forecast by the coarse prediction a further, or fine, prediction is made using the more up to date aircraft information. This process will be described later in more detail.

This prediction is used in the next process which is the time/position placement of the tracker gate in relative bearing and range. The hardware then gates in the radar returns.

The next process, the zone preprocessing, selects the return nearest to the centre of the tracker gate, allowing for scaling of the displacements.

The final process performs the tracking calculations to produce the estimates of the target position and velocity. The parameters FR and FT, known as error integrals, are updated together with the quality index N.

There are, as will be explained in the next section, two channels for each target and the tracker calculations select the appropriate channel to be used for the estimate of the target position and velocity. Another sub-process within the tracking calculations is to detect if the track has been lost and to warn the operator accordingly.

THE TWO CHANNELS AND THE BREAKPOINT TEST

As indicated in the previous section if a target is moving along a relatively straight line, the estimated target velocity will be sufficiently accurate to give good placement of the tracker gate and hence the weighting factors KP and KV will become small. This will produce a tight smoothing algorithm giving small corrections from the more recent plots. However, if the target then begins to turn, there will be the possibility of losing the target as a result of the tight smoothing.

There is therefore, in addition to the above main $\alpha\beta$ channel, a further channel that is non-adaptive and uses a lightly smoothed algorithm. This second, or idler, channel is operative in parallel with the main channel on every scan and uses a set of equations similar to those of the main channel. However, the values of the weighting factors KP and KV used in the idler channel are equivalent to using a fixed quality index N equal to six. Hence the idler channel gives more weight to the more recent target returns. As the estimated target velocity vector can be changed more rapidly, the idler channel is far less likely to lose track. The main channel has the advantage of giving a more accurate estimate of a target velocity and is normally used.

There remains the problem of determining when to use the idler channel. This is achieved by a set of further error integrals, OR and OT, for range and azimuth respectively. These latter integrals are running weighted averages of the displacements in range and azimuth of the target return in the tracker gate over successive scans. The following test, known as the breakpoint test, is made.

$$\left(\frac{OT}{FT}\right)^2 + \left(\frac{OR}{FR}\right)^2 > (KG)^2$$

where KG is a fixed threshold.

If the breakpoint test shows that the threshold has been exceeded, then there is an indication that the target is manoeuvring relatively out of the tracker gate. Clearly, the position and velocity estimates and the quality index N in the main channel are now invalid. They are therefore updated by copying their counterparts currently in the idler channel.

LOST TRACK

Occasional missed plots of target returns, either due to wrong placement of the tracker gate or complete radar fading of the target, are not usually enough to cause permanent loss of track.

To determine loss of target track, the error integrals FR and FT are incremented by a fill-in value for each scan period when there is a missed plot. Thus if there is a succession of a large number of empty tracker gates, the values of FR and FT will be forced to become artificially high.

Similarly, if the tracker gate moves away from the target but accidently moves into an area of high noise with false alarms appearing in the gate, then due to the random nature of the false alarms, the values of FR and FT will again increase to a level above that expected for a target.

The lost track test, then, is to compare FR and FT with a fixed threshold. If the threshold is exceeded, the operator is informed via the display that the track is lost.

THE PREDICTION PROCESS

The prediction process is much more complicated than that indicated by the earlier equation

$$F(r+1) = E(r) + V(r).T$$

As shown in figure 4, if the current target position is at P1 and the current aircraft position is at A1, then the target's relative bearing is in the direction A1-P1. One scan period later the aircraft and target will be at A2 and P2 respectively. However, the target will not lie in the line of sight since its relative bearing is now A2-P2 and the aerial will be pointing in the direction parallel to A1-P1 again. It is not sufficient to compute when the aerial will next be pointing in the direction A2-P2, since by that time the aircraft and target will have again moved on. The problem is to find when the target will next lie within the aerial beam A3-P3.

It is not possible to compute this with a simple equation and a complete iterative approach would be time consuming for the computer. Instead, the following process is used.

From the current aircraft and target positions and velocities, the angular velocity of the target relative to the aircraft is computed. This is subtracted from the aerial angular velocity to produce an effective angular closing rate between aerial and target. The time is then computed for the closing rate to move through 360 degrees. This time however is not quite sufficient to determine when the target is next in line of sight as the angular velocity of the target relative to the aircraft is changing as the relative position of the target changes.

The positions of the aircraft and the target are found at this first computed time to produce a relative bearing for the target. In addition, the aerial direction is found at that time and the residual error is found from the difference between the two directions. A new aerial closing rate is then found using the new relative angular velocity of the target for this new position. The time taken to correct the residual error is then found and added to or subtracted from the first computed time accordingly.

This process has been found to be highly accurate and does not warrant the computation and correction of any remaining error.

CONCLUSION

An operator may select a target of interest from the main PPI display and inject the target into the automatic tracking system using a free target file from within the computer. The use of the B scope allows accurate placement for the first tracker gate. The tracking process after this is fully automatic and the operator may display continuously updated data from the corresponding target file. Data from other files may also be displayed without interruption of other automatic tracks. Should the target cease to be of interest, the operator may at any time release the target from the system.

ACKNOWLEDGEMENTS

The author wishes to thank Mr. D.P. Franklin whose research, theoretical analysis and helpful suggestions were mainly responsible for the production of the tracker. This work has been carried out with the support of the Procurement Executive, Ministry of Defence.

REFERENCES

1. Sklansky, J., 1957, RCA Rev., pp 163-185.

2. Marks, B.L., Nov. 1961, RAE Tech Note Math 79.

3. Benedict, T.R., and Bordner, G.W., Oct. 1961, IRE Conf., Toronto.

4. Benedict, T.R., and Bordner, G.W., July 1962, Trans IRE AC-7, pp 27-32.

5. Quigley, A.L.C., 1973, IEE Radar Conf. Publn. No. 105.

6. Barham, P.M., and Humphries, D.E., May 1969, RAE Tech. Report No. 69095.

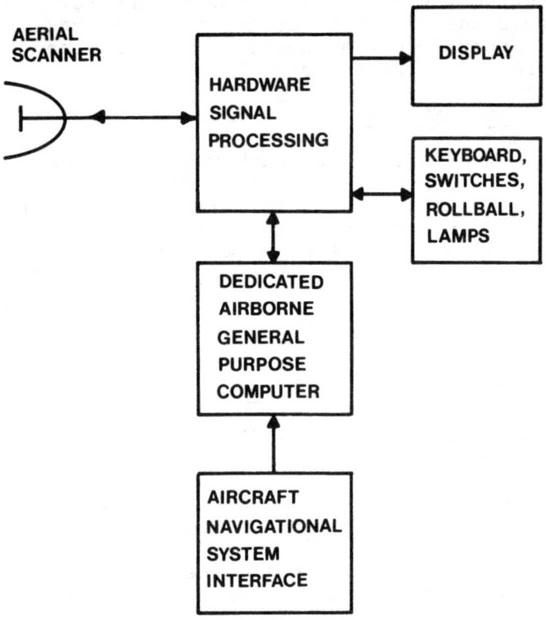

Figure 1 The Basic System Block Diagram

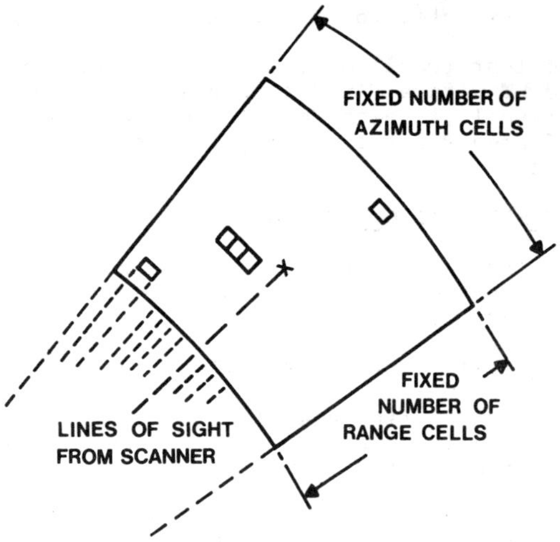

Figure 2 The Tracker Gate

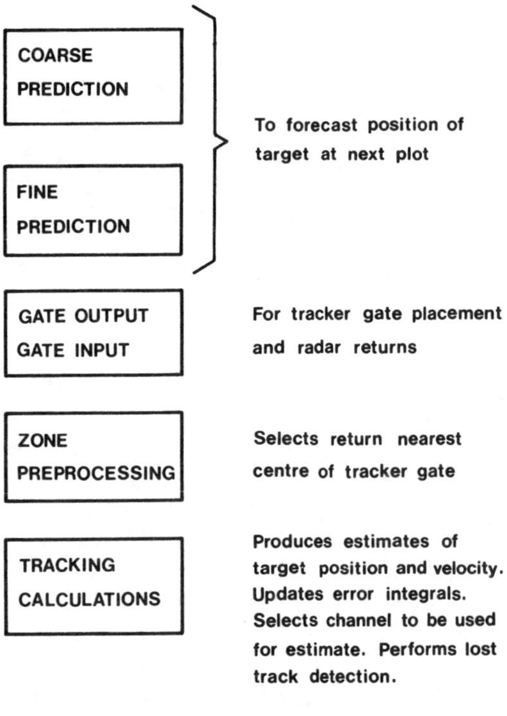

Figure 3 Auto Track While Scan Sequence of Processes

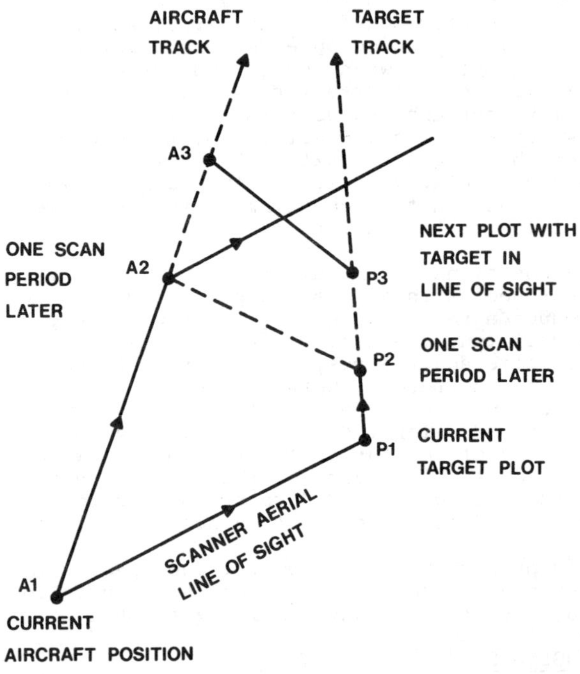

Figure 4 The Prediction Problem

AUTOMATIC INTEGRATION OF DATA FROM DISSIMILAR SENSORS

W.I. Citrin, R.W. Proue, and J.W. Thomas

The Johns Hopkins University Applied Physics Laboratory, USA

INTRODUCTION

Sensor systems have often been designed to detect specific target characteristics. As a result, each sensor has particular advantages and disadvantages. Radars can provide accurate position and velocity information on targets, but they may be degraded by clutter and electronic countermeasures and can be passively located by an enemy. Identification friend or foe (IFF) sensors can identify cooperative friendly targets as well as provide their position and velocity. Although clutter does not degrade IFF sensors, they too are susceptible to jamming and passive location. In addition, the enemy can passively locate units that are responding to IFF interrogations. Electronic support measures (ESM) can passively detect and identify emitting targets and determine their bearing but cannot provide range information. Infrared (IR) sensors can passively detect targets that have a sufficient IR signature. If the data from these different sensors are automatically integrated, the information needed to assess and engage a target is quickly available. Sensor operation may be managed so that the sensors complement one another in obtaining information while limiting enemy opportunities to counter them.

DISCUSSION

To integrate the data from any two sensors, the data from one are compared with the data from the other to determine which correspond to the same target and which to different targets. The decision process is called correlation.

The different sensors may provide different types of data (one may provide target range but not identification, the other may provide target identification but not range) with various accuracies (one may provide coarse bearing and the other very accurate bearing). A two-hypothesis log-likelihood ratio that uses selected sensor data as arguments can be used to combine the various data types and associated accuracies to obtain functions that are measures of the likelihood that the data correspond to the same or different targets. These functions are called discriminants. Having selected the parameter set from the available sensor data, the discriminant functions are obtained by forming the log likelihood ratio (also referred to as correlation likelihood) using the following expression:

$$LL(\bar{P}_A,\bar{P}_B) = \log_a \frac{P[\bar{P}_A,\bar{P}_B|H_0]}{P[P_A,P_B|H_1]}$$

where

$LL(\bar{P}_A,\bar{P}_B)$ = log likelihood ratio for the parameter sets $\bar{P}_A$ and $\bar{P}_B$,

$P[\bar{P}_A,\bar{P}_B|H_i]$ = conditional probability of observing the parameter sets $\bar{P}_A$ and $\bar{P}_B$, given that hypothesis H_i is true,

H_0 = hypothesis that the sensor A and sensor B tracks correspond to the same target,

H_1 = hypothesis that the sensor A and sensor B tracks correspond to different targets,

$\bar{P}_A = [P_{A1},P_{A2}, \ldots P_{AN}]$ = set of parameters obtained from the sensor A track file, and

$\bar{P}_B = [P_{B1},P_{B2}, \ldots P_{BN}]$ = set of parameters obtained from the sensor B track file.

If the parameters $[P_{Ai},P_{Bi}]$ are independent of the parameters $[P_{Aj},P_{Bj}]$ for all $i \neq j$, then the ratio may be rewritten as

$$LL(\bar{P}_A,\bar{P}_B) = \sum_{i=1}^{N} \log_a \frac{P[P_{Ai},P_{Bi}|H_0]}{P[P_{Ai},P_{Bi}|H_1]}$$

$$= \sum_{i=1}^{N} LL(P_{Ai},P_{Bi})$$

Thus the form of a discriminant may be obtained separately for each pair of parameters $[P_{Ai},P_{Bi}]$ from $LL(P_{Ai},P_{Bi})$. The discriminants are then summed to obtain the correlation likelihood.

Automatic Integration of Radar and ESM Sensor Data

Using the above approach, a real-time system that accepts actual radar and ESM sensor data, forms tracks from these data, and performs radar and ESM track correlation processing was designed and tested. Discriminants based on track bearing, identification, category, and heading were used in the correlation algorithm.

Figure 1 illustrates the major components of the real time system. A 2D short range radar and an ESM receiver are interfaced to an integration computer. The integration computer automatically accepts as inputs ESM tracks and radar target reports, forms radar tracks from the radar target reports, and correlates the radar and ESM tracks. The system also has the option of using simulated, rather than live sensor data.

Testing of the real time system was conducted using both live and simulated sensor data.

For both types of testing, the correlation algorithm performed very well and correctly correlated over 95% of the radar and ESM track pairs. The testing revealed some mis-correlations, most of which occurred when one target track crossed that of another or when targets were closely spaced in bearing. Since that time, several additional correlation discriminants designed to correct this problem and to make further improvements, have been tested by means of a computer simulation of the radar-ESM track correlation algorithm. The input to the simulation, consisting of tracked radar and ESM target reports versus time, is created by a scenario generator and tracking program. The output of the simulation (radar and ESM track correlation status versus time) is analyzed by a data reduction program. Figure 2 shows the key elements in the simulation. As a result of these tests, the likelihood calculations have been modified and extended, and improvements have been made to the algorithm, some of which are outlined briefly below.

The real-time correlation algorithm used discriminants based on the differences between radar and ESM bearing, identification, category, and heading, which can be found in or inferred from the radar and ESM track file data. The new discriminants are bearing history, used instead of bearing difference, and range, height, and speed. Bearing history uses current and past bearing information obtained from the radar and ESM track file data. Range and speed are determined directly from radar measurements and can be found in the radar track file. Height can be determined from radar measurements if a three-dimensional (3D) radar is used. These parameters can also be inferred from emitter characteristics found in the ESM track file data. In addition, association lists of correlation candidates are formed; they are composed of radar tracks that have remained within a specified bearing window about an ESM track throughout their history. The association lists offer the advantage of a very-long-term bearing history discriminant without the attendant problem of storing bearing samples. They also eliminate the need for the periodic searches for correlation candidates.

The performance of the algorithm was measurably improved by the additional discriminants, especially bearing history. Crossing targets no longer cause miscorrelations, and radial tracks that are close together in bearing but do not cross are correctly correlated a higher percentage of the time than with the previous algorithm.

The log likelihood approach to correlation permits each individual independent discriminant to be summed with the others to form an overall correlation likelihood. Consequently, it lends itself to the easy inclusion of new discriminants, even those that may not be formed from sensor-related parameters. This is done by adding the new discriminant, properly weighted, to the sum of the discriminants included previously. As an example, a "hysteresis" term, based on the previous correlation status for a radar-ESM track pair, was added to give extra weight to the pairs that were correlated in the past. This tended to stabilize correlations and prevent them from switching because of noise.

In instances where it was desirable to stabilize certain correlations, they were rendered permanent or "frozen" by adding a likelihood that outweighed all the other likelihoods combined. It was possible to add these likelihoods with minimal changes to the structure of the algorithm.

Because the correlation likelihoods are measures of the probability that the radar and ESM tracks represent the same target, the modified algorithm uses the likelihood sums to establish an ambiguity criterion for correlated track pairs. Track correlations with a likelihood greater, by a predetermined threshold, than the likelihood of any other pair involving either the radar or ESM tracks are now considered to be unambiguous correlations. Those that are not better by at least the threshold value are considered to be ambiguous with other potential pairings. This information may be useful to an operator or processing function assessing the quality of the track correlation data in cases where data must be used to make further engagement decisions.

Integration of Radar and IFF Sensors

The same likelihood approach to correlation was applied to the development of a method to automatically integrate IFF and radar sensor data in the AN/SYS-( ) Integrated Automatic Detection and Tracking (IADT) System, which was part of an experimental effort at APL. The integration design goals included improvements in shipboard identification, air traffic control, and Combat Air Patrol support functions.

An IFF sensor operates by transmitting a coded RF pulse train to which a cooperating airborne or surface unit automatically responds with a different coded pulse train. The pulse train transmitted by the IFF is referred to as an interrogation, and the reply transmission that it elicits from unit transponders is termed the interrogation response. The interrogation response is received and processed by the IFF sensor to provide target reports consisting of range, bearing, and code data, including altitude.

The SYS-1 IADT performs tracking for surveillance radars by forming and updating tracks in a single, unduplicated track file with target reports (contacts) from each radar. SYS-( ) refers to a generic form of SYS-1 that has been modified to accept IFF sensor data. Each radar video processor transforms the returned radar raw video into digital centroid reports, analogous to the IFF target reports formed from IFF response data. The IFF sensor and the SYS-( ) are integrated by performing the correlation process between the IFF target report data and the SYS-( ) radar track file data, using the log likelihood approach. Range, bearing, IFF-derived target altitude, 3D radar elevation, and IFF code data are used to evaluate discriminants.

As a first step toward demonstrating the feasibility of automatic IFF and radar integration, data from an IFF set and from two different radars were collected simultaneously at the Naval Tactical Data System Development and Evaluation Site, Mare Island, Calif.

The IFF set was an AIMS Mk XII with an experimental AN/UYK-20 computer-based beacon video processor (BVP). The BVP processes IFF data and outputs IFF target reports. The two radars were an AN/SPS-52B 3D air search radar and an AN/SPS-48C 3D air search radar. IFF/BVP and radar target reports for targets of opportunity were the primary data collected and evaluated. Figure 3 illustrates the data collection and processing schemes. Two radar digital data collectors developed by APL to interface with the SPS-52B and SPS-48C radars were used to collect digitized radar hits. The UYK-20 BVP computer program was modified to record digital IFF/BVP target reports. At APL, software processing was performed to centroid the recorded radar hits and to produce magnetic tapes of radar reports in a format accepted by the SYS-1 IADT software. The recorded IFF/BVP target report data were formatted identically to the radar reports, but IFF/BVP report range and bearing were used instead of radar range and bearing. An elevation angle derived from IFF-encoded altitude was used in place of the 3D radar elevation angle. Each target report tape could be played through the SYS-1 separately. Additional processing then merged the radar and IFF/BVP reports onto a single tape. After these modified target report tapes were used to drive the SYS-1 IADT program, existing data extraction and reduction software was used to analyze system performance.

The experimental results were encouraging. When the IFF/BVP report data were used separately, SYS-1 initiated and maintained tracks from the IFF/BVP-derived range, bearing, and elevation data. This indicated not only that IFF/BVP range and bearing accuracies are adequate to support automatic track initiation and maintenance but that a discriminant based on elevation derived from IFF code data is useful. When the merged tape containing IFF/BVP and radar reports was used, SYS-1 initiated and maintained integrated tracks that were updated asynchronously by both IFF/BVP and radar reports. The experiment was performed with no modifications to the SYS-1 IADT computer program.

SUMMARY

On the basis of the above two projects, significant advances have been made in the areas of sensor data integration. The log likelihood approach in sensor data correlation is appropriate for both similar and dissimilar sensor data. A future application involves the addition of an infrared search and track (IRST) set to the suite of existing shipboard sensors. In that case, the correlation may be performed between the IRST sensor data and a data base that has been formed from the integration of other multiple sensor data. The log likelihood approach and other previous correlation data processing techniques will be used to develop a similar integration and data correlation approach for the IRST sensor.

REFERENCE

1. W.G. Bath, "Automatic Tracking of Radar and ESM Returns for Point Defense Application", JHU/APL, FS-77-178

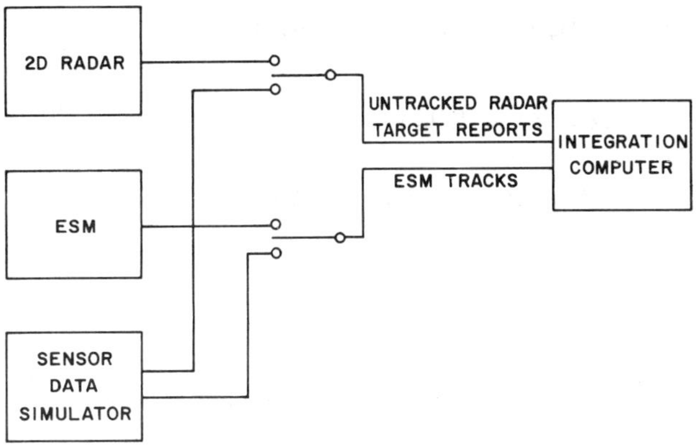

Figure 1 - Real Time Integration of Radar and ESM Sensor Data

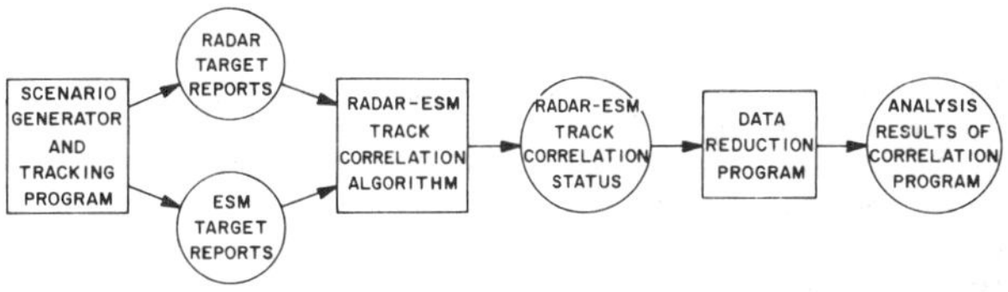

Figure 2 - Radar ESM Correlation Simulation Program

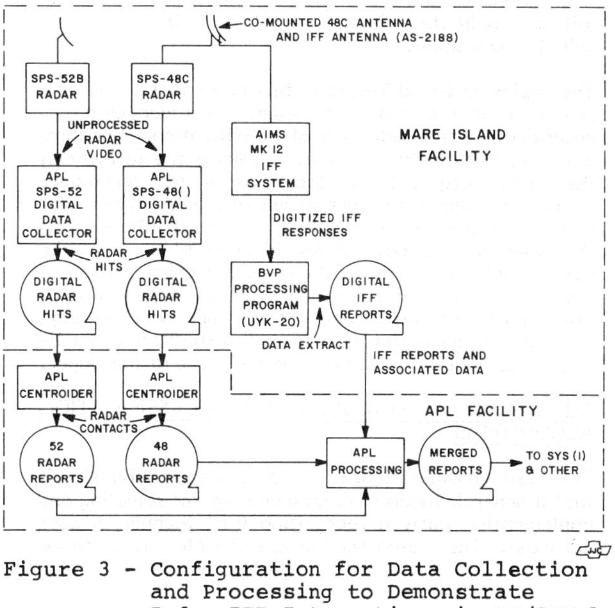

Figure 3 - Configuration for Data Collection and Processing to Demonstrate Radar-IFF Integration via AN/SYS-1

RECOGNITION OF TARGETS BY RADAR

N. F. Ezquerra and L. L. Harkness

Georgia Institute of Technology, USA

INTRODUCTION

The processing of radar signals, in conjunction with the use of pattern recognition techniques, has remained a subject of great interest over the past several years (1,2). Much of this interest may be due to the fact that the techniques used in radar target recognition have a wide range of areas of application. The interest in this field may also be partly due to the fact that several disciplines, such as information theory and pattern recognition methodology, can be combined with radar methods, thereby giving rise to challenging research areas amenable to new ideas and approaches. More specifically, pattern recognition theory provides a vast source of useful techniques with which to address problems such as the extraction of features from radar signatures (3) and the classification of specific target types (2).

There are several reasons for introducing the concept of target recognition techniques as part of a radar system. Firstly, it may be desirable to not only detect an object in the field of illumination of the radar beam, but also to know something more about the object than its mere presence (i.e., the identity of the target may be required). Secondly, the progress that has occurred in the field of signal processing has both nourished and driven the direction of research in radar data processing. The former is evidenced in the availability of fast, light-weight, and relatively inexpensive hardware which facilitates numerous radar tasks. On the other hand, at the same time that these advances in signal processing have allowed more tasks to be performed with greater efficiency, this very progress in signal processing has also made greater demands on radar operations in general. Thirdly, and perhaps most importantly, an increasingly large number of areas of application have placed great emphasis on this field of radar data processing. Among the various disciplines which employ pattern recognition techniques are geophysics and meteorology (e.g., weather radar), terrain and scene matching, civil engineering (e.g., detection and recognition of voids in highways), and military applications (e.g., recognition of aircraft, ground vehicles, and ships).

The topics to be addressed in this paper include: (1) the selection of features with which the radar data are described, (2) the selection of classification algorithms with which to make decisions regarding the identity of the radar target, (3) a discussion of the techniques employed to separate radar signals of interest from radar echoes produced by objects that are not of interest, and (4) major applications. It should be pointed out at the outset that such considerations as the enhancement of the signal with respect to noise, or the suppression of other forms of background interference sources (e.g., what is commonly referred to as "clutter"), will not be discussed, since these fall outside the scope of this paper.

SELECTION OF FEATURES AND CLASSIFICATION ALGORITHMS

Two basic considerations are emphasized initially. One is that a general, theoretical methodology for selecting and implementing pattern recognition (PR) techniques does not exist. The reason for this unfortunate circumstance is unclear, and two opposing schools of thought offer explanations: according to one, the theoretical formalism is not sufficiently mature or advanced, and thus a "set of rules" does not yet exist; according to the other, the nature of the subject is such that no single set of rules can be generalized to cover all applications under any given conditions. In either case, the fact remains that there is no clearly-defined, logical, analytic sequence which can be followed to determine an optimal set of target recognition techniques--i.e., there is no "recipe."

The second basic consideration is that, once a set of suitable pattern recognition techniques have been identified, the actual development, implementation, testing, and evaluation of these techniques are both time-consuming and iterative in character. It is the latter characteristic which should also be realized in applying PR techniques to radar data: the overall development of appropriate techniques is an inherently iterative process, meaning that identifying a set of candidate techniques is only the initial step in a long and demanding exericse.

Having pointed out some basic considerations regarding PR methodology, it would be instructive to begin with a discussion of feature selection. Generally speaking, it is desirable to extract features from the radar data with which to represent a target, and the reason for extracting features is threefold: to enhance performance, reduce the dimensionality of the data, and to ensure some degree of algorithm stability or robustness. A set of such features or target characteristics is called a feature vector.

Basically, there are three ways of constructing feature vectors using a set of data (4). The most obvious method is to simply use the complete information content of the collected data (i.e., the radar intensity profile) without introducing further corrections or transformations of the data. The chief advantage of this method is the fact that target information is retained in its entirety. Its main disadvantage is the usually high dimensionality of such a feature vector.

A second method of constructing features is by selecting certain characteristics from the data on the basis of a priori knowledge of the problem, or on the basis of intuitive judgement. This ad hoc method is commonly used for analysis; examples of representative "intuitive features" that may be extracted from radar data are the ratio of the highest to second highest peaks in an amplitude return, or the target's radar cross section. The primary motive behind the selection of these heuristic features is to reduce the dimensionality of the feature space; another advantage is its attempt at providing an understanding of the scattering phenomenon under investigation in an intuitively appealing manner.

A third way of constructing feature vectors is by deterministically extracting features, possibly non-linear functions of the original data, or by applying dimensionality-reduction procedures which estimate significant directions in the measurement space. This produces a subset of the original feature set which best represents the original pattern sample (5). These feature-extraction techniques are time-consuming, but the results are usually worth the effort. Two examples of feature-extraction techniques include the use of the

Karhunen-Loeve (K-L) expansion and algebraic moment invariants. The latter method, discussed in the Applications section of this paper, has been used primarily in image-processing. The K-L expansion has been used (3) in eigenanalysis of radar data with a two-fold purpose: to reduce dimensionality and to maximize the distance between classes of targets (in feature space). Both feature-selection procedures, based on invariant moments and K-L expansion, belong to the category of deterministically extracting useful information from radar data; both methods have yielded encouraging results (3, 7).

The task of selecting features is deeply interrelated with that of choosing a classifier. In order to develop a classifier, there are numerous techniques that may be considered as candidates for a classification algorithm. Depending on the nature of the problem and on the available information, a proper recognition algorithm can be selected. This may be accomplished by simply performing a literature search and finding a previously designed algorithm which will perform the required task, or it may require a novel formalism, to be completely developed and implemented as part of an automatic identification system. In either case, the algorithm will usually consist of a set of rules which will manipulate the input data (i.e., features) and produce some sort of decision as output; the form of output can be an electronic signal, visual display, or any other suitable action that rested on the outcome of this decision. The actual recognition algorithm should ideally be designed to be implemented in a few, simple steps in order to fully exploit the automaton character of the algorithm. On the other hand, the design and construction of the algorithm usually involves the development of sophisticated techniques, and this normally is a time-consuming task requiring a relatively large amount of data processing. The design and construction of the algorithm determines, to a great extent, the overall performance of the identification system.

The simplest algorithm is the linear machine, based on Fisher's linear discriminant (FLD) function (4). In this procedure, the feature vector is reduced from a multi-dimensional vector to a one-dimensional quantity by summing the weighted features to form one variable; the resultant variable is then compared with a threshold value which determines the classification decision. A second approach is based on the Nearest Neighbor (NN) rule (4). In this approach, feature vectors are stored such that the distances between these stored prototypes and a feature vector of an unknown origin can be calculated. The unknown is assigned to the class which contains the largest number of the k-nearest samples, where k takes on an odd value in order to implement a majority-rule decision. The FLD classifier is faster and simpler than the NN technique. In addition, the latter requires more memory (in order to store the prototypes for later comparison). On the other hand, the NN technique retains the full dimensionality of the data, thereby allowing the classifier to exploit the characteristics of the underlying probability density functions in feature space. The section entitled Applications includes a discussion regarding the use of different types of classifier.

To classify more than two radar targets simultaneously, the linear discriminant analysis can be extended to the multiple-category case. The extension of the formalism arises naturally for the case of R categories: a set of R discriminant functions is constructed, thereby partitioning feature space into R decision regions. The resulting classifier is a piecewise linear discriminant (PWLD) function, and an unknown feature vector is assigned to the class corresponding to the largest discriminant function. The PWLD function retains the desirable characteristics of the linear machine: simplicity, speed, and a fixed structure, thereby making it an attractive candidate for implementation as a computer.

In terms of preclassification, (i.e., the separation of alien targets from targets of interest), clustering techniques provide a valuable aid in investigating the inherent structure of the target classes. As an illustration of these techniques, Figure 1 shows a hypothetical, two-dimensional feature space in which three different classes are represented by clusters of data samples (differentiated by three kinds of symbols). The object is to design a clustering program which would correctly assign a given sample to its corresponding cluster, while at the same time declaring an input of unknown origin as an "alien" if the input sample is "sufficiently far" from any of the specified clusters. The decision of whether a sample is "sufficiently far" from a cluster is made by using any of a number of distance functions (e.g., Eucledian, Mahalanobis, etc.; see reference (4)).

Clustering techniques may be divided into two broad categories: hierarchical methods and non-hierarchical methods. In hierarchical clustering, the goal is the construction of a tree or diagram which depicts specified relationships among a set of data units. The tree begins with each data sample representing a single member cluster. At each step upward in the tree, the two closest clusters are merged to form a single cluster. The final merger results in a single cluster comprising the entire data set. In this way, the tree represents a hierarchy of clusters ranging from 1 to the number of data units. The analyst must then determine which partition best represents the data set. Once the proper number of clusters is found experimentally, an unknown data sample can once again be declared an alien or a recognizable sample, depending on its proximity to the nearest cluster. Figure 2 shows a display of the numerous levels of groupings, or clusters, which arise when actual radar data are processed through a hierarchical clustering program. The tree has been compressed somewhat in order to display all the data. As a result, each step in the tree actually represents several successive levels of clustering. To illustrate how this tree could then be used, consider a truncation at the first level down from the top. This would result in three clusters, with 8, 2, and 38 members, respectively. The specific members are then read off the bottom of the tree in the quantities listed above, and a decision is made experimentally whether adequate error rates are obtained with this partition. If the error rate is too high, a different truncation, at another level in the tree, is attempted.

APPLICATIONS

As previously pointed out, the fields of application of pattern recognition techniques to radar data processing are wide-ranging. One application is the classification of buried objects (6). Three techniques that have been considered are Fisher's Linear Discriminant (FLD), Nearest-Neighbor (NN), and Quadratic Discriminant Function (QDF). The targets were a metal plate, rocks, roots, and three non-meallic mines. It was found that, in many cases, classifiers using raw target data yielded relatively better results than classifiers that required other types of features (in particular, intuitively-derived features which were chosen on the basis of the degree to which these features seemed to represent the scattering characteristics of the targets). An overall comparison of the performances for all three classifiers indicated the 3-nearest neighbor algorithm to be as good as, or more often better than, other classifier techniques. The superior performance of the NN classifier suggests that the distributions for the two classes probably overlapped to a considerable extent in feature space. Under these circumstances it would have been very difficult to completely separate the clusters with either the FLD or QDF technique. However, better separation might have been achieved by using a set of different feature vectors which would yield a different distribution in feature space. Figure 3 shows a comparison between NN and FLD algorithms, as follows. The abcissa represents the probability $P_c(1)$ of correctly classifying one of the buried objects, and the ordinate represents $P_c(2)$ corresponding to a second buried object. Thus, Figure 3 shows the results of classifying an input by assigning it to one of two classes (i.e., a two-category case); the ideal performance level would be $P_c(1) = P_c(2)$, a point in the upper right hand corner of the graph. As the data points indicate, the NN technique (triangles) approaches this optimal performance more so than the FLD algorithm. As a result of these findings, the work of reference (6) concluded that the quadratic discriminant required relatively large amounts of processing time and storage memory capacity, and that improved performance with the NN technique may be achieved by either increasing the training set size, selecting different feature vectors, or both. This type of analysis can be extended to detection of highway voids, oil and gas pipes, or other sub-surface objects.

An interesting application of feature extraction techniques to two-dimensional radar imagery is provided by the construction of invariant moments (as features) in order to compare optical and radar images (7). The object of this exercise was to come up with features which are invariant with respect to rotations and scale variations of the images. By constructing a set of seven functions of algebraic invariants, the numerical values of the selected features did not appreciably vary when the images were rotated by 2° and 45°; the features were also invariant with respect to reduction to half-size and mirror-imaging of the images. These invariant moments are useful for cases where invariance with respect to such transformations is desired.

Pattern recognition techniques have also been applied to the recognition of aircraft by means of syntactic pattern recognition methods with relatively good success (8). The problem of distinguishing between re-entry vehicles and decoys has also been addressed via techniques similar to the ones described in this paper (9), as well as the identification of ballistic missiles (10.

Among geophysical and metorological applications, similar techniques have been employed to predict the motion of radar echoes in weather forecasting (11), using digitized weather radar data.

SUMMARY

Advances in signal processing, as well as progress in radar technology, have precipitated greater demands on the amount (and kind) of information to be extracted from radar targets. One manifestation of this can be found in the recent interest on target identification and the techniques which can be used for recognizing radar targets.

The methodology of pattern recognition offers a wide variety of techniques that can be used for recognizing targets by radar. These techniques generally are algorithmic in nature, and can be implemented as part of a computer- or processor-based, radar data-processing system. The techniques that may be used vary with respect to the demands which can be placed on the processor (both in terms of memory requirements and execution time), as well as with respect to the application. A discussion was presented of numerous methodologies employed in target recognition: several ways of extracting features from the radar data; nearest-neighbor, linear, and piecewise linear classification algorithms; and clustering techniques for separating alien targets from targets of interest. These techniques form the basis of target recognizers which may be used for various applications.

The types of data which may be processed can consist of range profiles and radar imagery, and the application may cover a wide spectrum: the recognition of ground vehicles, aircraft, and buried objects; terrain and scene matching also draw from the memory requirements and

execution time), as well as with respect to the application. A discussion was presented of numerous methodologies employed in target recognition: several ways of extracting features from the radar data; nearest-neighbor, linear, and piecewise linear classification algorithms; and clustering techniques for separating alien targets from targets of interest. These techniques form the basis of target recognizers which may be used for various applications.

The types of data which may be processed can consist of range profiles and radar imagery, and the application may cover a wide spectrum: the recognition of ground vehicles, aircraft, and buried objects; terrain and scene matching also draw from these techniques, as well as the problem of forecasting weather trends and distinguishing between re-entry vehicles and decoys.

The techniques described in this paper, used in conjunction with the references provided herein, should serve as an adequate starting point for those who are interested in further work in the development and implementation of pattern recognition techniques for applications to radar signal processing.

REFERENCES

1. F. Le Chevalier, "Radar Target and Aspect Angle Identification," (1978), Proceedings of the Fourth International Joint Conference on Pattern Recognition.
2. B. J. Burdick et. al., (1978), "Radar and Penetration Aid Design," Proceedings of the 1978 IEEE Computer Society Conference on Pattern Recognition and Image Processing.
3. C. W. Therrien et. al, (1974), "Application of Feature Extraction to Radar Signature Classification," Proceedings of the Second International Pattern Recognition Symposium.
4. R. Duda and P. Hart, (1973), Pattern Classification and Scene Analysis, J. Wiley and Sons
5. See, for instance, M. D. Levine, (1969), "Feature Extraction: A Survey," Proc. IEEE, Vol 57, No. 8.
6. J. D. Echard, et al, (1978), "Radar Detection, Discrimination, and Classification of buried, Non-Metallic Mines," Georgia Institute of Technology Final Report EES/GIT A-1828.
7. R. Y. Wong and E. L. Hall, (1978), "Scene Matching with Invariant Moments," Compt. Graphics Image Process., 8, pp. 16-24.
8. F. LeChevalier, G. Bohillot, and C. Fugier-Garrel, (1978), "Radar Target and Aspect Angle Identification," Proceedings IEEE International Association of Pattern Recognition Conference (Japan).
9. C. W. Therrien, W. H. Schoendorf, G. L. Carayannopoulus, and B. J. Burdick, (1974), "Application of Feature Extraction to Radar Signature Classification," Proceedings Second International Pattern Recognition Symposium.
10. B. J. Burdick, G. L. Carayannopoulus, A. A. Grometstein, W. H. Schoendorf, C. W. Therrien, and M. R. Williamson, (1978), "Radar and Penetration Aid Design," Proceedings of the 1978 Pattern Recognition and Image Processing Conference.
11. R. O. Duda and R. H. Blackmer, Jr., (1972), "Application of Pattern Recognition Techniques to Digitized Weather Radar Data," Final Report Contract 1-3692, SRI Project 1287, Stanford Research Institute, Menlo Park, Cal.

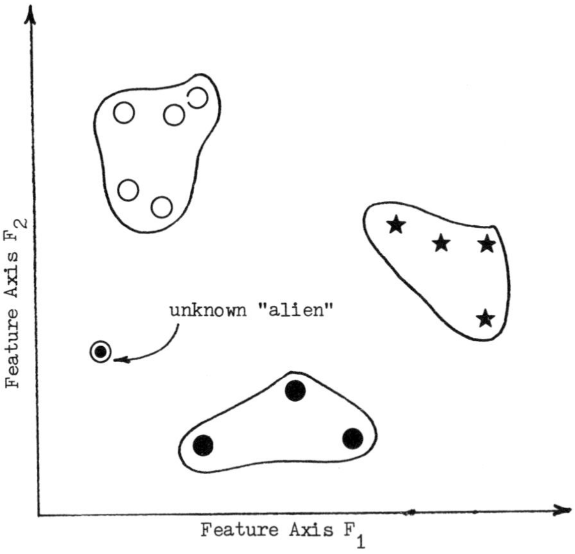

Figure 1 Clusters in Hypothetical Feature Space

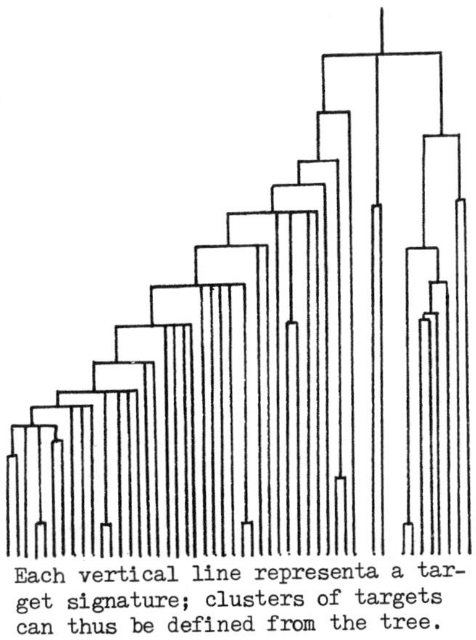

Each vertical line representa a target signature; clusters of targets can thus be defined from the tree.

Figure 2 Hierarchical Tree

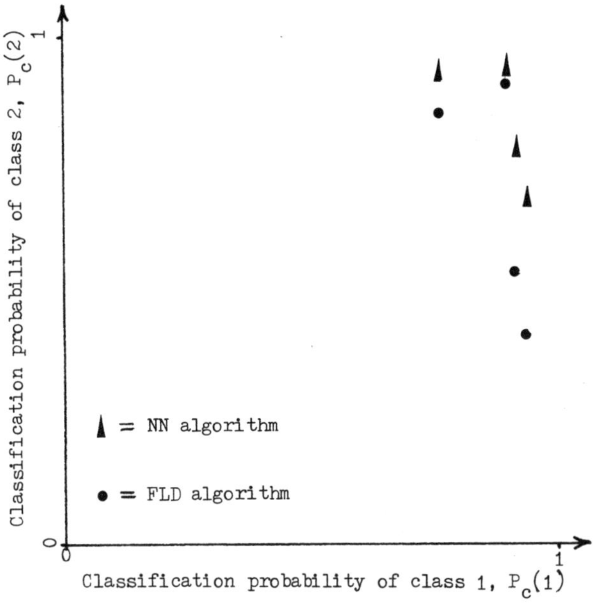

Figure 3 Comparison of FLD and NN classifiers

DIGITAL SIGNAL PROCESSING OF SCATTERING DATA FROM NONLINEAR TARGETS

Jae Y. Hong and Edward J. Powers

The University of Texas at Austin, U.S.A.

INTRODUCTION

A number of man-made objects, which are to be detected by radar, exhibit nonlinear effects which result in new frequency components (e.g., intermodulation products, harmonics, and "degenerate" frequencies) appearing in the backscattered field. Recently, Powers et al (1) presented a conceptual model which allows one to characterize nonlinear scatterers in terms of a hierarchy of linear, quadratic, cubic, etc. radar cross sections (RCS). The concept of nonlinear radar cross sections (NRCS) allows one to generalize the radar equation for a nonlinear target. The resulting equation may be regarded as a generalization of the harmonic radar equation discussed by Flemming et al (2). In reference (1), it was pointed out that the various nonlinear radar cross sections can, in principle, be computed, in terms of higher order spectral density functions, from the transmitted and scattered signals. The overall objective of this paper is to review how appropriate digital signal processing concepts may be used to estimate the nonlinear radar cross sections given the incident and scattered signals. Knowledge of such cross sections can provide additional signature information by which nonlinear targets may be identified and classified.

NONLINEAR RADAR CROSS SECTIONS

In this section we review some of the key ideas of NRCS as discussed in reference (1). Let $S_i(f)$ $[Wm^{-2}Hz^{-1}]$ denote the power flux spectral density incident upon a target, and let $P_s(f)[WHz^{-1}]$ denote the power spectral density of the scattered signal (measured at the target) which is assumed to be reradiated isotropically. Recalling that in general, radar cross sections relate the incident flux density to the power scattered isotropically and utilizing the input and output power spectrum relationship of the orthogonal nonlinear system model described by Hong et al (3), we have the following expressions relating $S_i(f)$ and $P_s(f)$:

$$P_s(f) = S_i(f)\sigma_1(f) + 2!\iint S_i(f_1)S_i(f_2) \times \sigma_2(f_1,f_2)\delta(f-f_1-f_2)df_1df_2 + 3!\iiint S_i(f_1)S_i(f_2)S_i(f_3)\sigma_3(f_1,f_2,f_3) \times \delta(f-f_1-f_2-f_3)df_1df_2df_3 + \cdots ,(1)$$

The quantities $\sigma_1(f)$, $\sigma_2(f_1,f_2)$, and $\sigma_3(f_1, f_2, f_3)$ are the radar cross sections representing the linear, quadratic, and cubic features of the target model. In general, the dimensions of the nth-order cross section $\sigma_n(f_1,f_2,\cdots,f_n)$ are $W^{1-n}m^{2n}$. The fact that the dimensions of the nonlinear cross section involves watts is a manifestation of the fact that the amount of power (at the harmonic and intermodulation frequencies) scattered by a nonlinear target is a nonlinear function of the incident power level.

Of particular importance is the fact that in reference (1) it was emphasized that the NRCS's can be computed directly from the time series of the transmitted and received signals. The approach is to compute, following the ideas in reference (3), the linear $H_1(f)$, quadratic $H_2(f_1,f_2)$, and cubic transfer functions $H_3(f_1,f_2,f_3)$, relating the transmitted and received signals. As discussed in reference (1), the hierarchy of cross sections can be expressed in terms of a hierarchy of transfer functions as follows:

$$\sigma_1(f) = \frac{|H_1(f)|^2}{\frac{\lambda^2}{4\pi}\frac{G_r(f)}{4\pi R^2}\frac{G_t(f)}{4\pi R^2}} , \qquad (2a)$$

$$\sigma(f_1,f_2) = \frac{|H_2(f_1,f_2)|^2}{\frac{\lambda^2}{4\pi}\frac{G_r(f)}{4\pi R^2}\frac{G_t(f_1)}{4\pi R^2}\frac{G_t(f_2)}{4\pi R^2}} ,$$

$$f = f_1+f_2 , \qquad (2b)$$

$$\sigma_3(f_1,f_2,f_3) = \frac{|H_3(f_1,f_2,f_3)|^2}{\frac{\lambda^2}{4\pi}\frac{G_r(f)}{4\pi R^2}\frac{G_t(f_1)}{4\pi R^2}\frac{G_t(f_2)}{4\pi R^2}\frac{G_t(f_3)}{4\pi R^2}},$$

$$f = f_1+f_2+f_3 , \qquad (2c)$$

where λ is the wavelength of an electromagnetic wave in free space and R represents the distance between a monostatic radar and a target. The receiver and transmitter antenna gain are denoted by G_r and G_t, respectively, which generally depend upon frequency. It is clear, if one is to successfully determine σ_1, σ_2, σ_3 etc. from transmitted and received signals, one must first demonstrate that the transfer function H_1, H_2, H_3 etc. can be successfully determined. This is the basic problem addressed in this paper with the aid of the nonlinear scattering simulation described in the next section.

SIMPLIFIED NONLINEAR SCATTERER MODEL

It is well known that one source of nonlinearity for metal targets is associated with various metal-to-metal contacts which have been observed to display nonlinear current-voltage characteristics. As a specific example, we consider the phenomenological model of a scattering nonlinear junction shown in Figure 1(a). In this model, which is basically the same one utilized by Harger (4), the voltage v_i is due to the incident field and the voltage v_r across R_r generates the scattered field. The current generator I is a nonlinear function of the voltage across it. The shunt admittance across the current generator is not considered in our discussion for simplicity since a major part of this admittance is the capacitance which exhibits memory characteristics. This results in the circuit equation being an algebraic one rather than a differential equation. The scattering properties of half-wave dipoles containing nonlinear metallic contacts were discussed by Misezhnikov et al (5), where it is pointed out that the following relationship, $I = a_1 v + a_3 v^3$, described the most frequently observed cubic I-V characteristic. Consequently, we shall use this same nonlinear voltage dependence for the nonlinear current source in Figure 1(a). The equations describing the phenomenological model of a scattering nonlinear junction are given by:

$$I = a_1(v_g - v_r) + a_3(v_g - v_r)^3 \ , \tag{3}$$

$$v_i = I \cdot R_i + v_g \ , \tag{4}$$

$$v_r = I \cdot R_r \ , \tag{5}$$

$$\mu = 1 + R_i/R_r \ . \tag{6}$$

From these equations, we find that the scattered voltage v_r satisfies a cubic algebraic equation, the solution of which is given by Hong (6) as

$$v_r = (W + \sqrt{D})^{\frac{1}{3}} - (\sqrt{D} - W)^{\frac{1}{3}} + \frac{v_i}{\mu} \ , \tag{7}$$

where D is the discriminant of cubic equation and W is given by $W = -0.5v_i/((1 + R_i/R_r)^4 a_3 R_r)^{-1}$. Equation (7) enables one to compute the "scattering" voltage v_r given "incident" v_i. Note that although the relationship between the diode current and voltage is of third order (i.e., cubic), the relationship between the scattering voltage v_r and incident v_i is of much higher order. This follows from the fact that a Taylor series expansion of v_r in terms of v_i will contain odd-order terms greater than three. Consequently, for a sinusoidal input, the output will contain, in addition to the fundamental and third harmonic, odd harmonics greater than three. We will refer to this observation when interpreting the simulation results.

DIGITAL RADAR SIGNAL PROCESSING

The actual computer simulations are performed without extraneous noise and with noise. When extraneous noise is included, Figure 1(a) is modified as shown in Figure 1(b).

In order to compute NRCS's of this model, we will utilize equations (2a) and (2c) since our cubic model does not generate a second harmonic response. Note that NRCS's in these equations are expressed in terms of nonlinear transfer functions. In reference (3), we have shown that these nonlinear transfer functions can be computed from higher-order cross-spectra between raw input and output time series data. Specifically, it is shown in this reference that the linear $H_1(f)$, and cubic transfer functions $H_3(f_1,f_2,f_3)$ can be expressed in terms of the cross-power spectrum (CPS) and the cross-trispectrum (CTS), respectively. As was mentioned in both references (1) and (3) the results relating the nonlinear transfer functions and the higher-order cross-spectra are valid for arbitrary Gaussian inputs only. Of particular practical importance is the fact that the higher-order cross-spectra can be computed directly from the Fourier transforms of the input (incident) and output (scattered) signals. This serves as the basis of the approach described in the following paragraphs.

The first computer simulation is performed on the configuration shown in Figure 1(a) without extraneous noise. In order to meet the input statistical requirement (i.e., arbitrary Gaussian), the input signal x(t) is chosen to be an ensemble of sinusoids with Rayleigh-distributed amplitudes and uniformly distributed phases. Note that each realization of this input is a pure sinusoid of frequency f_o with constant amplitude and phase determined by the above statistics. The digitized data sequence of the input $\{x[n]\}$, contains exactly one period of this sinusoid for computer memory efficiency since this model does not generate the subharmonic components. Furthermore, this choice gives no leakage of power spectra into adjacent frequencies. This input data sequence can be written as

$$x[n] = A_\kappa \cos(2\pi \ell n + \theta_\kappa) \ , \quad \ell = \frac{f}{f_o} \ , \tag{8}$$

where the subscript κ denotes each realization of an ensemble sinusoids. The input variance is chosen to be 2 in all the simulations since the effect of input power level is to be manifested. The resistances R_i and R_r are set to unit values. To show the effect of different degrees of cubic nonlinearity, the constant a_1 is set to unity and the constant a_3 is varied from 10^{-3} to 10^2 in (3).

Since the input spectrum for a given realization consists only of a single frequency component f_o, and since the nonlinear I-V model of junction contains only odd terms, the following cross-section are of principal interest: $\sigma_1(f_o)$, the linear RCS; $\sigma_3(f_o, f_o, f_o)$, the harmonic cubic RCS; and $\sigma_3(f_o,f_o,-f_o)$, the degenerate cubic RCS. There are actually three degenerate cubic RCS's $\sigma_3(f_o,f_o,-f_o)$, $\sigma_3(f_o,-f_o,f_o)$, and

$\sigma_3(-f_o,f_o,f_o)$. These cannot be distinguished given the input-output data. For uniqueness, we assumed they are equal by virtue of symmetry. Note that the degenerate RCS indicates that due to the cubic nature of the target there is additional contribution to the return at f_o. This is most easily seen from the following relationship for $(\cos\omega_o t)^3 = 3/4 \cos\omega_o t + 1/4 \cos 3\omega_o t$. Of particular importance is the fact that the CTS enables one to detect the degenerate return at f_o in the presence of the linear return at f_o. This has important implications, namely, that cubic targets may be identified by appropriately processing the return at f_o, rather than at $3f_o$.

As discussed in the preceding section, the output y(t) in Figure 1(b) will contain harmonics greater than three for a simple sinusoidal input. Therefore, in this simulation the Nyquist frequency is chosen to be $16f_o$ so that the higher harmonics in the output signal can be observed without serious aliasing problems.

The linear RCS, the degenerate cubic RCS and the harmonic cubic RCS are given in the digital form as follows (6):

$$\sigma_1(f_o) \propto |H_1[\ell]|^2 = \left|\frac{\langle Y[\ell]X^*[\ell]\rangle}{\langle |X[\ell]|^2\rangle}\right|^2 , \qquad (9a)$$

$$\sigma(f_o,f_o,-f_o) \propto |H_3[\ell,\ell,-\ell]|^2$$

$$= \left|\frac{1}{3!}\left\{\frac{\langle Y[\ell]X^*[\ell]X^*[\ell]X^*[-\ell]\rangle}{\langle |X[\ell]|^2\rangle^3} - 2\frac{H_1[\ell]}{\langle |X[\ell]|^2\rangle}\right\}\right|^2 , \qquad (9b)$$

$$\sigma_3(f_o,f_o,f_o) \propto |H_3[\ell,\ell,\ell]|^2$$

$$= \left|\frac{1}{3!}\,\frac{\langle Y[3\ell]X^*[\ell]X^*[\ell]X^*[\ell]\rangle}{\langle |X[\ell]|^2\rangle^3}\right|^2 , \qquad (9c)$$

where $X[\ell]$ and $Y[\ell]$ are the discrete Fourier transforms (DFT) of x[n] and y[n], i.e.,

$$X[\ell] = \frac{1}{N}\sum_{n=0}^{N-1} x[n]\cdot e^{-j2\pi\cdot\frac{n\ell}{N}} . \qquad (10)$$

The auto-power spectrum (APS) of the input, the CPS and CTS between input and output are given by: APS = $\langle |X[\ell]|^2\rangle$; CPS = $\langle Y[\ell]X^*[\ell]\rangle$; CTS = $\langle Y[\iota+\ell+\eta]X^*[\iota]X^*[\ell]X^*[\eta]\rangle$. The pointed brackets denote an ensemble average over the κ realizations. In this simulation the number of points in the input data sequence is given by N = 32.

In summary, one can estimate the linear and nonlinear cross sections of interest given the input and output sequence x[n] and y[n] by digitally implementing equations 9 and 10. This is a relatively straightforward and computationally efficient procedure since the fast-Fourier-transform algorithm may be used to transform the input and output sequences.

RESULTS

The results of the simulation without extraneous noise are shown in Figure 2 and with noise in Figure 3. As indicated in the previous section the principal parameter of interest is a_3 which was allowed to vary from 10^{-3} to 10^2. Since a_1 was set equal to 1, this range of a_3 spans the weakly nonlinear, moderately nonlinear, and strongly nonlinear cases.

The scattered power spectrum is plotted in Figure 2(b) as a function of a_3. Note that as a_3 is increased higher odd harmonics are introduced into the output power spectrum. The linear RCS σ_1 (computed from equation (9a)) is plotted in Figure 1(b). Since the orthogonal nonlinear system representation models a nonlinear target in the least-mean-square sense (3), the linear RCS represents the linear functional-like best-fit of a nonlinear target and the cubic RCS does the cubic functional-like best approximation. This is manifested by the increase in $\sigma_1(f_o)$ as a_3 increases. The increase in $\sigma_1(f_o)$ is due to the degenerate contributions to f_o from the higher-order nonlinearities, these nonlinearities becoming stronger as a_3 increases. The degenerate cubic RCS, $\sigma_3(f_o,f_o,-f_o)$ and the harmonic cubic RCS, $\sigma_3(f_o,f_o,f_o)$ are shown in Figure 2(c). Note that both RCS's have the same values since the model is independent of the frequency and the cubic nonlinearity is dominant with respect to the higher odd-order nonlinearities. This indicates that not only the cross-trispectrum is capable of identifying a degenerate contribution in the return at f_o, but also that one may obtain the same information concerning the nonlinear scatterer's cubic nonlinearity by appropriately digitally processing the return at f_o rather than observing the harmonic return at $3f_o$. In Figure 2(d) the scattered power spectrum at $3f_o$ is computed two ways. The "actual" value is computed directly from the actual output y[n]. The model output is computed via equation (1) where the digitally computed values of $\sigma_3(f_o,f_o,f_o)$ (shown in Figure 2(c)) are used for $\sigma_3(f_1,f_2,f_3)$. In this computation, the linear and quadratic terms in equation (1) do not contribute to the scattered power at $3f_o$. The agreement is quite good. At high values of a_3 the 5th, 7th, etc. order terms contribute to $3f_o$ (as well as f_o) and thus the model power spectrum computed using only $\sigma_3(f_o,f_o,f_o)$ is somewhat less than the actual power present at $3f_o$.

The simulation with extraneous noise is shown is Figure 3. In this example, the standard deviation of input noise is 0.01 and that of output noise is 0.2. Examination of the power spectrum in Figure 2(a) does not reveal the presence of any harmonics. this basically due to the fact that they are obscured by the relatively high noise level. It should be noted that the vertical scales used in Figure 2(a) and 3(a) are different. On the other hand the vertical scales used in the other corresponding plot are identical.

In Figure 3(b) we note the determination of $\sigma_1(f_o)$ agrees very well with the noise-free results shown in Figure 2(b). In Figure 3(c) the ability of the cross-trispectrum to detect the presence of the third harmonic buried in the noise allows one to determine $\sigma_3(f_o,f_o,f_o)$. The results of determining $\sigma_3(f_o,f_o,-f_o)$ are also shown. We note that when a_3 is small leading to a weak (relative to the noise) return at the degenerate frequency f_o and the third harmonic $3f_o$, statistical jitter characterizes the estimates. This jitter decreases as a_3 increases. These results indicate that, in spite of the high noise level, the presence of the nonlinearity can be detected from the return at f_o. In Figure 3(d) the scattered power spectrum at $3f_o$ is plotted. The actual (with noise) power spectrum was computed directly from the signal y[n]. Since the noise dominates the actual output, the output power spectrum is independent of a_3. The output power spectrum computed from the $\sigma_3(f_o,f_o,f_o)$ results shown in Figure 3(c) is denoted by the curve model w/n. For comparison purposes the actual scattered power spectrum without noise is repeated from Figure 2(d). The relatively good agreement between these later two curves provides further evidence of the ability of the cross-trispectrum to not only detect the presence of degenerate contributions to the return at f_o but also to detect harmonic contributions at $3f_o$ which are considerably below the noise level.

DISCUSSION

In conclusion, we observe on the basis of the simulation results that indeed it should be possible to estimate NRCS's from the raw time series data corresponding to the incident and scattered signals. In particular the results suggest that the cubic nature of a target may be identified by appropriately processing the return at f_o rather than $3f_o$. The success of this approach rests upon the ability of the cross-trispectrum to isolate the degenerate contribution to f_o and the insensitivity of the approach to relatively strong background noise. This insensitivity to noise is based upon the fact that the approach rests upon a novel method (based on the properties of the cross-trispectrum) to detect phase coherence, rather than on the absolute amplitude of the signals of interest. Lastly, it should be emphasized that although the simulation model used in this paper incorporated a memory-less nonlinearity, the approach is still applicable when the nonlinear junction is shunted by an admittance. In this latter case, the degenerate RCS and the cubic RCS will not necessarily be equal because of the frequency dependence of the admittance. In future studies we will consider the relationship between the quality of the NRCS estimates and the number κ of realizations to be included in the ensemble average.

ACKNOWLEDGEMENTS

This work was supported by the Department of Defense Joint Services Electronics Program through the Air Force Office of Scientific Research (AFSC) under contract F49620-77-C-0101. We wish to thank Dr. Young C. Kim for his many contributions in the early stages of this work, and Mr. James P. Coose for providing excellent computer facilities.

REFERENCES

1. Powers, E. J., Hong, J. Y. and Kim, Y. C., 1981, IEEE Trans. Aerosp. Electron. Syst., AES-17, 602-605.

2. Flemming, M. A., Mullins, F. H., and Watson, A. W. D., 1977, Proc. IEE Int. Conf. RADAR-77, 552-554.

3. Hong, J. Y., Kim, Y. C., and Powers, E. J., 1980, Proc. IEEE. 61, 1026-1027. See also Hong, J. Y., et al, 1982, Proc. IEEE, 70, 93.

4. Harger, R. O., 1976, IEEE Trans. Aerosp. Electron. Syst., AES-12, 242.

5. Misezhnikov, G. S., Mukhina, M. M., and Sel'skiy, A. G., and Shteynshleyger, V. B., 1979, Radio Engng. Electron Phys. (Radiotekhnika i Elektronika), 24, 117-119.

6. Hong, J. Y., 1982, "Nonlinear System Transfer Functions with Application to Nonlinear Electromagnetic Scatterers," Ph. D. Dissertation, Department of Electrical Engineering, University of Texas at Austin, Austin, Texas, U. S. A.

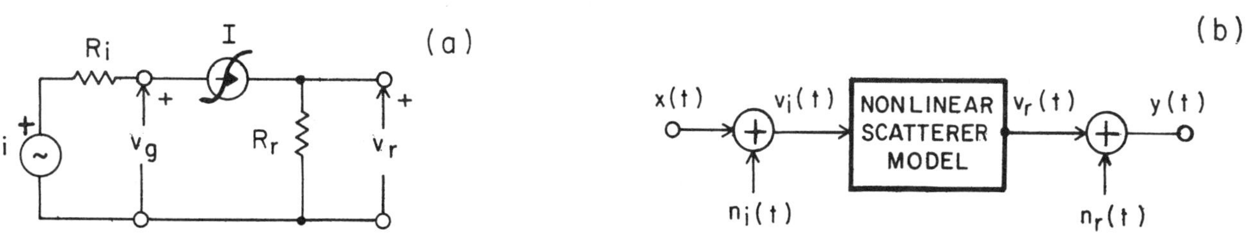

Figure 1. (a) phenomenological model of a nonlinear scattering junction, (b) phenomenological model of a nonlinear scattering junction with additive noise included.

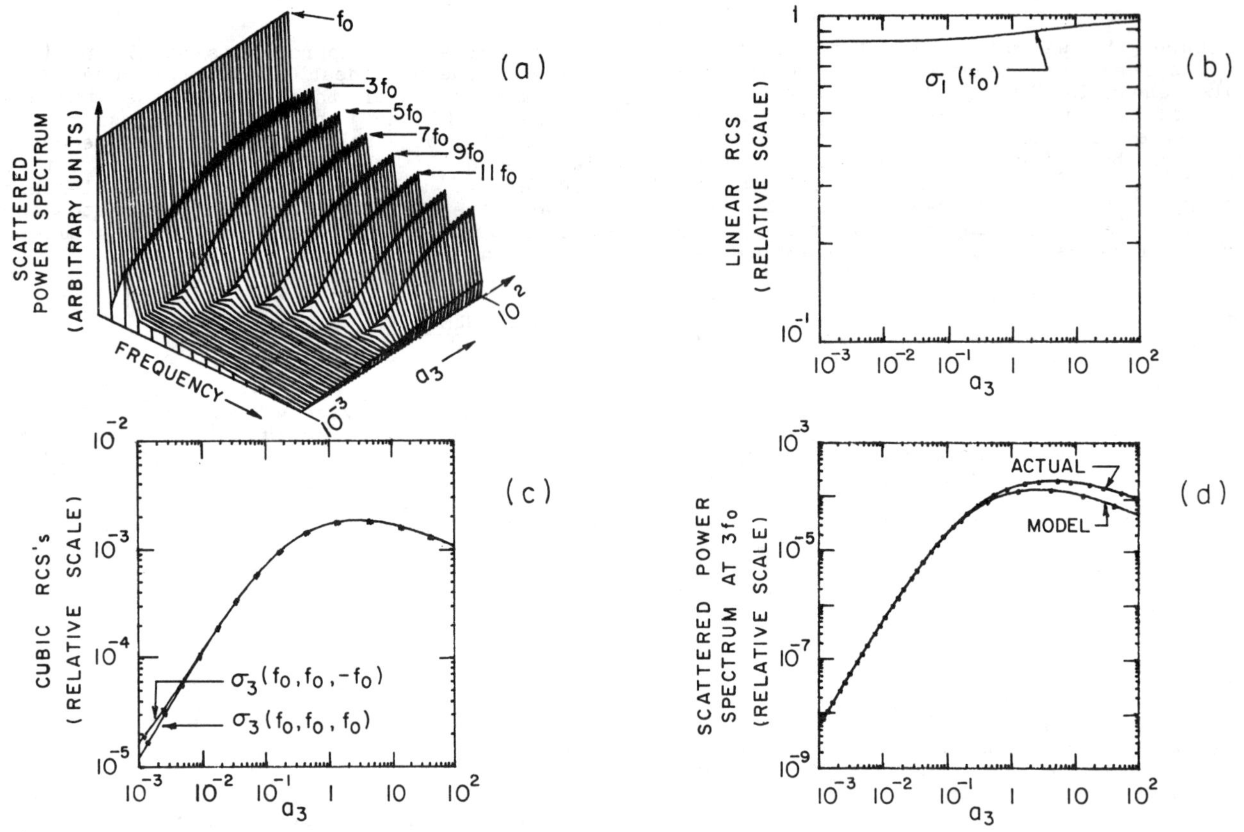

Figure 2. Simulation results with no additive noise: (a) power spectrum of scattered signal, (b) linear RCS, (c) cubic RCS's, and (d) power spectrum of scattered signal at $3f_o$.

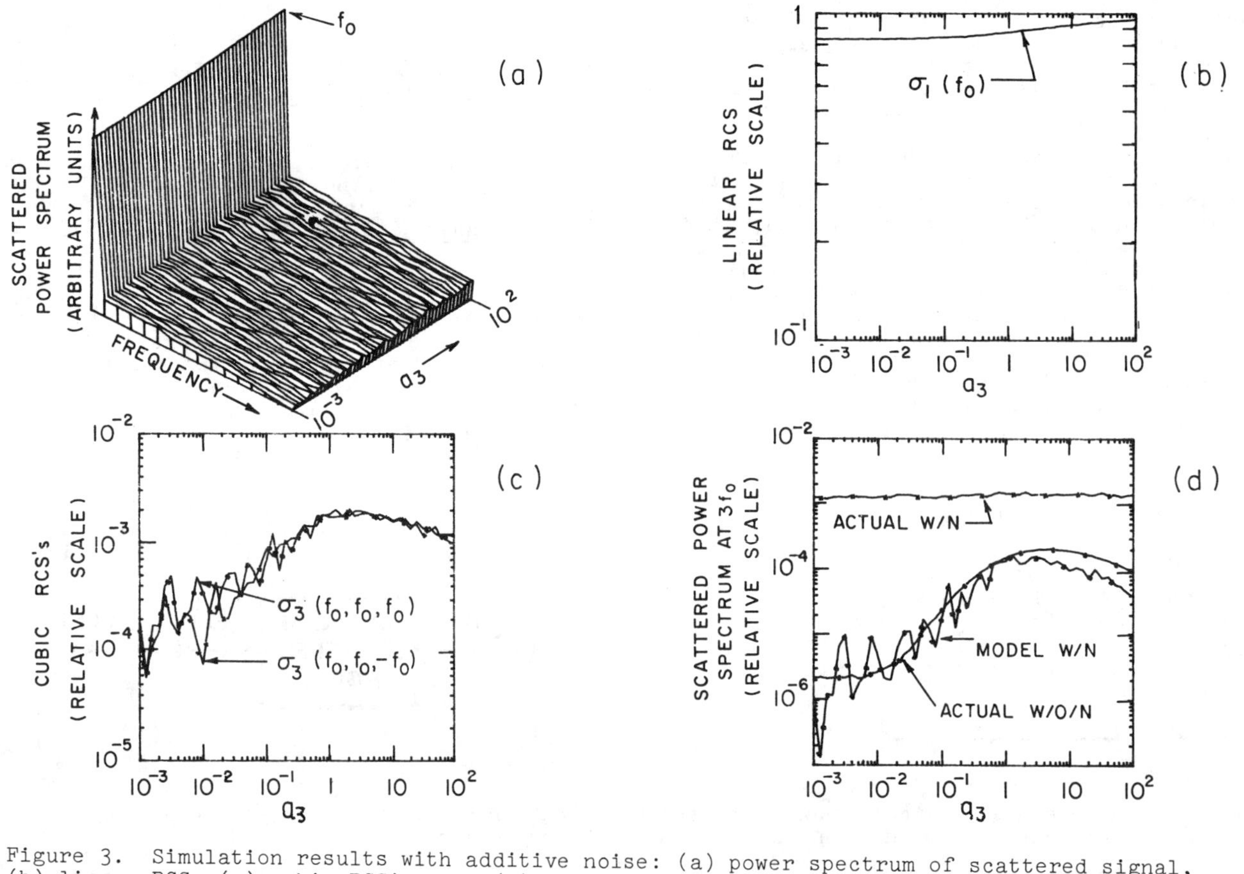

Figure 3. Simulation results with additive noise: (a) power spectrum of scattered signal, (b) linear RCS, (c) cubic RCS's, and (d) power spectrum of scattered signal at $3f_o$.

RADAR SPECTROSCOPY

P. J. Moser and H. Überall*

Naval Research Laboratory, Washington, D.C. 20375, USA
*Department of Physics, Catholic University, Washington, D.C. 20064, USA

The eigenfrequency spectrum of a radar target constitutes a code which can be used for target identification purposes. It corresponds to poles of the scattering amplitude in the complex frequency plane which can be obtained from observed frequencies and decay constants of ringing resonances (1). We obtain these poles for spheres, spheroids, and finite-lengths cylinders using Waterman's code (2), and study the "level" shifts and splittings as a function of target shape. The use of the radar spectrum of a target for characterizing target shape is thus indicated.

INTRODUCTION

In previous work (3) a physical interpretation of the Prony series solution for echoes of radar pulses from conducting spheres was given in terms of surface waves or "creeping waves." The sphere, however, is a very special example, as we shall see, but does demonstrate quite adequately the nature of the interaction of electromagnetic waves with a conducting target. The present study investigates the changes of a body's complex eigenfrequencies as the shape is deformed from the perfectly symmetrical sphere to axisymmetrical prolate spheroids and finite, short right circular cylinders.

PROCEDURE

Waterman's T-matrix solution (2) for the scattering cross sections of axisymmetric conducting bodies allows a secular equation to be written which determines the complex eigenfrequencies of those objects. Also, each value of the azimuth index, m, may be evaluated separately thus decoupling these modes. Hence, there will now be m=0,..., n secular equations to be solved for each order n.

The T-matrix is calculated from a preliminary Q-matrix (2) and for a given value of m

$$T = [Q]^{-1} \operatorname{Re}[Q] \qquad (1)$$

From equation 1 it is easy to see that the secular equation will be (4)

$$\det [Q] = 0 \qquad (2)$$

where det[ ] means "the determinant of." However, for values of $m \geq 2$, the elements of certain columns and rows of the Q-matrix will be all zeroes; therefore, a modification consisting of ignoring those columns and rows plus compressing the Q-matrix is required (4). A check of the multiplication to obtain the final T-matrix shows that this compression has no effect on the final resulting amplitudes. Also, the source order, n, and mode, TE or TM, can be obtained by an examination of the components of the imaginary J and K solutions, as defined by Waterman (2), which shows that the nth diagonal element passes through zero at or very near the nth order eigenfrequency (4). By inspection, the J element reflects the TE mode and the K element reflects the TM mode (1).

RESULTS

For the perfectly conducting sphere, the eigenfrequencies for each order and mode remained the same regardless of azimuth index chosen. However, for the prolate spheroids and right circular cylinders this is no longer true. Both real and imaginary components of the eigenfrequencies differed as the azimuth index changed. This is most dramatically seen as the real parts are plotted in an "energy level" type diagram. Figures 1 and 2 are these type of plots for the corresponding first TE and TM layers (4). The 100:1 cylinder results were extracted from one of Baum's articles (1). As can be seen, the sphere represents the equivalent most degenerate case of an optical spectroscopy energy diagram while the cylinders and spheroids show the first degeneracy removed. Hence, the symmetry of the levels is removed geometrically whereas, in optical spectroscopy, it is removed via the Zeeman effect.

The figures show clearly that the "level diagram" is characteristic for the shape of the conducting target. Therefore, an observed eigenfrequency spectrum may be used for target classification purposes, i.e., for the solution of the "inverse scattering problem".

In the Singularity Expansion Method (4), it is customary to plot the pole positions of a given target in the complex frequency plane; this procedure exhibits also the imaginary parts of the poles, thus providing a measure of the attenuation of the corresponding radar echoes. Figures 3 and 4 present the poles for the sphere (circles), and show (4) how they split into several components labeled by m as the sphere is deformed into a 2:1 prolate spheroid (Fig. 3) or into a 2:1 circular cylinder (Fig. 4).

REFERENCES

1. Baum, C. E., and Singaraju, B. J., 1980, "The Singularity and Eigenmode Expansion Methods with Application to Equivalent Circuits and Related Topics," Acoustic, Electromagnetic and Elastic Wave Scattering-Focus on the T-Matrix Approach, Varadan, V. K., and Varada, V. V., ed., Pergamon Press, New York.

2. Waterman, P. C., 1973, "Numerical Solution of Electromagnetic Scattering Problem," Computer Techniques for Electromagnetics, Vol. 7, Chapter 3, Mittra, R., ed., Pergamon Press, New York.

3. Überall, H., and Gaunaurd, G. C., 1981, "The Physical Meaning of the Singularity Expansion Method (SEM)", Appl. Phys. Lett. 39, 362-364.

4. Moser, P. J., 1982, "The Isolation, Identification and Interpretation of Resonances in the Radar Scattering Cross Section for Conducting Bodies of Finite General Shape", Ph D Dissertation, The Catholic University of America.

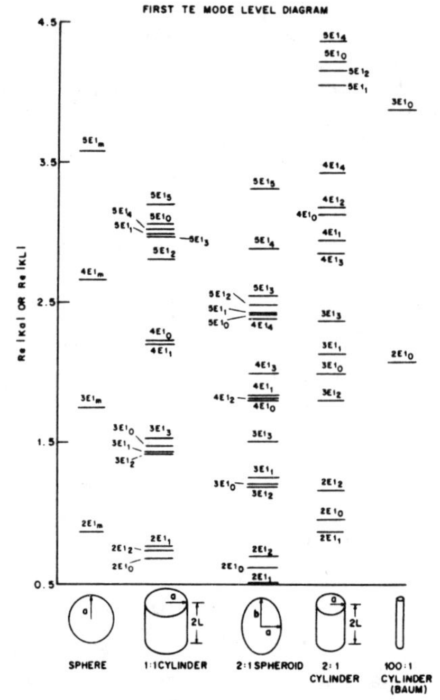

Fig. 1. Level Diagram for first TE mode eigenfrequencies only. Labeling scheme is $nE\ell_m$ where n=order, E denotes TE mode, ℓ= layer and m=azimuth index.

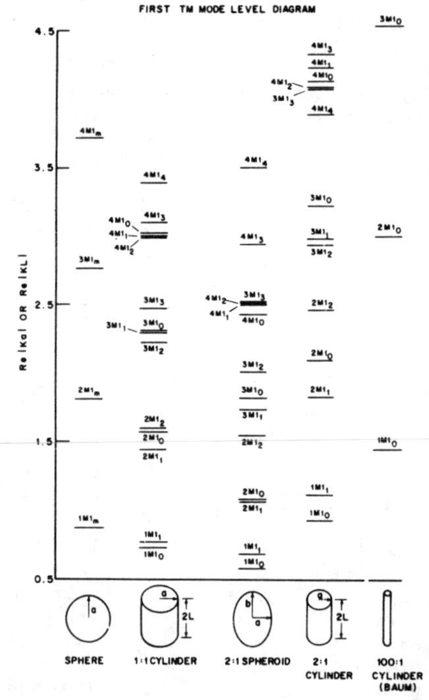

Fig. 2. Level Diagram for first TM mode eigenfrequencies only. Labeling is similar to Fig. 1 except M denotes TM mode.

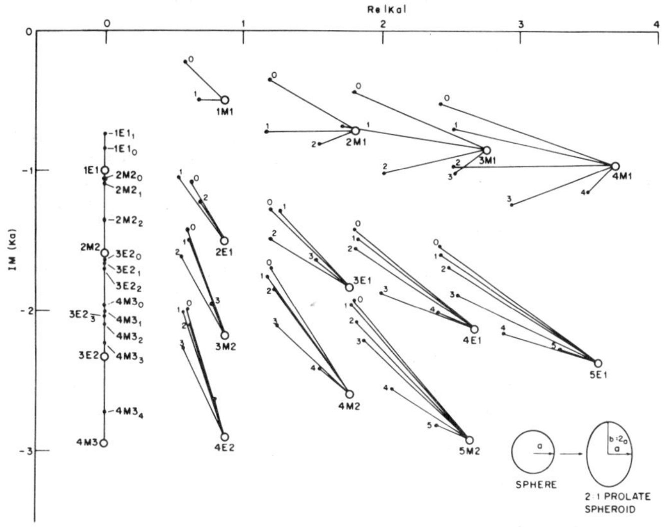

Fig. 3. Splitting and shifts of poles in the scattering amplitudes as a spherical target is deformed into a 2:1 prolate spheroid.

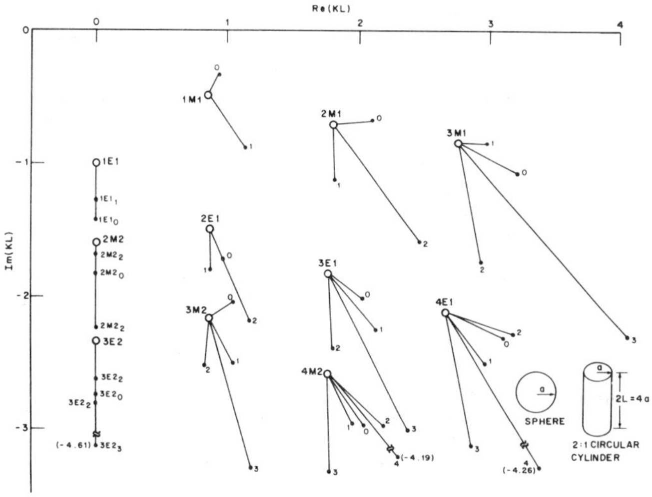

Fig. 4. Splitting and shifts of poles in the scattering amplitude as a spherical target is deformed into a 2:1 circular cylinder.

CLASSIFICATION OF SHIPS USING AN INCOHERENT MARINE RADAR

J. Maaløe

Electromagnetics Institute, Technical University of Denmark, DK-2800 Lyngby, Denmark

INTRODUCTION

Classification of ships on the basis of radar signals received from an ordinary marine radar is important for radar surveillance of harbours and coastal areas, where in addition to detecting and tracking of targets it is extremely desirable to maintain an identification of each ship detected. Further, in cases where the identification of a target has been lost due to clutter or because of a close passage of some other ship, it will be of great benefit to the surveillance if the targets can be reidentified. Since active identification information is normally not available in marine radar surveillance, the information must be extracted from the signals backscattered from the radar targets.

The classification method introduced in this paper only employs signatures which may be deduced from target returns using an incoherent radar, i.e. the method does not rely on measurement of doppler shift or target motion. Characteristics employed are for example the received amplitude level and the spatial target dimensions in both the range and azimuth. The classification procedure is performed using a stepwise discriminant analysis (1).

Stepwise Discriminant Analysis

The principle of the classification technique is to group the various target returns in a multidimensional observation space in such a way that all target returns falling into one specific group have some common characteristics which are in one way or the other different from the target returns falling into some other group. The axis of the multidimensional space represents the actual number of characteristics as calculated from the measured radar returns.

The classification technique employed in the analysis is the co-called stepwise discriminant analysis (1) which discriminates between any number of groups. The features used for classifying the target returns are chosen in a stepwise manner. At each step, the variable which causes the greatest separation of the groups is entered into the discriminant analysis, while the variable which has the least influence is removed from the analysis.

The results presented in this paper consist of a number of classification functions and a classification matrix. In addition, all the individual classifications are presented in a scatter plot, where also the mean of each group is shown. The axes of the scatter plots are the first two canonical variables (2).

The classification functions are linear combinations of radar return characteristics selected for the analysis. The aim of the classification functions is to separate the measurements into the classification groups selected. A new measurement of the characteristics is assigned to the group with the largest value of the classification functions.

Each target recorded is separated into a group according to the classification functions. The number of targets classified into each group and the percent of correct classifications are printed in a classification matrix.

The first two canonical variables define a new coordinate system in the multidimensional observation space. Each coordinate is a linear combination of the characteristics measured which maximizes the ratio between the variations among groups and inside groups. The two variables are determined in such a way that they become uncorrelated.

Measurements

The radar signals used for the classification trials are obtained from a Racal-Decca X-band marine radar situated at a coastal site overlooking the English Channel off Dungeness, England. The essential parameters of the radar as regard the classifications are

Antenna:		
-	Height	15 metres above mean sea level
-	Polarization	Horizontal
-	Azimuth resolution	0.8° at 3 dB
-	Revolution time	2.64 sec
-	Operating frequency	9410 MHz
-	Gain	33 dB
Transmitter		
-	PRF	1 KHz
-	Pulse length	250 nSec
Receiver		
-	Type	Superhet.-log IF.
-	Bandwidth	5 MHz
-	Noise factor	10 dB

During the measurements the weather was dry and the windspeed was about 5.7 m/s, corresponding to sea state 3.

The recording system used (3) stores the radar video signal together with the trigger pulse, the heading marker and the antenna angle as a composite video signal on a professional video recorder (IVC 601). When replaying, the signal is again separated into it's various parts. Subsequently, the data from a predetermined section of the PPI-picture is digitized and transferred into a computer (HP 21 MX). At present, the size of the window possible to store during an antenna revolution is limited to 2 NM in range and 200 A-scans in azimuth, corresponding to 24 degrees. The finite window size limits the number of targets which can be classified simultaneously. The limitation arises because of the finite memory size in the digital interface and the minicomputer and also because of the need of transmitting the digitized data from one A-scan during the subsequent interscan period.

The recordings available at present last for approximately ten minutes each.

Classification of Ships

The classification trial described in this paper includes three ships, all moving in the same direction at approximately equal speed. The image of the three ships shows three targets clearly separated in both range and azimuth, however close enough to be digitized within the same window. The distances between the radar and the targets are in the range interval from 4 NM to 6 NM. At these ranges, the clutter level at sea state 3 is negligible. The identities of the ships are completely unknown.

The initial classification of the three ships into there groups is based on various features of the radar returns over 80 antenna revolutions, corresponding to a time interval of 3.5 minutes. Five minutes and then ten minutes later, the same ships are reclassified into three similar groups, this time for 50 antenna revolutions, on each occasion corresponding to 2.2 minutes. The number of true classifications will then indicate the efficiency of the discriminant analysis.

The characteristic features selected for the classification purpose are:

NP, number of amplitudes exceeding a predetermined threshold

VA, variance of the amplitudes exceeding the fixed threshold

SM, spatial moment of inertia

PI, product of inertia

TM, template

In order to fully understand the classification procedure, some of the characteristics have to be further explained. The moments are calculated for an area consisting of 21×21 samples placed around the centre of the target, the weight being one if the amplitude exceeds the predetermined threshold and otherwise zero. The spatial moment of inertia (SM) is calculated relative to the centre of the target as

$$SM = \sum_{i=1}^{21} \cdot \sum_{J=1}^{21} d_{ij}^2 \cdot A_{iJ}$$

where d_{iJ} is the distance from the centre to the point (i,J), and A_{iJ} is 1 if the amplitude from the point (i,J) exceeds a predetermined threshold and otherwise 0. The product of inertia (PI) is calculated as

$$PI = \sum_{i=1}^{21} \sum_{J=1}^{21} X_{iJ} \cdot Y_{iJ} \cdot A_{iJ}$$

where X_{iJ} and Y_{iJ} are the distances from the two orthogonal axes parallel to the range direction and the azimuth direction, respectively, and going through the centre of the target. The product of inertia indicates the direction of the target.

The template (TM) is defined as the number of amplitudes backscattered from a target of which 10 out of 15 adjoining amplitudes (3 amplitudes in range and 5 amplitudes in azimuth) exceed a predetermined threshold. The reason for the name template is that the feature depends on the shape of the target (4).

The stepwise discriminant analysis produces the following classification functions

	Ship 1	Ship 2	Ship 3
NP	5.154	5.310	3.597
VA	0.046	0.035	0.020
SM	-3.443	-2.951	-2.168
PI	-3.027	-1.604	-0.168
TM	-3.988	-4.076	-2.924
Constant	-126.071	-108.970	-44.589

As already noted, a set of measured target characteristics is assigned to that of the groups with the largest value calculated from the classification functions.

According to the classification functions each measurement of the ships is classified. The results are listed in the classification matrix below, showing the number of ships classified into each group and the percentage of correct classifications. Each group is specified by the ship number and the time of the measurement.

Group	Percent correct	Number of cases classified into groups		
		Ship 1	Ship 2	Ship 3
Ship 1,0	91.3	73	7	0
Ship 2,0	92.5	5	74	1
Ship 3,0	100.0	0	0	80
Ship 1,5	74.0	37	13	0
Ship 2,5	94.0	2	47	1
Ship 3,5	100.0	0	0	50
Ship 1,10	80.0	40	7	3
Ship 2,10	100.0	0	50	0
Ship 3,10	98.0	0	1	49

The group means and all the measurements are plotted in figures 1, 2 and 3, showing respectively the initial classification as well as the classifications 5 minutes and 10 minutes later. In each figure, the group means from each classification trial are plotted. For example, the group means of ship 1 after 0,5 minutes and 10 minutes, respectively, are plotted as the numbers 1, 4 and 7.

The classification matrix and the scatter plots clearly show how ship 3 is easily distinguished from the other two ships. In addition, it should be noticed that the discrimination between ship 1 and ship 2 succeeds in approximately 94% of the times. Hence, it appears that the number of true classifications is high enough to indicate the efficiency of the classification method used.

CONCLUSION

The possibility of classifying ships on the basis of backscattered incoherent radar signals only, has been verified during several trials, one of which has been described in this paper. The parameters used for the classification are features characterizing the size of each target. The classification was performed by a stepwise discriminant analysis.

The classification experiments carried out as yet have only dealt with ships sailing in clutter-free environments and following a straight course. Future investigations will include ships following more complicated paths and passing through clutter areas. It will further be attempted to make an absolute identification of the individual ships so as to compare the response of various vessel types.

REFERENCES

1. Dixon, J.W. (ed.), Biomedical Computer Programs, University of California Press, Los Angeles 1973.

2. Andersen, T.W., An Introduction to Multivariate Statistical Analysis. John Wiley & Sons, New York 1958.

3. Maaløe, J, Radar Signal Recording and Analysing System, Release 1. Electromagnetics Institute, IR 253, October 1981.

4. Pratt, W.K., Digital Image Processing. John Wiley & Sons, 1978.

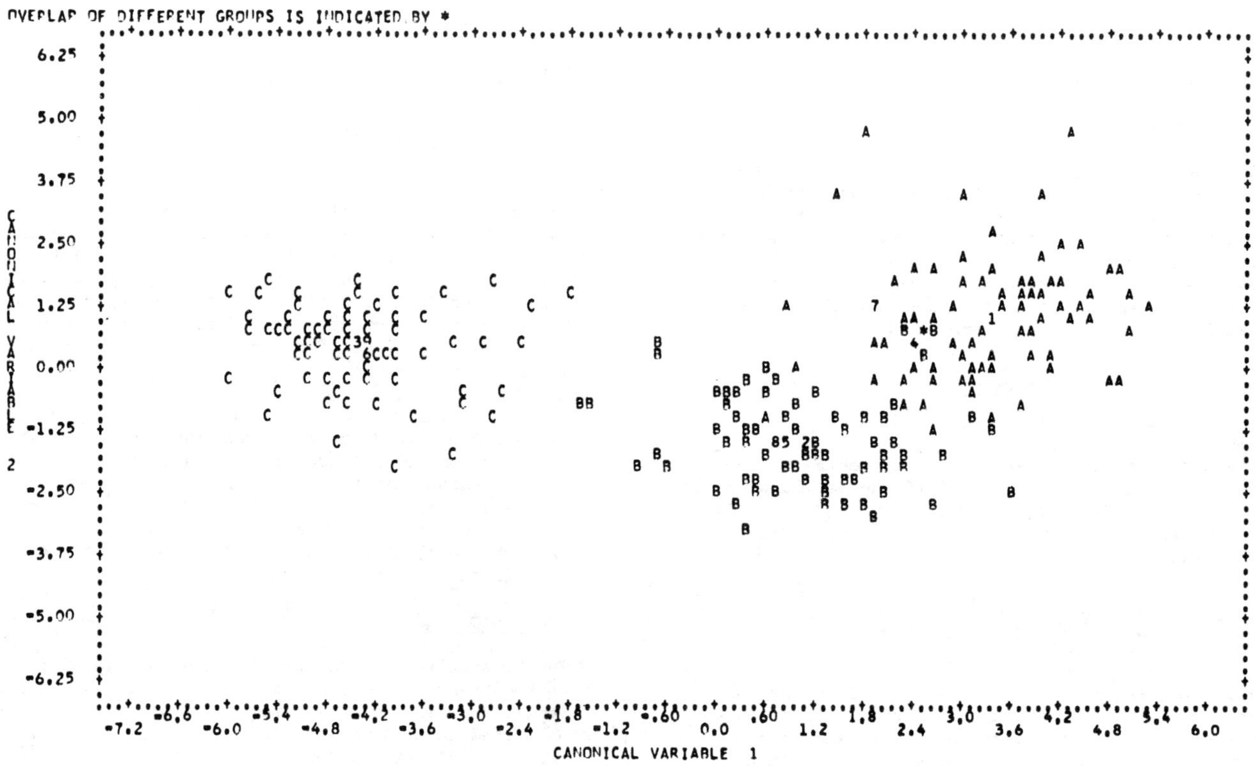

Figure 1 The initial classifications of three unidentified ships. The classifications of ship 1, 2 and 3 are plotted as A, B, and C.

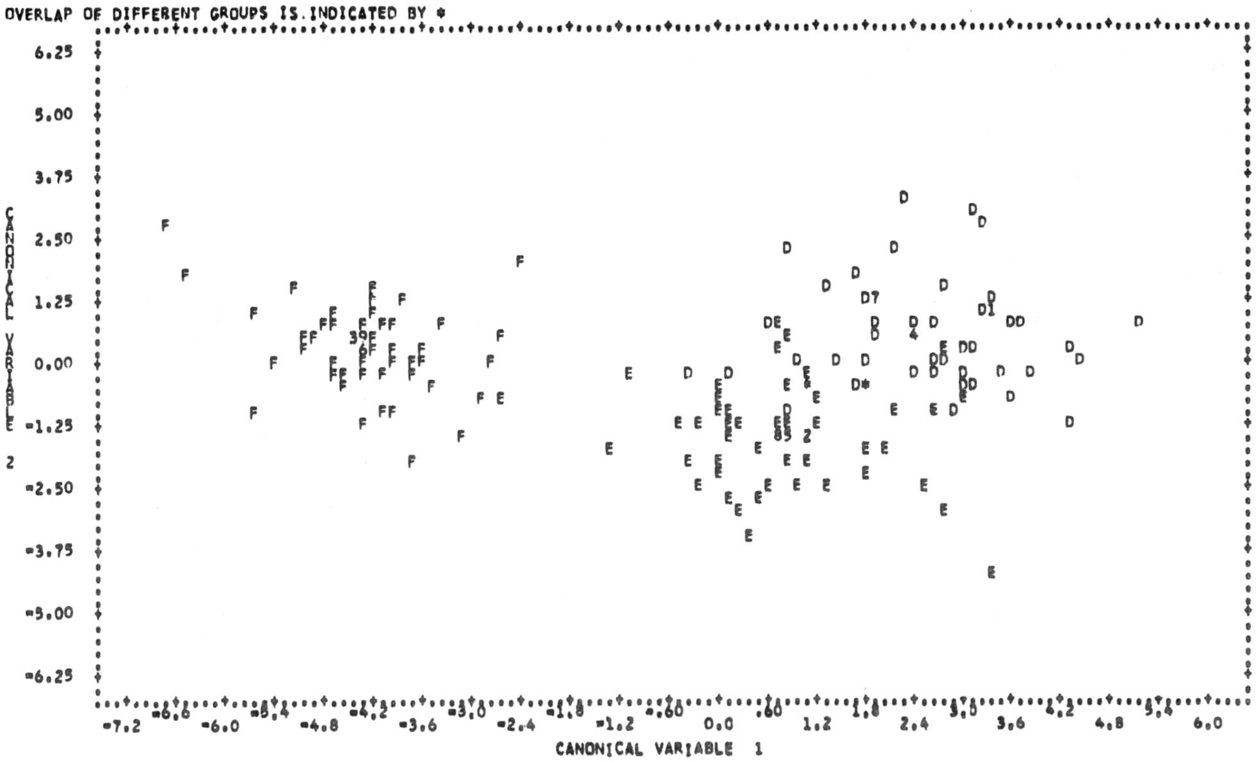

Figure 2 Reclassification of the three ships 5 minutes after the initial classification. The classifications of ship 1, 2 and 3 are plotted as D, E and F.

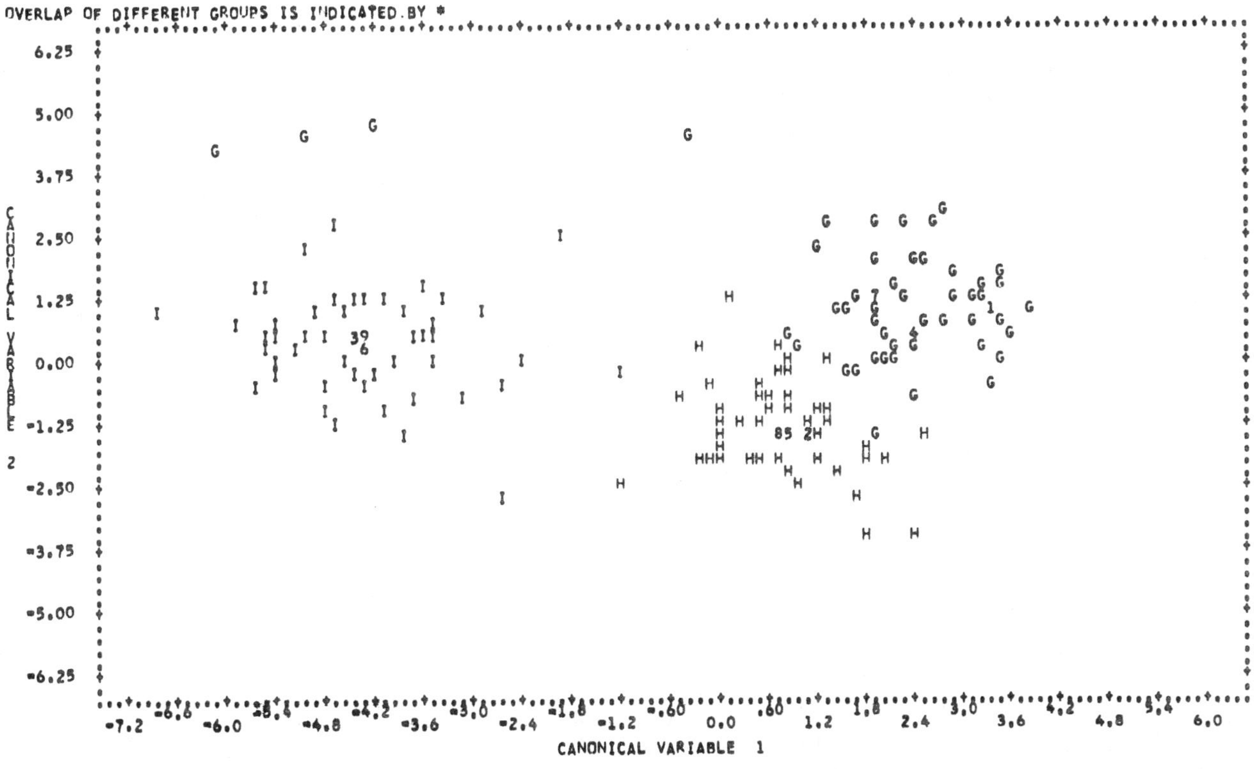

Figure 3 Reclassification of the three ships 10 minutes after the initial classification. The classifications of ship 1, 2 and 3 are plotted as D, E and F.

THE IMPACT OF WAVEFORM BANDWIDTH UPON TACTICAL RADAR DESIGN

Charles H. Gager

The MITRE Corporation, USA

INTRODUCTION

In the last several years a number of new radar designs have been implemented or proposed for tactical air surveillance and control functions. These radars use multibeam antennas or phased arrays with flexible beam control and beam shaping, operate over frequency bands of five to ten percent, and achieve high signal processing gain using digital processing. Typical examples are given in Cochran (1), Gostin (2), and Rosien (3) of radars that provide surveillance and target tracking to ranges of 300 to 400 kilometers through 360° azimuth coverage and an elevation coverage that extends between the horizon and the maximum likely aircraft altitudes. These designs illustrate that the development of modern components allow tactical radars to have significantly advanced capabilities.

Development of high-speed, integrated circuit digital processing and microwave components now permit the design of practical tactical radar that use wide bandwidth transmitted waveforms. These waveforms fall into two classes: wideband signals, in which each transmitted pulse has an instantaneous bandwidth that covers the radar's full operating frequency band, and frequency-agile signals, in which each narrow bandwidth transmitted pulse may have a different center frequency within a wide operating bandwidth. In this paper we compare the advantages and disadvantages for tactical radar of waveforms with wide instantaneous bandwidth, frequency agility, and conventional bandwidth. While discussing the unique capabilities of the wideband and frequency-agile waveforms, performance comparisons emphasize the resistance of each waveform to signal intercept and jamming during electronic warfare.

WAVEFORMS

Two illustrations define the frequency-agile and wideband signals. Figure 1 shows the frequency-versus-time plot of a waveform with pulse-to-pulse frequency agility. The time axis is expressed in units of the radar pulse repetition interval. The truncated frequency axis shows an interval which is about ten percent of the transmitter's operating frequency. In this frequency-agile waveform, the center frequency of each transmitted pulse is moved, in either a random or programmed schedule, between a large number of center frequencies on a pulse-to-pulse basis. The frequency of the next pulse cannot be generally predicted from the frequency of the current pulse. A variation of this frequency-agile waveform has a limited group of pulses at the same frequency before jumping to a new, randomly-selected, center frequency. Doppler processing and coherent integration may be achieved within each group of pulses at the same frequency.

Signals with wide instantaneous bandwidth are illustrated in the frequency-time plots shown in Figure 2. In these three plots, the time axis is expressed in units of the transmitted pulse width so that the plots show the frequency characteristics within each transmitted pulse. The frequency axis in this Figure again represents a spread of about ten percent of the transmitter center frequency. Three of the more common coded-pulse waveforms, as described by Cook and Bernfeld (4), are illustrated in Figure 2. Figure 2a shows a linear FM signal and Figure 2b shows a frequency-shift coded signal. Figure 2c shows the phase-time relationship for a phase coded signal in which the phase of the RF carrier is shifted at a chip rate equal to the bandwidth of the waveform.

The frequency-agile and wideband waveforms each have certain unique advantages that may cause the tactical radar designer to choose this waveform. Frequency-agile waveforms provide rapid target and clutter decorrelation, with a resultant improvement in target detection capability. Wideband signals permit measuring details of a target structure, measuring the range extent of a target, separating and counting the number of closely-spaced targets, resolving crossing target tracks and, together with coherent cross-range processing, providing a two-dimensional radar image of the target. The unique advantages of wideband waveforms are generally associated with the examination of a limited target volume or of tracked targets, while the advantages of frequency-agile waveforms are more generally associated with surveillance functions.

Each waveform has potential advantages for area surveillance during electronic warfare. Before making this comparison, let us first examine the characteristics of modern intercept receivers (often called elint or ESM equipment) and electronic jamming equipments that are the key adverse elements to a radar during electronic warfare.

RADAR SIGNAL INTERCEPTION AND JAMMING

Intercept receivers are used for electronic intelligence gathering, location of radars, recognition or classification of radars, and weapons guidance. An individual intercept receiver at a fixed location is only able to locate the angular coordinants of a radar emitter, but multiple receivers can use strobe triangulation, correlation, or time-difference-of-arrival for measurement of both range and angle location.

Figure 3 shows the functions of an intercept receiver. These include a wideband antenna and RF amplifier, a frequency separator, a pulse time-measurement processor, a signal identification processor, and an angle-of-arrival measurement unit. Characteristics of the frequency separator are of particular interest when comparing the susceptibility to interception of radar waveforms. Early intercept receivers used crystal video or scanning superheterodyne designs for frequency separation. More modern receivers are now being designed using channelized, compressive, and acousto-optical receivers (Rappolt (5)). While channelized, compressive, and acousto-optical receivers have specific differences and advantages, these modern receivers combine high sensitivitvy, wide instantaneous frequency coverage, and high frequency resolution. In this discussion, we will assume that the intercept receiver uses a channelized design. Parallel receiving channels, with offset center frequencies, divide the total intercept receiver frequency coverage into a number of equal bandwidth channels with continuous intercept ability. The number of narrow band channels, and the resultant sensitivity and frequency resolution, is limited only by hardware size, cost, and complexity.

The radar pulse's signal strength at the intercept receiver and the minimum signal detectable by the receiver are defined by:

$$S_r = \frac{P_R G_R g A_e}{4\pi R_e^2 L_e}$$

$$S_{min} = KTB_e F\ (S/N)_t$$

where

S_r	=	Signal Power at Receiver Input
P_r	=	Radar Transmitted Pulse Power
G_r	=	Radar Antenna Gain
g	=	Radar Antenna Sidelobe Ratio
A_e	=	Effective Receiving Aperture
R_e	=	Range from Radar to Receiver
L_e	=	Intercept Receiver Losses
S_{min}	=	Receiver Sensitivity
KTF	=	Noise Power per unit Bandwidth
B_e	=	Receiver Bandwidth
$(S/N)_t$	=	Detection Threshold

An intercept receiver at the same range as a radar target has a signal strength advantage, relative to the radar receiver, that is proportional to the range squared. Normally, this advantage is only partially offset by greater radar aperture and the processing gain obtained with complex radar waveforms. As a result, an intercept receiver is usually able to detect signals transmitted through sidelobes of the radar antenna.

Post World War II jamming equipments used either manual frequency tuning, slow-swept frequency, or broadband noise modulation. From these early equipments, modern jamming designs have advanced to capabilities for selecting victim radars from a complex ??remitter environment and for concentrating jamming power in the victim radar's frequency band. A functional block diagram of a modern jamming system is shown in Figure 4. The upper four boxes show receiving antenna, frequency separator, time measurement, and identification processing units that are similar to units of the intercept receiver previously described. The jammer control function sets the center frequency of the jamming signal to that of the radar and attempts to match the jammer modulation bandwidth to that of the radar signal.

The radar range equation, for a condition where the maximum radar detection range is limited by jamming noise rather than by radar receiver noise, can be expressed as:

$$R^4 = \frac{P_R G_R \sigma \tau R_J^2 B_J}{4\pi g (S/N)_T P_J G_J L_R}$$

where

R	=	Maximum Radar Detection Range
σ	=	Target Cross Section
τ	=	Transmitted Pulse Width
R_J	=	Jammer Range from Radar
B_J	=	Jammer Modulation Bandwidth
$P_J G_J$	=	Jammer Effective Radiated Power
L_R	=	Radar Receiver Loss Ratio-Radar Signals/Jamming Signals

This range equation formulation (Skolnik (6)), which uses jammer transmitted power and modulation bandwidth as parameters, emphasizes that the fourth power of radar detection range is directly proportional to the jammer modulation bandwidth. While a radar designer seeking to maximize radar performance against jamming can increase power aperture or reduce antenna sidelobes, these parameters are generally limited by practical considerations of cost, size, and mobility. An alternate approach, that may result in improved radar design, is to use a waveform that causes the jammer modulation bandwidth to be wide.

EXAMPLES OF TACTICAL RADAR PARAMETERS

Table 1 shows examples of tactical air surveillance radar parameters with three waveform cases: a simple pulse, a simple pulse with pulse-to-pulse frequency agility, and a wideband pulse. Examples of waveforms with a low pulse repetition rate are given in the first three columns and examples with a pulse doppler, high pulse repetition rate are given in the last two columns. The parameters are for an S-band tactical radar, with 40 db peak antenna gain, that has a nominal detection range of 350 km against a three square meter target. For the waveform comparison, we use an equal power aperture for each waveform, and assume an equal radar detection range. This assumption neglects fine differences in the detection performance of the three waveforms. The increased number of range cells and false alarm possibilities, and the possible reduction of target cross-section, with the wideband waveform tend to reduce detection range. Target decorrelation with the frequency-agile waveform tends to increase target detection range. In all cases we have kept a five microsecond transmitted pulse width for each waveform. Considerable pulse compression processing is required for the wideband waveform that has a time-bandwidth product of 1000.

Neither the frequency-agile nor the wideband low PRF waveforms are suitable for moving target doppler processing. Pulse-to-pulse frequency agility, of course, precludes coherent processing. The two foot range gate for the wideband waveform is so narrow that even point reflecting elements of a target complex will move from a range gate within a pulse repetition interval of 2.5 milliseconds, when the target radial velocity is greater than 500 knots.

TABLE 1 - Example of Radar Designs

	Low PRF Designs			High PRF Design	
	Simple Pulse	Frequency Agility	Wideband	Frequency Agility	Wideband
Frequency (GHz)	3.3	3.3	3.3	3.3	3.3
Peak Power (kW)	1000	1000	1000	83	83
Pulse Width (microsec)	5	5	5	5	5
Instrumented Range (km)	375	375	375	375	375
P.R.I. (microsec)	2500	2500	2500	50	50
Bandwidth per pulse (MHz)	0.2	0.2	200	0.2	200
Operating Band (MHz)	0.2	200	200	200	200
Range Gate Size (microsec)	5	5	0.005	5	0.005
Antenna Gain (dB)	40	40	40	40	40
Number of Range Gates	500	500	5×10^5	9	9×10^3
First Blend Speed (knots)	40	40	40	2000	2000
Pulses per Burst	1	1	1	12	12

The pulse doppler waveforms overcome these difficulties and permit coherent MTI processing. Bursts of twelve pulses, with pulse repetition intervals of 50 microseconds between pulses, are used for both the frequency-agile and the wideband waveforms. These twelve pulses are coherently integrated. In the frequency-agile waveform, subsequent groups of pulses are transmitted at new center frequencies. The peak transmitted power of each waveform in the pulse doppler mode is reduced by an amount equal to the number of pulses in the burst, thus neglecting processing losses.

WAVEFORM COMPARISONS

The resistance-to-intercept of these waveforms can be compared by referring to Figure 5. In this figure, the received signal strength at the intercept receiver is plotted as a function of range for cases where the waveform is transmitted through the radar antenna mainbeam and through radar antenna sidelobes whose response is 50 db below the peak of the mainbeam (-10 db isotropic). Plots are made for both the low PRF and the pulse doppler waveforms. For these plots, we use values of 0.5 square meters for the receiving aperture, A_e, and 4 db for intercept receiver losses, L_e. The values are conservative, since this size antenna can easily fit in even the smallest aircraft pods or mounting locations.

The minimum detectable signal (sensitivity) of the intercept receiver is also shown in Figure 5 for two channelized receiver bandwidths (200 MHz and 10 MHz). This sensitivity assumes a receiver noise figure of 6 db and a detection threshold setting 16 db above receiver noise. The receiver sensitivity with 200 MHz channelized receiver bandwidth is representative of the wideband signal detection capability by either a broadband intercept receiver or by a single 10 MHz bandwidth channel. In the latter case, each channel intercepts only a portion of the total pulse power spectrum.

While the detection range of the wideband signal is considerably reduced from that of either the frequency-agile or the simple pulse signal, the wideband signal is still detectable at ranges of 320 km, for the case of sidelobe transmission of the pulse doppler signal, and at considerable greater ranges for other cases. Differences between the intercept range of the different waveforms does not appear to be significant for most applications. Regarding more subtle factors, both the wideband waveform and the frequency agile waveform require greater processing complexities in the intercept receiver to recognize and classify the signals. While data processing power is available for this classification, additional size and speed are required. All of the waveforms can be implemented equally well with wave shapes that make time-of-arrival measurements difficult at the intercept receiver and therefore limit position measurement accuracies of time-of-arrival systems.

The capabilities of each waveform against jamming signals can be calculated for the case of a sidelobe-jamming source located at a standoff range of 180 km. Representative parameters used for this comparison include a jamming transmitted power (P_J) of 1 kw, an antenna gain (G_J) of 16 db, and modulation bandwidths (B_J) of 200 MHz or 1 MHz. For these parameters, which are typical of commercially available components, the effect of jammer modulation bandwidth is profound. When the jammer modulation bandwidth is 200 MHz, the jamming reduces the radar detection range from a normal value of 350 km to a value of 196 km. However, when the jammer is able to use a modulation bandwidth of 1 MHz, correctly centered at the radar receiver's center frequency, the jamming reduces the radar detection range to 49 km.

The merits of each radar waveform can be determined by its ability to dictate the effective modulation bandwidth of the jammer. The simple pulse waveform is vulnerable to frequency measurement and narrow band modulation by modern jammers. Use of this waveform is likely to result in the maximum detection range reduction to 49 km. Conversely, the wideband waveform forces the jammer to use broadband modulation and is therefore sure to be effective in limiting the radar range reduction produced by the jamming. The frequency agile waveform has a reasonable chance of forcing wideband jamming modulation. With pulse-to-pulse frequency agility, even the fastest jammer can only set-on to the correct frequency at times which correspond to ranges beyond its stand-off range (180 km in our example). The radar detection range is therefore limited to values similar to the 196 km detection range that is achieved against a broadband jammer. With pulse doppler waveforms, where frequency agility is only performed in pulse groups, a fast set-on jammer has a greater capability to inject narrowband modulation into the radar receiver before the integration interval is completed. However, this capability may be limited by pulse transit-time to the standoff range, the interval between pulse bursts, and the problem of fast recognition of the victim radar's signals in a dense pulse environment.

The wideband waveform gives greater assurance of improved radar detection range during jamming, but requires significantly increased equipment complexity for surveillance applications. The simple pulse and frequency-agile waveforms have a limited number of range gates (500 for the low pulse repetition frequency and 9 for the pulse doppler waveform) so that they do not require more components than current radar designs. Their signal processing bandwidths, analog-to-digital conversion speeds, and data processing bandwidths are also very tractable. The wideband waveform, however, requires 500,000 range gates to provide surveillance over all ranges with the low pulse repetition frequency and 9,000 range gates with the pulse doppler waveform. Range gate sampling and analog-to-digital conversion in each of these range gates must be performed at a 200 MHz rate. While large-scale integrated circuits make it possible to think of handling these numbers of range gates, the complexity and cost is high.

A second major complexity occurs when the wideband waveform is used with a phased array antenna. A wideband waveform requires that the array have an instantaneous bandwidth at least as wide as that of the wideband signal. With signal bandwidths of five to ten percent and the necessary array sizes, time-delay control of the array elements is required to steer the antenna beam from the axis of the array face (Skolnik (6)). Time delay steering, which compensates for the difference in signal propagation path to different elements across the face of the array, is more costly than phase control of the array elements.

The disadvantages of the wideband waveform for full-range surveillance are significantly reduced when this waveform is used only for target tracking, target examination, or jammer burn-through in limited areas. These functions infer some knowledge of the range to the target so that the number of range gates can be reduced by an amount equal to the ratio between the target's range position uncertainty and the total instrumented range of the radar. This ratio is typically several orders of magnitude. Similarly, the problem of the phased array bandwidth can be partially overcome when target position is approximately known. Techniques are available for compensation of the bandwidth limitation of phase-steered arrays over a limited target range interval, when a linear FM waveform is used (Pickens and Lyons (7)).

SUMMARY

In summary, our examples illustrate some tradeoffs between frequency-agile and wideband waveforms for tactical radar design. Selection between these waveforms must be made on the basis of detailed radar mission requirements. A wideband waveform gives the greatest assurance of maximum range performance against jamming, but at the expense of considerable additional radar receiver, processor, and phased array antenna hardware. The wideband waveform has special uses where its unique capabilities for target measurement and separation are desirable, but these uses often can be limited to narrow range windows so that the added complexity is low. A design with both frequency-agile and wideband waveforms may be advantageous for some applications if the wideband waveform is used for only limited functions.

ACKNOWLEDGMENT

This work was supported by the United States Air Force under Contract F19628-82-C-0001.

REFERENCES

1. Cochran, J. G., 1977, "Current Developments in the Design of Air Defense Ground Radar Systems," Radar 77 IEE Conference Publication, 16-19.

2. Gostin, J. J., 1980, "The GE592 Solid State Radar," Eascon '80 IEEE Conference Publication, 197-203.

3. Rosien, R. A., 1980, "The Series 320 Radar Family, A Cost Effective Approach to Inertialess Beam Steering," Eascon '80 IEEE Conference Publication, 182-189.

4. Cook, C. E. and Bernfeld, M., 1967, "Radar Signals," Academic Press, New York.

5. Rappolt, F., 1981, "Receiver Tools Improve Difficult RF Signal Acquisition," Microwave Systems News, 11, 62-86.

6. Skolnik, M. I., 1970, "Radar Handbook," McGraw-Hill Co., New York.

7. Pickens, R. A. and Lyons, B. J., 1963, "Compensation for the Bandwidth Limitation of a Wideband Phased Array Radar," US Air Force, Rome Air Development Center, Conference Publication RADC-TR-62-580.

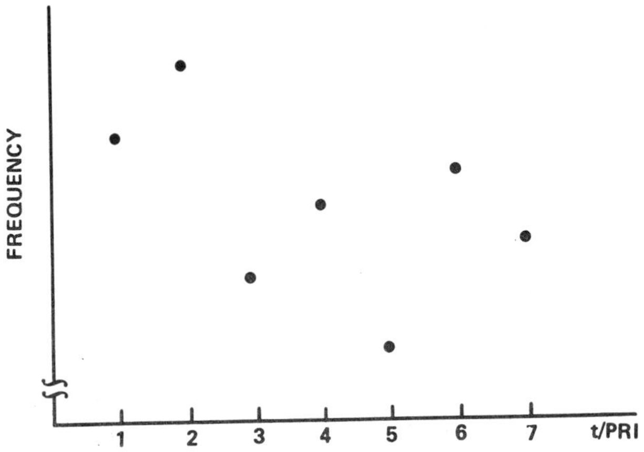

Figure 1. FREQUENCY-AGILE WAVEFORM

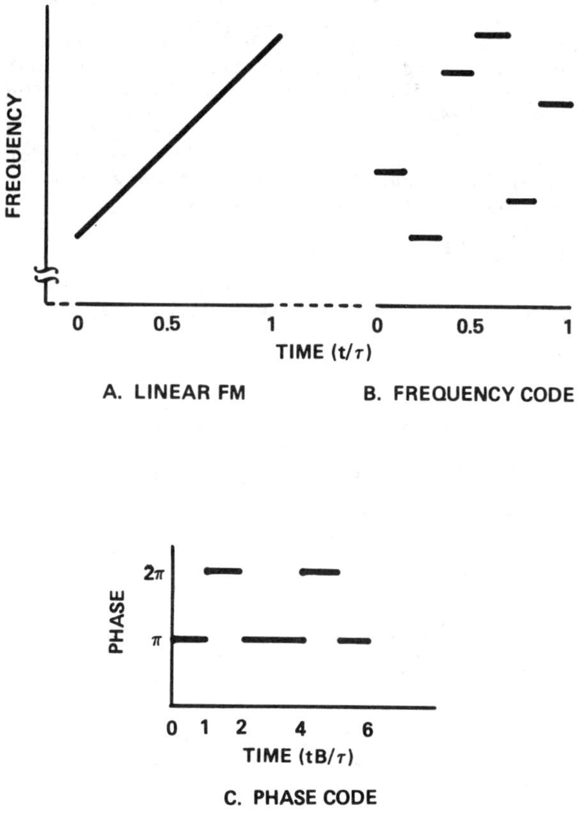

Figure 2. WIDEBAND WAVEFORMS

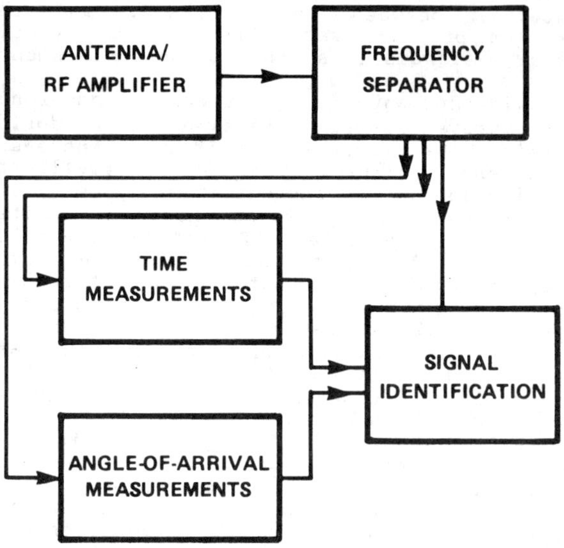

Figure 3. FUNCTIONAL DIAGRAM OF INTERCEPT RECEIVER

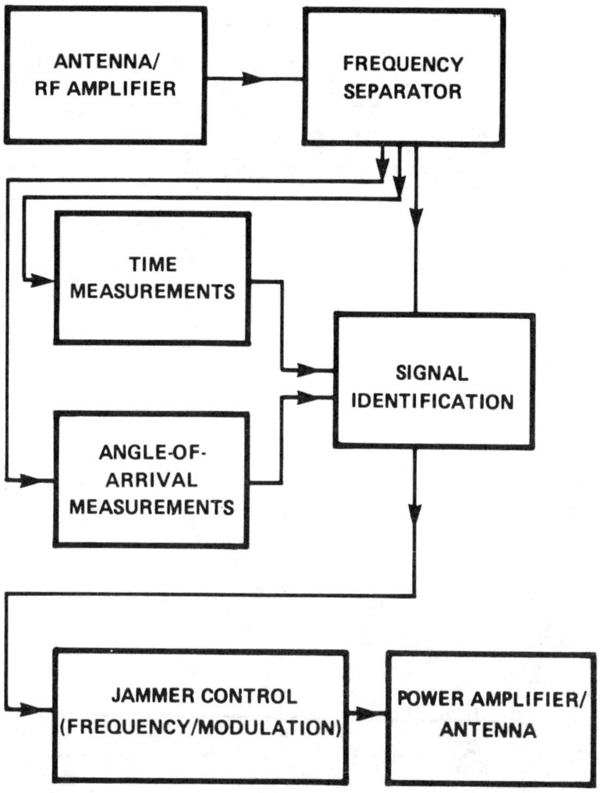

Figure 4. FUNCTIONAL DIAGRAM OF JAMMER

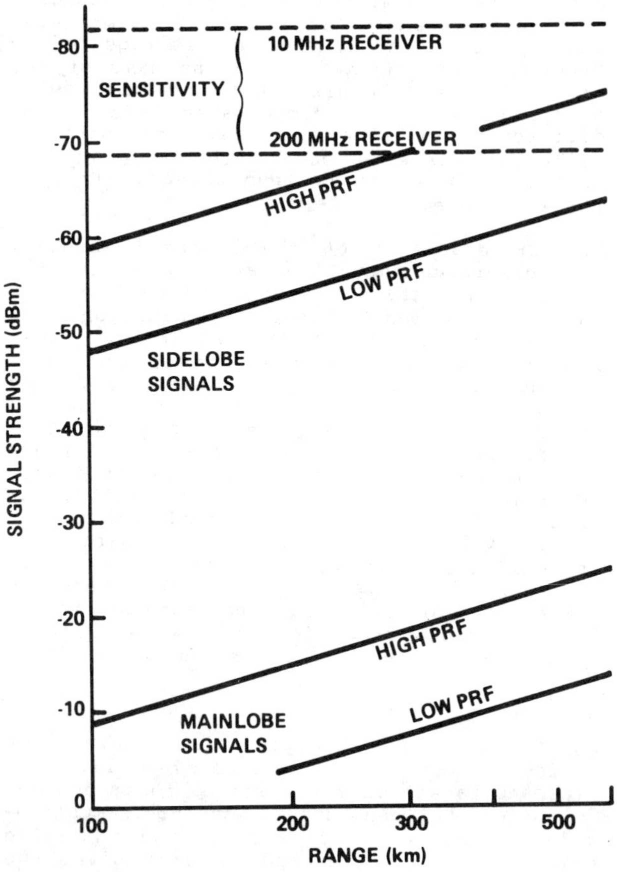

Figure 5. RECEIVED SIGNAL STRENGTH

RADAR ASSISTED PASSIVE DF TRACKING

R S Farrow

Admiralty Surface Weapons Establishment, UK.

INTRODUCTION

This paper describes the use of the time history of bearing data, obtained from microwave DF systems at separated sites, in order to calculate the position and velocity of a microwave emitter. Triangulation techniques appear to offer advantages over those techniques based on using data from a single site, eg a direct determination of range. The requirement for a large instantaneous azimuth field-of-view (generally 360°) in order to provide a high probability of signal intercept, however, leads to a relatively poor bearing accuracy when compared to that obtainable using radar systems. Due to this poor bearing measurement accuracy a series of position measurements calculated from the knowledge of bearing measurements from two separated platforms will not necessarily produce a consistent description of emitter motion. It is therefore necessary to use filtering techniques in order to generate a track which minimises the measurement errors and more accurately determines the emitter state vector.

For systems with poor bearing accuracy the filtered triangulation data may still not produce a track with a suitably high confidence level. In these cases it is possible to add radar information to the measurement vector and thereby improve the system tracking performance. Radar transmission must, of course, be minimised in order to prevent reverse interception of the beam, but knowledge of emitter bearing and approximate range will lead to illumination of the smallest possible volume of space and therefore, a lower probability of interception than conventional surveillance radar systems.

The performance of a Kalman filter algorithm is examined with respect to bearing measurement accuracy and the effect of radar updates. The algorithm comprises a 4-element, polar state vector, (bearing, course, range, speed) filter which is transformed to cartesian co-ordinates for the linear extrapolation stage. The data used in the analysis, including the 'measurement errors', have been generated synthetically.

THE DATA GENERATOR

The data that the track filter requires are:

(1) bearings from each surveillance receiver;

(2) knowledge of the relative position of the receivers.

These data should ideally be measured at the same time. In general, however, radar transmissions will not be intercepted at two widely separated sites at the same instant. If the radar is scanning there may be several seconds between intercepts. In a practical situation, therefore, the triangulated range calculation would need to use an extrapolated bearing measurement from one receiver in order to provide time coincident data. For this work it has been assumed that data is received by the DF sensors simultaneously - the bearing error introduced using this assumption is negligible for typical geometries and target radars.

A further error that will affect the range calculation will be the inexact knowledge of the relative position of the receivers. This has been simulated by assigning the arbitrary values of 1° in bearing standard deviation and 1 km in range standard deviation to the relative position error. The actual values of range and bearing are calculated using a truncated gaussian random number generator, the bearing and range variations are then added to the true bearing and range in order to generate a position for the secondary receiver. This type of error is particularly applicable to the naval scenario where shipborne receivers are probably in relative motion. For the tracks simulated in this work a baseline of 20 km has been assumed (~ horizon range).

The mechanism of the data generator is to:

(1) calculate the true range and bearing of the emitter from Receiver 1;

(2) calculate the true range and bearing of the emitter from Receiver 2;

(3) randomise the position of Receiver 2 with respect to Receiver 1;

(4) randomise the bearing measurement from each receiver;

(5) calculate the apparent range of the emitter from Receiver 1.

The error statistics of the corrupted measurements all have a gaussian form with the maximum errors limited to two times the standard deviation.

A further parameter required is that of the differential range error calculated using the measured bearings and separation. This is an approximate model of the variation in the accuracy of the range calculation according to geometry. Figure 1 indicates a typical disposition of the emitter and receivers and it is the size of the angle $\emptyset$ which controls the expected accuracy of the range measurement. As the emitter moves towards the receiver baseline the accuracy with which the range can be calculated increases and the track filter must be able to consider this variation in order to properly weight the range data.

In summary the data generator provides 'measurements' of bearing, range and expected range error. The sensor bearing error can be varied as can the receiver separation and associated errors. All measurement errors have a truncated gaussian form limited to two standard deviations. The effects of bias errors have not been modelled.

THE KALMAN FILTER ALGORITHM

The filter algorithm is based on the course and speed filter described by Clark (1). It has been modified in order to incorporate range and radar derived position measurements. The state vector A is defined by:

$$A = [B\ C\ S\ R]^T \qquad \ldots. (1)$$

where: B is the bearing of the emitter from the primary receiver;

C is the emitter course;

S is the emitter speed;

R is the range of the emitter from the primary receiver.

The superscript T denotes the matrix transpose.

In order to provide a linear extrapolation step for this basically polar filter it is necessary to carry out a polar to cartesian co-ordinate transformation followed by the extrapolation step and then the reverse transformation. This process is realised in a single step and the extrapolated state vector A' is defined by:

$$A' = [U\ C\ S\ \gamma R]^T \qquad \ldots. (2)$$

where:

$$U = \tan^{-1}\left[\frac{\sin (B + \epsilon \sin(C))}{\cos (B + \epsilon \cos(C))}\right]$$

$$\epsilon = \frac{S\Delta t}{R}$$

Δt = filter cycle time

$$\gamma = (1 + \epsilon^2 + \cos(C-B))^{1/2}$$

A linear covariance extrapolation is achieved using a similar technique and the transition matrix TM is defined by Equation 3.

This form of the Kalman filter although leading to a larger number of multiplications than the cartesian form, does avoid the biases that are present when the polar measurement errors are converted to cartesian form. It also has a state vector which comprises easily visualised emitter parameters, ie those generally used to describe emitter motion, and the inclusion of extra data in the measurement vector is thus simplified.

RESULTS FROM FILTER RUNS

The simulated emitter track used is shown in Figure 2 - a constant velocity track on a course of 105° at a speed of 900 km/hr respectively. The initial point for the tracks is at a range of 200 km on a bearing of 315°.

All runs of the track filter are initialised in the same way: range is set at 150 km; bearing is set to 315°; course is set to radial (135°); speed is set to 720 km/hr. The initial conditions are in error with respect to the true track thus simulating the live situation. The range standard deviation is the expected range error measurement; the standard deviation of the bearing is that of the data used; course standard deviation is set at 45°; and the speed standard deviation at 300 km/hr. The size of the latter two track initialisation parameters must be large in order to permit the velocity to be determined by the smoothing of the measured parameters rather than the initial estimate. Figure 3 illustrates the tracking performance of the filter against a non-manoeuvring target using the triangulated range measurement calculated using sensors with 1° standard deviation for bearing measurements. The true track is indicated by the heavy line, the filtered position is indicated by the cross (+). The velocity is indicated by the vector connected to the position indication. The positions/vectors are displayed at 30 second intervals and the vectors represent the distance that would be travelled in 30 seconds if the emitter remained on the smoothed course at the smoothed speed calculated for the position indicated.

As can be seen from the graph the filtered position estimates are in error by ~ 50 km in range until the range falls to ~ 100 km. Figure 4 shows graphs of true range error and measured range standard deviation (differential range error) vs time. The graphs illustrate that there is a critical range beyond which the filter has difficulty determining the true range. At this point it is worth examining the reasons for this feature. An analysis of the differential range error equation reveals that range error is highly dependent on the magnitude of the included angle Ø (see Figure 1) and until Ø increases to ~ 3-4 times the bearing standard deviation the magnitude of the range error will remain at an unacceptably high level. The range at which this magnitude is exceeded will depend on the disposition of emitter and sensors but a basic limit to the utility of triangulation for the direct determination of range is implied. This limit can be modified by extending the triangulation baseline or decreasing the bearing measurement error as can be seen from Figure 5 which illustrates the tracking performance of 0.1° bearing accuracy sensors in the triangulation mode.

$$TM = \begin{bmatrix} U/\gamma & 1 - U/\gamma & t \sin(C-B)/\gamma^2 R & -\epsilon \sin (C-B)/\gamma^2 R \\ 0 & 1 & 0 & 0 \\ 0 & 0 & 1 & 0 \\ \epsilon R \sin (C-B)/\gamma & -\epsilon R \sin(C-B)/\gamma & \Delta t\ (\epsilon + \cos(C-B)/\gamma & U \end{bmatrix} \qquad \ldots. (3)$$

Figure 6 illustrates tracking performance of the triangulation system when combined with an update from a radar system at one minute intervals and for comparison Figure 7 shows the performance of a bearing only filter with a radar update at the same rate.

This comparison illustrates a particular track filtering problem - that extra data is only of use if the errors can be modelled accurately. The range errors of Figure 6 are caused by the filter weighting the DF range measurements incorrectly because the 'measured error' is too small.

DISCUSSION

The use of two co-operating sites measuring bearing in order to calculate the range of a moving radar transmitter offers the possibility of direct range determination. It is apparent from an analysis of the performance of the tracking algorithm described in this paper, however, that the geometry of the tracking situation determines the utility of the technique. In particular the enclosed angle Ø must be greater than ~ 3 standard deviations of the bearing error before a reliable track can be formed.

The influence of Ø can be controlled by either the bearing measurement accuracy or by the triangulation baseline. (An examination of the differential range error expression reveals that a doubling of the baseline is equivalent to halving the bearing standard deviation over the majority of triangulation geometries.)

A comparison of Figures 3 and 6 indicates that utilising radar measurements improves the track. The improvement will not, however, be significant if the tracking geometry is such that Ø is always greater than the critical value. The rate of one radar update per minute is still rather high, but a radar system operating in the optimum mode, at this rate would only illuminate a small volume of space for a short time. Ideally the co-operating radar system would be able to be rapidly pointed in the required direction and transmit only when aligned with the target. It would also have a narrower beamwidth than a conventional continuously scanning surveillance radar in order to, again illuminate the minimum volume of space.

In this analysis a regular radar update has been used in order to improve the track quality and the quality of the track formed using the radar information only is worth discussion. For the cases examined the track might well be sufficiently accurate, however, the filtering technique should permit a much lower update rate than the once per minute used in this analysis, while still maintaining an accurate track in the non-transmitting intervals. The bearing measurements are also smoothed and therefore permit a smaller beamwidth for the associated radar than would be the case for unsmoothed bearing data.

The above work has simply analysed the problem of tracking single emitters. In a practical scenario it is likely that many emitters will be simultaneously within the field of view of the DF sensor and the association of the radar and DF data is necessary. The DF system will be able to group a time sequence of data from a particular emitter from its characteristic signature and problems of radar - DF association will only arise for targets in close proximity, ie targets which are separated by less than the beamwidth of the DF system. Three particular cases are envisaged: emitters travelling in consort; simultaneously crossing emitters; and emitters travelling on a radial course separated in range only. For emitters travelling in consort the correct association of radar and DF data is academic as it is the group track which is important. Simultaneously crossing emitters only present problems for the short time when all emitters are within the same beamwidth; simultaneous crossing is also a rare event. For emitters approaching on a radial course problems will occur if targets appear simultaneously at different ranges, however the secondary detector should permit some range differentiation as the emitters will not appear in the same beamwidth except for specific geometries.

CONCLUSIONS

A Kalman filter tracking algorithm has been presented and has been used for the derivation of smoothed position and velocity estimates from simulated bearing measurements received at separated DF sites. The error statistics of the bearing measurement data can be modified and non-manoeuvring tracks have been analysed. The inclusion of radar measurements in order to improve the tracking performance has shown that care must be exercised when combining two sets of range data unless the range data is accurately modelled.

Further work must be carried out in order to improve the estimate of the range error - it has been suggested that a running average of the differential error might be appropriate.

July 1982

REFERENCE

1. Clark B NSWC (USA) Report, 1980.

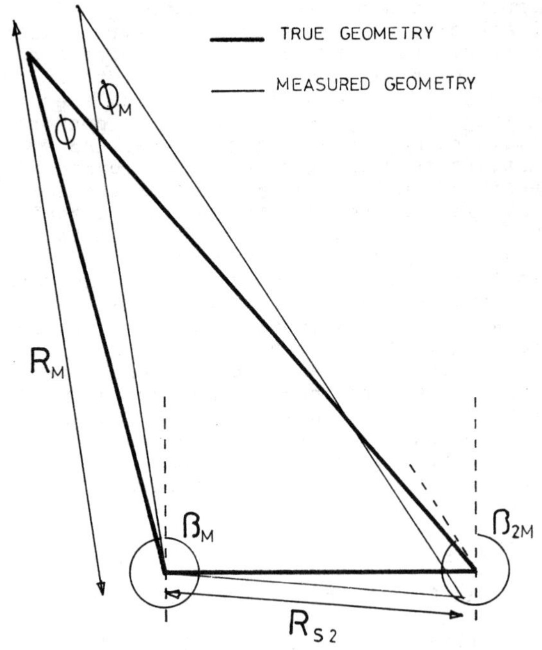

Figure 1 Range calculation geometry

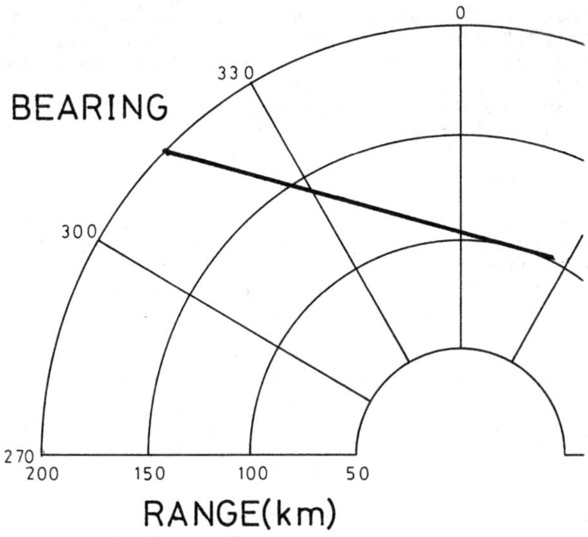

Figure 2 True emitter track

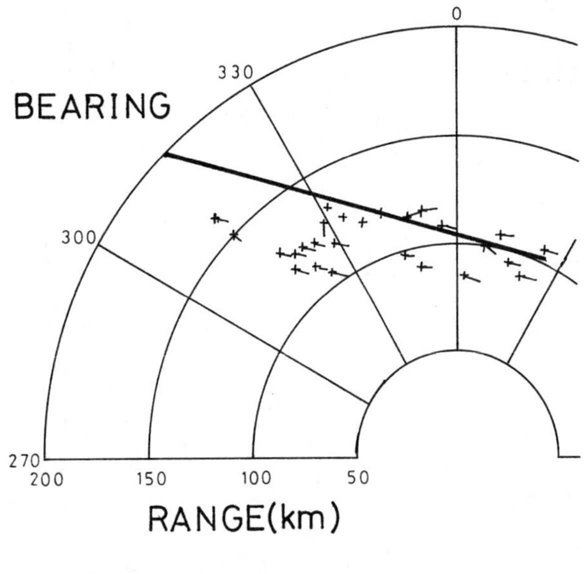

Figure 3 Filtered positions - 1.0° bearing accuracy sensors

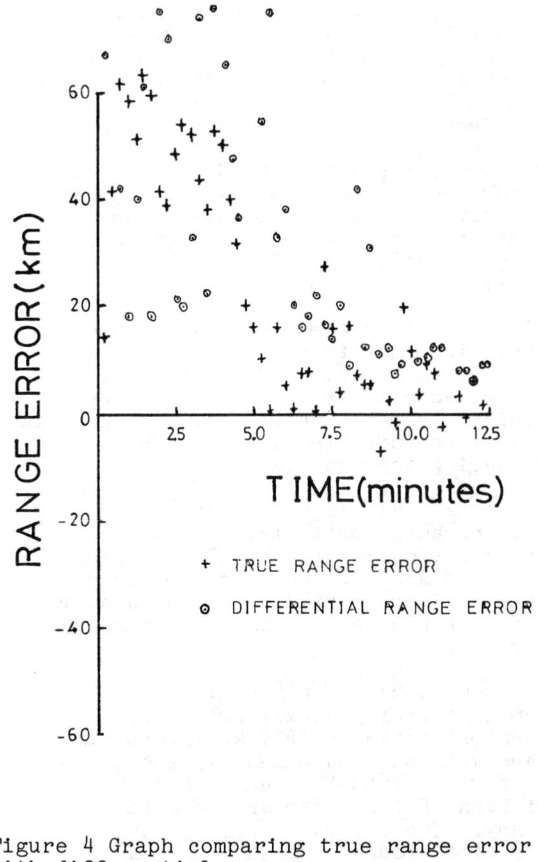

Figure 4 Graph comparing true range error with differential range error

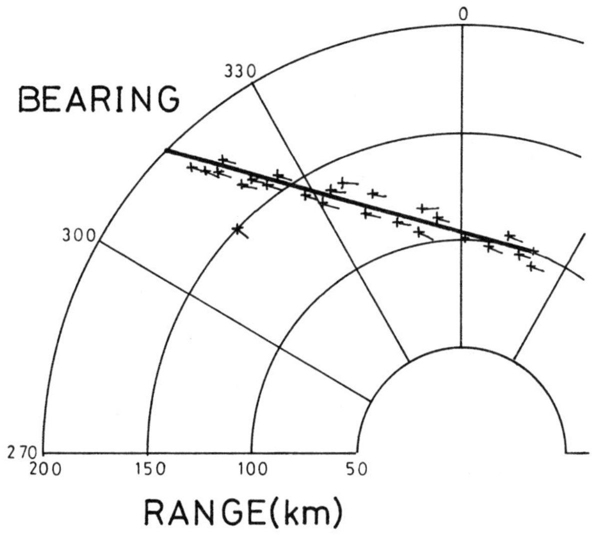

Figure 5 Filtered positions - 0.1° bearing accuracy sensor

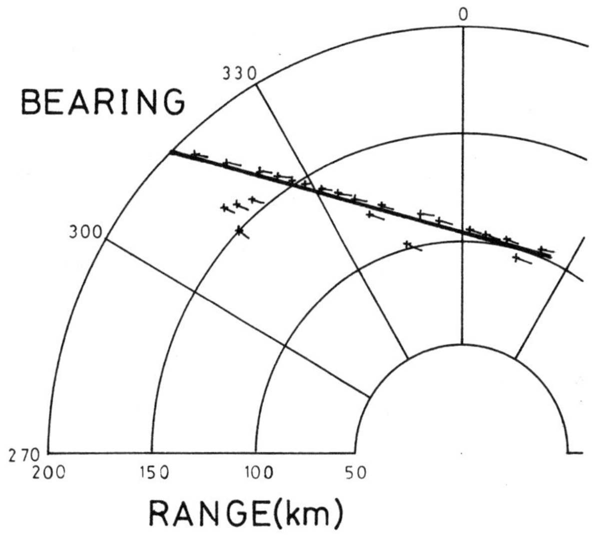

Figure 6 Filtered positions - 1.0° bearing accuracy plus one minute range update

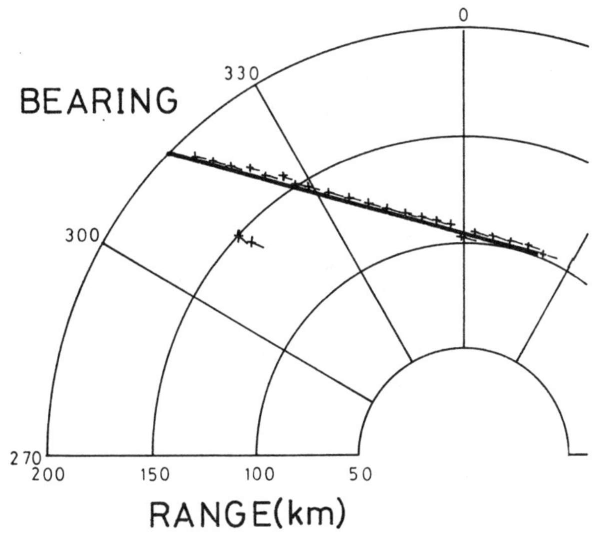

Figure 7 Filtered positions - 1.0° bearing accuracy (bearings only) one minute range update

A FILTERING TECHNIQUE OF PASSIVE RADAR IN HYPERBOLIC COORDINATE SYSTEM

Hungcun Chang, Zhuoying Wang, Yiyen Feng

Department of Radio and Electronic Engineering, Tsinghua University, Peking, PRC

INTRODUCTION

The signals from interfering sources are received by several radars. By crosscorrelation processing the signals of two radars, we can obtain the information of the distance difference between the target and two radars. The trace is a hyperbola. By intersecting two hyperbolic curves the target could be located. If there are several sources, then false location may be occured, but by appropriate pairing, we can eliminate these false targets. In order to increase the locating accuracy, the filtering techniques must be used. In this paper, alpha-beta filter in hyperbolic coordinate system is presented, it is shown that in some cases it is more efficient than conventional method.

THE INTERSECTING ERRORS

The measuring errors of the distance differences may result in the interseting error. Suppose there are three radar stations O,A,B, OA=OB=r. The distance differences between the target P(x,y) and the radar stations O,A or O,B are

$$\Delta_1 = \rho - \rho_A \qquad (1)$$

$$\Delta_2 = \rho - \rho_B \qquad (2)$$

where ρ, ρ_A, ρ_B are distance between target and stations O,A and B respectively. Suppose the measuring errors are $\delta\Delta_1$ and $\delta\Delta_2$ respectivly and the intersecting errors are δx and δy. We have

$$a\delta x + b\delta y = \delta\Delta_1 \qquad (3)$$

$$c\delta x + d\delta y = \delta\Delta_2 \qquad (4)$$

They are called the error intersecting equations. Where

$$a = \frac{x}{\rho} - \frac{x - r\cos\theta_1}{\rho_A} = \frac{\partial\Delta_1}{\partial x} \qquad (5)$$

$$b = \frac{y}{\rho} - \frac{y - r\sin\theta_1}{\rho_A} = \frac{\partial\Delta_1}{\partial y} \qquad (6)$$

$$c = \frac{x}{\rho} - \frac{x - r\cos\theta_2}{\rho_B} = \frac{\partial\Delta_2}{\partial x} \qquad (7)$$

$$d = \frac{y}{\rho} - \frac{y - r\sin\theta_2}{\rho_B} = \frac{\partial\Delta_2}{\partial y} \qquad (8)$$

Suppose

$$Var(\delta\Delta_1) = Var(\delta\Delta_2) = \sigma^2$$

$$Cov(\delta\Delta_1, \delta\Delta_2) = 0$$

Then the variances of the intersecting errors are

$$\sigma_x^2 = Var(\delta x) = f_x\sigma^2 \qquad (9)$$

$$\sigma_y^2 = Var(\delta y) = f_y\sigma^2 \qquad (10)$$

$$Cov(\delta x, \delta y) = f_{xy}\sigma^2$$

where

$$f_x = \frac{b^2 + d^2}{(ad - bc)^2} \qquad (11)$$

$$f_y = \frac{a^2 + c^2}{(ad - bc)^2} \qquad (12)$$

$$f_{xy} = \frac{ad + bc}{(ad - bc)^2}$$

FILTERING

The filtering is necessary in order to increase the locating accuracy. The requirement of the computation of Kalman Filter which can be found in reference (1) is so great that it is difficult to process in real time. But the computation of constant coefficient alpha-beta filter is much less than the former, real time processing can therefore easily realized, and by decreasing the sample period the higher filtering accuracy can be obtained.

Choosing The Filtering Parameters

The filtering equations and pridicting equation of the α-β filter are

$$\hat{x}_k = \hat{x}_{k-1} + \alpha(x_k - \hat{x}_{k/k-1})$$

$$\hat{\dot{x}}_k = \hat{\dot{x}}_{k-1} + \frac{\beta}{T}(x_k - \hat{x}_{k/k-1})$$

$$\hat{x}_{k/k-1} = \hat{x}_{k-1} + T\hat{\dot{x}}_{k-1}$$

Here, T is filtering period, x_k is measured value of the target position at moment kT, $\hat{x}_k$, $\hat{x}_{k-1}$, $\hat{\dot{x}}_k$, $\hat{\dot{x}}_{k-1}$ are estimating values of the position and velocity at the corresponding moments kT. Alpha and beta are constants.

If the input noise sequence is a zero mean, white Gaussian random one, and its variance is σ_i^2, then the steady variances of the random noises on the output are

$$\sigma_{\hat{x}}^2 = K_{\hat{x}}\sigma_i^2 \qquad (13)$$

$$\sigma_{\hat{\dot{x}}}^2 = K_{\hat{\dot{x}}}\sigma_i^2 \qquad (14)$$

$$\sigma_p^2 = K_p\sigma_i^2 \qquad (15)$$

where, $\sigma_{\hat{x}}^2$, $\sigma_{\hat{\dot{x}}}^2$. σ_p^2 represent the random noise variances of the estimates for the position, velocity and the prediction of position respectivly, and $K_{\hat{x}}$. $K_{\hat{\dot{x}}}$. Kp are the compressing coefficients of noise variance. They are

$$K_{\hat{x}} = \frac{2\beta - 3\alpha\beta + 2\alpha^2}{\alpha(4 - 2\alpha - \beta)} \qquad (16)$$

$$K_{\hat{\dot{x}}} = \frac{2\beta^2}{T^2\alpha(4 - 2\alpha - \beta)} \qquad (17)$$

$$K_p = \frac{2\beta + \alpha\beta + 2\alpha^2}{\alpha(4 - 2\alpha - \beta)} \qquad (18)$$

when the system enters in steady state, the position and velocity of the target will not result in any prediction error, the acceleration of the target will however cause prediction error, it can be found in reference (2)

$$\mathcal{E}_a(kT) = \frac{T^2}{\beta}a(kT) \qquad (19)$$

where a(kT) is the acceleration at moment kT. We find that the output error of the predicting filter consists in mainly the random error and the steady error by the acceleration. The reasonable rule, by means of which the filtering parameters are chosen, should be to minimize σ_s^2, the sum of the variances of the two kind of errors mentioned above.

$$\sigma_s^2 = \sigma_p^2 + \frac{T^4}{\beta^2} a^2(KT) \qquad (20)$$

In general, $0<\alpha\ll 1$, $0<\beta\ll 1$. From eqs. (13)-(20) it follows that

$$\beta \doteq \alpha^2 \qquad (21a)$$

$$\alpha \doteq \left(\frac{4T^4a^2}{\sigma_i^2}\right)^{\frac{1}{5}} \qquad (21b)$$

and

$$K_{\hat{x}} \doteq \alpha \doteq K_p \qquad (22)$$

$$K_{\hat{\dot{x}}} \doteq \frac{\alpha^3}{2T^2} \qquad (23)$$

σ_i^2, the variance of the measuring noise, may be determined according to a priori knowledge and a may be chosen as the maximum acceleration of the target. From eqs. (21) and (22) we see that the smaller T, the smaller α, and the higher the filtering accuracy. Accounting of the error caused by acceleration the equivalent coefficient in the prediction filter is

$$K_S = \frac{\sigma_s^2}{\sigma_i^2} = K_{\hat{x}} + \frac{T^4a^2}{\beta^2\sigma_i^2} \doteq 1.25\alpha \qquad (24)$$

when the acceleration is maximum, the error variance which results from it is a quarter of the error variance of the output random noise. The results have been verified with computer simulation.

The Filtering In Hyperbolic Coordinate System

When the filtering is performed in the rectangular coordinate, it is neccessary to intersect the target in every filtering period. The computation of the intersecting is much greater than that of the filtering itself. The speed of processing data is therefore greatly reduced. In fact, we find that it is unnecessary to know the state of the target in the rectangular coordinate in every filtering period when the period is very small. Therefore, it is possible to filter the time (or distance) difference signal, the target is then intersected at intervals of several filtering periods and not every period. Filtering is thus performed in parallel with intersecting, in consequence the processing speed is greatly increased. In addition, the sampling period can be decreased to increase filtering accuracy.

Relationship between the motion patterns in two coordinate systems. Differentiating the eqs. (1) and (2) with respect to time, we have

$$\dot{\Delta}_1 = a\dot{x} + b\dot{y} \qquad (25)$$

$$\dot{\Delta}_2 = c\dot{x} + d\dot{y} \qquad (26)$$

Here a, b, c, d can be computed from eqs. (5)-(8). Solving eqs. (25) and (26) it follows that

$$\dot{x} = \frac{d\dot{\Delta}_1 - b\dot{\Delta}_2}{ad-bc} \qquad (27)$$

$$\dot{y} = \frac{a\dot{\Delta}_2 - c\dot{\Delta}_1}{ad-bc} \qquad (28)$$

Using the above eqs., $\hat{X}$, $\hat{Y}$. $\hat{\dot{X}}$. $\hat{\dot{Y}}$ can be computed from $\hat{\Delta}_1$. $\hat{\Delta}_2$. $\hat{\dot{\Delta}}_1$. $\hat{\dot{\Delta}}_2$. Differentiating equations (25) and (26) again we have

$$\ddot{\Delta}_1 = a\ddot{x} + b\ddot{y} + \frac{\partial^2\Delta_1}{\partial x^2}\dot{x}^2 + \frac{\partial^2\Delta_1}{\partial y^2}\dot{y}^2 + 2\frac{\partial^2\Delta_1}{\partial x\partial y}\dot{x}\dot{y} \qquad (29)$$

$$\ddot{\Delta}_2 = c\ddot{x} + d\ddot{y} + \frac{\partial^2\Delta_2}{\partial x^2}\dot{x}^2 + \frac{\partial^2\Delta_2}{\partial y^2}\dot{y}^2 + 2\frac{\partial^2\Delta_2}{\partial x\partial y}\dot{x}\dot{y} \qquad (30)$$

where

$$\frac{\partial^2\Delta_1}{\partial x^2} = \frac{\rho^2-x^2}{\rho^3} - \frac{\rho_A^2-(x-r\cos\theta_1)^2}{\rho_A^3} \qquad (31)$$

$$\frac{\partial^2\Delta_1}{\partial y^2} = \frac{\rho^2-y^2}{\rho^3} - \frac{\rho_A^2-(y-r\sin\theta_1)^2}{\rho_A^3} \qquad (32)$$

$$\frac{\partial^2\Delta_1}{\partial x\partial y} = -\frac{xy}{\rho^3} + \frac{(x-r\cos\theta_1)(y-r\sin\theta_1)}{\rho_A^3} \qquad (33)$$

$$\frac{\partial^2\Delta_2}{\partial x^2} = \frac{\rho^2-x^2}{\rho^3} - \frac{\rho_B^2-(x-r\cos\theta_2)^2}{\rho_B^3} \qquad (34)$$

$$\frac{\partial^2\Delta_2}{\partial y^2} = \frac{\rho^2-y^2}{\rho^3} - \frac{\rho_B^2-(y-r\sin\theta_2)^2}{\rho_B^3} \qquad (35)$$

$$\frac{\partial^2\Delta_2}{\partial x\partial y} = -\frac{xy}{\rho^3} + \frac{(x-r\cos\theta_2)(y-r\sin\theta_2)}{\rho_B^3} \qquad (36)$$

From eqs. (29) and (30) we find that the acceleration in hyperbolic coordinate consists of two parts. One results from the accelerations of the target in rectangular coordinate:

$$\ddot{\Delta}_{1a} = a\ddot{x} + b\ddot{y} \qquad (37)$$

$$\ddot{\Delta}_{2a} = c\ddot{x} + d\ddot{y} \qquad (38)$$

The other from velocity:

$$\ddot{\Delta}_{1v} = \frac{\partial^2\Delta_1}{\partial x^2}\dot{x}^2 + \frac{\partial^2\Delta_1}{\partial y^2}\dot{y}^2 + 2\frac{\partial^2\Delta_1}{\partial x\partial y}\dot{x}\dot{y} \qquad (39)$$

$$\ddot{\Delta}_{2v} = \frac{\partial^2\Delta_2}{\partial x^2}\dot{x}^2 + \frac{\partial^2\Delta_2}{\partial y^2}\dot{y}^2 + 2\frac{\partial^2\Delta_2}{\partial x\partial y}\dot{x}\dot{y} \qquad (40)$$

It can be shown that the second order partial derivatives in the eqs. (39) and (40) are much less than the coefficients in the eqs. (37) and (38)(i.e. a. b. c. d.) it may be seen that $\ddot{\Delta}_1$ and $\ddot{\Delta}_2$ result mainly from $\ddot{X}$ and $\ddot{Y}$, and the effect of $\dot{X}$ and $\dot{Y}$ is so small that it can be neglected. As a consequence,

$$\ddot{\Delta}_1 = \ddot{\Delta}_{1a} + \ddot{\Delta}_{1v} \doteq \ddot{\Delta}_{1a} \qquad (41)$$

$$\ddot{\Delta}_2 = \ddot{\Delta}_{2a} + \ddot{\Delta}_{2v} \doteq \ddot{\Delta}_{2a} \qquad (42)$$

We can say accordingly that the uniform motion in rectangular coordinate system corresponds to approximately uniform motion in hypebolic one, particularly at the farther range.

Relationship Between The Filtering Effects In Two coordinate Systems.

Random error: Let the noise variance in measuring the distance difference is σ_Δ^2. The filtering effects in the rectangular coordinate system can be obtained from eqs. (13) (14) and (9) (10)

$$\sigma_{\hat{x}}^2 = K_{\hat{x}}\,\sigma_x^2 = K_{\hat{x}}\,f_x\,\sigma_\Delta^2 \qquad (43)$$

$$\sigma_{\hat{y}}^2 = K_{\hat{y}}\,\sigma_y^2 = K_{\hat{y}}\,f_y\,\sigma_\Delta^2 \qquad (44)$$

When the target is intersected after filtering in the hyperbolic coordinate system, it can be obtained from eqs. (9) (10) and (13) (14) that

$$\sigma_{\hat{x}}^2 = f_x\,\sigma_{\hat{\Delta}}^2 = f_x\,K_{\hat{\Delta}}\,\sigma_\Delta^2 \qquad (45)$$

$$\sigma_{\hat{y}}^2 = f_y\,\sigma_{\hat{\Delta}}^2 = f_y\,K_{\hat{\Delta}}\,\sigma_\Delta^2 \qquad (46)$$

With the same filtering parametes α and β, we have $K_{\hat{x}} = K_{\hat{y}} = K_{\hat{\Delta}} \doteq \alpha$. By comparing eqs. (43) and (44) to eqs. (45) and (46) we find that, as concerns the filtering random noise, the effects of the two methods are all the same.

The steady error caused by accelaration of the target: When filtering in hyperbolic coordinate system, the prediction errors caused by $\ddot{\Delta}_{1a}$ and $\ddot{\Delta}_{2a}$ can be obtained from eq. (19)

$$\varepsilon_{\Delta 1a} = \frac{T^2}{\beta} \ddot{\Delta}_{1a} \quad (47)$$

$$\varepsilon_{\Delta 2a} = \frac{T^2}{\beta} \ddot{\Delta}_{2a} \quad (48)$$

Substituting eqs. (47) and (48) into eqs. (3) and (4), we have

$$a\varepsilon_{xa} + b\,\varepsilon_{ya} = \varepsilon_{\Delta 1a} \quad (49)$$

$$c\,\varepsilon_{xa} + d\,\varepsilon_{ya} = \varepsilon_{\Delta 2a} \quad (50)$$

From eqs. (37). (38) and (47)-(50), it follows that

$$\varepsilon_{xa} = \frac{T^2}{\beta} \ddot{x} \quad (51)$$

$$\varepsilon_{ya} = \frac{T^2}{\beta} \ddot{y} \quad (52)$$

By comparing to eq. (19), we find that the steady errors caused by acceleration are just the same when filtering in these two different coordinate systems.

Error caused by the velocity of target: When filtering is performed in the rectangular coordinate system, the velocity of target does not result in any prediction error. But when filtering in the hyperbolic one, because the coordinate is uneven, the velocities $\dot{x}$ and $\dot{y}$ will thus result in the accelerations $\ddot{\Delta}_{1v}$ and $\ddot{\Delta}_{2v}$ in the hyperbolic coordinate, consequently result in the prediction errors:

$$\varepsilon_{\Delta 1v} = \frac{T^2}{\beta} \ddot{\Delta}_{1v} \quad (53)$$

$$\varepsilon_{\Delta 2v} = \frac{T^2}{\beta} \ddot{\Delta}_{2v} \quad (54)$$

Substituting them into eqs. (3) and (4) we have

$$a\varepsilon_{xv} + b\,\varepsilon_{yv} = \varepsilon_{\Delta 1v} \quad (55)$$

$$c\,\varepsilon_{xv} + d\,\varepsilon_{yv} = \varepsilon_{\Delta 2v} \quad (56)$$

Solving them we have

$$\varepsilon_{xv} = \frac{T^2}{\beta} \cdot \frac{d\ddot{\Delta}_{1v} - b\ddot{\Delta}_{2v}}{ad - bc} \quad (57)$$

$$\varepsilon_{yv} = \frac{T^2}{\beta} \cdot \frac{a\ddot{\Delta}_{2v} - c\ddot{\Delta}_{1v}}{ad - bc} \quad (58)$$

Because $\ddot{\Delta}_{1v}$ and $\ddot{\Delta}_{2v}$ are very small, the above errors are also small. The practical calculations have shown that they are smaller than the output random error for about two orders in magnitude. Therefore they may be neglected. The filtering effects in these two different coordinate systems are well closed. But the filtering method in the hyperbolic coordinate system can greatly increase the processing speed.

The above conclusions have been verified with the computer simulation. The filtering parameters are determined as $\alpha = 0.00757858$, $\beta = \alpha^2$, based on T = 0.01s, $\sigma_x^2 = (2\text{Km})^2$, a = 0.05Km/s². We suppose that the motion pattern of the target is $x_0 = y_0 = 300$Km $\dot{x}_0 = -0.1$Km/s, $\ddot{x} = -0.03$Km/s², $\dot{y}_0 = 0$, $\ddot{y} = 0$, the measured error variance of distance difference is $\sigma_\Delta^2 = (0.1\text{Km})^2$. Theoretical calculation shows that $K_{\hat{x}} = \sigma_{\hat{x}}^2 / \sigma_x^2 \doteq 0.00813$, $K_{\hat{y}} \doteq \alpha$, and the results of computer simulation are about 0.008 and 0.0075 respectively, whether in the rectangular coordinate system. or in the hyperbolic coordinate system. They are well closed to the previous theoretical analysis.

ACKNOWLEDGMENT

The authors wish to express our great indebtedness to Professor Wu You-Show, Lu Da-Jin, Mao Yu-Hai and Dr, Feng Shi-Zhang for their help.

REFERENCES

1. D. WILLNER, 1976, "Kalman filter configuration for multiple radar systems", AD-A026367.

2. Zhuoying Wang, 1977, "Synthesis and design about the Tracking loop in ranger".

DEGHOSTING IN AN AUTOMATIC TRIANGULATION SYSTEM

G. van Keuk

FGAN-Forschungsinstitut für Funk und Mathematik, Federal Republic of Germany

INTRODUCTION

With the advent of powerful computers for real-time real-data processing application automatic target tracking has become an important component in todays sensor-systems. In more general terms: techniques to integrate sensor outputs are used to improve the capability of mono-sensor-systems. Of particular importance, however, are methods for integrating distributed sensors in order to form a netted system. The aim is to process the individual sensor information to achieve more accurate, more reliable data of higher confidence. Excellent overview articles are available /1/. Data processing can be seen as a natural continuation of signal-processing work leading to a new level of integration. At this level we again, for instance, meet the conception of detection and false alarm probability, well known at the signal-processing level. But they now refer to the tracks formed from the output data of the sensors and not that much to the plots themselves.

Considering sensor nets we can generalize the concept of measurements (plots with radars) to include bearings also. Then automatic triangulation /2/ is the basic technique used to extract a complete air picture by data-processing means. If additional information is available for instance particular signal signatures, or precise time of arrival information, then the triangulation procedure can be generalized or replaced by other techniques. This, however, is out of the scope of this contribution.

As an example we confine on a multiple target (jammer) scene observed by a distributed net of conventional rotating passive radars measuring bearings only /2/. This contribution reports on a research study which is carried out at the FFM institute by analysis and computer simulation. To shorten this paper we restrict on a two-dimensional world and neglect all the screening (land masses) effects.

Typical problems which have been addressed are:
- trackers to process asynchronous bearings - only information from a net of mobile sensors /3/,
- bearings to track association (correlation) techniques /4/,
- deghosting techniques in redundant radar nets (at least three sites have contact to the targets in the common surveillance area) /5/,
- deghosting techniques in non-redundant nets /6/.

In triangulation systems ghosts accur from formal combination of bearings measurements in a sensor net. Deghosting means any process to automatically remove the ambiguities by analysis of the pattern of all the tentative tracks produced by processing the bearings. It sometimes is necessary to cancel many ghosts until the hidden targets can be detected. In a multiple target situation the average number of ghosts increases proportional to the square of the number of targets and sensor sites. So the deghosting process is extremly important to avoid targets to be hidden among their ghosts.

We here report on some general results on the last item only, without touching the questions of track-production and any tracking strategies. Techniques for deghosting as they are developed for redundant nets do not solve the problem in non-redundant ones. On the other hand: large nets temporarily may be broken up into smaller subnets. So the problem of the last item is important. Due to the complex geometrical description of an N target-2-sites environment, analytical results often are not realizable. Therefore some of our results are taken from Monte-Carlo simulations.

GEOMETRICAL CONSIDERATIONS ON GHOSTS

The following simple sketch shows two targets A, B observed by two sites 1,2.

Sketch 1

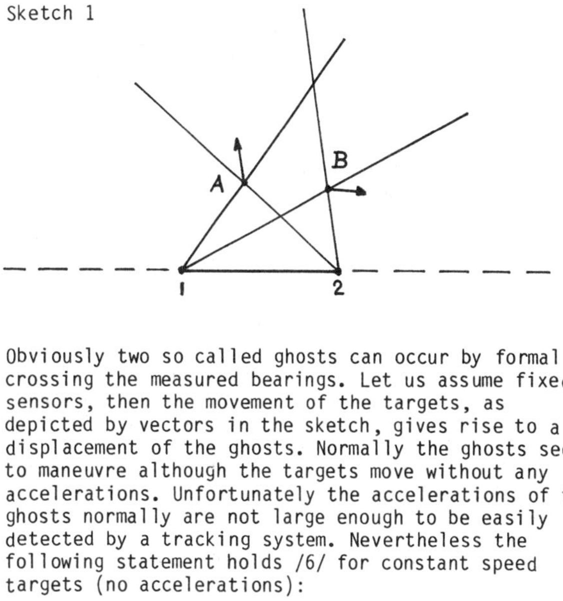

Obviously two so called ghosts can occur by formal crossing the measured bearings. Let us assume fixed sensors, then the movement of the targets, as depicted by vectors in the sketch, gives rise to a displacement of the ghosts. Normally the ghosts seem to maneuvre although the targets move without any accelerations. Unfortunately the accelerations of the ghosts normally are not large enough to be easily detected by a tracking system. Nevertheless the following statement holds /6/ for constant speed targets (no accelerations):

The ghosts move without any accelerations <u>iff</u> both the target speed vectors are parallel (including the sign) and
a) their velocities are proportional to the distance to the basisline B and they fly parallel to B, or
b) both the extrapolated target paths cross B at the same time.

Basing on this it can be reasonable to separate all the tentative tracks (including ghosts and true targets) into two classes depending on their estimated acceleration to be zero or not. If then the subset of constant speed targets is complete in the sense that the rest of tracks can be explained as their ghosts, we have detected the hidden targets among the ambiguities. This strategy of course can be deceived by co-operatively maneuvering targets because the roles of targets and ghosts can be exchanged. But this seems to be a purely academic case. Concluding from the above listed statement the probability of the strategy to be applicable will generally be large. We, however, learnt from our studies that the observation time necessary for a secure decision generally is very long due to the comparatively long time-basis needed for the separation into the above mentioned classes.

There are situations of two targets observed by two sites without any produced ghosts if the targets are separated by the basis-line. For simple geometrical reasons sometimes only one ghost exists and deghosting is possible by just detecting this pattern.

Sketch 2

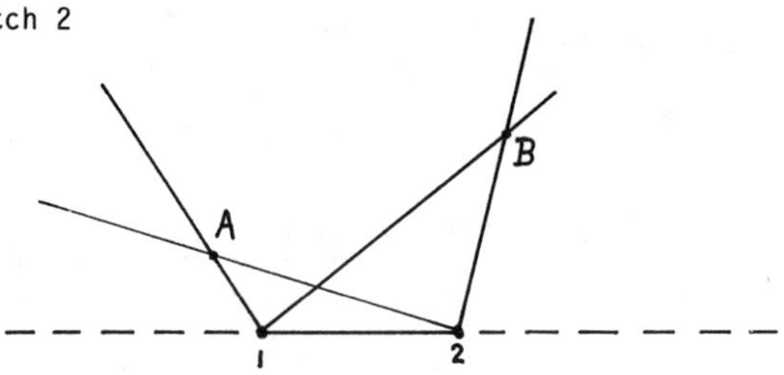

As depicted in the sketch the bearings are forming a W-like pattern. There exist bearings that can only be associated to exactly one tentative track. From this a deghosting strategy can be developed that holds for high detection and low false alarm probability. In addition the method can be generalized to be applicable in multiple target situations too /6/. We called this W-strategy.

DEGHOSTING BY ANALYSIS OF THE TARGET KINEMATICS

If the track production process leads to a number of tentative tracks an analysis at the estimated parameters

R (distance)
V (scalar speed)
Q (scalar acceleration)

can be helpful to detect ghosts. Obviously we have to check a constistency by proving the set of inequalities

$$R > R_0 \;\; \underline{or} \;\; V > V_0 \;\; \underline{or} \;\; Q > Q_0$$

R_0, V_0, Q_0 being reasonable limits derived from a priori knowledge on the sensor-range and the target capabilities. To avoid wrong decisions the estimation accuracy of the R, V, Q quantities have to be taken into account. This simply leads to increased limits R_0, V_0, Q_0 depending on the status of the tracks and the geometry of the whole configuration.

Considering ghost-tracks the quantities R, V, Q show a pronounced dependance. Ghosts which are far from the sensors are likely to show high velocities and strong accelerations. We followed from our study that a test on V is powerful enough.

If one of the ghosts as depicted in sketch 1 can be identified, the solution is unique and the tentative tracks A and B become established ones. They now represent the true targets. Particular flags are set in the corresponding segments of the track file to be taken along with the individual tracks.

It should be emphasized that this technique is of little power in multiple target environments, due to the comparatively large number of ghosts that have to be identified until a decision on the hidden objects becomes possible. Nevertheless this simple strategy (V strategy) should be applied. A combination of both the strategies W and V is easy as shall be explained along with another method.

TRACKING

Deghosting is part of the whole tracking procedure. If all the ambiguities have been solved the tracking reduces to a much simpler task of conducting established tracks by association and processing of bearings information. New targets entering the area of the triangulation system normally will not give rise to any new ambiguities, given that the new targets appear sufficiently separated in time.

Any error in deghosting that leads to wrong established tracks gives rise to consecutive wrong decisions and can have dramatic consequences on further deghosting and tracking. This principally is known in any tracking application but of primary importance here.

We here shall not consider the track-formation problem, but look onto the deghosting techniques from a more static point of view. This principally is correct as long as the interference between both the levels is not too tight. Nevertheless our numerical results represent upper limits of the deghosting probability.

THE DV-STRATEGY

One single strategy will not be powerful enough for deghosting, due to the fact that the deghosting capabilities strongly depend on the particular target-sensor geometry. We propose another technique again based on the pattern of all the tentative tracks. The new strategy considers the estimated scalar speeds and therefore can only be applied after the track formation process has been settled to give accurate speed estimates. We emphasize that it is not necessary to have precise estimates for all the tentative tracks due to the strong statistical dependence among them.

The idea is as follows: If all the N targets would move at the same but unknown speed V (possibly in independent directions) the estimated speeds v_i of all the conducted up to $N^2 - N$ tentative tracks generally differ from V. More precisely: if we think of N targets stochastically (uniformly and independently) distributed over an observation area as depicted in the following sketch

Sketch 3

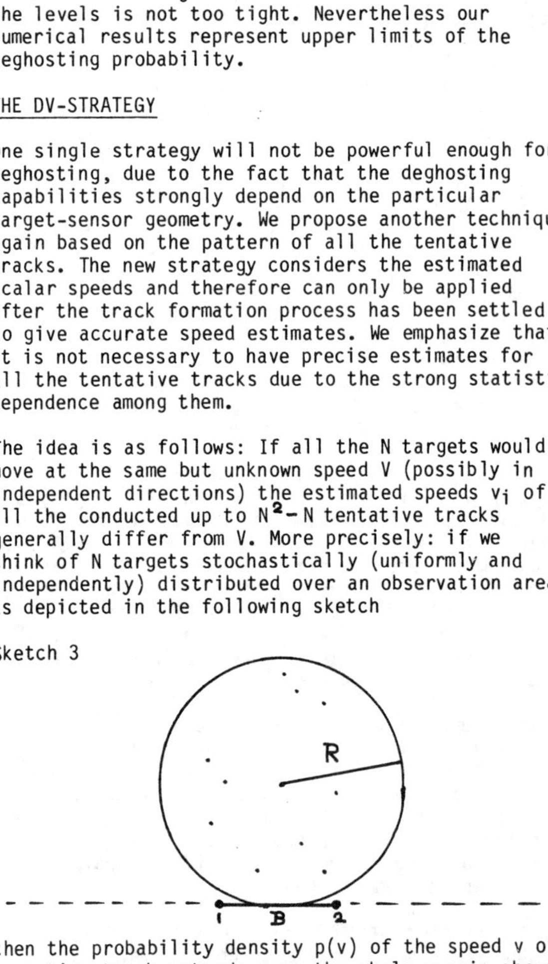

then the probability density p(v) of the speed v of a tentative track extends over the whole v-axis showing a peak around the exact V value. This pronounced peak structure is used to derive a non-parametric test for deghosting. Normally, however, the target speeds differ from each others. Let us assume a uniform distribution of the independent stochastic (constant) target speeds in the interval $V \pm DV$. As a consequence of this assumption the density p(v) is affected: the variance of v increases and the peak structure disappears along with increasing velocity spread DV. But the contrast between the "peaked-area", corresponding to the true targets, compared to the background of all the ghost-speeds remains significant enough to serve for deghosting. In cases of non negligible target spread DV a large fraction of ghost-speeds can be found in the "peaked-area" too.

The testprocedure /6/ will not be described in detail. All the up to N^2 tentative tracks formed from the N bearings observed by the two sites 1 and 2 (sketch 3) can be arranged in a NxN matrix showing the estimated speeds v_{ij} as elements. Some of them of course have to be particularly marked to represent no track. This will tell us whether a corresponding track could not been built due to geometrical reasons, or has been cancelled by the track-formation process or by applying the V-strategy.

Depending on the particular structure of the matrix we generally are faced with up to N! possible selections of candidates by looking for the hidden target-tracks. Each admissible selection can be represented as a sequence of corresponding matrix elements or as a N-vector of speeds $v_1, v_2, \ldots, v_N$. Consequently each candidate set allows to calculate the corresponding v-spread from the maximum and the minimum of the v_i. We look for a selection of minimum v-spread. Therefore it is not necessary to look up all possible selections.

Generally the set of minimal-solutions consists of more than one selection. The set will not be empty if the tracks corresponding to the true targets have not been lost.

The following sketch gives an example of a 4 targets situation in arbitrary speed units.

Sketch 4

	1	2	3	4
1	3	14	27	x
2	25	5	7	18
3	x	8	9	100
4	32	13	91	11

Two solutions (just represented by the matrix elements) give minimum v-spread: 3, 5, 9, 11 and 3, 8, 7, 11. If the solution is not unique we look for the common intersection to avoid fuzzy decisions. In this example 3 and 11 belong to the intersection. The corresponding tracks will then be established given that a particular measure of significance is fulfilled /6/. This measure depends on the observed v-spread itself and on the selection nearest to the minimum. By this measure the probability of wrong decisions can be controlled, but it will not be detailed here.

It should be emphasized that neither V nor DV is an input-parameter of the test. In addition the geometry given by R and B again is not part of the decision procedure, but they enter through the significance measure.

NUMERICAL RESULTS

To derive average deghosting probabilities we took a simulation approach. The N targets have been assumed to be uniformly distributed as explained in context with sketch 3. Parameters of the simulation have been R, B, V, DV and N apart from the above mentioned measure of significance. This parameter has been tuned to limit the probability of having a ghost among the established tracks below about one percent.

The speed estimates of course have to be derived from the trackprocedure. Noisy bearings generally will give noisy speed estimates. To incorporate this effect we corrupted all the bearings by adding pseudorandom $N(0,\sigma)$ distributed noise. The speed then has been estimated over a time interval of 40 sec assuming 5 to 10 sec rotating radars. It has been assumed that the targets undergo only weak maneuvres.

The W-strategy turned out to be insensitiv against small disturbances of $\sigma < 0.5°$ as long as the tracking process itself is not too much degraded. The V-strategy requires an increased limit V_0 depending on the achieved accuracy of speed estimates as derived from the tracking filter.

The DV-strategy, however, is sensitive against noise effects since the contrast between speeds of targets and ghosts becomes blurred and the testing becomes more difficult. Errors in speed-estimates are additiv. Therefore noise effects are of importance especially for slow targets (small V).

The figures 1 to 4 show the deghosting probabilities depending on the ratio of the triangulation basis B over the radius R of the observation area. The curves represent three different strategies as has been explained. The dashed curve demonstrates the influence of bearing noise on the DV-strategy. It should be considered that the DV-strategy is applied as the last deghosting technique, that means it is working on a pattern that could not be solved by the two preceding ones. Of course the deghosting capability is reduced if the target number N increases and seems to be more powerful if the B/R ratio remains small. Therefore the sensors should not be separated too far apart from each others. But we have to consider that the probability of track-formation has not been introduced. This probability vanishes due to nearly parallel bearings which often occur with small B/R values. So the numerical results for B/R values near or below 0.1 should be interpreted carefully.

The deghosting capability of the proposed methods of course depends severely on the geometrical assumptions. If the sensors would have been assumed to be localized more symmetrically near the origin of the circular observation area, the deghosting would have been improved considerably. This approximately doubles the number of targets to be deghosted at a given probability compared to our environment.

The last figure demonstrates the dependence of the deghosting probability on the assumed target's spread in speed (DV). As can be seen the power of the technique generally is decreased with increasing DV. This effect can easily be understood. But in the case of bearing noise small DV values will not considerably improve the deghosting due to unavoidable errors in the speed estimates.

SUMMARY

Deghosting in a two-site triangulation system, processing bearing information only, is possible but difficult. Precise measurements of high resolution are necessary. It is recommended to have the sites not too far apart from each others. The deghosting should be carried out basing on the whole pattern of all the tentative tracks.

The numerical results presented do not yet take the track-formation process into account. It has been reported on the TRIAS project which is going on at the FFM institute.

REFERENCES

1. A. Farina, S. Pardini, Survey of radar data-processing techniques in air traffic-control and surveillance systems, IEE Proc. 127, 6, 1980.

2. F.J. Berle, Multi-Sensor Concept for Future AADGE, Shape Technical Center, SN 17, 1978

3. G. v. Keuk, Extended Kalman Filter for Triangulation in a variable Multi-Sensor System, Regelungstechnik 10, 1978

4. G. v. Keuk, A method to correlate bearings in a computer aided multiple target triangulation system, FFM report 307, 2, 1981.

5. G. v. Keuk, Computer aided deghosting in a redundant triangulation system, FFM-report 294, 3, 1980

6. G. v. Keuk, Deghosting in non-redundant triangulation systems, FFM report 311, 12, 1981.

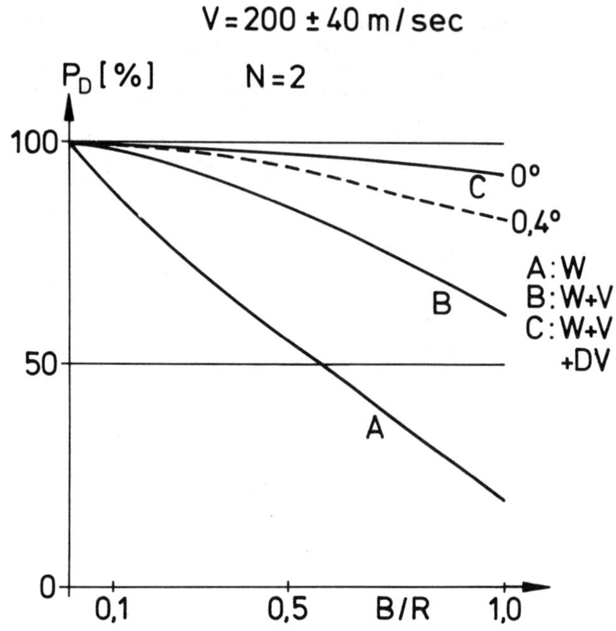

Fig. 1 Deghosting probability

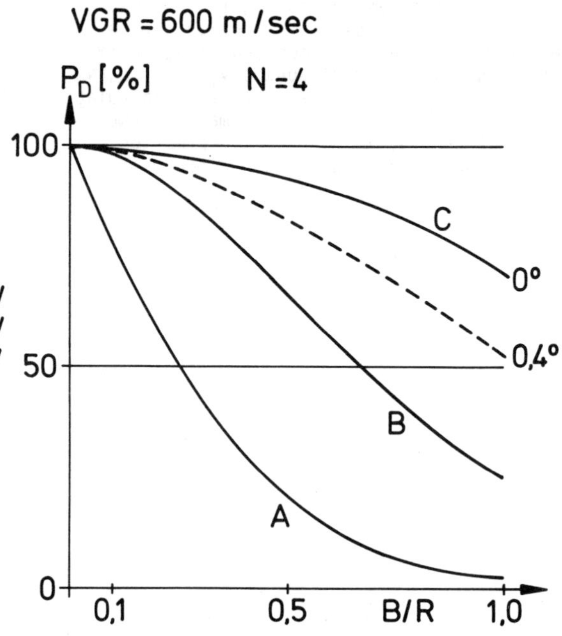

Fig. 2 Deghosting probability

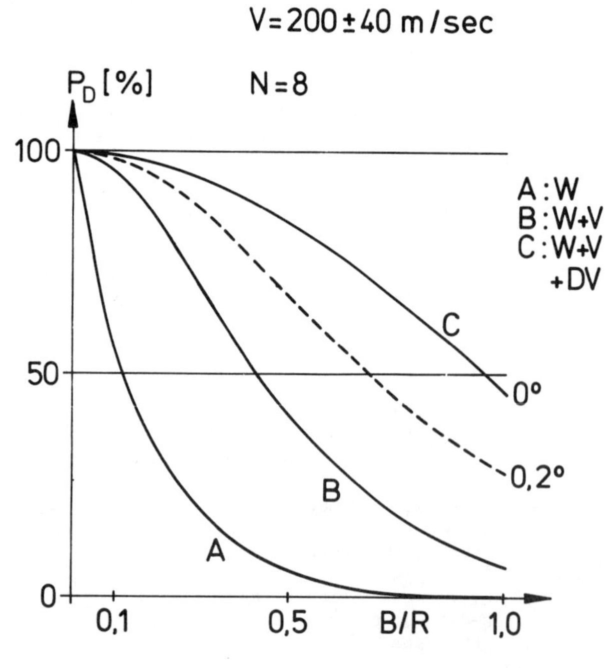

Fig. 3 Deghosting probability

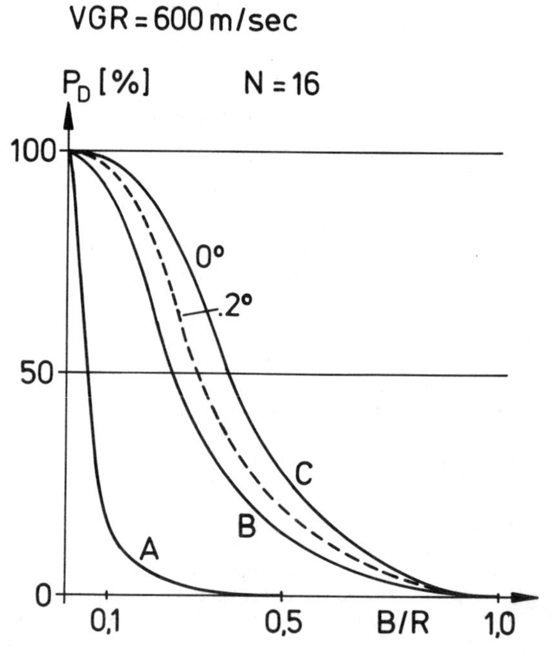

Fig. 4 Deghosting probability

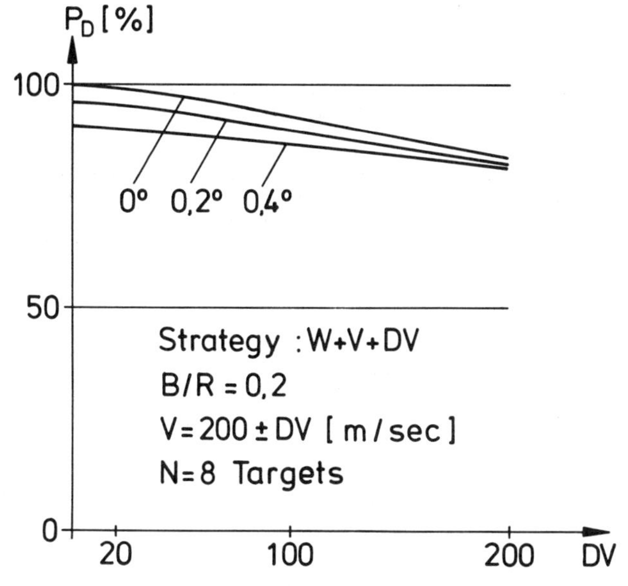

Fig. 5 Influence of speed spread on the deghosting probability

THE MULTIRADAR TRACKING IN THE ATC SYSTEM OF THE ROME FIR

G. Barale – G. Fraschetti – S. Pardini

Selenia S.p.A., I

INTRODUCTION

This paper deals with the multiradar tracking developed for the air-traffic-control system (ATCAS) of the Rome Flight Information Region (FIR).
The multiradar tracking (MRT) is, by definition, the target tracking performed by using measurements from two or more radars having overlapping coverages. Multiradar tracking systems are more and more extensively used in air traffic control and air defence applications due to their improved performance with respect to monoradar systems.

The most significant advantages of the MRT in comparison with a monoradar system are:

- continuity in the target tracking over an area wider than the coverage of each sensor;
- higher accuracy of the estimated track due to overall higher data rate;
- lower vulnerability to clutter and/or jamming effect because of the different geometry of the radar;
- gracefull degradation due to reconfiguration capability in case of failure of one or more radars.

Obviously these improvements in the performance imply some additional problems or difficulties, the most relevant of which are briefly resumed in the following:

- due to cooperating radars, the detections relevant to each target have variable accuracy and an apparent random data rate; as a consequence the simple filtering algorithm used for monoradar system should be modified;
- the position measurements should be referred to a single coordinate reference system;
- coordinate transformation from local site to a common centre should take into account the target altitude and the curvature of the earth;
- polar measurements performed by different radar have to be carefully aligned to geographic north so as to guarantee spatial congruence of plots concerning the same target;
- unlike the monoradar case, the data organization and the timing of the system cannot be derived from the scan rate.

CLASSIFICATION OF MRT SYSTEMS

Different types of MRT may be envisaged according to the spatial deployment of the cooperating radars and the integration level of information. A detailed analysis can be found in Farina and Pardini (1).

With respect to the spatial deployment the MRT may be classified as:

- colocated
- non-colocated

The former class is pertaining to applications having the antennas located in a rather limited area, as in shipboard case. In the latter class the distances between the different antennas are appreciable, as in ground based systems for air-traffic-control and air defence applications.

Depending on the degree of integration processing being used, MRT can be classified as:

- distributed
- centralized.

Block diagrams of these two classes of processing architectures are schematically shown in Figures 1 and 2.

The peculiar characteristic of the distributed architecture is the use of a computer in each site performing a monoradar tracking. The individual tracks are then sent to a common data processing centre, which combines them in order to determine a single multiradar track for each target.

The main characteristic of the centralized architecture is the use of a single data processing system to wich radar plots, instead of tracks, are transmitted from the sites. These measurement are processed to estimate a single track for each target.

The use of a MRT with distributed architectures implies the following items: prevalent use of monoradar type processing, capability of local operation, requirements of computing resources of limited power. For a MRT with centralized architecture the following items may be emphasized: full exploitation of the overall higher scan rate, requirement of processing resources of adequate power. A basic parameter to be considered in the selection of the architecture and in the comparison of the performance is the degree of the overlapping of the radar coverages. Obviously the performance of the two architectures tend to converge in the case of limited area of overlapping.

DESCRIPTION OF THE ATCAS MRT

The ATCAS (Air Traffic Control Automated System), which is schematically shown in Fig. 3, is devoted to the air traffic control of the Rome FIR.
The ATCAS MRT integrates five radar systems sited in the west side of the centre and south of Italy. In Figure 4 are indicated the sites and their qualitative coverages. They are located at Fiumicino (Rome), Monte Codi (East Sardinia), Poggio Lecceta (near Leghorn), Monte Stella (near Naples) and in the Island of Ustica (north Sicily).
The relative distance between adjacent sites is approximatively of 250 Km. The controlled area has an overlapping sufficient to guarantee the tracking continuity at the crossing of the coverages. The higher overlapping is in the Rome terminal area, which represents the most crowded one. In this way the best performance are guaranteed where higher requirements of accuracy and of redundancy capability are required.

The radar configuration of the integrated sites are as in the following:

- Site of Fiumicino-Rome
  - Primary radar Selenia ATCR-33A (approaching radar)
  - Primary radar Selenia ATCR-22 (radar for terminal area)
  - Secondary radar Selenia SIR-7
- Site of Monte Codi
  - Primary radar Selenia ATCR-2
  - Secondary radar SSI-70
- Site of Poggio Lecceta
  - Primary radar Selenia ATCR-2T
  - Secondary radar SSI-70

The sites of Monte Stella and Ustica have the same configuration of Poggio Lecceta.

The radar characteristics relevenat to this presentation are summarized in the following:

– Primary radar ATCR-3A

Range	70 NM
Beamwidth	1.5°
Scan period	4 s

– Primary radar ATCR-22

Range	150 NM
Beamwidth	1.2°
Scan period	6 s

– Primary radar ATCR-2T1

Range	170 NM
Beamwidth	1.2°
Scan period	6.5 s (13 s)

– Secondary radars SSI-70 or SIR-7

Range	170 NM
Beamwidth	2°

The error standard deviation in the range and azimuth is in the average respectively $\sigma_\rho = 100$ m and $\sigma_\Theta = 0.2°$. This values, theoretically estimated, have had an experimental validation.

Each site is mainly formed of a primary radar, a secondary radar, a combiner, a formatting device and a modem for the transmission of the plots to a common centre. Therefore the developed MRT is of the centralized type. This choice has been adopted in order to exploit at the best the pre-existing hardware. In fact, the initial design of the ATCAS (started in 1968) considered the use of raw video also for remote sensor, and therefore each displays was able to receive the radar information from a single sensor (channelized) display). Subsequent experience and the evolution in the field of synthetic radar presentation suggested the exclusion of raw video, thus allowing the possibility of the MRT in the present configuration.

It should be remarked that the selected solution (the centralized one) has been influenced by the pre-existing system and does not represent the unique Selenia approach. In fact for other systems Selenia is developing MRT with distributed architecture, as in the en-route system of Mazatlan ACC in Mexico.

Whichever is the type of MRT, special attention has to be paid to the north alignment of the various radar to properly integrate the position data coming from the different sensors. Various algorithms to reduce the effects of the misalignment errors for two or more radar can be applied. For our solution the use of some test targets in known position has been preferred. This way the azimuth alignment may be performed on each radar separately.

DATA PROCESSING AND TRACKING ALGORITHM

The sequence of the processing from the radar detection to the display is shown in Fig. 5 and listed in the following:

- detection
- primary/secondary radar combination
- formatting
- sending to the main centre
- stereographic projection
- polar to cartesian conversion
- correlation
- smoothing
- display

Each site has a primary and a secondary corotating radars which perform the detection of the plot and the estimation of their coordinates. After the combiner the plot is generally characterized by the following informations: range, azimuth, code and altitude (ρ, Θ, code, h). All the plots pertaining to the same angular sector are formatted and transmitted to the centre by means of duplicated telephone lines at 2400 bits/s. The transmission protocol is HDLC and is implemented with microprocessors.

At the centre the plots are processed according to their detection time. The plots coordinates are first transformed to the local stereographic plane and after to a common stereographic plan corresponding to the site of Fiumicino, in which the smoothing and the display will be performed. (A detailed analysis on the stereographic projection can be found in Mulholland and Stout (2)).
The plots are then correlated with candidate tracks, stored in a buffer. The correlation is performed according to the secondary code, if available. When a matching is verified between the track and plot codes a subsequent spatial validation is considered for the correlation. This requirement is usefull for a correct management of non-individualSSR codes and for discarding plots caused by reflection on obstacles. Obviously for plots without code the correlation is performed only on the basis of spatial congruence. Anyway the automatic initialization is performed only on plot with secondary code.

The correlated plot is used to perform the smoothing and the updating of the track, according to the well-known $\alpha - \beta$ algorithm. (A review of smoothing algorithm is done in Farina and Pardini (3)). The $\alpha - \beta$ algorithm works in two subsequent steps. The first refers to the filtering of the position and velocity through the following correction equations:

$$x\,(k/k) = x\,(k/k\text{-}1) + \alpha\,(\;x_m(k) - x\,(k/k-1))$$
$$\dot{x}\,(k/k) = \dot{x}\,(k/k-1) + \frac{\beta}{\Delta t(k)}(\;x_m(k) - x\,(k/k-1)) \qquad (1)$$

where:

$x\,(k/k)$	= x-component of the corrected position at the k scan
$x\,(k/k\text{-}1)$	= x-component of the forecast position at the k scan
$\dot{x}\,(k/k)$	= x-component of the corrected velocity
$\dot{x}\,(k/k\text{-}1)$	= x-component of the forecast velocity
$\alpha - \beta$	= smoothing parameters
$\Delta t(k)$	= time delay between scans k-1 and k

The second step performs the prediction of the same components

$$x\,(k+1/k) = x\,(k/k) + \dot{x}\,(k/k)\,\Delta\,t\,(k+1)$$
$$\dot{x}\,(k+1/k) = \dot{x}\,(k/k) \qquad (2)$$

Similar equations hold for the y-component. For the altitude, being transmitted as a code, the smoothing is not required; only a simple algorithm to estimate the trend is implemented. The smoothing is made on cartesian coordinate due to their inertial property. Thus a motion unaccelerated in nature remains unaccelerated when projected on the cartesian components.

The value of the $\alpha - \beta$ parameters determines the smoothing characteristics. Low values of $\alpha - \beta$ allow to obtain efficient smoothing of random measurement errors; high values of $\alpha - \beta$ guarantee promptness in the tracking of highly manoeuvring target. To adapt the behaviour to the target kinematic a manoeuvre sensor is used, which selects recoursively at each detection the values of $\alpha - \beta$ most suitable for the actual target path.

The manoeuvre sensor is based on the comparison of the spatial displacement between the track forecast position and the measured plot with a threshold selected from the standard deviation of the measurement errors. The radar measurement is polar in nature, and as a first approximation, the area of error around the true position of the target is an ellipse with one axis oriented toward the radar (Fig. 6). The axes of the ellipse are:

$$d_r = 2\,\sigma_\rho \quad \text{(radial axis)} \qquad (3)$$
$$d_t = 2\,\sigma_\Theta \quad \text{(transversal axis)}$$

where:

σ_ρ = rms error in the range measurement
σ_Θ = rms error in the azimuth measurement
R = target-radar range

Considering the following usual values σ_ρ = 100 m, σ_Θ = 0.2° and R = 200 Km it results for the error axes d_r = 200 m and d_t = 1360 m. Therefore the radar measurement error in the two components are higly asimmetrical. To exploit at the best the differential sensitivity of the manoeuvre sensor to the different components, the manoeuvre sensor has been implemented in polar coordinated oriented to the radar from which the plot is coming. Thus the equations for the manoeuvre sensor are:

$$|\rho_{track} - \rho_{plot}| < k\sigma_\rho$$
$$|\theta_{track} - \theta_{plot}| < k\sigma_\theta \qquad (4)$$

where k is a parameter of system (typically ranging from 2 to 3). If both of the previous relations hold, the target is considered moving along a straight line; in the other cases a manoeuvre is declared and the consequent variation in the smoothing procedure applies.

DISPLAYS

For what concerns the display different laws, with respect to the monoradar case, for the data renewal could be envisaged. In fact, the timing sequence of the plot detection does not involves a systematic sweeping of the radar coverage, as in the monoradar case. In any case the time frame for the scan of the display may be independent of the scan periods of the antennas. In addition, when tracks, rather than plots, are displayed, since the target velocity is available, a time phasing of the target position to the display time can be provided. For the renewal time of the display an interval of 4 s has been considered suitable.

Two main laws for the scanning of the display may be considered: a polar type, in order to reproduce the operation of a dummy monoradar, or a raster type. The former one is familiar to the operator, while the latter one has the advantages of implying a lower computation burden. Anyway with minor changes both solutions may be optionally implemented.

The most significant equipments in ATCAS subsystem presentation are:

– Consoles	n. 74
– PPI	n. 43
– Synthetic central unit	n. 15
– Video distribution units	n. 11
– Operational sectors	n. 15

EXPECTED RESULTS

The ATCAS system in the MRT configuration is now in the developing phase and a detailed analysis has been performed in order to determine the expected performance in terms of tracking accuracy. The performances strongly depend on the kinematic target condition. Therefore the two limit condition of straight line motion and accelerated path may be considered.

In the case of straight line motion the MRT tracking performance are very similar to the monoradar case. Thus, the error in the estimated position is comparable with the radar measurement error and the error in the velocity vector estimation is few knots (from 5 to 10 knots) for the speed and few degrees (from 2° to 6° depending on range and velocity condition) for the heading.

For accelerated path the MRT system has better performance than the monoradar one. Figure 7 shows the maximum error in the forecast position for accelerated path as a function of the distance from the radar. The values refers to the following operational conditions: two cooperating radars, probability of detection $P_D = 1$, scan period $T_s = 6$ s (for both radars), target acceleration 3 m/s^2. For the MRT system the two interleaved data rate has been assumed equal 1.5 s and 4.5 s. For the sake of comparison, a path has been considered for which the manoeuvre sensor has the lowest sensitivity.
For the multiradar we have considered the two limit cases. In the most critical case the overall geometry is so that both manoeuvre sensors are in the lowest sensitivity condition (target aligned with the two radar). In the most favourable case one of the manoeuvre sensor is in the highest sensitivity condition (geometry as in Fig. 6). Therefore in the multiradar system the performance may be widely spreaded depending on the overall geometrical conditions. In any case they are always better than for monoradar system.

The first experimental results will be available late this year. In order to evaluate the experimental performance in term of tracking accuracy, a recording system of plots and tracks has been implemented for an off-line performance analysis. The system will also be usefull to achieve an optimized design by means of a refinement of the system parameters.

CONCLUSION

The MRT system developed for the Rome air-traffic-control region has been described. Original solutions have been envisaged for the manoeuvre sensor, by exploiting at the best the different accuracy and data rate of each radar, and for the display renewal, by reproducing a scanning of an equivalent monoradar system. The first experimental results will be available in the late 1982.

REFERENCES

1 Farina, A., Pardini, S., 1982 "Introduction to multiradar tracking systems".
Selenia Technical Review, vol. 8, n. 1

2 Mulholland, R.G., Stout, D.W., 1982 "Stereographic projection in the National Airspace System".
Trans, IEEE on Aerospace and Electronic Systems, 18, n.1, 48-57.

3 Farina, A., Pardini, S., 1980 "Survey of radar data-processing techniques in air-traffic-control and surveillance systems".
Proc. IEE, 127, Pt. F, 190-204.

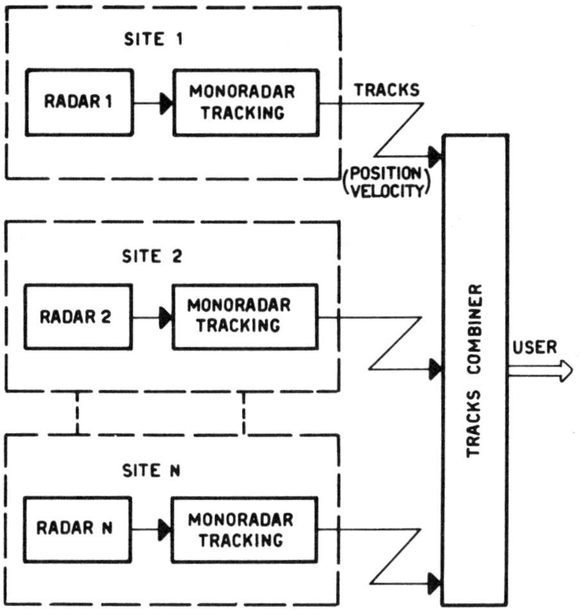

Fig. 1 — MRT with distributed architecture

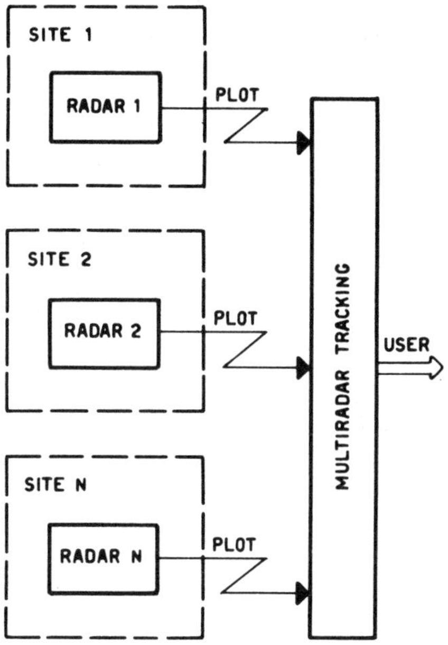

Fig. 2 — MRT with centralized architecture

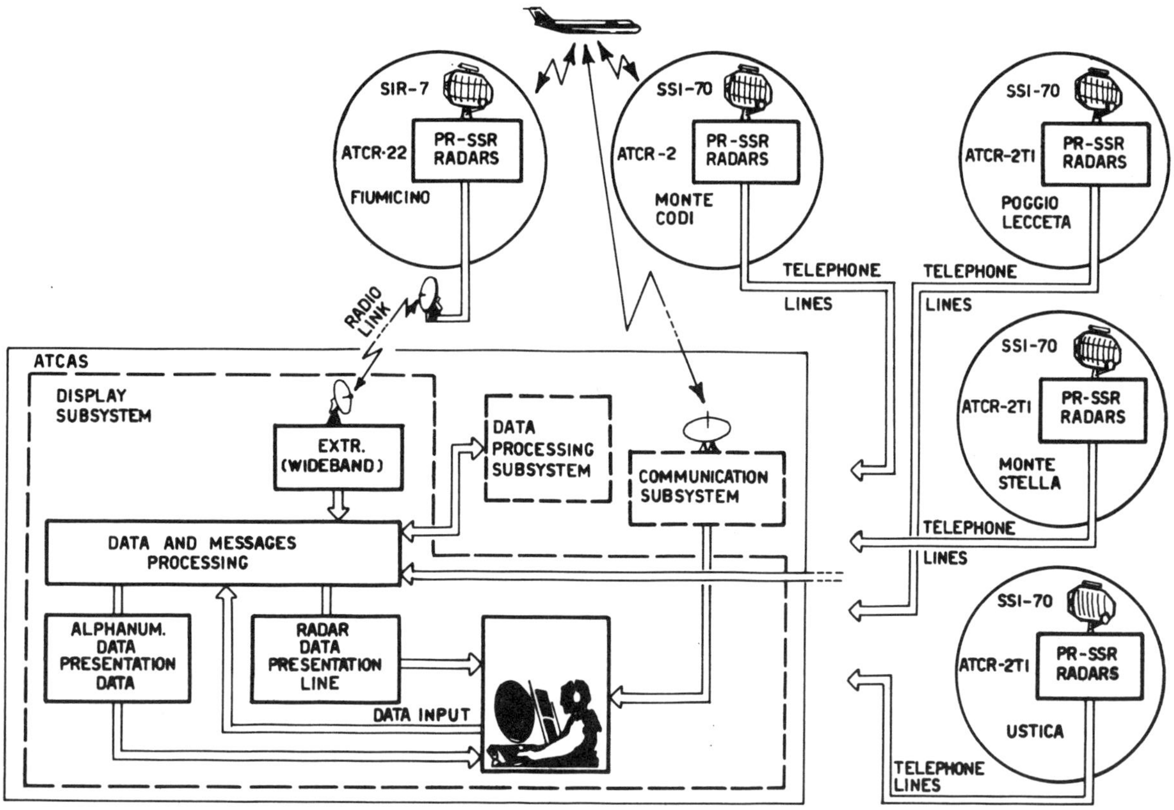

Fig. 3 — ATCAS multiradar tracking configuration

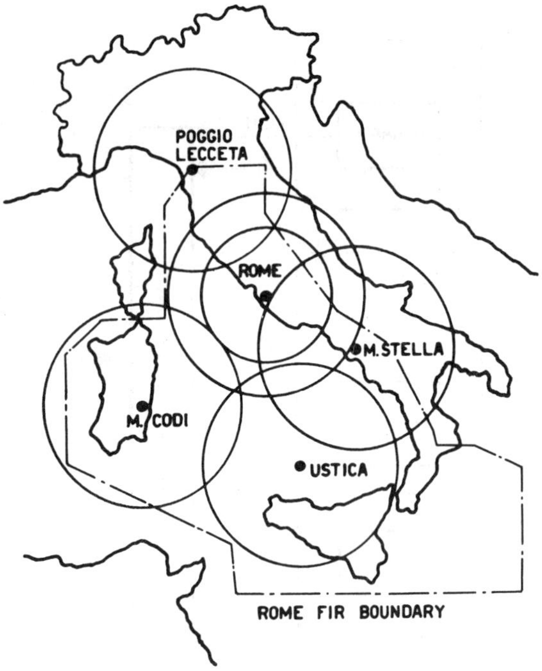

Fig. 4 — Schematic radar coverages

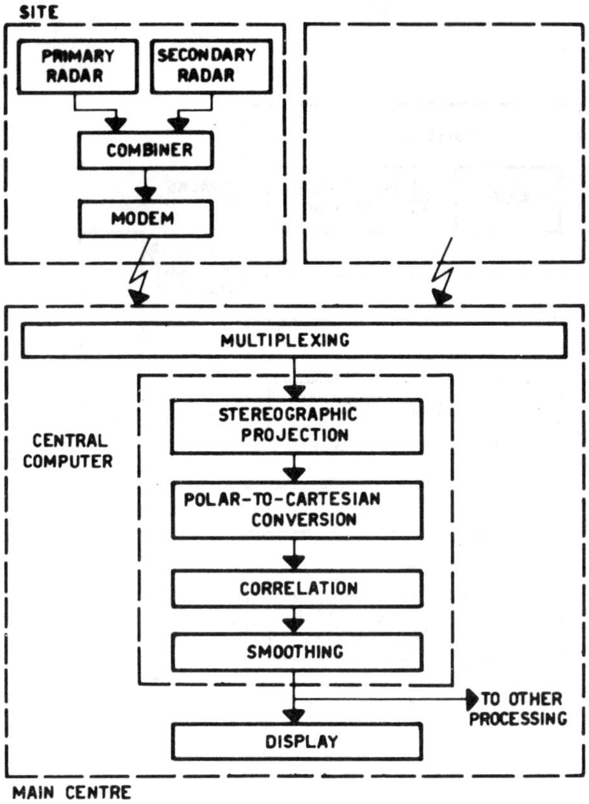

Fig. 5 — Structure of the ATCAS multiradar tracking processing

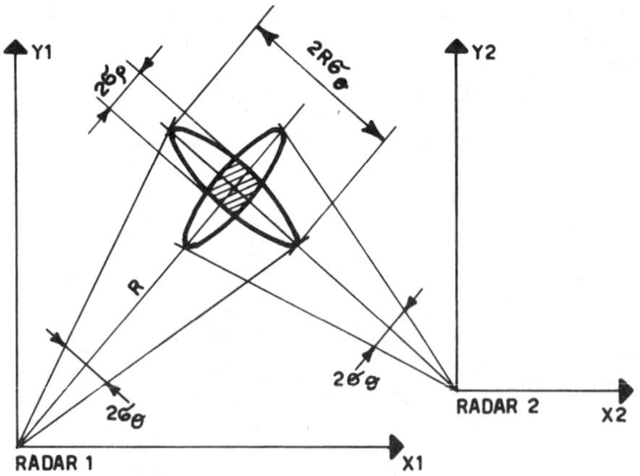

Fig. 6 — Influence of sensors relative position in the measurement accuracy

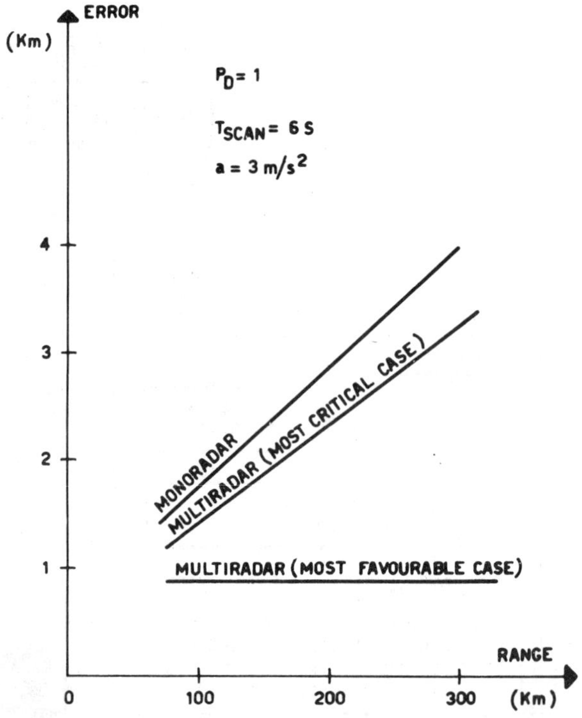

Fig. 7 — Maximum error on the forecast position

METHODS FOR RADAR DATA EXTRACTION AND FILTERING IN A FULLY AUTOMATIC ATC RADAR STATION

E. Giaccari

Selenia S.p.A., I

INTRODUCTION

In the past the relative low intensity of air traffic could leave time to the operator for an analysis of the raw radar data and for a correct discrimination between true and false targets. As soon as the intensity of air traffic increased and more information was necessary to assure a reliable control, an automatism in filtering, detection, extraction and tracking was required.

For this purpose adaptive processing has been applied to radar receivers to optimize detection of aircraft embedded in the clutter; extractors have been connected to the radars mainly for false alarm control and coordinate extraction; tracking and final filtering have been assigned to large computers operating on the basys of scan to scan correlation.

This system configuration (see Fig. 1-a), until now commonly applied, suffers of some limitations. In fact the main components of this system (radar, extractor, tracking) are normally independently procured and set, often requiring to each of them equivalent performances. Furthermore there is a limited link of information among the functions of the system and the absence of any feedback control.

In the next paragraph a method to obtain the maximum transfer and treatment of information with minimum impact in hardware is explained. This approach is made easy by the choice of parallel microprocessors which extends the concept of distributed intelligence to the radar data acquisition and evaluation. Figure 1-a is modified in figure 1-b where the already mentioned functions of detection, filtering, extraction and tracking are all carried out in the radar station. Full interaction among the functions is permitted by the feedback lines. Matching of the data flow to the properties of each block is realized with equivalent computing elements differently programmed.

Some preliminary results obtained with a practical implementation of the proposed method are given in the last part of the paper.

SYSTEM DESCRIPTION

The proposed architecture of the radar station consists in a sensor able to probe the environment and to select the proper processing section for a strict control of false alarms. The filtering capabilities of this sensor can be divided in three main cathegories (1). The first type is related to quasi-stationary interfering phenomena and consists in maps periodically updated whose boundaries define the areas where attenuation of the received signal and doppler filtering are applied. The second type works in real time for control of fast varying interference. This cathegory is represented by adaptive thresholding and correlators. The third type operates on a scan to scan basys and mantains the final false alarm probability within limits acceptable by the following processing devices.

The quality of the obtainable filtering reduces the amount of data to be examined and permits to adopt new criteria for plot extraction. Conventional assignement of range and azimuth coordinates can be substituted by a pattern recognition of the filtered data in order to improve accuracy and resolution (2). Furthermore strength and unambiguous radial velocity can be evaluated for a better characterization of the target. The inherent power of this part implemented with a powerful microprocessor is utilized to establish an interactive link with the radar receiver and processor. The double task to avoid unexpected instantaneous overflow of the input buffer and to supervise the radar for an automatic best setting of its functions is so achieved. This means that the adaptivity concept is applied to the radar functions for a control of the radar filtering activity proportional to the acceptable load.

The extracted output is not directly sent to the tracking algorithm, used for the estimation of the trajectories, but it passes through a post-extraction analysis and is filtered if the maximum number of plot, acceptable by the tracking function, is exceeded. This part, which applies in different ways for different plot distributions during the antenna scan, is based on the spatial density, quality, range relation and history of the incoming data. By this protection, the tracking function can operate for better performances in terms of association, and prediction because of the strictly controlled number of data to be processed. Furthermore the improved coordinates accuracy achieved by the extractor permits the reduction of the correlation window and of the smoothing coefficients, with improved performances in terms of resolution and responsiveness during maneuvring. By the way if for any reason the load for the tracking should become excessive a feedback control to the post-extraction filtering is available for applica -

tion of local harder criteria. Data from secondary radar can be directly accepted and processed according to conventional algorithm of secondary to primary data combination. The final part of the system is a narrow band link which transfers tracks coming out directly from the radar head to the ATC center.

Figure 1-b shows sinthetically the explained functions and their interaction. The feedback network for load control is evidenced. Except for the radar signal processing all the remaining functions are implemented in a parallel microprocessor structure that is part of radar receiver itself.

This architecture, a part from the already mentioned benefits, gives further advantages in terms of availability and maintenability as direct consequence of the limited amount of involved hardware. The optimization of the system is a built-in function and then requires minimum technical assistance both from local and from remote site. Furthermore the distributed intelligence and the reconfiguration capability reduces to the minimum the risk of complete loss of information.

METHODS AND ALGORITHMS OF VALUATION

A reliable knoledge of the performances of a radar system requires a large collection of data and powerful algorithms of valuation. In the following a method used for data analysis is described. The main functions to be valuated are the plot extractor as far as the coordinates accuracy is concerned, and the plot filtering for the load reduction.

Coordinates accuracy analysis

The accuracy evaluation of an extraction system is a complex problem because it involves not only the intrinsec characteristics of the extractor but those more general of the radar. In fact, in order to achieve a reliable accuracy measurement, it is necessary to consider the incidence of the radar and target parameters in a situation of operative use of the overall system. It is then necessary to apply dynamic methods such as scheduled flight tests or measurements carried out, in particular conditions, on flights of opportunity. In both cases it is necessary to define the motion characteristics, effective or presumable of the considered aircraft, so that, with simple calculation methods, the numerical values of accuracy of the overall system are obtained.
The basic hypothesis for the validity of the accuracy evaluation method are:

- the aircraft moves with straight and uniform motion;
- during the observation interval there is not any missing detection in N consecutive scans.

By these assumptions it is possible to evaluated the errors introduced by the system by solving the problem of true position estimation of an aircraft from the measured position. This problem has two aspects:

- The identification of the straight line that represents the estimated trajectory of the plane.
- The identification on the trajectory of the points that are the estimation of the true plane position at each scan.

These two aspects are jointly considered defining an optimization problem based on the least square method in the following way. From the measurement of coordinates in N consecutive scans:

$$(x_i, y_i) \qquad i = 1, 2, \dots N$$

minimize the function:

$$\sum_1^N {}_i\left[(x_i - X_i)^2 + (y_i - Y_i)^2\right]$$

on the coordinates (X_i, Y_i) that represent the true estimated positions in N consecutive scans, respecting the constraints of uniform and straight motion:

$$aX_i + bY_i + c = 0$$

$$(X_{i+1} - X_i)^2 + (Y_{i+1} - Y_i)^2 = K^2$$

where K can be evaluated as:

$$K = \frac{\sqrt{(x_N - x_1)^2 + (y_N - y_1)^2}}{N - 1}$$

With these assumptions the coefficients of the trajectory are given by:

$$a = -\sum_1^N {}_i (N+1-2i)\, y_i$$

$$b = \sum_1^N {}_i (N+1-2i)\, x_i$$

$$c = \frac{2}{N} \sum_1^N {}_i \sum_1^N {}_J (i - J)\, x_i y_J$$

and the true estimated positions:

$$X_i = \frac{1}{N} \sum_1^N {}_K x_K - (N+1-2i) \frac{Kb}{2\sqrt{a^2+b^2}}$$

$$Y_i = \frac{1}{N} \sum_1^N {}_K y_K - (N+1-2i) \frac{Ka}{2\sqrt{a^2+b^2}}$$

The measured position and the true position give rise to the azimuth and range errors through the following relations:

$$\Delta A_i = \cos^{-1}\left[\frac{r_i^2 + s_i^2 - t_i^2}{2\, r_i\, s_i}\right]$$

$$\Delta R_i = s_i - r_i$$

where

$$r_i = \sqrt{X_i^2 + Y_i^2}$$

$$s_i = \sqrt{x_i^2 + y_i^2}$$

$$t_i = \sqrt{(X_i - x_i)^2 + (Y_i - y_i)^2}$$

For each continuous interval of the tracjectory average values and r.m.s. values of the errors can be evaluated according to:

$$\overline{\Delta A} = \frac{1}{N} \sum_{1}^{N}{}_i \, \Delta A_i$$

$$\overline{\Delta R} = \frac{1}{N} \sum_{1}^{N}{}_i \, \Delta R_i$$

$$\sigma_A = \sqrt{\frac{1}{N-1} \sum_{1}^{N}{}_i \left[\Delta A_i^2 - (\overline{\Delta A})^2\right]}$$

$$\sigma_R = \sqrt{\frac{1}{N-1} \sum_{1}^{N}{}_i \left[\Delta R_i^2 - (\overline{\Delta R})^2\right]}$$

In fig. 2 an example of interpolation is shown; in fig. 3 and 4 the probability density function of range and azimuth as they are available directly from a computer program are depicted.
In fig. 5 a succession of plots relative to one flight is shown. It is evident the high level of coordinate accuracy obtained independently from the smoothing of the track.

Detection probability analysis

Automatic measurement of detection probability (P_D) can be obtained counting the number of presences and comparing them with the total number of presences and missing for each trajectory.
A practical result obtained dividing the radar coverage in sectors and ascribing to them the relevant P_D is given in fig. 6.

Plot filtering

The benefit obtainable by the proposed plot filtering system has been experimentally evaluated. This approach has been selected because of the difficulty to find a clutter model that could take into account a complex environment. The analysis has been carried out on recorded data whose content has been subdivided in true and false plots according to a tracking algorithm. On these two classes of data the statistical behaviour has been determined.
Fig. 7 shows the p.d.f. of the target quality while Fig. 8 is the correspondent p.d.f. for false plots. It appears evident the different polarization of the two functions that leads to a discrimination between true and false plots.Furthermore the two p.d.f. divided for range gates have shown a defined dependence of the quality from the range (R) and the possibility to establish some filtering thresholds (T_Q) for the best trade-off in terms of false plot reduction and true plot missing.
Table 1 summerizes the results obtained applying different T_Q (R) to the complete set of recorded data distinguished per range gates. The loss in detection probability L_{P_D} is a constant for every range gate while the percentuage of cancelled false plots(L_{FP}) shows a dependance from the range.
Besides the percentage of false plot filtered by the selected T_Q (R) is by far higher than the P_D reduction, thus permitting the use of this method for further improvement in radar processing.

CONCLUSIONS

The need for synthetic radar data presentation and the tendency for completely unattended radar stations lead to a solution where the radar head and the data processing are completely integrated.
The tight integration among the parts allows an optimization of the overall performances as far as coordinate accuracy and false plot control are concerned.

Table 1 - False plot reduction(L_{FP})and detection probability degradation (L_{P_D})in function of range (R) for different quality thresholds T_Q

R (nm)	T_{Q1}	L_{FP}(%)	L_{P_D}(%)	T_{Q2}	L_{FP}(%)	L_{P_D}(%)
0 - 10	14	60	20	17	80	30
10 - 20	18	80	20	22	85	30
20 - 30	18	90	20	20	90	30
30 - 40	16	90	20	19	90	30
40 - 50	14	90	20	16	94	30
50 - 60	14	95	20	16	97	30
60 - 70	14	97	20	15	97	30

AKNOWLEDGMENT
The author thanks Mr. Guido Grandi and Mr. Nicola Accarino for the cooperation given in evaluating the described system.

REFERENCES

1. E. Giaccari-G.Nucci, a Family of Air Traffic Control radars IEEE Transactions on AES Vol. AES-15 n° 3 May 1979.
2. N. Accarino-E.Giaccari, A new approach to radar plot extraction for ATC applications IEEE 1980 Int. Radar Conference.

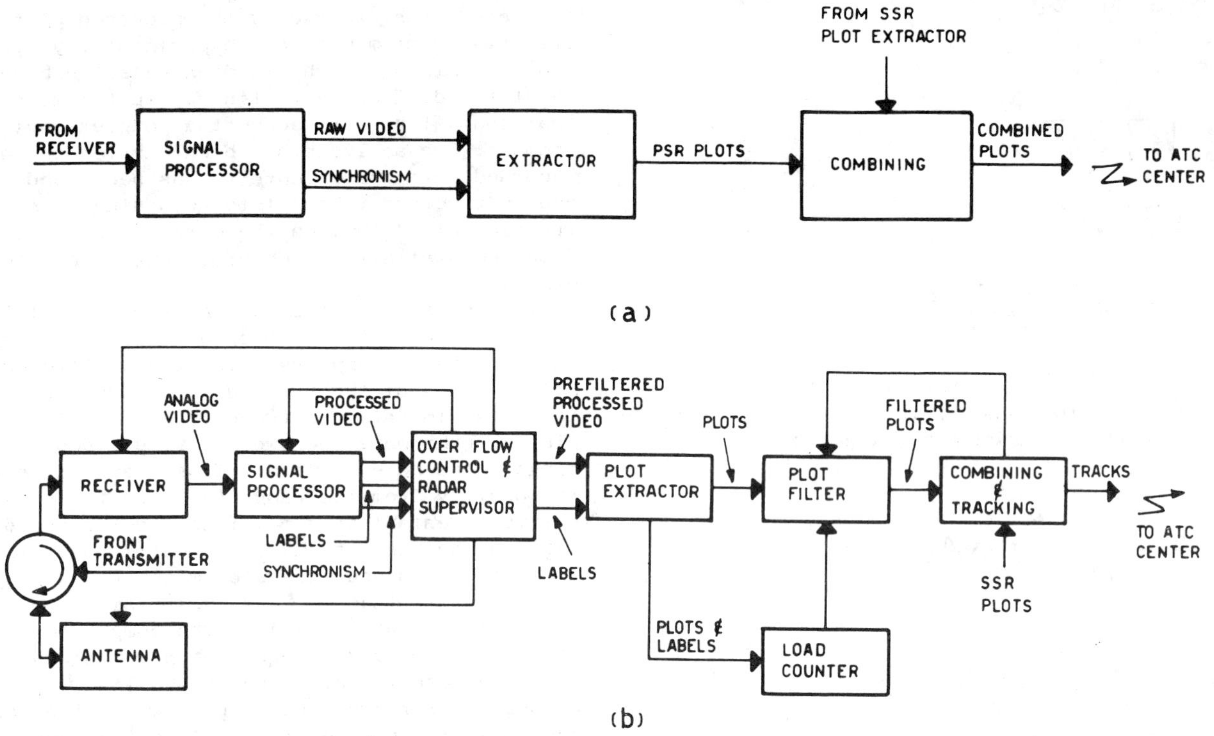

Figure 1 - Block diagram of an ATCR radar station a) conventional configuration; b) proposed configuration

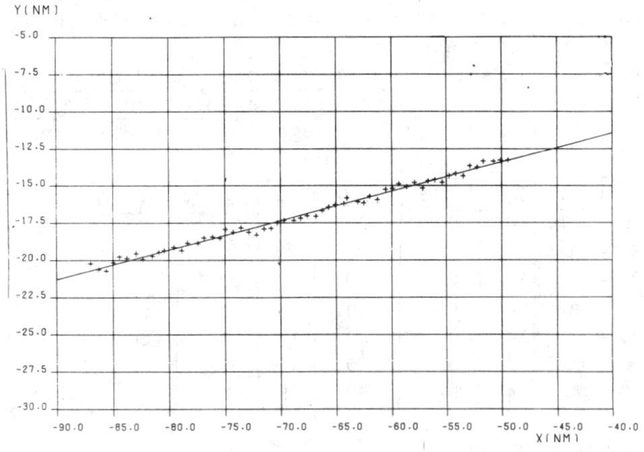

Figure 2 - Example of interpolation

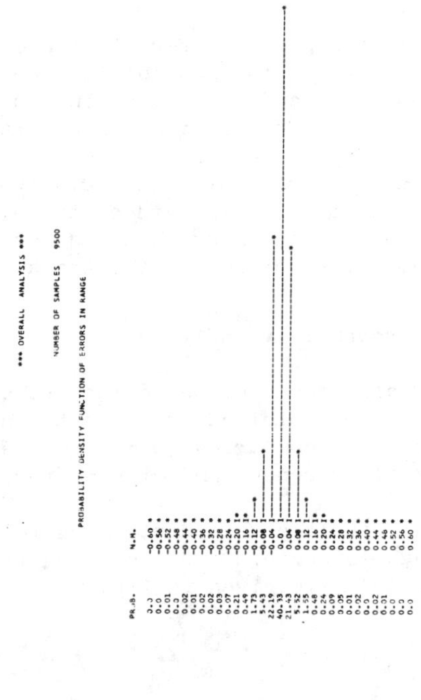

Figure 3 - Probability density function of er rors in range

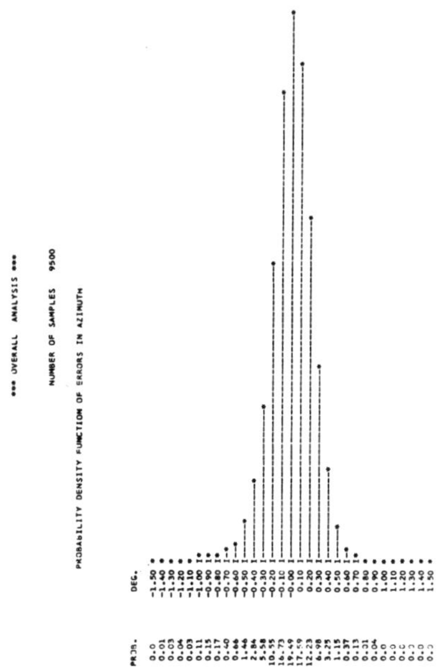

Figure 4 -Probability density function of errors in azimuth

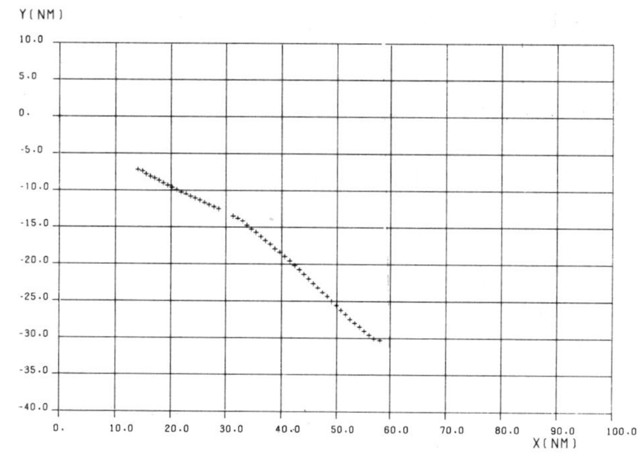

Figure 5 - Typical extractor output

PD VALUE FOR EACH SECTOR (PERCENT)

N.M.								
200-230	******	******	40.56	39.32	62.08	61.51	59.62	51.76
170-200	******	******	54.29	46.59	59.13	59.47	52.13	49.43
140-170	60.41	80.58	75.74	65.21	54.26	62.89	59.62	57.56
110-140	89.10	84.58	84.55	83.09	72.14	75.25	80.74	69.69
80-110	******	******	90.59	92.48	87.71	89.71	84.70	85.48
50-80	46.30	83.33	85.29	94.10	96.02	96.49	92.78	94.51
20-50	94.44	93.80	90.80	96.48	99.55	99.07	96.84	97.49
0-20	******	82.16	******	******	96.94	99.04	******	******
DEG.	0-45	45-90	90-135	135-180	180-225	225-270	270-315	315-360

Figure 6 - Example of automatic probability detection measurement

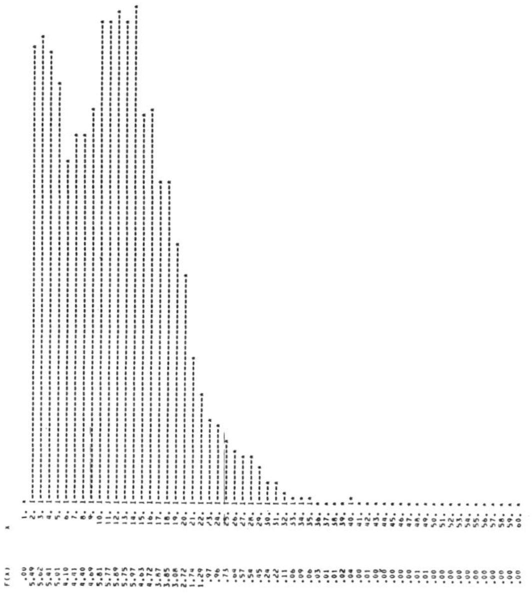

Figure 7 - Example of clutter probability density function

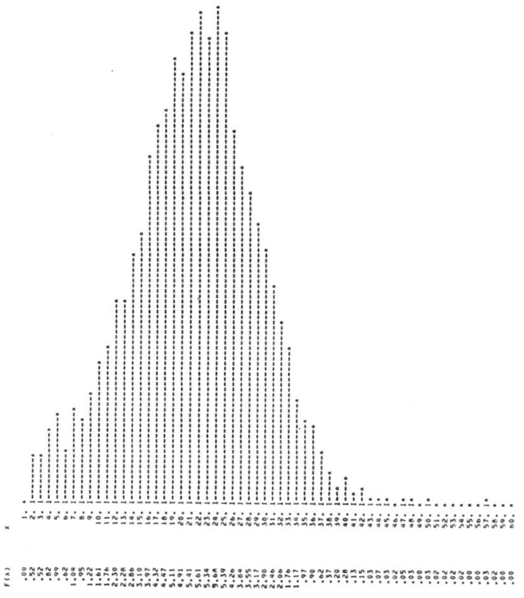

Figure 8 - Example of targets probability density function

PRESENTATION AND PROCESSING OF RADAR VIDEO MAP INFORMATION

R J G Edwards

Civil Aviation Division, Ministry of Transport, Wellington, New Zealand

INTRODUCTION

Five Air Traffic Control (ATC) centres at Auckland, Ohakea, Wellington, Christchurch and Dunedin are equipped with primary radars which provide a raw radar display of the air traffic situation over much of New Zealand's controlled airspace. At any given time a high proportion of traffic is involved in terminal area manoeuvring and it is likely that direct display of raw primary radar information will be required for some years yet.

As part of an ongoing radar modernisation program the radar displays and support systems are being replaced. The Civil Aviation Division's (CAD) radar engineering group have designed and are now building the Radar Display Support Systems (RDSS). The RDSS consists of a Maintrace Section which distributes the radar signals to the displays and an Intertrace Section which generates the video maps and provides interactive keyboard/tracker-ball facilities. Details of the RDSS design are contained in Baines (1) and Edwards (2). The first two RDSS being built by CAD are scheduled to enter service early in 1983.

A Radar Digital Video Map generator (RDVM) incorporated in the Intertrace Section enables vector drawn maps and also alpha-numeric text to be written on the radar display. A prototype RDVM has been built and following the completion of feasibility trials, is entering production. This paper, while being concerned primarily with the features of the RDVM, will make brief reference where necessary to other elements needed to form the complete RDSS.

THE RADAR VIDEO MAP

A video map superimposed on the target information presented by a radar display is intended to provide a fundamental set of geographically based references to which the air traffic controller may relate radar returns from aircraft targets. Map content is determined by the controllers task and this in turn may change with traffic requirements.

Map Content

Overlays. In general, more information must be made available than need be used at any one time. In order to reduce clutter on the display it has been suggested that the map be subdivided into several independent categories called overlays, any combination of which may be selected in order to perform a specific task. For example a set of overlays may be chosen so that the map may be configured for either Airways, Area or Approach control functions with a minimum of redundant data being displayed.

OVERLAY DATA	*ATC TASK*
Airways Reporting Points	Airways
Reporting Points Danger Areas Coastline Terminal Manoeuvring Area	Area
Terminal Manoeuvring Area Control Zones Approach Ladder RWY A Approach Ladder RWY B	Approach

A set of related overlays such as these is referred to as a map frame and the display system should make available a number of map frames according to the diversity of operation and flexibility required. The CAD RDVM provides up to eight overlays on each of up to three map frames.

Symbology. Display legibility can be assessed most objectively by considering the readability of groups of symbols, either words or map symbol groups. The legibility of individual characters is a function of lighting, size and shape while readability is affected by such factors as character spacing, line spacing and the relative shapes of characters when they are presented in groups. In order to preserve the optimum readability while at the same time reducing display clutter at short range settings, all symbols remain at constant size regardless of the range scale selected.

A repertoire of 64 symbols is preprogrammed into the RDVM and these may be drawn individually or in associated groups on text blocks.

Data Tables. There are advantages to be gained from presenting a limited amount of tabular data on the radar display. One must be careful however when suggesting the provision of such information. This may produce undesirable clutter if not suitably placed or in too great a volume and it may also confuse the existing natural division between the tactical and strategic information displays. The items listed in the table below have been carefully chosen and in general are items to which frequent reference is made or which relate directly to operations performed with the radar display. The Automatic Terminal Information Service (ATIS) status, barometric pressure at sea level (QNH) and time to the nearest minute are normally passed to pilots by radio telephone and require periodic updating.

DATA TABLE ENTRIES

ATIS status
QNH area value
TIME to nearest minute

RUNWAY in use
WIND direction/strength and gusts

BEARING/DISTANCE function.

Geographic Detail

The degree to which geographic detail may be displayed is related to the accuracy of the map generator. However there is often good reason to degrade the accuracy and hence detail of items such as coastline in the interest of display clarity. Coastline is useful for orientating the Controller and occasionally may be helpful typically in confirming the position of visual flight rule (VFR) traffic, military activities or in search and rescue operations. Most of the time however it is of little direct use and there is good reason in the interest of display clutter reduction, for simplifying the coastline to a form which is still aesthetically and operationally acceptable, but which involves minimum detail. Performing this task digitally rather than cartographically has the advantage that the accuracy of the resultant "smoothed" coastline may be specified quantitatively. Computer programs which have been developed to perform this task for the RDVM are described later.

Accuracy

The accuracy requirements of ATC radar users vary widely but are tending to become more demanding with the advent of extremely precise airborne navigation systems. The overall radar system accuracy is limited by the following factors:

intrinsic radar accuracy
slant range error
display resolution.

Producing a video map very much more accurate than the radar data is unnecessarily costly while less accuracy would degrade the overall radar system performance. The slant range error may be corrected if mode C Secondary Surveillance Radar (SSR) data is utilised. Most modern radar displays achieve positional accuracy approximately an order of magnitude better than the remainder of the radar system. The intrinsic radar accuracy is thus the primary limitation to radar system performance. The RDVM performance has been tailored therefore to match the intrinsic radar accuracy. Map generator accuracies are normally quoted in a way which does not incorporate any of the errors in the map data used. The RDVM accuracy calculations and measurements have by way of contrast been done so as to include all known errors from the initial map cartography through to the long term and temperature drift of the display generation hardware. The following positional accuracies have been achieved:

Airspace boundary and Coastline data	±0.4 NM
Reporting Points and symbols (including Approach Ladders)	±0.1 NM

THE VIDEO MAP DATA BASE

Digital Data

The accuracy and stability attainable by electronic data processing techniques far exceeds that available from traditional cartography. The New Zealand Department of Lands and Survey is compiling a digital data base containing the New Zealand airspace specifications. From this data base all aeronautical charts, including the radar video maps, will be produced. The radar video map digital data, instead of being plotted as would be the case for any other chart, is kept in digital form until processed for display by the RDVM.

Advantages. The main advantages of digital video map data are summarised below:

Fast, economical information update
Ease of data transformation and compilation
Security of map information
Compact, accurate, stable, inexpensive data storage
Reliable, economical hardware.

Data Categories

Coastline. The New Zealand master coastline file which includes major islands and lakes has been digitized from a number of cartographic stable base master charts. The resolution and accuracy requirements along with the data formats adopted for the master coastline file are detailed in Edwards (3). Coastline data for a particular radar video map is obtained by windowing the appropriate data from the master file and then transforming the projection to gnomonic (equidistant azimuthal) centred on the radar aerial site. The inclusion of cartographic and digitizing errors makes this the least accurate class of map data.

Airspace Boundaries. Unlike coastline data, airspace boundaries have precise mathematical definitions and thus may be generated without recourse to digitizing of cartographic charts. The boundaries rarely appear as straight lines on a gnomonic projection and are usually approximated to conserve memory space in the RDVM.

Point Data. Any data which can be completely specified by reference to one geographic location, such as reporting points for example, is referred to as point data. This is inherently the most accurate class of video map data as it may be entered into the data base directly as a geographic coordinate without recourse to digitizing of cartographic charts.

Data Compacting

The large volume of data required to store the entire New Zealand coastline definition to an accuracy suitable for the RDVM necessitated the development of a chain encoding technique optimised for the microprocessor based computing system employed by CAD for the video map project. Using the chain format provides a 4:1 data reduction and the 100,000 point coastline definition can be unpacked in 5 seconds.

The RDVM Map Memory Module

The link between the aeronautical digital data base and the RDVM generators at the radar sites is a reprogrammable map memory hardware module. The electronic memory components which store the video map information for display are contained on a single printed circuit board. These are programmed with the data from Lands and Survey in a few minutes by a computer at the CAD head office and are then sent to the appropriate radar station. Old data is erased from these memories and in this way they may be used over and over again. This is in contrast to the photographic process required by the map generators used at present for which a map update takes several weeks of preparation and costs several hundreds of dollars for time and materials.

Security

The possible technical and operational consequences of inaccurate or corrupted map data appearing on a radar screen are only too well understood. Manual and automatic cross-checking of the map data at various stages during the processing sequence ensures that any evident errors are removed. As a final confidence check the map data is read back from the programmed map memory module by a computer RDVM simulator and is plotted on paper for checking purposes and to provide a hard copy record. A different connotation is placed on security by military users and data encryption techniques can be employed to ensure that the map data is "unreadable" at all stages of processing from initial data entry until final data display.

ALTERNATIVE MAP GENERATION TECHNIQUES

Radar video map generators usually conform to one of three general classes of operation:

1. Real time maps (RTM)
2. Optically projected maps (OPM)
3. Vector drawn intertrace maps (VDM).

The CAD RDVM uses the VDM method for reasons outlined in (2). It must be emphasised that these techniques are only required when video map information must be superimposed on a raw radar display. If plot extracted primary radar target data is available then combined digital map and digital target information may be handled by a computer graphics display system.

RADAR INTERTRACE DIGITAL VIDEO MAP GENERATION

Radar Intertrace Writing

Radar intertrace time is the short period between the end of a radar sweep out to maximum range and the beginning of the next radar sweep. This "waiting period" is required to ensure that target returns beyond the maximum range do not appear as second-time-around returns at close range on the next radar sweep. During the intertrace period the display system is idle and can be employed on other tasks. The RDVM generator is able therefore to take control of the display system during this time in order to write map data on the screen. One intertrace period is too short to write the whole map so map writing is spread out over a number of successive intertrace periods. The map must be written on the screen many times a second in order to present a flicker free image and this results in a minimum refresh rate which limits the amount of data that may be drawn on a single map.

The main advantages of intertrace writing are that a steady bright image is produced which has reduced smear when moved across the screen. Also all symbols and text remain at a constant size regardless of the radar range being displayed thus reducing display clutter. A substantial amount of distracting glare is eliminated from the display because the map is no longer refreshed by the rotating maintrace sweep.

The RDVM Hardware

Vectors. The RDVM uses a computer graphics technique to write information on the screen whereby the picture is built up from a number of straight line segments of variable length and direction. The radar display may be expanded to show only one-eighth of the total picture and this has two important consequences for the vector generator. In order to achieve maximum capacity it is designed to write lines on the screen as fast as possible, yet the maximum writing rate must not be exceeded on eight times expansion. The vector generator is constrained therefore, to write at only one-eighth the maximum display writing speed. Secondly, any non-linearities or inaccuracies are magnified eight times so that the vector generator must be much more accurate than in a typical computer graphics display.

Symbols. The symbol generator has its own high speed memory in which are stored the descriptions of the entire 64 symbol repertoire. Each symbol may consist of up to 30 strokes and is drawn at a rate of 230 nsec per stroke with a 150 nsec setting up time per symbol. The average symbol generation time is 4.0 μsec. Symbols are drawn on an imaginary text block grid which is positioned anywhere on the radar screen by the vector generator. The text block dimensions are, +27,-22 symbols per line and +27,-22 lines, relative to the text block origin. Although the position of these text blocks may shift when the display is expanded, the block itself remains a constant size. This has the important effect that only a symbol drawn on the origin at the centre of any text block will maintain a constant geographic position as the display is expanded. Individual symbols and text blocks may be written at one of two software selectable sizes.

Variable data. The Map Memory module contains the preprogrammed fixed map data while a Random Access Memory (RAM) module within the RDVM may be loaded with locally variable data. Thus the RDVM has the ability to display information which requires to be changed frequently. Such information is generated locally by the RDSS or is obtained from the Aeronautical Fixed Telecommunications Network (AFTN) and may include the following items:

Interconsole marks
Data Table entries
SSR targets and labels
Simulated targets for training
Temporary map data.

Construction. The RDVM construction is modular thus helping to speed up maintenance procedures and permitting future expansion with minimal down time. Extensive use has been made of ground plane flat ribbon cable with mass termination connectors for both digital and analogue signal distribution within the equipment.

Maintenance. A monitor module built into each map generator allows the maintenance technician to relate digital data flowing through the system to the analogue waveforms and also to individual data points which may be identified on the radar display. This reduces the need for specialised test equipment and speeds up fault diagnosis.

Increasing the RDVM Display Capacity

The fact that only 20% of the total display time is available for intertrace writing means that the RDVM data capacity might be limited unless steps are taken to optimise the use of this time.

Intertrace Overrun. The map generator must halt during each maintrace period. However, when the signal indicating the end of an intertrace period is received by the map generator it is not always necessary for it to halt immediately. Should the function being performed by the map generator be of a non-visual variety, a transition between points on the screen for example, then it may continue to carry out this function even though it has been disconnected from the radar display and a maintrace period has commenced. In other words the map generator continues to run into the maintrace period until such time as a visual function is to be performed. The resulting overrun of inter-trace generation into maintrace time can result in an increased display capacity of up to 5%. The actual increase achieved is of course data dependent.

Continuous Integration. The vector generator hardware has been optimised to reduce the setting up time required for each vector. Even so, a complex curve such as coastline which is made up of many short vectors may accrue a significant time penalty due to the cumulative effect of the setting up time for each vector. A continuous vector integration facility has been developed which allows continuous sequences of vectors to be generated with setting up time being required for only the first vector in the sequence. The potential increase in data display capacity due to continuous integration is a further 5%. Continuous integration also produces an aesthetically pleasing curve because vectors in the sequence are connected by smooth radii.

Further increases in map data display capacity can be achieved by using techniques such as radar retiming (radar video time compression).

Accuracy and Resolution

In order to achieve the required accuracy level 12 bit coordinates are used to specify positions on the 4096 x 4096 display grid. For a radar with a maximum range of 160 NM the grid resolution is therefore 0.078 NM. The RDVM uses a 48 bit data word which is partitioned thus:

NO.OF BITS	*FUNCTION DESCRIPTION*
12	*X* position / integration rate
12	*Y* position / integration rate
12	vector integration/settling time
6	symbol/vector type
2	RDVM function and symbol size
3	overlay allocation
1	brightness level
48	

A / in the function description indicates that the interpretation of these bits by the RDVM depends upon the function being performed. The data is multiplexed within the 48 bit word to conserve memory.

Registration

Automatic Offset Correction. The map and the radar targets are drawn on the radar screen independently of each other. It is therefore important to ensure accurate registration of the map and radar images.This is aided by the use of automatic offset correction circuitry in the RDVM generator. For a brief period at the end of each map refresh cycle the Vector and Symbol Generators are placed in an offset nulling mode at which time any analogue output errors are removed. This process has the advantage that temperature and time dependent errors are reduced.

Intertrace Jitter. If the map refresh is asynchronous, that is, independent of the radar pulse frequency (PRF) a problem referred to as intertrace jitter may occur. This is a loss of map definition due to the map writing being interrupted by the maintrace periods at different points during each map refresh. This potential problem has been eliminated by synchronising the map refresh with the radar PRF thus ensuring that the same stopping points are used during each successive map refresh.

Limitations

Detail. The amount of map detail is related to the complexity of curves such as coastline and this is in turn dependent upon the number of vectors used to construct the curves. The use of continuous integration removes most of the time penalty from producing complex curves so that map detail is limited primarily by the data capacity of the RDVM memory module.

Line Length. Vector writing is the slowest task performed by the RDVM so that the display time used to generate the video map is directly related to the total length of vectors making up the map.

The above two statements may be summarised thus :

DETAIL	*relates to*	*MEMORY USAGE*
LINE LENGTH	*relates to*	*WRITING TIME*

The Operator's Interface

The Air Traffic Controller's interface with the RDVM takes the form of a microprocessor based Interactive Keyboard and Tracker-ball unit. The key-switches in the data area of the keyboard light up to give the status of the function which has been selected in the command area while at the same time allowing changes to be entered. This compact unit enables the following functions:

- Map and Overlay Selection
- Interconsole Identification of targets
- Bearing/Distance measurement
- Data table configuration and positioning
- Radar signal selection
- Display offset
- Range Ring selection.

The communications between the Interactive Keyboard/Tracker-ball units at each display position and the RDVM generator are supervised by a microprocessor based system controller unit.

VIDEO MAP DATA PROCESSING

Much of the flexibility and capacity of the RDVM is derived from the computer data processing facilities which support it. Combined software/hardware solutions have been applied to many of the problems encountered during development of the RDVM.

Data Reduction

For reasons of display clarity or memory conservation it is often necessary to reduce the amount and hence detail of coastline data. This is done by curve fitting with straight line or biarc approximations as described in Edwards (4). The accuracy of the resultant simplified coastline curve is specified as part of the approximation process. The biarc curve fit offers the advantage that even very coarse approximations result in a curved rather than straight line representation of the coastline thus reducing confusion with other straight line data on the map.

Data Editing

A syntactical map data editor has been developed to simplify the entry,formatting and modification of the video maps. The editor "knows" all the hardware limitations of the RDVM and radar display system and will not permit syntax errors. Although the operator must have some knowledge of the RDVM when formatting a map, he need not have any in-depth understanding of the RDVM generator. 'HELP' files may be invoked by the operator while using the editor.

Display Path Optimiser

This program is used to order the map data for display so as to minimise the display time used for transitions between data. As with continuous integration and intertrace overrun, the aim is to make optimum use of the limited intertrace display time.

Memory Imaging

The symbolic map data produced by the map editor must be converted to a form suitable for the RDVM generator map memory module. The map memory imaging program performs the following tasks.

Rate and time calculation. When vectors are required between points defined on the display grid, scaled 12 bit X and Y integrator rates and the integration time are generated.

Settling time. A minimum settling time for positioning moves across the screen is calculated based on the distance being moved.

Vector Sequences. Where a vector sequence is to use continuous integration an iterative algorithium is applied to the start coordinates of each vector based on the scaled and rounded 12 bit integrator rates and time calculated for the previous vector in the sequence. This prevents the build up of cumulative rounding errors along the vector sequence. There is however a remnant hardware rounding error in the RDVM which must be removed by inserting a transition to a known starting point every so often along the vector sequence. Sequences of between 10 and 20 vectors may be generated before the insertion of a transition is required. This process is carried out automatically by the memory imaging program.

Refresh Rate. The heading block attached to each video map data file carries the radar station parameters required for a refresh rate to be predicted and a warning is issued if this drops below the minimum refresh rate dictated by the Critical Flicker Function (CFF) for the radar display. Typical dual phosphor mix raw radar displays suitable for intertrace writing have a minimum refresh frequency of 16 Hz.

Memory Programming. The 48 bit map data words are shuffled into the appropriate bit order and then transferred into the map memory module along with a test map for RDVM alignment.

IMPLEMENTATION AND ENHANCEMENT

Although the first two RDSS are to enter service shortly,they will not initially have the locally variable data or tracker-ball functions.These will be retrofitted when they become available. The equipment has been designed so as to allow the introduction of additional features with a minimum disturbance to operational service.The flexibility of the microprocessor based RDVM controller and Interactive Keyboard means that new functions may be added with relative ease. Possible functions for future development are:

- SSR target labels
- Target simulation for training
- Interconsole marks between ATC centres
- Automatic update of the Data Table by the AFTN
- Temporary "special event" maps
- Display of Lateral Separation Charts
- Remote programming of new maps via the AFTN.

MOTIVATION

The limited engineering effort available within the Ministry for development work has necessarily led to a policy of purchasing ready built equipment wherever economically practicable. The reasons for undertaking the design and development of an RDSS were firstly,that the design approach taken could lead to the introduction of new features not readily available in typical medium cost raw primary radar display systems. Secondly, the aggregated savings in overseas expenditure are considerable.

ACKNOWLEDGMENTS

The author wishes to acknowledge the considerable encouragement and support provided by his colleagues in the Radar Engineering Section of the Civil Aviation Division in the development work and in the preparation of this paper.

REFERENCES

1. Baines,M J,June 1979,"A Radar Maintrace Display Backup System-Preliminary Investigation and Proposal",RER 45, CAD, MoT, Wellington, New Zealand

2. Edwards,R J G,October 1978,"A Radar Intertrace Digital Video Map Generator - Preliminary Investigation",RER 44, CAD, MoT, Wellington, New Zealand

3. Edwards,R J G,May 1979,"Master Coastline File Specifications",CAD,MoT,Wellington, New Zealand

4. Edwards,R J G,November 1977,"The Presentation of Static Information on Air Traffic Control Displays",PhD thesis, Cranfield Institute of Technology. U.K.

ASMI-18X AN AIRPORT SURFACE SURVEILLANCE RADAR

J.D. Holcroft, S.J. Martin

Racal MESL Radar Ltd.

INTRODUCTION

Airport Surface Surveillance Radars have been available for over 25 years*, yet only 34 are in use today world wide. Their importance in increasing safety and expediting ground movements has been recognised and is not questioned. It is the authors' opinion that, as might be expected, the most important reason for their failure to be more widely adopted has been the capital and running costs of equipment available. Therefore, the aim of our programme was, by re-examining the possible approach to the system implementation, to develop a system offering the necessary high level of performance whilst minimising total life cycle costs. A system has been produced which, as well as satisfying all the requirements of major international airports, may also be afforded by airports with movement rates substantially lower than have hitherto been considered.

OPERATING FREQUENCY

Because this class of radars is required to produce a high resolution display over a limited range (usually less than 4km), the use of high frequencies has generally been favoured. Previous systems have operated in the 35GHz, 24GHz and 16GHz bands. All of these systems have, from necessity, employed a narrow vertical beamwidth to avoid the effects of rain clutter. The vertical polar diagram has usually been an inverted cosecant^2 or cosecant pattern with a 'flat top'. However consideration of the operational requirement indicates that coverage above the height of the antenna (which might be as low as 20m (66ft)) has the following advantages:

a) Detection of aircraft on approach, on departure and during missed approach, is achieved. Flat top beams usually result in loss of a departing aircraft partway along the runway, and the assurance that it has completed a successful departure is not provided.
b) Light aircraft and helicopters manoeuvering over the aerodrome surface are detected.
c) Studies(1) have shown that in restricted visibility the effectiveness of an ASSR in maintaining a high level of movements is greatly improved when the ground movement controller can identify aircraft by means of labels. This will be facilitated by the extended vertical coverage above the antenna, to allow the early automatic hand-over of secondary radar derived labels for arriving aircraft to an ASSR plot extractor (and vice-versa on departure).

However a sufficiently low operating frequency must be chosen so that vertical beamwidth can be increased without suffering from an unacceptably restricted range in rainfall. X-Band was selected as the optimum compromise between improving rainfall performance and increasing antenna size.

In this band a fan beam elevation pattern can be employed, in combination with RF swept gain (s.t.c.), to obtain the overall sensitivity which is required. This approach simplifies the antenna structure, and because of the reduced vertical aperture leads to an antenna design with very low mass and wind drag, a most important factor when the total cost of an installation is considered.

Appendix 1 outlines the analysis which shows that, with the selected antenna parameters, good performance in rainfall is achieved. At a rainfall rate of 16mm/hr, a 2m^2 non-fluctuating target will be detected at 3km with a $P_d = 0.9$ and $P_{fa} = 10^{-6}$ (for a single pulse).

Component Costs

Having chosen X-band primarily for the above reasons, there are cost advantages too. Complete price comparisons are difficult but X-band components are generally cheaper than their higher frequency equivalents. This is partly due to the high volume of X-band radars in manufacture, notably in the marine radar market.

A particular case in point is the magnetron used in the ASMI-18X. Continued development has made available devices with lifetimes of over 10,000 hours, at a fraction of the cost of high frequency devices. The magnetron for the 35GHz system is at least ten times more expensive, and has a typical life of only 1,000 hrs.

Waveguide Losses

Another advantage offered by the lower operating frequency is reduced waveguide losses. A control tower roof installation will typically result in a waveguide run of 15m. If a remote tower is used it is desirable, especially where extremes of climate are experienced, to mount the Transceivers at ground level, requiring a waveguide run of 30m or more.

At X-band a continuous length of the readily available and easily installed elliptical flexible waveguide can be used. For longer runs use of overmoded waveguide(2) can be considered. Comparative figures for the two techniques are given in Table 1.

TABLE 1 Two Way Waveguide Loss

Path Length	X-Band 9.4 GHz	
	Elliptical EW85	Overmoded WG10(WR284)
15m	3.3 dB	N/A
30m	5.9 dB	3.1 dB

SYSTEM CONFIGURATION

A simplified block diagram of a dual channel

*The first system, which operated at 35GHz, was installed at London Heathrow in 1956.

system is shown in Fig. 1. In less critical applications a single channel can be employed. Salient characteristics of the radar are listed in Table 2.

TABLE 2 Radar Parameters

Transmitter Frequency	9.41GHz
Transmitter Power	20kW
Pulse Length	40ns
Antenna Gain	36dBi
Receiver Noise Figure	10dB max
IF Dynamic Range	80dB
IF Bandwidth	40MHz

Antenna

The antenna is a horizontally polarised non-resonant, end-fed slotted waveguide array. Characteristics of the complete antenna sub-system are shown in Table 3.

TABLE 3 Antenna Sub-System

Gain	36dBi
Aperture (horizontal)	5.4m
(vertical)	0.15m
Azimuth Beamwidth	0.4°
Vertical Beamwidth	15°
VSWR	1.18
Bandwidth	100MHz
Centre Frequency	9,400MHz
Turning Speed	60 r.p.m.
Weight (Antenna, Turning Unit and Pedestal)	265kg
Wind Drag @ 180km/hr	2.2kN
Overturning Moment @ 180km/hr	5.3kNm

The antenna is encased in an aerodynamic, electrically thin rotating radome, and thus avoids the well known problems attendant upon the use of fixed radomes. Because of the low weight of the antenna and turning unit combined with the low wind drag and overturning moment, the units can usually be mounted on existing control tower roofs without the need for structural modifications. If the control tower is not in a suitable location then a low cost steel lattice tower of small cross-section can be used.

Receiver

The system is required to provide high definition over the whole aerodrome surface including ranges down to about 50m. At these short ranges provision must be made to prevent overloading of the r.f. mixer, bearing in mind target echo area can be as high as $10^5 m^2$. Fig 2 shows how received power varies with range for a target near ground level. The graph shows three regions, namely:

a) A far field region beyond about 1km where the received power varies as R^{-4}.
b) Below 1km the azimuth beam can usefully be approximated as parallel sided, and the received power varies approximately as R^{-2}
c) At ranges less than 200m the target enters the skirts of the vertical polar diagram and the received power falls as range decreases. Phase 'error' across the antenna vertical aperture eliminates nulls and ensures complete elevation coverage.

It will be noted from Fig. 2 that the received power can be as high as +20dBm. This problem is overcome and the normal swept gain (s.t.c.) function implemented by a programmable attenuator prior to the r.f. mixer. This is achieved using a solid state pin diode/varactor limiter in place of the conventional T/R cell, in which the varactor bias current is modulated, under the control of a digitally stored waveform.

The IF Amplifier has a 3dB bandwidth of 40MHz. This overmatched bandwidth, defined by a linear phase filter, is used to ensure minimum pulse stretch. Loss of performance in comparison with a matched receiver is approximately 0.3dB. A logarithmic IF gain characteristic, obtained by "successive detection", and with a dynamic range of over 90dB is employed. A number of advantages over a linear receiver are achieved, namely:

i) IF and Video gain controls are eliminated (and thus cannot be miss-set).
ii) Adjustment of the RF swept gain control is not critical in that there is a large dynamic range window into which returns can be placed.
iii) Targets in close proximity and with widely differing radar cross sections are accommodated and their relative magnitudes are indicated to the operator.
iv) The well known ability of logarithmic receivers to reduce rain clutter to the receiver noise level applies[3]. This is achieved by differentiating the log video (f.t.c.); under these conditions performance equivalent to a linear receiver with perfectly adjusted s.t.c. can be obtained.

Bright Display System (BDS)

The main elements of the BDS are shown in Fig. 1. The Digital Scan-Converter (DSC) converts the radar video to TV raster format adding synthetic video maps for picture enhancement. The Control Processor supervises the transfer of data from the Map Store and interprets commands from the operator's Control Panel. The Map Store and Control Processor (MSCP) is a commercially available minicomputer. It has the capacity to control two DSCs and can accept data from external sources for presentation on the display. The features of the BDS are summarised in Table 4.

TABLE 4 Features of the Bright Display System

Video Inputs	2 alternative
Rain anti-clutter	FTC selectable
A/D conversion	4 bits (16 levels)
Max. sampling rate	75MHz (2 metres)
Range scales	20 (3 normally for ASSR)
Azimuth increments	4096 (0.088°)
Radar trigger	Asynchronous
Correlation	Pulse to pulse, 2 out of 2
Frame Store Size	Either (i) 512 lines (768 x 512 pixels) or (ii) 1024 lines (1024 x 1024 pixels)
Origin shift	Anywhere within frame store
Zoom/pan	Controlled by joystick
Display raster	1250 lines, 50 fields/sec., 2:1 interlace Programmable for other formats
VTR output	625 line CCIR standard
Cursor	Controlled by joystick
Video maps	4 independent maps
Alphanumerics	Full screen, 96 x 48 characters.

The Frame Store (FS) can be one of two sizes to meet customer requirements. Fig. 3 compares the resolution obtainable in each case for a 3:2 displayed area, and relates it to the radar resolution. Display resolution varies from P, the length of the pixel side, to $2\sqrt{2}P$, depending on the positions of the targets relative to each other and to the pixel grid. Typically, the 1024-line FS can show a whole airfield at a resolution 25% better than the 512-line FS, but on short range scales it can be up to 50% worse.

When the 512-line FS is displayed on the 1250-line raster, each line of the FS appears on two adjacent scan-lines; the effective 50Hz refresh makes flicker unnoticeable. Degradation of resolution by spot size is minimised because each pixel is two lines high and about two spot-diameters wide.

The FS comprises 4 memory planes allowing for 16 levels some of which may be reserved for synthetics. Data are decoded on output so that the brightnesses of all synthetics can be varied independently over the 16 levels while, at the same time, the radar video can span the full dynamic range. The algorithm that writes data to the FS is stored in PROM so that the number of levels for synthetics, and the persistence and other display characteristics, can be altered to suit the customer.

Video Maps

Provision is made for two outline maps, one blanking and one suppression map. Large scale maps are digitized and processed, using the same minicomputer as the MSCP, to produce the video map data in range-azimuth format. This format makes blanking easier and allows for elaborate variable suppression. Range-azimuth maps are independent of origin shift.

Outline maps are used to indicate runways, taxiways etc. Intricate outlines can be clearly shown because individual pixels are addressed by the line-drawing algorithms in the map preparation programs. Separate map data are used for each range scale so that the pixels which give the best line appearance can be chosen in each case. This also means that the features presented may be different on each range scale.

The blanking map completely suppresses the radar over selected areas of the display eliminating distractions. This will usually leave blank areas large enough to present useful alphanumeric information.

Suppression can be used in either SHIFT or CLIP mode. SHIFT reduces radar returns by a set amount so that areas of significant but lesser interest can be displayed at reduced brightness. CLIP completely suppresses returns that are below the suppression level while leaving large signals unaffected.

Distance from Threshold Indicator (DFTI)

The availability of suitable long range scales and separate video maps enables the provision of a DFTI display using the video from the approach radar. Alternative video inputs to the DSC can be selected by the MSCP so that either an ASSR or a DFTI picture may be shown on one display under operator control.

When both displays are required simultaneously, both DSCs may be controlled by one MSCP. A dual redundant system with ASSR and DFTI displays can be provided using only two MSCPs. Furthermore, the outputs of two DSCs can be synchronised to provide a composite ASSR/DFTI display.

TRIALS RESULTS

A trials and demonstration installation has been completed at Edinburgh Airport. The site is approximately 40 metres from the terminal building, an all metal clad structure. Antenna height is 30m, this being suitable for the majority of airports.

Standard $1m^2$ corner reflectors were placed in groups of three at different ranges on convenient grass areas on the aerodrome. All targets were at a height of 2m above the ground. Separations were adjusted to the minimum value at which the returns from all the groups of targets could simultaneously be clearly resolved on the PPI display. Results are shown in Table 5.

TABLE 5 Resolution

Range	Resolution	
	Range	Azimuth
300	9	6
1,000	9	6
2,000	10	15

Detection of vehicles was checked using a 'mini-van', the smallest target of that type which is likely to be encountered at an airport. Continuous detection of the moving vehicle without 'fading' was observed at all parts of the airfield including horizontal ranges down to less than 45 metres.

The use of the suppression feature described earlier has been found to be particularly valuable in enhancing the detection of men and vehicles on the grass areas. The suppression threshold can be set so that grass returns are completely eliminated whilst vehicles and men are seen at high contrast against the dark background thus created. Trials showed that a standing man whose r.e.a. at X-band is about $0.5\text{-}1m^2$ can be separated by this method from the grass returns down to approximately 200 metres range. Below this range the increasing grazing angle causes the returns from the grass to approach that from a man.

An important feature of an ASSR is its ability to allow the operator to identify aircraft by category (i.e. small, medium, heavy, extra heavy). Fig. 4 a photograph taken from the scan-converted display shows the complete airport, an area 3.4km x 2.3km. The large aircraft on the main runway is a B747, the group on the apron at the r.h.s. are medium size - Tridents and 1-11s. Although it is not always possible to distinguish between fixed and rotary wing aircraft from a single frame, observation of the display reveals a unique 'scintillating' effect from the latter type, presumably due to the variation of the rotor blade return from scan to scan.

ACKNOWLEDGEMENTS

The authors would like to acknowledge the early work carried out by personnel in a number of the Racal Decca companies, without whose large store of experience in ASSR systems the development of the new system would not have been so successful. We would also like to thank the Department of Industry for supporting the project.

REFERENCES

1. Peri, M., ICAO Surface Movement & Guidance Grp. WP/106.
2. Racal Decca, Patent Specification 905, 689 "Improvements in or relating to Waveguide Transmission Systems".
3. Croney, J., April 1956, Wireless Engineer, "Clutter on Radar Displays".

Appendix 1 Performance in Rainfall

ASSR systems use very short pulse lengths and direct energy towards the ground. This leads to the formation of three clutter volumes a, b and c, as shown in Fig. 5. Significant overlap between these volumes occurs at ranges of interest and the resulting interference effects must be considered. Fig. 5 identifies six regions within which the effective weighting factors relative to the free space case are as given in Table A1 (where ρ is the scalar ground reflection coefficient, assumed constant)

TABLE A1 Weighting Factors

W_1	=	1
W_2	=	ρ^4
W_3	=	$4\rho^4$
W_4	=	$\rho^4 + 4\rho^2 + 1$
W_5	=	$\rho^4 + 2\rho^2$
W_6	=	$2\rho^2 + 1$

The weighting factor for targets near the ground is W_4, provided their vertical extent h satisfies $h > \lambda R/2H$ (H = antenna ht).

Becuase the six regions are non-uniformly distributed in elevation the variation in antenna gain must be taken into account. The calculation can be effected with sufficient accuracy by segmenting the regions and applying a weighting factor Xi, proportional to the two-way gain at the elevation on which the segment is centred. Total clutter echoing area is then given by:

$$\sigma = \eta \sum_{j=1}^{6} W_j \sum_i V_{ji} X_i$$

where η = clutter reflectivity

V_{ji} = 1th element of volume j

The quoted result applies for $\rho = 1$, but it is found that performance is not sensitive to changes in ρ over the range of values likely to be encountered. The effect of path attenuation is negligible because of the high s/n ratio in clear weather (>30dB at 3km) together with the minimal path attenuation of 0.21dB/km which is suffered.

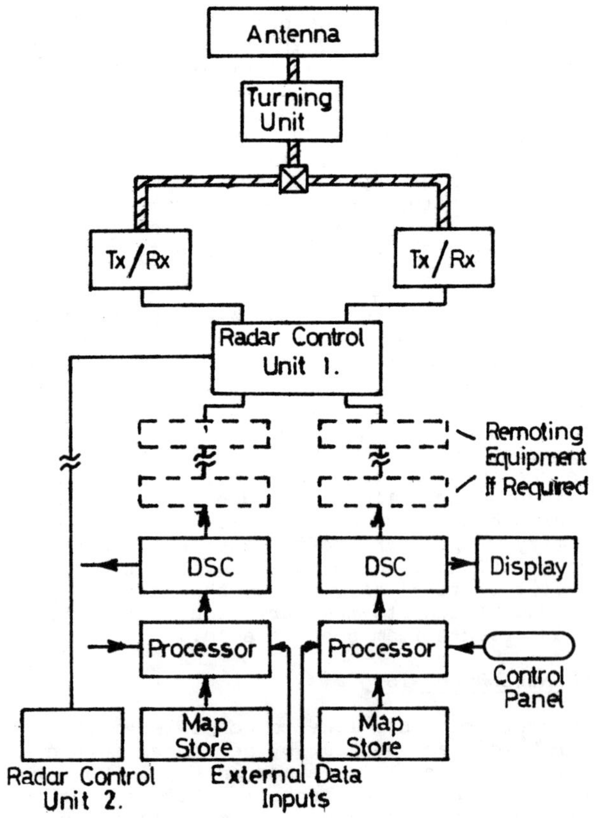

Fig. 1 Simplified System Block Diagram

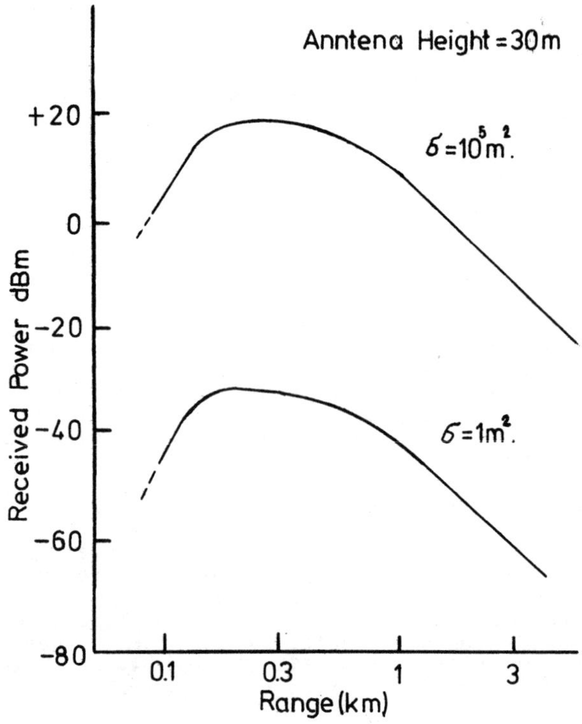

Fig. 2 Received Power

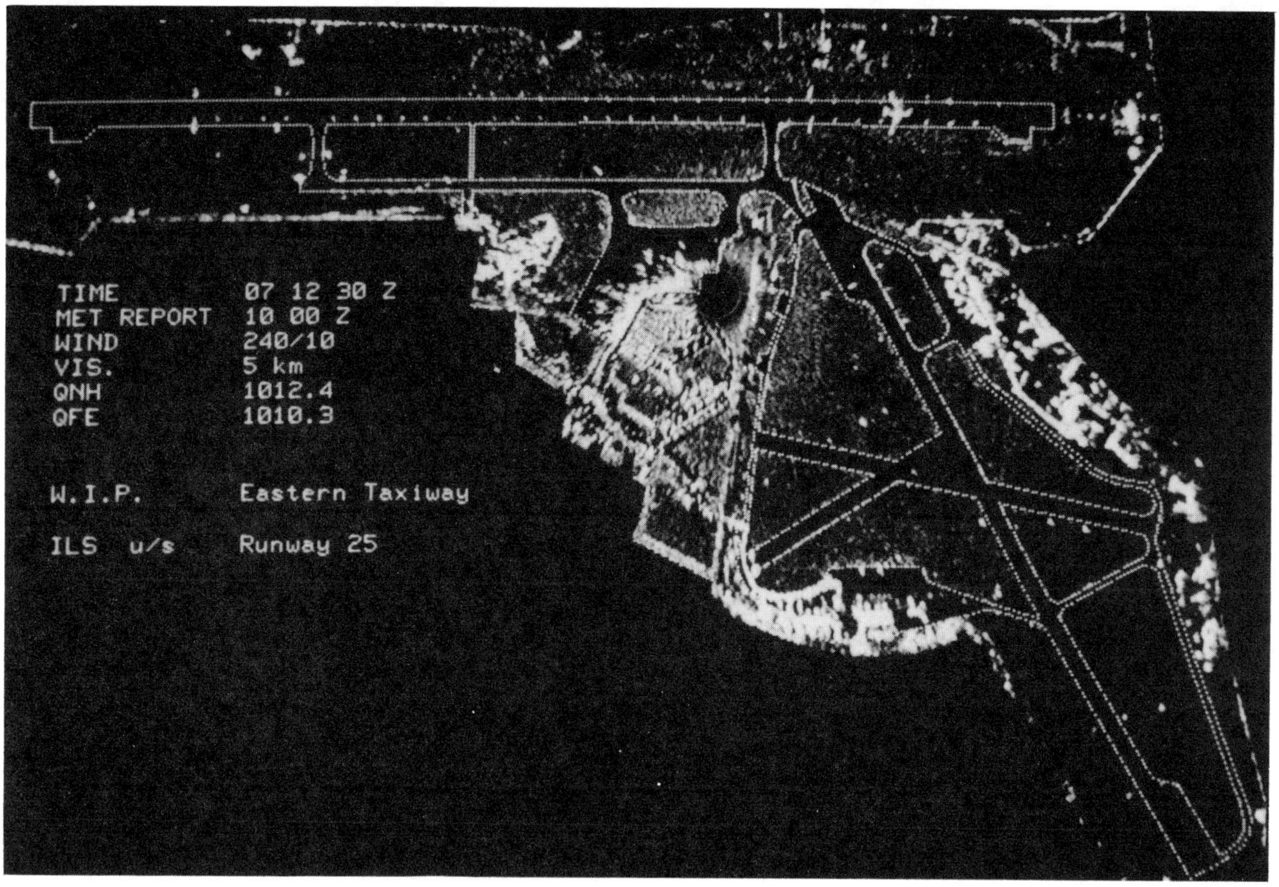

Fig. 4 Scan Converted Display

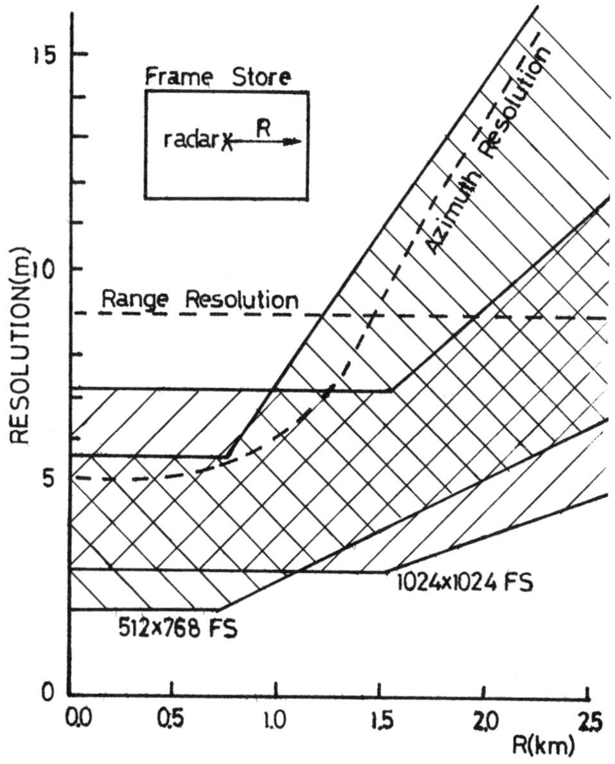

Fig. 3 Scan Converter Resolution

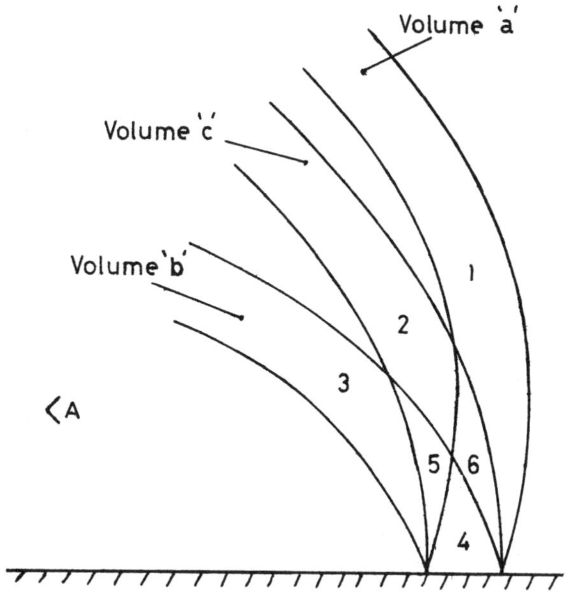

Fig. 5 Clutter Volumes

DETECTION OF HAZARDOUS METEOROLOGICAL AND CLEAR-AIR PHENOMENA WITH AN AIR TRAFFIC CONTROL RADAR

D.L. Offi, W. Lewis, T. Lee

Federal Aviation Administration Technical Center, USA

INTRODUCTION

Aircraft operating in and around the vicinity of airports may encounter a number of undetected hazards. Low-level wind shear and turbulence, associated both with thunderstorms and aircraft trailing wake-vortices, have long been indentified as problems confronting the Air Traffic Control (ATC) system. The reliable detection and measurement of these hazards in all weather regimes has been, and continues to be, a major goal of the Federal Aviation Administration (FAA).

Work performed by various researchers has shown that radar has the potential for making wind shear, turbulence, and wake vortex measurements. Doviak, et al. (reference 1) describe the origin and development of Doppler weather radar technology and present a thorough survey of significant efforts in this field. Strauch and Moninger (reference 2) have shown that pulse Doppler radar may provide wind shear warnings at all times, and Chadwick, et al. (reference 3) discuss the potential use of radar for trailing wake-vortex detection.

The FAA, in conjunction with the Wave Propagation Laboratory (WPL) and the Massachusetts Institute of Technology's Lincoln Laboratory (MIT/LL), developed modifications to the Airport Surveillance Radar (ASR)-8 test bed located at the FAA Technical Center, Atlantic City Airport, N.J., to observe the hazardous phenomena. The ASR-8 is a standard, dual-channel radar used for terminal area ATC purposes. The modification, consisting of a second antenna system and a separate receiver and Doppler processing chain, utilized the standby channel of the radar for these measurements. Initial tests have been completed and the results reported by D. Offi and W. Lewis, (references 4 through 6). This paper presents a summary of those results.

SYSTEM DESCRIPTION

A 15-foot parabolic antenna is interconnected with the ASR-8, operating with a peak transmitter power of 1 MW, a frequency of 2.79 GHz, a pulse repetition rate of 1030/second, and a pulse length of 0.6 μseconds. The separate receiver and Doppler processing chain includes inphase and quadrature (I&Q) detectors, analog/digital (A/D) converters, and a mini-computer with associated peripherals. A more complete description is provided by references 4 through 6).

RADAR WIND AND WIND SHEAR MEASUREMENTS

A 128-point Fast Fourier Transform (FFT) program produced the Doppler mean radial wind values. The radar's wind measurement capability was evaluated under various weather conditions by comparison of data with simultaneously recorded aircraft measurements. Potentially dangerous winds associated with thunderstorms were also detected and measured.

Flight Tests

Two different aircraft, equipped with meteorological sensors, flew various terminal area approach and departure paths. The radar was configured to obtain radial wind data with the antenna beam fixed in elevation and azimuth along a simulated glide slope profile for one set of measurements, and in a two-azimuth sampling sequence for another. Figure 1 is a sample of the radar/aircraft comparison for the simulated glide slope runs.

The Technical Center radar is located 2.5 km from the main airport runway. Wind components along the glide slope were derived using the two-azimuth pointing technique described by Strauch and Swezey (reference 7). This technique provides wind speed from the calculated longitudinal and transverse wind components. Wind direction is determined by appropriate trigonometry. Radar data were obtained for eight levels from 200 to 1,600 feet above ground level. The aircraft headwind/tailwind data were compared with the corresponding radar data. Table 1 is a statistical summary of the results for sets of approaches on separate days.

TABLE 1 - Comparison of Radar and Aircraft Headwind/Tailwind Components for Actual Approaches with Two-Azimuth Scan

Radar	A/C*	Difference*	Sigma*	Approaches	Total Comparisons
18.4	18.8	-0.4	4.7	6	38
21.9	22.9	-1.0	3.2	5	33

*Values in knots, headwind positive. Sigma is the standard deviation of differences.

It is important to emphasize that the two-azimuth pointing technique is applicable only to horizontally homogeneous winds and, as pointed out in reference 7, should not be used where complex three-dimensional wind fields are expected.

Thunderstorm Wind Detection

One of the causes of potentially dangerous, low level wind shear is the gust front, or wind outflow in advance of an approaching severe thunderstorm. This has been identified as a prime contributing factor in a number of aircraft accidents. Detection of the phenomenon has been a major goal of the FAA effort.

In one test sequence, radar data were obtained in the clear air as a fast-moving thunderstorm squall line approached from the West. Figure 2 shows profiles of radial wind velocities from two radar observations made just 3 minutes apart. The antenna was at +3° elevation and at 270° azimuth, pointing in the direction of the surface wind, as determined by a tower mounted anemometer. At the 300-foot level a distinct bulge, or nose, can be seen in the later profile, representing an increase of about 17 knots, going from an earlier reading of 8 knots to a sustained velocity of 25 knots.

On another occasion thunderstorm winds were measured to a higher level. Figure 3 is a radial wind profile from that storm, with the antenna in a +3° fixed glide slope configuration, looking into the direction of the surface wind. In this case the storm appeared to be producing an outflow at a higher altitude level. Note the potentially dangerous wind shear between 800- and 1,000-foot altitudes, with the radial wind velocity increasing from 26 to 44 knots.

THUNDERSTORM TURBULENCE MEASUREMENTS

ASR-8 Doppler Radar Measurement

Thunderstorm precipitation reflectivity, Doppler mean velocity, and turbulence values were achieved through spectral moment estimation using specific pulse-pair algorithms described by Labitt (reference 8). The zeroth moment of the Doppler spectrum provides the precipitation reflectivity factor in dBZ. The first moment provides the mean radial velocity in the pointing direction of the radar beam. The second moment (velocity spectral width) provides a measure of turbulence within the pulse volume. The spectral width is converted to $\epsilon^{1/3}$ (the cube root of the turbulence dissipation factor) by the following equation (reference 8):

$$\epsilon_r^{1/3} = \frac{\sigma_u}{1.352\ a^{1/3}} \quad (cm^{2/3}\ sec^{-1}) \qquad (1)$$

Where, σ_u is the spectral width (cm sec^{-1}), ϵ is the turbulence dissipation factor ($cm^2\ sec^{-3}$), "a" is the Gaussian half-power beamwidth (cm), and 1.352 is composed of known constants, assuming isotropic and homogeneous turbulence. Since "a" is a function of range, the cube root of ϵ is essentially a range-weighted spectral width and, as such, should be superior to spectral width as a turbulence measure. In subsequent text references, $\epsilon_r^{1/3}$ will be designated RADEP (radar epsilon).

Aircraft Turbulence Measurement

A Grumman Gulfstream-I turboprop aircraft was used to penetrate thunderstorms. The aircraft was equipped with a center-of-gravity (CG) accelerometer (20 samples/sec), a differential pressure transducer (100 samples/sec), an altitude transducer, a Rosemount temperature probe, a crystal-controlled clock, and a Litton inertial navigation system (INS), all sampling at 1/second. Data were digitized and recorded on magnetic tape.

Three measures of turbulence were derived from the aircraft instrumentation: (1) derived equivalent gust velocity (Ude) (Crooks, reference 9); peak Ude values were computed from incremental normal accelerations for running 7-second periods, updated each second; (2) $\epsilon^{1/3}$ from CG incremental normal accelerations (reference 8); and (3) $\epsilon^{1/3}$ from airspeed (Pitot pressure) fluctuations (reference 8). Both values were computed from running 7.5-second periods using a cosine-squared weighting, updated each second.

The two aircraft $\epsilon^{1/3}$'s are highly correlated. Linear coefficients of about 0.9 were obtained for several thunderstorm flights, with regression line slopes about 0.75 (airspeed derived values smaller). Therefore, they were combined into one value by computing the root mean square of paired one-second values, weighting the airspeed value with a four-thirds factor. In subsequent text reference the combined turbulence parameter ($\epsilon_{ap}^{1/3}$) will be designated ACEP (aircraft epsilon).

Aircraft Turbulence Scales

Peak Ude has been established to classify the degree of turbulence experienced by an aircraft (Lee, reference 10). This scale is: light ($\overline{>}$1.5 to 6.1 mps), moderate ($\overline{>}$6.1 to 10.7 mps), severe ($\overline{>}$10.7 to 15.2 mps), extreme ($\overline{>}$15.2 mps).

A corresponding scale for ACEP was determined but with the Ude light category further classified into negligible (0 to 3 mps) and light ($\overline{>}$3.0 to 6.1 mps). The ACEP scale was determined by defining an error function between ACEP and Ude which are highly correlated (~0.9). The square of this function (ACEP-λUde), was minimized over the data set. The factor λ was found to be 0.7 for several flights into thunderstorms. This was applied to the Ude category limits to determine the ACEP category limits, except that the upper limit of the severe category was set at 12.5, as established by MacCready (reference 11), for an aircraft traveling at 200 mph (Gulfstream speed). The ACEP turbulence scale (table 2) agreed well with recorded voice comments by the aircraft pilot for several thunderstorm flights.

Data Collection and Analysis

The aircraft was launched when thunderstorms were occurring within 40 nmi of the radar. National Weather Service WSR-57 radar video integrator and processor (VIP) level 3 and 4 storms (41-46 dBZ, heavy, and 46-50 dBZ, very heavy, respectively) were preferred. VIP level 5 and 6 (>50 dBZ) storms were avoided because of the high probability of damaging hail. The aircraft was flown in an altitude block of 3,000 to 5,000 feet to avoid encroaching on the New York Air Route Traffic Control Center airspace.

The beacon track of the aircraft was smoothed and interpolated by an FFT process to produce position points at approximately 1.2-second intervals. INS data were used to fill in missing points.

Comparison of Radar and Aircraft Data

The plots of ACEP and RADEP for the first half of a mission flown on July 17, 1980, into a very heavy (VIP level 4) thunderstorm are shown in figure 4. It is evident that there is good basic correspondence between aircraft and radar turbulence, although the radar curve is smoother and does not show the detailed turbulence fluctuations of the aircraft curve.

An analysis of some of the spectra showed that the very large RADEP peaks near 17:05 and 17:30 were due to markedly non-Gaussian spectra, which produced erroneously wide spectral widths. Also, the peak near 17:09 was caused by aircraft signal contamination in that range gate. This peak and and others contaminated by aircraft signal were removed in subsequent processing. However, peaks due to non-Gaussian spectra were not removed because a discriminating algorithm to correct such spectra has not been developed.

Correlation coefficients between ACEP and RADEP were computed. Data points were taken from consecutive independent 7-second periods, with the peak RADEP value for each period extracted for comparison with the aircraft values. RADEP values less than 2 were eliminated since there was insufficient signal strength to make a reliable width estimate.

The sequence from 17:09-15 showed the best agreement with a correlation coefficient between ACEP and RADEP of 0.81. The correlation coefficient for the entire mission was 0.51.

The low overall correlation between radar and aircraft turbulence is due to several factors. One is that the aircraft more realistically measures the actual turbulence fluctuations, as shown by figure 4. Other factors are the effect of the non-Gaussian spectra and deficiencies in the pulse-pair algorithm which tends to have increasing error with increasing spectral width. Wind shear could also broaden spectra but its contribution is generally thought to be small. (Doviak, et al., reference 12). Finally, errors may occur through interpolation of radar values along the aircraft track between the 80-second radar scans. (Computer capacity limits data acquisition to an approximately 10 x 10 nmi window at an 80-second update rate).

Radar/Aircraft Turbulence Categories

While the overall correlation between radar and aircraft turbulence is not high, it does contain useful information. This is shown by classifying the radar turbulence by category in relation to categorical aircraft turbulence. Table 2 shows the comparison between RADEP and ACEP for the entire July 17, 1980, flight.

The radar categories were chosen to provide a good discrimination between the classes of turbulence. Table 2 shows that this was essentially achieved up through the severe category, but at the expense of considerable mismatch in the corrresponding radar-aircraft categories.

Some of the "too-large" values of RADEP are due to non-Gaussian spectra. Others may be due to the pulse-pair method's deficiency in treating large spectral widths, which would contribute to the lower triangular distribution apprearance of table 2. Interpolation between scans is also a source of error, as noted previously.

Table 2 was simplified in order to better illustrate the radar/aircraft turbulence relationship. The five radar categories were reduced to three basic categories by combining N and L into "light" and combining S and E into "severe." The only change in the aircraft categories was to also combine S and E into "severe". The result is shown by the boxes in table 2 where the percentages on the right are occurrences of the indicated aircraft categories for the three resultant radar categories. The simplification shows that almost all aircraft turbulence occurrences with RADEP values less than 4.5 (light turbulence) were "N" or "L." For radar values in the 4.5-7 range (moderate turbulence), turbulence occurrences were concentrated in the aircraft "L" and "M" categories. When RADEP values were ≥7 (severe turbulence), aircraft turbulence was concentrated in the "M" and "severe" (S/E) categories, with the occurrence of severe increasing sharply.

The results for the combined radar severe turbulence category of table 2 can be improved by a screening based on the radar reflectivity factory (dBZ) in two categories, namely ≥35 and <35 dBZ. For RADEP values ≥7 and dBZ ≥35, there is an 86 percent association with moderate and severe aircraft turbulence (as opposed to 75 percent overall). However, for RADEP values ≥7 and dBZ <35, there is only a 45 percent association with moderate and severe aircraft turbulence. The association of high turbulence with high values of dBZ for flights into thunderstorms was also noted by Sand, W.R., et al (reference 13).

WAKE VORTEX MEASUREMENT

An analysis of factors involved in vortex detection by microwave radar indicated the possibility that a system such as the Technical Center's ASR-8 may also be able to detect vortices.

Fridman (reference 14) estimated that the radar cross-section of a vortex in clear air is between 10^{-13} m^2 and 10^{-9} m^2. Using the greatest estimate and the radar range equation below indicates that, in perfectly calm and homogeneous air, the ASR-8 and 15-foot parabolic antenna combination could detect a vortex on the basis of reflectivity at a distance of 2.2 kilometers (km).

$$R^4 = \left(\frac{P_t}{P_{rmin}}\right)\left(\frac{G^2 \lambda^2}{64\pi^3}\right) L\sigma \qquad (2)$$

where:

R is the radar range in meters
P_t is the transmitted power (1.4 x 10^6 MW)
P_{rmin} is the receiver minimum detectable signal (1.26 x 10^{-14} W, -109 dBm)
L is the system losses (2 dB)
G is the antenna gain (39 dB)
λ is the radar wavelength (10^{-1} meters)

Wake Vortex Detection

A limited number of tests were performed with the radar antenna pointing at +3° elevation and 90° elevation, while the radar system was configured to obtain data samples in the volume of space near a controlled B-727 jet aircraft flying in the airport vicinity.

TABLE 2 - $\epsilon_r^{1/3}$ Versus $\epsilon_{ap}^{1/3}$ for Flight of July 17, 1980

$\epsilon_r^{1/3}$		$\epsilon_{ap}^{1/3}$: 0 to 2.1	2.1 to 4.3	4.3 to 7.5	7.5 to 12.5	≥12.5	Total	Aircraft Cat.	Percent
		N	L	M	S	E			
2 to 3	N	10	10	0	0	0	20	N-L	90
3 to 4.5	L	10	25	6	0	0	41		
4.5 to 7	M	11	34	42	4	0	91	L-M	84
7 to 12.5	S	6	17	49	25	1	98	M-S	75
≥12.5	E	1	3	3	2	0	9		
Total		38	89	100	31	1	259		

Correlation coefficient = 0.51

Note: Turbulences categories: N = negligible, L = light, M = moderate, S = severe, and E = extreme.

There were indications of clear air signals in the data on a number of flights, but not in all cases. The exact nature of these signals and their relationship to aircraft wake vortices was not determined for these feasibility tests. A sample of the observed signals with the 90° elevation is shown in figure 5. In this case the aircraft was flying at 1,600 feet, on a path which was not directly over the site. The wind velocity, as measured by the aircraft, was approximately 15 knots at that altitude. Thus, the vortices would be expected to be in the antenna beam sampling volume approximately 1 minute after passage of the aircraft.

The system was configured to obtain data at one fixed point in altitude (range) at about 1,575 feet and recording was continuous. The resulting clear air signals associated with the aircraft's wake can be seen in the 8th, 9th, and 10th data samples, alternating from about -20 to +20 knots radial velocity as the vortices drifted across the radar antenna beam from 42 to 54 seconds after the aircraft's passage. Since the aircraft was traveling at a speed of 120 knots, it was approximately 2 nmi from the radar site when the signals, shown in figure 5, were detected. Therefore, they were not due to reflections from the aircraft. This is further verified by the change from negative to positive Doppler velocities associated with the signals as a function time.

For the test configuration used, there was very little radial velocity (vertical movement) associated with the aircraft wake signals being detected. The spectral signals centered about zero Doppler in all the data samples are due to side-lobe detected ground clutter and ambient clear air returns. The low amplitude signals at ±35 knots are caused by ground clutter signal saturation in the receiver.

REFERENCES

1. Doviak, R. J., Zrnic', D. S., and Sirmans, D.S., 1979, Doppler Weather Radar, Proceedings, IEEE, Vol. 67, No. 11, pp. 1522-1553.

2. Strauch, R. G. and Moninger, W. R., 1978, Radar Measurement of Windshear and Wind Profiles for Air Safety, Proc. of the 18th Radar Meteorology Conf., American Meteorological Society, Boston, Mass., pp. 432-436.

3. Chadwick, R. B., Moran, K. P., and Campbell, W. C., 1979, Design of a Wind Shear Detection Radar for Airports, IEEE Trans. on Geoscience Electronics, Vol. GE-17, No. 4, pp. 137-142.

4. Offi, D. L. and Lewis, W., 1981, Wind Shear Detection with a Modified Airport Surveillance Radar, Preprints, 20th Conf. on Radar Meteorology, American Meteorological Society, Boston, Mass., pp. 424-429.

5. Lewis, W., 1981, Doppler Radar and Aircraft Measurement of Thunderstorm Turbulence, Preprints, 20th Conf. on Radar Meteorology, American Meteorological Society, Boston, Mass., pp. 440-445.

6. Offi, D. L. 1982; Wake Vortex Detection with Pulse Dopppler Radar, FAA Technical Center Letter Report, No. CT-82-100-30LR, Atlantic City, N.J.

7. Strauch, R. G. and Swezey, W. B., 1980, Wind Shear Detection with Pulse Doppler Radar, FAA/SRDS, Final Report No. FAA-RD-80-26, Washington, D.C.

8. Labitt, M., 1981, Coordinated Radar and Aircraft Observations of Turbulence, ATC-108, MIT Lincoln Laboratory, Lexington, Mass, (Prepared for FAA Systems Research and Development Service.)

9. Crooks, W., 1965, High Altitude Clear Air Turbulence, Technical Report No. AFFDL-TR-65-144, AF Flight Dynamics Laboratory, Air Force Systems Command, Wright-Patterson AFB, Ohio.

10. Lee, J. T., 1971, Thunderstorm Turbulence, Proc. FAA Turbulence Symposium, Washington, D.C.

11. MacCready P. B., Jr., 1964, Standardization of Gustiness Values From Aircraft, J. Appl. Meteor, 3, pp. 439-449.

12. Doviak, R. J., Sirmans, D., Zrnic', D., and Walker, G. B., 1978, Considerations for Pulse-Doppler Radar Observations of Severe Thunderstorms, J. Appl. Meteor, 17, pp. 189-205.

13. Sand, W. R., Musil, D. J., and Kyle, T. G., 1974, Observations of Turbulence and Icing Inside Thunderstorms, Proc. 6th Conf. on Aerospace and Aeronautical Meteorology, Am. Met. Soc., Boston, Mass.

14. Fridman, J. D., 1974, Airborne Wake Vortex Detection, FAA, Report No. FAA-RD-74-46, Washington, D.C.

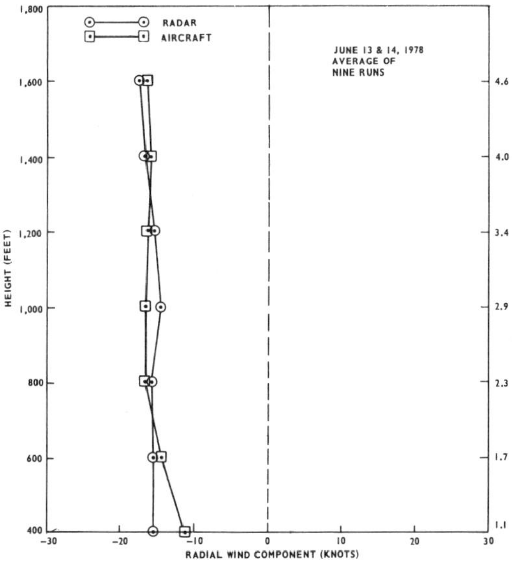

Figure 1 Radar/Aircraft Wind Comparison

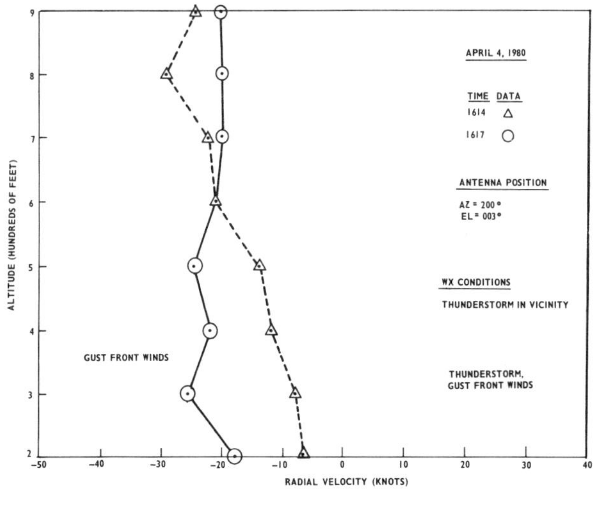

Figure 2 Thunderstorm Gust Front Winds

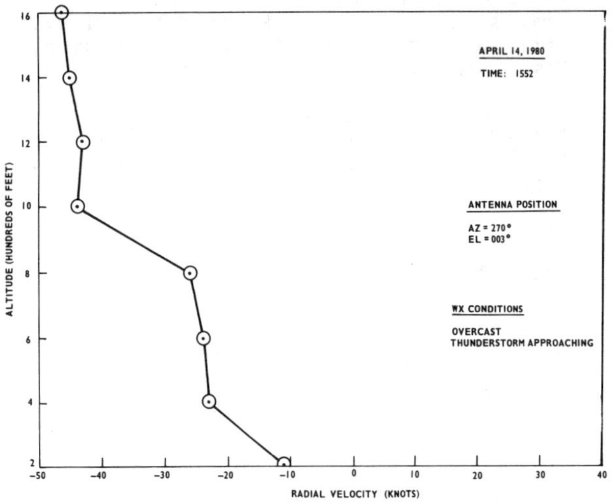

Figure 3 Thunderstorm Wind Outflow

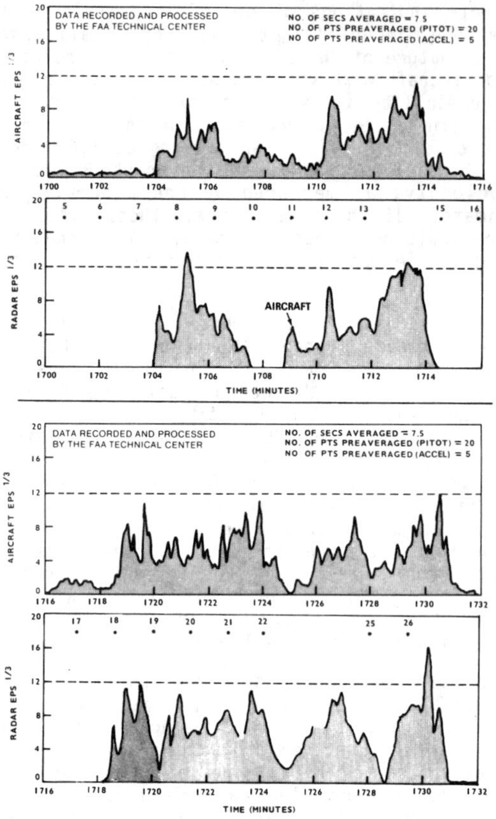

Figure 4 Comparison of ACEP and RADEP for July 17, 1980

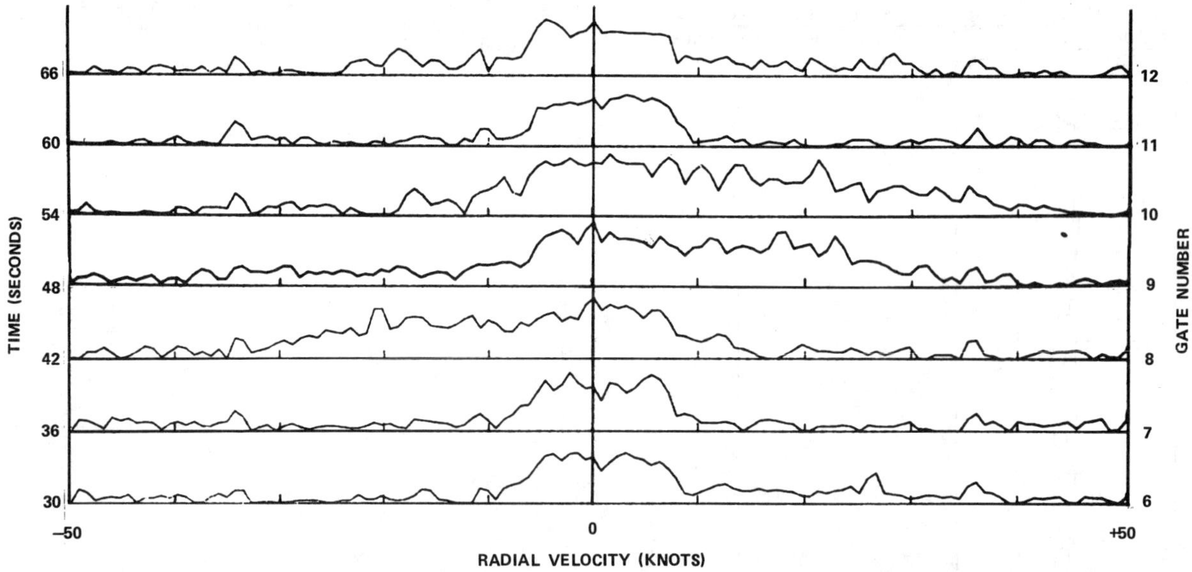

Figure 5 Wake-Associated Spectral Signals

MATCHED FILTERING USING SURFACE-ACOUSTIC-WAVE CONVOLVERS

D P Morgan, D R Selviah, D H Warne and J J Purcell

Plessey Research (Caswell) Ltd, Allen Clark Research Centre, Caswell, Towcester, Northants, UK

1. INTRODUCTION

Surface Acoustic Wave (SAW) devices have recently become widely used for analogue VHF and UHF signal processing applications in communications and radar systems. Particular examples are band pass filters, and dispersive filters for pulse compression radar. These devices employ acoustic waves propagating on the surface of a crystalline substrate, giving an accurately repeatable velocity, and hence the devices have excellent accuracy and reproducibility. The low velocity, typically 3mm/μsec, enables long delays to be obtained compactly.

The surface wave convolver is unusual in that it relies for its operation on a non-linear effect. Two counter-propagating surface waves are mixed, and an output given by the spatially integrated product is produced. With an appropriate reference input waveform, the device behaves as a matched filter, with potential applications in radar and spread spectrum communications, and with a high degree of programmability. Although this principle was first shown in 1970 (1), it is only recently that demanding standards of performance have been achieved, and devices have become commercially available. This paper reviews the principles and the performance of recent devices and then discusses some practical issues, particularly dynamic range, which need special consideration because the device is not linear.

2. DEVICE PRINCIPLES

As shown in Fig. 1A, the basic convolver is essentially a piezoelectric crystal with an evaporated metal film on the surface, etched to provide an interdigital transducer at each end and a uniform film, the parametric electrode, in between. The transducers generate surface waves when excited by input voltage waveforms S(t) and R(t), and the corresponding surface wave amplitudes are denoted $S_a(t,z)$ and $R_a(t,z)$. Assuming no distortion in the transducers, these are related to the applied voltages by

$$S_a(t,z) \propto S(t - z/v)$$

$$R_a(t,z) \propto R(t + z/v) \qquad (1)$$

where ±z are the propagation directions and v is the SAW velocity. The total acoustic disturbance would be the sum of equns (1) if the substrate behaved linearly. In practice, the substrate is slightly non-linear, so that in addition to the linear terms of equn (1) there will also be terms proportional to their squares, and a product term $R_a(t,z) \cdot S_a(t,z)$. Each term is accompanied by a corresponding electric field, which is sensed by the parametric electrode. The spatial uniformity of this electrode has the effect of suppressing all the terms except for the product (2).

The output waveform is thus primarily due to the product term, and for general input waveforms we have, ideally,

$$V_o(t) = K \int_{-d}^{d} S(t - z/v) \cdot R(t + z/v) \cdot dz \qquad (2)$$

where K is a constant and 2d is the length of the parametric electrode. The input waveforms are confined to a band of width Δw centred at w_o, say, this band being determined by the transducer responses. The required output is the product term, at the sum frequency, so it will be confined to a band of width 2Δw centred at $2w_o$. The input and output bands do not overlap, and it is common practice to add external bandpass filters to reject components outside these bands, as on Fig. 1A. These help to suppress the unwanted components of the output signal.

Now suppose that the input waveforms S(t) and R(t) have a finite length, less than the propagation delay 2d/v along the parametric electrode. If the inputs are appropriately timed the acoustic signals will overlap only in the parametric region, and the limits in equn (2) can be changed to ±∞. Using τ = t - z/v, we have

$$V_o(t) = v\,K \int_{-\infty}^{\infty} S(\tau) \cdot R(2t - \tau) \cdot d\tau \qquad (3)$$

This is the convolution of S(t) and R(t), except for the factor of 2 which causes the output waveform to have a contracted time-scale. Apart from this contraction, the output is the same as that of a linear filter, with impulse response R(t), and with an input waveform S(t). If R(t) is proportional to the time-reverse of S(t), the output will be the correlation function of S(t), so that the device behaves as a matched filter. For example, Fig. 1B illustrates the correlation of a 5-chip Barker-coded PSK signal; the output, of which the envelope is shown, gives the familiar correlation peak with small time-sidelobes on either side. Clearly, the convolver can correlate any waveform, subject to bandwidth and duration constraints, and is thus very versatile.

The device is ideally bilinear, with an output proportional to either input. Thus, multiple overlapping signals are effectively correlated independently. The processing gain, ie the enhancement of signal-to-noise ratio, is the same as that of a linear matched filter.

The efficiency of the convolver is quantified by a bilinearity factor, C(ω), measured by applying CW input waveforms at frequency ω and examining the output power at frequency 2ω. If P_1 and P_2 are the input powers available from 50 ohm sources and P_0 is the output power delivered to a 50 ohm load, we define

$$C(\omega) = P_0 - P_1 - P_2 \quad \text{(dBm)} \qquad (4)$$

where the powers are rms values measured in dBm. Because the device is bilinear, the value of $C(\omega)$ is independent of the input power levels, and can be interpreted as the output power for CW input signals with power 0 dBm. For an ideal device, obeying equn (2), $C(\omega)$ is independent of ω within the band of the input signals, and can be related to the constant K by using equn (2).

The convolvers discussed here are described as acoustic convolvers, since the non-linear effect is a property of the acoustic propagation medium. The substrate material is usually lithium niobate, since this material gives a relatively strong non-linear effect. Several other types of non-linear mechanism have been used, and are reviewed elsewhere (2). This review also shows that the convolver performance is essentially unaffected by attenuation, velocity changes or temperature changes, so that the device is more tolerant of errors than most linear filters.

3. CURRENT PERFORMANCE

Modern SAW convolvers are based on the principle illustrated in Fig. 1, though in recent work a variety of structural modifications have been made in order to improve the performance. We describe here a device with 120 MHz input bandwidth and 16 μsec interaction length, capable of correlating signals with time-bandwidth products up to 1920. The structure is illustrated in Fig. 2, which includes a photograph of a device. Technical details are given elsewhere (3,4). The non-linear interaction in lithium niobate is quite weak, so for good efficiency the SAW beams are confined to a narrow region, thus increasing the power density (5). The narrow parametric electrode, only 34 μm wide, performs this function because it acts as a waveguide. A wide output bus is required because of the resistance of the parametric electrode. To generate the narrow SAW beams efficiently, special chirp transducers are used; the dispersion produced by these transducers cancels out as far as the overall convolver response is concerned (3). Alternatively, a wider beam can be generated and then compressed using either a multistrip coupler (6) or a metal horn (7,8), all three methods giving comparable results. An important limitation of the basic structure, Fig. 1A, arises because the input transducers reflect surface waves. A wave launched at one end is reflected from the transducer at the other end, and the incident and reflected waves mix in the parametric region to give an unwanted output signal. This is known as fold-over convolution. To suppress this, the structure is duplicated and the electrode polarities of one of the transducers are reversed (Fig. 2). This reduces reflections because the waves generated at one end do not, ideally, excite the transducers at the other end. The input and output ports have impedances which are largely capacitive, so simple circuits are added to improve the matching to the 50 ohm sources and load.

The measured bilinearity factor is shown in the top curve of Fig. 3A. This was obtained by applying a CW waveform from a synthesiser to both inputs, and measuring the output level using a vector voltmeter. Bandpass filters were used, as shown in Fig. 1, and their frequency responses were measured separately and allowed for when calculating the bilinearity factor. The response shows a ripple of ±1½ dB, a centre frequency (300 MHz) bilinearity factor of -66 dBm, and a 3 dB bandwidth of 120 MHz. The two lower curves of Fig. 3A were each obtained by repeating the measurement with one of the inputs disconnected; the output in this case is a spurious unwanted signal due in part to the fold-over convolution mentioned above. The spurious suppression is better than 38 dB at most frequencies, reducing to 35 dB at the worst points. Figure 3B shows the phase of the output, after extracting a large linear term due to the delay. This was measured using a vector voltmeter, with the phase reference obtained from the CW input waveform by means of a frequency doubler. The phase error is less than ±25° over the 240-360 MHz band.

In addition to these frequency-domain measurements, it is also necessary to consider the spatial uniformity of the device. Ideally, the output signal should have an amplitude and phase independent of the location where the two surface waves interact, as implied by equn (2). A simple method of testing this is to apply a short RF pulse to each input. The two input pulses have the same carrier frequency and duration, and their relative timing is varied so that they overlap at varied locations. The output waveform is recorded photographically, with sufficient exposure time to show the output for all locations simultaneously. Fig. 3C shows such a result, where the input pulses were of 0.1 μsec duration, using a 300 MHz carrier and the output was synchronously detected. The sharp dips are associated with the tapping points (Fig. 2), while the slower double-humped curve arises from electromagnetic propagation effects on the parametric electrodes(9). Analysis of the effects of such variations shows that they are probably acceptable for most purposes (11).

A more rigorous test of the convolver is to measure its two-dimensional frequency response (10), though this is less convenient.

4. DYNAMIC RANGE

In practical applications, a matched filter is required to correlate signals with a range of power levels, in the presence of other signals, noise, and possibly jamming. Its ability to do this depends on various aspects of the device performance, and on the way in which the system uses it. In this context, the unusual nature of the convolver leads to special considerations which are discussed in this section. A full analysis would be very complex, so for brevity some simplifying approximations are used.

We first consider a noise-limited dynamic range, denoted DR_n. The input waveforms are taken to have flat envelopes, with input powers P_s and P_r for the signal and reference respectively, in dBm. Both waveforms have length T_w, and the convolver parametric length is $T_p = 2\ d/v$. It is assumed that $T_w \leqslant T_p$. The output power, at the correlation peak, is

$$\text{Output signal power} = C + P_s + P_r - L_t \text{ (dBm)} \tag{5}$$

where $L_t = 20 \log(T_p / T_w) \geqslant 0$ is a loss factor associated with the waveform length. This is valid irrespective of the waveform coding, provided the reference is the time-reverse of the signal, assuming that any distortion in the convolver does not affect the power level. In practice, there will be some maximum input power level. This is assumed to be the same for both input waveforms, and is denoted P_m. For the maximum output power we also take $T_w = T_p$, and hence the maximum output is C + $2P_m$ (dBm).

If the input signal level is reduced, the output reduces in proportion, and the minimum level is taken as that for which the output disappears into noise. The output noise power, denoted N_d, is of thermal origin and is given by $N_d = -109 + 10 \log B_o$, where B_o is the bandwidth of the output amplifier, in MHz, and a noise figure of 5 dB is assumed. The dynamic range is thus

$$DR_n = C + 2P_m - N_d \quad (dB) \qquad (6)$$

This is the maximum possible range of signal levels, and is equal to the maximum output signal-to-noise ratio obtainable. Quantitatively, P_m would be, say, + 20 dBm, allowing some margin below the transducer breakdown level of about 32 dBm (3,6). For the convolver of section 3, the output bandwidth is 240 MHz, so a reasonable amplifier bandwidth is, say, B_o = 300 MHz. This gives N_d = -84 dBm, and with the measured C = -66 dBm, equn (6) gives DR_n = 58 dB.

Now consider a signal accompanied by noise. In this case the output noise is mainly due to the noise applied at the input. If the input noise has power P_N (dBm) and a bandwidth compatible with the signal, the output SNR is $P_s - P_N + PG$, where PG is the processing gain in dB. Using equn (5) for the output signal power, the output noise is

$$\text{ideal output noise} = C + P_r + P_N - L_t - PG \ (dBm) \qquad (7)$$

The device thermal noise, N_d, should be lower than this so that it does not significantly reduce the output SNR. A reasonable criterion is that it should be at least 10 dB lower, so that the SNR degradation is less than 0.4 dB. Thus

$$C + P_r + P_N - L_t - PG \geqslant N_d + 10 \qquad (8)$$

Clearly it is best to maximise P_r and P_N. P_r can be set equal to the maximum input level P_m, and P_N can also be set equal to this if we assume that the input signal is buried in the noise. Using equn (6), the requirement becomes

$$DR_n \geqslant PG + L_t + 10 \quad (dB) \qquad (9)$$

This shows that there is a constraint on the dynamic range which must be satisfied if a given processing gain is to be useable in practice, as noted by Cafarella (12). For larger processing gain the dynamic range must be correspondingly larger. The processing gain, PG, is determined by the signal waveform, and is equal to 10 log (TB), where T and B are the signal duration and bandwidth, respectively. For the convolver of section 3, the processing gain would be maximised by using a signal with duration 16 µsec and bandwidth 120 MHz, giving a processing gain of 33 dB. Thus, since DR_n = 58 dB, equn (9) is readily satisfied.

Now consider the dynamic range for a signal accompanied by noise. It is assumed that the input noise power, P_N, is constant, while the signal level varies. The maximum input signal level, assumed to be above noise, is P_m, while the minimum is determined by setting the minimum output SNR (equns 5,7) to be, say, +10 dB. The range is optimised by using $P_r = P_m$ and setting P_N such that the equality of equn (8) is satisfied. This gives

$$\text{Dynamic Range} = DR_n - L_t - 20 \quad (dB) \qquad (10)$$

For the experimental device, with 16 µsec signals, the dynamic range is thus 38 dB , using P_N = + 5 dBm.

Spurious Signals

We consider here the spurious output signals which can be produced when a signal is applied to only one input port, as in Fig. 3. In practical usage, the output will include spurious components corresponding to both the signal input and the reference input and these are considered individually. In most convolvers, the spurious signals are primarily due to fold-over convolution.

To quantify the effects, we define spurious rejection factors R_s and R_r, in dB, referring to signals applied to the signal and reference inputs respectively. Suppose CW waveforms with equal power P_{in} (dBm) are applied to the two inputs, and the ideal output power is P_o (dBm). If the reference input is now removed, leaving the signal applied with power P_{in}, the spurious output has power P_{so}, say, and the spurious rejection is defined by

$$R_s = P_o - P_{so} \quad (dB) \qquad (11)$$

Similarly, if only the reference input is applied, with power P_{in}, and the output power is P_{ro}, the spurious rejection is defined by

$$R_r = P_o - P_{ro} \quad (dB) \qquad (12)$$

Since the spurious signals are in the same band as the ideal output signal, they must have powers proportional to the squares of the input powers, while the ideal output is proportional to the product of the two input powers. Hence R_r and R_s are independent of the input power level. They are functions of frequency (Fig. 3), but for simplicity we shall take them to be constants. The spurious output powers, for CW inputs, can be written

$$P_{so} = C + 2P_s - R_s \quad (dBm)$$

$$P_{ro} = C + 2P_r - R_r \quad (dBm) \qquad (13)$$

where now the input levels P_s and P_r do not have to be equal.

Consider first the spurious output generated by the reference. This is particularly important because, unlike linear matched filters, it gives an output which is present irrespective of the level of the signal. The spurious output power is given by equn (13) for CW waveforms, but in practice the reference waveform will be coded. This implies that the surface waves reflected by a transducer will not match the incident wave, so that the output power is reduced by a factor of about PG, the waveform processing gain. In practice, fold-over convolution is not the only spurious mechanism, but a reduction factor of about PG has been seen experimentally (13), so this is taken as a reasonable value. Assuming that P_r has its maximum value P_m, the spurious output power is thus $C + 2P_m - R_r - PG$. To avoid degrading the performance, this power should not exceed the thermal noise output N_d, and this requires

$$R_r \geqslant DR_n - PG \quad (dB) \qquad (14)$$

Experimentally, R_r was about 38 dB (Fig. 3) and DR_n was 58 dB; thus the spurious suppression is adequate for signals with processing gain 20 dB or more.

We now consider the spurious output due to the signal input. For a coded input signal, the requirement is that the spurious output component should be suppressed below the ideal time-sidelobe level, and this is easily satisfied. We must also consider an input noise waveform. The output power for wideband input noise is roughly the same as for an input coded waveform. Thus, if the input noise is allowed to take power levels up to the maximum, P_m, the requirement is the same as for the reference input, equn (14), that is, $R_s \geqslant DR_n - PG$.

Fig. 4 shows convolver output waveforms for correlation of a linear FM chirp signal. The input waveforms were obtained by impulsing a SAW expander and compressor, with 11 MHz bandwidth and 14 μsec pulse duration. From the above analysis, the output signal should exceed the output noise level for a range of signal levels DR_n = 58 dB, and the spurious due to the reference should not degrade this. The lower trace of Fig. 4 confirms this: for this trace one of the convolver inputs was reduced by 50 dB (and the amplitude and time scales were changed), but the output signal is still clearly above the noise.

5. ASYNCHRONOUS OPERATION

This topic is mentioned only briefly, since a complete analysis is given elsewhere (14). The convolver can correlate a signal only if a suitable reference waveform is also present in the device, and this generally implies some knowledge of the signal timing. However, if the signal has a finite duration T_w and the convolver parametric length is $T_p = 2T_w$, then the signal can be correlated correctly irrespective of its time of arrival, provided the reference waveform is applied repetitively. The reference repetition period is T_p. The output waveform can include some unwanted signals, which can be gated out. The output is then the same as would be produced by a linear matched filter, except for a time-scale distortion which arises because of the time-contraction in the convolver. This is true even if there are several signals overlapping, and even if noise is present. The method is applicable for a range of possible reference repetition periods; for a given repetition period the correct correlations can be produced for any waveform length up to a specified maximum. However, the length of convolver required is at least $2T_w$ in all cases.

6. CONCLUSIONS

The SAW acoustic convolver can serve as a versatile signal correlator with advantages of simple fabrication and notable tolerance to errors. Present devices, with 120 MHz bandwidth and 16 μsec length, can give output SNR's up to 58 dB, with spurious rejection of 38 dB. The amplitude and phase fidelity is adequate for most purposes, and some improvement can be expected.

Acknowledgements

Thanks are due to M F Lewis and C West for helpful discussions and for assistance with the measurements of Fig. 3B, and to R Gibbs for device fabrication. Part of this work was carried out with the support of the Procurement Executive, Ministry of Defence, sponsored by DCVD.

REFERENCES

1) M Luukkala and G S Kino, Appl Phys Lett, 18, 393-394 (1971)

2) D P Morgan, Ultrasonics, 12, 74-83 (1974)

3) D P Morgan, D H Warne, P N Naish and D R Selviah, IEEE Ultrasonics Symp, 1981, 186-191.

4) D P Morgan, D H Warne and D R Selviah, Electronics Lett, 18, 80-81 (1982)

5) P Defranould and C Maerfeld, Proc IEEE, 64, 748-751 (1976)

6) H Gautier and C Maerfeld, IEEE Ultrasonics Symp, 1980, 30-36

7) I Yao, ibid, 37-42.

8) B J Darby, D Gunton and M F Lewis, ibid, 53-58. Also B J Darby, D J Gunton, M F Lewis and C O Newton, Electronics Lett, 16, 726-728 (1980)

9) D P Morgan and J M Hannah, IEEE Ultrasonics Symp, 1974, 333-336.

10) D P Morgan, Electronics Lett, 17, 265-267 (1981)

11) D P Morgan, IEEE Ultrasonics Symp, 1981, 196-201.

12) J H Cafarella, Proc SPIE, 209, (Optical Signal Processing for C^3I, 1979) 53-56.

13) J H Cafarella, J A Alusow, W M Brown and E Stern, IEEE Ultrasonics Symp, 1975, 205-208.

14) D P Morgan, J H Collins and J G Sutherland, ibid, 1972, 296-299, and Proc IEEE, 60, 1556-1557 (1972).

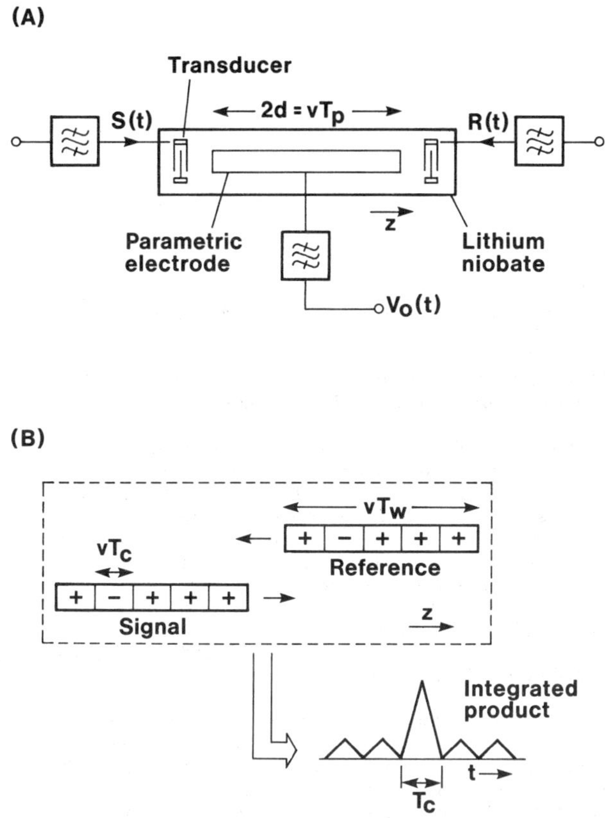

Figure 1 (A) Essential features of convolver. (B) Correlation of 5-chip Barker-coded signal.

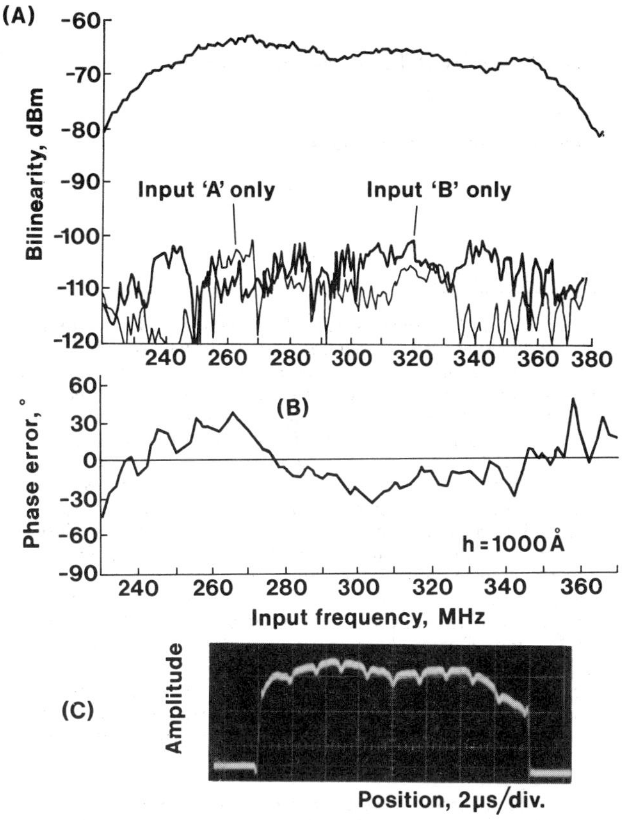

Figure 3 Performance. (A) Bilinearity and spurious levels. (B) Phase error. (C) Spatial uniformity.

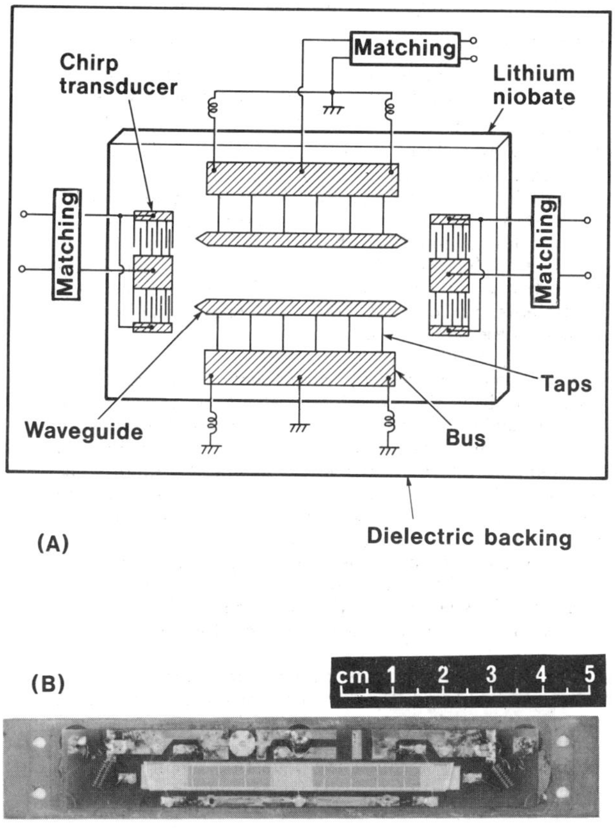

Figure 2 (A) Convolver structure (schematic). (B) Photograph of convolver.

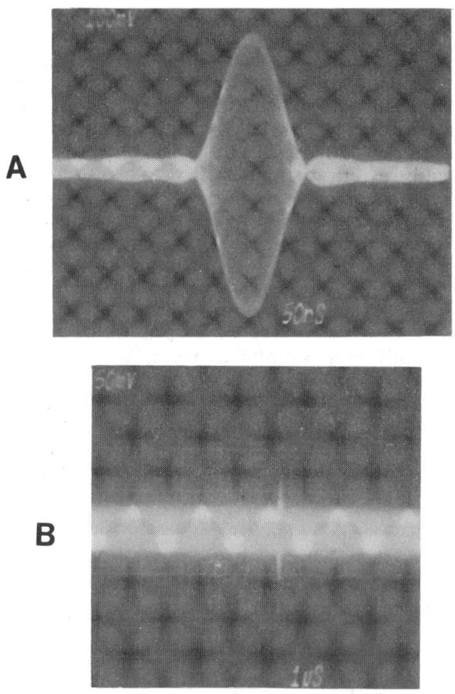

Figure 4 (A) Output for correlation of linear chirp waveform. (B) The same, with one input reduced 50 dB.

A FIBER OPTIC PULSE COMPRESSION DEVICE FOR HIGH RESOLUTION RADARS

E. O. Rausch, R. B. Efurd, and M. A. Corbin

Georgia Institute of Technology, USA

INTRODUCTION

Fiber optic (FO) cables are characterized by low loss (< 3dB/km), low dispersion (< 0.5 ns/km with a solid state laser), and wide bandwidth (> 1 GHz-km for multimode fibers). These characteristics make FO technology suitable for processing signals with bandwidths up to several GHz. A number of potential applications to radar signal processing have been reported previously by Dillard (1). The key element in each application is the FO transmission system which consists of a light transmitting device, a fiber cable, and a light receiving device. Of specific interest are those applications where the fiber optic transmission line is used as a precise time delay and where the fiber lengths are sufficiently short so that attenuation, dispersion, and temperature effects are minimized. One of these applications is associated with a stationary target identification scheme which uses pattern recognition theory and high range resolution "signatures" to classify targets. One way of achieving the appropriate range resolution is to employ a high resolution pulse compression device.

Pulse compression can be achieved by the correlation process. The transmitted pulse is divided into a number of subpulses such that the phase or polarization of each consecutive subpulse changes according to a binary code. The compressed range resolution is determined by the width of the subpulse, and the compression ratio is determined by the number of subpulses or, equivalently, by the number of elements in the binary code.

Correlators with bandwidths exceeding 1 GHz and a resolution $\leq$ 1 ns are difficult to implement. For example, Colvin (2) reported that digital correlators with 64 elements are presently limited to bandwidths of 20 MHz. Carr et al (3) demonstrated correlators utilizing surface acoustic wave (SAW) devices with a resolution of 5 ns (bandwidth of 200 MHz), and Berg et al (4) reported an acousto-optic correlator with a bandwidth of 250 MHz. This paper discusses a new fiber optic correlation device with a bandwidth greater than 1 GHz, a correlation rate equal to 1 GHz, and a resolution (or subpulse width) of 1 ns. A 1 ns subpulse implies a range resolution of approximately 15 cm (0.5 feet). A radar with such a high range resolution mode could resolve a number of scatterers on a target and acquire target "signatures" for target identification purposes.

FIBER OPTIC CORRELATOR CONCEPT

A simplified schematic of the fiber optic correlator is shown in Figure 1. Its basic elements are: a laser, fiber couplers, fiber delay lines, photodiodes, inverting amplifiers, non-inverting amplifiers and a power combiner. This system forms the cross correlation function C of the binary input signal A with the code B according to Equation (1)

$$C_j = \sum_{k=1}^{N} A_{j-N+k} B_k \qquad (1)$$

where

$k = 1, 2 .... N$

$j = 1, 2 .... M+N-1$

$M \geq N$

and

$$A_{j+k-N} = 0 \quad \text{if } j-N+k \leq 0 \quad \text{or} \quad j-N+k > M$$

N is the number of elements in the stored code, and M is the number of elements in the input data.

To illustrate this fact, consider the input signal in Figure 1 to consist of four sequential voltage pulses, each 1 ns wide. This signal modulates the output intensity of a solid state laser diode as shown in Figure 2. The modulations are superimposed on an optical d.c. bias level, such that the laser operates within the linear portion of its transfer function. The coded signal is then distributed by a fiber coupler to four delay lines which consist of coils of wideband optical fibers. Fiber A is a reference delay, whereas fibers B, C, and D delay the signal by 1ns, 2ns, and 3ns with respect to fiber A.

The purpose of the delay lines is to convert the input signal from a series sequence to a parallel sequence in time such that the input signal is aligned parallel to the code. The fiber delays in this case fulfill the same function as the shift register in a digital correlator.

The stored code, B_k, is generated by routing the output from the delay lines to either a non-inverting amplifier or an inverting amplifier. Fibers A, C, and D, which are linked to a non-inverting amplifier, are assigned a positive (+1) value. Thus, the output of the non-inverting amplifiers yield the sum of the correlation terms $A_{j-N+k} B_k$ where $B_k = +1$ and $A_{j-N+K} = \pm 1$. Fiber B is assigned a negative (-1) value, because it was routed to an inverting amplifier. Here, the output yields the correlation terms $A_{j-N+K} B_k$ where $B_k = -1$ and $A_{j-N+K} = \pm 1$. Summation of the output from both amplifiers results in the correlation function C_j given by Equation 1. In Figure 1, the code B and the input signal A are the same. Hence, the system, as shown will generate the aperiodic autocorrelation function of the ++-+ 4-element Barker code. This function, in an ideal form, is graphically demonstrated in Figure 3. Negative correlation coefficients can occur because the photodiodes are ac coupled to the amplifiers.

At this point a step by step analysis is useful to understand how the correlation function is generated. Let the numbers 1 through 4 mark the location of each element in delay lines A through D, such that symbol 1A, for example, refers to element 1 in line A. Line A has the shortest delay. Hence, element 1A is the first to emerge from the delay network. The 1A photon pulse will be detected by #1 photodiode, and enter the non-inverting amplifier as a positive voltage pulse. There, it will be amplified and routed through the power combiner. This pulse constitutes the first sidelobe of the autocorrelation function having an amplitude of 1 unit.

The next elements to emerge from the delay networks are elements 2A and 1B, which are coincident in time. 2A remains a positive voltage pulse because it is not inverted. 1B is transformed into a negative pulse, because it passes through the inverting amplifier. The positive and negative pulses cancel within the power combiner, thus, the output is zero. This summation constitutes the second sidelobe of the autocorrelation. The remaining sidelobes and the correlation peak can be traced in a similar manner.

The correlator schematic as illustrated in Figure 1 was purposely simplified for ease of understanding. In a practical design, photodiodes 1, 3, and 4 and the non-inverting amplifiers 1, 2, and 3 would be replaced by one photodiode and one amplifier. These would be preceeded by a fiber coupler which combines lines A, C, and D into one fiber. In a more general design, the routing of delay lines A, B, C, and D to a non-inverting or inverting amplifier would be accomplished by fiber optic switches, as shown in Figure 4. The sole function of the switches is to change the correlator code, B_k, as desired.9

EXPERIMENTAL

A correlator with a two-element Barker code (++) was constructed to demonstrate the fiber optic pulse compression concept. The fiber delay network, shown in Figure 5 was made from a Canstar two-to-two fiber optic coupler with one of the input fibers terminated, in effect making the coupler a one-to-two coupler. The output fibers were cut such that the propagation time through one fiber was 1 ns longer than the other.

The complete experimental setup is shown in Figure 6. The optical source was an injection laser diode, B&H model Oct-1000, with a 3 dB bandwidth of 1.5 GHz, wavelength of 850 nm, and rise and fall times less than 210 ps. The two avalanche photodiodes were B & H model OC-3002LN with 3 dB bandwidths of 4 GHz and rise and fall times of 80 ps. An SMA "T" served as the power combiner to sum the outputs of the APDs. The AVTECH model AVM-1-GT variable pulse generator produced the input signal. A Tektronix storage oscilloscope system, model 7613, sampling time base model 7T11, sampling unit model 7511, and sampling head model S-2 (DC to 4.6 GHz bandwidth), was used to view the signals.

The 2 ns pulse shown in Figure 7 was the input to the correlator. This pulse represents the ++ two-element Barker Code where each element is 1 ns wide. The reference code determined by the two noninverting preamplifiers at the output of the APDs was identical to the input code. Thus, the experimental correlator generated the autocorrelation function of the two-element code as shown in Figure 8. The main lobe has twice the amplitude of the two sidelobes on either side which corresponds to the expected 6 dB peak-sidelobe ratio for a two-element code shown in reference (5).

TEMPERATURE EFFECTS

Temperature variations affect the output intensity of injection lasers, the propagation times of optical signals in the delay network and the sensitivity of the photodiodes. In reference (6) APDs were shown to exhibit wavelength shifts of 0.4nm/°C to 0.5nm/°C at the peak of the response curve, and in reference (7) lasers show a shift toward higher threshold currents with increasing temperature accompanied by a change in the slope of the transfer curve. In commercial lasers both effects have been eliminated with temperature compensating circuity.

Temperature changes in the fiber delay network affect the performance of the FO correlator only if the relative transit time between the delay lines varies. The propagation delay change $\partial\tau/\partial T$ for a fiber is given by Cohen et al (8) in the form:

$$\frac{\partial \tau}{\partial T} = \frac{1}{c}\left[\frac{\partial n}{\partial T} L + n \frac{\partial L}{\partial T}\right] \qquad (2)$$

$\partial L/\partial T$ is the change in fiber length with temperature which can be expressed as $\partial L/\partial T = \alpha L$, where $\alpha \approx 8 \times 10^{-7}/°C$ is the linear thermal expansion coefficient for the combined fiber core and cladding and $\partial n/\partial T$ is the change in the index of refraction at the center of the fiber core. $\partial L/\partial T$ can be replaced with αL in Equation 3, and L can be factored out. The terms that remain within the brackets are $\partial n/\partial T$ and αn. For multimode silica fibers $\partial n/\partial T$ was measured by Malitson (9) to be 1×10^{-5} over a temperature range of -40°C to + 70°C which is much greater than $n\alpha(n \approx 1.2 \times 10^{-6})$. Therefore, propagation delay changes in multimode silica fibers are primarily due to refractive index variations with temperature.

Of importance here is the maximum allowable temperature variation δT assuming the differences in delay changes $\Delta\tau$ are restricted to $\leqslant \pm 0.01$ ns. Let $\delta\tau_1$ and $\delta\tau_2$ be the delay changes in fibers L_1 and L_2. Then $\Delta\tau$ is given by:

$$\Delta\tau = \delta\tau_1 - \delta\tau_2 = \left[\frac{L_1 - L_2}{c}\right]\left[\frac{\partial N}{\partial T}\right]\delta T \qquad (3)$$

where $L_1 - L_2$ is taken to be the difference (19.8m) between the longest (20m) and the shortest (0.2m) multimode fiber for a 100 element multimode delay network, and $\partial N/\partial T$ is $1 \times 10^{-5}/°C$ as before. Substituting these parameters into equation 4 and solving for δT yields a maximum allowable temperature change of about ±15°C.

CONCLUSIONS

Fiber optic technology offers new techniques for performing radar signal processing functions. Substantial performance improvements over conventional methods can be anticipated particularly where signal bandwidths exceed 500 MHz. In this paper we have described a new fiber optic pulse compression device with the potential of attaining range resolutions less than 15 cm (0.5 feet) and processing rates greater than 1 Gbit/sec. Pulse compression is achieved via the correlation process. The key element of the fiber optic correlator is a fiber network with a delay accuracy of ± 0.01 ns. For good performance, signal dispersion and attenuation within each fiber must be reduced to less than 0.01 ns and 0.1 dB, respectively. These specifications can be attained with a multimode fiber delay network if the fibers are restricted in length to 20m or less. This restriction implies a maximum code length of 100 elements. Single mode fibers offer the possibility of longer codes. The effects of temperature variations on the delay network have also been addressed. The maximum tolerable temperature change in a multimode fiber network with a maximum delay length of 20 m (100 ns) was calculated to be ± 15°C.

REFERENCES

1. G. M. Dillard, 1977, International Conference Radar 1977, London, England, 363.
2. R. Colvin, 1981, Microwaves 20, No. 2, 65.
3. P. H. Carr et. al., 1979, IEEE Ultrasonic Symposium Proceedings, 757.
4. N. J. Berg et al, 1979, SPIE Proceedings on Optical Signal Processing for C³I, 209, 57 .
5. M. I. Skolnik, 1970, "Radar Handbook," by McGraw-Hill Press, 20-19.
6. G. R. Elion and H. A. Elion, 1978, "Fiber Optics in Communications Systems", Marcel Dekker Press.
7. S. T. Eng and Bergman, 1980, Applied Optics, 19, No. 19, 3335.
8. L. G. Cohen and J. W. Fleming, 1979, Bell Systems Technical Journal, 58, No. 4, 945.
9. I. H. Malitson, 1965, J. Opt. Soc. Amer., 55, No. 10, 1205.

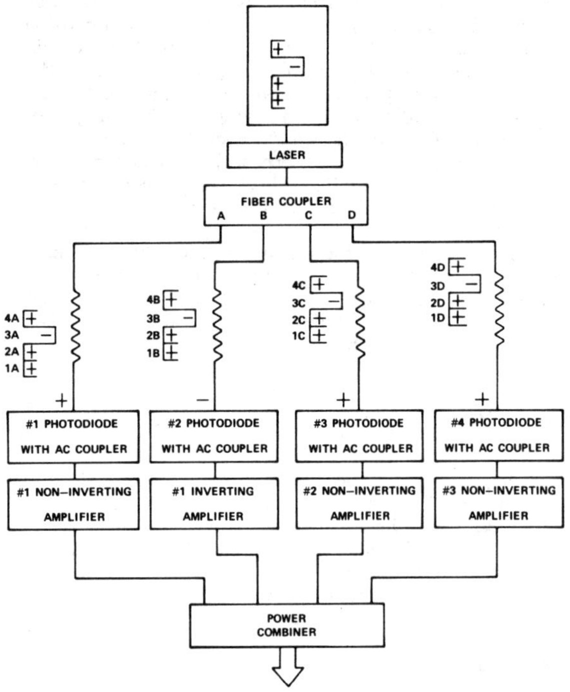

Figure 1 Simplified schematic of fiber optic correlator.

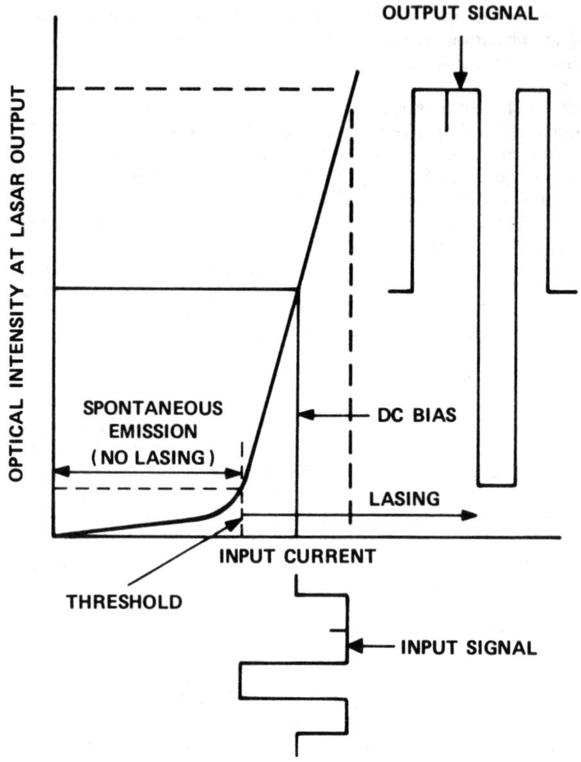

Figure 2 Laser transfer function.

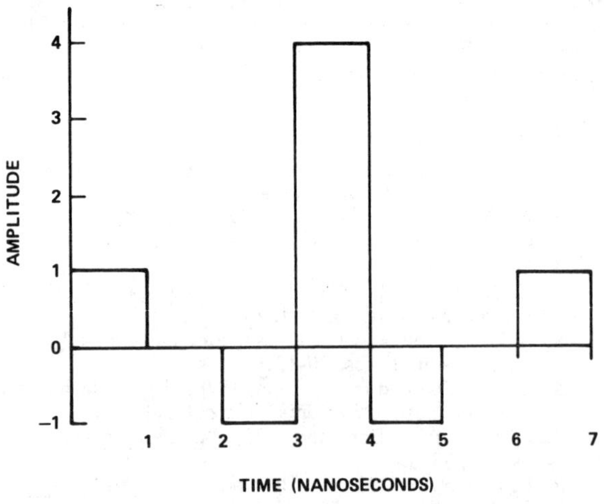

Figure 3 Ideal aperiodic autocorrelation function of 4-element Barker code.

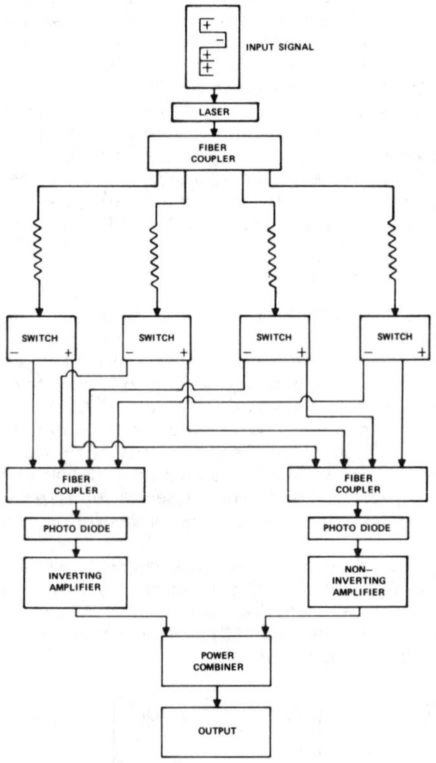

Figure 4 Schematic of practical fiber optic correlator.

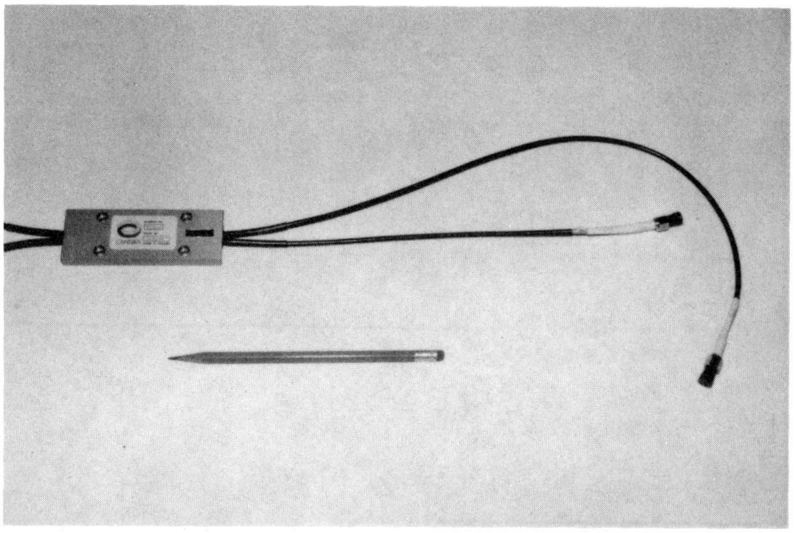

Figure 5 Fiber delay network for two-element correlator.

Figure 6 Experimental system of correlator with two-element Barker code.

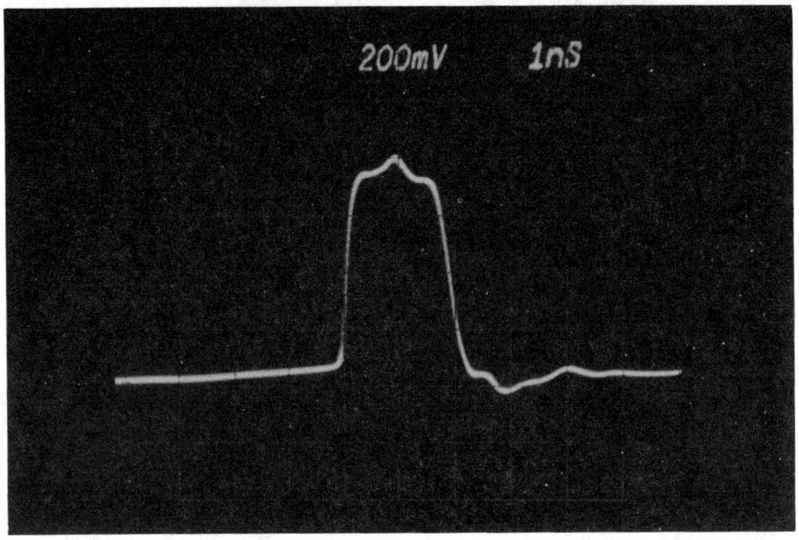

Figure 7 Input signal to correlator representing the ++ two-element Barker code.

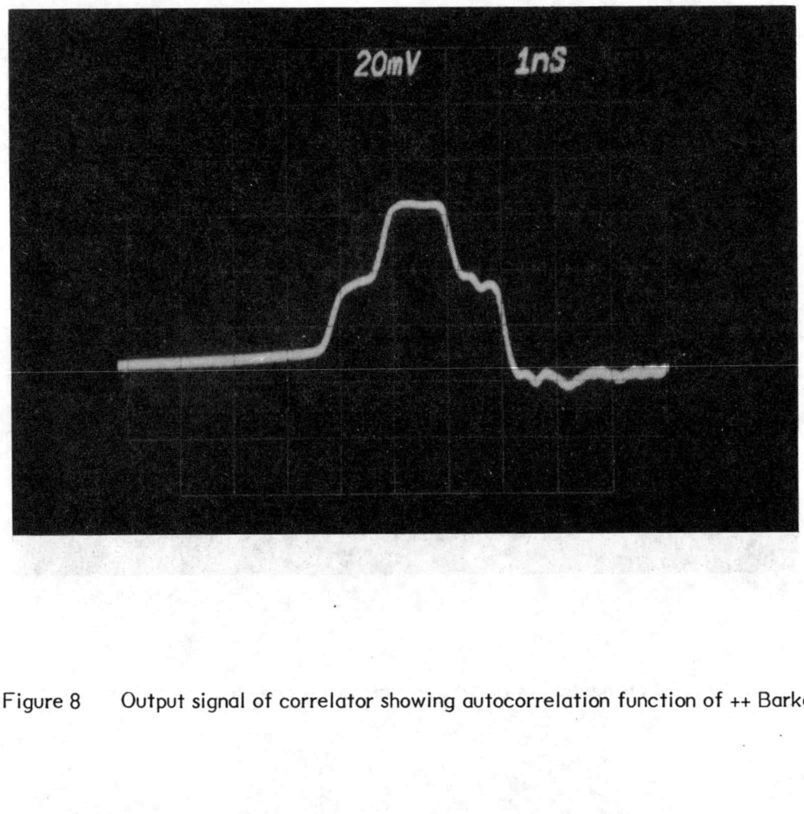

Figure 8 Output signal of correlator showing autocorrelation function of ++ Barker code.

NEW POLYPHASE PULSE COMPRESSION WAVEFORMS AND IMPLEMENTATION TECHNIQUES

Bernard L. Lewis and Frank F. Kretschmer, Jr.

Naval Research Laboratory, USA

INTRODUCTION

It is the purpose of this paper to introduce new polyphase pulse compression waveforms and efficient digital implementation techniques. The new waveforms provide previously unobtainable doppler and/or bandwidth limitation tolerance and the new implementation techniques permit systems to be implemented with cost effective available digital hardware.

Doppler tolerance in the paper will be taken to mean that the peak response and peak response to range-time-sidelobe ratio of compressed pulses do not degrade catastrophically with doppler.

Bandwidth limitation tolerance will be taken to mean that the peak response to range-time-sidelobe ratio of compressed pulses will not be degraded by receiver bandwidth limitations necessary to minimize the effect of receiver thermal noise.

POLYPHASE WAVEFORMS

The new waveforms to be discussed are carrier signals phase modulated with the P1, P2, P3, and P4 phase codes, Lewis and Kretschmer (1, 2). These codes are similar to the well known Frank polyphase code Frank (3), Cook and Bernfield (4) which will be labled PF in this paper. The PF code will be used as a standard for comparing the performances of the P codes.

The PF code can be derived as inphase I and quadrature Q samples taken at the Nyquist rate of a step approximation to a linear frequency modulation waveform coherently detected with a local oscillator of frequency equal to the lowest frequency of the waveform. This is illustrated by Fig. 1 in which a four frequency coherently detected step-chirp waveform that would yield a pulse compression ratio $\rho=4^2$ is sampled to obtain a 16 element PF code. A PF waveform is a carrier whose phase is changed every τ_c seconds by the phase difference between successive code elements of the PF code where τ_c is the uncompressed pulse length T divided by ρ. The ith code element of the jth frequency group of the PF code is defined by

$$\theta_{ij} = (2\pi/N)(i-1)(j-1) \qquad (1)$$

where i ranges from 1 to N for each j taken in ascending order from 1 to N.

It was noted Cantrell and Lewis (5) that (1) defines the complex conjugates of the steering phases used in an N point FFT to subdivide a bandwidth B into N subbands of width B/N where B is the I and Q sampling rate. This suggested that the PF code could be efficiently generated and compressed digitally as illustrated in Fig. 2. This implementation is the new form of implementation to be discussed in this paper.

Fig. 3 illustates the autocorrelation function of a 100 element PF code where sample number refers to resolvable time increments equal to the sampling period. Note the low range-time-sidelobes.

Experience with the PF waveform revealed that it was not doppler tolerant with high freqency carriers and that its peak to maximum range-time-sidelobe response ratio deteriorated with receiver bandwidth limitations. As a consequence, new codes and waveforms were sought. These statements will be justified in a following section of this paper.

NEW CODES AND IMPLEMENTATIONS

It was recognized that the intolerance of the PF waveform to bandwidth limitations was due to the fact that the code groups having the largest phase changes from code element to code element were in the center of the uncompressed pulse. This caused bandwidth limitations in a radar receiver to attenuate the center of the waveform more than the ends which adversely effected the peak to range-time-sidelobe ratio obtainable in the compressed pulse.

This suggested that the PF code groups representing the j frequencies be rearranged in time order of transmission to place the lowest phase increments from code element to code element in the center of the waveform. This triggered the invention of the P1 code whose code element phases are given by

$$\theta_{i,j} = -(\pi/N)[N-(2j-1)][(j-1)N+(i-1)] \qquad (2)$$

Note that when N is odd, the P1 code is the PF code rearranged to have conjugate symmetry about the D.C. term. The P1 code has an autocorrelation function nearly identical to that of the PF code (Fig. 3) with no bandwidth limitations.

Implementation of the P1 code for N odd simply requires changing the connections on the tapped delay line in Fig. 2 that delays the successive frequency groups. Implementation for N even is illustrated in Fig. 4.

In addition to the P1 code, the P2 code was developed whose successive code elements are defined by

$$\theta_{ij} = \{(\pi/2)[(N-1)/N]-(\pi/N)(i-1)\}\{N+1-2j\} \qquad (3)$$

This can be implemented as illustrated in Fig. 5 with N even. The P2 code is similar to the steering weights used on Butler matrices to steer antenna beams.

The P2 code has an autocorrelation function nearly identical in magnitude to that of the PF and P1 codes (Fig. 3) with no bandwidth limitations.

However, the P2 autocorrelation function is real rather than complex because the code is symmetrical.

The P1 and P2 waveforms were found to be much more bandwidth limitation tolerant than the PF waveform. However, they shared the PF doppler intolerance when used on high frequency carriers. This was recognized to be a characteristic of the analog step chirp waveform from which the phase codes were derived (4).

Recognizing that the linear chirp is much more doppler tolerant than the step chirp, a new phase code (the P3 code) was derived from the linear chirp waveform. This code consisted of Nyquist rate samples of a linear chirp waveform coherently detected using a local oscillator of frequency equal to the lowest frequency in the linear chirp waveform. The phases of successive code elements of the P3 code are given by

$$\theta_i = \pi(i-1)^2/\rho \qquad (4)$$

where i ranges from 1 to ρ and ρ is the pulse compression ratio. This code can also be implemented using a modified FFT as illustrated in Fig. 6.

The P3 code was found to be just as doppler tolerant as the analog linear chirp but it was found to be bandwidth limitation intolerant for the same reason as found for the PF code.

The same solution that worked for the PF code and transformed it to the P1 code was used on the P3 code to develop the P4 code which was implemented as illustrated in Fig. 7. The successive code elements of the P4 code have phases

$$\theta_i = [\pi(i-1)^2/\rho]-\pi(i-1) \qquad (5)$$

This code was found to be both doppler and bandwidth limitation tolerant.

BANDWIDTH LIMITATION EFFECTS

The effect of bandwidth limitations on the various polyphase coded compressors was investigated by bandlimiting the waveform to be compressed to the reciprocal of the length of a code element. The average response to waveforms with different times of arrival was then determined. Typical results are illustrated in Fig. 8 for the various polyphase codes. Note that the bandwidth limitation decreased the PF and P3 peak to range-time-sidelobe response ratios while it increased that of the P1, P2 and P4. Fig. 9 illustrates the effect of sliding window two sample adding the outputs of the various code compressors with ρ = 400.

DOPPLER EFFECTS

The effect of doppler on the various code peak to peak-sidelobe and image lobe responses is illustrated in Fig. 10. Note that grating lobes appear in the PF, P1 and P2 codes that are absent in the P3 and P4 codes. Note also the high peak to peak sidelobe ratio at zero doppler that these waveforms offer without amplitude weighting. This high ratio is characteristic of frequency derived polyphase coded compressors. It should be noted that bandlimiting suppresses the image lobe of the P1, P2 and P4 waveforms caused by doppler.

ADVANTAGES OF DIGITAL OVER ANALOG PROCESSING

Some advantages of digital over analog processing include the following:

1. Freedom from ringing produced by impedance mismatches or analog filters.
2. Low loss long delays.
3. Use of standard hardware.
4. Reproducibility of response.
5. High peak to peak-sidelobe ratios without weighting.
6. Ability to change bandwidth by changing clock frequency.
7. Ability to change pulse compression ratio by controlling number of time samples used and output tapped delay line delays.
8. Compatibility with other digital signal processors such as MTI.

CONCLUSIONS

At this point it can be concluded that doppler and bandwidth limitation tolerant digital pulse compressors are feasible for high carrier frequency radars. In addition, it can be concluded that such digital compressors have many advantages to offer over analog processors.

REFERENCES:

1. Lewis, B., and Kretschmer, F., May 1981, IEEE/AES, 17, 364-372.

2. Lewis, B., and Kretschmer, F., (scheduled for publication Sep, 1982), "Linear Frequency Modulation Derived Polyphase Pulse Compression Codes," IEEE/AES.

3. Frank, R., June 1963, IEEE/IT, 9, 43-45.

4. Cook, C., and Bernfield M., 1967, "Radar Signals, An Introduction to Theory and Applications," Academic Press, New York.

5. Cantrell, B., and Lewis, B., December 2, 1980, "High Speed Digital Pulse Compressor," U.S. Patent #4,237,461.

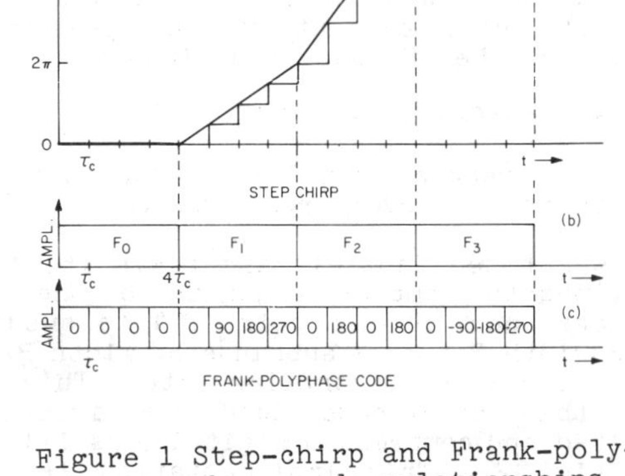

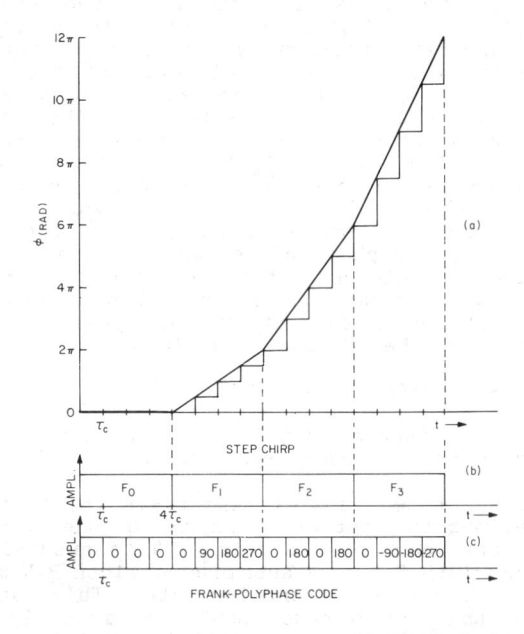

Figure 1 Step-chirp and Frank-polyphase-code relationships

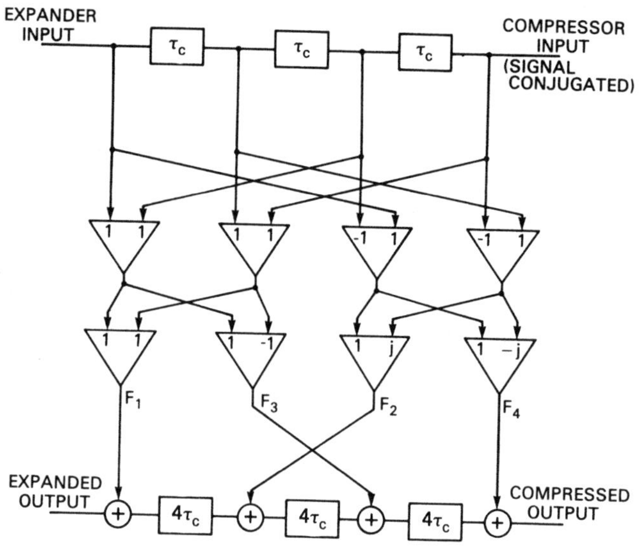

Figure 2 PF expander-compressor, $\rho=T/\tau_c=16$

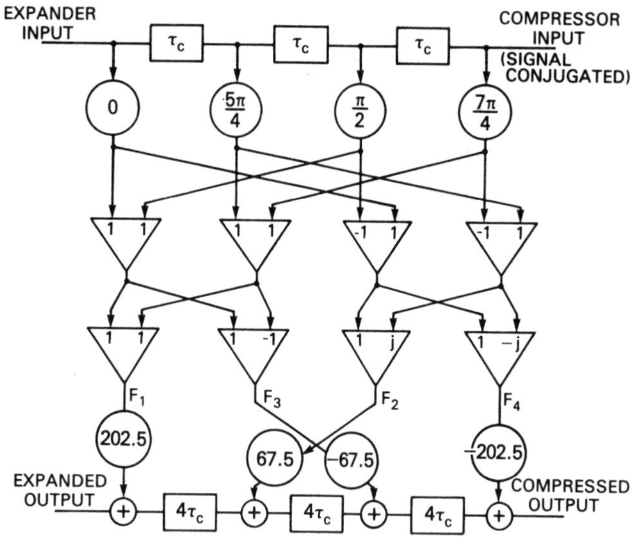

Figure 5 P2 expander-compressor, $\rho=T/\tau_c=16$

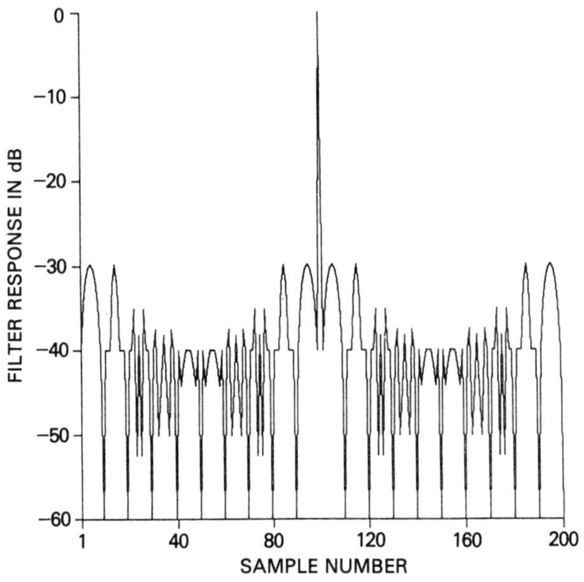

Figure 3 Compressed pulse of Frank code

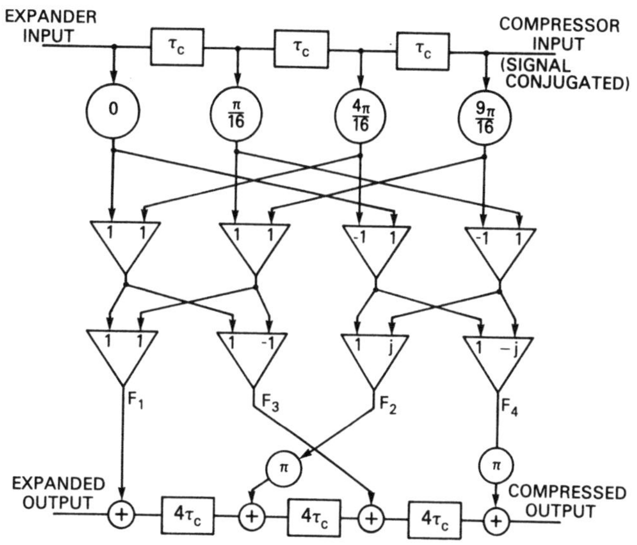

Figure 6 P3 expander-compressor, $\rho=T/\tau_c=16$

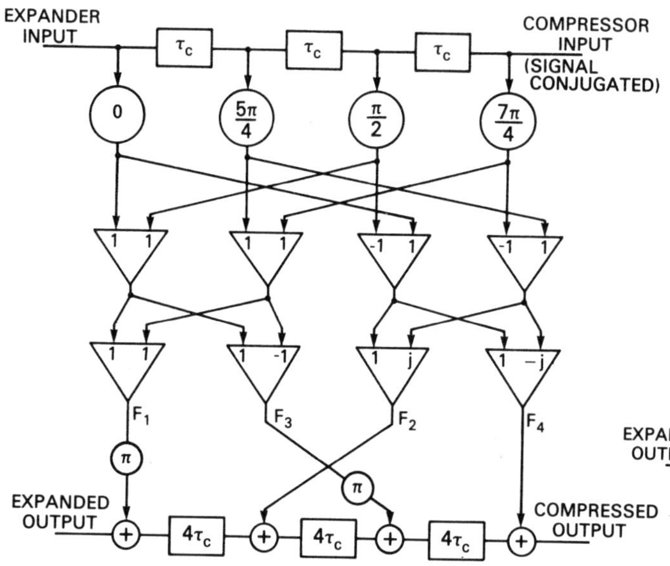

Figure 4 P1 expander-compressor, $\rho=T/\tau_c=16$

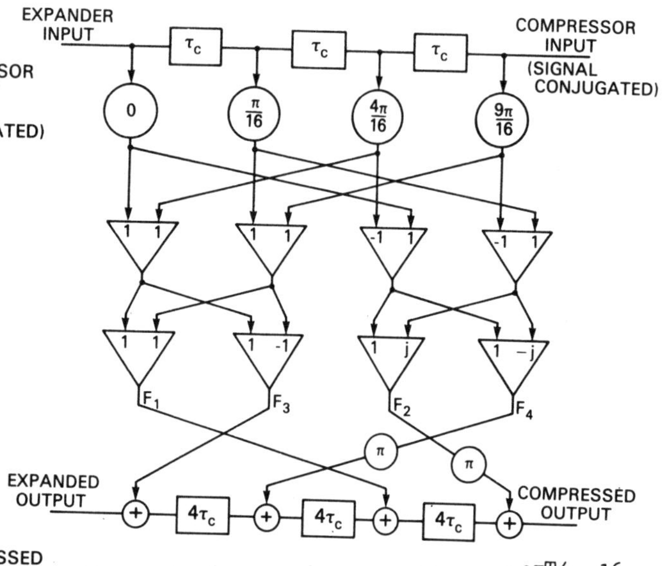

Figure 7 P4 expander-compressor, $\rho=T/\tau_c=16$

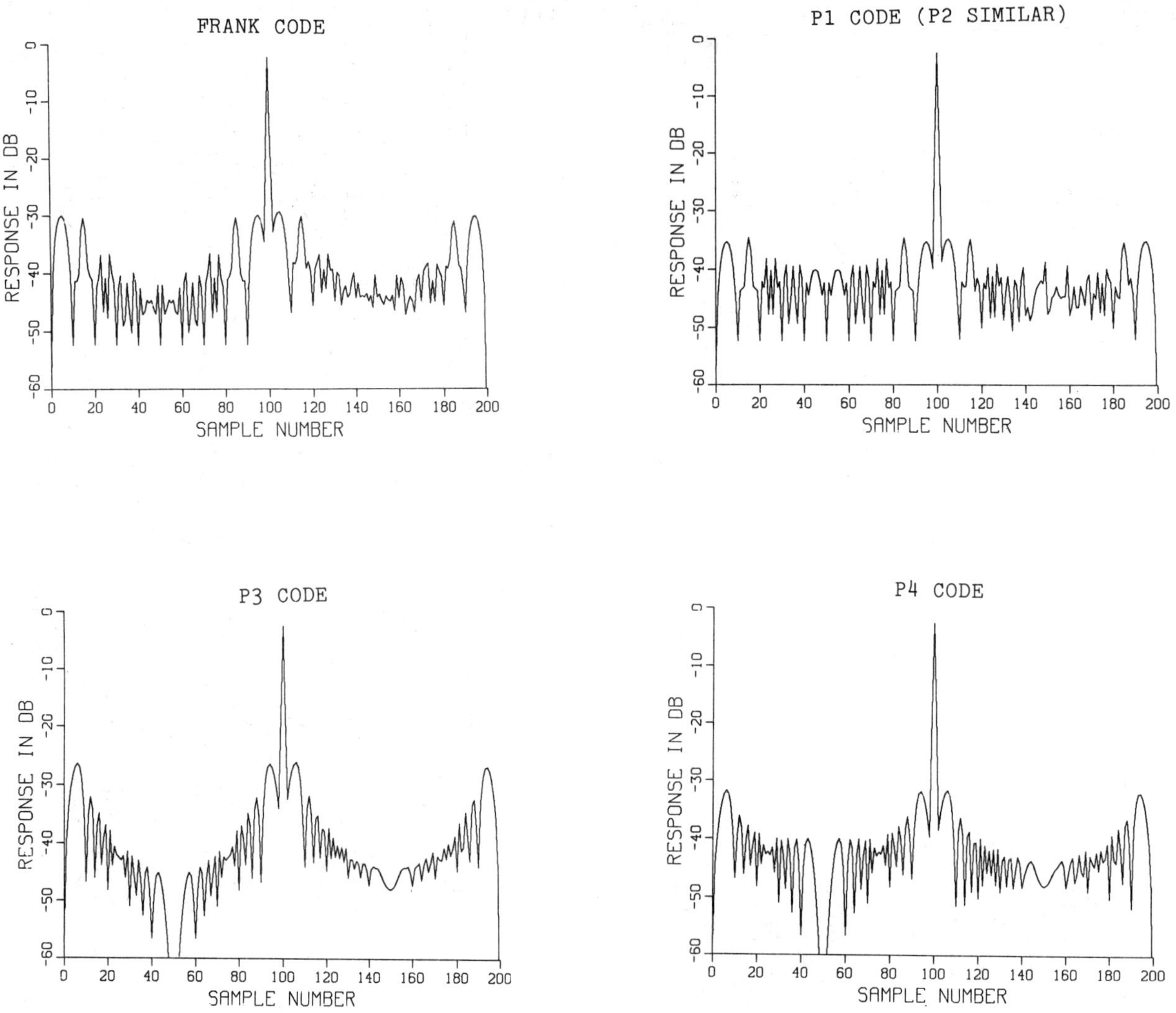

Figure 8 Effect of precompression bandlimiting on autocorrelation functions, ρ = 100. (Input oversampled by 5 and sliding window averaged by 5)

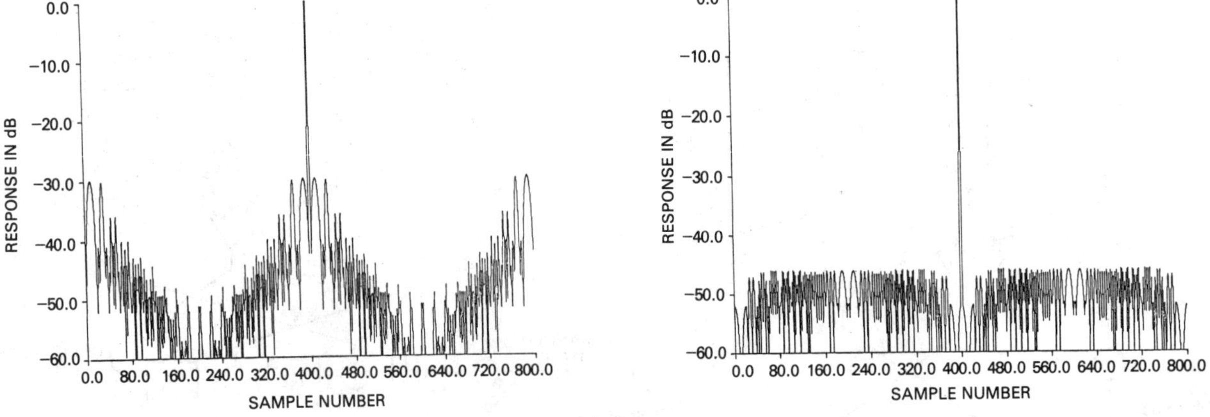

Figure 9(a) Effect of 2 sample adder on compressor output, ρ = 400

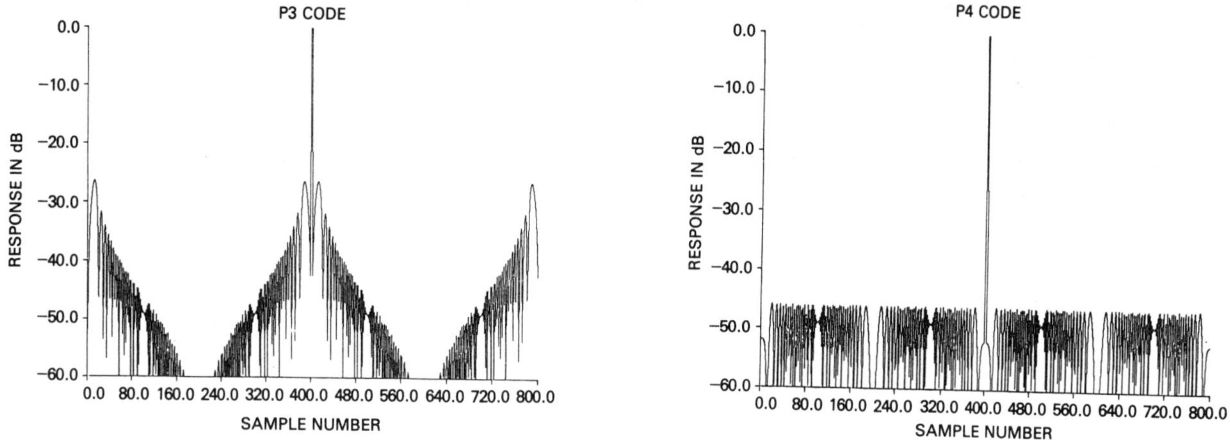

Figure 9(b) Effect of 2 sample adder on compressor output, ρ = 400

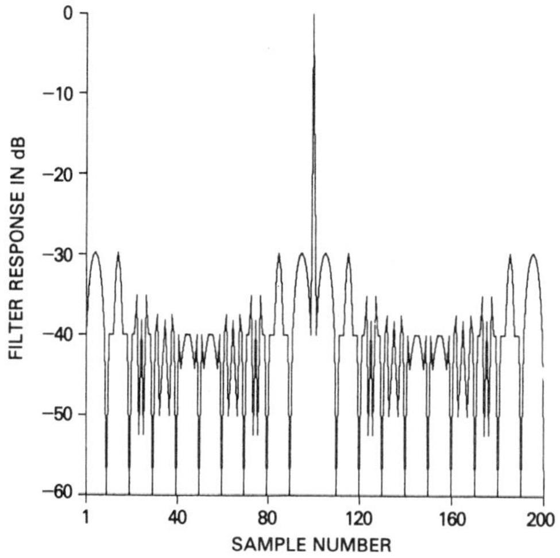

Figure 10(a) Frank code autocorrelation, ρ=100, zero doppler shift and no bandwidth limitation (P1 and P2 similar)

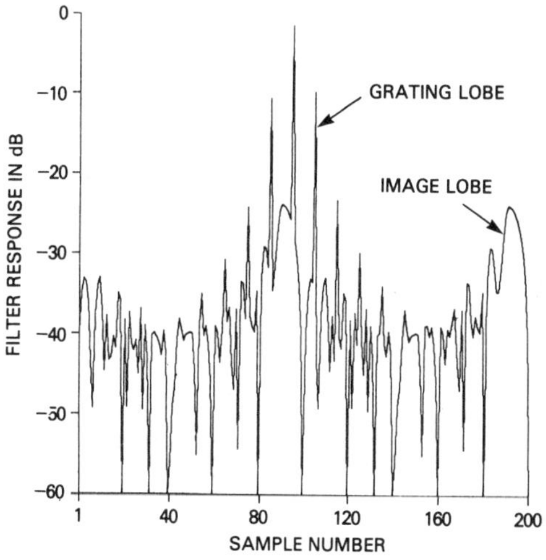

Figure 10(b) Frank code autocorrelation function, ρ=100, doppler =0.05B with no bandwidth limitation (P1 and P2 similar)

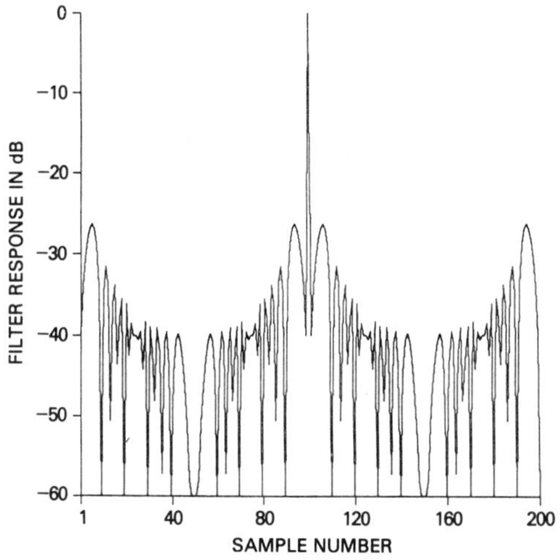

Figure 10(c) P3 or P4 autocorrelation function, ρ=100, zero doppler shift and no bandwidth limitation

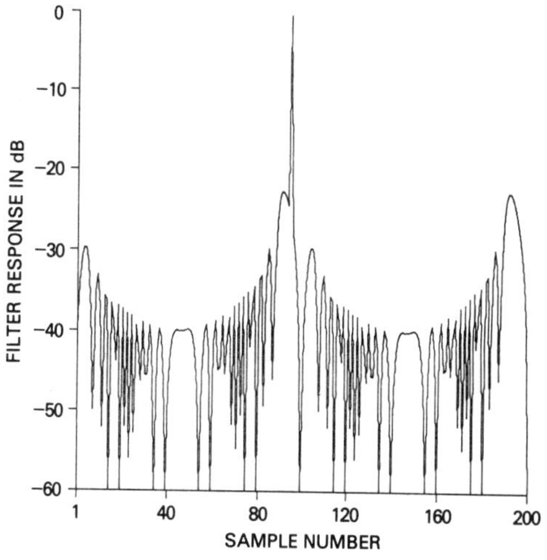

Figure 10(d) P3 and P4 autocorrelation function, ρ=100, doppler =0.05B with no bandwidth limitation

A DIGITAL HIGH-SPEED CORRELATOR FOR INCOHERENT-SCATTER RADAR EXPERIMENTS

H.J. Alker

Electronics Research Laboratory, University of Trondheim, Norway

INTRODUCTION

Radar observations of the ionosphere using the incoherent-scatter (IS) technique is a powerful ground-based method for probing the ionospheric plasma. Physical quantities including range-dependent plasma densities, temperatures, velocities and compositions, are deduced from the IS radar signal scattered from plasma fluctuations. At VHF/UHF radar frequencies, for which the IS technique is applicable, the effective scattering cross section of plasma volumes is extremly small as compared with normal radar operations. Observing the ionosphere as a "deep fluctuating target" requires IS radar facilities with high-power transmitters and low-noise receivers. The signal detection method is based on statistical estimation of the plasma autocorrelation function (ACF) or the spectral distribution of the plasma fluctuations. Special-purpose observational schemes and hardware have been developed for improving the detection performance.

The IS signal processing, after normal detection and amplification, consists of 2 parts: Real-time hardware preprocessing for data reduction purposes, and computer postprocessing for extracting pertinent physical information. The real-time preprocessing of IS data is essential for obtaining statistically accurate estimates of the ACF or the spectrum. Processing of data from the altitude range 80-2000 km is characterized by extreme low signal to noise ratios, demanding long integration times, typically 10 sec - 15 min. The computer postprocessing involves bias correction of the estimates and, by off-line analysis, the final evaluation of plasma parameters.

A basic problem in IS radar experiments is the limiting computational capacity of the on-line signal processing hardware. Particularly in monostatic (back-scatter) radar experiments for extracting plasma parameter variations as a function of altitude and time, the experimental temporal resolution is often set by the on-line equipment. Usually IS signal processing has been carried out by special-purpose digital autocorrelators with fixed hardwired constructions limiting the operational modes of the radar system.

This paper describes a novel design of a programmable, high-speed digital multiprocessor for IS radar experiments. The system is adaptable to different modulation modes of the transmitter. The design work, started as a feasibility study in 1977, was an integral part of a design concept for a European IS facility in the auroral zone (the EISCAT project). Hardware implementations of the multiprocessor were carried out in 1980/81, when the EISCAT radar system was brought into operation.

THE EISCAT RADAR SYSTEM

The EISCAT (European Incoherent SCATter) observatory located in the northern part of Scandinavia is an international research project with jointly scientific and monetary contributions from Finland, France, Norway, Sweden, the United Kingdom and West-Germany. The scientific justification for the EISCAT installation is described elsewhere (1), (2). EISCAT operates in a dual frequency configuration, consisting of a VHF radar located at Ramfjordmoen, near Tromsø in Norway, and a tristatic UHF system with the transmitter at Ramfjordmoen and remote receiver stations near Kiruna in Sweden and Sodankylä in Finland. The configuration is illustrated in Figure 1. The Ramfjordmoen site operates monostaticly.

The EISCAT system performs in the same range as the most powerful exsisting IS radars or beyond. System characteristics are given in Table 1. The transmitters are capable of providing a variety of modulation waveforms. The inherent flexibility of pulse modulation modes will enable the experimentators to match their techniques to the conditions encountered in the different altitude ranges of the ionosphere. The experimental requirements prescribe the use of single pulse trains as well as multipulse waveforms with groups of pulses in a specified pattern. For increasing the alti-

TABLE 1 - EISCAT radar system characteristics.

		VHF subsystem	UHF subsystem
Transmitters	Center frequency	224 ± 5 MHz	933.5 ± 1.5 MHz
	Peak power	5 MW	2 MW
	Mean power	625 kW	250 kW
	Duty cycle	12.5% max	12.5% max
	Pulse length	10 μsec - 1 msec	10 μsec - 10 msec
	Pulse rep. rate	0-1000 Hz	0-1000 Hz
	Bandwidth	3 MHz at 1 dB	10 MHz at 1 dB
	Polarization	Linear, L/R circular	linear, L/R circular, elliptical
Antennas	Type	Parabolic cylinder	Cassegrain parabol
	Dimensions	120 x 40 m	32 m diameter
	Effective area	3300 m^2	525 m^2
	Gain	44,5 dB	48 dB
Receivers	Structure	8 channels	8 channels
	Bandwidths	12.5, 25, 50, 100, 500 kHz	12.5, 25, 50, 100, 500 kHz
	Noise temperature	300 °K	120 °K (Tromsø) 40 °K (remote sites)

tude resolution the transmitted signal may be phase-coded. The transmitter average power capability is exploited using frequency commutation from pulse to pulse or within a modulation pulse embracing a group of RF subpulses. The circular polarization of the VHF system may also be altered. Both systems use high-power klystrons with the RF exciters controlled by computer.

The EISCAT UHF receiver and data handling system is sketched in Figure 2. The receiver front end consists of a dual-channel parametric preamplifier (20°K) and a polarizer which matches the system to the incoming wave. For calibration and test purposes the front end includes a noise source and a programmable attenuator. The front end components and the RF-to-IF conversion are located at the top of the antenna. The local oscillator frequencies of the first mixers can be switched for changing the position of upper and lower sidebands. Any asymmetry in filter or amplifier passband can be detected and corrected. The 120 MHz IF-section is split into 8 separate channels with different adjustable passband characteristics. The final baseband demodulation uses a quadrature detector generating the in-phase and quadrature (real and imaginary) components of the signal envelope. The digital data handling equipment comprises a multichannel AD-converter, a digital matched filter for 13-bit Barker code and a high-speed processor for ACF estimation.

The different parts of the EISCAT system including antennas, transmitters, receivers and data handling equipment are interfaced with local computers for on-line control and monitoring. Real-time control, synchronized to an accuracy of 1 μsec, is provided by local radar controllers and cesium clocks interfaced with the site computer. The site computers are linked by data lines enabling system supervision by an on-line experimental computer program.

THE ESTIMATION PROCESS

The stochastic scattering process may be described by a 3-dimensional autocorrelation function (ACF) (3), (4):

$$R_p(\tau,h,\lambda_p), \quad \lambda_p = \frac{\lambda_t}{2} \text{ for backscattering} \qquad (1)$$

The variables are the timelag τ, the altitude h, the plasma scale factor λ_p and the radar wavelength λ_t. The radar signals act as a Fourier analyzer of the plasma, picking out the component of the spatial fluctuation spectrum which has the correct scale for scattering. The backscatter spectrum has two basic components: The ionic component with Doppler shifts corresponding to the ion thermal velocities (bandwidth typically 25 kHz), centered at the radar frequency, and the electronic component appearing as up/down shifted spectral lines. The bandwidth defined by the lines is typically 20 MHz.

The ACF at the receiver front end is given by the received envelope v_s (3), (4):

$$R_s(\tau;t) \triangleq E[v_s^*(t-\tfrac{\tau}{2})\, v_s(t+\tfrac{\tau}{2})]$$

$$\approx R_p(\tau,h=\tfrac{ct}{2},\lambda_p)\cdot R_{tx}(\tau;t) \qquad (2)$$

where τ is the lag parameter, t transmission delay, h the altitude and c the velocity of light. E[·] denotes ensemble average and $(\cdot)^*$ complex conjugate. R_{tx} is defined by the transmitted envelope v_{tx}:

$$R_{tx}(\tau;t) \triangleq \int v_{tx}^*(t-\tfrac{\tau}{2}-\tfrac{2h}{c})\, v_{tx}(t+\tfrac{\tau}{2}-\tfrac{2h}{c})\, dh \qquad (3)$$

Defining a corresponding ACF for the receiver system, the ACF observed is (3), (4):

$$R_{out} = [R_p \cdot R_{tx}] \circledast R_{rx} \qquad (4)$$

where $\circledast$ denotes convolution. The functions R_{tx}, R_{rx} are deterministic, implying that these to some extent can be corrected for. The weight function R_{tx} of the plasma ACF (Eq. 3) may be chosen by proper signal-design to possess optimum detection performance. The truncation of the plasma ACF caused by the altitude-dependent R_{tx} demonstrates the well-known range/frequency resolution limitation. Figure 3 illustrates basic transmitter pulse modes. For single-pulse transmission, the weight function limits the range of the lag parameter to $|\tau| < \tau_{tx}$, the pulse duration, which also determines the range resolution. The 4-pulse pattern (Fig. 3) samples the plasma ACF at lags τ_1, $2\tau_1$, $3\tau_1$, ..., $6\tau_1$. The range resolution is determined by the element pulse duration $\tau_{tx,p}$.

Assuming an ideal receiver, the basic estimator equations for the real and imaginary part of R_s (Eq. 2) may be expressed by samples of the in-phase and quadrature components of the received signal:

$$R_{re,r,\ell} = x_r\, x_{r+\ell} + y_r y_{r+\ell}$$

$$R_{im,r,\ell} = x_r \cdot y_{r+\ell} - y_r \cdot x_{r+\ell} \qquad (5)$$

where the subscripts are the range-gate index r and lag parameter ℓ. The estimator given in Equation 5 is applicable for multiple-pulse transmission.

The received IS signal is embedded in background and receiver noise. Assuming the noise to be additive and independent, the noise bias in the estimates can be removed by subtraction. The noise ACF is computed at the end of the radar scan, when noise is received. The statistical errors of the estimators are reduced by averaging the estimator output (for a given range) over subsequent radar scans. It may be shown (4), assuming Gaussian processes, that the normalized estimator variance is:

$$\frac{VAR[R_{re,r,\ell}]}{P_s^2} \propto \frac{1}{N_{ip}} \{(n_p + SNR^{-1})^2 + \rho_{re,r,\ell}\} \qquad (6)$$

where P_s is the power of the received signal, SNR the signal to noise ratio, n_p the number of subpulses in the multiple-pulse pattern, N_{ip} the number of radar scan periods and ρ the normalized ACF at the receiver input (Eq. 2). For a given statistical estimator accuracy the integration time is strongly SNR-dependent. The SNR independent terms in Eq. 6 are caused by the stochastic nature of the IS signal and may be regarded as self-clutter.

In the single-pulse mode (Fig. 3) correlated returns from the same scatter volume may be averaged. Assuming N complex samples within the range-gate, the estimator equations become:

$$R_{re,r,\ell} = \frac{1}{N-\ell}\sum_{i=0}^{N-1-\ell}(x_{r,i}\cdot x_{r,i+\ell} + y_{r,i}\cdot y_{r,i+\ell})$$

$$R_{im,r,\ell} = \frac{1}{N-\ell}\sum_{i=0}^{N-1-\ell}(x_{r,i}\cdot y_{r,i+\ell} - y_{r,i}\cdot x_{r,i+\ell}) \qquad (7)$$

These estimates are then scan-to-scan integrated.

PROCESSOR ARCHITECTURE

The real-time data reduction requirements strongly affect the hardware design. Hardware programmability and remote control are ultimate design goals put forward by the variety of operational modes in IS signal processing. Design requirements for the EISCAT processor are: 1) Time domain ACF estimation, 2) Standarized interface with peripheral units, 3) Remote hardware check-out and fault detection, and 4) Remote control by on-site computer.

The digital processor developed combines high-speed, special-purpose hardware with standard elements for computer processing. The multi-processor structure, shown in Figure 4, consists of synchronized submodules controlled by a single microprogram execution. The system is fully microprogrammable from the on-site host computer by download of microinstructions to be stored in local memory. Real-time commands are generated by the on-site radar controller. The processor responsible for the program execution has an instruction set specially developed for multitask control. The instruction cycle is 250 nsec. A micro-instruction word of 126 bits controls all internal operations from cycle to cycle. The data processing section of the processor is implemented as a multi-stage pipelined arithmetical unit highly optimized for vector processing. The data pipeline includes a 2-port buffer memory with a high-speed bus interface (2 Mwords/sec. max) connecting the multi-channel data acquisition system to the processor. 8-bit in-phase and quadrature components are stored in the buffer memory as one word. A 16-bit address processor directs the range-gate data vectors as input operands to the data processing unit. After complex lagged product generation and digital integration, the ACF estimates are stored in the accumulator memory. The complex estimates are stored as one word controlled by a 12-bit address processor. The pre-processed data is transferred, using a high-speed direct memory access (DMA) channel, to the computer.

The 2-channel data processing unit, Figure 5, consists of distributed arithmetical elements implementing a 5-stage pipelined structure. In the first pipeline stage, 8-bit multiplier operands are selected from the buffer memory or from an external bus interface. The 8-by-8 bit multipliers and following adder-elements generate a 16-bit complex product of the operands. Scan-to-scan integration using 32-bit accuracy is performed by a read-modify-write cycle of the accumulator memory. The two channels are individually controlled by the local program execution.

The partitioning of processing tasks makes the system highly modular and sized to meet various real-time throughput requirements. One master module has the capacity of $40 \cdot 10^6$ arithmetical instructions per sec. Hardware flexibility for system expansion has been included. The 2-port buffer memory can accommodate 4 data pipelines in parallel. One master module, acting as system supervisor, is capable of controlling 3 additional slave processors as shown in Figure 6.

IMPLEMENTATION

The processor system, depicted in Figure 7, contains two mainframe modules housing the buffer memory and the master module (sub-assembly of the control section and data pipeline). 4 systems have been implemented as front-end pre-processors in the EISCAT radar facility. The pre-processor is configurated with 8K (16-bit words) of buffer memory, 2K(64-bit words) of accumulator memory and 128 words (128 bit) of program memory. The program memory is split into a RAM (Random-Access-Memory)-section storing computer generated programs, and a fixed PROM-section storing a library of IS operational tasks. The systems are interfaced with the on-site computer using CAMAC modules.

The processor hardware complexity is largely reduced exploiting commercially available TTL/NMOS circuit technology. VLSI-chips are used in time-critical parts of the processor: The control processor (bit-slice microcontroller), the address processors (bit-slice arithmetic/logic units) and as parts of the data pipeline (hardware multipliers). The system contains 13 boards (Eurostandard 233x160 mm) and the power dissipation is 275 W.

During the implementation phase, software tools have been developed. These include cross-assembler, simulator, microcode converter and a complete software system for remote testing and debugging of processor hardware. Standard microprograms for IS signal processing have been developed (5).

CONCLUSION

A description of the equipment used in performing IS radar observations of the ionosphere has been given. For the EISCAT radar facility, the measurement technique is based on estimating the plasma autocorrelation function. A digital hardware concept is described suitable for IS real-time signal processing. The high-speed correlator developed for the EISCAT system is microprogrammable and has a flexible adaption to multi-mission operations and experimentation with new algorithms. The developed real-time processor has an efficient hardware utilization with relaxed complexity as compared with conventional architectural designs based on hardwired implementations.

REFERENCES

1. du Castel, F., Holt, O., Hultqvist, B., Kohl, H. and Tiuri, M., 1971, "A European Incoherent Scatter Facility in the Auroral Zone, A feasibility Study", The Auroral Observatory, Tromsø, Norway.

2. du Castel, F. and Testud, J., 1974, Radio Science, 9, 113-119.

3. Hagfors, T., 1975, "Incoherent Scatter Radar Observations", in "Radar Probing of the Auroral Plasma", ISBN 82-00-02421-0, Universitetsforlaget, Norway, 75-101.

4. Alker, H.J., 1976, "A Design Study of a Multibit, Digital Correlator for the EISCAT Radar System", dr.thesis, The University of Trondheim, Norway.

5. Ho, T. and Alker, H.J., 1979, "Scientific Programming of the EISCAT Digital Correlator", MPAE-W-05-79-06, Max-Planck-Institut für Aeronomie, W-Germany.

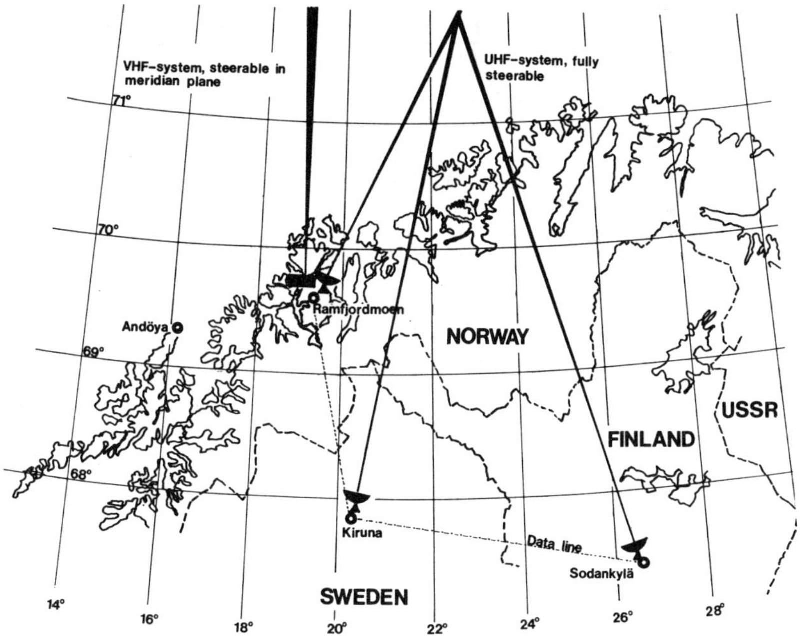

Figure 1. The EISCAT facility.

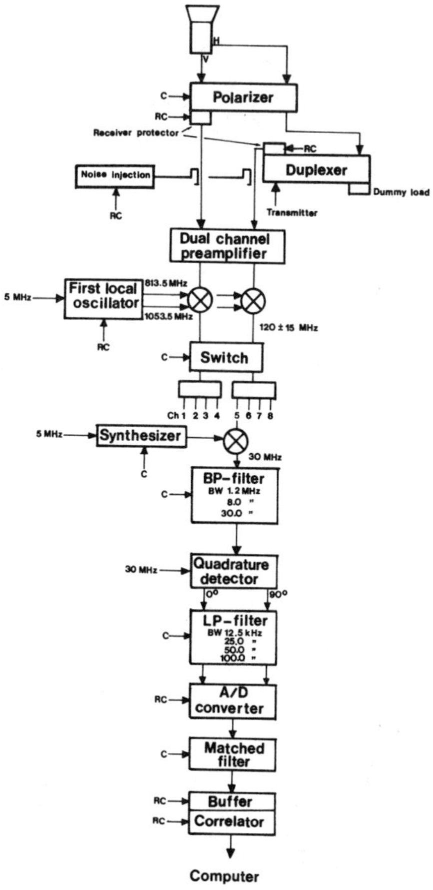

Figure 2. Block diagram of EISCAT UHF receiver system.

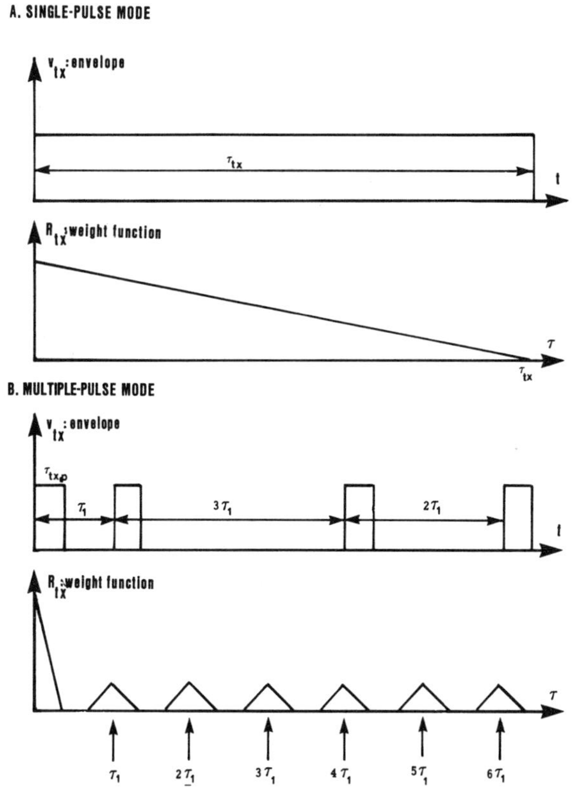

Figure 3. Basic modulation modes.

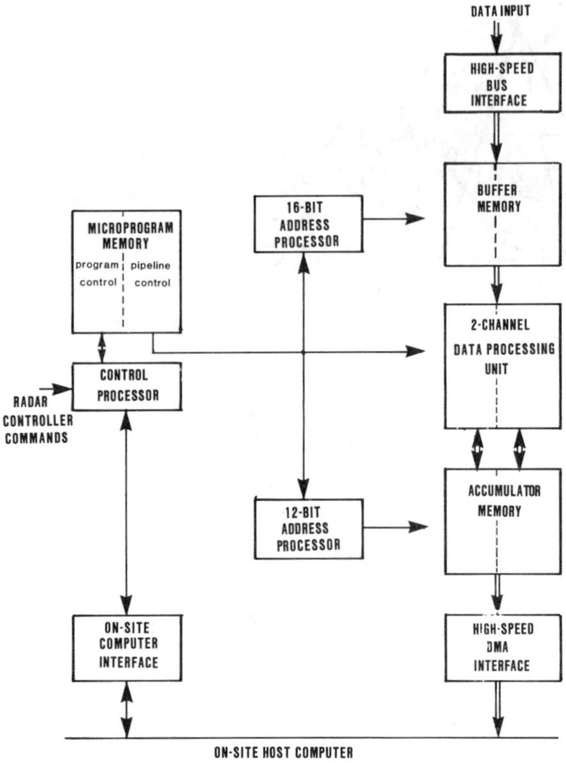

Figure 4. Block diagram of processor architecture.

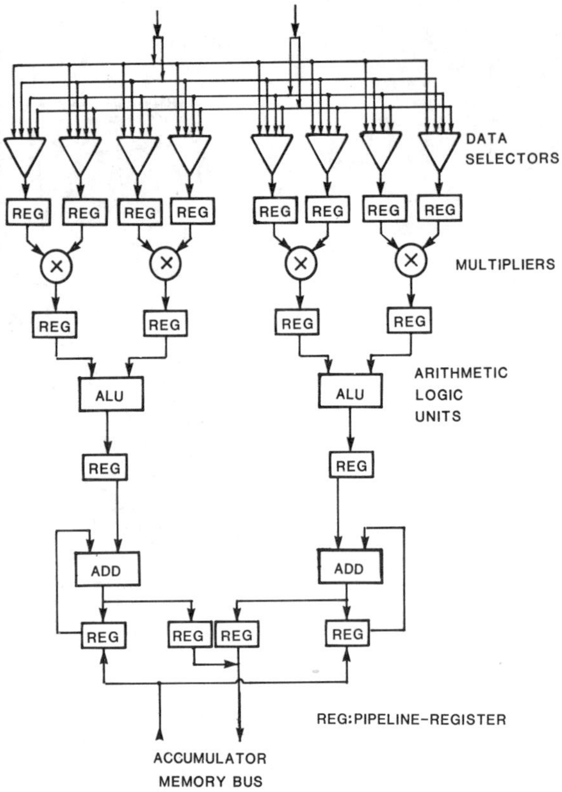

Figure 5. Block diagram of 2-channel data processing unit.

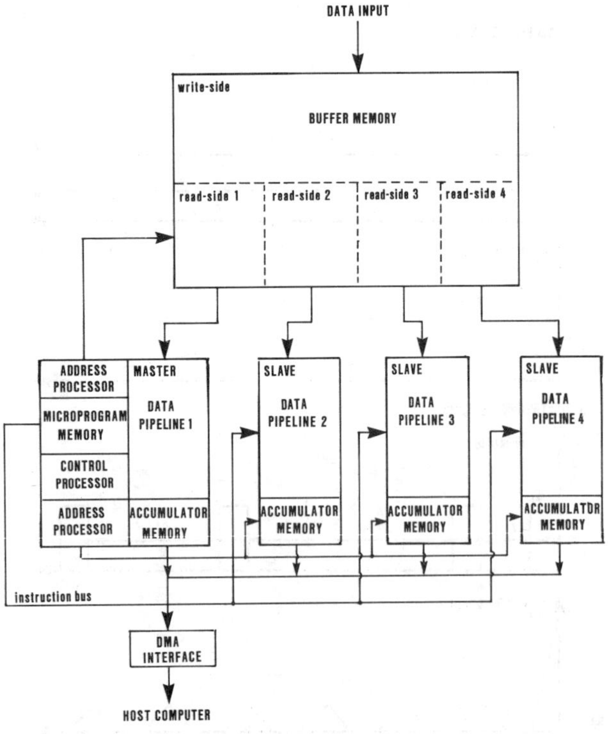

Figure 6. Parallel processing with master/ slaves.

Buffer memory

Master module

Figure 7. The processor mainframes.

A RETROSPECTIVE DETECTION ALGORITHM FOR EXTRACTION OF WEAK TARGETS IN CLUTTER AND INTERFERENCE ENVIRONMENTS

R. J. Prengaman, R. E. Thurber, W. G. Bath

The Johns Hopkins University Applied Physics Laboratory, Laurel, MD 20707 USA

INTRODUCTION

The ultimate use of all radar data depends on the ability to distinguish between signals returned from the environment or generated internally (noise) and those returned from desired targets. In the simplest systems, this target detection process is carried out by viewing the signals on a display scope. This is adequate when there are only a few targets in a reasonably clear environment; however, to obtain reliable detection of targets in dense or cluttered environments the process must be automated.

Automation of the detection process can take many forms, from simply setting a fixed threshold on the input signals to adaptively determining the detection threshold by measuring the statistics of signals returned from the local environment. This in turn establishes the signal-to-background ratio required for target detection in the given environment.

Due to the statistical nature of many types of clutter, a Constant False Alarm Rate (CFAR) detection device must set a high detection threshold to maintain a reasonable False Alarm Rate (FAR). A FAR of one in $10^5 - 10^8$ opportunities is typically required at the input to an automatic tracking system. By setting the threshold to obtain this FAR, detections of small and medium size targets can be missed.

A retrospective processor uses all contacts (or "plots") from several past radar scans, examining all possible target trajectories formed from stored contacts for each input detection. The retrospective processing architecture enables the processor to sort through many thousands of contacts efficiently, eliminating many false alarms and retaining those contacts describing reasonable trajectories. This approach allows a much lower detection threshold (higher FAR) to be set by the CFAR device resulting in large improvements in detection sensitivity in certain important clutter environments.

RETROSPECTIVE PROCESSING CONCEPT

Most theoretical studies of radar processors and most radar range equation predictions are based upon Gaussian noise and clutter assumptions. These assumptions are seldom valid in practice. As an example, Figure 1 illustrates measured cumulative probability distributions of the envelope of clutter returns from an S-band pencil beam 3D radar in three environments (sea clutter, land clutter, rain clutter). Also shown is the cumulative Rayleigh probability distribution which should describe the clutter envelope if the clutter or noise was in fact Gaussian. Three conclusions from this and similar clutter studies are: (1) Clutter statistics are not Gaussian; (2) No single non-Gaussian model accurately describes the variety of possible clutter types and situations; (3) The detection threshold required to maintain a low contact false alarm rate (e.g. $10^{-5} - 10^{-6}$) can be 10-20dB above that required with Gaussian noise (thus leading to a 10-20dB detection loss).

A retrospective processor overcomes these problems by allowing the contact detection process to operate at a much higher false alarm rate (e.g. 10^{-3}) at which the detection loss for non-Gaussian clutter is greatly reduced and the differences between different clutter types are less. A retrospective processor accepts as an input potential radar contacts from a radar input processor and produces filtered radar contacts for use as display data or data for automatic tracking systems.

The major elements of the process are: (1) Retrospective correlation using a bearing/range link structure; (2) Update of a velocity profile mask; (3) An output decision based on updates of the mask.

Linked list data structures are crucial to efficiently searching large collections of data. The bearing/range ordered linked list structure provides automatically the first dimension of a correlation process and allows a logical ordering of processes to follow a scanning radar. The major advantage of bearing/range linked correlation with multi-scan data is removal of old data and link restoration. Scan-to-scan linking is always newest to oldest, so that memory for the oldest scan can be reset or overwritten without disturbing the link for active scans.

To determine whether or not a target is present based on M scans of data, patterns of detection are evaluated for the "reasonableness" of these patterns under the assumption that the data represents true target returns. A velocity profile is constructed in the range-only dimension with data accumulated over bearing zones representative of the bearing resolution of the radar. Each profile consists of M bits indicating whether a detection occurred in a cell corresponding to that profile in each of the M scans preceding the contact of interest.

These velocity profile updates produce a hit/miss pattern for each velocity profile resulting in a multi-scan hit pattern rather than a simple count of detections associated with a profile. This allows a quality to be assigned to each pattern of hits and misses which is compared to a threshold. Only if the quality in any profile for a given contact exceeds the threshold is that contact output by the process.

As an example of this process, the various steps are described using three illustrations (Figure 2a-c) for the special case of a processor for surface target detections with a 7 scan correlation and 10 seven knot velocity profiles. This filter examines every contact and compares that contact with contacts from the previous 7 scans if they fall within a certain range and bearing window about the contact. Only if the contact of interest forms a probable track with contacts from the previous 7 scans is it output. The range and bearing windows are set to examine all prior contacts that could result from a target moving at up to 35 knots in any direction.

The top illustration is a window with a single scan worth of data shown. Contact number 1 is the contact to be examined with contact number 2 being another contact that falls in the range/bearing window. At the right of the range/bearing window, all contacts that appear in the window on this scan are plotted as a function of range.

The middle illustration is the same range/bearing window about contact 1 with all the contacts from the previous 7 scans shown. The false alarm rate used in this plot is 10^{-4}. It is clear from this 8 scan history that contact number 1 forms a target track with contacts 4, 6, 10, 12, and 14 from previous scans.

Here again, to the right of the range/bearing window the contacts within the window on each scan are plotted versus range. The function of the retrospective data processor (RDP) is to look at the contacts from the previous 7 scans and see if they form a probable track with the contact of interest. To do this the stored contacts are examined for patterns of detection data and the "reasonableness" of these patterns is evaluated under the assumption that the data represents a true target return.

The bottom illustration shows how the detection patterns are determined. A set of velocity profiles are constructed in the range-only dimension for contact data within the range/bearing window. These profiles are set up for target range rates in 7 knot increments for range rates of 0 to 35 knots inbound and outbound making a total of 10 profiles. For each profile the detection pattern is determined. There are 127 possible detection patterns that can occur, each one representing a certain probability the contact is real. This probability is computed assuming Poisson clutter spatial distributions. A threshold set on this probability determines whether a contact is to be output.

THEORETICAL ANALYSIS OF MULTISCAN DETECTION PROCESS

Suppose a design goal for an Automatic Detection and Tracking (ADT) system is to have the false track rate P_{fa} (say 10^{-8}) and that M scans of observation are allowed for target detection. It is clear there are several ways of doing this. One could set strict single scan detection criteria which might give a false alarm rate near $10^{-6}-10^{-8}$ on a single-scan basis. Relatively loose consistency and trajectory criteria could then be used in the multi-scan processing and still reach the ultimate 10^{-8} false track probability goal. Conversely one could set lenient single scan detection criteria which might give a much higher false alarm rate (10^{-3} to 10^{-4}) on a single scan basis. Strict consistency and trajectory criteria would then be used in multi-scan processing to reach the 10^{-8} false alarm track probability. Since almost any single scan false contact rate can be eventually reduced to the desired false track rate by using strict enough consistency/trajectory criteria, the relevant question is:

> What single-scan false alarm rate maximizes the probability of target detection in a given type of clutter, noise or jamming?

It appears that the answer to this question is often a much higher single-scan false alarm rate than is conventionally used in ADT systems.

A straightforward statistical analysis of this problem requires some simplifying assumptions. The multi-scan tracking process (including the RDP) is modeled as a bank of trajectory filters each basing final detection decisions on a simple count of the number of scans on which a contact falls in the trajectory window. This is somewhat oversimplified since the RDP uses a quality computation rather than a simple contact count. (A more precise analysis of the problem specifying the number of resolution cells in each trajectory window, the effects of the particular angle integration/centroiding technique used on single scan detection false alarm probabilities and multiplying the track false alarm rate by the total number of trajectory filters, [1] gives qualitatively very similar results).

The average single-to-noise required for a 95% probability of detection and a 10^{-8} probability of false alarm with M=8 scans of observation is shown in Figure 3 as a function of single scan false alarm rate, p_f, for the three clutter distributions shown in Figure 1.

The fundamental point illustrated by Figure 3 is that although performance depends upon the specific clutter distribution, detection sensitivity is maximized by selecting a single-scan false alarm rate in the 10^{-4} to 10^{-2} range. For all three long-tailed clutter distributions, the penalty for operating conventionally (i.e., with single scan false alarm rate set at 10^{-6}) is quite large (3-8dB). For the empirically fit mixture model there is a potential 12dB improvement in sensitivity when operating the single-scan false alarm rate in the range 10^{-3} to 10^{-4} as opposed to operation at 10^{-6}.

EXPERIMENTAL EVALUATION OF RETROSPECTIVE DATA FILTER

The initial thrust in the development of a retrospective processor was to enhance the signal processing capability of sea surface surveillance radars which are severely limited in detection performance by sea clutter returns.

Figure 4 shows the preprocessor and the associated data recording equipment which was used in initial retrospective processing experiments. The preprocessor consists of an amplitude processor where input video is digitized and filtered, a CFAR processing section where filtered data are detected with a well-regulated FAR in all environments, and data formatting circuits for interfacing with the digital magnetic tape system.

Digital recordings of the contacts are made and returned to the laboratory for reduction. These centroids of potential targets form the input to the retrospective processor. The initial retrospective processor was developed in software on a general purpose computer. This permitted the algorithms to be perfected without building hardware. The first version of the retrospective algorithm was used extensively to filter data collected in several sea surface surveillance environments. Once the basic technique was proven, a hardware design to operate in real time was developed. A software model that exactly emulated the hardware configuration has been used to prove the validity of the proposed hardware implementation and the actual hardware processor is currently being built and tested.

Figures 5-8 illustrate qualitatively the operation of the retrospective processor in a sea clutter environment. An AN/FPS-114 radar mounted on Laguna Peak on the California coast was instrumented as shown in Figure 4. This S-band radar performs sea surveillance for range-safety purposes. The radar has a high spatial resolution antenna (0.1μsec pulsewidth by 0.9° azimuth beamwidth) and a 4 second rotation period but no coherent doppler processing. The radar is sited approximately 1500 ft. above sea level. Figure 5 shows in PPI format a raw video recording from a single scan of the radar. After on site A/D conversion, the video signals pass through a digital range averaging logarithmic CFAR device producing the contacts shown in Figure 6. A high false alarm rate ($\sim 10^{-3}$) is maintained in the CFAR device resulting in approximately 2,000 false alarms per scan. False alarms from sea clutter are indistinguishable from small target returns when viewed on a single scan basis.

Figure 7 illustrates two effects of passing 100 scans of radar contacts through the retrospective processor. The false alarm rate has been reduced by at least 4 orders of magnitude and the ships and boats in the channel are clearly visible. The large reduction in false alarm rate which the retrospective processor produces is further illustrated by Figure 8 which shows the effects of passing 1000 scans (about one hour) of radar contacts through the processor. Only a few actual false alarms are visible. This ability to greatly reduce false alarm rates by exploiting the spatial and temporal correlation properties of sea clutter allows much greater target sensitivity to be achieved.

RETROSPECTIVE DATA FILTER HARDWARE IMPLEMENTATION

The hardware implementation of the Retrospective Data Filter was designed as a special purpose signal processor with limited programmability. The functions performed by this processor are shown in Figure 9.

The control function controls all the processes that take place during correlation of an input potential target contact with the stored contact data. Input contacts are stored in a buffer while waiting to be processed. First, input contacts are linked with contacts in memory that are located in the same range/bearing zone as the input contact and stored in memory. Then each contact in memory that falls in the range/bearing zone of the new contact is read out and compared with the new contact to see if the pair could form a target track. If they could, the appropriate velocity profile is updated and the next oldest contact in the range/bearing zone is read and compared to the new contact. This process continues, reading from newest contact in a zone to the oldest until all contacts have been compared with the input and rejected or used to update a velocity profile.

At this time the peak value of the velocity profiles is compared to a threshold. If a set threshold is exceeded in a velocity profile the input data is declared valid and sent to the output buffer along with the velocity of the profile in which a detection was made and the value of the detection data in that profile. Table I lists the performance characteristics of the hardware Retrospective Data Filter. The 2000 contact/scan processing capacity is currently limited only by memory size.

The construction technique used for this prototype unit is wirewrap on six inch by six inch circuit cards. The signal processor required six circuit cards containing 250 small and medium scale integrated circuits. The total power consumption is 30 watts.

CONCLUSION

Radar processing can be divided into two areas. First, radar detection devices which discriminate between targets and unwanted signals on the basis of the size of the echo, its doppler frequency and the shape of the returned pulse, and generally make decisions based on a single pulse or series of pulses spanning a relatively short period of time (i.e., several milliseconds). Second, track-while-scan techniques for target detection which examine these preliminary detections over a much longer period (i.e., several seconds) and check for consistency before a final detection decision is made.

Most automated radar systems rely primarily on the first type of processing to set a low false alarm rate. While this approach is valid in an idealized Gaussian clutter environment, in real-world clutter situations it often results in large target detection losses.

A retrospective processing approach appears to solve this problem by allowing a much higher false alarm rate to be set initially. The spatial decorrelation properties of the clutter or noise are then exploited to produce a low output false alarm rate. Both theoretical analysis and radar experiments indicate a substantial improvement in target detectability in clutter environments.

TABLE I Retrospective Data Processor Performance Characteristics

Number of Scans Correlated	8
Overwrite Adaptable Memory	
Contact Storage	16 K
Range Resolution	16 bits (24 ft)
Bearing Resolution	12 bits (0.088°)
Input Buffer Depth	64 Contacts
Velocity Coverage	± 126 knots
Number of Velocity Filters Used	36
Maximum Available	64
Velocity Filter Width	7 knots
Maximum Throughput Rate	500 contacts/sec
Detection Thresholds	selectable (128 discrete)
Link Sectors	
Range	64 (2 radar miles)
Bearing	16 (22.5°)
Typical Operating Characteristics	
Input Pfa $= 10^{-3}$ to 10^{-4}	(nominally 1000 false alarms/scan)
Output Pfa $= 10^{-8}$ to 10^{-11}	(1 false alarm per 10 to 10,000 sec)

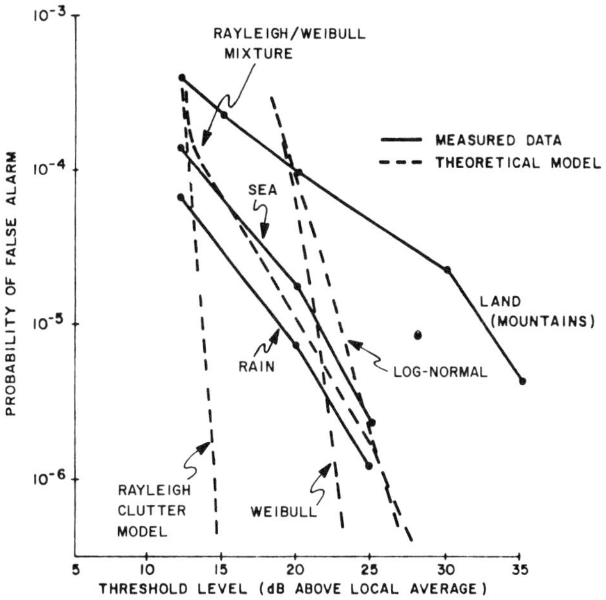

Figure 1 Measured and theoretical clutter probability distributions.

REFERENCES

1. Biddison, L. A., 1982, "Detection Performance of the Retrospective Data Filter for Fluctuating Targets in Gaussian Clutter". JHU/APL memorandum F3A-82-3-284.

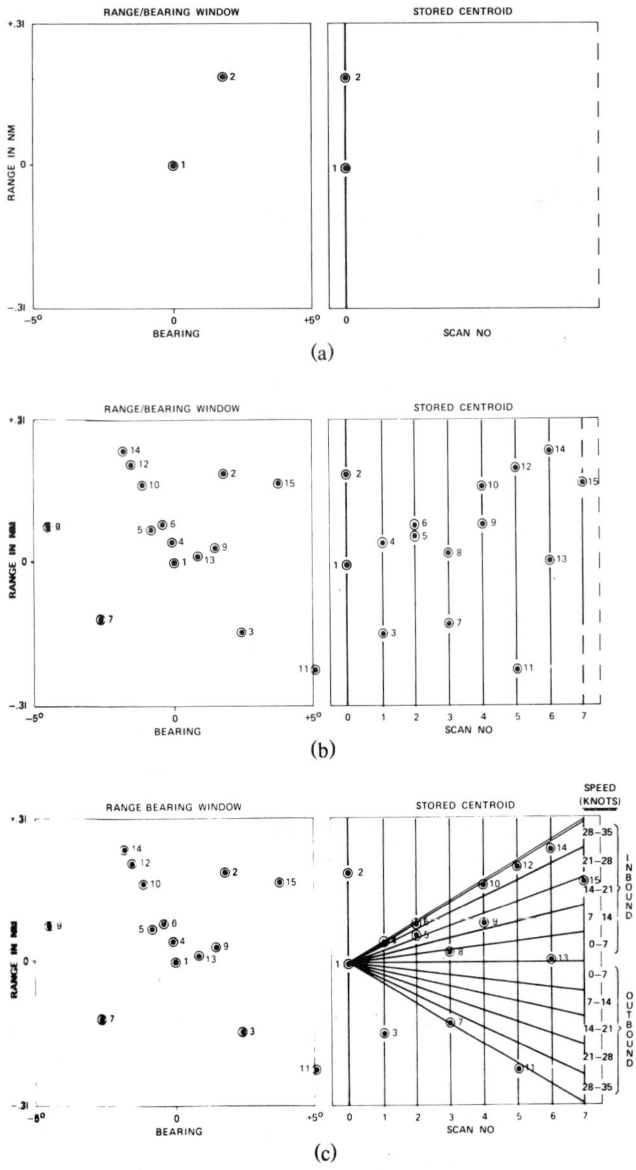

Figure 2 The retrospective process; (a) A single scan of data, (b) Eight scans of data, (c) Eight scans of data with trajectory filters applied.

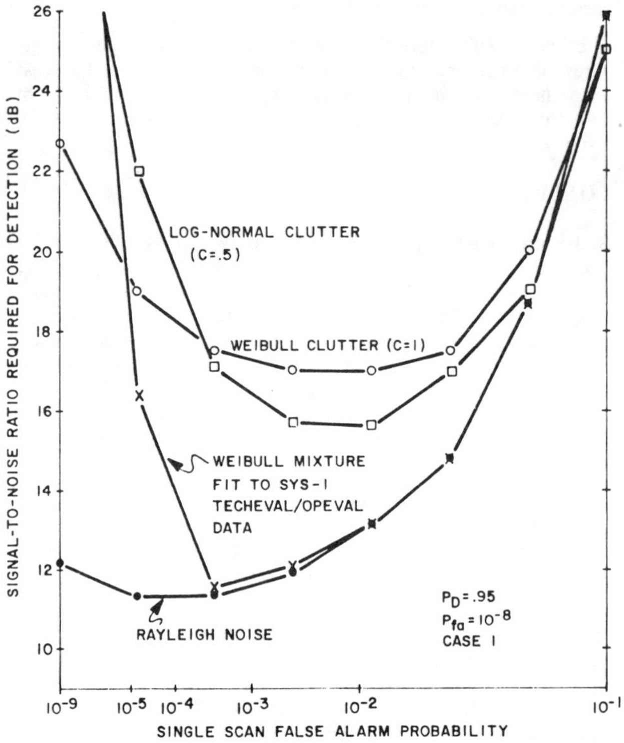

Figure 3 Detection performance of eight scan radar processor.

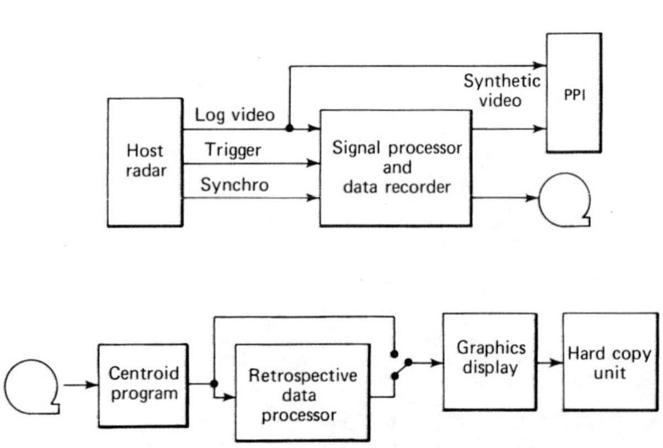

Figure 4 Data collection and analysis system employed in retrospective processing experiments.

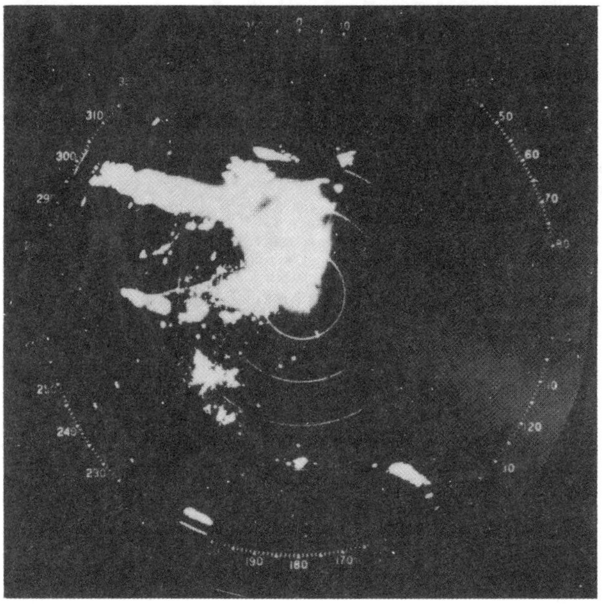

Figure 5 Plan position indicator display of raw radar video from sea surveillance experiment.

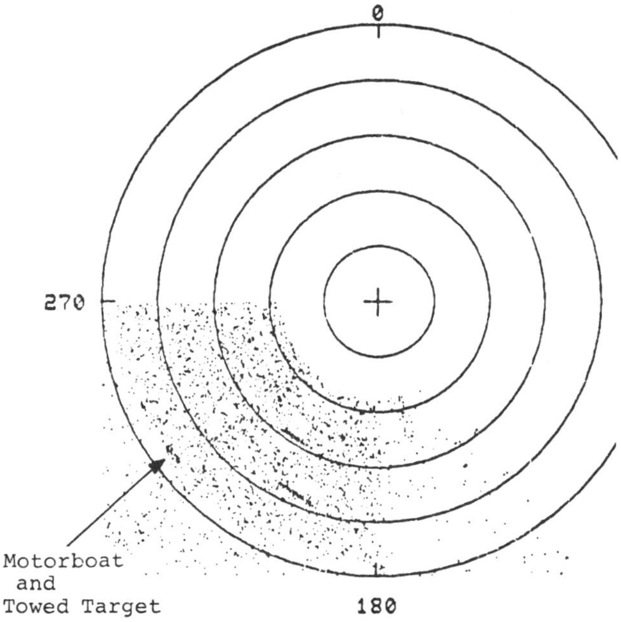

Figure 6 Raw contact data after CFAR processing (100 scans; 9 nm total range).

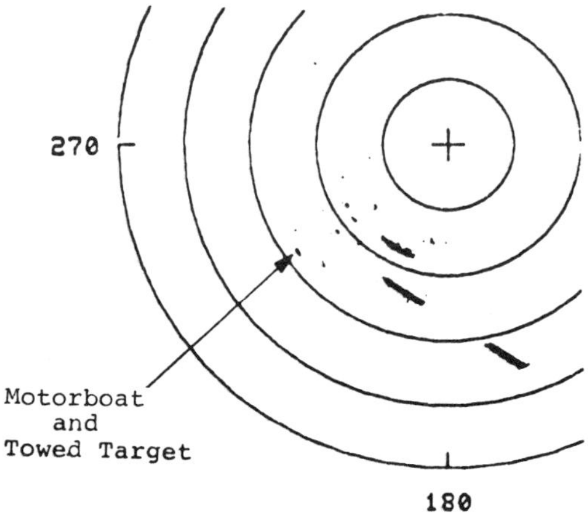

Figure 7 Contact data after retrospective processing (100 scans; 3 nm range rings).

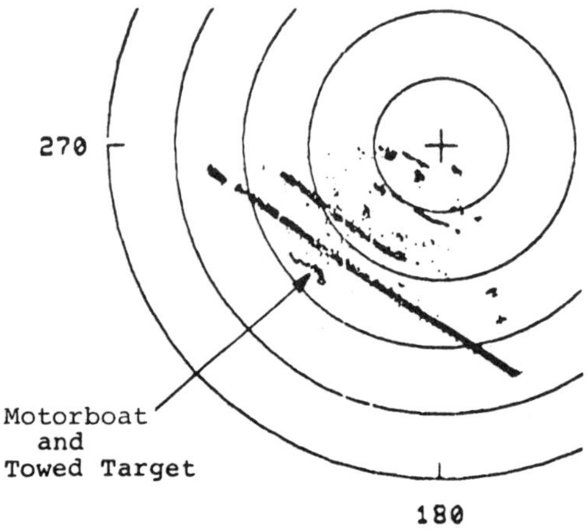

Figure 8 Contact data after retrospective processing (1000 scans; 3 nm range rings).

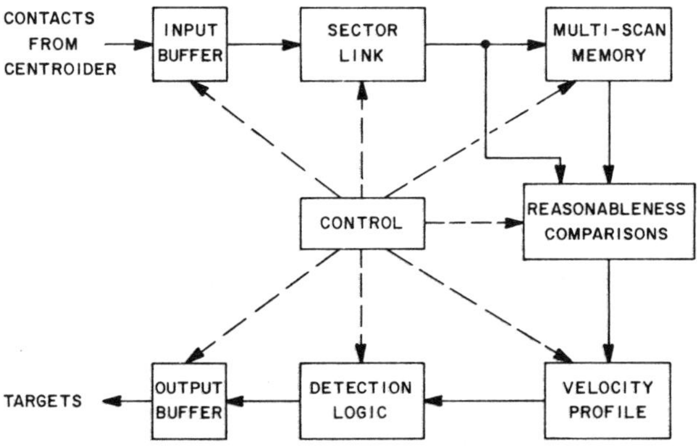

Figure 9 Retrospective processor architecture.

A KALMAN APPROACH TO IMPROVE ANGULAR RESOLUTION IN SEARCH RADARS

E. Dalle Mese*, G. De Fina*, V. Sacco**

*University of Pisa, Italy **S.M.A. S.p.A. - Florence, Italy

INTRODUCTION

In Surveillance Radar the problem of signal processing for improving azimuth resolution holds an important place. Generally, the antenna mainlobe width is a measure of the angular resolution of two targets. However two or more targets can exist at the same range from the radar, with an angular separation less than the antenna mainlobe width. In this case only one target is detected by conventional radar processing. However, there are many cases in which a better resolution is required. A way to obtain this goal is to use a suitable signal processing at the receiver. In fact, with a proper signal processing it is possible to obtain a pseudo-compression of the antenna pattern, without changes of the size of the antenna.

Let g(t) be the two ways (or power) antenna pattern and n(t) a white noise process. The received samples y_k can be written as:

$$y_k = g_k + n_k \quad , k = 1, 2, .. \quad (1)$$

where:

$$g_k = g(kT_R) = \text{sampled version of } g(t) \quad (2)$$

$$n_k = n(kT_R) = \text{additive sample of noise} \quad (3)$$

The sampling period T_R is, in our case, the pulse repetition period of the radar. Eq. (1) implies that the antenna beam interacts with the target from the instant k = 1. In this case the output sequence can be considered as the noisy output of a linear system whose discrete pulse response is $\{g(k), k = 1, 2, ... N\}$, where N is the number of considered antenna beam samples. The effect of this linear system is to spread the input signal energy, which is concetrated in a single azimuth resolution cell, in a number of consecutive resolution cells. So, two input pulses interact if their angle separation is less than $N \cdot \Delta\theta$, where $\Delta\theta$ is the angular width of a single cell.

A theoretical approach to this problem could be the utilization of an inverse filter, i.e. with transfer function given by 1/G(f), where G(f) is the Fourier Transform of g(t). Unfortunately this filter increases the noise power whenever the signal power is poor. So another approach to the problem must be devised.

Many authors treated the problem of optimum (in some sense) inverse filtering in connection with the sidelobe suppression problem in pulse compression radars. Among the solutions analysed we mention the technique of minimizing the sidelobe energy at the output of a mismatched filter via a least-square criterion by Ackroyd and Ghani (1), the technique of envelope-constrained filter design by Evans (2), the technique of inverse filtering using time reversal by Erikmats (3).

In this paper we analyse a Kalman approach to the problem, mainly for its simplicity in the analysis and implementation. A similar approach was used for the elimination of intersymbol interference in data transmission systems using dispersive channels, a problem which has many similarities with that treated in this paper.

In the following the structure of the optimum azimuthal compressor, some numerical results and some particular features relevant to practical implementation of the filter will be discussed.

This research activity is part of Italian Air Traffic Control (ATC) Project, supported by the Italian National Research Council (CNR).

A KALMAN APPROACH TO THE PROBLEM

In this section the problem of azimuth estimation is formulated as a problem of state estimation of a suitable linear system. Let us consider a fixed range resolution cell, and label with "Ø" an azimuth resolution cell which does not contain a target, and with "1" an azimuth resolution cell which contains a target. If N is the number of pulses on the target, we assume x_k as state vector at the time k, given by:

$$x_k = (u_k, u_{k-1}, \ldots, u_{k-n})^T \quad (4)$$

where $(\cdot)^T$ denotes transposition and the binary sequence (u_k, k = Ø, ±1, ±2, ...), u_k= Ø,1, characterizes the state of the k-th azimuth resolution cell. When the antenna rotates the state vector x_k varies accordingly to the following equation:

$$x_k = Fx_{k-1} + Bn_k \quad (5)$$

where F is the (NxN) shift matrix

$$F = \begin{pmatrix} 000..00 \\ 100..00 \\ 010..00 \\ .... \\ .... \\ 0....10 \end{pmatrix} \quad (6)$$

and B is the (Nx1) vector

$$B = (1, 0, 0, \ldots, 0)^T \quad (7)$$

Let us define the (Nx1) vector G as:

$$G = (g_1, g_2, \ldots g_N)^T \quad (8)$$

where the g_i, i = 1, ... N are samples of the antenna pattern. The received sequence y_k can be written as (see also eq. (1)):

$$y_k = G^T x_k + n_k \quad (9)$$

where n_k is a noise sequence, supposed Gaussian, with Ø mean value and variance σ_n. In addition, let us assume $\bar{u}$ and σ_s as the mean value and variance of the u_k sequence, respectively.

In the above framework, the angular position of the target is given by the position of the "1" in the sequence u_k, so an estimate of the state vector of the linear system described by equations (5) and (9) gives the solution of the problem (for reference see Meditch (4)).

As is well known, the best estimate $\hat{x}_k$, in the minimum mean square error sense, of the state is given by the Kalman solution:

$$\hat{x}_k = (F\cdot\hat{x}_{k-1}+B\cdot\bar{u})+K_k\cdot[y_k-G^T\cdot(F\cdot\hat{x}_{k-1}+B\cdot\bar{u})] \quad (10)$$

The (Nx1) gain vector K_k is obtained from the recursion:

$$\bar{P}_k = F\cdot P_{k-1}\cdot F^T+B\cdot B^T\cdot\sigma_s \quad (11)$$

$$K_k = \bar{P}_k\cdot G\cdot[G^T\cdot P_k\cdot G+\sigma_n]^{-1} \quad (12)$$

$$P_k = [I - K_k\cdot G^T]\cdot\bar{P}_k \quad (13)$$

The filter given in (10) is a time varying filter, so the practical implementation is quite difficult. However it is possible to show that when k tends to infinity the filter tends to a steady state solution which is quite simple. Fig. 1 shows the filter configuration in the steady state case. In addition it is simple to prove that, with the given model equations, the Kalman filter is stable. The (NxN) matrix P_k (in eq. (13)) is the covariance matrix of the estimation error, hence the diagonal terms are a measure of the goodness of the estimate.

A better estimate, i.e. with a lower value of the diagonal elements of the covariance matrix, can be obtained by means of smoothing techniques (Dalle Mese and Giuli (5)). The analysis is the same as Kalman filtering, if a suitable modifications of the matrices and vectors of the model is made. Particularly, the dimensions of all the matrices and vectors must be augmented by L which is the lag of the smoothed estimate, and the new antenna pattern vector is obtained from the vector G in eq. (8) by adding L components equal to zero. With these modifications the equations which give the smoothed estimate and the smoothing gains are just eqs. (10) ÷ (13). The structure of the smmothing filter is the same as in Fig. 1: the new dimensions is obviously (N+L), while the first N smoothing gains are just the filtering gains.

NUMERICAL RESULTS

In this section some numerical results relevant to the Kalman approach described in the preceding section will be discussed. In order to solve the recursion given by eqs. (11), (12) and (13), (in other words to find the steady state structure of the filter), it is necessary to fix the values of σ_s, σ_n and P_o (the initial value of the covariance matrix P). In addition the value of $\bar{u}$ is required to determine the state estimate in eq. (10). The variance σ_n depends on the value of the signal-to-noise ratio (SNR). We have assumed the following definition:

$$\mathrm{SNR} = 10\ \log_{10}(1/\sigma_n) \quad (14)$$

The matrix P_o is taken equal to $\sigma_s\cdot I$, where I is the identity matrix. Finally, a number of numerical tests suggested the values $\sigma_s=1$ and $\bar{u}=\emptyset$. A sample of the numerical results obtained by a computer calculation is reported in Figs. 2 ÷ 5. All the figures were obtained under the following assumptions:

- the antenna pattern is modeled by a (sin x/x) function;
- two targets with the same normalized power (=1) are supposed to be present within -3dB points of the antenna mainlobe;
- the number of pulses within -3db points of the antenna mainlobe is equal to 15;
- the dimension of the filter, i.e. N, is equal to 23 which is the number of pulses within -13dB points of the antenna mainlobe.

The last assumption was made because the Kalman filter is very sensitive to the model used to obtain the observed sequence. In our opinion, confirmed by a number of computer simulations, the number of samples considered is adequate to represent the antenna mainlobe.

The curves labeled a) represent the input sequence, while the curves labeled b) represent the processed output of the compressor. Fig. 2 was drawn for comparison purposes: it was obtained with SNR=50 dB (this practically corresponds to a noiseless case) and a separation Δ of the two targets equal to 7 azimuthal cells, i.e. equal to half the -3dB antenna mainlobe. The goodness of the proposed filter is clear; in this case it is possible to distinguish two targets with a separation of only two azimuthal cells. Fig. 3 is the same as Fig. 2, but with SNR=20 dB. The two targets are detected yet, but with an increase of the insertion loss and of the output noise. The separation capability, SC, of the filter can be evaluated by computing the ratio of the maximum output to the minimum output between the two targets. When SNR=20 dB, the value of SC is of the order of about 9 dB. In the case of SNR=15 dB the separation capability for Δ=7 is too small: Fig. 4 represents a typical result with Δ=10. In this case we have SC ≃ 15 dB. In the case of SNR=10 dB a filtering approach gives poor results in terms of both separation capability and output noise. In this case the smoothing technique seems to be the only practical approach. In Fig. 5 the result obtained with a lag equal to 15 is reported. It is noted that the output noise is quite large, while the separation capability is about 4.5 dB. This result is to be compared with curve a) which represents the input sequence at the compressor.

CONSIDERATION ON PRACTICAL IMPLEMENTATION

The implementation of a real-time digital filter is based on the following practical evaluations:

a) Total computing operations per iteration: this number is obtained by summing up the following contributions:
   - sum/subtraction operations;
   - multiplication operations;
   - division operations;
   - logical operations (ALU operation controls, memory addresses etc.);
b) Total computing time per iteration: this value is obtained by multiplying the preceding contributions by the single operation times;
c) Storage requirements are given as functions of both the filter parameters (final and temporary storage) and the program instructions;
d) Repetition period of the input pulses: this value must be compared with the evaluation at the item b) above;
e) Performance degradation is due to the quantization errors using fixed-point arithmetic.

TABLE 1 - Execution time, logic time and total computing time (in unit time)

Function	Execution Time	Logic Time	Total Time
$A_k = F \cdot X_{k-1} + B \cdot \bar{u}$	----	4·N	4·N
$B_k = G^T \cdot A_k$	MUL·N+SUM·(N-1)	47+6·N	45+8·N+MUL·N
$C_k = y_k - B_k$	SUM	6	8
$D_k = K_k \cdot C_k$	MUL·N	8·N	8·N+MUL·N
$\hat{S}_k = A_k + D_k$	SUM·N	MUL+27+5·N	MUL+27+7·N
Totals	2·N·(MUL+SUM)	MUL+23·N+80	MUL·(2·N+1)+2·N·SUM+23·N+80

The foregoing parameters depend on the hardware technology utilized (for example microprocessor and bit slice implementations). The starting point for the evaluation of the preceding parameters is the eq. (10), which is the mathematical expression of our filter. Note that in eq. (10) the $\{K_k\}$ coefficients are precomputed and stored.

Table I shows the execution time, the logic time and the total computing time. Note that:
- the times are indicated in unit times (i.e. the computing internal cycle);
- total computing time is the sum of the execution time and the logic time;
- N is the dimension of the filter;
- MUL and SUM are the single multiplication execution time and the single sum execution time, respectively. Also these values may be indicated in unit times;
- logic time is a purely indicative time: standard routines used by Mendel (6) are considered.

Figure 6 shows the plot of total computing time vs. the dimension of the filter, for representative values of MUL and SUM.
Note that generally MUL >> SUM and then the multiplication time is greater than the sum time. Besides, the MUL parameter has a little influence on the logic time.

In order to evaluate item c) above we must refer to the storage requirements which depend on the filter parameters. In fact the program and the temporary storage requirements are strongly determined by hardware configuration. Every way, see Table II for storage requirements of filter parameters.

TABLE 2 - Storage requirements

G^T	antenna mailobe coefficients	N words
K_k	filter gain coefficients	N words
x_k	input pulse	1 word
$\hat{S}_o$	initial state vector	N words
$\bar{u}$	mean value of U_k sequence	1 word

For clarity, let us illustrate the following example. Let us consider a microprocessor with the following characteristics:
- clock timing = 8 MHz
- internal cycle = 4 MHz
- MUL multiplication time = 37 internal cycles for 8÷16 bit word
- SUM sum/subtraction time = 4 internal cycles for 8÷16 bit word

Also, let us consider the following hypotesis:
- dimension of the filter (sample number) = 25
- Pulse Repetition Frequency = 1000 Hz

From the proceding figure and tables we have:
- total mul. time = 1850 int. cks.— 462.5 μs.;
- total sum time = 200 int. cks.— 50 μs.;
- total logic time = 692 int. cks.— 173 μs.;
- total comp. time = 2742 int. cks.— 685.5 μs.;
- max. sample repetition frequency ≃ 1460 Hz;
- storage req. of filter parameters = 77 words;

So, in this example there is a waiting time of about 300 μs between a filter output and the next input sample.

When item e) is considered, we observe that an implementation based on a fixed-point arithmetic gives a performance degradation which is function of:
- input sample quantization;
- arithmetic sample quantization.

The coefficient quantization are not considered because the gains of the filter are precomputed.
Stripad (7) suggests a mathematical model in which the quantizations generate error sequences. Particularly, Stripad (8) gives some relations in which the error sequences have the following statistical properties:
- the sequences are zero-mean;
- the variances of sequences are equal to $2^{-2t}/12$ where t is the bit number of fractional parts;
- the sequences are uncorrelated with input sequence and output sequence, respectively.

These hypotesis have been tested by comparisons between simulations and analytical predictions. So, one can evaluate the performance degradation in terms of error mean and error variance.

CONCLUSIONS

In this paper we analysed a spatial compressor obtained via a Kalman approach to improve angular resolution in search radars. The results obtained are generally satisfactory, when compared with the input sequence, which gives no possibilities to distinguish two or more targets within the

-3 dB antenna mainlobe. Some of the most important conclusions that can be drawn from this work are:

- the performance of the compressor decreases quickly when the SNR decreases. There is a threshold for SNR = 10 dB. Some improvement can be achieved by using smoothing techniques;
- in our opionion the poor results relevant to low values of SNR are mainly due to i) a statistical model of the sequence $\{u_k\}$ not reliable and to ii) the difficulty to invert the pulse response of the linear system (antenna pattern) which generates the input sequence;
- for a good target separation the thresholding of the filter output should be made over the samples within the antenna mainlobe. This means that a spatial filtering after the compressor is advisable. This spatial filtering could be accomplished by utilizing the output of the signal detector.

The insertion of the azimuthal compressor in a radar receiver needs a careful study. The insertion loss of the filter does not allow to eliminate the target detector. One possible solution is to put the compressor in parallel with the detection line. With this solution we have the possibility of processing the output of the compressor only in the angular interval where targets are declared.

This work is a preliminary study on the topic: future developments include a statistical analysis of the compressor and further investigations about the practical design and implementation of the filter.

REFERENCES

1. Ackroyd, M., and Ghani, F., 1973, IEEE Trans. AES, 9, 214 - 218.

2. Evans, R.J., 1977, "Design of robust sidelobe suppression filters", IEE Radar-77 Conference Proceedings, London, England.

3. Erikmats, O., 1978, "Range sidelobe elimination for discrete-coded pulse compression systems". International Conference on Radar Proceedings, Paris, France

4. Meditch, J.S., 1969, "Stochastic Optimal Linear Estimation and Control". Mc Graw Hill, New York, U.S.A.

5. Dalle Mese, E., Giuli, D., 1978, "On the use of smoothing technique for digital channel equalization". ICC International Conference on Communication Proceedings, Toronto, Canada.

6. Mendel, J.M., 1971, IEEE Trans. AC, 16, 748 - 758.

7. Stripad, A.B., 1981, IEEE Trans. AES, 17, 626 - 634.

8. Stripad, A.B., 1978, "Models for finite-precision arithmetic, with application to the digital implementation of Kalman filters", D. Sc. Thesis, Washington University, St. Louis, MO, U.S.A.

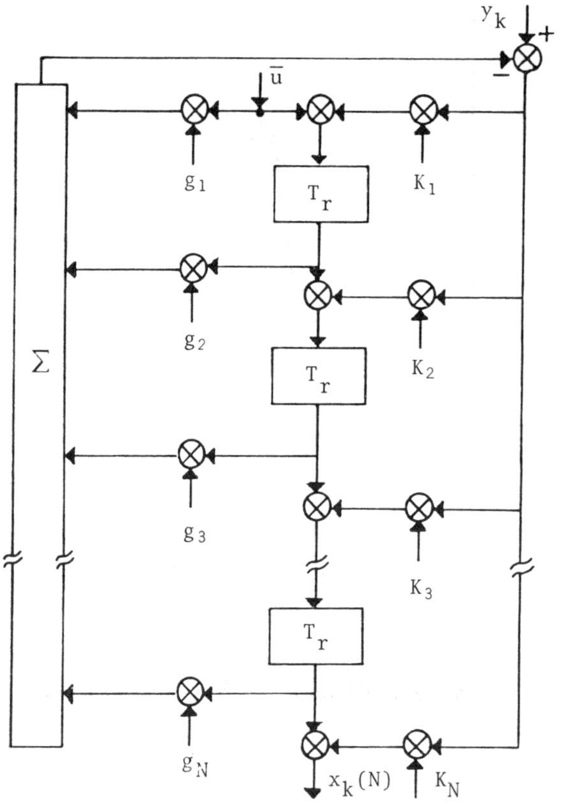

Figure 1 Block diagram of Kalman filter

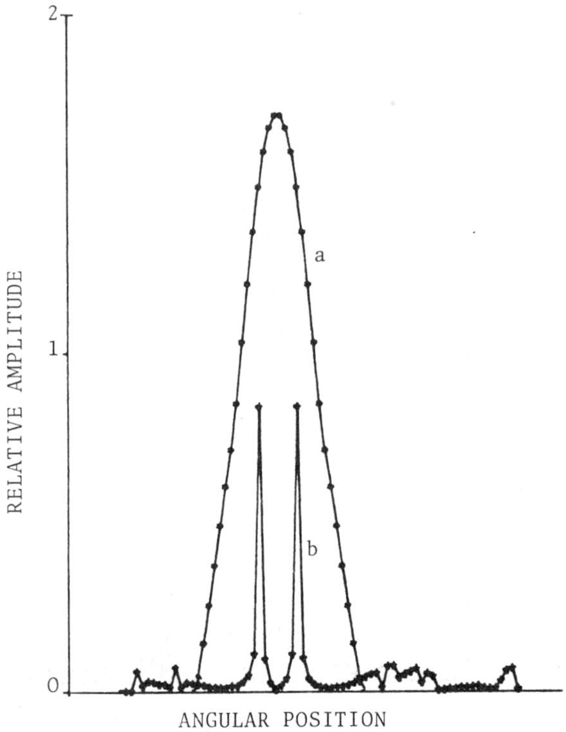

Figure 2 Input signal (a) and filter output (b) when SNR=50 dB and Δ=7 azimuthal cells

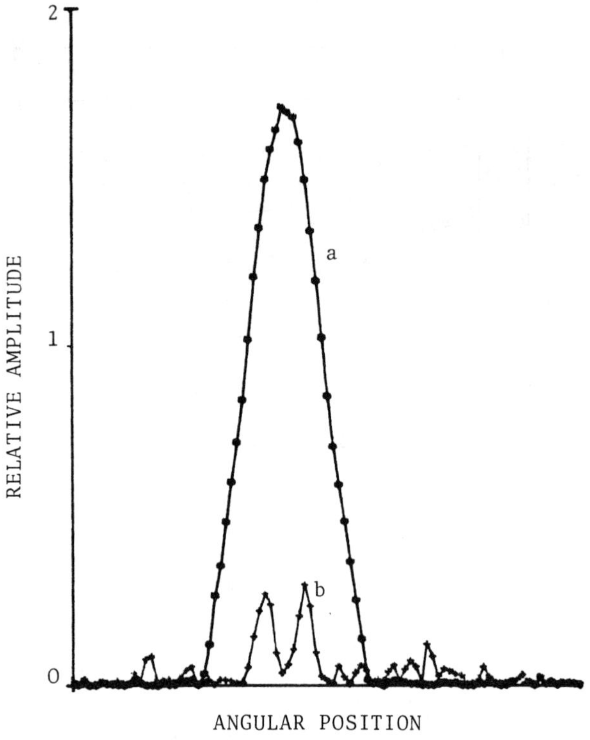

Figure 3 Input signal (a) and filter output (b) when SNR=20 dB and Δ=7 azimuthal cells

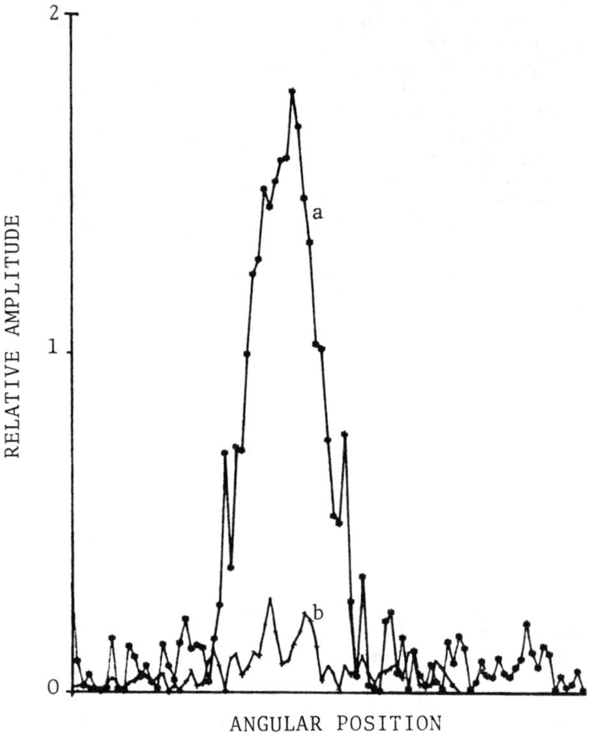

Figure 5 Input signal (a) and filter output (b) when SNR=10 dB, Δ=7 and L=15

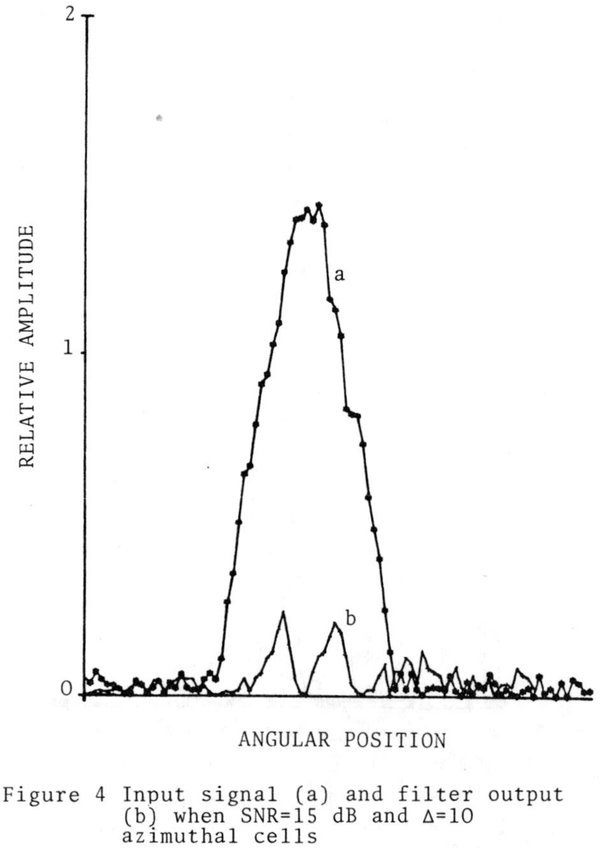

Figure 4 Input signal (a) and filter output (b) when SNR=15 dB and Δ=10 azimuthal cells

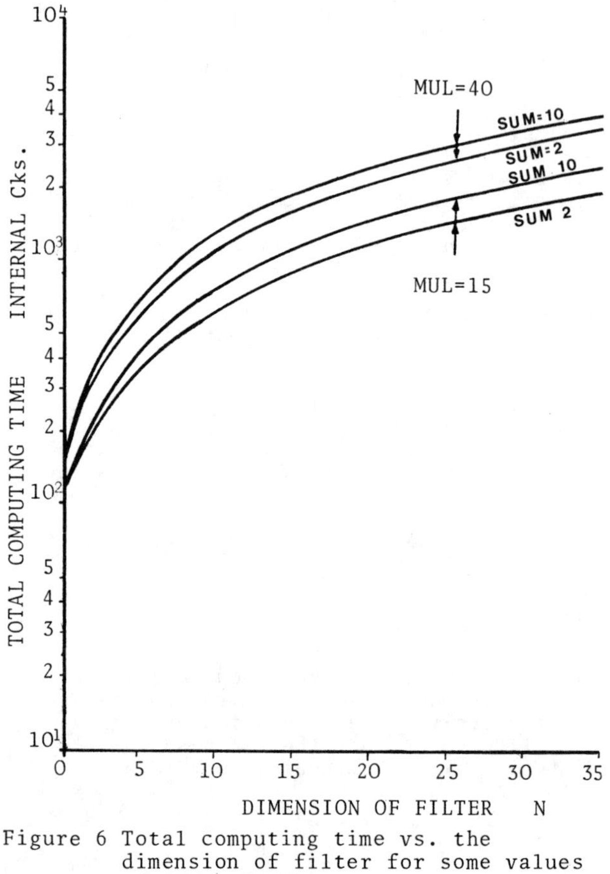

Figure 6 Total computing time vs. the dimension of filter for some values of SUM and MUL

REDUCED COST LOW SIDELOBE REFLECTOR ANTENNA SYSTEMS

N Williams*, P Varnish** and D J Browning*

ERA Technology Ltd* and ASWE** (Funtington) England

INTRODUCTION

In modern radar design, considerable emphasis is given to the suppression of sidelobe radiation. This involves careful control of the antenna aperture illumination and a number of theoretical distributions are available which allow a trade-off to be made between the aperture efficiency, the level of the first sidelobe and the rate of decay of sidelobes at wider angles. It was generally assumed that, to achieve the performance associated with these distributions, array techniques must be adopted. In surveillance systems, for example, consideration has been given to linear array feeds with cylindrical reflectors providing a shaped elevation plane coverage and, in stacked beam applications, planar arrays. Because of their complexity, they are usually singly polarised and often narrowband. Also, there are a number of applications for which they are too heavy and too expensive. In recent publications, it has been demonstrated that low sidelobe radiation patterns can be realised in the azimuth plane using simpler parabolic reflector systems (illuminated by point source feeds) which have an elliptical projected aperture, offset from the axis of the parent paraboloid. (Scudder (1), Rudge and Williams (2)). Consideration has also been given to shaped doubly curved reflectors which have essentially the same azimuth plane performance and a cosec$^2\theta$ elevation plane pattern (3). In this paper, we examine the design of a reflector system for surveillance applications which uses a vertical array of dual polarised feeds to produce two elliptical beams, stacked in the elevation plane. Emphasis is given to establishing the sidelobe performance over a range of elevation angles and the development of a relatively low cost, lightweight, reflector structure.

THEORETICAL DESIGN

Initial Studies

The basic antenna configuration is shown in Figure 1. The offset reflector parameters are similar to those which have been shown previously to provide low sidelobe characteristics in the azimuth plane. The aperture is elliptical with major and minor axes in the region of $45\lambda_o$ (where λ_o is the free space wavelength) and $15\lambda_o$, respectively, and the focal length is approximately $12\lambda_o$. The full illumination angle in the plane of offset is around 60° and the offset angle is chosen to avoid aperture blockage by the feed system. In the orthogonal plane, the full illumination angle is approximately 120° and, with feed apertures in the region of $1\lambda_o$ and the space taper produced by the deep reflector geometry, a highly tapered distribution can be realised in the projected aperture plane. In the studies reported here, conical horn feeds are employed because they provide a high level of polarisation purity over a reasonable bandwidth when the aperture diameter is of the order of $1\lambda_o$. To produce an acceptable illumination of the reflector in the plane of asymmetry and allow some flexibility in the elevation plane coverage, an array of four horns is used.

Before performing a detailed optimisation, an initial study was undertaken to examine the trade-offs between aperture efficiency and sidelobe performance. The far field patterns were computed using an aperture field integration technique and the computer program was structured such that the aperture fields could also be plotted and compared with idealised distributions. Typical co- and cross-polar plots of the aperture fields produced by a single feed horn are given in Figure 2; a smooth taper approaching 40dB is achieved in the azimuth plane. (The asymmetry in the elevation plane co-polar distribution and the cross-polar lobes in the azimuth plane result from the use of an offset geometry).

The azimuth plane distribution is plotted in Figure 3 and compared with that of a circular Gaussian distribution chosen to provide the same edge taper. The close similarity is confirmed by a comparison of their far field patterns, both producing sidelobes decaying away from the main beam region with a maximum value of approximately -50dB and a beam broadening factor (relative to that of a uniform circular distribution) of 1.47. This figure is considerably in excess of that for an ideal Tchebyscheff distribution producing uniform sidelobes of the same level. However, for many applications a fast decay in sidelobes is desirable. In this case, it is found that continuous distributions with a low edge value are required for which the aperture efficiency is closer to that achieved with reflector systems. For example, a Taylor distribution producing a first sidelobe level of -50dB and an $\bar{n}$ value of 2 (which results in a continuously decaying sidelobe envelope) has a beam broadening factor of 1.37.

Design optimisation

The geometry shown in Figure 1 was optimised for a vertically polarised surveillance beam with elevation and azimuth plane half power beamwidths of 5° and 1.7^o, respectively. Particular emphasis was given to the azimuth plane sidelobe performance over the complete range of elevation angles covered by the main beam. A predicted contour plot at the centre frequency is reproduced in Figure 4 and corresponds to excitation of two conical horns arranged about the focus of the reflector. Further computations indicated that, because of the space taper effect, the performance was relatively insensitive to frequency and polarisation changes.

The possibility of extending the elevation plane coverage was examined by computing the characteristics of a second beam produced with two additional feed horns, positioned beneath the principal pair. The corresponding contour plot is shown in Figure 5. It is seen that the coverage is extended by approximately 7.5°, although some broadening of the beam in the azimuth plane can be noted. This is due to phase errors introduced by the off-axis positioning of the feed horns.

EXPERIMENTAL STUDY

Antenna construction

The energy containment to the 40dB level implied in Figure 4 was found to be similar to that achieved with more complex array systems and an I band prototype model was fabricated and evaluated to confirm the theoretical results. Interest was shown in the system for naval applications (in which the antenna weight is a crucial parameter) and, to minimise the reflector mass, a construction using carbon fibre was sought. Conventional lay-up techniques using this material are generally expensive and tests were performed on a commercially available woven cloth which would allow processing in a manner similar to that of glass fibre reinforced plastics. The reflectivity of this material at I band (when used as a waveguide short circuit) is compared with that of a conventional carbon fibre reflector lay-up in Figure 6. It is seen that the difference is small, although further tests showed that it became more marked at higher frequencies.

The prototype reflector (approximately 1.4m x 0.47m in size) was fabricated over a high precision, NC turned, aluminium mould using carbon fibre cloth outer skins and a ⅜" Nomex honeycomb inner core to a tolerance of ± 0.15mm. The weight, including integral mounting ribs formed using the same construction on the rear surface, was 6.5 kilograms.

The feed system consisted of four electro-formed conical horns, bolted to an NC machined waveguide power splitting network.

Experimental results

The azimuth plane pattern measured through the peak of the beam produced by the upper feed horn only is plotted in Figure 7 for vertical polarisation. The main beam shape (including the broadening evident at the base) is in good agreement with theoretical predictions.

The first sidelobe level of -43dB is higher than the predicted level (-49dB) but this is attributable to scatter of spillover radiation from the feed support struts. The measured contour plot of the principal beam is given in Figure 8. The predicted -40dB contour is also shown and it is noted that, apart from some spurious lobes associated with spillover scatter, the correlation is generally good. Similar agreement was achieved with the secondary beam, as indicated in Figure 9. Further measurements confirmed that the performance was relatively broadband and similar for horizontal polarisation.

CONCLUSIONS

An antenna system has been described which is relatively lightweight and cheap to produce. The measured performance is generally in good agreement with theoretical predictions and demonstrates the potential for producing surveillance beams with good control of the sidelobe radiation. The simplicity of the feed system provides the possibility of dual-polarised operation and the incorporation of additional beams. Although the azimuth plane resolution of these is limited by phase error effects, reflector shaping techniques may be considered to provide a better compromise.

REFERENCES

1. Scudder, R.M., 1978 "Advanced antenna design reduces electronic countermeasures threat" RCA Engineer, 33, 61-65

2. Rudge, A.W. and Williams, N. "Low sidelobe reflector antennas" To be published.

3. Miersch, H.K., 1980, "Extremely low sidelobes with offset-fed doubly curved reflector antennas", IEEE International Symposium Digest - Antennas and Propagation, Quebec, Canada, 560-563.

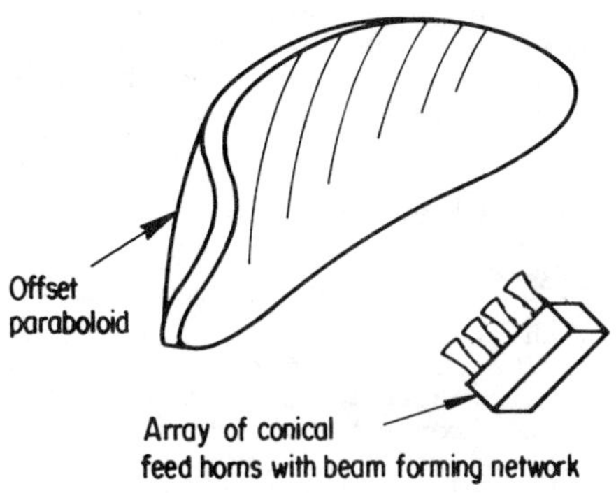

Figure 1 : Basic antenna configuration

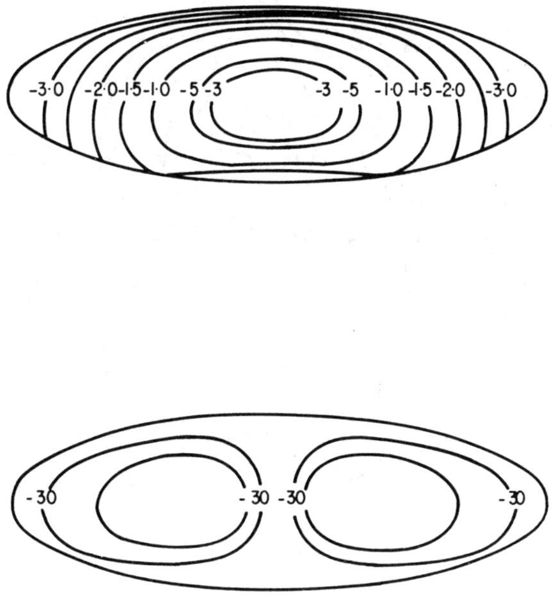

Figure 2 : Co- and cross-polar field contours in projected aperture

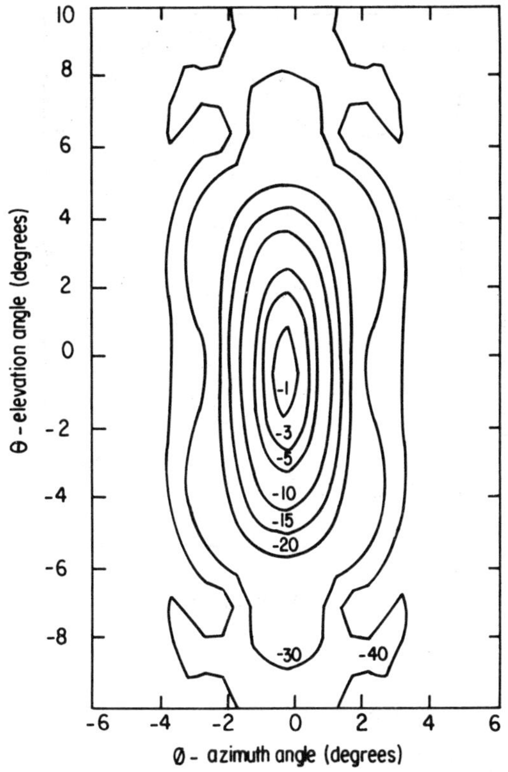

Figure 4 : Contour plot of principal beam at centre frequency

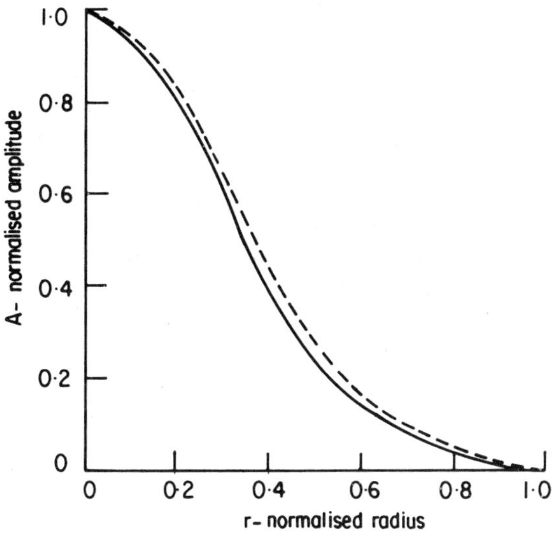

——— reflector system
------- Gaussian

Figure 3 : Aperture distribution in azimuth plane

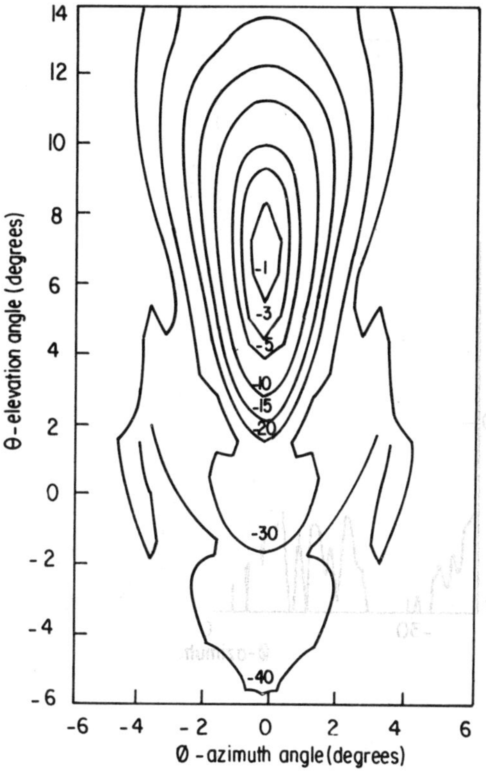

Figure 5 : Contour plot of secondary beam at centre frequency

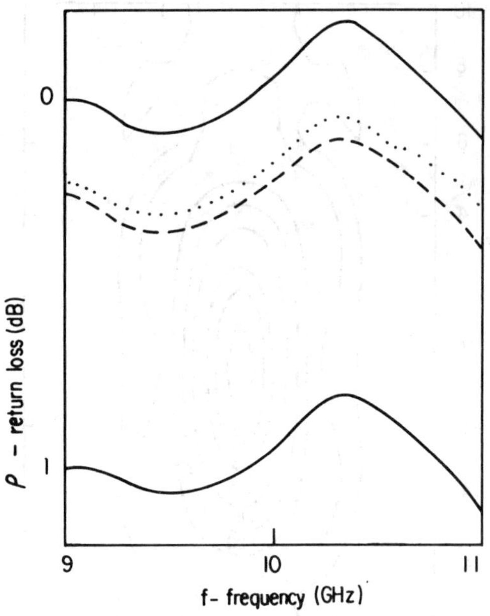

...... conventional lay-up
------ cloth

Figure 6 : Reflectivity of carbon fibre surfaces

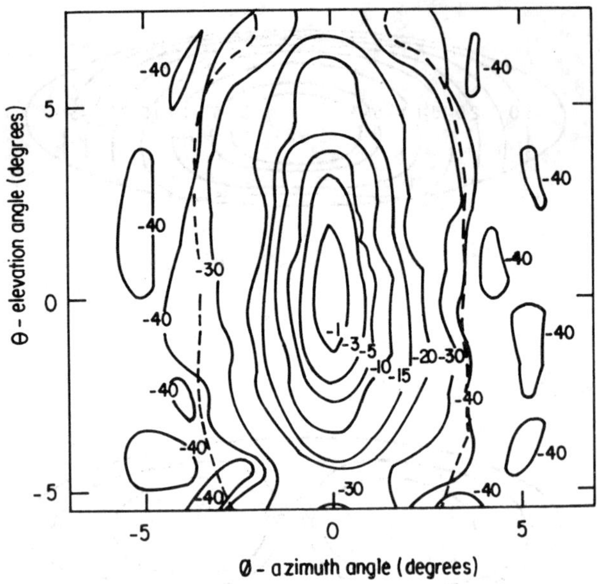

------ predicted -40dB contour

Figure 8 : Measured contour plot of principal beam

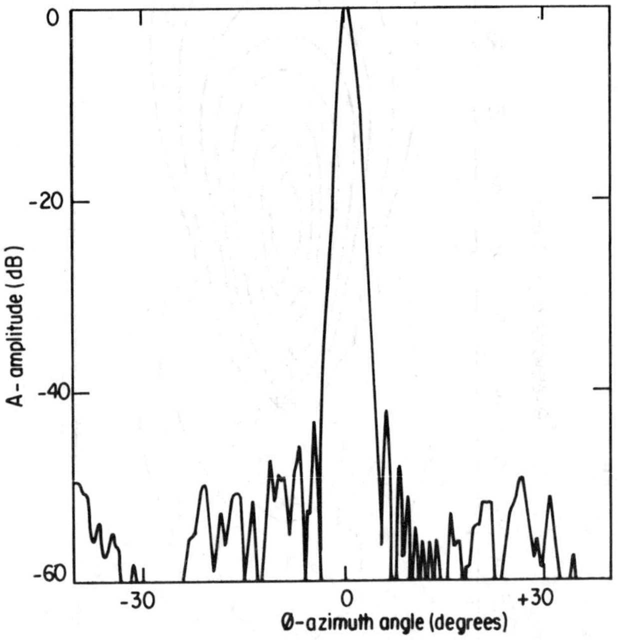

Figure 7 : Azimuth plane pattern of reflector illuminated by single feed

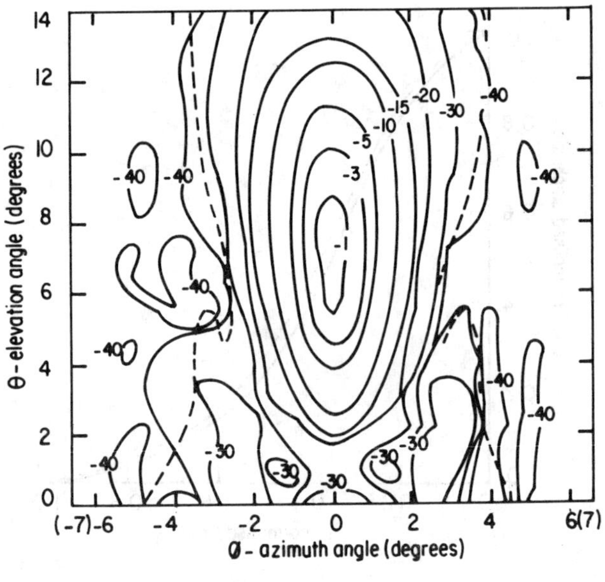

------ predicted -40dB contour

Figure 9 : Measured contour plot of secondary beam

ON THE PERFORMANCE DEGRADATION OF A LOW SIDELOBE PHASED ARRAY DUE TO CORRELATED AND UNCORRELATED ERRORS

James K. Hsiao

Naval Research Laboratory, Washington, D.C.

INTRODUCTION

Low-side-lobe array antennas have received wide interest in recent years. Theoretically, one can design an array antenna with any desired side-lobe level. However, in practice array errors and other imperfections limit the side-lobe level. One question is then how low a side-lobe level one may achieve in practice. In this paper, the relationship between the side-lobe level and the array random errors will be examined. In particular, the limitation on the side-lobe level as a function of the array errors will be presented. Intuitively, one may see that this limitation must be a function of the desired side-lobe level, the number of elements in the array, and the nature of these errors. Since these errors are generally random in nature, the results are in terms of probability distributions. It can be shown that all array errors that result from feed, phase shifters, mechanical location, and the orientation of radiating elements can be characterized by a phase error and an amplitude error for each element in the array. The results of this study, therefore, are in terms of these errors. In the past, these errors have generally been assumed to be statistically independent. However, there are cases for which errors in many elements are not necessarily independent. For example, the same phase and amplitude in a row or column feed network could feed to every element in a particular row or column, or in the case of subarray configuration, the same error may propagate to every element in a particular subarray. These errors are correlated in these groups of elements. The effects of these correlated errors will be also discussed.

STATISTICAL DISTRIBUTION OF ARRAY PATTERN

For simplification, linear arrays will be treated first. The array pattern of a linear array can be represented by

$$P(\theta) = \sum_n A_n \exp[-j(2\pi nd/\lambda)(\sin\theta - \sin\theta_o)], \quad (1)$$

where θ is the angle of incidence of a plane wave on the array and θ_o is the beam pointing angle. Element spacing d is assumed to be uniform. We further define

$$\mu = (2\pi d/\lambda)(\sin\theta - \sin\theta_o). \quad (2)$$

Equation (1) then becomes

$$P(\mu) = \sum_n A_n \exp(jn\mu). \quad (3)$$

For radiation in real space, μ is constrained so that

$$|\mu| \leq 2\pi. \quad (4)$$

Due to mechanical and electrical errors in the array, the array pattern becomes

$$G(\mu) = \sum_n A_n(1+\delta_n)\exp(j\phi_n)\exp(jn\mu). \quad (5)$$

where δ_n is the amplitude error and ϕ_n is the phase error. These errors vary from element to element and are random in nature. For simplification, we assume that these errors have a known probability density function. Because of randomness of these errors, $G(\mu)$ is a random complex function which is the sum of many random variables. Each of these random variables can be represented as

$$G_n(\mu) = (1+\delta_n)\exp(j\phi_n)\exp(jn\mu)$$
$$= X_n + jY_n. \quad (6)$$

These random variables are independent and have the same probability density function. The array pattern is hence a random function of the sum of many random variables, so that

$$G(\mu) = g_1(\mu) + jg_2(\mu) = \sum_n A_n X_n + j \sum_n A_n Y_n. \quad (7)$$

According to Lindenberg and Levy's central limit theorem [1], $g_1(\mu)$ and $g_2(\mu)$ are asymptotically normal. The means of $g_1(\mu)$ and $g_2(\mu)$ are the weighted sums of means and variances of X_n and Y_n. The means of $g_1(\mu)$ and $g_2(\mu)$ are then respectively

$$\overline{g_1(\mu)} = \Phi(1) \sum_n A_n \cos n\mu \quad (8a)$$

and

$$\overline{g_2(\mu)} = \Phi(1) \sum_n A_n \sin n\mu. \quad (8b)$$

where $\Phi(k)$ is the characteristic function of random variable x, defined as

$$\Phi(k) = \int g(x)\exp(jkx)dx,$$

where g(x) is the probability density function of random variable x. Furthermore, in deriving the above expression we have also assumed that the amplitude error δ_n has zero mean. If the linear array is symmetrically illuminated, such that $A_n = A_{-n}$, then $g_2(\mu) = 0$. The variances of $g_1(\mu)$ and $g_2(\mu)$ are, respectively,

$$\sigma_1^2 = 1/2 \sum_n A_n^2(F+B\cos 2n\mu) \quad (9a)$$

and

$$\sigma_2^2 = 1/2 \sum_n A_n^2(F-B\cos 2n\mu). \quad (9b)$$

where

$$F = 1 + \sigma_\delta^2 - \Phi^2(1) \quad (9c)$$

and

$$B = (1+\sigma_\delta^2)\Phi(2)-\Phi^2(1). \qquad (9d)$$

The quantity σ_δ^2 is the variance of the amplitude error δ. The covariance of $g_1(\mu)$ and $g_2(\mu)$ is

$$\sigma_{12} = 1/2 \sum_n A_n^2 B\sin 2n\mu. \qquad (10)$$

For a symmetrically illuminated array σ_{12} is zero. It can be shown [2] that in the sidelobe region $B\cos 2n\mu$ can be neglected; hence,

$$\sigma_1^2 = \sigma_2^2 = \sum_n A_n^2 F \qquad (11)$$

In the main-beam region where $\mu \approx 0$ and $\sigma_1 \neq \sigma_2$, the radiation level has a non-central chi-square density function. For this type of density function most of the probability mass concentrates within an ellipse with major and minor semiaxes of two times σ_1 and σ_2, centered at the mean value $\bar{g}_1$. Furthermore,

$$\bar{g}_1 >> \sigma_1 \text{ or } \sigma_2.$$

Therefore, for practical purposes one may assume that the radiation amplitude ($R_o = \sqrt{g_1^2 + g_2^2}$), is equal to $\bar{g}_1$ with a probability of unity,

$$R_o \approx \bar{g}_1 = \Phi(1) \sum_n A_n.$$

Since the illumination function A_n is normalized, one finds

$$R_o \approx \Phi(1). \qquad (12)$$

The probability density function in the side-lobe region has a Rician distribution. The side-lobe level is defined as the ratio of the main-beam level to the side-lobe level. Its probability function is the joint probability function of R_o and R. Since $R_o \approx \Phi(1)$ with a probability of unity, the probability density function of the ratio of R to R_o is the same as that of R with a scale factor $\phi(1)$; that is, $R'=R/\Phi(1)$, and the density function of R' is then

$$P(R') = \frac{R'\Phi^2(1)}{\sigma^2} \exp\{-[(R'\Phi(1))^2 - \bar{g}_1^2]/2\sigma^2\} I_o\left[\frac{R'\Phi(1)\bar{g}_1}{\sigma^2}\right], \qquad (13)$$

where

$$\sigma = \sigma_1 = \sigma_2$$

and

$$\bar{g}_1 = \Phi(1) \sum_n A_n \cos n\mu.$$

For convenience, we shall normalize this R' in such a way that

$$P(S) \approx \frac{S}{\sigma^2} \exp\left[-\frac{S^2+1}{2\sigma'^2}\right] I_o\left[\frac{S}{\sigma'^2}\right]. \qquad (14)$$

where

$$S = R'/\sum_n A_n \cos n\mu \qquad (15a)$$

and

$$\sigma' = \sigma/\left[\sum_n A_n \cos n\mu\right]\Phi(1) \qquad (15b)$$

We notice that $\sum_n A_n \cos n\mu$ represents the side-lobe level at the angle μ when there is no error present. Furthermore, in most cases when phase error is small, $\Phi(1)$ is close to unity. Therefore, both S and σ' in this equation are measured in terms of the designed side-lobe level. This more convenient to use than are values in terms of σ or $\bar{g}_1$.

The cumulative probability of S being less than S_L is then

$$P(S<S_L) + \int_0^{S_L} \frac{S}{\sigma'^2} \exp\left[-\frac{S^2+1}{2\sigma'^2}\right] I_0\left[\frac{S}{\sigma'^2}\right] dS. \qquad (16)$$

A family of such curves with σ' as parameter shown in Fig. 1. Each of these curves presents the cumulative probability that S (in terms of designed side lobe) is less than or equal to a level S_L for a given σ'. Since this surve is presented in such a way that it is not a function of the angle μ, these curves apply to all points in the side-lobe region. Secondly, these curves are normalized with respect to the ideal side-lobe level. It represents the probability of the deviation of side-lobe level from the designed value. Although they are not presented explicity as a function of μ, they are related to the side-lobe level. For example, at the peak of a side lobe, the normalized σ' may be only equal to 0.1, but at a point where the side-lobe level may be 10 times smaller, the normalized σ' then becomes 10 times larger. One can see the difference in the probability distribution for these two cases. This set of curves is universal. It applies to arrays with different illumination designs, different sizes, and different errors.

A set of constant probability curves is plotted in Fig. 2. These curves are plotted at a given cumulative probability with the deviation of the side-lobe level and σ from the designed values as abscissa and ordinate. Four such curves, each representing a constant cumulative probability, are shown. These curves are very useful in determining the required σ level to achieve a desired side-lobe degradation. For example, in order to assure that the side lobe not increase more than 3 dB for 90 percent of the time, the required σ^2 has to have a value of -11 dB relative to the designed side-lobe value. If the designed side lobe is -40 dB, in order to keep the side lobe below -37 dB, the required σ^2 value must be less than -51 dB. These curves are very convenient to use. Once this σ is known, the allowable array amplitude and phase errors can be determined. In the next section we shall discuss how these errors are related to the σ value.

In general, it is difficult to achieve a smaller σ value. One may, therefore attempt to design an array with lower side-lobe level and let it deteriorate more in order to achieve the same overall side-lobe level. However, the requirement on the value of σ in this case may not be as tight. For instance, in the previous example, if one designs a -45 dB array instead of -40 dB, to achieve the same -37 dB side lobe, the required σ value becomes -46 dB instead of -51 dB. This may significantly ease the component tolerance requirement. However, other factors such as illumination loss must be taken into account. Furthermore, σ is also a function of array illumination. Adjusting the illumination function to achieve a lower design side-lobe level may at the same time inadvertently increase the σ value. This will be discussed in the next section.

σ VALUE OF UNCORRELATED ERRORS

As shown in equation (11), the σ value can be approximated

$$\sigma^2 = \sum_n A_n^2 F$$

Each σ consists of two parts. The first part is $\sum_n A_n^2$, which is a function of the number of elements in the array and the illumination function of the array. The second part F is determined by random array errors.

In finding the first part of σ, for convenience of comparison, let us normalize the array illumination function A_n in such a way that

$$\sum A_n = 1.$$

This implies that at the peak of the main beam the radiated field has unit strength (or zero dB). The summation of A_n^2 is then always less than unity. In the case of a uniformly illuminated array,

$$\sum_n A_n^2 = \frac{1}{N}, \tag{17}$$

where N is the total number of antenna elements. In the event that the array illumination is not uniform, it can be shown [2] that

$$\sum_n A_n^2 > \frac{1}{N}. \tag{18}$$

This means that when the array illumination function is tapered to achieve low side lobe, the sum of A_n^2 becomes larger and hence σ^2 also becomes larger. However, to a first order estimation, one may assume that $\sum_n A_n^2 = \frac{1}{N}$.

The function F, according to equation (9c), is

$$F = 1+\sigma_\delta^2-\Phi^2(1),$$

where σ_δ^2 is the variance of the amplitude error and

$$\Phi(1) = \int p(x)\exp(jx)dx .$$

If p(x) has zero mean and its probability mass concentrates in the vicinity of the mean, the above equation can be approximated by

$$\Phi(1) \approx \int p(x)(1-jx-\frac{x^2}{2}...)dx \tag{19}$$

$$= 1-\frac{\sigma_\phi^2}{2},$$

where σ_ϕ^2 is the variance of the phase error and

$$\Phi^2(1) \approx 1-\sigma_\phi^2,$$

when high order terms are neglected, one has

$$F \approx \sigma_\delta^2+\sigma_\phi^2 \tag{20}$$

The F function is then the sum of the variances of both the amplitude error and phase error.

For a planar array, if errors in each element are independent, the results of the linear array analyzed above can be applied directly. In the next section, we shall discuss the case of correlated errors in a planar array.

PLANAR ARRAYS WITH CORRELATED ERRORS

In this section we shall discuss the effects of correlated errors in a planar array. For simplification and without loss generality, we shall discuss the case for which a planar array is fed by rows and columns. In this case, the same error from the feed network may appear in every element in an entire row (or column) or in many cases all elements of a subarray may contain the same error due to mechanical reasons. Besides this correlated error, there are also independent errors in every element. The array pattern can them be assumed to have the following form:

$$G(\mu,v) = \sum_n(1+\delta_n)\exp(j\phi_n)\sum_m A_{nm}(1+\delta_{nm})\exp$$

$$(j\phi_{nm})\exp[(m\mu+nv)] . \tag{21}$$

where δ_n and ϕ_n are, respectively, the amplitude and phase errors which appear in the nth row. And,

$$\mu = (2\pi d_x/\lambda)\ (\sin\theta\ \cos\phi - \sin\theta_o\ \cos\phi_o),$$

$$v = (2\pi d_y/\lambda)\ (\sin\theta\ \sin\phi - \sin\theta_o\ \sin\phi_o).$$

It can be shown [2] that

$\sigma = \sigma_1 = \sigma_2$, and

$$\sigma^2 = \tfrac{1}{2}\sum_n\sum_m\left[(1+\sigma_n^2)(1+\sigma_{nm}^2)-\Phi_n^2(1)\Phi_{nm}^2(1)\right]A_{nm}^2$$

$$+\ \tfrac{1}{2}\sum_n\left[(1+\sigma_n^2)-\Phi_n^2(1)\right]\Phi_{nm}^2(1)\sum_m\sum_s A_{nm}A_{ns}\cos(m-s)\mu \tag{22}$$

In the derivation of equation (22), it is assumed that the amplitude errors δ_n and δ_{nm} have zero mean and their respective variances are σ_n^2 and σ_{nm}^2. In the formulation, we also assumed that the amplitude error consists of two levels δ_n and δ_{nm}, and the total error is $(1+\delta_n)$ $(1+\delta_{nm})$. The mean of this error is zero; however, the composite variance is given by

$$(1+\sigma_n^2)(1+\sigma_{nm}^2)=1+\sigma_n^2+\sigma_{nm}^2+\sigma_n^2\sigma_{nm}^2 . \tag{23}$$

Let the total amplitude variance σ be given by

$$\sigma_\delta^2 = \sigma_n^2+\sigma_{nm}^2+\sigma_n^2\sigma_{nm}^2. \tag{24}$$

In both σ_n and σ_{nm} are small, the total variance σ_δ^2 can be viewed as the sum of the individual variances σ_n^2 and σ_{nm}^2. Let the total phase error ϕ be the sum of the two phase errors ϕ_n and ϕ_{nm}, then the characteristic function of Φ is

$$\Phi(1) = \Phi_n(1)\Phi_{nm}(1). \tag{25}$$

$$\sigma^2=\frac{1}{2}\sum_n\sum_m[1+\sigma_\delta^2-\phi^2(1)]A_{nm}^2 + \frac{1}{2}\sum_n[(1+\sigma_n^2)-\Phi_n^2(1)]|G_n(\mu)|^2. \qquad (26)$$

where

$$|G_n|^2=\Phi_{nm}^2(1)\sum_m\sum_s A_{nm}A_{ns}\cos(m-s)\mu. \qquad (27)$$

and σ_n, σ_{nm}, Φ_n and Φ_{nm} are replaced by σ and Φ as shown in Eqs. (24) and (25).

The mean radiation pattern of the linear array of the nth row is

$$G_n(\mu) = \Phi_{nm}(1)\sum_m A_{nm}\exp(jm\mu). \qquad (28)$$

therefore $|G_n(\mu)|^2$ is its power pattern. The σ consists of two parts. The first part $\frac{1}{2}\sum_n\sum_m [1+\sigma_\delta^2-\phi^2(1)]A_{nm}^2$, is identical to that of an array with uncorrelated errors when both the correlated and uncorrelated errors in each element are taken into account. The second part $\frac{1}{2}\sum_n [(1+\sigma_n^2-\Phi_n^2(1)] |G_n(\mu)|^2$, has a value similar to that of a linear array with the illumination weights A_n^2 replaced by the power radiation pattern function $|G_n(\mu)|^2$. It is interesting to note that when $\mu = 0$, $|G_n(\mu)|^2$ has its maximum value. In other words, at $\mu = 0$ the correlated errors have the strongest effect on σ and the array patterns would most probably deteriorate more at $\mu = 0$ in the μ, v domain. For example, a column and row fed array, as discussed in this section, has rows in the x direction and columns in the y direction. Elements in each column have a common error. If the array beam is steered at the broadside ($\theta_o=0$), then $\mu = \sin\theta\cos\phi$, and the maximum degradation will occur in a plane along the y direction (the E plane), when $\phi = 90^o$. However, if elements in each row have a common error, the worst degradation may occur at v = 0, along the x direction (the H plane). Under this condition,

$$|G_n(\mu)|^2 = \left[\sum_m A_{nm}\right]^2. \qquad (29)$$

If the illumination coefficient A_{nm} values are normalized such that

$$\sum_n\sum_m A_{nm} = 1. \qquad (30)$$

one can show that

$$|G_n(\mu)|^2 \geq \frac{1}{N}, \qquad (31)$$

where N is the total number of rows which have correlated errors. One may hence conclude that:

- σ consists of two parts, with the first part due to total errors (including both correlated and uncorrelated) at each array element (σ_μ) and the second part due to the correlated error fed to each row (or column) (σ_c). The σ is the root sum square (RSS) of these two parts.

- Both σ_μ and σ_c are functions of both the array element errors and the illuminations. The effects of amplitude and phase errors on the σ value are identical to that of the case of a linear array. The array illumination can be approximated by the number of elements in the array (for σ_μ) and the number of rows (for σ_c) for the correlated errors.

When the array is symmetrically illuminated and the phase center of the array is taken at the center of the array, one may shown that $\sigma_{12} = 0$. Therefore, the distribution of array patterns of a planar array is Rician as shown in Fig. 1. The value of σ is the RSS of σ_μ and σ_c. The above results can be applied to an array which is fed by subarrays and the correlated error appears in each subarray. In this case, one has to replace the parameters μ and v with functions of μ and v.

CONCLUSIONS

In this paper we have shown the following:

1. The amplitude distribution of the radiation pattern for both a linear array and a planar array is Rician. A set of universal curves for such distributions is shown in Fig. 1 and Fig. 2. The curves are presented in different σ values and are normalized to the designed side-lobe level.

2. The σ value for both linear arrays and planar arrays has a similar form, which is a function of errors and array illuminations. It shows that the illumination function of the array can be approximated by the number of elements, and it also shows that σ^2 is the sum of the variances of both phase and amplitude errors.

3. For a planar array with correlated errors the variance σ is the RSS of σ_μ and σ_c are functions of the errors and the illuminations. However, σ_c is also a function of the pattern of the subarray in which the errors are correlated.

REFERENCES

1. Harald Cramer, Mathematical Methods of Statistics, Princeton University Press, Princeton, N.J., 1946, p. 45.

2. J.K. Hsiao, "Effect of Errors on the Side-lobe Level of a Low-side-lobe Array Antenna", NRL Report 8485, June 28, 1981.

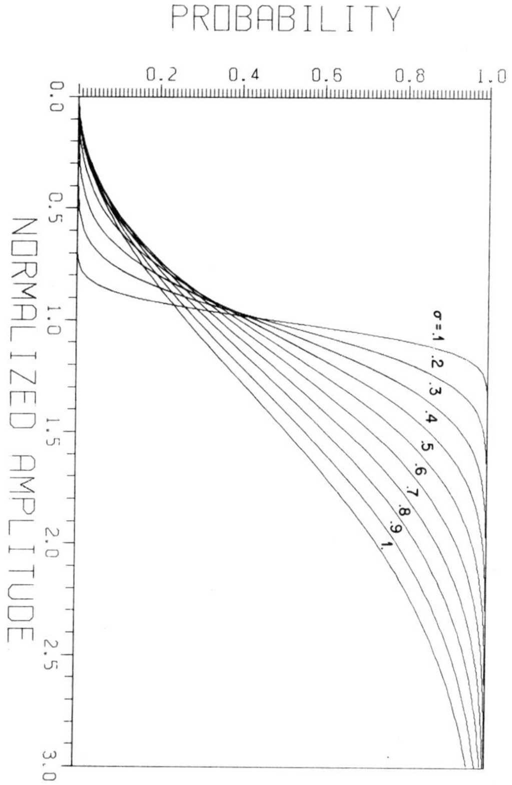

Fig. 1 Cumulative probability distribution of normalized sidelobe level (normalized to design value)

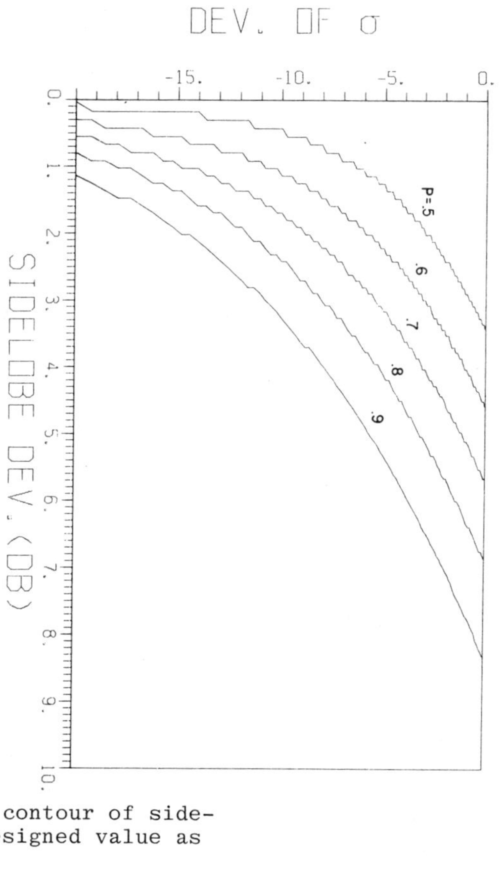

Fig. 2 Constant probability contour of sidelobe level deviation from designed value as function of normalized σ

MINIMISATION OF SIDELOBES FROM A PLANAR ARRAY OF UNIFORM ELEMENTS

G J Halford and W J McCullagh

ASWE, Portsdown, Cosham, Portsmouth, PO6 4AA

1. INTRODUCTION

The use of active arrays employing distributed transmitting and receiving elements is becoming more feasible with the development of cheaper and better solid-state devices. In general there should be no great difficulty in obtaining low sidelobes on reception from a large array of such elements, since any desired weighting can be applied to the outputs of the individual receivers. On transmission, however, the desire to employ identical elements operating at maximum efficiency may preclude any possibility of amplitude tapering. This could result in an unacceptably high level of near-in sidelobes, and a severe mismatch between the beamwidths on transmission and reception.

This paper investigates whether density tapering - the selective omission of elements - could be of value in this context. The reduction of near-in sidelobes and widening of beamwidth are offset by losses in gain and total power output, and by an increase in the background of random sidelobes above that already caused by phase quantization and random errors. An expression for the sidelobe level due to thinning is first derived, and applied to a variety of circular aperture distributions. Their parameters are compared with those for a uniformly filled aperture to show what compromises and trade-offs are possible. A specific example is then given showing how conclusions might be altered if the beamwidths were normalised to match that of a -40 dB Taylor distribution used for reception.

The reduction of random sidelobes which can be achieved with transmitting elements of two different power levels is considered theoretically. The optimum ratio of these levels, and the resulting sidelobe reductions, are then evaluated for two particular aperture distributions, and some conclusions are drawn as to the results for more elaborate schemes. The analogous situation for rectangular arrays is briefly discussed, and finally some examples are given of radiation patterns.

2. SIDELOBES DUE TO THINNING

A previous paper (1) quoted the RMS sidelobe level of a thinned array as (1-k)/N, where k is the filling factor N/M, and N elements are present compared with M in a fully filled array. The term 'RMS sidelobe level' referred to the square of the RMS sidelobe level expressed in voltage; it will be more correctly described here as 'mean sidelobe level', in decibels below main beam peak. The value (1-k)/N is correct if k is constant over the array or a good approximation if the local value of k does not vary greatly. However this paper investigates arrays where k may vary from 1 to 0 with radius. In this case it turns out that k must be replaced by k′, derived as sketched below.

Any desired circular aperture distribution f(r), Fig 1, where r and f(r) are ≤ 1, is simulated by assigning weights of 1 or 0 to each element such that the probability of 1 occurring for elements at radius r is f(r). If the array has n elements per unit area, spaced distances δr apart, a ring of radius r and width δr will contain approximately $2\pi n r \delta r$ elements.

Approximately a proportion f(r) of these will have weight 1, (1-f(r)) too high, and a proportion (1-f(r)) will have weight 0, f(r) too low. Thus besides the desired pattern due to the distribution f(r), there will be sidelobes due to the random combination of the errors in weights. The RMS field at distant points due to these errors, relative to that of a single element of unit weight, will be given by summing their squares ie

$$E^2 = 2\pi n \sum_{r=\frac{\delta r}{2}}^{r=1-\frac{\delta r}{2}} r\left[\left(f(r)+\alpha_r\right)\left(1-f(r)\right)^2+\left(1-f(r)+\beta_r\right)\left(f(r)\right)^2\right]\delta r \qquad -(1)$$

where α_r and β_r are corrections due to the limited integral numbers of elements in each ring. As the number of elements in the array increases, $\delta r, \alpha_r$, and β_r all tend to zero and the series attains a constant value $\int_0^1 r\left[f(r)\left(1-f(r)\right)^2+\left(1-f(r)\right)\left(f(r)\right)^2\right]dr.$

Thus with further simplification

$$E^2 = 2\pi n\int_0^1 r f(r)\,dr - 2\pi n\int_0^1 r\left[f(r)\right]^2 dr \qquad -(2)$$

This is exact when $n\to\infty$, and a good approximation if n is large. At the peak of the main beam, the field, E_{max} times that of a single element, is given by

$$E_{max} = 2\pi n\int_0^1 r f(r)\,dr = kM = N. \qquad -(3)$$

Hence the mean sidelobe level E^2/E^2_{max} is given by

$$\frac{E^2}{E^2_{max}} = \frac{\int_0^1 r f(r)\,dr - \int_0^1 r\left[f(r)\right]^2 dr}{\left[\int_0^1 r f(r)\,dr\right]^2} = \frac{1-K'}{N} \qquad -(4)$$

$$\text{where } K' = \frac{\int_0^1 r\left[f(r)\right]^2 dr}{\int_0^1 r f(r)\,dr} \qquad -(5)$$

Using the value for aperture efficiency obtained later, equation (4) is found to be identical to that given by the statistical approach of Skolnik et al (2), whose value $(1-\bar{N}_E/N\rho_a)/\bar{N}_E$ corresponds to $(1-k/\eta)/N$ in our notation.

In general the radiation pattern will have a region close to the main beam where the sidelobes are given by the Fourier transform of the distribution f(r), a region far out where the random sidelobes dominate, and an intermediate region where the two effects are comparable and additive. The exact boundaries will depend on the aperture size and distributions used and on the constraints (such as those in reference 1) placed on the random choice of elements.

3. COMPARISON OF DISTRIBUTIONS FILLING THE APERTURE

Various aperture distributions implemented by thinning are compared in Table 1 with a fully filled circular array of M similar elements.

The filling factor, k, determines gain and power output. Each thinned array has k times the gain, k times the total power output, and hence k^2 times the effective radiated power of the filled array.

The mean random sidelobe level $\frac{1-k'}{N}$ or $\frac{1-k'}{kM}$ will normally dominate the effects of phase quantization or random errors (3-bit quantization corresponds to 0.051/N.) As it will vary with the number of elements it is convenient to quote it relative to $\frac{1}{M}$.

Some of the more interesting aperture distributions are plotted in Figure 2. The Hansen one-parameter distributions (3) have a sidelobe level which falls off similarly to a uniform distribution but starting at -20 or -25 dB instead of -17.6 dB. Both these and the circular Taylor distributions (4,5) have been integrated numerically to obtain k and k'.
The Taylor -25 dB distributions reach an optimum at $\bar{n}$=8 when the edge illumination is close to that at the centre (Figure 2). Higher values of $\bar{n}$ result in a rapid fall in k and k' as the edge illumination increases above that of the centre.
The $1-r^2$ distribution is fairly well known, and has the advantage of a considerably more rapid fall-off of sidelobes than the other distributions (Figure 3). However, it has a filling factor of only 0.5 and hence $\sqrt{1-r^2}$ (k=.667) could have advantages whilst still providing a rate of sidelobe fall-off greater than for the uniform distribution. Patterns for these two distributions have been evaluated using Ramsay's collected data on Lambda functions (6), whilst values of k and k' have been obtained by exact integration.

It is interesting to compare the $\sqrt{1-r^2}$ and the -25 dB $\bar{n}$=8 Taylor distributions since they have closely similar values for filling factor, gain and random sidelobe level.
However, the Taylor distribution gives a far wider solid angle of pattern sidelobes at the -25 dB level, whilst there is less power in the narrower main beam. This latter fact is also likely to be a disadvantage, as will be evident in the next section. Thus Taylor patterns may well be non-optimum for this application, unless very low first sidelobes are of paramount importance.

At first sight the conclusion from Table 1 is that for many applications there is little to be gained by departing from the filled array, since small improvements in near-in sidelobes are more than offset by losses in ERP and by increases in far-out sidelobes. However, Table 1 is pessimistic to the extent that changes in beamwidth are not allowed for.

TABLE 2

Comparison of Distributions of Equal Beamwidth (75.02°xD/λ)

DISTRIBUTION	No OF ELEMENTS (x M)	REL POWER (dB)	REL RMS S/L Level (dB)
UNIFORM	.617	0	-13 (3-Bit)
HANSEN (-20dB)	.535	-1.25	-6.50
$\sqrt{1-r^2}$	.520	-1.50	-5.27
$1-r^2$	.470	-2.36	-3.59
HANSEN (-25dB)	.456	-2.63	-3.27
TAYLOR 5,-25 dB	.426	-3.23	-2.76
$\bar{n}$ = 8,-25 dB	.439	-2.97	-4.28
$\bar{n}$ = 9,-25 dB	.384	-4.13	-2.76
$\bar{n}$ = 5,-30 dB	.414	-3.48	-2.13
$\bar{n}$ = 10,-30 dB	.419	-3.37	-2.75
$\bar{n}$ = 5,-35 dB	.403	-3.70	-1.81
$\bar{n}$ = 10,-35 dB	.405	-3.65	-2.02
$\bar{n}$ = 5,-40 dB	.395	-3.89	-1.61
$\bar{n}$ = 10,-40 dB	.394	-3.90	-1.66

4. COMPARISON OF DISTRIBUTIONS WITH EQUAL BEAMWIDTH

For an active array used for surveillance there is likely to be a requirement that the beamwidth on transmission should match that on reception, or bear some fixed relation to it. The figures given in Table 2 have therefore been normalised for a more extreme case, where the beamwidth is to correspond to that of a -40 dB $\bar{n}$=5 Taylor distribution. In particular, the uniform distribution has to be considerably reduced in size, and hence its advantages in radiated power and random sidelobe level over the other distributions are smaller. In practice better performance might be achieved using slightly narrower beamwidths for transmission than reception, in order that more power can be radiated. However the relative figures in Table 2 will still hold provided that any restrictions on diameter are not exceeded.

TABLE 1

Comparison of Distributions Filling the Aperture

DISTRIBUTION	BEAMWIDTH (Deg)	FIRST S/L (dB)	k	k'	REL POWER k^2 (dB)	REL MEAN S/L (dB) $\left(\frac{1-k'}{k}\right)$
UNIFORM	58.9	17.6	1	1	0	(-13 3 BIT)
HANSEN (-20 dB)	61.8	20	.789	.806	- 2.06	- 6.09
$\sqrt{1-r^2}$	66.2	21.3	.667	.750	- 3.52	- 4.26
$1-r^2$	72.7	24.6	.500	.667	- 6.02	- 1.76
HANSEN (- 25 dB)	67.2	25	.568	.652	- 4.92	- 2.12
TAYLOR (- 25 dB)						
$\bar{n}$ = 5	63.3	(25)	.598	.636	- 4.47	- 2.15
$\bar{n}$ = 8	60.9	(25)	.666	.735	- 3.53	- 4.00
$\bar{n}$ = 9	60.4	(25)	.592	.671	- 4.56	- 2.55
TAYLOR (- 30 dB)						
$\bar{n}$ = 5	67.7	(30)	.508	.589	- 5.88	- 0.92
$\bar{n}$ = 10	64.7	(30)	.563	.640	- 4.99	- 1.94
TAYLOR (- 35 dB)						
$\bar{n}$ = 5	71.6	(35)	.443	.569	- 7.07	- 0.12
$\bar{n}$ = 10	69.0	(35)	.479	.588	- 6.39	- 0.65
TAYLOR (-40 dB)						
$\bar{n}$ = 5	75.0	(40)	.395	.559	- 8.08	+ 0.49
$\bar{n}$ = 10	73.0	(40)	.416	.565	- 7.62	+ 0.20

It will be seen that the Hansen, $\sqrt{1-r^2}$ and $1-r^2$ distributions all provide greater radiated power than the -25 dB Taylors, whilst having lower second and subsequent sidelobes. The Hansen -20 dB and $\sqrt{1-r^2}$ distributions also provide lower random sidelobes than any of the Taylors. However, their sole advantage over the uniform distribution lies in their lower pattern sidelobes: this would be particularly marked for $\sqrt{1-r^2}$ in the case of large arrays with low random sidelobe levels.

5. INTER RELATION OF k′ AND OTHER PARAMETERS

For the types of thinned arrays already discussed, the total radiated power is the product of the number of elements and their output power P ie $2\pi n P \int_0^1 r f(r)\, dr$, whereas a fully filled amplitude tapered array with the same gain would radiate only $2\pi n P \int_0^1 r [f(r)]^2 dr$.

The difference is, of course, the power radiated in the random sidelobes.

Thus $k' = \dfrac{\text{power in pattern}}{\text{total radiated power}}$ ie: k′ can be thought of as the relative efficiency of a thinned array with respect to a filled one having the same aperture distribution.

The aperture efficiency η of an array is defined as the ratio of its gain to that of one with a uniform distribution.

For the present case

$$\eta = \frac{\left[\int_0^1 r f(r)\, dr\right]^2 \int_0^1 r (1)^2 dr}{\int_0^1 r [f(r)]^2 dr \left[\int_0^1 r\, dr\right]^2} = \frac{k}{k'} \qquad (6)$$

Thus $k = \eta k'$ can be considered the product of aperture efficiency and thinned array efficiency.

6. DISTRIBUTIONS WITH TWO OR MORE POWER LEVELS

A uniformly filled circular array transmits 16.7% of its total power in the pattern sidelobes. As might be expected, any tapered thinned array which reduces this, gives a larger increase in the random sidelobe power. This limitation, however, can be overcome if two or more fixed levels of weighting can be used in the thinned array.

It is convenient to start by deriving a general formula. Consider part of an array (Figure 4) between radii p and q where the function f(r) is simulated by elements of weights f(p) and f(q).

At radii p and q there will be 100% of the elements with these respective weights.

At an intermediate radius r there will be a proportion α of weight f(p) and 1-α of weight f(q) where $\alpha = \dfrac{f(r)-f(q)}{f(p)-f(q)}$

with errors equal to f(p)-f(r) and f(r)-f(q) respectively. Summing the squares of the errors as in Eqn 1, assuming n is large, and simplifying,

$$E^2_{p,q} = 2\pi n \int_p^q r\,(f(r)-f(q))(f(p)-f(r))\, dr \qquad (7)$$

This will now be applied to the two-level case where the weights are 1 and f(a), together with missing elements of weight zero. Integrating for the regions $0 \leqslant r \leqslant a$ and $a \leqslant r \leqslant 1$, and adding,

$$E^2_{0,1} = E^2_{0,a} + E^2_{a,1} = 2\pi n \int_0^a r\,(f(r)-f(a))(1-f(r))\, dr + 2\pi n \int_1^a r\,(f(a)-f(r)) f(r)\, dr \qquad (8)$$

$$= 2\pi n \left[\int_0^a r\,(f(r)-f(a))\, dr + \int_0^1 r f(r)(f(a)-f(r))\, dr\right] \qquad (9)$$

In the case of the distributions $1-r^2$ and $\sqrt{1-r^2}$ it is possible to find a closed solution of these integrals so that an optimum value of f(a) can be obtained, as in Table 3.

The $1-r^2$ distribution is unique in that the optimum weight for f(a) is exactly 0.5, and the error contributions from the regions where r is greater or less than a are equal. Thus if the weighting of the elements were continuously variable between 1 and 0.5 a further 3 dB reduction of random sidelobes would be possible.

It will also be found for this distribution that the sidelobe power will fall off as $1/\chi^2$ where χ multiple levels of weights $\frac{1}{\chi}, \frac{2}{\chi}, \ldots\ldots, 1$ are employed.

At least three such levels would be needed to reduce random sidelobe power to a level comparable with that resulting from 3-bit quantization.

Roughly similar results would be expected with the other distributions, but the optimum weights would not be linearly distributed.

TABLE 3

Optimum values for Second Level of Weighting

DISTRIBUTION	$1-r^2$	$\sqrt{1-r^2}$
ΣE^2 (see eqn 7)	$\frac{\pi n}{6}(3a^4-3a^2+1)$	$\frac{\pi n}{6}(1-2a^2\sqrt{1-a^2})$
Minimum value $\left(\frac{\pi n}{6}\right.$ (Single level $\left.\frac{\pi n}{6}\right)$	$\frac{\pi n}{24}$	$\frac{0.2302\pi n}{6}$
Reduction on single level	-6.02 dB	-6.38 dB
Optimum value of a	$1/\sqrt{2}$	$\sqrt{2/3}$
Optimum value of f(a)	0.5	0.577
Optimum relative to 1	-6.02 dB	-4.77 dB
% of elements of weight 1	25	36.31
% of elements of weight f(a)	50	52.58
% of elements of weight 0	25	11.11

7. RECTANGULAR ARRAYS

If a uniform rectangular distribution is employed it will have sidelobes which fall-off as $\frac{\sin u}{u}$ along the principal axes.

Between these, the sidelobes will be much lower - falling off as $\left(\frac{\sin u}{u}\right)^2$ along the normals to the diagonals.

If the high level of principal axes sidelobes, and their slow fall-off are unacceptable, it is less easy to implement low sidelobe distributions in each plane because of the low values of weights required at the corners of the array. Once thinning is introduced, the spread of random sidelobes will largely negate the advantage of the low sidelobes in the inter-cardinal planes. It is more likely better in such cases to simulate circular or elliptical distributions in the ways that have already been described.

8. DEMONSTRATION AND VERIFICATION

As a demonstration, Figures 5-6 show how a filled array of 2004 elements has been thinned to provide $1-r^2$ distributions, using single level and two levels respectively. This particular distribution was chosen because of the rapid fall-off of its sidelobes which enables them to be distinguished from those due to the thinning. The radiation patterns of these distributions, also shown, have been computed by a two-dimensional Fast Fourier Transform program providing 256 x 256 points uniformly covering sin θ space. With half-wave spacing of elements and the main beam at boresight, only the region within the circle will be in visible space.

However, as the main beam is deflected, or the element spacing becomes greater, the remaining portions of the pattern, and their repetition, become visible.
In practical cases, of course, the effect of element pattern must be added. The two levels of shading represent sidelobes above -30 dB or -40 dB respectively. Worst random sidelobes in the two cases are below -23.5 dB and -29.5 dB respectively.

As all the sidelobe levels are known it becomes possible to calculate the mean sidelobe level due to thinning, and hence check predictions based on the earlier part of this paper. It is first necessary to exclude from the average the power in the main beam and any significant pattern sidelobes. This was done progressively correcting for 1st, 2nd and 3rd sidelobes as shown in Table 4. The results are in excellent agreement with theory.

TABLE 4

Comparison with Theory

	Mean Random Sidelobe level (dB) 1 level	2 levels
Predicted $\frac{(1-k')}{N}$ N = 1003	- 34.78	- 40.80
Corrected for Sidelobe 1	- 34.78	- 40.67
Sidelobes 1,2	- 34.81	- 40.77
Sidelobes 1,2,3	- 34.81	- 40.79

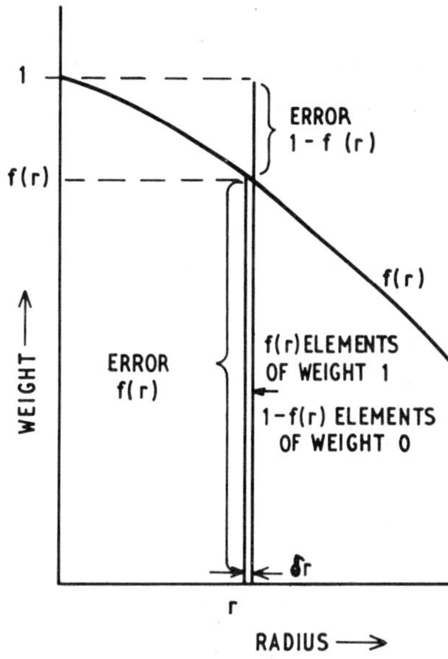

FIGURE 1

SIMULATION OF APERTURE DISTRIBUTIONS BY THINNING

9. CONCLUSIONS

Compared with a filled array of the same beamwidth, density tapered arrays have been shown to employ less elements, and hence have lower gain and output power. A fraction 1-k′ of the total output power is radiated as random sidelobes. The advantage of density tapering is the reduction of near-in sidelobes, and with large arrays lower peak sidelobes can be obtained over the whole radiation pattern.
If a reduction of total sidelobe power is required, at least two levels of amplitude quantization must be used. Finally, methods for calculating k′ and sidelobe levels are shown to give accurate results for two particular cases.

REFERENCES

1. Halford, G J and McCullagh, W J 1981 "Design Studies for a thinned active planar array", Second I.C.A.P. proceedings, York.

2. Skolnik, M I , Sherman, J W , and Ogg, F C 1964, "Statistically Designed Density-Tapered Arrays", IEEE Trans AP-12, 408-417.

3. Hansen, R C, 1976, "A one-parameter circular aperture distribution with narrow beamwidth and low sidelobes" IEEE Trans AP-24, 477-480.

4. Taylor, T T, 1960, "Design of circular apertures for narrow beamwidth and low sidelobes", IRE Trans AP-8, 17-22.

5. Hansen, R C, 1960 "Tables of Taylor distributions for circular aperture antennas", IRE Trans AP-8, 23-26.

6. Ramsey, J F, 1967 "Antenna Design Supplement" Microwaves 6.6 69-107.

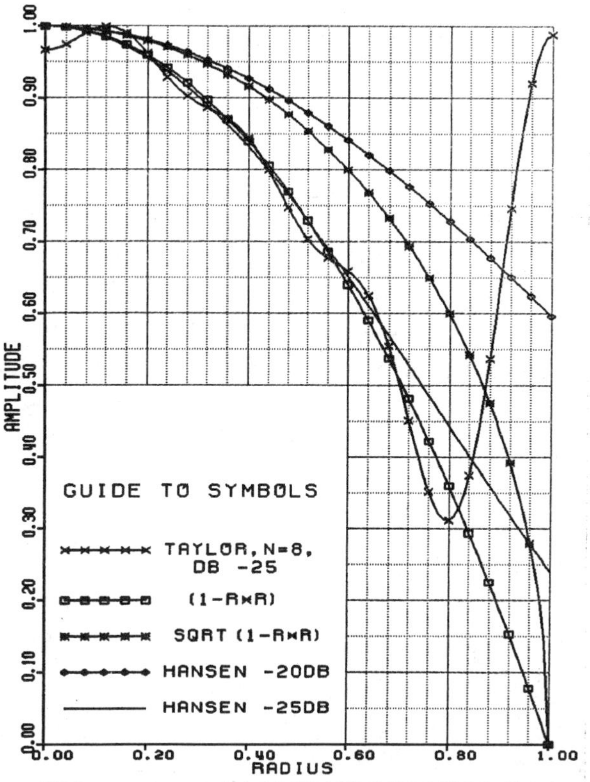

FIGURE 2

VARIOUS APERTURE DISTRIBUTIONS

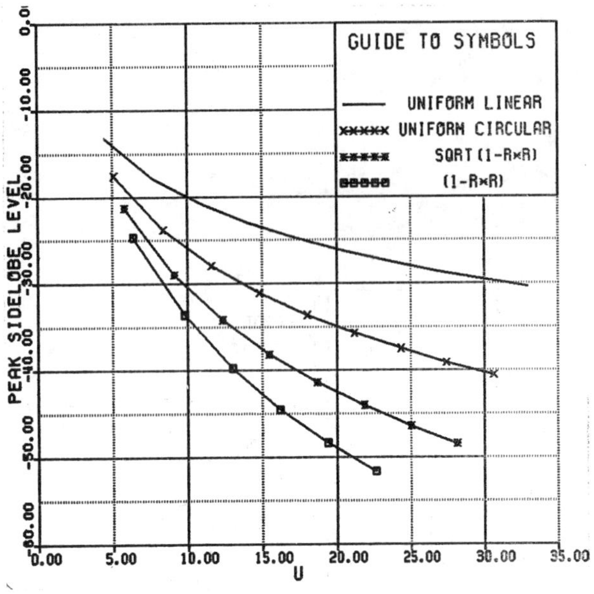

FIGURE 3

FALL-OFF OF SIDELOBES FOR VARIOUS DISTRIBUTIONS

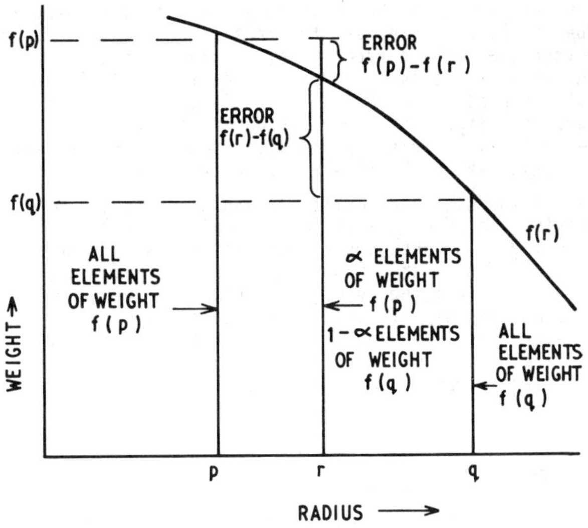

FIGURE 4
SIMULATION OF APERTURE DISTRIBUTIONS USING MULTIPLE WEIGHTS

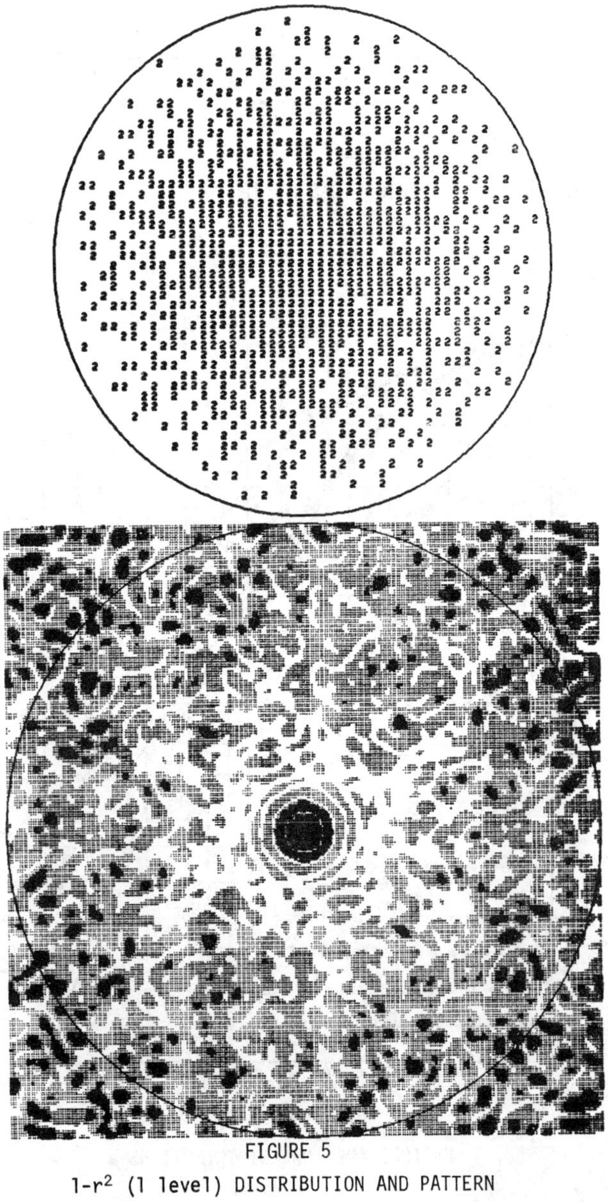

FIGURE 5
$1-r^2$ (1 level) DISTRIBUTION AND PATTERN

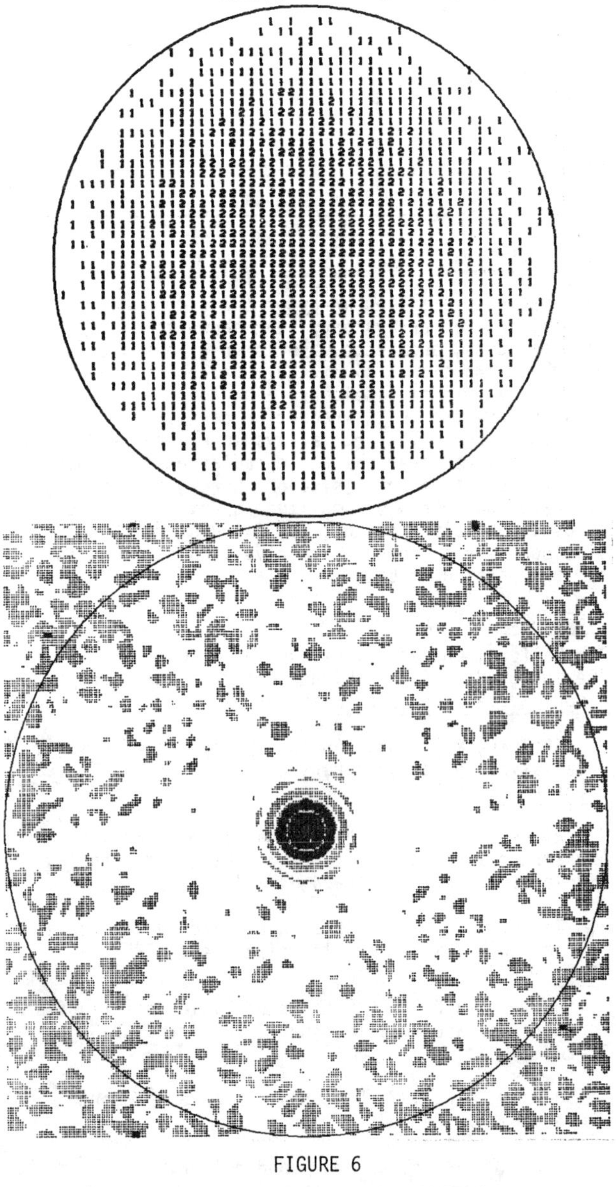

FIGURE 6
$1-r^2$ (2 LEVEL) DISTRIBUTION AND PATTERN

THE IN-SITU CALIBRATION OF A RECIPROCAL SPACE-FED PHASED ARRAY ANTENNA

Eric K.L. Hung, N. Ross Fines, and Ross M. Turner

Department of Communications, Canada

1. INTRODUCTION

A space-fed phased array antenna can be calibrated through the measurement of the following parameters associated with each array element: (a) relative current, (b) selectable phase-shifts, and (c) phase adjustment to theoretically calculated beamforming phase shifts. The relative current at an array element is the effective current through this element compared with the average of the effective currents through all the working elements. Selectable phase-shifts are the allowable modifications on the phase of the elemental radar signal received or transmitted by the array element. Phase adjustment to theoretically calculated phase shifts is an overall correction to suppress errors which affect the phase of the elemental signal transmitted or received by the feed horn through the array element. These errors include antenna distortion, misalignment of antenna boresight direction, and inaccurate knowledge of array element insertion phases.

Presented here is a technique for the in-situ calibration of the space-fed phased-array antenna at the Communications Research Centre (CRC) in Ottawa, Canada. A brief description of this CRC antenna is given in the next section. The equations used to estimate the array calibration parameters are derived in Section 3. Section 4 is concerned with the measurement of radar signals. Some results of antenna calibration are given in Section 5.

A technique for the in-situ calibration of a space-fed phased array antenna has been described by Hüschelrath and Sander (1). This technique uses auxiliary horn antennas in the near-field zone and is restricted to antennas in which the actual antenna configuration is very accurately known. The CRC calibration technique uses a remote calibration source and does not have the above restrictions. Here, the actual antenna configuration is taken into account automatically in the calculation of calibration parameters.

2. THE CRC PHASED ARRAY ANTENNA

The CRC space-fed phased array antenna is a reciprocal S-band antenna. It consists of a rectangular feed horn mounted 111 cm behind a vertical planar array as shown in Figs. 1 and 2. The planar array contains 295 elements configured into an elliptical aperture with a 167 cm horizontal axis and a 105 cm vertical axis. Each array element is constructed as a 3-bit PIN-diode phase-shift assembly with flared waveguide horns at the ends. The PIN-diodes are used to switch three transmission line segments in and out of the signal path, thereby producing $2^3=8$ selectable phase-shift states. The nominal lengths of the line segments are $\lambda/8$, $\lambda/4$, and $\lambda/2$, where λ is the signal wavelength at the desired radar frequency. The selectable phase-shift states are identified by a 3-bit integer ℓ with ℓ = (0,0,1)=1, ℓ = (0,1,0)=2, and ℓ = (1,0,0)=4 indicating the bypassing of line segments $\lambda/8$, $\lambda/4$, and $\lambda/2$, respectively.

The selectable phase shifts at the kth array element are denoted by $\{\beta_{k\ell};\ \ell=0,1,2,\ldots,7\}$. They satisfy relations

$$\beta_{k\ell} = 0\,, \qquad \ell=0, \tag{1}$$

$$\beta_{k\ell} = \frac{\ell\pi}{4}\,, \qquad \ell=1,2,4 \tag{2}$$

$$\beta_{k\ell} = \beta_{k1} + \beta_{k2}\,, \qquad \ell=3 \tag{3}$$

and

$$\beta_{k\ell} = \beta_{k4} + \beta_{k(\ell-4)} \qquad \ell=5,6,7 \tag{4}$$

The independent selectable phase shifts are β_{k1}, β_{k2}, and β_{k4}. Their values depend on the actual lengths of the nominally $\lambda/8$, $\lambda/4$, and $\lambda/2$ line segments, respectively, in the kth phase shift assembly.

3. EQUATIONS

Because the CRC antenna is a reciprocal antenna, it is sufficient to calibrate this antenna with the radar in the receive mode of operation only. In this section, an expression for the radar signal due to a remote calibration source is given. The conditions for radar signal measurement are stated. Equations for the estimation of calibration parameters from measured radar signals are derived. A summary is presented at the end.

A. Expression for Radar Signal

The antenna output measured at the radar receiver can be written as a sum of the contributions from a specified array element plus other contributions,

$$A_{k\ell} = a_{k\ell}\, e^{j(\alpha_k+\beta_{k\ell})} + B_k \tag{5}$$

Here, k signifies the quantities associated with the specified element, and ℓ signifies the phase-shift state of this element. The elemental radar signal from this element has an effective amplitude $a_{k\ell}$. Its phase is α_k before phase-shifting. Phasor B_k represents the sum of all other contributions to $A_{k\ell}$. It includes the sum of elemental radar signals from all other array elements, mutual coupling, and background noises.

Because the sum of α_k and $\beta_{k\ell}$ appears in the expression for $A_{k\ell}$. It is convenient to write

$$\gamma_{k\ell} = \alpha_k + \beta_{k\ell} \tag{6}$$

From the properties of the $\beta_{k\ell}$s in (1) to (4), one has

$$\gamma_{k0} = \alpha_k \tag{7}$$

$$\gamma_{k\ell}-\gamma_{k0} = \beta_{k\ell}\,, \qquad k=0,1,2,\ldots,7 \tag{8}$$

$$\gamma_{k\ell}-\gamma_{k(\ell-4)} = \beta_{k4}, \qquad \ell=4,5,6,7 \tag{9}$$

and

$$\gamma_{k1}+\gamma_{k2} = \gamma_{k0}+\gamma_{k3} \tag{10}$$

In the rest of this paper, it shall be assumed that $a_{k\ell}$, α_k, and B_k are constants. This means that the experiment is designed so that the following conditions are satisfied during calibration:

1. The output power of the signal source used to calibrate the antenna is extremely stable.

2. The phase coherency between the signal source and the reference signal at the radar receiver is very high.

3. The magnitude of fluctuations in the noise background is small compared with $a_{k\ell}$.

Phasor B_k includes the contribution due to mutual coupling among the array elements. Strictly speaking, it is also dependent on the phase-shift state ℓ of the kth array element. However, this dependence on ℓ is very weak and can be ignored completely.

B. Estimation of Relative Current

The value of $a_{k\ell}$ in (5) is dependent on both k and ℓ. The average value of the $a_{k\ell}$s over all the phase-shift states of the kth array element is calculated as

$$a_k = \text{radius of the best-fit circle through the set of radar signals } \{A_{k\ell};\ \ell=0,1,2,\ldots,7\} \text{ in the Argand diagram.} \quad (11)$$

The relative current at the kth array element is defined here as

$$b_k = c\ a_k, \quad k=1,2,3,\ldots,295\ , \quad (12)$$

where

$$c = \frac{K'}{\Sigma'\ a_k} \quad (13)$$

is a normalization constant, K' is the number of functioning elements, and Σ' is a summation over the functioning elements. The b_ks defined here satisfy the condition

$$\Sigma' b_k = K' \quad (14)$$

An array element is identified here as a functioning element if its average current amplitude is larger than 0.01 a_{max}, where a_{max} is the largest of the a_ks.

The position and radius of the best-fit circle through $\{A_{k\ell};\ \ell=0,1,2,\ldots,7\}$ is defined here as the circle which minimizes the sum

$$s = \sum_{\ell=0}^{7} (d_\ell^2 - a_k^2)^2 \quad (15)$$

where d is the distance of $A_{k\ell}$ from the circle centre C_k. It can be shown that the coordinates of C_k can be calculated in closed form as

$$\text{Re}\{C_k\} = [f(x,y)-g(x,y)]h^{-1}(x,y) \quad (16)$$

$$\text{Im}\{C_k\} = [f(y,x)-g(y,x)]h^{-1}(x,y) \quad (17)$$

where

$$f(x,y) = 2(\overline{y^2}-\overline{y}\overline{y})(\overline{x^3}-\overline{x}\overline{x^2}+\overline{xy^2}-\overline{x}\overline{y^2}) \quad (18)$$

$$g(x,y) = 2(\overline{xy}-\overline{x}\overline{y})(\overline{x^2y}-\overline{x^2}\overline{y}+\overline{y^2}-\overline{y}\overline{y^2}) \quad (19)$$

$$h(x,y) = 4[(\overline{x^2}-\overline{x}\overline{x})(\overline{y^2}-\overline{y}\overline{y})-(\overline{xy}-\overline{x}\overline{y})^2] \quad (20)$$

and

$$\overline{x^u y^v} = \frac{1}{8}\sum_{\ell=0}^{7} [\text{re}(A_{k\ell})]^u[\text{Im}(A_{k\ell})]^v \quad (21)$$

The radius of the circle is given by

$$a_k = [\overline{x^2}+\overline{y^2}-2\overline{x}\ \text{Re}(C_k)-2\overline{y}\ \text{Im}(C_k)+|C_k|^2]^{\frac{1}{2}} \quad (22)$$

C. Estimation of Selectable Phase Shifts

The independently selectable phase shifts at the kth array element are β_{k1}, β_{k2}, and β_{k4}. In the following, these independently selectable phase shifts are estimated through the estimation of the $\gamma_{k\ell}$s given by (6). There are five steps in the procedure. First, a set of initial estimates, denoted by $\{\gamma_{k\ell}^{(1)};\ \ell=0,1,2,\ldots,7\}$ is constructed. Second, the estimate of β_{k4} is calculated. Third, condition (9) is imposed to produce a set of improved estimates $\{\gamma_{k\ell}^{(2)};\ \ell=0,1,2,\ldots,7\}$. Fourth, condition (10) is imposed to produce the final estimates $\{\hat{\gamma}_{k\ell};\ \ell=0,1,2,\ldots,7\}$. Finally β_{k1} and β_{k2} are calculated.

The circle centre C_k is an estimate of the tip of phasor B_k in (5). Therefore, the phase of $(A_{k\ell}-C_k)$ is an estimate of $\gamma_{k\ell}$, i.e.,

$$\hat{\gamma}_{k\ell}^{(1)} = \text{phase of } (A_{k\ell}-C_k), \quad k=0,1,2,\ldots,7 \quad (23)$$

There are four values of phase difference given by $\{[\hat{\gamma}_{k(\ell+4)}^{(1)}-\hat{\gamma}_{k\ell}^{(1)}];\ \ell=0,1,2,3\}$. According to (9), each phase difference can be used to obtain an estimate of β_{k4}. Because the nominal value of β_{k4} is π radians, four initial estimates of β_{k4} are calculated as

$$\beta_{k4}^{(1)}(m) = \hat{\gamma}_{k(\ell+4)}^{(1)} - \gamma_{k\ell}^{(1)} + 2\pi n_{km} \quad (24)$$

$$m = 1,2,3,4; \quad \ell=m-1.$$

where n_{km} is an integer which restricts $\hat{\beta}_{k4}^{(1)}(m)$ to the range [0,2) radians, a range which is centred at the nominal value of β_{k4}. The final estimate of β_{k4} is defined as the average of the initial estimates,

$$\hat{\beta}_{k4} = \frac{1}{4}\sum_{m=1}^{4} \hat{\beta}_{k4}^{(1)}(m) \quad (25)$$

The manner condition (9) is imposed on the estimates of the $\gamma_{k\ell}$s is illustrated in Fig. 3. Initially, the phasors $\{(A_{k(\ell+4)}-C_k),\ (A_{k\ell}-C_k)\}$ are separated by an angle of $\hat{\beta}_{k4}^{(1)}(m)$, where $m=\ell+1$. The phasors in this pair are rotated towards each other by the same angle $\frac{1}{2}[\hat{\beta}_{k4}^{(1)}(m)-\hat{\beta}_{k4}]$ to form a new pair $\{(A'_{k(\ell+4)}-C_k),(A'_{k\ell}-C_k)\}$ so that they are separated by an angle of β_{k4} after rotation. Estimates $\gamma_{k(\ell+4)}^{(2)}$ and $\gamma_{k\ell}^{(2)}$ are defined as the phases of $(A'_{k(\ell+4)}-C_k)$ and $(A'_{k\ell}-C_k)$, respectively. The improved estimates of the $\gamma_{k\ell}$s are calculated as

$$\hat{\gamma}_{k\ell}^{(2)} = \begin{cases} \hat{\gamma}_{k\ell}^{(1)} + 0.5\left[\hat{\beta}_{k4}^{(1)}(\ell+1)-\hat{\beta}_{k4}\right]; & \ell=0,1,2,3 \\ \hat{\gamma}_{k\ell}^{(1)} - 0.5\left[\hat{\beta}_{k4}^{(1)}(\ell-3)-\hat{\beta}_{k4}\right], & \ell=4,5,6,7 \end{cases} \quad (26)$$

By construction, these improved estimates satisfy the relation

$$\hat{\gamma}_{k\ell}^{(2)} - \hat{\gamma}_{k(\ell-4)}^{(2)} = \hat{\beta}_{k4}\ , \quad \ell=4,5,6,7 \quad (27)$$

Relation (10) is imposed in a two-step calculation. Initially, a dummy phase angle is calculated as

$$\delta_k = \hat{\gamma}_{k1}^{(2)} + \hat{\gamma}_{k2}^{(2)} - \hat{\gamma}_{k0}^{(2)} - \hat{\gamma}_{k3}^{(2)} + 2\pi p_k \quad (28)$$

where p_k is an integer which restricts δ_k to the range $[-\pi,\pi)$. This range is chosen because δ_k is zero if (10) is satisfied. The second step calculates the final estimates $\{\hat{\gamma}_{k\ell};\ \ell=0,1,2,\ldots,7\}$ as

$$\hat{\gamma}_{k\ell} = \hat{\gamma}^{(2)}_{k\ell}+e_\ell\delta_k, \quad \ell=0,1,2,\ldots,7. \quad (29)$$

It can be shown that the e_ℓs are given by

$$e_\ell = \begin{cases} 0.25, & \ell=0,3,4,7 \\ -0.25, & \ell=1,2,5,6 \end{cases} \quad (30)$$

Estimates of selectable phase shifts β_{k1} and β_{k2} are calculated from the final estimates of the $\gamma_{k\ell}$s as

$$\hat{\beta}_{k\ell} = \hat{\gamma}_{k\ell}-\hat{\gamma}_{k0}+2\pi q_{k\ell}, \quad \ell=1,2, \quad (31)$$

where $q_{k\ell}$ is an integer which restricted $\hat{\beta}_{k\ell}$ to the range $[(0.25\ell-1)\pi,\ (0.25\ell+1)\pi)$, which is centred at the nominal value of $\beta_{k\ell}$.

It may be noted that (31) can also be used to calculate the other estimates of $\beta_{k\ell}$. The estimates thus calculated satisfy relations [1], (3), and (4), i.e.,

$$\hat{\beta}_{k\ell} = 0, \quad \ell=0 \quad (32)$$

$$\hat{\beta}_{k\ell} = \hat{\beta}_{k1}+\hat{\beta}_{k2}, \quad \ell=3 \quad (33)$$

$$\hat{\beta}_{k\ell} = \hat{\beta}_{k4}+\hat{\beta}_{k(\ell-4)}, \quad \ell=5,6,7 \quad (34)$$

D. Estimation of Phase Adjustment

Angle α_k in (5) is the phase of the kth elemental signal at the radar receiver output. A phase reduction of α_k, equivalent to a phase advance of $(2\pi-\alpha_k)$, applied to the kth elemental signal would force it to have a zero phase at the radar receiver output. If all the elemental signals have zero phase at the radar receiver output, the radar receiver output signal has the largest possible amplitude. Hence, apart from a phase constant common to all array elements, α_k is the phase reduction which must be applied to the kth elemental signal to point the main beam of the antenna at the calibration signal source.

The theoretical value of α_k, denoted by α^T_k, can be calculated from antenna geometry, signal direction relative to antenna boresight direction and array element insertion phases. The difference

$$\Delta_k = \alpha_k-\alpha^T_k \quad (35)$$

is the phase adjustment which must be added to the theoretically calculated α^T_k to point the antenna beam at the signal source.

From (7), an estimate of α_k is given by:

$$\hat{\alpha}_k = \hat{\gamma}_{k0} \quad (36)$$

Therefore, an estimate of phase adjustment is:

$$\hat{\Delta}_k = \hat{\alpha}_k-\alpha^T_k = \hat{\gamma}_{k0}-\alpha^T_k \quad (37)$$

Summary

The conditions for radar signal measurement are stated in Section 3A. The equations for the calculation of antenna calibration parameters are derived in Sections 3B to 3D. There are eight steps in the calculation of these calibation parameters:

- Step 1 Set k=0
- Step 2 Replace k by k+1
- Step 3 Calculate circle radius a_k and circle centre C_k with (16) to (22).
- Step 4 Calculate $\{\hat{\gamma}^{(1)}_k;\ \ell=0,1,2,\ldots,7\}$ with (23). Calculate $\{\hat{\gamma}^{(2)}_k;\ \ell=0,1,2,\ldots,7\}$ with (24) to (26). Calculate $\{\hat{\gamma}_{k\ell};\ \ell=0,1,2,\ldots,7\}$ with (28) to (30). Calculate estimates of selectable phase shifts $\{\hat{\beta}_{k\ell};\ \ell=0,1,2,\ldots,7\}$ with (31).
- Step 5 Calculate phase adjustment $\hat{\Delta}_k$ with (37)
- Step 6 If k=295 proceed direction to Step 8
- Step 7 Return to Step 2
- Step 8 Calculate relative currents $\{b_k;\ k=1,2,3,\ldots,295\}$ with (12) and (13).

4. RADAR SIGNAL MEASUREMENT

The experimental set-up for radar signal measurement is shown in Fig. 4. The horn antenna was mounted on a tower located at a distance of slightly over one kilometre in front of the space-fed phased array antenna. It had a gain of about 17 dB in the pointing direction and was used to transmit a calibration signal towards the phased array at a power level of slightly less than 2 watts and a frequency of 2.97 GHz. An auxiliary dish antenna was mounted on another tower located at a distance of about 7 metres from the phased array antenna. This dish had a diameter of 183 cm and was pointed at the source. Its output was used as a reference signal for the quadrature sampling of the output from the phased array antenna. The digitized in-phase and quadrature components were identified as the real and imaginary components, respectively, of the antenna output signal.

The use of the auxiliary antenna output as a reference signal for radar signal measurement was an important feature of this array calibration. With this arrangement, slow drifts in signal source frequency produce almost identical frequency shifts in the output signals from the dish and array antennas. Because the dish antenna output was used as a reference signal in the quadrature sampling of the array antenna output signal, drifts in signal source frequency were expected to have negligible effects on the I and Q components of the measured radar signal.

Preliminary measurements showed that the magnitude and phase of antenna output fluctuated slightly about their respective mean values. Two precautions were taken to suppress errors caused by these fluctuations. First, a null of the array beam pattern was pointed at the calibration source to ensure that the magnitude of B_k in (5) was as small as practically possible. Second, each $A_{k\ell}$ used in the calculation of calibration parameters was identified as the arithmetic mean of a large number of individual signals.

The details in the construction of $A_{k\ell}$ are given below. There are eight steps.

- Step 1 Set up the experiment as shown in Fig. 4. Point the signal source at the phased array antenna. Point the dish antenna at the source. Point a null (not necessarily a perfect null) of the phased array beam pattern at the source.
- Step 2 Set k=0.
- Step 3 Replace k by k+1
- Step 4 Step the kth array element through all its eight phase-shift states. Take M individual antenna output signals at each phase-shift state. Identify the mth signal at the ℓth phase shift state as $D_{k\ell}(m)$.

Step 5 Return the kth array element to its original phase-shift state, i.e., the phase-shift state at the end of Step 1.
Step 6 If k=295, proceed directly to Step 8.
Step 7 Return to Step 3.
Step 8 Calculate $A_{k\ell}$ as

$$A_{k\ell} = \frac{1}{M} \sum_{m=1}^{M} D_{k\ell}(m) , \qquad (38)$$

$k = 1,2,3,\ldots,295; \quad \ell=0,1,2,\ldots 7.$

A choice of M=2000 was made. This choice produced phase shift estimates which had a standard deviation of less than 0.25 degrees.

5. RESULTS

The results of two array calibrations are presented here. The first was carried out with a distortion-free antenna. The second calibration was carried out with a rotated feed horn and was an example of the calibration of a distorted antenna.

A. Distortion-Free Antenna

In Fig. 5 is the best-fit circle through the set of phasors $\{A_{k\ell};\ k=148;\ \ell=0,1,2,\ldots,7\}$ measured with centre element k=148 stepped through all its phase-shift states. Here, 0 is the origin and C_k is the circle centre. The positions of the $A_{k\ell}$s are marked by crosses labelled with the value of ℓ.

Listed in Table I are some selectable phase shifts obtained by estimation and by bench measurements. Array elements k=128, 148, and 175 were in the centre of the middle column. Elements k=141 and 168 were the two elements in the right-most column when the array was viewed from the back. The differences between the estimated and the bench-measured values were smaller than expected, because the bench measurements were carried out with a different signal source about twenty months earlier.

In Fig. 6 are the array beam patterns measured by electronically sweeping a beam across the calibration source. The solid curve was constructed with estimated values of $\beta_{k\ell}$ and Δ_k. The broken curve was constructed with ideal values defined as $\beta_{k\ell}=0.25\ell\pi$ and $\Delta_k=0$. There was no significant difference between the beam patterns.

B. Distorted Antenna

The distorted antenna is shown in Fig. 7. The feed horn was pointed at the left of the array. The antenna was recalibrated, and Fig. 8 was produced. In this figure the uncalibrated beam was on the left of the correct position. The widths of both the calibrated and uncalibrated beams were also larger than those obtained with a distortion-free antenna.

6. REMARKS

The CRC calibration technique can be modified to calibrate other array antennas with phase-shifter beam steering. The phase-shifters need not be restricted to three bits. The antenna need not be in the receive mode of operation. The details of this modification are straightforward and shall not be presented here.

ACKNOWLEDGEMENT

The authors are grateful to D.J. Mabey and A.L. Poirier for their suggestions and assistance during the development of this calibration technique. This work was supported by the Canadian Department of National Defence under Research and Development Branch Project 33C69.

REFERENCE

1. Hüschelrath, G. and Sander, W., 1976, "The ELRA Phased-Array Radar with Automatic Phase Adjustment in Practice", AGARD Conference Proceedings, 197, on New Devices, Techniques and Systems in Radar, Reference No. 36.

TABLE I: Selectable phase-shifts obtained by estimation and by bench measurements

ARRAY ELEMENT IDENTIFICATION		SELECTABLE PHASE-SHIFT $\beta_{k\ell}$ (degrees)					
		ESTIMATED			BENCH MEASUREMENT		
k	(ROW,COL)	ℓ=1	2	4	ℓ=1	2	4
121	(2, 0)	47	89	166	49	94	170
148	(0, 0)	49	84	175	48	88	173
175	(-2, 0)	48	85	167	49	90	169
141	(1, 13)	48	104	189	49	104	194
168	(-1, 13)	45	104	196	44	103	193

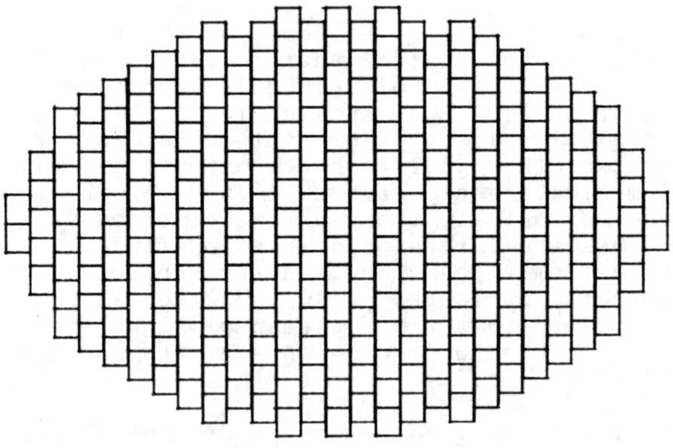

Figure 1 Layout of array elements

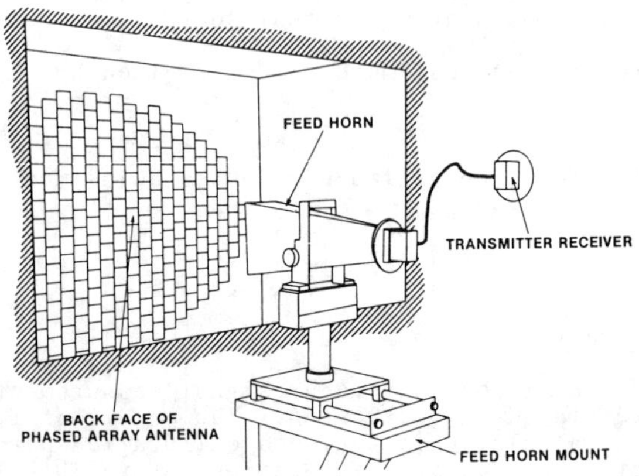

Figure 2 Rear view of array antenna with feed horn in position

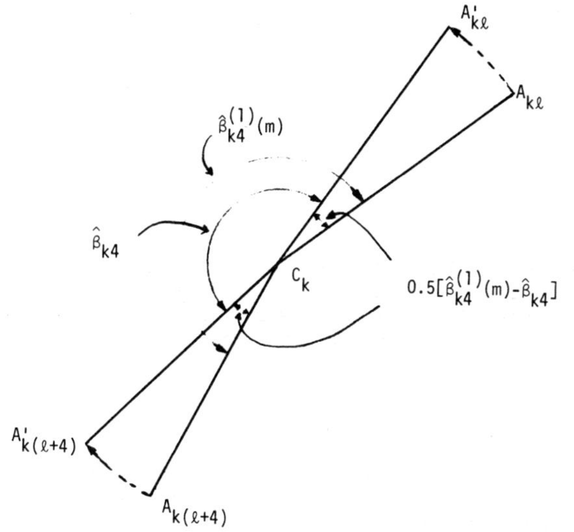

Figure 3 Relationship between phase pairs $\{A_{k(\ell+4)}-C_k\}$, $(A_{k\ell}-C_k)\}$ and $\{(A'_{k(\ell+4)}-C_k)\}$ in the construction of estimates $\hat{\gamma}^{(2)}_{k(\ell+4)}$ and $\hat{\gamma}^{(2)}_{k\ell}$.

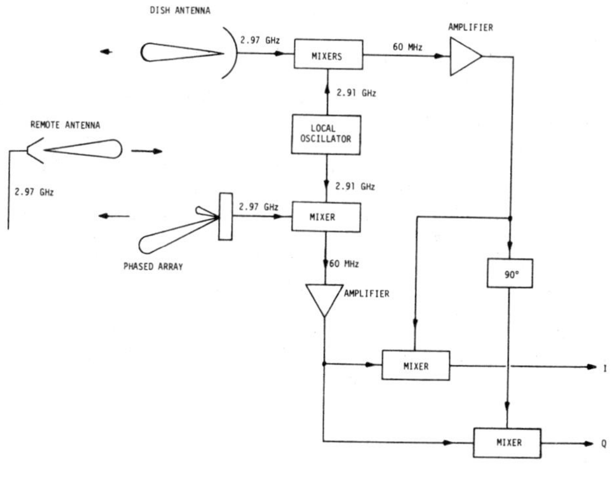

Figure 4 Experimental set-up for radar signal measurement

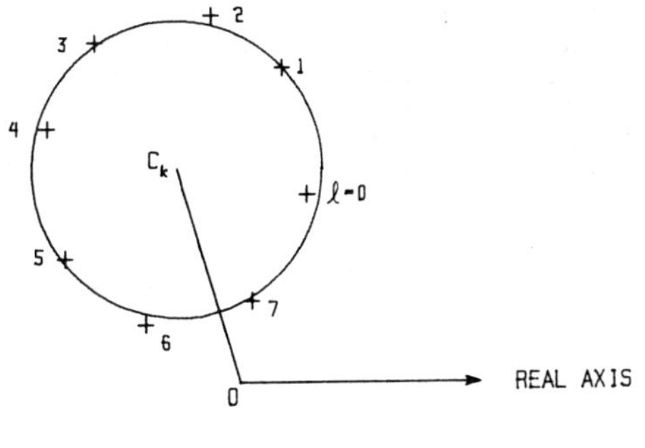

Figure 5 The best circle through the set of signals $\{A_{k\ell};\ k=148;\ \ell=0,1,2,\ldots,7\}$

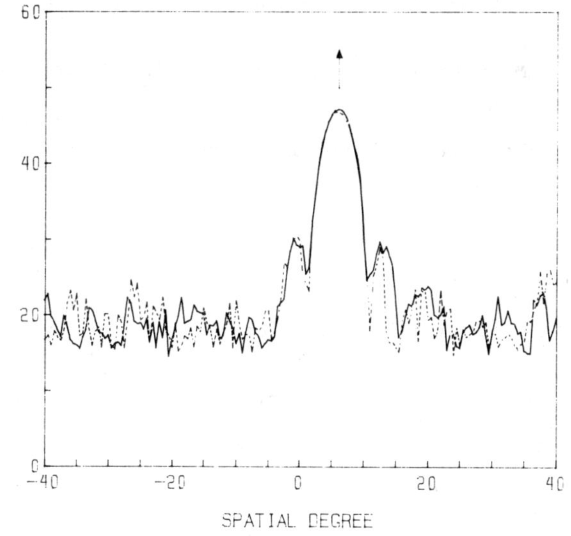

Figure 6 Array beam patterns before and after the calibration of the distortion-free antenna. The solid curve was constructed with estimates of $\beta_{k\ell}$ and Δ_k. The azimuth of the arrow is the direction of the signal source.

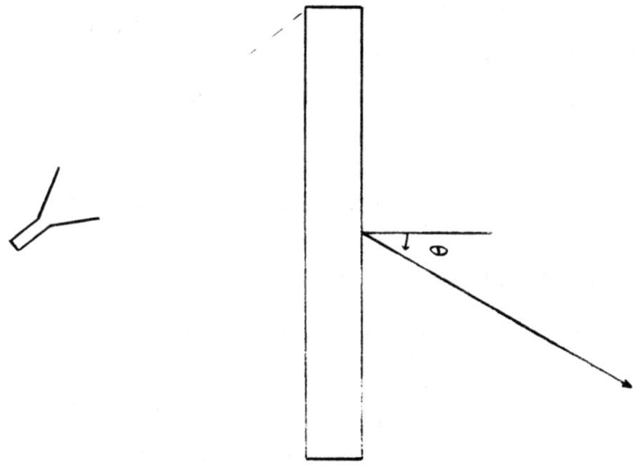

Figure 7 A diagram of the distorted antenna. The feed horn is pointed at the left edge of the array

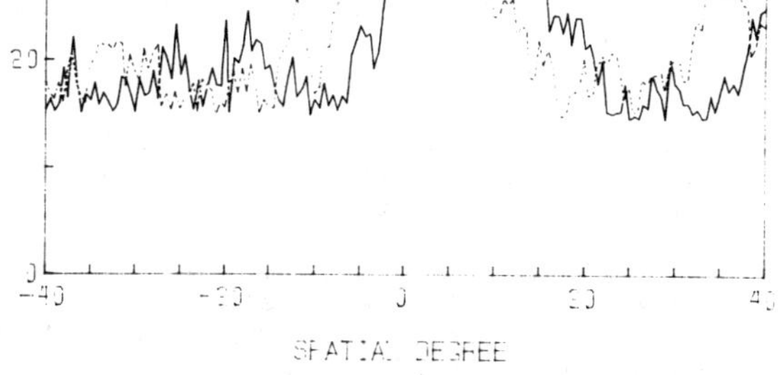

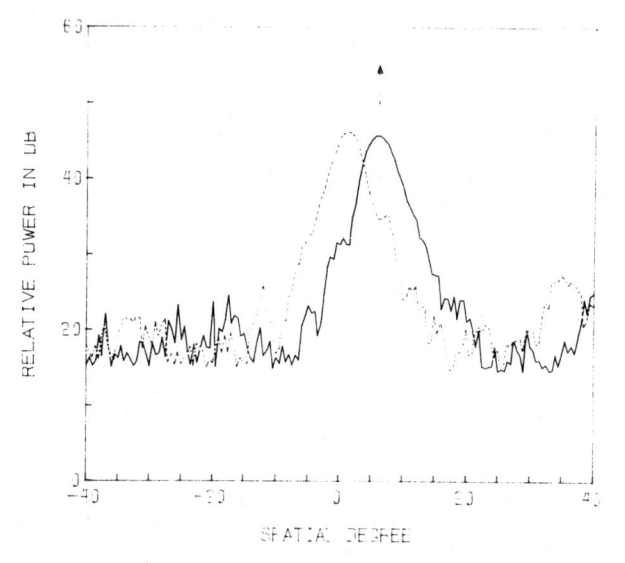

Figure 8 Array beam patterns before and after the calibration of the distorted antenna

MULTIPLE-PARAMETER-RADAR TECHNIQUES AND APPLICATIONS FOR PRECIPITATION MEASUREMENTS: A REVIEW

S. M. Cherry, M.P.M. Hall and J.W.F. Goddard

Rutherford Appleton Laboratory, U.K.

INTRODUCTION

Meteorological radar provides a useful tool for investigating the structure and effects of raincells. The type of investigation may be divided into two categories; the determination of the nature and abundance of the hydrometeors, and the measurement of the dynamics of hydrometeors in terms of wind velocities and turbulence. This review is restricted to the first category except where the dynamic behaviour of the hydrometeors is used to infer the nature of them.

The measurement of only one backscatter property of precipitation is often inadequate. For example, the ability to accurately estimate rainfall-rate from the intensity of the echo from rain has proved erratic (Battan (1)). One reason for this is that the radar reflectivity and rainfall-rate are linked by the statistical distribution of drop sizes in rain (1), and a well-known model for the distribution is given by

$$N(D) = N_o \exp(-\Lambda D) \qquad (1)$$

where N(D) is the number of drops per unit volume per unit interval of the drop diameter, D. This model and other proposed models contain at least two parameters, and therefore to estimate rainfall-rate from one radar observable requires that either N_O, Λ or that a relationship between them is assumed. The nature of rain has proved sufficiently variable that these assumptions can lead to substantial errors. Similar multiple-parameter problems exist when trying to predict other properties of precipitation such as the microwave specific attenuation or depolarization effects. To remove or at least reduce the uncertainties in interpreting the radar measurements it is necessary to measure more backscatter properties of the precipitation. The radar techniques considered in this review are:-

1. Circular depolarization ratio radars (CDRR), which measure the depolarization of the backscatter signal when using circularly polarized transmissions, and so investigate the anisotropy of hydrometeors.

2. Dual linear polarization radars (DLPR), which measure the reflectivity of precipitation with horizontally and vertically polarized transmissions, the reception being copolar.

3. Multiple wavelength radars (MWR) which measure the reflectivity of precipitation at more than one wavelength.

4. Vertical incidence Doppler radars (VIDR) which measure the Doppler spectrum to deduce the fall-speed of hydrometeors and so estimate their drop-size distribution.

An URSI symposium on 'multiple-parameter radar measurements of precipitation' was held in Bournemouth, U.K. in August 1982, and papers from that symposium were used as a source for the latest developments in multiple-parameter radar techniques and applications.

TECHNIQUES

Circular depolarization radars

In these radars right- or left-hand circular (RHC or LHC) polarization is transmitted, and a dual-channel receiver measures simultaneously the amplitude of the echoes with RHC and LHC polarizations so that their relative phase is available. This radar technique has been developed considerably by McCormick and Hendry (2), and the theory relating the radar observables to the properties of the hydrometeors has been described by them.

CDR is defined as

$$CDR = 10 \log_{10} (P_{RHC}/P_{LHC}) \qquad (2)$$

where P_{RHC} and P_{LHC} are the powers received in the RHC and LHC channels respectively when RHC polarization is transmitted (RHC and LHC being interchanged for LHC transmission). CDR gives an indication of hydrometeor shapes, while the complex correlation between received amplitudes for the RHC and LHC channels gives information about the degree of orientation of the particles, ρ, and the mean canting angle, α, of the particles when depolarization exists. The estimate of α is affected if the particles exhibit non-Rayleigh scattering, but this can be resolved if estimates are made with RHC and LHC transmissions, and the difference in α so determined indicates non-Rayleigh scattering.

The propagation of the radar transmission through rain affects the measured values of CDR, and especially ρ and α, through the effects of differential attenuation and differential phase shift. Thus at short wavelengths (below 5 cm) the intrinsic properties of the precipitation can only be measured when the path through the rain is short or when the rain is weak. Within large storms the transmission is so depolarized that radar measurements characterise not the target but the intervening precipitation, which in some cases can be a useful measurement in itself (Hendry and Antar (3)).

Measurements (3) have shown that CDR varies from -30 dB in light rain to -10 dB for large hail, while the percentage of aligned particles varies from 60 to 90% in rain to less than 30% for hail and particles in the melting region.

Depolarization effects can be measured with other pairs of orthogonal polarizations than RHC and LHC, and Boerner (4) proposes that an optimum pair of polarizations can be chosen to determine particular properties of the hydrometeors. The use of coherent radars with polarization diversity receivers has been proposed (Metcalf and Echard (5)) as a means of acquiring additional information. Pasqualucci et al (6) describe a 35 GHz radar with this capability while Schrott et al (7) present the design of a 6 GHz coherent radar with polarization agility.

The main technical requirement of the CDRR technique is an antenna system with excellent polarization characteristics. That is the polar diagram should be matched between RHC and LHC polarizations, and the isolation between the channels should be 40 dB or better when integrated over the beamwidth. Further, if the polarization is not exactly circular then this needs to be allowed for by a calibration which measures the ellipticity.

Linear dual-polarization radar

This radar measures the ratio of the radar reflectivity for horizontal polarization, Z_H, to that for vertical polarization, Z_H. The ratio is known as the differential reflectivity, Z_{DR}. That is Z_{DR} expressed in decibels is

$$Z_{DR} = 10 \log_{10} (Z_H/Z_V) \quad (3)$$

This type of radar was primarily developed to reduce the uncertainty through assumptions about the drop-size distribution when predicting rainfall-rate or specific attenuation. The technique uses the phenomena that raindrops tend to have oblate spheroidal shapes with their axes of symmetry near vertical, and the larger the drop the more oblate it is. Consequently for radars observing with low elevation angles Z_H is larger than Z_V, and for rain with a drop-size distribution described by equation (1), Z_{DR} is a function of Λ alone (Seliga and Bringi (8)). N_O can then be determined from Λ and Z_H so giving both parameters of the distribution. Values of Z_{DR} expected for rain are 0 to 4 dB (8).

Experiments have been performed to test the physical assumptions behind the technique. Predictions of Z_{DR} from drop-size distributions measured by ground-based distrometers 200 m below the radar beam have been compared with the radar measurements, by Goddard et al (9), and Goddard and Cherry (10). The results showed that Z_{DR} did reflect the size of the raindrops, but small drops appeared more spherical than predicted by the model of Pruppacher and Pitter (11). Limited comparisons between the parameters N_O and Λ measured by an airborne distrometer compared with predictions by LDPR support this finding (Cherry et al (12)). Seliga et al (13) have compared rainfall-rates measured by DLPR and a network of raingauges, and found substantial improvements over normal rainfall-rate-reflectivity relationships. Cherry et al (14) have compared attenuation on a 11 GHz satellite-to-earth path predicted by DLPR with that directly measured, and found a marked improvement over methods where N_O had to be assumed.

Seliga et al (15) and Dissanayake et al (16) have investigated theoretically the effect of raindrop oscillation on the expected values of Z_{DR}. The results are contradictory in that (15) predict Z_{DR} is reduced progressively for large raindrops while (16) show Z_{DR} is reduced for small drops, in line with the findings of (7), (8) and (12).

Ulbrich and Atlas (17) and (10) have investigated the effects of non-exponential drop-size distributions on the relationships between Z_{DR} and other parameters. Both found that data from ground-based distrometers predicted that rainfall-rate would be typically overestimated by 30% when using DLPR, and assuming an exponential distribution.
Z_{DR} is a good indicator of ice/water boundaries in raincells (Hall et al (18)). Aydin et al (19) propose that values of Z_{DR} uncharacteristic of rain (near zero or negative) in storms can be used to detect the presence of hail, while Leitao and Watson (20) describe how the poor correlation of Z_{DR} with Z_H within a storm may be used as a hail indicator.

Propagation of the radar transmission through rain can reduce the measured values of Z_{DR} from the true one through the effect of differential attenuation, but at 10 cm wavelength the effects are negligible except for long paths through very heavy rain (Cherry and Goddard (21)), and McGuinness et al (22). However for 5 cm and below the effects are significant (21).

The ability to use DLPR for the purposes described above depends on measuring Z_{DR} to a few tenths of a decibel (for example see (8) and (14)). This can only be done reliably and without excessive integration time by sampling Z_H and Z_V rapidly on alternate transmitted pulses. This technique eliminates biases which might be introduced by Z_H and Z_V changing during the integration time, and the effects of slow changes in the sensitivity of the radar. Further, because samples of Z_H and Z_V taken in this way are highly correlated (Bringi et al (23)) sampling errors in Z_{DR} due to the noise-like nature of echoes from precipitation are greatly reduced. The standard error of Z_{DR} for the 10 cm wavelength, Chilbolton DLPR is only 0.1 dB for 0.2 s integration time (21). The Chilbolton radar uses a mechanical waveguide switch to achieve the alternate sampling with a pulse recurrence frequency (prf) of 610 Hz, while recently an electronic switch has been used on the CHILL radar (Seliga and Mueller (24)) with a prf of 950 Hz.

As for CDRR, the polarization performance of the antenna for DLPR must be good in that the polar diagram must not depend on the polarization, but very good isolation between polarizations is not necessary and 25 dB is adequate (21).

Stapor and Pratt (25) propose that the linear depolarization ratio (LDR) be measured by observing the cross-polar signals received by DLPR, and this would indicate the canting angle of the hydrometeors. They also suggest that DLPR could be used with polarization 45° to the vertical to increase the cross-polar signal so making it easier to measure. Obviously LDR measurements would require better isolation in the antenna between the polarization channels than Z_{DR} measurements.

Dual wavelength radars

Measuring the differential reflectivity of precipitation with wavelength is proposed as a means of detecting hail shafts in rain storms (Eccles and Atlas (26)). The backscatter cross-section of small hydrometeors such as rain is reasonably well approximated by Rayleigh's theory for wavelengths as low as about 3 cm. While Rayleigh theory applies, the effective radar reflectivity Z_e is independent of frequency. Large hydrometeors such as hail tend to have smaller cross-sections than predicted by Rayleigh theory, especially for wavelengths less than 5 cm. Consequently

$Z_1/Z_2>1$ where Z_1 is the reflectivity for wavelength λ_1 and Z_2 is the reflectivity for a shorter wavelength λ_2. Z_1/Z_2 can be as large as 18 dB for hail with an exponential drop-size distribution when λ_1 = 10 cm and λ_2 = 3 cm (Eccles and Atlas (27)).

The measured value of Z_2 can be reduced by attenuation caused by rain along the radar beam to the measurement point. This effect shows as a gradual apparent increase in Z_1/Z_2 with range, but should not be confused with $Z_1/Z_2>1$ produced by hail because this would eventually decrease with increasing range once out of the hail shaft into rain.

Eccles and Mueller (28) have proposed that the attenuation indicated by the measured values of $Z_1 \neq Z_2$ in the absence of non-Rayleigh scatterers be used as an improved estimator of rainfall-rate or liquid water content. Eccles (29) describes an experiment where rainfall measured this way was better than estimates computed directly from reflectivity, when compared with raingauges. However, hail was present and treated as rain in the determination of rainfall-rate from reflectivity, which resulted in a severe overestimation of rainfall-rate.

The range resolution of measurements of rainfall rate by DWR is limited because the attenuation has to be sufficiently large (i.e. it requires a long path-length between measurement points) compared with the measurement errors. Measurement errors come from the sampling of the "noisy" rain signal, and Eccles (29) has described a means of reducing these errors by applying a curve fitting process to the way the apparent Z_1/Z_2 varies with range, so making use of many data points to average out fluctuations. Attenuations are then deduced from the slope of the curve. The assumption $Z_1 = Z_2$ for rain also leads to errors. For example, at 3 cm, heavy rain of 50 mmh^{-1} at a temperature of 20°C causes a two-way attenuation of about 1 $dBkm^{-1}$. This has to be compared with the errors in the Rayleigh approximation which would typically underestimate Z_2 by 1.4 dB and overestimate Z_1 by 0.4 dB for this rain. The errors are a function of the drop-size distribution parameter Λ, so unless rain with small drops is around the heavy rain reliable measurement points will not be available. Use of shorter wavelengths to produce larger more easily measured attenuations aggravate the problem of non-Rayleigh scatter.

The major technical problem of DWR is producing an antenna system with the same polar diagram, including the major sidelobes, for both wavelengths. Rinehart and Tuttle (30) discuss the effect of mis-matched beams and show that to keep spurious values of Z_1/Z_2, through reflectivity gradients across the beam, to less than ±3 dB requires the beamwidths to be matched to 10%. Eccles (31) also comments on these effects and suggests a design goal of 5% and that sidelobes should be kept 30 dB below the main lobe.

Barge and Humphries (32) describe the use of a 10 cm wavelength CDRR with a 5 cm wavelength radar to distinguish between mixtures of small hail and light rain from mixtures of small hail and heavy rain. Because the two radars do not have synchronised antennas with exactly the same beamwidth, they rely on spatial averaging of the reflectivity to reduce errors, and base their deductions on large ensembles of data to avoid the spurious effects which can happen with mis-matched beams.

Vertical incidence Doppler radar

The spectrum from VIDR allows the measurement of the fall-speed of hydrometeors relative to the radar. If the vertical wind speed, w, is zero then the drop-size distribution can be computed through the relationship of terminal velocity with drop-size. Vertical winds shift the spectrum and this effect has to be determined. Methods (Atlas (33)) exist for predicting the mean reflectivity weighted Doppler velocity as a function of reflectivity, and comparison with the measured mean Doppler velocity, v, allows w to be determined. However, the prediction relies on the drop-size distribution parameters (as in equation (1)), and is therefore underdetermined by v alone.

Hauser and Amayenc (34) describe a means of comparing N points of a spectrum derived as a function of N_o, Λ and w, with the measured spectrum. Then for N>3 the predicted and measured spectra can be matched by varying N_o, Λ and w, when these parameters have their correct values.

The technique is affected by vertical air velocity turbulence and uncertainties in the terminal-velocity size relationships which cause spectrum broadening (34). To keep errors within a reasonable range the broadening must not exceed 25% of the total width (34).

Hauser and Amayenc (35) report a comparison of VIDR with an airborne distrometer, and found that although the VIDR predicted larger N_o and Λ than the distrometer, the estimates of liquid water content agreed well. A similar observation has been seen when comparing DLPR estimates of N_o, Λ, and rainfall-rate with those of an airborne distrometer (12), where the effect was ascribed to non-exponential drop size distributions.

Pasqualucci (36) describes a VIDR which is bistatic and can observe echoes as close as 19 m. At this height above ground the vertical wind speed is assumed to be zero and hence needs no determination before drop-size distributions can be deduced.

APPLICATIONS

Rainfall-rate measurement

Many radars used operationally for the prediction of rainfall-rates for short-term weather forecasting are often required to work at ranges where the beam is either not filled with rain, or contains a mixture of ice, melting hydrometeors and rain is present. Such radars (for example, Browning (37)) are calibrated with relationships derived empirically, and it is doubtful if multiple-parameter radar (MPR) techniques would offer any significant improvement. However, different prevailing weather conditions will affect these considerations. For example, the USA compared with the UK typically has greater rain height, tornadic weather and areas where large hail is common. Under these conditions MPR would have much to offer.

Spurious ground-clutter caused by anomalous propagation is a general problem to weather radar and may be usefully detected by DLPR and CDRR through its different polarization characteristics. For example, ground-clutter with DLPR gives values of Z_{DR} which are very variable and predominantly negative (known through the authors' connection with the Chilbolton DLPR).

Water resources management usually requires rainfall prediction for a large catchement area and the averaging inherent in this means improved estimates of localised rainfall-rate by DLPR may not be necessary. However, if hail is common then DWR could offer improved estimates as should DLPR and CDRR by eliminating the confusion between hail and rain echoes.

Early warnings of flash-floods require rainfall -rate predictions with 10 to 30% accuracy, a spatial resolution of a few kilometres, and with only 10 minutes delay (Atlas and Thiele (38)). The improved accuracy of rainfall-rate by DLPR should be useful for this purpose.

The possible development of rainfall-rate measurement from satellites, giving global coverage has produced a demand for a ground-truth precipitation calibration facility (PCF) with a spatial resolution of 1 km, and DLPR has been suggested as a suitable technique (38). It has also been suggested that large numbers of low-power inexpensive CW VIDR be used as remote 'raingauges' for this purpose when area coverage radars are not possible (38).

Hydrometeor identification

The ability of MPR to distinguish between different types of hydrometeors lends itself to many practical and pure research applications. CDRR, DLPR and MWR can be used to detect hail shafts in storms, and MWR has been used (29) to estimate the size and flux of hail. Here the applications are warnings of possible hail damage and investigations into weather modification.

At low heights in hail free areas large raindrops can cause erosion to high-speed aircraft, and the knowledge of the drop size distribution through DLPR or VIDR would help estimate the effects.

Super-cooled water which may exist in columns above the normal freezing level in raincells is a danger to aircraft, especially helicopters because of the rapid icing it can cause. For this reason its detection is important, in addition to the cloud physics interest. The ability of CDRR and DLPR to distinguish between rain and small hail or graupel should therefore be useful for this purpose.

Clearly hydrometeor identification can contribute much to cloud physics. For example, Carbone et al (39) describe an experimental programme using DLPR which includes LDR measurement, MWR, and Doppler facilities to investigate the origin and evolution of ice in convective clouds, drop breakup and coalescence in rain and snow, and the effects of evaporation and sedimentation on drop-size distributions. Hendry and McCormick (40) have used CDRR to investiage the alignment of ice crystals by electric fields in clouds.

Modelling of radio propagation effects

This application requires hydrometeor identification, especially between ice and water, and a knowledge of the drop size distribution in rain. Goddard and Cherry (41) report the use of DLPR to collect data systematically to predict attenuation at 11 GHz on earth-space paths with various elevation angles. The data were used to predict cumulative statistics of attenuation, and the gain achieved by the use of space diversity reception. Although the cumulative statistics were based on only a 10% sample in time, they were consistent with previous statistics from satellite ground-stations.

Models for predicting attenuation from point rainfall-rates, for which there is much data, require the knowledge of the ratio of rainfall-rates averaged over a path to the point value. Data from DLPR has been used to evaluate these ratios for path lengths between 1 and 12 km (20).

Depolarization is another problem facing engineers of microwave communications and radar. Here the shape, degree of orientation and alignment of the hydrometeors is important. Hence CDRR is particularly suited to investigating these problems. Radars for air surveillance frequently use circular polarization to reject clutter from rain, and depolarization of the radar transmission can limit the effectiveness (Hendry and McCormick (42)).

DLPR, although it does not determine as many parameters as CDRR, may also be used to investigate depolarization effects qualitatively by detecting highly distorted horizontally aligned hydrometeors.

SUMMARY

This review has attempted to cover the wide range of multiple-parameter radar techniques which are currently in varying stages of development. It is inevitable that some aspects of MPR have been omitted, but it is hoped that the extent of the activity and interest in this developing field has been conveyed.

REFERENCES

*Note SMPRMP refers to the Pre-Prints of the URSI-IEE Symposium on Multiple-Parameter Radar Measurements of Precipitation, Bournemouth, England.

1. Battan, L.J., 1973, "Radar Observations of the Atmosphere", University of Chicago Press.

2. McCormick, G.G., and Hendry, A., 1975, Radio Sci., 10, 421-434.

3. Hendry, A., and Antar, Y.M.M., 1982, "Precipitation particle identification with centimetric-wavelength dual-polarization radars", SMPRMP.

4. Boerner, W-M, Al-Arini, M.B., 1981, "Utilization of the optimal polarization concept in radar meteorology", Pre-Prints 20th Conference on Radar Meteorology, Amer. Met. Soc., Boston, U.S.A..

5. Metcalf, J.I., and Echard, J.D., 1978, J. Atmos. Sci., 35, 2010-2019.

6. Pasqualucci, F., Kropfli, R.A. and Moniger, W.R., 1982, "A polarization-diversity, 35 GHz meteorological Doppler radar, SMPRMP.

7. Schrott, A., Meischner, P., and Schuster, H., 1982, "Technical concept of the planned agile coherent polarization-diversity radar for the DFVLR", SMPRMP.

8. Seliga, T.A., and Bringi, V.N., 1976, J. Appl. Met., 15, 69-76.

9. Goddard, J.W.F., Cherry, S.M., and Bringi, V.N., 1982, J. Appl. Met., 21, 252-256.

10. Goddard, J.W.F., and Cherry, S.M., 1982, "The ability of dual-polarization radar (co-polar linear) to predict rainfall rate and microwave attenuation", SMPRMP.

11. Pruppacher, H.R., and Pitter, R.L., 1971, J. Atmos. Sci., 28, 86-94.

12. Cherry, S.M., Goddard, J.W.F., and Ouldridge, M., 1982, "Simultaneous measurements of rain by airborne distrometer and dual-polarization radar", SMPRMP.

13. Seliga, T.A., Bringi, V.N. and Al-Khatib, H.H., 1981, J. Appl. Met., 20, 1362-1368.

14. Cherry, S.M., Goddard, J.W.F., and Hall, M.P.M., 1981, Ann. des Telecom., 36, 33-39.

15. Seliga, T.A., Aydin, K., and Bringi, V.N., 1982, "Behaviour of the differential reflectivity and circular depolarization ratio radar signals and related propagation effects in rainfall", SMPRMP.

16. Dissanayake, A.W., Chandra, M. and Watson, P.A., 1982, "Theoretical prediction of differential reflectivity using dual-polarisation radar for various types of hydrometeors", SMPRMP.

17. Ulbrich, C.W., and Atlas, D., 1982, "Assessment of the contribution of differential polarization to improved radar rainfall measurements", SMPRMP.

18. Hall, M.P.M., Cherry, S.M., Goddard, J.W.F., and Kennedy, G.R., 1980, Nature, 285, 195-198.

19. Aydin, K., Seliga, T.A., and Bringi, V.N., 1982, "Differential radar scattering properties of hail and mixed phase hydrometeors", SMPRMP.

20. Leitao, M.J., and Watson, P.A., 1982, "Application of dual linearly polarized radar data to prediction of slant-path attenuation at 10-30 GHz", SMPRMP.

21. Cherry, S.M., and Goddard, J.W.F., 1982, "The design features of dual-polarization radar which affect the accuracy of measuring differential reflectivity", SMPRMP.

22. McGuinness, R., Holt, A.R., and Evans, B.G., 1982, "Some comments on the accuracy of using dual-polarization radar to estimate raindrop size distributions", SMPRMP.

23. Bringi, V.N., Cherry, S.M., Hall, M.P.M., and Seliaa, T.A., 1978, "A new accuracy in determining rainfall rates and attenuation due to rain by means of dual-polarization radar measurements", Antennas and Propagation, IEE Conference publication, 169.

24. Seliga, T.A., and Mueller, E.A., 1982, "Implementation of a fast-switching differential reflectivity dual-polarisation capability on the CHILL radar: first observations", SMPRMP.

25. Stapor, D.P., and Pratt, T., 1982, "Dual-polarized radar for propagation research", SMPRMP.

26. Eccles, P.J., and Atlas, D., 1970, "A new method of hail detection by dual-wavelength radar", Pre-Prints, 14th Radar Meteorology Conf., Amer. Met. Soc., Boston.

27. Eccles, P.J., and Atlas, D., 1973, J. Appl. Met., 12, 847-856.

28. Eccles, P.J., and Mueller, E.A., 1971, J. Appl. Met., 10, 1252-1259.

29. Eccles, P.J., 1979, IEEE Trans. Geos. Elect., GE-17, 205-218.

30. Rinehart, R.E., Tuttle, J.D., 1982, "Dual-wavelength processing - the effects of mismatched antenna beam patterns", SMPRMP.

31. Eccles, P.J., 1982, "The accuracy required for dual-wavelength radar analysis of observations", SMPRMP.

32. Barge, B.L., and Humphries, R.G., 1980, "Identification of rain and hail with polarization and dual-wavelength radar", Pre-Prints, 19th Conf. on Radar Meteorology, Amer. Met. Soc., Boston.

33. Atlas, D., Srivastara, R.C., and Sekhon, R.S., 1973, "Doppler radar characteristics of precipitation at vertical incidence", Rev. Geophys. Space Phys., 11, 1-35.

34. Hauser, D., and Amayenc, P., 1982, J. Appl. Met., 20, 547- 555.

35. Hauser, D., and Amayenc, P., 1982, "Raindrop size-distributions and vertical air motions as inferred from zenith pointing Doppler radar with the RONSARD system", SMPRMP.

36. Pasqualucci, F., 1982, Drop-size distribution measurements in convective storms with a vertically pointing 35 GHz Doppler radar", SMPRMP.

37. Browning, K.A., 1981, "Maximising the usefulness of rainfall data from radars", Papers of the Workshop/Seminar on Weather radar at Reading, U.K., COST-project 72, Brussels.

38. Atlas, D., and Thiele, O.W., (Editors), 1981, "Precipitation Measurements from Space", Report of Workshop, NASA Goddard Space Flight Center.

39. Carbone, R.E., Serafin, R.J., Frush, C., Bringi, V.N., and Seliga, T.A., 1982, "CP-2 radar developments", SMPRMP.

40. Hendry, A., and McCormick, G.C., 1976, J. Geophys., 81, 5353-5357.

41. Goddard, J.W.F., and Cherry, S.M., 1982, "Site diversity advantage as a function of spacing and satellite elevation angle, derived from dual-polarization radar data", SMPRMP.

42. Hendry, A., and McCormick, G.C., 1974, Electron. Lett., 10, 165-166.

THE FEDERAL AVIATION ADMINISTRATION WEATHER RADAR RESEARCH AND DEVELOPMENT PROGRAM

D.E. Johnson

Federal Aviation Administration, USA

INTRODUCTION

Real time weather information for Federal Aviation Administration (FAA) air traffic control (ATC) uses currently comes from ATC surveillance radars. Existing surveillance radars have three severe limitations for weather detection uses. First, the radars are optimized for aircraft detection. This degrades the weather detection capability due to the fact that: (a) the sensitivity time control is set for aircraft detection instead of weather, (b) the moving target indicator eliminates the low radial velocity weather returns, and (c) use of circular polarization reduces the level of the weather signals detected by the radar.

Second, the fan beam antenna (4° to 5° elevation, 1° to 1.4° azimuth) of the surveillance radars causes inaccuracies in weather indications due to lack of beam filling by the weather cells and because no altitude information can be derived from the radar returns. Third, the surveillance radars provide only reflectivity information whereas Doppler information has been shown to provide significant improvement in hazardous weather detection capability.

In order to overcome these limitations, the FAA is conducting research and development activity focusing on the use of Doppler radars for weather detection and is also planning enhanced weather detection capabilities for future ATC surveillance radars.

FAA WEATHER RADAR REQUIREMENTS

The FAA's weather radar requirements tend to be very demanding in terms of basic radar technology, radar signal processing, and weather product generation and display. This is true because of both the pilot's needs and the air traffic controller's needs.

The pilot's needs are most critical in the terminal areas. The airport terminal area and its approach and departure routes are relatively small, confined areas and offer limited opportunity for avoiding hazardous weather. This is particularly true for the larger, extremely busy terminals where the airspace is quite crowded. Furthermore, the aircraft operating in such airspace are often operating in modes (slow speed, reduced power, landing gear down, etc.) which do not lend themselves to recovering from the effects of encountering severe weather.

With respect to the controller's needs, the weather radar products are used by air traffic control personnel who are not meteorologists and who have limited training in using meteorological information. Furthermore, the air traffic control specialist's primary function is air traffic control, not meteorology. The result is that he has little time to spend analyzing weather information.

These factors taken together require that for the weather products to be useful to the air traffic control specialists, they must meet several critical criteria: They must be presented automatically in a simple clearly understandable format. They must present basically yes or no type information which the controller at a glance can understand. They must be available essentially in real time especially for airport terminal applications. They must be easily put on the controller's display along with other ATC information, meaning they must not clutter the display or obscure aircraft target information. Finally, they must be easily correlatable with the aircraft position information on the display and must be in a form that is easily describable to pilots when the need arises.

NEXRAD

The majority of the FAA's weather radar research and development (R&D) activities are in support of the Next Generation Weather Radar (NEXRAD). NEXRAD is a joint FAA, National Weather Service (NWS), and U.S. Air Force Air Weather Service (AWS) Program to develop a weather radar with the goal of satisfying the weather radar needs of all three participants. The NEXRAD Program is being managed by a Joint System Program Office (JSPO) which is headed by a program manager and three deputy program managers representing FAA, NWS, and AWS (Bonewitz (1)).

The NEXRAD JSPO, with inputs from the three participating agencies, has identified 26 technical areas which require some development before the specific capabilities for NEXRAD can be established (1) (JSPO (2)). Of those 26 areas, the nine listed in table 1 are the ones believed critical to meeting the FAA's weather needs. Of these nine critical areas, the first four are related to either basic radar technology or the mode of operation of the NEXRAD radars while the remaining five areas are related to the processing and use of the weather information.

The FAA's weather radar R&D is concentrated in these nine critical areas because it is believed such concentration will provide the maximum benefit both to the technology requirements of the NEXRAD program and to the specific weather radar needs of the FAA.

TABLE 1 - Technical areas critical to FAA weather needs

Basic Radar Technology/Mode of Operation
- Data Acquisition Rate Needs and Strategies
- Clutter Suppression Techniques
- Data Compaction and Transmission Techniques
- Multiple Radar Mosaicking Techniques

Processing/Use of Weather Information
- Turbulence Analysis Techniques
- Thunderstorm Analysis Techniques
- Interpretive Techniques
- Operator/Machine Interface Techniques
- Fine Line Analysis Techniques

RESEARCH AND DEVELOPMENT ACTIVITIES

The FAA weather radar research and development activities are managed by the Systems Research and Development Service (SRDS) located at FAA headquarters in Washington, D.C. Because of the number and complexity of the areas of investigation, and due also to the particular expertise of the performing organizations, five separate groups are involved in the weather radar research and development. While each organization generally is performing separate activities, there is some overlapping of efforts where warranted due to the difficulty of the task or for efficiency reasons. Specific examples of overlapping activities include data collection, ground clutter filtering, and product display. The activities of each organization will be briefly discussed.

FAA Technical Center

Data Collection. The general approach used by the FAA to support weather radar research and development is to collect live data on severe weather and to use that data, supplemented by existing data bases to explore and characterize those weather radar parameters necessary to meet FAA needs.

The primary data gathering activities are taking place at the FAA Technical Center near Atlantic City, New Jersey, and at the National Severe Storms Laboratory (NSSL) in Norman, Oklahoma. The normal procedure is to fly instrumented aircraft through areas of turbulence while simultaneously observing the same turbulence area with a Doppler radar. Data is recorded on the aircraft for later correlation with the data recorded at the same time from the radar on the ground. Table 2 lists the ground instrumentation at the Technical Center (Jelatis and Drury (3)).

The FAA instrumented a Grumman Gulfstream turbo prop aircraft in 1980 for use in the severe weather data collection activities at the Technical Center. Table 3 is a complete listing of the aircraft instrumentation (3). The first use of the collected data is to positively establish the correlation between the Doppler radar signals returned from storm cells and the actual turbulence as experienced by the penetrating aircraft. Data gathered during the 1977 through 1979 storm seasons at NSSL indicate that there is correlation between Doppler radar returns and turbulence; however, before dependence can be put upon these returns for use in critical air traffic control (ATC) applications it is essential that the exact relationship between the turbulence and the Doppler returns be defined (JDOP Staff (4)). Lincoln Laboratory (LL) has performed an analysis of this relationship (Labitt (5)). The second use of the data is to provide a data base to support the other activities listed in Table 1.

Wind Shear and Wake Vortex Detection. Using basically the same pulsed Doppler radar test bed as is being used for the data collection described in the previous section, the Technical Center has also been conducting experiments designed to detect low level hazards in the airport environment. Specifically, wind shear and aircraft generated vortex detection activities have been undertaken.

For the wind shear investigations, instrumented aircraft were flown on simulated approach and departure paths while the radar antenna remained at a fixed azimuth and elevation for an extended dwell time in order to provide inputs to a 128 point Fast Fourier Transform (FFT) program. Data recorded on the aircraft was compared to that obtained from the radar. A limited amount of wake vortex investigations have been conducted by attempting to observe vortices in the proximity of aircraft flying controlled patterns near the radar. The results of these wind shear and wake vortex experiments have been reported by Offi, et.al. (6).

TABLE 2 - Doppler radar data collection ground instrumentation

Turbulence Sensing Radar	Station-Keeping Radar
ASR-8 Transmitter	ASR-7 Transmitter
Pencil-Beam Antenna	Fan-Beam ASR Antenna
1.6° Beamwidth	1.2° AZ Beamwidth
15-foot Diameter Paraboloid	csc^2 EL Beam
Selectable Antenna Elevation	
AZ Scan - Variable	14.5 RPM AZ Scan - Fixed
0.6 us Pulselength	0.833 us Pulselength
1.4 Mw Peak Power	0.5 Mw Peak Power
Receiver modified to be "Coherent-on-Receive"	Receiver: "Coherent-on-Receive"
Display: PPI format with selectable parameter alphanumeric readout	Display: PPI Analog for storm and aircraft (skin and beacon) returns

TABLE 3 - Aircraft instrumentation

Parameter	Sensor	Readout
DP - Delta Pitot	Pitot tube to differential pressure transducer	100/second
IAS - Indicated Air Speed	Pitot tube through low-pass filter	1/second
TAS - True Air Speed	Air-data computer with various sensor inputs	1/second
BALT - Barometric Altitude	Ambient pressure transducer	1/second
OAT - Outside Air Temperature	Resistance-wire probe/bridge	1/second
DP - Dew Point	Probe/computer	1/second
CG - Center of Gravity Vertical Motion	Vertical accelerometer	20/second
ITS - Universal Indicated Turbulence Sensor	CG sensor/computer	1/second
TOD - Time-of-day	Digital clock	1/second
Lat/Long	Inertial Navigation System	1/second

Surveillance Radar Separate Weather Channel. The FAA currently uses weather radar information from surveillance radars. Since these radars are optimized for aircraft detection, the weather information is degraded by the use of the sensitivity time control (STC) setting, the moving target indicator (MTI), and the use of circular polarization (CP) all of which reduce the weather detection capability of the radar. The weather detection capability is further reduced by the fan beam antenna pattern which causes inaccuracies due to lack of beam filling by weather cells and due to lack of altitude information on the weather. Finally, these radars provide only reflectivity information.

In recognition of the probability that dedicated weather radars covering all desired airspace will not be economically feasible, the FAA is planning to specify a separate receive-only weather channel for the newest terminal surveillance radar, the ASR-9. This will permit the STC to be optimized for weather detection. This will considerably improve the weather detection capability of the ASR-9, but will leave the deficiencies due to the fan beam and the lack of Doppler information.

Consideration is also being given to retrofitting some existing terminal surveillance radars with a separate weather channel. It is probable that weather information from surveillance radars will be used to supplement NEXRAD weather information for FAA uses, in some cases being mosaicked with the NEXRAD information and in other cases, being used instead of NEXRAD information.

MITRE Corporation

Mosaicking. The preferred method of displaying weather radar information for air traffic control uses is to have it on the same display as surveillance radar information. Obviously, the weather information must neither degrade the surveillance information nor interfere with the air traffic control function. The task of weather radar display is complicated by the fact that weather data from several different weather radars and possibly from surveillance radars as well must be mosaicked onto the display while maintaining the proper registration and alignment with video maps and aircraft position information.

MITRE Corporation is using data remoted from NWS radars at Atlantic City, New Jersey, and Patuxent River, Maryland, to display with surveillance radar data from the Washington Air Route Traffic Control Center. The effort includes data compaction, contouring, mosaicking, and merging of the weather radar data and the surveillance data onto a single display. The display work is being done using both separate weather displays and plan view displays of the type used by air traffic controllers (Gasper (7)).

National Severe Storms Laboratory

Data Collection. The data collection activity at NSSL is similar to that described previously at the FAA Technical Center. The FAA sponsored the use of an instrumented T-28 aircraft for storm penetration flights in the Spring of 1980 and has also participated in data collection activities using the National Aeronautics and Space Administration's lightning research aircraft, an F-106.

Among the advantages of data collection at NSSL is the fact that dual Doppler data can be collected at different aspect angles from two 10 cm radars, one located at Norman, Oklahoma, and one at Cimmaron, Oklahoma.

Scan Strategies. In order to provide adequate warning of any threatening weather in the airport environment, the FAA believes that the full volume coverage in the airport terminal area (e.g.: 0 to 20°) must be updated approximately every 2.5 minutes. This is based on data gathered by NSSL and others which show that thunderstorm cells can develop in intensity as rapidly as 10 dBZ per minute and that severe turbulence may be encountered at dBZs of 30 and greater (Rempfer, et. al. (8)).

NSSL has been analyzing Doppler weather radar data on severe storms to document weather cell profiles to determine weather radar update rate requirements. These analyses include growth rates, lifetime, and decay rates for severe weather cells.

Reflectivity, velocity, and spectrum width information will be examined to determine storm lifetime characteristics and the data analyzed will be both from single Doppler and dual Doppler data bases.

The FAA need to update this terminal area volume on the order of every 2.5 minutes would restrict the operational parameters of the radar. For example, to sample the full volume in 2.5 minutes would place some restrictions on such parameters as the antenna rotation rate, the maximum range, the Pulse Repetition Frequency (PRF), etc. To reduce the impact on radar operational parameters, the FAA has proposed that the full 0-20° volume be sampled each 2.5 minutes. This would be accomplished through an interlaced scanning mode whereby selected elevation angles would be alternately scanned each 2.5 minutes (for example, odd elevation angles on one scan sequence and even elevation angles on the next scan sequence, etc.). The weather presentation could then be updated with each elevation scan, at the end of 2.5 minutes, and/or at the end of 5 minutes as desired by the user. In this manner, the FAA believes the early detection of hazardous weather necessary for ATC purposes can be accomplished. NSSL is collecting data using various scan modes and is examining the data to determine the adequacy of these modes for detecting hazardous weather and for providing the weather updates the FAA requires. NSSL will also be examining the collected data to ensure that the scan modes used will not adversely impact any of the algorithms being used to process data and derive weather products for the FAA, NWS, and AWS.

Ground Clutter. NSSL is examining alternative digital ground clutter filtering techniques to determine the levels of clutter cancellation that can be expected for various types of clutter cancellers. These examinations include such parameters as the suitability of filter techniques for use with batch and uniform pulse train transmissions, the impact of the number of samples on filter performance, the impact of filtering on reflectivity and Doppler products, the impact of antenna rotation rate on filter capability, the depth and width of the filter notch, and the hardware and software required to implement clutter cancellers (Zrnic' and Hamidi (9)).

NEXRAD Interim Operational Test Facility (IOTF)

Product Development/Display. The IOTF is staffed by personnel from the NEXRAD JSPO and from NWS, AWS, and FAA. The primary responsibilities of the IOTF are related to the testing of proposed data processing algorithms, and the generation, display and evaluation of candidate weather products. The IOTF works closely with the three sponsoring agencies to develop and test products which are of common use to all three and also products of special interest to the individual agencies (2).

Lincoln Laboratory

Signal Processing. The most difficult technical challenges to the use of Doppler radar products for aviation weather purposes lie in the area of the signal processing techniques that are required for automatically extracting meaningful information from the Doppler radar returns. Some of these challenges arise from the fact that various operational requirements for the radar conflict with each other while other difficulties are due to natural phenomena, such as ground clutter.

Among the conflicting operational requirements for NEXRAD, for example, are those of maximum Doppler velocity measurement versus maximum range. Since the Doppler velocity,

$$V = \pm \frac{\lambda \, PRF}{4} \quad \ldots\ldots\ldots\ldots\ldots (1)$$

is directly proportional to the pulse repetition frequency (PRF) whereas the range,

$$R = \frac{C}{2 \, PRF} \quad \ldots\ldots\ldots\ldots\ldots\ldots (2)$$

is inversely proportional to the PRF, any maximization of range comes at the expense of Doppler velocity measurements and vice versa. If, for example, a uniform pulse train is employed with a PRF of 1000 providing a maximum unambiguous Doppler velocity of ±30 m/sec, the maximum range is 125 Km. If measurements are required at longer ranges, then multiple trip returns must be used. This requires some method of unfolding the second, and higher order trip returns so that accurate information can be extracted. Another complicating factor with the use of the multiple trip returns is that any ground clutter near the radar which shows up in the first return period would also show up in each of the multiple trip returns, occupying ever larger portions of the area for each higher period return. The ground clutter would also remain at a constant level in each return period and would be competing with relatively much weaker weather signals in each succeeding period due to the longer ranges.

Lincoln Laboratory (LL) has established a 10 cm Doppler radar test bed and is attempting to resolve the velocity and range ambiguities by investigating techniques for automatically and accurately measuring parameters (reflectivity, velocity, spectrum width) in one interval while eliminating interference from other ambiguous intervals (Laird (10)). Also, along with NSSL, LL is investigating techniques for separating the reflectivity and Doppler information from the contaminating ground clutter returns.

Data on different types of clutter environments are being collected and analyzed to determine the unique spectral and amplitude profiles. Various types of filters are being evaluated to determine losses introduced into the desired weather information (e.g.: the reflectivity, velocity, and spectrum width of weather returns) (Anderson (11)).

The evaluations of techniques for dealing with the velocity and range ambiguities as well as the ground clutter filtering are being carried out both through simulations and through actual hardware and software implementations. The most promising of these techniques will be incorporated into LL's Doppler weather radar test bed for further testing, evaluation, and refinement.

Subject to funding availability, the FAA plans to have LL build a transportable Doppler radar test bed which can be used to investigate and resolve FAA-peculiar weather problems for both terminal and en route environments. Following operation in the Boston environment, the test bed will be moved to two or more other locations so that operation can be conducted under varying geographic and meteorological conditions and so that various ATC environments can be experienced along with exposure to as many ATC personnel as is practical.

As experience is gained with the test bed in the various environments, the radar signal processing parameters will be refined to optimize the test bed performance and to identify those signal processing characteristics best suited for weather radar operation. Continuing feedback of technical information will be made from the test bed activities to the NEXRAD project to assist in the joint radar development activity.

It is envisioned that the test bed will eventually be installed at the FAA Technical Center where it will continue to support FAA hazardous weather research activity.

SUMMARY

The FAA weather radar research and development program is oriented toward providing real time, accurate weather radar information for air traffic control uses. Development activities are being conducted by multiple organizations and concentrate on those technical needs: (a) which are considered most critical to the FAA, (b) which have been identified by the NEXRAD JSPO as most needing research and development, and (c) which the FAA feels most capable of investigating. In parallel with the NEXRAD related research and development, the FAA is improving the weather detection capability of ATC surveillance radars to supplement the information from the network of NEXRAD radars.

REFERENCES

1. Bonewitz, D.J., 1981: The NEXRAD Program -- An Overview. 20th Conference Radar Meteorology, Boston, Amer. Meteor. Soc.

2. NEXRAD Joint System Program Office, 1981: Next Generation Weather Radar Research and Development Plan, December 1981.

3. Jelatis, J.G., and W.H. Drury, 1982: Storm Turbulence Measurement Program Rationale and Instrumentation, ATC Project Memorandum No. 46PM-WX-0004, Lincoln Laboratory, Massachusetts, Institute of Technology, March 1982.

4. Staff of JDOP, 1979: Final Report on Joint Doppler Operational Project (JDOP) 1976-1978. NOAA Tech. Memo. NSSL-86, March 1979.

5. Labitt, M., 1981: Coordinated Radar and Aircraft Observations of Turbulence, Report No. FAA-RD-81-44, Lincoln Laboratory, May 1981

6. Offi, D.L., W. Lewis, T. Lee, 1982: Detection of Hazardous Meteorological and Clear-Air Phenomena with an Air Traffic Control Radar. International Conference Radar-82, London, Institution of Electrical Engineers.

7. Gasper, M.I., 1982: Weather Radar Line Contours, Report No. MTR-81W304, MITRE Corporation, March 1982.

8. Rempfer, P., J. Kuhn, and L. Stevenson, 1981: NEXRAD Radar Scanning Strategy Study, Report No. DOT-TSC-FAA 288-PM-81-67, U.S. D.O.T., Transportation Systems Center, December 1981.

9. Zrnic', D.S., and S. Hamidi, 1981: Consideratons for the Design of Ground Clutter Cancellers for Weather Radar, Interim Report DOT/FAA/RD-81/72, U.S. D.O.T., Federal Aviation Administration, February 1981.

10. Laird, B.G., 1981: On Ambiguity Resolution by Random Phase Processing. 20th Conference Radar Meteorology, Boston, Amer. Meteor. Soc.

11. Anderson, J.R., 1981: Evaluating Ground Clutter Filters for Weather Radars. 20th Conference Radar Meteorology, Boston, Amer. Meteor. Soc.

LAND CLUTTER STUDY: LOW GRAZING ANGLES (BACKSCATTERING)

J.W. Henn*, D.H. Pictor*, A. Webb**

*British Aerospace Dynamics Group, UK, ** RSRE, Malvern, UK

List of Symbols

σ_o (backscatter) cross-section per unit area

θ_d Depression angle of the radar

θ_g Grazing angle at the surface, viz., the angle between the incident ray and a plane surface

θ_i Angle of the incident wave as measured from the vertical

λ Wavelength of incident radiation

f Frequency of incident radiation.

1. INTRODUCTION

One recent trend in modern missile technology is towards low flying low radar cross-section weapon systems. In order to obtain the maximum probability of detecting these weapons by a ground based radar, the signal processing associated with any radar system must be designed to give a large clutter rejection factor, and, the radar must be positioned to give the best possible performance. In order to achieve these objectives knowledge of clutter characteristics is essential, and, as the particular threat under consideration is low flying, of particular relevance is the clutter characteristics near grazing incidence.

One of the important parameters used in clutter modelling is the radar cross-section. Often used to describe clutter returns is the normalised cross-section per unit area, σ_o. The use of σ_o tacitly implies that the returns received are caused by a large number of scattering mechanisms distributed uniformly throughout the physical area illuminated by the radar. Ground terrain is by no means uniform. Undulations in the terrain, trees and buildings lead to shadowing effects. Buildings and other man-made effects act as discrete sources. The effects of non-uniformities become more pronounced as the grazing angle is decreased and the cell to cell variation in the observed radar cross-section increases (Linell (1)). In spite of these problems σ_o is widely used (e.g. Hayes and Dyer (2)) and the results presented in this paper concentrate on this parameter.

Section 2 summarises surveyed results (both theoretical and empirical) whilst Section 3 discusses a promising way to model ground clutter using a digital data base.

2. SURVEY OF PAPERS ON LOW GRAZING ANGLE CLUTTER

2.1 Different Clutter Models

One of the fundamental problems when dealing with clutter modelling is the definition of what comprises a clutter model. Table 1 summarises some possible clutter models.

2.2 Simple Clutter Models

Based upon empirical measurements the model is presented either as a set of curves or as a mathematical formula obtained from the empirical data. The average value of σ_o, obtained by Hayes and Dyer (2) is given by:

$$\sigma_o = -c_1 + c_2 \log_{10}\left(\frac{\theta_g}{\theta_o}\right) - c_3 \log_{10}\left(\frac{\lambda}{\lambda_o}\right) \qquad (1)$$

$$0.3 \text{ cm} \leq \lambda \leq 3 \text{ cm}$$

and $$0^o \leq \theta_g \leq 45^o$$

The constants used are shown in Table 2.

Trebits (3) also produced a mathematical relationship for dry trees and grass fields.

$$\sigma_o = -20 + 10 \log_{10}\left(\frac{\theta_d}{25}\right) - 15 \log_{10}\lambda \qquad (2)$$

Equation used between 0.3 cm $\leq \lambda \leq$ 3 cms.

Equation is valid for angles $\theta_d \leq 45^o$.

Where σ_o is in dB
θ_d is the depression angle in degrees below the horizon
λ is the wavelength in centimetres.

Vertical polarised backscatter is usually 3 to 4 dB higher than horizontally polarised backscatter. The return from wet foliage is about 5 dB stronger than the return from dry foliage. Figure 1 shows results obtained at 9 GHz (∿3 cms) by Hayes and Dyer (2) plus results obtained using equations (1) and (2). Also plotted are results by Krason and Randig (4) at 9 GHz which show a rise in σ_o as grazing is approached. This anomalous behaviour has been noticed by a number of researchers (1, 4, 5) and has been attributed to reflection from vertical objects such as trees or buildings. Figure 2 are results obtained by Hayes and Dyer (2) at 35 GHz. Also plotted on this figure are results obtained by DeLoor et al (6) for crops, which do not show the rise in cross-section as zero grazing is approached. The frequency dependency of σ_o is difficult to define. It depends on incidence angle as well as terrain type. A surface could behave as a smooth surface at 'low' frequencies and a rough surface at 'higher' frequencies. Table 3 shows the frequency dependencies obtained from some of the theoretical scattering models developed.

2.3 Statistical Models

Statistical models are based on measured returns from the countryside. In these models Lognormal or Weibull distributions are fitted to low angle measurements as the radar scans across the countryside. Rayleigh distributions are fitted to the returns from higher angles of incidence or to the returns from one particular cell. (At greater incidence angles the terrain appears more homogeneous). These models produce spatial and temporal probability density functions, spatial and temporal autocorrelation functions and spectral information.

In statistical models the average or median value of σ_o is not usually taken as a function of incidence angle (the angle subtended at the ground varies as the array is scanned). Because of the existence of discrete clutter sources it was found (7, 8) that σ_o varied with cell size. Allan (8) quotes the following values (assuming a lognormal distribution):

a) Average Model

1) Pulse duration $\leqslant 1\,\mu s$ or azimuthal beamwidth $\leqslant$ 0.5 degrees

σ_{omed} = -38 dB standard deviation 12 dB

2) Pulse duration $> 1\,\mu s$ or azimuthal beamwidth > 0.5 degrees

σ_{omed} = -34 dB standard deviation 8 dB

b) Severe Model

σ_{omed} = -25 dB standard deviation = 10 dB

The frequency range where these values apply is 3 to 17 GHz.

Barton (9) quotes σ_o values for severe clutter of -25 dB at 3 GHz and -20 dB at 9 GHz.

The statistical distribution observed will change as the range changes. This is associated with the change in resolution cell size and with the effects of shadowing.

In simple models the average normalised cross-section is kept constant to the radar horizon and then rapidly decreases. The smooth spherical earth model gives the maximum range of clutter as:

$$R_{emax} = (2R_e h_r)^{\frac{1}{2}} \qquad (3)$$

R_e is the 4/3 earth radius (8500 km)
h_r is the height of the radar

A continuous clutter curve developed for mountainous terrain in the United States (Barton (10)) gives σ_o ($\sim$3 GHz) as:

$$\sigma_o = k \frac{R_h^{\,4}}{R_h^{\,4} + R^4} \qquad (4)$$

Where k is a constant $\simeq$ -25 dB worst-case, -30 dB heavy clutter and -45 dB average clutter.
R_h is the radar horizon which varied from 17 km to 35 km
R is the range of the clutter cell.

This model is further refined by Barton (10) using a diffraction approach. For standard deviations use Allan's values above.

2.4 Scattering Models

These models are generally used to calculate the average value of the normalised cross-section, σ_o. They are usually divided into three classes.

a) Semi-Empirical - generally simple models which use empirical constants. These constants are functions of the properties of the surfaces considered.

b) Geometrical Models - employ a surface made up of deterministic or simple shapes. The models are usually analysed by multiple scattering techniques and are chosen to ease the analytical solution of difficult problems.

c) Statistical Models - treats the surface height above the xy plane as a random variable. In order to analyse the problem one must choose a form for the statistical properties of the surface height random variable (eg Gaussian). Results are written explicitly in terms of average surface properties rather than arbritrary constants.

The wavelength and angle dependencies of σ_o for various models is summarised in Table 3. The angular dependency shown in this Table should be compared with the dependencies shown in Figures 1 and 2.

3. CLUTTER MODELLING

3.1 Proposed Method

One of the major problems involved when comparing results is the adequate definition of the 'target' (ground truth) (11). The terrain definitions used to date have tended to be very generalised, for instance (12, 13):

- rural countryside with undulating farmland and villages, rocky mountains, low rolling wooded hills, grassland, forest, urban.

Further unknowns are introduced because the varying weather conditions are not usually taken into account. Even when the clutter cell is well specified (14) there are significant variations in the returns associated with plant and soil moisture content, angle of incidence, row direction (of crops) and even time of the day. Thus the returns from the generalised categories can only be used as a broad estimate of the average σ_o value and the statistical distribution.

It is suggested that potentially the most accurate method of modelling would be using a computer simulation. The terrain would be described by a detailed digital data base; atmospheric and ground conditions would be given by a well defined set of rules/states; scattering models would be used to predict the backscatter from the empirical formula because the theoretical scattering models attempt to take into account the physical processes that are taking place. For instance scattering models will predict how backscatter changes with angle of incidence.

The terrain, meteorological and ground conditions must be matched to a particular scattering model and to the coefficients of that model.

Different scattering models will describe the scatter from different types of terrain surface. The same clutter patch may be described by different scatter models due perhaps to change in plant height (eg corn field) or change in radar frequency. It can therefore be seen that the previous clutter measurements will probably not be directly relevant due to inadequate definition of ground truth and meteorological conditions.

Problems involved in using the proposed method:

- Most scattering models give the average value of σ_0, $\overline{\sigma_0}$, as a function of surface characteristics and incidence angle. Temporal statistics will have to be added to the values obtained.
- Some of the models were not developed for use at very low grazing angles (eg Spetner and Katz (15)) but could be modified comparatively easily (by means of a shadowing function).

3.2 Data Bases Available

3.2.1 Digital Landmass System (DLMS) Data Base (16). Digital maps (Level 1) have been produced covering most of the USA and Europe. Level 1 information is equivalent to maps having scales of 1:150000 to 1:200000. There are limited areas which have been covered to Level 2 data, e.g. Dorset in England. Level 2 data is equivalent to 1:50000 maps.

3.2.2 Second Land Utilisation Survey. The Second Land Utilisation Survey was directed by Miss Alice Coleman of Kings College London during the nineteen sixties. The mainland of Britain was completely covered. This survey involved marking down what was growing in every field and wood in the country. Maps based on the 1:10000 scale Ordnance Survey Maps were used. Figure 3 shows the amount of detail that is available and how the data is presented.

3.3 Computer Simulation

INPUT DATA

General Information

Detailed Digital Map of area under consideration	including geological and cultural features
Scattering Models	basic theoretical information required to calculate clutter characteristics

Specific Information

Radar characteristics	position and height beam characteristics, frequency, radar type etc.
Weather conditions	past and present rainfall wind velocity etc.
Time	season, day.

METHOD

a) Analyse the landshapes of map - establish moisture retention, plant growth and shadowing characteristics - sun and wind.

b) Divide area into rectangular grid - probably 15 to 25 metre spacing.

c) Establish visibility of grid squares to radar.

d) For each grid square observable choose appropriate mathematical model(s) and assign relevant constants.

e) Produce polar plot with radar at centre taking into account beamwidth and pulse length characteristics. Each clutter cell so produced is the sum of the constituent grid squares.

OUTPUT

a) Radar Detection Simulation.

b) Generate accurate statistics - spatial probability distribution functions autocorrelation functions etc.

4. DISCUSSION

Theoretical considerations and empirical measurements must be carried out to test the feasibility of the proposed model. The scattering models used to describe clutter must be fully developed and the terrain and meteorological parameters which influence clutter returns must be identified before proceeding with a large scale measurement programme.

Recent trials by RSRE, Malvern set out to measure different areas of land clutter at low grazing angles using high resolution radars (∿1° azimuth beamwidth, ∿100 ns pulse length) at three frequencies (I, K and M band). The data base is by no means complete due to the limited nature of the trials. The results from the trials together with previous published data may be used with a general terrain description to predict the spatial distribution of clutter. RSRE are also investigating the feasibility of using a digital data base together with either trials information or statistical models with the aim of providing a terrain description sufficient to assess the performance of ground-based radars.

ACKNOWLEDGEMENT

This work was supported in part under contract by the Ministry of Defence.

REFERENCES

1. Linell, T., 1963, "An Experimental Investigation of the Amplitude Distribution of Radar Terrain Return". 6th Conf. Swedish Comm. Sci. Radio Inst. of National Defence, Stockholm, Sweden. Translated paper 1966.

2. Hayes, R.D., and Dyer, F.C.,1973, "Land Clutter Characteristics for Computer Modelling of Fire Control Radar Systems" Engineering Experiments Station, Georgia Inst. of Technology, Contract No. DAAA 25-73-C-0256 AD 912490.

3. Trebits, R.N., Hayes, R.D. and Bomar, L.C., "Millimetre Wave Reflectivity of Land and Sea", Microwave Journal, August 1978, pp 49-53, 83.

4. Krason, H., and Randig, G., 1966, "Terrain Backscattering Chracteristics at Low Grazing Angles for X- and S-band", Proc. IEEE (letters) Vol. 54, pp 1964-1965.

5. Katzin, M., Wolff, E.A. and Katzin, J.C., 1960, "Investigations of Ground Clutter and Ground Scattering", Final Report on Contract AF19 (604)-5198, AFCRC-TR-60-127 Report No. CRC-5198-4 Electromagnetic Research Corporation.

6. DeLoor, G.P., Jurriens, A.A. and Gravesteijn, H., 1974, "The Radar

Backscatter from Selected Agricultural Crops", IEEE Trans., Vol GE.12, No.2 pp 70 to 77.

7. Riley, J.H., 1971, "An Investigation into the Spatial Characteristics of Land clutter at C-band", ASWE Tech. Report TR-71-6.

8. Allan, D., 1980, "Predicting the Performance of Surface Radars in Land Clutter", Aerosystems Engineering Course M.Sc. Report, Loughborough University of Technology and Airforce College Cranwell.

9. Barton, D.K., 1967, "Radar Equations for Jamming and Clutter" Eastcon 1967 Rec. pp 340-55. Reprinted by Barton, D.K. Radars Vol.2, Artech House, Inc. 1977.

10. Barton, D.K., 1980, "Ground Clutter Model", Raytheon Company Missile Systems Division, Bedford Ma, Memo No. 7101-80-159 File No. DKB.80.02. Feb 1980.

11. Currie, N.C., Dyer, F.B. and Hayes R.D., 1973, "Radar Land Clutter Measurements at Frequencies of 9.5, 16, 35 and 95 GHz", Georgia Institute of Technology, Tech Report No.3, Contract DAA-25-73C-0256 AD A012 709.

12. Rigden, C.J., 1973, "High Resolution Land Clutter Characteristics", ASWE Technical Report, TR-73-6.

13. Boothe, R.R., 1969, "The Weibull Distribution Applied to the Ground Clutter Backscatter Coefficient", US Army Missile Command, Technical Report, RE-TR-69-15, AD691109.

14. Ulaby, F.T., 1980, "Vegetation Clutter Model", IEEE Trans. Vol. AP.28, No.4. July 1980.

15. Spetner, L.J., and Katz, I., 1960, "Two Statistical Models for Radar Terrain Return", IRE Trans. Vol. AP-8 May 1960. pp 242-6.

16. 1978, "Digital Landmass System (DLMS) Data Base Definition", Issued by MCERE, Elmwood Avenue, Feltham, U.K.

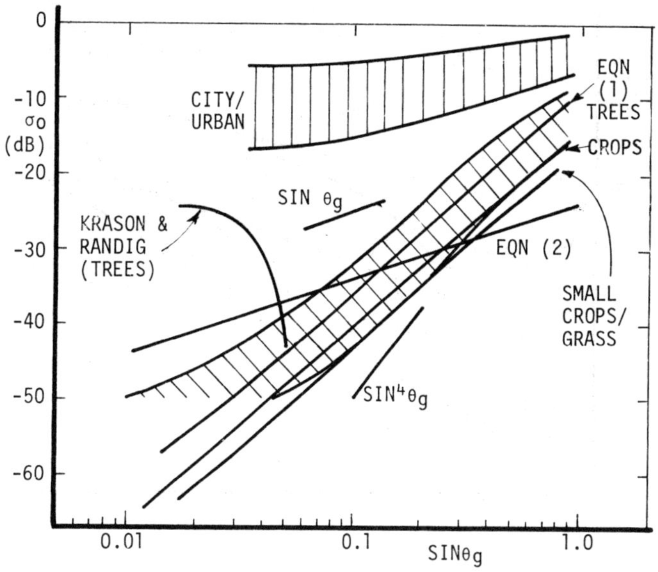

Figure 1 Results obtained at 9 GHz (∿3 cm)

TABLE 2 Table of constants used in equation 1

CLUTTER TYPE (DESCRIPTION)	VALUES OF CONSTANTS c_1 (dB)	c_2 (dB)	c_3 (dB)	θ_0 (deg)	λ_0 (cm)
Trees	11.5	26	8	35	1.0
Crops	16.3	26	8	35	1.0
Grass/small crops	20	26	10	35	1.5
Ploughed ground	31	18	15	25	1.5
Cereal	28	18	15	25	1.5
Snow	25	25	15	30	1.5
Concrete	39.1	32	20	35	2.2
City/urban	6	5	3	30	1.0

TABLE 1 Summary of clutter models

CLUTTER MODEL	CLUTTER CHARACTERISTIC CALCULATED	USED FOR	TYPE OF ANALYSIS
Simple	Average backscatter value of σ_0. Perhaps different value of σ_0 for each type of terrain (measured)	Signal to clutter calculations	Mathematical analysis
	Variation of σ_0 with frequency, angle of incidence and terrain type	Signal to clutter calculations	Mathematical analysis
Statistical	i) Variation of mean or median value (σ_0) with frequency, incidence angle and terrain type ii) Single cell pdf (σ_0) iii) Spatial pdf (σ_0) iv) Range extent of model v) Spatial autocorrelation function	Signal to clutter calculations Detection probability Size of clutter patches	Mathematical modelling Computer simulation
	vi) Temporal characteristics Time autocorrelation spectra	MTI processing pulsed doppler	Mathematical modelling Computer simulation
Scattering	Average backscatter σ_0 calculated as a function of surface characteristics. Surface assumed of a definite form or some sort of statistical distribution.	Signal to clutter calculations	Mathematical/computer analysis

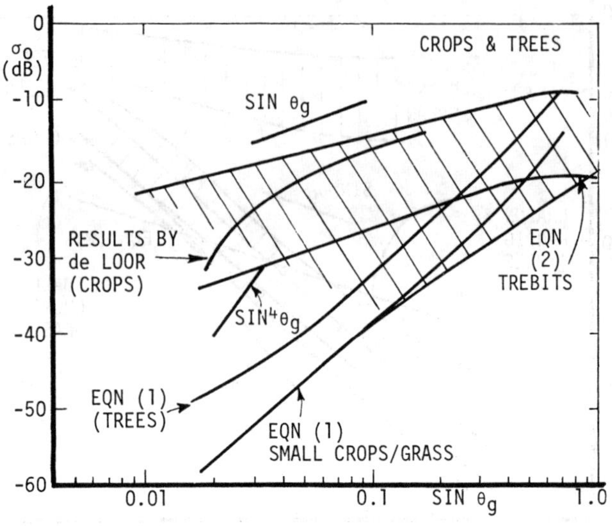

Figure 2 Results obtained at 35 GHz (∿1 cm)

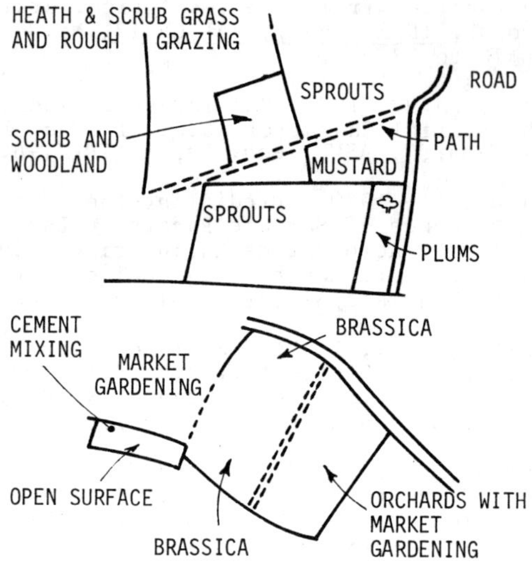

Figure 3 Amount of detail that can be extracted from the second Utilisation Land Survey

TABLE 3 Summary of results

MODEL	ANGLE DEPENDENCE	WAVELENGTH DEPENDENCE	COMMENTS
Clapp 1 - Lambert Law	$\sin^2\theta_g$		
Clapp 2	-		
Clapp 3	$\sin\theta_g$		
Ament (L)	${\theta_g}^2$	λ^{-2}	Semi-infinite vertical wires
Ament (L)	${\theta_g}^2$	λ^{-1}	Wires grouped into halfcylinders
Spetner & Katz - Random Scatterer	-	$\lambda^{-2},\lambda^{-4},\lambda^{-6}$	Angular dependency could be introduced
Spetner & Katz - Specular Point	$\exp(-\cot^2\theta_g/2s^2)$	$\lambda^{-2},\lambda^{-6}$	Depends on spectrum assumed for slopes of surface
Twersky's Semicylinder (L)	${\theta_g}^4$, ${\theta_g}^2$	$\lambda^{-2},\lambda^{0},\lambda^{-3},\lambda^{-1}$	Depends on polarisation and relative radius of hemisphere and λ
Twersky's Hemispheres (L)	${\theta_g}^4$, ${\theta_g}^2$	$\lambda^{-4},\lambda^{-1},\lambda^{-2},\lambda^{0}$	Depends on polarisation and relative radius of hemisphere and λ
Peake's Cylindrical Vegetation	$\cos^2\theta_i$	-	λ dependence not easy to see
Fung & Ulaby Vegetation	-	-	Dependences not straightforward
Statistical Models Peake's Slightly Rough Surface	$\approx\cos^4\theta_i$	λ^{-4}	Incoherent cross-section *Depends on statistics of surface
Rough Surface	$\exp((\frac{1}{s^2}\tan^2\theta_i)/\cos^4\theta_i)$ $\exp((-\frac{\sqrt{6}}{s}\tan\theta_i)/\cos^4\theta_i)$	- -	Depends on statistics of surface independent of λ
Composite Surface Beckmann	$\exp((-\tan^2\theta_i)/\Sigma_j s_j^2)/\cos^4\theta_i$ $\approx\cos^{-4}\theta_i$	λ^2 -	Gaussian statistics Exponential statistics
Brown			Complicated formulations

(L) - Low Grazing Angle Approximations.

MILLIMETER WAVE LAND CLUTTER MODEL

N. C. Currie and S. P. Zehner

Georgia Institute of Technology, USA

INTRODUCTION

As a result of the interest in millimeter wave technology in recent years, a number of measurement programs to characterize targets and clutter have been conducted and a considerable number of reports and papers are in the open literature, particularly for millimeter wave clutter characteristics. However, few attempts have been made to develop a realistic millimeter wave clutter model which could be used for evaluating radar system performance. Another area of interest that has not been addressed, even at microwave frequencies, is the effects of low angle clutter on system performance.

A program to develop a semiempirical model for low angle clutter (< 45° depression angle) at microwave frequencies for use in computer modeling was completed at Georgia Tech in 1979.[1] This paper will extend that model, which was developed for the 1 to 16 GHz frequency regime, to 95 GHz

BASIC CLUTTER MODEL

The clutter model developed at Georgia Tech was based on general assumptions about the dependence of the clutter response on depression angle and surface roughness as derived from previous clutter models. Least square fit curves were generated for primarily 10, 35, and 95 GHz data, and a model equation was developed of the form:

$$\sigma^o = A\,(\gamma + C)^B \text{Exp}\left[-D\,\left(1.0 + \frac{0.1 * \sigma_h}{\lambda}\right)\right] \quad (1)$$

where: γ is the depression angle in radians
σ_h is the RMS surface roughness
A,B,C,D are empirically determined constants

For very rough surfaces ($\frac{\sigma_h}{\lambda} >> 1$) , Equation (1) becomes

$$\sigma^o = A\,(\gamma + C)^B \quad (2)$$

For very smooth surfaces ($\frac{\sigma_h}{\lambda} << 1$) , Equation (1) becomes

$$\sigma^o = A(\gamma + C)^B \text{Exp}\left[-D\right] \quad (3)$$

Thus, by appropriately determining constants A, B, C, and D, the average radar cross section per unit area (σ^o) can be predicted as a function of depression angle γ. In fitting the model to available data, we chose the easier approach of developing a different set of constants (A,B,C,D,) for each frequency band, rather than adding a frequency dependent term to the model. For the clutter types chosen, D was essentially 0 so that the form of the model given by Equation (2) was used to calculate "fits" to the data.

Table 1 summarizes the calculated coefficients for three types of clutter (snow, trees, and grass/crops) for 10, 35, and 95 GHz. These coefficients were calculated from clutter data available primarily from Georgia Tech, the University of Kansas Center for Research, Goodyear Aircraft Corporation, and the Rome Air Development Center.

COMPARISON OF MODEL WITH CLUTTER DATA

The results of the model equation using the constants from Table 1 are plotted in Figures 1 through 8 with the actual data values used to generate the Table 1 coefficients. The data sets were obtained from different sources and include dependencies not accounted for in the experiments; thus, there is considerable spread between the individual points. Thus, the model attempts to describe the "average" behavior of the data. The model results for the three different types of clutter considered here are discussed below.

Snow Model

Figures 1 through 3 show the model results as compared to snow data. Note that two snow conditions are plotted, wet snow and dry snow. It was discovered in 1976 that the snow reflectivity is highly dependent on the amount of free water present in the snow.[2] In particular, when the free water is 2% or less, the reflectivity at low depression angles is 5 to 10 dB higher than for conditions of higher free water content. The effect is most noticeable at 35 GHz, but the effect is evident at both higher and lower frequencies as can be seen in Figures 1 and 3. At 10 GHz the reflectivity for dry snow is approximately 9 dB higher than for wet snow, at 35 GHz the reflectivity for dry snow is approximately 12 dB higher than for wet snow, and at 95 GHz the reflectivity for dry snow is approximately 5 dB higher than for wet snow.

Tree Clutter

Figures 4 through 6 show the model equation prediction plotted with tree clutter data at 10 GHz, 35 GHz, and 95 GHz. The 35 GHz data are plotted for wet (water present on leaves) and dry conditions. The wet data are 2 to 3 dB higher than the dry data. The 35 GHz data are approximately 5 to 10 dB higher than the 10 GHz data at low depression angles but are equal at the plateau region. The 95 GHz data are approximately 8 to 10 dB higher than the 35 GHz data. Also, the 10 GHz and 95 GHz data exhibit a stronger dependence on depression angle than the the 35 GHz data.

Grass Clutter

Figure 7 and 8 show the model equation predictions compared with grass and crop data at 10 GHz and 35 GHz. The data for tall grass are very similar to the tree data in average values, while the short grass data are 5 to 10 dB lower in average values. (No 95 GHz grass data were available for use in generating a model during the development of the coefficients.)

GENERATION OF A STATISTICAL CLUTTER MODEL

The curves for clutter backscatter discussed so far allow one to generate an "average" value for clutter, but are not helpful in evaluating radar performance on a statistical basis. However, a statistical model can be generated by assuming that the average value of the reflectivity is given by Equation (1) and that the variations in the data are due to: 1) a temporal variation which is Rayleigh distributed and 2) a spatial variation (as the radar antenna scans the ground) which is Weibull distributed. The Weibull distribution is given by

$$P(\sigma)\ d\sigma = \frac{b\sigma^{b-1}}{\alpha}\ \text{Exp}\left[\frac{-\sigma^b}{\alpha}\right] d\sigma \quad (4)$$

where $\alpha = \frac{\sigma m^b}{\ln 2}$ (σm = median value)

b = Weibull slope Parameter
(b = 1 is Rayleigh Distribution)

Data at 10 GHz have indicated that the appropriate slope parameter b for low angle clutter is as given in Table 2.[11] If one then picks an average value from Equation (1) and substitutes the value as σ_m in Equation (4) with the appropriate Weibull slope parameter from Table 2, then a set of clutter statistics can be generated. If statistical values were generated for 10 GHz tree clutter, summarized statistically, and overlaid on actual data, the results would be as shown in Figure 9. Figure 9 gives the 10, 30 50, 70, and 90 percentile levels of the cumulative distribution of the statistical model prediction overlaid on tree clutter at 10 GHz. It can be seen that the model-generated data match the actual data very well. Finally, Figure 10 illustrates the effects on the predicted clutter distribution as the depression angle becomes small and the Weibull slope parameter becomes large. The high level tails remain large even though the median values fall off rapidly with depression angle. This phenomenon has been observed in actual data.[12]

SUMMARY

This paper has summarized the characteristics of an average clutter model for snow, trees, and grass baskcatter over the frequency range of 10 GHz to 95 GHz and compared the model predictions to actual data available in the open literature. In addition, a brief description was given of a method for generating statistical data from the average model thus simulating "actual" data that would be viewed by a scanning radar. Comparison of the statistical model results with X-band data show excellent correspondence.

REFERENCES

1. Zehner, S. P. and M. T. Tuley, "Development and Validation of Multipath and Clutter Models for TAC ZINGER in Low Altitude Scenarios," Final Technical Report on Contract F49620-78-C-0121, Georgia Institute of Technology, March 1979.
2. Currie, N. C., F. B. Dyer, and G. W. Ewell, "Radar Millimeter Backscatter Measurements on Snow," 1976 IEEE-APS International Symposium, Amhurst, Massachusetts, October 1976.
3. Stiles, William H. and Fuwwaz T. Ulaby, "Microwave Remote Sensing of Snowpacks," Final Technical Report on Contract NAS5-23777, University of Kansas Center for Research, Inc., June 1980.
4. Hayes, D. T., et al, "Millimeter Wave Backscatter from Snow," 1978 IEEE-APS International Symposium, Seattle, Washington, June 1978.
5. Hayes, R. D., N. C. Currie, and J. A. Scheer, "Reflectivity and Emissivity of Snow and Ground at MM Waves," 1980 IEEE International Radar Conference, Washington D.C., April 1980.
6. Currie, N. C., F. B. Dyer, and R. D. Hayes, "Radar Land Clutter Measurements at Frequencies of 9.5, 16, 35, and 95 GHz," Technical Report No. 3 on Contract DAAA25-C-73-0256, Georgia Institute of Technology, April 1975.
7. Bush, T., et al, "Seasonal Variations of the Microwave Scattering Properties of the Deciduous Trees as Measured in the 1-18 GHz Spectral Range," Final Technical Report on Contract NAS9-10261, The University of Kansas Center for Research, June, 1976.
8. "Radar Terrain Return Study," Goodyear Aircraft Corporation, Final Report on Contract NOas-59-6186-C, GERA-463, September 1959.
9. Currie, N. C., "Characteristics of Millimeter Radar Backscatter From Wet/Dry Foliage," 1979 IEEE-APS International Symposium, Seattle, Washington, June 1979.
10. Stiles, W. H., F. T. Ulaby, and E. Wilson, "Backscatter Response of Roads and Roadside Surfaces," Final Technical Report on Contract DE-AC04-76DPOD789, The University of Kansas Center for Research, January 1979
11. Boothe, R. R., "The Weibull Distribution Applied to Ground Clutter Backscatter Coefficient," Report No. RE-TR-69-15, U.S. Army Missile Command, Huntsville, Alabama, June 1969.
12. Currie, N. C., and F. B. Dyer, "Methods for Comparison of Clutter Processing Techniques," Technical Report No. 5 on Contract No. N00024-68-C-1125, Georgia Institute of Technology, February 1971.

TABLE 1. CALCULATED CLUTTER MODEL COEFFICIENTS

CLUTTER TYPE	FREQUENCY	A	B	C
Wet Snow	10 GHz	0.0246	1.7	0.0016
	35 GHz	0.195	1.7	0.0016
	95 GHz	1.138	0.83	0.0016
Dry Snow	10 GHz	0.195	1.7	0.0016
	35 GHz	2.45	1.7	0.0016
	95 GHz	3.60	0.83	0.0016
Trees	10 GHz	0.062	1.7	0.012
	35 GHz	0.036	0.59	0.001
	95 GHz	3.60	1.7	0.012
Grass/Crops	10 GHz	0.06	1.5	0.012
	35 GHz	0.301	1.5	0.012
Short Grass	10 GHz	0.023	2.2	0.012
	35 GHz	0.125	1.5	0.012

TABLE 2. STATISTICAL MODEL

	WEIBULL SLOPE PARAMETER		
CLUTTER TYPE	TEMPORAL	SPATIAL	
		Above 5°	Below 5°
Snow	0	2	1.5
Grass Crops	1	0	1.5
Trees	1	1	2

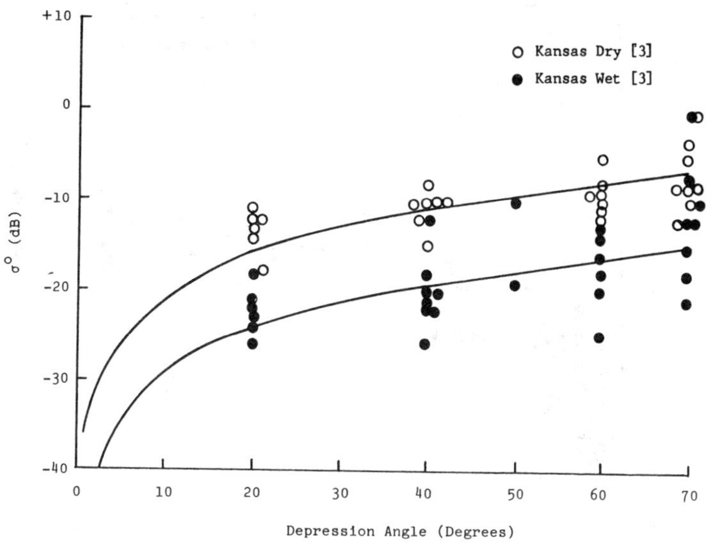

Figure 1. Comparison of Wet/Dry Snow Backscatter Data With Model at X-Band.

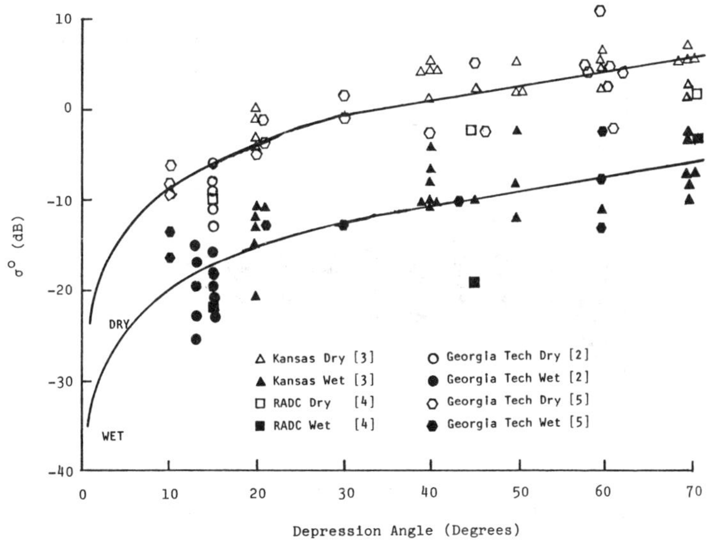

Figure 2. Comparison of Wet/Dry Snow Backscatter Data with Model at 35 GHz.

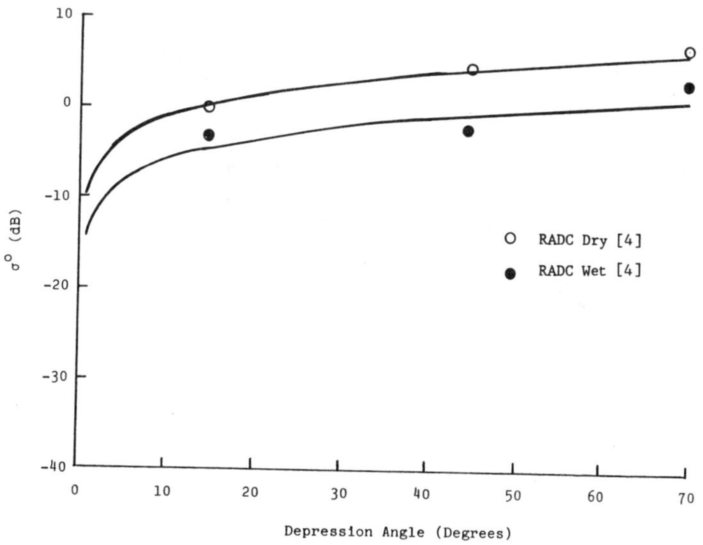

Figure 3. Comparison of Wet/Dry Snow Backscatter Data With Model at 95 GHz.

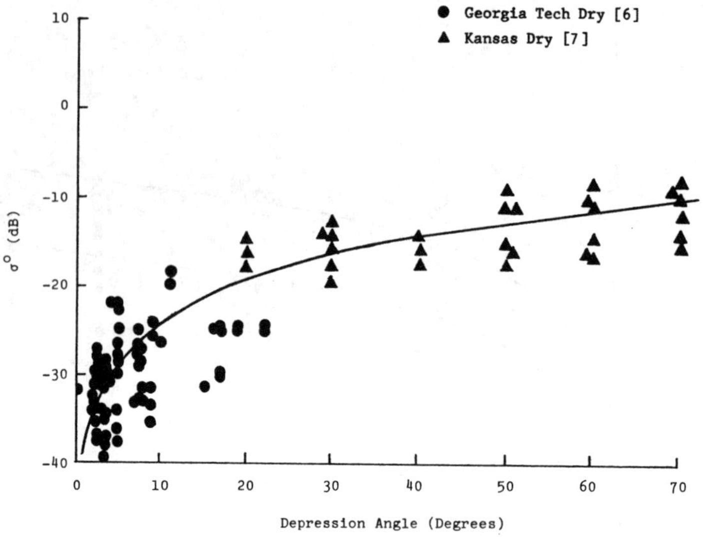

Figure 4. Comparison of Tree Backscatter Data With Model at X-Band.

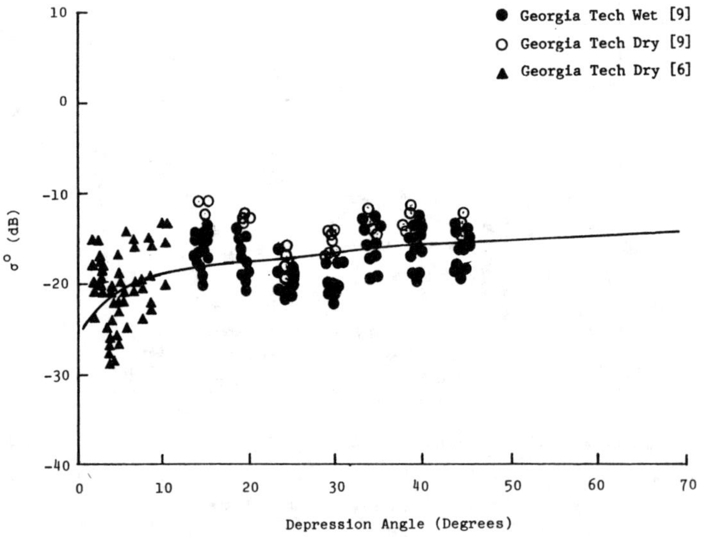

Figure 5. Comparison of Tree Backscatter Data with Model at 35 GHz.

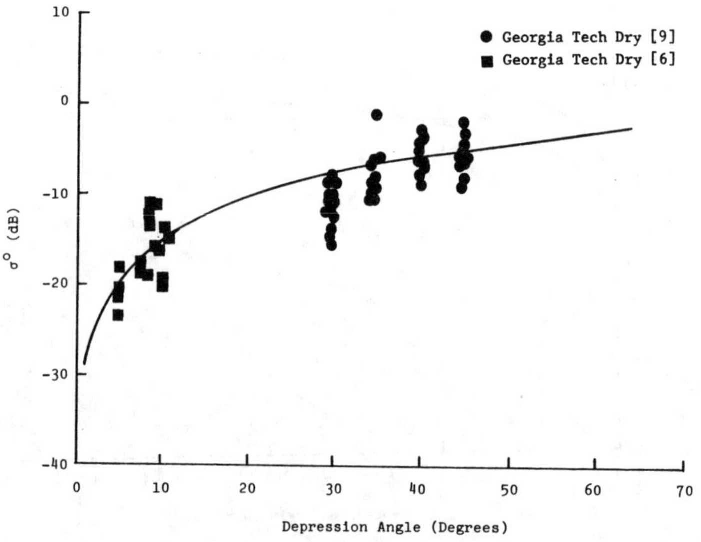

Figure 6. Comparison of Tree Backscatter Data With Model at 95 GHz.

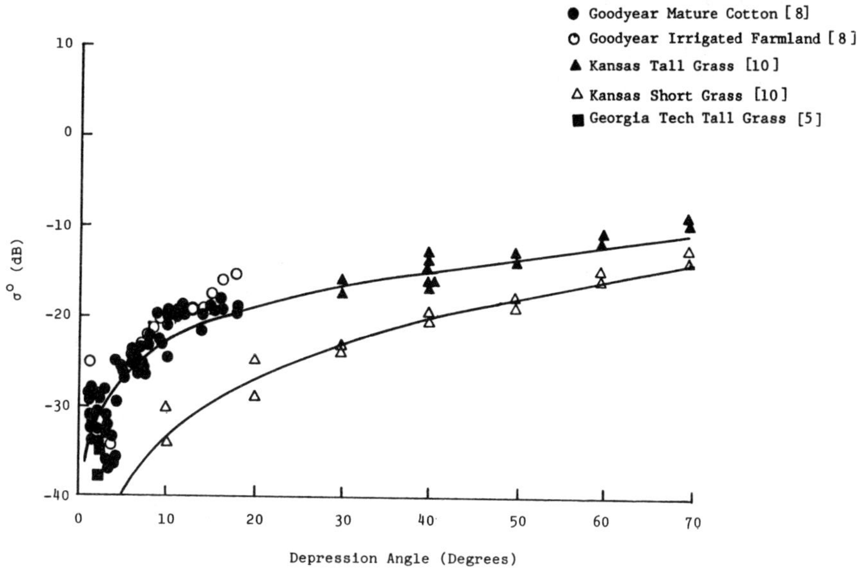

Figure 7. Comparison of Grass/Crops Backscatter Data With Model at X-Band.

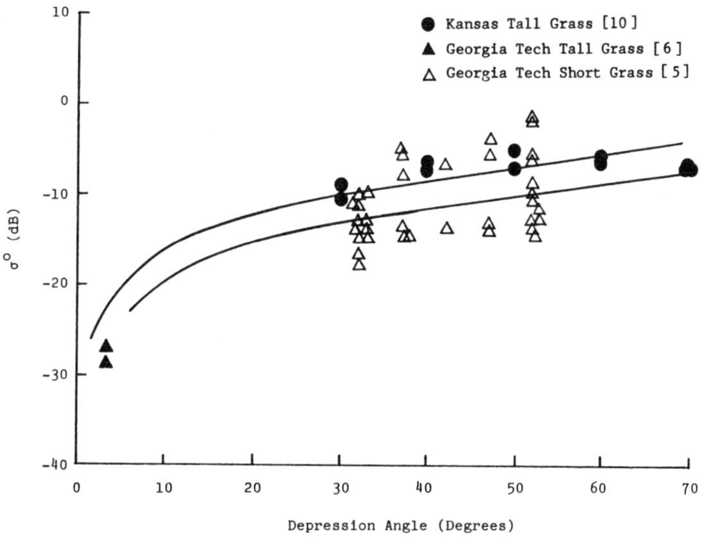

Figure 8. Comparison of Grass Backscatter Data With Model at 35 GHz.

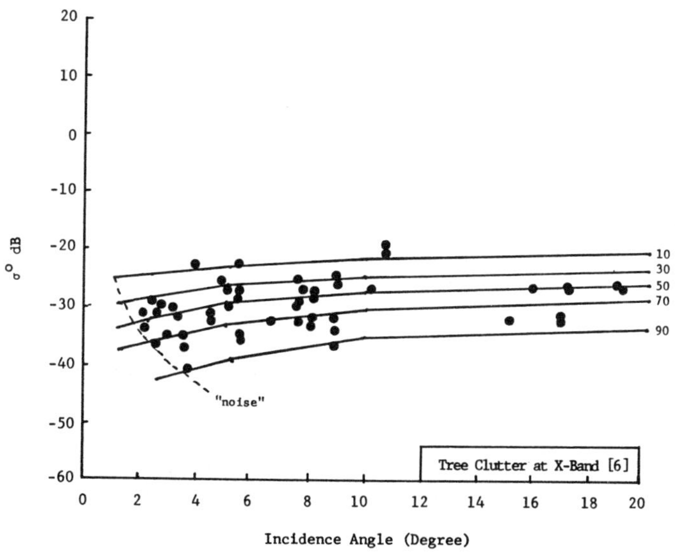

Figure 9. Statistical Model Output Values for 10, 30, 50, 70, and 90% Levels Compared to Tree Clutter Data at X-Band.

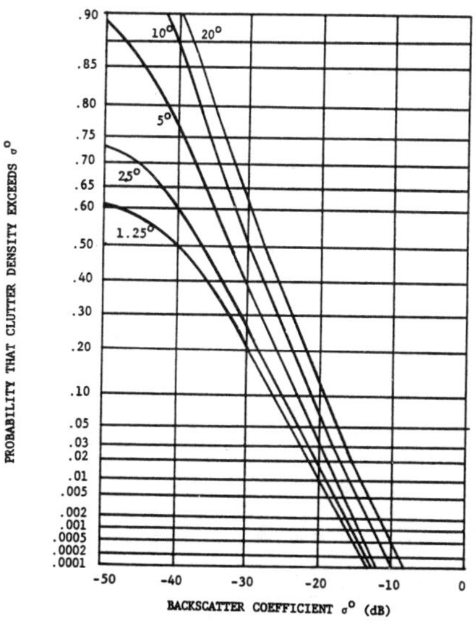

Figure 10. Distribution of σ° Due to Scanning Over Trees at Depression Angles as Indicated (1.25° to 20°) As Predicted by the Statistical Model.

BEAMFORMING FOR A MULTI-BEAM RADAR

J.M. Chambers, R. Passmore, J. Ladbrooke

The Plessey Company, UK

INTRODUCTION

A surveillance radar must be able to resolve targets and measure their angular co-ordinates throughout the desired volumetric coverage. Furthermore the radar may have to operate in heavy clutter and jamming environments. Good angular discrimination is therefore essential and this can be achieved by directing narrow receiving beams, with low sidelobes, into the volume of space illuminated by the radar transmitter. For optimum performance it is important to have flexible management of these beams thus ensuring that each can be matched to its particular part of the coverage. For example, beamwidths may not necessarily be the same at all parts of the coverage, then again sidelobes may be made lower in some sectors at the expense of others less critical.

FORMING BEAMS

Partitioning the radar coverage by generating beams is the function of the antenna and its associated processing networks. Beams may either be formed at the radar's radio frequency, by microwave networks (r.f. beam forming), or by processing samples of the antenna aperture after amplification and frequency down conversion to an intermediate frequency (i.f. beam forming). The simplest example of r.f. beam forming is the parabolic reflector with one or more feeds placed close to the focus. More elaborate versions are well known including linear and planar arrays with microwave beam forming networks of varying complexities.

CHOICE OF TECHNIQUE

The most serious disadvantage of any direct r.f. beam forming matrix is the fundamental restriction on beam overlap. Generally a target will be between the peaks of two adjacent beams and its exact angle must be determined by interpolation. This need, and importance of maintaining a substantially even coverage, forces the beams close together. A cross-over level of -3dB from the beam peaks is typical. Avoiding significant coupling losses and consequent reduction in radar sensitivity requires that the beams be relatively far apart resulting in lower cross-over levels.

Beams generated by a uniform amplitude distribution and therefore with -13dB first sidelobes call for a -4dB cross-over point. In practice beams with distributions giving much lower sidelobes are used in radar applications and these demand much lower cross-over levels.

A practical disadvantage of r.f. multi-beam forming networks is that the required beam format is rigidly designed in with fixed phase and amplitude distributions set by the microwave hardware. The expense, and additional signal loss, of variable weighting elements are both prohibitive, particularly if wide bandwidths or very high microwave frequencies are in use.

I.f. beam forming offers a cost effective means of flexible beam management while making optimum use of the antenna aperture. Essentially each element of a linear or planar array has its own microwave receiver which down converts the received signal to a convenient i.f. typically at a few tens of MHz. Since the signal-to-noise ratio of every antenna sample is set by its own receiver each i.f. output signal may be divided, weighted, and combined with the others to form beams without incurring high losses due to overlap, moreover variable weighting coefficients are cheaper at i.f. than at r.f. Thus relatively inexpensive and flexible electronic beam steering and shaping is possible although close control of the phase and amplitude error distribution of the receivers must be maintained.

EXAMPLE OF I.F. BEAMFORMING

Multiple receiving beams may be formed at i.f. in either or both principal planes of a radar antenna. The following example, used to demonstrate the technique, operates in the elevation plane on a planar antenna having fixed linear feeds in the horizontal plane.

The particular application calls for coverage from 0 to 30° with varying beamwidth, narrow at low angles and wider at the higher angles and peak sidelobes lower than -30dB with respect to peak beam gain.

Achieving the required minimum beamwidth of 3° with a low sidelobe aperture distribution dictates an antenna aperture of around 22 wavelengths. Spacing of the antenna receiving elements was chosen to ensure that grating lobes are suppressed by the receiving pattern of the array-embodied element.

Thirty-two elements 0.7 wavelengths apart were chosen. This spacing satisfies the grating lobe criterion and the quantity 32 allows convenient binary power divider design.

THEORETICAL CONSIDERATIONS

Given the beam formation requirements a suite of computer software was generated to compare Cosine^2 on a pedestal with Chebychev and Taylor illumination functions in the presence of randomly distributed amplitude and phase errors for a broadside beam and a beam squinted to 13.5°.

Beam pointing angle, worst sidelobe level, and 3 dB beamwidth degradation were found to be relatively independent of the chosen illumination function in the presence of errors, and an error budget of 3° phase (r.m.s.) on 0.5 dB amplitude (mean absolute) per channel was established as a resonable design target for a prototype beamformer. It was possible to generate all the beams using the cosine^2 on a pedestal function. In this application the dynamic range of the amplitude weights was lower than that required by the other two functions.

It was proposed that i.f. amplitude weighting be established using π resistive attenuators with phase weights implemented using either:

(1) miniature coaxial delay lines in conjunction with 90 or 180 degree transformers, to minimise frequency dependence to less than that of half a wavelength of line, or

(2) via quadrature splitters using resistive pads to attenuate the in phase (I) and quadrature (Q) components of the signal before recombining them in phase to give the required overall phase shift to the channel.

A second suite of software was developed to model beams by simulating amplitude and phase weights resulting from the techniques described above.

In operation, the aperture excitation coefficients are entered into computer memory and the phase shifters defined either by choosing I and Q pad attenuators from nominal E24 series resistors or deriving the minature coaxial linelength/transformer combinations required. The relative loss per channel due to the phase shifters is then estimated. Amplitude pads are then chosen using the required channel excitation coefficients modified by the phase shifter losses established above. All the component values are printed out, together with amplitude and phase coefficients, and the systematic errors calculated assuming perfect components. The array factor is then calculated and multipled by an externally defined element factor to give graphical output of the beam, and an estimate of the pointing angle and 3 dB beamwidth of the beam printed out.

The program was run for i.f.'s across 10% bandwidth and it was found that although the miniature coaxial line lengths were able to supply nominally better beams than those generated by the I-Q system (typical phase tolerance approximately 1° across 10% bandwidth) at i.f. band centre, beam degradation at the band edges was intolerable due to phase shift with frequency in the coaxial lines.

Although the practical problem of interfacing the subminiature coaxial cable (type UT-34) with the stripline propagation system used in the rest of the beamformer was adequately resolved, it would have been necessary to trim at least 256 delay lines to within ± .5mm to achieve the necessary phase tolerance.

Given the formidable prospect of trying to maintain these tolerances in the production environment and the systematic phase shift versus frequency inherent in the line length technique, it was decided that this phase shift scheme be rejected in favour of the I-Q system for the development of an experimental beamformer.

PRACTICAL CONSIDERATIONS

The requirement for the Beamformer may be summarised as follows. From 32 available receiver signals it is necessary to derive, from a preferably passive compact network, 8 separate fixed beams of specified beam-width, squint and sidelobe level. In addition the required performance should be achieved over a bandwidth of approximately 10%.

The operation of the Beamformer will first be discussed in general terms, followed by the practical considerations which resulted in the final design.

If it is assumed that each of the available inputs contributes to each of the beams then it is necessary to split these inputs into at least as many paths as there are beams. The outputs from this process form samples, and each of these, together with those from all the other input divisions, are essentially passed through parallel channel phase shifting networks. These networks, each one of which corresponds to a particular beam, impress a linearly progressing phase shift (ramp) across their sample signals. The slope of this ramp is appropriate to the squint of the beam which is formed when these signals are combined. Incorporated in the phase shifting channels are attenuators or pads which apply the required amplitude distribution function. Alternatively, the attenuators may be located at each of the input dividers, using either a single unit at the divider input or several at each of the outputs. This first alternative is inflexible and has application only when all beams are formed using identical taper distributions, which naturally implies that all inputs contribute to all beams. This however is not always the case, and the second alternative must be used to give one pad/sample input. Conceptually N input signals, one from each receiver, are sampled, weighted, phase shifted (with a linear phase ramp) and recombined in an N-way summing network. Now in order to generate M beams, the input sampling divider must generate at least M outputs. Neglecting for the moment the phase shifting operation, (N-1) x M + (M-1) x N is the irreducible minimum number of discrete two-way divide/combine operations necessary to achieve formation of all the beams. This is equal to the number of two-way divider/combiner devices required to realise the co-axial delay line Beamformer solution. To form all beams, (and it is assumed again that all inputs contribute to each of them), it is necessary to weight and phase shift all M x N input samples. The I/Q technique has been adopted as the phase shifting method. Each I/Q channel comprises a quadrature divider (or combiner), an in-phase combiner (or divider) and weighting attenuators appropriate to the phase shift required. To achieve all the necessary phase shifts throughout the Beamformer then an additonal 2x M x N devices are required, some of which are quadrature divider/combiners. Although this derivation has really assumed a particular Beamformer arrangement, it can be shown that, whatever the final configuration, the total number of two-way operations (in-phase or quadrature) remains constant. In practice, the cost of quadrature dividers is twice that of the in-phase types, and the material cost of a passive beamformer (of the scale under consideration) is likely to be mostly that of the discrete divider/combiner units used in it. The minimum number of quadrature devices is equal to M, the number of beams, whilst the maximum is M x N, the product of the number of beams and available inputs. Consequently, even at the preliminary stages of design, it is possible to estimate what the minimum cost of an adequate system would be, i.e., that system with the minimum number of quadrature dividers.

Of necessity any system of this kind must incorporate a large number of phase matched signal path interconnections, each one routing sample signals (or resultants of them) to the inputs of successive two-way combiners to form the required beams. The interconnections will be by a combination printed and coaxial line. The latter is unavoidable since somewhere in the system signal paths must physically cross-over. Each screened interconnection must be fabricated separately and is, therefore, a likely source of manufacturing error, (c.f. printed line, say). Clearly then, a preferred beamformer configuration would attempt to use the minimum number of coaxial links, not only to reduce potential errors, but also to ease problems in system manufacture, and reduce size.

The selection of the divider/combiner components, be they quadrature or in-phase, is governed by device performance, system cost and system weight. The relative numbers of each depends on the configuration used, and this is influenced by the need to minimise channel errors, particularly in phase. The performance of individual devices, in terms of amplitude and phase balance, isolation and terminal match are not strictly under the control of the system designers. Quoted figures are usually (necessarily) typical and whilst such qualities as output balance are specified, little, if any information is available relating to, say, insertion phase. This parameter is of particular significant here, since it directly contributes to the absolute phase shift in any one signal path, any one of which will contain ($\log_2 N + \log_2 M + 2$) quadrature and in-phase devices. In practice it has been found that a greater variation in insertion phase occurs with the quadrature devices. These have a greater internal complexity than the in-phase types, and are more liable to the slight variations in construction which result in effects on electrical characteristics.

RECEIVER PHASE AND GAIN CONTROL

During the course of preliminary investigation it became evident that automatic gain and phase control of the receivers would be necessary to maintain the total system errors within the limits derived in the theoretical analysis. Here amplitude and phase error signals are derived at the input to the beamformer, are fed back to voltage controlled attenuators and phase shifters. Thus the amplitude is corrected at the beamformer input and phase in the receiver second local oscillator.

For ease of manufacture it was decided to use a microstrip configuration through the i.f. circuitry, with copper clad fibre-glass board as the substrate material and standard printed circuit photo-etching processes as the method of fabrication. The relatively low cost microstrip/p.c. approach permits the generation of accurately controlled highly repeatable circuit boards. The fibre glass substrate, although not particularly rigid, requires no special drilling or device mounting techniques. A further advantage of this arrangement is that it can be readily developed to allow circuit hybridization, giving significant weight and size reduction.

The basic I/Q phase shifting network comprising the quadrature divider/combiner, in-phase combiner/divider and sine and cosine weighting attenuators permits phase control in one quadrant only. If however, the I and Q channels, and their inversions, are available then four quadrant control is established. This may be achieved by introducing 1:1 isolation transformers in each channel. By suitably grounding one side of the output windings, it is possible to select the quadrant appropriate to the phase shift required. To improve the impedance match of the transformers, shunt connected ceramic capacitors are used. These compensate finite primary and leakage inductances of the windings so reducing the in-channel return loss from 14 dB to about 32 dB. Excess insertion phase of the inverting transformers necessitates the use of non-inverting transformers in the complementary channels where no inversion is required. Failure to redress the quadrature balance in this or a similar way can, in some cases, result in the generation of 10-15° phase errors.

The resistive pads used for the I/Q and taper weighting functions, were formed from π connections of 1 or 2% metal film resistors, derived from E24 preferred value series, and selected to provide a match (return loss > 32 dB) to the nominal 50 ohm system impedance.

The insertion loss of the selected in-phase and quadrature dividers was $\leqslant$ 0.5 dB, with phase balance and quadrature deviations $\leqslant 0.5^\circ$. Amplitude balance of both devices was $\leqslant$ 0.3 dB. The insertion phase variation was 2° for the quadrature devices and negligible for the in-phase types.

For mechanical reasons the beamformer is built in two units, each processing 16 alternate receiver signals, 1 to 31, and 2 to 32. Each unit contains 16 x 8-way input divider boards (incorporating AGC, APC), and 8 x 16-way (half beamforming) boards. The latter comprises 16 parallel I/Q phase shifting channels, each one of which contains a discrete taper pad. The resultant of the 16 weighted, phase shifted channels is formed in a 16-way in-phase combiner, the output of which forms one "½ beam". The input divider boards (16) are mounted horizontally and stacked vertically, whilst the beamforming boards (8) are mounted vertically and 'stacked' horizontally. Coaxial interconnection between the divider/beamforming boards is by means of coaxial connector pairs (128 in each unit). Slide mating connectors are used in preference to the Snap or Coupling Nut types, to reduce mating/unmating forces and to ease assembly. The use of these connectors with this particular circuit configuration provides a simple means of achieving the coaxial signal path interconnections discussed earlier. The mechanical arrangement is also quite compact. One set of beamforming boards (in one of the units) is fitted with an additional in-phase combiner to add corresponding "½ beam" outputs together. The physical links between the two units are by means of pre-formed coaxial cables, the additional phase shifts of which are balanced by compensating delay lines mounted on those beamforming boards fitted with the additional ½ beam combiners.

A MORE COMPACT NETWORK

The resistive weighting networks described earlier can be reduced in size by the use of thick film hybrid technology. A more significant reduction in volume can be achieved by reverting to a 'line-length' network in which the incremental time delays are realised by a series of Surface Acoustic Wave (S.A.W.) tapped delay lines. A practical S.A.W. beamforming network has a configuration similar to that of Maxson and Blass (1). Signals originating from the separate elements of the antenna are launched from a rank of input transducers ranged along one edge of the S.A.W. material. In their parallel propagation across the surface these signals are intercepted and collected by a rank of lightly coupled transducers on a line set at an angle to the input rank. This angle is chosen to import a phase tilt appropriate to the required beam angle. Many such output ranks may be introduced simultaneously at different angles to provide multiple beams.

CONCLUSION

Angular discrimination in a surveillance radar can be achieved by versatile beamforming at the i.f. stage of the receiver.

Beam position, beamwidth and sidelobe levels can be accurately predicted by the design theory even when readily available standard components are used in manufacturing the weighting networks. In practice a beamformer may be modular allowing simple amendments to the beam format by substitution of plug-in circuit cards. Moreover the beam outputs are phase coherent which allows optimum use to be made of the information in the radar's subsequent signal processing.

ACKNOWLEDGEMENTS

The author wishes to thank the Directors of The Plessey Company for their kind permission to publish this paper.

REFERENCES

1. E. Shaw, February 1969, "The Maxson Multi-Beam Antenna: Theory and Design for Non-Interacting Beams", The Radio and Electronic Engineer, 37 No.2, 117-129.

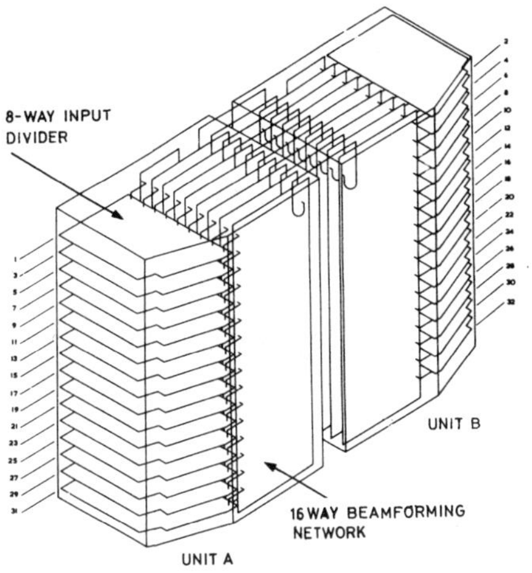

Mechanical form of an 8 beam i.f. beam former

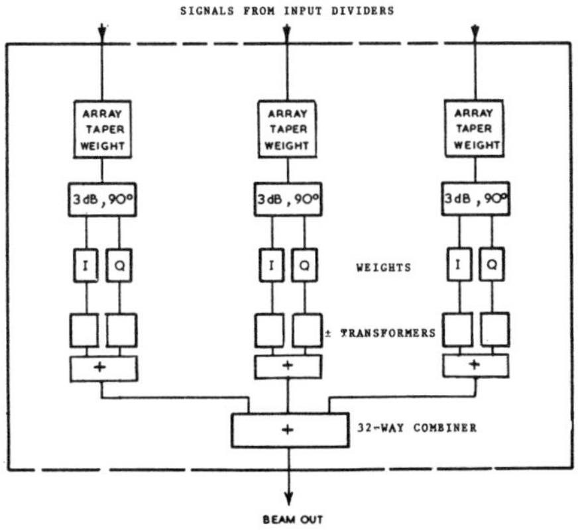

Circuit of an I/Q beamforming network

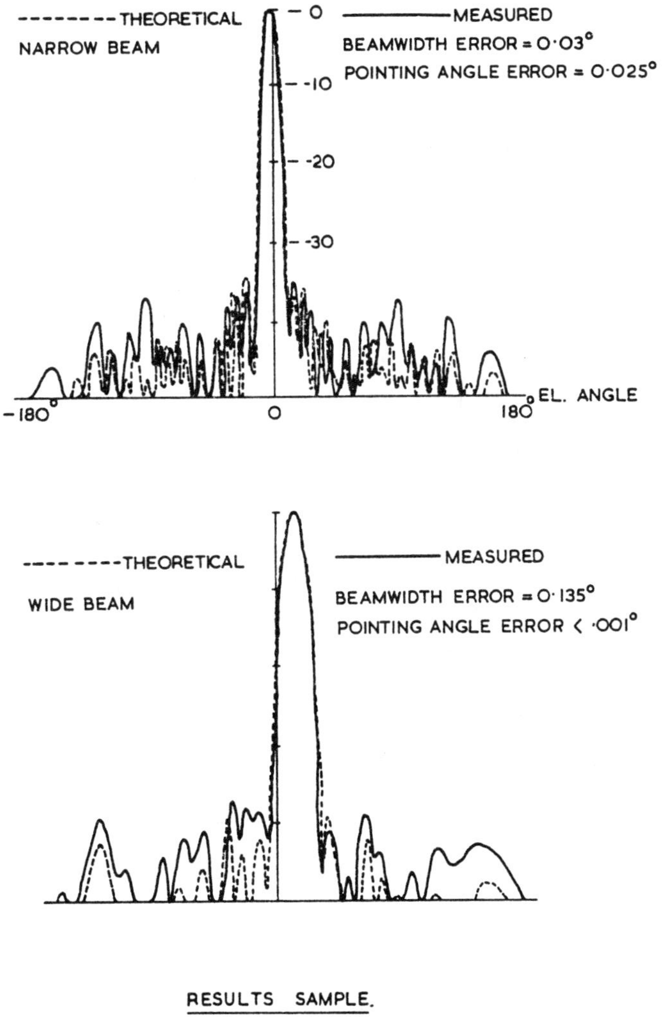

Typical Performance

AN X-BAND MICROSTRIP PHASED-ARRAY ANTENNA WITH ELECTRONIC POLARIZATION CONTROL

C H Hamilton

AEG-Telefunken, Federal Republic of Germany

INTRODUCTION

Electronically scanned phased-array-antennas can significantly improve the performance of airborne radar systems, but their high cost and complexity have so far prevented their widespread introduction. In this paper, the results of a study into possible configurations for such an antenna are summarized. The study requirements called for an X-band monopulse antenna with a circular aperture of about 25λ in diameter. Electronic scanning to ±50° was required in at least one plane, with the option of mechanical scanning in the other. Other requirements were wideband operation and polarization agility.

Four different configurations were considered at the outset of the study;

- Two-axis reflection array
- Two-axis transmission array
- Single-axis constrained-fed transmission array with electronic scanning in AZ plane and mechanical EL scanning
- Azimuth phased-array line feed combined with cylinder parabolic reflector for EL focussing

Of these, the two-axis arrays were considered unsuitable due to the high phase shifter cost. Since many of the missions carried out by airborne radars involve frame scanning where one axis is scanned much more slowly than the other, a hybrid solution in which electronic scanning is used in one plane and mechanical scanning in the other may be almost as effective as a fully agile antenna.

Of the two single-axis alternatives, the line feed plus reflector is the least expensive; it does, however, suffer from the disadvantages of aperture blockage, spillover and high inertia. To avoid these problems, the more complex transmission array with electronic AZ scanning was chosen as the subject for this study. In the following sections, the basic antenna and its component parts will be described. Experimental results will also be given for a model array with 192 radiators which was built to test the concept.

ANTENNA CONFIGURATION

Fig. 1 shows a schematic diagram of the antenna. It consists of five different sections;

- the radiating aperture
- the equiphase elevation feeds
- phase shifters
- azimuth feeds
- monopulse comparator

Polarization control is achieved by using radiators with two orthogonal inputs; by suitably controlling the phase difference between the inputs any desired polarization can be generated. In order to meet the monopulse requirement the radiating aperture is divided into four quadrants. Each quadrant consists of 24 radiator columns in which the elements are spaced 0.9λ apart; between 4 and 14 elements per column are used depending on the position in the quadrant. Each radiator column is fed by two identical triplate feeds which are in turn connected to two sets of 4-bit diode phase shifters. The phase shifters are supplied by a total of eight waveguide series feeds the outputs of which are combined in a suitable monopulse comparator. Note that this arrangement may be regarded as two individual phased arrays which are orthogonally polarized but which share the same radiating aperture. This approach was chosen since, apart from polarization control, it also allows two independent orthogonal beams to be generated.

RADIATING ELEMENTS

Because of the large number of radiators required - some 1100, each with two inputs - the only type of element seriously considered was a planar microstrip antenna. The element used is shown in Fig 2. It consists of a circular microstrip antenna etched on to 1/8" thick Duroid. Sheets of about 150 elements were printed and attached to an aluminium support plate with dielectric screws. The two orthogonal inputs were polarized at +45° and -45°. The position of the feed points was experimentally varied to obtain the lowest VSWR and highest decoupling over the operating band. Coaxial lines made from the inner conductor and teflon insulator taken from semi-rigid cable were soldered to the surface of the discs. Holes drilled in the support plate formed the outer conductors. At the rear of the support plate, the dielectric was cut flush with the surface, while the inner conductor was allowed to protrude about 3mm to permit mating with the EL feeds.

ELEVATION FEEDS

In order to maximize bandwidth, parallel triplate feeds with a Taylor amplitude distribution were used for EL power feeding. Here, the main problem was one of connecting to over 2000 antenna inputs without resorting to cables or connectors. Soldering would have been a possibility, but it was rejected since it is both difficult to implement and bad for repair work. Instead, a plug system employing gold-plated sockets of the type used in low-frequency connectors was devised (Fig 3). The socket diameter was chosen to securely grip the inner conductor of the antenna cables. The edges of the triplate feeds were suitably drilled and the sockets inserted and soldered to the centre conductors. Pairs of feeds were riveted together and fitted at the base with two aluminium strips.Semi-rigid cables were used to connect the EL feeds to the phase shifters.

PHASE SHIFTERS

Both ferrite and diode phase shifters were considered for use in the antenna. Although ferrite devices have the advantages of lower loss and higher power handling capacity, diode phase shifters are faster, less temperature sensitive and potentially much cheaper.

In this work, 4-bit diode phase shifters employing 12 diodes in a loaded-line circuit were developed. Eighteen ohm coaxial line was used to keep the RF voltage low and to allow high capacitance, high power diodes to be used. The diode impedances are suitably transformed by stub lines, and bias is introduced via 3-section $\lambda/4$ filters. Fig 4 shows a partially opened prototype in which the main line and some of the stubs with their diodes can be seen. Note the block was anodized to allow metallic bias filters to be used. In order to reduce unit costs, blocks of 24 phase shifters were manufactured on an NC mill. The filters and stub lines were made on automatic lathes. Loss and phase shift for a typical phase shifter are shown in Fig 5. Catalogue diodes were used here; by using special diodes losses of about 1dB should be possible.

The outputs of the phase shifter blocks were spaced in such a way as to allow direct mating with the outputs of the azimuth feeds.

AZIMUTH FEEDS

Because of the high power level at the antenna input, it was decided that some form of waveguide feed was required in the AZ plane. Here a series feed can be used since the frequency-dependent output phase can be corrected for by the phase shifters. (Out-of-phase outputs are actually advantageous since they break up quantization sidelobes). Eight identical feed halves are needed to satisfy the monopulse and polarization agility requirements.

Fig 6 shows a cross-section of the series feed developed for this antenna. It is a travelling wave feed in which power is coupled out by probes mounted at intervals of $\lambda/4$ along the guide. By suitably selecting the length of the probe, which is formed by the centre conductor of an SMA connector, any coupling factor between about 4 and 20dB can be set up. Provided the coupling factor is looser than about 8dB, the reflective loss is small and the arrangement may then be regarded as a reactive power divider. When several probes with approximately the same coupling factor are mounted $\lambda/4$ apart in the guide, the reflections at the input cancel leading to a low VSWR over a wide frequency range. A simple theoretical and experimental procedure was devised to determine the probe lengths for a particular amplitude distribution. The feed was then built and the probes experimentally trimmed to set the coupling factors as accurately as possible. Total losses, including termination loss, amounted to about 0.5dB. Fig 7 shows the theoretical and experimental coupling values for a pair of series feeds centre-fed via a magic Tee.

The outputs of the series feeds are combined in a conventional monopulse comparator.

EXPERIMENTAL ANTENNA

In order to keep the cost of the experimental antenna low, the aperture was limited in the mechanically scanned EL plane and EL monopulse was omitted. Fourty-eight columns of 4 radiators were used which were fed by 48 pairs of triplate feeds and 96 phase shifters. Azimuth feeding was carried out by two pairs of waveguide feeds. Fig 8 shows a rear view of the model during assembly. On each side the phase shifter blocks can be seen connected to their series feeds. The magic Tee outputs correspond to the sum and difference patterns for the two 45° polarization vectors. In the centre of the model, the aluminium support plate is visible with five EL feed pairs shown mounted. The microstrip radiators are mounted on the front side.

Before measurements were started, the insertion phase from the antenna input to the outputs of the phase shifters was measured at several frequencies and all 16 phase states. These measurements were stored in a minicomputer and later used to calculate the phase shifter settings required to produce a particular phase front with the minimum error. The minicomputer calculated the phase front for a given pointing angle and polarization and set up the phase shifters via an IEC bus interface.

MEASURED RESULTS

Computer simulations using measured errors in the feeding system ($\pm$ 1.5dB amplitude and $\pm$ 14° phase) indicated sidelobe levels of 20-25dB (depending on beam angle) could be expected. This was in agreement with measured values which were slightly better due to phase error optimization. Fig 9 shows a broadside beam, horizontally polarized, with a sidelobe level of 25dB. Fig 10 shows a set of beams scanned from -50 to +50°. Measured sidelobe level at 50° was about 20dB. Cross-polar attenuation (VP to HP) was about 20-40dB depending on scan angle. Beamwidth at boresight was 3°, increasing to 4.5° at 50°. Difference pattern nulls of over 40dB at boresight and about 30dB at 50° scan were measured.

CONCLUSIONS

The concept described above allows a single-axis, polarization-agile phased array to be realized at an acceptable cost for airborne applications. Use of microstrip radiators, pluggable triplate feeds and series waveguide feeds leads to a compact, low weight structure. Better sidelobe performance could be obtained by optimizing the phase shifter design to reduce incidental amplitude modulation. Cost could also be reduced, at the expense of higher insertion loss, by using printed triplate phase shifters integrated into the EL feeds.

AKNOWLEDGEMENT

The author would like to thank the German Ministry of Defense for permission to publish this work.

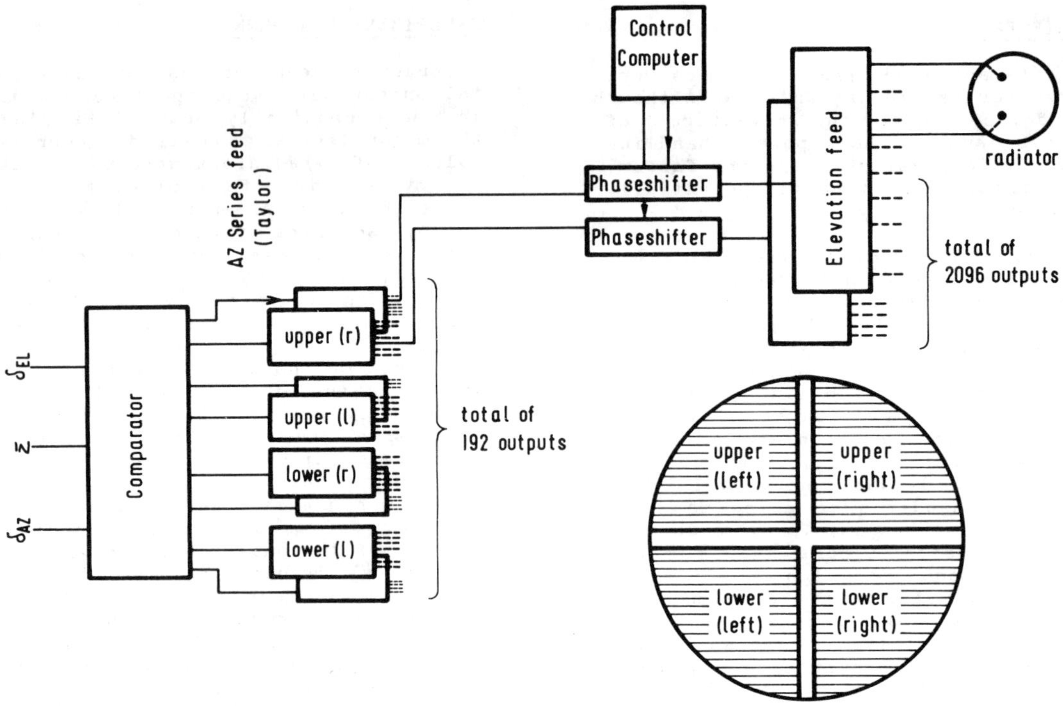

Fig 1 Antenna block diagram

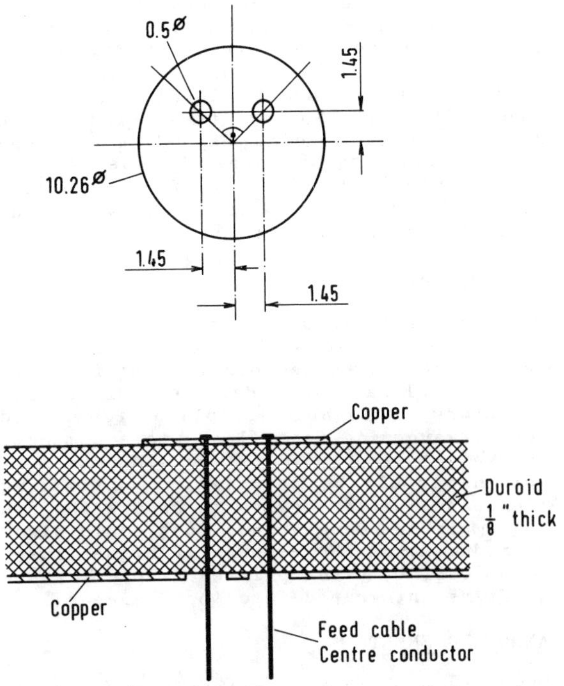

Fig 2 Radiating element

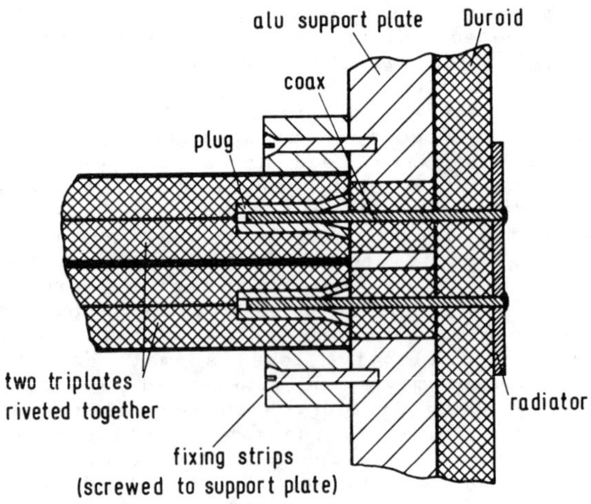

Fig 3 Section through EL feed to element plug system

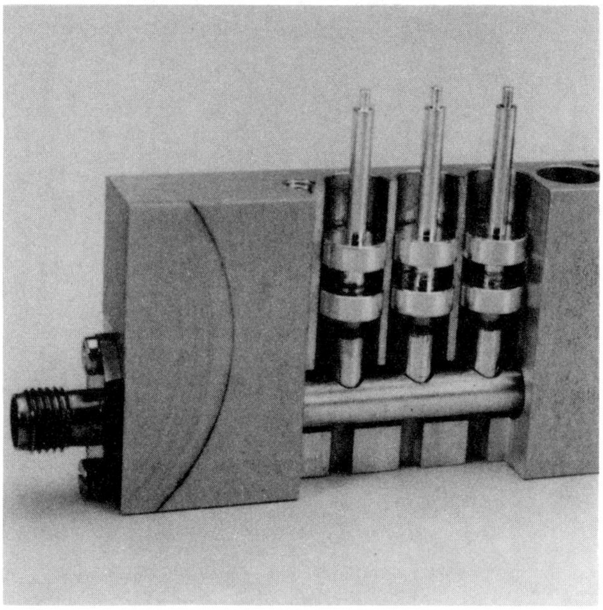

Fig 4 Prototype 4-Bit phase shifter showing diode and filter arrangement

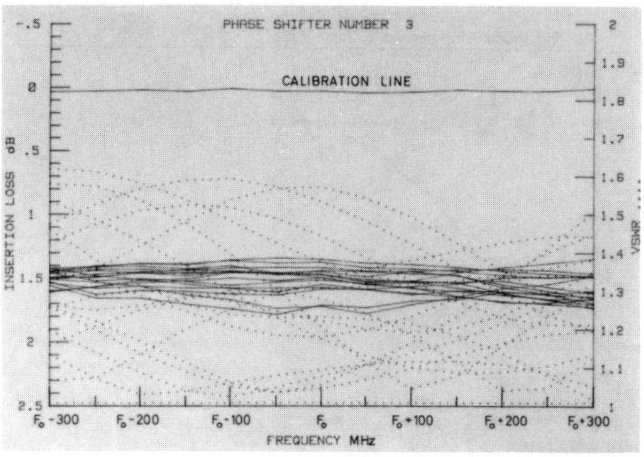

Fig 5(a) Insertion loss response in all 16 states

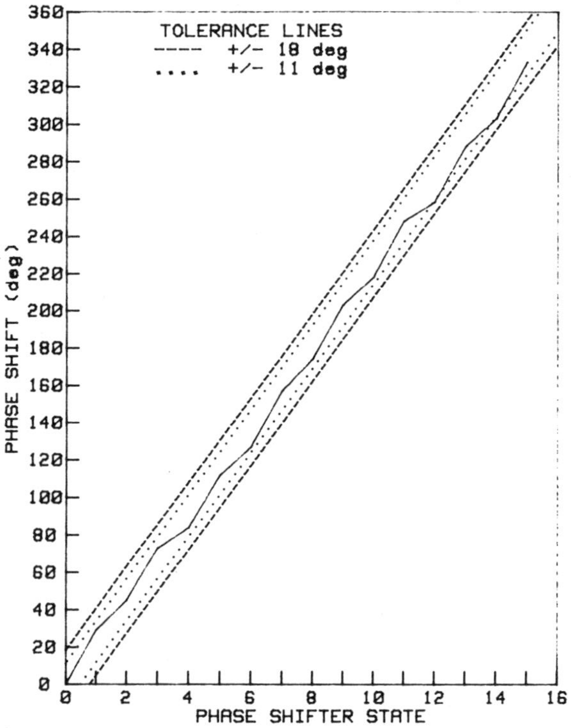

Fig 5(b) Insertion phase as function of set phase

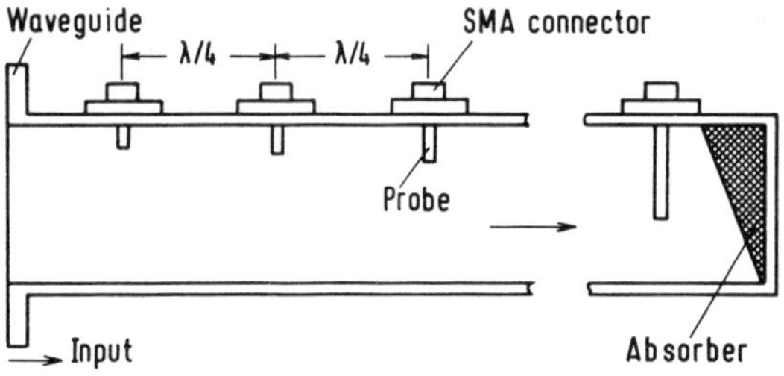

Fig 6 Schematic diagram of series feed

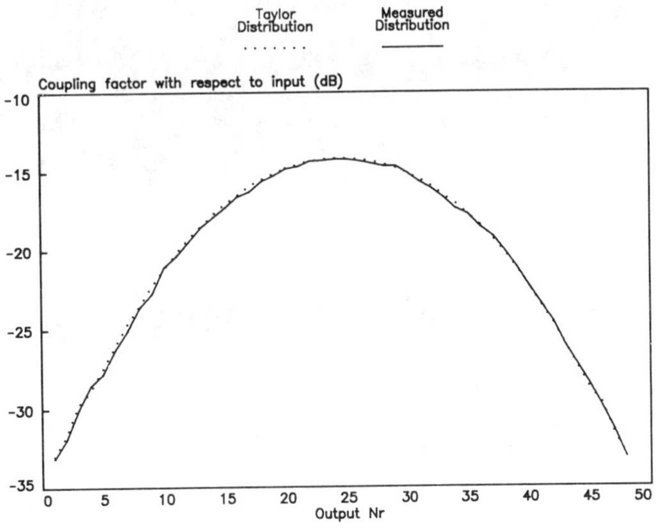

Fig 7 Required and measured response of series waveguide feed

Fig 8 Rear view of antenna during assembly

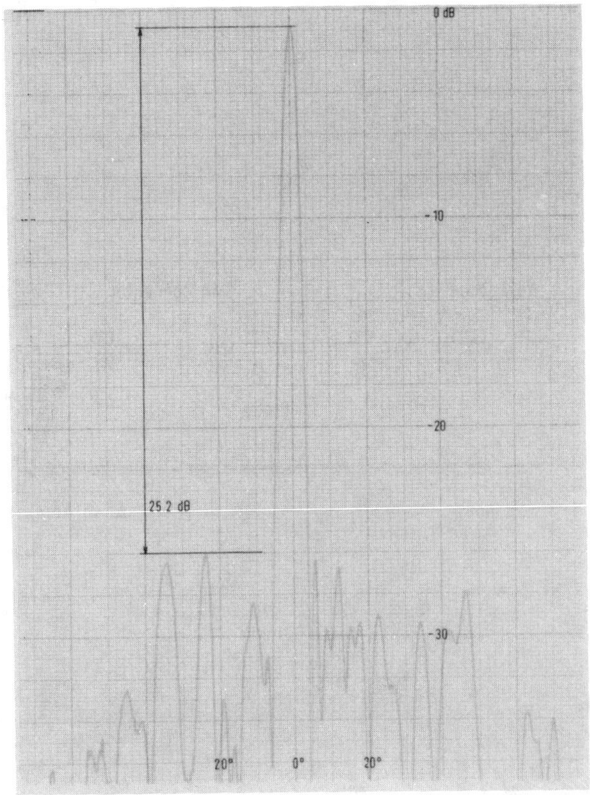

Fig 9 Horizontally polarized broadside beam

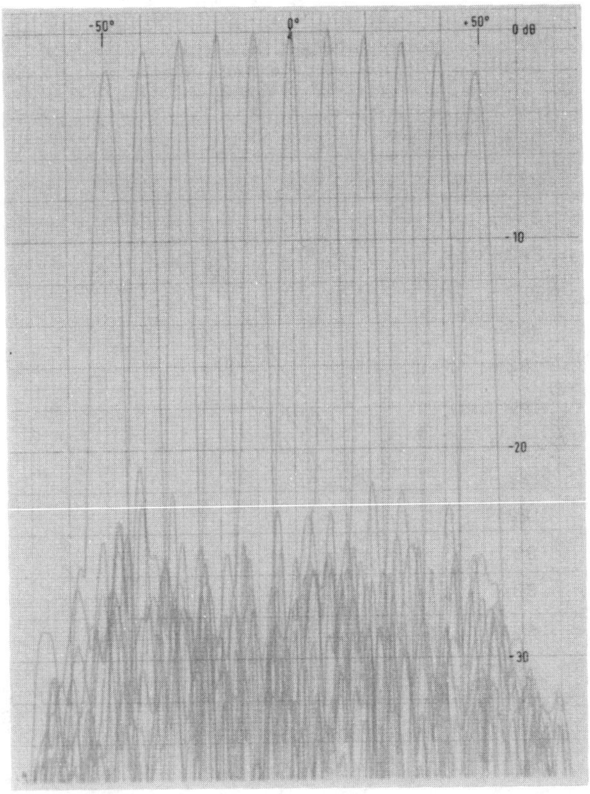

Fig 10 Horizontally polarized pattern electronically scanned to ± 50°

RESULTS FROM AN EXPERIMENTAL RECEIVING ARRAY ANTENNA WITH DIGITAL BEAMFORMING

U. Petri

AEG-Telefunken, Federal Republic of Germany

INTRODUCTION

Radar antennas with electronic beam steering and forming offer the advantage of nearly instantaneous adaptation of the antenna diagramm to the requirements of the system strategy. At present nearly all antennas of this kind are realised as phased arrays in which the individual antenna elements are connected to the transmitter or receiver via phase shifters. The necessary phase shift to appropriately direct a transmitted or received plane wave is calculated by a computer. The amplitudes of the elements have fixed values (Fig. 1). Beam forming may be achieved within certain limits by additional curvature of the wavefront (beam spoiling). Phase errors in the distribution network and, in the case of spoiling, even amplitude errors (1), may be compensated for by appropriate setting the phase shifters.

The constraint of phase-only setting is a practical one since electronic amplitude control is more complicated and generally implies a loss in gain. As a result, beam shaping possibilities are limited (as required by adaptive sidelobe suppression for example) and amplitude error corrections are limited. Amplitude control is, however, extremely desirable for multibeam receive - only arrays if the restriction of mathematical orthogonality imposed by lossless feeding networks is to be avoided. The alternative - lossy feed networks - requires amplifiers at the element level which must be carefully matched to avoid introducing additional amplitude errors.

After a comparision of different methods of electronically steering phase and amplitude (2) digital video beamforming, i.e. generation of the antenna space factor by numerically weighted superposition of the digitized complex video outputs of the single antenna elements has turned out to be the most promising method. An experimental receiving array with 20 active elements has been realized in order to study the feasibility of this method and the behaviour of critical components. The following paper will give a short insight into the theory and illustrate some important results.

THEORY AND MODE OF OPERATION

The principle of digital beamforming shall be explained by regarding a linear equally spaced array (Fig. 1) neglecting the element space factor and polarisation. In this case the complex signal S_n at the output of the n`th element of a plane wave from the direction ϑ will be

$$S_n = S_o \, e^{j\, u(\vartheta) \cdot n} \qquad (1)$$

with $u(\vartheta) = 2\pi \sin(\vartheta)\, d/\lambda_o$ and S_o being a complex constant proportional to the amplitude of the received wave. In a phased array the signals S_n will be submitted to phase shifts φ_n, resulting from the transmission phase of the distribution network and from the phase set by the phase shifter, and to weighting A_n before being superposed at the output of the network. Mathematically speaking phase shifting and weighting is the same as the multiplication of the complex value of S_n with a complex factor W_n

$$W_n = A_n \, e^{j\varphi_n} \qquad (2)$$

and the complex output of the array G is the sum over the complex products of element signals and weights

$$G = \sum_{n=1}^{N} S_n \, W_n \qquad (3)$$

If one is able to obtain the digital values of the complex signals at the element outputs then the following weighting and summation could be performed by a computer instead of by HF components. One means to do this is the demodulation by a quadrature detector (Fig. 2) followed by a A/D-convertor. Neglecting amplification and phaseshift within the detector, the output of the I- and Q-channel will equal the real and the imaginary part of S_n. After digitizing, the complex multiplication could be performed in the subsequent "beam forming processor" by 4 real multiplications:

$$S_n W_n = I_n \operatorname{Re}(W_n) - Q_n \operatorname{Im}(W_n) + j\{I_n \operatorname{Im}(W_n) + Q_n \operatorname{Re}(W_n)\} \qquad (4)$$

Real receivers generally will show slightly different amplifications and transmission phases as well as different gains within I and Q channels. Thus I and Q output no longer will equal the real and imaginary part of S_n. In order to compensate for this error I and Q signals must be submitted to a linear transform before applying equ. (4). This transform can be combined with the complex multiplication and will result in a multiplication with generally 4 different coefficients

$$S_n W_n = I_n W_{RI_n} + Q_n W_{RQ_n} + j\{I_n W_{JI_n} + Q_n W_{JQ_n}\} \qquad (5)$$

Another error to be compensated for is the offset error resulting from improper biasing of the A/D-convertors. It can be corrected by subtraction of a constant from the output G (equ. (3)).

The quantisation error however cannot be corrected. It should be decorrelated from the signal by choosing the size of the least significant bit (LSB) to 1 or 2 times the RMS value of the noise and will then behave as additional noise.
The dynamic range of the A/D-convertors equals the ratio of the maximum radar return to the LSB and therefore to the receiver noise. Under the assumption that noise of all receivers is decorrelated the signal to noise ratio will be improved by the superposition of the element signals proportional to the gain of the group. Thus with the minimal detectable signal given, the tolerable receiver noise will increase with the gain, i.e. approximately with the number of elements, and the necessary dynamic range of the A/D-convertors will shrink by the same amount. For larger arrays the element signals may even be smaller than the LSB. Because of the high sampling rate defined by the video bandwidth parallel multiplications in the processor must be executed. The summation following is performed in steps of two in a "pipeline" configuration. Fig. 3 shows the principle structure of the processor. Multiple beams may be generated by several parallel processors of this kind. The complex output signals can immediately be accessed by the following signal processors (MTI, pulse compression, etc.)

THE EXPERIMENTAL MODEL CONCEPT AND RESULTS

The model realized consists of 20 active array elements each with a quadrature detector and a pair of A/D-convertors, one beam forming processor and a desktop computer for control and evaluation. The array elements are columns of 4 open circular waveguides with fixed EL distribution network in order to achieve a smaller EL beamwidth. The receivers are conventional receivers taken from an existing radar system. They were tuned only for about equal noise power. The "critical components" A/D-convertors and processor, were chosen for a sampling rate of more then 10 MSPS. Resolution for converter and weighting factors is 7 bits (plus sign.). The processor is built up according to the scheme of Fig. 4. The controller (desktop computer) calculates the weights, records and evaluates the received digital signal and drives pulse generator and pedestal. When computing the complex weights the errors previously discussed can be compensated for. The size of the deviations is calculated from the results of a calibration measurement with 4 different phases during which successively all elements besides the one to be measured receive zero weighting.
The following diagrams show typical results. Fig. 5 shows the antenna diagram with equal weights with and without error compensation. Digital beamforming enables easy realisation of any weighting commonly known from theory. As an example, Fig. 6 shows results for Chebyshev-weighting. Deviations from the theoretical curves result from further errors not compensated for. The main reason for the coarse sidelobe structure is the difference between the elements`space factors (only 20 elements) and errors in the calibration measurement. The fine structure is due to receiver noise and quantisation.

Another example for the correction of deviations is given in Fig. 7. Two columns had been half covered by absorbers in order to simulate a degradation of amplification. This would result in sidelobes rising about 8 dB. After a new calibration measurement and calculation of new weights, sidelobes will come down again. However there is a loss in gain.
Fig. 8 shows the possibility to get fairly good diagrams even when the element signal level is down below the least significant bit as long as the array will have enough elements. Since the model only consists of 20 elements, which would not give enough improvement of the S/N ratio, the average was taken from 400 measurements which would lead to the same results as an array with 8000 elements.

PROSPECT

The experience with the model proved the feasibility of digital beamforming for radar. Digital beamforming enables flexible shaping of the antenna diagram as well as the correction of amplitude and phase deviations. However its practical use is still restricted by the high costs of the components used. But since these components (A/D-convertors multipliers etc.), unlike phaseshifters, find increasing usage within consumer electronics, it is possible that this technology will be attractive for future radar systems.

REFERENCES

1. Petri,U.: Kompensation von Amplitudenfehlern durch Phaseneinstellung. Wiss.Ber. AEG-Telefunken 54 (1981) 1-2

2. Borgmann D.: Steuerung und Formung von Strahlungscharakteristiken mit Gruppenantennen. Wiss.Ber.AEG-Telefunken 54 (1981) 1-2

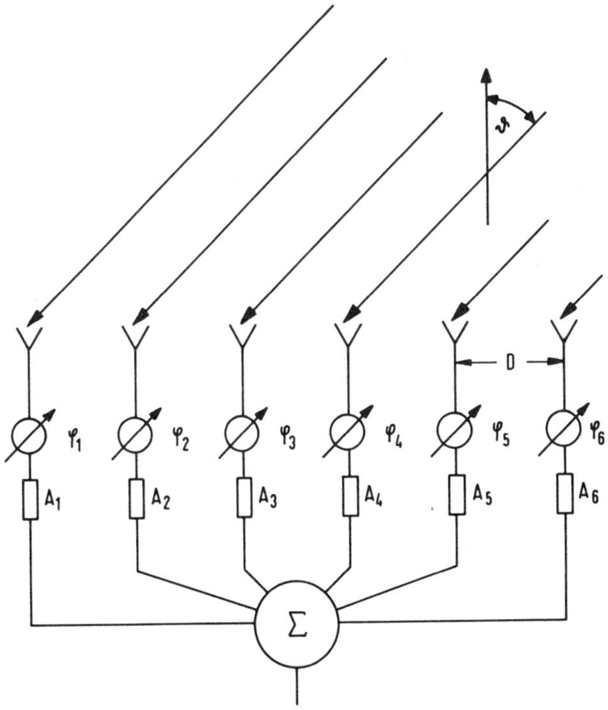

Figure 1 Principle of array beamforming

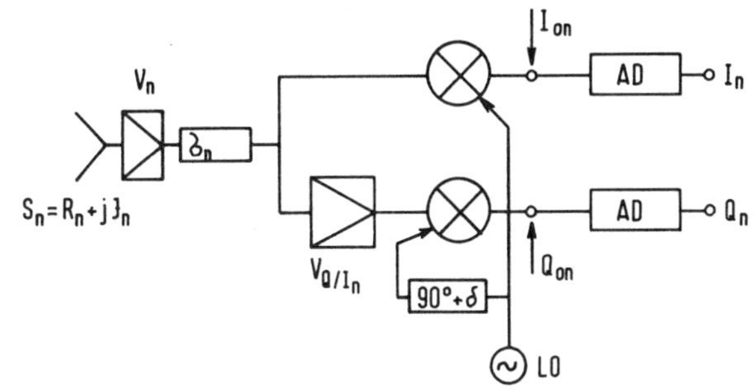

Figure 2 Quadrature detector with different error sources

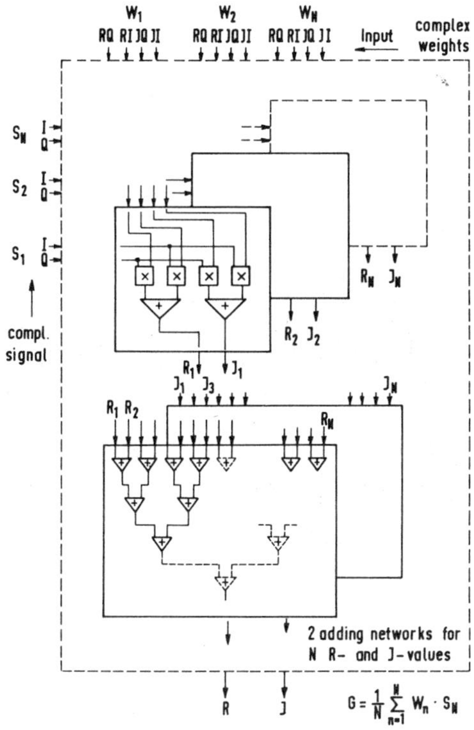

Figure 3 Concept of the "Beam Forming Processor"

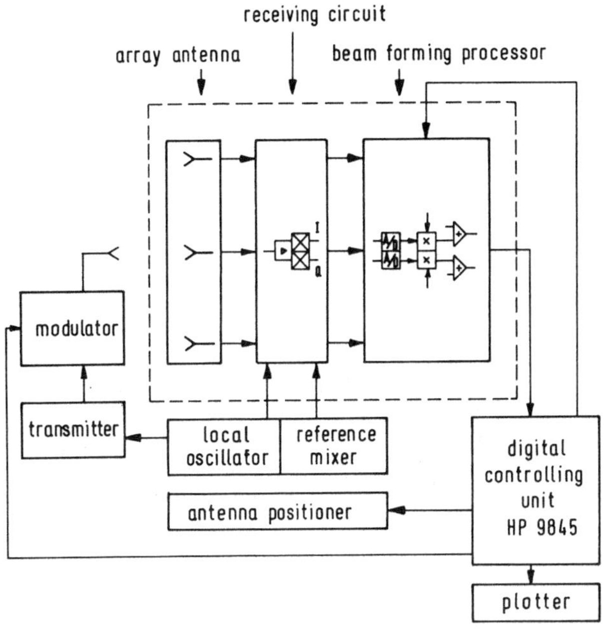

Figure 4 Scheme of the experimental model

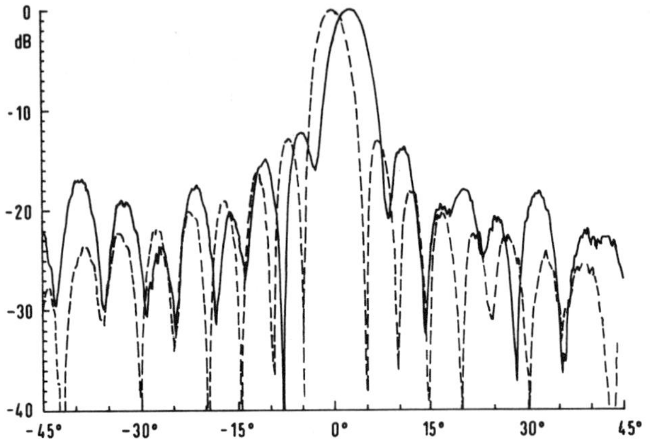

Figure 5 Measured antenna diagram for uniform weighting
no calibration-with calibration---

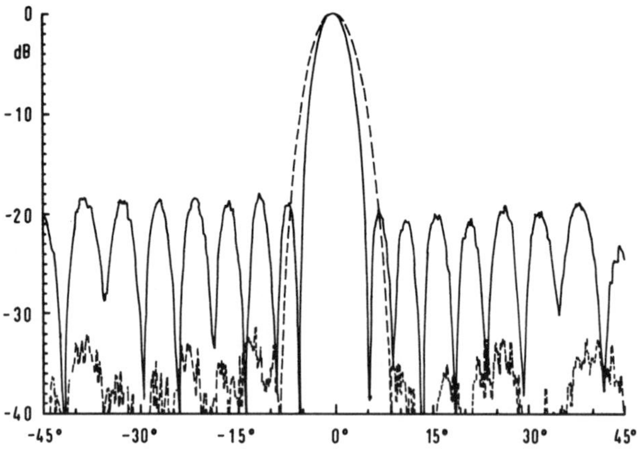

Figure 6 Measured diagram for Chebyshev weighting 20 dB sidelobe level—
40 dB sidelobe level---

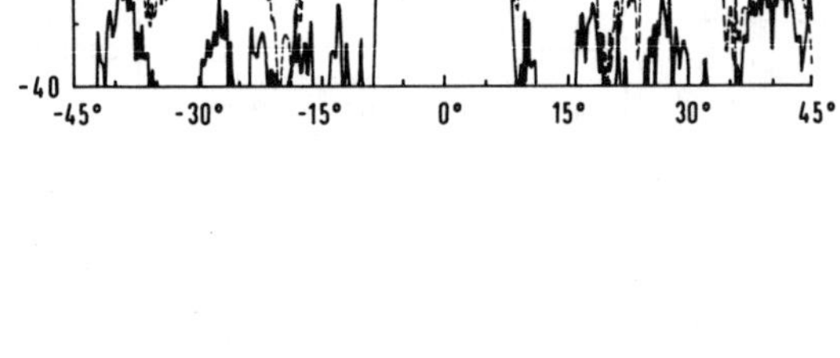

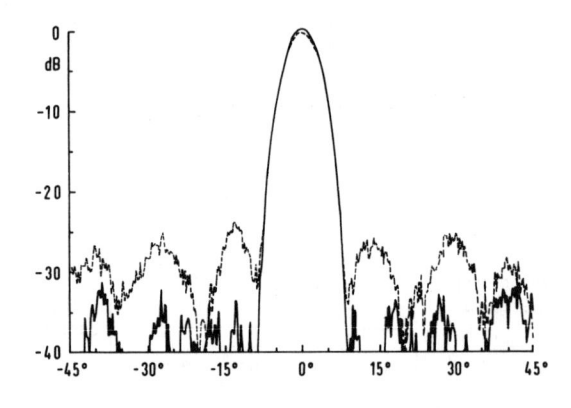

Figure 7 Measured diagram for additional attenuation in two antenna elements
no compensation---with compensation—

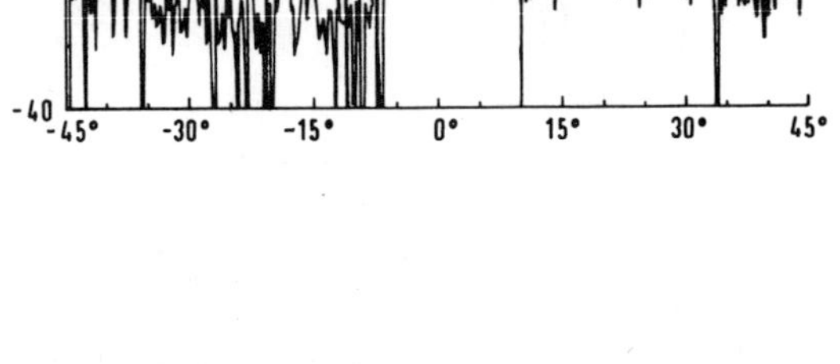

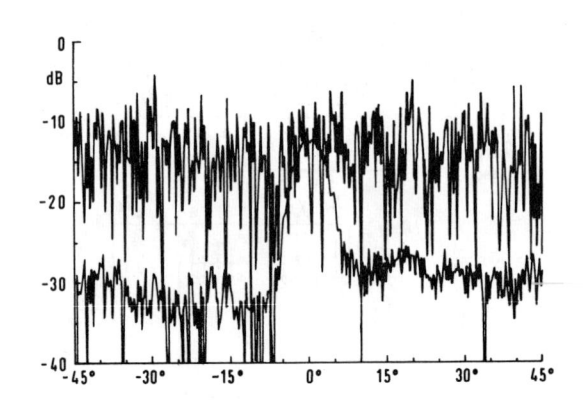

Figure 8 Measured diagram at a signal level of about.5LSB. One measurement and average from 400 measurements.

BEAM FORMING WITH PHASED ARRAY ANTENNAS

W. Sander

Forschungsinstitut für Funk und Mathematik (FFM) of FGAN F. R. Germany

INTRODUCTION

Phased array antennas with many active receiving modules sample the electromagnetic field at the radiator locations. By an exact analysis of the received wavefront pattern across the array a great deal of information on targets and interferences can be extracted. For example, an array with N elements can in principle form N simultaneous beams and thus, signals from all directions can be discriminated. This case of multi-beam forming is highly efficient with multistatic radar systems or passive direction-finders. Moreover, the fine structure of the wavefront contains information on multipath or multi-target situations. Though the necessary signal processing involves theoretical difficulties, greater problems arise with the practical implementation of processing the high number of samples in real-time. In radar systems time constraints are especially severe, more so than e.g. in sonar systems. Thus, parallel processing cannot generally be replaced by serial processing, which entails less effort. An optimum processing which retains control of each of say some hundred element signals, requires an immense expense. Therefore, in most practical systems signal combining is restricted to conventional beam forming which assumes an undistorted planar wavefront produced by an ideal target. In most cases more refined systems which form several or all possible beams, have only few elements or do not work in real-time. The principle process of beam forming requires the individual multiplication of all signals with a complex factor and then the sum of these weighted values. Different beams can be formed by changing the weighting factors which generally depend on the element location of the aperture. Fixed weights can be easily implemented by analog devices such as resistive networks and operational amplifiers. Difficulties arise when beams with a variable characteristic are to be formed by analogue means. This application aims for digital processing which is additionally characterized by high stability and reproducibility. Swartzlander (1) has given an analysis of a multiple beam former with a pure digital processor. The feasibility of his proposal requires the development of special integrated circuits which are not yet available. In future, this solution will offer a high technical performance, compared with analogue implementations, at reduced costs. Similiar results have been obtained in a detailed analysis of digital beam forming by Barton (2). He estimates the limits of practical beam formers which are feasible with modern technology to about 100 inputs at sample rates of 100 ns. A much larger number of inputs and smaller sampling intervals are expected from anticipated advances in digital technology within a few years.

SUBARRAY TECHNIQUE

The discrepancy between the desirable optimum processing and the present limitations of practical implementation must be decreased by a suboptimal compromise, e.g. by subarraying. This procedure yields a reduction in dimensionality by cascading two steps of signal processing. In the first stage, neighbouring elements are grouped in subarrays. Their signals are combined by fixed weights, forming (broad) subarray beams. These beams must be combined in a second beam former which may be more versatile and comfortable, since the number of its inputs is considerably reduced by the preceding subarraying. An obvious disadvantage is, that the second beam former can only generate beams within the beamwidth of the subarray beams. Nevertheless, in most applications progress is achievable, especially if the subarray beams can be steered by phase shifters. The practical implementation may be performed by analogue or digital devices or by a combination of both these types. The weights for processing the subarray signals must be determined by the following four steps:

1) Given the weighting function for the desired beam shape, the optimum weight for each element is calculated.
2) The average of all element weights belonging to the subarray under consideration is the preliminary subarray weight.
3) A normalization of all subarray weights may become necessary. The results should be that the power of the thermal noise at the output of the beam former is equal for all beams.
4) In digital systems, a quantisation, resp. rounding to a suitable word size, yields the final weights.

The smaller the subarray size, the better the approximation of the optimum weighting function and the smaller the gain in reducing the complexity. Therefore, one has to prove, how far the subarrays may be enlarged for a given application. This should be explained by the example of forming difference beams.

Figure 1 shows three weighting functions for the generation of difference beams. The abscisse is normalized to the aperture length. Usually, separate sums of the left and right half of the aperture are formed. The sum and the difference of these partial sums yield the final sum beam, resp. difference beam. The corresponding weighting function is a staircase with two steps at + 1 and - 1, and is mostly applied because of its easy implementation. However, with active arrays the optimum weighting function, which yields the highest angular sensitivity factor, is a linear function that is, the contribution of elements has to be decreased the more they approach the centre of the aperture. With subarrays the linear function must be replaced by a staircase function. The angular sensitivity depends on the slope S_D of the array factor at zero angle:

$$S_D = \frac{\pi N^2}{M}\sqrt{\frac{M^2-1}{12}} \qquad (1)$$

where M = number of subarrays
N = total number of elements

and a linear array with $\lambda/2$ spacing is assumed. Equation (1) is valid in the case of ± 1 weights (M=2) and in the optimum case (M=N), too. A comparison between the simplest and optimum case shows an improvement factor of 1.15 which corresponds to an increase of 1.25 dB in the signal-to-noise ratio. However, with only a few subarrays, e.g. four or eight, an improvement of 0.9 dB, resp. 1.2 dB is achievable. More subarrays only increase the complexity, whereas the further improvement is marginal. In figure 2 the effect of proper weighting is shown with an 80 element linear array. The sum beam pattern has been drawn for comparison whereas a difference beam with 8 subarrays has been omitted since it approximates the optimum difference pattern

almost ideally. This example should demonstrate that in some cases a weighting function may be roughly approximated by only a few subarrays without a heavy loss.
The pattern of a shifted beam can be written as the product of the fixed subarray pattern and the shifted array factor of an array with wide element distances corresponding to the subarray spacing. The second term has grating lobes which coincide in the non-shifted case with zeroes of the subarray pattern and are therefore suppressed. When the pattern is shifted, the coincidence disappears, and the side-lobes increase at these points as shown in figure 3. Here, five sum beams have been calculated with an incremental shift of half of the beamwidth. Together with the decrease of the main beam an increase of the sidelobes can be observed. To achieve a broader main beam and a lower sidelobe level, the subarray pattern should be tapered by a suitable function such as sin x/x or cos-squared. Thus, the amount of shift could be enlarged. But this advantage must be paid for by a loss of signal-to-noise ratio. This loss depends on the tapering function used, the amount of beam broadening and the reduction in sidelobe level. For example, a beam broadening of 1.8 and a sidelobe reduction of about 16 dB is possible with a loss of 2.2 dB. Other parameters achieve less sidelobe reduction but broader main beams. Similiar results are obtainable with irregular spaced arrays in which subarray patterns can be improved by space tapering.

BEAM FORMING WITH ELRA

An experimental electronic steerable array radar ELRA has been developed enabling various research projects to be performed on array control, signal processing and data management with phased arrays. Sander has given an overview on the whole antenna system (3). Some theoretical and experimental studies in connection with ELRA have been reported by Wirth (4) who also gives some results on signal processing in (5). Here only those parts of the radar system which deal with the beam forming are being described.
Figure 4 shows the element distribution of the receiving array with triangles indicating radiator positions. The planar aperture is filled by 768 radiating elements in a space-tapered configuration. Since the aperture has a diameter of 39 wavelengths a completely filled array with a regular grid of 0.5 wavelengths would comprise about 4800 radiating elements. The resulting thinning factor of 16% increases the average element distance and therefore reduces mutual coupling. This coupling is additionally decreased by special radiators which consist of inexpensive etched dipoles with tilted arms. 192 of them are crossed dipoles which occupy a horizontal strip of 5 by 39 wavelengths in the middle of the aperture. Their linear polarisation can be switched from vertical to horizontal orientation enabling differences in received signal polarisation to be studied. The element distribution shown yields a pencil beam of about 2° beamwidth and an average sidelobe level of about -29 dB.
The radiating elements are separated into 48 subarrays of 16 elements each as indicated in figure 4, in which the outermost elements of a subarray are connected by straight lines. The division into non-overlapping subarrays is arbitrary and can be easily changed e.g. to get mutually overlapping subarrays of more equal dimensions. Since the element distribution and the size of these subarrays differ, the radiation patterns obtained by summing up all the individual element contributions of a subarray are different, too. In figure 5 the calculated antenna patterns of 12 subarrays located on a horizontal strip of the array have been drawn. The direction cosine has been normalized to the half-power beamwidth of the whole array. While three groups of beamwidth can clearly be recognized and attached to the subarray size, no classification of the sidelobe region is possible. This stochastic behaviour avoids the increase in sidelobe level when shifted beams are to be created, which is not the case with regular spaced arrays.
Actual subarraying is done at baseband (figure 6). Each radiating element is followed by an individual receiving module consisting of a low-noise preamplifier in S-band, an image rejection mixer, a 30-MHz-IF-amplifier with two selectable bandwidths of 0.1 resp. 1 MHz, and a second mixer which produces coherently the I- and Q-components of the received signal at baseband. The 3-bit phaseshifter is located in the reference path of the second mixer. Two summing operational amplifiers combine separately the I-components, resp. Q-components of the 16 elements belonging to a subarray. The weights, easily attained with resistors, are chosen equal for all elements, that is, no beam shaping at subarray level occurs. The output signals of all summing amplifiers are sampled, held, and digitised every 500 ns by analogue-to-digital-convertors (ADC). For reasons of cost ADC's of 6 bit wordsize, produced in our own laboratory have been employed. Module gain is levelled so that root-mean-squared thermal noise equals one quantisation unit. Thus the dynamic range of subarray signals is constraint to 36 dB, but can be increased by reducing module gain by 20 dB. Final beam forming increases the dynamic range by a factor of $\sqrt{48} \simeq 17$ dB. In our experimental system this range seems to be sufficiently high, but operational systems should use ADC's of higher wordsize, available nowadays at lower cost. The 48 pairs of digitised subarray signals are processed by 4 parallel beam formers.

Fixed beam former (FBF)

All beam formers generally consist of three major parts; a) the weighting section b) the adding network and c) the output stage. Three of the beam formers generate "standard" beams for normal monopulse operation, namely a sum beam and two difference beams for azimuth and elevation. Beam shapes and their mutual locations are fixed, that is, the maximum of the sum beam coincides with the zeroes of the difference beams. The three beams can be commonly scanned by phase shifter beam steering. Weights are hardwired and cannot be changed easily. The sum beam former has no weighting section. The weights for the difference beams are determined by calculating first the optimum weight for each element as described previously. The quantisation procedure used here chooses the nearest power of 2, e.g. 2, 1, 0.5, 0.25 and so on. These weights obtained can easily be realized by shifting the signal before adding resp. by appropiate wiring. Negative values are realized by inverting the signal. Weights are real, that is, we use the same set of weights for I- and Q-components. Though the optimum weighting function is roughly approximated by a staircase with different step heights, the slope of the difference beam is 1 dB steeper than with usual ± 1 weights.
The adding network has seven cascaded adder stages with growing wordsize. The settling time is small enough for the signal rate of 2 MHz.
If only thermal noise is present, the output components should have a zero mean. In the output stage this is measured and, if an offset occurs, automatically compensated for. Further tasks of the output stage are code conversion, truncation of word length and distribution to various signal processors.
These beamformers are fully integrated into the ELRA system and have been in operation for some years.

Variable beam former (VBF)

The fourth beam former is in the experimental stage but offers many research projects to be performed in the future. The main difference between the VBF and the FBF's described above is in the weighting section. Each of the 48 subarrays has a complex binary multiplier for generating all possible 8 bit weights Thus, the shape of the beams can be changed, and their location can be shifted from the location given by the settings of phase shifter. However, the amount of shift is limited by the beam-width of the subarrays. Figure 7 shows a block diagram of the complex weighting and storage unit. In baseband a complex multiplication requires four real multiplications and two additions. These arithmetic operations could be performed serially by one multiplier or simultaneously by four multipliers. The latter version is distinguished by its higher cost since the multiplier is the most expensive part of the circuit. Nevertheless, the processing speed is higher, the design is more "straightforward" and therefore control is simpler. Our implementation is a compromise between the two versions. It is able to perform two partial 8 x 8 - bit products and its addition by two separate multipliers and one adder.

$$F_I = W_I \ S_I + W_Q \ S_Q$$

where W is the weight and S is the signal. The subscripts I and Q denote the orthogonal in-phase and quadrature components of either the signal or the weight or the output. The weights are stored in a random-access memory. For the calculation of the second component F_Q, the weights W_I and W_Q must be replaced by - W_Q, resp. W_I. This is easily performed by storing the weights in successive memory cells and by increasing the memory address by one. Thus, the weights for one beam occupy two addresses of the memory. 128 different sets of weights can be stored in advance and used at real-time speed. The processing rate of 250 ns per component, resp. 500 ns per beam is limited mainly by the cycling time of the memory types used. Another limiting factor is the settling time of the adder network. Since seven adding stages with a word length of up to 24 bits are cascaded, the ripplethrough time prohibits faster rates. Therefore, carry-look-ahead adders or registers switched in between the adding stages should be used for increased processing speed.

The number of different beams generated in real-time (synchronous mode) depends on the sampling interval of the signal. If the minimum sampling interval is extended, e.g. to 5 µs - allowable with the transmitted search pulse of 10 µs length - up to ten different beams can be generated serially. If one wishes to have more beams than allowed by the sampling rate, up to 256 signals out of successive range bins can be stored in a second memory and may be processed off-line (asynchronous mode).

The universal design of the weighting section offers a variety of potential modes of operation, but requires extensive control circuits. The control section of the VBF can be divided into two main parts. A fast control section delivers the appropriate pulses, timing signals and memory addresses for the on-line operation mode. A slower section is driven by a microprocessor INTEL 8086 via an IEC-bus, and serves mainly for the preparation of the real-time processing, e.g. transmitting the calculated weights to the memory, setting the mode of operation, sampling rate, number of beams required or the location of the required beams in the memory. In the asynchronous mode output signals can be fed by the IEC-bus to the computer. For the fast synchronous mode a multiplexer is being built which injects the signal processed by the VBF back to signal processors of the ELRA-system. In order to detect faulty circuits a test signal can be injected into each of the subarray inputs. Picture 8 gives an impression of the amount of hardware used for the FBF in the right rack and the VBF in the left rack. The uppermost cardholder of the VBF contains the input section, the next two cardholders are filled with the weighting section, the adding network and a part of the control circuit, whereas the lowest cardholder takes the output section and the rest of the control circuit.

APPLICATIONS OF THE VARIABLE BEAM FORMER

The main aim of the VBF will be to prove the practicability of results obtained by theoretical analysis and computer simulation which can never take into account all the imponderables of the real environment. Some of these research activities will deal with

a) difference beams with optimum angular sensitivity
b) symmetrical difference beams for better recognition of multipath or multi-target situations
c) a cluster of shifted sum beams for amplitude comparison monopulse
d) difference beams with unsymmetrical sidelobe behaviour against multipath
e) partial sum beams, e.g. two beams with different polarisation planes for studying the influence of targets on signal polarisation
f) sharpened beams for higher angular resolution
g) difference beams with sidelobes above the sidelobe of the sum beam for discriminating signal arriving from sidelobe directions
h) pattern variability as a function of range for reducing clutter

CONCLUSIONS

An experimental digital beam former has been described which processes the signals of an active phased array with some hundred receivers. A reduction of dimensionality has been achieved by forming subarrays. This suboptimal procedure is sufficient for many applications and is easier to implement than the optimum case where all elements must be separately accessible. A variety of different beams can be calculated in advance, stored in memories and called up in real-time. The maximum processing speed of 2 MHz enables multi-beam operation if the sampling interval exceeds 0.5 µs. With the same devices a doubling of the processing speed seems possible. Further increase of speed will require faster devices and a higher degree of chip integration.

REFERENCES

1. Swartzlander, E.E., Jr., 1980, IEEE EASCON 80 Record, 234 - 238
2. Barton, P., 1980, IEE PROC., Vol. 127, Pt F, 266 - 277
3. Sander, W., 1980, IEE PROC., Vol. 127, Pt F, 285 - 289
4. Wirth, W.D., 1978, Proc. of the 1. Mil. Microwave Conf. 379 - 390
5. Wirth, W.D., 1981, IEE PROC., Vol. 128 Pt F, 311 - 315

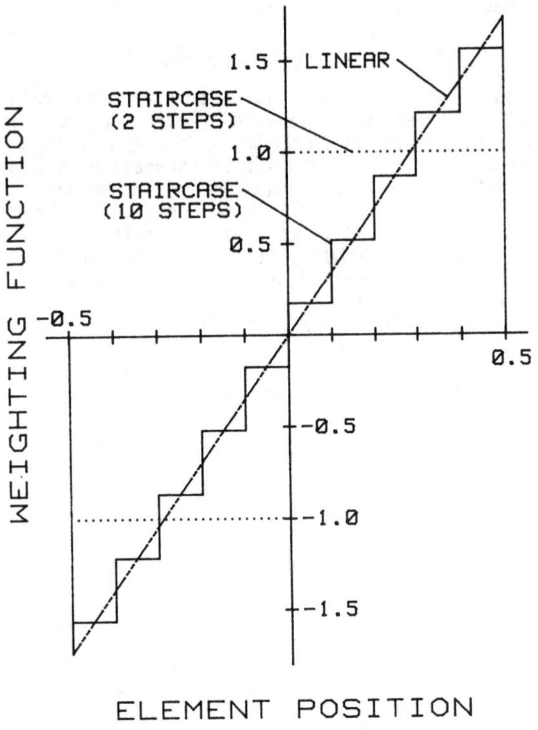

figure 1: weighting functions for generating difference beams

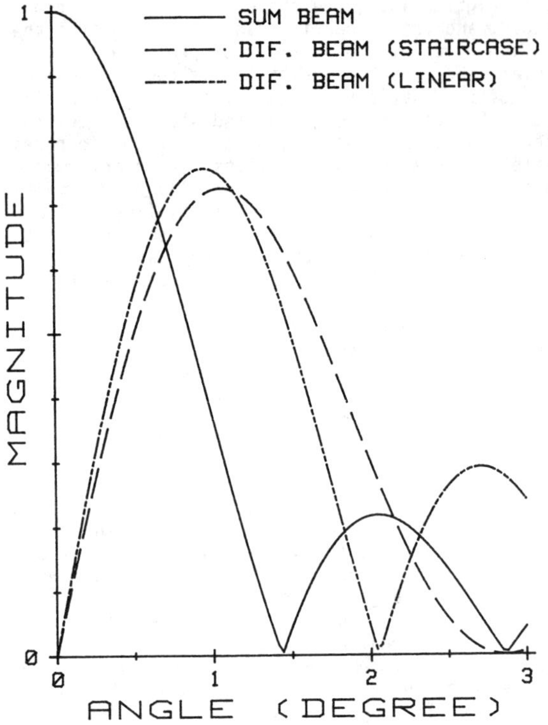

figure 2: sum and two difference patterns of a regular spaced linear array

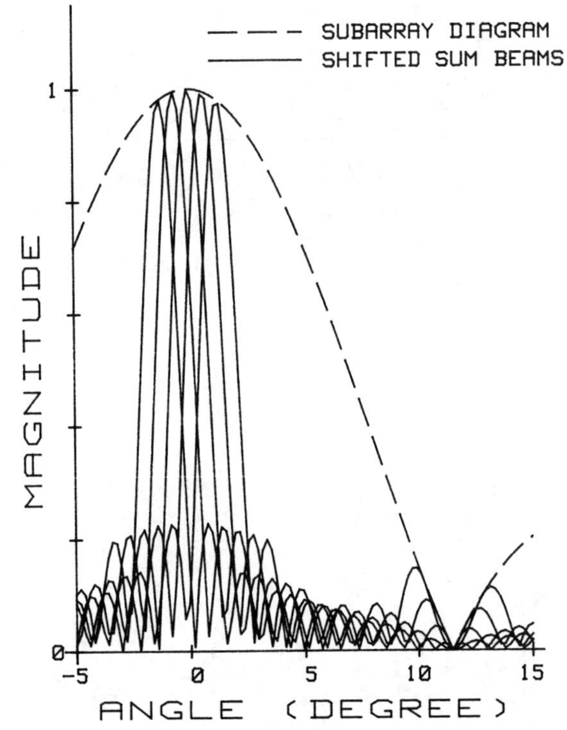

figure 3: a cluster of 5 shifted sum beams together with the subarray pattern

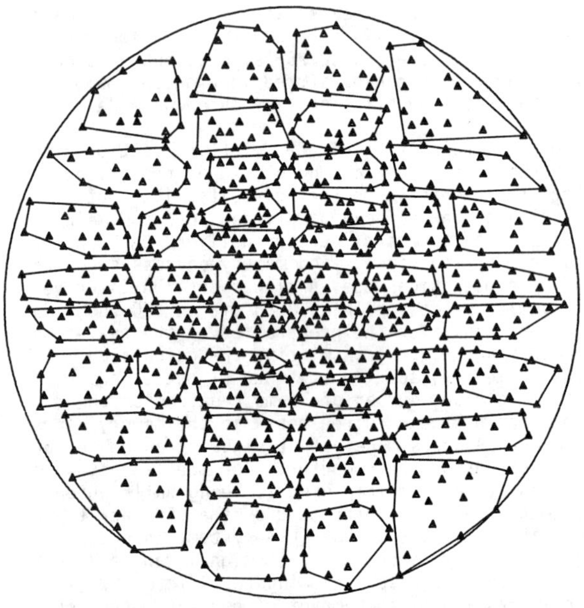

figure 4: ELRA receiving array: element distribution and division into subarrays

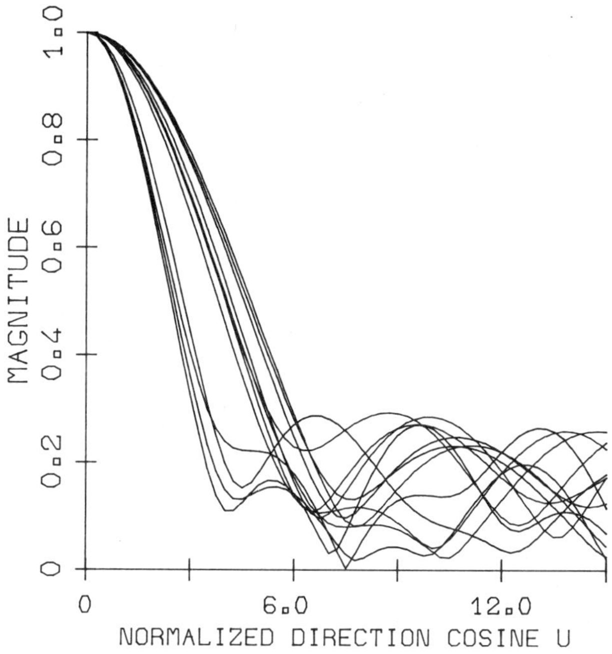

figure 5: calculated antenna patterns of 12 selected subarrays

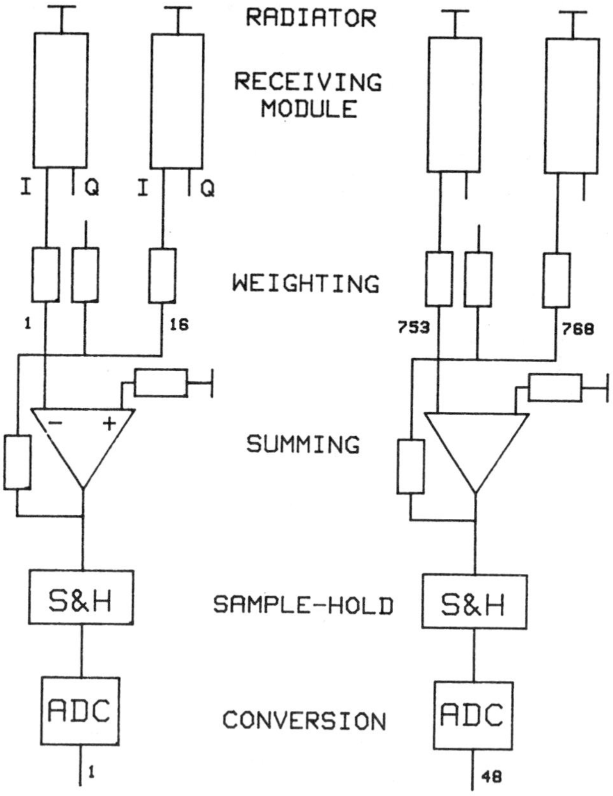

figure 6: block diagram of the receiving system and subarray beam forming

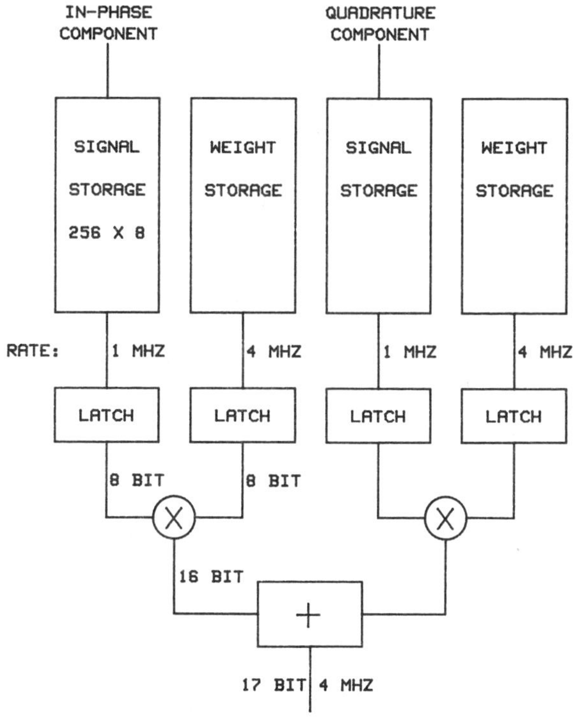

figure 7: subarray weighting unit with a complex multiplier and associated memories

figure 8: Fixed (on the right) and Variable (on the left) Beam Former

OPTICAL FIBRE NETWORKS FOR SIGNAL DISTRIBUTION AND CONTROL IN PHASED ARRAY RADARS

J.R.Forrest, F.P.Richards, A.A.Salles

Microwave Research Unit, University College London,
Torrington Place, LONDON WC1E 7JE, U.K.

P.Varnish

Ministry of Defence (P.E.), A.S.W.E.,
Funtington, Chichester, Sussex, U.K.

INTRODUCTION

A recent trend in phased array radars(1), such as the GE592 and Marconi Martello, is towards an active sub-array approach. This allows one-dimensional electronic beam steering or multiple beam formation and flexibility in aperture illumination control for low side-lobes; equally important, it allows solid-state technology to be used in distributed microwave power amplifiers and brings advantages in reliability.

Systems under development in the U.K.(2,3) and the U.S.A.(4) are now going to the next stage in an individual active element modular approach. This enables a relatively low r.f. power per element to be used, since the number of elements is high, and is well-suited to the evolving monolithic microwave integrated circuit (MMIC) technology.

Although it is difficult to predict exactly when a mass-produced active array element in MMIC form will be viable, it is certain that this will be the way of the future high performance radars. The time is therefore right to address the problems of signal interfaces with such MMIC modules. In the past, the complexity of the active module has been so great that this has dominated the attention of research and development teams. Inadequate study has been devoted to the r.f., i.f., and control signal manifolds that are required in active arrays as regards phase and amplitude tolerances, cross-talk and interconnection reliability. Unless careful thought is given at the present stage, such manifolds could become the major cost element and problem area in future arrays.

The simplest active array element (Fig. 1) would require, apart from power supply lines, connection only to an r.f. manifold and digital control bus structure for phase shifter beam steering commands. A more complex active array element (Fig. 2) for highly demanding future needs could require an r.f. manifold, an i.f. manifold, and a baseband digital bus structure. The i.f. manifold might be used to provide an i.f. reference to each array module for downconversion of radar returns to baseband and digitisation; alternatively, it might be used to carry downconverted radar i.f. signals from each module back to an i.f. beam former.

The most severe tolerances usually occur in the r.f. manifold. Manifolds for future systems are likely to require insertion phase accuracy of approximately 1° and insertion loss accuracy of 0.1dB unless some overall closed loop adaptivity to errors is included. Such tolerances are difficult to achieve since the manifold is a structure many wavelengths in dimension. Past systems have not required quite this degree of accuracy, but even so have needed very carefully matched waveguides or semi-rigid cables. Insertion phase accuracy and loss are easier to control with waveguides, but the resulting structure is bulky, which introduces problems in cooling of active array modules, and it is heavy. With cables, reproducible insertion phase and loss are more difficult to achieve particularly at the connector interfaces.

The i.f. manifold is not usually a difficult structure since the wavelength is longer than, or of the same order as, the structure dimension and losses at i.f. are relatively low. An important consideration, however, may be the ability of such a manifold to cope with a wide dynamic range of signal level and not be vulnerable to noise or electromagnetic interference.

The digital bus structure again is fairly straightforward, but careful attention needs to be directed to elimination of cross-talk between control lines and to careful screening from the active module d.c. supply lines. Each module will contain active r.f. devices drawing currents of several amperes in turn-on times of nanoseconds and this poses a considerable EMC task.

The obvious solution to the whole signal distribution problem is to use fibre optics. Fibres have key advantages in their light weight and flexibility, their high immunity to electromagnetic interference, their very large bandwidth capability and the ease of multiplexing different signals through optical wavelength diversity. In principle therefore, all signal interfaces with an active array module could be made using a single fibre and perhaps the power supply line realised with a conducting braid on that fibre. This paper describes some of the constraints that define such an optical fibre distribution network.

COMPONENTS FOR THE FIBRE OPTIC DISTRIBUTION NETWORK

Any fibre optic distribution network must comprise a light source, some means of introducing modulation on the optical carrier, a fibre, and a demodulating detector. The performance of these components will now be considered.

Optical Source

The spectral line width, $\Delta\nu$, is an important parameter of the source; it determines the coherence time, t_c, or coherence length, l_c , of the light:

$$t_c = 1/2\pi\Delta\nu \qquad l_c = t_c c/n$$

where c/n is the velocity of light in a medium of refractive index n.

For most optical links, unless heterodyne detection is involved, the coherence length can be less than the link length; it must, however, be longer than any paths over which interference effects are used, such as in some types of modulator.

For propagation over a path length L in a dispersive medium such as glass, the finite line width limits the signal bandwidth, B, on the optical carrier:

$$B_{max} \sim 10^3/DL\Delta\lambda \qquad ......(1)$$

where D is the medium dispersion (ps/nm.km), L is in km. and $\Delta\lambda$ is in nm.

Gas lasers have very narrow line widths and provide optical output power in the visible and infra-red regions at up to the watt level or more; however, they are prone to low frequency instabilities, are bulky, and require an external modulator. A solid-state optical source is preferable and this gives the choice of a light emitting diode (LED) (5) or semiconductor laser. The linewidth of an LED is very broad ($>$10nm), its output power rather low ($\sim$1mW), and its modulation frequency response limited (usually less than 200MHz), so the most attractive source for the present application is the semiconductor laser; these now provide output power, P_0, in the region of 10mW at a range of wavelengths such as 850nm (GaAlAs), 1300 and 1500nm (InGaAsP). Multiple stripe lasers producing powers of 150mW and pulse power up to 1.5W (6,7) have also been reported.

The first semiconductor lasers were multi-moded with a broad spectrum (e.g. 2nm) of emission frequencies, but it is now possible to obtain single-mode index-guided (channelled substrate planar,CSP) lasers (8) which provide only one narrow spectral line of width $\Delta\lambda$ 10^{-5}nm (~5MHz); the use of optical feedback has been shown to narrow the line further to some tens of kHz in width (9). The use of a single-mode laser avoids the problem of partition noise - the fluctuations of energy between laser modes; in a dispersive propagation path, or through connectors whose coupling efficiency is frequency dependent, such frequency fluctuations are converted to a very troublesome source of a.m. noise.

Semiconductor lasers may be directly modulated through their bias current and show a very linear dependence of optical output power on bias current I_0 above the threshold bias I_{th}:

$$P_0 = K_L\, I_0 \qquad \ldots\ldots \qquad (2)$$

In general, a mixture of a.m. and f.m. is produced (10,11); the modulation sensitivity passes through a resonance and then falls off rapidly above a critical frequency

$$\omega_c \approx \left\{\frac{1}{\tau_s \tau_p}\left[\frac{I_0}{I_{th}} - 1\right]\right\}^{\frac{1}{2}} \qquad \ldots\ldots \qquad (3)$$

where τ_s and τ_p are the electron and photon lifetimes, typically 2ns and 2ps respectively. Since $I_0 \sim 1.2 I_{th}$, it follows that such lasers may be directly modulated at frequencies up to a few GHz. A possible extension of this range to almost 10GHz might occur in future with the newly-developed low-threshold current devices which can operate at many times the threshold current. However, microwave modulation of semiconductor lasers at appreciable modulation depths frequently causes the single mode to break into a multi-mode structure; modulation at frequencies beyond 1-2GHz is not well under-stood and can present difficulties. Matching of the microwave modulation drive into the laser chip is also not an easy matter since the chip presents a very low and capacitive impedance at these frequencies (12). For these reasons, there is considerable interest in the development of external electro-optic modulators. Research versions of such modulators in integrated-optic form are now demonstrating modulation rates up to at least 10GHz with relatively modest drive input, such as 100mW.

Semiconductor laser noise has been extensively studied in optical communication applications. The noise arises from a variety of causes, such as shot and recombination effects. A typical noise characteristic (8,13) is shown in Fig.3. Far from carrier signal to noise (S/N) levels greater than 130dB (1Hz bandwidth) may be obtained at typical bias currents and this satisfies the usual radar reference frequency source requirements; close to carrier noise shows a typical 1/f character (in terms of detector output power) with a corner in the region of 10kHz. Phase noise is only important in systems which involve interference or heterodyne detection unless the system contains elements which can introduce p.m./a.m. conversion.

Considerable care must be taken to avoid reflections in the optical link. If energy is reflected back into the laser, an external resonant cavity may be formed and the interaction of this with the laser cavity can give rise to a splitting of the laser line into external cavity modes; reflexions close to the laser at the coupling point to the fibre can introduce low frequency noise due to random changes in the mechanical format.

Optical Detector

The most common optical detector is the p-i-n diode. The incident photon flux produces carrier pairs which are swept to the electrodes under a d.c. bias field and produce a photocurrent in the external circuit. The photocurrent, I_p, is directly related to the incident optical power, P_d, through the equation :

$$I_p = q\eta P_d/h\nu \;\; = \mathcal{R}\, P_d \qquad \ldots\ldots \qquad (4)$$

where q is the electronic charge (1.6×10^{-19}C), h is Planck's constant (6.626×10^{-34} J.s), ν is the optical frequency (3.53×10^{14}Hz at 850nm) and η is the quantum efficiency (14). η here takes account of light lost by reflexion at the detector surface and various mechanisms of carrier loss, a typical value for η being 0.3 - 0.5. $\mathcal{R}$ is the detector responsivity, an ideal detector at 850nm wavelength having a value $\mathcal{R} \sim 0.68$A/W.

Noise in the detector arises from any dark current, I_D, or background illumination current, I_B, shot noise, together with the thermal noise contribution of the detector circuit load, R_L, and following amplifier.

The minimum optical power, $P_{d,min}$, necessary to achieve a given (S/N) is :

$$P_{d,min} = \frac{2h\nu B}{\eta}\left(\frac{S}{N}\right)\left\{1 + \left[1 + \frac{I_B + I_D + 2kT/qR_L}{qB(S/N)}\right]^{\frac{1}{2}}\right\} \qquad (5)$$

where k is Boltzmann's constant (1.38×10^{-23} J/K) and T is the noise temperature of the detector and following amplifier (a value of 1000-3000K is typical). (S/N) represents the d.c. carrier to noise ratio and is replaced by $(S/N)\sqrt{2}/m$ for the signal to noise ratio of a sinusoidal modulation with modulation index m. Since $2kT/qR_L \gg I_B + I_D$ under most conditions,

$$P_{d,min} \approx \frac{2h\nu}{\eta q}\left[\frac{2kTB}{R_L}\left(\frac{S}{N}\right)\right] \qquad \ldots\ldots \qquad (6)$$

Taking a rather severe requirement of (S/N) = 130dB in 1Hz bandwidth, with $h\nu = 2.34\times10^{-19}$, $\eta = 0.5$, $kT = 3.2\times10^{-20}$ and a value $R_L = 50\Omega$ (a usual value for detection of high frequency modulation), then:

$$P_{d,min} \sim 6.6\times10^{-4}\ \text{W} \qquad \ldots\ldots \qquad (7)$$

Under this requirement, relatively few detectors (less than 10) could be fed from one laser, taking into account only very modest coupling and transmission losses (e.g. < 3dB). To feed more would require a relaxation of the (S/N) specification or the use of an avalanche p-i-n photodetector which provides gain. At an optimum multiplication factor with avalanche detectors (14), the value of $P_{d,min}$ may be reduced over that for the p-i-n detector. However, for large (S/N) the advantage may be very slight (e.g. a factor 2-5) and not outweigh the disadvantage that the avalanche detector imposes in the need for a well-stabilised d.c. bias supply of approximately 100V; this would require an extra power supply line to each module and is a complexity to be avoided if possible. Future progress with FET photodetectors (15), which can also provide gain, may result in an acceptable low voltage alternative to the avalanche detector.

From this analysis, it is seen also that signal dynamic ranges of approximately 70dB are possible with signal bandwidths of 1MHz, but if radar return signals were to use the optical network, considerable amplification in each element prior to modulation on to the optical carrier would be necessary.

Optical Fibre

Optical fibres may be of either single mode (core diameter ~ 5μm) or multimode type (core diameter ~ 50μm).

The multi-mode fibre is easily joined and interfaced with active devices because of its larger core diameter and is frequently used in lower data rate communication links. The different modes, propagating a slightly different velocities in the fibre, interfere with each other and give a characteristic speckle pattern when the fibre output is imaged on to a detector. This speckle pattern is very sensitive to mechanical distortions or vibration of the fibre and to variations in the laser emission wavelength. A degradation of (S/N) to some tens of dB on the optical link can result and this is unacceptable in the analogue modulation application.

It therefore appears that to preserve the high (S/N) of the source, a single mode fibre is required. The single mode fibre can occasionally give rise to problems with two polarisation states (16) and it may be necessary to place polarisation filters at either end of the fibre.

Single mode fibre connectors having an insertion loss of less than 1dB have now been developed (17) and it is clear that the use of single mode fibre is not as daunting as was at first imagined.

The choice of optical wavelength for the fibre path must also be considered. Currently available fibres show zero dispersion for a wavelength around 1300nm, and maximum bandwidth.length products in excess of 100GHz.km are possible. For a wavelength of 850nm, the dispersion $D \sim 50$ps/nm.km, which, in Eqn(1) yields a bandwidth.length product of some 10GHz.km. Since the phased array application is unlikely to involve propagation paths of more than some tens of metres maximum, it appears that optical wavelength is not an important consideration. With attenuation in the fibre at values of 1-2dB/km for 850nm and less than 1dB/km for 1300nm, the choice of wavelength is again unimportant. Moreover, attenuation due to the fibre may effectively be neglected in an overall system analysis.

Changes in temperature of a fibre affect the signal propagation delay through a thermal expansion in length L, and a change in refractive index, n. The overall path delay $\tau = Ln/c$ and has a value of approximately 5ns/m for typical fibres with $n \sim 1.5$. Then:

$$\frac{d\tau}{dT} = \frac{1}{c}\left[n\frac{dL}{dT} + L\frac{dn}{dT}\right] \quad \ldots\ldots \quad (8)$$

For a modulation frequency f_m on the optical carrier, the change in phase of the received modulation signal with temperature is:

$$\frac{d\phi}{dT} \approx \frac{2\pi f_m}{c}\left[n\frac{dL}{dT} + L\frac{dn}{dT}\right] = \frac{2\pi L f_m}{c}\left[n\alpha + \frac{dn}{dT}\right] \quad (9)$$

The values of thermal expansion coefficient α and $\frac{dn}{dT}$ vary significantly between fibre types, but for order of magnitude estimation may both be taken to have values of about $10^{-6}/^oC$. For L=10m and f_m =3GHz, $d\phi/dT \sim 0.1deg/^oC$. Unless different fibre lengths to array elements are used, it is only temperature differences between fibres that are relevant. Clearly, temperature effects may be neglected in the example chosen.

Fibres can handle relatively high optical power and a single mode fibre will carry approximately 1W of optical power before the onset of non-linear effects.

Optical Distribution Network

In order to divide the optical signal to individual modules, an optical manifold is required. Such networks, for single mode fibres, have been constructed and can take the form of "T" or "star" couplers (18). A network of T-couplers is the direct analogue of the microwave corporate feed, whereas the star coupler has the fibres joined essentially at one point. Star couplers for up to about 10-way division have been realised with losses of approximately 1dB at the junction. For a very large distribution network, it is likely that the choice would be to cascade star couplers, each providing maybe 10-way division and each manufactured in highly reproducible integrated-optic form.

Because of the rather high optical signal level needed at each photodetector for high (S/N) requirements, a logical structure for the optical distribution network would be to use a laser amplifier, acting in a similar way to a repeater in a communications link, at each star coupler (Fig.4). Injection locked lasers (19,20) have been used with considerable success in the telecommunications application with high frequency f.m. signals. Conventional locking theory applies and locking gains in the region of 20dB have produced locking ranges of several GHz.

SYSTEM LOSS BUDGET

For a fibre system consisting of a source, N-way fibre distribution network and detectors, the relationship between input r.f. or i.f. modulation power, p_{in}, and output r.f. or i.f. power from a detector, p_{out}, may be derived.

$$p_{in} = i_m^2 r = \left(\frac{p_o}{K_L}\right)^2 r \quad \ldots\ldots \quad (10)$$

where p_o is the r.f. or i.f. component of the optical output power, i_m is the laser modulation current and r is the slope resistance of the laser. K_L is also frequency dependent. The r.f. or i.f. optical power reaching a detector, p_d, is:

$$p_d = \frac{\alpha_c \alpha_t \alpha_b}{N} p_o \quad \ldots\ldots \quad (11)$$

where α_c, α_t, and α_b are optical transmission coefficients representing losses due to coupling in and out of the fibre, and in the N-way distribution network. The i.f. or r.f. power from the detector, p_L, with a load R_L is then:

$$p_{out} = i_p^2 R_L = (\mathcal{R} p_d)^2 R_L \quad \ldots\ldots \quad (12)$$

$\mathcal{R}$ is also frequency dependent.

Hence,

$$\frac{p_{out}}{p_{in}} = \left(\frac{\mathcal{R}\alpha_c\alpha_t\alpha_b K_L}{N}\right)^2 \frac{R_L}{r} \quad \ldots . \quad (13)$$

With typical values of $\mathcal{R}$ =0.25A/W (p-i-n detector), α_c=0.5 (3dB), α_t=1, α_b=0.7, K_L=0.2W/A, R_L=50Ω, r=4Ω

$$\frac{p_{out}}{p_{in}} = \frac{3.83\times10^{-3}}{N^2} = -(24+20\log N) \text{ dB} \quad (14)$$

For N=1 (which approximates a single optical path), the figure of -24dB is close to that found by Kiehl (21).

This needs to be compared with a conventional r.f. microwave corporate feed structure; in such a feed the number, N_d, of T 2-way power dividers is given by $N_d = \log N/\log 2$. If each splitter has a through loss α_d (dB), the total loss between source and each port is 3.322 $\alpha_d$log N. To this must be added the loss in the waveguide, cable or stripline connecting the splitters. Thus, the total insertion loss becomes:

$$\frac{P_{out}}{P_{in}} = -(3.322\,\alpha_d \log N + \alpha_t' \ell_t)\ \text{dB} \qquad (15)$$

where α_t' is the transmission line loss/unit length in dB/m and ℓ_t is the path length (m) from source to array element. Typically, ℓ_t will be of the order of the array face dimension. For values α_d =0.5dB, α_t =3dB/m (cable), and ℓ_t =2m,

$$\frac{P_{out}}{P_{in}} = -(6 + 1.6\log N)\ \text{dB} \qquad (16)$$

It is clear that there is a considerable loss penalty paid in the use of the optical system, which must be set against the advantages given earlier. However, in an active phased array, the distribution network operates at very low power level and losses are insignificant in the overall array power budget.

EXPERIMENTAL STUDIES

Several research establishments are already reporting preliminary results in research programmes on optical control of phased arrays (2,22,23). A 40W S-Band active array module has been demonstrated with the r.f reference signal provided by an optical link (2), the optical detection being performed by an injection-locked MESFET oscillator.

Because of the modulation frequency limitations with semiconductor lasers and external modulators, recent work at U.C.L. is concentrating on optimising MESFET photodetectors and on a scheme to provide the r.f. reference signal as a difference frequency between the optical frequencies of two single mode c.w. lasers (Fig.5); one laser is locked to the other at a constant microwave frequency offset by means of a heterodyne phase locked loop. Such a technique offers the prospect of avoiding any direct or external modulation on lasers.

CONCLUSIONS

Some preliminary considerations for optical fibre distribution networks in active phased arrays have been outlined. It has been shown that optical fibre distribution networks can provide the required signal to noise ratio for r.f. signal distribution, though the number of active modules that can be supplied from a single semiconductor laser is rather limited if high (S/N) ratios are required. An active distribution network using star couplers and laser amplifiers is proposed for large arrays.

Optical fibre networks show more power loss than their microwave transmission line equivalents, but bring many advantages that outweigh this disadvantage. However, to justify their use it is important that one fibre should replace all the signal interfaces to a module and thus simplify the overall interconnection network. The high bandwidth of fibres and the ease of wavelength diversity permits this.

The topic is still in its infancy and there are many possibilities yet to be investigated for the use of fibre optics and optical signal processing in arrays.

ACKNOWLEDGEMENT

Support for this work is being provided under a research agreement between UCL and the Ministry of Defence (PE), A.S.W.E., together with personal support grants (F.P.R. & A.A.S.) from CNPq, Brazil.

REFERENCES

1. J.R.Forrest :'Phased arrays-current technology and future prospects'. Proc. 11th European Microwave Conf., Amsterdam, 1981, pp.81-90.
2. C.J.Ward et al.:'High phase accuracy active phased array module for multifunction radars'. IEEE MTT-S Symp, Dallas, 1982, Tech. Digest, pp.179-181.
3. R.S.Pengelly:'GaAs monolithic microwave circuits for phased array applications'. Proc. IEE, Part F, 127, 301-311 (1980).
4. D.N.McQuiddy:'Solid state radar's path to GaAs'.IEEE MTT-S Symp., Dallas, 1982, Tech. Digest, pp.176-178.
5. R.Plastow et al.:'Light emitting diodes for optical fibre systems'. Proc. Int. Conf. 'Communications 82' IEE Conf. Publ. No. 209, pp.276-279.
6. D.E.Ackley and R.W.H.Engelmann:'Twin stripe injection laser with leaky mode coupling'.Appl. Phys. Lett. 37, pp.866-868 (1980).
7. D.E.Ackley and R.W.H.Engelmann:'High power leaky mode multiple stripe laser'. Appl. Phys. Lett., 39, pp.27-29 (1981).
8. K.Peterman and G.Arnold:'Noise and distortion characteristics of semiconductor lasers in optical fibre communication systems'. IEEE Trans. MTT-30, pp.389-401 (1982).
9. L.Goldberg et al.:'Spectral characteristics of semiconductor lasers with optical feedback'. IEEE Trans. MTT-30, pp.401-410 (1982).
10. S.Kobayashi et al.:'Direct frequency modulation in AlGaAs semiconductor lasers'. IEEE Trans. MTT-30, pp.428-441 (1982).
11. A.J.Seeds and J.R.Forrest:'High rate amplitude and frequency modulation of semiconductor lasers'. To be published in Proc. IEE, Solid State & Electron Devices, Dec. 1982.
12. R.S.Tucker:'Microwave circuit models of semiconductor injection lasers'. IEEE MTT-S Symp., Dallas, 1982, Tech. Digest, pp.104-106.
13. T.G.Giallorenzi et al:'Optical fibre sensor technology'. IEEE Trans. MTT-30, pp.472-511 (1982).
14. S.M.Sze:'Physics of Semiconductor Devices'. 2nd. Ed. J.Wiley & Sons, 1981.
15. R.I.MacDonald:'High gain optical detection with GaAs field effect transistors'. Appl. Optics, 20, pp.591-594 (1981).
16. D.N.Payne et al.:'Development of low and high birefringence optical fibres'. IEEE Trans. MTT-30, pp.323 -334 (1982).
17. J.I.Minowa et al.:'Optical componentry used in field trial of single mode fibre long haul transmission'. IEEE Trans. MTT-30, pp.551-563 (1982).
18. T.G.Giallorenzi:'Optical communications research and technology:fiber optics' Proc. IEEE, 66, pp.744-780 (1978).
19. S.Kobayashi and T.Kimura:'Optical FM signal amplification by injection locked and resonant type semiconductor laser amplifiers'. IEEE Trans. MTT-30, pp.421-427 (1982).
20. D.J.Malyan and D.W.Smith:'The application of injection locked semiconductor lasers for optical communications in the wavelength range 0.8um to 1.6um'. Proc. Int. Conf. 'Communications 82'. IEE Conf. Publ. No.209, pp.285-290.
21. R.A.Kiehl and D.M.Drury:'Performance of optically coupled microwave switching devices'. IEEE Trans. MTT-29, pp.1004-1010 (1981).
22. A.M.Levine:'Fiber optics for radar and data systems' SPIE, 150, Laser and Fiber Optics Communications, pp.185-192 (1978).

23. R.Wolfson et al.:'Coherent R.F. modulation of an optical carrier for application to phased array antennas'. IEEE AP-S Symposium, Albuquerque, 1982, Tech. Digest.

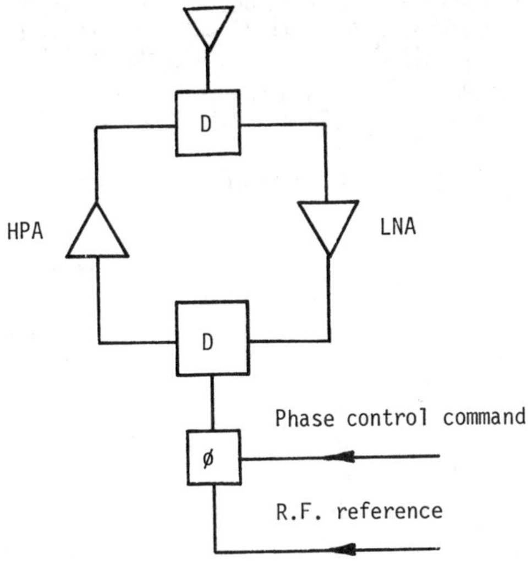

Fig.1: Simple Active Array Element.
(HPA: High power amplifier; LNA: Low Noise Amplifier, D: Duplexer)

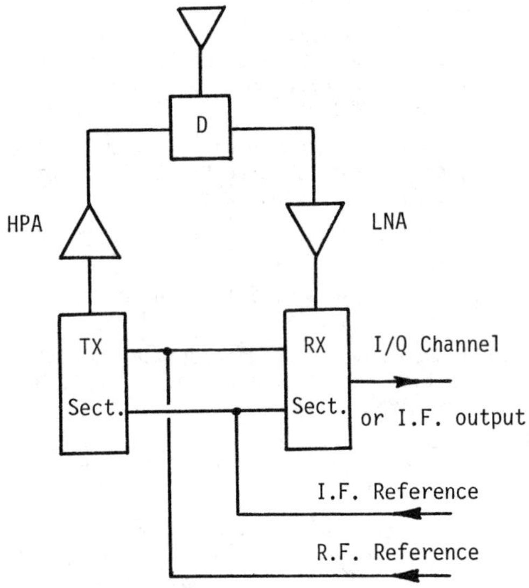

Fig.2: More Complex Active Array Element

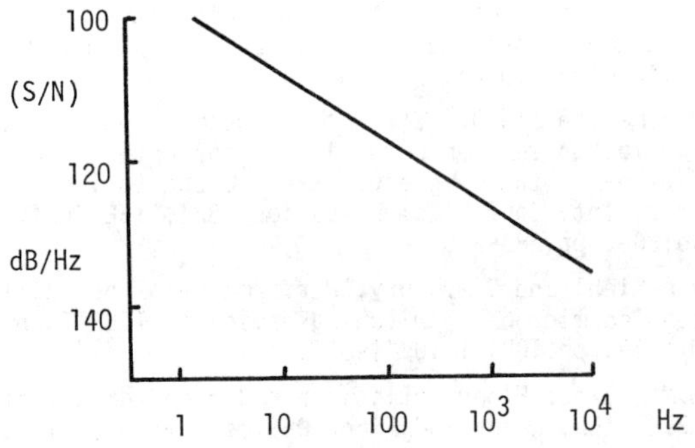

Fig.3(a): Close-to-Carrier Laser A.M. Noise.(13)

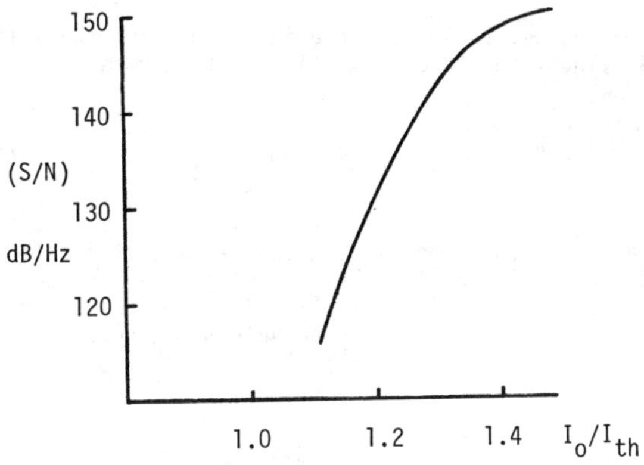

Fig.3(b): Laser A.M. Signal-to-Noise Ratio as a Function of Bias Current. (8)

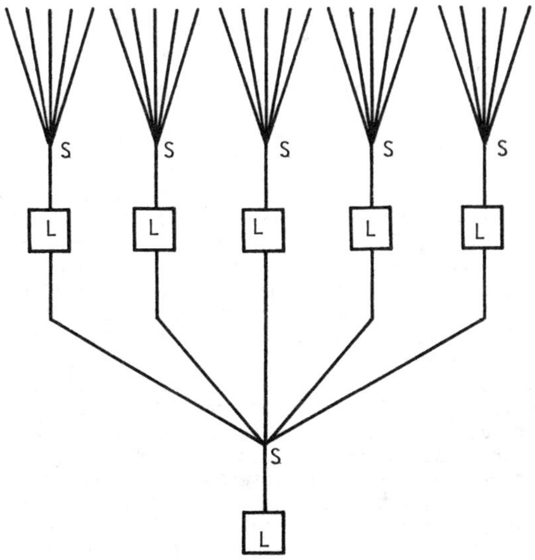

Fig.4: Active Optical Fibre Distribution Network using 5-way Star Couplers (S) and Laser Amplifiers (L).

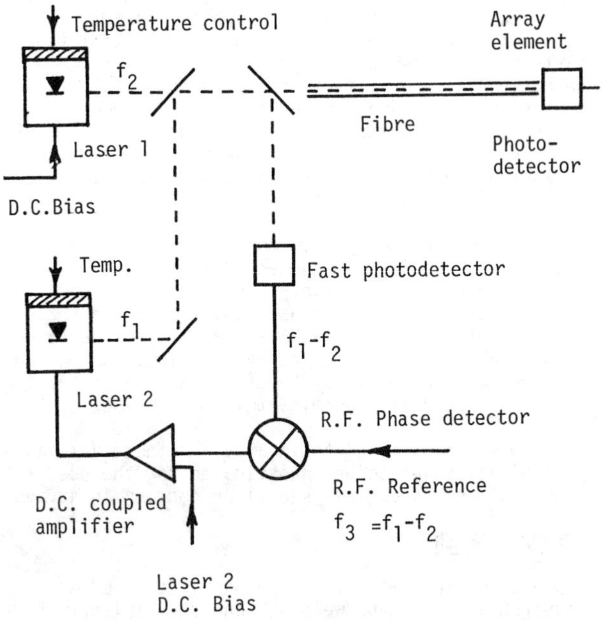

Fig.5: Proposed R.F. Reference Distribution.

PORTABLE FMCW RADAR FOR LOCATING BURIED PIPES

A. D. Olver, L. G. Cuthbert, M. Nicolaides and A. G. Carr

Queen Mary College, University of London, UK

INTRODUCTION

The development of a portable FMCW radar is described. The aim of the development was to assess the potential of this type of radar for locating metallic or plastic pipes buried up to about one metre in urban environments. The radar had to be portable, non-contacting with the ground, capable of being operated by one person and able to provide an almost instantaneous display showing the presence or absence of public-utility service-pipes; it also had to be a relatively low cost system.

The radar described is a development of an earlier laboratory system (Clarricoats (1) and Clarricoats et al (2)) which had shown that the detection of shallow buried objects was feasible using an FMCW radar. A schematic diagram is shown in Figure 1. The transmitted signal, from a solid state microwave source, is focussed onto the ground by means of an off-set paraboloidal reflector fed by a ridged horn feed. The beam is incident at an oblique angle on the ground so that most of the strong ground-return-signal is reflected forwards and does not interfere with any return signal from a target. The relatively high dielectric constant of the ground means that the portion of the transmitted signal which enters the ground propagates down nearly vertically. A photograph of the radar mounted on a portable trolley is shown in Figure 2. The solid state microwave source and control components together with associated power supplies are at the base of the trolley. In the version used for evaluation of system performance on site the received signal is processed by a Hewlett Packard Spectrum Analyser, but in any future versions this would be replaced by a micro-computer which would obtain the frequency spectrum by means of the Fast Fourier Transform and also perform the analysis necessary to obtain range information and to enhance the probability of detecting a target (Carr et al (3)).

FMCW RADAR

The QMC radar uses the FMCW principle, transmitting a burst of energy which sweeps from 1.0 GHz to 2.0 GHz. The basic repetition rate (which determines the range resolution) is about 10 Hz and the sweep time is about 10 ms. Neither of these two latter parameters has been found to be critical and acceptable performance is obtained over a range of values.

The use of FMCW arose after a detailed study had been made of the relative merits of FM and pulse-radar systems. Pulse systems are usually used for space radars but the major difference between space radars and radars for locating buried objects is the distance the transmitted signal has to travel before being reflected by a target. In a ground-probing radar the distance is very short, necessitating a short pulse. Short pulses can be generated without too much difficulty but, to be useful, they must be faithfully transmitted into the air, and it is here that the problem arises. A short pulse consisting of a single cycle has a frequency spectrum which extends from DC to a high frequency. In the pipe-locating radar the bandwidth to obtain good reproduction needs to be about 8:1. Very few antennas are capable of transmitting or receiving this bandwidth. In addition to the broad bandwidth, the antenna for this application needs to satisfy two other criteria: it must have a constant beam-width throughout this broad bandwidth of operation; it must be highly directive. Much of the energy in a short pulse is transmitted at frequencies below 1 GHz where highly directive antennas are not easily portable.

The problems of constructing a good antenna for a pulse system led to the conclusion that an FMCW system is preferable as the antenna requirements, although still stringent, can be met with modifications to available designs.

The FMCW radar has the disadvantage that the detected signal must be transformed into a frequency spectrum in order to yield the range information. In the past this has required a Spectrum Analyser; it is now possible to do the transformation digitally as part of the signal processing system. The waveforms for the FMCW radar are shown in Figure 3. Although the output spectrum has a sin x/x type response the sidelobes can be reduced by modifying the envelope of the transmitted signal.

The basic FMCW radar equation relating the difference frequency f_d to the range R (in metres) is:

$$f_d = \frac{R\sqrt{\varepsilon}\,\Delta f}{1.5\times10^{8}T_s}$$

where ε is the relative dielectric constant of the ground, Δf the frequency sweep and T_s the sweep time. Substitution of typical values for the QMC radar (Δf = 1.0 GHz, T_s = 10 ms) gives values of f_d in the audio range for values of R of a few metres.

CONSIDERATIONS GOVERNING CHOICE OF FREQUENCY BAND

The frequency band used is a compromise between a number of factors which are detailed below. They may be summarised as a compromise between the requirement for a physically small antenna, high resolution, and the need to overcome the loss in the ground.

Horizontal resolution

The horizontal resolution depends on the antenna: the larger the size of the antenna, in wavelengths, the more concentrated is the transmitted beam and the smaller is the size of the object that can be resolved in the horizontal plane. This is particularly important in regard to distinguishing two or more targets spaced close together and is illustrated in Figure 4. The estimated resolution (3 dB radius) of the QMC radar using a 600 mm diameter reflector is 110 mm at 3 GHz and 280 mm at 0.75 GHz. This relatively good resolution is due to the choice of antenna, in this case a focussed reflector system, which gives a particularly directive pattern at the focal plane on the ground.

A higher equivalent resolution could be obtained by making use of polarisation discrimination, transmitting on one polarisation and receiving on the orthogonal polarisation. This requires a more complicated feed antenna but should be practical and would be worth investigating in the future. Polarisation discrimination is particularly suitable for pipe location because the direction of the pipes is usually known.

Antenna size

For portable operation, the antenna should not be more than 1.0 m in diameter and even this is larger than ideal. This size of the antenna limits its ability to produce a highly directive beam. Since, for a given antenna diameter, a higher frequency produces higher directivity it is desirable in this application to use the highest practicable frequency.

Non-contacting antenna

An important principle in the success of the QMC radar is that the antenna makes no contact with the ground. This is easier to achieve at a high frequency. At low frequency the ground is relatively very close to the antenna and probably in the inductive near-field region. This means that the parameters of the ground surface considerably influence the antenna performance and the true antenna is actually the combination of above-ground antenna plus ground plane. In these circumstances a directive beam would be very difficult to achieve.

Vertical resolution

The range resolution, or resolution to which two objects can be resolved vertically, is inversely proportional to the frequency bandwidth. In typical soils a resolution of about 100 mm requires a frequency bandwidth of 1 GHz and 200 mm requires 0.5 GHz. If the presence of only a single object is required, the frequency bandwidth need only be a few hundred MHz. This can be a significant feature of FMCW radar in contrast to pulse radar where the frequency spectrum is necessarily very broad.

Attenuation of the radar signal in the ground

The attenuation of electromagnetic waves in the ground increases approximately linearly with both increasing frequency and increasing soil-volumetric-water-content. This is the most important feature in deciding the choice of operating frequency. An investigation of published papers has shown that practical soils vary considerably in their water content, and the electrical properties of a vertical section of soil are unlikely to be uniform. The worst type of soil is very wet clay, the best, dry sand. Figure 5 shows the approximate attenuation to be expected from various soils and water contents. There is some indication that these values may be optimistic for UK soil conditions and more work needs to be done to quantify the expected losses.

Assuming a transmitted power of about +5 dBm and a receiver sensitivity of about -85 dBm it is clear from the figure that detection of an object buried 1 m in very wet soils will not be feasible at frequencies above about 200 MHz. At such a frequency the production of a compact portable high resolution radar of any type will not be easy to achieve.

Choice of operating frequency

Bearing in mind all the considerations described above, the frequency range for the QMC radar was chosen to be 1.0 GHz to 2.0 GHz because the antenna could be made compact and reasonable resolution might be expected. For assessment purposes the restrictions imposed by very wet soils were accepted.

PORTABLE ANTENNA SYSTEM

The 600 mm diameter, fibreglass reflector used in the QMC radar is a portion of a 1.2 m symmetric paraboloidal reflector with an f/D ratio of 0.25, the offset configuration being used in order to avoid the effects of feed blockage. The feed is a ridged horn which has a broadband, nearly constant pattern, characteristic. The crucial design area, since it controls the VSWR of the horn, is the junction between the coaxial line input and the feed plates; this was optimised by trial and error.

The antenna is directed so that the ground-reflected signal does not return into the antenna. This improves the detection capability, but has the disadvantage that the sidelobes of the antenna radiation pattern become more significant. If a sidelobe is large and directed vertically at the ground then the ground return can give a false target. This is more of a problem as the frequency is reduced and at 1.0 GHz to 2.0 GHz it was found that the forward look angle should be kept below 30° to minimise the sidelobe ground return.

TRANSMITTER

FMCW systems have not been widely used in the past because of the problems of generating an accurate, linear swept-frequency signal. Voltage-controlled transistor-oscillators now make it feasible to make a low cost, low power, portable FMCW source. The frequency-time characteristic needs to be linear in order to obtain sensitive detection of targets. This was found to be a considerable problem at one stage in the project, but was overcome by using digital techniques to store and reconstruct the correct-shape voltage waveform needed to obtain an accurate frequency sweep. The block diagram of the transmitter and control electronics is shown in Figure 6. The voltage values required to drive the VCO to produce a linear frequency sweep are stored in 256 bytes of an EPROM addressed by a counter controlling the sweep. Data values read from the EPROM are converted to a voltage by a D/A converter and filtered before being fed to the control input of the VCO. The VCO is followed by a voltage-controlled attenuator

(VCA). The output from the VCA is sampled and compared with the data stored in a second EPROM. The difference voltage controls the VCA and enables the amplitude of the FM signal to be shaped. A uniform rectangular shape gives a sin x/x response of the frequency spectrum of the detected signal (Figure 3). However, this is undesirable because the high range sidelobes (-13.2 dB below peak) can be mistaken for a false target, particularly in situations where two or more targets are present and intermodulation terms are generated. The range sidelobe level can be reduced, at the expense of main beam broadening, by using a tapered amplitude shape which has a low value at the minimum and maximum frequencies and a peak value at the centre frequency. The effect is similar to the way an illumination taper across an antenna aperture alters the radiation pattern. The amplitude control EPROM contains a number of weighting functions which can be switch selected. Experience indicated that a Taylor type function, giving -40 dB sidelobes is the optimum choice.

SYSTEM CONFIGURATION

The received signal is fed to one port of a mixer where it is mixed with a sample of the transmitted signal. Since the received signal is delayed in time with respect to the transmitted signal, the output from the mixer is a difference frequency, proportional to range, which is fed to the Spectrum Analyser. The layout of the receiver microwave components was found to be important in determining the self clutter of the radar. Discrete components are used, but in a production version it would be better to use a custom-built integrated front-end. A number of configurations were evaluated. The most obvious choice is to use a bistatic system with separate transmitting and receiving antennas. However, this was not found to be successful because it is extremely difficult to electromagnetically isolate the two antennas and because, with two horns feeding the same reflector, neither horn can be at the focal point, (their "footprints" on the ground do not then coincide). Monostatic systems were found to be better, even though signal is lost in the return directional coupler. The arrangement shown in Figure 7 was found to give a self-clutter level of about -90 dBm, monitored by removing the antenna and connecting a matched load (Figure 8). With the antenna connected and facing RAM, the clutter levels for the frequency range 2-10 kHz are between -83 and -87 dBm. A strong target return (metal plate or ground) provides almost 30 dB of signal to clutter ratio.

TESTS WITH RADAR

A typical radar return from the QMC radar is shown in Figure 9. This shows the amplitude against range output from the Spectrum Analyser for a 150 mm diameter ceramic pipe buried 1.0 m below the surface. The Figure shows the radar return when the radar is both over the pipe and not over the pipe, clearly indicating the target at a free space range of 4 m. Since the depth of the target is known in this case the average relative dielectric constant of the ground can be deduced as 4.8. Here the ground was covered by concrete which kept the soil dry; hence the attenuation could be expected to be low.

Early experience gained in the laboratory showed that a trained operator is needed to interpret the display. When the ground is rough the radar return can vary from position to position because of the changing antenna-ground interaction. In these circumstances it is often quite difficult to detect a change in the radar return when a pipe is being illuminated. The addition of digital processing eases this problem. Even the simple subtraction of the two radar returns in Figure 9 would considerably improve the display.

A series of field trials was conducted at locations in South East England. The ground at these locations contained a number of buried public-utility-pipes. Overall a detection capability of about four out of five pipes was recorded.

However, a second set of measurements with a modified system at the same sites recorded a detection capability of only one out of five. There is some uncertainty as to the differences between the two sets of measurements but one explanation is changes in the ground attenuation. As was shown earlier, any microwave radar is dependent on the loss in the ground being low enough to ensure that a transmitted signal, after having travelled down through the ground, and been reflected by a pipe and travelled back up through the ground, is strong enough to be detected. Unfortunately it is impossible to see from a ground-probing radar return whether the absence of a target implies no target or whether the soil attenuation is too high. The only way of resolving the problem is to operate at a lower frequency. However, as shown above, that inevitably means less resolution with a portable antenna system. Probably the best compromise would be to use a frequency sweep of 0.5 GHz to 1.0 GHz or 0.25 GHz to 0.5 GHz and redesign the antenna.

The tests have shown that, without signal processing, the QMC radar will detect the presence of pipes buried up to 1 m, provided the ground attenuation is acceptable.

ENHANCED DETECTION CAPABILITY WITH DIGITAL SIGNAL PROCESSING

The results discussed above had no additional processing after the Fourier transform was obtained using the Spectrum Analyser. Studies on digital signal processing to enhance the target detection capability have been reported by the authors (3). This processing is particularly useful in situations where the radar return from a pipe is lost in the clutter of the radar. The output from the mixer is digitised, then low pass filtered using a digital filter. Filtering "smooths" the data and helps detection because unwanted signals at high frequencies are attenuated. The conversion to a frequency spectrum is then performed with an FFT algorithm and the resulting signal is correlated with a template corresponding to the return from a known target. The process minimises the error between the unknown radar return and up to five templates using a non-linear numerical optimisation technique.

The digital processing has been implemented on a Motorola 6800 microprocessor with a hard-wired multiplier and the output displayed on a VDU. An example of processing is shown in Figure 10 where two 70 mm diameter plastic pipes buried on top of each other in a box of sand are detected. The method has been shown to enhance the detection capability but its limits have not yet been assessed. The main disadvantage of the numerical optimisation approach is the large amount of computational time required, amounting to several minutes on the 6800 prototype system. Faster and more advanced microprocessors will reduce this time substantially so that real-time processing may be almost achievable.

ACKNOWLEDGEMENTS

The authors would like to thank the British Gas Corporation for permission to publish this paper. The views expressed in this paper are solely those of the authors and do not represent those of the British Gas Corporation.

One of the authors (A.G. Carr) is indebted to the UK Science & Engineering Research Council for a studentship during the period of research.

REFERENCES

1. Clarricoats, P.J.B., 1977, "Portable radar for the detection of buried objects," IEE Conference Proceedings, "Radar 77", London, England.

2. Clarricoats, P.J.B., Kularajah, R., Lentz, R.R. and Poulton, G.T., 1977, "Detection of buried objects by microwave means", Microwave Exhibitions and Publications Ltd Conference Proceedings, "7th European Microwave Conference", Copenhagen, Denmark.

3. Carr, A.G., Cuthbert, L.G. and Olver, A.D., 1981, IEE Proceedings Part F, 128, 331-336.

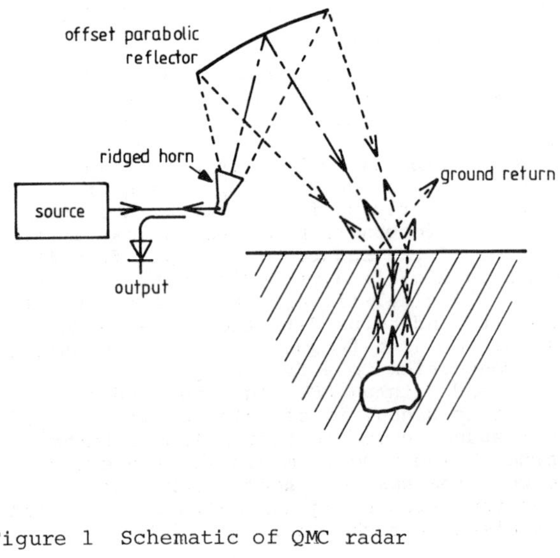

Figure 1 Schematic of QMC radar

Figure 2 Portable radar

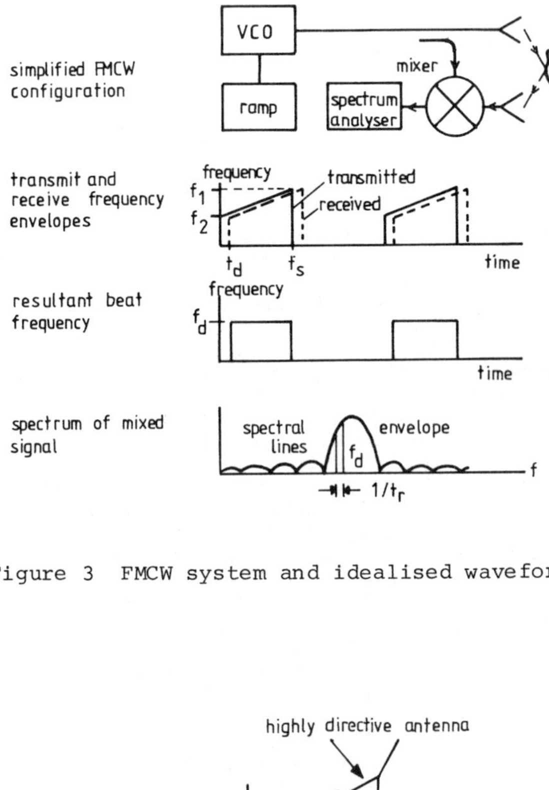

Figure 3 FMCW system and idealised waveforms

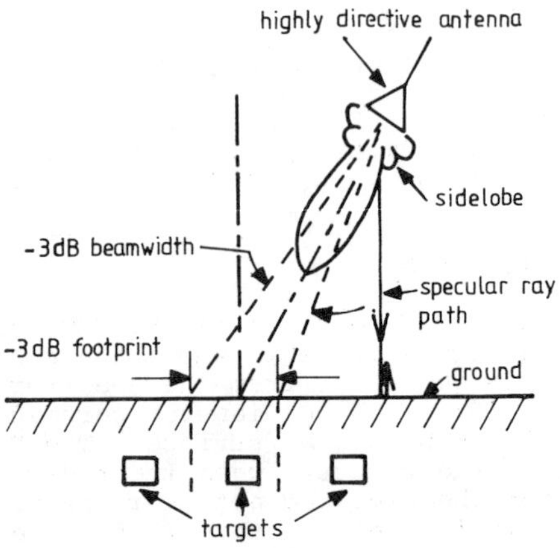

Figure 4 Antenna directivity requirement for satisfactory horizontal resolution

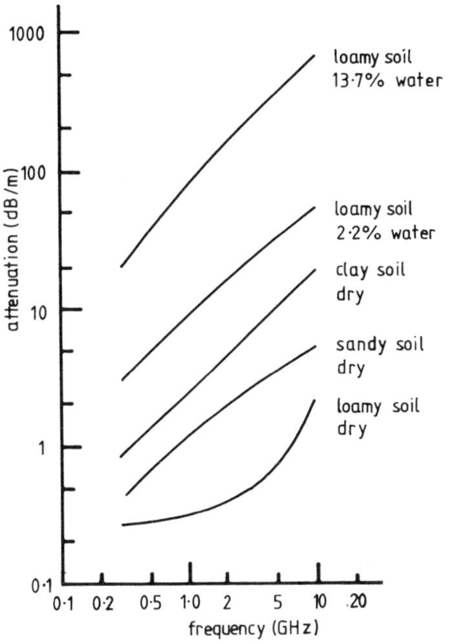

Figure 5 Approximate attenuation as a function of frequency for different soil types

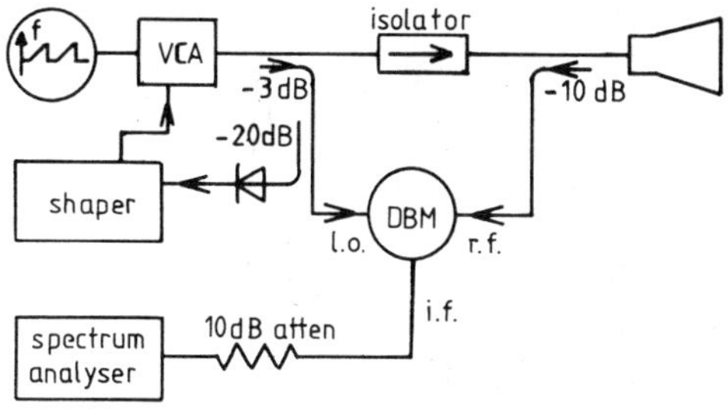

Figure 7 Receiver block diagram

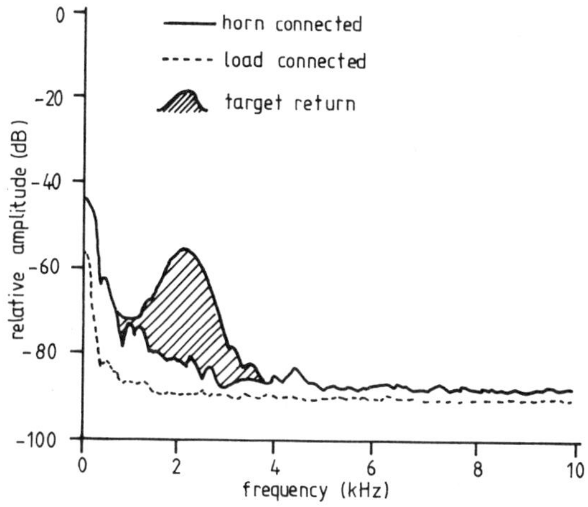

Figure 8 Radar return using configuration of Figure 7

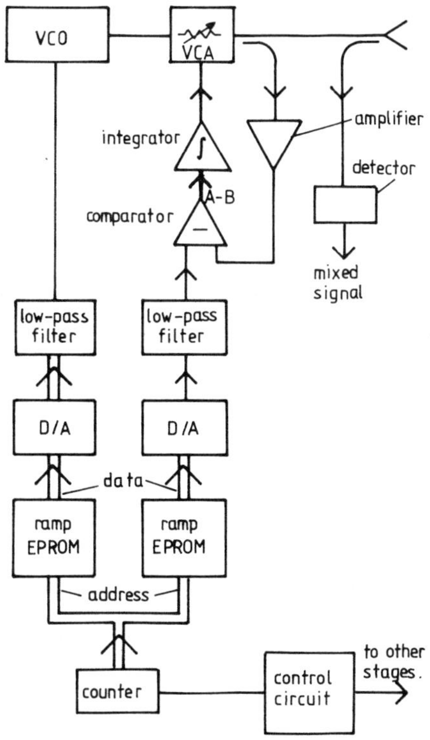

Figure 6 Simplified block diagram of transmitter and control electronics

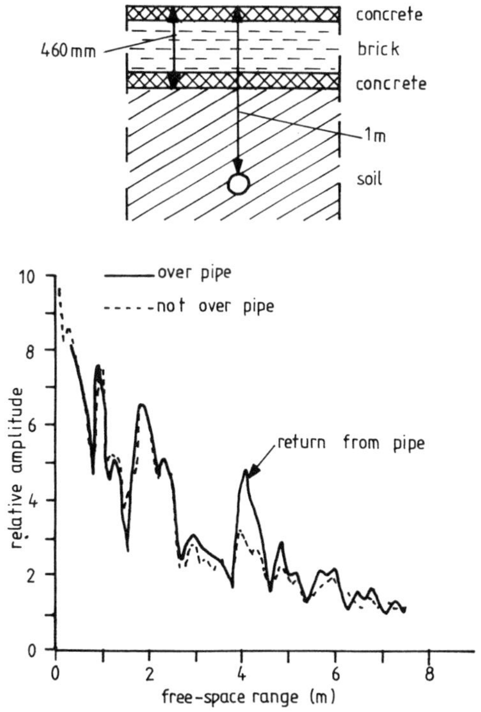

Figure 9 Radar return from a ceramic pipe

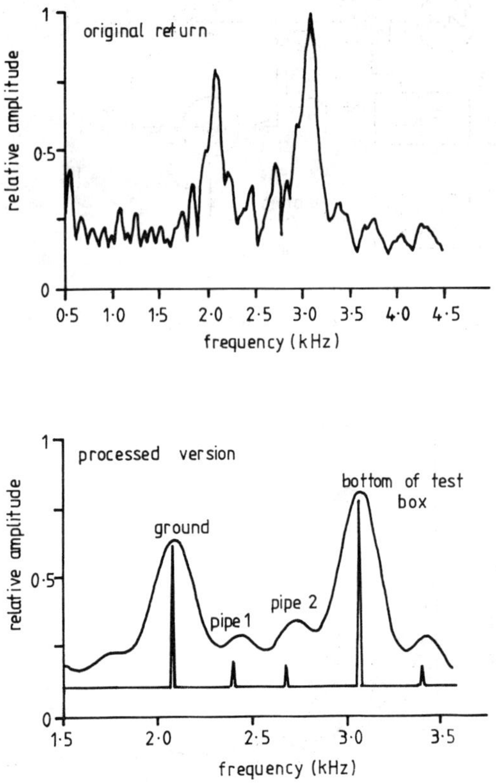

Figure 10 Radar return from two 70 mm plastic pipes

A NOVEL METHOD OF SUPPRESSING CLUTTER IN VERY SHORT RANGE RADARS

A. Al-Attar, D.J. Daniels and H.F. Scott

British Gas, Engineering Research Station, UK

INTRODUCTION

Considerable efforts have been made by a number of workers to design and build a microwave radar system capable of detecting and accurately locating various classes of buried objects, from buried pipes and cables to explosive mines. However most systems have failed to overcome the major problem of clutter which limits the detection capability of any ground probing radar. The visibility of the target returns over the clutter profile is the key factor in determining operational success. This paper describes one technique which is capable of reducing the clutter profile. A block diagram of a radar is shown in Figure 1 and a typical signal return, in dB, plotted against range, is shown in Figure 5a. The operational requirements define the operating parameters of the radar. Depth of detection governs overall dynamic range and subsequently frequency of operation. Plan resolution governs antenna characteristics and depth resolution governs modulation bandwidth. From extensive measurements we have concluded that successful penetration of more than 1 m in the UK cannot be achieved by frequencies over 1 GHz except in a minority of situations.

Various techniques to overcome clutter have been proposed and implemented. In the case of linear targets, vis pipes, polarisation discrimination provides a useful clutter suppression and has the extra benefit of improving plan resolution. Forward look (angled antenna) systems can be effective on flat ground but ground roughness causes significant backscatter, in addition sidelobe interference limits clutter suppression. For most types of systems the most significant cause of clutter is interaction between the ground surface and the antenna. This effect can be reduced by eliminating the transmission path between the antenna and the ground surface and matching the antenna characteristic impedance to that of the ground. However, because of the variability of the latter it is difficult to achieve a satisfactory design compromise. In addition operational deployment becomes severely limited by the necessity for intimate ground contact.

Therefore a filtering technique which does not rely on physical stability of the antenna - ground gap is the major requirement for successful operation of a ground probing radar. If the complete system is modelled as shown in figure 2, where the scattering parameters are complex functions of frequency, then it can be seen that the reflection from the target ρ_{TARGET} becomes transformed and multiplied by the scattering parameters of the radar system and the ground. Because of the variability of the characteristics of the ground it is not sensible to attempt to model the latter in detail. Therefore the front surface reflection is regarded as a composite containing the target return. Removal of the radar system parameters by means of an inverse transformation of the S matrix can suppress the clutter caused by interaction of the target return with the system parameters. The limitations to this approach are the variation of the frequency characteristics of the front surface reflections and the open structure of the antenna-ground transmission path, however the effect of the latter can be minimised by careful antenna design.

There are a number of possible signal processing techniques which can be employed to reduce clutter; subtraction, deconvolution, cross correlation, cepstrum filtering and S parameter filtering. Subtraction in either the time or frequency domain, of a stored version of a scan without a target, from a version of a scan with a target present has been found unsatisfactory because the target signal is swamped by variations caused by differences originating from causes earlier in time. Deconvolution filtering is a partial solution because it only removes the effect of the impulse response of the radar system and leaves multiple interaction from real interfaces unaffected. Cross correlation in either frequency or time domain is limited because of the unpredictability of the dispersive characteristics of the ground. Cepstrum filtering appears to hold some promise for this application in that multiple antenna-ground interactions in the cepstrum provide a series of delta functions at known locations. An adaptive comb filter in the cepstrum could be used to remove these echoes. However a signal processing technique based on an S parameter analysis of the radar system (including the open structure of the antenna and ground) can be used to suppress clutter. A computer model was used to show that 50dB of clutter suppression could be achieved in an equivalent time of 12nS. This is described in a later section. Following the work on modelling various practical measurements were made and the results of these are given later in the paper. It became evident that the processing imparted strict requirements on sweep stability for an FM radar and the system used for measurement was not capable of achieving adequate stability for full implementation of the processing. However the promise of the technique is such that further work is planned to improve the physical deficiencies of the experimental radar system used for measurements.

S PARAMETERS PROCESSING-THEORY

A block diagram of an FMCW radar used to locate buried pipes is shown in Figure 3. The signals entering and leaving different parts of the system are described by their frequency domain representations a_1, b_1, a_2, b_2, a_3 and b_3. The reflection coefficient is the reflectivity looking into the ground (including the front surface reflection) and its Fourier transform gives the bandlimited impulse response of the ground which represents the reflectivity as a function of distance into the ground. The system shown in figure 3 measures b_3 which is the ground reflectivity transformed through the system. This transformation distorts the impulse response of the ground. The system inside the dotted box can be represented by a three port S matrix relating the signals in different parts of the system as shown below.

$$\begin{bmatrix} b_1 \\ b_2 \\ b_3 \end{bmatrix} = \begin{bmatrix} S_{11} & S_{12} & S_{13} \\ S_{21} & S_{22} & S_{23} \\ S_{31} & S_{32} & S_{33} \end{bmatrix} \begin{bmatrix} a_1 \\ a_2 \\ a_3 \end{bmatrix}$$

The delay lines L1 and L3 are made long enough to shift b_1 and a_3 outside the desired time window so that they can be ignored. Under these circumstances the ground reflectivity can be derived from the measured signal b_3 and the S parameters of the network as shown below.

$$\frac{a_2}{b_2} = \frac{(b_3 - S_{13}\, a_1)}{S_{12}\, S_{23}\, a_1 + S_{22}\,(b_3 - S_{13}\, a_1)} \quad ..(1)$$

As can be seen from (1) only three network parameters are needed to remove the effect of the system on the ground reflectivity. S_{13} represents the leakage signal between input and output when the system is looking into free space. $S_{12}\, S_{23}$ represents the transfer function of the system while S_{22} represents the reflectivity looking back into the system from the ground interface. The system parameters are calculated by measuring b_3 when looking into three known values of which results in three equations in three unknowns. The reference reflections used were free space ($\rho = \emptyset$), a metallic ground plane ($\rho = 1$) and a wire grid ($\rho = RW$)(1). If the corresponding values of b_3 for these references are FS, GP and GW then the network parameters are given by:

$$S_{13}\, a_1 = FS \quad ......(2)$$

$$S_{22} = \frac{RW\,(GP - FS) + (GW - FS)}{RW\,(GW - GP)} \quad ......(3)$$

$$S_{12}\, S_{23}\, a_1 = (1 + S_{22})\,(FS - GP) \quad ......(4)$$

In the configuration of Figure 3 the signal at the IF output of the mixer represents the real part of the complex signal b_3 and a method of obtaining the imaginary part has been derived, as the above analysis requires phase and amplitude information.

GROUND PROBING RADAR COMPUTER MODEL

In order to predict the effects of multiple interaction between the antenna and the ground, a computer model was constructed. Network calculations were carried out in the frequency domain and the impulse response of the system reflection coefficient obtained by means of a discrete Fourier transform. Having obtained the overall system response and the network parameters of the antenna, system effects were removed to demonstrate the recovery of targets buried in clutter.

Antenna. The model for the antenna consisted of two impedance mismatches to represent the reflections from the feed point and the aperture. Between these two mismatches was placed a lossless transmission line which simulated the physical length of the antenna. To complete the model, account was taken of the 'scattering aperture' of the antenna. The scattering aperture allows for that portion of the power incident upon the antenna which is absorbed in the radiation resistance and is hence re-radiated. For conditions of maximum power transfer ie a perfectly matched antenna and zero antenna losses, half of the incident power is re-radiated. In order to simulate this effect, a network with the scattering matrix shown below was placed in cascade with the aperture mismatch network:

$$[S] = \begin{bmatrix} 0 & 1 \\ 1 & .707 \end{bmatrix}$$

The antenna model is shown in Figure 4a. The individual networks were represented by their respective A B C D transmission matrix parameters and the overall system parameters obtained by matrix multiplication in the usual manner.

Free space. For free space the following scattering parameter representation can be applied where β = Phase constant of free space.

$$[S] = \begin{bmatrix} 0 & \frac{Gt\lambda}{8\pi R} e^{-j\beta R} \\ \frac{Gt\lambda}{8\pi R} e^{-j\beta R} & 0 \end{bmatrix}$$

From which the corresponding transmission matrix can be calculated.

Lossy ground. The ground interface mismatch can be represented as a transformer of turns ratio η, where η represents the ratio of impedances given by

$$\eta = \sqrt{\frac{\varepsilon_1}{\varepsilon_2}} \quad \text{where } \varepsilon_1 = \text{relative dielectric constant of free space} = 1.0$$

$$\varepsilon_2 = \text{relative dielectric constant of the ground}$$

Transmission through the ground can be represented by a lossy transmission line and for this it will be assumed that S11 and S22 are zero and that:-

$$S_{12} = S_{21} = \sqrt{\frac{Gt\lambda}{8\pi R}}\ \sqrt{\varepsilon_2}\ e^{-\alpha}\ e^{-j\beta R\sqrt{\varepsilon_2}}$$

where $e^{-\alpha}$ = the absorption coefficient of the ground and, in this case R = the depth of ground to the target. See Figure 4b.

Underground target. This is represented as a shunt susceptance whose return loss can be varied. It acts, in the model, as the terminating impedance. See Figure 4c.

Results. Figures 5 a, b and c show the results for a typical simulation where a target is buried at a depth of 37.5 cm in ground of relative dielectric constant 16 and loss of 30dB/metre increasing at the rate of 10dB/metre/GHz. The response of the target is well below the system generated clutter levels and can only be detected after their removal by the signal processing method. Some 49 dB improvement in clutter is shown in this calculation and is only limited by the modulation bandwidth chosen. As shown later, in the practical case, other considerations degrade this potential improvement.

Signal processing. The signal processing used is similar to that described for the practical work except that the analysis is for two port networks, for which it can be shown that:-

$$\rho_{ground} = \frac{S_{11} - \rho_{in}}{S_{22}\,(S_{11} - \rho_{in}) - S_{12},\, S_{21}}$$

Where ρ_{ground} = frequency response of ground reflection coefficient = a_2/b_2

ρ_{in} = frequency response of measured reflection coefficient = b_3/a_3

From the calculated frequency response i.e. the total reflection coefficient, which in the real world is the only quantity which can be measured, the reflection coefficient of the ground can be calculated. The Fourier transform of ground gives the impulse response which is free from the effects of system clutter.

SYSTEM IMPLEMENTATION

A monostatic FMCW radar was implemented as shown in figure 3 and the important points to be noted are detailed below. Measurement of the reflected signal from the antenna was by means of a high directivity SWR bridge which was used in preference to a coupler because of its superior directivity (better than 55dB) and its inherent broadband properties (0.1 to 2 GHz). Delay lines, L1 and L2, were used in order to shift the effects of source and mixer mismatch outside the measurement time (range) window and their difference in length was used to shift the I.F. output of the mixer away from d.c. A ridged horn antenna designed to cover the frequency range 0.7 GHz to 1.7 GHz with a return loss of better than -10 dB and an average gain across the band of 9 dB; was used for the measurements. Control of the sweep generator and data acquisition sub-systems was implemented by means of a microcomputer. A/D conversion, signal averaging, Fourier transforms and S parameter processing were all carried out by the microprocessor driving software. The S parameters of the system were measured for the case where the antenna is to be operated 10 cm above the ground interface. A ground plane and a wire grid were used as references and equations (2) - (4) were used to calculate the S parameters. The calculated time domain S parameters are shown in Figures 6 and 7.

The signal processing method was used to detect a one inch diameter metal pipe placed behind a dielectric backed grid. As can be seen from Figure 8a the pipe is buried in the clutter caused by multiple reflections between the grid and the antenna. Figure 8b shows the processed radar return which clearly indicates the presence of the pipe. The processing is seen to be successful in removing multiple bounces. The processed radar return when looking into the dielectric backed grid alone is shown in Figure 8b as the dotted line. To test the radar in a practical situation a metal pipe buried 35 cm in sand was used as a target and Figures 9a and 9b show the processed and unprocessed results respectively. The processing technique is seen to provide an improvement and clearly identify the target. The degree of clutter suppression achieved practically, demonstrates the viability of the method of processing, but falls short of the predicted improvement; in particular the rate of fall off is significantly degraded.

Investigations into the causes revealed that the actual S parameters of the system drifted with time, two likely causes have been identified; firstly the frequency and amplitude stability of the swept frequency source used was inadequate for the required precision, secondly, mechanical flexure of the antenna caused a variation in its transfer function. Steps are being taken to improve the performance of both elements of the system.

CONCLUSIONS

The technique of S parameter filtering has been shown by modelling to be an effective method of clutter suppression. The degree of clutter with time suppression is primarily limited by the system bandwidth. The variation of the surface reflection coefficient with frequency is a further limiting factor and has not yet been fully investigated. It is evident that achieving the full benefit of clutter suppression; in the order of 40 dB - 50 dB per 10 nS, imposes severe constraints on system stability and our initial measurements highlighted this factor. It is anticipated that appropriate improvements will allow the practical clutter suppression to approach that predicted by the computer model.

ACKNOWLEDGEMENT

The authors wish to thank the British Gas Corporation for permission to publish this paper and comment that the views expressed are solely those of the authors. The authors also gratefully acknowledge the assistance of Mr Graham Sexton with the practical work.

REFERENCE

1. Larsen, T., May 1962, 'A Survey of the Theory of Wire Grids', IRE Transactions on Microwave Theory and Techniques, pp191-201.

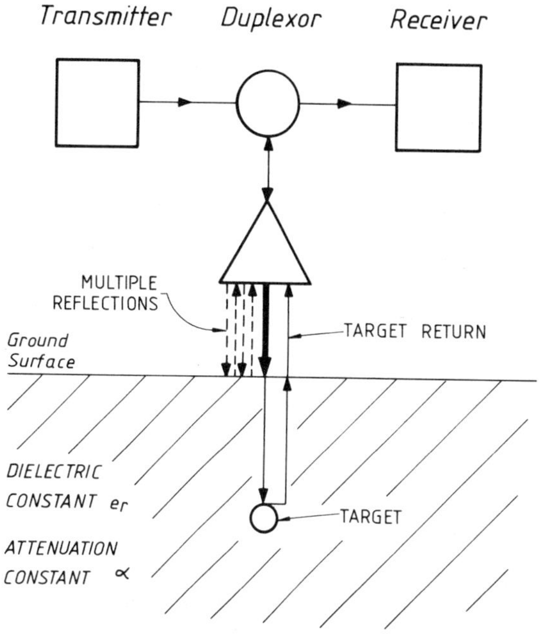

Figure 1 Block diagram of radar.

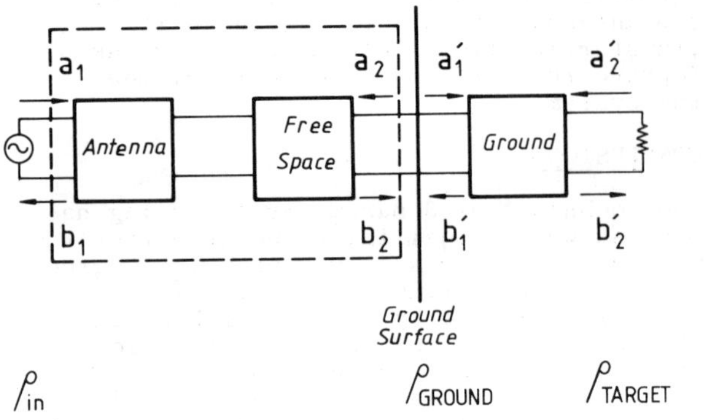

$$\rho_{in} = \frac{b_1}{a_1} = \frac{S_{11}(1 - S_{22}\rho_L^*) + S_{12}S_{22}\rho_L^*}{1 - S_{22}\rho_L^*}$$

$$\text{where } \rho^* = \frac{b_1'}{a_1'} = \frac{S_{11}'(1 - S_{22}'\rho_{TARGET}) + S_{12}'S_{21}'\rho_{TARGET}}{1 - S_{22}\rho_{TARGET}}$$

Figure 2 Equivalent circuit model of radar.

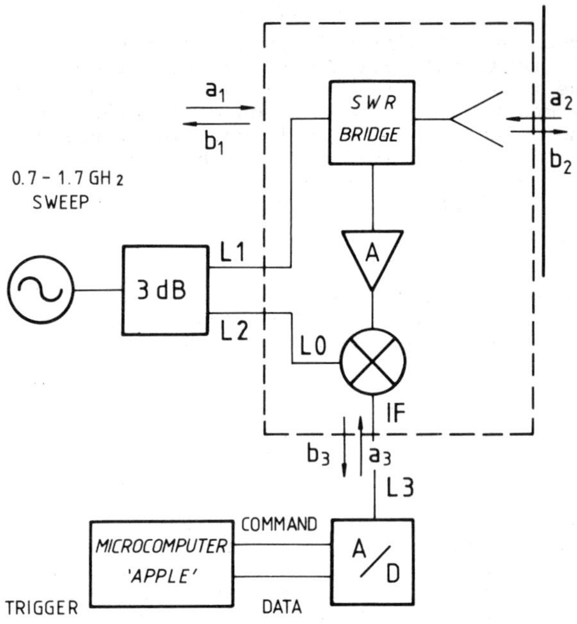

Figure 3 FMCW radar block diagram.

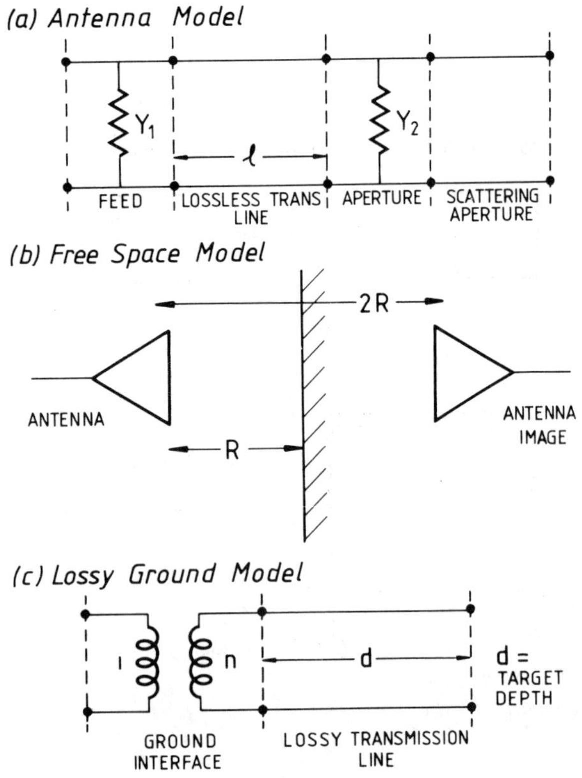

Figure 4 Equivalent models.

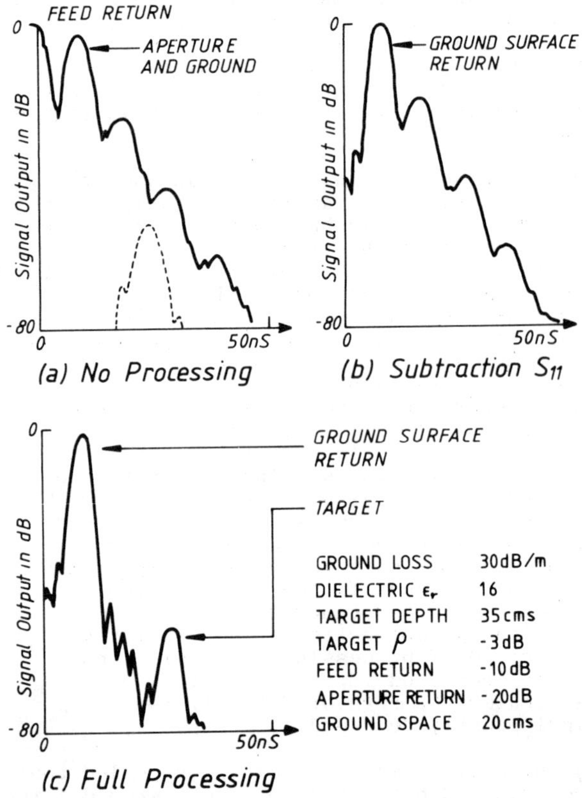

Figure 5 Model output results.

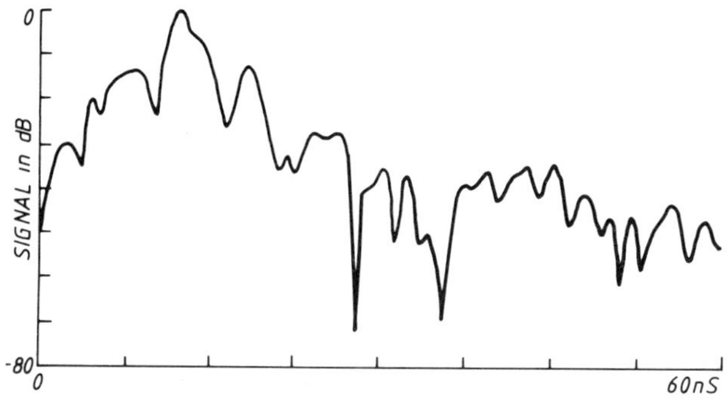

(a) Scattering Parameter S_{11}

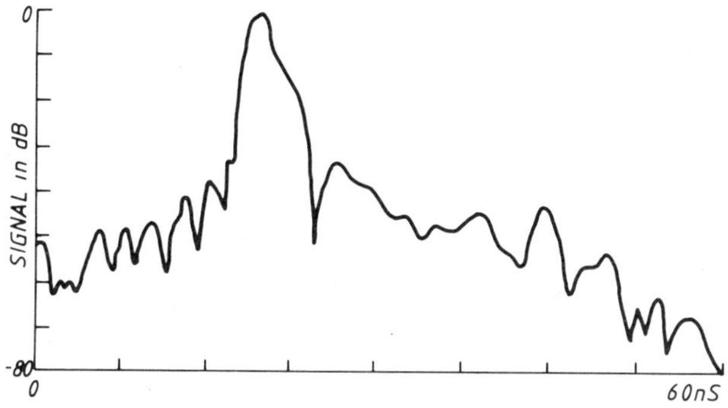

(b) Scattering Parameter $S_{12}S_{21}$

Figure 6 Measured scattering parameters.

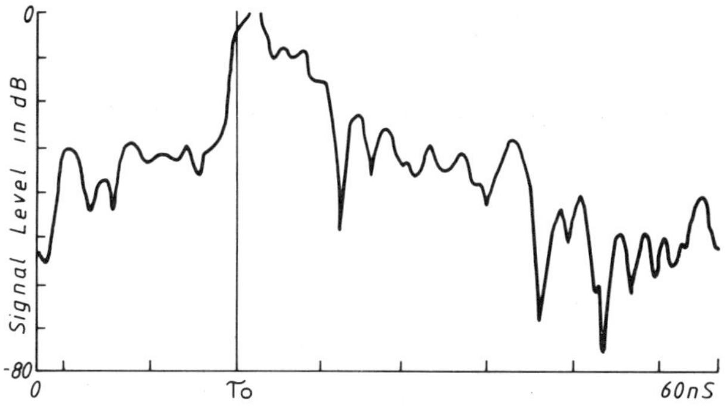

Scattering Parameter S_{22}

Notes :- THE EFFECTIVE TIME POSITION OF MEASUREMENT IS NORMALLY AT To

Figure 7 Measured scattering parameters.

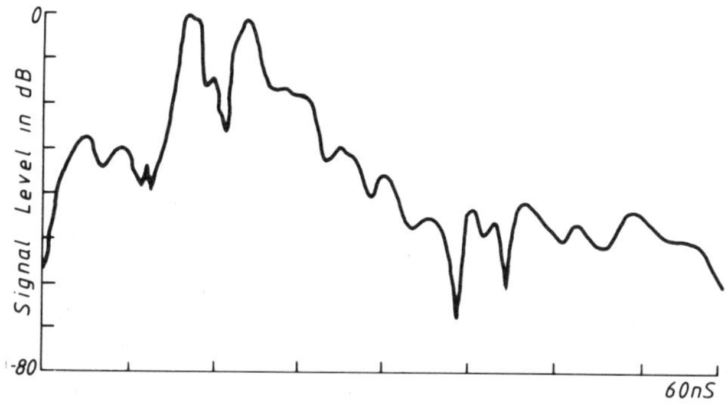

(a) Unprocessed Return from Pipe behind Grid

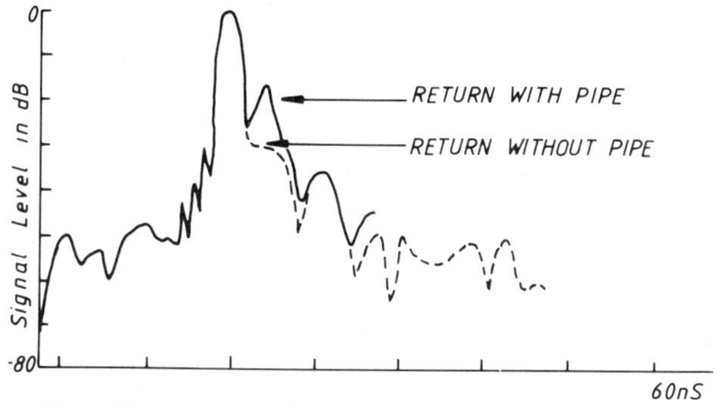

(b) Processed Return with and without Pipe

Figure 8 Processed and unprocessed returns, from pipe behind wire grid.

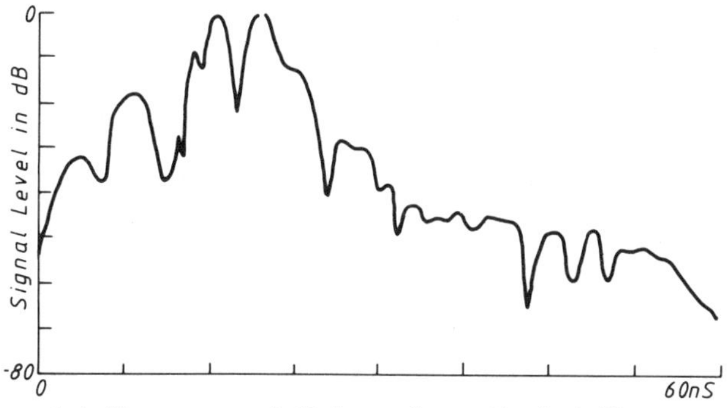

(a) Unprocessed Return from Buried Pipe

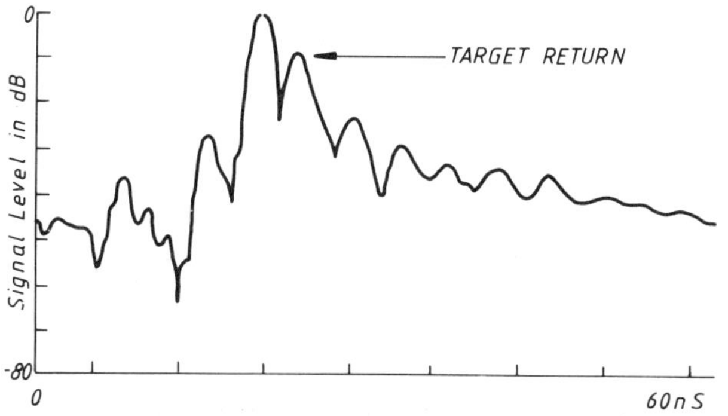

(b) Processed Return from Buried Pipe

Figure 9 Processed and unprocessed returns from buried pipe.

CABLE RADAR FOR INTRUDER DETECTION

A. C. C. Wong and P. K. Blair

Standard Telecommunication Laboratories, UK

BACKGROUND

In recent years, growing interest has been aroused in the use of radiating cables as sensors in cable radar systems for intruder detection (Patterson and Mackay (1)). Radiating cables, also known as 'leaky' or 'ported' cables, were first developed for use in local communication networks where normal propagation is difficult, and have been applied successfully to radio links in mines, tunnels, motorways, and inside buildings (Johannessen (2)). Such a cable is usually coaxial in construction, either made with a loose outer braid, or with a slot or a number of slots along the outer conductor, so that part of the internal field is coupled to the outside.

In the early 1970's, STL carried out extensive work on radiating cables including characterisation of their performance in a realistic environment, covering a range of mounting conditions including burial underground.

Some of the advantages in applying radiating cables as the basis of a perimeter protection system are quite apparent. For example, the cables can be laid out according to the contour of the perimeter and thus could provide overall coverage without the usual line-of-sight limitations of microwave or optical systems. Furthermore, when cables are buried underground there are no visible signs of the existence of the sensors, and the system can be shielded to some extent from some of the variations in the environment. They also have the useful property of being able to carry wide bandwidth signals which are required for low intercept probability and accurate definition of the point of intrusion.

Recognising the potential of these sensors, especially in combination with modern digital signal processing techniques, Standard Telecommunication Laboratories have built an experimental intruder detection system based on these ideas, to demonstrate the feasibility of the concept, and to allow realistic field trials to be carried out.

PRINCIPLE OF OPERATION

As shown in Figure 1, the system works on the pulse-doppler radar principle, with the pulses guided along the length of the radiating cable instead of propagating in free-space as in a conventional radar. The key components of the system are two cables laid out in parallel, one for transmission and one for reception, a transmitter unit, a receiver unit, and a signal processor. The transmitter unit sends pulses into the transmitting cable at pre-determined intervals and, as a pulse travels along its length, part of the energy leaks into the surrounding space. A small amount of this energy is picked up by the nearby receiving cable, and registered in the receiving unit. This forms the direct coupling between the cables. If now an intruder is within the zone of influence set up by these two cables, part of this field will be perturbed and the variation in the received signal can be detected by the signal processor which will, in turn, indicate the presence of an intruder on a display or trigger off an alarm.

The distance to the target along the cable is measured by round-trip timing in the same way as in a conventional radar. Spread spectrum pulse compression can also be applied, not only to match the resolution of the range-cell to the target size for improving its signal-to-clutter ratio, but also to make the transmission more noise-like and of lower peak power to reduce its chance of interception.

The main problem from the detection point of view is the large direct coupling component with respect to the small variation due to the intruder. Doppler filtering will help to separate these two components. Since an intruder may try to beat the system by creeping across the cables as slowly as possible, extremely high resolution is required, probably down to 0.1 Hz. The dynamic range of the input is also large. As such, a digital Doppler filter is the best practical approach, usually taking the form of a Fast Fourier Transform.

We shall first describe the characteristics of our experimental hardware, and then go on to discuss the main signal processing issues.

HARDWARE CONFIGURATION

The radiating cables used in the experimental system are of the continuous slot type made by Standard Telephones and Cables, with a longitudinal attenuation of 30 dB/Km. The cables are laid out in parallel at 1 meter separation, and buried at 7 to 10 cm below the earth's surface in a test field. Part of the cable run is in close proximity to a protecting wire fence, so that its effect on the system can be investigated. From previous work, we found that no particular attention needed to be paid to the orientation of the slot in the cables during installation, the cables were simply "mole drilled" into the ground.

In general, radiating cables are intended for use in the VHF spectrum. The system, accordingly, is designed to operate at selected frequencies between 70 and 200 MHz. Figure 2 shows the system block diagram. Since the purpose of the system is for experimental research and measurement rather than operational, such features as alarm indicators or multiple range-cell capability have not been incorporated. Instead, the display is a spectrum of the target return, and an approach is adopted whereby a single range-cell can be examined at one time through manual selection.

A 28-bit pseudo-random spread spectrum code is generated in the logic control unit at a selectable pulse repetition frequency (p.r.f.), and a chipping rate of 20 M-bits/ sec. is used, corresponding to a range-cell size of around 5m. In order to maximise the effective signal power, the p.r.f. should be run as fast as the returns can be handled by the processor, up to a limit fixed by range ambiguity. For a range of 500m, this upper limit is 200 KHz.

The transmitting spread spectrum code is bi-phase modulated onto the r.f. carrier in the transmitter unit, amplified to a peak power of 3W, and injected into the transmitting cable.

On its return, the signal is filtered and amplified in the receiver front-end unit, stripped of its code, formed into an i.f. pulse of 1.4 μS duration on a 40 MHz carrier, and passed on to the zero i.f. unit for further processing. The demodulation process is performed with an exact replica of the transmitting code generated in the logic control unit, but delayed in time corresponding to the expected delay for the range-cell of interest.

The zero i.f. unit has three inputs: the 40 MHz demodulated signal pulse, and two local oscillator (L.O.) inputs, both at 40 MHz but at 90° apart from one another in phase, generated in the clock and synthesiser unit. The L.O. inputs are mixed with the signal pulse in two separate channels, called the I and Q channels, to produce d.c. pulses in quadrature. These pulses are then further filtered, sampled-and-held, and converted into two digital words. These are the in-phase and quadrature components representing the incoming signal vector for passing on to the microprocessor for subsequent analysis.

THE SIGNAL PROCESSING ENVIRONMENT

The heart of the signal processor is a floating point Fast Fourier Transform (FFT) executed in a microprocessor. For practical reasons, the size of the FFT is limited to 1024 points, with an update time including the subsequent spectral display of around 2 secs. Since a resolution of 0.1 Hz is necessary for the successful discrimination of slow moving targets, this implies a total sampling time per FFT of 10 secs. In the hardware implementation, therefore, the 10 secs. interval will form a sliding sampling window within which one-fifth of the samples will be renewed every update.

A fast runner will probably be able to cross the zone of influence of the cables in about a second, perturbing only a tenth of the total samples in the integration process at any one time. This will induce a degradation of the processing gain of moving targets amounting to 10 dB in the worst case, referred to as sampling window loss in the following calculations.

The problem confronting the signal processor is thus as follows. Using the parameters of the experimental system as a basis, first consider the noise in the receiving channel. Starting with a thermal noise of -174 dBm/Hz and a 40 MHz channel bandwidth, the thermal noise floor is -98 dBm. The receiver front-end probably contributes a further 4 dB, and noise from the surroundings picked up by the cable amounts to, by measurement, 6 dB. The noise input to the system is therefore:

$$-98 + 4 + 6 = -88 \text{ dBm}.$$

Consider now the signal. We start with a peak transmitting power of 35 dBm and a coupling loss which, in the type of cable employed, is around 140 dB from transmitting cable to receiving cable. The human target tends to provide a slight increase in the coupling between the cables, but not much, of the order of 3 dB when mid-way between the cables. Consider a modest case of receiving a signal coupled across the cables at 500m from the source. With a cable attenuation of 30 dB/Km each way, a sampling window loss of 10 dB, and a gain of 14 dB due to pulse compression, the signal input to the system before any FFT processing is:

$$35 - 140 + 3 - 30 - 10 + 14 = -128 \text{ dBm}.$$

As can be seen, the required signal is some 40 dB below visibility, and a large amount of coherent integration is hence required to pull it above the noise floor for positive detection. Taking a signal-to-noise ratio of 13 dB for a 90% probability of detection on a first look basis and a false alarm rate of 1 in 10^6 (Nathanson (3)), a processing gain of 53 dB is required in the subsequent processing, implying the need for a 200,000 point FFT!

Other alternatives also seem to be unpromising. Increasing the peak transmitting power, even to ten-fold or a hundred-fold, is quite ineffective; and increasing the spread spectrum code-length to the order required would drastically increase the circuit complexity. The best solution still lies in increasing the p.r.f. as much as possible, and to somehow enhance the throughput of the microprocessor to enable more pulses to be coherently integrated.

THE PRE-FFT ACCUMULATOR

Consider an N-point discrete Fourier transform with each of the input points made up of n time-samples x_i added together:

$$F(w) = \sum_{k=0}^{N-1} \left(\sum_{i=0}^{n} x_i \right) e^{-jwnkT}$$

The result is a spectral output with a $\frac{\sin x}{x}$ magnitude weighting superimposed on the display of the corresponding nN point to N point transform, as shown in Figure 3.

The first weighting nulls are at $\pm\frac{1}{nT}$, where T is the pulse repetition time. The maximum signal loss for the N-point FFT will be at the two end points of the display, amounting to 4 dB. This loss, analogous to the Doppler loss of a pulse compression radar with long codes, is insignificant in the present context, as the spectral lines of a human target are usually concentrated around the zero line, and hardly exceed ±10 Hz.

Thus, if a number of samples are added up vectorially before the transform, either internally by software within the micro-processor, or better still, by an external accumulator inserted between the A to D convertors and the FFT processor, coherent integration of an extremely large number of points can be effected without any increase on the complexity of the FFT itself.

In the hardware described, some initial experiments were done with this pre-FFT accumulation implemented in software in the microprocessor. Subsequently, an external accumulator was built to maximise the throughput. With a p.r.f. of 200 KHz, 4097 samples could be accumulated in this unit over a 20 ms period, giving a sampling window of 10.49 sec. duration for a 512-point FFT. The total number of individual returns in each FFT is therefore 2,097,664, with a processing gain of 63 dB, 10 dB above the minimum required value.

FIELD TRIALS

Figures 4, 5 and 6 show snapshots of spectrum outputs obtained when an intruder is crossing the cables at right-angles, respectively in walking, running and slow creeping movements. The characteristic shapes of these displays are all very similar, as can be expected. The peak of the running spectrum is lower than that of the walking one, as less pulses within the some 2 million returns are perturbed, but the energy is spread out further in frequency. The spectrum of the creeping movement, on the other hand, is concentrated around the centre.

The d.c. lines in these displays are suppressed to avoid confusion. An adaptive threshold can thus be set according to the prevailing noise floor for target detection. Scope also exists for performing target recognition within the microprocessor.

In addition to the runs above, experiments have also been carried out with the cables:

(i) buried near a wire fence which vibrates in the wind,

(ii) laid out in a wood with multiple reflections from trees,

(iii) covered over with a wire mesh acting as a screen and with the intruder crossing the mesh,

(iv) covered as in (iii), but with the intruder crossing in an adjacent range cell.

In all cases, adequate signal strengths are obtained for positive detection. The combination of the accumulator followed by the FFT is seen to be the key factor in providing this encouraging result.

Being a system that works in the VHF band, our approach can reap an additional advantage: the human target, as a radar reflector, has a response that rises to a maximum at around 70 MHz. Radiating cable losses tend to increase substantially as frequency is increased. Our opinion, as borne out by Poirier (4), is that a frequency of around 60 to 70 MHz is the optimum to use. At this frequency, anything below the size of a small human, e.g. a stray rabbit or dog, is suppressed far below the ratio which one would expect from the size proportionality, and this will help to reduce the false alarm rate due to such causes. Occasional observations in the field tend to substantiate this, but controlled experiments have yet to be done for its confirmation.

Using an artificial target device which reradiates a received signal with a defined level and Doppler shift, the sensitivity profile around the cable has been explored. This shows that the zone of detection extends about a meter on either side of each cable and about 2 meters above the centre line. This shape can be influenced by the cable spacing.

One of the less desirable features of radiating cables used for narrow band VHF communication purposes has been the existence of standing wave patterns producing large variations in the signal strength along the cable. The use of a wide bandwidth waveform tends to reduce the variation by a frequency diversity process, although there will obviously be some distortion of the waveform due to the frequency selective fading. The sensitivity variation along the cable has been investigated and a figure of about 10 dB observed.

The radar system sensitivity and given cable parameters enable the maximum perimeter length to be estimated. This length can be maximised by the use of sections of cables with lower coupling factor close to the transmitter and sections of higher coupling, to compensate for the preceding cable loss, at the remote end. Further extension can be achieved by means of repeaters in the receive path, the feasibility of units with gains of about 20 dB having previously been demonstrated in communication applications.

The use of an equipment driving two cable sections in a centre fed arrangement can also be contemplated as a means of extending the perimeter. Either time or frequency sharing can be used.

CONCLUSIONS

The experiments have shown the feasibility of using a spread spectrum pulse-doppler radar signal format with radiating cable sensors in a practical covert perimeter protection system. Such a format has advantages over a more conventional narrow band system in that it provides:

(i) high levels of coherent integration allowing operation with low peak power,

(ii) lower probability of intercept and system discovery,

(iii) low level of interference to other services,

(iv) potential for target classification by its spectral characteristics, and

(v) high resolution of the point of intrusion.

We conclude that a design of this type has high potential for use in advanced perimeter protection systems.

REFERENCES

1. Patterson, R.E. and Mackay, N.A.M. "A Guided Radar for Obstacle Detection" IEEE Transactions on Instrumentation and Measurement, IM-26, No.2, 137-143, June 1977.

2. Johannessen, R. "Radiating Cables" Wireless World, June 1978.

3. Nathanson, F.E. "Radar Design Principles" McGraw Hill Book Company, New York, 1969.

4. Poirier, J.L. "An Evaluation of Some Factors Affecting the Choice of Operating Frequency of a Guided Wave Radar for Intruder Detection", Rome Air Development Centre, In-House Report, RADC-TR-79-83, March 1979.

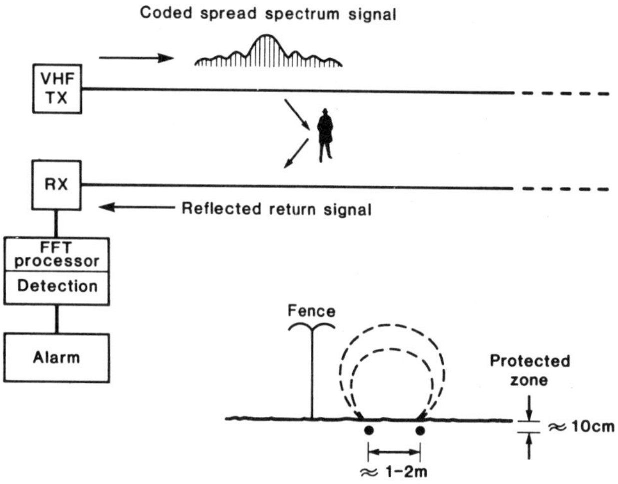

FIGURE 1. Intruder Detection Radar Principle

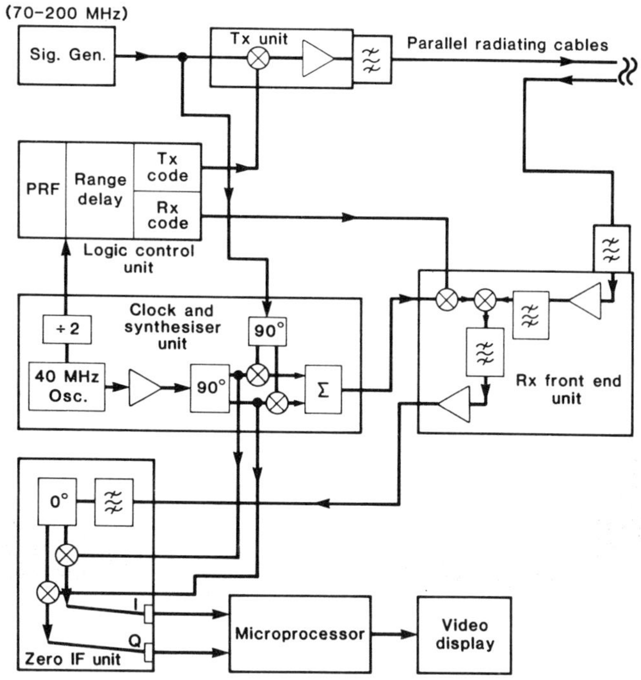

FIGURE 2. Block Diagram of Intruder Detection System

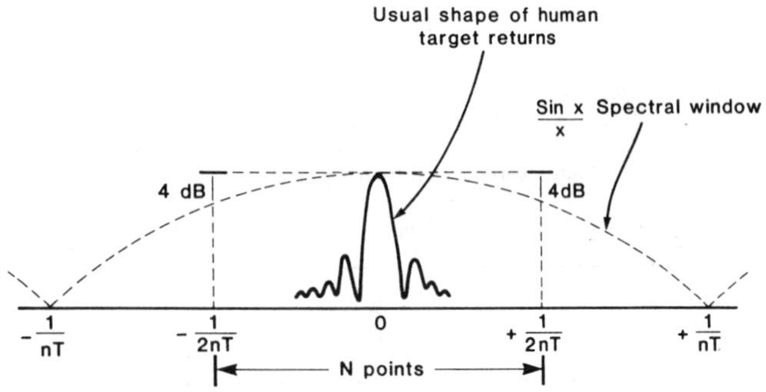

Number of points in discrete fourier transform	= N
Number of pulses accumulated for 1 point of transform	= n
Total number of pulses coherently integrated	= nN
Pulse repetition time	= T sec.
Interval between successive DFT input points	= nT sec.
Duration of total sampling time	= nNT sec.
Range of coverage of DFT	$= \pm \frac{1}{2nT}$ HZ
Resolution of discrete fourier transform lines	$= \frac{1}{nNT}$ HZ
Position of first nulls of spectral window	$= \pm \frac{1}{nT}$ HZ

FIGURE 3. Effect of Pre-FFT Accumulation

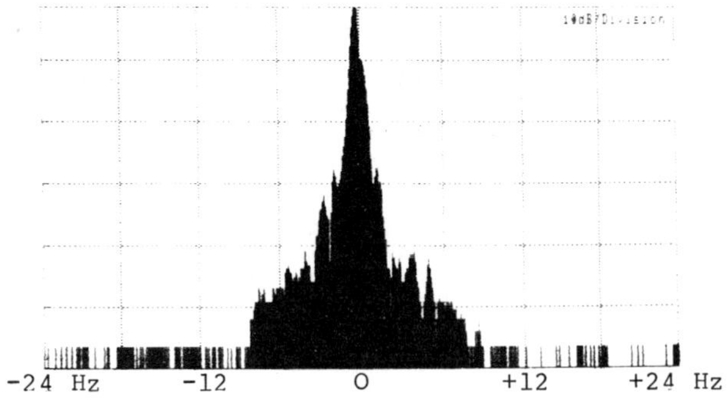

FIGURE 4. Spectrum of Intruder Walking Across Cables.

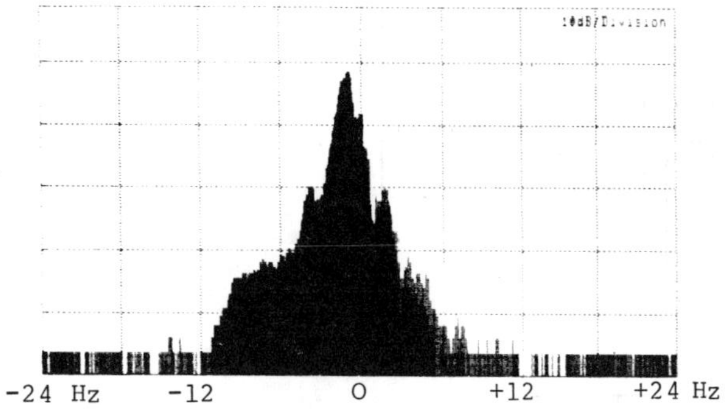

FIGURE 5. Spectrum of Intruder Running Across Cables.

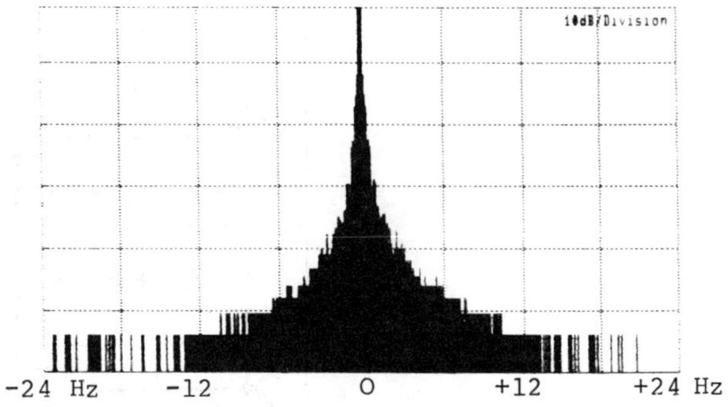

FIGURE 6. Spectrum of Intruder Creeping Slowly Across Cables.

COUPLING MECHANISM FOR GUIDED RADAR

P. W. Chen (1), G. O. Young (2) and R. K. Harman (3)

(1) ESL Incorporated, U.S.A., (2) TRW Incorporated, U.S.A.,
(3) Computing Devices Company, Canada (Now with Senstar Corporation, Canada)

INTRODUCTION.

Leaky coaxial cables are similar to conventional open wave guides that enable the internal guided signals to be coupled into the external environment. Because of this unique coupling scheme, these leaky cables have found wide applications in continuous-assess guided communication and guided radar (Harman and MacKay (1), Beal et al. (2), Farmer and Shephard (3), Mackay and Mason (4)). For a typical guided radar the leaky cable(s) is (are) mounted along the perimeter of the protected area either above or underneath the ground surface. The RF signal with either continuous wave (CW) or pulsed modulation is sent through the leaky cable. Energy is then guided along the cable through two separate guided modes: one is the internal mode, which propagates inside the cable, and the other is the external mode, which is bounded by the outer conductor of the cable and the surrounding media. The external wave interacts with the internal wave through leaky holes. In the meantime, the external guided wave continuously interacts with the surrounding environment. A target object located in the path of propagation will cause the external wave or energy to be reflected. The reflected energy will be received through coupling via reciprocal process by the same or the other adjacent leaky cable. In the receiver, MTI techniques are used to process the received signal and detect the target.

This paper describes the basic mechanism of generating the external mode and the coupling mechanism between the internal and external modes. In the past the mutual coupling between these two modes has been studied extensively (Delogne (5), Delogne and Deryck (6), Deryck (7), Martin (8), and Delogne and Safak (9)). However, most theories are complicated and are not developed for designing guided radar system. This paper presents a simple formulation of the basic coupling mechanism based upon normal mode transmission line theory. Instead of using conventional line voltage and current as variables, the normal mode approach uses a linear combination of voltage and current. This new variable permits separation of the propagation modes traveling in positive and negative directions. This approach can simplify the mode coupling expression if the transmission lines are properly terminated, because only the modes traveling in one direction need to be considered.

MATHEMATICAL DESCRIPTION OF MODE COUPLING ALONG A LEAKY CABLE

Leaky coax can support two propagation modes: an internal mode and an external mode. The internal mode is basically a TEM wave whose propagation velocity is equal to the speed of light divided by the square root of the relative dielectric constant of the insulator. The external mode is determined by the properties of the outer conductor of the coax and the surrounding medium. For coax laid horizontally above the ground, the external mode is a quasi-TEM mode propagating with the speed of light (Carson (10)). When the coax is buried in the ground, the external mode is a hybrid mode that experiences considerable attenuation (Wait (11)).

If there is no leakage in the coax, the internal and external modes propagate independently, that is, no coupling takes place. However, if there is leakage, the internal and external modes will couple. Both cases can be approximated by the following general coupled transmission-line equations (Johnson (12)).

$$(\partial V_n/\partial z) = -j\beta_n z_n I_n - jk \sum_m z_{mn} I_m \quad (1)$$

$$(\partial I_n/\partial z) = -j\beta_n y_n V_n - jk \sum_m y_{mn} V_m \quad (2)$$

where

V_n and I_n are the voltage and current of mode n, and vary as $\exp[j(\omega t-\beta_n z)]$

β_n is the uncoupled propagation constant of mode n.

ω is the angular frequency.

k is the propagation constant in free space $(\omega^2\mu_0\varepsilon_0)^{1/2}$.

z_n and y_n are the characteristic impedance and admittance of mode n.

z_{mn} and y_{mn} are the normalized coupling impedance and admittance of mode m to mode n.

Equations (1) and (2) reduce to the ideal uncoupled transmission-line equations if z_{mn} and y_{mn} equal zero.

The transmission mode can be described by the normal mode functions.

$$a_n = (8z_n)^{-1/2} (V_n + z_n I_n) \quad (3)$$

$$a_n' = (8z_n)^{-1/2} (V_n - z_n I_n) \quad (4)$$

and conversely

$$V_n = (2z_n)^{1/2} (a_n + a_n') \quad (5)$$

$$I_n = (2/z_n)^{1/2} (a_n - a_n') \quad (6)$$

where a_n and a_n' describe the propagation wave traveling in the +z and -z directions, respectively.

For simplicity, assume that the transmission lines are properly terminated in order to avoid excessive reflections from the end. Then the normalized coupled transmission-line equation can be written as

$$j(\partial a_n/\partial z) = H_{nn}a_n + \sum_m H_{mn}a_m \tag{7}$$

where

$$H_{nn} = \beta_n \tag{8}$$

$$H_{mn} = (k/2)[y_{mn}(z_m z_n)^{1/2} + z_{mn}(z_m z_n)^{-1/2}] \tag{9}$$

and

$$H_{mn} = H_{nm},$$

since the coupling between modes m and n is passive.

For ideal termination of modes m and n, with $a_n' = 0$, and $a_m' = 0$, it is required that

$$y_{mn}(z_m z_n)^{1/2} = z_{mn}(z_m z_n)^{-1/2} \tag{10}$$

If only two modes are considered, the coupled normal mode equations become:

$$j(\partial a_1/\partial z) = H_{11}a_1 + H_{12}a_2 \tag{11}$$

$$j(\partial a_2/\partial z) = H_{21}a_1 + H_{22}a_2 \tag{12}$$

where subscripts 1 and 2 refer to the internal and external modes, respectively. Equations (11) and (12) are typical eigenvalue problems.

Let a_1 and a_2 be proportional to $\exp(-j\beta z)$. Then the propagation constants or eigenvalues for the two coupled modes are

$$\beta_{1,2} = (H_{11} + H_{22})/2 \pm [(H_{11} - H_{22})^2/4 + H_{12}^2]. \tag{13}$$

Further assume that

$$H_{11} = \beta_{10} \tag{14}$$

$$H_{22} = \beta_{20} \tag{15}$$

$$H_{11} - H_{22} = \beta_{10} - \beta_{20} = \Delta\beta \tag{16}$$

and the coupling is weak such that

$$H_{12}^2 << \left(\frac{\Delta\beta}{2}\right)^2 \tag{17}$$

then

$$\beta_{1,2} = (\beta_{10} + \beta_{20})/2 \pm (\Delta\beta/2)[1 + 2\,(H_{12}/\Delta\beta)^2] \tag{18}$$

$$\text{or } \beta_1 = \beta_{10} + H_{12}^2/\Delta\beta \tag{19}$$

$$\beta_2 = \beta_{20} - H_{12}^2/\Delta\beta \tag{20}$$

consequently

$$a_1(z) = A_{11}\exp(-j\beta_1 z) - (H_{12}/\Delta\beta)\,A_{22}\,\exp(-j\beta_2 z) \tag{21}$$

$$a_2(z) = (H_{12}/\Delta\beta)A_{11}\exp(-j\beta_1 z) + A_{22}\,\exp(-j\beta_2 z) \tag{22}$$

where A_{11} and A_{22} are dependent upon the boundary condition.

For instance, let $a_1(z) = A$ and $a_2(z) = 0$ at z=0. Then

$$A_{11} = A \tag{23}$$

$$A_{22} = -\,AH_{12}/\Delta\beta \tag{24}$$

Inside the coax,

$$a_1(z) = A\,[\exp(-j\beta_1 z) + (H_{12}/\Delta\beta)^2\,\exp(-j\beta_2 z)] \tag{25}$$

Outside the coax,

$$a_2(z) = A\,(H_{12}/\Delta\beta)[\,\exp(-j\beta_1 z) - \exp(-j\beta_2 z)] \tag{26}$$

It should be noted that the powers carried by the internal and external modes are

$$P_1(z) = |a_1(z)|^2 \tag{27}$$

$$P_2(z) = |a_2(z)|^2 \tag{28}$$

and the total power flow along the cable is

$$P(z) = P_1(z) + P_2(z)\quad . \tag{29}$$

CHARACTERISTICS OF EXTERNAL PROPAGATION MODES

The coupling between the internal and external modes results in some changes in the propagation constants as shown in Equations (19) and (20) and, consequently, causes power interchange between these two modes as a function of position.

Assume leaky coax is laid horizontally above the ground. The propagation constants for the internal and external modes are

$$\beta_{10} = \omega/v_p - j\alpha_1 \tag{30}$$

$$\beta_{20} = \omega/c - j\alpha_2 \tag{31}$$

where ω is the operating frequency in radians per second, and α_1 and α_2 are the attenuation constants associated with the internal coaxial mode and external mode, respectively.

v_p and c are the propagation velocities of the uncoupled internal and external modes. Consequently, Equation (26) becomes

$$a_2(z) = b\,\{\exp[-\alpha_1 z - j(\omega z/v_p - \phi)] - \exp[-\alpha_2 z - j(\omega z/c - \phi)]\} \tag{32}$$

where

$$b = A\,H_{12}[(\alpha_1 - \alpha_2)^2 + (\,\omega/v_p - \omega/c)^2]^{-1/2} \tag{33}$$

where

$$\phi = \tan^{-1}[(\alpha_1 - \alpha_2)/(\omega/v_p - \omega/c)] \tag{34}$$

The in- and quadrature- phase components of the external air mode are

$$I_2(z) = b\,[\exp(-\alpha_1 z)\,\cos(\omega z/v_p - \phi) - \exp(-\alpha_2 z)\,\cos(\omega z/c - \phi)] \tag{35}$$

$$Q_2(z) = b\,[\,-\exp(-\alpha_1 z)\,\sin(\omega z/v_p - \phi) + \exp(-\alpha_2 z)\,\sin(\omega z/c - \phi)] \tag{36}$$

and the amplitude and phase are

$$M_2(z) = [\ I_2^2(z) + Q_2^2(z)]^{1/2} \quad (37)$$

$$\phi_2(z) = \tan^{-1}[Q_2(z)/I_2(z)] \quad (38)$$

The power carried by the outer mode is

$$P_2(z) = |a_2(z)|^2 = M_2{}^2(z) \quad . \quad (39)$$

$P_2(z)$ is proportional to the coupling coefficient H_{12}^2 and inversely proportional to $|\Delta\beta|^2$, in general. Note that $\Delta\beta$ cannot be too small; otherwise, the weak coupling assumption as shown in Equation (17) will be invalid.

The total power flow along the leaky coax is

$$P(z) = P_1(z) + P_2(z) \cong A^2 \exp(-2\alpha_1 z) \quad (40)$$

The power loss is primarily due to the conducting loss associated with the coaxial mode. Equation (40) shows that

$$P(z) = A^2 \qquad \text{if} \quad \alpha_1, \alpha_2 = 0 \quad . \quad (41)$$

This equation represents ideal passive coupling between the two propagation modes.

In general, the attenuation constant α_2 is larger than α_1, since the ground is much lossier than the cable dielectric material. If $\exp(-\alpha_2 z)$ is not small, the sinusoidal term in Equations (37) will be visible. This represents the energy coupling back and forth between the internal and external coaxial modes along the cable. According to Equations (32) - (38), the received field intensity and phase are strongly dependent upon the position along the coax. The minimum amplitude occurs at $[2m\pi/(\omega/v_p - \omega/c)]$ and the maximum at $[(2m+1)\pi/(\omega/v_p - \omega/c)]$, where m is an arbitrary integer. The phase varies by approximately $(\omega/v_p + \omega/c)$ radians per unit length.

When $\exp(-\alpha_2 z)$ is significantly smaller than $\exp(-\alpha_2 z)$, the sinusoidal term in Equation (37) will not be visible and the mutual coupling between modes becomes weak. Equations (37) and (38) then become approximately,

$$M_2(z) = b \exp(-\alpha_1 z) \quad (42)$$

$$\phi_2(z) = -\ \omega z/v_p \quad . \quad (43)$$

These equations also represent the asymptotic field strength and phase variation at far distance, i.e., at $z \to \infty$.

Any object in the propagation path of the external air mode will generate a reflected wave. This wave can be thought of as the reflection caused by the resulting impedance mismatch in the transmission line. According to the transmission line theory, the reflection coefficient is

$$\Gamma = (z_\ell - z_2)/(z_\ell + z_2) \quad (44)$$

where

z_ℓ is the load impedance at the object position.

z_2 is the characteristic impedance of the transmission line.

The load impedance is

$$z_\ell = (z_t - z_2)/(z_t + z_2) \quad . \quad (45)$$

z_t is the object impedance, which depends upon the object size, shape, material, position, and orientation with respect to the transmission line.

Usually, $1 \leq |\Gamma| \leq 0$.

The power reflected by the object at position z is

$$P_r = |\Gamma|^2\ P_2(z) = |\Gamma\ a_2(z)|^2 \quad . \quad (46)$$

This reflected external wave will in turn couple back to the internal coaxial mode. The field strength r(z) received inside the coax at z = 0 can be obtained by using the reciprocity principle; thus,

$$r(z) = (\Gamma/A)[a_2(z)]^2 \quad (47)$$

The received field strength and phase due to a specific object depend upon its position. As the object moves parallel to the leaky coax, both the field strength and phase received at z=0 vary accordingly. By measuring the phase and amplitude variation of the received signal, it is possible to estimate the direction of movement and the size of the object (as is done in conventional radar systems). However, the effective detection area is limited to that area close to the leaky coax where the field intensity is large.

To illustrate quantitatively how the external propagation mode varies as a function of distance, assume that the leaky coax is laid horizontally and the coupling is weak, and

f = 60 MHz

α_1 = 0.00226 neper/meter

v_p = 0.79c

$c = 3 \times 10^8$ meter/sec

α_2 = 0.01417 neper/meter.

Figure 1 shows the values of M_2 and ϕ_2 as a function of z. M_2 and ϕ_2 are the amplitude and phase of the signal incident upon the target object at position z along the cable. Note that the signal amplitude reaches a minimum at intervals of approximately 19 meters, and the phase varies 360 degrees per 4.6 meters. These values can be measured by using a voltage probe moving along the coax.

When an object moves along the coax, the received signals at z=0 due to this object can be uniquely obtained by special processing. An MTI processing technique can remove the stationary signals from the irregular background. The reflected signal amplitude and phase from the object are plotted in Figure 2. The in- and quadrature-phase components of the reflected signal are shown in Figure 3. The mutual coupling between the internal and external modes is clearly seen in Figures 1 and 2.

If, in the above example, the attenuation constant α_2 equals 0.1417 neper/meter, the values of I_2, Q_2, M_2, and ϕ_2 can be calculated using Equations (35) to (38), consecutively. If an object moves along the cable, the reflected signal amplitude and phase received at z = 0 are shown in Figure 4. The mutual coupling is visible only at near range (z < 35 meters). The received signal amplitude decreases exponentially at far range (z > 35 meters).

Comparison of some of the theoretical results with experimental data generated by the Computing Devices Company of Canada resulted in excellent agreement (Harman (13)).

SUMMARY AND RECOMMENDATIONS

In this paper, the coupling mechanism between the internal and external modes is formulated using a simple normal mode approach. The propagation constants are affected slightly by the coupling. The magnitude of the effect depends upon the difference between the propagation constants and the coupling coefficient.

The field strength and phase for the external mode are dependent upon the position along the cable. The power carried by the external mode varies with position. However, the total power carried by the internal and external modes remains constant if the transmission is lossless.

The signal reflected from an object is formulated. The expression can be used to aid the design of a guided coax radar system.

This paper assumes that the propagation constants for the uncoupled modes and the coupling coefficient between modes are known. However, the derivation of these constants is not straightforward. The propagation characteristics of an uncoupled system are dependent upon the configuration of the cable. Placement of the leaky coax above or below the ground results in different propagation characteristics. Much effort has been devoted to studying these propagation problems. The problem still needs more study and experimental work. The dependence of field strength upon the distance away from the coax also needs to be studied. This field strength profile defines the effective detection area for the guided radar system.

ACKNOWLEDGEMENT

The authors would like to thank Computing Devices Company of Canada for providing experimental data and useful comments. This work was supported partially by the United States Air Force under Contract F04804-81-C-0022.

REFERENCES

1. Harman, R.K. and Mackay, N.A.M., 1976, "Guidar: An Intrusion Detection System for Perimeter Protection", Conference on Crime Countermeasures, 155-159.

2. Beal, J.C., Josiak, J., Mahmoud, S.F. and Rawat, V., 1972, "Continuous-Assess Guided Communication for Ground-Transportation Systems", Proc. of IEEE, 61, 562-568.

3. Farmer, R.A. and Shephard, N.H., 1962, "Guided Radiation, the Key to Tunnel Talking", IEEE Trans. Veh. Commun., VC-14, 93-102.

4. Mackay, N.A.M. and Mason, J.L., 1975, "A Guided-Radar Technique for Vehicle Detection", IEEE International Radar Conf., 123-127.

5. Delogne, P.P., 1976, "Basic Mechanisms of Tunnel Communication", Radio Science, 11, 295-303.

6. Delogne, P.P. and Deryck, L., 1980, "Underground Use of a Coaxial Cable with Leaky Section", IEEE, Trans. Ant. Prop., AP-28, 875-883.

7. Deryck, L., 1975, "Control of Mode Conversions on Bifilar Line in Tunnels", Radio and Elect. Engr., 45, 241-247.

8. Martin, D.J.R., 1975, "A General Study of the Leaky-Feeder Principle", Radio and Elect. Engr., 45, 205-214.

9. Delogne, P.P. and Safak, S., 1975, "Electromagnectic Theory of the Leaky Coaxial Cable", Radio and Elect. Engr., 45, 233-240.

10. Carson, J.R., 1926, "Wave Propagation in Overhead Wires with Ground Return", Bell System Tech Journal, 5, 539-554.

11. Wait, J.R., 1972, "Theory of Wave Propagation Along A Thin Wire Parallel to an Interface", Radio Science, 7, 675-679.

12. Johnson, C.C., 1965, "Field and Wave Electrodynamics", McGraw-Hill, New York.

13. Harman, R.K., 1978, "Free Space Mode Cancellation in GUIDAR", Tech. Memo., Computing Devices Co., Ottawa, Canada.

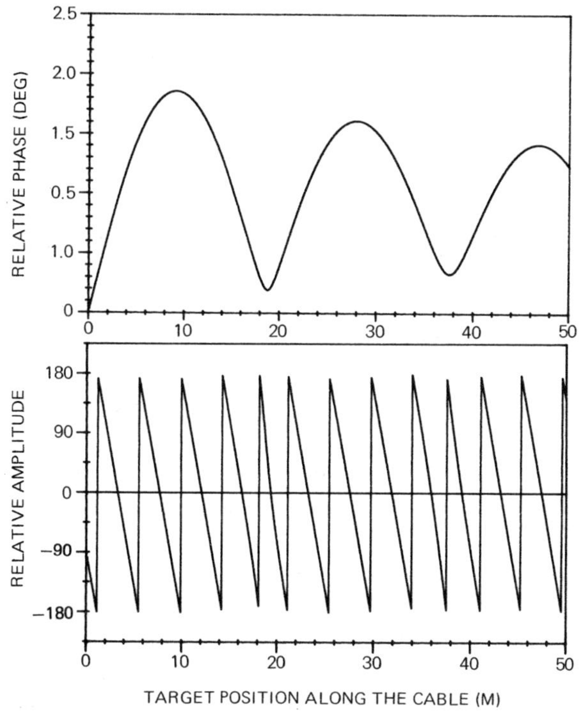

Figure 1. Relative amplitude and phase of the incident signal $a_2(z)$ on the target

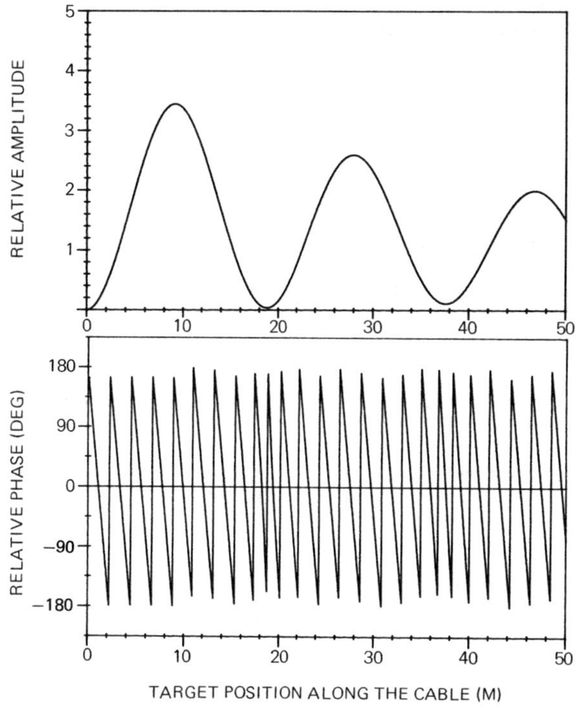

Figure 2. Relative amplitude and phase of the reflected signal r(z)

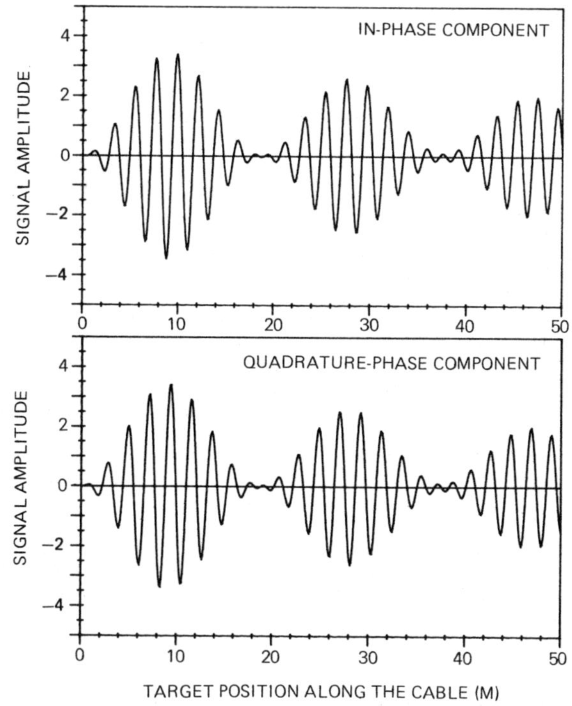

Figure 3. In- and quadrature-phase component of the reflected signal r(z)

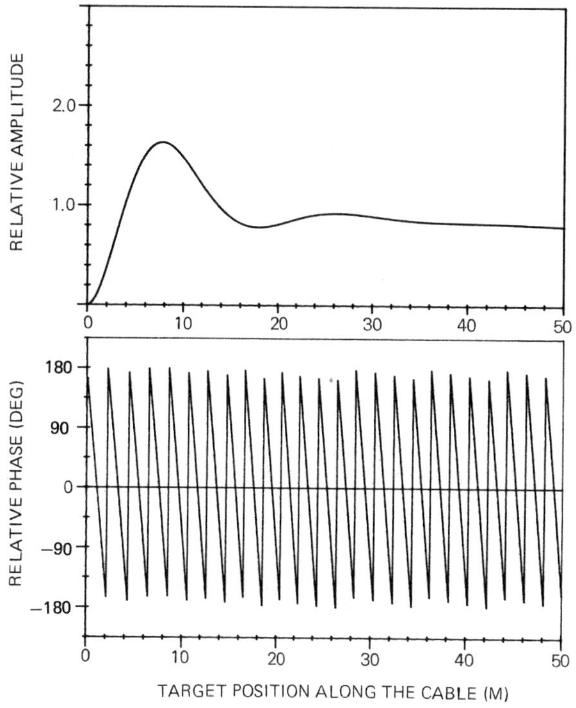

Figure 4. Relative amplitude and phase of the reflected signal r(z) with α_2 = 0.1417 neper/meter

A NEW FAMILY OF SELENIA TRACKING RADARS; SYSTEM SOLUTIONS AND EXPERIMENTAL RESULTS

T. Bucciarelli (*) — U. Carletti (*) — M. D'Avanzo (*) — G. Picardi (**)

(*) Selenia S.p.A., I/(**) University of Rome, I

1. INTRODUCTION

The new family of tracking radars, developted and tested during last years, represents the natural evolution of the Selenia presence in the weapon systems field.
The older Selenia tracking radar solution, with its simple signal processing due to the use of a non-coherent single pulse transmission, doesn't allow to reach the present system requirements especially in a very sophisticated environment. Today the evolution of the threat, including the target ability to fly at very low levels in a ECM (electronic counter measures) environment, requires the use of a more complex waveform design, able to give a better degree of interference suppression.
An up-to-date technology, especially in the receiver area, has been introduced having the purpose to reduce radar dimensions and production costs.

2. SYSTEM REQUIREMENTS AND RADAR DESCRIPTION

Although the required performances vary with the system in which the radar is inserted, a series of common aspects have advised the development of only one radar design, having a good versatility to match the various uses.
The basic requirement of a modern radar is the ability to operate both in medium range mode (conventional mode) against fast maneuvring targets and in short range mode, against small low altitude targets.
The capability to give some additive informations, such as projectiles or missiles trajectory is another important requirement. Due to the kinematics of the targest (speed, low altitude) a fast reaction time is needed, so automatic operation during search, acquisition and tracking modes is mandatory.
To reach these requirements the Selenia X-band monopulse fully-coherent tracking radar makes use of:

- pulse-coded waveform, for medium range mode (~ 50 Km unambiguous range);
- high p.r.f., single-pulse waveform, for short range mode (~ 10 Km);
- adaptive MTI signal processing, with a closed loop doppler correction;
- adaptive anti-multipath angular error processing.

The radar solution is characterized by the following features:

- monopulse antenna head, with twist of polarization to reduce sidelobes levels;
- wide band, low-noise RF receiver, including built-in system for monopulse channels matching;
- coherent transmitter type using a TWT amplifier final stage and a solid state frequency generation;
- combined analogical and digital signal processing, using SAW, CCD and μprocessor technologies.

In the following paragraph the receiver will be described in more detail due to the new technologies involved.

3. DESCRIPTION OF THE RECEIVER AND ITS FUNCTIONS

As previously noted, the receiver is the part of these new radars in which new techniques and technologies have been adopted; its block diagram is shown in fig. 1 where it is easy to realize that two different blocks act to implement the required functions:

- a high speed multi range bin channel (for search and acquisition modes)
- a high dynamic four samples channel (for tracking).

- Search and acquisition

As already shown, two different pulses are transmitted: a single short pulse, in short range mode, and a 13 bit Barker code in the medium range mode.
In the high speed channel the received pulses are filtered in the IF amplifier then coherently detected, using a phase modulated coherent oscillator to account for the relative speed of clutter and platform (see later), obtaining the in phase (I) and quadrature (Q) components of the narrow band signal.
The I and Q signals are then cancelled in a four pulses canceller (implemented analogically with CCD shift-registers) able to adapt the lenght of the delay lines to the two different modes of the radar.
The cancelled pulses are then up converted to allow an IF hard-limiting and a code SAW compression in long range mode. In short range mode obviously these functions are bypassed.
The cancelled and compressed pulses are then converted to a digital form; the used small number of bit is clearly due to the limited dynamic range of the cancellation output.
To get CFAR (Constant False Alarm Rate) an adaptive circuit is used in short range mode (a mean disturb estimator) while a fixed threshold is used in the other mode for the presence of a hard limiter.
A single pole integration is performed in either cases, changing the circuit parameters to match the different pulse repetition frequencies and a second adaptive circuit is used to obtain the plot information in CFAR conditions. This second adaptivity is necessary for the high integration gain which could allow the clutter residues, or the code sidelobes, to be detected if a fixed threshold would be used.
The required performances of this search channel are optimized for a probability of detection equal to 0,8 with a probability of false alarm of 10^{-6} in heavy clutter conditions.
The improvement factor which can be obtained using a 4 pulses binomial canceller is greater than 40 dB for ground and sea clutter and of 32 dB for the hypothised rain clutter. These figures are obtained using Charge Coupled Devices as it will be explained later.
The code compression is implemented in Selenia designed S.A.W. matched filters.
The digital part of the channel is implemented using up to date digital devices thus obtaining a very small, very little dissipative search signal processor.
One simple more circuit (acquisition logic: s. fig. 1) is used in acquisition mode, which could be considered as a first step in the tracking function of the radar: a large window (64 pulses) is opened before and later a range which can be either evaluated by the radar itself or can be transmitted by another measuring instrument.
Following proper procedures the window can be narrowed and the range tracking can begin.
During search and acquisition, one of the two difference channels can be used to detect the presence of a jammer; as a point of fact some "delta" samples are processed in a way similar to that of the sum channel; the special design of the monopulse antenna allows to obtain the information of a jammer in the sidelobes comparing the amplitude of the sum and of the delta channel; so, if wanted, a tracking procedure can begin on the jamming source after a proper acquisition.

- Tracking

An automatic gain control (AGC) acts to keep the useful signal level of the output of IF stages constant.
Due to the high clutter level a heavy cancellation must be performed on tracking signals.
As the difference channels dynamic range is large (greater than 65 dB) a 12 bit convertion is performed; the present characteristics of CCD devices don't allow to use the search cancellers and a digital cancellation must be implemented.

As the rate of the tracked signals is slow (only two targets are sampled and tracked in each sweep) a single time shared circuit is used; as a point of fact the outputs of the coherent detectors are sampled and then stored analogically; an analog switch allows to serialize the signals and only one converter and one MTI are used. The accuracy of each processing stage must be consistent with the wanted dynamic range. Obviously, due to the changing characteristics of the radar in the long range and in the short range modes the parameters of these circuits must be varied. This fact and the relatively slow rate of the output have suggested to implement a microprogrammable unit able to adapt itself to the different constraints present in the processing.
As the IF stages are matched to the signal either in the short or in the long range modes, a high resolution timing unit (it works with a 64 MHz clock) is able to create proper pulses which can gate and sample the incoming signals with the required precision for range tracking.
To allow the reconfigurability of the receiver, these circuits (video section) have been doubled; usually they are able to track two targets different in range, but if one of them cannot work the tracking function can go on with only one circuit even if some information is lost. In fig. 2 the sampling portion of the tracking circuit is better detailed.
Besides the already considered automatic gain control the microprogrammed unit can perform the range tracking filters computations (the range tracker uses a technique similar to the split gate technique) and the angle tracking filters computations.
Also the nodding computations and correction are performed in this unit. During search the programmable unit performs the side lobe jamming control and the faults control function.
Proper signals are inserted in the receiver during "dead times" (intervals in which there is no processing) under the control of this programmable unit which is able to test the output of each board and strip of the receiver, thus detecting faults in real time. The family used to design this unit is the AMD 2900, allowing to reach a great computation power (4 Mips in the implemented solution), needed for the high prf (about 10 KHz) used in the short range mode.
It is clear the great flexibility of this approach.

— Clutter speed estimation

One of the greatest problems, if not the greatest, in tracking radars mounted on a mobile platform (ships, aircrafts.....) is the presence of strong clutter echoes having a doppler carrier. This problem is present also in fixed radars when a rain or a chaff are to be cancelled, but in X band radars on mobile platform it is more difficult to be solved with non adaptive MTI.
In Selenia tracking radars a circuit has been designed to estimate the doppler carrier; it uses a closed loop able to null the error between clutter speed (V) and its estimation (V_R). In fig. 3 the theoretical loop (used for transient and sensitivity analysis) is shown and in fig. 4 its hardware implementation is detailed.
The gain H present in fig. 3 is the speed (V) to phase transducer gain; it is due to the prf and to the transmission frequency (f_t). Ho is its value averaged on all prf and f_t.
The input signal (s. fig. 4) is amplitude normalized by a hard limiter so it is described by the equation

$$x(t) = \cos\left[\omega_c t + \phi_d(t) + \phi_o\right] \qquad (1)$$

being f_c the carrier frequency and ϕ_d (t) the doppler shift.
If ϕ_R (t) is the feedback phase, the signals at the input of the phase detector are

$$\underline{s}(t) = \exp\left[j\left(\phi_d(t) + \phi_o - \phi_R(t)\right) + \omega_c t\right] \qquad (2)$$

and

$$\underline{s}(t-T_R) = \exp\left[j\left(\phi_d(t-T_R) + \phi_o - \phi_R(t-T_R) + \omega_c t\right)\right] \qquad (3)$$

The delay lines and the mixers can be thought as a IF delay.
The output of the phase detector is an error which in the case of a single doppler frequency is

$$\varepsilon = 2\pi\left(\omega_d - \omega_R\right) T_R \qquad (4)$$

f_r being the feedback frequency. Eq. (4) can be rewritten

$$\varepsilon = \frac{4\pi}{c} f_t T_R \left(V_d - V_R\right) \qquad c = \text{velocity of light} \qquad (5)$$

thus obtaining the expression of gain H (fig. 3).
To avoid the dependance on prf and transmitted frequency a gain factor is introduced before and after the loop integrator, whose output is the estimated frequency.

$$y = 4\pi/c\ \overline{f_t}\ \overline{T_R}\ V_R \qquad (6)$$

The output of the phase accumulator is used to phase shift the coherent oscillator used as a reference in the input mixers. The phase modulator must be fast enough to follow the doppler changes along the sweep.
The gain control can be inserted to speed up the circuit during its transient time.
The "presence control" is intended to avoid the influence of code sidelobes; when the amplitude of the signal in the reference range cell is more than 20 dB larger than those of adjacent cells no information is inserted in the estimation circuit.
The improvement factor loss due to the performance of this clutter speed estimator is very small.

4. TECHNOLOGICAL PROBLEMS

The early design of these new radars receivers was intended to adopt a full conventional digital solution; it was soon realized that this approach would have determined an unaccettable hardware, a great power dissipation, a longer development time. Then analogical technologies were studied to implement the processing functions in a reduced hardware, devoting a bit slice microprocessor digital hardware to the tracking units and a pipelined distributed hardware to the thresholding and integrating units of the high speed search processor.
The functions analogically implemented are three:

- the code compressor and the related compensating delay when the radars are in the short range mode
- the four pulses search cancellation
- the doppler carrier estimator circuits.

The code compressor is implemented using the known SAW technique [1]. On the same substrate (ST quartz for high temperature stability and matching of the three IF channels) simply using an electronic switch, it is possible to match either a 13 bit Barker code or the single pulse used in the short range mode.
The peak to sidelobe ratio obtained is larger then 21.5 dB while the spurious effects (direct coupling, triple transit, rigeneration and so on) are under a - 40 dB level related to the peak of the compressed signal.
Twelve of the thirteen compressor taps are connected together; the thirteenth is electronically connected only in the long range mode. This last tooth is the output of the device; in this way an equal delay is obtained in the two modes avoiding changes in the timing unit.
The canceller is implemented using CCD shift registers [2]. As used in the search channel the distortion effects of the device are negligible and due to the absence of dead time the spatial noise has no pratical effect on the cancellation performances. To have the best possible results the canceller is implemented cascading three single zero stages and properly modifying the dynamic range between the stages by a discrete limiting differential amplifier. Output sampling is adopted to avoid the residues of shift clock, large enough if not compensated.
The lenghts of the delay-lines in the two modes are changed using an analog low coupling switch.
The measured D.C. cancellation is larger then 60 dB and the filter mask match the theoretical one.
The circuit is temperature stable.
The delay lines in the doppler carrier estimation circuits are implemented like the MTI delay lines and the other circuits of this part of the receiver are designed using up to date compact analogic devices to obtain temperature stability, great linearity and precision in the phase detector and phase modulator.
The final hardware of the analog/digital design is one third of the estimated initial hardware of the fully digital design.

5. EXPERIMENTAL TESTS

Many tests were done to investigate the radar response to different types of target, such as aircraft, missiles, gun projectiles and helicopters in presence of sea and ground echoes effects.

- Sea tests

The effect on radar accuracy due to the sea reflections (multipath) are well known and analized in literature [3].
Although a general solution of the accuracy problem (excluding the obvious use of a very narrow beam), is not available, an implementation of an effective anti-multipath technique was done and tested during the past years.
The errors reduction effect, based on the adaptive smoothing of the radar errors measured in frequency diversity [4], is shown in the figures 5 and 6.
In fact these figures show the elevation error measured on a target flying at a very low levels with and without the filter; by comparing the two figures, a reduction of the error of about 3 times is evaluated.

- Ground tests

Due to the clutter residues after the MTI filtering two problems arise in ground applications:the growth of probability of false alarm and the birth of an angle error bias.
In the region selected for the tests, with clutter echoes levels well in accordance with the theoretical models, no target acquisition was possible without MTI operation; moreover no clutter acquisition was recorded in MTI operation.
Figure 7 shows the azimuth and elevation errors measured on an aircraft flying at an elevation less than 1/2 the antenna beam; the analysis of these and other similar records, shows a little bias angular error due to the back-scatter and, perhaps, to the diffuse reflection of the ground.

REFERENCE

[1] Bucciarelli T., Fazio C., Picardi G., 1980, "An Analog (SAW-CCD) Radar Processor", EUSIPCO 80, Lausanne.

[2] Bucciarelli T., 1979, "CCD non linearities in MTI receivers", 5.th International Conference on CCD. Edin burgh 1979.

[3] Barton D., "Radars" Vol. 4.

[4] Selenia Internal Classified Report.

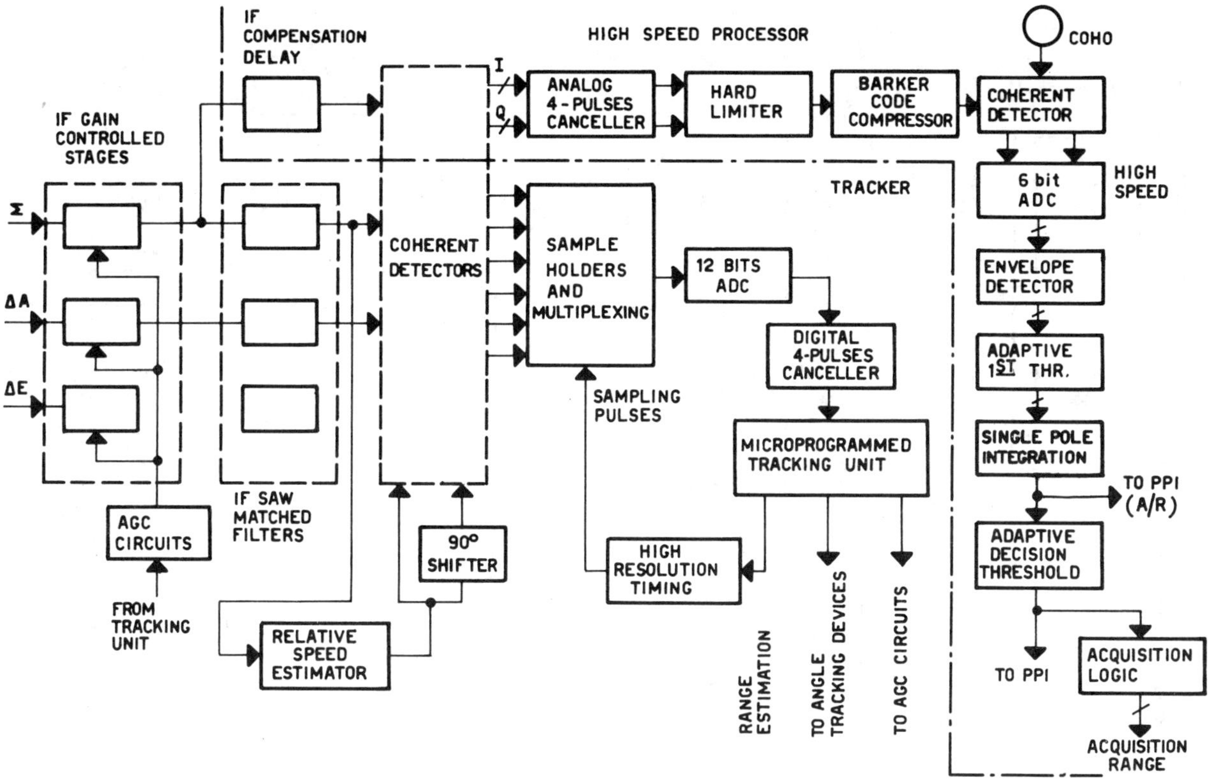

Fig. 1 — Schematic block diagram of the receiver

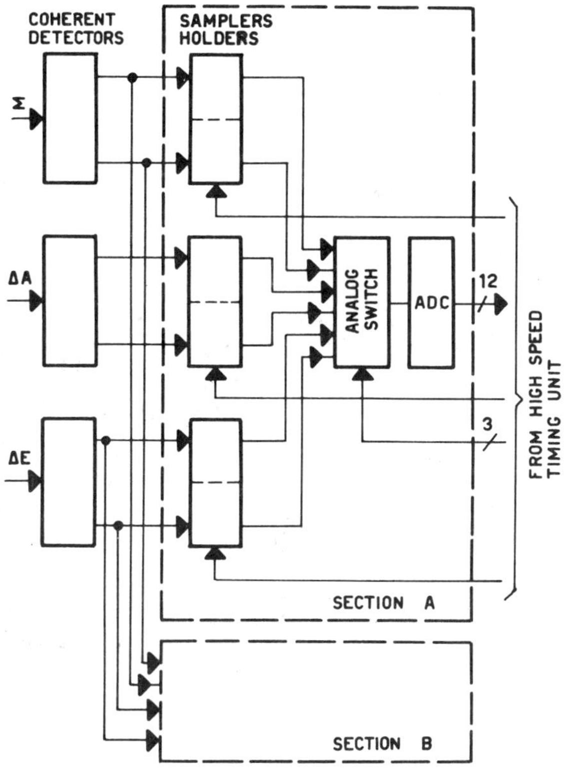

Fig. 2 — Sketch of the tracking sampling and conversion circuit

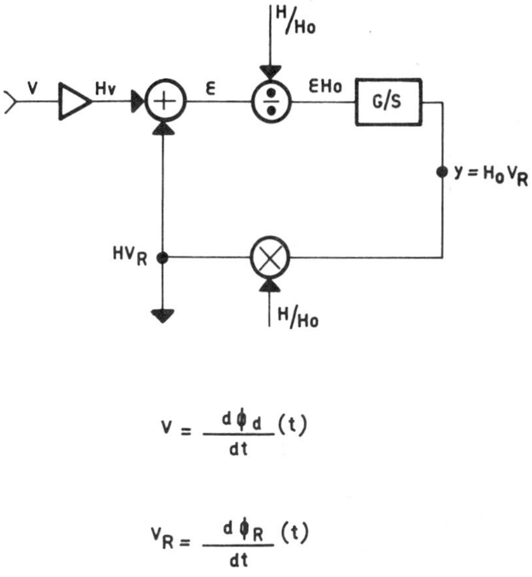

Fig. 3 — Explicative block diagram of speed estimation loop

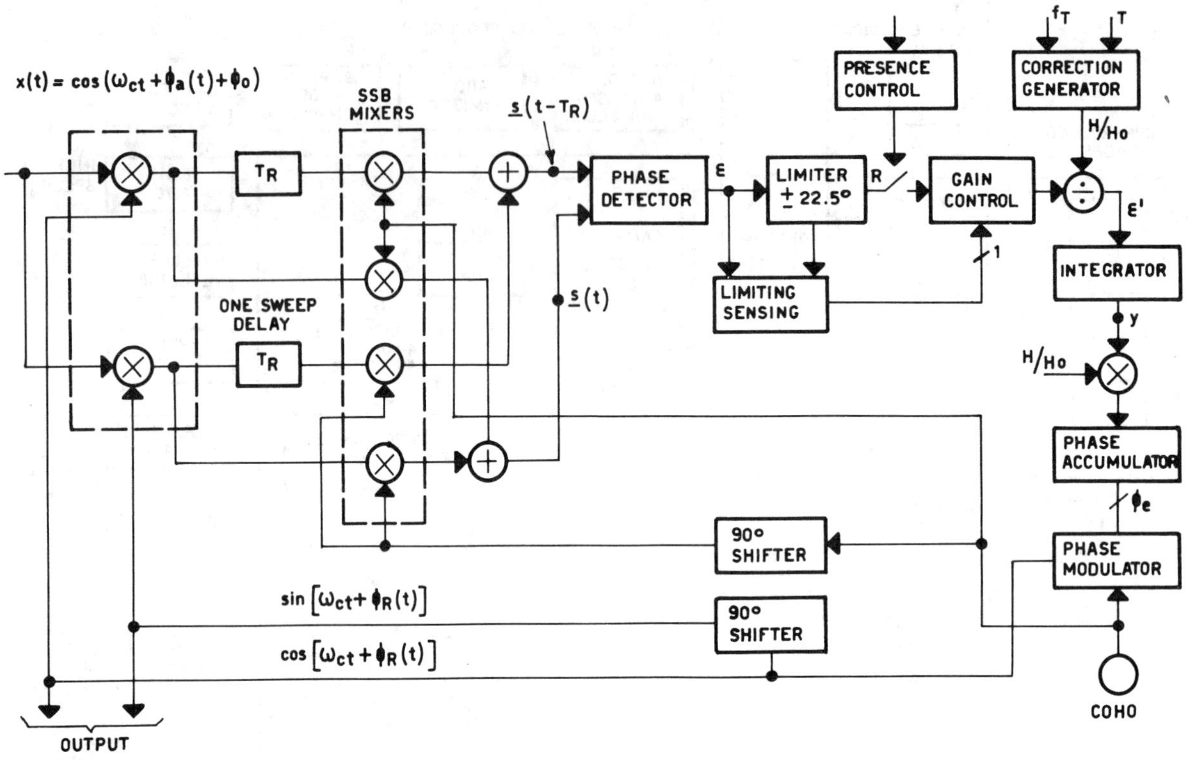

Fig. 4 — Detailed block diagram of the estimation loop

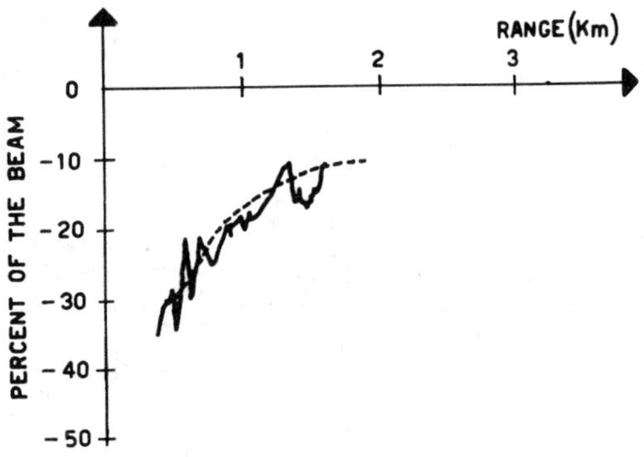

Fig. 5 — Measured target elevation

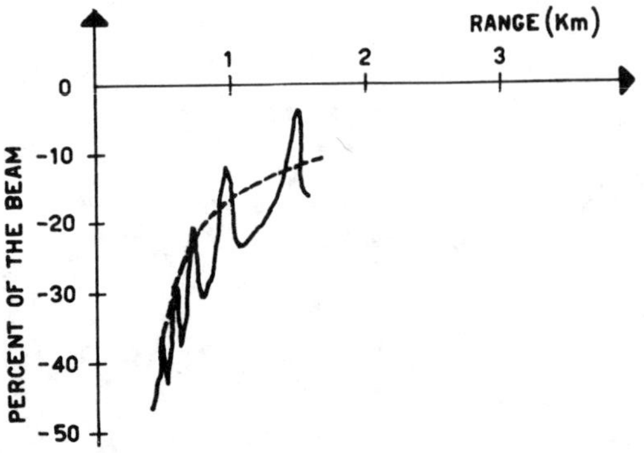

Fig. 6 — Measured target elevation

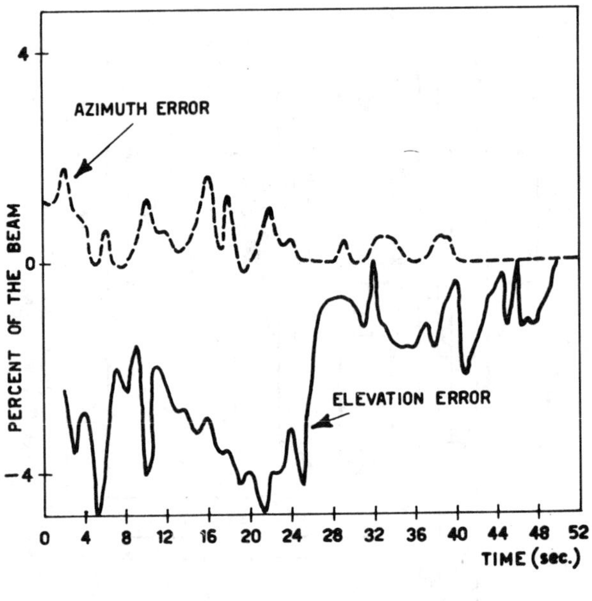

Fig. 7 — Azimuth and elevation errors, as a function of the time

AN X-BAND ARRAY SIGNAL PROCESSING RADAR FOR TRACKING TARGETS AT LOW ELEVATION ANGLES

A. Pearson[+], P. Barton[+], W. D. Waddoup[+], R. J. Sherwell[*]

+ Standard Telecommunication Laboratories, UK

* Admiralty Surface Weapons Establishment, UK

INTRODUCTION

The problem of tracking low flying targets is well known. When the separation between a target and its multipath image becomes sufficiently small, signals from both may be within the main beam of an antenna at the same time and a conventional tracking radar will have difficulty in resolving them and thus become subject to substantial errors. Various methods of reducing these errors have been proposed and this paper deals with a particularly promising approach which uses spatial filtering to suppress the multipath signals. This technique has been investigated jointly by STL and ASWE for a number of years and the results of some earlier 'one-way' trials have already been reported (Barton et al (1) and Barton and Waddoup (2)). More recently an experimental radar facility has been established on the coast near Portsmouth in order to assess the effectiveness of spatial filtering under realistic conditions and a large number of elevation measurements have been made of aircraft flying low over the sea surface. In the following sections the spatial filtering technique is discussed and details of the experimental radar are given together with an assessment of the results obtained so far.

SPATIAL FILTERING

If an antenna array is divided into N rows which are connected to N well matched receiver channels, then a plane wave narrow-band signal arriving at an elevation angle θ above the array normal will give rise to a signal vector $\underline{V}$ at the receiver outputs of the form

$$\underline{V} = a \begin{bmatrix} \exp j\psi \\ \exp j\left(\psi - \frac{2\pi d}{\lambda}\sin\theta\right) \\ \cdot \\ \cdot \\ \exp j\left(\psi - \frac{2(N-1)\pi d}{\lambda}\sin\theta\right) \end{bmatrix} \qquad ..(1)$$

Discrete Fourier transformation (DFT) of $\underline{V}$ yields Fourier coefficients from which the angular response in elevation of the antenna to the plane wave can be reconstructed. If the DFT is represented by the matrix $\underline{F}$ and the resultant coefficients by the vector $\underline{Y}$ then

$$\underline{Y} = \underline{F}\,\underline{V} \qquad ..(2)$$

It will be advantageous in practice to obtain more than N Fourier coefficients because, even though the outputs will be correlated, their close spacing (relative to the natural beamwidth of the antenna) will aid the accurate determination of the beam peak by simple interpolation methods. Use of this arrangement at low angles is subject to the same limitations as conventional monopulse tracking since the angular response to multipath overlays the response to the direct signal and the peak of the composite beam can be at a significantly different angle from that of the direct signal beam by itself.

The process of spatial filtering is one in which the individual receiver channel (or element) responses represented by $\underline{V}$ are modified, prior to beam formation by the DFT, so that the discrimination of the system against unwanted signal directions can be improved. The finite extent of the aperture will of course mean that the transition from the passband to the stopband will take place over an angular region proportional to the natural beamwidth of the aperture. Several outputs with different effective phase centres are to be provided, spanning as far as possible the original extent of the aperture. In general, the number of useful outputs (M) which can be synthesised is less than the number of inputs (N). The spatial filter is therefore represented by a transfer matrix $\underline{H}$ of dimensions (M,N) and is cascaded with the Fourier transformation so that modified Fourier coefficients $\underline{Y}'$ are obtained, i.e.

$$\underline{Y}' = \underline{F}\,\underline{H}\,\underline{V} \qquad ..(3)$$

Various response profiles can be suggested for the filter. If the multipath reflection is specular (as is generally the case for very low grazing angles over the sea) the provision of a correctly positioned notch filter characteristic can provide a relatively sharp recovery of gain away from the notch direction in terms of the natural beamwidth of the array. On the other hand, if the multipath is expected to be angularly spread, e.g. a combination of a specular component and diffuse reflections from rough ground or sea surfaces, then the filter's stop band should be spread to suppress all components. The filter shape used for the trials reported in this paper is a bandpass type with the stopband on one side associated with directions of arrival below the horizon. Regardless of the multipath distribution some benefit will always be obtained from the use of this filter.

The most plausible design synthesis requires that the element h_k of $\underline{H}$ is given by

$$h_k = T\{\ell-k\}\,\frac{\sin\left(\pi x_b \frac{\ell-k}{N}\right)}{\pi x_b \frac{\ell-k}{N}}\exp\left[j2\pi x_o \frac{\ell-k-\frac{1}{2}}{N}\right] \qquad ..(4)$$

where $T\{\ell-k\}$ is a windowing function, x_o defines the centre of the bandpass region in beamwidths and x_b is the width in beamwidths of the bandpass region. A simple case is to choose $T\{\ell-k\} = 1$ and obtain angular responses at 6 outputs of the matrix $\underline{H}$ as illustrated in Fig. 1 for N = 8, $x_b = 4$ and

$x_o = 2.25$. Note that an angle of arrival (θ) with respect to the boresight of a line array of N elements having an inter-element spacing of d can be represented as an angle shift (x) in units of standard beamwidths given by

$$x = Nd\lambda^{-1} \sin \theta.$$

It should be noted that, in the transition region between the pass and stop band of the filter, phase distortion arises which desensitises the change in beam centre position after the spatial filter for a given change in signal direction prior to the filter by about 30 per cent in the angular region of most interest; this will somewhat counterbalance the increased discrimination against multipath provided by the spatial filter. The effect, however, is largely independent of the multipath distribution and is corrected for in the processor by a simple look-up table procedure.

EXPERIMENTAL HARDWARE

The radar (Fig. 2) operates at 9.6 GHz and employs separate transmit and receive antennas fixed side by side on a common mount. A TV camera situated between them forms part of an optical tracking facility for controlling the azimuth position of the mount and provides elevation data which is used for comparison with that obtained by the radar. The antennas do not move in elevation. Each consists of a vertical array of 16 horizontally mounted slotted waveguides and associated flares enclosed within a rectangular structure some 1.1m high by 0.8m wide by 0.2m deep. The waveguide slots are designed to give a 3.5 degree azimuth beam with vertical polarization and -28 dB side-lobes. The transmit array is fed by a high power divider network giving an elevation beam which has a sharp cut-off at the horizon and provides adequate illumination up to 8°. In the receive antenna the waveguides are coupled in pairs through 3 dB hybrids to 8 front-end units consisting of a mixer and modular amplifiers. The first IF frequency is 100 MHz and the overall noise figure and gain of these units are 9 dB and 37 dB respectively. A stalo-coho unit mounted behind the antennas contains the 9.5 GHz stable source for the front-end mixers and a 100 MHz source for driving base-band mixers. This source is phase locked to a 100 MHz version of each transmitted pulse in order to allow coherent (Doppler) processing of the base-band signals. The transmitter is a standard unit from a Type 903 naval radar and consists of a CV 2261 magnetron with associated modulator and power supplies. The peak power is approximately 40 kW, the pulse width 0.15 μS and the PRF used for the recent trials was 2.5 KHz.

The 100 MHz signals from the receive antenna are fed to a 'zero IF' unit where they are split and suitably mixed with the coho signal to provide in-phase (I) and quadrature (Q) components of the 8 base-band signals. After further amplification and filtering these 16 components go to sample and hold (SH) circuits which at the appropriate time are accessed sequentially by a high speed analogue to digital converter (ADC) which transfers the data to a PDP 11-34 mini computer. The 8 receiver channels have been balanced to within ±0.6 dB in amplitude and $\pm 10^{o}$ in phase. The zero IF unit also contains a timing board which is linked to the computer and provides the appropriate timing signals for the various devices (magnetron, SH, ADC, etc.). The computer has a VT 55 visual display unit (VDU) and two RL01 hard disc storage units.

SIGNAL PROCESSING

For reasons of cost and flexibility, all the digital signal processing is carried out in software in the PDP 11-34 mini computer. Assembler language is used to achieve real-time tracking of fast moving aircraft. Apart from elevation angle estimation, the processing is not particularly optimised. An outline is given with reference to Figure 3. The first function of the radar processor is to acquire the target in range (acquisition in azimuth having been achieved optically). The range gate is made to scan from 2 to 10 Km in 30m steps. The returned signals of each range-gate are processed in Doppler over 8 returns and summed over the 8 array channels. The range-normalised responses from such processing are plotted versus range on the VDU to give an A-scan display. The range cells with the largest and next largest responses above a given threshold are identified as possible targets. The Solent is a very busy waterway and this facility was necessary to prevent acquisition on marine-vessels. Operator intervention determines which of the two will be tracked, or, by default, whether reacquisition is to occur.

Limited acquisition cycles, over decreasing range spans, now occur on the target chosen, to provide an up-to-date range estimate prior to tracking.

In track mode, range tracking is accomplished by performing the same processing on returns from 5 range cells centred around the range cell with the maximum response in the previous cycle. Range up-dating is carried out by noting which of the current 5 range cells has maximum response. Range cells of 15m are employed in tracking mode to enable closer sampling to the peaks of the return pulses. Elevation tracking is then carried out on the Doppler-filtered signal vectors from the up-dated range cell. This consists of spatial-filtering followed by a 32 point FFT acting as an overlapping beamformer giving outputs every 1/4 BW over the range ±4 BW. A peak of beam tracker incorporating an interpolation routine scans the beamformer outputs to determine the target elevation. This is output on a new display of elevation versus range on the VDU. The cycle-time in track-mode is 144 ms, giving a data rate of ≃ 7 Hz. Execution of the spatial filtering takes only 4 mS of this total. The relatively slow data rate is due to the use of a general purpose computer and does not represent what might be achieved with a radar of this type. A consequence of the slow data rate is that, since only 8 pulse returns are processed every 144 mS, full use is not made of the available mean power.

Concurrently with the range and elevation processing, the computer outputs data to hard-disc from the previous tracking cycle. This consists of signal vectors (after Doppler-filtering), range, elevation, Doppler and amplitude of the target. At any stage in the tracking process, a return to the acquisition mode can be made. Alternatively, tracking can be discontinued and if this occurs the data-file is closed and a hard-copy of the elevation versus range display currently on the VDU is made. The original A-scan display is then reconstituted from memory and also copied. The process then terminates and can be rerun as required.

FIELD TRIALS

The trials site is on the coast near Portsmouth and the radar is some 15 metres from the edge of the sea and about 7 metres above mean sea level. The antennas face out to sea in a southerly direction with an azimuth coverage of about 135°. Within this arc there are various fixed objects such as buoys and towers and usually a steady traffic of small and large boats and an occasional search and rescue helicopter. These have acted as useful targets for calibration and operator confidence purposes during the setting up of the radar and subsequent trials. In addition, some initial tests were also made using model aircraft and towed kites with radar reflectors attached to them. However, the trials themselves were carried out using military aircraft flying at sub-sonic speeds. The general procedure was for the aircraft to fly a 'race track' circuit, approaching the radar on a constant bearing and at constant height from a range of about 12 Km to 2 Km. Typically an aircraft would carry enough fuel to make 12 approach runs at different heights as requested by the trials team. The targets were acquired initially by the optical tracker which then maintained the radar on the correct azimuth bearing. The radar operator observed a conventional A-scope display on an oscilloscope or the computer generated equivalent on the VDU. When the target signal was seen above the noise level the operator could instruct the computer to enter the tracking routine and the VDU would convert to an E-scope display (Fig. 4) which plots a number indicating the current Doppler cell on a display of elevation against range. During tracking, all the complex signal voltages on the 8 channels (after Doppler filtering) were stored on hard disc together with results of the real-time processing and data from the optical tracker. It is by off-line analysis of this stored data that the accuracy of the elevation measurements has been determined.

Over 100 data files have been created representing about 10 hours of tracking data on a wide variety of targets. So far, it has only been possible to analyse a small quantity of these data in any detail and the results presented here concentrate on 6 separate aircraft runs when the targets were tracked at between 0.1 and 0.5 beamwidths elevation. Beamwidth here is 1.79° which is $\sin^{-1}(\lambda/L)$, L being the vertical extent of the antenna and λ the operating wavelength. Figure 5 shows a typical plot of elevation as a function of range and allows a comparison between optical tracking and radar tracking, both with and without spatial filtering. The benefits of spatial filtering are immediately apparent. There is no tendency for the radar to lock onto the multipath image and there is a substantial reduction in errors. As expected, it can be deduced from the geometry of the trials site that the ranges at which peak positive and negative errors occur correspond to conditions of anti-phase and in-phase multipath respectively. In Figure 6 these peak errors have been plotted for all six aircraft runs together with theoretical curves for the spatially filtered ASP case and for a conventional off-set monopulse radar. The theoretical curves are for a specular reflection coefficient of 0.95 which is the value expected at 0.15 beamwidths elevation for vertically polarized signals. The very low measured errors at higher elevation angles will correspond to a lower value for this coefficient.

CONCLUDING COMMENTS

A comprehensive series of trials, involving typical targets flying under realistic conditions, has confirmed that the spatial filtering technique gives a substantial reduction in elevation measurement errors resulting from multipath interference. The hardware required to implement the technique is quite straightforward and has proved realiable in practice. During the development of the radar and the subsequent trials the advantages of digital array signal processing have been clearly apparent. System modifications and improvements have been rapidly achieved by changes in software. The filter algorithm employed so far has proved to be quite robust and not excessively critical of the phase and amplitude balance of the receiver channels.

ACKNOWLEDGEMENT

This work has been carried out with the support of the Ministry of Defence, Procurement Executive.

REFERENCES

1. Barton, P., Wong, A., Kelly, K., and Gwynn, P. 1977 "Array Signal Processing for Tracking Targets at Low Elevation Angles". IEE Conference Proceedings, "Radar 77", London, England.

2. Barton, P., and Waddoup, W.D., 1978 "Array Signal Processing Techniques for Low Angle Tracking", Conference Proceedings, "Military Microwaves", London, England.

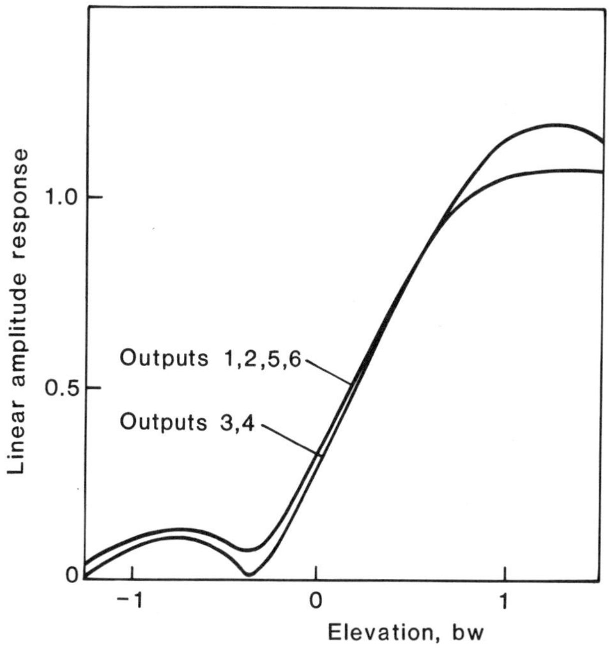

Figure 1 Bandpass Spatial Filter Responses ($N = 8$, $x_b = 4$ and $x_o = 2.25$)

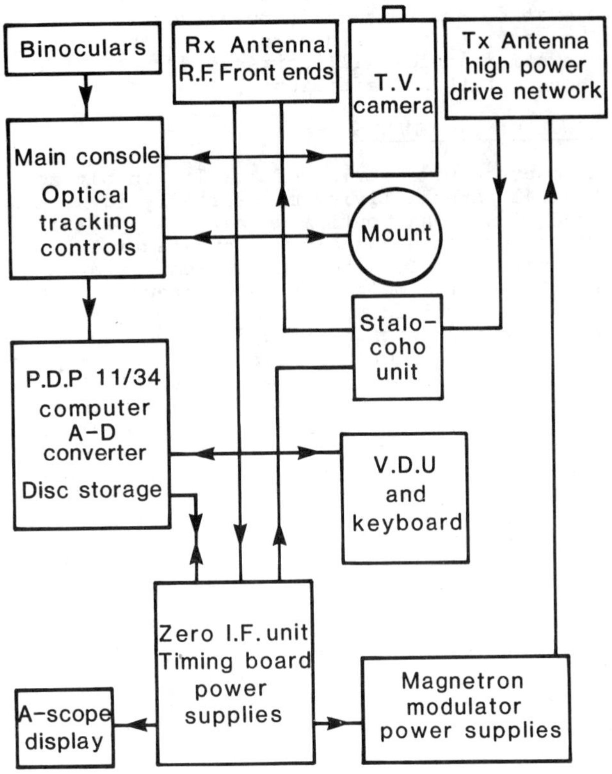

Figure 2 Block Diagram of the Radar

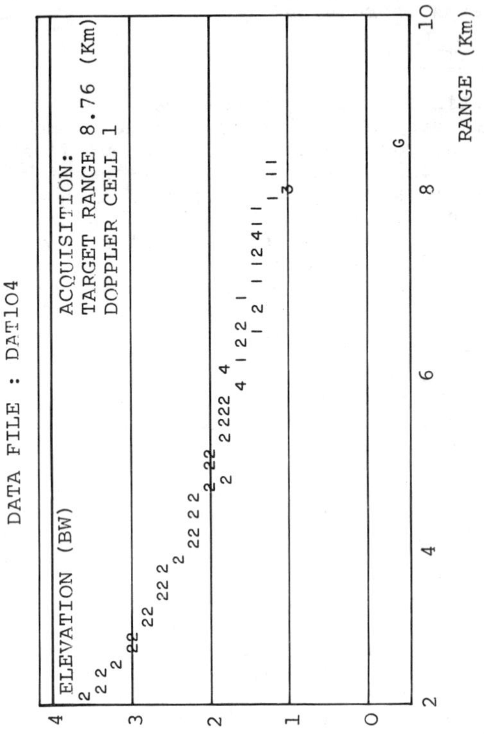

Figure 4 Real-time Display of Elevation against Range

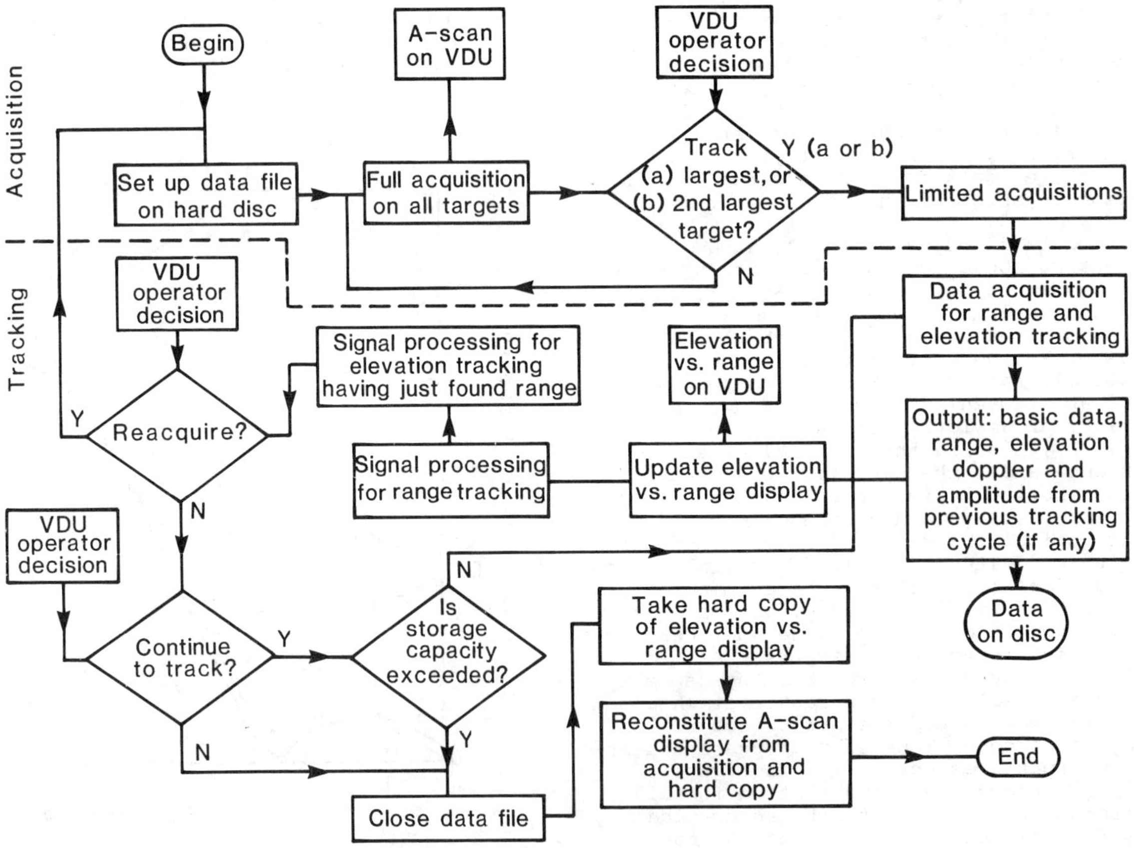

Figure 3 Flow Diagram for the Radar Software

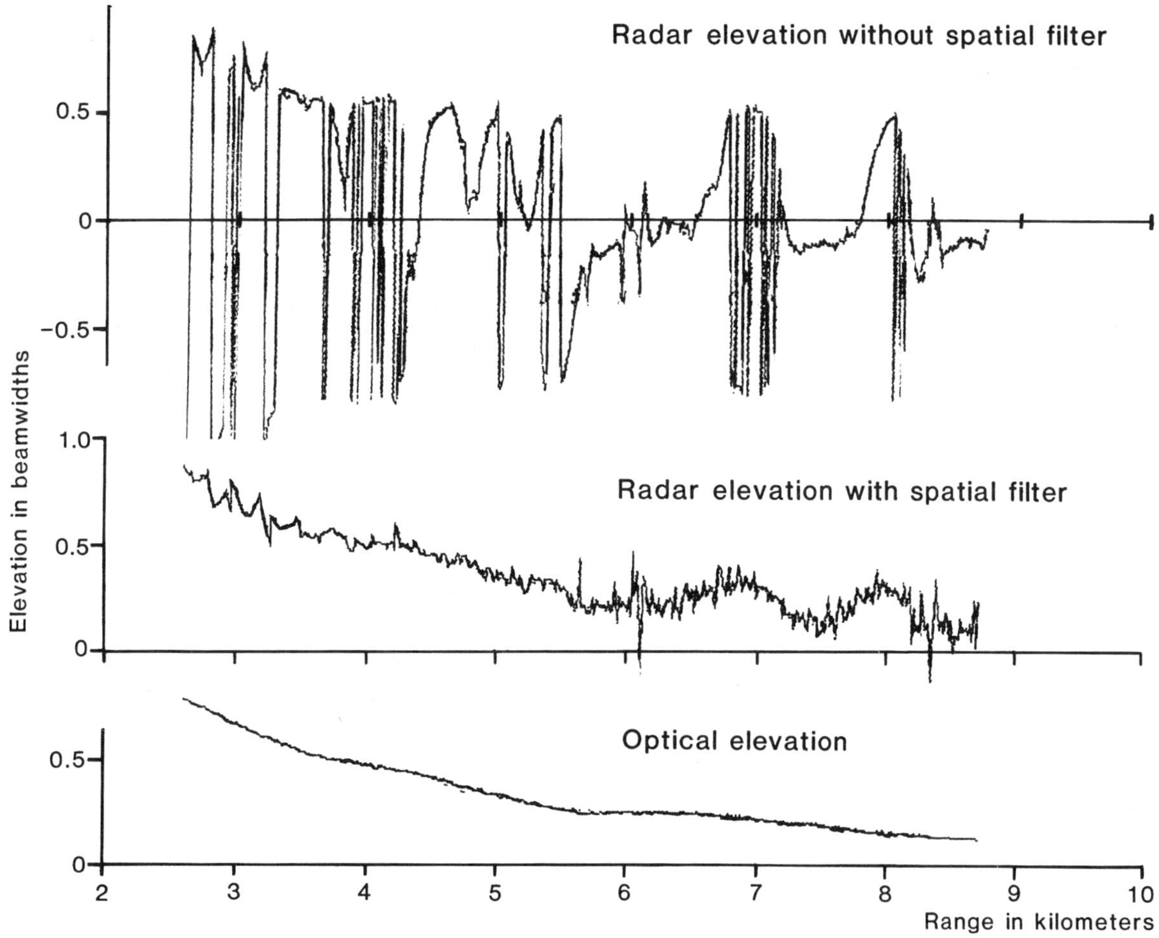

Figure 5 Elevation as Function of Range for a Low-Flying Aircraft

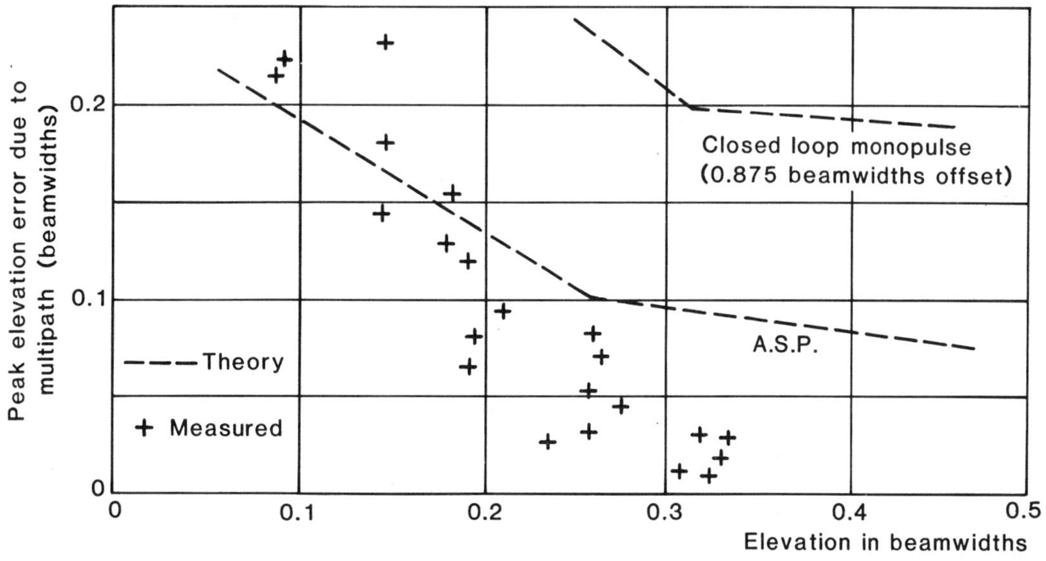

Figure 6 Peak Errors due to Multipath

TRACKING RADAR ELECTRONIC COUNTER-COUNTERMEASURES AGAINST INVERSE GAIN JAMMERS

Stephen L. Johnston

International Radar Directory, USA

INTRODUCTION

Conical scanning is a form of angular tracking which has been widely used in automatic angle tracking radars. In conical scanning a pencil beam whose direction of maximum radiation is offset from the axis of the antenna is rotated or nutated about that axis. Variation of signal amplitude as the beam scans provides information on the amount and direction of displacement of the target from the axis of rotation (1). This technique was a key to the success of the widely used World War II SCR 584 radar (2). Identity of the inventor(s) of conical scanning is unclear (2) although the principles of conical scanning were described in 1927 by Whitford and Kron (3) for another application. That early work was cited by this author almost a quarter of a century ago (4)!

Conical scanning has been used in military systems for both tracking radars, homing guidance systems and in infrared homing missiles. The last application is a close relation to the work of Whitford and Kron (automatic guidance of astronomical telescopes). During World War II the SCR 584 was not jammed by Germany since the Germans did not develop microwaves until they discovered Allied use of this frequency region (5). It is well known that performance of conical scan radars in a benign environment can be degraded by target amplitude fluctuations at or near the radar conical scan frequency. This deficiency of conical scan is a principal advantage of monopulse over conical scan. Even so, the relative simplicity of conical scan over the complexity and cost of monopulse had made conical scan very attractive until advent of inverse gain jammers.

A major disadvantage of conical scan for military applications is vulnerability of "conventional" conical scan to the inverse gain jammer. This angle deception ECM uses a receiver to detect the tracking radar signal and re-transmits a return signal with tracking scan modulation shifted in phase so as to cause the tracking radar to move the antenna in opposition to the direction of the true target position. Performance of an inverse gain jammer against a conical scan radar was vividly shown in the simulation results of Cikalo and Greenbaum (6). A "conventional" inverse gain jammer is effective only at ranges greater than crossover range, i.e. where the jammer signal is greater than the skin echo signal. Kline (7) described a controlled deception device wherein the inverse gain jammer transmits at either of two power levels to cause the radar to track by a side lobe. This device is claimed to be effective at all ranges.

Defeat of the inverse gain jammer is essential if conical scanning is to remain a viable technique for military applications. Review of the Western and Russian books on radar ECCM (8, 9) and their voluminous references revealed only limited information in the public domain on anti-inverse gain jammer techniques. In the process of expansion of the author's short course on Radar ECCM (10), an extensive search was made of U S patents on radar ECM and ECCM. That search identified a number of techniques for defeating inverse gain jammers. This paper will describe some techniques with their advantages and disadvantages. These have not been previously treated in the professional literature.

Due to space limitations, descriptions will be very brief. Additional explanations together with timing/waveform/block diagrams are contained in the patents*. Two dates will be given for patents in the bibliography for this paper: issue date and filing date. The former indicates when information on the patent was made public, while the latter approximates when the invention was conceived.

ANTI INVERSE GAIN JAMMER TECHNIQUES

While monopulse is indeed an excellent electronic countermeasure against the inverse gain jammer, there are in fact a number of ECCM techniques which may be applied to conical scan radars to defeat this ECM. Anti inverse gain jammer techniques may be placed into four groups: a. "Natural" lobe on receive only; b. True lobe on receive (LORO)/ Conical scan on receive only (COSRO); C. Variable conical scan frequency; d. Hybrid monopulse-conical scan systems. Each group has a number of interesting methods of implementation.

"Natural" Lobe On Receive Only

Inverse gain jammers e.g. (7) usually derive the scan modulation frequency from the radar transmitted signal. Accordingly, a good anti inverse gain jammer technique is to deny the radar scan modulation frequency to the jammer receiver. N B: It is not necessary that the radar transmitted signal be conically scanned; conical scan of only the received signal is adequate. In some radars, conical scan of only the received signal occurs as a "natural" part of the radar. Such radars generally use separate transmitting and receiving antennas, with conical scan being applied to the receiving antenna only. Three such examples are: bistatic radars, CW tracking radars, passive radar homing systems, and receivers for beam rider missiles. N B: A bistatic radar has its transmitting and receiving antennas at different locations (1). A semi-active radar-homing system is actually a combined monostatic and bistatic radar. Even though the missile receiver (bistatic receiver) may use COSRO, the tracking illuminating radar must use some form of COSRO. A CW tracking radar normally uses separate side-by-side transmitting antennas. A passive homing system is inher-

*Unfortunately the figures in these patents are not of adequate quality for reproduction in this paper.

ently COSRO. As a historical item, it is interesting to note that scan on transmit only was proposed in World War II as a means of improving tracking accuracy (5)!

True LORO/COSRO techniques

Pulsed tracking radars generally use a common antenna aperture for transmitting and receiving in order to minimize radar size and cost. In such radars e.g. SCR 584, conical scanning is applied on both transmission and reception. A number of schemes for applying scan on receive only have been devised. These will be described in chronological order.

Prichard (11) reported a method for reducing the effectiveness of jamming in a single axis lobe switching on receive only radar. An artificial locally generated signal is injected into the receiver in the time vicinity of the reception of each echo pulse. Through an elaborate set of switches and bridges, effects of jamming in one lobe are mitigated. The intent of this invention is to handle off-axis jamming although it could handle on-axis switched jamming if the jamming were switched at the radar lobe switching frequency.

Adam (12) described a single antenna COSRO system which he claimed can be used with either CW or pulsed radars. This invention utilizes wave polarization properties. A reflective wire grid placed at the focal plane of the antenna radiates only one component of the circular polarization from the transmitter. A grid of conductive strips on the concave side of the parabolic reflector converts the linear polarization to a radiated circular polarization. On reception, the reflected circular polarized wave is converted into a plane wave by the grid strips on the reflector. It is of orthogonal polarization to the transmitted wave which was radiated by the grid at the antenna focus. Being of opposite polarization, it passes thru that grid to a second grid of conducting strips which performs a conical scanning action thru its off-axis rotation as it reflects received energy back towards the receiver. The specific construction actually imparts nutation rather than rotation to retain a constant polarization in scanning. Conventional conical scan reference generator and angle error detectors are used.

The invention of Tellier et al (13) pertains to an improved Lobe-on-Receive-on-Track-Only (LORTO) antenna. This antenna is used for a dual function radar, search/track. In the search mode the radar functions as a normal search radar. Conical scan is utilized in receiving only when the radar is in the track mode. In the prior invention an ATR device and a phase shift element couple all of the received energy into a special track receive wave guide thru a spinning trislot junction which creates the conical scan action. On transmit, a pre-TR would fire so that the trislot junction would appear as a solid cylinder and thus no lobing would be affected.

This prior invention required two waveguide feeds. The Tellier et al invention uses only one waveguide feed. A push-pull solenoid which is activated to change from search to track moves a short tuning stub with a polarization sensitive grid. In the search mode, this grid shorts the energy from the trislot, preventing lobing. For tracking, the solenoid rotates the grid by 90° permitting trislot lobing on receive only. The patent also indicates that this device will also cancel circularly polarized jamming.

Holman (14) proposed a method of achieving the appearance of COSRO to an ECM receiver on board a target while actually using con scan on both tracking and receiving. A transmitted signal (either CW or pulsed) is divided (actually modulated) by the received angle error signal from the target being tracked. Let the received angle error signal be K cos wt where K is the conical scan error coefficient and wt is the conical scan angular lobing frequency. The modulated wave is now of the form $A(1 - K \cos wt)$ where A is the magnitude of the transmitter unmodulated output. Upon radiation, the signal becomes $A(1 + K \cos wt)(1 - K \cos wt)$ or $A(1 - K^2 \cos^2 wt)$ due to the lobing on transmission. Since the quantity $K^2 \cos^2 wt$ is a second order term, it may be disregarded making the apparent radiated signal a constant. On receiving, the signal now becomes $A(1 + K \cos wt)$. The receiver angle error detector recovers the component K cos wt. Although separate transmitting and receiving antennas are shown in the patent a common antenna could be used for a pulsed radar.

The Schmidt (15) COSRO system can be thought of as being the combination of a conventional non-con scan radar and a separate additional COSRO receiver. Outputs of the conventional radar receiver and the COSRO receiver are subtracted to cancel jamming modulation. The first embodiment of Schmidt uses a common parabolic antenna with two feeds. One feed has mechanical conical scanning and is used for receiving only. The other non-scanning feed is used for both transmission and for jamming reception. Output of the COSRO receiver will consist of con scan error signals and jamming modulation. Output of the jamming receiver will be jamming modulation only. This latter signal is inverted and then used to modulate the COSRO receiving channel to effectively cancel the jamming modulation. Another embodiment uses only one antenna feed in a four-horn monopulse arrangement. Equivalent conical scanning is electrically imparted to the monopulse elevation and azimuth difference signals. The Holman invention will be recognized as that of a cancellation system. Described in reference (9) are two CW jamming cancellation schemes. The Holman invention resembles the Russian cancellation systems.

Variable Con Scan Frequency

Variation of the radar con scan frequency has also been proposed to defeat inverse gain jamming. Simple con scan frequency variation would provide only limited benefit. The optimum frequency variation method is variation in a pseudo-random manner. Reid (16) has discussed properties of pseudo-random codes and methods for their generation. Two patents for pseudo-random variation of con scan frequency have been described. Rabow (17) achieved this variation by pseudo-randomly applying either of two voltages to the radar nutating motor.

Neri and Albrande (18) proposed the same concept as that of reference (17) but with digital techniques used to either control the voltage applied to the motor or to control a braking magnetic field. Their scheme uses the reference generator in a feedback loop to monitor actual nutation motor speed.

Hybrid Monopulse-Conical Scan Systems

The last group of ECCM techniques against inverse gain techniques is that of hybrid monopulse-conical scan systems. Three such techniques - CONOPULSE (20), Scan With Compensation (Russian) (21), and the Two Channel Monopulse System (22) have been reported elsewhere and will not so be discussed individually here. CONOPULSE is similar to scan with compensation. CONOPULSE uses two antenna feeds rotating 180 degrees apart. Two different methods of processing the received signals were described by Sakamoto and Peebles - MOCO and COMO. One uses monopulse processing up front and conical scan processing at the end. The other scheme reverses the sequence.

Generally, hybrid monopulse-conical scan systems use a conventional four horn (two axis) monopulse feed followed by four hybrid junctions to provide sum, azimuth difference and elevation difference signals. Hybrid systems then use some form of amplitude lobing modulation to create a conical scan effect. Several hybrid monopulse-conical scan patents have addressed the inverse gain jammer problem.

Butler (23) described three schemes, two of which somewhat resemble reference (20); in fact, Butler states that his invention functions "as a conically scanning monopulse radar". In the first embodiment, a circularly polarized non-scanned beam is radiated. Upon reception, two separate conically scanned beams spaced 180 degrees apart are formed by the antenna. These signals are separately fed to the two channels of a conventional 2-channel monopulse receiver. The antenna uses a gaseous discharge detuning device which, when activated by the transmitted signal effectively short circuits the slots of the conical scanning radiator.

In his second embodiment, Butler uses a single channel monopulse receiver which is alternately connected to either of two receiving waveguides. Antenna is similar to that in the first scheme. The third scheme uses the same antenna in a "conventional" COSRO system. A polarization gyrator rotates the electric field at one half the desired scan rate producing a rotating circular polarization. As before, on reception, the gaseous discharge detuning device becomes inactive with the slots producing conical scanning for reception at twice the rate of gyrator polarization rotation.

Schöneborn (24) proposed a switched single axis monopulse system, useful for homing missiles. The RF portion differs from conventional monopulse in that it uses a separate transmitting antenna feed and that it has a phase shifter in one antenna feed. A common receiving channel is alternately switched between the two outputs of the hybrid junction.

Longuemare and Moody (25) described a modified monopulse system which uses only two IF amplifiers in contrast to the usual three. This is made possible by employment of a lobing frequency for amplitude modulating the az error and el error signals. A composite sum of these two signals and a portion of the sum signal are amplified in one IF amplifier. This system resembles that of Noblit (22). Inventors claim that this simplified system avoids amplitude effects from inverse gain jammers and does not radiate its internal lobing frequency. This scheme is also claimed to be suitable for use with Doppler radars which use highly selective filters for target/clutter signal separation. Such filters generally introduce large phase shifts making matching of multiple channels of conventional monopulse systems difficult.

Felsenthal (26) used a system similar to that of (25) except that the lobing reference signals for the azimuth and elevation signals are periodically reversed in phase. Reversal of the direction of nutation in synchronism with the radar pulse repetition frequency reduces effectiveness of inverse gain jammers.

Gulick (27) invented a scheme similar to that of (25). Notable differences include application to CW semiactive missile seekers and inclusion of a rear reference antenna with Doppler tracking loop. The IF amplifier bandwidth is about 1 KHz and the modulating oscillator frequency is on the order of 5 KHz in contrast to the usual 100 Hz or less lobing frequency. Reference (27) also cited ability of this hybrid monopulse system to defeat inverse gain jammers. Use of a rear reference antenna and rear reference receiver has now been eliminated by the inverse receiver described by Ivanov (28). This could be adapted to the invention of (27).

Further ECM Considerations

The thrust of this paper is defeat of inverse gain repeaters against tracking radars. Inverse gain repeaters can also be used against search radars (29, 30). Some search radars employ monopulse techniques to improve angular accuracy of the radar. This is also known as "beam sharpening". Boucher (31) has described methods for defeating inverse gain jammers, sidelobe repeaters and standoff jammers against such search radars. His techniques resemble those of the above described hybrid monopulse-conical scan techniques for tracking radars.

Reference (19) pointed out that interfering amplitude modulation by repeater jammers which is slightly different from the actual radar lobing frequency can cause the radar boresight to spiral around the target at the difference frequency. Accordingly, schemes which merely avoid transmission of the lobing frequency can be defeated. In fact, Simonaire (32) has invented an ECM against LORO/COSRO radars. This is sometimes called a "jog detector", or swept audio modulated repeater. In one implementation the jammer transmitter radiates pulses on the victim radar frequency at a repetition rate which repeatedly sweeps over a range selected to include the unknown lobing frequency of the radar. A receiver stops the PRF sweep when the signal from the radar drops to a low level.

In view of the Simonaire invention, techniques such as the cancellation scheme of (15), random con scan frequency methods of (17) or (18), or the reference phase switching of (19) take on special importance. Some of the hybrid monopulse-conical scan systems are also very attractive.

CONCLUDING REMARKS

A number of techniques for defeating the inverse gain jammer against tracking radars have been described, largely based on U S patents. Indeed there are viable alternatives to monopulse radar such that conical scan can still be useful for military applications. The techniques described have varying complexities, capabilities, and limitations. Selection of an ECCM technique to be used can be influenced by the maturity of the environment in which the radar is to be used. Certain ECCMs which can defeat a simple inverse gain jammer can be defeated by a more sophisticated jammer. Specification of the detailed implementation of a deception jammer is thus of great importance in selection of ECCM techniques.

Although not specifically addressed in this paper, use of other types of ECCM techniques in the radar must also be investigated when considering inverse gain jammers. Frequency agility and jittered PRF are two very powerful ECCMs (8-10). Additionally, antenna cross polarization characteristics must be determined (10).

REFERENCES

1. "IEEE Standard Radar Definitions", 1982 (in press), IEEE Std 686, IEEE, USA.

2. Getting, I. A., 1975, IEEE Trans Vol AES-11, No. 5, 922-36.

3. Whitford, A. E. and Kron, G. E., 1937, Rev. Sci Instr, 8, 78-82.

4. Johnston, S. L., 1956, IRE Trans PGI, 60-5.

5. Johnston, S. L., 1980, "Radar ECCM History", IEEE NAECON Rec, 1210-14.

6. Cikalo, J. and Greenbaum, M., 1975, Radar/ECM Computer Modeling", IEEE NAECON Rec, 431-7; reptinted in (8) below.

7. Kline, C. R., 1977, "Controlled Deception Jamming Device", U S pat. 4,037,227, (filed Apr 11, 1966).

8. Johnston, S. L., 1979, "Radar Electronic Counter-Countermeasures", Artech House, Dedham, Mass, USA.

9. Maksimov, M. V. et al, 1979, "Radar Anti-jamming Techniques", Moscow (in Russian), English language translation: Artech House, Dedham, Mass, USA

10. Johnston, S. L., 1980 ff, Lecture Notes for Course 686, George Washington University, USA, Continuing Engineering Education Program.

11. Prichard, A. C., 1949, "Radio Apparatus", U S pat. 2,464,258 (filed Jan 4, 1945).

12. Adam, W. B., 1967, "Conical Scan Radar System and Antenna", U S pat 3,307,183, (filed March 11, 1957).

13. Tellier, J. et al, 1974, "Antenna With Short Line Tuning", U S pat 3,858,213, (filed Oct 18, 1965).

14. Holman, J. G., 1980, "Anti-jam Device for a Conical Scan Tracking Radar", U S pat. 4,183,023, (filed Apr 20, 1962).

15. Schmidt, J. D., 1980, "Apparatus for Eliminating Amplitude Modulation Interference in Conically Scanning Radars", U S pat. 4,224,622, (filed Apr 20, 1962).

16. Reid, M. S., 1969, IEEE Trans GE-7, No 3, 146-56, reprinted in Johnston, S. L., 1980, "Millimeter Wave Radar", Artech House, Dedham, Mass, USA.

17. Rabow, G., 1975, "Conical Scan Tracking System", U S pat. 3,859,658, (filed Oct. 16, 1972).

18. Neri, F. and Albrande, A., 1978, "Device For Controlling the Conical Scanning Frequency for Conical Scanning Radar Systems", U S pat. 4,107,683 (filed Mar 15, 1977).

19. Felsenthal, H. D., Jr., 1976, "Radar System", U S pat 3,947,847 (filed Mar 24, 1969).

20. Sakamoto, H., and Peebles, P. Z., Jr., 1978, IEEE Trans AES-14 No. 1, 199-208; AES-14 No. 4, 673, reprinted in (8).

21. Bakut, P. A., and Bol'shakov, I. S., 1963, Questions of the Statistical Theory of Radar Vol II, Chaps 10, 11, Moscow, Sovetskoye Radio, English language translation, NTIS, USA, AD 645775.

22. Noblit, R. S., 1967, Microwaves, 6, 59-60.

23. Butler, J. L., 1971, "Conical Antenna System", U S pat 3,618,091 (filed Nov. 10, 1961).

24. Schöneborn, H., 1970, "Antenna Energizing Arrangement for Direction Finding Utilizing Amplitude Comparison", U S pat. 3,495,246 (filed Dec 5, 1967).

25. Longuemare, R. N., Jr., 1973, "Guarded Monopulse Radar System", U S pat 3,778,829 (filed Feb. 18, 1972).

26. Felsenthal, J. D., Jr., 1976, "Radar Systems", U S pat 3,947,847 (filed Mar 24, 1969).

27. Gulick, J. R., Jr., 1977, "Phase Modulated Monopulse System", U S pat. 4,011,864, (filed July 7, 1966).

28. Ivanov, A., 1976, Microwaves Magazine, 28, 54ff; reprinted in (8) above.

29. Steer, D. J., 1978, Proc Military Microwaves Conf, MM-78, London, 29-38, reprinted in (8) above.

30. Prior, J. R., and Woodward, N., 1978, Proc Mil. Microwaves Conf, MM-78, London, 39-50; reprinted in (8) above.

31. Boucher, R. J., and Brackney, R. L.,Jr., 1978, "System for Overcoming The Effect of Electronic Counter-Countermeasures", U S pat 4,107,682, (filed Apr. 5, 1967).

32. Simonaire, F. M., 1978, "Countermeasure for LORO Radar", U S pat. 4,126,862, (filed Apr 23, 1968).

A NEW BROADBAND ARRAY PROCESSOR

K.M. Ahmed and R.J. Evans

Department of Electrical & Computer Engineering, University of Newcastle, New South Wales 2308, Australia

ABSTRACT

This paper presents a new optimization procedure which allows the designer to impose specified bandwidth constraints on an array output response pattern. In many situations the new technique has similar broadband performance to the Frost algorithm yet does not require broadband pre-steering delays. The method can be implemented adaptively with a structure moderately more complex than the Frost processor. Several simulations are presented.

INTRODUCTION

There is currently considerable knowledge concerning the design of narrowband array processors [5], however, there remain very few substantial results on the design of broadband systems. The major practical result, due to Frost [1], employs frequency dependent weights (tapped-delay-lines) selected to achieve a specified frequency response in a look direction determined by broadband pre-steering time delays. Broadband interference from other directions is reduced by minimizing the array output power. This approach works quite well, however the need for broadband presteering delays and the lack of ability to precisely control the array response makes the technique unsatisfactory for many applications. An attempt to alleviate the need for presteering, investigated by Takao [3], uses the partial equivalence between bandwidth and angular spread to justify the use of an angular derivative constraint in the look direction for improving array bandwidth. The look direction is controlled by a constraint rather than presteering. The method works well only for low bandwidth signals. More recently, Mayhan [2, 4] has studied certain fundamental aspects of array systems in an endeavour to determine the achievable broadband performance of an array.

Our approach is conceptually different to the above techniques in that we attempt at the outset to synthesize a specified array response over a prescribed bandwidth. Via this approach we are able to design arrays with a broadband look direction response and acceptable broadband nuls without the need for presteering delays. If the interference directions are known we can ensure truly broadband nulls in these directions, via a new constraint function, provided of course, that the imposed constraints are feasible.

Below we formulate the broadband design problem and develop an optimization procedure for solving this problem and determining the array weights. Section III following presents and discusses simulations and comparisons with other techniques. Finally, in Section IV we conclude with a discussion of an adaptive implementation of our algorithm, and the application of our results to beamspace processors.

BROADBAND ARRAY DESIGN

For simplicity of notation we consider a linear array of K isotropic receiving antennas separated by ℓ, each connected to 2 real weights in a tapped delay line configuration with delay T. The complex array output for a frequency f from direction θ, is well known to be [5],

$$y(\theta,f) = \underline{c}'\underline{w} + j\underline{d}'\underline{w}$$

where $\underline{w}$ is the weight vector, and $\underline{c}$ and $\underline{d}$ are 2K steering vectors

$$\underline{c}' = [\cos\psi_1(\theta),..\cos\psi_K(\theta),\cos(\psi_1(\theta)+2\pi fT),.. ..\cos(\psi_K(\theta)+2\pi fT)]$$

$\underline{d}$ is similarly defined with cos replaced by sin, and $_k = \frac{2\pi f\ell}{v}(k - \frac{K+1}{2})\cos\theta$, k = 1,...K, v being the speed of light.

The broadband design problem we would like to solve is:-

> determine a set of array weights $\underline{w}^*$ for which the complex array output voltages satisfy
>
> $y(\theta_0,f) = 1 \quad \forall f \in F$
>
> $y(\theta_i,f) = 0 \quad \forall f \in F, \quad i = 1,2, \ldots M$

where θ_0 is the known look direction, θ_i are M unknown interference directions, and F is the set of frequencies over which the array must operate. Unfortunately this problem is ill-posed and $\underline{w}^*$ may not exist. A more realistic problem, is to find the weight vector $\underline{w}$ which minimizes the array output power while giving satisfactory performance in the look and interference directions over the frequency band of interest, i.e.

$$\text{minimize} \quad \underline{w}'\underline{R}\,\underline{w}$$

subject to $y_i^- \le y(\theta_i,f) \le y_i^+$, $\forall f \in F$, i = 0,1, ... M where $\underline{R}$ is the broadband signal correlation matrix [5], and y_i^+, y_i^- are complex upper and lower limits. This problem as it stands is not amenable to any known optimization procedure because of the compact set F. However, via the result outlined in Appendix A we can show that for an F of the form

$$f_0 - \Delta f \le f \le f_0 + \Delta f$$

then any weight vector $\hat{\underline{w}}$ which minimizes $\underline{w}'\underline{R}\,\underline{w}$ subject to

$$|y(\theta_i,f_0)-d_i| + \xi_i'|\underline{w}| \le \varepsilon_i, \quad i=0,1, \ldots u$$

also satisfies the problem above, where

$$\xi_i = [|\gamma_{1_i}\Delta f|\sin(\gamma_{1_i}f_0), \ldots, |\gamma_{K_i}\Delta f|\sin(\gamma_{K_i}f_0,+ 2\pi f_0T)]$$

$$+j[|\gamma_{1_i}\Delta f|\cos(\gamma_{1_i} f_0),\ \ldots,$$

$$|\gamma_{K_i}\Delta f|\cos(\gamma_{K_i} f_0 + 2\pi f_0 T)]$$

and $\gamma_{k_i} = \frac{\psi_k(\theta_i)}{f}$, k=1, ... K, when the array element spacing $\ell = {}^{v}/_{2f_0} = {}^{\lambda_0}/_{2}$, and $d_i + \varepsilon_i = y^+$, $d_i - \varepsilon_i = y^-$. Following [6] a convergent algorithm for solving this new problem can be developed. The algorithm is briefly described in Appendix B.

REMARKS

(1) It is, of course, easy to select a set of voltage and frequency constraints ($d_i, \varepsilon_i, \Delta f$) for which there is no feasible solution given the chosen array structure. However, our experience with the algorithm has shown that in such cases the algorithm behaves very robustly and tends to least squares approximate the constraints.

(2) The extension of the results above to more than 2 array weights per receiving element is straightforward.

(3) The simplestway of using the above algorithm is to impose one broadband constraint (in the desired look direction) and let the minimization of $\underline{w}'R\underline{w}$ null the interferences. In this way no broadband presteering delays are needed to establish a broadband look direction. The resulting broadband response patterns tend to have a broad (angular) mainlobe in the constrained look direction and nulls only in the interference directions. If the beam is steered via an ordinary narrowband constraint [3], then the interference is not always cancelled and extra nulls appear.

(4) If broadband constraints are imposed in other directions (i.e. known interference directions) the resulting response pattern certainly contains truly broadband nulls in those directions whenever this is possible. These constraints are satisfied at the expense of null depth in unconstrained directions.

The simulations following further illustrate these observations.

SIMULATION RESULTS

The simulation results below all refer to a standard narrowband 4 element linear array each with 2 weights. The array elements are spacially separated by ${}^{\lambda_0}/_{2}$ where λ_0 is the centre frequency wavelength of some broadband signal. Similarly the weights are separated by $\frac{1}{4f_0}$. The aim of the simulations is to indicate that the new broadband algorithm can (subject to fundamental limitations) steer a specified bandwidth broadband mainlobe and satisfactorily cancel broadband interference without the need for broadband presteering delays. More extensive simulations on these and other aspects of the algorithm will be presented during the conference.

Note that in nearly all the cases presented below the new broadband algorithm converges in about 200 iterations. Also, all array patterns below are patterns averagedover the operating bandwidth.

CASE 1. When the look direction is broadside our technique performs similarly to a presteered Frost system. For example, with 3 equal power sources each of 30% bandwidth (i.e. 30% deviation either side of f_0) good nulls (-34dB) are produced for the 2 broadband interference sources (30° and 155°)

CASE 2. When the look direction is changed to 85°, and the interference sources remain at 30 and 155° our algorithm still provides nulls around -28dB without presteering. In contrast however if the array is constrained to look in the 85° direction via a standard algorithm [3] the nulls produced are only about -15dB. Figure 1 shows the array patterns averaged over the 30% (2 octave) frequency interwave.

CASE 3. (Figure 2) The look direction has now been rotated another 5° to 80° and both interference sources have also been rotated by 5° in the same direction (i.e. 25° and 150°). Comparing this case with Figure 1 and Case 1 we see that the nulls have increased to about -25dB. This is worse than a presteered Frost system but still considerably better than the standard constraint steered system which only achieves nulls around 9dB down.

CASE 4. For this case the look direction is rotated to 70° and interferences to 15° and 140°. With a bandwidth of 15% the standard constraint steered array achieves only 8dB down nulls whereas the new broadband processor achieves 15dB nulls (see Figure 3). Note that in this and every other case the standard algorithm achieve its look direction constraint and possesses nulls in the interference direction. However the pattern also often contains an extra mainlobe and an extra null.

Figure 3(a) also shows averaged patterns (over a 15% bandwidth) for different levels of white noise in the $\underline{R}$ matrix. Clearly, the greater the noise, the better the pattern for the ordinary constraint steered system. For our system, there is very little difference. Part of the explanation for this is that there is already a type of $\|w\|$ constraint on our weight vector through the $\xi|w|$ term. The full relationship between this constraint, bandwidth and $\|w\|$ is not yet understood, however we believe that the connection may be important to a further understanding of broadband systems. Currently we are able to show that $|y(\theta,f_0)-y(\theta,f)|^2 \le (f-f_0)^2\|w\|^2\|\gamma\|^2$ where γ is a vector of relative delays between weights. This result holds for moderately broadband systems with not too many weights or elements. Figure 3(b) shows the weight vectors for the cases in Figure 3(a).

CASE 5. Next, we rotate the look direction to 60° and the interference source to 5° and 130°, still with a 15% bandwidth (1 octave). Again the new system is considerably better than the ordinary constraint steered system which has nulls about 8dB down yet not as good as the presteered Frost system. Our system achieves nulls averaged over the full 15% bandwidth of approximately 20dB down. See Figure 4 for details.

CASE 6. Finally we consider the case of a 15% bandwidth system with a 60° look direction, a noise source at 120° and an imposed broadband sidelobe constraint at 145°. The resulting response pattern is shown in Figure 5 where the ability of the algorithm to achieve constraints over a repspecified bandwidth is clearly shown. Note however that the constraints are satisfied at the expense of performance in the unknown interference direction. A standard narrowband constraint 145° for this problem [6] only satisfies the constraints for a 2% bandwidth.

REMARKS

These simulations, although brief, clearly indicate the ability of the new technique to improve the bandwidth performance of a typical narrowband array structure. Near the broadside direction the system performs similarly to a pre-steered Frost system. However away from broadside where the narrowband array is notoriously sensitive to bandwidth (typically only a 1-2% bandwidth is achievable) our approach achieves satisfactory performance for up to 10-15% bandwidth. Our approach is of course limited by the fundamental performance limits of the array, however it appears that by imposing a broadband inequality constraint in the look direction we are able to make maximum use of the available degrees of freedom.

We are currently investigating in detail the effect on our algorithm of noise in the R matrix, the performance with larger tapped delay lines and the performance comparison of our technique with various derivative constrained systems.

CONCLUSION

This paper has presented a synthesis procedure for designing controlled broadband array response patterns, and demonstrated its application to some simple linear array space problems. The procedure also has applications to beamspace problems where multiple beams are created from a common set of array elements. These beams are then weighted and combined to effect interference cancellation. For such systems the design of prescribed bandwidth specified beams is very important [7].

Finally, it is interesting to observe that when only a mainlobe broadband constraint is imposed our algorithm can be implemented adaptively. Each update requires approximately 10NK real multiplications and 20NK real additions for a K element, N taps per element array. The adaptive algorithm can be shown to converge with probability one for noise corrupted inputs [7].

APPENDIX A

The set of array weights $\underline{w}$ for which $|y(\theta_i,f_0) - d_i| + \xi_i|\underline{w}| \le \bar{\varepsilon}_i$, i=0,1, ... M also ensure that $|y(\theta_i,f)-d_i| \le \varepsilon_i$, i=0,1,... u for all frequencies $f_0-\Delta f \le f \le f_0+\Delta f$. The proof of this statement follows from simple trigonometric manipulations as briefly outlined below. Consider the real part of $y(\theta_i,f)$ first.

$$B \triangleq |\sum_k w_k\,(\cos\psi_k \cos(\psi_k f + \phi_k) - d|$$

$$= |\sum_k w_k\,(\cos\Psi_k f \cos\Phi_k - \sin\Psi_k f \sin\Phi_k) - d|$$

But $f = f_0 - \tilde{f}$ for some $|\tilde{f}| \le \Delta f$

$$\therefore B = |\sum_k w_k[\cos\psi_k f_0 \cos\psi_k f \cos\phi_k + \sin\psi_k f_0 \sin\psi_k \tilde{f} \cos\phi_k - \sin\psi_k f_0 \cos\psi_k f \sin\phi_k + \cos\psi_k f_0 \sin\psi_k f \sin\phi_k\,] - d|$$

$$\le |\sum w_k[\cos\psi_k f_0 \cos\phi_k + \psi_k \tilde{f}_0 \cos\psi_k f_0 \sin\phi_k + \psi_k f_0 \sin\psi_k f_0 \cos\phi_k - \sin\psi_k f_0 \sin\phi_k] - d|$$

$$= |\sum w_k[\cos(\psi_k f_0+\phi_k) + \gamma_k \tilde{f}\, \sin(\psi_k f_0+\phi_k) - d|$$

$$\le |\sum w_k \cos(\psi_k f_0+\phi_k) d| + |\sum w_k \gamma_k \tilde{f}\, \sin(\Psi_k f + \Phi_k)|$$

$$\le |y(\theta,f_0)-d| + |\sum w_k|\gamma_k \tilde{f}||\sin(\psi_k f_0 + \;_k||$$

But when $|\tilde{f}| \le \Delta f$

$$\le |y(\theta,f_0)-d| + |\sum w_k|\gamma_k \Delta f||\sin(\psi_k f_0+\phi_k)||$$

$$= |y(\theta,f_0)-d| + \xi'|w|$$

A similar proof follows for the imaginary part of $y(\theta,f)$.

APPENDIX B

The weight update algorithm for solving the broadband design problem with constraints in M+1 directions is given by

$$\underline{w}(k+1) = \underline{w}(k)\ \alpha \frac{\partial L}{\partial \underline{w}}$$

$$\underline{\lambda}_1(k+1) = \max\ [0, \underline{\lambda}_1(k) + \alpha \frac{\partial L}{\partial \lambda_1}\ ]$$

$$\underline{\lambda}_2(k+1) = \max\ [0, \underline{\lambda}_2(k) + \alpha \frac{\partial}{\partial \lambda_2}$$

where α is a fixed step size and

$$L \triangleq \underline{w}'\underline{R}\,\underline{w} + \sum_{i=0}^{m} \lambda_{1_i}\ (y(\theta_k,f_0)-d_i-\varepsilon_i+\underline{\xi}|\underline{w}|) + \sum_{i=0}^{m} \lambda_{2_i}\ (y(\theta_i,f_0)-d_i+\ _i+\xi|\underline{w}|)$$

The major difficulty is determining $\frac{\partial L}{\partial \underline{w}}$ since L is not differentiable w.r.t. w. A directional derivative must be used here [7].

REFERENCES

1. O.L. Frost III, "An Algorithm for Linearly Constrained Adaptive Array Processing", Proc. IEEE, Vol. 60, pp.926-935, August 1972.
2. J.T. Mayhan, "Some Techniques for Evaluating the Bandwidth Characteristics of Adaptive Nulling Systems", IEEE Trans. Antennas and Propagat. Vol.AP-27, pp.363-373, May 1979.
3. K. Takao & K. Komiyama, "An Adaptive Antenna for Rejection of Wideband Interference", IEEE Trans. Aerosp. & Elect. Syst., Vol. AES-16, pp. 425-459, July 1980.
4. J.T. Mayhan, A.J. Summons & W.C. Cummings, "Wide-band Adaptive Antenna Nulling Using Tapped Delay Lines", IEEE Trans. Antennas & Propagat. Vol. AP-29, pp. 923-936, Nov. 1981
5. J. Hudson, Principles of Array Processing, IEE Monographs 1982.
6. R.J. Evans & K.M. Ahmed, "Robust Adaptive Array Antennas", Jrnl. Acoust.Soc.America, February 1982.
7. K.M. Ahmed, Robust Signal and Array Processing, Ph.D. Thesis, University of Newcastle (In preparation).

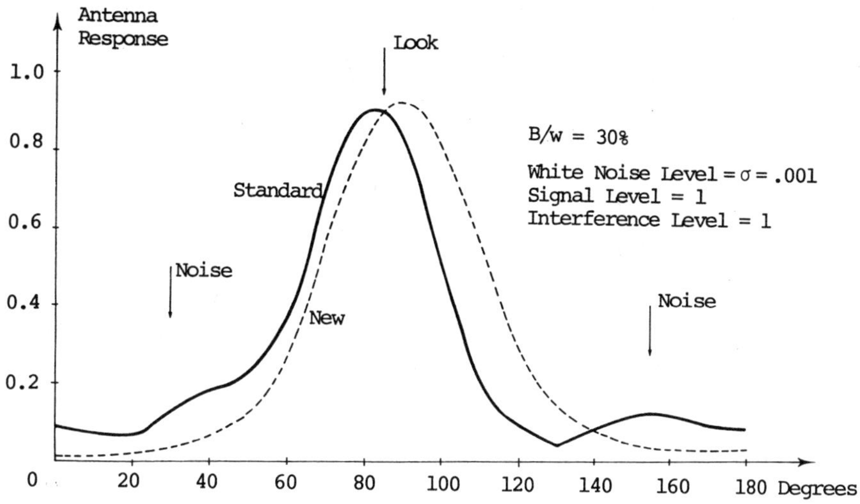

Figure 1 Averaged patterns over 30% bandwidth for signal from 85° with two interferences.

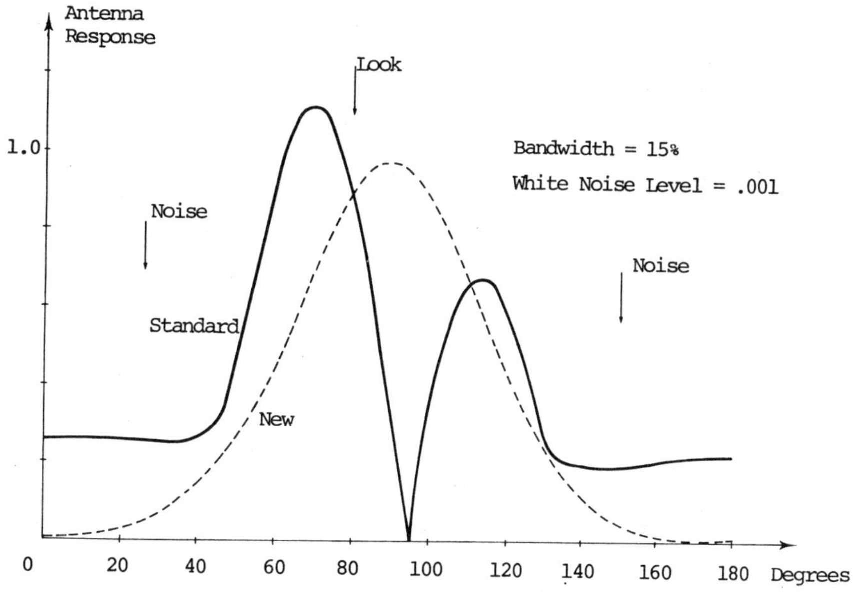

Figure 2 Averaged patterns over 15% bandwidth for signal from 80° with two interferences.

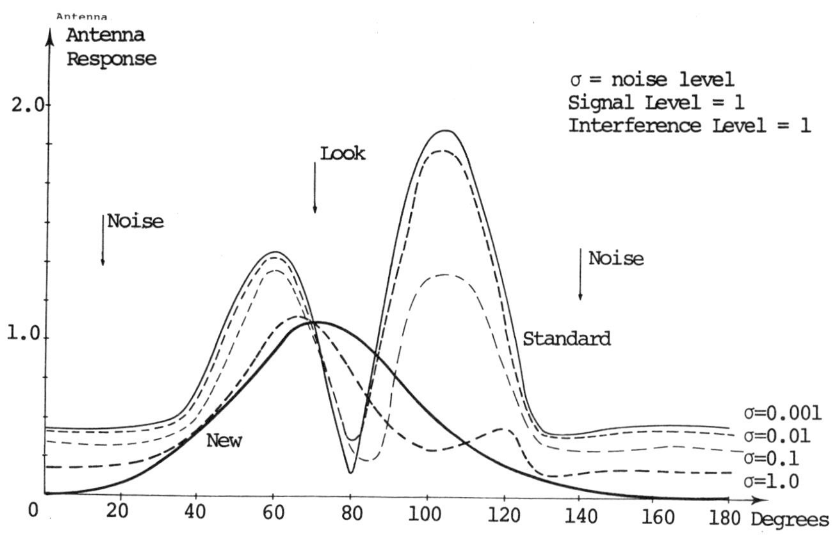

Figure 3(a) Averaged patterns for different white noise levels for signal from 70° with two interferences.

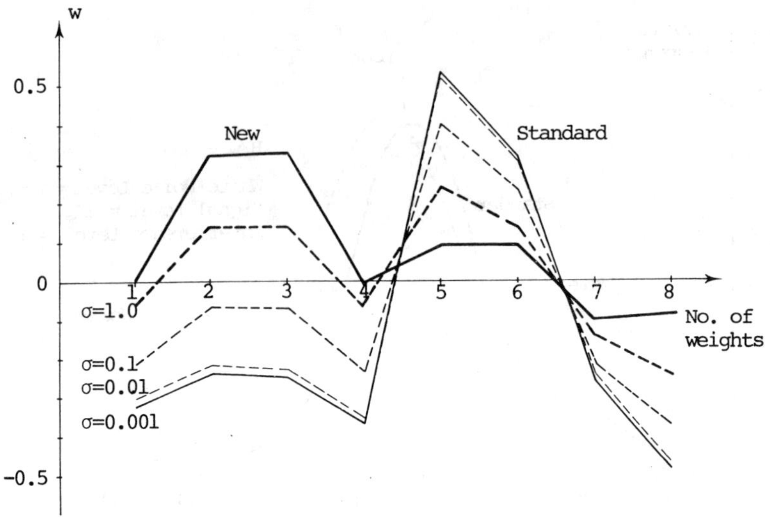

Figure 3(b) Weights for different white noise levels for the example in Figure 3(a).

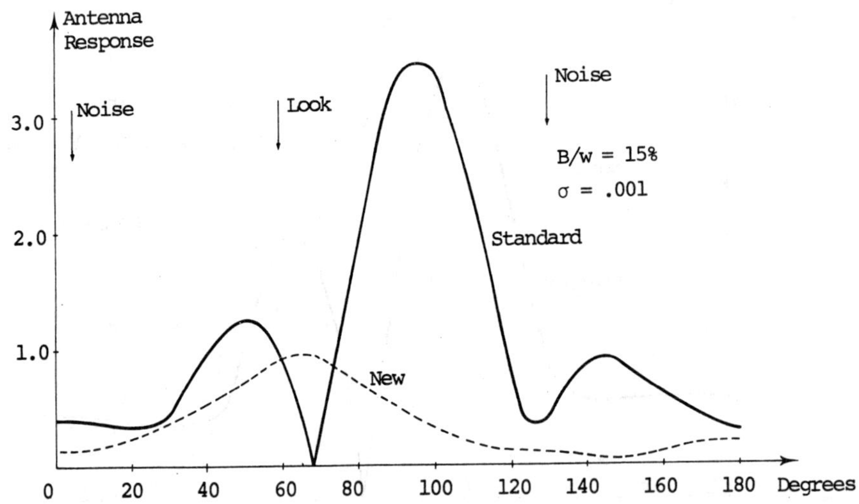

Figure 4 Averaged patterns over 15% bandwidth for signal from 60° with one interference.

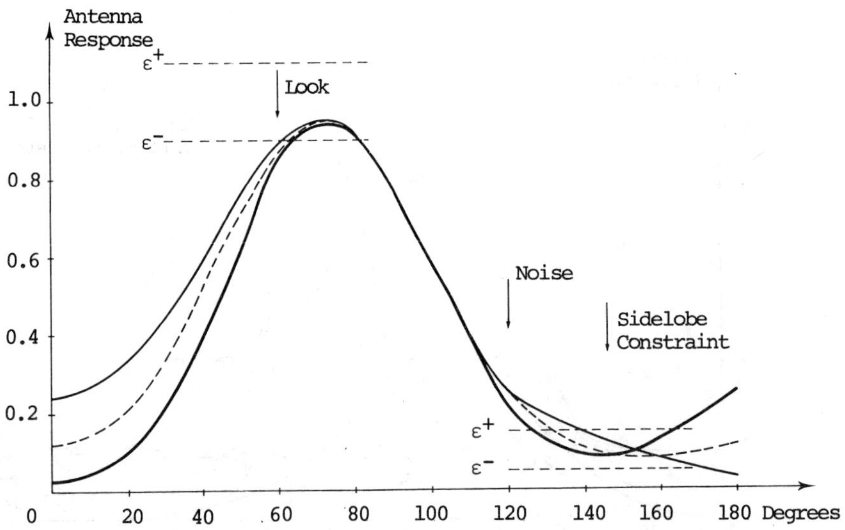

Figure 5 Patterns drawn at f_L, f_0 and f_u for signal from 60° with one interference and one sidelobe constraint at 145°.

ESTIMATION OF SHIP´S MANOEUVRES WITH A NAVIGATION RADAR

G.F. Lind

Lund Institute of Technology/University of Lund, Sweden

INTRODUCTION

One use of navigation radars onboard ships is to closely watch the behaviour of all other ships in a surrounding area so that no collision risk occurs. This is done manually or automatically by plotting the other ship´s positions, extracting in the plotting procedure their courses and speeds, from which the closest-point-of-approach (CPA) and time to CPA (TCPA) can be calculated. One more function is necessary, however, to observe if one nearby ship is at a constant course and speed or if it is manoeuvring, i e decreasing or increasing its speed or turning. Even if ship´s manoeuvres in the above sense are rare, it is important to observe the few that occur. The reason for this is that the collision avoidance situation suddenly becomes more complex, especially if the other ship´s manoeuvre occurs at close range, and it may take all possible alertness to decide on the own ship´s action under the new circumstances.

Technically, course and speed are given by the first derivatives of position from radar data. Ship´s manoeuvres show up in the second derivative.

The purpose of this paper is to study the possibilities of observing this second derivative and thereby decide if an observed ship is manoeuvring or not. One difficulty is that the position measurements in themselves have random errors, causing random errors in the acceleration estimate and therefore require averaging or filtering to bring those errors down to an acceptable level. This averaging or filtering takes time, however, and if this time is excessive it might endanger navigation safety by making the reports on nearby ship´s acceleration come in too late.

This is precisely the situation reported by practical navigators. Given only radar as the source of information, which is the case in heavy fog, Curtis (1) has reported that it takes 2-3 minutes before a manoeuvre can be detected. Since this is a rather long time, it is important to find out if this delay is necessary. Therefore, this paper will study a typical case to see how long time after the decision to turn rather sharply onboard one ship this is positively detected using a navigation radar onboard another ship.

The first section following this introduction defines the typical test case. The next section discusses the maximum random errors in the acceleration measurements that can be accepted in order to make adequate navigation decisions.

The following section analyses the noise or perturbation that cause errors in the position measurements like glint, receiver noise, sea and rain clutter.

Thereafter a suitable filter is selected to calculate acceleration data from position data given by the radar. The filter used here is optimum in the sense of giving the least random error for a certain finite length of filtering time.

The necessary filter length and the "lost time" are calculated by adding to the filter response a model of a ship´s dynamics after a certain rudder has been ordered.

A discussion is then given of the different properties of the today standard navigation radars which are either X-band or S-band, and what radar measures are possible to improve the present situation. The possible influence of automatic plotting aids (ARPA) or collision avoidance systems is also treated.

TEST CASE DATA

Own ship.	speed 12 knots rolling 5°, period 10 s
Observed ship.	displacement 10000 tons length 150 m width 20 m speed 12 knots rate of turn 0 or $1^{o}/s$ yawing $0,1^{o}$, period 12 s radar cross section 5000 m^2
Situation geometry.	Meeting, parallell courses CPA 1 nm
Radar data.	band X frequency fixed 9300-9500 MHz polarization horisontal horisontal beam width 1° vertical beam width 20° antenna rotation 24 rpm antenna height 15 m peak power 25 kW pulse width 0,5 µs pulse repetition frequency 2000 Hz
Weather data.	rain 1 mm/h sea state 4

SHIP´S ACCELERATION AND ESTIMATION ERRORS

The navigation radar produces position data, one measurement or sample each time the antenna scans the beam across the ship which is measured. The position data are then used to estimate acceleration. Random errors in position therefore will cause random errors in the acceleration estimate. Here we will address the question of how much random errors in the acceleration estimate that can be tolerated.

A first orientation is given by the actual limitation in acceleration that typical ships have. This limit is generally quite low. A few examples are given in table 1.

TABLE 1 - Acceleration capacity of ships

Type of ship	Acceleration capacity m/s^2
Fast patrol boat	1-2
Passenger ferry	0,1-0,2
Supertanker (VLCC)	0,01-0,03

The test case ship has been given a capacity of 0,11 m/s^2 across (1°/s at 12 knots) and is assumed to use at least 0,06 m/s^2 in a turn during fog conditions. Here we use the fact that during fog or other conditions of limited visibility, all turns must be forceful.

The threshold acceleration above which we should declare a turn in our test case, and the random errors that can be tolerated in the accelerations estimate can all be calculated through the theoretical framework of Decision Theory or Statistical Hypothesis Testing. For the complete Bayes criterion there is the usual difficulty of finding the costs, and the a priori probabilities. In spite of these problems a sketch of an analysis using Bayes criterion will be made. The final result of this paper, the delay time of detection, is not very sensitive when the assumptions used here are varied within reasonable limits.

Only 10 nm before the meeting are essential to the development and perhaps only the final 5 nm are critical, so we assume 25 minutes of time important for each encounter.

The filter time will be in the order of 2 minutes, which we assume for the moment. The acceleration signal will be strongly correlated during the filter time, which means that during 25 minutes only about 12 decisions on manoeuvre/no manoeuvre are made.

Now, we assume that the a priori probability for a manoeuvre is 10^{-3}, and the cost ratio between non detection of a manoeuvre and a falsely declared manoeuvre is taken as 10. This last assumption is based on the argument that not detecting a manoeuvre has a higher risk of causing a collision, since a false alarm in many cases leads to increased alertness only.

For a complete solution we need an absolute accepted cost, which can be studied in the following manner. Assume false alarm leads to one collision in 100 cases, and one false non-detection in 10 cases, according to the cost-ratios above. Further assume that the number of collisions tolerated for this reason is lower than one in 10^5 encounters, which is approximately what Lewison (2) reports for clear weather encounters. The number of decisions per encounter is 12 as given above. We then have

$$P_{FA} + 10^{-2}(1 - P_D) < 8 \cdot 10^{-5} \qquad .....(1)$$

where P_{FA} is the false alarm rate and
P_D the probability of detection

The result is then

$\sigma = 0,009\ m/s^2$ (acceleration error standard deviation)
$a_T = 0,37\ m/s^2$ (detection threshold acceleration)

It must be noted that the above analysis is only a sketch, with many approximations involved. All in all, the author feels that a correct order of magnitude analysis has been made, and hopes that this approach will sometimes be widened to a more complete analysis, including an operational analysis of the collision process.

RANDOM POSITION NOISE

The inherent errors, which will constitute the best possible performance of any system of this kind, include radar glint, radar fading, receiver noise, sea clutter and rain clutter. They have been calculated in detail for the test case in angle and range. Only a brief outline of the calculations is presented here.

Radar glint is a target induced tracking error, basically a function of target size. Using the methods of Barton and Ward (3), ch 6.7, with the geometry and the size of the test case ship we get the glint standard deviation from

$$\sigma_\theta \approx 0,35\ L_x \qquad .....(2)$$

$$\sigma_r \approx 0,25\ L_r \qquad .....(3)$$

where σ_θ is glint standard deviation in angle
σ_r is glint standard deviation in range
L_x is target size perpendicular to line of sight
L_r is target size along line of sight

The glint varies in time, with a correlation time given by

$$\tau_c = \frac{\lambda}{2\dot{\phi}L_x} \qquad .....(4)$$

where τ_c is the correlation time
λ is the radar wavelength, here 3 cm
$\dot{\phi}$ is angular rate of change

The angular rate of change has three sources, aspect change from geometry, yawing and turning. With aspect change from the geometrical situation, yawing $0,1^o$ with a period of 12 s and turning rate either 0 or 1°/s, the correlation time can be calculated.

The effective number of independent samples is

$$n_e = 1 + \frac{\tau_o}{\tau_c} \qquad .....(5)$$

where τ_o is the observation time. Antenna rotation 24 rpm and 1° azimuth beamwidth gives $\tau_o \approx 7$ ms for small targets. This has to be corrected for the ship's size, which is a noticeable part of the radar beam at short range.

There is no correlation between measurements and a slight decorrelation during the illumination time. Results, see table 2.

TABLE 2 - Glint errors.

Range	Glint, angular direction, $\sigma_{\theta eff}$, m		Glint in range σ_{reff}, m	
nm	no turn	turn 1°/s	no turn	turn 1°/s
2	31	25	34	28
4	20	18	37	34
6	16	14	38	35

As can be seen, when the observed ship is turned, glint is somewhat reduced.

Radar fading can also be seen as a target induced error, but it is uniquely connected with the method of angle measurement, scanning one radar beam at a regular rate across the target and observing the amplitude. Amplitude variations from other sources than scanning the antenna diagram, such as target induced fading, will lead to measurement errors. Fading errors in range are not considered, since range can be measured without scanning methods.

The fading error has been estimated using (3), especially fig 6.10. Here also the observation time has been corrected for the target extension. Results are in table 3.

TABLE 3 - Fading errors.

Range	Error in angular direction, $\sigma_S R$, m	
nm	no turn	turn 1°/s
2	2,9	7
4	2,6	10
6	2,9	14

As can be seen, and opposite to the effect on glint, the 1°/s turn here leads to increased errors.

Receiver noise. At the short ranges in question and with a ship of actual size normal navigation radars have very high signal - to - noise ratio (S/N) and consequently very small position measurement errors caused by noise. These errors are briefly calculated below. The method used is Urkowitz (4) formula for the high S/N case and nonfluctuating targets, corrected according to Avdeyev (5) with a factor 1,8 to include the effect of a fluctuating target and also corrected for target finite size relative the radar resolution. The formula is then for error in the angular direction

$$\sigma_\theta = \frac{0,54 \cdot \theta' \cdot 1,8}{\sqrt{n}\ \sqrt{S/N}} \qquad \ldots\ldots (6)$$

where σ_θ is the rms angular error
n is number of pulses
S/N is signal-to-noise ratio
θ' is lobewidth corrected for target size

S/N, the single pulse signal-to-noise ratio is calculated from the standard radar equation with the following extra radar data assumed

Antenna Gain 31 dB (beam width $20^o x 1^o$)
Radar cross section 5000 m^2 according to Skolnik (6)
Receiver bandwidth 3 MHz
Receiver noise factor 12 dB
Miscellaneous losses 3 dB

The resulting errors in the angular direction in meters, and range errors calculated in a similar way, are shown i table 4.

TABLE 4 - Receiver noise errors.

Range nm	S/N dB	Error in angular direction, m	Error in range m
2	60	0,02	0,01
4	48	0,2	0,06
6	41	0,5	0,1

Sea clutter causes errors in position measurements in the same way as noise, and the same formula may be used with two changes. One is a substitution of S/N with the signal-to-clutter ratio S/C. The second is the use of the equivalent number of degrees of freedom of the clutter n_e instead of the number of pulses.

Using the data and method of Nathanson (7) the sea clutter is calculated for sea-state 4 for X-band, horisontal polarisation and the grazing angles from 15 m antenna height, see table 5.

TABLE 5 - Sea clutter data.

Range nm	Sea clutter echo σ_{sea}, m^2	S/C dB
2	0,24	43
4	0,16	45
6	0,12	46

The number of degrees of freedom are also calculated according to (7), where the standard deviation of internal movement in the sea clutter is seen to be 1,4 m/s for sea state 4, giving a correlation time of 1,7 ms for X-band. See table 6.

TABLE 6 - Sea clutter errors.

Range nm	n_e	Error in angular direction, m	Error in range m
2	8,1	0,3	0,1
4	5,5	0,3	0,1
6	5,2	0,4	0,1

The errors are obviously insignificant compared to the glint and fading errors.

Rain clutter. The calculation procedure is similar to the sea clutter case, and differs only by the resolution cell beeing three-dimensional and in the mechanisms for decorrelation. Rain data are from Hawkins and LaPlant (8). See table 7.

TABLE 7 - Rain clutter data.

Range nm	Rain clutter echo, σ_{rain}, m^2	S/C dB
2	0,6	39
4	2,5	33
6	5,6	29,5

The correlation time is calculated according to (7). There are two sources of internal movement within the clutter, turbulence which can be assumed to be 1 m/s and wind shear, taken to be 4 m/s/km in the main wind direction or a projected average 2,8 m/s/km. The result is then as in table 8.

TABLE 8 - Rain clutter errors.

Range nm	n_e	Error in angular direction, m	Error in range m
2	11	0,4	0,2
4	12	0,9	0,4
6	14*	1,7	0,5

*limited by number of pulses during the dwell time.

Also here the errors are smaller than the dominant errors from glint and fading.

Total error. The errors are independent and are quadratically added. Result see table 9.

TABLE 9 - Total errors.

Range	Error in angular direction, m		Error in range, m	
	no turn	1°/s turn	no turn	1°/s turn
2	31	26	34	28
4	20	21	37	34
6	16	20	38	35

It seems safe to conclude that a standard deviation of the position measurement from 20 to 40 meters, typically 30, can be used for further calculations. The errors are random and independent from scan to scan.

FILTER FOR ACCELERATION ESTIMATION

We have selected the filter with limited length in time, which estimates acceleration with the least influence of random and sample independent input noise. Following the procedure of Monroe (9) ch. 9, we have the following results. The filter is defined as a series of weight factors a_k and the optimal filter is defined by

$$a_k = \mu_1 + \mu_2 k + \mu_3 k^2 \qquad .....(7)$$

$$\mu_1 = 60/N(N + 1)(N + 2)T^2$$

$$\mu_2 = -360/N(N - 2)(N + 1)(N + 2)T^2$$

$$\mu_3 = 360/N(N - 1)(N - 2)(N + 1)(N + 2)T^2$$

The variance propagation factor, relation output variance to input variance is

$$\sigma_{out}^2/\sigma_{in}^2 = \mu_3 \cdot 2/T^2$$

Fig 1 shows the weight factors of the filter for the case N = 10, T = 2,5 s. As can be seen the weight factors form a symmetrical parabola, with average zero.

How long will the filter be? Using eq 8, with σ_{out} = 0,009 m/s^2, σ_{in} = 30 m, T = 2,5 s we get N <45,95, select N = 46, filter length (N - 1)T = 112,5 s. For such high values of N we can note the approximation.

$$\sigma_{out}^2/\sigma_{in}^2 \approx 720/N^5T^4 \qquad .....(8)$$

DYNAMICS AND TIME DELAY

The time lost in identifying a manoeuvre of another ship is ideally the time from the moment the manoeuvre is decided till the moment it is detected. The reason for this is that other means of communication, e g VHF radio, which has been discussed by Cahill (10), can transfer the manoeuvre when it is decided and a rudder command is given. This means that the dynamics involved are of two kinds, the dynamics of the manoeuvring ship and the dynamics of the filter used to estimate acceleration.

The ship´s dynamics is approximated using the Nomoto model as described by Åström (11) which is a first order low-pass filter between rudder angle and acceleration with a time constant T_D.

For the test case ship T_D is approximately 75 seconds. The rudder is assumed to set in 5 seconds. Since this is small relative other times involved, the rudder set time is approximated by adding half of it or 2,5 seconds to the final time delay.

The total dynamic time function from a rudder angle step to the acceleration estimation filter has been calculated and is shown in normalized variables in fig 2.

In the test case, the threshold was 0,037 m/s^2 and the acceleration 0,06 m/s^2, or Y(z) = 0,62. T_{tot} = 112,5 s and T_D = 75 s; giving k = 0,67. Interpolation in fig 2 gives Z = 1,18. The total response time is then 1,18 x 112,5 + 2,5 = 135 s.

X-BAND OR S-BAND.OTHER TECHNICAL ALTERNATIVES.

Until now, an X-band navigation radar has been assumed. Many ships are equipped with both X-band and S-band navigation radars, and both are used for the navigational purposes here discussed.

The difference in the random position error is not great. Calculations give approximately 10% higher S-band radar errors than X-band radar errors, which does not seem important.

What can be done to reduce the errors in position measurement using radar technology in different ways than present-time navigation radars do? Since glint dominates, the introduction of frequency agility is a natural choice. Glint can then be reduced $\sqrt{14}$ times in our test case. Fading errors are, however, increased. The total result is an improvement of approximately 3 times in accuracy, reducing filter time to about 70 s. Further improvements can be made by introducing monopulse technology or transponders, but this would mean rather drastic changes from present day navigation radar technology.

ARPA, automatic radar plotting aid, is certainly of value when it comes to reducing the workload on the bridge, but it must be clearly stated that an automatic tracker in an ARPA cannot perform any better than the data here presented, which are valid for the best automatic tracker that can be designed, since all the errors are basic and common to all navigation radars of today.

CONCLUSIONS

With the above analysis, in a representative test example, it has been shown that the time delay in detecting a ship´s maneuver with radar only is inherently long. The delay calculated for the test case, 135 s, agrees well with the 2-3 minutes reported by practical navigators. No simple change in radar technology seems able to reduce the delay appreciably. If such a reduction is necessary for an enhancement of navigational safety, other methods must be used.

ACKNOWLEDGEMENTS

The author would like to thank professor Karl Johan Åström of the Department of Automatic Control, Lund Institute of Technology, and lecturer Göran Nydell, the Merchant Marine Academy in Malmö, for providing data of ship´s dynamics and data for the test case.

REFERENCES

1. Curtis, R.G., 1980, Journal of Navigation, 33, 329-340.

2. Lewison, G.R.G., 1978, Journal of Navigation, 31, 384-406

3. Barton, D.K. and Ward, H.R., 1969, "Handbook of Radar Measurements", Prentice-Hall, Englewood Cliffs.

4. Urkowitz, H., 1981, IEEE Trans AES, AES-17, 156-158.

5. Avdeyev, V.V., 1965, Radio Engineering and Electronic Physics, 10, 125-126.

6. Skolnik, M.I., 1980, "Introduction to Radar Systems", Second Edition, McGraw Hill Kogakusha Ltd, Tokyo.

7. Nathanson, F.E., 1969, "Radar Design Principles", McGrawhill, New York.

8. Hawkins, H.E. and LaPlant, O., 1959, IRE Trans on Aeronautical and Navigational Electronics, 26-30.

9. Monroe, A.J., 1962, "Digital Processes for Sampled Data Systems", John Wiley, New York.

10. Cahill, R.A., 1979, Journal of Navigation, 32, 259-265.

11. Åström, K.J., 1981, "Ship Steering, a Test Example for Robust Regulator Design" Report, Department of Automatic Control, Lund Institute of Technology, Sweden.

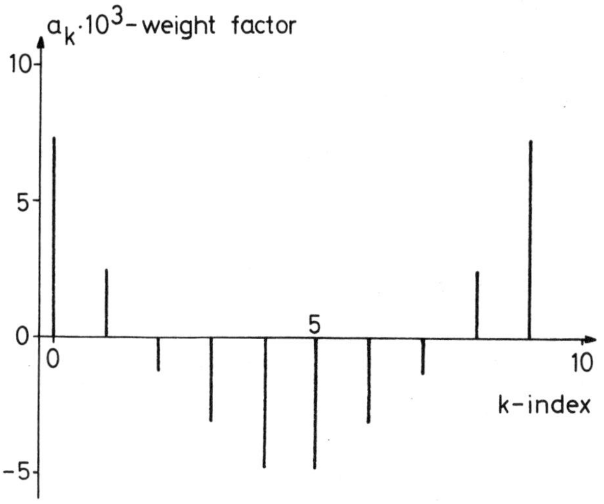

Figure 1. Filter for acceleration estimation, weight factors a_k, example N=0.

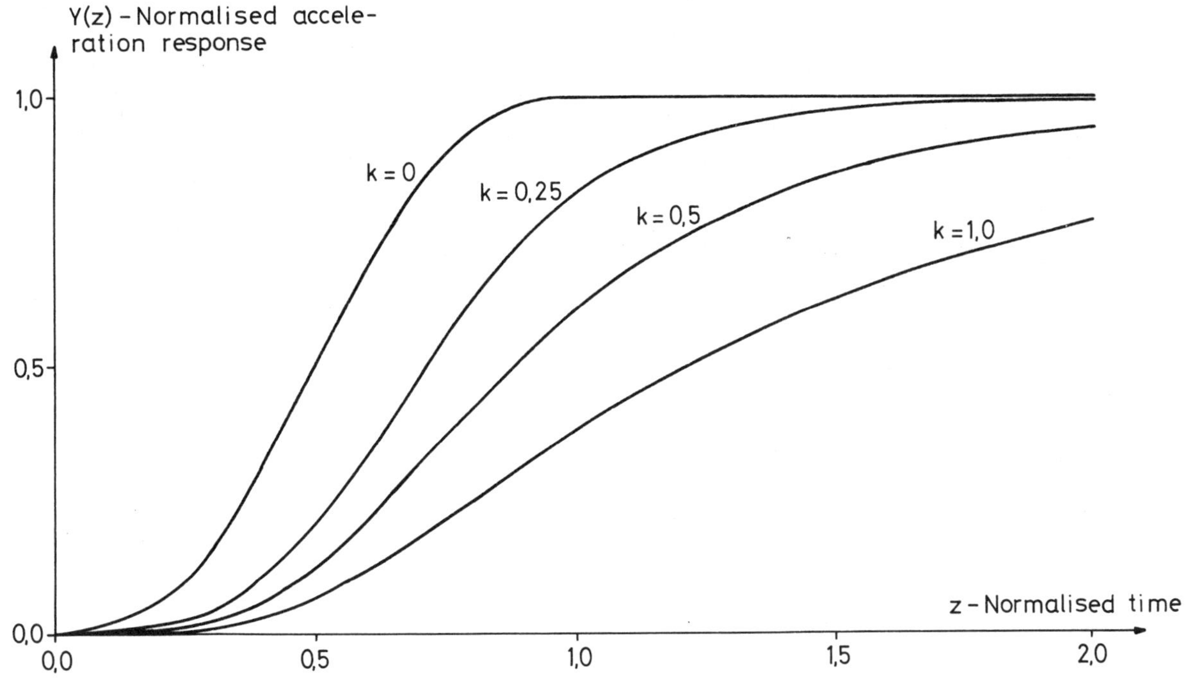

$$k = \frac{T_D}{T_{total}} = \frac{\text{ship's dynamic time constant}}{\text{acceleration estimation filter time}}$$

Figure 2. Normalised response of ship's dynamics and acceleration estimation filter to rudder angle step.

MEASURING TARGET POSITION WITH A PHASED-ARRAY RADAR SYSTEM

Gerard A. van der Spek

Physics Laboratory TNO, the Hague, Netherlands

1. INTRODUCTION

In a multi-function phased-array radar system great emphasis should be placed on an efficient use of radar time and -energy. For the search process this implies minimizing the dwell time and -energy for each antenna beam position, given a required detection quality, and an efficient transition to the succeeding echo confirmation and track initiation process for each detected alarm. Such an efficient transition is obtained when the search process delivers for each detected echo an accurate estimation of the target parameters viz. its position in 3 co-ordinates and its radar cross-section. The radar cross-section is useful to determine the required dwell energy for a confirmation transmission. The position is needed to steer the antenna beam and to set a video gate of minimal width over the echo to be confirmed. The estimation of radial velocity during search may be useful for echo confirmation, provided that the required longer dwell time is already needed for clutter suppression. The accurate position calls for a true monopulse capability, i.e. the direction of a target somewhere inside the antenna beam is obtained from one single radar transmission. This is accomplished by using three simultaneous antenna patterns on reception: one sum pattern, which might be identical to the antenna pattern used for radiating the radar transmission, and two difference patterns (corresponding, roughly, to azimuth and elevation).

FUCAS, an experimental phased-array radar system (van der Spek (1)), is provided with such a monopulse facility. The phased-array antenna (CAISSA, Snieder (2)), is a space-fed lens-type planar circular array with 850 non-reciprocal 4-bits phase shifters. The antenna covers a frequency band from 5400-5900 MHz and has a useful scan angle of 60 degrees off broadside. At broadside its pencil beam has a half-power width of 4 degrees (70 mrad). In 1976 antenna measurements have been made to determine a suitable monopulse processing algorithm and to calculate its expected performance (Rijsdijk and van der Spek (3)). It was found that one single relation could be used to map the monopulse ratios obtained from both difference channel outputs and the sum channel output on the two directional co-ordinates relative to the beam's boresight axis. Since then, this algorithm has been implemented in the radar system and its operation has been evaluated in the complete system.

In order to evaluate realistically the overall performance, measurements have been made for which all relevant system parts were in their normal operating mode. As a target a Westland Lynx helicopter of the Royal Netherlands Navy was used. A fire-control radar of the Netherlands Army was used to get reference position data of the helicopter with an accuracy which was sufficiently superior to that expected of the FUCAS system. This paper describes the measuring arrangements, summarizes the results and discusses several of their interesting aspects.

2. MONOPULSE AND RANGE PROCESSING

According to the earlier antenna measurements (3) the relation between the monopulse ratio r and the echo direction relative to the boresight axis can be determined by:

$$\Delta \sin\,(\text{angle}) = -0.011\; r^3 + 0.048\; r \qquad (1)$$

This relation is applicable in both phased-array co-ordinates (sinA and sinB). The monopulse ratios are derived from the echo signals which are received via the sum and both difference channels. These channels originate in the multi-mode horn of the phased-array antenna and are further composed of 3 identical mixer- and subsequent IF amplifier stages, synchronous detectors, matched filters, and multiplex combined A/D convertors (7 bits plus sign). The resulting digital I and Q signal samples for the 3 channels are processed for MTI, if applicable, and envelope detected (digital equivalent). If applicable, incoherent integration over successive echoes is the last operation before the actual monopulse ratio is formed by taking the quotients of each of the difference channel outputs and the sum channel output. This division process is combined with the range estimation process in the following way. The envelope samples of the sum channel, which appear at an interval corresponding to ΔR = 75 m i.e. half a range resolution cell, are compared to a detection threshold. Whenever two successive samples are above this threshold an interpolation procedure determines the position of the maximum of the sum channel response (equivalent to the echo range) from the largest samples s_m and its two neighbours s_f and s_l. If R_m is the range corresponding to s_m, the estimated echo range is:

$$R = R_m + \Delta R \frac{s_l - s_f}{s_m + \frac{|s_l - s_f|}{2}} \qquad (2)$$

The monopulse ratios are obtained by using the difference channel outputs corresponding to s_f, s_m and s_l. For both difference channels the monopulse ratio is formed:

$$r = \frac{0.5\; d_m + 0.25\; (d_l + d_f)}{0.5\; s_m + 0.25\; (s_l + s_f)} \qquad (3)$$

and the results are substituted in (1). The estimated echo direction in sine-space is obtained by adding (or subtracting) the outcomes of (1) to the sine co-ordinates of the beam axis. The signs (addition or subtraction) are derived by determining the larger of the two components of the sum channel (I or Q) at the sample instant of s_m and comparing its sign with that of the other branch of the difference channels (Q or I). If incoherent integration is applied the selected signed I and Q samples are added and the signs of their sums are compared. This processing is based on the monopulse property that, for a single point target, the noiseless echo vectors in sum and difference channels have a phase difference of + or -90 degrees, depending on the echo direction relative to boresight.

The described monopulse processing was chosen for its convenient implementation. When the case without incoherent integration is considered it corresponds to the following simplified one dimensional model. Suppose the relevant echo vectors in sum and difference channel are

$$S = u + j\,v \qquad D = x + j\,y \tag{4}$$

For FUCAS we form the monopulse ratio

$$r = \left|\frac{D}{S}\right| \cdot \mathrm{sgn}\left\{\frac{\max\,(u,v)}{y \text{ if } u,\ x \text{ if } v}\right\} = \sqrt{\frac{x^2 + y^2}{u^2 + v^2}} \cdot \mathrm{sgn}\left\{\ \right\} \tag{5}$$

where $\max\,(u,v) = u$ if $|u| > |v|$
$= v$ if $|u| < |v|$

and the result of the sgn-operation is either +1 or -1.
This monopulse ratio does approximate the proper expression:

$$r = \mathrm{Im}\left(\frac{D}{S}\right) = \frac{yu - xv}{u^2 + v^2} \tag{6}$$

For a point target (5) and (6) become asymptotically equal with increasing signal-to-noise ratio. Since target detection ensures a sufficient signal-to-noise ratio (> 13 dB) the performance of the FUCAS implementation will be close to the optimum performance. Moreover, the measurements have been arranged such that the signal-to-noise ratio was quite large (> 30 dB).

To conclude this section it is observed that the position accuracy to be obtained should be expressed in terms of beamwidth and range resolution cell. Unlike the direction and range accuracy, the cross range accuracy expressed in meters will deteriorate with range, even when the signal-to-noise ratio were constant.

3. MEASURING ARRANGEMENT

The initial monopulse measurements (3) were done with the antenna proper. Here the complete system is involved. For conventional antennas the monopulse performance does not depend on the direction to the target. A phased-array antenna, however, produces different antenna patterns (sum and both difference patterns) for each beam direction. For this reason a selection of 8 different scan directions was used, suitably spaced over the accessible observation cone of the antenna (see figure 1).
Since the antenna, when coupled to the FUCAS system, is fixed (at a tilt angle of 20 degrees), isolated reflecting objects had to be provided at various locations off the ground. For this purpose a helicopter was used which was positioned above a number of suitable landmarks at various heights. The corresponding slant distances (up to 8 km) guaranteed a high echo signal-to-noise ratio so that receiver noise could hardly contribute to the variability of the monopulse results. For each of the indicated helicopter positions a number of measurement cycles was made. Each cycle consisted of 160 x 100 x 3 measurements corresponding to 160 beam positions, for each of which a series of 100 pulses was transmitted at 3 different radar carrier frequencies. One cycle was completed in about 200 seconds.

For 90 percent of all transmissions the beam axis was directed to the target centre. Every 10th transmission the beam axis, on both transmission and reception, was put in an offset direction (figure 2). The offset directions were arranged according to a cross-formed pattern, the centre of which corresponds to the target centre. Both axes of the cross extended over an angle corresponding to a bit more than the half-power beamwidth of the sum pattern, the horizontal arm over $\Delta\sin A = \pm 0.040$, the vertical arm over $\Delta\sin B = \pm 0.040$.

By feeding the measured helicopter positions to a simple smoothing filter (α = 0.01) the FUCAS-system kept track of the target centre while the target might gradually drift away from its intended position. The tracking was hardly affected by measurement errors during beam offset because of the low duty cycle of these offset measurements (1 in 10). For each of the 160 offset directions 10 single pulse position measurements were made per carrier frequency, respectively on frequency 0 (5500 MHz), 1 (5650 MHz) and 2 (5800 MHz). In another mode incoherent integration was applied over 3 pulse transmissions on respectively frequency 0, frequencies 0, 1 and 2 (frequency agility), and frequency 2. In addition instead of single pulses, 3-pulse staggered MTI transmissions could be chosen on frequencies 0, 1 and 2 (pulse interval ≈ 250 μs). Incoherent integration over 3 MTI-series combined with frequency agility has been used as well.

The FUCAS-measurements were directly combined with those of an X-band fire-control radar (KL/MSS-3012, made by Signaal). This system has an elliptical antenna beam (az x el: 2.4 x 1.2 degrees) which is mechanically scanned around the direction of the target which is being tracked, at a rate of 50 Hz. Its pulse repetition frequency is 3800 Hz and it produces an estimated target position every 0.5 sec which is based on the received echoes during the last 3.5 sec and on the estimated state of motion of the target.

4. SOURCES OF ERROR

4.1 Receiver noise

The aim was to restrict the influence of receiver noise to a minimum. This was done by selecting a target of sufficient cross-section and keeping the slant range small. The obtained signal-to-noise ratio will typically be larger than 30 dB. Only in the case of MTI the signal-to-noise ratio may become small.

4.2 Quantization noise

By adaptively controlling the receiver gain the output of the sum channel was kept within 30 to 40 units (least significant bit=1) for the beam axis centred on the target. This corresponds to a r.m.s. value of the quantization noise of 42 dB below the echo strength (for MTI the surplus in receiver gain was occasionally not sufficient).

4.3 Target glint

Target glint was the major source of error. It was caused by the continuous change in aspect of the target and, most significantly, by the fluctuating echo contributions of the rotor blades of the helicopter. The last effect is responsible for the remaining echoes when MTI was employed; without the rotating blades no noticeable MTI echoes can be expected of a nearly stationary object. The target glint has not only a definite effect on the FUCAS-data but is also visible in some of the reference data obtained from the fire-control radar, which could not suppress glint frequencies below 2 Hz.

4.4 Multipath

The terrain over which the experiments have been conducted consists of grass-land and dunes with a sparse vegetation. Especially at longer ranges and lower target heights main beam multipath might have taken place, for instance via the elevation difference channel. Measurements below or close to the boresight axis might have been affected by multipath reception via a sidelobe of either difference pattern. Due to its smaller beam size no multipath was to be expected for the fire-control radar.

4.5 Imperfections in antenna and receiver

The characteristics of the sum- and difference antenna patterns as a function of scan direction and radar frequency may depend on minor imperfections in antenna phase steering, illumination and geometry. The monopulse performance depends also on the linearity of the 3 receiver channels and the difference in gain between the channels. During the design of the system great care has been taken to reduce these imperfections.

5. RESULTS

The presented results correspond to measurements carried out for 8 distinct helicopter positions. The angular co-ordinates of these positions are shown in figure 1 together with the directions used for the initial antenna measurements (3).

5.1 Discussion of some examples

In figures 3-9 some results are shown for one helicopter position. Each figure shows the average direction error (central line) plus and minus the standard deviation of the error as a function of the offset position of the beam axis relative to the centre of the helicopter. The half-power beamwidth of the antenna corresponds in sine-space to the range -0.035 to 0.035, which is a bit less than the length of the x-axis (which corresponds to the horizontal arm of the cross pattern for sinA and to the vertical arm for sinB).

Figure 3 shows the difference between the sinA estimate of FUCAS and that of the reference radar. Every ΔsinA offset corresponds to 10 measurements, each resulting from the incoherent integration of 3 radar echoes. The average sinA error varies from point to point due to target glint. The value of the standard deviation tends to grow towards the edge of the beam.

Figure 4 shows the results when frequency agility over three radar frequencies is applied. Its use reduces the standard deviation of the sinA error, not the effect of glint on the average of the 10 estimates of sinA.

In figure 5 the reference for the sinA error is obtained by using the average of the 18 surrounding axis-on-target measurements of FUCAS instead of the filtered output of the fire-control radar. It shows that the reference data of figure 4 is suffering from glint around ΔsinA = 0.016.

Figures 6 and 7 give the sinB error as a function of ΔsinB. The results are similar to those for the sinA error in figs 3 and 4.

Figures 8 and 9 illustrate that the sinA error does not depend on ΔsinB and vice versa: both the average (bias) and the standard deviation do not vary significantly with position along the other arm of the cross pattern.

5.2 Discussion of summarized results

The single frequency direction error results are summarized in figure 10. The error data from 8 helicopter positions are reduced to the average bias at the beam edge and the beam centre plus and minus the average standard deviation of the direction estimates. The sinA and sinB bias each follow a specific pattern, irrespective of frequency. This bias pattern should be corrected by modifying the monopulse algorithm based on the initial antenna measurements. The discrepancy with the initial results is presumably due to the fact that the monopulse algorithm was based on the sum and difference patterns measured with the array in the transmit condition, which is not identical to the receive condition obtained by a different setting of the non-reciprocal phase-shifters.

The standard deviation of the sin-errors is smaller in the beam centre than at the beam edge and is smaller for sinB than for sinA. These trends may be related to the increased influence of quantization errors at the beam edge and to the predominance of the horizontal (sinA) component of the glint. The average standard deviation of the error in range near boresight and near the beam edge is 3.1 m and 3.8 m, irrespective of frequency. Frequency agility gives both in range and angle a reduction in standard deviation to 70 percent.

A small set of MTI measurements indicates that the standard deviation of the range and direction errors are roughly doubled by applying MTI. It can be inferred that the position of a hovering helicopter can reasonably well be estimated even when MTI is used, provided that the echoes are sufficiently strong.

6. CONCLUSIONS

The following conclusions are not necessarily valid for all kinds of phased-array antennas.

1. A single target echo from within the half-power beam contour is sufficient to provide an accurate position estimate (a few percent of the beamwidth/range cell).
2. When more than one transmission is used for one position estimate the application of frequency agility will reduce the effect of target glint.

3. Although a single monopulse algorithm may be adequate, the monopulse performance can be enhanced by using a set of monopulse relations in order to cope with differences related to direction co-ordinate, radar frequency and scan direction.

ACKNOWLEDGEMENT

This work has been a joint effort of a large number of people at the Physics Laboratory and LEOK-TNO. The co-operation of several services of the armed forces is gratefully acknowledged.

REFERENCES

1. van der Spek, G.A., "FUCAS, an experimental phased-array radar system", Int. Conf. RADAR-77, IEE Conference Publication 155.

2. Snieder, J., "The design, construction and test results of the CAISSA space-fed phased-array antenna", Military Microwaves Proceedings, pp. 391-402, October 1978.

3. Rijsdijk, F.B., and van der Spek, G.A., 1978, IEEE Trans. AES, 14, 226-236.

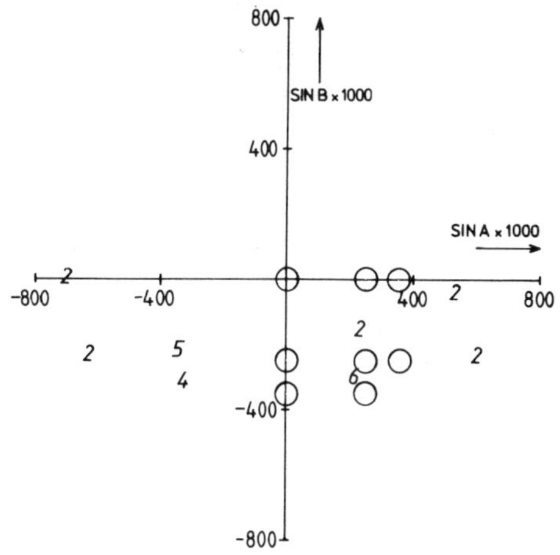

Fig 1 Direction centres of the initial monopulse measurements (o) and of the cross-measurements (number of cycles).

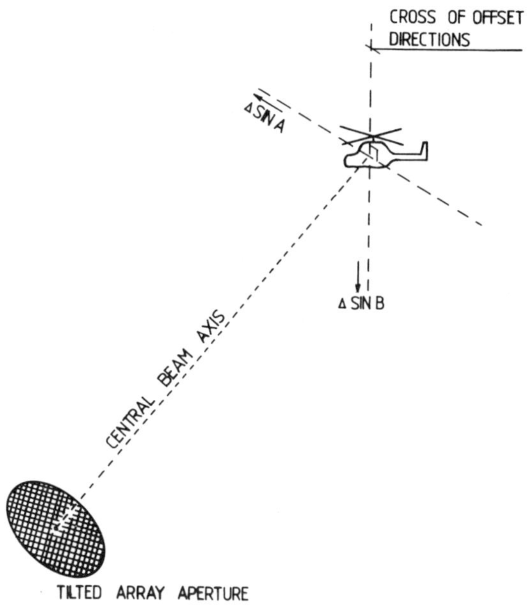

Fig 2 Cross pattern to measure monopulse errors for 80 sinA and 80 sinB offsets between target direction and beam axis.

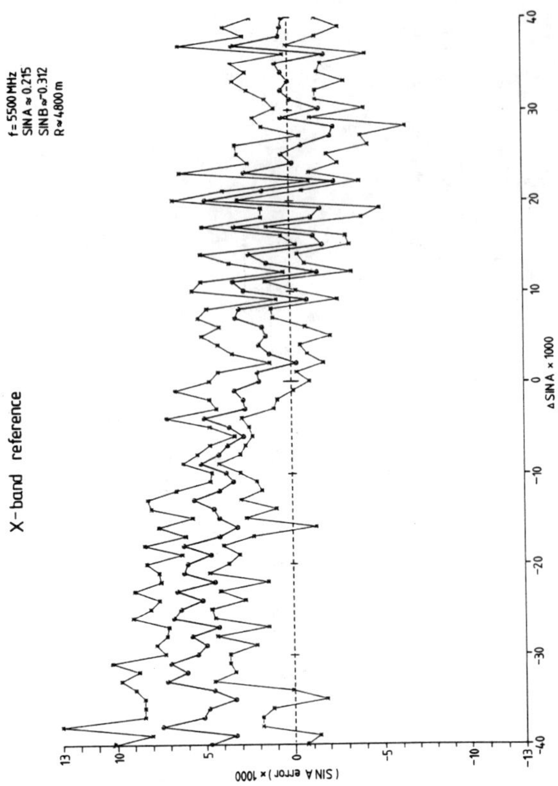

Fig 3 SinA error, averaged over 10 measurements (central line) plus and minus the standard deviation of the error, as a function of the sinA offset between target and beam axis. Single frequency case.

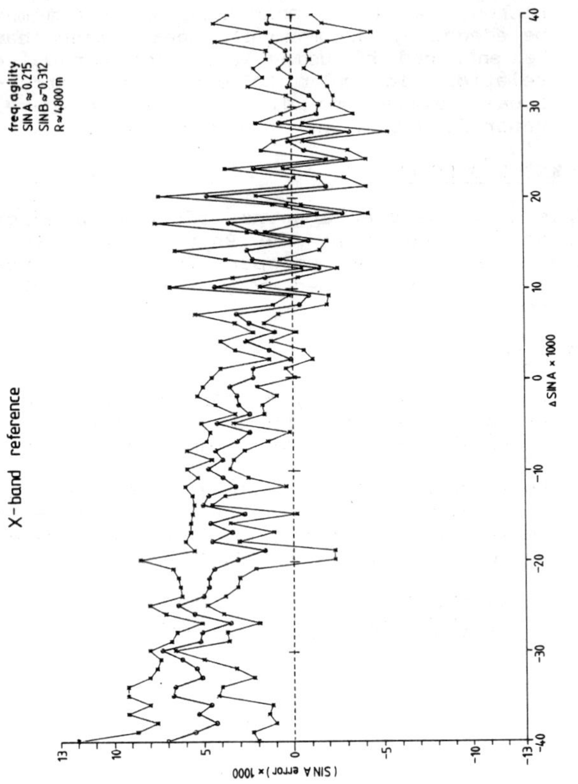

Fig 4 Same as fig 3 with frequency agility over 3 radar frequencies.

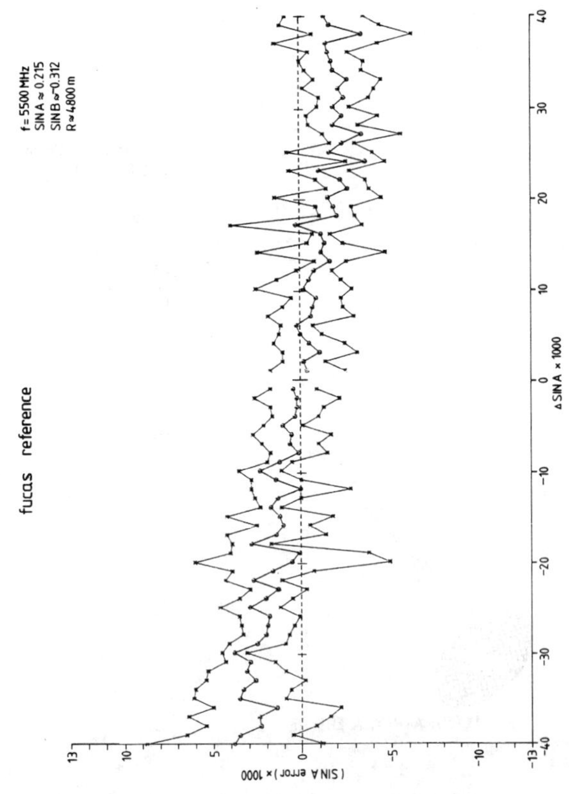

Fig 5 Same as fig 3 but reference data obtained by averaging over 18 neighbouring non-offset FUCAS measurements.

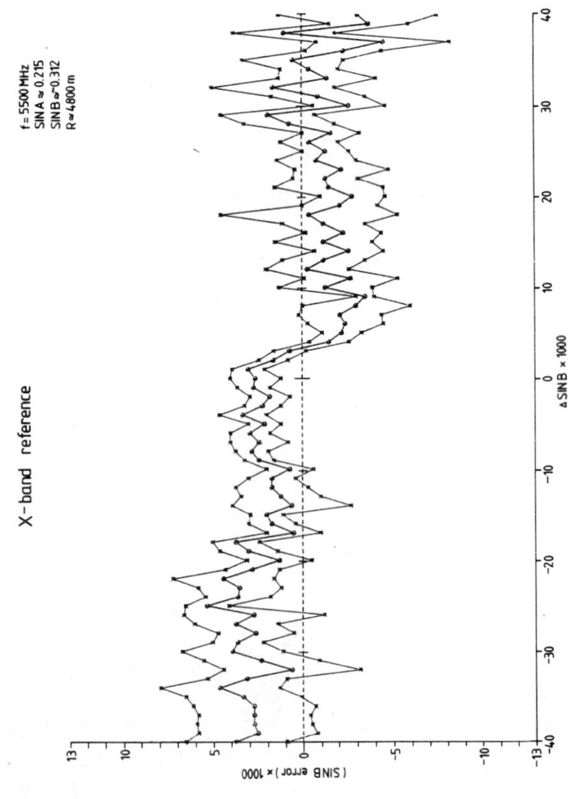

Fig 6 Same as fig 3 for sinB error versus sinB offset.

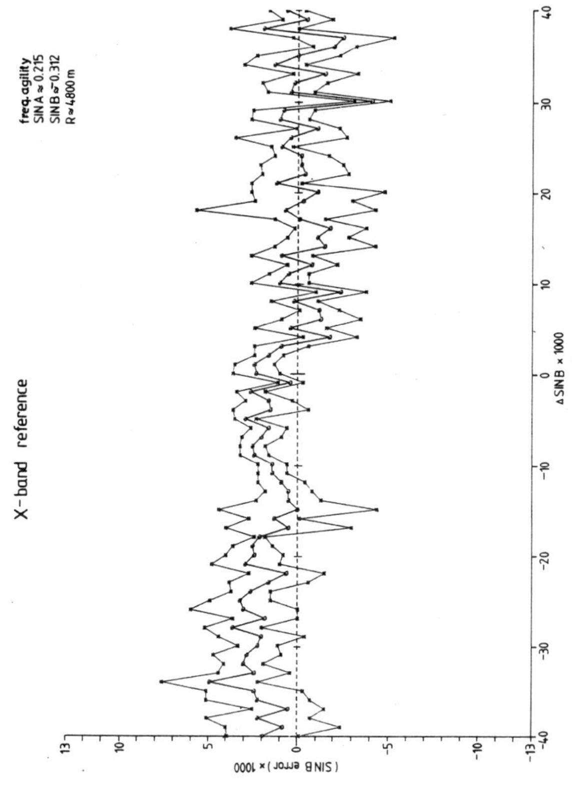

Fig 7 Same as fig 6 with frequency agility over 3 radar frequencies.

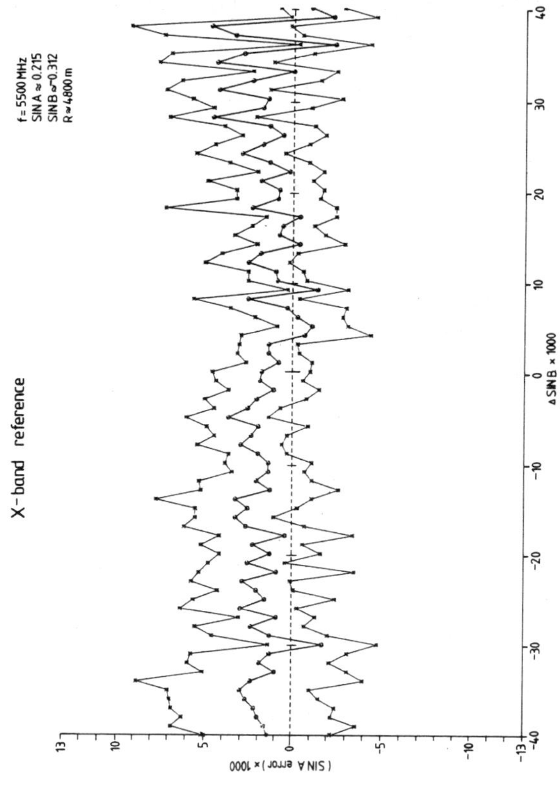

Fig 8 Same as fig 3 for sinA error versus sinB offset.

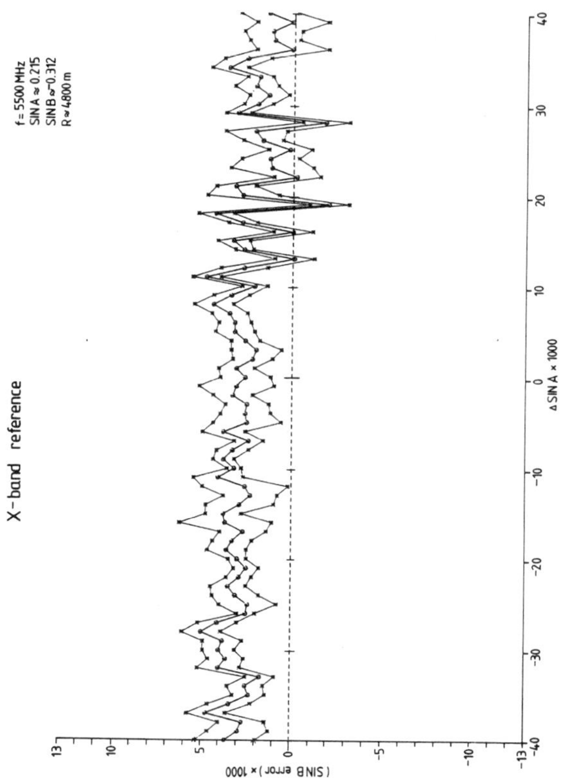

Fig 9 Same as fig 8 for sinB error versus sinA offset.

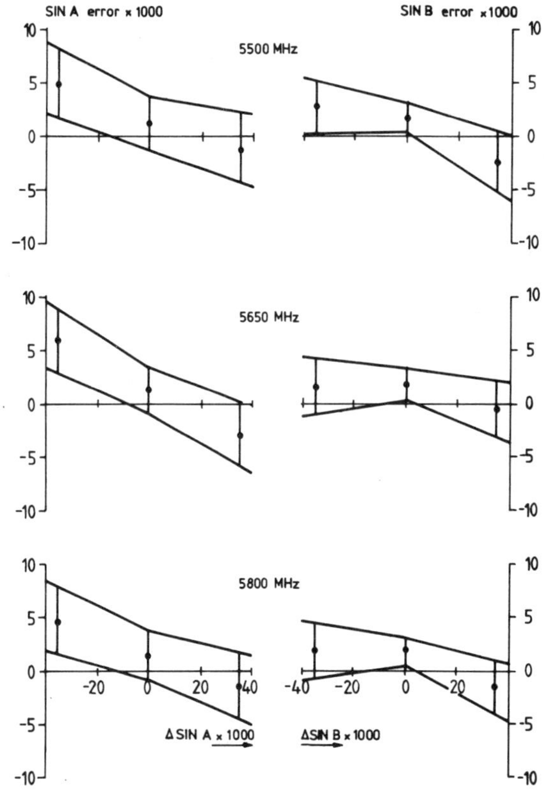

Fig 10 Bias plus and minus standard deviation of the measured angular coordinates sinA and sinB as a function of target direction offset. Average over all measured target positions.

AUTOMATIC DETECTORS FOR FREQUENCY-AGILE RADAR

Gerard V. Trunk and Paul K. Hughes II

Naval Research Laboratory, Washington, D.C. USA

INTRODUCTION

Modern, long-range surveillance radars can detect 1-m^2 targets at ranges of 200 nmi or longer. However, if the radar transmits at a constant frequency, a jammer needs only to jam a narrow bandwidth to reduce significantly the radar detection range. To force the jammer to do broadband jamming, and consequently to reduce its effectiveness, most modern radars employ pulse-to-pulse frequency agility. When frequency agility is used, the received sidelobe jamming power can vary by as much as 20 dB pulse-to-pulse even though the transmitted jamming power is constant over the radar bandwidth. To see why this is true, consider Fig. 1, which shows the measured jamming power received by the SPS-39 radar as it scans by a jammer [1]. In a space of several degrees the jamming power usually varies by 15 dB and sometimes varies by 20 or more dB. These data were recorded at a constant frequency. Thus, if the frequency is varied pulse-to-pulse, the sidelobes pointed in the direction of the jammer will shift by several degrees; therefore the received sidelobe jamming power will vary pulse-to-pulse by as much as 20 dB. The purpose of this paper is to investigate the detection performance of various detectors for a radar employing frequency agility in the presence of broadband sidelobe jamming.

AUTOMATIC DETECTORS

If the jamming is white Gaussian noise, the density of the ith output pulse x_i from an envelope detector is

$$P(x_i|A_i) = \frac{x_i}{\sigma_i^2} \exp[-(x_i^2+A_i^2)/2\sigma_i^2] I_0(A_i x_i/\sigma_i^2)$$

where A_i is the signal amplitude, the signal-to-noise ratio (S/N) is 10 log $(A_i^2/2\sigma_i^2)$, and the noise power σ_i^2 varies pulse-to-pulse because frequency agility is being used in the presence of sidelobe jamming. The detection problem of interest is

$$H_0:\quad A_i \equiv 0,\ \text{any } \sigma_i\ ;$$

$$H_1:\quad A_i > 0,\ \text{any } \sigma_i\ .$$

Since σ_i is unknown, no uniformly most powerful test exists. That is, no optimal (in the sense of maximizing the probability of detection for a given false-alarm probability) test exists, and suboptimal tests must be used.

We have shown [2] that good detection performance can be obtained using

$$\sum_{i=1}^{n} \frac{x_i^2(j)}{\frac{1}{2m}\sum_{k=1}^{m}\left[x_i^2(j+l+k) + x_i^2(j-l-k)\right]}$$

where $x_i(j)$ is the ith envelope-detected pulse in the jth range cell and 2m is the number of reference cells. The denominator is the maximum likelihood estimate of σ_i^2 and essentially, the detector sums signal-to-noise power ratios. This detector, which we will call the ratio detector, is shown in Fig. 2.

The ratio detector has good performance because it sums power ratios. Thus, it will detect targets even though only a few returned pulses have a high signal-to-noise ratio. Unfortunately, this will also cause the ratio detector to declare false alarms in the presence of short-pulse interference. Consequently, to reduce the number of false alarms when short-pulse interference is present, the individual power ratios will be soft-limited to a small enough value so that interference will only cause a few false alarms.

In some cases, limiting will have very little effect on the detection performance of the ratio detector. However, since we know nothing about the detection properties of the soft-limiting ratio detector, we will compare its detection performance with the more commonly used detectors, such as the cell averaging CFAR, the log integrator, and the binary integrator. These detectors can be described mathematically as follows: The cell-averaging CFAR is given by

$$\frac{\sum_{i=1}^{n} x_i^2(j)}{\frac{1}{2m}\sum_{k=1}^{m}\sum_{i=1}^{n}\left[x_i^2(j+l+k) + x_i^2(j-l-k)\right]}$$

the log integrator is given by

$$\sum_{i=1}^{n}\left[\ln x_i(j) - \overline{\ln x_i(j)}\right]$$

where

$$\overline{\ln x_i(j)} = \frac{1}{2m}\sum_{k=1}^{m}\left[\ln x_i(j+l+k) + \ln x_i(j-l-k)\right]$$

and the binary integrator is given by

$$\sum_{i=1}^{n} y_i(j) \text{ where}$$

$$y_i(j) = \begin{cases} 0, & \text{if } \left[\ln x_i(j) - \overline{\ln x_i(j)}\right] < T \\ 1, & \text{if } \left[\ln x_i(j) - \overline{\ln x_i(j)}\right] \geq T \end{cases}$$

We will compare these detectors by first calculating the appropriate detection thresholds using the importance-sampling technique and then calculating the detection performance using a Monte Carlo simulation.

DETECTION THRESHOLDS

As noted previously [1], the ratio detector is the sum of samples from an F-distribution, and its density function can only be represented in terms of an integral. To avoid numerical integrations, the detection threshold will be calculated using the importance-sampling technique [3]. Since many simulations were run, we chose to use only n = 6 pulses integrated and 2m = 16 reference cells. These numbers correspond to those of the NRL radar which will eventually be used to record data to validate the simulation results. The results for the ratio detector are shown in Fig. 3. As mentioned previously, if short-pulse interference is present, the ratio detector will have too many false alarms. To relieve this problem the individual ratio was limited to a maximum value of 10, the simulation was repeated, and the results are shown in Fig. 4.

To show that limiting will yield a low false-alarm rate in the presence of nonsynchronous interference, we will calculate the false-alarm rate when the probability of an interference spike is 10^{-3} in each range cell. The probability of false alarm is

$$P_{FA} = \sum_{i=0}^{n} P_{FA}(i)\ P(i) \tag{1}$$

where $P_{FA}(i)$ is the conditional probability of false alarm given that i interference spikes are present in the n pulses and P(i) is the binominal probability that there are exactly i interference spikes present. The detection threshold is set so that $P_{FA} = 10^{-6}$ when there are no spikes present. Thus, from Fig. 4, the threshold is set equal to 29.2. The calculation of P_{FA} proceeds as follows: for i = 0 (i.e., no spikes present) P(0) = 0.994 and $P_{FA}(0) = 10^{-6}$. For i = 1 (one spike present), P(1) = 0.006, and $P_{FA}(1)$ equals the probability that the sum of the other five ratios exceeds 19.2 (here we are assuming that the ratio containing the spike obtains the limiting value; hence 29.2 - 10.0 = 19.2). From Fig. 4, $P_{FA}(1) = 2.0 \times 10^{-4}$. Similarly, for i = 2, $P(2) = 15 \times 10^{-6}$ and $P_{FA}(2) = 0.036$. Since $P(3) = 20 \times 10^{-9}$, the probabilities associated with i = 3,4,5, and 6 can be neglected. Substituting the previously calculated probabilities into Eq. (1) yields $P_{FA} = 2.7 \times 10^{-6}$. Thus, limiting can be used to obtain a low false-alarm rate when short-pulse interference is present. The limiting value of 10 was found by trying various values. For instance, if the limiting value is 12, $P_{FA} = 6.5 \times 10^{-6}$.

The importance-sampling technique was used to calculate the thresholds for the cell-averaging CFAR, the log integrator with a limiting value of 1.5, and the single-pulse log detector which is needed for the binary integrator. The threshold curves are given in Figs. 5 to 7. Using Fig. 6, the P_{FA} in the presence of interference is calculated to be 1.7×10^{-6} when the log integrator is limited to a value of 1.5. Using Fig. 7, the appropriate thresholds yielding $P_{FA}(0) = 10^{-6}$ for 2 out of 6, 3 out of 6, and 4 out of 6 detectors are found to be 1.51, 1.27, and 1.10, respectively.

DETECTION RESULTS

To compare the performance of the various detectors, probability-of-detection vs signal-to-noise-ratio curves were generated using simulation techniques. All results are for the case of six pulses integrated, 16 reference cells, and a false-alarm probability of 10^{-6} in thermal noise. The detection performance for nonfluctuating and pulse-to-pulse Rayleigh fluctuating targets in thermal noise (i.e., no jamming) of the five detectors discussed is shown in Figs. 8 and 9, respectively. For nonfluctuating targets, the cell-averaging CFAR is the best detector; however, the ratio detectors and log integrator are within a few tenths of a decibel of each other. The binary integrator is 1 to 1.5 dB worse than the cell-averaging CFAR. On the other hand, for Rayleigh fluctuations, the variation is between 0 and 3 dB. The cell-averaging CFAR is still the best detector. The ratio detector is better than the log integrator, which is better than the binary integrator.

To compare the various detectors in jamming it was first necessary to generate the variation of jamming power when frequency agility is employed. The received jamming power can be written as $J = CG^2$ where C is an appropriate constant (a function of jammer power, jammer antenna gain, and range from jammer to radar) and G is the radar-antenna voltage gain in the direction of the jammer. In the far-out sidelobe region, it is a reasonable assumption that the gain varies as (sin x)/x. Therefore, when one changes the radar frequency, the variable x, which is proportional to frequency, changes and consequently the received jamming power changes. If one assumes that x is uniformly distributed from X_0 to $X_0 + 2\pi$ (where X_0 is any angle greater then π), then the jamming J is approximately proportional to $\sin^2 x$, and the probability density of J (normalized to a maximum value of 1) is given by $P(J) = 1/(J-J^2)^{1/2}$ for $0 \leq J \leq 1$. Thus, the total noise-plus-jamming power is given by $\sigma^2 = 1 + J_0 J$, where J_0 is the maximum received jamming power, J is a random variable between 0 and 1 whose density is given above, and the thermal noise power has been set to 1. The signal-to-noise ratio is defined as S/N(dB) = 10 log $(A^2/2)$, where A is the nonfluctuating signal amplitude, and the maximum jamming-to-noise ratio is given by J/N(dB) = 10 log (J_0).

Detection curves for nonfluctuating targets in jamming are shown in Fig. 10. Now the ratio detector is better than the log integrator, which is better than the cell-averaging CFAR, which is better than the binary integrator. The results for Rayleigh fluctuating targets are shown in Fig. 11. The ratio detector is still the best detector. However, the next-best detector is the cell-averaging CFAR. The log integrator and binary integrator are very similar in performance.

In the jamming cases considered, the ratio detector without limiting was the best detector and the ratio detector with limiting was the next-best detector. In thermal noise, the ratio detectors are only several tenths of a dB worse than the cell-averaging CFAR.

Therefore, if one does not have any short-pulse interference problems, one should use the ratio detector; and if one does have short-pulse interference problems, one should use the ratio detector with limiting. Since the ratio detector without limiting can be several dB better than the ratio detector with limiting, one possibility is to use the ratio detector without limiting and then decide whether the received signal is from a target or is interference.

INTERFERENCE DETECTION

We will now consider a second test used after a detection is declared by the ratio detector without limiting. The purpose of this second test is to detect the presence of short-pulse interference (SPI). Again, if the jamming is white Gaussian noise, the density of the ith output pulse, x_i, from an envelope detector in the presence of only SPI is

$$P(x_i|B_i) = \frac{x_i}{\sigma_i^2} \exp[-(x_i^2+B_i^2)/2\sigma_i^2] I_0(B_i x_i/\sigma_i^2)$$

where B_i is the interference amplitude, and the interference-to-noise ratio is 10 log $(B_i^2/2\sigma_i^2)$. The binary hypothesis of the second test is

$$H_0: A_i = 0, \; B_i > 0, \quad \text{any } \sigma_i$$

$$H_1: A_i > 0, \; B_i = 0, \quad \text{any } \sigma_i$$

The interference signal amplitude was assumed to be Rayleigh in one case and one-dominant plus Rayleigh in another case. The results of both cases are similar so only the latter is discussed.

A likelihood ratio is formed assuming that at most only one SPI pulse is present in N pulses. The likelihood ratio

$$L = \frac{\prod_{i=1}^{N} \int P(x_i|A_i)\, dA_i}{\int P(x_j|B_j)P(B_j)dB_j \prod_{\substack{i=1\\ i\neq j}}^{N} \frac{x_i}{\sigma_i^2} \exp\left[-x_i^2/2\sigma_i^2\right]}$$

where x_j is the pulse containing the SPI (assumed to be the ratio with the largest value). The above equation simplifies to

$$L = \sum_{i=1}^{N} \left[\frac{S_i\, x_i^2}{2\sigma_i^2(S_i+1)} + \ln\left(\frac{1}{S_i+1}\right)\right] - \frac{S_j\, x_j^2}{2\sigma_j^2(S_j+2)}$$

$$- \ln\left[1 + \frac{S_j\, x_j^2}{2\sigma_j^2(S_j+2)}\right] - 2 \ln\left(\frac{2}{S_j+2}\right)$$

where S_i is the MLE estimate of the signal-to-noise ratio of the ith pulse. This estimate is given by

$$\frac{I_1\left(\sqrt{2S_i\, x_i^2/\sigma_i^2}\right)}{I_0\left(\sqrt{2S_i\, x_i^2/\sigma_i^2}\right)} = \sqrt{\frac{2S_i\, \sigma_i^2}{x_i^2}}$$

which is difficult to solve for S_i. An approximation to the above was used to find S_i given x_i^2/σ_i^2.

The threshold for this second detector was determined using Monte Carlo techniques with a single 30 dB interference-to-noise spike for a probability of false alarm of 2.4 x 10^{-4}. The overall probability of false alarm is calculated using Eq. (1). When the probability of an interference spike is 10^{-4} in each range cell, the over-all probability of false alarm becomes 1.4 x 10^{-6}. The ratio detector with limiting has a probability of false alarm of 1.14 x 10^{-6} in this case.

Fig. 12 shows the probability of detection curve for the second detector compared to the ratio detector, with and without limiting, in the presence of 20 dB jamming and a Rayleigh fluctuating target. The second detector gets back about half of the lost detection performance of the ratio detector with limiting. As the probability of the interference spike increases, maintaining a low false alarm rate decreases the detection performance of the second detector faster than that for the ratio detector with limiting. This is largely due to the increased probability that two of the six pulses will contain an interference spike. As the probability of an interference spike increases to and above 10^{-3}, little difference is seen between the ratio detector with limiting and the second detector.

SUMMARY

When pulse-to-pulse frequency agility is employed, the received sidelobe jamming power can vary by as much as 20 dB. The best detector for this situation is the ratio detector, which normalizes the received power on every pulse (using the neighboring reference cells) and then sums these normalized power ratios. In the presence of short-pulse interference, a ratio detector using limiting, or an interference detector should be employed.

There are two areas which require further work. First, pulse-to-pulse frequency agility should be used in the presence of broadband sidelobe jamming, and the recorded data should be used to test the behavior of the various detectors mentioned in this paper. Second, one should continue investigating whether the ratio detector without limiting can be used in the presence of short-pulse interference by performing further testing to discriminate between target and interference using the individual ratios, not their sum.

REFERENCES

1. G.V. Trunk, "Comparison of Detectors in the Presence of Sidelobe Jamming," NRL Report 8449, Oct. 23, 1980.

2. G.V. Trunk, "Automatic Detectors for Frequency-Agile Radars," NRL Report 8571, April 24, 1982.

3. R.L. Mitchell, "Importance-Sampling Applied to Simulation of False Alarm Statistics," IEEE Trans. Aerosp. Electron. Syst. AES 17 (1), 17-24 (Jan. 1981).

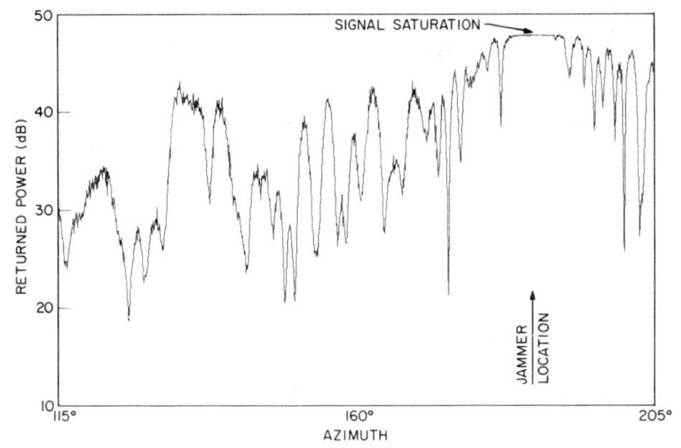

Fig. 1 - Received jamming power as the SPS-39 sweeps over a jammer

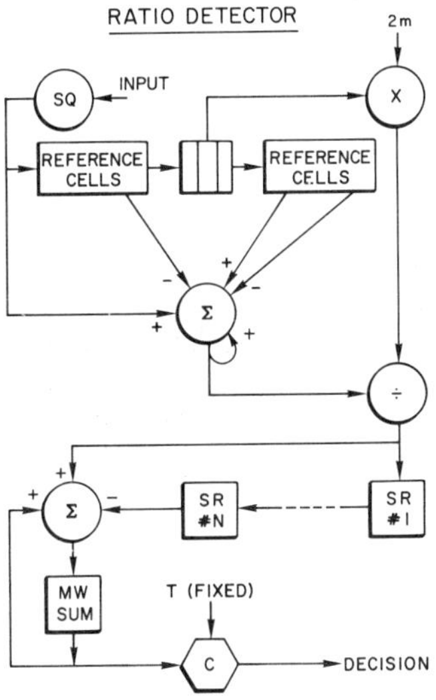

Fig. 2 - Ratio detector: SR is a shift register, MW is a moving window, 2m is the number of reference cells, and C is a comparator

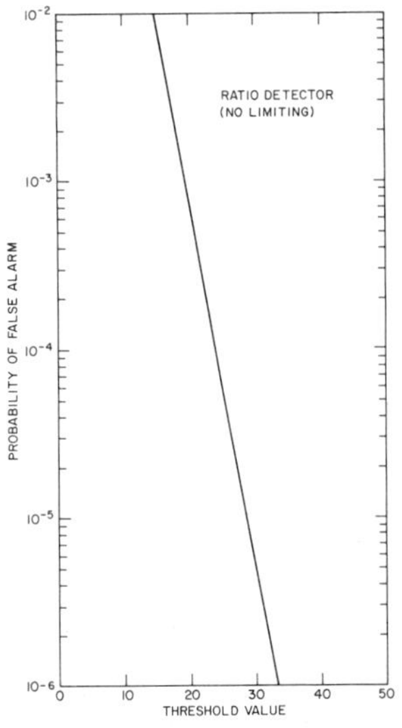

Fig. 3 - Threshold values for the ratio detector: six pulses integrated and 16 reference cells used

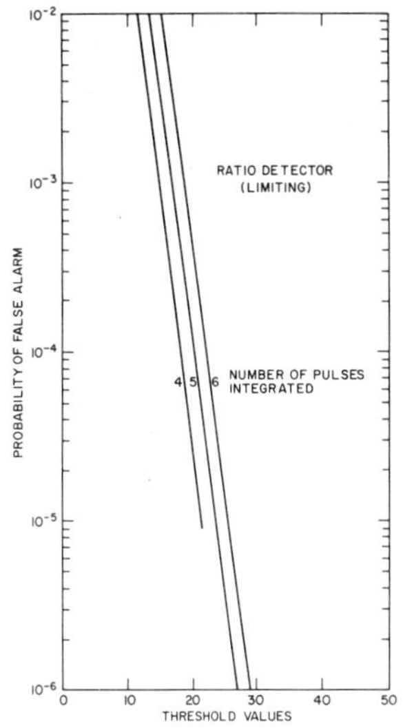

Fig. 4 - Threshold values for the ratio detector: maximum value of each ratio limited to a value of 10

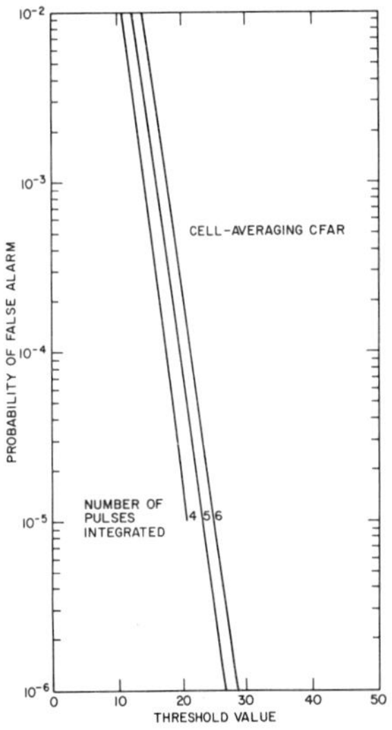

Fig. 5 - Threshold values for the cell-averaging CFAR

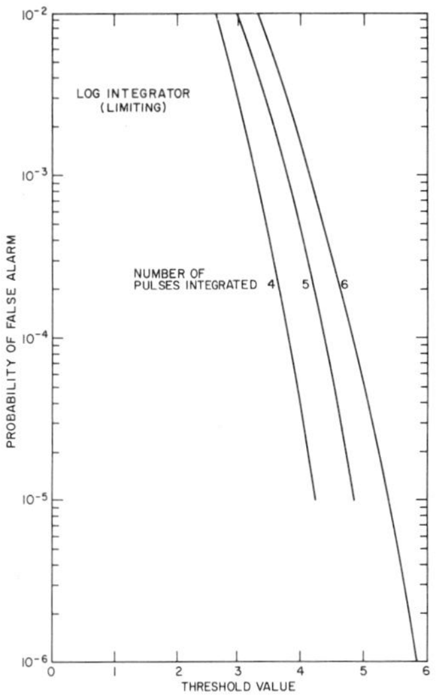

Fig. 6 - Threshold values for the log integrator: maximum value of each normalized log term limited to a value of 1.5.

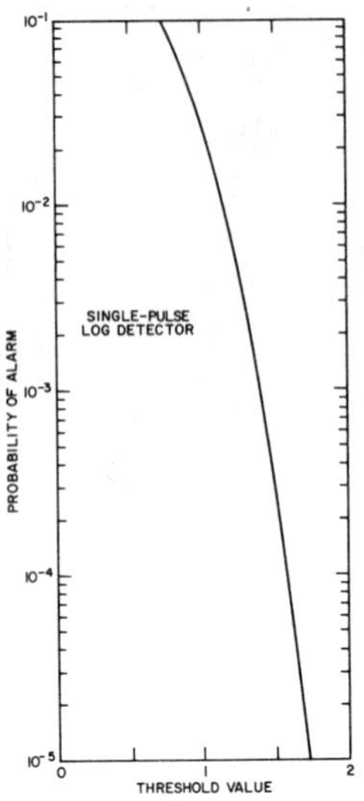

Fig. 7 - Threshold values for single-pulse log detector: 16 reference cells used

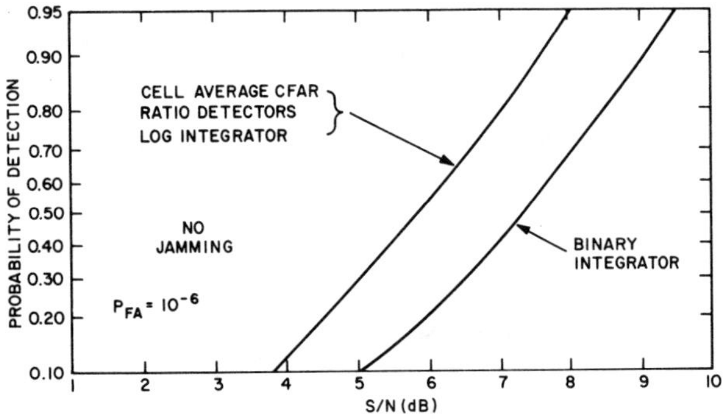

Fig. 8 - Curves of probability of detection vs signal-to-noise ratio for the cell-averaging CFAR, ratio detectors, log integrator, and binary integrator: nonfluctuating target and probability of false alarm = 10^{-6}

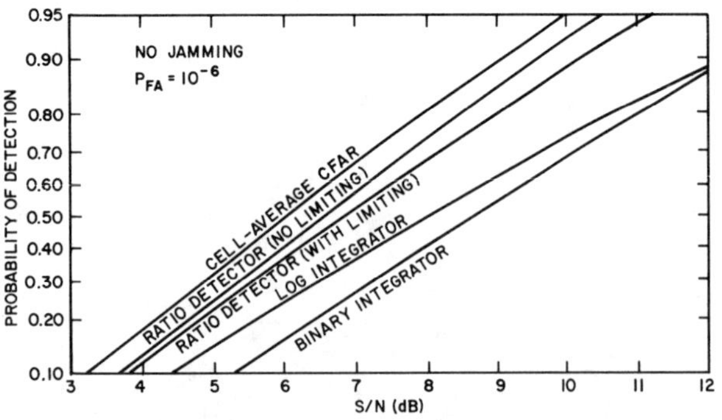

Fig. 9 - Curves of probability of detection vs signal-to-noise ratio for the cell-averaging CFAR, ratio detectors, log integrator, and binary integrator: Rayleigh fluctuating target and probability of false alarm = 10^{-6}

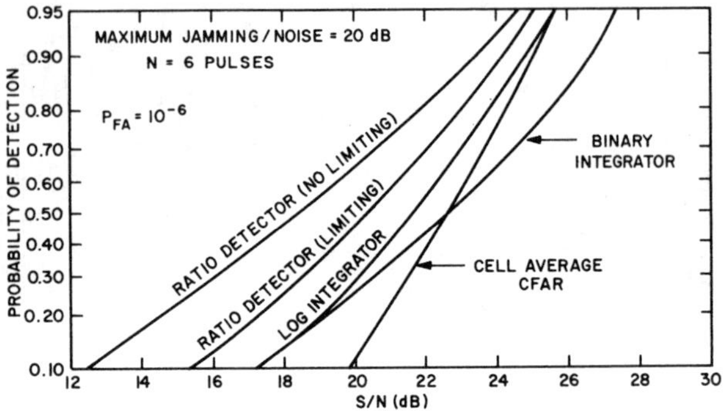

Fig. 10 - Curves of probability of detection vs signal-to-noise ratio for the cell-averaging CFAR, ratio detectors, log integrator, and binary integrator: nonfluctuating target, probability of false alarm = 10^{-6}, and maximum jamming-to-noise ratio = 20 dB

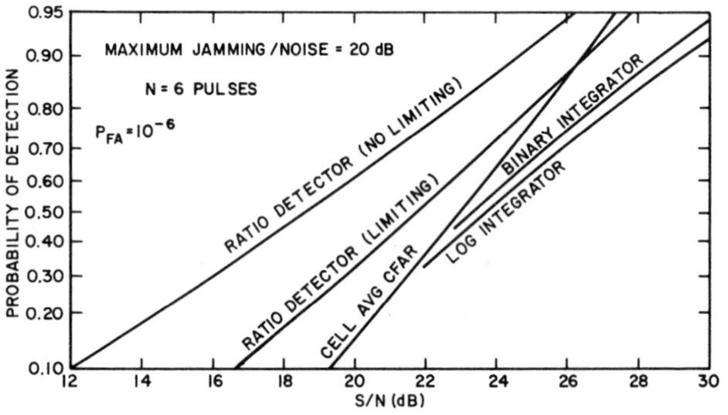

Fig. 11 - Curves of probability of detection vs signal-to-noise ratio for the cell-averaging CFAR, ratio detectors, log integrator, and binary integrator: Rayleigh fluctuations, probability of false alarm = 10^{-6}, and maximum jamming-to-noise ratio = 20 dB

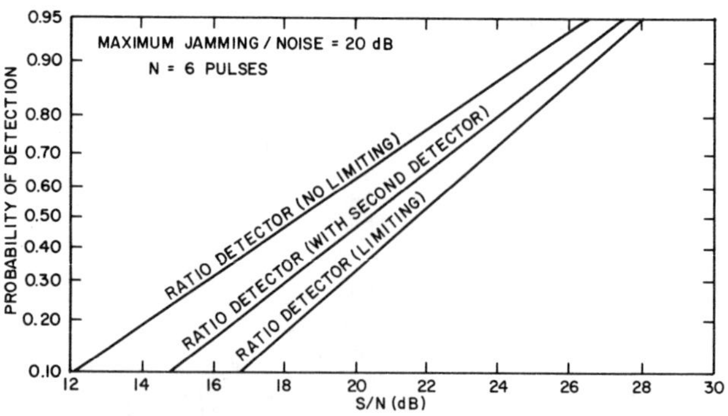

Fig. 12 - Curves of probability of detection vs signal-to-noise ratio for ratio detectors and ratio detector with a second detector: Rayleigh fluctuations, and maximum jamming-to-noise ratio = 20 dB

A NOVEL 35 GHZ 3-D RADAR FOR FLIGHT ASSISTANCE

G.M. Ritter

Siemens AG, Munich, Germany

INTRODUCTION

This paper describes some of the work carried out at SIEMENS AG in Munich to evaluate a 35 GHz 3-D radar for flight assistance. One application of this radar system is the generation of 3-D images of the terrain to assist the pilot of a helicopter or an aircraft in low-level missions. To form a 3-D image it is necessary that each point of the terrain can be measured in two angles, that is azimuth and elevation. Normally this requires a flat antenna having good resolution in both azimuth and elevation. The disadvantages of such an approach are dimensions, weight, cost and scan-time. The investigated radar system avoids these problems by the application of fan-beam antennas having high resolution in azimuth only. The missing resolution in elevation is achieved by interferometer and doppler processing.

ELEVATION RESOLUTION BY INTERFEROMETER PROCESSING

If two fan-beam antennas are vertically displaced by d, the echoes returned from an elevation ε will produce a phase difference φ between both antennas:

$$\varphi = 2\pi \cdot \frac{d}{\lambda_0} \cdot \sin \varepsilon \qquad (1)$$

λ_0 = free space wavelength

With the measured phase difference φ the elevation ε may be determined. There has been some theoretical work in the past by Potter (1) on a similar application, restricted to a small unambiguous displacement. The sensitivity may be increased essentially with increasing displacement. Our system uses a vertical distance of approximately 6 wavelengths. The measurement then will be ambiguous, for there may be a path difference of several wavelengths between both antennas. This ambiguity causes several phase changes of 2π for the full range of elevation. It may be resolved if the phase difference is tracked for all range cells. The achieved resolution is best in the boresight direction of the antenna system, while there is a poor resolution in the perpendicular directions. This means good resolution in front of the aircraft, while there is poor resolution underneath the aircraft. The ambiguity resolution may be aided by an additional unambiguous interferometer measurement with a small displacement, requiring a third receiver. However it is much more efficient to combine the high-resolution interferometer system with doppler processing.

ELEVATION RESOLUTION BY DOPPLER PROCESSING

The doppler shift f_D of echo returns obtained by a radar system on board an aircraft is given by:

$$f_D = 2 \cdot \frac{v_0 \cdot f_0}{c} \cdot \cos \Delta_{FV} \qquad (2)$$

with v_0 = speed in the direction of the flight vector
f_0 = transmitted frequency
Δ_{FV} = angle relative to the flight vector

Targets in the direction of the flight vector will produce the highest possible doppler shift, while echoes from perpendicular directions will show no doppler shift. The measured doppler shift allows determination of the angle relative to the flight vector, if the highest possible doppler shift is known. If the direction of the flight vector is also known, this information may be used to achieve the missing resolution in elevation when using fan-beam antennas. There is a rotational ambiguity centered on the flight vector and also a poor resolution in this region. The best resolution is given perpendicular to the flight vector, this normally means underneath the aircraft. The resolution is a direct function of speed and operating frequency. An existing ambiguity with the PRF is easily resolved by tracking the doppler shift of the range cells.

COMBINATION OF INTERFEROMETER AND DOPPLER PROCESSING

A comparison of both methods shows that in regions where one method gives a poor resolution the other method leads to a high resolution and vice versa. Furthermore the combined information of both methods allows determination of the flight vector in magnitude and direction, this means speed, azimuth and elevation.

THE EXPERIMENTAL RADAR SYSTEM

Flight tests have been performed using a 35 GHz pulsed radar system. The system consists of a magnetron transmitter of 20 kW peak power and a pulse width of 100 nsec. The transmitter feeds the center antenna of three vertically stacked slotted waveguide antennas, having a beamwidth of .7° in azimuth. The beamwidth in elevation is approximately 60°. The antenna system is mechanically scanned over usually ± 20°. The upper and lower antenna are spaced approximately 6 wavelengths apart and form the interferometer system. The antenna system is depressed 20° from the horizontal direction. Both receivers deliver the logarithmic detected amplitude information, and after IF limiting, the detected quadrature components to obtain the phase information. The following data processor reduces the data rate to match the capability of a high-speed tape recorder. This data processing is done for each range cell. The necessary speed of 100 nsec per range cell is achieved by pipelining and look-up tables. The amplitude information is simply low-pass filtered. The quadrature components delivered by the receivers are converted to polar components. This is realized by look-up tables. Now the phase difference may be formed by a simple subtraction. The low-pass filtering must be done in rectangular components. A final conversion to polar representation is required to obtain the averaged phase differences for each range cell. The doppler shift is processed by computing the phase difference of the echoes from consecutive transmitting pulses and low-pass filtering as described above.

ALGORITHM FOR 3-D IMAGE GENERATION

The generation of 3-D images of the terrain in the experimental system is done off-line using an 8086 microprocessor system and a colour display. The complete information of one antenna scan is read from the magnetic tape to the memory of the microprocessor system. First the flight vector is calculated using the combined information from interferometer and doppler measurement. This determines the three different areas, where only the doppler information is taken, or combined doppler and interferometer processing is done, or the interferometer information only is processed. Usually underneath the aircraft doppler processing is applied. With increasing elevation it is changed to combined processing and then to interferometer measurement. The amplitude information is used to control the brightness of each range cell after a range dependent correction factor has been applied, similar to the operation of an STC. The range information is used to control the colour, changing from green to yellow then blue with increasing distance. Special processing is required for regions with no echo returns, e.g. hidden valleys, lakes and artificial plane areas like runways. Also moving targets like cars require special attention.

RESULTS OF FLIGHT TESTS

Several images taken during flight tests are shown in Figures 4 - 7. They all suffer loss of range information contained in the colour, due to the black and white reproduction. Figure 4 shows the mountain Waxenstein close to Garmisch, as seen by the radar system. The dots from the range cells line up to lines of constant distance from the radar system. The increment is 15 m. The spacing of the lines indicates the steepness of the terrain. A close spacing means a smooth terrain; where the lines are further apart there is a steeper change of the terrain. The two crossed bars indicate the axis of the aircraft, acting similarly to a pitch and roll indicator. At the upper right the ground speed is indicated, as determined from the doppler measurement. In the background there are some incomplete echoes returned from mountains at the range limit of the radar system.

For flights over relatively smooth terrain the algorithm is changed somewhat to eliminate the roughness of the lines of constant distance. Figure 5 shows the junction of two highways. Some details of the different tracks are visible. Figure 6 is taken somewhat later during a left turn leaving the highway. The radar image clearly shows the roll angle of the left turn. Finally Figure 7 was generated during the approach to an airfield. The runway, taxiways and free areas in front of buildings are characterized by practically no echo returns. Also this image gives a clear indication of the aircraft attitude.

ADDITIONAL CAPABILITIES

The capabilities of this radar system reach beyond the generation of 3-D images of the terrain. First the flight vector may be determined in magnitude (speed) and direction (azimuth and elevation). Second the ground speed may be approximated by the highest doppler shift received from the azimuth of the flight vector. A better estimate is obtained taking into account more data from different directions similarly to a doppler navigation system. Third the flight level above ground may be simply calculated from the time required for the first echo after transmitting a pulse. Finally the position of the aircraft relative to the ground may be obtained. An example is Figure 6, where the roll angle of a left turn is obvious. All this additional information may be very helpful for navigation purposes. The available range information and the ability to detect moving targets may proove useful in some other applications.

SUMMARY

The capabilities of our novel system to provide 3-D images of the terrain from an aircraft has been demonstrated. The additional information offered by this system may assist in navigation. Especially for helicopters such a system may reduce the problem of ground avoidance for low-level flights in darkness and poor visibility.

REFERENCES

1. Potter K.E., "Experimental Design Study of an Airborne Interferometer for Terrain Avoidance", 1977, Radar-77 pp. 508 - 512

2. Leysieffer H.,"In Fahrzeugen, insbesondere Luftfahrzeugen, angebrachte Einrichtung zur perspektivischen Darstellung eines Geländeausschnittes auf einem Radar-Bildschirm", 1978, patent DE 2543373

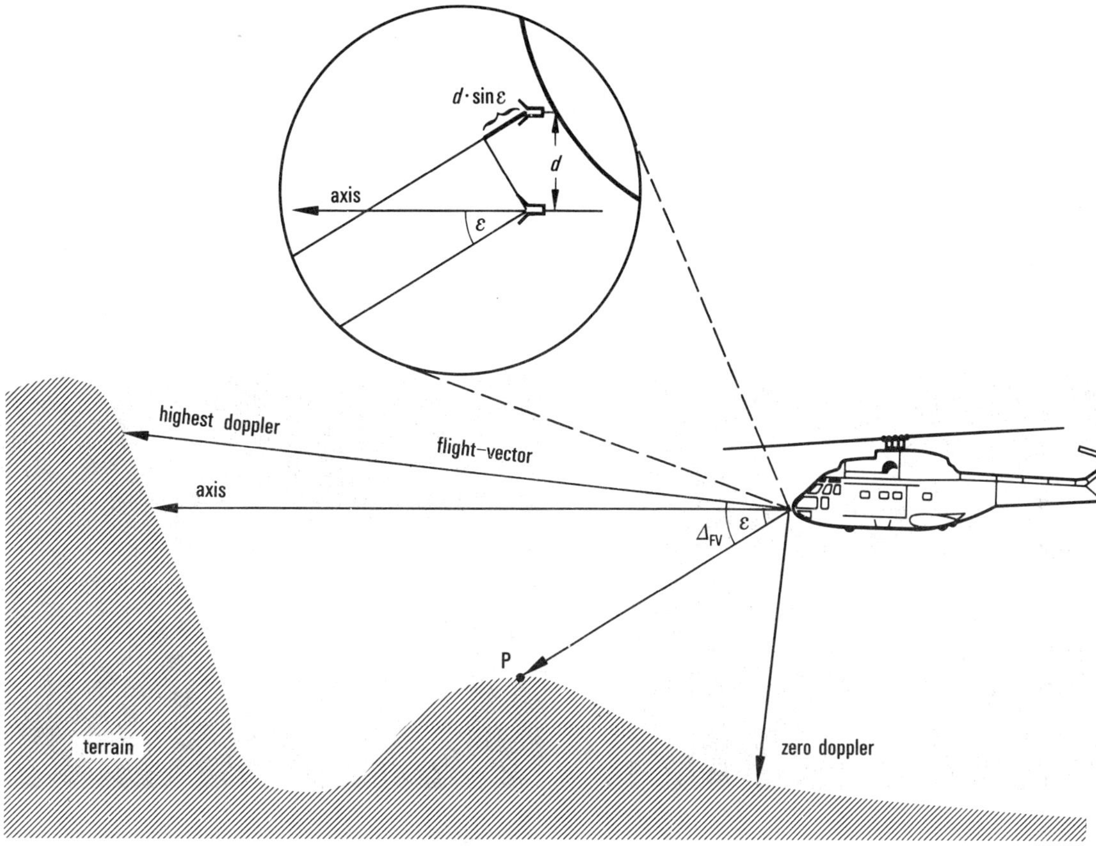

Figure 1: Elevation-resolution by interferometer-and doppler-processing

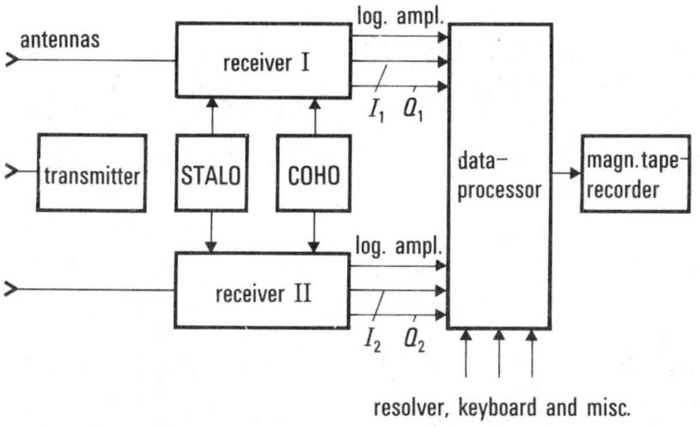

Figure 2: Block-diagramm of the flight-equipment

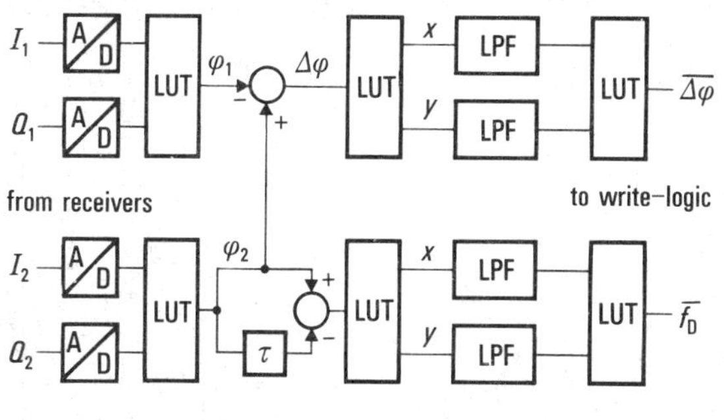

Figure 3: Processing of phase-difference and doppler-shift
LUT=look-up table
LPF=low-pass filter

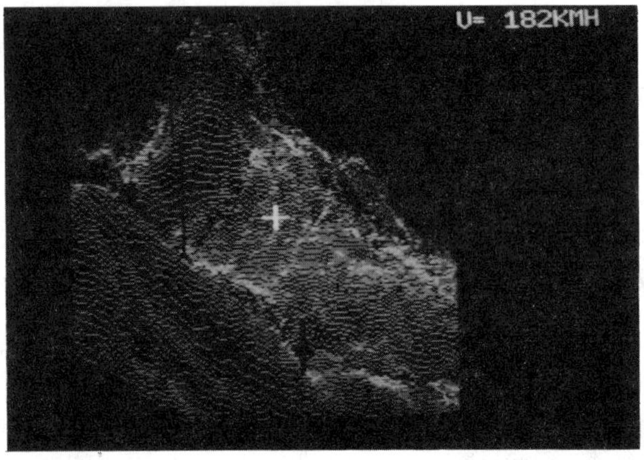

Figure 4

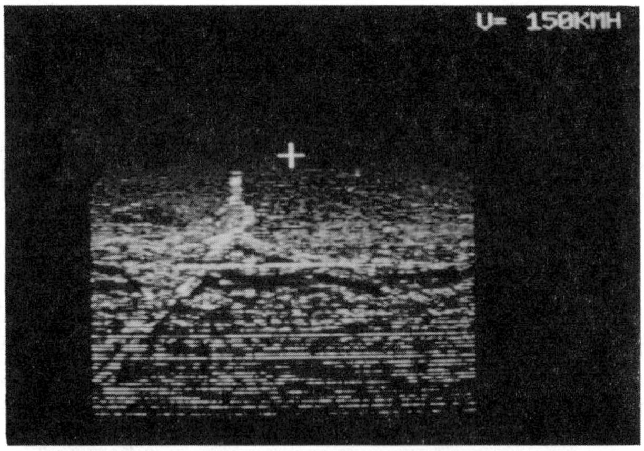

Figure 5

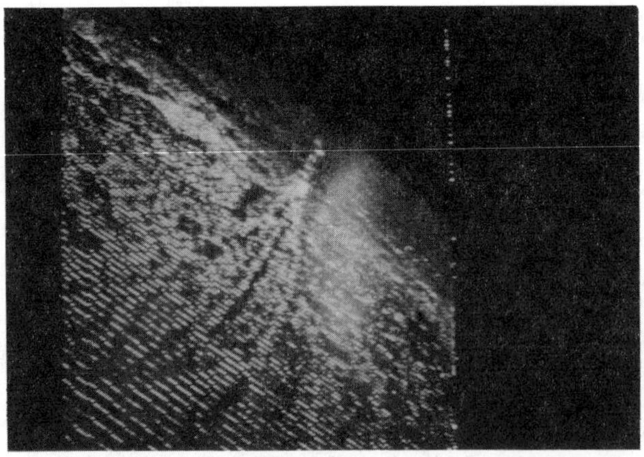

Figure 6

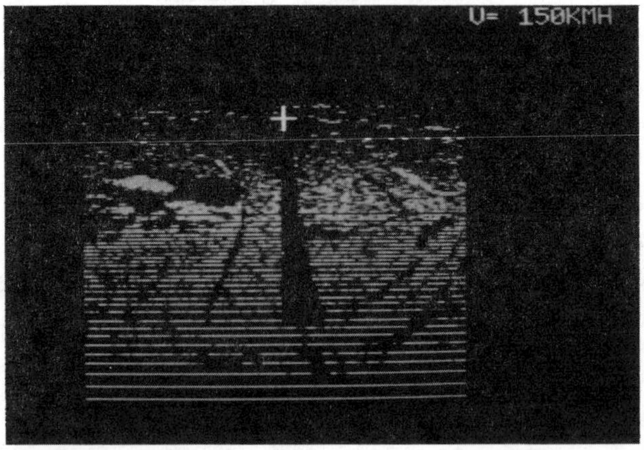

Figure 7

SUBOPTIMUM CLUTTER SUPPRESSION FOR AIRBORNE PHASED ARRAY RADARS

R. Klemm

Forschungsinstitut für Funk und Mathematik, D-5307 Wachtberg-Werthhoven

INTRODUCTION

General remarks

There are several motivations for carrying radar by an aircraft, one of them being the raised position which enables the radar to look from above. Thus detection of low flying aircraft and vehicles in a hilly landscape may be significantly improved. By doing so, one encounters two serious problems. First, the level of clutter returns will be much higher than for a ground based radar because the clutter aspect angles are larger. Secondly, the clutter returns are doppler-shifted due to the aircraft motion. The doppler shift is proportional to the directional cosine between the aircraft's velocity vector and the direction of the individual clutter element. Therefore, two-dimensional sampling of the reflected field (space-time) has to be carried out. This can be realized by using an active phased array antenna and coherent pulse trains. A variety of different space-time filters are discussed briefly in the paper. Suboptimum solutions for real-time processing are of special interest.

Literature

In the open literature a few proposals for solving the airborne clutter problem have been made. Apart from the well-known TACCAR /7/ and related approaches (Voles(1)) several authors (Andrews (2), Zeger and Burgess (3), Brennan et al. (4)) have been working on adaptive time-space filters for suppression of moving clutter. Brennan et al. (4) discussed the performance of the optimum two-dimensional clutter filter and its realisation by iterative algorithms. A similar approach has been made by Andrews (2). He optimises so called correction patterns to minimise the clutter power. Another wellknown technique called DPCA (displaced phase centre antenna) compensates for the platform motion by a shift of the array's aperture against the flight direction between every two pulses. Zeger and Burgess made some investigation and experiments on this technique in conjunction with an adaptive space-time filter. There is one disadvantage common to all of the quoted papers: They are influenced by the existing MTI-technique in that they consider only the case of two temporal samples, thus putting up with the unfavourable resolution properties of two-pulse clutter cancellers.

Optimum Clutter Suppression

Several aspects of optimum clutter suppression for airborne array antennas have been discussed by Klemm (5, 6). One major result of this investigation was that 2-dimensional filters operating on 2-dimensional clutter spectra are as efficient as one-dimensional filters operating on one-dimensional spectra. More specifically, appropriate signal processing applied to a moving array yields basically the same gain in signal-to-clutter ratio as if the platform were not moving. As a second achievement, it was found that the numbers of spatial and temporal samples should be roughly equal in order to maximise the gain. This is in contradiction to the papers quoted before. Fig. 1 shows the gain obtained by optimum processing vs doppler frequency plotted for four different look directions (0°, 30°, 60°, 90°; 0° being the flight direction, 90° abreast of it). As can be seen there are minima only at the frequencies belonging to the look directions, i.e., where the optimum processing is perfectly matched to the clutter. Everywhere else the white noise limited gain is reached. This result encourages us to look for suboptimum processing schemes which approximate the optimum gain curves but are much less expensive and time consuming than the optimum processor. The latter one is used in the following for comparison.

CLUTTER MODEL

Covariance matrix

The clutter model used in this investigation is given by the covariance matrix of spatial and temporal clutter samples of a moving array antenna.

Let $\underline{c}^* = (c^*_1\,(t_1)\dots c^*_N(t_1),\ c^*_1\,(t_2)\dots$
$\dots,\ c^*_1\,(t_M)\dots c^*_N(t_M))$

be the conjugate complex transpose vector of clutter samples, then the corresponding covariance matrix becomes

$$\underline{Q}_c = E\{\underline{c}\ \underline{c}^*\} = \begin{pmatrix} \underline{C}(0) & \underline{C}(\tau_0) & \dots & \underline{C}(\tau_0(M-1)) \\ \underline{C}^*(\tau_0) & \underline{C}(0) & \cdot & \\ \vdots & & \ddots & \vdots \\ \underline{C}^*(\tau_0(M-1)) & \dots & \dots & \underline{C}(0) \end{pmatrix}$$

with the elements

$$q_c(i,k,\tau) = q_0(\tau)\int_{\varphi=0}^{\pi} F(\varphi,\tau)L(\varphi)C_{ik}(\varphi,\tau)C_D(\varphi,\tau,v)\,d\varphi \quad (1)$$

where

- τ_0 = $t_n - t_{n-1}$ (equidistant pulses)
- $L(\varphi)$ = inhomogeneity function
- $F(\varphi,\tau)$ = $G(\varphi, 0)G(\varphi,\tau)$
- $G(\varphi,\tau)$ = transmit pattern moved during time τ
- $q_0(\tau)$ = correlation of clutter fluctuations
- $c_{ik}(\varphi,\tau)$ = phase relations between the i-th and k-th sensor and two pulses τ ms apart
- $C_D(\varphi,\tau,v)$ = doppler phase factor.

The submatrices $\underline{C}(\tau)$ contain the spatial information wheres the temporal information lies between these matrices.

In addition white sensor noise with power P_W is assumed. Thus we get

$$\underline{Q} = \underline{Q}_c + P_W\,\underline{I}\,. \quad (2)$$

A few assumptions have been made in order to obtain just a single integral in Eg. (1):

- the clutter returns are range gated
- the range gates are so small that the doppler frequency is constant with range
- clutter scatterers are uncorrelated.

Throughout the paper a line array along the aircraft's axis is assumed. The following parameters have been kept constant if not denoted separately:

clutter-to-noise ratio at each sensor	20 dB
frequency	3 GHZ
platform velocity	200 km/h
platform height	500 m
range	1000 m
number of sensors (N)	16
number of pulses (M), also length of doppler filters	8
spacing of sensors	$\lambda/2$
pulse rate	2 f_{max}(maximum clutter frequency)
transmit aperture (line source)	(N-1) λ /2
correlation of clutter fluctuations	1
$L(\varphi)$	1,φ = 0...180°

Eigenspectra

There are several possibilities of spectral representation of moving clutter returns, such as Fourier, maximum entropy, maximum likelihood a.s.f. (Klemm (6)). For the signal processor the most meaningful spectrum is the distribution of eigenvalues of the covariance matrix. It reveals the interior of the matrix much better than other spectral representations and tells something about the possibilities of signal processing. In particular it shows the number of degrees of freedom of the covariance matrix which is a measure for the minimum rank of the clutter filter. Fig. 2 and Fig. 3 give some more insight in the relations between spatial and temporal samples (sensors and pulses). The transient from the clutter eigenvalues to the noise part is seen clearly. The number of clutter eigenvalues never exeeds N+M, i.e. the sum of numbers of spatial and temporal samples. The product NM is always 128. In Fig. 2 we have the cases M> N, in Fig. 3 M< N. As a further result one can conclude that it is basically better to increase the number of sensors than the number of pulses (NM being constant). Experience has shown, however, that generating coherent pulses is much easier than implementing coherent receive channels. Fig. 4 shows the influence of the transmit aperture on the eigenspectrum (receiver aperture: 0.8 m). As can be seen the number of clutter eigenvalues becomes smaller with increasing transmit aperture. This is because the transmit beam acts as a spatial filter and, simultaneously, as a doppler filter. Therefore, clutter returns tend to become narrowband with increasing aperture which results in a smaller number of clutter eigenvalues.

SUBOPTIMUM APPROACHES

Auxiliary channel concepts

The gain of any linear processor $\underline{h}$ in signal-to-clutter ratio is given by

$$G = \frac{|\underline{h}^* \underline{s}|^2 \, P_c}{\underline{h}^* \, \underline{Q} \, \underline{h} \, P_s} \qquad (3)$$

with $\underline{s}$ being the space-time vector of a target signal and P_c and P_s the powers of clutter and signal respectively. In particular, the optimum processor is given by $\underline{h}_{opt} = \underline{Q}^{-1}\underline{b}$, with $\underline{b}$ being the azimuth-doppler matched filter vector. For $\underline{b} = \underline{s}$ (target match) the optimum gain is obtained:

$$G_{opt} = \underline{s}^* \, \underline{Q}^{-1} \, \underline{s} \, \frac{P_c}{P_s} \qquad (4)$$

This form has been used in Fig. 1. Applying a suitable pre-transform $\underline{T}$ to the data we get $\underline{g} = \underline{T}\,\underline{b}$, $\underline{N} = \underline{T}\,\underline{Q}\,\underline{T}^*$ and $\underline{z} = \underline{T}\,\underline{s}$. One may get a good approximation to the optimum gain by using the processor $\underline{y} = \underline{N}\,\underline{g}$ if the pre-transform $\underline{T}$ is properly chosen. $\underline{T}$ may contain auxiliary sensors, beams or any other linearly independent sets of weights /8/. The method is useful if the number of rows of $\underline{T}$ is small compared to the rank of $\underline{Q}$ (expense), but large compared to the number of clutter eigenvalues of $\underline{Q}$. It has been proven to be useful if there are only one or two eigenvalues (example: one interfering source radiating on an array with narrowband channels). However, as follows from the eigenspectra, the two conditions mentioned cannot be satisfied by moving clutter. We drop therefore any further consideration of these methods.

Cascading adaptive space and time processors

The idea is to cascade a purely spatial adaptive filter with a purely temporal one. The spatial filter is based on the spatial covariance matrix $\underline{C}$ (0) as described above: $\underline{k}(\varphi) = \underline{C}^{-1}(0) \cdot \underline{b}(\varphi)$, with $\underline{b}(\varphi)$ being a conventional beamformer. Then the elements of the temporal covariance matrix $\underline{Q}_t$ become $q_t(\tau) = \underline{k}^*(\varphi)\,\underline{C}(\tau)\,\underline{k}(\varphi)$. The adaptive temporal filter is $\underline{d} = \underline{Q}_t^{-1}\,\underline{f}$, $\underline{f}$ being a doppler filter vector. This sort of processing cannot be expected to work very well because the spatial adaptive filter perceives the clutter environment as omnidirectional due to a lack of spectral resolution. There is only some directivity in the field caused by the transmit antenna pattern. This, however, leads to a minimum in the response of the adaptive spatial processor in the look direction, thus reducing the power of the expected target signal. The spatial gain is therefore rather little. Furthermore, the temporal filter receives a broadband clutter signal because the spatial filter hasn't any selectivity which in turn leads to poor temporal gain. Fig. 5 shows a corresponding example (compare with Fig. 1). This approach is of no use.

Beamforming and temporal filtering

This is the conventional approach in that the spatial filter part is reduced to the beamformer operation. These results are, therefore, also valid for reflector antennas. On the contrary to the purely spatial adaptive filter the beamformer preserves the target signal power. Fig. 6 shows the gain curves of six different ways of processing. For comparison, two curves due to the optimum space-time processor (16/8,64/2) are plotted, the numbers denoting the numbers of spatial/temporal samples. Notice the lit similarity between BF + MTI (complex 2-pulse canceller) and 64/2. 64/2 is the gain curve attempted by the adaptive methods of several authors /2,3/. In conclusion one may say that all the effort put in adaptive algorithms and DPCA techniques is not worth doing if the clutter filter is based on two pulses only. BF + optimum temporal clutter suppression is almost identical to BF + MTI. The narrow margin between these two curves is all one can expect by any sort of clutter canceller after beamforming. BF + doppler filter without any clutter cancellation doesn't work well. BF + doppler nulling (orthogonal projection of received signals such that a doppler null at the beam frequency is created) generates nulls that are too narrow to cope with the doppler bandwidth cut out of the clutter spectrum by the beamformer. Therefore, no significant gain is obtained.

Space-time projection methods

The idea of these methods is to project the received data vector onto a clutter-free vector subspace. This is done by the operation $\underline{y} = \underline{P}\,\underline{x}$, with $\underline{P}$ being a projection matrix:

$$\underline{P} = \underline{I} - \underline{C}\,(\underline{C}^*\underline{C})^{-1}\,\underline{C}. \qquad (5)$$

$\underline{P}$ is orthogonal to $\underline{C}$. If $\underline{C}$ describes the subspace of clutter only then $\underline{y} = \underline{P}\,\underline{x}$ is a clutter-free projection of the data. The problem is how to choose $\underline{C}$. One possibility might be a set of azimuth-doppler vectors (combination of beamformer and doppler filters) matched to a number of azimuth angles and to the corresponding doppler frequencies. Then the projection matrix provides a set of nulls along the diagonal of the 2-dimensional φ-v-spectrum. As can be seen from Fig. 7 this kind of azimuth-doppler nulling is not satisfactory. This can be interpreted in several ways. If the clutter background is homogeneous as assumed in this paper the φ-v-spectrum is continuous; a continuous spectrum, however, can hardly be suppressed by setting single point shaped nulls. Another interpretation is that the subspace given by C which was found by physical intuition does not agree well enough with the actual clutter subspace of Q. Then P is not perfectly orthogonal to the clutter part of Q, thus letting some clutter remainders pass through and causing attenuation of target signals.

A much more accurate separation of the subspaces of clutter and signals is obtained if C contains the eigenvectors belonging to the clutter eigenvalues. In Fig. 7, there are four curves showing the performance of the eigenvector method (EVM). The corresponding eigenspectrum was shown in Fig. 2 (16/8). Best performance is obtained if the number of eigenvectors used is roughly equal to the number of significant eigenvalues (N_e = 16). If N_e is too large (N_e = 32) then the signal subspace is reduced, thus causing loss in signal power. If N_e is too small a part of the clutter power remains in the signal subspace.

CONCLUSIONS

A model study has been carried out in order to investigate the performance of a variety of subobtimum space-time filters for clutter suppression in moving phased array radars. Some results can be summarized as follows:

Adaptive space-time filters are approaches to the optimum filter (Q^{-1}) as far as a reasonable estimate of Q can be obtained (spatial variability negligible against processing speed). Adaptive filtering is worth doing only if the numbers of spatial and temporal samples (sensors and pulses) are of the same order of magnitude. Adaptive filters are insensitive to any unknown distribution of clutter scatterers and, in particular, to sidelobe clutter.

The performance of the optimum processor (Q^{-1}) can be approximated very well by a space-time projection matrix P based on the clutter eigenvectors of Q if P is properly matched, i.e. if Q is known and the correct number of significant eigenvectors is chosen. The number of clutter eigenvalues depends on several parameters, such as transmit aperture, clutter fluctuations and inhomogeneities. Filtering as such needs comparatively few operations. However, the computation of the eigenvectors is fairly complicated and hardly of use for real-time operations. The weighting coefficients of P might be calculated on a model basis and be stored, but then the filter is very sensitive to any wrong assumption on the clutter environment.
There are several possible ways of clutter filtering that do not work sufficiently and should be rejected:
- cascading a d a p t i v e spatial and temporal filters
- azimuth-doppler nulling
- beamforming and doppler filters only (no clutter cancellation)

Conventional beamforming followed by a complex 2-pulse canceller and a doppler filter bank (BF + MTI) is a well known fair approach that works rather outside a relatively broad stop band. The advantage of adaptive temporal filtering after beamforming compared with BF + MTI is very small. Therefore, clutter cancellers using more than 2 pulses are not worth considering. All beamformer + temporal filter concepts are sensitive to inhomogeneities of the reflecting background.

The paper dealt with a survey over several possible processor structures. However, a final recipe for the design of an "ideal" processor having the nice properties of the optimum clutter filter, but at much lesser expense, has not been presented and, very likely, doesn't exist. So the designer is left with the choice between the expensive fully adaptive solution, which possibly can be implemented for small arrays, the nonadaptive eigenvector projection method and, for large arrays and reflector antennas, the simple BF + 2-pulse canceller concept. A final decision can be taken only on the basis of experiental data which so far we do not have.

REFERENCES

1. R. Voles, 1973, "New approach to MTI clutter locking", Proc. IEE, Vol. 120, No 11, pp. 1383 - 1390

2. G.A. Andrews, 1978, "Radar Antenna Pattern Design for Platform Motion Compensation", IEEE Trans. AP - 26, No. 4, pp. 566 - 571.

3. A.E. Zeger, L.R. Burgess, 1974, "An adaptive AMTI Radar Antenna Array", NAECON '74 Record, pp. 126 - 133.

4. L.E. Brennan, J.D. Mallett, I.S. Reed, 1976, "Adaptive Arrays in Airborne MTI Radar", IEEE Trans. AP - 24, No. 5, pp. 607 - 615.

5. R. Klemm 1982, "Problems of clutter suppression in moving array antennas", Proc. of the NTG-meeting on Antennas, Baden-Baden, Germany

6. R. Klemm, 1982, "Optimum Clutter Suppression in Airborne Phased Array Radars. Proc. of IEEE-ICASSP 82, Paris, France.

7. M. Skolnik, 1970, Radar Handbook. McGraw-Hill pp. 17 - 32.

8. R. Klemm, 1980, "Horizontal Array Gain in Shallow Water. Signal Processing 2, pp. 347 - 360.

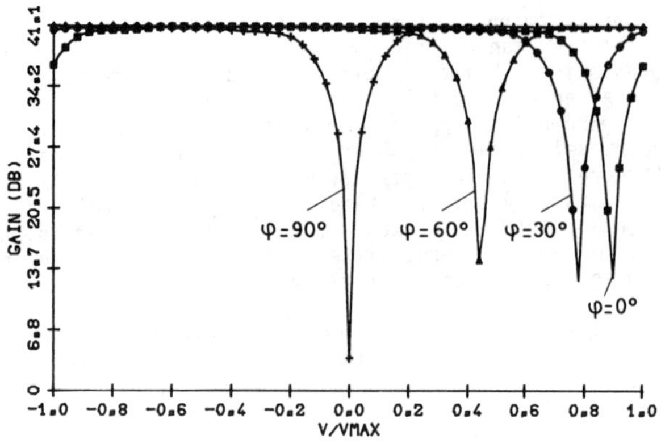

Fig. 1: Optimum clutter suppression

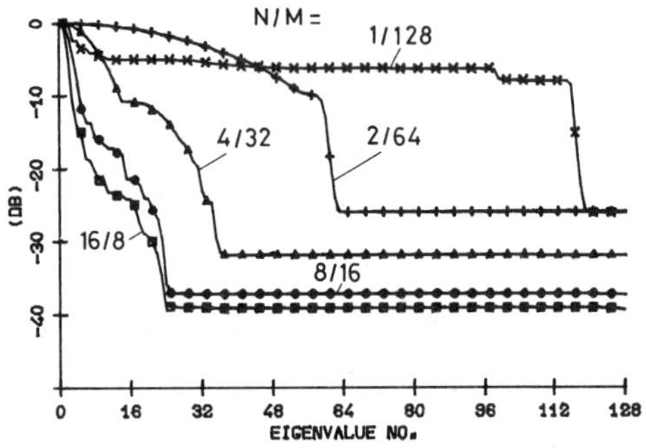

Fig. 2: Distribution of samples (M>N)

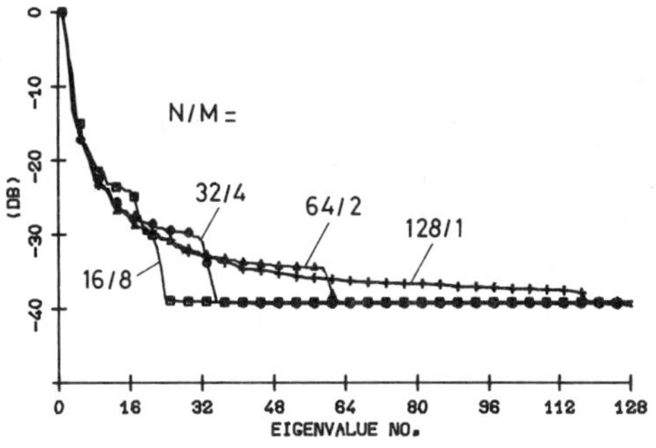

Fig. 3: Distribution of samples (M<N)

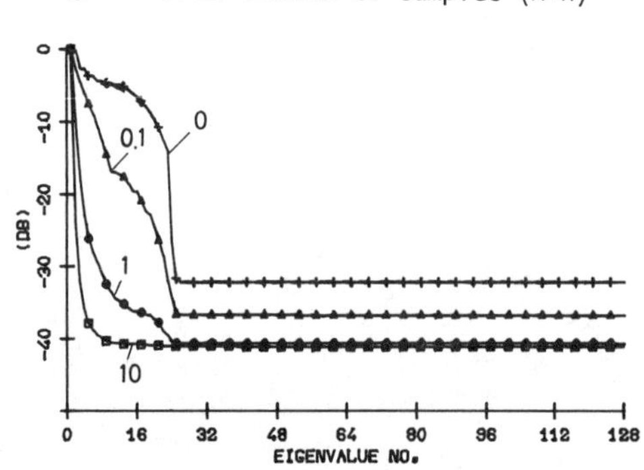

Fig. 4: Influence of transmit aperture [m]

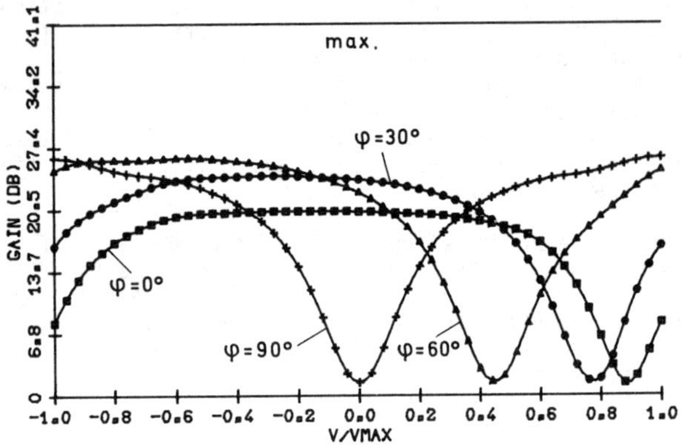

Fig. 5: Adaptive spatial and temporal filtering

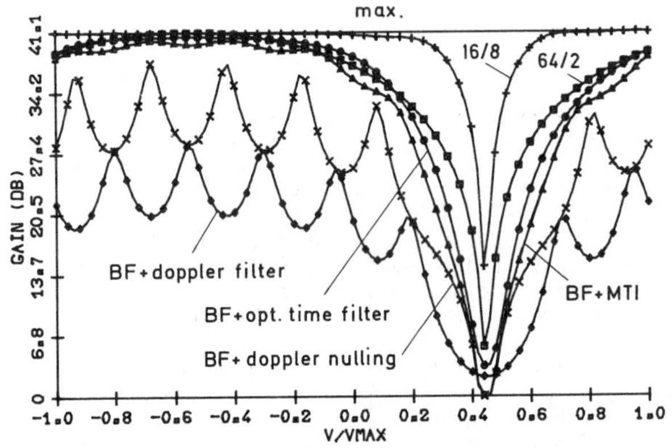

Fig. 6: Pre-beamforming and temporal filtering

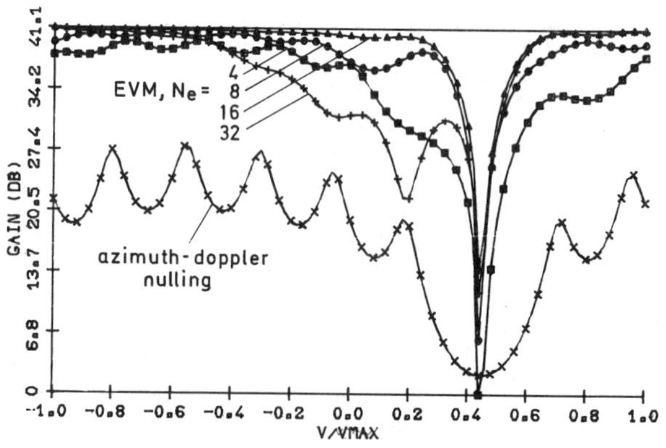

Fig. 7: Projection methods

AMBIGUITY FUNCTIONS OF COMPLEMENTARY SERIES

J A Cloke

Ministry of Defence, UK

INTRODUCTION

A pair of complementary series has the property that the range sidelobes in the autocorrelation function (ACF) for one member of the pair exactly cancel those of the other member. If the two ACFs are added, a new ACF is produced with identically zero sidelobes. The questions which arise are thus what are the drawbacks to such an attractive scheme (both in theoretical and practical terms), and also what properties do the ambiguity functions of such series display. This paper examines the ambiguity functions of complementary series and thus gives a more complete view of the possible application of such series to particular radar and communication systems.

Autocorrelation and Ambiguity Functions. It is well known that the so called "matched" filter is the optimum linear filter for extracting a known signal from a background of white noise. Even in situations where a matched filter is not used, or where the noise background is not white, the matched filter is a standard against which other processing schemes can be compared. The output from a matched filter is the autocorrelation function of the input signal (ie the signal to which it is matched) provided there is no noise present and no doppler shift. The ambiguity function is a description of the output from a matched filter in the presence of a doppler frequency shift.

Complementary Series. Complementary series can produce ACFs with zero (time) sidelobes. This means that a large target will never mask an adjacent smaller target, or produce spurious targets at other ranges. For communications applications this means that there will be no ambiguous points of synchronism for a correlation receiver. A very simple example of complementary series is shown in Fig 1. Such series were first reported by Golay (1) and are sometimes known as Golay codes. Golay gives the theoretical bases of complementary series and also methods of constructing such series, confining his attention to the ACF. Later authors have expanded Golay's work but none deal in detail with the ambiguity functions of either the original Golay series or of the extensions to the theory which they report. (Tseng and Lui (2), Sivaswamy (3, 4), Hollis (5)). The allowable lengths and codes are shown in Fig 2.

Digital Phase Coded Waveforms

Two groups of binary codes which have been extensively studied are the Barker Codes and Maximal length (or M) sequences. The Barker codes will now be briefly reviewed in order to demonstrate the techniques used later and to obtain ambiguity function diagrams for comparison with those of the complementary series.

Barker Codes. The Barker Codes are optimum in the sense that the peak of the ACF is of height N (where N is the number of sub pulses) whilst all the sidelobes of the ACF are $\leqslant 1$. Only a small number of these codes exist, and none with a length greater than 13 is known. The ambiguity functions for the 5, 11 and 13 bit Barker Codes are shown in Figs 3 to 5. The point to notice from these figs is that although the sidelobes along the wd = 0 axis are only of height 1/N (ie 20%, 9%, 7.7% of the main lobe height) there are very large sidelobes further out in the doppler plane. In many cases these sidelobes are comparable to the main lobe height and could give rise to ambiguity problems.

AMBIGUITY FUNCTIONS OF COMPLEMENTARY SERIES

The allowable lengths and codes are shown in Fig 2. Ambiguity functions for the 2 10-bit codes are at Figs 6 and 7. It is clear that the overall sidelobe level is substantially lower than for the equivalent Barker code, and that the true zero along the time axis can be seen. There is a very clear central peak.

16-bit Complementary Series: 1. The 16-bit codes are constructed from shorter codes as described by Golay (1). Fig 2 shows the "original" 16 bit codes constructed from the fundamental 2 bit codes (++,+-) by way of the 4- and 8- bit codes. The ambiguity function for this complementary pair is shown at Fig 8. There is complete cancellation over the entire τ, wd plane for all $\tau > 8$ bits (ie half the code length). Such a cancellation was predicted by Turyn (6), but its importance does not seem to have been emphasised either by Turyn or later authors who have considered the properties of complementary series. We already have complete cancellation along the time axis; we now see that the ambiguity function is entirely confined to the area it would occupy if it were only half its length (ie 8 bits in this case). The cancellation is only achieved at the expense of a somewhat higher average sidelobe level in the central region, however, and since the total volume cannot be reduced the effect of truly evenly distributed (in the τ, wd plane) clutter will be unaffected. The fact remains, however, that the ambiguity diagram has a very desirable appearance which approximates both the "ideal thumbtack" and "ideal impulse" function (see Skolnik (7)).

16-bit Complementary Series: 2. Golay (1) shows that by means of suitable transformations, any pair of complementary codes may form the basis of 2^6 different pairs of the same length. The question therefore arises as to the effect of these transformations on the ambiguity function. Because of the symmetry of the ambiguity function, it is clear that identical reversals or inversions of both codes of a pair will have no effect on the ambiguity function. Also, repeated application of the transformations often leads back to a pair of codes obtained earlier. The 2 transformations which appeared to offer the greatest differences from the original codes were (a) inversion of only one of the pair of codes and (b) inversion of alternate elements of each code. The original series and the results of these transformations are shown below:

a. Original series	+++-++-++++---+- +++-++-+---+++-+	16 bit (1)
b. One inversion	+++-++-++++---+- ---+--+-+++---+-	16 bit (2)
c. Alternate bits inverted	+-+++---+-++-+++ +-+++----+--+---	16 bit (3)

To the casual observer these pairs of codes are extremely dissimilar, especially a and c. Surprisingly the ambiguity diagrams are absolutely identical to that for the original series (fig 8). Since the 16 bit complementary pairs (and all complementary pairs of length 2^N) are constructed from shorter series and may be transformed to produce whole families of complementary pairs it was assumed that a fruitful area of study would be the precise affect of the construction details and transformations on the ambiguity diagram. The discovery that the transformations have no effect whatever means that there is in fact only one 16-bit complementary series ambiguity function.

SUGGESTED IMPLEMENTATION SCHEMES FOR COMPLEMENTARY SERIES

The justification for employing complementary series is that the time sidelobes of the ambiguity function (and ACF) are zero. This effect is achieved with a matched filter operation, and only two implementations are able to achieve perfect sidelobe cancellation:

Alternate PRI Code Transmission. This scheme (which is only suitable for radar use) involves transmitting each of the pairs of codes on alternate pri and switching in the appropriate matched filter on reception.

Two Frequency Operation. Here each of the codes is modulated onto a different carrier. The scheme would be suitable for either radar or communications use.

Each of the above schemes requires a correlation receiver or a matched filter for each of the pair of complementary codes being used. The matched filters will be the same as those for any phase coded waveform and may have one of the general forms shown in Fig 9. Because binary phase coding is employed, a digital implementation using shift registers or charge coupled devices (CCD) would be appropriate.

The combination of appropriate matched filters to give a full complementary series decoder is shown at Fig 10. There are advantages and disadvantages with each system, both as regards complexity of implementation and performance. The main problem with the alternate pri system is the need for an extra delay element of 1 pri, whilst the problem with the 2-frequency method is the need to use 2 frequencies (although this may be an advantage from a security or counter-measures viewpoint). From a performance point of view both systems may suffer from difficulties with correlating the two independent ambiguity functions: the clutter returns may be decorrelated when 2 frequencies are used, leading to imperfect cancellation; in the alternate pri system the target may move significantly between pri's, again leading to imperfect cancellation.

Typical System Parameters. For a typical radar application at either microwave or HF frequencies, the maximum doppler frequency encountered will only be a fraction of the frequency axis shown in figs 3-8. A sub-pulse length of 1 μS, for example, will generate an ambiguity diagram frequency axis of length 1 MHz, although the maximum doppler encountered in practice will be of the order of 10 KHz.

Figure 11 shows the 16-bit complementary series ambiguity diagram with lines drawn at $\pm$ 1.5% of the doppler domain. The scale of the diagram is not appropriate for investigating the off-axis sidelobes in this very narrow region, and further study would be needed on this point. Some rough estimates indicate that the off-axis sidelobes will be about 20dB down on the main lobe at 1.5% of the maximum doppler.

In the two-frequency implementation, which would be applicable to a communications system, the 2 codes are modulated onto different carriers; the doppler shifts will not be equal and so the individual ambiguity diagrams will not cancel. They will cancel exactly on the zero doppler axis, however, and there will be good cancellation close to this axis. Now at 5 GHz, a reasonable frequency separation would be say 30 MHz. This figure leads to a doppler difference of only 0.6%; a frequency separation of less than 30 MHz would give a correspondingly lower doppler difference. Thus the two ambiguity diagrams are not only restricted to the region very close to the time axis ($\pm$ 0.4%), but their individual doppler shifts will be within 0.6% of one another. It appears, therefore, that such a 2-frequency implementation is feasible for a satellite communication system, and that very good sidelobe cancellation should be achieved. Rihaczek (8) shows that the effect of adding ambiguity functions with different doppler shifts, in the general case, is to cause a smearing of the off-axis sidelobes, and results in a less spiky, more noiselike ambiguity surface.

PERFORMANCE OF COMPLEMENTARY SERIES WAVEFORMS

From the foregoing investigations we may approximate the ambiguity surface of a complementary series as shown in fig 12. Over the doppler region of interest we see that the function has a "ridge" appearance giving excellent range resolution with poor doppler resolution, but with no ambiguity. The sidelobes are very low, and are zero on the time axis and over the whole τ, wd plane for time delays greater than half the pulse length. Such an ambiguity function will give excellent clutter rejection due to the low sidelobe level and very narrow central ridge. In addition the complete absence of time sidelobes will enable closely spaced targets to be distinguished even when there is a large disparity in their reflectivity. There is no range-doppler coupling, so range can be determined accurately at all times, at the expense of doppler resolution. For the range of doppler frequencies encountered there will be no need to implement separate doppler channels, so the additional complexity of using complementary series will not be exacerbated by such a requirement. The ambiguity diagram shown has been achieved with no additional weighting networks.

REFERENCES

1. Golay, M.J.E., 1961, "Complementary Series", IRE Trans. Inf. Th.

2. Tseng and Lui, 1972, "Complementary Sets of Sequences", IEEE Trans. Inf. Th. IT18.

3. Sivaswamy, R., 1978, "Digital and Analogue Sub-complementary Series for Pulse Compression", IEEE Trans. Aerospace & Electronics. AES 14, 343-350 (correction p 692).

4. Sivaswamy, R., 1978, "Multiphase Complementary Codes", IEEE Trans. Inf. Th. Vol 24, 546-552.

5. Hollis, E.E., 1967, "Constructing Broad Sense Complementary Series of Length 4R", IEEE Trans. Aerospace & Electronics AES 3.

6. Turyn, R., 1973, "Ambiguity Functions of Complementary Series", IEEE Trans. Inf. Th. IT9.

7. Skolnik, M.I., 1970, "Radar Handbook", McGraw Hill.

8. Rihaczek, A.W., 1969, "Principles of High Resolution Radar", McGraw Hill.

ACKNOWLEDGEMENT. The research for this paper was carried out at the University of Birmingham and the Royal Signals and Radar Establishment as part of the author's work for the degree of Master of Science in Electronic and Electrical Systems (Radar Specialisation) in 1979.

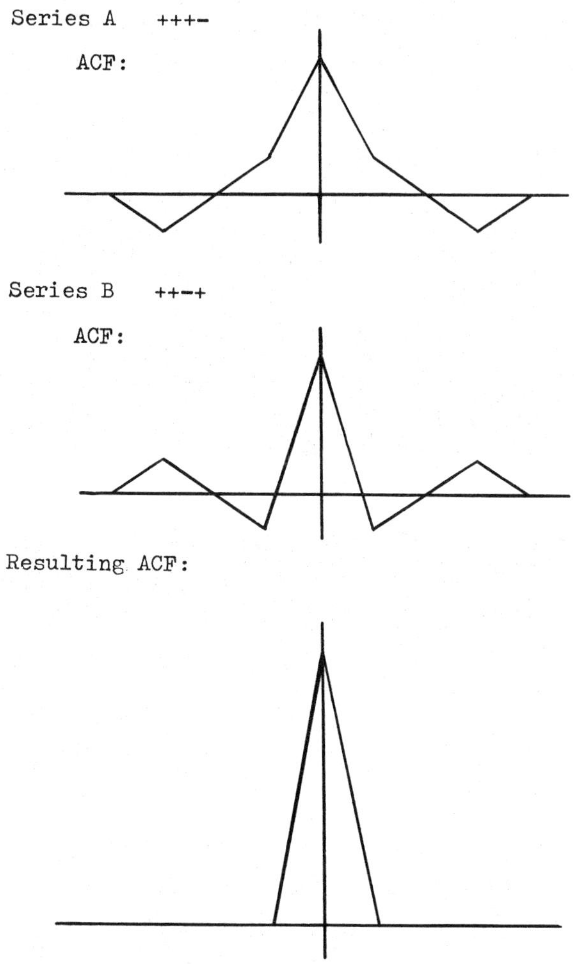

Figure 1 Simple Example of Complementary Series Autocorrelation Function

2 ++
+-

4 +++-
++-+ (CONSTRUCTED FROM 2 BIT)

8 +++-++-+
+++---+- (CONSTRUCTED FROM 4 BIT)

10 (1) +--+-+----+
+--------++- (2) +-+-++++--
++++-++--+

16 +++-++-++++---+-
+++-++-+---++++-+ (CONSTRUCTED FROM 8 BIT)

18 NONE EXIST

20 CONSTRUCTED FROM 10 BIT.
ALSO FOR 40, 80, 160 ETC

26 NONE FOUND. ALSO TRUE OF 34, 36, 50 ETC

32 CONSTRUCTED FROM 16 BIT.
ALSO FOR ALL HIGHER POWERS OF 2

Figure 2 Allowable Lengths and Codes

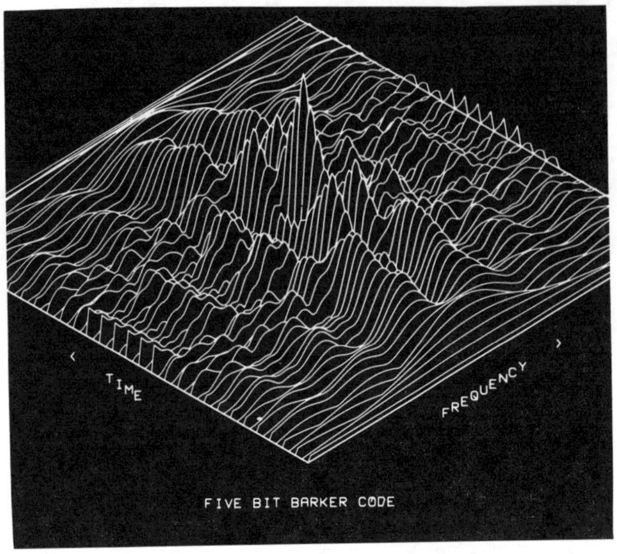

Figure 3. Five Bit Barker Code

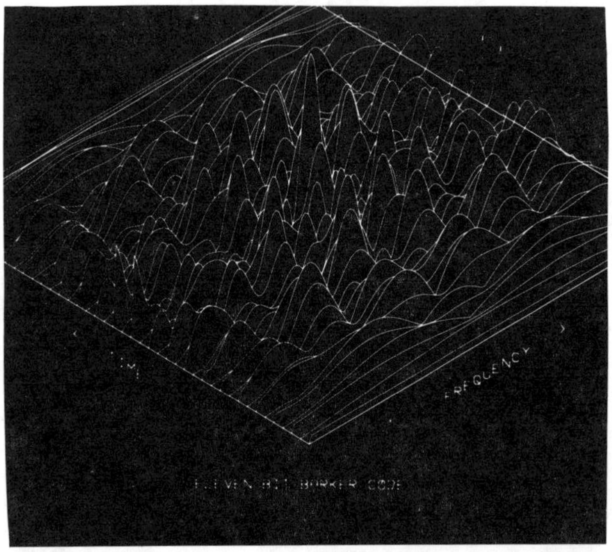

Figure 4. Eleven Bit Barker Code

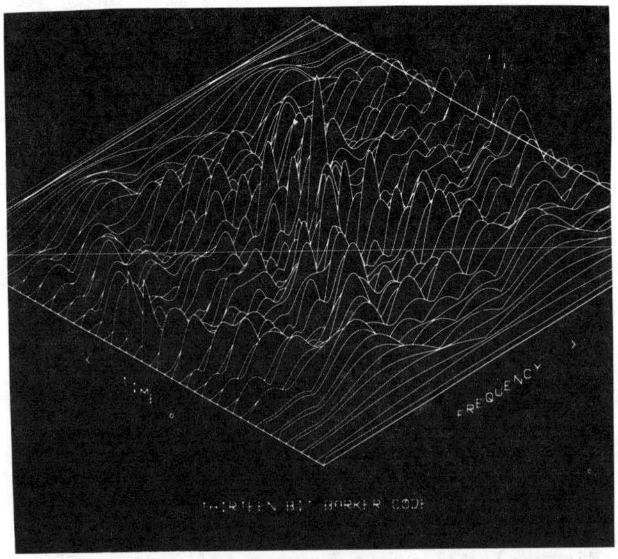

Figure 5. Thirteen Bit Barker Code

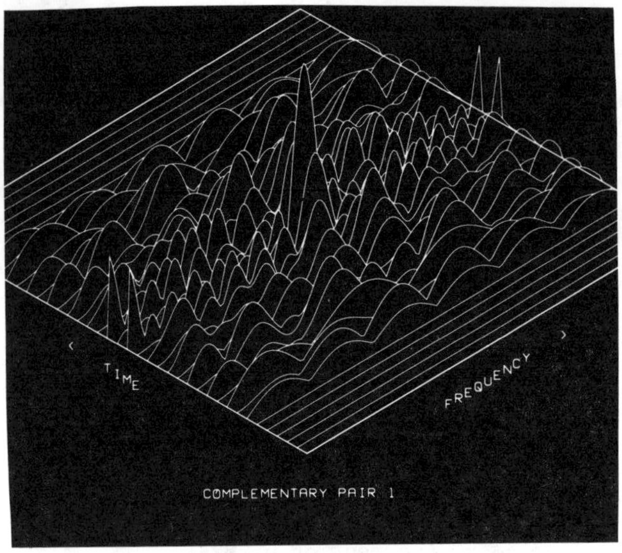

Figure 6. 10 Bit Complementary Pair (A)

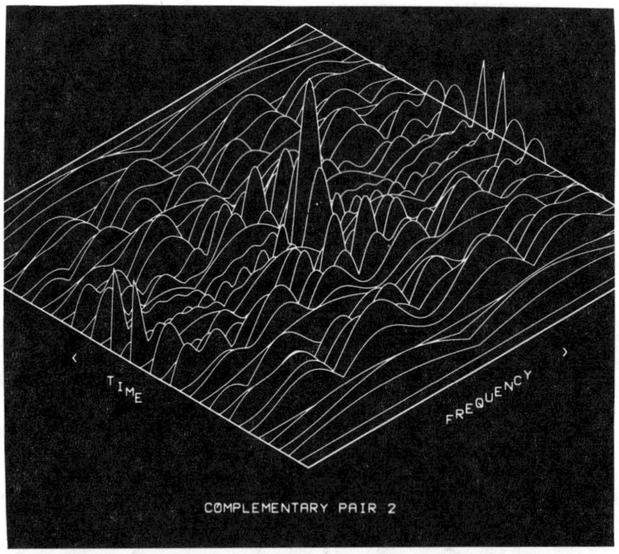

Figure 7. 10 Bit Complementary Pair (B)

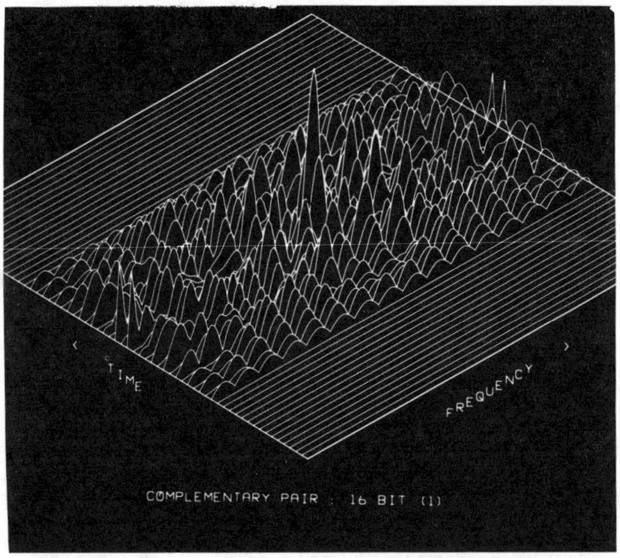

Figure 8. 16 Bit Complementary Pair

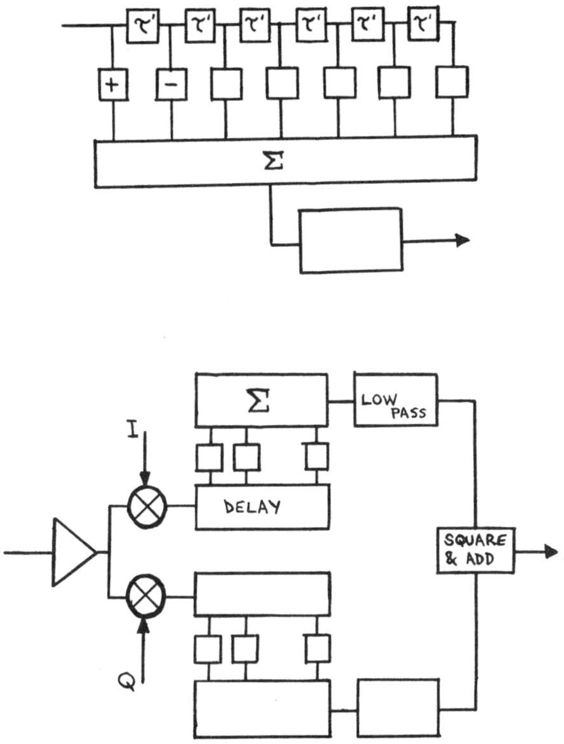

Figure 9. Matched Filters for Codes

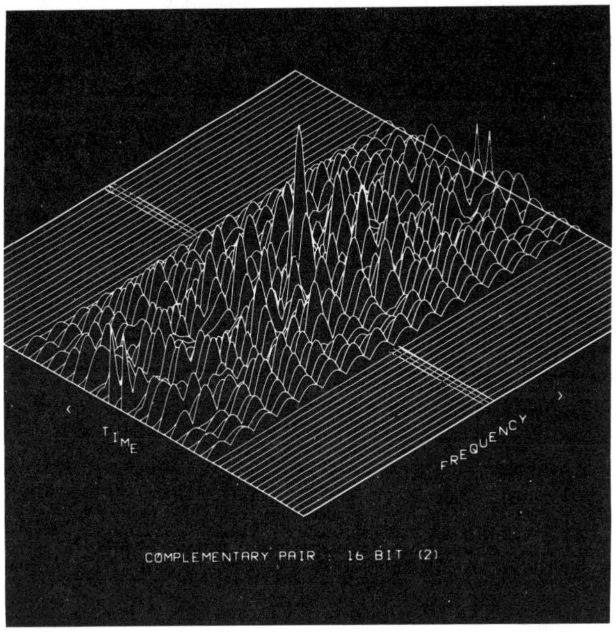

Figure 11. Practical Doppler Width

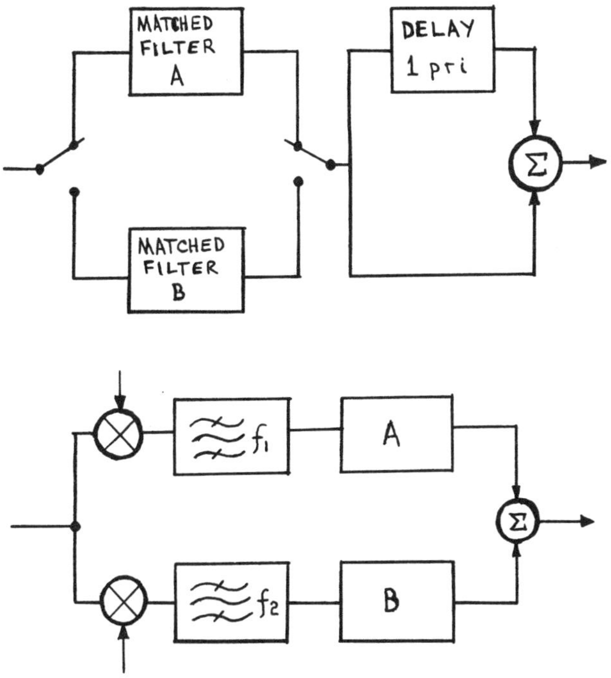

Figure 10. Complementary Code Receivers

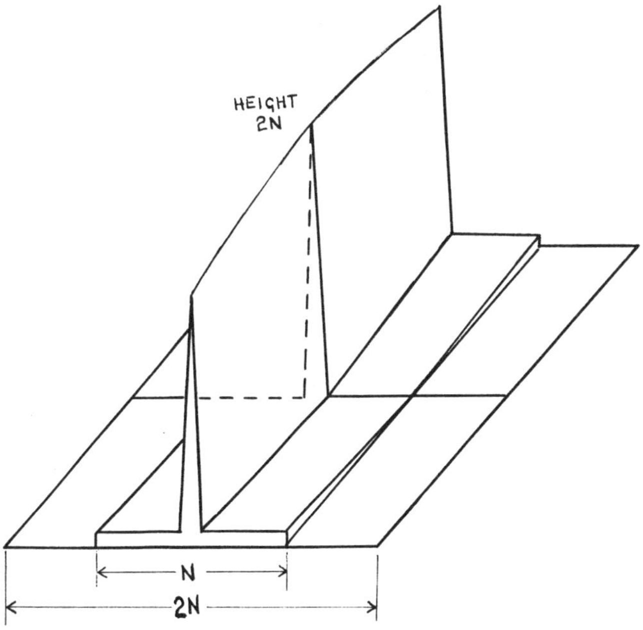

Figure 12. Approximate Ambiguity Surface

MTI-FILTERING FOR MULTIPLE TIME AROUND CLUTTER SUPPRESSION IN COHERENT ON RECEIVE RADARS

Stefan Carlsson

Dept. of Telecommunication Theory, Royal Institute of Technology, Stockholm, Sweden

INTRODUCTION

In a radar equipped with a transmitter that is not coherent from pulse to pulse, e.g a magnetron, coherence on receive can be acheived by locking a coherent oscillator to the phase of the tranmitted pulse. Coherent signal processing,MTI-filtering, can then be applied to the returned signals. Since the phase of each transmitted pulse is random, echoes from targets outside the 1:st ambiguous range interval will be mixed with the wrong phase in the receiver. These echoes will then be randomly phase modulated and appear as white noise. A conventional MTI-filter will suppress clutter from the 1:st ambiguous range interval only, while so called multiple time around clutter from outside the 1:st ambiguous range interval will appear in the passband of the filter. This inability to suppress multiple time around clutter has been cited as the main difference in performance between a coherent on receive radar and a radar that is fully coherent, i.e where there is no phase-change between transmitted pulses. Barton (1).

In this paper however, we will show that it is in fact possible to acheive 1:st and multiple time around clutter suppression simultaneously in a radar that is coherent on receive only, by measuring the phase-shift between succesively transmitted pulses and using this information in the calculation of the MTI-filter weights.

MTI-filtering

Let $S_n^{(i)}$ be the complex amplitude of the echo returned from the i:th ambiguous range interval after transmission of the n:th pulse, using a fully coherent radar. Let p_n be the phase of the n:th transmitted pulse in a coherent on receive radar. Echoes from the i:th ambiguous range interval will then have the transmitted phase p_{n-i+1} and will be mixed in the receiver with the phase p_n. They will then be phase-shifted $p_n - p_{n-i+1}$ relative to the echo returned using the fully coherent radar. The complex amplitude of the echo from the i:th ambiguous range interval using a coherent on receive radar can then be written:

$$e^{j(p_n - p_{n-i+1})} S_n^{(i)}$$

For simplicity we will consider echoes from the 1:st and 2:nd range interval only. Defining:

$$q_n = e^{j(p_n - p_{n-1})}$$

the sum of the 1:st and 2:nd time around echoes can be written:

$$S_n^{(1)} + q_n S_n^{(2)}$$

If this signal is filtered in an MTI-filter with L delays and filter weights $c_0, c_1, \dots c_L$ the output will be:

$$\sum_{k=0}^{L} c_k S_{n-k}^{(1)} + c_k q_{n-k} S_{n-k}^{(2)}$$

The frequency response for 1:st and 2:nd time around echoes will be:

$$H^{(1)}(e^{j\omega}) = \sum_{k=0}^{L} c_k e^{-j\omega k} \qquad (1)$$

$$H^{(2)}(e^{j\omega}) = \sum_{k=0}^{L} c_k q_{n-k} e^{-j\omega k} \qquad (2)$$

From these equations we see that in order to specify the filter response for both 1:st and 2:nd time around echoes we need information about the phase shifts q_{n-k}. This information can be obtained through the use of an extra oscillator that measures the phase shift of the coherent oscillator in the receiver.

Note that since the coefficents c_k depends on the time-varying phase shifts q_n we will get time varying frequency responses.

Given information about the phaseshifts q_n, we can design the responses $H^{(1)}$ and $H^{(2)}$ by proper location of the zeroes of the z-transforms:

$$H^{(1)}(z) = \sum_{k=0}^{L} c_k z^{-k} \qquad (3)$$

$$H^{(2)}(z) = \sum_{k=0}^{L} c_k q_{n-k} z^{-k} \qquad (4)$$

In order to completely specify the responses $H^{(1)}$ and $H^{(2)}$ we would like to choose L zeroes of $H^{(1)}$ and $H^{(2)}$ independently. This can however not be done since both depends on the same available weights c_k. We can only specify L1 zeroes for $H^{(1)}$ and L-L1 zeroes for $H^{(2)}$. The remaining L - L1 zeroes for $H^{(1)}$ and L1 zeroes for $H^{(2)}$ will depend on the values of q_{n-k}. The generalization to the case of any number of range ambiguities is quite straightforward. Defining:

$$q_n^{(i)} = e^{j(p_n - p_{n-i+1})}$$

The transfer function for echoes from the i:th ambiguous range interval will be:

$$H^{(i)}(z) = \sum_{k=0}^{L} c_k q_{n-k}^{(i)} z^{-k}$$

The main difference compared to the fully coherent case is that we need longer delays L in order to specify the zeroes of the $H^{(i)}$:s since they are independent. Because of the independence of the transfer functions $H^{(i)}$ however, we are completely free to choose very broad stopbands for multiple time around clutter without affecting the frequency response for 1:st time around echoes. This effect will be seen in the evaluations in the last section.

Cascaded 1:st and 2:nd time around filters

Specification of zeroes for $H^{(1)}$ and $H^{(2)}$ leads to a system of linear equations that have to be solved at each time n. In this section we will show that it is possible to recursively compute the filter weights that assigns specified zeroes to $H^{(1)}$ and $H^{(2)}$.

We will consider a cascaded structure for 1:st and 2:nd time around filters as shown in fig,(1). Suppose we have determined a filter with weights $c_0 \ldots c_{L1}$ that assigns zeroes to $H^{(1)}$. These weights will be kept constant for all times. In cascade with this filter we put single delay filters with weights $a_{1,n}\ b_{1,n} \ldots a_{L2,n}\ b_{L2,n}$. These time varying weights will now be chosen in order to assign the specified zeroes for $H^{(2)}$.

The output from the 1:st time around filter is

$$U_{0,n} = \sum_{k=0}^{L1} c_k S_{n-k} \quad (5)$$

The outputs from the single delay filters are

$$U_{i,n} = a_{i,n} U_{i-1,n} + b_{i,n} U_{i-1,n-1} \quad (6)$$

$$i = 1 \ldots L2$$

Denoting by $H^{(1)}_{i,n}$ and $H^{(2)}_{i,n}$ the transfer functions between signals S_n and $U_{i,n}$ for 1:st and 2:nd time around echoes respectively, we get:

$$H^{(1)}_{0,n}(z) = \sum_{k=0}^{L1} c_k z^{-k} \quad (7)$$

$$H^{(2)}_{0,n}(z) = \sum_{k=0}^{L1} c_k q_{n-k} z^{-k} \quad (8)$$

From eq. (6) we get:

$$H^{(1)}_{i,n}(z) = a_{i,n} H^{(1)}_{i-1,n}(z) + b_{i,n} z^{-1} H^{(1)}_{i-1,n-1}(z) \quad (9)$$

$$H^{(2)}_{i,n}(z) = a_{i,n} H^{(2)}_{i-1,n}(z) + b_{i,n} z^{-1} H^{(2)}_{i-1,n-1}(z) \quad (10)$$

From these relations we see that if z_0 is a zero of $H^{(1)}_{i-1,n}$ or $H^{(2)}_{i-1,n}$ for all n, it will also be a zero of $H^{(1)}_{i,n}$ or $H^{(2)}_{i,n}$.

Continuing the recursion we see that it will also be a zero of $H^{(1)}_{L2,n}$ or $H^{(2)}_{L2,n}$. This means that the fixed L1 zeroes for $H^{(1)}_{0,n}$ that were chosen with the time invariant weights c_k, will also be zeroes of the total transfer function $H^{(1)}_{L2,n}$. We also conclude that a sufficent condition for $z_1 \ldots z_{L2}$ to be zeroes of $H^{(2)}_{L2,n}$ is:

$$H^{(2)}_{i,n}(z_i) = 0 \qquad \text{for all n and all } i=1,,,L2 \quad (11)$$

Using (10) we get the conditions on $a_{i,n}$ and $b_{i,n}$:

$$a_{i,n} H^{(2)}_{i-1,n}(z_i) + b_{i,n} z_i^{-1} H^{(2)}_{i-1,n-1}(z_i) = 0 \quad (12)$$

This relation will be fulfilled if:

$$a_{i,n} = -H^{(2)}_{i-1,n-1}(z_i) \quad (13)$$

$$b_{i,n} = z_i H^{(2)}_{i-1,n}(z_i) \quad (14)$$

Using (8) and (10) we then have a recursion for computing the weights $a_{i,n}\ b_{i,n}$ $i = 1 \ldots L2$.

Performance with linear prediction filtering

In order to compare the performance of the coherent on receive radar with that of the fully coherent radar we need some kind of optimum filter. The conventional way is to choose the filter that optimizes the so called MTI-improvement factor, which is defined as the average signal/clutter improvement assuming that target doppler has uniform probability distribution. The filter weights in this case are obtained as the eigenvector belonging to the minimum eigenvalue of the clutter covariance matrix. A simpler way which has been shown to give approximately the same performance as the eigenvector filter, Chiuppesi et. al. (2), is to use linear prediction filtering. The weights $c_0 \ldots c_L$ of the linear prediction filter are chosen in order to minimize the so called prediction residual:

$$E\left(|c_1 S_{n-1} + c_2 S_{n-2} + \ldots + c_L S_{n-L} - S_n|^2\right)$$

where $c_0 = -1$ and S_n is the clutter signal. E means that we take the expected value. Equating the expression and minimizing with respect to the c_i:s we get the system of linear equations:

$$\begin{matrix} c_1 R_{n-1,n-1} + c_2 R_{n-1,n-2} + \ldots + c_L R_{n-1,n-L} = R_{n-1,n} \\ \vdots \\ c_1 R_{n-L,n-1} + c_2 R_{n-L,n-2} + \ldots + c_L R_{n-L,n-L} = R_{n-L,n} \end{matrix} \quad (15)$$

Where $R_{i,j} = E(S_j \bar{S}_i)$ is the i,j element of the clutter covariance matrix and $\bar{S}$ denotes complex conjugation. In the coherent on receive radar when clutter is composed of 1:st and 2:nd time around components we get for the clutter covariance matrix:

$$R_{i,j} = E\left((S^{(1)}_j + q_j S^{(2)}_j)(\bar{S}^{(1)}_i + \bar{q}_i \bar{S}^{(2)}_i)\right) =$$

$$= R^{(1)}_{i,j} + \bar{q}_i q_j R^{(2)}_{i,j} \quad (16)$$

where $R^{(1)}$ and $R^{(2)}$ are the covariance matrixes for 1:st and 2:nd time around clutter respectively. In (16) we assume uncorrelated 1:st and 2:nd time around clutter, which means the cross terms are zero.

Solving the system of linear equations (15) for the coherent on receive case, with 1:st and 2:nd time around clutter, we get weights c_k depending on the phase shifts q_n. With signal and clutter power at the filter input normalized to unity we get for the signal to clutter improvement at the normalized doppler shift ω

$$(S/C)_i(\omega) = \frac{\left|\sum_{k=0}^{L} c_k e^{-j\omega k}\right|^2}{\sum_{k=0}^{L}\sum_{i=0}^{L} \bar{c}_k c_i R_{n-k,n-i}} \qquad (17)$$

where the doppler shift ω is normalized with respect to the pulse repetition frequency.
Since we have time varying filter weights c_k and a time varying clutter covariance matrix R, we will get a time varying signal/clutter improvement.
We have therefore computed an average signal/clutter improvement by randomly choosing 10 different sets of phase shifts $q_0 \ldots q_L$, which gives the signal/clutter improvement at frequency ω :

$$(S/C)_i(\omega,k) \qquad k = 1 \ldots 10$$

The average signal/clutter improvement was then computed through:

$$(S/C)_{ave}(\omega) = \frac{1}{10}\sum_{k=1}^{10} \log\left[(S/C)_i(\omega,k)\right] \qquad (18)$$

In the evaluation we assumed the following parameters for 1:st and 2:nd time around clutter:

Spectral density: Gaussian

	1:st	2:nd
Standard deviation:	0.02	0.06
Mean doppler:	0	0; 0.25; 0.5

The relative clutter level between 1:st and 2:nd time around clutter was assumed to 10 dB.

Performance was also computed for the case of 2:nd time around clutter only.

The number of delays L in the filter were in the case of combined 1:st and 2:nd time around clutter chosen to 9. In the case of 2:nd time around clutter only 4 delays were used. This means that the filter uses 10 and 5 pulses respectively.

The signal to noise ratio was in all cases set to 50 dB. As can be seen in fig (2) this limits the improvement in the fully coherent case to 50 dB.

The chosen parameters roughly corresponds to a case of rain clutter at 0.5 and 1.5 of the unambiguous range with spectral spread due to wind shear varying linearly with height.

The resulting signal/clutter improvement as a function of normalized doppler is illustrated in fig (2) in the left column. The right column illustrates the normalized MTI-response for 2:nd time around echoes. In the coherent on receive case the normalized MTI-response was computed by the same averaging procedure as in the signal/clutter improvement computations.

The first 3 diagrams in fig.(2) shows the effect of varying the mean doppler of 2:nd time around clutter. As can be seen this greatly affects the fully coherent case since, due to the coherence 1:st and 2:nd time around echoes will have the same transfer function. In the coherent on receive case however, the mean doppler of 2:nd time around clutter does not affect the signal/clutter improvement of 1:st time around echoes, due to the random phase shifts. The effect of suppressing 2:nd time around clutter is spread out over the whole doppler interval for 1:st time around echoes. The last diagram in fig. (2) shows the case of 2:nd time around clutter only. For the coherent on receive case there is no stopband in the 1:st time around transfer function while the normalized response for 2:nd time around echoes will be the same as in the fully coherent case.

The performance acheived with the coherent on receive radar could of course also be acheived with the fully coherent radar by deliberately introducing pseudo-random phaseshifts in the transmitted signal. The MTI-filter could then use a set of precomputed filter weights which would simplify the signal processing.

(The main part of this work was carried out while the author was still employed by Philips Elektronikindustrier AB, Järfälla, Sweden)

REFERENCES.

1. Barton, D.K, 1975, "Comment on "Comparision of Two major Classes of Coherent Pulsed Radar Systems" ", IEEE Trans. Aerospace and Electronic Systems, vol AES-11, no 5, 920

2. Chiuppesi, F , et.al., 1980, "Optimisation of Rejection Filters" IEE Proc. 127, pt.F, 354-360

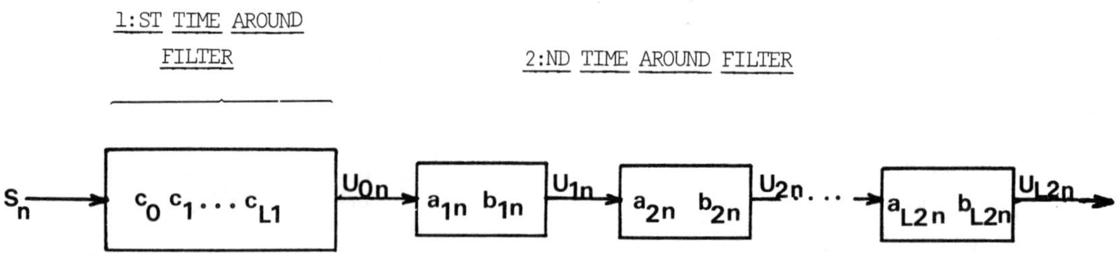

FIG.1 CASCADED 1:ST AND 2:ND TIME AROUND FILTERS

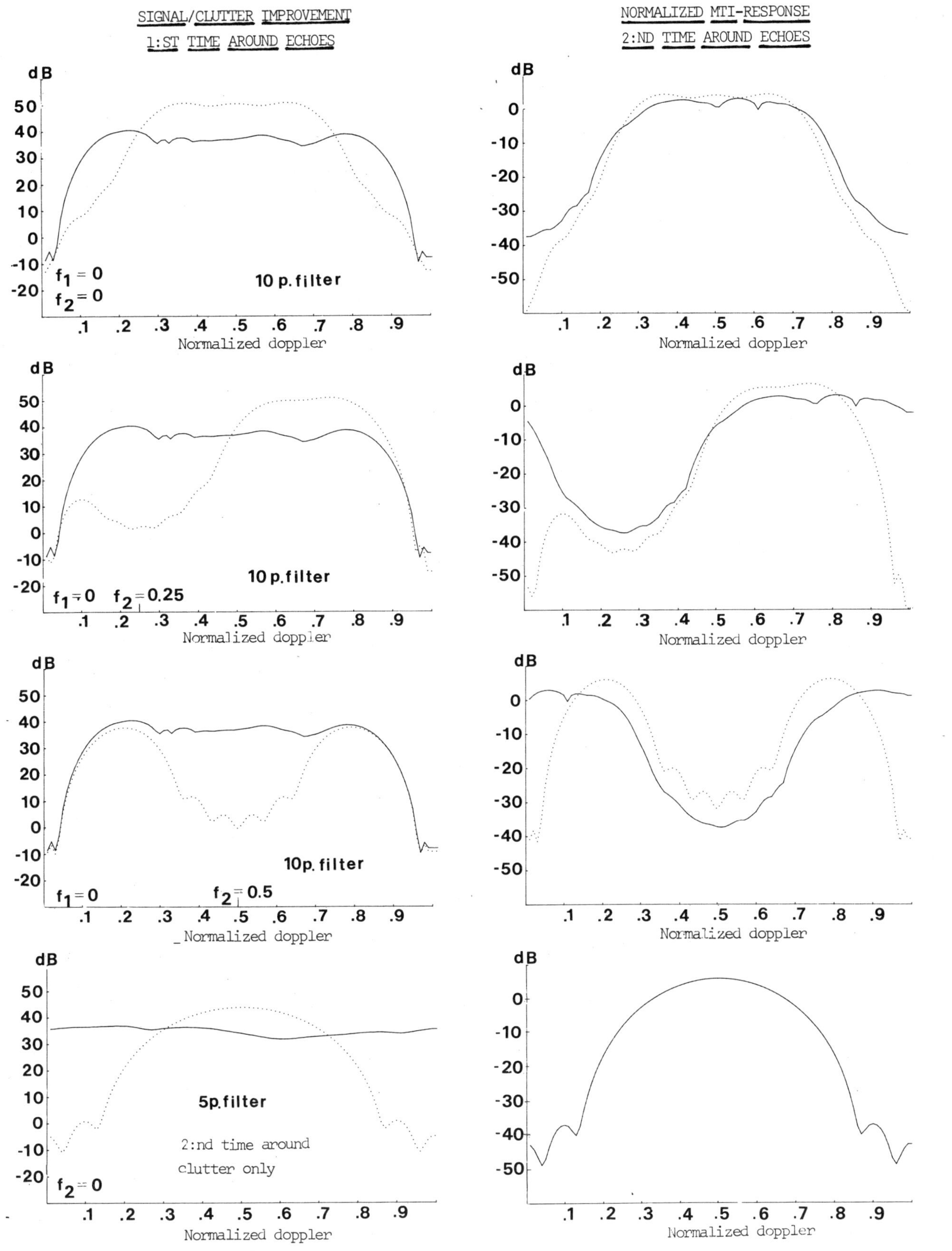

FIG.2 LINEAR PREDICTION FILTERS

——— Coherent on receive (average)

.... Fully coherent

f_1 Mean doppler 1:st time around clutter

f_2 Mean doppler 2:nd time around clutter

THE GRAM–SCHMIDT SIDELOBE CANCELLER

T. Bucciarelli, M. Esposito, A. Farina, G. Losquadro

Selenia S.p.A., I

INTRODUCTION

A sidelobe canceller can be efficiently employed to reduce the effect of jammers received through the sidelobe of a radar system. A well established technique refers to the SideLobe-Canceller (SLC) approach, in which external aerials (called "auxiliary antennas") placed around the radar antenna (called "main antenna") are subject to control. The signals received through the auxiliary antennas are multiplied by proper weights and then summed obtaining an estimate of the jammer received through the radar sidelobes. The cancellation is performed by subtracting the jammer estimate from the radar output. The weights are usually obtained by evaluating the correlation coefficients between each auxiliary signal and the residue of cancellation. The processing is performed by an adaptive loop, Howells-Applebaum technique, consisting of a multiplier and a low pass filter. The number of auxiliary antennas determines the degree of freedom on the sidelobe structure of the overall system and, thereby, the number of jammers which can be cancelled. Fuller details of this technique can be found in Monzingo and Miller (1), Hudson (2). Two figures of merit define the SLC system performance: the power cancellation ratio and the time required for adaption of all the loops. Unfortunately these figures are in contrast to each-other in the Howells-Applebaum implementation. In fact, the greater the loop bandwidth, the faster its response to a non-stationary jamming situation; however, a wider bandwidth reduces the filtering effect on the input jamming process. A detailed analysis of these conflicting effects on the jammer cancellation, for a SLC having two auxiliary aerials, have been described by Farina and Studer (3).

One way to speed convergence is based on the Gram-Schmidt orthogonalization procedure which maintains the same steady state cancellation of the standard loop and it is easily implemented. This paper gives the performance evaluation of this canceller. Mathematical expression of the steady-state cancellation as a function of the ratio between the adaptive circuit bandwidth to the radar receiver bandwidth is given. A comparison, in terms of performance and implementation complexity is made with the canceller based on the Howells-Applebaum technique.

PERFORMANCE EVALUATION OF THE OPTIMUM CANCELLER

The optimum weights control law can be obtained resorting to the linear prediction theory. Indicate with the N-dimensional vector $\underline{X}$, the set of the complex signals received through the auxiliary aerials and with X_M the signal received through the main antenna (a coherent signal processing at a convenient frequency band is assumed).
The useful target signal, if present, is contained only in X_M while the vector $\underline{X}$ contains only jamming due to the low gain of auxiliary aerials, set to match the sidelobe average level of the main antenna. Assume the jamming source to be a zero mean white Gaussian noise producing correlated processes when received through the main and auxiliary antenna, and filtered through the receiving channels.
Indicate with $\underline{M}$ (N, N) the covariance matrix of the process $\underline{X}$ and with $\underline{R}$ the N-dimensional vector containing the covariance of the signal X_M with the auxiliary signals X_i (i = 1,2,........N):

$$\underline{M}(i,j) = E\{x_i\, x_j^*\} \qquad \underline{M}(i,i) = \sigma^2 \tag{1}$$

$$\underline{R}(i,1) = E\{x_i\, x_M^*\} \tag{2}$$

In the hypothesis of an interference having much more power than the useful signal, the rejection of the disturbance in the main channels is achieved by subtracting from X_M the estimation of the jamming signal. The estimation is performed through a linear prediction of the jamming X_M on the basis of the auxiliary signals X_i. The optimum vector $\underline{W}$ (N, 1) containing the weights W_i of the linear combination of the data X_i is determined minimizing the mean square prediction error. Reference (1) shows that the following equation holds:

$$\underline{W} = \underline{M}^{-1}\,\underline{R} \tag{3}$$

The prediction error is:

$$V(t) = X_M - \underline{W}^T\,\underline{X} \tag{4}$$

therefore the power of jamming residue is:

$$E\{|V(t)|^2\} = E\{|X_M - \underline{W}^T\underline{X}|^2\} = \\ = E\{(X_M - \underline{W}^T\underline{X})\, X_M^*\} = \sigma^2 - \underline{W}^T\underline{R} \tag{5}$$

the second equality follows from the orthogonality of the data X_i to the prediction error. The jammer cancellation g, defined as the ratio of input jamming power σ^2 to output residual power $E\{|V(t)|^2\}$ is:

$$g = \sigma^2/(\sigma^2 - \underline{W}^T\underline{R}) \tag{6}$$

Equation (6) has been evaluated assuming that the processes X_M and X_i (i = 1,2, ...N) have the following autocorrelation of Gaussian shape:

$$E\{x_M(t)\,x_M^*(t+\tau)\} = E\{x_i(t)\,x_i^*(t+\tau)\} = \sigma^2 e^{-(\tau/\tau_c)^2} \tag{7}$$

where the correlation time τ_c depends on the receiving channel bandwidth. If a perfect amplitude and phase equalization is assumed, between the main and auxiliary receiving channels, X_i (i = 1,2,....N) and X_M are delayed version of a same process. The covariance matrix $\underline{M}$ and the cross-correlation vector $\underline{R}$ can be evaluated by means of the equation (7) and taking into account the geometrical displacement of the auxiliary aerials around the main antenna. As an example consider the auxiliary and main antennas set in a line and regularly spaced at a constant distance d among them. The matrix $\underline{M}$ and the vector $\underline{R}$, relevant to jammer having an incidence angle θ with respect to the array, are as follows:

$$\underline{M}(i,j) = \rho^{(i-j)^2}\sigma^2$$

$$\underline{R}(i,1) = \rho^{i^2}\sigma^2 \tag{8}$$

where the coefficient ρ is:

$$\rho = \exp[-(\Delta t/\tau_c)^2] \tag{9}$$

being Δt the time difference with which the jamming signal is received by two contiguous antenna. Indicating with c the velocity of light the following equation holds:

$$\Delta t = d \,\mathrm{sen}\, \theta / c \tag{10}$$

Fig. 1 shows the jammer cancellation g versus the number of auxiliary aerials and with different values of parameter ρ. Two sets of curves are drawn, those with solid lines refer to jammer only as disturbance source while those dashed refer to jammer plus thermal noise having a variance σ_n^2. In the following section two methods will be described to implement the optimum sidelobe canceller weights (3).

PERFORMANCE EVALUATION OF THE HOWELLS-APPLEBAUM METHOD

The Howells–Applebaum method is a well known embodiment of the equation (3). The functional diagram of this standard technique is depicted in fig. 2 where only two auxialiary aerials are considered for sake of example.
The integrator to correlate X_1 and V (t) is digitally performed through a single pole filter with a constant:

$$\beta \cong 1 - T_S/\tau \qquad (11)$$

where T_S is the sampling interval of the signals and τ the constant time of the integrator. This method which attracts for its simple implementation, has a slow convergence of the adaptation weights. The problem of slow convergence arises whenever there is a wide spread in the eigenvalues of the input signal correlation matrix. See references (1) and (2) to have more details on this problem.

To show the mutual influence between the steady-state and transient behaviours, a digital simulation of the system shown in fig. 2 has been performed in order to evaluate the steady-state jammer cancellation and the number of samples needed for the adaptation of all the loops. These two figures have been obtained as a function of the parameter α :

$$\alpha = \tau_C / 2\,\tau_L = B_L / 2\,B_C \qquad (12)$$

equal to the ratio of the bandwidth B_L of a single closed loop to twice the channel bandwidth B_C.
The parameter τ_L :

$$\tau_L = \tau/(1 + G\sigma^2) \qquad (13)$$

is the closed loop time constant of a canceller having a single aerial and can be assumed as mean to represent the parameters of each identical loop in a canceller having more than one auxiliary antenna. The greater α the faster the adaptation of the weights; however, a wider system bandwidth reduces the accuracy of the jammer signal estimation in the main channel and the attainable jammer cancellation.

In fig. 3 the cancellation of a single jammer is drawn versus α and for several cases of interest. It can be noted that when α is small (slow adaptation of weights) the steady state jammer cancellation is equal to the optimum value shown in fig. 1;on the other hand, when a faster reaction time is required,the cancellation can dramatically decrease due to the strong fluctuation of the weight estimate. The limit value of α corresponding to the cancellation increases with the number of auxiliary aerials but it is independent of the correlation value ρ . Fig. 4 shows instead the number of samples needed to reach the steady state cancellation value with a 10% error. It can be seen that this number of samples decreases with the value of α and increases with the number of auxiliary aerials and with the correlation coefficient ρ . It is an easy matter to recognize that non-stationary disturbances (e.g.: sidelobes variation, along a certain direction, in a rotating antenna) cannot be afforded with the Howell-Applebaum system when high cancellation values should be obtained this requiring a high value of ρ and a sufficient number of auxiliary antennas. In the next section it will be described a canceller with better transient response characteristics and equal steady-state cancellation values.

THE GRAM-SCHMIDT JAMMING CANCELLER

A way to overcome the problem of slow convergence is to reduce the eigenvalue spread of the input signal correlation matrix through the introduction of an appropriate transformation which resolves the input signals into their eigenvector components. Then equalizing in power the resolved signals,the eigenvalue spread may be considerably reduced. An orthogonal signal set can be obtained by a transformation based on the Gram-Schmidt orthogonalization procedure, introduced by Giraudon (4) and White (5). To understand the coordinate transformation based on the Gram-Schmidt orthogonalization procedure, consider the case of two auxiliary aerials and a main antenna, as shown in fig. 5, which receive the signals X_1, X_2 and X_M, respectively. Generalization of this scheme to more aerials is straightforward; transformations S_{ij} each having two input signals, and giving one output uncorrelated with the left side input , are used in the canceller. Three blocks of this type need to orthogonalize the input signals thus obtaining the new signal set X_1, Y_1 including the useful output V(t).
A possible implementation of the decorrelating blocks S_{ij} is represented by the standard SLC with a single auxiliary signal that is the same block shown in fig. 2. Assuming that each block S_{ij} have the same parameters G (amplifier gain) and τ (integrator time constant), each S_{ij} has a time constant

$$\tau_{Li} = \tau/(1+G\sigma_i^2) \qquad (14)$$

where σ_i^2 is the power of the auxiliary signals (left side input), in the block S_{ij}. The two blocks S_{11} and S_{12} have the same time constant and S_{22} too, due to the power equalizer inserted after Y_1 to restore the input jamming power σ^2 . Therefore, the overall transient time is the loop time constant τ_L times the number of loops in the main channel. Instead,in the case of the Howells-Applebaum implementation, it can be noted (3) that the transient time duration depends on ρ and increases indefinitely when the correlation coefficient goes to unity. In other words, the main reason of the fast response of the Gram-Schmidt canceller with respect to the standard SLC system implementation of fig. 2, is the transversal architecture of the former while the latter has a feedback architecture and therefore a slower dynamic. In the next section the performance evaluation of the Gram-Schmidt canceller will be evaluated and compared with those of the Howells-Applebaum system.

PERFORMANCE EVALUATION OF THE GRAM-SCHMIDT CANCELLER AND COMPARISON WITH THE HOWELLS-APPLEBAUM SYSTEM

The Gram-Schmidt and the Howells-Applebaum systems have the same steady-state cancellation. To demonstrate this statement the useful output of the two compared systems are made equal and a condition is found on the weights of the two systems. Indicate with Wij the weight of the Gram-Schmidt block S_{ij} and with W_j the weight of the Howell-Applebaum block S_j. It can be shown that the following equation holds between the two sets of weights:

$$W_{N-i} = \sum_{j=1}^{i} W_{N-i,\,N-j}\; W_{N-j+1} \qquad (15)$$

with the inizial condition $W_N = 1$, where N is the maximum number of auxiliary aerials. As an example consider the relationship existing between the systems of figs. 2 and 5, that is for two auxiliary aerials, from equation (15) follows:

$$W_2 = W_{22}$$

$$W_1 = W_{12} + W_{11}\,W_{22} \qquad (16)$$

which are the same found in reference (1) p. 368. A further confirmation of this statement is given by the evaluation of the Gram-Schmidt jammer cancellation as a function of α , obtained by means of a digital simulation. Results very close to those shown in fig. 3 are obtained.

The main difference between the two compared systems refers to their transient behaviours. Fig. 6 shows the number of samples, needed to reach the steady-state in the Gram-Schmidt case. These results refer to a number of auxiliary aerials up to nine and a correlation coefficient ρ = 0.9. Comparison can be made with the curves of fig. 4 for a number N_A of auxiliary aerials up to three. This comparison cannot be extended to a greater number of N_A due to the slow reaction time of the Howells-Applebaum system which involves very long computer time in the Montecarlo simulation of the system. This comparison shows that the Gram-Schmidt canceller has an excellent transient response. If the same transient time is required to the Howells-Applebaum system a reduction of steady state cancellation is paid.

IMPLEMENTATIONS COMPARISONS AND CONCLUSIONS

From the previous considerations the Howells-Applebaum solution seems to be smaller than the Gram-Schmidt canceller; in fact the basic block of both solutions is similar and while in Applebaum approach only N_A blocks must be used, in Gram-Schmidt canceller 0.5 N_A (N_A + 1) circuits are needed, the difference so increasing with the number of auxiliary aerials. Moreover additional hardware must be inserted to performe the power equalization between successive stages. This is not true indeed; the short transient time of the Gram-Schmidt system allows to employ a single arithmetic unit thus obtaining a great reduction of the hardware to be used, as it is explained in the following.
In fig. 7 a detailed block diagram of the Gram-Schmidt canceller is drawn and two sections can be evidenced: the former in which a real time processing is performed; the latter where, after buffering the real time results, a single arithmentic unit is used to obtain the

weights. The signals coming from the main and the auxiliary antennas, coherently detected, are multiplexed in an analog switch and serialized allowing the use of a single analog to digital converter. In the case of two auxiliary aerials a 6 input multiplexer must be used and, in the hypothesis of a 1 μsec sampling, a 166 nsec conversion time must be obtained. The output of the converter (a TDC 1019J can be adopted able to digitize an analog signal at a rate of 20 M samples per second) is stored in a proper buffer. This buffer is RAM implemented and is very small due to the limited number N_S of samples (e.g. : 30 - 40) to be stored.
After loading all the converted samples, each S_{ij} block processing is performed.
The stored data are multiplied and accumulated in a proper device (TDC 1009J) able to perform each step of computation every 125 nsec. Each 500 nsec the complex multiplication of inputs and the previously stored weights is performed and after about 1000 N_S nsec the updated weights are obtained and stored in the weights memory (also the dimension of this buffer is very small as it must be designed to store only a number of data equal to twice the number of S_{ij} blocks). After performing the evaluation of the outputs of each stage of computation, the power of the decorrelated samples is measured thus allowing the equalizations of the following stage of computation.
The circuits adopted are speedy enough to allow the real time cancellation of the incoming jammers and are easily programmable thus enabling to change the parameter (time constants) of the design. A proper timing and control unit must be used. One important parameter influencing the hardware complexity is the number of bits; simulations have been performed and some special results are shown in fig. 8; for a high number of auxiliary antennas at least 8 bits must be used to avoid cancellation losses.

From the previous analysis and hardware considerations it is possible to select the suitable architecture and parameters to obtain the required performances in a specified jamming environment; this paper is intended to contribute to further investigations of a fast suppressor of active interferences which is an important problem in the field of continuous scanning and phased array radars.

REFERENCES

1. Monzingo, R.A., and Miller, T.W., 1980, "Adaptive Arrays". John Wiley and Sons, N.Y.

2. Hudson, J.E., 1981, "Adaptive Array Principles". IEE Electromagnetic Waves Series 11, Peter Peregrinus LTD.

3. Farina A., and Studer F.A., 1982, "Evaluation of sidelobe-canceller performance". Proc. IEE , 129, Pt. F, 52-58.

4. Giraudon, C., 1976, "Optimum antenna processing: a modular approach". Proc. of NATO Advanced Study Institute on signal processing and underwater acoustic, Porto Venere, Italy.

5. White, W.D., 1978, "Adaptive cascate networks for deep nulling". "Trans IEEE on Antennas and Propagation, 26, 396-402.

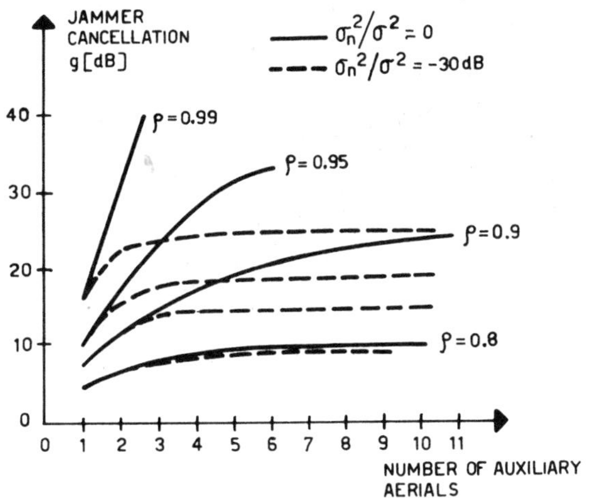

Fig. 1 — Jammer cancellation of the optimum coherent sidelobe canceller

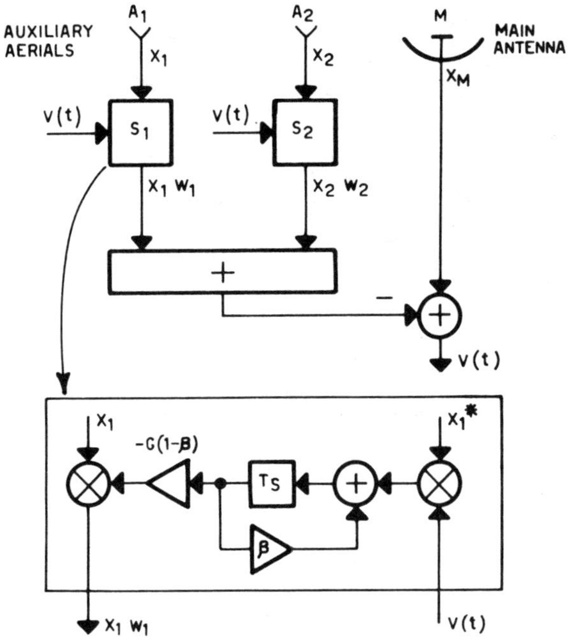

Fig. 2 — Standard sidelobe canceller

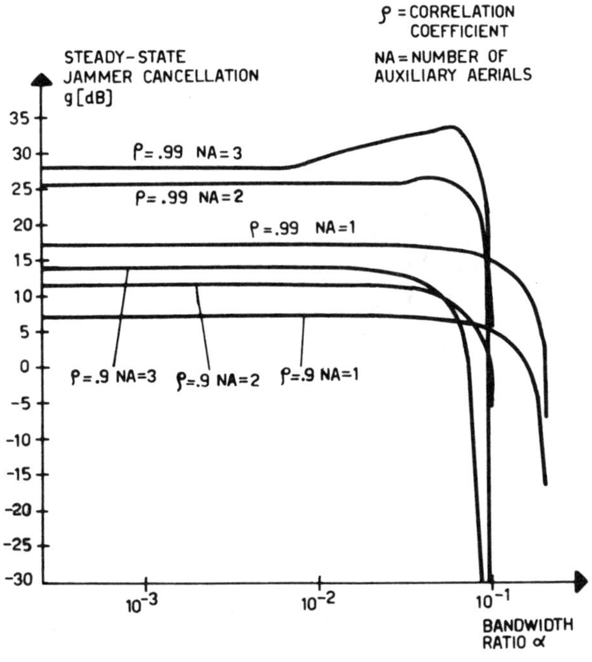

Fig. 3 — Steady-state cancellation of the Howells-Applebaum system

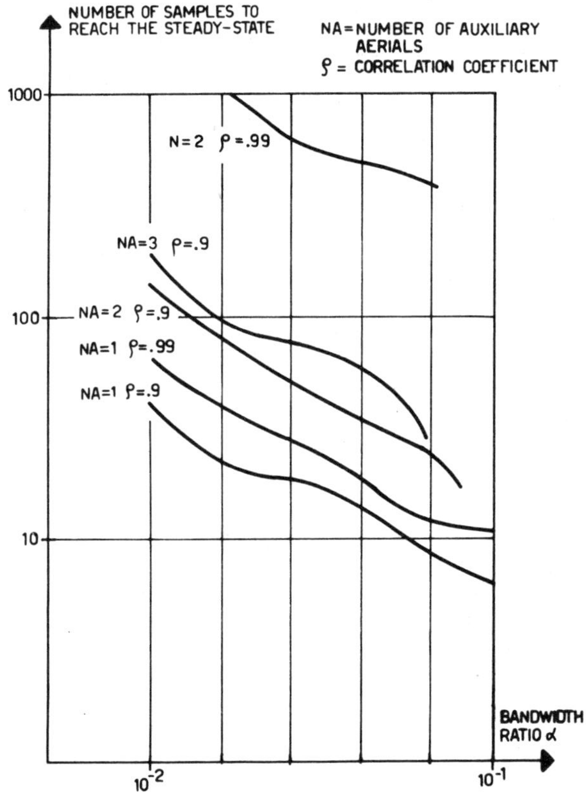

Fig. 4 — Number of samples to reach the steady-state in the Howells-Applebaum canceller

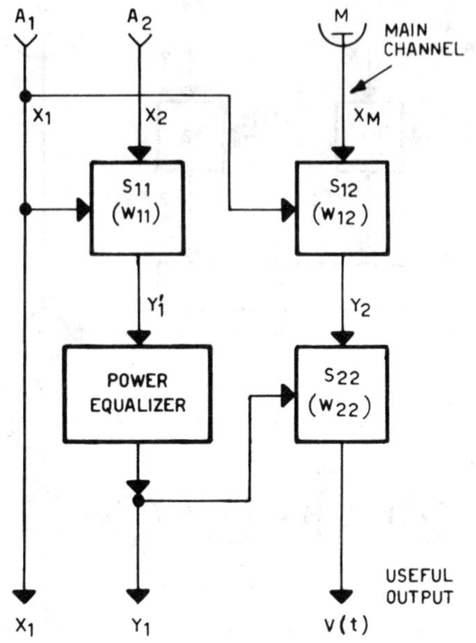

Fig. 5 — Gram-Schmidt canceller with two auxiliary aerials

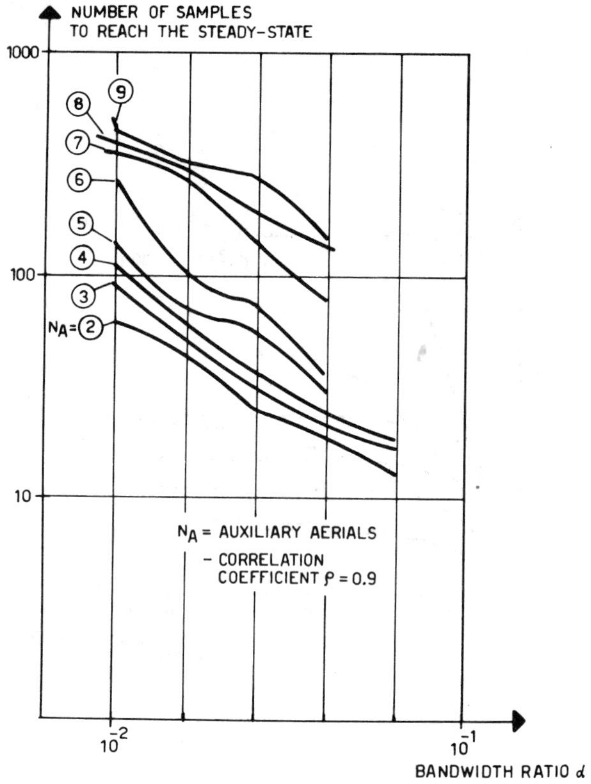

Fig. 6 — Number of samples to reach the steady-state in the Gram-Schmidt canceller

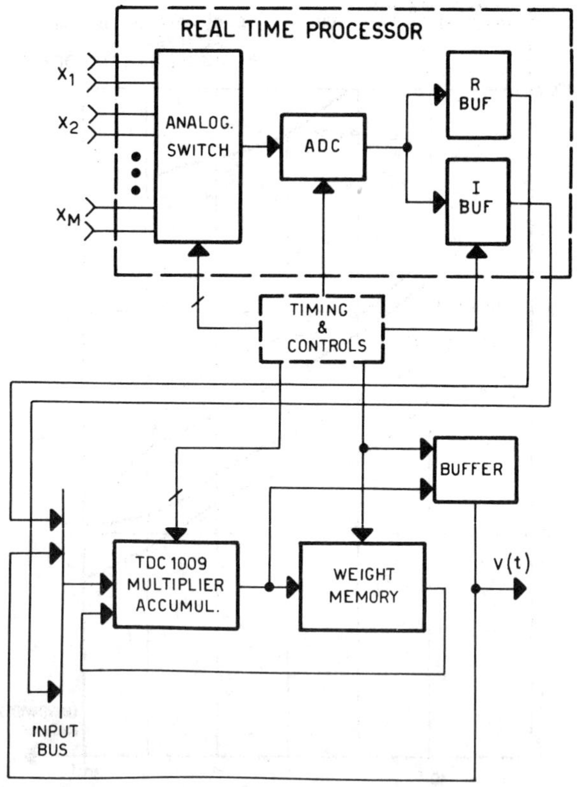

Fig. 7 — Hardware implementation of the Gram-Schmidt canceller

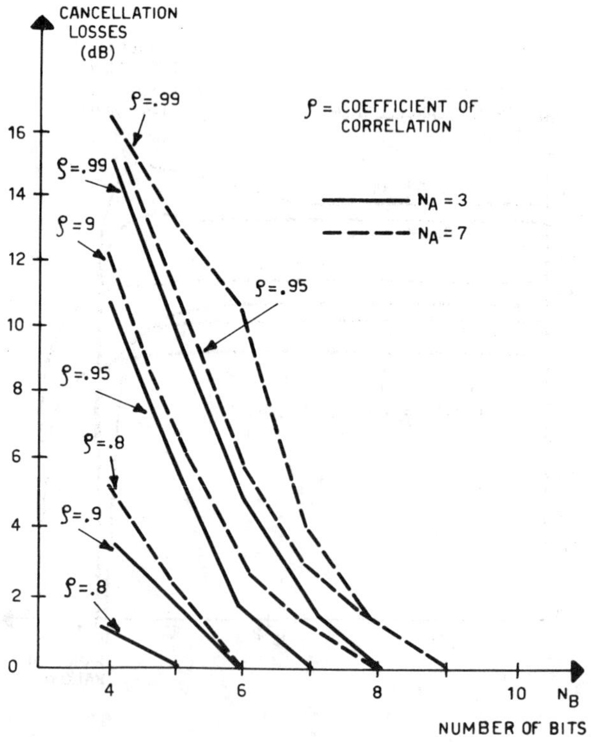

Fig. 8 — Conversion losses vs. bit number

GROUND CLUTTER SUPPRESSION USING A COHERENT CLUTTER MAP

John S. Bird

Department of Communications, Canada

INTRODUCTION

A chronic problem for most surveillance radar systems is the detection of radially stationary targets in the presence of ground clutter. Returns from aircraft flying tangential to the radar or from a hovering helicopter can be buried in the clutter and hence the target lost to the radar operator (or system). In addition targets that are not radially stationary can be lost because of Doppler filter leakage from large ground clutter returns. Many schemes for handling ground clutter have been proposed, analysed and implemented (Dax (1), Sutherland (2), Muehe et al (3), O'Donnell et al (4), O'Donnell and Muehe (5)). Generally they fall into two broad classes: variable threshold techniques (noncoherent clutter maps and sliding window normalizers), and pulse cancellation. Variable thresholding is used when the target and clutter cannot be separated by Doppler filtering and pulse cancellation is used when separation is possible. These traditional methods for handling ground clutter have been developed under constraints imposed by continuous-scan antennas. For step-scan phased array antennas the constraints are removed, permitting the use of a coherent clutter map which allows the radar to operate as if the clutter were almost nonexistent.

A typical processing scheme for a step-scan phased array radar is shown in Figure 1. It consists of a receiver, Doppler filter and detector. Pulse cancellers have been used in such systems to reduce dynamic range and filter leakage from high level ground clutter returns (5). This however renders the radar totally blind to targets flying at the so-called blind speeds, which include the important zero Doppler region. Variable thresholding has been used to combat clutter in the low Doppler regions; however detection performance is poor even for moderate clutter levels and decreases as the clutter increases. Conversely, the performance of a coherent clutter map (as will be demonstrated in this paper) does not decrease with increasing clutter level and there are no blind speeds.

The key to the utility of the coherent clutter map lies in the very nature of the clutter returns. These returns must be coherent over the long term (minutes, hours, days, etc.). It is shown here that at least some ground clutter does possess this long term coherency and this characteristic is used to establish the merits of a coherent clutter map. Since we are primarily concerned with detecting low Doppler targets the comparison will be between variable thresholding techniques and the coherent clutter map. Pulse cancellers will be included in a brief discussion of the Doppler responses of the various ground clutter suppression techniques. The paper concludes by proposing several modes of operating a coherent clutter map that depend on system size and mission objective.

GROUND CLUTTER

To establish the long term coherency of ground clutter as seen by a step-scan phased array radar, returns from an isolated island (Aylmer Island) in the Ottawa River were examined. This island is 5.7 km from the phased array and at the time the data were recorded it was snow covered and the river was frozen. Bursts of 512 1 μs pulses at a repetition rate of 1 kHz were transmitted every 8 seconds. Returns from 16 range bins spread across the island were recorded. The total elapsed time from the first burst to the last was 20 minutes and the antenna look direction was held constant for this time.

A summary of the experimental data is shown in Figure 2. For each range bin the maximum and minimum values for the real (I) channel and the imaginary (Q) channel over the 20 minutes are plotted. Also plotted is the average echo power returned on a relative decibel scale. The average power plot shows the peak return (range bin 6) from the island was approximately 30 dB above the background noise and returns from the ice. The quadrature components of the ground clutter were remarkably stable during the 20 minutes of recording indicating that the clutter returns are coherent over the long term.

A closer look at the nature of the long term stability is provided by the phase diagram shown in Figure 3 where the samples for range bin number 6 are shown in time sequence. The full scale plot indicates the significance of the variations relative to the clutter amplitude. The small scale plot describes the actual nature of the variations by illustrating the drift with time of the average for each burst (solid line) and the envelope of the standard deviations (dashed line). The return vectors have a relatively stable mean position that drifts slowly with time and small random components that distribute the vector about the mean. It is not known at this time whether the drift is due to system instabilities or to slow changes in the clutter but the significant fact is that it is small and trackable.

DETECTION PERFORMANCE AND THRESHOLDING

It will now be shown that the stable coherent component of the clutter has a very significant effect on the detection performance of a radar system. Consider the three sets of phase diagrams shown in Figure 4. The first set of two phase diagrams shows the probability distributions of the received signals under the target absent and target present hypotheses (a Rayleigh target was assumed). In the absence of a target the distribution is an offset circularly symmetric Gaussian distribution. The offset represents the phase stable part of the clutter and the circular symmetric distribution the combination of the diffuse part of the clutter return and receiver noise. With the Rayleigh target present the variance of the circular symmetric part of the distribution grows according to the signal strength of the target. The problem is to test the null hypothesis (distribution is offset Gaussian with diffuse variance σ_N^2) against an alternative (distribution is offset Gaussian with diffuse variance $\sigma^2 > \sigma_N^2$). Various types of thresholding procedures can be used to make this test.

The first set of thresholding procedures considered are the standard CFAR and noncoherent clutter map thresholds shown in the second set of phase diagrams. The first phase diagram of this set shows the threshold "T_G" established assuming the clutter and noise are circular symmetric and Gaussian. Since this is not the true situation the probability of false alarm is not what is expected (it can be shown to be too low (Bird (6)) and hence detection performance is poor. To improve performance somewhat it is possible, but difficult in practice, to adapt the threshold "T_A" until the desired probability of false alarm is reached (second phase diagram of the second set). This improves the detection performance but still is not optimum.

The final set of phase diagrams shows the optimum thresholds for testing the two hypotheses. One can either establish a circular symmetric threshold around the end point of the stable clutter vector (Diagram 1) or remove the stable clutter vector coherently and establish a circular symmetric threshold around the origin (Diagram 2). Both techniques perform identically but the latter is easier to implement and leads to the concept of a coherent clutter map.

Before proceeding to the coherent clutter map, it is instructive to quantify the degradation in performance expected by using the standard variable thresholds T_G and T_A. Figure 5 shows the probability of detection curves for the different thresholding techniques for an assumed probability of false alarm of 10^{-6}. A clutter-to-noise ratio (CNR) of 21 dB was chosen for the purpose of illustration (CNR is the power ratio between the stable coherent part of the clutter and the diffuse random part, which of course includes the noise). The best performance is obtained when the clutter-to-noise ratio is 0 (or $-\infty$ dB) in which case all the distributions are circular symmetric and hence the different thresholding techniques are equivalent. As the coherent component of the clutter grows however the performance of the noncoherent clutter map thresholding techniques decreases rapidly. (It should be noted that the calculation of these curves requires the evaluation of the Marcum Q function and establishing T_A requires an accuracy of about 10 significant figures. See Bird and George (7) for a technique to accomplish this). At a clutter-to-noise ratio of 21 dB the threshold under the Gaussian assumption "T_G" suffers a 21 dB signal-to-noise ratio (SNR) penalty (penalty $\simeq$ CNR) and the adaptively set threshold "T_A" an 11.2 dB penalty (measured at a probability of detection of 0.8). The performance of the threshold associated with the coherent clutter map however is unchanged (i.e. is the same as that for no clutter) as there is no offset of the distribution and the threshold remains optimum. Therefore there is quite an advantage to be gained in removing the clutter coherently even for a moderate clutter-to-noise ratio of 21 dB. For our experimental data the maximum clutter-to-noise ratio was in excess of 30 dB indicating a very substantial potential for a coherent clutter map processor.

THE COHERENT CLUTTER MAP

A conceptual description of the coherent clutter map processor is given in Figure 6. This processor is identical to that shown in Figure 1 except that a recursive mean estimator has been added at the end of the Doppler filter. (All signal paths are assumed to be complex). The principle of operation is as follows. The radar is directed to the desired location and the processing system is allowed to reach a steady state. In this condition the response to the coherent component of the clutter is zero. This is known as the learning phase. During subsequent operation the radar is periodically directed to the location to provide an update for the estimate. When testing for a target the update on the clutter estimate is inhibited. The current estimate is subtracted from the returns and then they are processed by the Dopper filter. If no target is detected the returns are permitted to enter the mean estimator for updating.

The performance of this system depends on how well the clutter estimator can track the clutter. A single stage recursive mean estimator was chosen because of its simplicity, low memory requirements, and an ability to track clutter variations. In addition this estimator gives near optimum performance for the assumed ground clutter model. The estimator parameter "n" may be adjusted to suit the clutter drift rate which thereby tunes the system to the particular clutter at hand.

To compare the performance of the coherent clutter map to variable threshold techniques an expression for the coherent clutter map system will be derived. As stated before, once the estimator has reached steady state the response to the constant clutter is zero. Since the rest of the inputs are circular symmetric Gaussian and the system is linear the output is circular symmetric Gaussian. Therefore to determine the SNR penalty only the single component variance for the noise response with and without the estimator need be found. Without the estimator the output due to the noise in the delay line has a single component variance of

$$\sigma_{no}^2 = N\sigma_n^2 \quad \ldots\ldots\ldots\ldots (1)$$

where N is the filter length and σ_n^2 is the single component variance of each independent return. Since the noise output of the estimator is made up of returns previous to those in the delay line it is independent of the filter noise. Hence the single component variance at the output with the estimator present is

$$\sigma_{no}^2 + \sigma_e^2 \quad \ldots\ldots\ldots\ldots (2)$$

where σ_e^2 is the output variance due to the estimator. If the filter coefficients have a constant amplitude of 1.0 and linear phasing then

$$\sigma_e^2 = \frac{\sigma_n^2}{2n-1}\left(\frac{\sin\left(\frac{N\theta}{2}\right)}{\sin\frac{\theta}{2}}\right)^2 \quad \ldots\ldots (3)$$

where n is the estimator parameter and θ is the normalized Doppler frequency and we have the following expression for the SNR penalty (SNR_{pm}, subscript m referring to coherent clutter map).

$$SNR_{pm} = \frac{\sigma_{no}^2+\sigma_e^2}{\sigma_{no}^2} = 1+\frac{1}{N(2n-1)}\left\{\frac{\sin\left(\frac{N\theta}{2}\right)}{\sin\frac{\theta}{2}}\right\}^2 \quad ....(4)$$

In order for this penalty to be meaningful it must be compared with those for the variable threshold techniques and also for completeness the pulse canceller. The penalties for the variable thresholding techniques are not easily determined because the distributions involved are not circular symmetric and Gaussian. However expressions have been derived for high signal-to-noise ratios and have been shown to be valid to within 1 dB for SNRs as low as are required for a probability of detection of 0.8 in the absence of clutter (6). These results are quoted to emphasize some of the positive qualities of the coherent clutter map. Similarly, the SNR penalty for the pulse canceller is also quoted from (6) for the purpose of comparison.

The expressions for the various SNR penalties discussed above are shown in Figure 7. As was mentioned earlier the expressions corresponding to T_G and T_A are valid only for SNRs such that the probability of detection is greater than 0.8 when no clutter is present. In addition the penalty for T_A can only be bounded as there is no closed form expression of this threshold. These expressions include the effect of the Doppler filter and hence the clutter-to-noise ratio has been modified (CNR replaced by CNR_1) for T_G and T_A to reflect this. In other words CNR_1 is the constant clutter power to diffuse clutter and noise power ratio at the output of the filter for a given Doppler frequency θ.

There are several conclusions to be drawn from these expressions. First, the penalties for T_G and T_A increase with increasing clutter strength; however they are finite for all Doppler frequencies and hence radially slow or stationary targets can be detected provided the target return is strong enough. Secondly, the penalty for the pulse canceller is independent of the clutter strength but goes to infinity for the blind speeds ($\cos\theta=1$) which includes the important case of radially stationary targets. Finally, the penalty for the coherent clutter map processor is both independent of clutter strength and finite for all velocities. Therefore it possesses the positive properties of both the variable threshold and pulse canceller techniques.

To better illustrate the conclusions drawn above a specific example of the SNR penalties is shown in Figure 8. In this figure N (the filter length) is 8 and CNR is 15 dB. Only the upper bound for T_A is shown; the lower bound is within 1 dB of this. All the penalties are periodic in θ (period = 1) and symmetric about zero; therefore only one-half the cycle is shown.

Again the basic principles are evident: the penalties for the thresholds T_G and T_A are high but finite so detection in the zero Doppler region is possible, the penalty for the pulse canceller is relatively low over most Doppler frequencies, but infinite in the zero Doppler region. The penalty for the coherent clutter map on the other hand is relatively low for all Doppler frequencies. Even when only one previous return is used for the clutter estimate (n=1) there is a significant improvement over the other techniques. For an n of 10 the situation improves considerably to where the penalty never exceeds 1.6 dB and for 80% of the Doppler frequency region it is below 0.25 dB: for all practical purposes this is optimum performance. In addition it still permits the clutter estimator to respond fairly quickly to clutter drifts and in fact practical values for n may be considerably higher, thereby lowering the signal-to-noise ratio penalty even farther.

COMMENTS AND CONCLUSIONS

In practice the coherent clutter map can be implemented in many forms depending upon the complexity and resources of the radar system. A map could be built that covers the whole radar space keeping track of which detection bins require correction and what that correction should be. This type of system would require a considerable amount of memory - twice that of a noncoherent clutter map. Alternatively the map could be restricted in its area of operation to reduce the memory requirements. For example, a radar tracking an aircraft can predict the next detection bin that the target is most likely to appear in. Since the target has not arrived there yet a small clutter map could be built up around the expected location, thus effectively preparing the area for target detection. This kind of clutter mapping would not be of much help in the initial detection stages but after a track has been established it would be of great value in following it through a heavy clutter region.

In conclusion, it has been shown that at least some ground clutter, as seen by a step-scan phased array radar, possesses long term coherency. This coherency permits the use of a coherent clutter map which significantly outperforms traditional methods for handling such clutter.

REFERENCES

1. Dax, P.R., April 1975, Microwaves, page 34.
2. Sutherland, J.W., 1977, "World Market Trends in Radar for Defence and Air Traffic Control", IEE Radar-77, pp. 1-2.
3. Muehe, C.E., Cartledge, L., Drury, W.H., Hofstetter, E.W., Labitt, M., McCorison, P.B., and Sferrino, V.J., 1974, Proc. of the IEEE, 62, 716-723.
4. O'Donnell, R.M., Muehe, C.E., Labitt, M., Drury, W.H., and Cartledge, L., 1974, "Advanced Signal Processing for Airport Surveillance Radars", EASCON '74, 71-71F.
5. O'Donnell, R.M., and Muehe, C.E., 1979, IEEE Trans. Aero. and Elect. Syst., AES-15, 508-517.
6. Bird, J.S., 1980, "The Application of Circular Symmetric Signal Theory to the Detection of Signals in Noise and Clutter", Ph.D. dissertation, Carleton Univ., Ottawa, Ont., Canada.
7. Bird, J.S., and George, D.A., 1981, IEEE Trans. on Com., COM-29, 1357-1365.

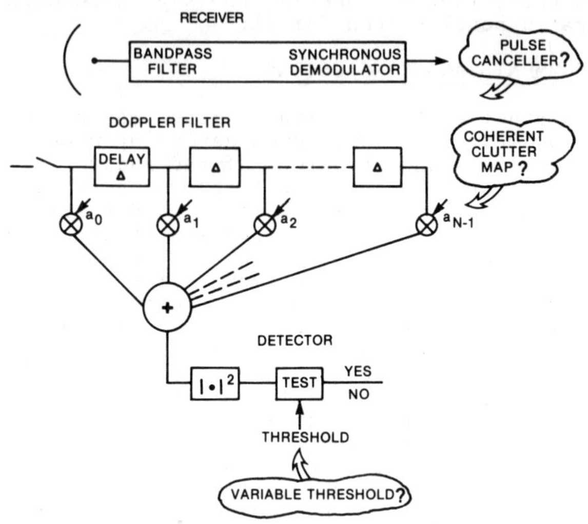

Figure 1 A typical signal processing scheme for a step-scan phased array radar

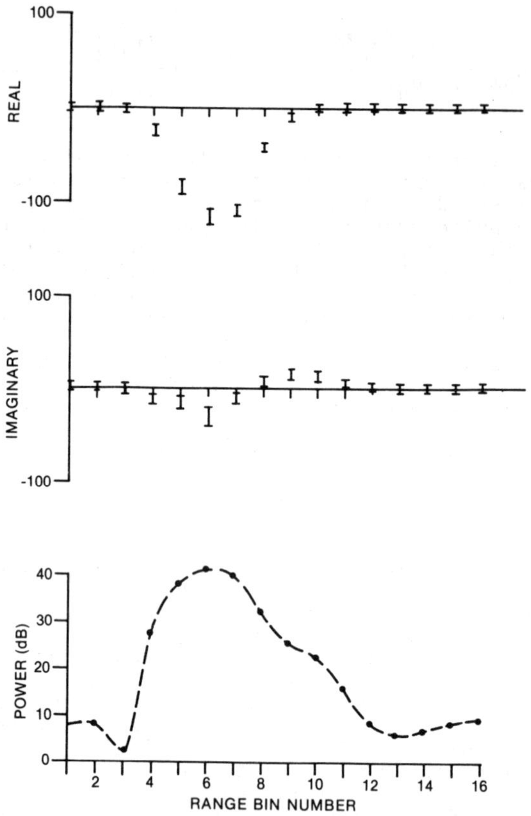

Figure 2 Summary of returns from Aylmer Island

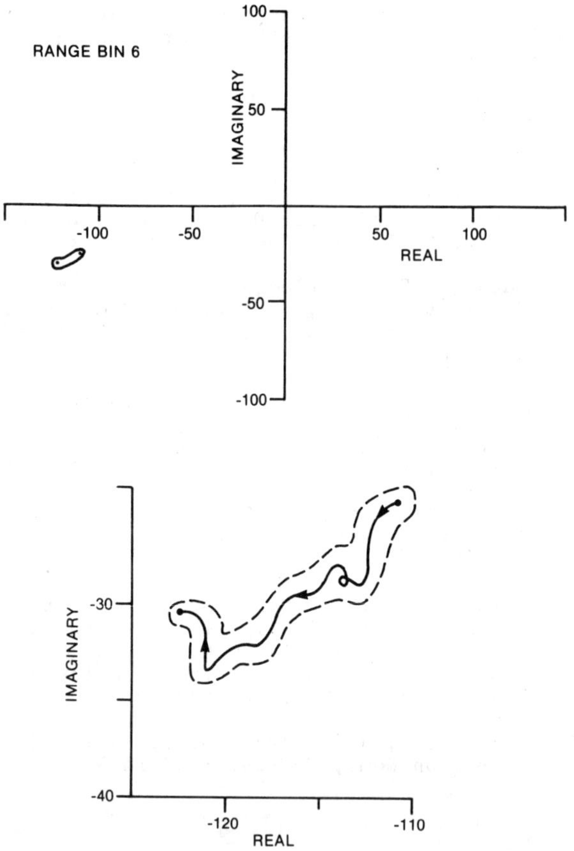

Figure 3 Phase diagram showing returns for range bin number 6

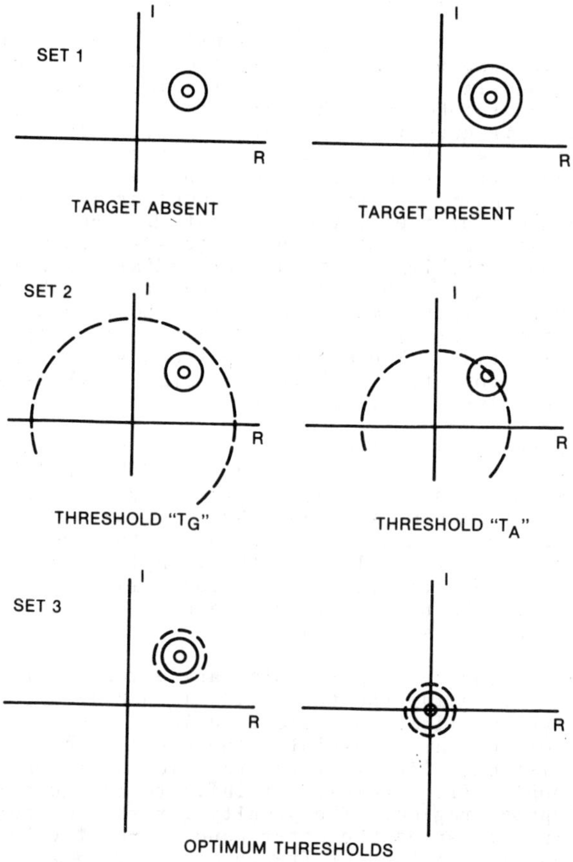

Figure 4 Phase diagrams and detection threshold

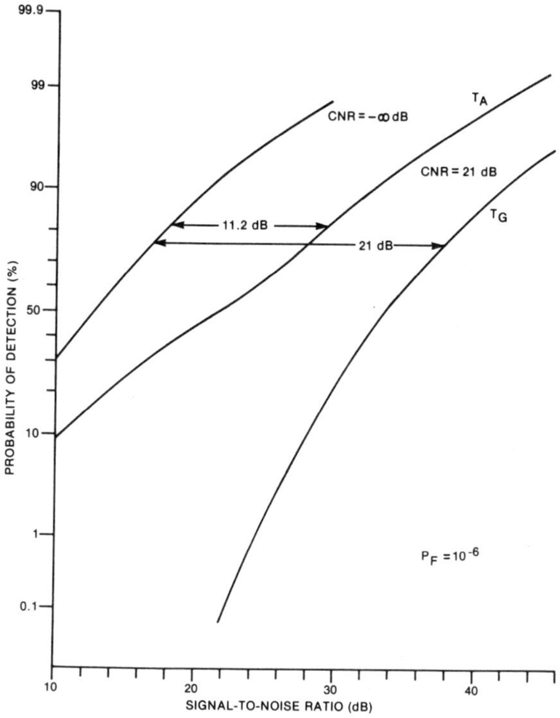

Figure 5 Probability of detection curves for the different thresholding techniques

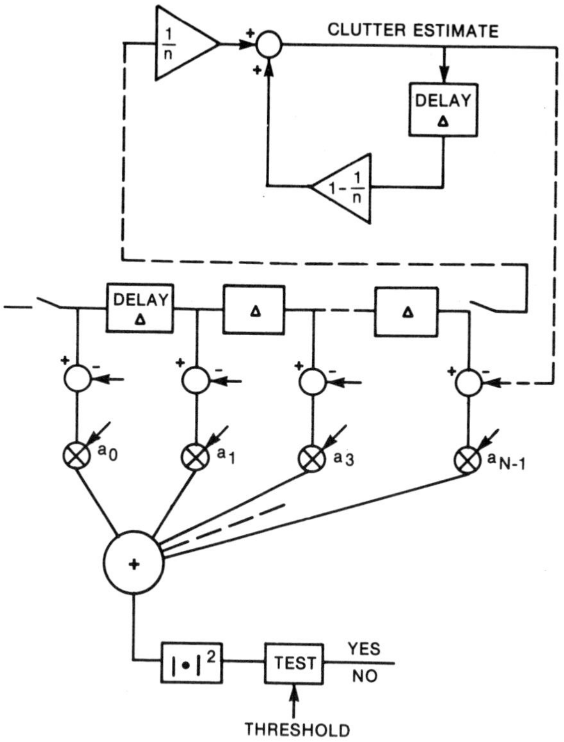

Figure 6 Coherent clutter map signal processor

Signal-to-Noise Ratio Penalties

1) Variable thresholding

$$SNR_{PG} = CNR_1 + 1$$

$$\max\left[\begin{array}{c}\dfrac{[CNR_1^{\frac{1}{2}} + \frac{1}{\sqrt{2}}\mathrm{erfc}^{-1}(P_F)]^2}{\ell n\frac{1}{P_F}} \\ 1\end{array}\right] < SNR_{PA} < \frac{[CNR_1^{\frac{1}{2}} + (\ell n(\frac{1}{P_F}))^{\frac{1}{2}}]^2}{\ell n\frac{1}{P_F}}$$

where

$$CNR_1 = \frac{CNR}{N}\left(\frac{\sin\left(\frac{N\theta}{2}\right)}{\sin\left(\frac{\theta}{2}\right)}\right)^2$$

2) 2 Pulse canceller

$$SNR_{PP} = 1 + \frac{N + \cos\theta}{N^2(1-\cos\theta)}$$

3) Coherent clutter map

$$SNR_{PM} = 1 + \frac{1}{N(2n-1)}\left(\frac{\sin\left(\frac{N\theta}{2}\right)}{\sin\left(\frac{\theta}{2}\right)}\right)^2$$

Figure 7 Signal-to-noise ratio penalties

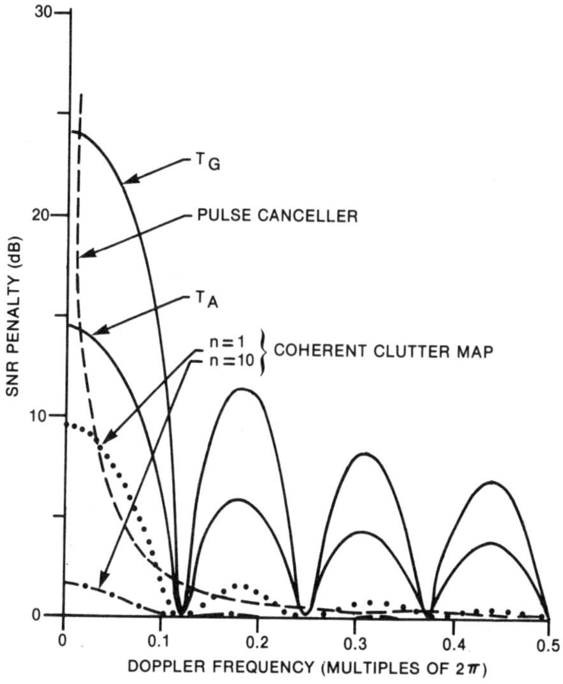

Figure 8 Signal-to-noise ratio penalties for N=8 and CNR=15 dB. A 2 pulse canceller was used.

AN EXPERIMENTAL ADAPTIVE RADAR MTI FILTER

Y.H. Gong and J.E. Cooling*

Chengdu Radio Engineering Institute, The People's Republic of China
*Loughborough University, UK

INTRODUCTION

Experience has shown that the performance of Moving Target Indicator (MTI) radars can be seriously affected by noise sources such as chaff, moving rain clouds, and similar effects. This is due to the fact that:

(a) The signal to noise ratio at the receiver input is often very low.

(b) The noise ("clutter") spectrum is not fixed but varies randomly with time.

As the signal and clutter spectra do not usually overlap the noise can normally be filtered out. For best performance, however, the filter stop band should automatically track the spectrum of the clutter. In earlier designs analogue filters were used, but these are limited and allow only simple structures to be realised in practice. Digital methods, however, enable complex self-adaptive filters to be implemented. Their drawbacks previously have been those of cost and size. This is now changing with the advent of relatively low cost VLSI devices, and so digital signal processing systems are likely to become much more commonplace in the near future. This paper describes the theoretical and practical features of a self-adaptive filter which is designed to remove clutter noise from a radar signal. The hardware uses an 8 bit microprocessor/fast hardware multiplier combination together with analogue-digital and digital-analogue interfaces. The software is implemented in assembler language.

DIGITAL FILTERING - THEORETICAL ASPECTS

General

There are two aspects of the filter which need to be decided upon

(a) The required digital transfer function of the filter system.

(b) The self-adaption method used by the filter.

These are described below. Two methods of self adaption are evaluated for the selected filter algorithm though the basic concept is the same in each case.

The Digital Transfer Function Algorithm

It is possible to implement very complex filters using digital techniques. However in many applications it has been found that a 2nd order cancellation filter (which is relatively simple) gives a satisfactory performance. Its transfer function is (fig. 1):

$$H(z) = 1-2WZ^{-\frac{Tr}{Tsam}} + Z^{-2\frac{Tr}{Tsam}} \quad \ldots\ldots\ldots (1)$$

where Tr is the delay period
Tsam is the sampling period
W is the weighting coefficient.

Its frequency response is:

$$H(f) = 2(\mathrm{Cos}\ 2\ f\pi Tr-W)e^{-j2\pi f\ Tr} \quad \ldots\ldots\ldots (2)$$

The filter output y(t) is:

$$y(t) = X(t)-2\ WX(t-Tr) + X(t-2Tr) \quad \ldots\ldots (3)$$

The average output power $\overline{y^2(t)}$ is:

$$\overline{y^2(t)} = \left[2R(0) + 2R(2Tr)\right] - 8R(Tr)W + 4R(0)W^2. \quad (4)$$

where $R(n) = \overline{X(t)\ X(t-n)}$

R(n) can be seen to be the autocorrelation function of the input signal X(t).

Self Adaption Techniques

The concept of the self-adaption methods described here is straightforward. It assumes basically that the noise power is greater than the signal power and that their spectra are non-overlapping. Thus if the total radar signal is filtered until it reaches a minimum value then the filter is mainly attenuating the noise (ideally the signal is unaffected). The self-adaption process operates by varying the filter stop band until the condition of minimum output power is achieved. Thus, provided the filter can be varied sufficiently quickly, the signal can usually be extracted from the noise even if the noise spectrum is time varying.

MTI performance is generally described in terms of an Improvement Factor (I) (Skolnik (1)), defined as

$$I = \left(\frac{\overline{So}}{\overline{Si}}\right)\left(\frac{Ci}{Co}\right) \quad \ldots\ldots\ldots\ldots\ldots (5)$$

where $\overline{Si}$, $\overline{So}$ are the input and output powers averaged over all possible Doppler frequencies and Ci, Co are the input and output clutter powers. Thus for a (virtually) constant $\overline{So/Si}$ during adaption I is maximising by minimising Co. Therefore it is required to minimise $\overline{y^2(t)}$ of eqn. (4) (a mean-square error criterion).

It can be seen from eqn. (4) that $\overline{y^2(t)}$ is a parabolic function which is minimised by varying either W or Tr. Both methods are described below.

W Adaption Algorithm (Coefficient Method)

This is a standard case as described by Widrow (2) where W is adapted until $\overline{y^2(t)}$ is a minimum. The gradient of the function w.r.t. W is zero at this point and W has its optimum value (W opt).

$$\frac{\partial(\overline{y^2(t)})}{\partial W} = \nabla W(\overline{y^2(t)}) = -\ 8R(Tr) + 8R(0)W \quad \ldots (6)$$

For an optimum value of W (i.e. W opt), $\nabla W(\overline{y^2(t)}) = 0$

$$\text{Thus } W\ opt = R(Tr)/R(0) \quad \ldots\ldots\ldots\ldots (7)$$

It is also required to adapt as fast as possible, and the "steepest descent" algorithm is

$$W(n+1) = W(n) - K\nabla W(\overline{y^2(t)}) \quad \ldots\ldots\ldots (8)$$

Writing Ks = 4K (loop gain)
and R(O) = Pin (input power, $X^2(t)$),

$$W(n+1) = W(n)(1-2Ks\ Pin) + 2Ks\ Pin\ W\ opt \quad . . . (9)$$

When using Widrows Least Mean Square (LMS) algorithm (2) an estimated gradient of $\hat{\nabla y}^2(t)$ is used instead of $\nabla Wy^2(t)$ where

$$\nabla^2 y^2(t) = \frac{\partial(y^2(t))}{\partial W}$$

For this the adaption equation (8) becomes

$$W(n+1) = W(n) + Ksy(n)\ X(n-Tr/Tsam) \quad . . . . . (10)$$

which is a stochastic approximation of (8). It can be shown that provided Ks < 1/Pin then W(n) will converge exponentially to its optimum value (W opt) with a time constant (τ) of

$$\tau = \frac{-1}{\log_e (1-2\ Ks\ Pin} \quad . . . . (11)$$

The block diagram of this adaptive filter is shown in fig. 1. The practical implementation of this filter (described later) uses the W algorithm only.

Tr Adaptive Algorithm (Delay Method)

In this case Tr is adapted until $\overline{y^2(t)}$ is a minimum and Tr attains the value of Tr opt. The steepest descent algorithm is

$$Tr(n+1) = Tr(n) - K\nabla Tr\ \overline{y^2(t)} \quad . . . . (12)$$

(Compare with eqn. (8)).

It can be shown that as

$$\overline{-y(t)\ X(t-Tr)} \simeq \nabla Tr\ \overline{y^2(t)} \quad . . . . (13)$$

and using the stochastic approximation (valid for O<f Tr<½) of

$$y(n)\ X(n-Tr/Tsam) = \overline{y(n)\ X(n-Tr/Tsam)} . . . . . (14)$$

Eqn. (12) becomes

$$Tr(n+1) = Tr(n) + Ks\ Y(n)\ X(n-Tr/Tsam) \quad . . . . (15)$$

For narrow band clutter with centre frequency fco and input power Pin the optimal value of Tr is;

$$Tr(opt) = \frac{Cos^{-1} W}{2\pi\ fco} \quad . . . . (16)$$

with an adaption time constant of

$$\tau = \frac{1}{\log_e (1-Ks\ 2\ fco\ Pin\ \sqrt{1-W^2})}$$

It can be seen from eqn. 16 that as fco tends to zero then Tr (opt) tends to infinity. This sets an additional constraint on the Tr algorithm.

Computer simulation results for the Tr algorithm were in close agreement with these theoretical deductions. However this filter was not implemented on the microcomputer system.

Computational Problems

Two specific questions have to be considered in practical digital systems:

(a) Which particular number representation system should be used in the filter ? and

(b) What errors are introduced by the finite word length of the computer ?

Number Representation

For this work fixed point 2's complement numbering was used. This method minimises computation times and allows the same software routines to be used for addition and subtraction. Because of the circular nature of this code partial sums generated during computation can overflow without causing problems (provided that the final sum does not exceed the defined maximum value of the number system). This last problem, the overflow one, can be catered for by using scaling factors before each summation node. In the adaptive filter of fig. 1 node N1 has the scaling factor of

$$\frac{1}{|H(f)m|} = \frac{1}{4}$$

where H(f)m is the maximum value of transfer function of the filter (eqn. (2)).

For node N2 Ks itself serves as the scaling factor.

Quantisation Noise Effects

Both analytical estimation and computer simulation of the quantisation noise effects were carried out for this filter. Three main sources of noise were identified. These are

(a) A-D conversion quantisation noise (e_{AD})

(b) Round-off noise due to the multiplication process $[(-2W) . X(t-Tr)]$, denoted as e_{R1}

(c) Round-off noise due to the multiplication $[(y) . X(t-Tr)]$, denoted as e_{R2}.

Computer simulation results are given in fig. 2 which show the effects of quantisation noise on the filter performance. From this it can be seen that for a performance specification of I>35 dB the word length requirements are:

ADC Wordlength; > 7 bits
Multiplication at Node 1; > 8 bits
Multiplication at Node 2; >12 bits.

The word lengths on the micro system were selected to fit in with the computing structure of the filter. For ADC and node 1 multiplication 8 bits is used whilst 16 bit working is used at node 2.

MICROCOMPUTER SYSTEM

The purpose of this work was to devise the adaptive filter of fig. 1 using a microprocessor based system. Fig. 3(a) illustrates the hardware system, which uses an Intel 8085 microprocessor, a fast hardware multiplier (TRW MPY 8HJ), A-D and D-A converters and associated circuitry. A 5th order Butterworth network was used for the input (anti-aliasing) analogue filter and also for the output reconstitution filter.

A hardware multiplier is used in order to speed up the computational time. It also has the advantage of simplifying the software. Although it is an 8x8 bit multiplier 16x8 bit operations are easily implemented. A 2K Byte EPROM is used for programme memory and 256 bytes of RAM is available for use as the data store. The complete unit is packaged on a single printed circuit board.

The working cycle of the software is shown in fig. 3(b). The programme is written in 8085 assembler code, occupies 200 bytes of store, and has an approximate cycle time of 580 μSec.

EXPERIMENTAL RESULTS

General Test Procedure

The filter performance was evaluated by measuring its output response to an input consisting of signal and noise. The parameters of importance are the adaption time (the "transient time") and the steady state noise reduction factor (the improvement factor I). The input signal was a pulse-modulated c.w. one whilst the noise has a defined centre frequency (fco) and relative bandwidth ($\Delta f/fco$). A white noise source was used in conjunction with a tuneable bandpass filter to generate the noise signal.

Transient (Adaption) Time

In practice the transient time is dependent on;

(a) The input power to the filter (Pin)

(b) The initial value of weighting coefficient (W)

(c) The filter loop gain (Ks)

(d) The centre frequency and bandwidth of the clutter.

Specimen results are given in fig. 4, whilst a picture of the transient (adaption) process is shown in fig. 5. These are given for a signal pulse width of 10 mSec (signal frequency of 500 Hz) having a repetition rate of 10 Hz. The clutter has a centre frequency of 100 Hz, and the input signal to noise ratio is -20 dB. Sets of values are given for four different clutter bandwidths. It can be clearly seen that fast adaption is achieved when Ks has a high value.

A problem was found in handling wide dynamic range input powers. This can be overcome by using $[X(t-Tr)/|X(t-Tr)|]$ instead of $[X(t-Tr)]$ in the correlation operation (Gabrial (3)). Subsequent computer simulation work produced good results and validated this technique.

Steady State Improvement Performance (I)

This depends on the centre frequency and bandwidth of both the signal and the clutter and also on the input signal to noise ratio. Specimen results for the adaptive filter are given in fig. 6 and the actual circuit waveforms are shown in fig. 7. In fig. 6 the performance of a 2nd order non adaptive filter is also given (for comparison purposes). This was obtained by using the approximate expression;

$$I = \frac{1.5}{[\mathrm{Cos}(2\ fcoTsam)-1]^2 + [\mathrm{Sin}(2\ fcoTsam)]^2\ [2\pi\sigma cTsam]^2}$$

where σc is the standard deviation of the clutter.

This expression is valid only when the clutter bandwidth is very small.

CONCLUSIONS

The performance of an MTI radar which is subject to clutter noise can be significantly improved by using adaptive filtering methods. A second order filter which uses output power minimisation techniques has been shown to satisfy this requirement. This assumes, however that there is little overlap between the signal and noise spectra, and that the noise power is much greater than that of the signal.

One of the most important factors to be considered when designing digital filters is the quantisation noise. This degrades the steady state performance from that of the ideal (infinite word length) filter. Similation tests which were carried out to evaluate quantisation noise effects provided a valuable basis for hardware design decisions.

The major limitation of the filter described here is low sampling rate (1.72 kHz) due mainly to the time spent on the multiplication routines. However the methods discussed in this paper are general purpose and can be applied to both traditional and more complex radar MTI systems provided that the filter sampling frequency is increased. The most promising way forward (at the present time) appears to be through the use of dedicated VLSI signal processors such as the NEC 7720 I.C.

REFERENCES

1. Skolnik, M.I., "Radar Handbook", McGraw-Hill, 1970.

2. Widrow, B., et al, Dec. 1967, "Adaptive Antenna Systems", Proc. IEEE, Vol. 55, 2143-2159.

3. Gabrial, W.F., Feb. 1976, "Adaptive Arrays - An Introduction", Proc. IEEE, Vol. 64, 239-272.

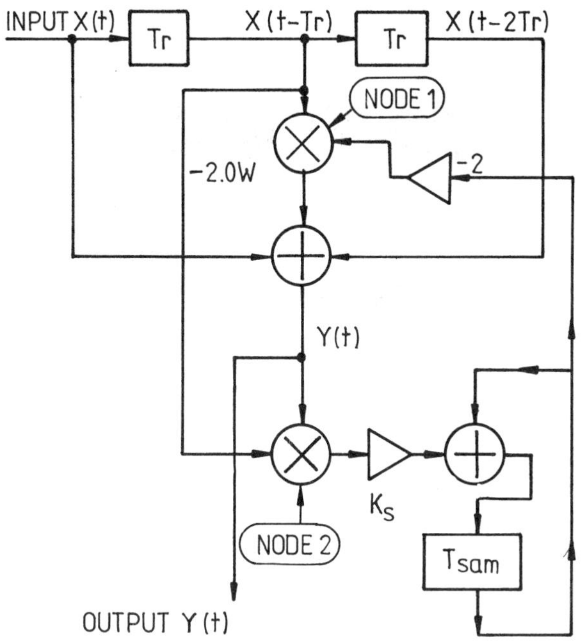

FIG. 1. Second-Order Radar MTI Adaptive Filter (Coefficient Adaption).

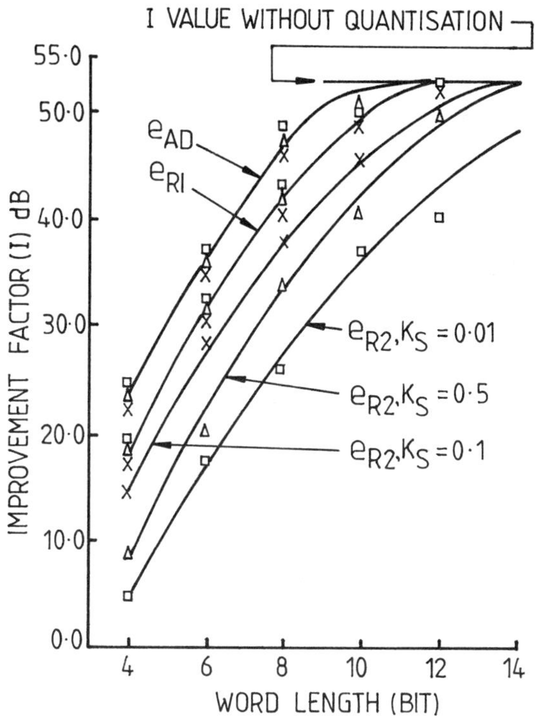

FIG. 2. Degradation of Filter Performance Due to Quantisation Noise.

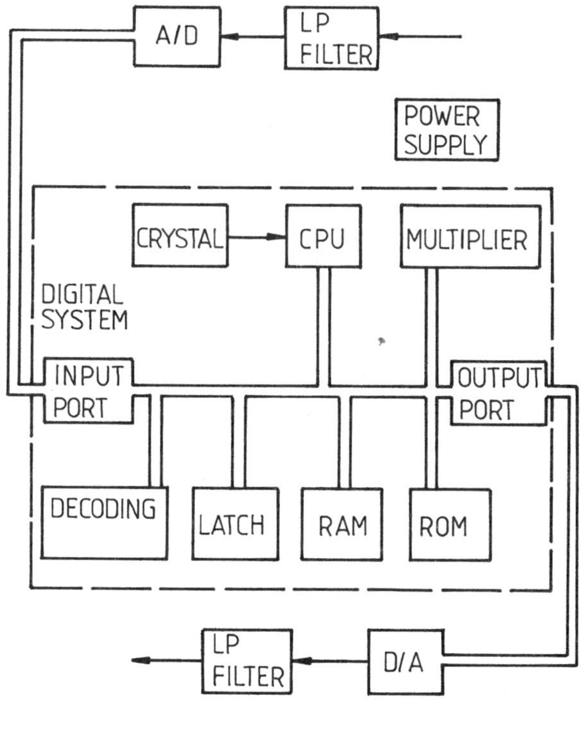

FIG. 3(a). Block Diagram of the Adaptive Filter.

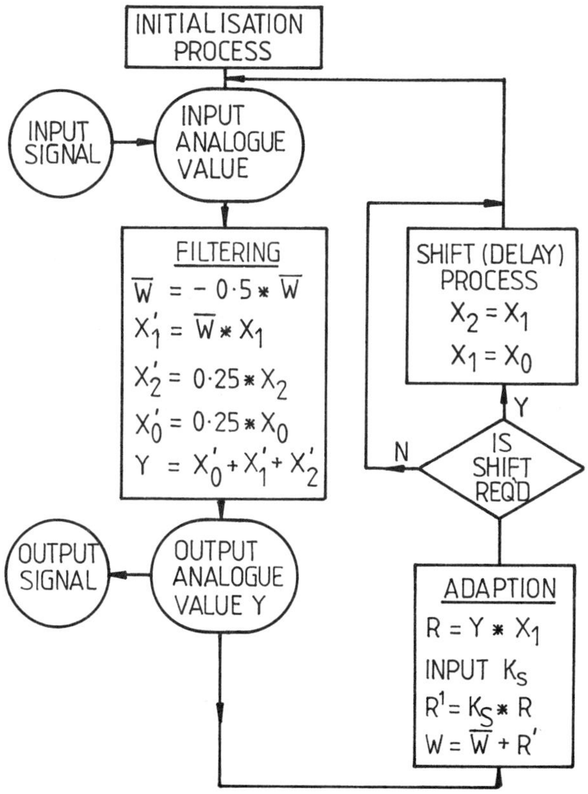

FIG. 3(b). Software Working Cycle.

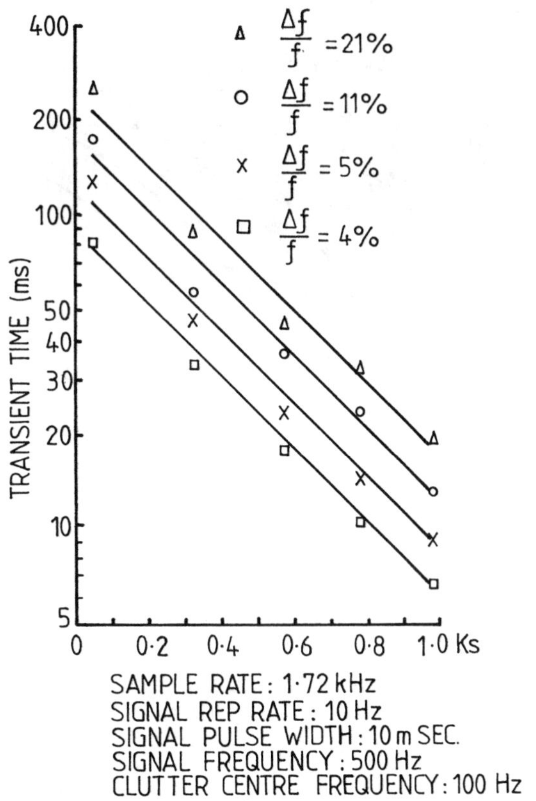

FIG. 4. Filter Transient (Adaption) Time.

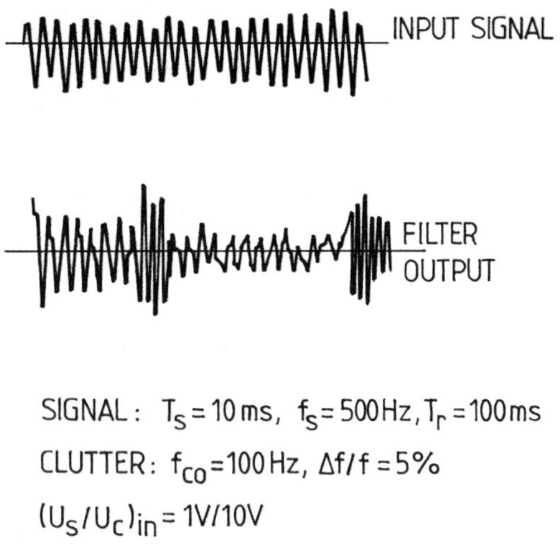

FIG. 5. The Transient Process of the Adaptive Filter.

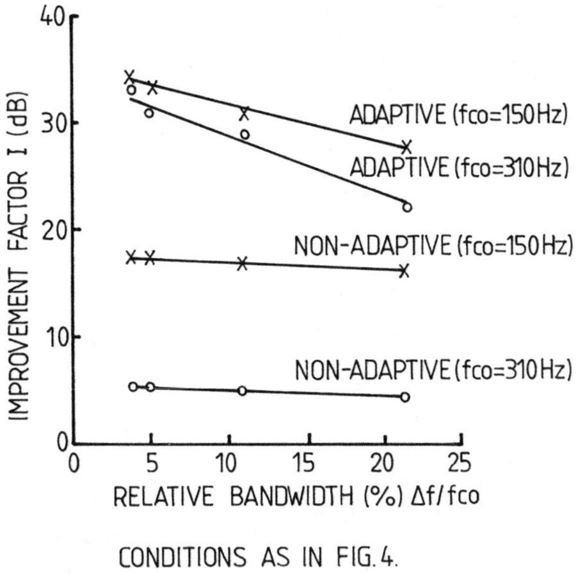

FIG. 6. Improvement Factor of the Adaptive and Non-Adaptive Filters (Conditions as in fig. 4).

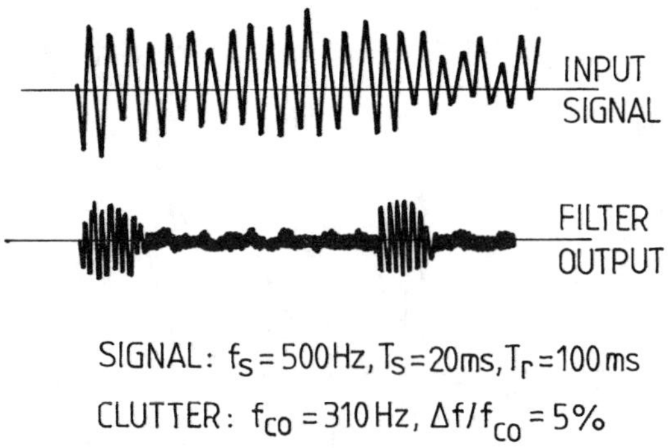

FIG. 7. Steady State Performance of the Adaptive Filter.

THE USE OF A MULTI-LEVEL QUANTISER IN PLOT EXTRACTION

P. N. G. Knowles

Plessey Electronic Systems Research Ltd., Southleigh Park House, Havant.

INTRODUCTION

One of the principal functions of a surveillance radar plot extractor is to integrate the video returns in each range cell across the azimuthal beamwidth. It is common practice to use a binary digital integrator, which must be preceded by some form of quantiser to select, as single-level hits, only those video samples which exceed an arbitrary threshold. The level to which this threshold is set is clearly a critical compromise between the loss of weak targets and the acceptance of excessive numbers of noise hits.

Many plot extractors based on this principle are in current operational service, providing excellent detectability of targets in a thermal noise background. In regions of ground or weather clutter, however, their performance is less acceptable due to the rather large numbers of unwanted clutter plots and the relatively poor detection of aircraft.

The Clutter Problem

This paper is concerned only with non-coherent, post-detection processing, where clutter presents the following difficulties:-

a) Non-uniformity over the coverage area.

b) Non-Rayleigh amplitude distribution (1)

c) High azimuthal correlation (2)

Non-uniformity can be dealt with by a fast-acting threshold device such as a background averager (BA), albeit with some detection loss in the clear. Some of these devices are, however, highly distribution dependent and fail on problem (b). Even where good distribution independence is achieved, as, for instance in the rank-ordered quantiser, problem (c) remains a serious obstacle.

Area MTI exploits the velocity difference between aircraft and clutter by comparing the video level in each resolution cell with that on previous rotations. Further, since it averages in time rather than space, it tends to decorrelate adjacent azimuth samples and reduces problem (c) (3). Consequently it is a highly effective process (4). However, the cost of the multi-level store needed for every resolution cell over the whole coverage area becomes unacceptable for high-resolution pulse compression radars, even using the latest devices.

PRINCIPLES OF MLQ OPERATION

Multi-level (8-bit) storage is provided for every range cell on every pulse-recurrence interval over an azimuth extent equal to, or somewhat greater than, the aerial beamwidth. The availability of this stored data avoids the need to make immediate and irrevocable threshold crossing decisions on the current PRI, as is the case with binary systems. It allows the thresholding of a given cell to be determined with hindsight, taking account of a variety of characteristics of the data surrounding it, both in range and azimuth.

One of the most powerful features of MLQ is its ability to measure the azimuth correlation coefficient in range cells surrounding the current sample. A high density of high correlation values is taken to indicate a clutter background and is used to raise the threshold of the current cell proportionately. The generation of clutter plots is consequently reduced, with little loss of detectability in the clear. Super-clutter visibility of strong targets is retained.

The storage requirements are moderate even for a high resolution radar.

IMPLEMENTATION

The basic functions of an experimental plot-extractor using MLQ principles are shown in figure 1. The equipment is described in reference 5.

A threshold is established as a function of the following measures:-

a) A long-term noise meter (NM), sampling at long range and effective in thermal noise regions.

b) A short-term background averager (BA), sampling in range azimuth cells surrounding the current cell of interest and effective in localised areas of high-amplitude noise or clutter. Its degree of control is preset by a multiplier for linear video and an offset for logarithmic video.

c) An azimuth correlation estimator (ACE), sampling the density of high correlation values and of nonstationarities in range in the area surrounding the current cell and effective in clutter and in noise discontinuities. Its degree of control is preset by a multiplier and a function selector.

Video samples which exceed the threshold are passed to a conventional binary moving-window integrator (BMWI).

As a separate processing chain, the pattern of video amplitudes as a function of azimuth in the current range cell is compared with the known 2-way beamshape of the aerial. When the cross-correlation value exceeds a preset threshold, indicating the centre of a point source target, the AND gate is enabled, allowing BMWI plots to be declared.

The equipment has the following main characteristics.

a)	Pulse length	0.1 μsec min.
b)	Range cells per PRI	8192 max.
c)	No. of PRIs stored	14
d)	Dynamic range of store	256 levels
e)	Total MLQ storage	0.92 Mbits

RESULTS

The MLQ plot extractor was evaluated using video tape recordings of an AR15 radar, having a pulse length of 1.0 μsec, located at Cowes (6). The VTR allows the same radar situation to be analysed with different extractor parameters. A test target, controllable in amplitude and beamshape, is mixed with the radar data with a pseudo-random phase relationship, representative of a true radar target. The radar data was unprocessed logarithmic video, with no MTI, swept gain or other fixes. There was extensive ground clutter from the Isle of Wight and the South Downs, but virtually no weather returns.

Figure 2 shows the total number of plots and the detection performance in the clear for operation with fixed threshold, noise meter, background averager alone and background averager plus ACE, the latter having multiplier settings of x1, x2, and x4. In each case the threshold additive offset was set to give a 5% probability of noise hit in the clear, equivalent to a false plot probability of 10^{-6}. The datum (OdB) for detection loss was the test target amplitude required to give an 80% detection probability with a fixed threshold. (Note: The number of clutter plots with FT and NM was so high as to approach the limits of resolution and is consequently understated).

The effectiveness of BA plus ACE in reducing clutter plots with little or no detection loss in the clear is particularly striking.

Figure 3 shows a similar comparison, but in this case the total number of plots was held at 70 (of which some 20 were aircraft).

The effectiveness of BA plus ACE is again well illustrated.

Measurements of super-clutter detection with BA plus ACE x1 were closely similar to those with BA alone. However, ACE multiplier settings of x2 and above led to significant loss of super-clutter visibility.

Figures 4 and 5 show flags from the BMWI (before being collapsed to single plot statements) over 30 aerial rotations using BA alone and BA plus ACE x2. In the latter case, the reduction in clutter plots and the improvement in the detectability of targets in the clear is quite apparent.

Tests of the beamshape cross correlator, used alone, confirmed simulation predictions of an unacceptably high probability of false plot from noise when its threshold is set low enough to accept noise-contaminated targets. However, used in conjunction with the BMWI, it gives a substantial improvement in azimuth resolution. As an example, 2 targets, differing in amplitude by 20 dB, were readily resolved when separated by 1 beamwidth. Moreover, unlike conventional plot extractors, the resolution capability increases with target amplitude. The BCC also provides an improvement in interclutter visibility. It is expected to provide protection against some forms of diffuse weather clutter, and non-point-source ground clutter, but this has not yet been measured.

CONCLUSIONS

The multi-level quantiser provides substantial improvements in plot extractor performance in the presence of clutter. This results from its ability to defer threshold decisions until the region surrounding the cell of interest is available for examination.

The azimuth correlation estimator is particularly effective in reducing clutter plots without losing target detectability in the clear.

The beamshape cross correlator gives a substantial improvement in azimuth resolution.

The storage and processing requirements are practicable, even for high-resolution long-range radars.

REFERENCES

1 Fay F.A. Clarke J. and Peters R.S. 1977 "Weibull distribution applied to sea Clutter" IEE Conference publication "Radar-77".

2 Hansen V.G. 1973 "Constant false alarm rate processing in search radars" IEE Conference publication "Radar - present and future".

3 Blythe J.H., Treciokas R. 1977 "The application of temporal integration to plot extraction ". IEE conference publication "Radar - 77".

4 Hinson N.R., Knowles P.N.G., Selwyn N.A. and Sheppard A.G. 1979 "Report of the evaluation of the experimental Area MTI Equipment" Plessey Report No. 17/79/R101U.

5 Hinson N.R. 1977 "Final report on a study into plot extractor improvements" Plessey Report No. 17/79/R073C.

6 Knowles P.N.G. and Selwyn N.A. 1982 "Performance evaluation of the improved plot extractor" . Plessey Report No. 17/82/RO91U.

ACKNOWLEDGEMENTS

This work has been carried out with the support of the Procurement Executive, Ministry of Defence.

The author wishes to thank the Directors of the Plessey Company plc for permission to publish this paper. Special thanks are due to Mr. N.R. Hinson who designed the equipment.

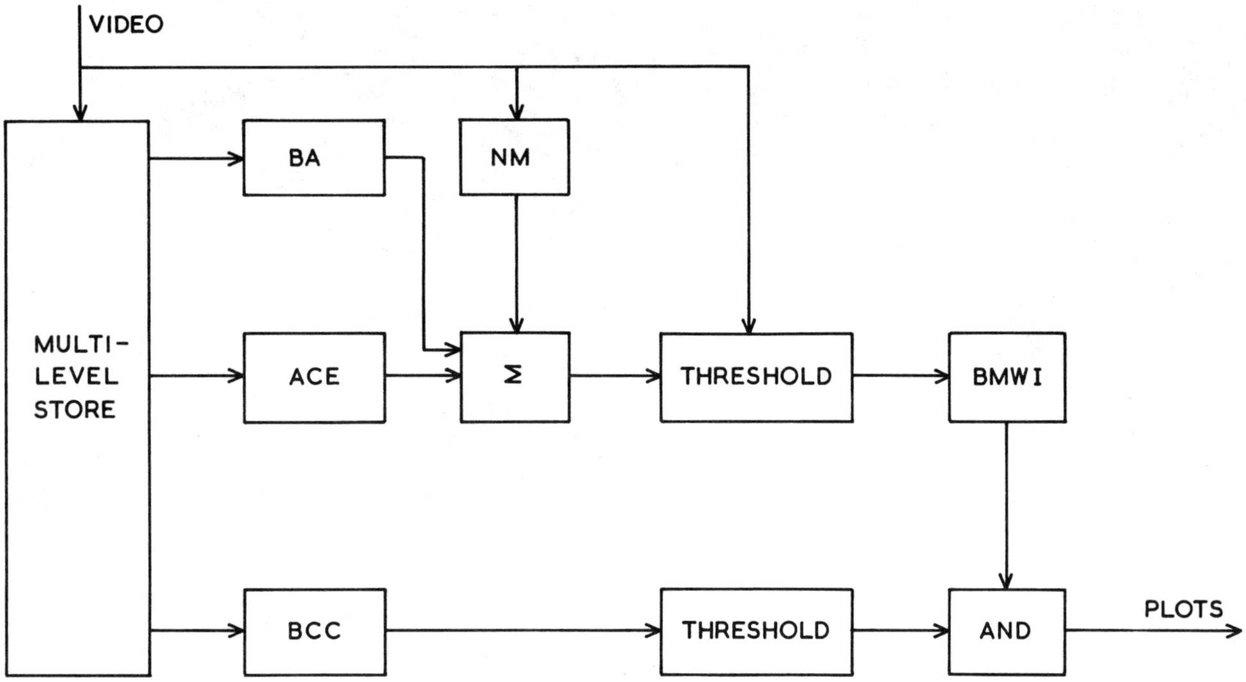

FIGURE 1 : MLQ PLOT EXTRACTOR

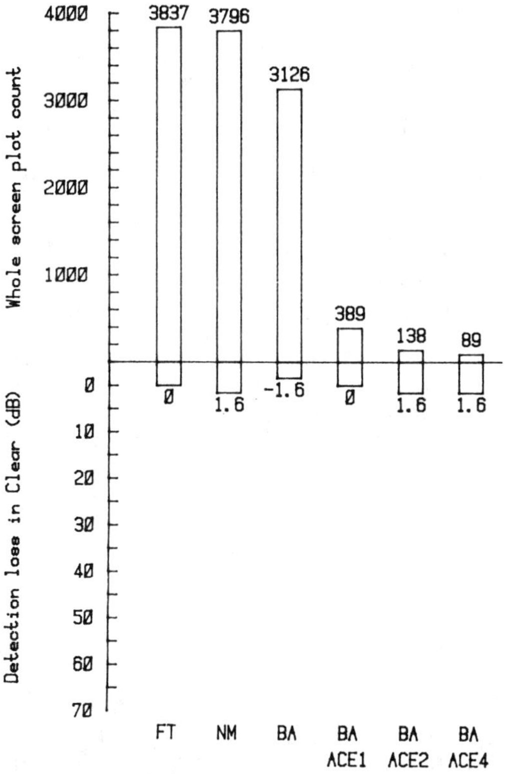

FIGURE 2 : PLOT COUNT AND DETECTION LOSS
FOR 5% Pn IN CLEAR

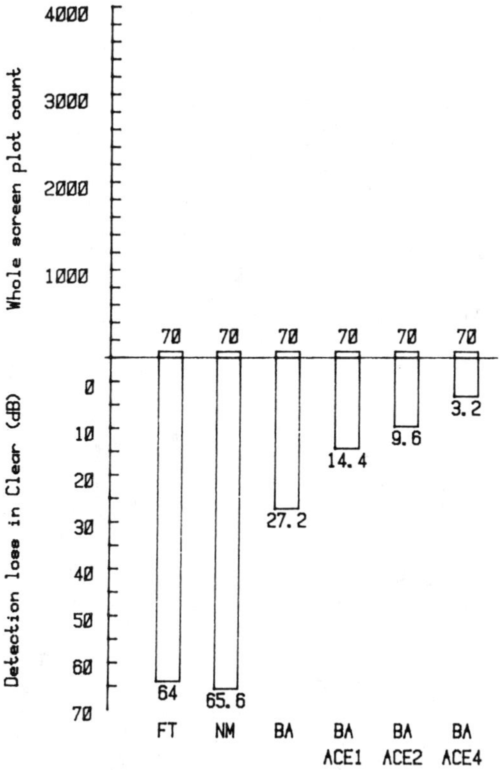

FIGURE 3 : DETECTION LOSS FOR 70 PLOTS
PER ROTATION

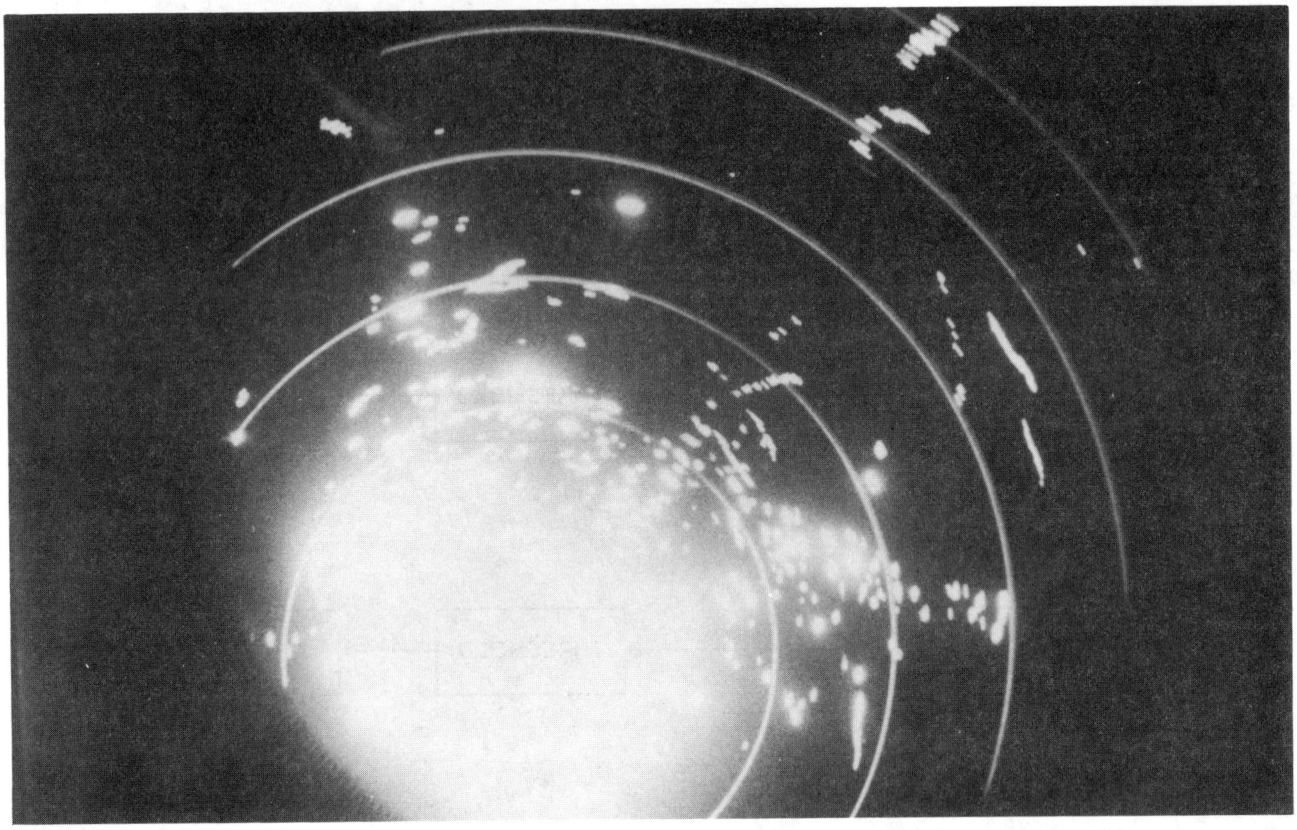

FIGURE 4 30 revs, BA alone

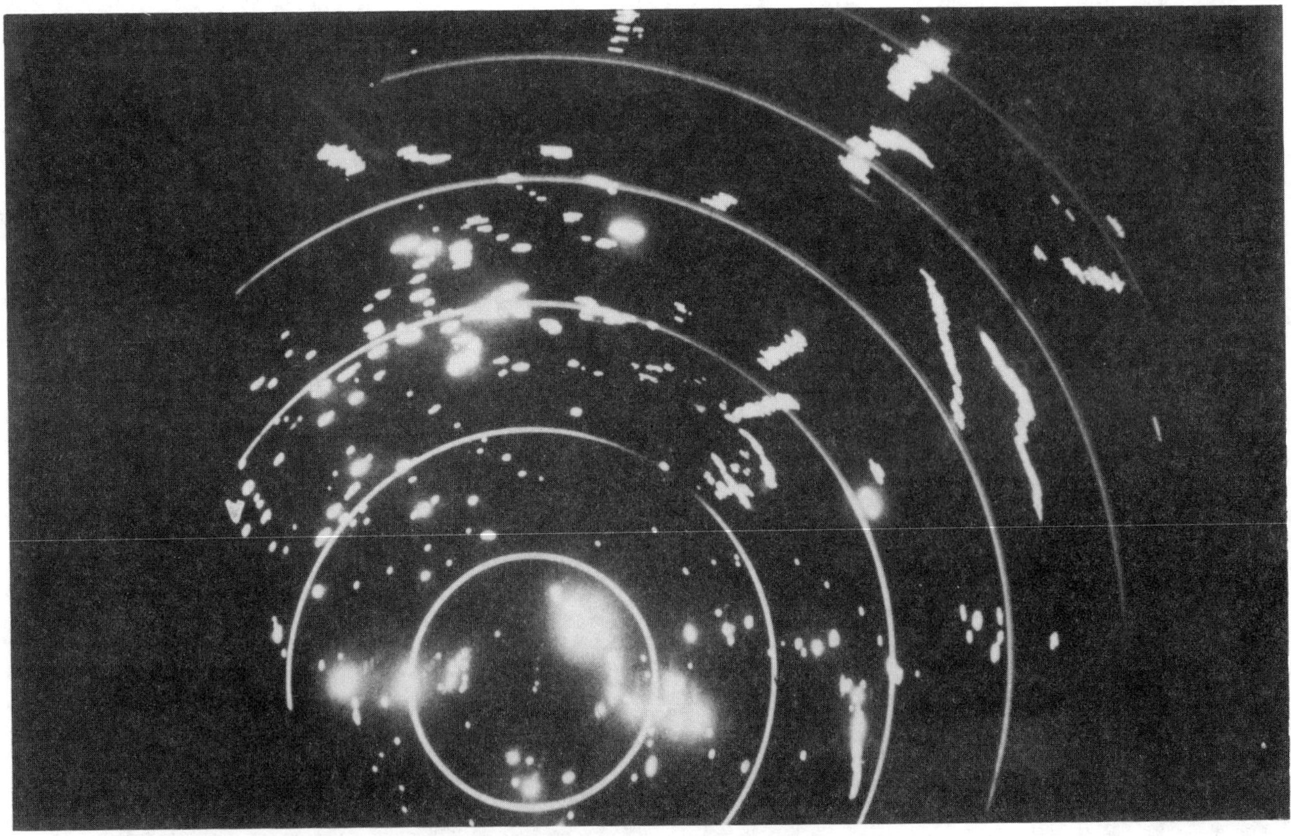

FIGURE 5 The same 30 revs, BA + ACE

IMPROVED COHERENT-ON-RECEIVE RADAR PROCESSING WITH DYNAMIC TRANSVERSAL FILTERS

R. L. Trapp

The Johns Hopkins University Applied Physics Laboratory Laurel, MD 20707 USA

INTRODUCTION

Radar returns originating from a pulsed noncoherent transmitter can be processed coherently by altering the return signal phase as a function of the corresponding transmit pulse phase [1]. Typically, a local oscillator is phase shifted a constant amount during a given pulse repetition interval and mixed with the return signal to achieve coherency. Figure 1 is a simplified block diagram of such a coherent-on-receive radar. Since the transmitter is noncoherent, the phase shift value changes from pulse to pulse; the phase shift required for a given interval is derived from a measurement of pulse phase during transmission.

Coherent-on-receive radars that rely on single-value estimates of transmit pulse phase can have degraded performance if the phase across a transmit pulse changes significantly. Measurements of magnetron transmitter intrapulse phase indicate considerable phase deviation across a short duration pulse [2]. Figure 2 is an example of the measured intrapulse characteristics of a Ka-band magnetron. Although the phase is constant near the pulse center for this example, it changes rapidly near the pulse edges and traverses more than 120° between the pulse half-amplitude points. Previous measurements also indicate a relationship between intrapulse phase behavior and transmitter load characteristics [2].

A coherent-on-receive approach that effectively uses multiple intrapulse phase estimates can compensate for the potentially large intrapulse phase variations in a noncoherent transmitter. This paper describes a coherent-on-receive configuration that embodies such an approach, and includes the results of system simulations that show the potential for improvement. The simulations were based on actual magnetron intrapulse data collected in previous tests [2].

COHERENT-ON-RECEIVE RADAR PROCESSING WITH TRANSVERSAL FILTERS

Matched filtering in both magnitude and phase not only maximizes the signal-to-thermal-noise ratio but also makes coherent an otherwise noncoherent signal [3]. A transversal filter approximates a matched filter when the impulse response is a time-reversed approximation of the transmitted signal. Noncoherent transmitters such as magnetrons are random in starting phase from pulse to pulse. Therefore, pulse-to-pulse adaption of transversal filter response enables coherent-on-receive operation [4].

Figure 3 illustrates a coherent-on-receive radar with dynamic response transversal filters. Coincident with the start of a transmit pulse, in-phase and quadrature (I and Q) components of the transmit signal are sampled and shifted into shift registers. The trans-

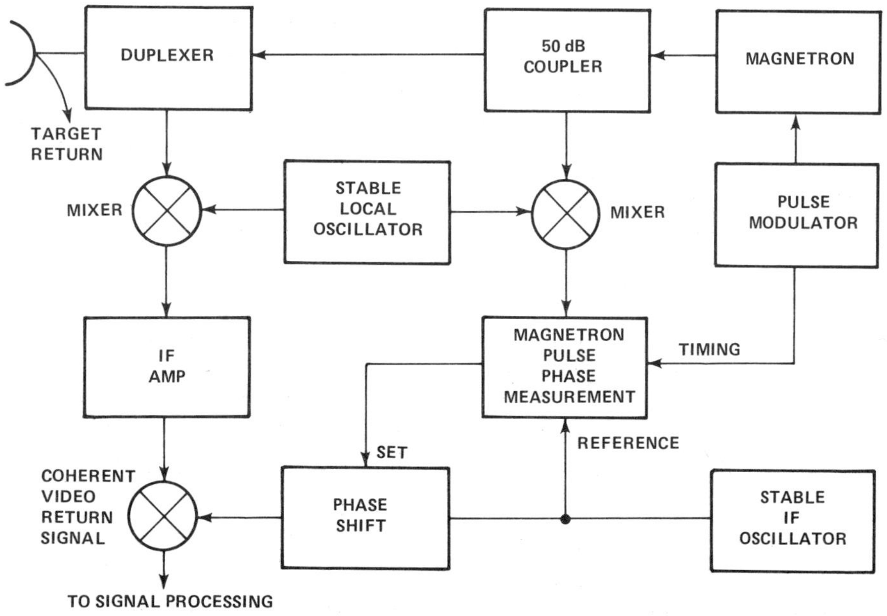

Figure 1 Typical coherent-on-receive radar.

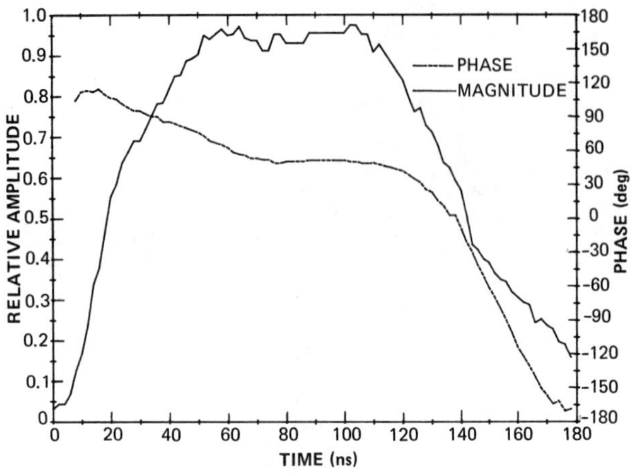

Figure 2 Example of the magnitude and phase of a Ka-band magnetron pulse.

mit samples in the shift registers serve as transversal filter tap weights. The desired transversal filter accuracy (i.e., the number of taps) establishes the number of intrapulse transmit samples required. The return signal, after translation to an IF frequency and amplification, is separated into I and Q components. Each of the quadrature return signals is directed into a series of delays with weighted taps that are summed to form a transversal filter. The maximum delay through the filter is nearly equal to the transmit pulse width. A filter delay element is equal in duration to the time between intrapulse transmit samples. Since transmit intrapulse samples are tap weights, the transversal filter impulse response changes from pulse to pulse and is a time-reversed replica of the most recent I or Q transmit pulse. The summed output of the I and Q channel transversal filters forms a filter that is approximately matched in magnitude and phase. The resulting video output is a coherent return signal.

A coherent-on-receive radar implemented with transversal filters uses multiple intrapulse samples. It can have performance superior to a coherent-on-receive radar that phase shifts a local oscillator on the basis of a single-valued estimate of transmit pulse phase. The transversal filter processing more closely approximates a matched filter. In addition, there can be less residual noise when using transversal filters, since the local oscillator in effect more closely follows the phase variations within a pulse.

COHERENT-ON-RECEIVE SIMULATIONS

Computer simulations of the dynamic transversal filter processing in a coherent-on-receive radar provided the primary means for evaluating performance. The simulations were based on the processing of actual magnetron pulse data collected in a series of measurements [2]. The data available consisted of 512 pulses with intrapulse sampling intervals of 2 ns from each of two Ka-band magnetrons. The magnetrons had pulse widths of approxi-

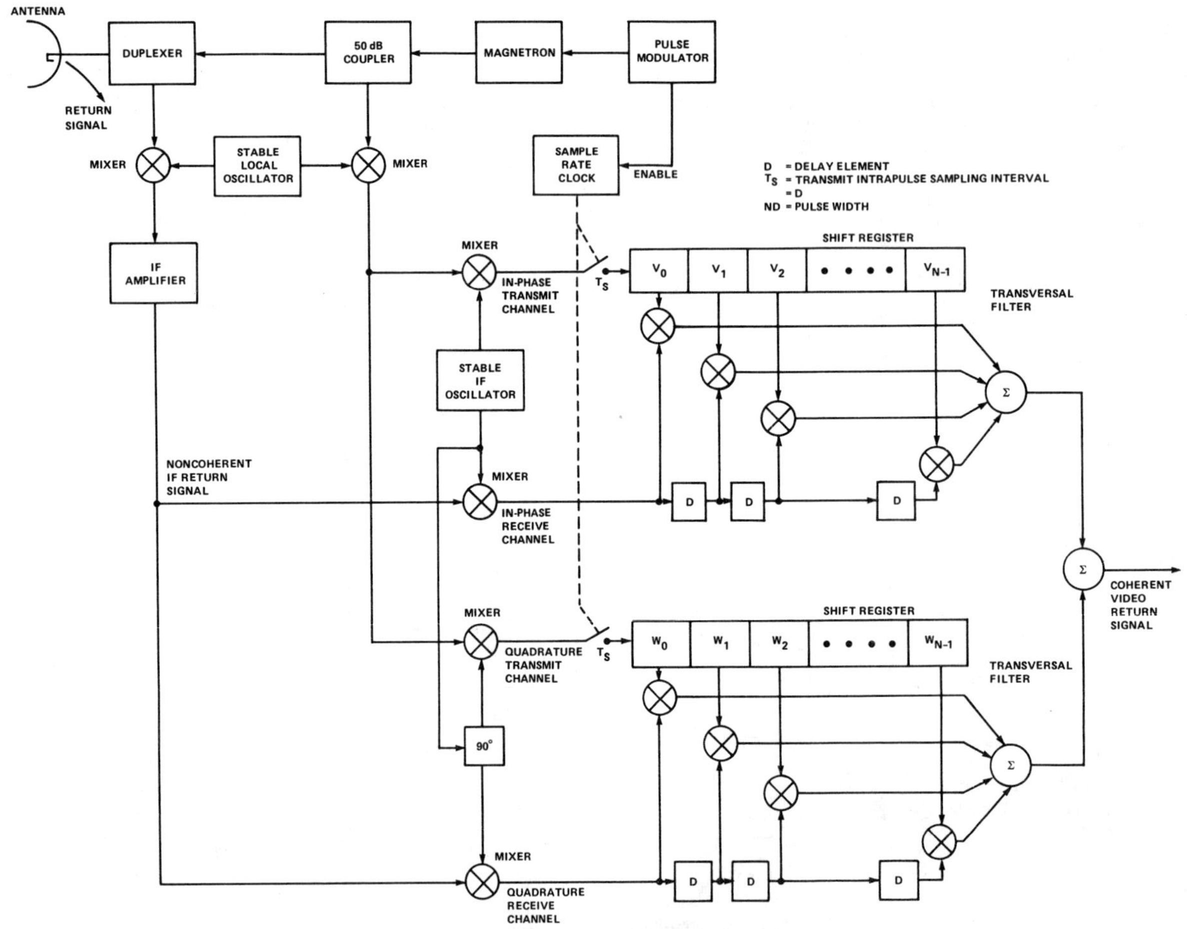

Figure 3 Coherent-on-receive radar with dynamic transversal filters.

mately 130 ns. Within the simulation, a point target return signal is generated from the magnetron data with the addition of wideband Gaussian noise and the Doppler effect. The transversal filters are simulated with a selectable number of taps. The wideband I and Q target return components (2 ns intrapulse spacing) are initially reduced in bandwidth by prefiltering with a moving average. The moving average time extent is set to equal the required intrapulse sampling interval for the transversal filters (i.e., pulse width/tap number). The transversal filters process the I and Q channel signals as described in the previous section. After rangegating the coherent video return signal, a 512 point fast Fourier transform (FFT) is performed. The evaluation of individual simulations is based on the frequency domain results.

The wideband Gaussian noise density was held constant throughout the simulations. Odd numbers of transversal filter taps from 1 to 9 were selected. A one-tap transversal filter is equivalent to coherent-on-receive processing with a local oscillator phase shift and a single-valued estimate of transmit phase (Figure 1). Therefore, the transversal filter approach to coherent-on-receive that uses multiple intrapulse transmit samples can be compared with the local oscillator phase shift method by comparing the multi-tap transversal filter results with the one-tap results.

Signal-to-noise (S/N) ratios computed in a 512 point FFT resolution bandwidth (e.g., Doppler filter) are graphed in Figure 4 versus transversal filter tap number for two magnetrons. The improvement in signal-to-noise ratio achievable with dynamic transversal filters is determined by comparing the multi-tap results with the single-tap results. As expected, signal-to-noise ratio increases with the number of transversal filter taps. The most significant S/N ratio improvement (almost 1 dB for magnetron 2 and 2.8 dB for magnetron 1) occurs when using only 3 taps. Simulations with magnetron 2 data show an increase in signal-to-noise ratio only slightly more for tap numbers greater than 3, to an improvement of 1.2 dB for a 9-tap transversal filter. However, the magnetron 1 simulations show a continuing improvement, reaching 3.9 dB when 7 transversal filter taps are used. There is obviously a point of diminishing return, where an increased number of transversal filter taps yields little improvement in signal-to-noise ratio. However, that point is dependent on the characteristics of the particular magnetron being used.

The magnetron 2 simulations increase in S/N less drastically than magnetron 1, but have a consistently higher absolute S/N ratio. The differing results between the two magnetrons can be correlated with their respective intrapulse phase characteristics. Typical intrapulse phase responses for magnetrons 1 and 2 are shown in Figure 5. The magnetron 1 intrapulse phase changes more rapidly and deviates over a greater phase extent than does magnetron 2. Therefore, a greater number of transversal filter taps are needed for magnetron 1 to establish a given signal-to-noise performance than with magnetron 2. It should be noted that although intrapulse phase characteristics vary from magnetron to magnetron, they also vary with load characteristics (e.g., standing wave ratio) [2]. A coherent-on-receive system should therefore be configured with transversal filters that are adequate for the more severe intrapulse variations that result from the possible load conditions.

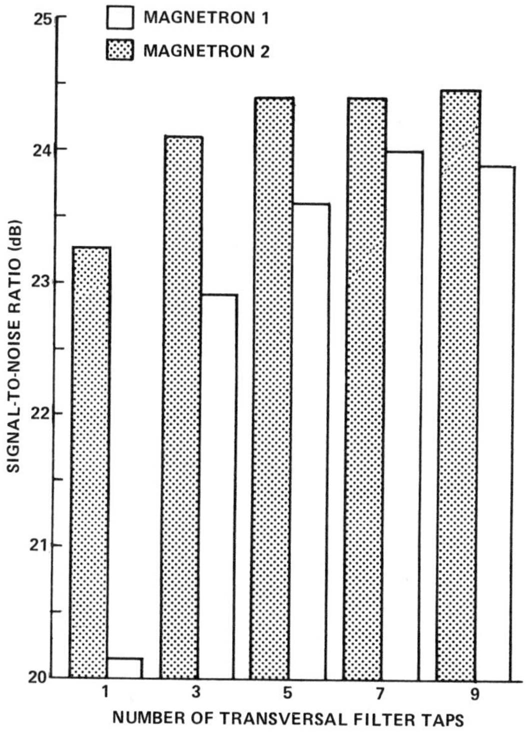

Figure 4 Coherent-on-receive signal-to-Guassian-noise ratio when using dynamic transversal filters.

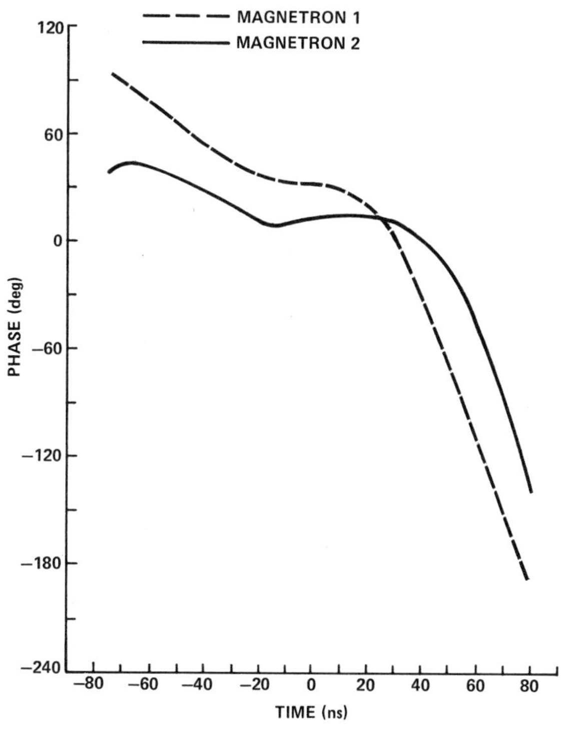

Figure 5. Typical phase responses of two Ka-band magnetrons.

Although not indicated in the S/N ratio results, dynamic transversal filter processing of the magnetron return pulses results in some degree of pulse compression. This is clearly evident when viewing the time domain output of the transversal filters. Since the transversal filters approximate a matched filter, pulse compression is an expected result in view of the significant phase deviations seen across magnetron pulses.

Additional coherent-on-receive simulations were run without wideband Gaussian noise to assess the effect of dynamic transversal filters on subclutter visibility. Limitations on subclutter visibility are typically established by equipment instabilities and signal processing. Imperfect coherent-on-receive processing can cause residual components of a return signal to be present across the Doppler band. This residual noise can be reduced by processing with dynamic transversal filters, so as to improve subclutter visibility.

The mean signal-to-residual-noise ratios that resulted from the Gaussian noise-free simulations are graphed in Figure 6. Magnetron 1 simulations exhibited higher signal-to-residual-noise ratios and greater improvement with tap number then did magnetron 2 simulations. Improvement in residual noise suppression of 6.2 dB and 2.5 dB are indicated with 9-tap transversal filters for magnetrons 1 and 2, respectively. However, a minimal 3-tap transversal filter improves residual noise suppression over 5 dB for magnetron 1 and 2 dB for magnetron 2.

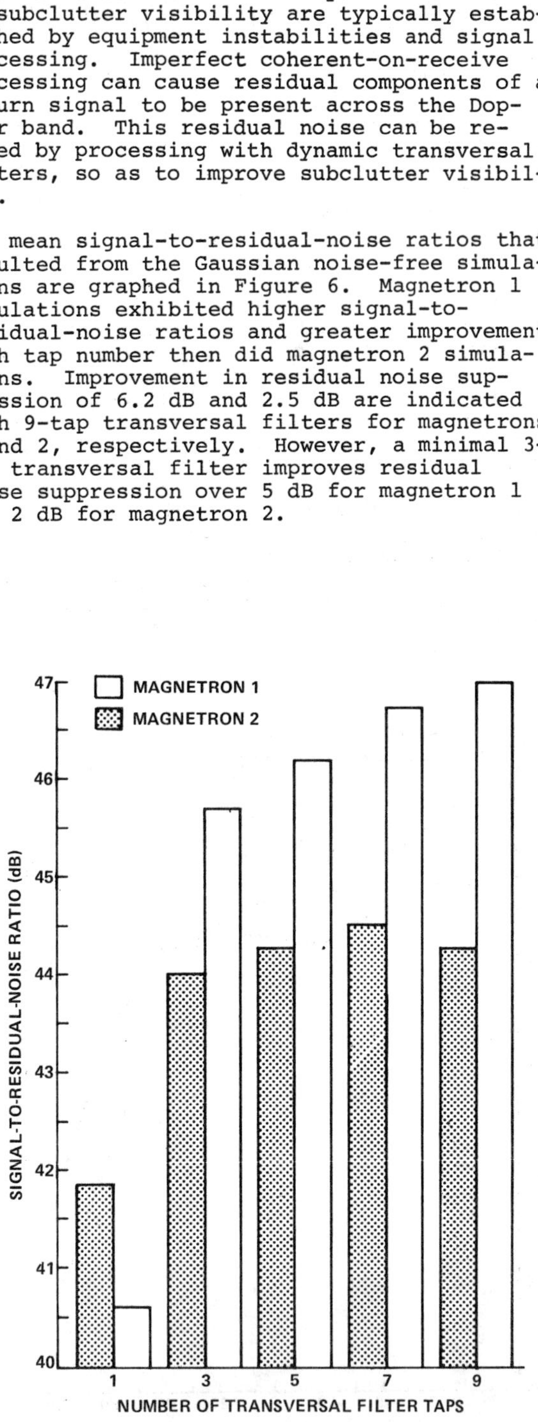

Figure 6 Coherent-on-receive signal-to-residual-noise ratio when using dynamic transversal filters.

CONCLUSIONS

Processing noncoherent radar returns with dynamic transversal filters is a technique that can effectively form a coherent return signal. With the use of multiple I and Q transmit samples across a pulse, transversal filters can yield significant improvements in signal-to-thermal-noise ratio and subclutter visibility over more typical coherent-on-receive approaches, which use only one measurement of the phase of a transmitted pulse. Simulations based on extensive and detailed magnetron pulse data indicate improvement of the signal-to-Gaussian-noise ratio of 0.9 to 3.9 dB. And although the realizable improvement correlates directly with the intrapulse phase response of the transmitter, a simple 3-tap transversal filter configuration yielded significant S/N ratio improvements of 0.9 to 2.8 dB that would be adequate for many systems. A coherent-on-receive radar configured with dynamic transversal filters yields improved subclutter visibility by reducing the residual signal components caused by either imperfect processing or equipment instabilities. Residual noise reductions of 2 to over 6 dB were indicated in simulation results.

Dynamic transversal filter processing can approximate matched filtering. Therefore, pulse compression of the return signal can result, depending on transmitter intrapulse phase deviation. The pulse compression capability could be exploited by purposely tuning the magnetron during transmission to force a larger deviation phase response [4].

ACKNOWLEDGMENTS

The author gratefully acknowledges the extensive magnetron data collection and simulation work of Ms. C. D. Sayles. In addition, the author is indebted to E. C. Jarrell, D. R. Marlow, and P. R. Sodergren for their encouragement, review, and ideas.

This work was supported by the Department of the Navy under general contract N00024-81-C-5301.

REFERENCES

1. M. I. Skolnik, Introduction to Radar Systems, Second Edition, McGraw-Hill Book Co., 1980, p. 106.

2. S. P. Williams and R. L. Trapp, "Characterizing and Modeling the Intrapulse Magnitude and Phase Responses of Two Ka-Band Magnetrons," JHU/APL F1B81U-077, 15 Jul 1981.

3. R. L. Trapp, "Coherent-on-Receive System Feasibility: Initial Analytical Approach," JHU/APL F1B79U-054, 23 Apr 1979.

4. D. E. N. Davies, "Signal Processing for Radar Applications," Case Studies in Advanced Signal Processing, IEE Conference Publication No. 180, Sep 1979, pp. 145-150.

MODULAR SURVIVABLE RADAR FOR BATTLEFIELD SURVEILLANCE APPLICATIONS

E. L. Hofmeister, W. E. Szczepanski, R. F. Oot, D. C. Dalpe and M. E. Davis

General Electric Company, Aircraft Equipment Division, Utica, NY, USA 13503

SUMMARY

This paper describes a flexible, computer-controlled multimode radar developed to exploit advanced air-to-ground radar system techniques for battlefield applications. The radar features an electronically scanned phased array and an airborne LSI radar signal processor that provides considerable mode flexibility due to the reconfigurable processor architecture. Multiple waveforms (linear FM, nonlinear FM, binary phase codes) are digitally generated to increase the overall system flexibility.

The radar mode capabilities include slow and fast moving target (MTI) wide area or sector search, MTI area track and track update, doppler beam sharpening (DBS) to provide a clutter reference background map for MTI detections, and SAR spotlight to image fixed target installations. Mode switching is rapidly performed by an airborne radar controller (ARC) which uses an LSI-11M for mode and parameter flexibility. A separate LSI-11M in the ARC provides real time motion compensation for MTI, DBS, and SAR modes.

The MSR system operates in the K_u band region with system bandwidths of 5 and 20 MHz. The system is currently in the demonstration flight test phase in the central New York area. The demonstration flight tests are to provide a quantitative evaluation of MTI and SAR imaging modes of operation. The flight tests use the Rome Air Development Center, Griffiss Air Base Stockbridge and Verona test sites. Both test sites can support calibrated corner reflector arrays, moving targets, and have a significant variety of terrain features including grasses, bush, wooded areas, buildings, and vehicles.

TACTICAL REQUIREMENTS CONSIDERATIONS

Long range, wide area Tactical Battlefield Surveillance (TBS) for the detection, precision location and tracking of both fixed and moving ground targets in dense target and electronic warfare (EW) environments in near all weather conditions is one of the most challenging radar and radar data processing design problems. Such an operational capability is required for the effective command and control of simultaneous, multiple target engagements with limited and diversified strike resources. Emerging sensor technologies will support this basic system concept but the indicated operational requirements for system survivability, simultaneous target detection and track performance, and command and control interfacing will dictate the radar performance and data processing complexities and overall system architecture.

One proposed system configuration would include one or more self-contained airborne radar platforms providing processed, prefiltered real time target track data directly to airborne weapon systems, ground based fire control centers, and a Tactical Air Control Center (TACC) via data link.

The data link requirements and ground data processing burden need to be addressed. The hardware for systems under consideration for tactical battlefield surveillance consists of an airborne radar with a phased array antenna coherent radar, data link, and a ground signal and data processing system. The disadvantage of ground processing is indicated by the wide bandwidth required for the data link. For an operational system, it is advisable to have much of the signal and data processing accomplished in the air to provide wide area data transmission and to provide the potential for autonomous operation.

MSR SYSTEM DESCRIPTION

General

The objective of the Modular Survivable Radar (MSR) Program was to design, develop, and demonstrate in flight test a mature advanced development level modular radar capability that will maximally satisfy the Tactical Battlefield Surveillance mission. The critical design requirements for this modular radar that must be considered to insure that this hardware can be transitioned into an Engineering Development program with minimal changes in the manufacturing approach are summarized in Table 1.

TABLE 1 - DESIGN REQUIREMENTS FOR MSR

- Incorporate advance technology in waveform and signal processing via LSI/VLSI development
- Develop computer controlled, real time, reconfigurable multimode capable radar
- Implement Standardized Electronic Module (SEM) design throughout in microwave, analog, and digital circuit areas
- Partition into Line Replaceable Units (LRUs) consistent with the building block approach

The integrated system configuration is illustrated in Figure 1 and consists of functional units which include a coherent frequency-agile transmitter-receiver, electronically-agile phased array antenna, coherent digital signal processor, data converter, airborne radar control group, exciter/synchronizer, display, and instrumentation. The overall design objectives for the totally integrated system configuration are shown in Table 2.

TABLE 2 - MSR SYSTEM OBJECTIVES

Radar Modes	Wide Area MTI Search MTI Track Clutter Reference Map SAR Spotlight
Receiver Channels	Σ , Δ
Airborne Signal Processing	
Feasibility Demonstration Tests	Antenna Range Test Laboratory Test & Calibration Flight Test

The MSR emphasized mode and airborne processing flexibility to support all aspects of the tactical battlefield surveillance issues. The result is a considerable degree of airborne processor configurability in a pipeline architecture. This permits a wide range of processing alternatives to fit various mode requirements. For example, the coherent processing intervals (CPIs) from an FFT-32 (search) to 1024 for SAR spotlight operation.

A displaced phase center antenna (DPCA) capability is incorporated to null the effects of platform motion clutter spread. DPCA is selectable through the control computer and uses sum and difference patterns of the antenna beams to synthesize two receive phase centers - one leading and one lagging. With appropriate scaling, platform motion noise is nulled on a two pulse basis.

Functional System Description

The flight test configuration consists of nine core radar units (see Figure 3) and 26 units of support equipment including recording instrumentation. Debug hardware, and an airborne display are system support units. Generically the support units include the following subsystems

Inertial Navigation	Control and Display
Data Recording	Power Conversion

There are three LSI-11M microcomputers in the MSR system. Two LSI-11s are located in the Airborne Radar Controller. One LSI-11 performs radar mode, mode parameter selection, beam steering computations, and peripheral control functions. The second LSI-11 provides motion compensation computation using the INS outputs. The third LSI-11 is in the Recorder Computer and controls all of the data recording control functions.

Transmit pulses are digitally generated in the Exciter/Synchronizer, heterodyned to K_u band, amplified, and fed to the transmitter. Flexibility of transmit waveform is achieved merely by loading different phase controls in the digital pulse generator memory. The transmitter amplifies the transmit waveform to the 1.5 kW peak power level and feeds it to the transmit corporate feed in the antenna.

In area search, only a single channel is used in the Converter. Digital data is buffered out of the converter unit, fed to the preprocessor, through the pulse compressor, and stored in the spectrum analyzer main memory (capacity 4 megabits). At the end of each coherent processing interval (CPI), the data collected for each beam position is read from memory (using a corner turn), and FFT is performed, the appropriate filters are applied to CFAR circuitry, and the data thresholded to the target detect and data report unit. A display interface is provided through the recorder computer.

The Carousel IV INS provides the following data to the Airborne Radar Controller (ARC):

Latitude	Velocity North	Velocity Vertical	Roll
Longitude	Velocity East	Pitch	Yaw
			Heading

The INS interface is digital and provides updates at a 20 Hz rate. The motion computer uses this information to provide linear motion corrections in MTI modes to center the clutter return in the MTI canceller notch (and the zero spectral filter). For SAR modes (area doppler beam sharpening for clutter reference and spotlight), the motion computer provides the quadratic phase correction (focus) and corrects any line of sight aircraft motions which would (otherwise) perturb the map.

The signal processor utilizes custom LSI devices. The majority of these devices (developed by General Electric) are contained in the radix-4 board which is used 15 times accounting for a total of 600 devices. The processor, shown in Figure 1, consists of four LRUs: Preprocessor, Range Compressor, Spectral Analyzer, and Target Detect and Data Report. The LRUs are functionally chained together providing pipeline signal processing starting with raw digital radar data collected in the converter and resulting in target detections or clutter map data displayed in real time.

The pulse compression function is implemented using a dual transform technique with 50% overlap and add and can be programmed to process waveforms with TW products from 32 through 512. The FFT operations are based on a radix-4 node. Dynamic range is preserved through the pulse compressor by using vector floating point arithmetic. At the output of the pulse compressor, the exponent is fixed based upon the local operational environment encountered.

Maintenance of antenna sidelobes requires both an aperture amplitude taper and control of phase and amplitude errors throughout the antenna system. In all cases, the array scans the azimuth sector from -60 degrees to +60 degrees, utilizing element phase shifters.

The MSR system controllable parameters are listed in Table 3.

All of the parameters listed in Table 3 are digitally controlled by system software through the ARC control computer and the 1553A data bus.

TABLE 3 - MSR SYSTEM CONTROLLABLE PARAMETERS

Antenna
- Beam Steering Angle
- Collimation
- Switch Matrix Control

Exciter/Synchronizer
- PRI Control
- Pulsewidth Controls
- A/D Pulse Rate/ Delay
- LFM Waveform Select
- Pulse Compression Select
- Frequency Select
- Transmit Gate Controls

Transmitter
- Warm-Up Controls
- High Voltage Control
- Transmit
- Dummy Load Select
- Power Management

Converter
- Range Gate Controls
  - Start
  - Width
- Channel Controls
- Bandwidth Controls
- Gain Controls
- BITE Test Controls

Signal Processor
- MTI/SAR Control
- SAR Prefilter Select
- FFT Control
- Size
- Fixed/Floating Point Control
- CFAR Control
- Window Size
- Thresholds
- Monopulse Control

MSR FLIGHT TESTS

The radar system is installed on a DC-4 aircraft. Figure 2 shows the antenna installation on the DC-4 aircraft with the antenna pod mounted beneath the aircraft. The INS system inertial platform is mounted just above the antenna to ease measurement of antenna motions in SAR modes.

Installation of the Core radar and support subsystem inside the aircraft is illustrated in Figure 3.

The flight test program has the basic objectives of demonstrating wide area MTI search, selected area MTI precision track and SAR spotlight imaging. Emphasis will be placed upon slow moving target detection in a tactical clutter environment.

Figure 4 illustrates MTI data taken near the New York State thruway exit 32. Thruway traffic detected on a given scan is illustrated in Figure 4a against a real beam clutter reference map. The MTI detections are the rectangular highlights in the radar image. Figure 4b illustrates the road map configuration of the area. Figure 5 illustrates MTI detections in the vicinity of the Verona test facility against a digitized cartographic road underlay.

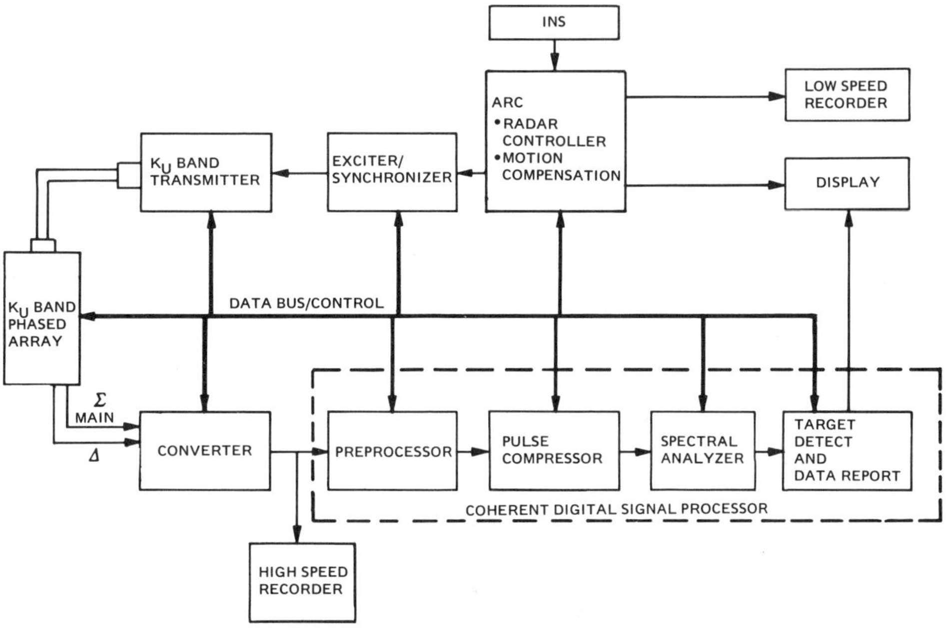

Figure 1. MSR System

AIRCRAFT - DOUGLAS DC-4 (C-54)
SPEED - 100-200 KNOTS
ALTITUDE - 30,000 FT MAX
PAYLOAD - 22,000 LBS MAX

Figure 2. MSR Antenna Pod and Test Flight Aircraft

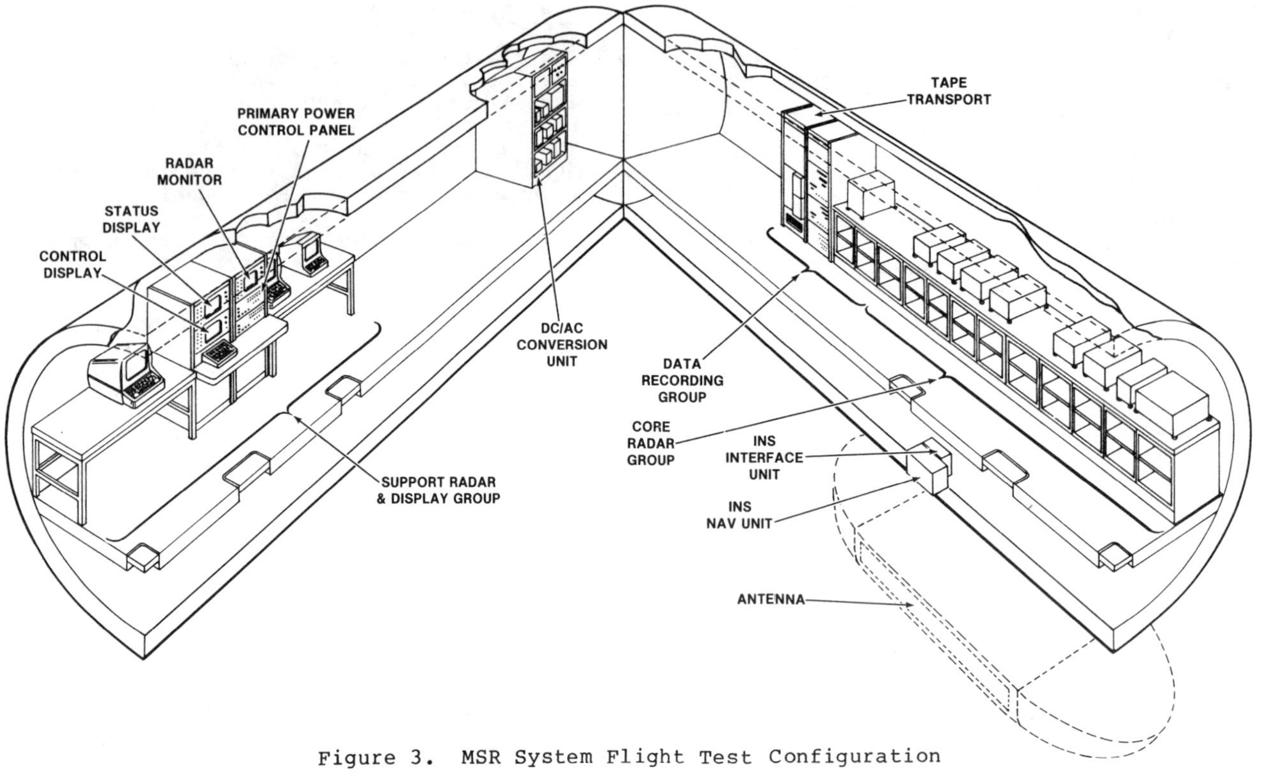

Figure 3. MSR System Flight Test Configuration

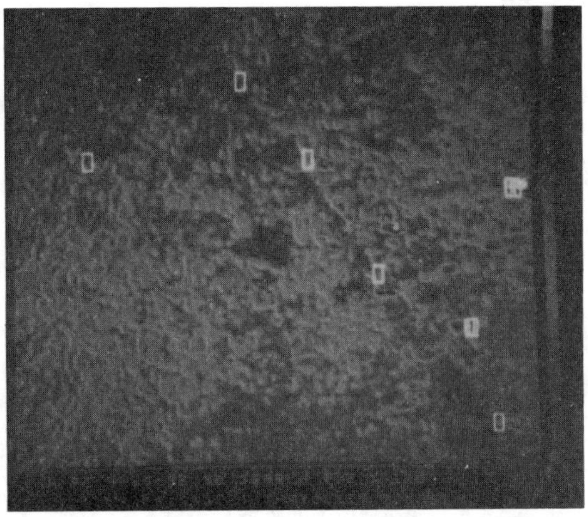

a) GROUND PROCESSED REAL
BEAM MAP SHOWING DETECTIONS

TARGET RANGE: 3 km
MAP AREA: 3 km x 4 km

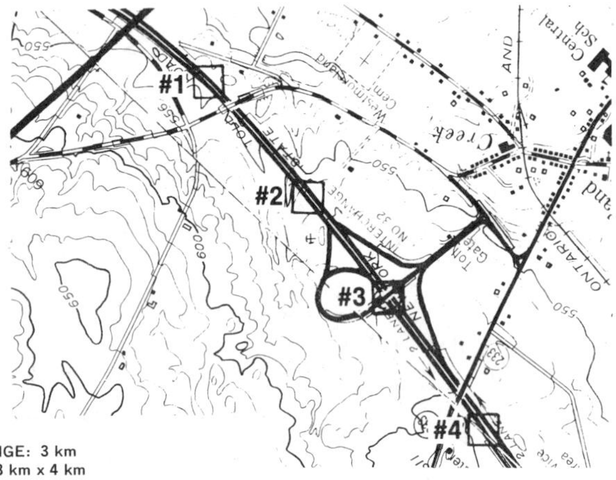

b) GEODETIC MAP OF THRUWAY AREA

Figure 4. MSR Ground Processed Search Data

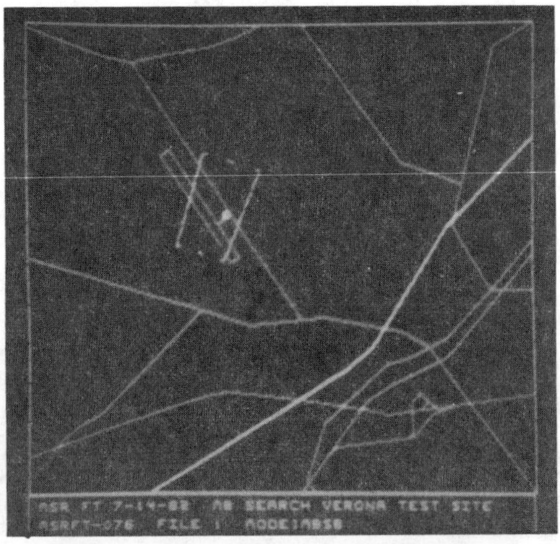

Figure 5. MTI Detection - Verona Test Site